COMPREHENSIVE ORGANOMETALLIC ANALYSIS

To
Elisabeth
on our
Silver Wedding Anniversary

PREFACE

It is now some sixteen years since the author's first series of books on the analysis of organometallic compounds. Many developments in the subject have occurred since that time and a new book on the subject is now overdue.

The present book aims to provide a comprehensive review of the subject. It covers not only all aspects of the analysis of organometallic compounds but also contains two additional chapters, dealing with environmental analysis and the use of chelates of metals in the determination of very low concentrations of organic metals.

Whilst reviewing the literature for the present book, it was observed that whereas papers published prior to 1973 dealt almost exclusively with various forms of analysis, a high proportion of those published during the past ten years were concerned with the application of proven or newly developed methods to the determination of organometallic compounds in environmental samples such as water, air, soil, river and ocean sediments, fish life and biota samples. An increasing range of elements including mercury, lead, arsenic, tin, antimony, selenium and manganese are now being found in organically bound forms in the environment, some resulting from pollution, others formed in nature by bacterial processes. As many of these substances have appreciable implications to human and animal health and the ecosystem as a whole, it was considered that it would be timely to include a separate chapter in the book devoted entirely to this subject.

Many elements, upon reaction with particular chelating agents, produce chelates which are amenable to chromatographic analysis and, in many cases, are sufficiently volatile to be gas chromatographed. This has opened up a whole new area of analysis of mixtures of metals at very low concentrations.

In many instances chelate formation-gas chromatographic methods have absolute detection limits several orders of magnitude lower than those achieved by competing techniques such as atomic absorption, neutron activation analysis, emission spectrography and spark-source mass spectrometry.

The purpose of this separate chapter is to gather together the world literature on this new subject so as to enable analytical chemists to take a strong interest in and develop further, the technique which, in the opinion of the author, has even greater potential in the future.

The first seven chapters of the book cover each of the major analytical techniques that can be applied to the analysis of organometallic compounds. Within each chapter, the elements are discussed in alphabetical order. The first two chapters cover the determination of elements and functional groups. Suceeding chapters, respectively, cover the applications of titration techniques, visible and ultra-violet spectroscopic techniques, other spectroscopic techniques particularly infrared, NMR, PMR, etc., polarography, gas chromatography and finally a chapter covering other chromatographic techniques.

Errors are inevitable in a work of this size; the author would be grateful to receive notification of any errors so that they can be rectified in future editions.

It is hoped that this book will serve its aim of being a source-book of all aspects of the analysis of organometallic compounds, their occurrence in the environment and their uses in analytical chemistry.

The volumes will interest workers in a wide variety of fields both industrial and academic, at both the pure and the applied ends of the subject. In addition to analytical chemists the book will be of interest to organic chemists and those concerned with the environment and public health. Organometallic chemistry is a field which continues to grow and to which students should direct their interest and it is hoped that they will find much to interest them.

T. R. Crompton

ACKNOWLEDGEMENTS

The author wishes to express his gratitude to the publishers of various journals for permission to reproduce the following illustrations.

FIGURE

1,2,5,6,7 T.R. Crompton, Chemical Analysis of Organometallic Compounds, Vol. V, Academic Press, London, New York, San Francisco, 1977, Chapter 20, pp. 115-235.

3 R. Dijkstra and E.A.M. Dahmen, Z. Anal. Chem., 181, 399 (1961).

4 B.J. Phillip, W.L. Mundry, and S.C. Watson, Anal. Chem., 45, 2298 (1973).

8 I. Dunstan and J.V. Griffiths, Anal. Chem. 33, 1598 (1961).

9 H. Allen and S. Tannenbaum, Anal. Chem., 31, 265 (1959).

10 S.A. Greene and H. Pust, Anal. Chem., 30, 1039 (1958).

11 J.C. Boaker and T.L. Isenhour, Anal. Chem., 41, 1705 (1969).

12 H. Pieters and W.J. Buis, Mickrochem. J., 8, 383 (1964).

13,14 R.A. Mostyn and A.E. Cunningham, J. Inst. Pet., 53, 101 (1967).

15 B.C. Southworth, J.H. Hodecker, and K.D. Fleischer, Anal. Chem., 30, 1152 (1958).

16 I. Lysyj and J.E. Zarembo, Microchem. J., 2, 245 (1958).

17,18 B.D. Holt, Anal. Chem., 37, 751 (1965).

19 O. Meier and N. Shaltiel, Mikrochim Acta 580 (1960).

20 K. Ziegler H. Gellert and Justus Liebigs, Ann. Chem., 629, 20 (1960).

21 T.R. Crompton, Analyst (London), 91, 374 (1966).

22 D.F. Hagen and W.D. Leslie, Anal. Chem., 35, 814 (1963).

23 E. Bonitz, Chem. Ber., 88, 742 (1955).

24 M. Farina, M. Donati and H. Ragzzini, Ann. Chim., (Rome), 48, 501 (1958).

25 L. Nebbia and B. Pagani, Chim. Ind. (Milan), 44, 383 (1962).

26 E.H. Hoffman and W. Tornau, Z. Anal. Chem. 186, 231 (1962).

27,28 M. Dimbat and G.A. Harlow, Anal. Chem., 34, 450 (1962).

29,30,34 W.L. Everson, Anal. Chem., 36, 854 (1964).

31 E.G. Hoffman and W. Tornau, Z., Anal. Chem., 188, 321 (1962).

32 G. Pilloni and G. Plazzogna, Anal. Chim. Acta, 35, 325 (1966).

33 A.F. Clifford and R.R. Olsen, Anal. Chem., 32, 544 (1960).

35 S.C. Watson and J.F. Eastham, Anal. Chem., 39, 171 (1967).

36 M. Dimbat and G.R. Harlow, Anal. Chem., 34, 450 (1962).

37 C. Jolibois, R. Acad. Sci., 155, 213 (1912).

38,39 G. Tagliavini, Anal. Chim. Acta, 34, 24 (1966).

40,41 W.L. Everson and E.M. Ramirez, Anal. Chem., 37, 812 (1965).

42,43 E. Bonitz, Chem. Ber., 88, 742 (1955).

44 W.P. Neumann, Angew. Chem., 69, 730 (1957).

45 J.H. Mitchen, Anal. Chem., 33, 1331 (1961).

46,47,48 B.T. Commins and P.J. Lawther, Br. J. Ind. Med., 22, 139 (1965).

49 J. Stary, K. Kratzer and K.J. Prasilova, Anal. Chim. Acta, 100,627 (1978).

50 S. Sass, W.D. Ludemann, B. Witten, V. Fishen, A.J. Sisti, and J.I. Miller, Anal. Chem., 29, 1346 (1957).

51 K.S. Pitzer and R.J. Sheline, J. Chem. Phys., 16, 552 (1948).

52,53,61 E.G. Hoffman, Z. Elektrochem., 64, 616 (1960).

54,155 E.G. Hoffman and G. Schomerg, Z. fur Elecktrochemie 61, 1101 (1957).

56 E.G. Hoffman and G. Schomberg, Z. fur Elecktrochemie, 61, 1110 (1957).

57 J.V. Bell, J. Heisler, H. Tannenbaum, and J. Goldenson, Anal. Chem., 25, 1720 (1953).

58 D.V. Guertin, S.E. Wiberley, W.H. Bauer, and J. Golderson, J. Phys. Chem., 60, 1018 (1956).

59 V. Yamamoto, Bull. Chem. Soc. Japan, 35, 619 (1962).

60 J. Smidt, M.P. Gruenewage, and H. de Vries, Rec. Irav. Chim, 81, 729 (1962).

62 J.J. Kaufman, W.S. Koski, L.T. Kuhns, and S.S. Wright, J. Amer. Chem. Soc., 85, 1369 (1963).

63,64 W.S. Koski, J.J. Kaufman, and P.C. Lanterbur, J. Amer. Chem. Soc., 79, 2382 (1957).

65 A.F. Reid, D.E. Scaife, and P.C. Wailes, Spectrochim. Acta, 20, 1257 (1964).

66 J.R. Urwin and P.J. Reed, J. Organometal Chem., 15, 1 (1968).

67,68 H.Susi and H.E. Rector, Anal. Chem., 30, 1933 (1958).

69 J.H. Lowry, R.B. Smart, and K.H. Mancy, Anal. Chim., 50, 1303 (1978).

70 J.E. de Vries, A. Lauw-Zecha, and A. Pellecer, Anal. Chem., 31, 1995, (1959).

71 M. Mehner, H. Jehring, and H. Kriegsmann, 3rd Analytical Conference, Budapest, 24-29 August, 1970.

72,73 R. Geyer and H.T. Seidlitz, Z. Chem., 4, 468 (1964).

74 L.M. Brown and K.S. Mazdiynasi, Anal. Chem., 41, 1243 (1969).

75,76,79 B.J. Gudzinowicz and J.L. Driscoll, J. Gas Chromatogn, 1, 25 (1963).

77,78 G.E. Parris, W.R. Blair, and J.E. Brinkman, Anal. Chem., 49, 2215 (1977).

80,81,82 C. Feldman and D.A. Batistoni, Anal. Chem. 49, 2215. (1977).

83 G. Schomberg, R. Koster, and D. Henneberg, Z. Anal. Chem., 170, 285 (1959).

84 G.R. Seely, J.P. Oliver, and D.M. Ritter, Anal. Chem., 31, 1993 (1959).

85 T.D. Parsons, M.B. Silverman, and D.M. Ritter, J. Am. Chem. Soc., 79, 5091 (1957).

86 J.A. Semlyen and C.S.G. Phillips, J. Chromatography, 18, 1 (1965).

87,88 J.J. Kaufman, J.E. Todd, and W.S. Koski, Anal. Chem., 29, 1032 (1957).

89,90 H.W. Myers and R.F. Putman, Anal. Chem., 34, 664 (1962).

91,92 H. Veening, J. Graver, D.B. Clark, and B.R. Willeford, Anal. Chem., 41, 1655 (1969).

93 H. Veening, J.S. Keller, and B.R. Willeford, Anal. Chem., 43, 1516 (1971).

94 W.J.A. Van der Heuvel, J.S. Keller, H. Veening, and B.R. Willeford, Analyt. Lett., 3, 279 (1970).

95 C.S.G. Phillips and P.L. Timms, Anal. Chem., 35, 505 (1963).

96 B. Iatridis and G. Parissakis, Anal. Chem., 49, 909 (1977).

97 O.E. Ayers, T.C. Smith, J.D. Burnett, and B.W. Pouder, Anal. Chem. 38, 1606 (1966).

98,99,100 J.E. Lovelock and A. Zlatkis, Anal. Chem. 33, 1958 (1961).

101,102 H.J. Dawson, Anal. Chem., 35, 542 (1963).

103 E. Barrall and P. Ballinger, J. Gas Chromatography, 1, 7 (1963).

104 E. Bonnelli and H. Hartmann, Anal. Chem., 35, 1980 (1963).

105,106 N.L. Soulages, Anal. Chem., 38, 28 (1966).

107 W.S. Leonhardt, R.C. Morrison, and W.C. Kamicnski, Anal. Chem., 38, 466 (1966).

108 L.V. Giold, C.A. Hollingsworth, D.H. McDaniel, and J.H. Wotiz, Anal. Chem., 33, 1156 (1961).

109 A. Wowk and S. Di. Giovanni, Anal. Chem., 38, 742 (1966).

110,111 P.C. Uden, R.M. Barnes, and P. Di. Sanzo, Anal. Chem., 50, 852, (1978).

112,113 B.D. Quimby and P.C. Uden, Anal. Chem., 50, 2112 (1978).

114 A. Apelblat and A. Hornik, J. Chromatography, 24, 175 (1966).

115 A. Apelblat and A. Hornik, Trans Faraday Soc. No., 529, 63, 185 (1967).

116 A. Apelblat, J. Inorg. Nucl. Chem., 31, 483 (1969).

117 R.L. Grob and G.L. McCrea, Anal. Lett., 1, 55 (1967).

118 K.P. Berlin, T.H. Austin, M.E. Nagahushanam, J. Peterson, J. Calvert, W.A. Wilson, and D. Hopper, J. Gas Chromatogr., 3, 256 (1965).

119 A. Davis, A. Roadi, J.G. Michalovic, and A.M. Joseph, J. Gas Chromatography, 1, 23 (1963).

120 C.S. Evans and C.M. Johnson, J. Chromatography, 21, 202 (1966).

121,122 C.S.G. Phillips, P.L. Timms, Anal. Chem., 35, 505 (1963).

123 F.H. Pollard, G. Nickless, and P.C. Uden, J. Chromatogr., 19, 28 (1965).

124 H. Rotzsche, Z. Anorg. Chem., 324, 197 (1963).

125 H. Rotzsche, Z. fur Anorg. und Allgemeine. Chemie., 328, 79 (1964).

126 M. Wurst, Coll. Czech. Chem. Commun., 30, 2038 (1965).

127 J.B. Carmichael, D.J. Gordon, and C.E. Ferguson, J. Gas Chromatography, 4, 347 (1966).

128,129,130 M.A. Osman, H.H. Hill, M.W. Holdren, and H.H. Wetberg, Anal. Chem., 51, 1286 (1979).

131,132,133 G. Garzo and F. Till, Talanta, 10, 583 (1963).

134 D. Thrash L. Viosinet and K.E. Williams, J. Gas Chromatography July 248 (1965).

135,136 F.H. Pollard, G. Nickless, and D.B. Thomas, J. Chromatogr., 22, 286 (1966).

137 D.D. Schlenter and S. Siggia, Anal. Chem., 49, 2343 (1977).

138 D.D. Schleuter, Ph.D. Dissertation, University of Massachusetts, Amherst, Massachusetts (1976).

139,140 F.H. Pollard, G. Nicklett, and D.J. Cooke, J. Chromatography, 13, 48 (1964).

141 R.C. Putnam and H. Put, J. Gas Chromatography, 3, 2 (1965) and 3, 2 (1965).

142 H. Geissler and H. Kriegsmann, Z. Chemie, Lpz., 4, 354 (1964).

143,144 B.L. Tonge, J. Chromatography, 19, 182 (1965).

145,146,147 S.G. Perry, J. Gas Chromatography, 93 March (1964).

148 J.E. Schwarberg, R.W. Moshier, and J.H. Walsh, Talanta, 11, 1213 (1964).

149 J.E. Schwarberg, Master's thesis, University of Dayton, 1964.

150 P. Jacquelot and G. Thomas, Bull Soc. Chim. Fr. 702 (1971).

151 H. Veening and J.F.K. Huber, J. Gas Chromatog., 6, 326 (1968).

152,153 J.A. Stokeley, Master's thesis, Oak Ridge University, Diss. Abstr., 27, 1388 B (1966).

154 K.J. Eisentraut and R.E. Sievers, J. Am. Chem. Soc., 87, 5254 (1965).

155 Anonymous, Chem. Eng. News, 43, November 22, 39 (1965).

156 T. Shigematsu, M. Matsui, and K. Utsunomiya, Bull. Inst. Chem. Res., Kyoto Univ., 46, 256 (1968).

157 W.I. Stephen, I.J. Thompson and P.C. Uden, Chem. Commun., pp. 269-270 (1969).

158 A. Khalique, W.I. Stephen, P.E. Henderson, and P.C. Uden, Anal. Chim., Acta, 101, 117 (1978).

159 T. Fujinaga and Y. Ogino, Bull. Chem. Soc., Japan, 40, 434 (1967).

160 J.K. Foreman, T.A. Gough, and E.A. Walker, Analyst (London), 95, 797 (1970).

161 R. Belcher, C.R. Jenkins, W.I. Stephen, and P.C. Uden, Talanta, 17, 455 (1970).

162,163 T.J. Cardwell, D.J. Resarro, and P.C. Uden, Anal. Chim. Acta, 85, 415 (1976).

164 Y. Shimoishi and K. Toei, Anal. Chim. Acta, 100, 65 (1978).

165 E.G. Gaetani, C.F. Laureri, A. Magnia, and G. Parolari, Anal. Chem., 48, 1725 (1976).

166,167 T.J. Cardwell, D. Caridi, and M.S. Loa, J. Chromatography, 351, 331 (1986).

168 M. Saitoh, R. Kurada, and M. Shibukawa, Anal. Chem., 55, 1025 (1983).

169 K. Saitoh, M. Kabayashi, and N. Suzuki, Anal. Chem. 53, 2309 (1981).

170 G.R. Ricci, L.S. Shepard, G. Colovos, and N.H. Hester, Anal. Chem., 53, 610 (1981).

171 J.D. Messman and T.C. Rains, Anal. Chem., 53, 1632 (1981).

172 G. Misson, Chem. Ztg., 32, 633 (1908).

173 R.B. Lew and F. Jakob, Talanta, 10, 322 (1963).

174 Department of the Environment and National Water Council (U.K.) H.M. Stationary Office, London, 23 pp (pt 23 Abenv) (1978).

175 K. Minagawa, Y. Takizawa, and I. Kufune, Anal. Chim. Acta., 103, 115 (1980).

176,177 K. Chiba, K. Yoshida, K. Tanabe, H. Horaguchi, and K. Fuwa, Anal. Chem., 55, 450 (1983).

178 G.A. Hambrick, P.N. Froebich, O.A. Meirate, and B.L. Lewis, Anal. Chem., 56, 421 (1984).

179 Y.K. Chau, P.T.S. Wong,and G.A. Bengert, Anal. Chem., 54, 246 (1982).

180 C.J. Soderquist and D.G. Crosby, Anal. Chem. 50, 1435 (1978).

181,182,183 R.S. Raman and M.A. Tomkins, Anal. Chem., 51, 12 (1979).

184 Y.K. Chau, P.T.B. Wong, and O. Kramar, Anal. Chem. Acta., 146, 211 (1983).

185,186 M.O. Andreae, Anal. Chem., 49, 820 (1977).

187 A.A. Grabinski, Anal. Chem., 53, 966 (1981).

188,189 J. Aggett, R. Kadwani, Analyst (London), 108, 1495 (1983).

190 C.E. Stringer and M. Attrep, Anal. Chem., 51, 731 (1979).

191,192 L. Brown, S.J. Haswell, H.M. Rhead, P.O'Neill, and C.C. Bancroft, Analyst, 108, 1511 (1983).

193,194,195 Y.K. Chau, P.T.S. Wong, and H. Saitoh, J. Chromatogr. Sci., 14, 162 (1976).

196,197,198 S. Hanamura, B.W. Smith, and J.D. Winefordner, Anal. Chem., 55, 2026 (1983).

199,200,201 J.G. Gonzales, and R.T. Ross, Anal. Lett., 5, 683 (1972).

202 W.A. MacCrehan, R.A. Durst, and J.M. Bellama, Anal. Lett., 10, 1175 (1977).

203,204 C.J. Cappon and V. Crispin Smith, J. Anal. Chem., 49, 365 (1977).

205 M. Morita, T. Uehiro, K. Fuwa, Anal. Chem., 53, 1806 (1981).

206 M. Loe, R. Cruz, and J.C. Van Loon, Anal. Chim. Acta, 120, 171 (1980).

207 G. Torsi, F. Palmisano, Analyst (London), 108, 1318 (1983).

208 G. Torsi, E. Desimoni, F. Palmisano, Analyst (London), 107, 96 (1982).

209 J.W. Robinson, E.L. Kicsel, J.P. Goodbread, R. Bliss, and R. Marshall, Anal. Chim. Acta, 92, 321 (1977).

210,211 A.J. McCormack, S.C. Tong, and W.D. Cooke, Anal. Chem., 37, 1470 (1965).

212 W.R.A. De. Jonghe, D. Chakraborti, and F.C. Adams, Anal. Chem., 52, 1974 (1980).

213 H. Koizumi, R.D. McLaughlin, and T. Hadeishi, Anal. Chem., 51, 387 (1979).

214,215 R. Moss and E.V. Browett, Analyst (London), 91, 428 (1966).

CONTENTS

CHAPTER 1 DETERMINATION OF ELEMENTS AND FUNCTIONAL GROUPS

CHAPTER 2 TITRATION PROCEDURES

CHAPTER 3 SPECTROSCOPIC TECHNIQUES, VISIBLE AND ULTRAVIOLET SPECTROSCOPY

CHAPTER 4 OTHER SPECTROSCOPIC TECHNIQUES

CHAPTER 5 POLAROGRAPHIC TECHNIQUES

CHAPTER 6 GAS CHROMATOGRAPHY

CHAPTER 8 OTHER CHROMATOGRAPHIC TECHNIQUES

CHAPTER 9 ORGANOMETALLIC COMPOUNDS IN THE ENVIRONMENT

CHAPTER 1

DETERMINATION OF ELEMENTS AND FUNCTIONAL GROUPS

ORGANOALUMINIUM COMPOUNDS

A. Determination of Aluminium

Many organoaluminium compounds are spontaneously pyrophoric upon contact with air and, as such, are extremely hazardous and should be handled with care. Even contact with traces of oxygen will alter the composition of the sample during analysis and vitiate the analytical results obtained. A sampling procedure for organoaluminium compounds is illustrated in Figure 1.

To determine aluminium, the organoaluminium sample must first be decomposed by the addition of an aqueous reagent in order to provide an aqueous extract in which the aluminium is quantitatively recovered. A cooled hydrocarbon solution of the organoaluminium sample is hydrolysed by the gradual addition of aqueous hydrochloric acid in an inert atmosphere. Aluminium is quantitatively recovered in the aqueous extract and is then determined in this extract complexiometrically using disodium EDTA. This gives satisfactory aluminium recoveries from all types of organoaluminium compounds, from the most reactive types such as neat triethylaluminium to the less reactive higher molecular weight compounds containing alkyl groups up to C_{18}[2]. Sample decomposition is performed in a specially designed flask of the type illustrated in Figure 2. In order to obtain 100% recovery of aluminium in this method, it is necessary for a 20% excess of EDTA over the amount of aluminium present to be added. For this reason a trial titration is carried out, the data obtained being used to calculate the volume of disodium EDTA to be added in the final titration.

Hagen et al[1] have described a method for the determination of aluminium in organoaluminium compounds based on controlled deactivation hydrolysis followed by chelatometric titration with cyclohexane-diamine tetraacetate (Wannier and Ringbom[3], Nydahl[4]).

Shvindlerman and Zavadovskaya[5] described a method for the determination of aluminium in organoaluminium compounds based on hydrolysis with dilute nitric acid followed by addition of excess EDTA to the aqueous extract and estimation of unused EDTA by titration with standard lead solution. Chlorine could also be determined volumetrically in the hydrolysed extract of the sample[6].

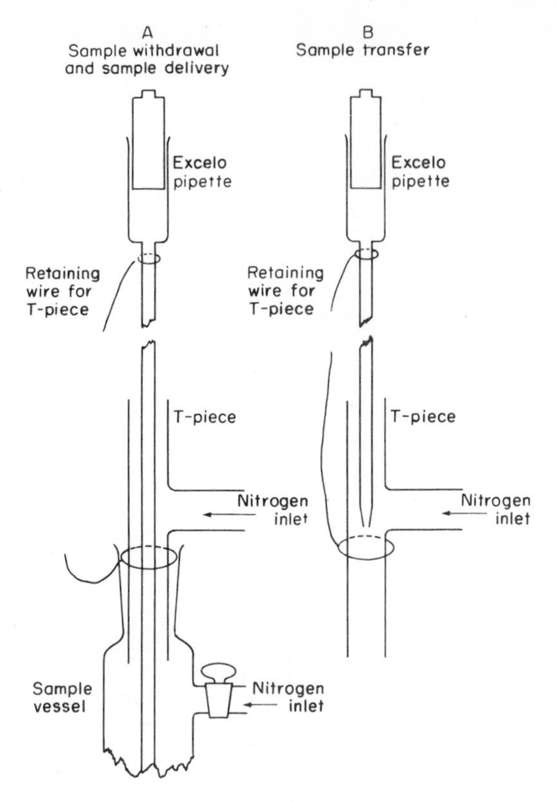

Fig. 1. Sampling procedure for organoaluminium compounds.

Hennart[6] described a similar procedure involving decomposition of the sample with aqueous hydrochloric acid, followed by titration of the excess EDTA with standard copper sulphate solution to the catechol-violet end-point. Alternatively, a xylene solution of the organoaluminium sample is transferred into a Erlenmeyer flask[7] to which is added aqueous acetone containing hydrochloric acid. An excess of 0.05 M cyclohexanediaminetetraacetic acid (CDTA) solution and 200 ml of isopropanol is added and the solution heated nearly to boiling and buffered. Excess of CDTA is titrated with standard 0.05 M zinc sulphate solution to the dithizone end-point. The end-point is very sharp from green to red. Excellent accuracy and precision are claimed for this method. It has been applied to very pure and to very complex mixtures without interference or matrix effects.

Organoaluminium compounds can be analysed by digestion with a nitric-sulphuric acid mixture in a sealed tube, followed by a spectrophotometric determination of aluminium as the 8-hydroxyquinoline complex.[8]

An amount of 20-40 mg of organoaluminium samples that also contain silicon and/or phosphorus can be fused with sodium peroxide in a bomb under an atmosphere of oxygen. After dissolving the product in sulphuric acid, excess of EDTA is added and the excess is determined by titration with standard copper sulphate solution to the catechol violet end-point[9]. An absolute error of less than 0.3% was claimed at the 14% aluminium level in the sample.

X-ray fluoroescence can be used for determining aluminium in organo-

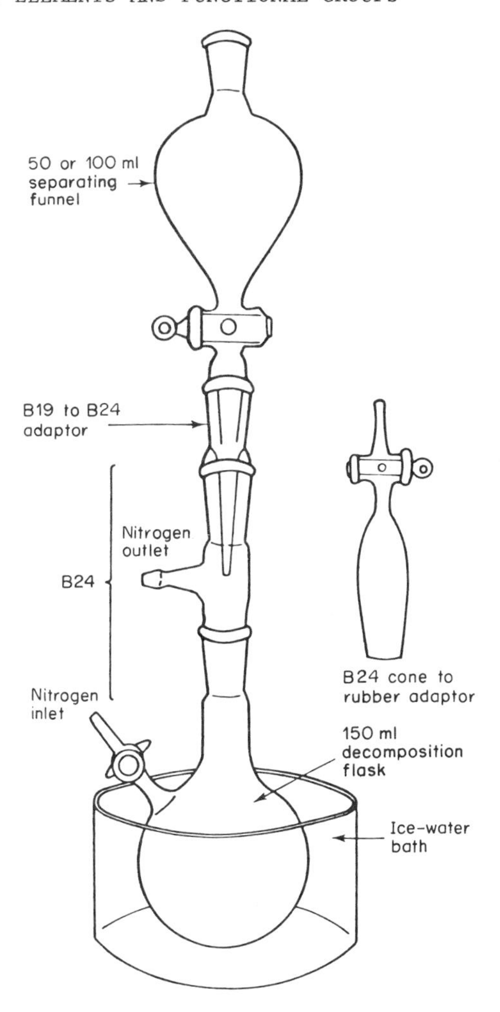

Fig. 2. Sample decomposition flask for organoaluminium compounds.

aluminium compounds[10]. The use of a chromium target tube, a pulse height discriminator, and a modified sample chamber for sample cooling together with nylon sample cells, permits the determination of aluminium over the range 0.05 - 10% in organoaluminium compounds.

B. Determination of Carbon and Hydrogen

Gilimon and Bryushkova[11] have described a procedure for the elementary analysis of carbon, hydrogen, halogens and aluminium in reactive or non-reactive organoaluminium compounds. They burn 3-12 mg of the sample in an open capillary tube in an atmosphere or argon or nitrogen using apparatus which they describe in detail. Aluminium is determined gravimetrically as alumina in an oxygen medium and carbon, hydrogen and halogens are determined by the procedure described by Korshun[12].An error of less than ±0.5% is claimed with trialkylaluminium compounds. The method is also suitable for unstable compounds of boron, bismuth, antimony and phosphorus.

Head and Holley[13] modified the conventional combustion method for carbon and hydrogen for the analysis of lithium and aluminium hydrides. Recoveries of carbon and hydrogen were 99% or better and agreed closely with results obtained by hydrolysis procedures.

C. Determination of Aluminium-Bound Halogens

Aluminium-bound halogens may be determined by addition of an aqueous solution of nitric acid to a cooled hydrocarbon solution of the organo-aluminium[14]. This converts aluminium-bound halogens into the halogen hydracid which is extracted into the aqueous phase and titrated potentiometrically or directly with N/15 or N/100 silver nitrate:

$$R_2AlX + 3H_2O \rightarrow HX + 2RH + Al(NO_3)_3$$

$$X = \text{halogen}$$

If a low concentration of aluminium-bound bromine is to be determined, an alternative procedure is available[14] for the determination of this element in amounts down to 50 ppm with an accuracy of ±1% of the determined value. Aluminium-bound iodine interferes in the determination of bromine but chlorine does not. The sample, diluted with isooctane, is quantitatively decomposed at 50°C with aqueous sulphuric acid, converting bromine to the ionic form, which is then extracted from the organic phase with dilute aqueous sulphuric acid:

$$2R_2AlBr + 3H_2SO_4 \rightarrow 2HBr + 4RH + Al_2(SO_4)_3$$

The bromide content of the extract is then determined by a volumetric procedure in which the buffered bromide solution is treated with excess of sodium hypochlorite to oxidize bromide to bromate. Excess of hypochlorite is destroyed with sodium formate. Acidification of the test solution followed by addition of excess potassium iodide liberates an amount of iodine equivalent to the bromide content of the sample. The iodine is titrated with sodium thiosulphate. A method for the specific determination of iodine is also available[14]. It is applicable at concentrations as low as 40 ppm of iodine and also for iodine at the macro level. In the procedure for determining iodine the diluted sample is decomposed quantitatively at 50°C by the addition of hydrochloric acid. To the acid extract is added standard potassium iodate solution which converts iodide to iodine monochloride:

$$R_2AlI + 3HCl \rightarrow HI + 2RH + AlCl_3$$

$$KIO_3 + 2KI + 6HCl \rightarrow 3KCl + 3ICl + 3H_2O$$

The end-point, which occurs with the complete conversion of iodide to iodine monochloride, is indicated by the disappearance of the violet iodine colour from a chloroform layer present in the titration flask. The silver nitrate titration obtained in the chlorine determination would include iodide if any were present. If the iodine analysis of the material is available, the iodine-corrected chlorine analysis may be calculated from the following equations:

$$\% \text{ chlorine (wt./wt.) (corrected for iodine)} = \left(\frac{T_A \times f_A}{w} - \frac{2 \times T_B \times f_a}{W}\right) \times 3.5456$$

$$\text{equivalent chlorine/100 g alkyl} = \left(\frac{T_A \times f_A}{w} - \frac{2 \times T_B \times f_B}{W}\right) \times 0.100 \text{ g}$$

where:

Chlorine determination:

T_A = titre of silver nitrate;
f_A = normality of silver nitrate;
w = grams of alkyl represented by the aliquot of decomposed alkyl solution employed in a silver nitrate titration.

Iodine determination:

T_B = titre of f_a molar potassium iodate;
f_a = molarity of potassium iodate;
W = grams of alkyl employed per iodine determination.

N.B. These corrections can be ignored if the iodine content of the sample is less than 0.5%.

A typical analysis of an approximately 4:1 mixture of trimethylaluminium and dimethylaluminium iodide is given below. It is seen that the determined elements and functional groups add up to near 99% of the sample:

		% w/w
aluminium	*	18.6
methyl	+	25.2
iodine		55.1
		98.9

* determined by the EDTA procedure described previously.
+ determined by the alcoholysis-hydrolysis procedure described later in the section on functional groups.

D. Determination of Aluminium-bound Alkyl Groups up to Butyl and Hydride

Lower alkyl and hydride groups in organoaluminium compounds may be determined by reacting a known weight of sample at a low temperature with 2-ethylhexanol in a specially constructed nitrogen- or helium-filled gasometric system[15,16]. Upon alcoholysis, each alkyl group liberates 1 mol of an alkane gas and each hydride group liberates 1 mol of hydrogen, as follows:

$$\gt AlC_nH_{2n+1} + HOCH_2-\underset{\substack{|\\C_2H_5}}{CH}-CH_2CH_3 \longrightarrow \gt AlOCH_2-\underset{\substack{|\\C_2H_5}}{CHCH_2CH_3} + C_nH_{2n+2}$$

$$\gt Al-H + HO-CH_2-\underset{\substack{|\\C_2H_5}}{CH}-CH_2-CH_3 \longrightarrow \gt Al-OCH_2-\underset{\substack{|\\C_2H_5}}{CHCH_2CH_3} + H_2$$

The alkyl and hydride contents of the samples are then calculated from the amount of gas evolved from a known weight of sample and from the composition of the gas withdrawn from the system at the end of the analysis, obtained by mass spectrometry and other methods. Gas recoveries obtained by this procedure were lower than expected from the composition of the samples analysed. Crompton and Reid[17] and Crompton[18] have tested this procedure against pure redistilled samples of triethylaluminium and triisobutylaluminium and confirmed that lower than expected gas yields were obtained when either 2-ethyl hexanol or n-hexanol were used as alcoholysis reagents. It was evident, however, from their results that incomplete reaction of alkyl and hydride groups with the alcoholic reagent was the principal cause of the low gas recoveries obtained. Thus, appreciably higher gas yields were obtained when sample decomposition was effected using a mixture of n-hexanol and monoethylene glycol or a mixture of water and monoethylene glycol than when 2-ethyl hexanol was used. Crompton and Reid[17] studied the reaction of lower alkyl groups up to butyl and hydride groups with a wide range of hydroxylic reagents (alcohols, glycols, water) to find a suitable reagent as described below, for the quantitative decomposition of each type of organoaluminium compound. These workers used gas chromatography for the analysis of the evolved gas mixtures. Typical analyses obtained by this procedure are shown in Tables 1 - 4.

Analysis of methyl aluminium and ethyl aluminium compounds:

It has been stated already that incomplete decomposition of the alkyl and hydride groups in triethylaluminium samples occurs upon reaction with 2-ethyl hexanol or with n-hexanol. Table 5 shows (experiments 1 and 2) that a higher gas yield is obtained when a 4:1 mixture of anhydrous n-hexanol and anhydrous monoethylene glycol is used to decompose triethylaluminium, instead of anhydrous n-hexanol alone. Aluminium-bound ethyl and hydride groups react very vigorously with water. It is not feasible to add water directly to triethylaluminium as the ensuing reaction is extremely vigorous, even when carried out at - 70°C. Also, an undesirable "fissioning" side-reaction, which converts alkyl groups to ethylene and hydrogen instead of ethane, occurs to some extent when aqueous reagents or aqueous monoethylene glycol reagents are added directly to neat organoaluminium compounds of low molecular weight. (Crompton and Reid[17]).

Fission does not occur, however, when anhydrous n-hexanol is added to neat ethylaluminium compounds (Table 5, experiment 1) the trace of hydrogen in the gas obtained in this experiment being due to a small amount of aluminium hydride in the sample. When 20% aqueous sulphuric acid was added to the reaction product obtained in experiment 1, a further appreciable liberation of gas occurs (Table 5, experiments 3 and 4). This additional liberation of gas takes place very smoothly and, as can be seen from the results in Table 5, no fissioning occurs. Thus, by successively reacting triethylaluminium with anhydrous n-hexanol and then with an aqueous reagent a maximum gas yield is achieved smoothly and without the "fissioning" side-reaction.

The recovery obtained by this alcoholysis-hydrolysis procedure was tested for very pure triethylaluminium and, also for a sample of trimethylaluminium. The triethylaluminium, in addition to ethyl groups, also contains a small amount of hydride groups and some butyl groups. The results in Table 6 show that almost quantitative recovery is obtained in the determination of the compounds using the alcoholysis-hydrolysis procedure and the determination is satisfactorily reproducible. The small percentage of these samples unaccounted for (up to 1.7%) included alkoxide groups which were not determined. This procedure is suitable therefore, for the analysis of trimethylaluminium and triethylaluminium preparations. The same procedure can be applied to the analysis of chloroethylaluminium and alkoxide ethylaluminium preparations, e.g. diethylaluminium chloride and diethylaluminium ethoxide.

Table 1

Alcoholysis/hydrolysis of triethylaluminium and triisobutylaluminium preparations

Sample No.	Determined constituent, % wt						
	Triethylaluminium compounds						
	$Al(C_2H_5)_3$	$Al(C_2H_5)_2H$	$Al(C_2H_5)_2(OC_2H_5)$	$Al(C_4H_9)_3$	$Al(C_6H_{13})_3$	Total determined	Unaccounted
1	82.1	1.4	6.9,6.9[a]	5.3	Not determined	95.7	4.3
2	82.8,83.6	1.1,1.1	2.9,2.9[a]	7.4,7.7	0.5,0.5	94.7,95.8	5.3,4.2
3	76.5	9.9	2.6[a]	5.9	Not determined	94.9	5.1
4	75.4	15.8	1.7[a]	3.0	1.1	97.0	3.0
5	71.8	17.7	4.3[a]	4.3	Not determined	98.1	1.9
6	66.9	20.3	2.3[a]	5.4	Not determined	94.9	5.1
7	65.2	26.0	2.0[a]	3.2	Not determined	96.4	3.6
	Triisobutylaluminium compounds						
	$Al(C_4H_9)_3$	$Al(C_4H_9)_2H$	$Al(C_4H_9)_2(OC_4H_9)$	$Al(C_8H_{17})_3$	-	Total determined	Unaccounted
1	95.8	4.6	Nil[a]	Nil	-	100.2	Nil
2	95.4	2.5	Not determined	Nil	-	97.9	2.1[b]
3	90.0	8.0	Not determined	Nil	-	98.0	2.0[b]
4	87.6,87.1	6.6,6.6	6.0,6.0[a]	Nil	-	100.2,99.7	Nil,0.3

[a] Determined by procedures described in this section.
[b] Includes butoxide groups which were not determined in these samples.

Table 2

Analysis of crude and distilled tri-n-propylaluminium

Sample description	Aluminium (%wt)	n-Propyl (%wt)	n-Propoxide (%wt)	Hydride (%wt)	Hexyl (%wt)	Total (%wt)	2-Methylpentene-1 (by difference) (%wt)
Crude tri-n-propylaluminium	15.73	69.3	1.5	0.027	7.5	94.1	5.9
	Empirical formula: $Al_{1.00}Pr_{2.76}OPr_{0.04}H_{0.05}(Hexyl)_{0.15}$						
Distilled tri-n-propylaluminium	16.12	76.8	1.7	0.025	Nil	94.6	5.4
	Empirical formula: $Al_{1.00}Pr_{2.98}OPr_{0.05}H_{0.04}$						

Table 3

Alcoholysis/hydrolysis of diethylaluminium ethoxide and di-n-propylaluminium isopropoxide samples

Determined constituents and empirical formula - Diethylaluminium ethoxides

Sample No.	Aluminium, (%wt)	Ethyl, (%wt)	Butyl, (%wt)	Hydride, (%wt)	Ethoxide, (%wt)	Total (%wt)	Empirical formula	Sum of subscripts in empirical formula	Departure of sum of subscripts from 3.00 wt (stoichiometric AlR_3), (%)
1	20.1	41.3	1.6	0.001	35.3	98.3	$Al_{1.00}Et_{1.91}Bu_{0.04}H_{0.001}(OEt)_{1.05}$	3.00	0.0
2	20.0	41.9	3.2	Nil	32.8	97.9	$Al_{1.00}Et_{1.94}Bu_{0.07}(OEt)_{0.99}$	3.00	0.0

Determined constituents and empirical formula - di-n-propylaluminium isopropoxides

Sample No.	Aluminium, (%wt)	n-Propyl, (%wt)	Hydride, (%wt)	Isopropoxide, (%wt)	Total (%wt)	Empirical formula	Sum of subscripts of empirical formula	Departure of sum of subscripts from 3.00 wt (stoichiometric AlR_3), (%)
1	15.4	45.1	Nil	39.7	100.2	$Al_{1.00}Pr_{1.84}(OPr)_{1.18}$	3.02	+0.7
2	15.6	51.9	Nil	32.0	99.5	$Al_{1.90}Pr_{2.08}(OPr)_{0.94}$	3.02	+0.7
3	15.4	52.9	0.001	31.1	99.4	$Al_{1.00}Pr_{2.14}H_{0.002}(OPr)_{0.92}$	3.06	+2.0

Table 4

Alcoholysis/hydrolysis of diethylaluminium chloride batches

Sample No.	$Al(C_2H_5)_2Cl$, (%wt)	$Al(C_4H_9)_2Cl$, (%wt)	$AlC_2H_5Cl(OC_2H_5)$, (%wt)	Iodine[a] (%wt)	Total determined (%wt)	Chlorine, calculated from compound analyses "A" (%wt)	Chlorine[b] (determined), "B" (%wt)	Chlorine believed present as alkylaluminium dichloride "A" - "B" (%wt)
1	87.9	4.5	4.9	0.15	97.5	28.0	29.3	1.3
2	88.3	1.8	10.6	0.08	100.8	29.1	29.2	0.1
3	88.6	0.7	9.2	0.03	98.5	28.6	30.0	1.4
4	90.5	1.7	6.9	0.08	99.2	28.7	29.2	0.5
5	90.0	2.2	5.7	0.07	98.9	28.7	29.2,29.4	0.6
6	92.3	0.6	6.0	0.03	98.9	28.8	29.8	1.0
7	92.5	0.8	6.6	0.06	100.0	29.1	29.5	0.4

[a] Elementary iodine added during preparation of diethylaluminium chloride-iodine determined by methods described in this section.

[b] Chlorine determined by method described in this section.

Table 5 - Gas yields obtained by the use of various reagents in the alcoholysis and/or hydrolysis of triethylaluminium

Experiment No.	Alcoholysis and/or hydrolysis reagent and decomposition technique employed	Gas yield at S.T.P. Mls. evolved/g sample				
		Hydrogen	Ethane	Ethyllene	Butane	Total (ml/g)
1	Added 3 ml anhydrous n-hexanol to sample at -30°C, then slowly heated to 100°C	0.5	324	Nil	34	358.5
2	As above, using 3 ml n-hexanol (80% vol) - monoethylene glycol (20% vol) mixture	0.5	337	Nil	32	369.5
3	Added 1.5 ml n-hexanol to sample at - 30°C then 1 ml 20% wt aqueous sulphuric acid at - 30°C. Slowly heated to 100°C.	0.4	351	Nil	36	387.4
4	Added 1.5 ml n-hexanol to sample at - 30°C then heated to 100°C. Added 1 ml 20% aqueous sulphuric acid at 100°C.	0.4	351	Nil	36	387.4

Normal reaction

$$''AlC_2H_5 + H_2O = C_2H_6 + ''AlOH$$

Fissioning reaction

$$''AlC_2H_5 + H_2O = C_2H_4 + H_2 + ''AlOH$$

Table 6 - Analysis of pure triethylaluminium and trimethylaluminium samples

	Distilled triethylaluminium (%wt)	
$Al(C_2H_5)_3$	94.1	94.6
$Al(C_2H_5)_2H$	2.2	2.2
$Al(C_2H_5)_2(C_4H_9)$	2.0	2.1
Total	98.3	98.9

	Distilled trimethylaluminium (%wt)
$Al(CH_3)_3$	99.5
$Al(CH_3)_2H$	nil
Total	99.5

Analysis of Propyl Aluminium Compounds - A "fissioning" side-reaction occurs when aqueous reagents or aqueous monoethylene glycol reagents are added directly to ethylaluminium compounds. Preliminary reaction of the sample with anhydrous n-hexanol followed by reaction with an aqueous reagent prevents this side-reaction from occurring. Propyl aluminium compounds are, however, less reactive than the ethyl compounds.

A 3:7 mixture, by volume, of monoethylene glycol and water was added to a sample of di-n-propylaluminium isopropoxide cooled to - 30°C. Extensive "fissioning" of propyl groups to propylene and hydrogen occurred under these conditions:

$$"Al - CH_2 - CH_2CH_3 + H_2O \rightarrow "AlOH + CH_3 - CH = CH_2 + H_2$$

Next, a 3:7 mixture by volume of monoethylene glycol and 20% aqueous sulphuric acid was added to di-n-propylaluminium isopropoxide. Rather surprisingly, propylene was completely absent in the gas generated in this experiment indicating that "fissioning" did not occur when the acidic aqueous reagent was used.

The acidic-glycol decomposition procedure was used to determine n-propyl and hydride groups in samples in di-n-propylaluminium isopropoxide. The results obtained in determination of propyl and hydride groups, n-propoxide groups and aluminium are shown in Table 7. It is seen that the total determined constituents add up to about 95% and that the empirical formulae is satisfactory.

Analysis of Butylaluminium Compounds. In the alcoholysis-hydrolysis of organoaluminium compounds containing methyl, ethyl and propyl groups, two main types of decomposition procedure have been evolved. In one the sample first reacts with anhydrous n-hexanol and then with aqueous sulphuric acid. The second procedure employs a solution of monoethylene glycol dissolved in aqueous sulphuric acid for decomposition of alkyl and hydride groups.

The results obtained by applying these procedures to a sample of triisobutylaluminium and to a sample of diisobutylaluminium ethoxide are shown in Table 8. In the case of both butyl containing compounds, gas yields about 22% higher are obtained using single phase mixtures of aqueous sulphuric acid -30-50% and monoethylene glycol 70 - 50% instead of the n-hexanol-aqueous sulphuric reagent system. There is a small tendency for isobutyl groups to "fission" to isobutylene and hydrogen when the acidic ethylene glycol reagent is used. "Fissioning" does not occur with this reagent, however, if the sample is cooled to - 65°C and the addition of reagent made slowly. As with the propylaluminium compounds, a small correction is applied to results for the saturation vapour pressure exerted by the aqueous glycol reagent.

The composition of a sample of triisobutylaluminium, based on an analysis made using acidic-ethylene glycol reagent is shown in Table 9. It is seen that the determined constituents add up satisfactorily to near 100% with good reproducibility.

A one-stage procedure for determining alkyl groups up to butyl and hydride groups has been described[19]. A cyclohexane solution of the organoaluminium sample is injected directly on to a small reagent-containing pre-column, usually containing lauric acid on a porous carrier (Sil-0-1,50-80 mesh), which is connected directly prior to the gas chromatographic column (Figure 3). Alkyl and hydride groups are decomposed by lauric acid as follows:

$$AlC_2H_5 + RCOOH \rightarrow AlOOCR + C_2H_6 \qquad AlC_4H_9 + RCOOH \rightarrow AlOOCR + C_4H_{10}$$

$$AlH + RCOOH \rightarrow AlOOCR + H_2$$

Table 7

Analysis of propylaluminium compounds

n-Propyl % wt	Hydride % wt	n-Propoxide % wt	Aluminium % wt	Not determined by difference % wt	Empirical formula	Sum of subscripts in empirical formula	Departure sum of subscripts from 3(i.e. stoichiometric AIR_3), %
di-n-propylaluminium isoproxide							
48.8	0.01	3.01	14.7	6.4[a]	$Al_{1.00}Pr_{2.08}H_{0.01}(OPr)_{0.94}$	3.02	+ 1
tri-n-propylaluminium							
78.7	0.01	0.7	16.7	3.9[b]	$Al_{1.00}Pr_{2.96}H_{0.02}(OPr)_{0.02}$	3.00	0

[a] These samples are known to contain a small amount of a hydrocarbon diluent.
[b] Shown to contain some 2-methylpentene-1 impurity produced during manufacture by dimerization of propylene.

Table 8 - Gas yield obtained by the use of various reagents in the alcoholysis-hydrolysis of butylaluminium compounds

Experiment No.	Alcoholysis-hydrolysis reagent and decomposition procedures used	Gas yields, ml gas evolved at STP/g sample			
		Hydrogen	Isobutane	n-Butane	Total
	Triisobutylaluminium				
1 (0.1 g sample taken)	Added 1 ml n-hexanol to sample at - 60°C, then heated to 50°C, and slowly added 1.5 ml 20% aqueous sulphuric acid. Finally heated to 100°C.	7.5	242	20	269.5
2 (0.2 g sample taken)	Added 1 ml n-hexanol to sample at - 60°C, then heated to 50°C, and slowly added 1.5 ml 20% aqueous sulphuric acid. Finally heated to 100°C.	7.5	238	24	269.5
			Total isobutene plus isobutene		
3	Slowly added 3 ml 1:1 (v/v) 20% aqueous sulphuric acid: monoethylene glycol to sample at - 65°C and then heated to 100°C.	10.3	322		332.3
4	Slowly added 4 ml 3:7 (v/v) 20% aqueous sulphuric acid; monoethylene glycol to sample at - 65°C and then heated to 100°C.	10.2	320		330.2
	Diisobutylaluminium ethoxide				
5	Added 1 ml n-hexanol to sample at - 60°C, then heated to 50°C and slowly added 1.5 ml 20% aqueous sulphuric acid. Finally heated to 100°C.	Nil	178		178
6	Slowly added 4 ml 2:3 (v/v) 20% aqueous sulphuric acid: monoethylene glycol to sample at - 65°C and then heated to 100°C.	Nil	220 220 216		220 220 216

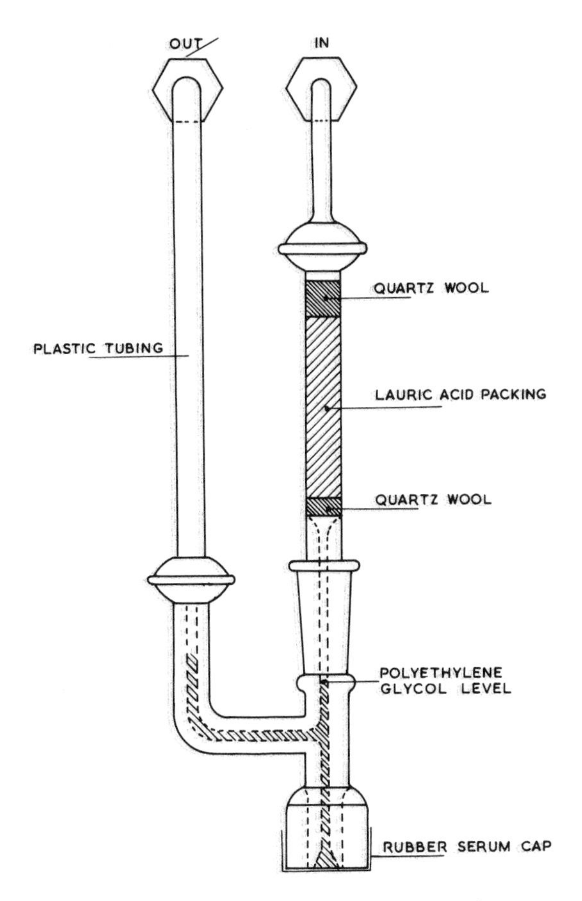

Fig 3. Dijkstra and Dahmen procedure for analysis of organoaluminium compounds. Details of lauric acid decomposition column.

Table 9 - Analysis of triisobutylaluminium

Determined	% wt/wt	
$Al(C_4H_9)_3$	87.6	87.1
$Al(C_4H_9)_2H$	6.6	6.6
$Al(C_4H_9)_2(OC_4H_9)$[a]	6.0	6.0
Total	100.2	99.7

[a] Isobutoxide groups determined by the procedure described in the section on determination of alkoxide groups.

The alkane gases and hydrogen are then swept on to the chromatographic column by the carrier gas and are resolved and determined. To determine the total volume of gas evolved a suitable marker compound (n-pentane) is added to the organoaluminium sample prior to injection into the gas chromatograph.

The Stauffer Chemical Company have developed a procedure for the determination of alkyl groups up to butyl and hydride groups, similar to those described by Dijkstra and Dahmen[19] and by Crompton and Reid[17] based on hydrolysis and gas analysis. In this procedure the sample is exposed to water vapour at sub atmospheric pressures and the generated gases analysed. Philipp et al[20] point out that one of the disadvantages of the Stauffer method[21] ia that elaborate glassware is required and extended periods of time ranging from 45 to 120 min are required to ensure that hydrolysis is completed by water vapour-metal alkyl contact in such a manner as to avoid cracking as referred to previously[17,18].

Philipp et al[21] have developed a simple more rapid hydrolysis method to meet the requirements of a routine quality control check and yet retain the same degree of precision and accuracy of the more time consuming procedures. The method employs a small, crown-capped, thick-walled borosilicate glass bottle, shown in Figure 4. In the pre-evacuated bottle, a measured amount of paraffin oil is first added to serve as a diluent and reaction medium for the metal alkyl sample. The metal alkyl is then injected by a syringe through the rubber liner of the crown cap, thoroughly mixed with the oil, and immediately hydrolyzed to its corresponding alkanes and hydrogen by addition of dilute hydrochloric acid at room temperature according to the following expression:

$$R_yAlX_{3-y} + 3HOH \rightarrow Al(OH)_3 + yRH + (3-y)HX$$

where X = Cl,Br,I,H and y = 1,2,3. With proper technique, the degree of alkene formation with this method is minimized to the same degree as the previous methods.

The major advantages that this new technique provides over the former methods are: The hydrolysis is carried out at room temperature in a simply designed hydrolysis flask; a fourfold less sample size is required; and the time required to complete the hydrolysis is reduced tenfold for the hydrolysis of any aluminium (or zinc) alkyl containing up to five carbons with no loss of accuracy or precision.

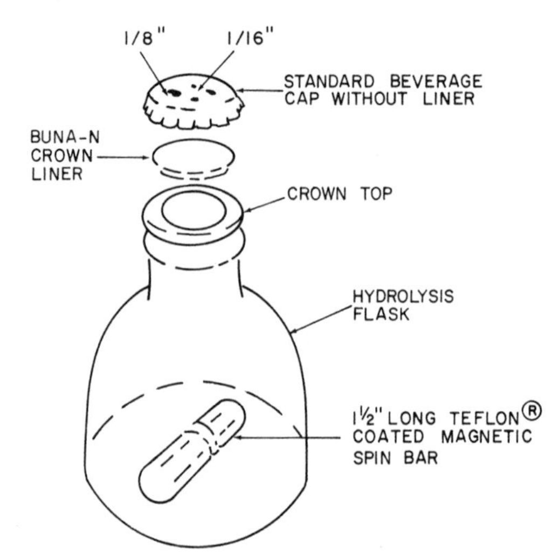

Fig 4. Rapid Philipp method. Hydrolysis bottle.

Cracking can result in varying degrees if the hydrolyzing agent is added to the non-uniformly mixed metal alkyl-paraffin oil blend. In this case, it is believed to be caused by areas of concentrated metal alkyls reacting with the hydrolyzing reagent, resulting in the evolution of excessive localized heat.

$$(C_2H_5)_3Al + 3HOH \rightarrow 3C_2H_4 + 3/2H_2 + Al(OH)_3$$

The resulting hydrolysis gas products from aluminium alkyls were quantified by a gas chromatographic system calibrated in such a manner that a peak height for a component represents the partial pressure of that component. The partial pressure of each component - i.e., the peak height of each component multiplied by its gas calibration factor - equalled the total absolute pressure, which is obtained by a manometer reading at the time of sample injection. The resulting values of each component were obtained in units of absolute mole per cent. For alkylaluminium halides, the hydrolysis products were reported in units of normalized mole per cent. The final calculation involved normalization of the absolute values.

Hydrocarbon hydrolysis products and halogen produced from trialkylaluminiums and dialkylaluminium hydrides are reported in weight per cent values as the corresponding trialkylaluminium and aluminium hydride.

The calculation to accomplish this involves multiplying the absolute mole per cent values of each component by a factor equal to 1/3 the molecular weight of the corresponding trialkylaluminium for each hydrocarbon present and 1/3 the molecular weight of AlH_3 for hydrogen. The values are normalized to 100 wt %.

An estimate of the wt% aluminium in the sample could also be derived directly from the gas analysis. This could be performed by multiplying the wt% of each trialkylaluminium by the theoretical wt% aluminium of that component, and summing over all components. A sample calculation for wt% aluminium in triethylaluminium is shown in Table 10.

Table 10 - Estimation of wt% aluminium in a sample of triethylaluminium

Hydrolysis gas products distribution as	R_2Al	Wt%	x	Atomic wt of Al / Formula wt of R_3Al	=	Wt % Al
Methan	Me_3Al	0.3	x	0.3743	=	0.1
Ethane	Et_3Al	95.5	x	0.2363	=	22.8
Propane	Pr_3Al	0.1	x	0.1727	=	0.01
Isobutane	iBu_3Al	0,9	x	0.1360	=	0.1
n-Butane	$n\text{-}Bu_3Al$	1.8	x	0.1360	=	0.2
Hydrogen	AlH_3	1.4	x	0.8993	=	1.3
		Calcd wt% aluminium			=	24.5
	Found wt% aluminium (by EDTA)				=	24.1

The calculated wt% aluminium determined from the gas distribution could be compared with the experimentally found wt% aluminium of a given product, determined by EDTA titration of an aliquot from a separate hydrolysis solution. This comparison can be used to check the approximate accuracy of the analysis as well as an estimation of non-aluminium containing diluent levels in products, provided there are relatively small amounts of C_6 and above aluminium alkyls present in the sample.

An estimated aluminium content could also be derived for the alkyl aluminium halides if the wt% halide is known. The total estimated wt% aluminium is the weighted combination of the theoretical wt% aluminium derived from the hydrocarbon gas analysis and the theoretical wt% aluminium in the trihalide calculated to be present. The latter can be computed from the found wt% halide values obtained by titrimetric procedures. A sample calculation for wt% aluminium in diethylaluminium chloride as used by Philipp et al[21] is shown below:

It is assumed that for the purposes of the calculation alkylaluminium halides are simple mixtures of R_3Al and AlX_3. The contribution of the aluminium in the R_3Al components can be obtained from the hydrolysis gas analysis by the method shown in Table 10. This value is called the "uncorrected calculated wt% aluminium" and given the symbol(A_1).

The aluminium in the AlX component (A_2) is assumed to be theoretical for AlX_3 (e.g. 20.238 for $AlCl_3$).

Then the "calculated wt% aluminium" (A_T) equals the weighted average of A_1 and A_2.

$$A_T = A_1\ (1 - F) + A_2(F)$$

where F = the wt fraction of AlX_3 in the total sample.

$$F = B\ \frac{(\text{mol wt of } AlX_3)}{3(\text{atomic wt of X})}\ 100$$

For example, a sample of diethylaluminium chloride gave the following results;

A_1 = 23.4 = Uncorrected wt% Al from hydrolysis gas analysis
B = 29.01 = wt% chloride (experimentally determined from triplicate analysis)

$$F = (29.02) \times \frac{(133.346)}{3(35.453) \times 100} = (29.01) \times (0.012537)$$

$= 0.3637.$

Therefore,

$$\begin{aligned} A_r &= 23.4\ (1 - 0.3637) + 20.238\ (0.3637) \\ &= 14.89 + 7.36 \\ &= 22.2 = \text{calculated wt\% aluminium in diethylaluminium chloride.} \end{aligned}$$

The wt% aluminium in this sample of diethylaluminium chloride found by EDTA titration was 22.0.

In Table 11 are compared results obtained for a range of aluminium alkyls using both the Stauffer[20] and the rapid Philipp et al[21] methods. The results for comparison using some of the products are in Tables 11a, b and c. As can be seen, the methods agree well within experimental error. The agreement achieved using the wide variety of products demonstrates that the rapid method is a completely suitable replacement of the Stauffer method, as none of the values fell out of the expected range of values for each component. Philipp et al[21] also compared their method with the alcoholysis-hydrolysis procedure described by Crompton and Reid[17] and Crompton[18]. Again good to excellent agreement is claimed, and in much less time.

Lioznova and Genusov[22] have also described a hydrolysis procedure for the determination of ethyl groups in triethylaluminium. In this procedure a correction is made for the solubility of ethane in the solvent.

Hagen et al[1] have discussed the gasometric determination of alkyl groups in organoaluminium compounds.

Table 11 - Comparison of hydrolysis methods using hydrolysis gas products

a. From the hydrolysis of R_3Al.

	Trimethylaluminium		Triethylaluminium		Triisobutylaluminium	
Component wt%	Stauffer	Rapid	Stauffer	Rapid	Stauffer	Rapid
Methane as Me_3Al	99.9	99.9	0.3	0.3	0.1	0.1
Ethane as Et_3Al	0.0	0.0	93.5	93.5	0.2	0.2
Propane as Pr_3Al	0.0	0.0	0.4	0.4	0.2	0.2
Isobutane as $isoBu_3Al$	0.0	0.0	0.7	0.8	96.0	96.1
n-Butane as $n\text{-}Bu_3Al$	0.0	0.0	4.5	4.4	0.2	0.2
Isobutylene as isobutylene	0.0	0.0	0.0	0.0	2.5[a]	2.4[a]
Hydrogen as AlH_3	0.1	0.1	0.6	0.6	0.8	0.9
Number of runs averaged	2	3	2	6	2	6
Wt% Al, calculated	37.5	37.5	23.5	23.5	13.9	13.9
Wt% Al, found (by EDTA titration)	37.4		23.4		13.8	

[a] Isobutylene actually exists in samples of triisobutylaluminium.

b. From the hydrolysis of R_yAlX_{3-y}

	Diethylaluminium chloride		Diethylaluminium iodide		Ethylaluminium dichloride	
Component, mole %	Stauffer	Rapid	Stauffer	Rapid	Stauffer	Rapid
Methane	0.3	0.3	0.5	0.6	0.0	0.1
Ethane	97.7	97.8	97.1	97.0	99.5	99.4
Propane	0.1	0.1	0.2	0.2	0.0	0.0
Isobutane	0.3	0.2	0.2	0.2	0.4	0.4
n-Butane	1.5	1.5	1.8	1.8	0.1	0.1

Table 11 (cont.)

b. From the hydrolysis of R_yAlX_{3-y}

	Diethylaluminium chloride		Diethylaluminium iodide		Ethylaluminium dichloride	
Component, mole %	Stauffer	Rapid	Stauffer	Rapid	Stauffer	Rapid
Hydrogen	0.1	0.1	0.2	0.2	0.0	0.0
Number of runs averaged	6	6	3	3	3	3
Wt% Al, calculated	22.2	22.2	12.5	12.5	21.3	21.3
Wt% Al, found	22.0		12.5		20.9	

c. From the hydrolysis of miscellaneous metal alkylis

	Diethylzinc		Diisobutylaluminium hydride		Ethylaluminium sesquichloride	
Component mole %	Stauffer	Rapid	Stauffer	Rapid	Stauffer	Rapid
Methane	0.5	0.5	0.4	0.4	0.1	0.1
Ethane	97.1	97.1	0.3	0.3	99.4	99.4
Propane	0.2	0.2	0.3	0.3	0.0	0.0
Isobutane	0.3	0.3	65.9	65.7	0.1	0.1
n-Butane	1.9	1.8	0.2	0.2	0.4	0.4
Hydrogen	0.0	0.1	32.9	33.1	0.0	0.0
Number of runs averaged	2	4	9	24	3	3
Wt% metal, calculated	53.1	53.1	19.0	19.0	21.8	21.8
Wt% metal, found	52.6		18.8		21.6	

An N-methylaniline method for the rapid, accurate determination of aluminium-bound hydride groups in organoaluminium compounds is particularly useful in the case of organoaluminium samples which contain colloidal aluminium that cannot be removed from the sample by filtration or centrifugation[23]. Aluminium metal interferes in the alcoholysis/hydrolysis and the lauric and decomposition methods by reacting with these reagents to produce hydrogen, thereby leading to falsely high aluminium-bound hydride determinations. In this procedure an excess of anhydrous N-methylaniline is added to a known weight of the organoaluminium compound at -40°C, in a nitrogen-filled gasometric apparatus. The temperature of the reaction mixture is then slowly increased to 20°C. The volume of hydrogen evolved under these conditions is equivalent to the dialkylaluminium hydride content of the sample.

$$\mathrm{AlR_2H} + \mathrm{HN}\begin{matrix} \diagup \mathrm{CH_3} \\ \diagdown \mathrm{C_6H_5} \end{matrix} \rightarrow \mathrm{H_2} + \begin{matrix} \mathrm{CH_3} \diagdown \\ \mathrm{C_6H_5} \diagup \end{matrix} \mathrm{N - AlR_2}$$

The method can be used to determine the concentration of hydride groups in all types of dialkylaluminium hydride-containing samples. Dialkylaluminium hydride contents between 1 and 60% may be determined by this procedure. Halogen and alkoxide substituents and higher alkyl groups do not interfere. The accuracy of individual determinations is of the order of ±5% of the determined value.

Table 12 shows results obtained by applying this method to a range of triethylaluminium-diethylaluminium hydride mixtures containing between 10 and 60% of the latter component. These results are in good agreement with dialkylaluminium hydride contents obtained by the alcoholysis-hydrolysis procedure described earlier in this section. The N-methylanlinie method takes less than 1 h, compared with 4 - 5 h for the alcoholysis-hydrolysis procedure.

Table 12 - Comparison of the alcoholysis-alcoholysis procedure and the N-methylaniline procedure for the determination of hydride groups in triethylaluminium

Sample No.	$AlEt_2H$(%wt./wt.) N-methylaniline procedure	Alcoholysis-hydrolysis procedure
1	1.2	1.1, 1.2
2	2.2	1.2, 1.6
3	7.7	8.2, 8.3
4	12.4	12.3
5	13.0	13.0
6	22.6	21.3, 21.3
7	24.3	23.5
8	32.1	28.9, 31.9
9	60.7	57.5

E. Determination of Higher Aluminium-Bound Alkyl Groups

The gasometric alcoholysis-hydrolysis procedure[17] described earlier for the determination of alkyl groups up to butyl is not applicable to the determination of aluminium-bound alkyl groups in the C_5 - C_{10} range which, upon hydrolysis, produce liquid alkanes. Hydride groups in these compounds can, of course, be determined by alcoholysis-hydrolysis or lauric acid decomposition[19] or N-methylaniline procedures.

C_5 - C_{10} alkvl groups can be determined by conversion to the corresponding alkane with a proton-donating reagent[24]. The organoaluminium sample is decomposed smoothly without loss of liquid paraffins at - 60°C under nitrogen, by the addition of glacial acetic acid dissolved in ethylbenzene, (Figure 5).

$$AlR + CH_3COOH \rightarrow AlOOCR + RH$$

The cold ethylbenzene solution is then contacted with aqueous sodium hydroxide to extract aluminium acetate and excess of acetic acid and provide an ethylbenzene solution of the C_5 - C_{10} alkanes, which can be directly

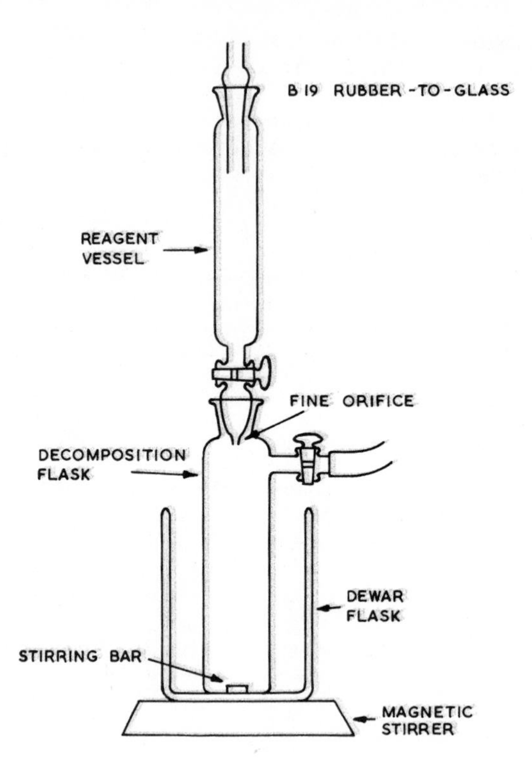

Fig. 5 - Determination of higher alkyl and alkoxide groups in organo-aluminium compounds. Apparatus for decomposition of sample.

injected into a gas chromatographic column.

F. Determination of Aluminium-Bound Alkyl and Alkoxide Groups up to C_{20}

Higher molecular weight trialkylaluminium compounds are of interest commercially. Ziegler has described a process for the manufacture of higher molecular weight organoaluminium compounds based on the reaction of higher olefins with triisobutylaluminium.

$$Al(C_4H_9)_3 + 3C_nH_{2n} = Al(C_nH_{2n+1})_3 + 3C_4H_8.$$

Higher organoaluminium compounds can also be synthesized by the reaction of ethylene with triethylaluminium under appropriate conditions, (i.e. the "Aufbau" route).

$$\text{"}AlC_2H_5 + nC_2H_4 \xrightarrow[100/120°C]{100\ at} \text{"}Al(C_2H_4)nCH_2CH_3$$

Products obtained in these reactions, in addition to higher alkyl groups, might contain the following components: unreacted higher olefins, higher aluminium-bound alkoxide groups (which are produced by the reaction of traces of atmospheric oxygen with aluminium-bound higher alkyl groups) and unreacted aluminium-bound alkyl and alkoxide groups (up to C_4) and hydride groups.

Higher molecular weight aluminium alkoxides are usually made by oxidation of the trialkylaluminium compound using air and/or oxygen. Hydrolysis of these compounds is a useful method of preparing fatty alcohols:

$$2Al(C_nH_{2n+1})_3 + 3O_2 = 2Al(OC_nH_{2n+1})_3.$$

In the method for the determination of aluminium-bound alkyl and alkoxide groups up to C_{20}, these groups are hydrolyzed to produce the corresponding paraffins and alcohols respectively.

$$''AlR + H_2O = ''AlOH + RH$$

$$''AlOR + N_2O = ''AlOH + ROH$$

This is achieved by diluting the organoaluminium sample with petroleum ether and refluxing the solution with an aqueous solution of sulphuric acid and sodium sulphate. The paraffins and/or alcohols obtained as hydrolysis products are recovered for analysis by evaporating the light petroleum extract to dryness. The mixture so obtained usually has a fairly wide carbon number range and it is convenient to carry out a detailed analysis of the individual alcohols and hydrocarbons in these mixtures by gas chromatography and of total alcohols by catalysed acetylation. Figure 6 shows a typical GLC trace obtained in the analysis of a mixture of partially oxidized C_6 to C_{16} aluminium alkyls. In the catalysed acetylation procedure[25] for the determination of higher alcohols in the alcohol-hydrocarbon sample obtained by hydrolysis of the organoaluminium sample, a portion of the extract is reacted with a reagent consisting of acetic anhydride and a perchloric acid acetylation catalyst dissolved in pyridine. Following the reaction period the mixture is diluted with aqueous pyridine and titrated with standard sodium hydroxide solution to the cresol red-thymol blue mixed indicator end-point. Alternatively, the mixture may be titrated potentiometrically using an electrode system. The analytical results are calculated from the measured consumption of acetic anhydride, after correcting for any small amount of 'free organic acidity' in the sample. The accuracy of the procedure is of the order of ±1.5% of the determined result.

G. Determination of Aluminium-Bound Alkoxide Groups up to Butoxy

Trialkylaluminium or dialkylaluminium halide compounds containing alkyl groups up to butyl are often contaminated with small amounts of alkoxide groups produced by oxygen contamination during manufacture or subsequently. Also dialkylaluminium alkoxide procatalysts containing alkyl and alkoxide groups up to C_4 are themselves used as catalysts and may be required to determine the alkoxide content of these procatalysts.

The method devised by Crompton[26] for the determination of alkoxide groups in reactive low molecular weight organoaluminium compounds is described below. Dilute aqueous acids react with aluminium trialkoxides to produce the corresponding alcohols in quantitative yield. However, this method of decomposing the sample cannot satisfactorily be appliwed to lower molecular weight alkylaluminium compounds or to their solutions in hydrocarbons because of the highly exothermic nature of the reactions involved; recoveries of the alcohols are variable even when the reaction mixture is cooled during decomposition. However, even the most reactive organoaluminium compounds can be smoothly decomposed without loss of the alcohol, when a dilute solution of glacial acetic acid dissolved in toluene is slowly added under nitrogen to the diluted sample maintained at a temperature of - 60°C in the apparatus shown in Figure 5. The subsequent addition of an aqueous solution of sodium hydroxide to this mixture dissolves the precipitated aluminium salts and extracts the alcohol quantitatively into the aqueous

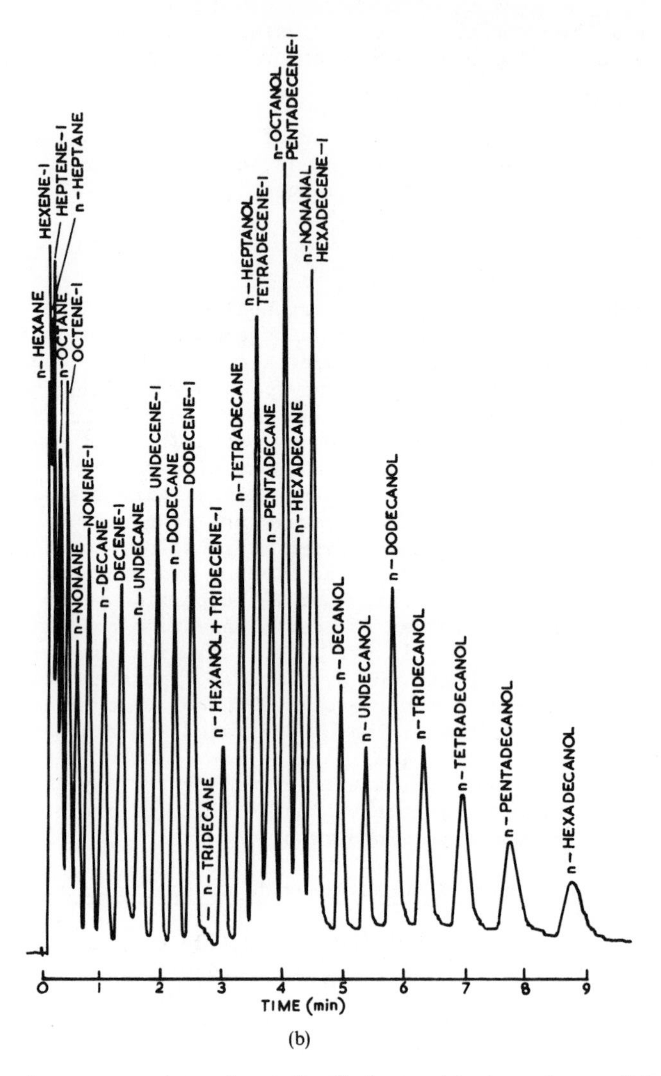

Fig. 6. Gas chromatography using polyethylene adipate column with temperature programming and flame ionization detector, (a) mixture of C_6 to C_{16} alkanes and n-alcohols; (b) mixture of C_6 to C_{16} alkanes, alkenes and n-alcohols.

phase, from which it can be isolated from electrolytes by steam distillation using an apparatus of the type recommended by Vogel[27] and then determined colorimetrically in the distillate by the ceric ammonium nitrate method (Reid and Truelove[28] and Reid and Salmon[29]). The method determines alkoxide groups up to butoxide in reactive organoaluminium compounds containing a high percentage of aluminium bound alkyl groups up to butyl or higher. At the 0.1% alkoxide level accuracy is within ±5% of the actual alkoxide content of the sample. Higher concentrations of alkoxide in the 10-60% range can be determined with an accuracy of ±1%.

In Table 13 are shown the results obtained for full analyses and empirical formulae of several samples of di-n-propyl aluminium isopropoxide.

Table 13 - Analysis and empirical formulae of di-n-propylaluminium isopropoxide

Sample No.	1	2	3
Isopropoxide content found, %	39.7	32.0	31.1
n-Propyl content found, %	45.1	51.9	52.9
Aluminium, content found, %	15.4	15.6	15.4
Sum of contents, %	100.2	99.5	99.4
Empirical formula	$Al_{1.00}Pr_{1.84}(OPr)_{1.19}$	$Al_{1.00}Pr_{2.08}(OPr)_{0.94}$	$Al_{1.00}Pr_{2.14}(OPr)_{0.92}$
Sum of subscripts in empirical formula	3.02	3.02	3.06
Theoretical sum of subscripts	3.00	3.00	3.00
Difference from theoretical, %	+0.6	+0.6	+2.0

A procedure described by Bondarevskaya et al[30] is concerned with the determination of ethoxide groups in triethylaluminium. They do not discuss the application of this method to other types of organoaluminium compounds although it may well have a wider range of application.

In the procedure an ampoule containing about 0.1 g of organoaluminium compound is crushed under a mixture of 10 ml of 5% aqueous potassium dichromate solution and 5 ml 1:1 aqueous sulphuric acid. The mixture is refluxed for 30 min and then cooled. Unconsumed potassium dichromate is then determined by addition of 25 ml 10% iodine solution, followed by titration of excess iodine with 0.1N sodium thiosulphate. The alcohol (i.e. alkoxide) content of the sample can then be calculated from the amount of potassium dichromate consumed during the analysis.

Bondarevskaya et al[31] have also described alternative procedures for the determination of isobutoxide groups in triisobutylaluminium. In one method the sample is hydrolysed under an inert gas to produce isobutyl alcohol which is then reacted with sodium nitrite to produce isobutyl nitrite. This is then diazotized with sulphanilic acid and coupled with 1-naphthylamine which can be estimated spectrophotometrically. Their second procedure for estimating isobutoxide groups utilizes the ceric ammonium nitrate reagent. It is claimed that the error in this method is less than ±3% when determining 0.09 - 1% of isobutoxide groups in triisobutylaluminium.

Belova et al[32] described a procedure for determination of isobutoxide groups in toluene solutions for triisobutylaluminium based on reaction with ethyl alcohol under an inert atmosphere. The ethyl alcohol-isobutyl alcohol mixture thus obtained is centrifuged to remove alumina and analysed by gas chromatography on a column comprising 20% tritolyl phosphate on Celite 545 at 80°C, using helium carrier gas. The error is claimed to be ±10%.

Mitev et al[33] have described a method for the coulometric determination of ethoxydiethylaluminium based on hydrolysis to ethanol, followed by oxidation with excess potassium dichromate in sulphuric acid medium and subsequent coulometric titration of the unconsumed dichromate with ferrous iron electrogenerated from ferrous ammonium sulphate with ferroin as visual end-point indicator. Mitev et al[23] determined ethoxydiethylaluminium in triethylaluminium by this method and obtained a coefficient of variation for 0.04 - 0.16 mg of ethanol in aqueous solution of 2.4% or less.

H. Determination of Aluminium-Bound Amino Groups

Aluminium-bound amino groups can be determined by procedures based on hydrolysis or ethanolysis as described below. Both of these methods are free from interference by aluminium-bound alkyl, alkoxide, hydride, halogen or S-alkyl groups in the sample.

In the hydrolysis method for determining Al-NH_2 groups[34] a suitable weight of the neat or dilute sample is dissolved in isooctane under nitrogen and cooled to 50°C and then decomposed with dilute hydrochloric acid:

$$R_2AlNH_2 + 4HCl(aq) \rightarrow NH_4Cl + 2RH + AlCl_3$$

The ammonium chloride is extracted with water from the isooctane phase. Steam distillation of this extract in the presence of excess of sodium hydroxide provides an amount of ammonia proportional in amount to the $AlNH_2$ content of the original organoaluminium sample. Higher concentrations of ammonia in the steam distillate are determined by a titrimetric procedure. Traces of ammonia are determined by the colorimetric indophenol blue method. This procedure is more suitable for higher molecular weight organoaluminium compounds which do not react too vigorously with water. An alternative method[35] for determining aluminium-bound amino groups utilizes ethanol as the reagent:

$$R_2AlNH_2 + C_2H_5OH \rightarrow R_2AlOC_2H_5 + NH_3$$

This procedure is suitable for the analysis of more reactive types of organoaluminium compounds. The ammonia produced is determined by conventional spectrophotometric or titrimetric procedures. The apparatus used to carry out these determinations is shown in Figure 7. The weighed sample in the reaction tube under dry oxygen free nitrogen is reacted with 4 N hydrochloric acid or 10:1 ethanol benzene contained in the separatory funnel. The evolved ammonia is collected in the conical flask and determined.

I. Determination of Aluminium-Bound Thioalkoxide Groups

SR groups containing alkyl groups up to $C_{20}H_{41}$ can be determined in amounts down to 1% in the sample by a procedure[36] involving hydrolytic decomposition of the sample at - 60°C followed by argentimetric determination of the thiol produced.

$$R_2AlSR + 3HNO_{3aq} \rightarrow RSH + 2RH + Al(NO_3)_3$$

Especially in the case of lower molecular weight mercaptans (e.g. ethyl mercaptan), there exists a great danger of losses of these substances from the decomposition flask during sample decomposition: Low S-alkyl analysis will then be obtained. The occurrence of losses of mercaptan will certainly be evident due to their highly characteristic odour.

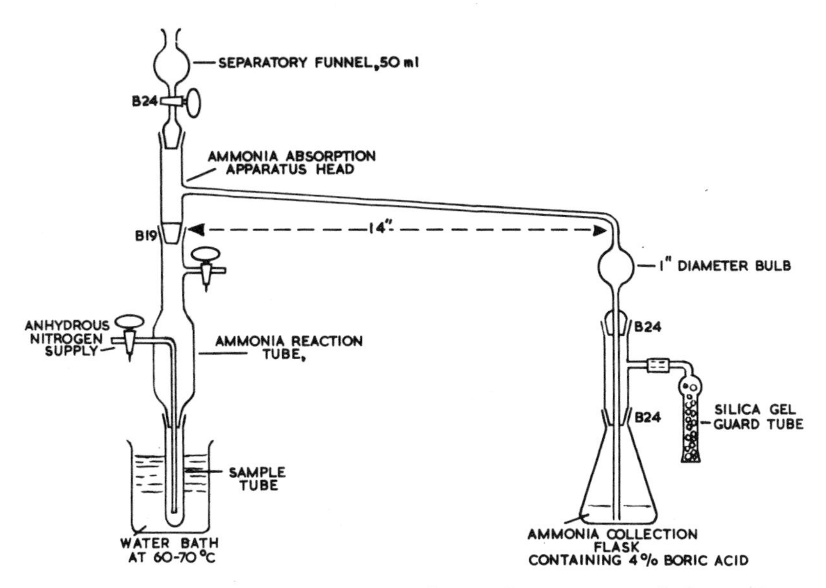

Fig. 7. Ammonia method. Alumnium amide decomposition tube.

To overcome the danger of losses of the more volatile alkyl mercaptans it is necessary to carry out the hydrolysis at the temperature of an acetone-cardice mixture, i.e. at about -60 to -70°C. Negligible losses of mercaptan occur at this temperature. Any losses of mercaptan will be indicated by the formation of a white precipitate of silver mercaptide in the silver nitrate containing bubbler attached to the decomposition apparatus.

ORGANOANTIMONY COMPOUNDS

A. Determination of Antimony

Several methods for the determination of antimony in organoantimony compounds have been described[37-41]. In one the organoantimony compound is burnt with metallic magnesium to convert antimony to magnesium antimonide [37,38]. This is then decomposed with dilute sulphuric acid to produce stibine, which is absorbed in 6 N hydrochloric acid containing sodium nitrite. The resulting hexachloroantimonic acid can be determined colorimetrically after having been converted to a blue coloured compound by reaction with methyl violet. The blue colour is extracted with toluene for spectophotometric evaluation. Nitrogen, phosphorus, and arsenic do not interfere. An alternative procedure uses Rhodamine B (C.I. Basic Violet 10)[39]. The decomposition of the sample is best effected by the use of concentrated sulphuric acid and potassium sulphate, followed by the addition of 30% hydrogen peroxide. The solution is adjusted to 6 N with respect to hydrochloric acid and antimony is extracted with diisopropyl ether. Colour is developed by the addition to the ether phase of 0.02% Rhodamine B in 1 N hydrochloric acid. Compounds containing trivalent antimony which are unstable in oxygen or moist air may be stabilized by exposure to sulphur for 8-48 h in vacuo and the resulting compounds are examined by pyrolysis[41]. Marr and Sithole[42] have described a procedure for the microdetermination of antimony in organoantimony compounds by atomic absorption spectrophotometry.

ORGANOARSENIC COMPOUNDS

A. Determination of Arsenic

Various techniques have been applied to the determination of arsenic in organoarsenic compounds. These include oxygen flask combustion, digestion with mineral acids, and fusion with magnesium.

Oxygen Flask Combustion. Corner[43] was the first to recommend the determination of arsenic in organic materials by oxygen flask combustion. Arsenic attacks the platinium sample holder, hence many workers recommend a silica spiral. Use of a quartz spiral was not satisfactory because the quartz devitrified during even a single combustion. Spectrographic analysis has shown that arsenic forms alloys with platinium even under the strongly oxidizing conditions obtained when paper impregnated with potassium nitrate is burned (Belcher et al[44]). The following is a fairly conventional procedure for the determination of arsenic. After combustion the arsenite and arsenate formed are absorbed in a suitable solution; commonly a solution of sodium hydroxide is used. Following conversion to arsenic trichloride, this is distilled into sodium bicarbonate solution and is estimated iodometrically. In the procedure described by Merz[45], discussed below, it is pointed out that "phosphorous-resistant" platinum can be used as the sample holder for certain materials, but in general a silica spiral is recommended. A modified silica holder has been described (Eder[46]) which, it is claimed, is less prone to dropping the sample into the absorption solution. When a silica spiral is used combustion is never as satisfactory as with platinum. For example, Forner[43] was unable to achieve proper and immediate combustion of resinous materials. Some workers (Belcher et al[44]; Tuckerman et al[47]) are of the opinion that wet combustion methods are preferable to the oxygen flask method for the determination of organic arsenic.

The oxygen flask method described by Merz[45] mentioned above for the determination of arsenic involves burning the sample in an oxygen-filled flask, and the combustion products are absorbed in dilute iodine solution in which trivalent arsenic is quantitatively oxidized to pentavalent arsenic. The arsenic is determined by the molybdoarsenate blue reaction, with hydrazine sulphate as reductant; the excess of iodine from the absorber does not interfere. Concentrations of arsenic greater than 10μg are measured on a filter photometer and those of less than 10μg with a spectrophotometer. The mean error for a single determination is ±0.2%.

Pushchel and Stefanac[48] use alkaline hydrogen peroxide instead of iodine solution in the oxygen flask method to oxidize arsenic to arsenate. The arsenate is titrated directly with standard lead nitrate solution with 4-(2-pyridylazo) resorcinol or 8-hydroxy-7-(4-sulpho-1-naphthylazo) quinoline-5-sulphonic acid as indicator. Phosphorus interferes in this method. The precision at the 99% confidence limit is within ±0.67% for a 3 mg sample. In a further finish to the oxygen flask method Stefanac[59] used sodium acetate as the absorbing liquid. The arsenite and arsenate so produced are precipitated with silver nitrate solution. The precipitate is dissolved in potassium nickel cyanide ($K_2Ni(CN)_4$) solution and the displaced nickel is titrated with EDTA solution, with murexide as indicator. The average error is within ±0.19% for a 3 mg sample, Halogens and phosphate interfere in the procedure.

Belcher et al[50] described a method for arsenic in which the sample (30 - 60 mg) is burnt by a modified oxygen flask procedure and arsenic is determined by precipitation as quinoline molybdoarsenate which is then reduced with hydrazine sulphate and determined spectrophotometrically at 840 nm. The absolute accuracy is within ±1% for arsenic. Phosphorus interferes in this procedure.

Wilson and Lewis[51] developed a method for arsenic, which includes precipitation as ammonium uranyl arsenate and subsequent ignition under controlled conditions to triuranium octaoxide. Phosphate and vanadate form similar insoluble ammonium uranyl salts and the method is not applicable to organic compounds containing these elements without prior separation.

Various sample supports have been tried in the hope that they might prove more satisfactory than platinum [51]. For this purpose the stopper of the oxygen flask was fitted with a glass hook, so that spirals and gauzes of various materials could be hung from it. Oxidized copper spirals were first used in the hope that the oxide film would assist combustion and prevent reduction to arsenic but, despite excellent combustions, poor results were obtained, again, presumably, because of alloy formation. The use of steel gauzes and aluminium spirals gave much improved results; there were no signs of attack on these materials (although the aluminium tended to melt) and combustion was adequate. Steel spirals gave good combustions and aluminium less so; with aluminium it was necessary to use tissue-paper rather than filter-paper and to get the combustion going well initially. After dissolution of the white sublimate of arsenic oxides, which were formed on the sides of the flask, and subsequent acidification and oxidation to arsenate with bromine water, the arsenic was determined as previously described. The results of these preliminary investigations are given in Tables 14 and 15. Further results obtained with steel and aluminium supports gave satisfactory recoveries of arsenic from o-arsanilic acid, arsenious oxide, and acetarsol.

Digestion Procedures. Organic arsenicals may be analysed by heating them in a Kjeldahl flask with a mixture of concentrated nitric and sulphuric acids and hydrogen peroxide[52]. After cooling, the mixture is diluted with water and an aliquot is treated with concentrated sulphuric acid and zinc. The arsine evolved passes into a solution in pyridine of silver diethyldiocarbamate, and the molar absorptivity of the resulting solution is measured at 540 nm. Halogens and sulphur do not interfere. Arsenic in 10-mg amounts of organic compounds may be determined iodimetrically[53]. The method is suitable for the determination of all types of organic arsenic compounds, including those which give low results by the classical wet oxidation methods using sulphuric and nitric acids. The sample is weighed out into a Kjeldahl flask, concentrated sulphuric acid is added, and the solution set aside for 5 min, warming it if necessary to dissolve any solids. Small amounts of potassium permanganate are added, warming gently after each addition until the dirty green of the solution disappears and the precipitated manganese no longer dissolves. Water and 30% hydrogen peroxide are added and the solution is heated until a clear solution is obtained. Arsenic is then determined iodimetrically.

Pietsch[54] described a digestion procedure for arsenic which is not interfered with by the following elements - magnesium, calcium, strontium, barium, cobalt, nickel, zinc, manganese, cadmium, copper, and halogens. In this procedure the sample (50 - 200 mg) is transferred to a Kjeldahl flask containing concentrated sulphuric acid (1 ml)copper sulphate (1 crystal) and concentrated nitric acid (1 ml). The solution is heated to fuming and then cooled prior to the addition of further 1 ml portions of nitric acid

Table 14 - Recovery of arsenic after oxygen flask combustion with various supporting materials. o-Arsanilic acid was the organic compound used in these recovery experiments

Support Material	Arsenic added (mg)	Arsenic recovered (mg)	Observations
Platinum gauze	11.16	7.52	Platinum pitted and swollen. Good combustions
Glass spiral	11.02	9.98)	Poor combustion, smoke and
	14.65	13.26)	carbon deposits
Copper spiral	17.32	13.75)	
Copper gauze	14.03	7.12)	Good combustions. Copper
	14.01	6.91)	melts
Aluminium spiral	14.29	14.28)	Moderate combustions
	14.11	13.77)	Aluminium melts
Stainless-steel gauze	16.02	15.90)	
	13.92	13.53)	Good combustions
	14.04	13.74)	

Table 15 - Recovery or arsenic from various compounds when steel or aluminium supports were used

Support material	Arsenical oxide	Arsenic calculated (mg)	Arsenic found (mg)
Aluminium spiral	Arsenious oxide	(16.04	16.08
		(15.50	15.60
Stainless-steel gauze	Arsenious oxide	15.76	15.88
Mild-steel gauze	Arsenious oxide	15.71	15.74
		(14.29	14.28
		(14.11	13.77
Aluminium spiral	O-Arsanilic acid	(14.98	14.10
		(13.87	13.70
		(16.02	15.90
Stainless-steel gauze	o-Arsanilic acid	(13.92	13.53
		(14.04	13.74
		(14.49	14.49
Mild steel gauze	o-Arsanilic acid	(14.13	13.85
		(15.63	15.23
		(11.29	11.28
Aluminium spiral	Acetarsol	(12.06	12.04
		(11.48	11.04
Mild steel gauze	Acetarsol	(11.20	11.18
		(11.18	10.81

and then again heated to fuming until colourless or pale blue. It is then transferred to a beaker, diluted to 300 ml, heated to boiling point, and a solution of 7 g barium nitrate in 50 ml water added. The solution is then cooled, filtered, and the filtrate neutralized to methyl orange with sodium hydroxide solution. Aqueous 25% nitric acid (1 drop) is added and excess 0.1 N silver nitrate solution added followed by dropwise addition of a concentrated solution of sodium acetate until precipitation is complete. The solution is then made up to volume, filtered, and the excess silver nitrate back-titrated with 0.1 N ammonium thiocyanate solution.

DiPietro and Sassaman[55] modified the micro Carius procedure for arsenic originally described by Steyermark[56] in which the sample is digested with fuming nitric acid to form quinquevalent arsenic. The arsenic is then determined by iodine titration. DiPietro and Sassaman[55] evaluated this procedure by checking it against several organic arsenic compounds whose purity had been checked by determination of carbon and hydrogen. None of these compounds yielded correct values for arsenic using the Steyermark procedure[56] - Table 16. Because incomplete destruction of the arsenoorganic compound was suspected, the digestion temperature and time were increased to 300°C for 10 hours. Following this relatively more severe treatment, all solutions remained clear upon the addition of potassium iodide and the end-points became sharp. Results obtained by the modified procedure (DiPietro and Sassaman[55]) are shown in Table 16.

Table 16 - Comparison of percentage arsenic found by micro Carius method at different temperatures, with and without potassium chloride

Sample	Digestion temp: Digestion time: Theory	I 250°C, 8 h, K Cl absent	II 300°C 10 h, K Cl present
As_2O_3	75.73	71.99 72.01	76.04 75.66 75.59 75.66
$C_6H_5CH_2AsO_3H_2$	34.67	31.12 31.63	34.50 34.49
$C_6H_5AsO_3H_2$	37.08	36.62 34.06 35.26 35.21	36.91 36.88 36.97
$HOC_6H_4AsO(OH)_2$	34.36	30.76 31.70	34.46 34.42
$(C_6H_5)_3As$	24.46	turbid	24.40 24.46 24.51 24.49
$((C_6H_5)_3AsCH_3)I$	16.72	turbid	16.83 16.72

Tuckerman et al[57] state that because chloric acid oxidizes organic matter more smoothly and rapidly at 180°C, it is to be preferred to the more widely used sulphuric acid or sulphuric-nitric acid digestions or alkaline fusions which have been recommended by various workers for the determination of arsenic in organic compounds. Excess chloric acid is easily removed by boiling to leave a perchloric acid solution of inorganic arsenic present in a higher valency state. Rapid micro and semi-micro methods for the determination of arsenic based on chloric acid digestion are described.

Samples of p-arsanilic acid ranging from 1.6 - 4.0 mg in weight were run by the micro-method. The absorbance per mg of arsenic in 10 ml of final dilution were: 15.8, 15.1, 15.4, 15.0, 15.4, 15.3, 15.2: average: 15.3. The 99% confidence interval is ±0.4 or 26 parts/10^3.

Using the average absorbance per mg of arsenic found for p-arsanilic acid, samples of 4- (2-hydroxyethylureido) -phenylarsonic acid were found to contain the theoretical 24.6% of arsenic by the micro method. Results are shown in Table 17. Other compounds not containing their theoretical amount of arsenic are also shown in Table 17.

Phosphorus interferes in the micro method by formation of a heteropoly blue similar to that formed by pentavalent arsenic. Phosphorus and arsenic, if present together, may be separated and determined, after digestion of the sample with chloric acid, by the method given by Je n[58].

Magnesium Fusion. The sample is heated in a sealed tube for 5 min with a mixture (3 + 1) of magnesium and magnesium oxide, which converts all the arsenic into magnesium arsenide. The tube is opened and the contents are decomposed by dilute sulphuric acid. Arsine is evolved and is absorbed in a 0.5% solution of silver diethyldithiocarbamate in pyridine[59,60]. The colour produced has an absorption maximum at 560 nm and is proportional to concentration up to 20 μg of arsenic in 3 ml of solution. Alternatively, the arsine is oxidized by bromine and determined iodimetrically[61]. Methylated arsenicals have been determined by vapour generation atomic-absorption spectrometry[62].

Table 17 - Assays by micro method

Substance	Sample weight (mg)	Absorbance	% as found	% error
Glycobiarsol, U.S.P.(14.78%As) (Milibis^R)	5.605	0.253	14.7	-0.6
o-Arsanilic acid(33.7%As)	4.860	0.495	33.3	-1.2
	4.205	0.423	32.8	-2.7
	2.224	0.235	34.5	+2.4
4-(2-Hydroxyethylureido)-phenylarsonic acid (24.6%As)	2.897	0.221	25.9	+5.3
	2.395	0.186	25.4	+3.3
	2.346	0.180	25.0	+1.6
	2.446	0.186	24.8	+0.8
	2.421	0.186	25.1	+2.0

Atomic Absorption Spectrophotometry. Fleming and Taylor[63] described a method for the determination of total arsenic in organoarsenic compounds by arsine generation and atomic absorption spectrophotometry using a flame heated silica furnace. Denyszyn et al[64] collected arsine at the 2 μg/ m^3 level produced from organoarsene compounds, on charcoal, then desorbed the arsine in acid and analysed it by electrothermal atomic absorption spectrophotometry using a nickel pretreatment. Mean percentage recovery and standard deviation were respectively, 89.1% and ±0.10 . Havard and Arhab-Zavar[65] determined inorganic, arsenic III, arsenic V, methylarsenic and dimethylarsenic species by selective hydride evolution atomic absorption spectrometry.

B. Determination of Carbon

The determination of carbon and arsenic in organic compounds using magnesium fusion and the elemental analysis of organoarsines and organobromarsines have been reviewed[66,67].

C. Determination of Fluorine

A volumetric microdetermination based on oxygen flask combustion for 5 - 40% of fluorine in organic compounds containing arsenic has been described[68]. The sample (containing 0.2 - 0.6 mg of fluorine) is burnt in an oxygen combustion flask and the combustion products are absorbed in 5 ml of water. Hexamine and sufficient murexide-naphthol green B (C.I. Acid Green 1)-hexamine (1:3:100) are added and the flask contents are titrated with 0.005 M cerium (IV) sulphate to a green end-point. Methods have been reported for the determination of arsenic in organoarsenic compounds with halogen attached to the arsenic, organoarsenic compounds with halogen attached to the carbon, and in organoarsenic compounds with halogen in the anion[69].

D. Determination of Sulphur

Compounds containing arsenic which are unstable in oxygen or moist air can be stabilized by exposure to sulphur for 8 - 48 h in vacuo. The resulting compounds may be analysed by pyrolysis for sulphur and arsenic[70].

ORGANOBERYLLIUM COMPOUNDS

A. Determination of Carbon and Hydrogen

Carbon and hydrogen in dimethylberyllium and in beryllium hydride can be determined by combustion[71]. After weighing into a small tin capsule, the sample is ignited in a tube furnace at 600°C under a stream of pure helium to destroy any stable carbonates and then under a stream of pure oxygen at 1050°C. The resulting carbon-dioxide and water are weighed off.

Because of the highly reactive nature of dimethylberyllium, and beryllium hydride, a method is needed whereby the material can be easily weighed and placed into the combustion system. This is accomplished by using a small tin capsule which can be introduced with the sample and completely oxidized in the course of the combustion. These capsules are made from pieces of tin 4 cm square; about 3 mm thick, weighing approx. 0.6 g. After a sheet has been rolled on a mandril, the side and one end are sealed by crimping. The open capsule is placed in a weighing bottle, taken into a drybox, filled with the inert gas of the box, taken out, and weighed using another bottle as a tare. The capsule and bottle are taken back into the drybox; the capsule is loaded, sealed by crimping, taken out of the drybox, and reweighed. Samples of 50 mg are convenient. The relatively large quantity of ether liberated from materials such as beryllium hydride etherate is burned completely upon its passage through the combustion furnace. This was tested by burning a sample of purified ether in the apparatus. The ether was introduced from a tube equipped with a breakoff seal, which was broken after the tube had been sealed into the system. The ether was then vaporized and carried into the combustion region by means of helium passing over the opening of the tube. By this method, a sample of ether weighing 63.3 mg gave 99.1% of the expected amount of carbon dioxide and 99.3% of the expected amount of water.

Table 18 shows some results obtained in an analysis of dimethylberyllium by combustion and by a procedure involving hydrolysis of the sample and subsequent mass spectrometric measurement of the quantity of methane produced.

Table 18 - Determination of carbon and hydrogen in dimethylberyllium

	Sample weight (mg)	%C	%H
Combustion	46.3	61.04	15.27
Hydrolysis/mass spectroscopy	77.8	60.19	15.05

ORGANOBISMUTH COMPOUNDS

A. Determination of Bismuth

Bismuth in pharmaceuticals such as bismuth tribromophenoxide, bismuth salicylate, and bismuth subgallate has been determined complexiometrically [72]. The sample is shaken with dilute nitric acid for 5 min and then diethyl ether added with gentle swirling followed by water, and the solution is heated to 60°C. The solution is then cooled to 50°C, ignoring the precipitate and a 1% titration of methyl thymol blue in potassium nitrate is added and the solution titrated immediately with 0.1 M EDTA disodium salt to a pale red-violet colour. Then 10% aqueous ammonia is added and this solution titrated further until it becomes yellow.

ORGANOBORON COMPOUNDS

A. Determination of Boron, Carbon and Hydrogen

Rittner and Culmo[73] have described a Pregl type combustion procedure for the micro-determination of carbon, hydrogen and boron in organoboranes. In this procedure samples containing 1 to 4 mg of boron are combusted by the standard Pregl-type procedure, using the Brinkman-Heraeus microcombustion assembly. Carbon and hydrogen values are obtained in the normal fashion. The residue which remains behind in the boat is transferred to a beaker, dissolved in water, and the boric acid is titrated by the identical pH method using 7.10 as the critical pH.

Typical results obtained by this procedure for a range of organo-boron compounds, some of which contain phosphorus are tabulated in Table 19. The effects of adding oxidizing agents and of changes in heating temperature in the microdetermination by furnace methods of carbon and hydrogen in boroxins and other boron-containing compounds have been studied [74]. The best results were obtained with an oxygen flow-rate of 10 ml/min, a stationary furnace temperature of 900°C, and with the addition of about 100 mg of tungstic oxide to the sample, which was then heated to a final temperature of 1000°C. With these conditions, errors in the determination of carbon and hydrogen were reduced to ±0.28% and ±0.18% respectively, for the compounds studied. When organoboron compounds are oxidized by conventional methods low results are obtained for carbon and low or high results are obtained for nitrogen[75]. To reduce errors in the determination of organoboron compounds, oxidation conditions which are satisfactory for oxidizing methane should be used.

Boron has been determined by pyrolysing the organoboron compound in a stream of pure hydrogen containing methanol vapour, and burning the gaseous mixture at the end of a quartz tube (acting as a blowpipe) in a stream of oxygen[76]. The water-vapour formed contains the bulk of the boron and, after condensation, is collected in a flask containing a known volume of water that will absorb the boron. The Carius nitric acid oxidation procedure for the micro-determination of boron in organoboranes has been studied (Tables 20 and 21).[77] The standard deviation is ±0.33% absolute. A majority of the analytical results were slightly above the theoretical values, indicating a possible bias in the method. This may be due to leaching of boron from the borosilicate glass Carius tubes during the oxidation.

Dunstan and Griffiths[78] describe in detail a method for the determination of boron and carbon in alkyldecaboranes and related compounds, based on oxidation by alkaline potassium persulphate, followed by potentio-metric titration of boric acid in the presence of mannitol. Carbon contents are determined by a modification of the Van Slyke wet combustion technique. (Figure 8). Although these methods are limited to compounds which are nonvolatile at room temperature, they are rapid and reliable and do not require elaborate apparatus, and is usually accurate within 1.0 and 0.5% absolute for boron and carbon contents, respectively. The methods are readily applicable to the analysis of organoboron compounds containing nitrogen. The chief limitation of the potassium persulphate oxidation method is that it cannot be used to analyse steam-volatile compounds. Thus, whereas decaborane and its alkyl derivatives dissolve rapidly during the oxidation, more highly alkylated derivatives of pentaborane, for example, tetraethylpentaborane, are streamvolatile and remain unchanged. None of the compounds examined resisted oxidation by Van Slyke combustion fluid, but the method is limited to the analysis of compounds with a low vapour pressure at room temperature. A more complicated apparatus would be required to prevent loss of volatile material during preliminary evacuation of the system. Methods for determining boron in organoboron compounds based on oxidation with alkaline potassium persulphate have been compared with methods based on oxidation using (a) sodium peroxide, (b) alkaline hydrogen peroxide, and (c) trifluoroacetatic acid[79,80]. Oxidation with sodium peroxide followed by titration with sodium hydroxide using mannitol as indicator gave an error not exceeding 1%. Oxidation with sodium peroxide or with alkaline hydrogen peroxide, followed by titration with barium hydroxide, gave errors as low as 0.1%. Analyses obtained by this procedure for decaborane $B_{10}H_{14}$ and a number of its alkyl derivatives are presented in Table 22.

Table 19 - Determination of Carbon, Hydrogen and Boron

Compound	Calculated	Found	Mean value and precision	Calculated	Found	Mean value and precision	Calculated	Found	Mean value and precision
Diphenylphosphinedecaborane, $C_{12}H_{23}B_{10}P$	47.02	46.99 46.92 47.00	46.97 ± 0.02	7.56	7.40 7.70 7.45	7.52 ± 0.07	35.30	35.38 35.32 35.38	35.36 ± 0.02
Bis(aminodiphenylphosphine)decaborane, $C_{24}H_{33}B_{10}N_2P_2$	55.14	55.11 55.30 55.14	55.18 ± 0.04	6.94	7.00 7.05 7.06	7.04 ± 0.01	20.70	20.84 20.83 20.60	20.76 ± 0.06
Bis(ethoxydiphenylphosphine)decaborane $C_{28}H_{42}B_{10}O_2P_2$	57.90	58.10 57.82 57.90	57.94 ± 0.06	7.29	7.40 7.24 7.40	7.35 ± 0.04	18.63	18.88 18.56 18.68	18.71 ± 0.07
Bis(methylaminodiphenylphosphine)-decaborane, $C_{26}H_{40}B_{10}N_2P_2$	56.70	56.90 56.58 56.82	56.77 ± 0.07	7.32	7.40 7.45 7.30	7.38 ± 0.03	19.65	19.40 19.71 19.51	19.54 ± 0.06
Bis(hydroxydinphenylphosphine)decaborane, $C_{24}H_{34}B_{10}O_2P_2$	54.93	54.77 54.60	54.69 ± 0.06	6.53	6.67 6.70	6.68 ± 0.01	20.63	20.98 20.72	20.85 ± 0.09
Bis(hydroxydiphenylphosphine)decaborane, $C_{23}H_{48}B_{10}N_2O_2P_2$,bis dimethylami e salt	54.69	55.08 55.14	55.11 ± 0.02	7.87	8.26 8.37	8.32 ± 0.04	17.60	17.98 17.66	17.82 ± 0.11
Bis(triethylammonium)perhydrodecaborane, $C_{12}H_{42}B_{10}N_2$	44.90	44.70 44.89 44.78 44.82	44.80 ± 0.03	13.08	13.20 13.10 13.15 13.10	13.14 ± 0.02	35.50	33.57 33.59 33.35 33.61	33.53 ± 0.04
Trimethylamine-dimethylphenylphosphine decaborane, $C_{11}H_{32}B_{10}NP$	41.60	41.65 41.81 41.72	41.73 ± 0.03	10.16	10.23 10.30 10.01	10.18 ± 0.06	34.07	34.10 34.25 33.94	34.10 ± 0.06
Trimethylamineborane, $C_3H_{12}BN$	49.45	49.30 49.35	49.33 ± 0.02	16.48	16.18 ±6.20	16.19 ± 0.01	14.84	14.95 14.72	14.84 ± 0.08

Table 20 - Microdetermination of boron by the Carius method

Compound	Boron, % Calcd.	Boron, % Found	Detns.
Dimethylamineborane, $(CH_3)_2NHBH_3$	18.36	18.66 ± 0.09	4
Sodium tetraborate, $Na_2B_4O_7$	21.50	21.69 ± 0.13	2
Decaborane, $B_{10}H_{14}$	88.46	89.49 ± 0.23	2
Isopropylamineborane, $C_3H_7NH_2BH_3$	14.83	15.24 ± 0.04	2
Trimethylamineborane, $(CH_3)_3NBH_3$	14.83	15.25 ± 0.02	2
Morpholineborane, $C_4H_9NOBH_3$	10.72	10.76 ± 0.02	2
Chlorodecaborane, $B_{10}H_{13}Cl$	69.02	69.74 ± 0.33	3
bis-Acetonitriledecaborane, $B_{10}H_{12}$-$(CH_3CN)_2$	53.46	53.17 ± 0.02	2

Table 21 - Analysis of phosphorus and fluorine compounds. (Carius tubes appeared to be etched after oxidation of samples containing fluorine).

Compound	Boron, % Calcd.	Boron, % Found
$(C_4H_9)_3PBH_3$	5.01	3.26
		1.88
$(C_4H_9)_3PB(C_2H_5)_3$	3.60	3.96
		4.46
$(C_6H_5)_3PBH_3$	3.92	2.87
$B_{10}H_{14}$(NaF added)	88.46	147.98
		120.08
$(CH_3)_3NBF_3$	8.52	10.16
		12.01

Strahm and Hawthorne[81] have studied the determination of boron in borohydrides and in organoboron compounds by oxidation with trifluoroperoxy acetic acid. The sample (containing 2 to 6 mg boron) is placed in a test-tube, methyl cyanide (1 ml) is added, a reflux-condenser is fitted, and the tube is cooled in an ice bath for a few minutes. Trifluoroperoxyacetic acid (2 ml) is cautiously added through the top of the condenser and, after any initial reaction has subsided, the tube is heated in a boiling-water bath for 5 - 60 min, according to the stability of the sample. For samples requiring long heating a further 1 ml of trifluoroperoxyacetic acid is added after 30 min. When decomposition is complete the solution is washed into 50 ml water and boiled gently for a few min, before titrating the boric acid in the usual manner, after its conversion into mannitoboric acid. The reagent is prepared by adding hydrogen peroxide (90%) (1 ml) to methyl cyanide (9 ml), cooling in ice and cautiously adding, while swirling, trifluoroacetic anhydride (6.2 ml).

Pierson[80] has examined six simple oxidation procedures for the determination of 1 to 90% boron in compounds also containing carbon, hydrogen, nitrogen, oxygen, bromine and phosphorus. In all cases the analysis is completed by titration of the solution with 0.05 - 0.1 N sodium hydroxide

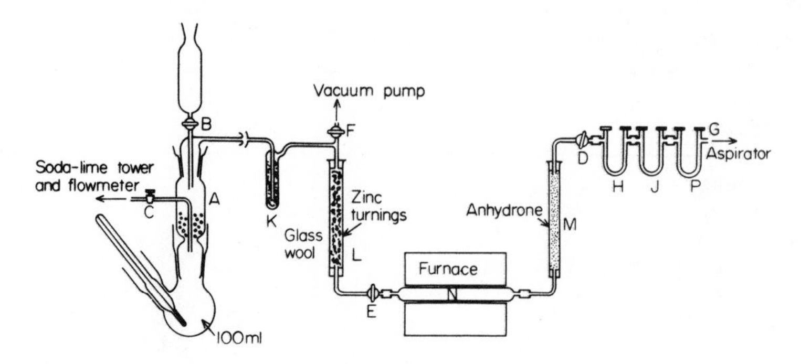

Fig. 8. Wet combustion apparatus.
K. Trap, 150 x 20 mm.
L.M. Tubes, 300 x 25 mm.
N. Silica tube, filled with copper oxide.
H,J. Weighed U-tubes. filled with soda-lime and Anhydrone, respectively.
P. Guard tube, filled with Anhydrone.

in the presence of a large excess of mannitol. Sharp beginning- and end-points are achieved by controlling the type and degree of initial oxidation according to the specific compound or group of compounds of similar structure. For the 27 different compounds analysed the average relative standard deviation was 0.29% with a range from 0.06% to 0.57% boron.

Arthur and Donahoo[82] have developed a procedure for the combustion of micro samples of organoboron compounds. The boric oxide formed by the combustion is quantitatively removed by refluxing water in the combustion tube, and the resulting boric acid solution is titrated with standard sodium hydroxide with mannitol as a complexing agent with an initial pH of 8.4. Best results are obtained when the boron content of the solution is between 0.8 and 1.5 mg.

Kuck and Grim[83] have described a semi-micro procedure for the determination of boron in organic compounds. In this procedure, the sample (0.1 - 0.2 g) is covered with 10 -15 drops of liquid paraffin and burnt in a Parr high-pressure oxygen bomb containing 25 ml water, and the solution is made up to 100 ml with water. An aliquot (10 ml) is brought to pH 4.8 with 0.01 N hydrochloric acid and heated under reflux for 10 min, to remove carbon dioxide. While the solution is cooling a stream of pre-purified nitrogen is passed through by means of a filter-stick. The stick is washed down with water, the pH adjusted to 5.15 and 10 ml of 10% aqueous mannitol solution added. The mixture is titrated with 0.05 N sodium hydroxide, a pH meter and glass electrode being used, with magnetic stirring. The exact end-point (pH 7.5 to 8.0) is located by the method of differencing described by Kolthoff[84]. The result is calculated from the following equation:

$\%B = 100(V_N F)/W$, where V is the volume (ml) of sodium hydroxide, N its normality, F the factor applicable to mannitol titrations (10.82) and W the weight (mg) of sample. Results of analyses of 10 compounds (1.69% - 19.7%B) showed a good agreement with theoretical values. Replicate analyses showed a relative accuracy of 0.4%.

Table 22 - Determination of boron and carbon in organoboron compounds

Compound	Boron, % Calcd.	Boron, % Found	Carbon, % Calcd.	Carbon, % Found
Boric acid	17.5	17.6		
Decaborane	88.5	88.5		
Decaborane derivatives				
2-Ethvl	71.9	70.9	16.0	16.3
5-Ethyl	71.9	68.7	16.0	16.3
6-Ethyl	71.9	71.0	16.0	16.4
2-Methyl	79.4	77.0	8.8	8.9
5-Methyl			8.8	8.9
6-Methyl	79.4	77.9	8.8	9.3
Diethyl	60.0	59.4	26.9	26.8
Triethyl			34.9	36.7
Tetraethyl			41.0	41.4
Mixed ethyl *	63.1	62.4	24.6	24.0
	66.7	65.0	21.1	21.2
	70.2	69.3		
Bisdiethylsulphide	36.0	35.7	32.0	32.3

* Calculated analyses for these mixtures are based on gas chromatography

Allen and Tannenbaum[85] described methods based on combustion in an Inconel bomb at high temperatures and pressures for the determination of carbon, hydrogen and boron - Figure 9. The combustion products (boric oxide, carbon dioxide and water) are separated by conventional techniques and determined. The method was applied successfully to several trialkyl-borates and-boranes. For the carbon, hydrogen and boron determinations, respectively, the precisions are ±0.11%, ±0.07% and ±0.07% absolute, and the accuracies ±0.22%, ±0.12% and ±0.1% absolute, in the ranges 60% - 88% 11% - 16% and 3% - 11% respectively.

Allen and Tannenbaum[85] claim that their method accurately determines all three elements and is not, therefore, subject to the errors involved in only analysing two of the three elements and obtaining the concentration of the third element by a difference calculation. The novelty of this procedure arises from the technique employed to burn to completion a group of elements which are difficult to burn completely by standard methods. The organoboron compound is oxidized quantitatively with oxygen in an Inconel bomb at high temperature and pressure. Upon completion of the oxidation process, the combustion products - carbon dioxide, water and hydrated boric oxide - are separated by appropriate techniques and their quantities are determined subsequently. The percentages of carbon, hydrogen and boron in the original compound are then calculated in the usual manner.

Results obtained by applying the Allen and Tannenbaum[85] method to the analyses of several trialkylborates, trialkyboranes, and hydrocarbons are shown in Table 23. The method gives results which usually are slighly lower than theoretical. This trend is believed to represent the summation of losses during the transfer operations employed in the procedure. The error is small, however, and the analyses are sufficiently precise and accurate for most analytical purposes, as indicated by the standard deviations in Table 23.

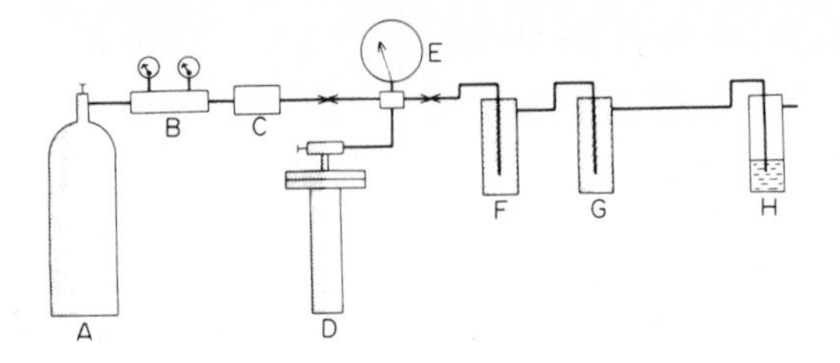

Fig. 9. Oxidation bomb connected to pressurization and absorption systems.
A. Oxygen tank. E. Regulator.
B. Regulator. F. Anhydrone bulb.
C. Check valve. G. Ascarite bulb.
D. Bomb. H. Mineral oil.

These workers state that the procedure might prove applicable to the analysis of all types of compounds containing boron, carbon, and hydrogen, if sufficient hydrogen is present to form water of hydration for all the boric oxide produced in the combustion reaction. This procedure offers an easy and safe method of analysis for spontaneously flammable compounds, such as boron hydrides and their hydrocarbon derivatives. The technique also is applicable to the analysis of hydrocarbons and offers possible advantages over the train combustion method in the analysis of volatile hydrocarbons, in which problems are encountered in weighing and transferring the sample to a combustion train, and to polynuclear aromatic compounds which are sometimes difficult to oxidize completely.

Bailey and Gehring[86] have described a procedure for the determination of boron in organic compounds based on oxygen bomb combustion using a platinum-lined Parr bomb. The sample (up to 1 g) is placed in the sample holder, 10 ml of water is added to the bomb, and the bomb is fired in the inverted position. The combustion products are absorbed in water, (sometimes containing, if desired, 1 mequiv. of base to neutralize any nitric acid formed during the combustion), and the solution is analysed for the required cation or anion by standard micro-chemical methods.

Shanina et al[87] describe a procedure for the determination of boron based on decomposition by fusion with sodium hydroxide in a nickel bomb followed by a spectrophotometric determination of boron in the combustion residue. They applied the procedure to the determination of between 3% and 75% of boron in milligram samples of barenes (carboranes), dicarba-unde caboranes and their related compounds (containing as well as C, H and B, appreciable amounts of N, Na, Si, P, Cl, Fe, Ni, Ge, As, Br, Sn, Cs or Hg). The sample is decomposed by fusion with potassium hydroxide in a nickel bomb at 800° - 850°, the melt is extracted with water, and boron in the extract is determined spectrophotometrically at 415 nm with azomethine H(4-hydroxy-5-(salicylideneamino)naphthalene-2,7-disulphonic acid). The amount of boron is calculated from the extinctions of the complexes formed from the sample solution and from two standard solutions containing boron in amounts slightly different from that in the sample solution. The coefficient of variation of the method is less than 0.4%.

The oxygen flask combustion technique has been successfully applied to the determination of boron in organoboron compounds. Yasuda and

Rogers[88] describe a technique wherein the sample (0.5 - 1 mg of boron), together with 10 mg of sucrose, is burned in oxygen and the products are absorbed in 10 ml of water. The solution is transferred with 5 ml of methanol to the cathode compartment of a coulometric titration cell, 1 ml of saturated sodium nitrate solution is added to both cathode and anode compartments, and the liquid level in the latter is adjusted with water. Any strong acid is pre-titrated to pH 7, 1.6 - 1.7 g mannitol is added, and the boric acid is titrated to the potentiometric end-point. The curve of pH v time interval is plotted, and the end-point is established by the method of concentric arcs. If chlorine is present, a little 30% hydrogen peroxide solution is added to the flask before combustion.. Recoveries of boron are good except with very refractory materials such as boron carbide.

In a further oxygen flask combustion technique, Obtemperanskaya and Likhosherstove[89] fold the organoboron sample inside a filter paper, leaving a strip of paper free, and attach this to a platinum wire fixed to the ground glass stopper of a 250 ml flask, 5 ml of sodium hydroxide (0.5 N) and 6 ml of water are transferred to the oxygen filled flask and the free strip of paper ignited and the stopper placed firmly in the flask. The flask is then set aside for 15 - 20 min to cool. A drop of methyl red and then 0.5 N sulphuric acid is added until the solution is slighly acidic. The solution is then boiled for a few minutes to remove carbon dioxide, neutralized with 0.02 N sodium hydroxide. Mannitol (0.5 g) and phenolphthalein (3 drops) are then added and the boric acid titrated with standard 0.02 N sodium hydroxide.

Corner[90] described a rapid micro-determination of organically bound boron using the oxygen flask technique. The sample is mixed with finely powdered potassium hydroxide before combustion, the products are collected in water. Boron is determined using the mannitol procedure.

The aspirating burner of Wickbold (Wickbold and Nagel[91]) (Wickbold[92]) has been applied to the determination of boron in organoboron compounds. The sample is placed in a flask containing sulphuric acid through which passes a rapid stream of nitrogen which leads into an oxy-hydrogen flame. When the sample is decomposed and any water present driven off, methanol is added and the flask is heated to 70° - 80°. On passage through the flame, methyl borate and other compounds are decomposed into boric oxide, water, and carbon dioxide. The boric oxide is absorbed in water and determined volumetrically. Recoveries of boron were within ±0.1 mg of the theoretical for H_3BO_3, $Na_2B_4O_7$, and sodium and potassium tetraphenylboron.

Flame photometry has been applied to the determination of boron in organoboron compounds. Buell[93] describes a procedure for the determination of boron in amounts down to 0.1% at 519.5 nm by volatilizing the compound in an organic solvent. The accuracy and precision are of the order of 1% - 2% of the amount present. The method has been applied to the determination of boron in lubricating oils.

Yoshizaki[94] points out that boron emission in flame photometry depends not only on boron content, but also on the molecular structure of the sample. It was therefore necessary to decompose the sample with nitric acid to boric oxide, which was used as the standard substance. The coefficient of variation was 0.67%.

Shah et al[95] heat the sample (5 - 10 mg) with a length of sodium wire (0.4 m x 0.1 m) in a nickel crucible at 400°C for 30 minutes. The residue

Table 23 - Application of combustion bomb method to determination of carbon, hydrogen and boron in alkylboron compounds

Compound	Carbon, %			Hydrogen, %			Boron, %		
	Theoretical	Found	Dev.	Theoretical	Found	Dev.	Theoretical	Found	Dev.
$(C_5H_{11})_3BO_3$, triamylborate	66.17	65.93	0.24	12.22	12.13	0.09	3.97	3.85	0.12
		65.94	0.23		12.10	0.12		3.74	0.23
		66.10	0.07		12.20	0.02		3.93	0.04
		66.08	0.09		12.09	0.13		3.92	0.05
		65.89	0.28		12.30	-0.08		3.83	0.14
		66.14	0.03		12.14	0.08		3.84	0.13
		66.00	0.07		12.23	-0.01		3.84	0.13
		65.93	0.24		12.17	0.05		3.75	0.22
		65.91	0.26		12.10	0.12		3.81	0.16
Av.		65.99	0.18		12.16	0.06		3.83	0.14
$(C_4H_9)BO_3$, tributylborate	62.62	62.46	0.16	11.83	11.75	0.08	4.70	4.69	0.01
		62.50	0.12		11.81	0.02		4.64	0.06
Av.		62.48	0.14		11.78	0.05		4.56	0.04
$(CH_3)_3BO_3$, trimethylborate	34.67	34.59	0.08	8.73	8.60	0.13	10.41	10.28	0.13
		34.59	0.08		8.74	-0.01		10.35	0.06
Av.		34.59	0.08		8.67	0.06		10.32	0.10
$(C_2H_5)_3$, triethylborane	73.53	73.21	0.32	15.43	15.22	0.21	11.04	10.91	0.13
		73.36	0.17		15.34	0.09		11.09	-0.05
		73.57	-0.04		15.39	0.04		10.86	0.18
Av.		73.38	0.15		15.32	0.11		10.95	0.09
$(C_3H_7)_3$, tripropylborane	77.16	76.80	0.36	15.11	14.86	0.25	7.72	7.54	0.18
		76.80	0.36		14.87	0.24		7.56	0.16
Av.		76.80	0.36		14.86	0.24		7.55	0.17

C_6H_{14}HBA,2,3-di-methylbutane	83.62	83.80	-0.18	16.38	16.36	0.02	
		83.53	0.09		16.31	0.07	
Av.		83.66	0.04		16.34	0.04	
C_8H_{18}NBS 2,2-di-methylhexane	84.12	84.33	-0.21	15.88	15.83	0.05	
		84.19	-0.07		15.80	0.08	
Av.		84.26	-0.14		15.82	0.06	
C_5H_8, vinyl cyclopropane	88.16	87.96	0.20	11.84	11.73	0.11	
		87.95	0.21		11.76	0.08	
Av.		87.96	0.20		11.74	0.10	
C_6H_{12}, isopropylcyclopropane	85.62	85.28	0.34	14.38	14.22	0.16	
		85.35	0.27		14.28	0.10	
Av.		85.32	0.30		14.25	0.13	
Std. dev.							
Precision			0.11			0.07	0.07
Accuracy			0.22			0.12	0.1

is then made up to 5 ml with water and the solution passed down a column of Amberlite 1R - 120 (H^+ form). The percolate is diluted to 100 ml with water and its boron content determined flame photometrically at 518 nm. All types of organoboron compound could be determined by this method. No interference occurs from any nitrogen present in the sample.

Giorgini and Lucches[96] have devised an absolute spectrographic method for the determination of small amounts of organic boron in mixtures of terphenyls. No preliminary treatment of the sample is required, and analyses are carried out using a large quartz spectrograph in a matrix of graphite and terphenyls (9:1). The accuracy of the method was examined against boric oxide and phenylboronic anhydride standards.

Erdev et al[97] studied the spectrochemical analysis of methyl borate. This substance was led from a micro-distillation apparatus directly through a bored electrode into the spark or arc gap. Solutions of methyl borate were sprayed directly into the arc or spark gaps. He describes, in detail, a suitable spraying apparatus.

Malysheva et al[98] have applied neutron absorption to the determination of boron in nitrogen-containing organoboron compounds. In this method a stream of neutrons from a polonium-beryllium source (90 mc) is used, with a paraffin-wave cube surrounded by cadmium sheets to slow down the neutrons; neutrons passing through the sample are counted. Standards are prepared by mixing an aqueous solution of boric acid with triethylamine or (if the sample contains more than 4% boron) with tris-dimethylaminoboron. The presence of the amino-compound is necessary, as the samples used by these workers contained 20% - 30% of nitrogen. Calibration graphs were prepared by plotting the number of slow neutrons passing through the sample against boron content. Down to 0.05% boron could be determined with a relative error not exceeding ±8% in about 20 minutes.

B. Determination of Chlorine

No doubt oxygen flask combustion techniques[99], which are capable of determining organically bound boron, can be applied with suitable modifications, to the determination of halogens in organoboron compounds, as indeed can other types of combustion techniques described earlier. An accurate and reproducible method has been described for the determination of chlorine bound to boron in organoboron compounds containing boron - chlorine and carbon - chlorine bonding[100]. Analysis for chlorine bound to boron mixtures containing compounds which possess both chlorine bound to boron and to carbon necessitates a method which distinguishes each type of chlorine. Although procedures are available for the determination of chlorine in organoboron compounds (Parr bomb - sodium peroxide fusion, Volhard, and Carius methods), the values obtained give the total chlorine content. Aqueous hydrolysis of chloro-2-chlorovinylboranes, with subsequent titration of hydrochloric acid with either base or silver nitrate, fails to differentiate between chlorine bound to boron and to carbon, because these compounds break down in aqueous medium (above pH 3) to form hydrochloric acid, orthoboric acid, and acetylene. The hydrolysis of chlorine bound to carbon proceeds at a much slower rate than that for the chlorine bound to boron. Various attempts were made to stabilize the 2-chlorovinylboric acid produced upon hydrolysis of the carbon - chlorine bond[100]. When it was dissolved in benzene and titrated with a solution of potassium methylate in benzene containing a small amount of methanol, only chlorine bound to boron was determined. The titration was carried out to a thymol blue endpoint. The analysis must be rapid since on standing some breakdown of the 2-chlorovinylboric acid occurred.

This procedure was found to be effective for chlorovinylboranes, but if sample mixtures contained chloroethylboranes, the method was inaccurate. Under these conditions all of the chlorine bound to boron in the chloroethylboranes did not dissociate. It is known[100] that chloroethylboranes can be hydrolysed quantitatively. The poor recovery of chlorine obtained when dealing with mixtures containing alkylboron chlorides by the benzene-potassium methylate titration technique is related to the low polarity of this solvent[99].

A polar solvent system enables the chlorine - carbon bonds to be stabilized. Accordingly, the compounds, or mixture of compounds containing both types of chlorine, are dissolved in a strong nitric acid - methanol solution. Hydrochloric acid formed through esterification of the chlorine bound to boron is titrated potentiometrically with standard silver nitrate. Chlorine bound to boron in the chloro-2-chlorovinylboranes and in the chloroethylboranes esterifies quantitatively; further the 2-chlorovinyldimethoxyborane is found to be very stable in this solvent system. Some typical results obtained by this procedure are given in Table 24.

Table 24 - Determination of chlorine bound to boron in various organoboron compounds

Compound	Analysis of Reference Sample %		% Chlorine Bound to Boron Present	Found	Average
$ClCH = CHBCl_2$	$ClCH = CHBCl_2$	92.50			
	$(ClCH = CH)_2BCl$	3.60	48.84	47.84	47.77
	BCl_3	2.00			
	Inert	1.90		47.69	
$(ClCH = CH)_2BCl$	$(ClCH = CH)_2BCl$	89.70			
	$ClCH = CHBCl_2$	7.40	22.46	24.30	24.34
	Boron oxides	3.00		24.37	
$(ClCH = CH)_3B$	$(ClCH = CH)_3B$	85.40			
	$(ClCH = CH)_2BCl$	5.00	1.07	1.31	1.32
	Boron oxides	8.60			
	Inert	1.00		1.33	
$C_2H_5BCl_2$	C_2H_5BCl	91.90			
	$ClCH = CHBCl_2$	5.40	61.97	61.58	61.58
	BCl_3	0.50			
	Inert	2.10		61.58	
$(C_2H_5)_2BCl$	$(C_2H_5)_2BCl$	95.70		32.40	
	Boron oxides	4.30	32.54	32.20	32.30
$C_2H_5(ClCH = CH)_2$ BCl	$C_2H_5(ClCH = CH)BCl$	84.30			
	$C_2H_5BCl_2$	6.00	27.71	27.92	
	$(ClCH = CH)_2BCl$	9.70		28.00	27.96

C. Determination of Nitrogen and Boron in Amineboranes

Noth and Beyer[101] and Ryschkewitsh and Birnbaum[102] have described procedures for the estimation of boron and nitrogen in amineboranes. They hydrolyse the amineborane with acidified methanol to produce trimethylborate in quantity equivalent to the boron content of the original sample. The trimethylborate is determined by conventional procedures using mannitol or glycerol. They determined nitrogen on a separate sample of the amineborane by the Kjeldahl or the Dumas procedures.

Melly[103] has described a method for the consecutive titrimetric determination of boron and nitrogen in amineboranes. He points out that difficulties are sometimes encountered in the determination of boron by procedures based on the use of mannitol or glycerol when the sample contains nitrogen. In his procedure the amineborane is first hydrolysed to produce a hydrolysate in which the amine fragment has not been degraded - consecutive titrations with a standard base are then carried out in aqueous solution using a pH meter, boron being determined by conventional titration in the presence of mannitol, and nitrogen by titration of the corresponding ammonium ion. Typical results are presented in Table 25. In this method, for the titration of ammonium ions having pK_a values greater than about 9.5, points of inflection are slight and, consequently, the accuracy of the nitrogen determination is diminished. Amine salts having pK_a values less than about 4 are sufficiently acidic to be titrated with the excess hydronium ion introduced as hydrochloric acid in the hydrolysis of the amineborane. Nevertheless, the method may be used in the analysis of diaminebisboranes, where hydrolysis produces a diammonium ion, even when only one of the corresponding acid functions can be titrated conveniently. For example, in the analysis of ethylenediaminebisborane (ethane-1,2-diamineborane), the end-point in the neutralization of the first - NH_3^+ group of the ethylenediammonium ion is obtained readily, but the -anionethylammonium ion is too weakly acidic (pK_a = 9.93) to be determined accurately. For m- and p-phenylenediaminebisboranes and for triethylenediaminebisborane, the first acid dissociation constants of the respective diammonium ions are high and a separation of the first ammonium ion from excess hydronium ion is not obtained. The second dissociation constants, however, are in a range amenable for titration of the corresponding monoammonium ions. The aminophenylammonium ions are relatively acidic and, therefore, are titrated prior to the determination of boron while the reverse is true for the monoammonium ion of triethylenediamine. The determination of only one of the acid functions in a diammonium ion (and estimation of the pH_a for that ion) is usually sufficient for characterization and analysis of the bisborane.

Rittner and Culmo[104] have described a simple method for the microdetermination of boron, either alone or simultaneously with nitrogen and/or phosphorus, in organoboranes. They digest the sample with sulphuric acid in the presence of selenium powder and copper sulphate-potassium sulphate (3:1). The solution is diluted, and boric acid is titrated with sodium hydroxide solution by the fixed-pH method. If nitrogen and phosphorus are to be determined simultaneously, a solution of hydrogen peroxide is added to the digestion reagents, and suitable aliquots are taken for measurement of the nitrogen by a micro-Kjeldahl method and of the phosphorus by a spectrophotometric method. The relative errors are ±0.05, ±1.34 and ±0.3% for boron, nitrogen and phosphorus respectively.

Table 25 - Analysis of Amineboranes

		Calcd.				Found			
Amineborane	R_3NH^+	pK*	H^-	B	N	pK*	H^-	B	N
$H_3BH_2NCH_2CH_2NH_2BH_3$	$H_3NCH_2CH_2NH_3^{2+}$	7.31 †	6.89	24.6	31.9	7.31	6.82	24.3	32.0
O NHBH$_3$	O NH$_2^+$	8.7	3.00	10.72	13.87	8.60	2.97	10.71	13.97
H$_3$BN NBH$_3$	N NH$_2^+$	8.98 ‡	4.32	15.47	20.03	8.98	4.30	15.44	19.94
$HOCH_2CH_2NH_2BH_3$	$HOCH_2CH_2NH_3^+$	9.44 §	4.04	14.44	18.69	9.6	4.00	14.15	18.32
NBH$_3$	NH$^+$	10.58	2.42	8.65	11.21		2.34	8.63	
$p\text{-}CH_2C_6H_4NH_2BH_3$	$p\text{-}CH_3C_6H_4NH^+$	5.30 ‖	2.50	8.94	11.58	5.22	2.40	8.98 ¶	11.78
$m\text{-}CH_3C_6H_4NH_2BH_3$	$m\text{-}CH_3C_6H_4NH_3^+$	4.74 ‖	2.50	8.94	11.58	4.8	2.39	9.07 ¶	11.36
$m\text{-}C_6H_4(NH_2BH_3)_2$	$m\text{-}H_2NC_6H_4NH_3^+$	5.0 ‡	4.45	15.9	20.6	5.04	4.34	15.5	20.5
$p\text{-}C_6H_4(NH_2BH_3)_2$	$p\text{-}H_2NC_6H_4NH_3$	6.23 ‡ 6.04 ‖	4.45	15.9	20.6	6.18	4.20	15.3	20.1
$CH_3NH_2BH_3$	$CH_3NH_3^+$	10.64 §	6.73	24.10	31.20		6.61	23.58	
$(CH_3)_3CNH_2BH_3$	$(CH_3)_3CNH_3^+$	10.45 §	3.48	12.44	16.10		3.45	12.7	

* Lit. value at 25°C.
† At 30°C, 0.2 M HCl, 0.1 M $BaCl_2$, 0.1 M KCl.
‡ Obtained from titration of separate sample of the diaminedihydrochloride.
§ Calculated from pK_a.
‖ Calculated from pK_a.
¶ No mannitol added.

Arthur et al[105] describe a combustion technique for the determination of carbon, hydrogen and nitrogen in organoboron compounds. When applied to the determination of organoboron compounds, conventional combustion techniques yield low results. They showed that by employing quartz combustion tubes and by heating to between 1000° and 1100° in the unpacked section around the combustion boat, both carbon and hydrogen were determined to within about ±0.2% and ±4% respectively. Determinations of nitrogen in acetanilide - BF_3 were improved by restricting the sample sizes to 7 mg or less by employing a longer combustion tube with two successive packings.

D. Determination of Hydride and Active Hydrogen

Boranes and chlorinated boranes may be determined gas chromatographically by passing them, with argon as carrier gas, through a column of molecular sieve 5A moistened with water, then through a column of dry molecular sieve 5A to determine the amount of liberated hydrogen produced by hydrolysis of the B - H bonds in the sample[106]. The results are within 5% of the known values. Some typical results are presented in Table 26. If the original sample contains hydrogen as an impurity, it is frozen out in a cold trap at - 78°C. Gaseous hydrogen is swept out of the trap with a current of argon, then the residual boranes in the trap are allowed to warm up the room temperature and are collected in a gas burette. The sample is now ready for analysis by chemically active gas chromatography[107] - Figure 10. A very detailed study of the gas chromatographic technique for the determination of hydridic and active hydrogen in borane compounds has been reported [109]. This technique is based on the liberation of active or hydridic hydrogen in a micro-reaction cell incorporated in a gas chromatographic flow system and measurement of the hydrogen gas band in nitrogen gas carrier by thermal conductivity detection. The difference in thermal conductivity between hydrogen and nitrogen is detected with a Teflon-coated hot wire. The hydrogen gas is formed from the borane compounds by acid hydrolysis using 10% hydrochloric acid. Application of heat to the reaction cell accelerated the reaction rate and provided rapid and stoichiometric release of hydridic hydrogen from the compounds investigated. The main advantage of this method is that it can be applied to the analysis of a variety of borane compounds using a conventional commercially available gas chromatograph.

This procedure was applied to the determination of boron-hydrogen bonds in trimethylaminomonoborane, ($(CH_3)_3N - BH_3$) and borazole ($B_3N_3H_6$).

In the procedure, 10% aqueous hydrochloric acid was used to liberate hydrogen from these compounds. Application of heat to the reaction cell accelerated the reaction rate and provided rapid and stoichiometric release of hydridic hydrogen from the compounds investigated. Determinations of hydridic hydrogen in trimethylaminomonoborane and borazole (Table 27) give satisfactory recoveries.

ORGANOCALCIUM COMPOUNDS

A. Determination of Carbon and Hydrogen

Carbon and hydrogen in alkali and alkaline earth metal compounds can be determined by igniting the sample at 900°C mixed with eight times its weight of tungstic oxide in oxygen[110].

Table 26 - Accuracy of the chemically active gas chromatography method in determining known samples

Standard samples introduced (ml of gas, STP)					Analytical results (ml of gas, STP)				
$BHCl_2$	B_2H_5Cl	B_2H_6	B_5H_9	H_2	$BHCl_2$	$B_2H_5Cl_2$	B_2H_6	B_5H_9	H_2
0.182	—	—	—	—	0.177	—	—	—	—
0.182	—	—	—	—	0.181	—	—	—	—
0.181	—	—	—	—	0.175	—	—	—	—
—	—	0.126	—	0.095	—	—	0.125	—	—
—	—	0.125	—	0.095	—	—	0.119	—	0.095
—	—	0.066	—	—	—	—	0.066	—	—
—	—	0.331	—	—	—	—	0.323	—	—
—	0.062	—	—	—	—	0.060	—	—	—
—	—	—	0.191	—	—	—	—	0.188	—
—	—	—	0.195	—	—	—	—	0.185	—
—	—	—	1.196	—	—	—	—	1.162	—
—	—	—	1.120	—	—	—	—	1.124	—

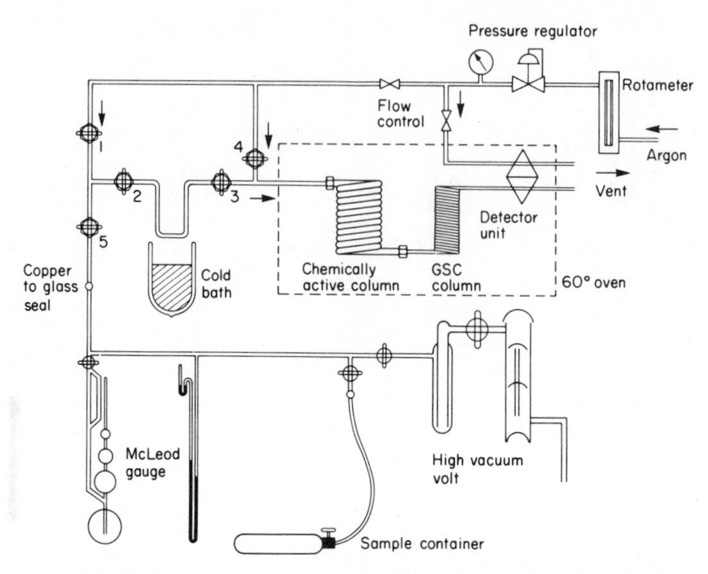

Fig. 10 - Schematic diagram of apparatus for chemically active gas chromatographic method for determination of boron-hydrogen banding.

Table 27 - Determination of hydridic hydrogen in organoboron compounds

	Sample size (mg)	Recovery (%)
Trimethylaminoborane, $(CH_3)_3N\ BH_3$	1.623	95.43
	1.2984	88.65
	0.9738	90.63
	0.6492	99.73
	0.3246	94.43
	0.1623	108.71
Av. and s.d.		96.3 ± 5.3
Borazole, $B_3N_3H_6$	2.1312	90.43
	1.5984	87.48
	1.0656	96.88
	0.5328	97.07
	0.2664	95.87
Av. and s.d.		93.6 ± 3.5

ORGANOCHROMIUM COMPOUNDS

A. Determination of Chromium

Chromium in organochromium complexes used as additives in drilling fluids can be determined by evaporating a sample to dryness on a water-bath mixing the residue with nitric acid and potassium chlorate solution, and again evaporating to dryness[111]. The residue is dissolved in water, the solution is boiled with 2.5 M sulphuric acid and silver nitrate solution,

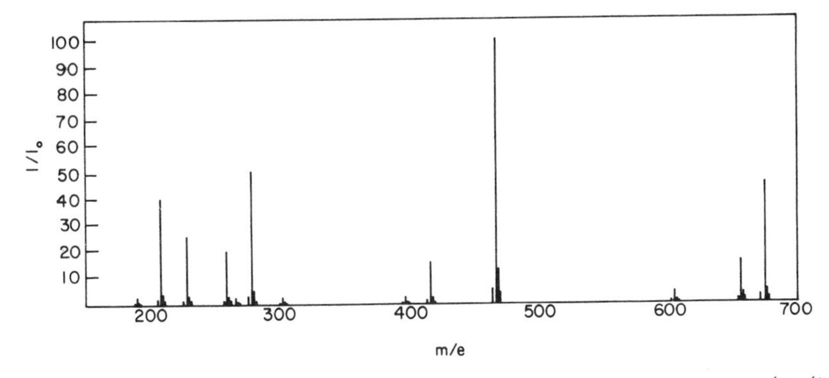

Fig. 11 - Mass spectrum of chromium hexafluoroacetylacetonate ($Cr(hfa)_3$) above mass 200.

then gently boiled for 30 min with ammonium persulphate, sodium chloride solution is added, the mixture is filtered, and chromium in the filtrate is determined with diphenylcarbazide.

Brooker and Isenhour[112] have described a rapid mass spectrometric determination of chromium as chromium(III)hexafluoroacetylacetonate. The determination can either be directly applied to nanogram samples or to larger samples such as stainless steels containing several percent chromium by aliquoting. In this procedure about 25 mg of the chromium-containing sample is quantitatively oxidized with perchloric acid in a sealed tube and then reacted with hexafluoroacetylacetone to produce chromium (III) hexafluoroacetylacetonate. This chelate is then quantitatively transferred as an ether or benzene solution to the inlet block of an MS9 mass spectrometer. Figure 11 shows a mass spectrum of the compound above mass 200. This spectrum is characteristic of chromium hexafluoroacetylacetonate for ionizing voltages greater than 25 V. The most prominent peak is that assigned to $_{52}CR(hfa)_2$ (I/Io = 100) and this peak is used in the analysis.

ORGANOCOBALT COMPOUNDS

A. Determination of Cobalt

Palade[115] has described a method for the determination of cobalt in its complexes with 1,10 phenanthroline and 2,2'-bipyridyl. To 0.2-0.3 g of sample in a Kjeldahl flask is added 3-4 g of potassium perchlorate and 4-5 ml of concentrated sulphuric acid. The mixture is heated until it becomes blue and is then diluted with 20-30 ml of water. Solid or aqueous saturated sodium carbonate is then added until basic cobalt salts are precipitated. The precipitate is then dissolved by adding a few drops of acetic acid and cobalt is determined gravimetrically by the standard pyridine-thiocyanate method.

A further method involving X-fluorescence spectroscopy has been described by Bartkiewicz and Hammatt[114] for the determination of cobalt in amounts between 10 ppm and 1% in organic substances. Only a minimum of sample preparation is required, and cobalt (also iron and zinc) can be determined in 10 - 15 min. Matrix effects and minor instrumental variables are compensated for by using a solution of copper octanoate or naphthenate in 2-ethylhexan-1-ol (containing 0.2% of copper) as internal standard. The lower limit of detection is about 10 ppm of cobalt for concentrations of up to about 0.5% of cobalt; the precision (95% confidence level) is about 0.1%.

Sharif[115] estimated cyanocobolamin by estimating cobalt. The cobalt content of a solution of cyanocobolamin was determined by means of nitroso-R salt reagent using the method of Kidson and Eskew[116]. The cobalt complex obtained was measured colorimetrically at 550 mm. For a 2% accuracy, 100μg of cyanocobolamin is required in the sample weight taken.

A method described by Mitra et al[117] is suitable for the determination of cyanocobolamin in injections, tablets and liver extracts. The method depends upon the colour produced with Nitroso-R salt at controlled pH after oxidation of the sample with hydrogen peroxide. Any colour due to metals other than cobalt is destroyed with nitric acid. The extinction of the final solution is measured at 420 mm. Beer's law is obeyed in the range 100 - 600 μg cyanocobolamin and recovery is between 100 and 104%.

Kinnory et al[118] have described a method for the determination of urinary excretion of radiocobalt labelled cyanocobolamin by cobalt sulphide precipitation. A tracer close of 0.05 μg C of 60Co, labelled cyanocobolamin given orally is followed after 1 hour by intra-muscular injection of 1 μg of non-radioactive cyanocobolamin. The total urine during the next 24 hours is collected, cobaltic chloride is added as carrier and cobalt is precipitated with ammonium sulphide in the presence of ammonium chloride. The cobalt sulphide is collected on glass fibre filter paper and counted using a well type scintillation counter.

Several workers have described methods for the determination of cyanocobolamin based on preliminary decomposition to produce cobalt ions followed by spectrophotometric determination of cobalt using a suitable range.

Abd El Raheem and Dokhana[119] report that the 3:4 complex formed by cyanocobolamin with Fast Navy 2R (C.I. Mordant Blue 9) has a maximum at 550 nm and an inflection at 620 nm; the latter point is used for the spectrophotometric determination of cobalt. Beer's law is obeyed for 0.2 - 5 μg of cobalt per 25 ml, and the colour is stable at 15 - 80°C for 3 h. The presence of calcium, lead, manganese and zinc does not cause interference if EDTA is used as a masking agent; intereference from lead can also be suppressed by the use of tartaric acid; from manganese by the use of ascorbic acid; and from nickel by treatment with EDTA at pH 8.5 for 15 min.

Menyharth[120] has reported a method in which the cyanocobolamin containing sample is decomposed with sulphuric acid and hydrogen peroxide and cobaltous ion is oxidized and precipitated as cobaltic hydroxide by addition of sodium hydroxide and hydrogen peroxide solution and the dissolved precipitate is titrated iodometrically. The error in this method is not greater than ±0.5% of the determined result.

Brustie et al[121] have described methods for the determination of cyanocobolamin in river extracts, tablets etc. in which the sample is digested to dryness with perchloric acid. Cobalt in the digest is determined spectrophotometrically using Nitroso-R-salt.

Parissakis and Issopoulos[122] have described a method for the determination of cyanocobolamin as cobalt in pharmaceutical products in the presence of liver extract and salts of cobalt, iron and copper. The buffered sample (pH 8.0) is treated with 8-hydroxyquinoline in chloroform to remove salts of copper, cobalt and iron. The acidic aqueous phase containing cyanocobolamin is treated with hydrogen peroxide and buffered to pH 4.65 prior to extraction with a light petroleum solution of o-nitrosophenol. The determination is completed by spectrophotometric evaluation of the extracted cyanocobolamin complex at 570 nm.

Konechy and Talgyessy[123] used ozone to decompose cyanocobolamin to cobalt. They followed the course of the decomposition by determining cobalt by an isotope - dilution method or by radiometric titration with 10^{-5} M EDTA. From the amount of cobaltic ions liberated the amount of cyanocobolamin present could be calculated.

ORGANOCOPPER COMPOUNDS

A. Determination of Copper

Copper in organocopper compounds may be determined by first oxidizing the sample by heating it with concentrated sulphuric acid in a Kjeldahl flask for 2-3.5 h[124]. Copper in the cooled and diluted residue is titrated with 5 mM EDTA (pH 4-4.5; disodium ethylbis (5-tetrazolylozo) acetate as indicator) or iodimetrically after the addition of potassium iodide solution. Alternatively, digestion may be followed by spectrophotometric determination using 1(2-pyridylazo)naphthol[125]. A polarographic method has been used for the determination of ionic copper in copper chlorophyllins[126]. In this method known amounts of dried copper sulphate are dissolved in 1.5 M aqueous ammonia solution and the solution is electrolysed at the dropping electrode.

Sebon et al[127] have studied the influence of the type of organocopper (and organoiron) compounds on the determination of these elements by flame atomic absorption spectroscopy.

ORGANOGERMANIUM COMPOUNDS

A. Determination of Germanium

Germanium in organogermanium compounds can be determined by mixing the sample with 100-fold amount of chromium(III) oxide and heating in a tube at 900°C under a current of oxygen[128,129]. Carbon dioxide and water are collected and weighed by conventional procedures to provide estimates for the carbon and hydrogen contents of the sample. The content of germanium is obtained from the change in weight of the ignition tube before and after the ignition. In a tube combustion procedure, which is applicable to the determination of germanium in volatile organometallic compounds, oxygen is bubbled through a weighed portion of the sample until volatilization is complete[130]. The vapours are passed into a weighed silica tube and ignited in a plug of prepared asbestos. The silica tube is then disconnected and ignited to constant weight at 800°C. The germanium content of the sample can then be calculated from the residual weight of germanium oxide found in the silica tube. The method may be modified to accommodate samples that hydrolyse rapidly to non-volatile products[131]. It is accurate to ±0.5% and takes 3 - 4 h. In a bomb combustion procedure for the determination of organically bound germanium, the sample is fused with sodium hydroxide and sodium carbonate for 1.5 h in a sealed nickel bomb at 920 - 940°C[131]. After cooling, the residue is dissolved in water and ice-cold concentrated sulphuric acid added. The mixture is then heated almost to boiling point and diluted prior to determination of germanium and halogens.

A direct spectrographic procedure has been developed for the determination of germanium in organogermanium compounds[132]. A solution of the sample in cumene containing polymethylphenylsiloxane as internal standard is transferred to a fulgurator cooled to solid carbon dioxide - acetone temperature. Spectra are excited by a condensed spark discharge between the graphite rod in the fulgurator and graphite counterelectrode. X-ray fluorescence spectroscopy may also be used[133]. The sample is dissolved in dioxane. Water-soluble species are dissolved in water and insoluble compounds are powdered and pressed into discs with $Na_2B_4O_7 10H_2O$. Arsenic is added as an internal standard. An instrument with a lithium fluoride analysing crystal is suitable. Measurements are made of germanium and arsenic K radiation. The relative error for 0.15 - 0.30% of germanium was within ± 3.3%. Atomic-emission spectrography has a detection limit for hydride of 0.4 µg[134].

Burns and Dadgar[135] investigated the determination of germanium in organogermanium compounds using carbon furnace atomization - atomic absorption spectroscopy.

B. Determination of Carbon and Hydrogen

Pieters and Buis[136] have pointed out that they could not obtain reproducible results when carrying out tube combustions on organogermanium compounds either alone or mixed with tungstic oxide or other oxidants. Liquids were particularly difficult to analyse. Pieters and Buis[136] overcame the problem by developing a modified method using very slow combustion (Figure 12). The success of this procedure depends strongly on the experience and attention of the operator. Apart from this each new type of compound should be analysed critically and should be analysed in duplicate to be sure of reproducible results. As can be seen from Table 28 the method gives results which are reliable and compatible with the standards of normal microanalytical practice.

C. Determination of Alkoxygermanes and Mercaptogermanes

Anderson[137] determined mercaptogermanes indirectly with iodine. Klimova et al[138] used a modification of the Ziesel procedure for the microdetermination of alkoxygermanes. GeOC and GeSC linkages can be determined by the perchloric acid catalysed acetylation method of Fritz and Schlenk[139]. The mechanism for the reaction of an alkoxy or mercaptogermane with acetic anhydride is presumably the same as that proposed for the acetylation of SiOC (Dostal et al[140], Magnuson[141], and SiSC linkages (Berger and Magnuson [142]). The acetylation method depends on the reaction of the alkoxy or mercaptogermane with excess acetic anhydride to form the corresponding acetate or thiolacetate and acetoxygermane. The acetoxygermane and remaining acetic anhydride are rapidly hydrolysed to acetic acid which is then titrated with standard base. The alkoxy or mercapto content is calculated from the difference in volume of base between the sample and blank titrations.

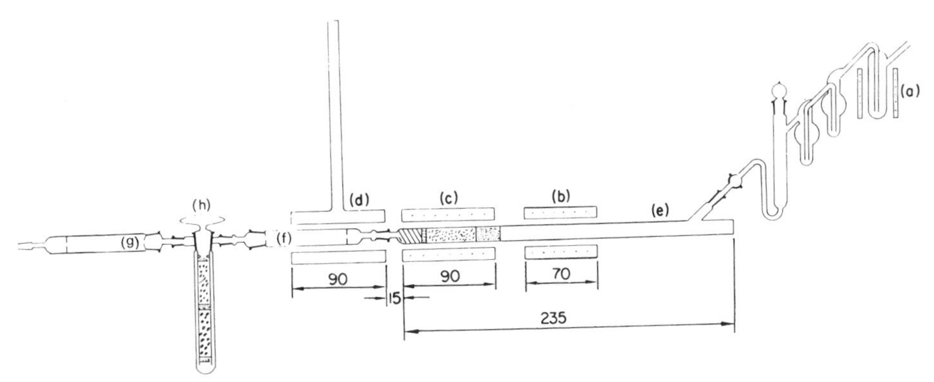

Fig. 12. Determination of carbon and hydrogen in organogermanium compounds. Combustion trains, all dimensions in millimeters (a) Preheater; (b) burner; (c) furnace; (d) mortar; (e) combustion tube (quartz wool, catalyst, quartz wool, silver wool); (f) water absorption tube (with a capillary 0.4 mm diameter); (g) carbon dioxide absorption tube (h) manganese dioxide tube.

Magnuson and Knaub[143] have described a further method for the assay of GeOC linkages based on reaction with in situ generated hydrogen bromide. Under anhydrous conditions, standard perchloric acid is titrated into an acetic acid solution of tetraethylammonium bromide and the sample. The hydrogen bromide which is formed reacts with the alkoxygermane to form a bromide and the corresponding parent alcohol. The first excess of hydrogen bromide is detected by a Blue BZL (Ciba 22062S) indicator end-point. This method allows a quantitative determination of alkoxygermanes in the presence of alkoxysilanes and alcohols.

Magnuson and Knaub[143] described procedures based on acetylation and on reaction with hydrogen bromide for the determination of alkoxy and mercapto groups and compared the results obtained by these two procedures for a range of organogermanium compounds which by elemental analyse and gas chromatography were known to have a purity between 98.7 and 99.9%.

The results in Table 29 suggest that most alkoxy and mercaptogermanes may be quantitatively determined by either of the above methods. Acetylation however, is preferred for mercaptogermane determinations.

Results obtained for dimethylphenoxygermane and dimethyldibutyl mercaptogermane by the acetylation procedure following a 5 minute acetylation time were low. There was no increase in recovery when the acetylation time was increased to 15 minutes. Magnusson and Knaub[143] found that analysis of alkoxygermanes by the hydrogem bromide method suffers interference by alkoxysilanes, lower alcohols or phenols.

ORGANOIRON COMPOUNDS

A. Determination of Iron

Iron in ferrocene derivatives has been determined by first digesting 25 - 400 µg in a mixture of concentrated nitric acid and concentrated sulphuric acid in a sealed glass tube and heating to 300°C[144]. After the addition of hydroquinone and suitable buffering the iron is determined spectrophotometrically using 1.10-phenanthroline (Table 30).

Table 28 - Determination of carbon and hydrogen in organogermanium compounds

		Sample weight	Carbon theory	Carbon found	Carbon diff.	Hydrogen theory	Hydrogen found	Hydrogen diff.
$C_{20}H_{33}NGe$	$(C_4H_9)_3Ge-N$ (indole ring)	4.162	66.71	66.62	-0.09	9.24	9.38	+0.14
		6.334		66.43	-0.28		9.36	+0.12
$C_{20}H_{45}NGe$	$(C_4H_9)_3Ge-NH-C_8H_{17}$	3.402	64.54	64.65	+0.11	12.19	12.04	-0.15
		2.902		64.67	+0.13		12.11	-0.08
$C_{24}H_{55}NGe_2$	$((C_4H_9)_3GE)_2NH$	5.885	57.32	57.36	+0.04	11.02	10.86	-0.16
		2.498		57.35	+0.03		11.02	0.00
$C_{16}H_{31}O_2NGe$	$(C_4H_9)_3GeN(CO.CH_2)_2$	1.963	56.19	56.30	+0.11	9.14	9.01	-0.13
		2.196		56.42	+0.23		9.17	+0.03
$C_{20}H_{31}O_2NGe$	$(C_4H_9)_3Ge-N(C=O)_2C_6H_4$	6.392	61.58	61.43	-0.15	8.01	8.09	+0.08
		5.090		61.60	+0.02		8.13	+0.12
$C_9H_{21}NS_2Ge$	$(C_2H_5)_3Ge-S-\underset{S}{C}-N(CH_3)_2$	4.289	38.60	38.62	+0.02	7.56	7.57	+0.01
		3.701		38.78	+0.18		7.50	-0.06
$C_4H_9Cl_3Ge$	$C_4H_9GeCl_3$	3.150	20.35	20.45	+0.10	3.84	3.94	+0.10
		3.458		20.36	+0.01		3.95	+0.11
$C_{12}H_{27}BrGe$	$(C_4H_9)_3GeBr$	2.718	33.50	44.38	-0.12	8.40	8.44	+0.04
		2.840		44.30	-0.20		8.35	-0.05

Table 29 - Determination of Alkoxy and Mercaptogermanes

Compound	% of theory* Perchloric acid catalysed acetylation	in situ HBr	% Purity + by gas chromatography
$(CH_3)_2(OCH_3)_2Ge$	98.9 ± 0.3	97.1 ± 0.4	99.9
$(C_6H_5)_2(OCH_2)_2Ge$	99.5 ± 0.1	——	98.8
$(C_4H_9)_3(OC_2H_5)Ge$	99.3 ± 0.1	100.3,101.1	99.8
$(CH_3)_2(isoC_3H_7)_2Ge$	100.0 ± 0.4	98.3 ± ±.±	98.7
$(CH_3)_2(OC_6H_5)_2Ge$	98.0 ± 0.2	100.1 ± 0.9	98.8
	98.3 ± 0.3‡	101.0	
$(OH_3)(SC_4H_9)_2Ge$	98.4 ± 0.3	101.5 ± 0.9#	99.5
	98.4 ± 0.1‡		

* Average and average deviation of triplicate determinations.
+ Based on area % using a 4 ft stainless steel column packed with G.E. SE-30 on Chromosorb D.
‡ 15 min. reaction time.
# End-point slow to attain and poorly defined.

Table 30 - Determination of iron in ferrocene derivatives

Compound	Range of sample weight (μg)	Iron (%) Calc.	Iron (%) Found (mean)	No. of determinations	Standard deviation (%)	Range of errors (%)
Ferrocene	29.4-157.8	30.02	30.11	8	0.27	-0.34 to ±0.43
1.1-Dibenzoyl-ferrocene	44.0-158.0	14.17	14.22	8	0.12	-0.15 to -0.23
Ferrocene 1,1-dicarboxylic acid	63.4-99.3	20.38	20.13	8	0.14	-0.45 to -0.10
Ferrocene 1,1'-dicarboxy-anilide	50.5-179.6	13.16	13.21	10	0.21	-0.30 to ±0.40

In a polarographic method, the sample is decomposed with concentrated sulphuric acid in the presence of hydrogen peroxide and the unused hydrogen peroxide is destroyed by boiling[145]. The iron is then absorbed on KU-2 cationite, from which it is eluted with 4N hydrochloric acid. The pH of the eluate is adjusted to 9, the iron-catecholdisulphonic acid complex is formed, and the solution of this complex is analysed polarographically for iron. An absolute error of approximately ±0.5% is claimed for this method. Iron can be determined spectrophotometrically following digestion with concentrated nitric acid and hydrogen peroxide[146]. Silicon and fluorine do not interfere in the procedure.

Simultaneous determination of iron and titanium in donor-acceptor complexes of the ferrocene bases can be effected by decomposing the sample with a mixture of nitric acid, anhydrous acetic acid, and aqueous bromine [147]. The iron is titrated complexometrically, with sulphosalicyclic acid as indicator, and the titanium is determined colorimetrically by extraction of its 8-hydroxyquinoline complex into chloroform. The standard deviation for each analysis is not greater than ±0.3%.

Ignition and mineralization techniques involve a risk of inflammation and sublimation of ferrocene derivatives. In a procedure which avoids these difficulties the sample is treated in a Kjeldhal flask with concentrated hydrochloric acid, followed by the addition of concentrated nitric acid[148]. After a few minutes the solution is heated to gentle boiling and, after 30 min, concentrated nitric acid is added and the solution boiled to expel nitrous fumes. After cooling, 110-volume hydrogen peroxide is added and the iron is determined spectrophotometrically using 1,10-phenanthroline. Iron in ferrocene and its organosilicon derivatives can be determined by dissolving the sample in carbon tetrachloride[149]. Hydrochloric acid and ammonium persulphate are added and the solution is stirred, whilst heating, until the blue colour of the solution changes to yellow. Boiling is continued until chlorine evolution ceases and then the solution is diluted with water. Potassium iodide solution is then added, the pH is adjusted to 4.5 and the solution titrated with standard EDTA, using sulphosalicylic acid as indicator, to a golden yellow colour change.

Renger and Jenik[150] have described a volumetric micro-determination of iron in ferrocene and its derivatives. The sample (2-4 mg) is shaken with acetone or anhydrous acetic acid and aqueous bromine (25 ml) added followed by shaking for 1 min and repeat shaking at 5 min intervals over the next 30 min. Then, to the aqueous layer, is added M sodium acetate (4 ml), formic acid (0.5 ml) and sulphosalicylic acid indicator (3-5 mg). The solution is then titrated to the colourless end-point with 0.01 M EDTA.

Iron has been determined spectrographically in ferrocene and its derivatives. Kuchkarev[151] atomized a cooled solution of the sample containing a known concentration of methylphenylpolysiloxane or of tributyl phosphate in cyclohexanone or kerosene into a spark between graphite electrodes. For standardisation a similar solution was used containing known concentrations of pure ferrocene. The analytical line pairs used were either Fe 2525·39 and Si 2524·12 or Fe 2493·18 and P 2553·28.

Fischl[152] proposed a method for the determination of iron in haemoglobin. The blood sample (0.02 ml) is pipetted into 5 drops of water contained in a 15 ml centrifuge tube. Concentrated sulphuric acid (0.3 ml) aqueous potassium persulphate (0.5 ml) were added and the solution cooled and 5 ml of a reagent comprising, 30 g potassium thiocyanate dissolved in 100 ml, 4 ml acetone, and 500 ml isobutyl alcohol added and the mixture shaken and centrifuged at high speed for 10 min. The ferric thiocyanate colour of the supernatant solution is then evaluated spectrophotometrically under standard conditions. Alternatively, there is a simple reproducible method involving oxygen flask combustion which does not need internal standards, for the liquid scintillation counting of haemoglobin and haemin, labelled with carbon-14[153]. In a further method the blood sample, saturated potassium chlorate solution, and concentrated sulphuric acid are heated in boiling water[154]. After cooling, sodium tungstate solution is added and the mixture is centrifuged. Iron in the supernatant solution is determined using potassium thiocyanate. X-ray spectroscopy can be used to determine total blood iron[155]. Samples of serum are placed directly on to confined spots on paper, dried, and passed through the X-ray field. Results for total phosphorus in serum and total iron in whole blood showed no significant difference from those given by wet-washing procedures, except that the total iron is more precisely determined by wet-washing.

ORGANOLEAD COMPOUNDS

A. Determination of Lead

Lead may be determined by first decomposing the organolead compound with a mixture of 1:1 fuming sulphuric acid (25% SO_3) and fuming nitric acid (d = 1.52)[156]. The mixture is carefully heated until all of the sulphuric acid has been evaporated. The residue is dissolved in glacial acetic acid and 25% ammonia and diluted. After suitable buffering, lead is determined by titration with 0.05 M disodium EDTA to the Eiochrome Black T end-point. In a direct spectrographic procedure, a cumene solution of the sample and polymethylphenylsiloxane internal standard is poured into the inner vessel of a fulgurator, the outer vessel of which contains 50% aqueous monoethylene glycol cooled to - 70°C in solid carbon dioxide-acetone[157]. Spectra are excited by a condensed spark discharge between the graphite rod in the fulgurator and a graphite counter electrode. Tetraethyllead has been analysed by beta-ray back-scattering using ^{90}SR as a source[158]. Good agreement with chemical analyses for lead content is reported by this method.

B. Determination of Organolead Compounds in Petroleum via the determination of Lead

Atomic absorption spectroscopy has been used to determine organolead compounds in petroleum[159-170]. The method is rapid, reproducible and remarkably free from interferences by other elements present in the petroleum. The results obtained agree favourably with those obtained by X-ray fluorescence, wet chemical methods, and flame photometry. Lead absorbs strongly at 283.3 nm. At this wavelength the degree of absorption is so intense that the useful range of analysis is limited to between 0 and 70 ppm so that sample dilution may be necessary. Tin, sodium, bismuth, copper, zinc, chromium, iron and nickel do not interfere if an oxygen-hydrogen flame is used when present at about 1% of the concentration of lead in the sample[154]. There appeared to be no interference from sulphur, halogen or nitrogen compounds. Isoctane has also been used as the solvent[161]. A precision of about 1% for the determination of either tetraethyllead or tetramethyllead in petroleum using atomic-absorption spectroscopy is claimed. Using an oxy-hydrogen flame and a standard Beckman flame atomizer burner, Robinson[160] showed that the following elements did not interfere in the procedure when present at about 1% of the concentration of lead in the sample; tin, sodium, bismuth, copper, zinc, chromium, iron, and nickel.

The sulphur and nitrogen content of gasoline vary significantly, depending on the source of the crude oil. To study the effect of interference by sulphur and nitrogen, Robinson mixed equal quantities of leaded petroleum and the interfering sulphur or nitrogen compound, diluted them 10-fold with isooctane, and determined the apparent lead content.

As shown in Table 31, there appeared to be no interference from sulphur or nitrogen compounds, even though the compounds were present in an equal volume to gasoline. If these elements do not interfere at these high concentrations, they would not interfere at the lower concentrations

Table 31 - Interference by sulphur and nitrogen on lead determination

Lead standard parts/10^6	Apparent analyses in presence of interfering compound Carbon disulphide parts/10^6	Thiocresol parts/10^6	Diethylamine parts/10^6
50	50	51	51
125	127	127	123
220	220	220	220
325	325	320	320

Sample contained 50% interfering compound, 50% petroleum. The mixture was diluted ten-fold with iso-octane.

normally encountered in gasoline analysis. Other studies indicated freedom from interference from halides providing precipitation of the lead did not occur.

Dagnall and West[162] have reported their observations on the atomic absorption spectroscopy of lead at 283.3 nm in aqueous solutions in organic extracts and in petroleum. They found in their preliminary work that atomic absorption spectroscopy using an air-propane flame and the Hilger and Watts "Uvispek" with atomic absorption attachments and a perforated plate type burner lead, 10 cm long x 1.5 cm wide, gave low results for tetraethyllead in petroleum. As a consequence, they undertook a detailed investigation including a study of the effect of various extraneous ions, extraction procedures and solvents on the determination of lead.

These workers examined several solvents for dilution of the petroleum sample prior to analysis, e.g. ethanol, methyl ethyl ketone, isopropanol, isobutyl methyl ketone, hexane, isooctane, but mostly they were either too volatile (hexane and iso-octane), or slowly increasing reading were obtained upon spraying the tetraethyllead samples (tetraethyllead was very difficult to wash out from the burner head with many solvents). The best solvent appeared to be methyl ethyl ketone.

Calibration curves were prepared from two standard substances, lead nitrate and lead 8-hydroxyquinolate dissolved in methyl ethyl ketone. In each instance 5.18 - 25.9 parts/10^6 of lead gave absorbances of 0.092 - 0.455. Tetraethyllead diluted with the same solvent gave a straight line calibration curve, but with a 3-fold decrease in absorbance for the same lead content (11.2 - 67.7 parts/10^6 of lead gave absorbances of 0.067 - 0.402).

A similar series of results was obtained using ethyl acetate as solvent. Standard additions of lead as its nitrate were also made, but the same conclusions were reached, i.e. disagreement between lead nitrate and lead 8-hydroxyquinolate on the one hand and tetraethyllead on the other.

They showed that the position of measurement in the flame was a vital factor. When the absorbance measurements were made at progressively higher positions in the flame relative to its base, the absorbance from lead nitrate or 8-hydroxyquinolate both containing 21 parts/10^6 lead in methyl ethyl ketone slowly increased. However, under the same conditions, the absorbance from tetraethyllead in methyl ethyl ketone decreased significantly (Table 32).

Table 32 - Effect of position of absorbance path in flame in relation to absorbance by lead nitrate and tetraethyllead (TEL)(Pb = 21 parts/10^6)

Position of absorption path in flame	Absorbance (283.3 nm) in methyl ethyl ketone	
	$Pb(NO_3)_2$	$(C_2H_5)_4Pb$
Top (most suitable for $Pb(NO_3)_2$ and $Pb(C_9H_6ON)_2$)	0.350	0.130
Middle	0.326	0.156
Base	0.289	0.203

Dagnall and West[162] concluded that satisfactory results for the determination of tetraethyllead in petroleum will only be obtained if the standard solutions are prepared from tetraethyllead. Their experiments on the position of the absorption path in the air-propane flame suggests that this is caused by a different "burn-off" rate or easier atomisation of tetraethyllead compared with lead nitrate or lead 8-hydroxyquinolate. Consequently, the population of ground state atoms is highest at the base of the flame when tetraethyllead is involved and measurements are best made at this point. The maximum population of ground state atoms of lead is obtained higher up in the flame when less readily atomised compounds of lead are present. It is probable that for the majority of lead compounds the upper regions of an air-propane flame may be preferable for absorption measurements.

It is to be noted that Robinson[160] also recommends the use of tetraethyllead rather than the more conventional lead standards. He used an oxygen-hydrogen flame and did not comment on the position of the absorbance path in the flame. Difficulties have been reported in the determination of lead in petroleum when using a high-efficiency burner of the total combustion type or a pre-mix burner[162,163]. The problem may be solved by using 1:1 v/v acetone-isooctane as the solvent for petroleum sample dilutions and the preparation of standard lead solutions rather than isooctane alone[163].

Table 33 compares results, obtained on leaded petroleum samples by atomic-absorption spectroscopy[163] using isooctane alone and mixed isooctane acetone as solvents with results obtained by various other established gravimetric and polarographic[171] techniques. It is seen that much better agreement with these latter techniques is obtained when the mixed solvent is used in atomic-absorption spectroscopy. The problem seems to be due to the complex phenomena associated with the vaporization and burning of the lead solutions at the burner tip and in the flame itself[163] and the success of the mixed solvent may be due to its increased vapour pressure[160,164,165], brought about by its acetone content relative to that of raw petroleum. Alternatively, it may be merely that the presence of acetone in the mixed solvent increases the solubility of tetraethyllead.

A typical atomic absorption method for the determination of lead in petroleum is that of Mostyn and Cunningham[166], quoted below:

Reference standards were prepared from tetraethyllead and lead-free petroleum covering the concentration range 0.6 ml tetraethyllead per imperial gallon. These were analysed by the Institute of Petroleum Chemical Method[172] and the results used for the calibration of the atomic absorption method. Operating conditions are shown in Table 34.

Table 33 - Mean lead content values of selected petroleum samples

Lead content (ml/Imperial gallon)		
	Atomic-absorption spectroscopy	
Gravimetry and/or polarography[b]	Isooctane	Acetone-isooctane[c]
2.15[d]	2.75	2.57
1.42	1.55	1.37
1.26	1.42	1.24
3.28	3.70	3.32

[b] Ref. 171
[c] 1:1 mixture of acetone and isooctane used as a sample diluent instead of isooctane alone.
[d] The experimentally determined values for all three methods were found to vary by less than 0.025 ml/Imperial gallon from the mean values listed.

Table 34 - Atomic Absorption Determination of Lead, operating conditions

Instrument	Peakin Elmer Model 303 double beam
Wavelength	283.2 nm
Lamp current	20 mA
Air flow supply at	25 psi; Flowmeter at 8½ divisions
Acetylene flow supply at	3 psi; Flowmeter at 2½ divisions
Burner height	3 divisions
Response time	3
Noise supression	x 2 (recorder read-out)
Scale expansion	x 3
Slit width	position 4 (3 mm)
Sample flow	3.5 ml/min
Spraying time	15 sec. *

* vital for good precision

The normal operating conditions for the determination of lead by atomic absorption are far too sensitive for undiluted petroleum samples to be used and dilution with an organic solvent is necessary to reduce the lead content of the test solution. The choice of dilution is closely linked with the very distinctive "memory" effect obtained when solutions of tetraethyllead are sprayed and measured. A dilution ratio of at least 50:1 is desirable to minimise the inconvenience, i.e. 2 ml of petroleum sample diluted to 100 ml with isooctane.

The memory effect is illustrated in Fig. 13. Recorder trace A is from a gasoline containing 3.21 ml tetraethyllead per imperial gallon diluted 50 times with isooctane. After 90 s spraying, no clear absorption plateau has been reached and over 4 min. washing-out with isooctane is required to return to zero absorption. The two traces, (B), from solutions of lead acetate in methylated spirit, show that the "memory" effect is not present in this case; the absorption reading levels off after 20 - 30 s and returns rapidly to zero when spraying is discontinued.

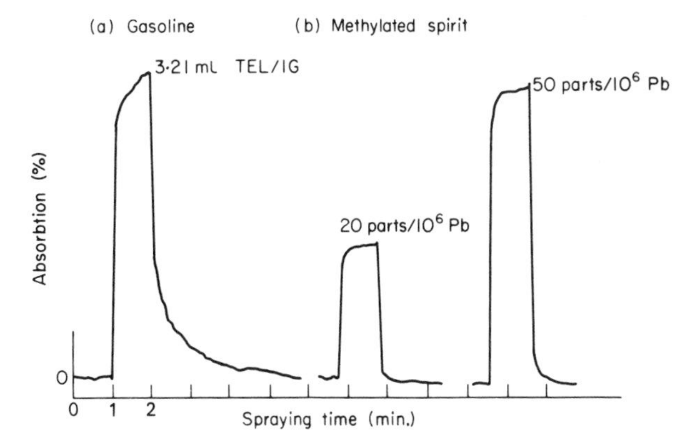

Fig 13. - Atomic absorption of spectroscopy of lead alkyls. The "memory" effect. Recorder trace of lead in petroleum (as tetraethyllead) in lead (as acetate in methylated spirits).

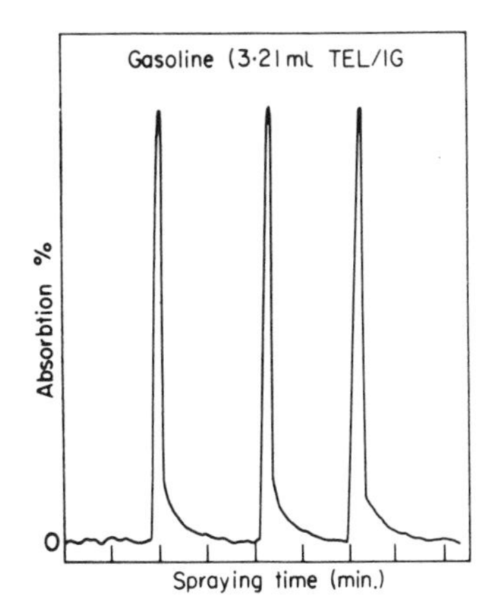

Fig. 14. - Atomic absorption spectroscopy of lead alkyls. Recorder trace of successive 15 second sprayings of a standard petroleum.

Fig. 14 shows three successive 15 s sprayings of a petroleum sample, each followed by a $1\frac{1}{2}$ - 2 min. washing-out spray. This technique gives very reproducible measurements and it is possible to reduce the washing-out period to about 1 min. for tetraethyllead contents below about 2.5 ml per imperial gallon. Hence it is recommended to wash out until the absorption falls to 1% or less, as shown by the null meter.

In Table 35 are given a comparison of results for tetraethyllead in petroleum obtained by Mostyn and Cunningham[166]. The Table lists respectively the tetraethyllead contents of synthetic standards given by chemical analysis, and the tetraethyllead contents indicated by drawing a smooth curve through the absorbance readings.

Table 35 - Lead content of synthetic reference standards ml TEL per imperial gallon

Standard	Chemical (1)	Atomic absorption (2)	Difference (1) - (2)
1	0.00	0.00	0.00
2	0.49	0.52	-0.03
3	1.10	1.04	0.06
4	1.55	1.54	0.02
5	2.02	2.02	0.00
6	2.48	2.53	-0.05
7	3.23	3.19	0.04
8	3.52	3.48	0.04
9	4.19	4.16	0.03
10	4.58	4.58	0.00
11	5.11	5.13	0.02
12	5.56	5.56	0.00
13	6.24	6.17	0.07

The standard deviation of the method is claimed to be 0.056 ml tetraethyllead per imperial gallon at a concentration level of 3.46 ml tetraethyllead corresponding to a coefficient variation of 1.6%. The method is precise over the range 0 - 0.6 ml tetraethyllead per imperial gallon.

The determination of organolead compounds by direct atomic absorption spectroscopy has also been discussed by Moore et al[167].

The workers point out that in spite of the number of attempts made during the past decade[160-163,167] to exploit the obvious advantages of atomic absorption spectroscopy to the direct determination of organolead compounds, in general, these have not been successful. The main problem is that tetramethyllead and tetraethyllead give rise to difference absorption coefficients for lead in the flame, necessitating very carefully matched standards. As the proportions of tetramethyllead and tetraethyllead may well vary over a given batch of samples, a method is required which is independent of the alkyl type ratio.

In this connection, Campbell and Moss[168] showed that lead alkyls may be efficiently extracted into aqueous iodine monochloride at room temperature. They applied the system to the extraction of trace amounts of lead from a variety of petroleum products excluding, however, gasoline. The reagent reacts rapidly with lead alkyls, converting them to water soluble dialkyl lead compounds. Inorganic lead compounds are also soluble in iodine monochloride, enabling the total lead content to be extracted from the gasoline. It was considered probable that after extraction and suitable dilution the aqueous iodine monochloride solution containing the lead could be sprayed directly into an air acetylene flame, and calibration effected using aqueous solutions prepared from a lead salt. The initial experiments examined the recovery of lead from synthetic blends of tetramethyllead and tetraethyllead in isooctane, covering the concentration range normally found in commercial gasolines. 5 ml of each mixture were pipetted into a 100 ml separating funnel containing 10 ml of 1 M iodine monochloride. After the addition of 25 ml of isooctane the mixture was shaken for 3 min, and the layers allowed to separate. The lower aqueous phase was transferred to a

100 ml volumetric flask, 10 ml of water was added to each funnel and the mixture shaken for 1 min after which the aqueous phase was transferred to the respective flask and then diluted to the mark with water. Standard solutions were prepared from aqueous lead nitrate to cover the range 5 - 50 mg/1 lead.

The instrument used for this investigation was a Unicam SP90 Series II Atomic Absorption Spectrophotometer, fitted with a standard 10 cm air acetylene burner. Absorption measurements were made at 283.3 nm, aspirating water between each sample and standard solution. A lean air-acetylene flame was used.

Table 36 shows the extraction efficiency for the series of tetramethyllead and tetraethyllead nominal concentrations and ratios. An extraction efficiency of better than 96% was indicated and it was felt that this might well be improved if a second wash with water was included in the extraction procedure. In Table 37 are compared results obtained on petroleum samples by this method using a single water extraction and by X-ray fluorescence analysis. The standard deviation found at the 3 g lead per imperial gallon level was 0.024 with a coefficient of variation of 0.87%.

Atomic absorption spectroscopy has been applied to the determination of organolead compounds following their separation on a gas chromatographic column. Ballinger and Whittemore[169] separated tetramethyllead, trimethylethyllead, dimethyldiethyllead, methyltriethyllead and tetraethyllead on a column comprising 60/80 mesh Chromosorb P coated with 1% potassium hydroxide operated at 85°C and determined lead in quantities down to 20 ng in the effluent by the absorption of the lead 283.3 nm emission line. Kolb et al 170 and Ballinger and Whittemore[169] applied a similar technique to the analysis of the five methylethyllead compounds.

Table 36 - Analysis of tetraethyllead/tetramethyllead mixtures by iodine monochloride/atomic absorption procedure

Nominal Tetraethyllead	Composition (gPb/Imperial Gallon Tetramethyllead	Total lead	Recovery (%)
0	0.06	0.06	99.5
0.6	0	0.06	99.5
0.6	0.6	1.2	96.5
0	1.2	1.2	96.0
1.2	0	1.2	99.0
1.2	1.2	2.4	97.3
1.8	0	1.8	98.1
0	1.8	1.8	99.0
1.8	1.8	3.6	96.7

Vickrey et al[798] have studied the effect of the surface treatment of graphite furnace tubes for the analysis of lead in organolead compounds by graphite furnace atomic absorption spectrometry.

Scott and Halbale[799] determined organolead in gasoline by Zeeman atomic absorption spectrometry.

Table 37 - Comparison of determination of lead in petroleums by iodine monochloride/atomic absorption (AAS) and by X-ray fluorescence (XRF) methods

	Lead (g)/Imperial gallon	
	AAS	XFR
Petrol A	2.80	2.72
Petrol B	2.82	2.82
Petrol C	2.76	2.80

Torsi and Palmisano[800] have recently studied the electrostatic capture of gaseous tetralkyllead compounds and their determination by electrothermal atomic absorption spectrometry.

The determination of organolead compounds in gasoline has been recently reviewed by Holding and Palmer[801].

Flame Photometry. Procedures for the determination of tetraethyllead in petroleum by flame photometry have been described[173-176]. The sample is burned in a flame fed with oxygen and hydrogen at 293 and 14 nmHg pressure, respectively. The flame is measured with a monochromator at either 406 or 402 nm by means of a photocell previously calibrated with petroleum of known lead content. Emission spectrography using a medium-dispersion quartz spectrograph and a high-voltage spark discharge in conjunction with a rotating double-disc electrode has been used to determine lead[177]. The film of leaded petrol is transferred by contact onto spectrally pure carbon. The internal standard consists of a cobalt-pentanol complex. Standards are analysed by using the following line pairs: Pb 257.73 - Co 276.42 nm; Pb 282.32 - Co 301.76 or 276-42 nm; and Pb 287.33 - Co 301.76 or 307.23 nm. The method permits the determination of 0.005 - 1% of lead with a relative error of ±3%.

X-Ray Fluorescence. X-Ray fluorescent permits the determination of tetraethyllead in the concentration range 0.1 - 6 ml/Imperial gallon to be determined with a standard deviation of ±0.28 ml/Imperial gallon[178]. The error caused by sulphur in the sample is very small and that due to possible petrol additives such as phosphorus is negligible. The time required for one analysis is about 5 - 10 min. For samples of petroleum containing about 0.1 ml/1 of tetraethyllead a molybdenum anti-cathode (50 kV, 12 mA), a curved crystal of lithium fluoride (r = 110 cm), and a scintillation counter to ensure maximum sensitivity have been used[179]. The lines $L_1 + L_2$ of lead and K (second order) of molybdenum are most suitable, but an internal standard must be used if other elements, e.g. 0.05 - 0.1% of sulphur, are at present. The standard deviation is 0.85%; the lower limit of determination is 25 ppm of lead. About 0.25% of bromine (as bromobenzene in tetralin) must be present as internal standard to compensate for the interference by sulphur and the fluorescence should be excited at 50 k V and 8 mA. The standard deviation is ca. 0.8%. Petroleum manufactured from different crude oils derived from different oil fields does not greatly effect the results in the determinations of tetraethyllead by X-ray fluorescence[180]. The organic bromine scavenger compound added to the petroleum with the tetraethyllead also has little effect, provided that the bromine to lead ratio in the sample is constant, which is the case for any particular additive composition. Good overall agreement in the determinations of lead in petrol between X-ray fluorescence using platinum metal as internal standard and a variety of chemical, gravimetric and X-ray absorption procedures has been reported[181]. Other references on the use of X-ray fluorescence for the determination of lead in petroleum include refs 176, and 182 - 184.

Tetraethyllead in petroleum has been determined by X-ray absorption methods [185-189]. One method consists of measuring the absorption increment corresponding to the sublevel L_{111} of the tetraethyllead present in the petrol[185]. An apparatus of the General Electric type XRD^3 was used with an anti-cathode tube of molybdenum (18 kV). The emitted radiation, rendered monochromatic by a crystal of sodium chloride, is passed through the cell containing the petroleum sample and its intensity is measured for various angular positions with a Geiger counter connected to a scaler. Constructional details of an X-ray absorptiometer and its application to the determination of tetraethyllead in petroleum over the range 0.0 - 1.5 ml/1 with an accuracy of ±0.005 ml/1 have been described[186]. Results obtained by the tritium Bremsstrahlung technique agree to within ±0.02% with those by the gravimetric procedure for concentrations of 0.5 - 2% of tetraethyllead in petroleum[189]. Absorption measurements were made on a sample dissolved in heptane. Equations were derived for calculating the content of tetraethyllead in the presence of dibromoethane and 1-chloronaphthalene.

β-Ray Back Scattering. Ashbel et al [190] give a brief description for the determination of tetraethyllead using this technique with Sr as a source.

Mass Spectrometry. Howard et al[191] have described an accurate mass spectrometric technique which permits the determination in petroleum of tetramethyllead, tetraethyllead or mixtures of these two compounds. A specific advantage of this technique, is that it is capable of rapidly determining in petroleum the mixtures of tetraethyllead and tetramethyllead which are now becoming increasingly used by the petroleum industry, instead of tetraethyllead alone. The concentration of tetramethyllead and tetraethyllead fluid in gasolines is determined in ml per gallon by comparing the mass spectrum of the sample with mass spectra of two synthetic gasoline blends, one having a known concentration of tetramethyllead and the other a known concentration of tetraethyllead.

Howard et al[191] analysed a range of synthetic mixtures of tetramethyllead fluid, tetraethyllead fluid and petroleum by their mass spectrometric method and found good agreement with the expected values (Table 38).

They analysed ASTM cooperative samples of tetraethyllead in petroleum and service station petroleum of various brands by their method and by an ASTM method. These results, shown in Table 39, are in good agreement, illustrating the compatibility of two methods where only tetraethyllead is present.

Howard et al [191] pointed out that a possible limitation regarding the ability of the mass spectrometric method to differentiate between tetraethyllead and tetramethyllead could be the use of some additive which might mark the presence of tetramethyllead at m/e + if 253 or tetraethyllead at m/e + if 295. Even if an additive of this sort were used, its presence would be easily recognised and appropriate corrections could be made.

They attempted analysis of petroleum containing tetramethyllead and tetrathyllead with a high molecular weight mass spectrometer. The sample was inserted by capillary dipper through a liquid-gallium valve into an inlet system heated to 300°C. No tetraethyllead or tetramethyllead peaks were observed. The lead compounds either decomposed under these conditions or did not pass through the gallium.

Methods for the determination of lead anti-knock compounds in petroleum have been reviewed[193-197]. Results from ten cooperating laboratories indicate that the limits of reproducibility for compounds other than tetraethyllead are not as precise as those quoted in the ASTM Standard Method[193]. Consequently, wider limits are quoted in the revised standard for petroleum containing the more volatile anti-knock agents[193]. Critical comparisons have been given of flame photometric and complexometric methods [195] and of various polarographic, colorimetric, and gravimetric methods[196] for the determination of tetraethyllead.

Table 38 - Mass spectrometric results on known blends of tetramethyllead tetraethyllead in petroleum. (All values in ml of tetramethyllead and tetraethyllead fluid per gallon).

Blend	Tetramethyllead		Tetraethyllead	
	M.S. *	Synthetic	M.S.	Synthetic
1	0.14	1.14	2.92	2.86
2	0.50	0.50	2.51	2.50
3	0.74	0.75	2.21	2.25
4	1.18	1.20	1.88	1.80
5	1.52	1.50	1.52	1.50
6	2.27	2.25	0.75	0.75
7	2.52	2.50	0.51	0.50
8	2.91	2.86	0.17	0.14

* Mass spectrometry

Table 39 - Tetraethyllead content of ASTM cooperative samples and commercial gasolines. (All values in ml of tetraethyllead per gallon; samples 4 to 10 are commercial gasolines).

	M.S.	ASTM #
ASTM-1	4.68	4.75*
ASTM-2	0.49	0.46*
ASTM-3	2.97	2.84*
4	0.54	0.52
	0.55	0.55
5	2.75	1.78
	2.82	2.94
6	2.48	2.75
	2.50	2.77
7	0.44	0.45
	0.44	0.46
8	2.84	3.02
	2.91	
9	0.94	0.91
	0.98	
10	3.14	3.28
	3.30	3.39

* Average of 22 separate determinations on each sample by 6 different laboratories.

(American Society Testing Materials 1956)

C. Determination of Carbon and Hydrogen

The general opinion in the limited amount of published work[156,198,199] is that the presence of lead does not offer any serious difficulties in the determination of these elements in organolead compounds. About 4 mg of the sample may conveniently be burnt in a Heraeus furnace at 850°C and the combustion gases conducted in a stream of pure oxygen at 500°C over platinum gauze, silver permanganate, and silver gauze, respectively[156]. To avoid the risk of explosion the organolead sample is covered with about 20 mg of tungsten oxide.

D. Determination of Halogen

Organolead compounds containing ionic halogen can be titrated directly in ethanol or acetone solution with standardized silver nitrate [156]. If the halogen is covalently bonded, or if the ionic halide cannot be titrated directly, the sample can be completely decomposed by Parr bomb combustion in the presence of sodium perioxide. In the case of a bromine or iodine determination it is recommended that the decomposed sample is reduced with hydrazine in order to avoid losses through free halogen formation. It is also necessary to employ this reduction when determining chlorine in the presence of lead, otherwise some chlorine gas is evolved causing the chlorine content found to be low. The chlorine is formed probably through oxidation by tetravalent lead. This possibly accounts for the low chlorine recoveries reported[200].

ORGANOLITHIUM COMPOUNDS

A. Determination of Lithium, Carbon, Hydrogen and Oxygen

A microcombustion procedure has been described for the determination of lithium, carbon and hydrogen in organolithium compounds[201]. The sample is intimately mixed with finely ground quartz in an empty tube and combusted in a stream of oxygen. Combustion in an empty tube prevents absorption of carbon dioxide by the lithium residue and permits carbon and hydrogen to be determined by standard methods. The amount of alkali metal can then be obtained from the weight increase of the ignition tube after the combustion.

Head and Holley[202] have described a procedure for the determination of carbon and hydrogen in lithium hydrides and it is possibly also suitable for the analysis of organolithium compounds. This is a modified conventional combustion procedure which gave recoveries exceeding 99% for hydrogen and exceeding 99.5% for carbon.

Honeycutt[203] Johnson and Clark[204] and Juenger and Seyfirth[205] have described methods for the determination of carbon and hydrogen in vinyllithium.

Very low levels of oxygen in butyllithium can be determined by using butylllithium as a source of tritons in the determination of oxygen by ^{18}F counting after activation according to the reaction $^{6}Li(n,\ )t$; $^{16}O(t,n)$ ^{18}F [206]. The sample is mixed with butyllithium; if monomers are present, trimethylaluminium is added to overcome the polymerizing effect of butyllithium. The solvent or monomer is distilled in a high vacuum. The residue of butyllithium is irradiated with a neutron flux of $5 = 10^{12}$ neutrons $cm^{-2}s^{-1}$. After addition of fluoride ion as carrier, the total fluorine is distilled as hydrofluosilicic acid; this is hydrolysed, and the fluoride ions are precipitated as lead chlorofluoride. The ^{18}F is determined by counting the positron-destroying radiation.

ORGANOMAGNESIUM COMPOUNDS

A. Determination of Carbon and Hydrogen

A combustion procedure which could probably be applied to the analysis of organomagnesium compounds has been described.[202]

B. Determination of Alkyl Groups

The determination of alkyl groups in Grignard compounds is based on hydrolysis with dilute sulphuric acid to produce a hydrocarbon which can then be determined gas volumetrically[207].

$$RMgX + H_2O \rightarrow RH + Mg(OH)X$$

The method, as used, is restricted to those RMgX compounds and their etherates which give a hydrocarbon that is gaseous at ordinary temperatures.

ORGANOMANGANESE COMPOUNDS

A. Determination of Manganese

In one method for the determination of manganese in organic compounds, the sample is digested with nitric acid or aqua regia in a long-necked flask, the excess of acid is evaporated off, and dilute sulphuric acid containing hydrogen peroxide is added[208]. The solution of manganese(II) sulphate is then titrated with standard potassium permanganate. The error does not exceed ± 0.3% (absolute).

In a further method, the sample is heated with potassium hydrogen sulphate and mercury(II) oxide moistened with 98% sulphuric acid[209]. The neutralized solution is titrated with 0.05 N potassium permanganate. This method was applied successfully to a wide range of cyclopentadienylmanganese tricarbonyl derivatives containing between 14 and 20% of manganese without interference by various elements present in the sample as major constituents including chlorine, iodine, nitrogen, sulphur and mercury.

ORGANOMERCURY COMPOUNDS

A. Determination of Mercury

Mercury in organic compounds can be determined by burning the sample at 750 - 800°C in a standard automatic combustion furnace[210]. The carrier gas (nitrogen mixed with a small amount of oxygen) sweeps the combustion products into a stationary furnace containing reduced copper gauze and combustion catalyst (CuO, MnO_2 and Co_3O_4). Sulphur and halogens are absorbed in an auxiliary furnace in a tube containing MnO_2, Co_3O_4, and silver granules at 550°C. The absorption tube for the mercury is placed close to the last furnace so that its temperature remains between 40 and 100°C by radiation. Under these conditions mercury is quantitatively absorbed on silver granules and may be weighed. The results show a deviation of the mean from theoretical values of -0.06% to +0.08% with a standard deviation ranging from 0.10% to 0.17%. Tube combustion and wet oxidation methods for the determination of mercury in organic material are also available[212,213]. If halogens are present, loss of mercury as halide can be avoided if the sample is placed in a combustion tube filled with calcium oxide and burnt in a current of air at 370°C [214]. Mercury is then collected in a bubbler containing concentrated nitric acid, and determined by titration with 0.005 N potassium thiocyanate in the presence of hydrogen peroxide with iron alum as indicator.

In a further method the sample is heated in a quartz tube in a stream of pure dry nitrogen[215]. Nitrogen oxides, halogens, sulphur, phosphorus, and arsenic compounds are absorbed by a 6-cm layer of a decomposition product of potassium permanganate. The mercury is subsequently absorbed on silver sponge and weighed. The results quoted show an error of ±0.2%.

Organomercury compounds may be analysed by placing them between layers of calcium culphide and heating in a slow stream of air[216]. The vapours are passed through successive layers of granular calcium oxide and silver pumice at 750 - 800°C and the mercury is absorbed in a tube containing gold leaf. The method is suitable for 1 - 20 mg of mercury. In a further method, the sample is pyrolysed in a stream of hydrogen and burnt in an oxy-hydrogen flame, ensuring that a portion of the oxygen feeding the flame is bubbled through saturated bromine water[209]. The mercury(II)bromide so formed is collected in water. The excess of bromine in the condensate is removed by adding hydrazinium chloride in small portions until the color is discharged. After stirring, 0.1 M EDTA (disodium salt) is added, followed by pyridine, and the mercury is determined by potentionmetric titration with 0.01 M sodium diethyldithiocarbamate. The results by these last two methods range from 98 to 101% of the theoretical.

Several workers have described methods based on the Schoniger oxygen flask combustion technique for the determination of mercury in organomercury compounds[171,212,230]. The sample may be burned in a flask containing nitric acid in which the mercury is absorbed. Following adjustment of pH the mercury is titrated amperometrically using ethylenedinitriloacetic acid. The only commonly encountered interference comes from chloride ion, which stablizes the mercury as mercury (I) chloride. It is then necessary to reflux the sample in the nitric acid to oxidize the mercury to the divalent form. For o(3-hydroxymercuri-2-methoxy-2-propylcarbamyl)phenoxyacetic acid, the 95% confidence interval is 15 parts per 1000 for both the micro and the semi-micro determinations. Alternatively, after oxygen flask combustion a simple visual titration method with sodium diethyldithiocarbamate may be used[220-222]. Chlorine and bromine, if present in the sample, can be titrated immediately after the mercury determination in the same combustion run. Iodine cannot, however, be determined as it leads to indistinct end-points and low recoveries (80 - 90%). Other procedures using an oxygen flask include (i) the use of 8 M nitric acid as an absorbent solution and determining mercury gravimetrically as $[Co(NH_3)_6][Hg)S_2O_3)_3]_3 \cdot IOH_2O$ [223] and (ii) applying the sample to a strip of filter-paper and when burning it, using saturated aqueous bromine water as the absorbing liquid[224,225]. After combustion, excess of bromine is removed by aspiration. A measured volume of 0.005 M EDTA is then added and the buffered solution titrated with 0.01 M zinc chloride to the 1-(2-pyridylazo)-2-naphthol end-point in the presence of potassium iodide.

Southworth[219] burn the sample in an oxygen filled flask containing nitric acid in which the mercury is absorbed. Following pH adjustment to

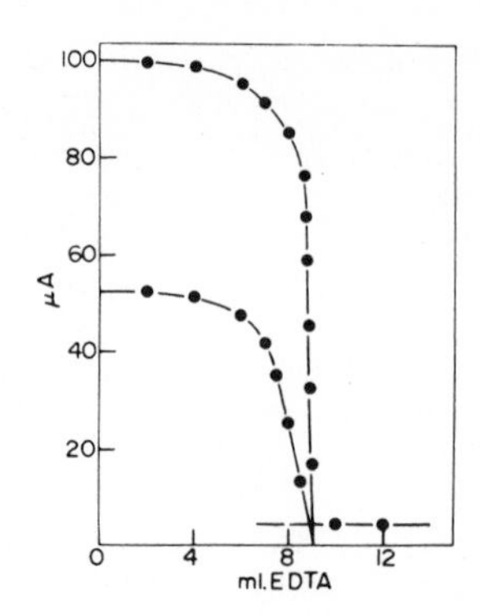

Fig. 15 - Titration of mercuric nitrate 0.00104 M with 0.003 M EDTA.

7.5 the mercury is titrated amperometrically using ethylenediamine tetra-acetic acid. A typical titration curve is shown in Figure 15. The only commonly encountered interference comes from chloride ion, which stabilizes the mercury as mercurous chloride. It is then necessary to reflux the sample in the nitric acid to oxidize the mercury to the divalent form. Volatile organic mercury compounds can be treated by the procedure outlined by Schoniger for liquid samples. The results of determination of mercury by this method are tabulated in Table 40. Using the results for o-(3-hydroxy mercuri-2-methoxy-2-propyl)(carbamyl) phenoxy acetic acid, the 95% confidence interval is 15 parts per 1000 for both the micro and the semi-micro determinations.

Gouveneur and Hoedeman[220] confirmed the suitability of the oxygen flask technique for mercury determinations in organomercury compounds. For the analysis of mercury II ions formed after flask combustion they used the simple visual titration method with sodium diethyldithiocarbamate, based on work by Wickbold[221] as elaborated by Roth and Beck[222], Gouveneur and Hoedeman[221], found that chlorine and bromine, if present in the sample, can be titrated immediately after the mercury determination in the same combustion run. Iodine could not, however, be estimated as it led to indistinct end-points and low recoveries (80 - 90%). Some results obtained by this procedure are given in Table 41.

The titration method of Roth and Beck[222] is based on the fact that on addition of sodium diethyldithiocarbamate to an ammoniacal solution (pH c 9) containing both mercury(II) and copper(II) ions, first the white mercury(II) carbamate complex is precipitated quantitatively, followed by the yellow-brown copper(II) carbamate complex. In actual titrations a copper(II) solution is added as the indicator. The colour change from white to yellow-brown copper (II) carbamate complex. In actual titrations a copper (II) solution is added as the indicator. The colour change from white to yellow-brown is particularly sharp, even with a 0.005 N carbamate solution, if a few ml of chloroform are present to concentrate the colour.

The carbamate solution is not stable and should be standardized every day against standard mercuric sulphate solution. The normality of the solution was found to decrease by about 0.4% per day; this value was also observed by Wickbold[221].

Several procedures based on digestion with mineral acids have been published for the determination of mercury in organic mercurials[226-234]. The sample may be digested by treatment with 75% sulphuric acid and then with

Table 40 - Titration of mercury with (ethylenedinitrilo) tetraacetic Acid

Sample	% Hg Calcd.	% Hg Found*	Sample size, Mg.	No. of Detns.	% Hg found Av.	Max.	Min.
o-(3-Hydroxymercuri-(2-methoxypropyl) carbamyl)phenoxy-acetic acid	43.06	42.5	10.40	11	42.5	43.0	42.4
			2-4	5	42.6	43.0	42.2
Anhydro-N-(-methoxy- -hydroxymercuri-proply)-2-pyridone-5-carboxylic acid	48.9	49.4	10.40	6	49.1	49.4	48.7
			2-3	2	49.1	49.4	48.7
3-Chloromercuri-2-methoxypropylurea	54.63	54.5	15-30	2	54.3	54.4	54.2
3-Carbethoxymethyl-mercaptomercuri-2-methoxypropylurea	47.44	—	15	1	47.6	—	—
3-Acetoxymercuri-8-carbethoxymethyl-4-methoxy-1,6,8-triazabicyclo-(4,3,0) nonane-7,9 dione	37.86	37.8	20	1	37.6		

Table 41 - Analysis of organic compounds containing mercury and chlorine or bormine

Compound	Mercury % Calcd.	Found
Phenylmercuric nitrate with added p-chlorobenzoic acid	59.27	59.86
		59.43
P-chlorobenzoic acid	56.16	56.33
		56.01
Phenylmercuric acetate (with added bromobenzoic acid)	59.57	59.86
		59.74
P-mercuricbenzoic acid	56.16	56.01
		56.33

potassium permanganate, heating until the odour of bromine has disappeared 226. After the addition of acidic hydrogen peroxide the solution is boiled to destroy excess of peroxide and titrated with 0.1 N ammonium thiocyanate to the ferric alum end-point. The sample may also be digested with a mixture of sulphuric acid (60%) - nitric acid (70%) - perchloric acid (concentrated) (3:3:1) in an ignition tube for 30 min[227]. The mixture is washed into a beaker with water and excess of 0.005 M EDTA added. The pH is adjusted to 10 and excess of EDTA titrated with 0.005 M magnesium sulphate to the blue to purple Eriochrome Black T colour change. An error not exceeding ±0.4% is claimed for this method. In a further method the sample is wet oxidized with nitric acid, boiled until colourless, and mercury is determined by EDTA titration[228].

A procedure which determines total mercury, mercury(II) acetate, triacetoxymercuribenzene, and diacetoxymercuribenzene in crude phenylmercury (II) acetate involves heating the sample to fuming with concentrated sulphuric acid, diluting, and boiling with bromine water[229]. After making the solution alkaline with sodium hydroxide and aqueous ammonia, mercury(I) is titrated with 0.05 N sodium thioglycollate, with thiofluorescein as indicator. Mercury(II) acetate is determined in an aqueous extract by titration before and after decomposition with bromine water. The residue insoluble in hot 80% acetic acid is filtered off and weighed as triacetoxy-mercuribenzene; after dilution of the filtrate with water the precipitated diacetoxymercuribenzene is separated and the total mercury remaining in the clear solution is again determined.

A further digestion procedure for the determination of 2 - 20 mg amounts to mercury in organic matter depends on digestion of the sample with a mixture of nitric acid, sulphuric acid, and perchloric acids in a Kjeldahl flask fitted with a separating funnel, condenser, and receiver[230]. After digestion, a mixture of hydrochloric and hydrobromic acids is added to isolate the mercury in the residue, which is then extracted as the tetraiodo complex using ethyl acetate at pH 2 - 3 and finally determined colorimetrically using dithizone. If the sample does not contain antimony and tin, then the formation of the tetraiodide complex can be omitted. The recovery of mercury is between 87 and 100%. Mercury in plant protective substances such as phenylmercury(II) acetate, phenylmercury(II) chloride, and methoxyethylmercury(II) silicate may be determined by boiling the sample with an aqueous mixture of potassium iodide and iodine and concentrated sulphuric acid[231-233]. Excess of iodine is removed, the filtered residue is washed with boiling water, and enough EDTA (disodium salt) is added to complex contaminating metals, such as iron, aluminium, zinc, magnesium, manganese, nickel and cobalt, followed by the dropwise addition of concentrated aqueous ammonia to pH 7. The solution is boiled and treated with a boiling saturated solution of the reagent $[Cu(en)_2](NO_3)_2.2H_2O$(en = ethylenedimine) and mercury is determined gravimetrically as $[HgI_4][Cu(en)_2]$. The accuracy is about ±1%.

Precipitation of mercury with Reinecke salt has been used for the determination of mercury in organomercurial fungicides such as phenyl-mercury(II) acetate, nitrate, or borate, diphenylmercury(II) and methoxy-ethylmercury(II) silicate[235] and drugs[236].

A method for the identification and determination of mercury in N-organomercury compounds is based on the reaction of N-organomercury compounds with thiols, whereby S-aryl(alkyl)mercury compounds are quantitatively produced[237]. To determine N-organomercury compounds, the sample is treated with excess of an ethanolic sodium salt of 2-mercaptobenzothiazole. The precipitate is filtered off and excess of 2-mercaptobenzothiazole in the filtrate is titrated with 0.1 N iodine solution, with starch as indicator. The error of this determination if ±0.3%.

In an ignition method for the determination of mercury in phenylmercury (II) acetate and in mercury(II) acetate the sample, in a nickel crucible, is covered with layers of copper oxide, copper, iron, and calcium oxide and heated at 500 - 750°C[238]. The crucible is covered with a gold plate of known weight. The mercury evolved is retained by the gold plate and weighed. A photometric method for the determination of mercury in ethylmercury(II) chloride, phenylmercury(II) acetate, and phenylmercury(II) chloride uses copper diethyldithiocarbamate as the chromogenic reagent[239]. A very simple method for the determination of total mercury in organomercurials involves reduction of the sample with zinc amalgam in glacial acetic acid followed by dissolution of the filtered off amalgam in nitric acid and titration of the solution with standard ammonium thiocyanate[240]. A review has discussed the differential determination in organomercurials of phenylmercury(II) acetate and metallic mercury[241].

A semi-automated procedure has been developed for the determination of mercury in fish and animal tissue based on digestion with concentrated nitric acid and sulphuric acids at 58°C followed by flameless atomic-absorption spectroscopy of the treated extract[242-244]. Using Technicon Auto Analyzer Equipment, samples can be analysed at the rate of 30 samples per hour with a recovery of 95% (standard deviation ±3 - 8%). This digestion has been found to be satisfactory for all types of fish meat and other food products. Mercury in silicon-containing organomercurials can be determined by atomic-absorption spectroscopy[245]. Cold vapour atomic-absorption spectroscopy has been used for the determination of methylmercury in muscle of marine fish[246] and to determine low levels of organic mercury in natural waters after pre-concentration on a chelating resin[247]. Volatile mercury compounds in air have been determined with a Coleman mercury analyser system[249]. Methylmercury compounds in fish have been analysed using a graphite furnace atomic-absorption spectrometer to analyse a dithizone-treated toluene extract of the sample[250]. A detection limit of 0.08 µg/g of mercury is claimed.

Oda and Ingle[277] have described a procedure for identifying inorganic and organic mercury species in water by cold vapour atomic absorption spectrometry using selective reduction. Inorganic and organic mercury are selectively reduced by stannous chloride and sodium borohydride respectively. Detection limits are in the 0.003 - 0.005 ppb range and both types of mercury are determined in - 3 min in a 1 ml sample.

Burns et al[295] have a method using atomic absorption spectrometry for the determination of mercury in silicon containing organomercury compounds.

Langmyhr and Aarmodt[893] applied atomic absorption spectrophotometry to the determination of organic mercury in sediments and aquatic organisms [894]. Hanamura et al[895] applied thermal vapourization and plasma emission spectrometry to the determination of inorganic and mercury and organomercury compounds in solid biological samples. Mercury compounds in air using a Coleman mercury analyser system have been determined (see section on air analyses).

Neutron activation analysis[896] has been very successfully applied to the determination of mercury in the parts per billion range. This technique does not, of course, distinguish between inorganic mercury and organically bound mercury.

B. Determination of Carbon, Hydrogen, Sulphur, Halogens and Oxygen

Lebedeva and Kramer[251] used the apparatus developed by Korshun and Sheveleva[252] for the simultaneous micro-determination of carbon, hydrogen and mercury in samples free from halogen. The sample (3-7 mg) is decomposed in a stream of oxygen. (18 - 20 ml/min) at 900°- 950°C. The products of decomposition are burned at 600°C - 650°C over Co_3O_4, and the mercury formed is collected on silver-impregnated pumice. The water and carbon dioxide are collected in anhydrite and Ascarite, respectively. The accuracy of the determination is within ±0.3% for carbon and hydrogen and within ±0.5% for mercury.

Lebedeva and Fedorova[253] developed a further method for the micro-determination of carbon, hydrogen and mercury in organomercury compounds that do not contain halogen. In this method the organomercury compound is burnt in a current of oxygen in a quartz tube containing the product of the thermal decomposition of silver permanganate heated at 60°. Carbon and hydrogen are determined with errors not greater than ±0.3%. The rapid micromethod for the determination of mercury in organomercury compounds that do not contain halogens involves burning them, absorbing the reaction products in concentrated nitric acid and titrating the mercuric ions formed with 0.1 N ammonium thiocyanate using ferric alum as indicator.

A further combustion procedure for the determination of carbon and hydrogen in mercury-containing organic compounds has been described by Abramyan and Kocharyan[254]. The sample (3 - 5 mg) is burnt in oxygen at 850° - 900° and the combustion products are passed through a catalyst at 400° - 450°. The catalyst consists of the thermal-decomposition products of potassium permanganate on asbestos, and is satisfactory for about 60 analyses; it must then be regenerated by being heated at 800° - 900°. Mercury in the sample is oxidized and retained on the catalyst, as are halogens and their compounds. Carbon and hydrogen are determined gravimetrically by this procedure with an accuracy within ±0.25%.

Holmes and Lander[255] point out that most methods reported for the determination of carbon and hydrogen in organic compounds containing mercury are based on slow combustion procedures in which, during the combustion, elemental mercury is retained temporarily by fillings of the Pregl universal type. Such fillings will, in subsequent determinations, pass on to the absorption train, resulting in high hydrogen values. Various workers have overcome this effect by using gold wire in the beak-end of the combustion tube[256-260]. A boat containing ceric oxide, litharge, silver dichromate, silver oxide and lead chromate placed immediately after the ceria-copper oxide-lead chromate combustion catalyst was used[261]. A rapid method based on the cobalto-cobaltic oxide method of Vecera[262,263] has been developed by Garwagious and Macdonald[264]. In this method the exit tube and beak-end are packed with tightly coiled gold wire to remove the mercury. It is necessary to regenerate the gold after 5 - 6 determinations. Holmes and Lander[255] adopted the "rapid" empty-tube method as described by Belcher and Ingram[265] for the determination of carbon and hydrogen.

Initially they attempted to remove the mercury by packing the beak-end of the combustion tube with gold foil. This failed, presumably owing to the short contact time. Correct carbon and hydrogen values were obtained, however, when gold deposited on Gooch crucible asbestos, loosely packed into a Flaschentrager tube, was placed between the beak-end of the main combustion tube and the water-absorption tube.

Numerous determinations can be carried out by this procedure without it being necessary to regenerate the gold-impregnated asbestos. Although it might be expected that there would be some loss of water in the tube of gold-impregnated asbestos, as this is not heated, in fact with prolonged flushing at the flow-rate specified such a discrepancy does not occur. Some typical results are given in Table 42.

Holmes and Lander[255] conclude that this method gives satisfactory results that are within the accepted limits of accuracy It permits the analysis of mercury compounds to be carried out relatively rapidly and only necessitates one combustion apparatus, for the determination of carbon, hydrogen and mercury.

Pechanec and co-workers have described, in a series of papers, methods based on the Dumas procedure for the micro or semi-micro determination in organomercury compounds of carbon and hydrogen (Pechanec and Horacek[266], mercury and halogens[267], carbon, hydrogen and mercury[268], sulphur[269] and halogens[270].

Pechanec and Horacek[271] claim that these methods, especially those for the determination of sulphur (Pechanec[269]) and halogens (Pechanec[267,270]) are applicable to all solid and liquid organomercury compounds with a maximum error of ±0.3% and a maximum analysis time of 50 minutes. In all the procedures the usual Duman combustion apparatus is used, but because of the presence of mercury vapour, the tube is packed either with a layer of the decomposition product of potassium permanganate or of silver permanganate as a combustion catalyst and absorbent for sulphur and halogens, or copper granules to decompose nitrogen oxides, or a layer of silver sponge to absorb mercury vapour which causes errors in the conventional Dumas procedure. These procedures are described below:

Determination of Carbon and Hydrogen (Pechanec and Horacek[268]) A layer of silver sponge is used to absorb the mercury in the cooler part of the combustion tube. Conditions recommended are an oxygen flow rate of 15 ml/min a layer of Co_3O_4 to catalyse the combustion, and silver in the hot zone to remove halogens and sulphur oxides. A 2.5 cm layer of silver sponge in a combustion tube of 11 mm diameter then suffices for 200 determinations.

Determination of Carbon, Hydrogen and Mercury (Pechanec[268]). The combustion tube (heated at 550° - 600°) is packed with a layer of the decomposition product of silver permanganate (prepared by heating $AgMnO_4$ at 90° - 95° for 24 h), a layer of copper granules, and a further layer of the decomposition product. Mercury is trapped in an absorption tube packed with silver sponge and silver wool; carbon dioxide and water are absorbed in the usual manner.

Determination of Mercury and Halogens (Pechanec[267]). The combustion of the sample is carried out at 700 - 750°C in an atmosphere of nitrogen in the presence of granular decomposition products of potassium permanganate on a support of glass splinters. Mercury is absorbed in a layer of silver sponge (obtained by reduction of silver nitrate by acetaldehyde) and determined gravimetrically. The halogens are washed out of the combustion tube with water and titrated.

Determination of Sulphur (Pechanec[269]). The sample is burnt in oxygen in the presence of Mn_2O_3; any $MnSO_4$ formed is extracted with boiling water and Mn^{2+} is determined complexometrically with EDTA. The error for determinations of sulphur in di(phenylthio)mercury varied from 0.11 to +0.09%.

Abramyan and Megroyan[272] have described the micro-determination of carbon, hydrogen, mercury and chlorine or bromine in organomercury compounds.

Table 42 - Results of the determination of carbon and hydrogen in various compounds.

Compound *	Carbon Content % calculated	Carbon Content % found	Hydrogen Content % calculated	Hydrogen Content % found
$(MeHgPMe_3)Cl$	14.5	14.7	3.70	3.73
$(MeHgPMe_2Ph)Cl$	27.8	27.5	3.60	3.66
$(MeHgPEt_3)Br$	20.3	20.5	4.35	4.50
$(MeHgPMe_2Ph)I$	22.5	22.5	2.94	2.92
$(etHgPMe_3)Cl$	17.6	17.4	4.13	4.23
$(EtHgPEt_3)Br$	22.5	22.5	4.71	4.75
$(nPrHgPMe_3)Cl$	20.3	20.2	4.54	4.60
$(nPrHgPMe_2Ph)Cl$	31.7	31.5	4.35	4.30
$(nBuHgPMe_3)Cl$	22.8	22.7	4.91	4.81
$(nBuHgPMe_2Ph)Cl$	33.4	33.4	4.67	4.74
$(nBuHgPMe_3)I$	18.25	18.1	3.91	3.86
$(MeHgdipy)NO_3$	30.45	30.4	2.55	2.55
$(MeHgAsPh_3)ClO_4$	36.7	36.95	2.90	2.90
$(MeHgPPh_3)(Cr(NH_3)_2(SCN)_4)$	34.8	34.9	3.05	3.15
$(PEt_3)_2HgBr_2$(12 results)	24.15	24.2±0.1	5.07	5.13±0.14

* This compound has been used as a standard for the determination of carbon and hydrogen.

The water and carbon dioxide produced by pyrolytic oxidation of the sample are absorbed in the conventional way and determined gravimetrically. Mercury is absorbed on fine-grain metallic bismuth and determined gravimetrically. To absorb halogens, a boat containing a product of the thermal decomposition of potassium permanganate is inserted in the combustion tube: the halogens are then determined by conventional procedures.

Korshun et al[273] describe rapid methods for the micro-determination of carbon, hydrogen, mercury and halogen in a single combustion of the organomercury compound. The method is based on igniting the substance in a stream of oxygen and gravimetric determination of the four elements. Carbon, hydrogen and halogen are determined by previously described procedures (Korshun and Gel'man[274]), and mercury is collected in the combustion tube on silver and determined by the increase in weight. The error for mercury is not greater than 0.7% absolute. Mercury nitrate is not formed during the determination of mercury in substances containing nitrogen. Heavy metals that do not form volatile salts with the halogen can also be determined by this procedure.

C. Determination of Carbon, Hydrogen and Mercury

Farag et al[278] point out that mercury produced during the combustion of organomercury compounds, frequently causes troubles in the determination of carbon and hydrogen owing to its high volatility. The combustion products either poison the tube fillings or are swept into the water and carbon dioxide absorption tubes[279]. For this reason, various reagents have been used in the combustion tube for the collection of mercury combustion products. The commonly used reagents include gold in different forms, e.g. gold foil[280-283], gold wire[284], and gold deposited on asbestos[285]. Also, different forms of metallic silver[286-288] have been recommended for the retention of mercury produced during the combustion of organomercury compounds. Finely granulated metallic bismuth has also been employed for the same purpose.

Although many of these reagents[289-291] have been adopted for the simultaneous determination of mercury together with carbon and hydrogen, the dependence of mercury retention on the temperature of most of absorbents makes these methods less popular.

Farag et al[278] tested polyurethane foam as a new trapping material for the recovery of mercury vapors from the gaseous combustion mixture at room temperature. The carried out a detailed study of the various factors influencing the complete retention of mercury on the foam material using the following different combustion techniques: (1) the rapid straight empty tube method of Korshun-Klimova[292]; (2) the rapid empty tube method of Belcher-Ingram[286]; (3) the rapid flash-combustion method[293]; (4) the cobaltocobaltic oxide method[294]. The carbon dioxide and water produced were absorbed in the usual way in soda asbestos and anhydrone. The mercury vapor is trapped in polyurethane foam packed in a standard Pregi absorption tube, connected externally between the combustion and water absorption tubes, and determined gravimetrically.

For determination of the ability of polyurethane foam to trap mercury vapors, a layer of about 7 cm length packed in the cooler part of the combustion tube was tested by repeatedly analyzing phenylmercuric acetate employing the rapid straight empty tube combustion method. Correct carbon and hydrogen figures were obtained in more than 20 runs by using the same combustion tube. Also, quantitative recoveries for carbon and hydrogen were recorded in analyzing a wide range of organomercury compounds using this modification.

The above results indicate that mercury produced during the decomposition of organomercury compounds is completely retained by the foam layer under these rapid combustion conditions. This suggests the possibility of using polyurethane foam, packed in a suitable absorption tube, i.e. outside the combustion tube, for the quantitative retention and determination of mercury. Obviously, the success of this work allows extending the application of the inexpensive polyurethane foam to the simultaneous gravimetric micro-determination of carbon, hydrogen, and mercury in organomercury compounds. A standard Pregl absorption tube packed with 0.5 g of dry foam and connected between the combustion and water absorption tubes was examined for this purpose.

On employing the proposed foam tube, it was observed that, during some analyses, small amounts of mercury vapors were condensed in the capillary of the foam-packed tube. Warming the end of combustion tube and the front of the foam tube by means of an extension piece on the furnace or by employing a microburner prevented this condensation. This warming not only facilitated quantitative displacement of mercury vapors to the foam tube but also prevented any probable condensation of water vapors in this tube.

When these modifications were introduced, correct and reproducible carbon, hydrogen, and mercury results (Table 43) were obtained.

The foam tube was tested with other combustion methods, namely, the rapid empty tube method of Belcher-Ingram, the rapid flash-combustion method, and the cobaltocobaltic oxide method. It was found that the foam tube functions efficiently with all the combustion procedures examined.

The results obtained for carbon, hydrogen, and mercury in various organomercury compounds using the above mentioned combustion methods are presented in Tables 44 - 46.

Table 43 - Simultaneous Microdetermination of Carbon, Hydrogen, and Mercury in Organomercury compounds by the rapid straight empty tube method of Korshun and Klimova

compound	element	% calcd.	av found [a](x)	std dev(s)	confidence limits x ±ts/ n[b] (t = 0.95)
phenylmercuric acetate	C	28.54	28.52	0.212	±0.25
	H	2.39	2.50	0.089	±0.10
	Hg	59.57	59.48	0.222	±0.26
mercuric benzoate	C	36.32	36.43	0.230	±0.27
	H	3.04	2.93	0.103	±0.12
	Hg	43.34	43.38	0.279	±0.32
mercuric acetate	C	15.07	14.86	0.112	±0.13
	H	1.90	2.16	0.055	±0.06
	Hg	62.94	62.88	0.218	±0.25
mercuric oxalate	C	8.32	8,34	0.185	±0.21
	H	0.00			
	Hg	69.50	69.54	0.279	±0.32
mercuric tartrate	C	13.77	13.70	0.233	±0.27
	H	1.15	1.29	0.148	±0.17
	Hg	57.53	57.56	0.259	±0.29
fluoresceinmercuric acetate	C	33.92	33.93	0.215	±0.25
	H	1.89	1.98	0.126	±0.15
	Hg	47.22	47.23	0.235	±0.27
research I $C_{10}H_{26}N_2O_5Cl_4Hg$	C	20.75	20.64	0.145	±0.17
	H	4.17	4.30	0.211	±0.24
	Hg	34.66	34.64	0.307	±0.35
research II $C_9H_{14}NO_3Cl_2BrHg$	C	20.17	20.22	0.258	±0.29
	H	2.63	2.75	0.231	±0.27
	Hg	37.45	37.57	0.227	±0.26

a Average of four determinations.
b n is the number of determinations.

Organomercury compounds containing nitrogen together with carbon, hydrogen, and oxygen were successfully analyzed by using the proposed foam tube (cf. Table 43). A manganese dioxide tube or silica gel (suitably treated with a sulfuric acid-chromic acid mixture) tube connected as usual between the water and carbon dioxide absorption tubes was employed for the elimination of nitrogen combustion products produced from the combustion of organic compounds containing nitrogen.

Organomercury compounds containing halogen have been successfully analyzed (cf Table 43). Halogen (chlorine or bromine) was absorbed on a silver wool layer placed in the combustion tube after the main combustion furnace and kept at 550°C by means of a short furnace.

Table 44 - Simultaneous microdetermination of carbon, hydrogen and mercury by the Belcher-Ingram Method

compound	element	% calcd	av found[a] (x)	std dev (s)	confidence limits x ± ts/n[b] (t = 0.95)
Phenylmercuric acetate	C	28.54	28.64	0.122	±0.14
	H	2.39	2.42	0.232	±0.27
	Hg	59.57	59.59	0.269	±0.31
Mercuric benzoate	C	36.32	36.42	0.311	±0.35
	H	3.04	3.04	0.197	±0.23
	Hg	43.34	43.39	0.315	±0.36
Mercuric acetate	C	15.07	15.17	0.174	±0.20
	H	1.90	1.95	0.217	±0.25
	Hg	62.94	63.03	0.296	±0.34
Mercuric tartrate	C	13.77	13.98	0.108	±0.12
	H	1.15	1.22	0.129	±0.15
	Hg	57.53	57.50	0.121	±0.14

[a] Average of four determinations.
[b] n is the number of determinations.

Table 45 - Simultaneous microdetermination of carbon, hydrogen and mercury by the Flask Combustion Method

Phenylmercuric acetate	C	28.54	28.57	0.227	±0.26
	H	2.39	2.50	0.261	±0.30
	Hg	59.57	59.63	0.291	±0.34
Mercuric benzoate	C	36.32	36.37	0.276	±0.32
	H	3.04	3.18	0.139	±0.16
	Hg	43.34	43.38	0.252	±0.29
Mercuric acetate	C	15.07	15.15	0.195	±0.23
	H	1.90	1.91	0.235	±0.27
	Hg	62.94	62.81	0.185	±0.21
Research compund	C	20.75	20.69	0.209	±0.24
	H	4.17	4.24	0.199	±0.23
	Hg	34.66	34.62	0.274	±0.32
Mercuric tartrate	C	13.77	13.69	0.221	±0.26
	H	1.15	1.32	0.061	±0.07
	Hg	57.53	57.62	0.315	±0.36

[a] Average of four determinations.
[b] n is the number of determinations.

D. Determination of oxygen

Labedeva and Nikolaeva used a conventional combustion train (Lebedeva and Nikolaeva[276]) to determine oxygen in organomercury compounds. The sample (4 - 10 mg) is decomposed at 1000° in a stream of argon (9 - 12 ml/min) and passed through a combustion tube heated at 1120° ± 10°. The tube is packed with a piece of platinum mesh, then asbestos (3 cm), anthracene carbon black (12 cm) and asbestos (1 cm). Mercury vapour is adsorbed on a layer of pumice (8 cm x 8 mm in diameter) at room temperature, and the carbon monoxide formed is oxidized to carbon dioxide cupric oxide at 300°C and absorbed in Ascarite.

The method has been used for about 80 determinations on samples containing up to 60% of mercury without replacement of the pumice. A single determination takes about 45 min and the error is less than ±0.3%.

Table 46 - Simultaneous microdetermination of carbon, hydrogen and mercury by the Cobaltcobaltic Oxide Method

compound	element	% calcd (x)	av found [a]	std dev	confidence limits x ± ts/n[b] (t = 0.95)
Phenylmercuric acetate	C	28.54	28.65	0.242	±0.28
	H	2.39	2.45	0.194	±0.22
	Hg	59.57	59.56	0.218	±0.25
Mercuric benzoic	C	36.32	36.43	0.115	±0.13
	H	3.04	3.06	0.176	±0.20
	Hg	43.34	43.41	0.294	±0.34
Mercuric acetate	C	15.07	15.19	0.201	±0.23
	H	1.90	1.97	0.224	±0.26
	Hg	62.94	62.93	0.198	±0.23
Mercuric tartrate	C	13.77	13.78	0.129	±0.15
	H	1.15	1.31	0.055	±0.06
	Hg	57.53	57.56	0.197	±0.23

[a] Average of four determinations.
[b] n is the number of determinations.

ORGANONICKEL COMPOUNDS

A. Determination of Nickel

The quantitive determination of traces of nickel in oils has been discussed[296,297].

ORGANOPALLADIUM COMPOUNDS

A. Palladium and Chlorine

Palladium and chlorine in organopalladium complexes may be determined by decomposing the sample with sodium peroxide in an atmosphere of oxygen in a bomb, and then reducing the divalent palladium with sodium formate to palladium metal, which is determined gravimetrically[298]. The chloride ion in the filtrate is determined by potentiometric titration. The gravimetric determination of palladium with dimethylglyoxime[299] tends to give low results[300], so that gravimetric analyses based on 2-thiophene-trans-aldoxime[300] or precipitation of palladium as PdI_2 [301] are preferred when only limited amounts of palladium are present.

ORGANOPHOSPHORUS COMPOUNDS

A. Determination of Phosphorus

As some types of organometallic compounds contain phosphorus in addition to a metal, the determination of phosphorus is discussed here. Various procedures have been described for the determination of organically bound phosphorus. The main procedures for sample decomposition which have been described, and which are discussed in the following sections, involve digestion with mixtures of sulphuric and perchloric acids, or with mixtures of nitric and perchloric acids, or with fuming nitric acid; fusion in a bomb with sodium peroxide and oxygen flask combustion have also been used extensively[302-304]. It is claimed that for mineralization, the open-tube wet-combustion method is the fastest, simplest, and most convenient for samples containing down to a few micrograms of phosphorus[303]. The sealed tube method permits the analysis of both volatile compounds and aqueous solutions but takes longer, and large samples cannot be analysed. The flask combustion method works satisfactorily, but there is a slight tendency towards low results. The spectrophotometric methods investigated were the yellow molybdophosphoric acid and the molybdenum blue procedures. Measurement of the molybdophosphoric acid colour at 400 or 430 nm is the fastest, simplest, and most accurate of the colorimetric methods tested; 460 nm is preferable if a lower sensitivity is desired. Amyl acetate is an excellent extractant for separating molybdophosphoric acid completely from molybdosilicic acid and most other interfering substances. Molybdenum blue methods should be used when high sensitivities are required. The various sample digestion procedures that have been employed in the determination of phosphorus are discussed below.

Digestion with mixtures of sulphuric acid, perchloric acid and nitric acid. Salvage and Dixon [305] have determined phosphorus at the microgram level (30 - 500 μg) by a procedure involving preliminary digestion of the sample with a mixture of sulphuric acid and perchloric acid followed by spectrophotometric evaluation of the yellow molybdovanadophosphate complex at 430 nm. These workers developed the method from a similar semi-micro procedure described by Saliman[306], for which the detection limit was 3 - 50 mg of phosphorus. Some typical results obtained by this procedure are given in Tables 47 and 48.

Apodacu[307] has discussed the use of perchloric acid - sulphuric acid mixtures for the digestion of organophosphorous compounds in colouring matters, plastics, insecticides and pharmaceutical products.

Christopher and Fennell[308] and Fennell and Webb[309] compared digestion procedures using three different mineral acid systems and open tube digestion for the determination of phosphorus in organic compounds, some of which contained fluorine. They determined the resulting phosphate spectrophotometrically, either as molybdenum blue at 735 nm or as phosphovanadomolybdate at 315 nm. Christopher and Fennell[308] concluded (i) that the most sensitive simple method available for the determination of microgram amounts of phosphorous measured the absorption of the phosphovanadomolybdate complex at 315 nm. Only one reagent is added; the complex forms rapidly and is stable for a considerable time; the sensitivity is ten to fifteen times that at 430 nm. On the other hand, measurement in the ultraviolet renders the method liable to interference from "colourless" molecules or ions. (ii) For absorption in the visible region of the spectrum, the molybdenum blue method described by Levine et al[310] is to be recommended.

It is less sensitive than the phosphovanadomolybdate method, involves the addition of more reagents and requires 30 min for colour development. However, measurement at 735 nm should be liable to less interference than measurement at 315 nm.

The various acid digestion and alkali fusion procedures investigated by Christopher and Fennell[308] are discussed below.

Table 47 - Phosphorus contents of solid compounds

Compound	Phosphorus (% ww) Theory	Found
Triphenylphosphine $(C_6H_5)_3P$	11.8	11.6 - 11.9
Diethyl-1-hydroxycyclohexyl phosphonate $C_6H_{10}(OH)PO(OC_2H_5)_2$	13.1	12.9 - 13.7
Diphenylphosphorochloridothionate $(C_6H_5O)_2PSCl$	10.9	10.5 - 10.8
Phenylphosphinic acid $(C_6H_5)HPO.OH$	21.8	20.9 - 21.4
1-Phenyl-1-thiophosphorane $C_5H_{10}PSC_6H_5$	14.7	14.3 - 14.7
Vinyldiphenylphosphine oxide $(CH_2 = CH)(CH_6H_5)_2$ PO	13.6	13.1 - 13.9

Table 48 - Phosphorus contents of liquid samples

Compound	Phosphorous (% w/w) Theory	Found
Tri-o-tolyl phosphate $\left(C_6H_4(CH_3)-O-\right)_3PO$	8.4	8.4 - 8.8
Di(2,6-xylyl)phosphorochloridate $\left(C_6H_3(CH_3)_2-O\right)_2POCl$	9.6	9.4 - 9.5
Diphenyl lauryl phosphorothionate $(C_6H_5O)_2(PS[O(CH_2)_{11}CH_3]$	7.1	7.2 - 7.3
Phenyl bix-(n-dibutyl)-phosphono-diamidothionate $C_6H_5PS[N(C_4H_9)_2]_2$	7.8	7.7 - 7.9
Diphenyl-(2-ethylhexyl)phosphorothionate $(C_6H_5O)_2PS[OC_6H_{10}C_2H_5]$	8.2	7.9 - 8.1
Diphenyl tridecyl phosphorothionate $(C_6H_5O)_2PS[OC_{13}H_{27}]$	6.9	6.6 - 7.2

Table 48 (cont.)

Tri-m-tolyl phosphate[a] (CH_3-C_6H_4-O-)$_3$ PO	8.4	8.0 - 8.1

[a] Commercial grade

Open Tube, Acid Digestion. These workers confirmed that the perchloric-sulphuric acid mixture was satisfactory although, in order to keep the final acidity down to the level recommended by Michelsen[312] for the phosphovanadomolybdate finish, they added only half the quantity of sulphuric acid used by Salvage and Dixon[305]. They established, using triphenylphosphine as a test material, that heating for 15 min at 275°C gave complete digestion.

The results obtained using this procedure for mineralization of a number of solid and liquid compounds, with either the ultraviolet or visible spectrophotometric finish, are given in Table 49. These results are satisfactory for non-fluorinated compounds; it is seen, however, that recoveries tended to be slighly low for some fluoro-compounds and very low for others (cf. Fennell et al[311]).

Table 49 - Phosphorus analysis of organic materials: open tube perchloric acid-sulphuric acid digestion

Compound	Ultraviolet finish			Visible finish		
	Wt. taken (μg)	Phosphorus (%) Found	Error	Wt. taken (μg)	Phosphorus (%) Found	Error
Triphenylphosphine 11.81%P	35.12	11.70	-0.11	42.85	11.62	-0.19
	43.91	11.77	-0.04	43.79	11.81	0.00
	62.59	11.66	-0.15	47.51	11.79	-0.02
	78.21	11.87	+0.06	49.14	11.88	+0.07
	83.57	11.80	-0.01	50.79	11.73	-0.08
	92.94	11.75	-0.06	55.07	11.46	-0.35
	95.64	11.73	-0.08	61.84	11.85	+0.04
$C_{20}H_{20}NO_3P$ 8.77%P	42.47	8.78	+0.01	69.62	8.99	+0.22
	51.41	8.83	+0.06	81.08	8.84	+0.07
$C_{19}H_{19}N_2O_2P$ 9.15%P	44.71	8.97	-0.18	69.54	9.49	+0.34
	68.30	9.14	-0.01	86.57	9.32	+0.17

Open Tube, Hydrogen Peroxide Digestion. The use of 50% hydrogen peroxide for mineralisation as advocated by Whalley[313] and by Taubinger and Wilson[314] for the decomposition of organic materials is generally satisfactory. Results for triphenylphosphine tend to be low, and fluorophosphorus compounds are not completely mineralized. The results given by this method of decomposition are shown in Table 50.

Sealed Tube, Acid Digestion. Using their previously published technique for sealed tube digestion with a mixture of nitric and sulphuric acids (95 + 5) with both fluorometric and visible finishes, and making a correction for the reagent blank in both cases, Fennell and Webb[309] obtained the results shown in Table 51. Results for triphenylphosphine tend to be low but two of the fluorocompounds which had not been amenable to mineralization by the open tube methods gave reasonable recoveries after sealed tube digestion.

Sodium Peroxide Fusion. Previous work by Fennell et al[311] has shown that, on the semimicro scale, fusion with sodium peroxide resulted in complete decomposition of "difficult" fluorocompounds. Attempts to apply this technique on a centimilligram scale were unsuccessful, however; severe interference was found with both spectrophotometric finishes. It is suspected that this interference was caused by impurities in the sodium peroxide, and by random adsorption and desorption of phosphorus in the nickel capsules.

Table 50 - Analysis of organic materials: open tube H_2O_2-H_2SO_4 digestion

Compound	Wt. taken (μg)	Phosphorous (%) Found	Error
Triphenylphosphine 11.81%P	51.50	11.57	-0.24
	53.37	11.60	-0.21
	58.51	11.45	-0.36
$C_{19}H_{19}N_2O_2P$ 9.15%P	54.46	9.14	-0.01
	54.59	9.25	+0.10
	58.16	9.20	+0.05
	61.63	9.02	-0.13
	63.77	9.20	+0.05
	74.53	9.11	-0.04

Hot Flask Combustion. Kirsten's hot flask technique[315] using sodium hypochlorite as absorbent was applied successfully to liquid and solid compounds, including "difficult" fluorocompounds (Table 52). No blank was found, but it was essential to heat the absorbent, after acidification, presumably to convert meta- into orthophosphate.

Results obtained by the various methods for the analysis of triphenylphosphine are summarized in Table 53. From these data, it can be concluded that (a) the phosphovanadomolybdate finish is significantly more precise than the molybdenum blue finish, (b) decomposition involving open tube digestion with hydrogen peroxide and sealed tube digestion with acid have a significant negative bias and (c) open tube acid decomposition leads to significantly higher precision than hot flask combustion with sodium hypochlorite. Christopher and Fennell[308] concluded that open tube digestion with nitric and perchloric acids followed by a phosphovanadomolybdate finish at 315 nm is the preferred method for estimating phosphorus in organic compounds which do not contain fluorine. Optimum reaction times for triphenylphosphine appear to be about 30 min (sample decomposition process from completion of weighing and transfer of digest to volumetric flask) and 3 min (standing time between making up to volume and measurement of absorbance) They compared the results obtained by this method with those obtained by

and perchloric acids have been used for the decomposition of organophosphorus compounds followed by colorimetric determination of the phosphate produced as phosphovanadomolybdate[321]. (vi) Heating the sample for 2 h with concentrated sulphuric acid, iron (III) chloride and perchloric acid, followed by measuring the molar absorptivity of the molybdenum blue complex at 700 or 840 nm [322]. (vii) Digestion of the sample with fuming nitric acid in a sealed tube (Carius)[323]. (viii) Sealed tube digestion procedures using either concentrated sulphuric acid at 460°C or fuming nitric acid[325] as the digestion reagent. (ix) Carius methods have been used for the determination of phosphorus in organic fluorine compounds[326]. Low recoveries have been reported for this technique and ascribed to the adsorption of phosphate on the walls of the tube[325]. An effective way of avoiding the decomposition of the tube of a white insoluble material is to add a small amount of an alkali metal salt to the acid mixture in the decomposition tube[323]. Potassium chloride is the most effective salt. Arsenic, tungsten, tin, titanium, vanadium, and zirconium interfere in this procedure.

Table 53 - Comparison of methods of phosphorus analysis of triphenylphosphine (11.81% P)

Method	Reagent	Finish	No.of results	Phosphorus, %			
				mean	SD	Bias	SD if Bias
Open tube	acid	A	7	11.75	0.07	-0.06	0.03
Open tube	acid	B	8	11.77	0.18	-0.04	0.06
Open tube	H_2O_2	A	3	11.54	0.08	-0.17	0.05
Sealed tube	acid	A	5	11.37	0.17	-0.44	0.08
Sealed tube	acid	B	2	11.62	-	-0.19	-
Hot flask	NaO1	A	5	11.60	0.44	-0.21	0.20

Combustion A - phosphavanadomalybdate at 315 nm
B - molybdenum blue at 753 nm

Table 54 - Determination of phosphorus in organic compounds (fluorine absent). Comparison of methods.

Method	No.of results	No.of compounds	Precision (% P) pooled SD	Accuracy % P	
				Bias	SD if bias
Salvage and Dixon[305]	32	14	0.30	-0.11	0.06
Belcher et al[315]	6	2	0.24	-0.14	0.10
Belcher et al[316]	44[a]	7	0.23	+0.07	0.07
	14[b]	3	0.38	-0.13	0.05
Christopher and Fennell[308]	24	7	0.07	-0.02	0.02

[a] Titrimetric finish.
[b] Colorimetric finish.

In Table 55 the results using micro-Carius combustions, with and without the addition of potassium chloride, are compared with results using Kjeldahl flask digestions.

Results obtained when the sample was destroyed in the presence of potassium chloride indicate no loss of phosphorus. The method is as accurate as the Kjeldahl flask procedure. Furthermore, precision is excellent, and lies within the ±0.2% absolute error generally acceptable for elemental analysis.

Following the micro-Carius digestion of organophosphorus compounds, the tubes which had contained an alkali metal salt were perfectly clear. Those tubes from which the salt had been omitted invariably exhibited a residue on the inner wall similar to lightly frosted glass. This residue was enhanced greatly by heating the tube to a light red and allowing it to cool, whereupon a heavily frosted spot remained.

Results obtained by the procedure of DiPietro et al[323] for the determination of phosphorus in tri-α-naphthyl phosphate using various alkali metal salts are given in Table 56. The agreement of these with the theoretical values appears to indicate that the formation of a deposit is prevented by the presence of an alkali metal salt.

Table 55 - Comparison of phosphorus analyses by the micro-Carius method, with and without alkali salts, with the Kjeldahl flask method

Sample	% Phosphorus			
	Theory	Carius combustion: Alkali metal salt absent	Carius combustion: KCl present	Kjeldahl flask digestion
Tri-α-naphthyl phosphate	6.50	2.81	6.49	6.45
		4.33	6.61	6.57
		3.49	6.55	6.48
		4.76[a]	6.53	6.50
5-Chloro-4-hydroxy-3-methoxy benzyl isothiourea phosphate	8.99	6.47	9.14	
			9.12	
Bis,β-chloroethyl vinyl phosphonate	13.30	10.42	13.42	13.38
			13.54	13.63
Chloromethyl phosphoric dichloride	18.50	15.07	18.58	18.91
			18.60	18.72
Tris(aziridinyl)phosphine oxide (85% in methanol)		11.91	14.93	14.55
			14.66	14.76
Treated textile		1.17	3.03	3.02
			3.00	2.99

[a] Shorter period of combustion, and using only 0.3 ml of fuming nitric acid.

Sodium Peroxide Fusion. Following the finding that wet digestion with mixtures of concentrated nitric and sulphuric acids did not give reliable phosphorus determinations in fluorinated compounds, the applicability of fusion with sodium peroxide in a Parr bomb followed by determination of phosphate by Wilson's method[327] was examined[311] for the determination of down to 2 - 3 mg of phosphorus in fluorinated organic compounds. Relatively large and variable blank values were obtained when a semi-micro (8.5 ml capacity) nickel bomb and the usual amounts[328] of reagents for semi-micro operation were used, i.e. 4 g of sodium peroxide, 200 mg of potassium nitrate, and the organic material made up to 200 mg with sucrose. These blank values were caused by silica picked up from the glass apparatus used during the weighing of the sodium peroxide and during the leaching of the bomb, and could be eliminated by using platinum apparatus for these operations, the bomb leachings being transferred to glass apparatus only after acidification. Tests carried out with a 8-ml micro-bomb with 25 mg of standard compounds, 25 mg of sucrose, 50 mg of potassium nitrate, and 1 g of sodium peroxide were satisfactory, as shown in Table 57.

The presence of fluoride in the bomb leachings give rise to positive errors in determination of phosphorus, presumably by attack on the glass flask. This interference by fluoride was overcome by evaporating the bomb leachings, acidified with hydrochloric acid to dryness two or three times, or by adding boric acid.

Table 56 - Analysis of combustion products of tri-α-naphthyl phosphate

Salt Used	% Phosphorus	Comments
KCl	6.55	Average %,Table 55, 20 mg salt used
KHphthalate	6.67	20 mg
NaBr	6.61	20 mg (light frosted spot on tube)
NaOAc	6.64	20 mg
NaCl	6.54	40 mg
NaCl	6.49	60 mg
LiCl	6.53	20 mg
RbCl	6.49	20 mg

Table 57 - Phosphorus analysis of standard compounds by the micro-bomb fusion method

Compound	No. of determinations	Phosphorus calculated (%)	Phosphorus found (%)	Deviation from mean (%)
Triphenylphosphine	6	11.81	11.80	±0.14
Tri-n-butyl phosphate	4	11.63	11.69	±0.14
Tri-m-cresyl phosphate	5	8.41	8.42	±0.04

The following comment has been made about the method[311]:

(a) Over 50 mg of material can be decomposed by using 1 g of sodium peroxide and 50 mg of potassium nitrate. When these amounts are used, no addition of sucrose is made.

(b) The use of platinum apparatus for weighing the sodium peroxide and in leaching the residue from the bomb assists materially in eliminating the blank values caused by the pick-up of silica. Easily measurable contamination by silica was found when the peroxide was weighed on a watch-glass. In the presence of fluoride, the leaching of the residue from the bomb, acidification, and evaporation must be carried out in platinum apparatus or very serious errors caused by high blank values will result. On the other hand, when platinum apparatus is used, the presence of fluorine

assists in reducing the blank values by removing silica as silicon tetrafluoride during evaporation of the acidified leachings.
(c) Wilson's method for the determination of phosphate[327] has proved, with modification, to be suitable for the determination of 1 - 3 mg of phosphorus.
(d) Results should be within ±0.15% (absolute) and the majority of the results obtained on fluorinated materials are within ±0.5%.
(e) A single determination can be completed in about 2 h.
(f) It is thought that arsenic will interfere, although the method was not tested in the presence of arsenic. Silicon does not interfere if sufficient fluoride is present to remove silicon as the tetrafluoride or if the bomb leachings are evaporated to dryness twice after acidification with hydrochloric acid.

Table 58 - Phosphorus analysis of fluorinated compounds

Compound	Sample weight (mg)	Phosphorus calcd. (mg)	Phosphorus found (mg)	Phosphorus calcd. (%)	Phosphorus found (%)	Absolute error (%)
$(C_3F_7CH_2O)_3PO$	57.22	2.75	2.78	4.81	4.87	±0.06
	49.40	2.38	2.39	-	4.84	±0.03
$(C_3F_7CH_2O)_2P(O)-$	49.57	2.79	2.81	5.62	5.67	±0.05
$CH.C_6H_4.CH_3$	51.31	2.88	2.86	-	5.58	-0.04
$(C_3F_7CH_2.O)_2P(O)-$	49.55	2.86	2.87	5.77	5.80	+0.03
$NH.C_6H_5$	38.44	1.11	1.10	-	5.72	-0.05
$C_3F_7CH_2.O.P(O)-$	23.51	1.59	1.58	6.76	6.74	-0.02
$(NH.C_6H_4.CH_3)_2$	21.60	1.46	1.46	-	6.76	0.00
$CF_3.CH_2O.P(O)-$	18.98	1.78	1.75	9.38	9.24	-0.14
$(NH.C_6H_5)_2$	18.49	1.73	1.71	-	9.24	-0.14
$(CF_3.CH_2.O)_2P(O)NH_2$	24.31	2.88	2.86	11.87	11.77	-0.12
	19.36	2.30	2.27	-	11.75	-0.12

The determination of phosphorus in organophosphorus compounds by fusion with sodium peroxide in a Parr bomb has been studied by other workers[302,320]. Sodium peroxide fusion has been compared with two other procedures for the determination of phosphorus in glycerophosphates[330]. The three methods involved are: (i) carbonization in a porcelain crucible in a gas flame (10 min), dissolution of the residue in concentrated nitric acid and evaporation; (ii) evaporation on a sand bath in perchloric acid-nitric acid mixture followed by dissolution in 10% nitric acid and evaporation; (iii) heating for 15 min in a Parr-Wurzschmitt bomb with sodium peroxide by a gas flame, and dissolution of the residue in water to decompose the excess of sodium peroxide. In each method phosphate was determined by precipitation with excess of bismuth nitrate in nitric acid and titration of the excess of bismuth with EDTA in the presence of catechol violet. The results were the same by all three methods, but method (ii) was of advantage in the presence of chloride ion which is eliminated as hydrochloric acid. The following ions do not interfere: NH_4^+, Li^+, Mg^{2+}, Ca^{2+}, Sr^{2+}, Ba^{2+}, Al^{3+}, Zn^{2+}, Ce^{3+}, Mn^{2+}, Co^{2+}, Ni^{2+}, Cd^{2+}, Cu^{2+}, Pb^{2+}, UO_2^{2+}, and Ag^+; however, Fe^{3+}, Ga^{2+}, In^{3+}, Zr^{4+}, Th^{4+}, Hg^{2+}, SO_4^{2-}, AsO_4^{3-}, $Cr_2O_7^{2-}$, and Cl^- should be absent. Method (ii) had the advantage of not being subject to interference by the presence of chlorine in the sample. Another group compared three methods of sample decomposition involving (i) fusion with sodium peroxide in a steel bomb in a burner flame, (ii) fusion with sodium peroxide in a calorimeter bomb, and (iii) heating with concentrated sulphuric acid-nitric acid mixture[331]. They recommend method (ii) for a variety of organophosphorus compounds, including polymers.

Buss et al[332] decompose the organophosphorus or ferrophosphorus compound (5 - 60 mg P_2O_5 equivalent) with sodium peroxide and precipitate the resulting phosphate as zinc ammonium phosphate, ($ZnNH_4PO_4$). The precipitate is dissolved and zinc determined chelatometrically, masking iron, if present, with triethanolamine.

Other Digestion Reagents. Saliman[306] has described a digestion reagent comprising hydriodic acid, calcium hydroxide, water, phenol and acetic acid for the determination of microgram quantities of phosphorus in organic compounds. During removal of solvent and excess reagent by volatilization and combustion, phosphorus is converted to orthophosphate. The molybdenum blue colour is developed. The procedure can be adapted to either ultramicro or trace analysis, and is applicable to the determination of organic phosphorus in a wide variety of solvents. They also describe a rapid semimicro procedure which utilizes digestion in sulphuric and perchloric acids, followed by formation of the phosphovanadomolybdate complex.

Saliman[306] points out that wet digestion using strong oxidizing reagents, are sometimes difficult to apply to solutions of relatively volatile phosphorus compounds such as 2.2-dichlorovinyldimethyl phosphate or O.O.S-trimethyldithiophosphate or to samples which are relatively non-volatile such as glycol or grease. The use of strong oxidizing reagents with some solvents may be hazardous. Because of the limitations of strong oxidizing procedures and because of the solvents, concentrations, and compounds encountered particularly in biological investigations, they decided that a method of wider application than those available at the time was needed.

All the organic phosphorus compounds tested by Saliman[306] yielded their phosphorus to this digestion (Table 59). A very few volatile compounds required a hydrolysis period at room temperature in the stoppered flask, or the addition of bromine water to ensure complete recovery of phosphorus.

Using the hydriodic acid-phenol reagent described above, phosphorus was determined in such compounds as 2,2-dichlorovinyl dimethylphosphate, trimethylphosphate, methylparathion, 0,0,0-tri-p-tolylphosphorothioate, 0,0,0-triethylphosphorothioate, tricyclohexylphosphine oxide, phosphrin and others (Table 60).

In addition to being applicable to these materials as an ultramicro method for phosphorus determination, the method was applicable as a trace method for phosphorus at the level of 1 ppm or less in various media, such as organic matter, water, lubricating oils, carbon tetrachloride, acetone, acetic acid, xylene, glycol and mineral oil. Recovery of phosphorus from compounds dissolved in methanol, ethanol, and isopropanol, all of which might be expected to react with the hydriodic acid, was also quantitative. Apparently the reaction of the hydriodic acid with the phosphate is much more rapid than with the alcohol solvent. The reagent reacted rapidly with most phosphate insecticides, requiring no digestion other than volatilization of solvent and excess of reagent on a stream bath and/or a hot-plate, a process which usually required about 10 min. However, some phosphorus-containing materials, including aryl esters of phosphates, phosphonamides, and dithiophosphate esters such as OOS-triethylphosphorodithioate, required 30 min or more of digestion on a steam bath. Triphenylphosphine required an even longer digestion. Thiophosphate esters required no special treatment. Dithiophosphate esters, in aqueous or non-aqueous solution, were pretreated with aqueous bromine solution. The excess of bromine was boiled off, reagent was added, and the procedure was continued as described. This allowed the determination of phosphorus without further digestion.

Persulphate[333] and hydrogen peroxide[334] oxidations and sodium carbonate potassium nitrate fusion (1 g; 2:1) in a platinum crucible[335] have also been used for the determination of phosphorus in organic compounds. Sulphuric acid and potassium permanganate have been mentioned for the decomposition of organophosphorus compounds[336]. Following decomposition of various substituted phosphonic and phosphonothionic acids by this methed, the determination of phosphate by the molybdenum blue method gave poor results. However, good results were obtained when phosphorus was determined in these compounds by a semi-micro method involving pyrolytic decomposition of the sample in a silica tube[337]. In a further method the organophosphorus sample was heated with potassium permanganate in a sealed glass tube at 400 - 500°C, thus oxidizing phosphorus to phosphate, which was then determined titrimetrically[338].

Decomposition with magnesium has been used for the microdetermination of phosphorus[339]. When organic compounds containing phosphorus are burned with metallic magnesium, the phosphorus is converted into magnesium phosphide, which can be decomposed in bromine water and oxidized to phosphoric acid, which can be determined photometrically as molybdophosphoric acid after having been extracted with ethyl acetate. The presence of nitrogen, sulphur, and halides does not cause interference. The method is suitable for 1.0 - 2.5 mg of sample, and the error is ±0.4%.

Table 59 - Phosphorus analyses of typical compounds, semi-micro scale; KH_2PO_4 standard

Sample	Phosphorus, (mg) Added	Found
Trimethyl phosphite	1.95	1.89
	1.92	1.90
	3.16	3.13
	2.83	2.82
p.p-dibutyl-N,N-diisopropylphosphinic amide	2.08	2.10
	1.16	1.18
	0.70	0.72
Triphenylphosphine	0.92	0.93
	2.22	2.22
	1.00	0.99
	1.92	1.93
Methyl parathion insecticide	1.83	1.85
	1.29	1.31
Phosdrin insecticide	2.88	2.86
	2.22	2.19
Phosdrin insecticide,)	2.09	2.99
1% on Pyrex dust)	2.62	2.49
1-(Dichlorophosphinyl) piperidine	2.90	2.87

Table 59 (cont)

Sample	Phosphorus, (mg) Added	Found
5 empty gelatine capsules, size 5	0	0
Silicic acid, silica gel, 1 g	0	0

Table 60 - Micro phosphorus determination

Compound	Solvent	Solution (ml)	Phosphorus, (μg)	
			Added	Found
Methyl parathion	chloroform	5	8.25	8.3
	none	-	5.2	5.1
Triphenylphosphine	chloroform	5	6.3	3.2
$(CH_3O)_2P(=O)CH_3$ (with =O above and below P and C as printed)	chloroform	5	15.8	15.3
$(CH_3O)_2PH$	chloroform	1	6.0	5.9
Phosdrin insecticide	isopropanol	1	5.0	5.9
	xylene	1	9.8	9.7
	methanol	5	7.3	7.3
	ethanol	5	7.3	7.5
	chloroform	25	7.1	7.0
	chloroform	50	1.0	1.05
	chloroform	55	1.1	1.3
	glycol	50	2.9	2.9
	mineral oil	50	2.9	3.0
	water	5	5.0	5.0
	water	25	2.5	2.4
Tri-p-tolyl phosphate	chloroform	5	5.1	5.1
	toluene	1	9.5	9.6
	toluene	1	9.5	9.2
	toluene	0.5	4.7	4.7
	toluene	0.5	4.7	4.4
	gasoline[a]	0.5	4.75	5.0
	gasoline[a]	0.5	4.75	4.8
	gasoline[a]	2.5	4.75	4.5

[a] Mercaptobenzoic acid added at start of procedure to prevent interference from lead.

Oxygen Flask Combustion. This technique has been extensively studied for the determination of phosphorus in organophosphorus compounds[331,340-352]. One procedure involves ignition over dilute nitric acid followed by reaction of the combustion products with magnesia mixture, which is filtered off and determined by reaction with ethylenediaminetetracetic acid to the Eriochrome Black T end-point (semi-micro method) or are reacted with molybdate reagent for a spectrophotometric finish (micro method)[341]. Other elements that form heteropoly acids with molybdates which are reducible to molybdenum blue (in the micro method) are silicon, arsenic and germanium. Silicon, as silicate, does not interfere with the colorimetric method. Arsenic may be separated from the phosphate and determined quantitatively[342], but arsenic occurs rarely with phosphorus in organic compounds. Germanium is encountered infrequently in organic analysis. Complete transformation of phosphorus pentoxide to phosphoric acid requires boiling with dilute nitric acid for 10 - 15 min prior to application of the molybdate procedure. A further oxygen flask combustion uses 1 N sulphuric acid as the absorbent and a molybdate finish[343]. The analysis should be completed soon after the combustion, since phosphorus pentoxide may be lost from the solution (about 5% loss in 24 h, probably through adsorption by the glass).

Table 61 indicates the accuracy and applicability of the method. The first two samples are standard samples from a collaborative study of phosphorus analyses under the auspices of the Association of Official Agricultural Chemists. The other two samples are at least 97% pure based on elemental analysis for carbon, hydrogen, and nitrogen or chlorine.

A comparative study has been made of the methods available for the semi-micro determination of phosphorus in fabrics flame-proofed with organic phosphorus compounds[344]. In one method the products from oxygen flask combustions are absorbed in aqueous hydrogen peroxide solution[345]. After combustion of the sample over dilute hydrogen peroxide solution, the solution is boiled for 30 min and made slighly acidic. Eriochrome Black T solution and excess of 0.01 N lead nitrate are added, the excess being back-titrate with 0.01 N potassium dihydrogen phosphate.

In another very simple method the combustion products were absorbed in water and, after boiling, the resulting phosphoric acid was titrated with 0.1 N sodium hydroxide to the thymolphthalein end-point[346]. This method is obviously subject to interferences. A mixture of 0.4 N perchloric acid and 0.4 N nitric acid has been used as absorbent[321]. Phosphate can then be determined by the ammonium molybdovanadate method. In another procedure the combustion products were absorbed in 0.4 N sulphuric acid and the solution was boiled prior to adding ammonium persulphate[347]. An aliquot was diluted, treated with acidic $(NH_4)_2[MoO_4]$, bismuthyl carbonate, and ascorbic acid (as reducing agent), and the molybdenum blue measured at 710 nm. Sodium carbonate was used with silicon compounds to facilitate dissolution and boric acid was used with fluorine compounds to prevent etching of the glass. In the presence of arsenic and divalent nickel high and low results, respectively, were obtained.

Organophosphorus compounds can be burned over a solution of ammonia[348]. After boiling the resulting solution to remove excess of ammonia it is passed through a column of Amberlite-IR-120(H^+ form, 20 - 50 mesh). The percolate is acidified with nitric acid and pyridine, acetone, and a 0.1% dithizone solution in acetone added, and the solution is titrated with 0.01 M lead nitrate to a red colour. If fluoride is present, the absorber solution after removal of ammonia is acidified with nitric acid and evaporated to dryness; the residue is dissolved in water and then treated as before. Technicon Auto Analyzer system has been adopted to the analysis of solutions of organophosphorus compounds decomposed by the oxygen flask technique[349]. This procedure is capable of determining less than 0.1µg of phosphate per millilitre of test solution. A modification of the oxygen flask technique provides a flow of oxygen to the combustion bottle, the products from which pass into a Wickbold absorber[352]. The technique was applied successfully to the determination of phosphorus in solids and liquids (e.g. lubricating oil); recoveries from 92 - 106% were obtained for the range 10 - 1000 ppm of phosphorus.

Table 61 - Accuracy of semi-micro and micro oxygen flask method for phosphorus

Compound	P Found (%)			Error, Parts/1000	
	Calcd.	Micro	Semi-micro	Micro	Semi-micro
6-Allyl-6,7-dihydro-5H-dibenz-(c,e)azepine-phosphate(Ilidar Phosphate[a])	9.20	9.16	9.29	-17	0
			9.33		+4
			9.30		+1
			9.29		0
			9.25		-4
3-Diethylamino-2,2-dimethylpropropyl tropate phosphate (Syntropan Phosphate[a])	7.64	7.73	7.61	+9	-4
3,4-Dichlorobenzyl-triphenyl-phosphonium chloride	6.77	6.84	6.86	+10	-12
Ethylenediamine salt of monoguaiacylphosphoric acid	11.73	11.59	12.08	-14	+29

[a] Registered trade marks, Hoffman-LaRoche, Nutley, N.J.

Atomic Absorption Spectrometry. Driscoll et al[400] have described a direct determination of phosphorus in gasoline by flameless atomic absorption spectrometry. Phosphorus is determined by direct injection into the graphite furnace. The detection limit is 20 μg P and the precision at 80 μg is 20% measured as relative standard deviation.

Fourier Transform NMR. Kaslen and Tierney[401] have described the non-destructive determination of phosphorus by this technique. Results are accurate to ±0.4% with a relative error of 2 - 3%.

Miscellaneous Methods. The continuous band spectrum of phosphorus in ethanol solutions of organophosphorus compounds has been investigated[353]. For concentrations between 0.01 and 0.03 M the average error was 0.0006 M; sodium and calcium ions cause positive errors whereas nitrogen, iodine, sulphur, and chlorine do not interfere in amounts equivalent to that of phosphorus. Flame emission measurements at 540 nm were made on a variety of organophosphorus compounds. The standard deviation for tributyl phosphate in paraffin was ±0.88% at the 46% level and ±0.015% at the 1.7% level.

These methods capable of detecting 0.2 μg of combined phosphorus are applicable to acids, esters, acyl halides, and anhydrides, together with their thio analogues[354]. They involve degradation by refluxing with concentrated sulphuric acid. Two of these methods then involve the use of o-dianisidine molybdate reagent and indicate phosphorus by the formation of a reddish brown precipitate. The third is based on the production of molybdenum blue, and hydrazine hydrate is used as the reducing agent. Methods of sampling and the application to air analysis were also discussed. Unstable organophosphorus compounds may be stabilized by exposure to sulphur for 8 - 48 h in vacuo, and the resulting compounds are analysed by pyrolysis[355].

B. Determination of Phosphorus, Iron, Silicon, Titanium and Arsenic

To determine phosphorus and iron in organophosphorus compounds, Kotova[356] fused the sample (13 - 30 mg) with sodium peroxide in a bomb; the phosphate in an aliquot of a solution of the melt in nitric acid (adjusted to pH7 with sodium hydroxide solution) is titrated potentiometrically (glass indicator electrode) with standard lanthanum nitrate-ammonium chloride solution of pH8; sulphosalicyclic acid is added to mask iron. In another aliquot of the solution, iron is determined photometrically with sulphosalicylic acid in aqueous ammonia medium. The absolute error is less than 0.3% for each element.

Phosphorus and silicon occur together in certain types of organometallic compounds. To determine these, Luskina et al[357] decompose the sample by heating it with potassium persulphate or hydrogen peroxide solution dissolved in concentrated sulphuric acid. The silicic acid formed is determined separately[358]. The filtrate is used for the photometric determination of phosphorus as molybdenum blue. For samples containing 10 - 20% each of silicon and phosphorus, differences between the calculated and determined contents were about 0.2% in single determinations. When organic phosphorus silicon compounds are fused with potassium metal, phosphorus is reduced to potassium phosphide, which does not interfere with the amperometric titration of silicon; similarly, the potassium silicate formed does not interfere with the determination of phosphorus by amperometric titration with uranyl acetate solution[359]. Methods for the determination of free phosphorus, combined phosphorus, and silicon in reaction products of tetraalkoxysilanes with potassium halides are given in refs 360 and 361. To determine titanium, phosphorus, and silicon in organic compounds that are difficult to decompose the sample may be fused with ammonium fluoride and potassium pyrosulphate and the cooled melt treated with concentrated sulphuric acid, followed by evaporation in an air-bath until fumes appear[361]. The residue is dissolved in 70% sulphuric acid and titanium, phosphorus, and silicon are determined by standard procedures.

Phosphorus and arsenic in organic compounds may be determined by first burning the sample by a modified oxygen flask method[325]. Phosphorus is determined by precipitation as quinoline molybdophosphate and titration with sodium hydroxide solution, or spectrophotometrically at 750 nm as molybdenum blue, with iron (II) ammonium sulphate as the reductant. Arsenic is determined spectrophotometrically at 840 nm by a similar method, with hydrazine sulphate as reductant. The absolute accuracy is within ±0.5% for phosphorus by either method, and within ±1% for arsenic.

C. Determination of Carbon and Hydrogen

The determination of carbon and hydrogen in organic compounds containing phosphorus and sulphur is difficult, especially in compounds containing phosphorus[362]. This is due mainly to the formation of a phosphorus pentoxide carbon film inside the combustion tube. The phosphorus pentoxide crystals surround the particle of carbon, making them thermoresistant, and a very high temperature (around 900 - 1000°C) is necessary to destroy this complex. The heaters of a standard Pregl combustion unit are capable only of temperatures around 700 - 800°C and it requires an extremely long time to decompose phosphorus pentoxide residue left inside the tube using such a technique. In order to provide a simple, fast, and economical method for the analysis of materials that are difficult to combust the Korbl method[363-366] is recommended[362]. This uses a packing of thermally decomposed

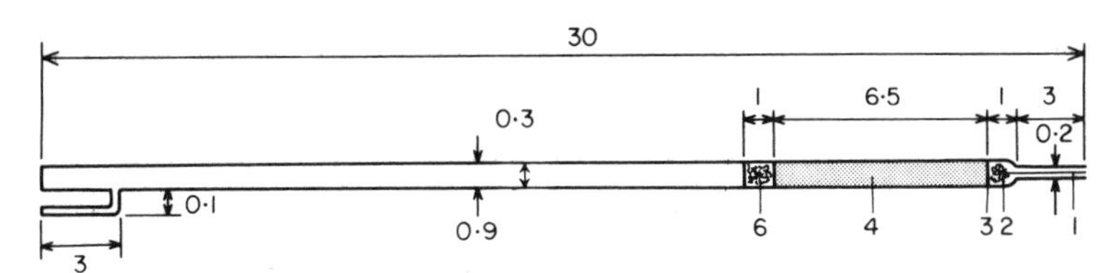

Fig. 16 - Determination of carbon and hydrogen in organophosphorus compounds. Combustion tube (all dimensions in centimetres). (1) Silver wire; (2) silver wool; (3) asbestos; (4) thermally decomposed silver permanganate; (5) silver wool.

silver permanganate without a buffer zone of asbestos (Figure 16). In order to effect complete combustion of phosphorus compounds, it is necessary to use a Fischer blast burner and to heat the sample vigorously for 5 min. The special heating element used by Korbl to prevent condensation of water in the capillary end of the absorption tube (Anhydrone) may be replaced by a simple steel hook connected to the stationary heater[362]. The total time required for one carbon and hydrogen determination is about 40 min. The principle of the method is that the organic material is burned at 600 - 700°C in an oxygen atmosphere. The conversion of carbon to carbon dioxide and of hydrogen to water is accelerated by passing the combustion products over decomposed silver permanganate. Sulphur and phosphorus oxides combine directly with silver wool and the contact mass of decomposed silver permanganate. Water produced from the combustion is absorbed in Anhydrone, and carbon dioxide is Ascarite tubes.

Lysyj and Zarembo (1958) found that the filling of the combustion tube as used by Korbl for elementary analysis was unsatisfactory for the determination of organophosphorus compounds unless some modifications were made.

In the combustion of organophosphorus compounds, phosphorus pentoxide forms a film which surrounds and entraps particles of carbon. This makes it impossible for an ordinary heating element to decompose the mixture of phosphorus pentoxide and carbon.

The asbestos plug in front of the decomposed silver permanganate filling, as used by Korbl, to some degree prevented contact between the partially decomposed organophosphorus material and the oxidizing media of silver permanganate. There is also the possibility of chemical reaction between asbestos and the phosphorus pentoxide carbon mixture with the formation of an extremely thermally resistant compound. This difficulty was overcome by eliminating the asbestos plug. The complete decomposition of the phosphorus pentoxide-carbon residue was achieved by the use of an air blast burner.

The results (Table 62) for carbon-hydrogen analyses using these modifications were of good precision and accuracy. The greatest advantages of this method are the simplicity of the apparatus, the ease in the working technique, and the speed of analysis. The preparation and filling of the combustion tube is very simple when compared to the regular Pregl tube. No break-in period is necessary for the combustion tube, and the train is ready for operation 30 min after it has been assembled.

Table 62 - Semi-microdetermination of carbon and hydrogen

Compound	Formula	Sample (mg)	Found		Theory		Difference	
			% C	% C	% C	% H	% C	% H
Bis,beta-chloro ethyl vinyl phosphonate	CH_2=CH.PO($OCH_2CH_2Cl)_2$	24.89 24.08	30.70 30.65	4.92 4.80	30.92	4.76	-0.22 -0.27	+0.16 +0.04
Diethyl phosphoro-chloride	$(C_2H_5)_2POCl$	17.54 19.14	27.80 27.75	6.01 5.97	27.83	5.84	-0.03 -0.08	+0.17 +0.13
Tris-benzoyloxymethyl phosphine	$(C_6H_5.COOCH_2)_3PO$	21.51 20.05	63.68 63.79	4.71 4.87	63.71	4.64	-0.03 +0.08	+0.07 +0.23
Dimethyl hydrogen phosphite	$(CH_3O)_2POH$	24.91 22.12	21.69 21.72	6.56 6.45	21.83	6.40	-0.14 -0.11	+0.16 +0.05
Trimethyl phosphite	$(CH_3O)_3P$	14.99	29.03	7.09	29.03	7.25	0.00	-0.16

Carbon in organophosphorus compounds may be determined by wet combustion using a modified Van Slyke method[367,368]. For the micro-determination of carbon and hydrogen a standard combustion train with cobalt(III) oxide as the oxidizing agent may be used[369]. The combustion products from the sample are passed through finely divided silver supported on pumice to remove oxides of phosphorus, and the carbon dioxide and water are then determined by conventional gravimetry.

Other procedures involve combustion on pumice for the determination of carbon, hydrogen, and phosphorus[337], and for the simultaneous micro-determination of phosphorus, sulphur, carbon, and hydrogen in compounds containing these elements plus nitrogen pyrolysis of the sample followed by combustion with a large excess of hydrogen[370]. Phosphorus pentoxide produced in the combustion is absorbed by powdered quartz, which has been etched with caustic alkali. The method is claimed to be particularly suited to compounds with the C - P linkage.

D. Determination of Nitrogen

Nitrogen and phosphorus in organophosphorus compounds may be determined by first heating the sample with 70% perchloric acid[371]. In separate aliquots of the dilute solutions nitrogen is determined spectrophotometrically at 420 or 500 nm with Nessler reagent and phosphorus is determined at 830 nm as molybdenum blue. This method has been applied to organic compounds and various natural products, such as egg and milk lipids, urine, meat extract and some amino acids.

E. Determination of Oxygen

A novel approach to the carbon reduction method permits the direct determination of oxygen in organophosphorus compounds with an average recovery of 100.0% and a relative standard deviation of ±0.05%[372]. The method uses a carbon reduction bed contained in an induction-heated graphite pipe. The silica of the quartz reaction chamber is not directly exposed either to the corrosive vapours of the sample or to the hot reducing carbon of the graphite. This permits considerably more latitude in the operating temperatures of the carbon bed than was found necessary by earlier workers 310,373-376. Table 63 gives some results obtained by the Holt[372] procedure. The normal blank value that persisted with these analyses was about 0.025 mg of oxygen. This amounted to about 1% of the total oxygen measured per analysis.

In Unterzaucher's original apparatus for the determination of oxygen the temperature of the specially prepared carbon bed had to be held fairly close to 1120°C. At higher temperatures the carbon monoxide blank, resulting from the interaction of carbon in contact with the quartz container, became significant, whereas at lower temperatures the conversion of oxygen-bearing vapours to carbon monoxide tended to be incomplete. Since the development of this method, platinum-catalysed carbon has come into use with which the carbon bed can be operated at lower temperatures. The induction heating unit used by Holt[372] provided graphite-pipe temperatures from about 600°C to 2000 C. No special treatment went into the preparation of the chip bed except that of crushing up AUC-grade graphite, sifting it on a 20-mesh sieve, and preheating it in the graphite pipe in an inert gas stream at 1800°C. The minimum operating temperature of about 1400°C can be conveniently set (Figures 17 and 18). In a fluorination method for the determination of oxygen in organophosphorus compounds the sample is placed in a nickel

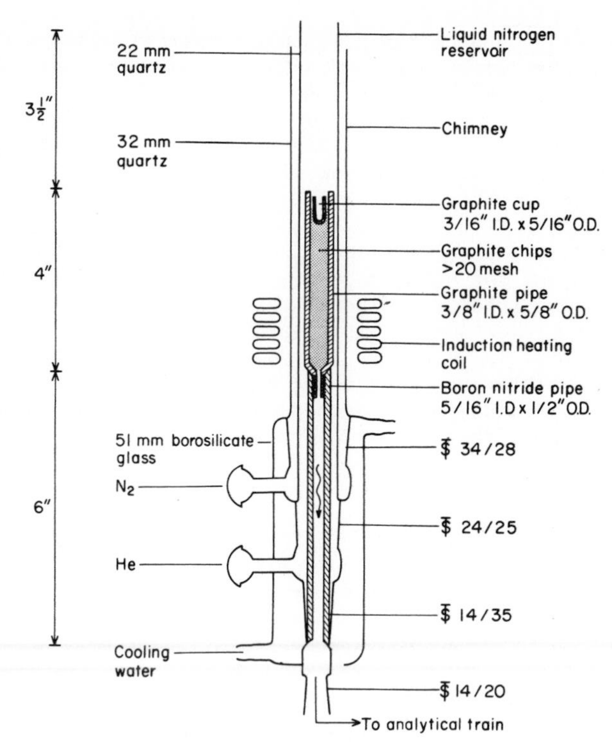

Fig 17. Determination of oxygen in organophosphorus compounds. Reaction chamber.

vessel containing $BrF_2.SbF_6$, which is then evacuated and heated at 500°C to fluorinate the sample and convert the oxygen to oxygen gas[377]. The latter is determined by mass spectrometric analysis. This procedure is tedious and lengthy.

F. Determination of Halogens

In an ultramicro method for the determination of fluorine in 1 - 20±g samples of volatile organophosphorus compounds the sample is decomposed by combustion in an oxygen-filled flask and oxidized with a mixture of nitric and perchloric acids[378]. Fluorine is then determined photometrically with the zirconium-cyanine or the zirconiun-norin complex. Another method depends upon the bleaching action of fluoride ions on the iron(III)-sulphosalicylic acid complex at pH 2.85 - 2.90[379]. The determination of fluorine and phosphorus in organic compounds using the Parr bomb technique followed by spectrophotometric determination has been discussed[380]. Phosphate present in the alkaline melt in the bomb after the fusion interferes in the determination of fluorine. Phosphate and fluoride are therefore separated by ion exchange. Other methods for determining fluorine involving precipitation of lead chlorofluoride[381,382], and the use of the thorium-alizarin lake complex[383-386]. For the latter, distillation of fluoride ions from the phosphate residue prior to titration is recommended because of the interference of orthophosphate with the fluorine determination[384].

Another method for total fluorine uses sodium ethylate in ethanol to

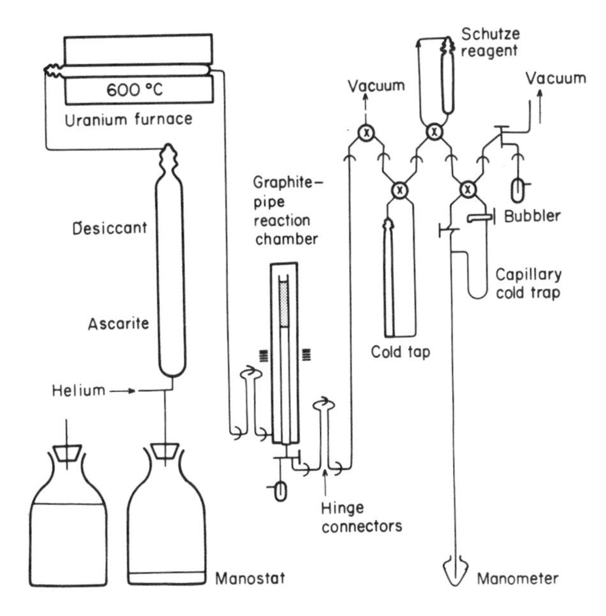

Fig. 18 - Determination of oxygen in organophosphorus compounds. Analytical train.

form fluoride ion from phosphorus fluoridate while esterifying the phosphorus moiety[387]. This precludes the formation of orthophosphate or alkyl phosphonic acid during the conversion of fluoridate to fluoride. The fluoride is titrated using a thorium-alizarin method without prior distillation of fluoride from the phosphorus residue and in this sense is superior to the method described above[383].

Owing to the hydrolytic stability of certain fluoridates in water at pH 3, ionic and hydrolysable fluorine can be determined in the presence of ester phosphorus monofluoridates by direct titration using the thorium-alizarin method. Ionic fluorine (hydrogen fluoride) can be distinguished quantitatively from hydrolysable fluorine (phosphonic difluoridate) by two acidity titrations - one with aqueous sodium hydroxide, and the other with an ethanolic solution of a tertiary amine.

In the determination of total fluorine, sodium ethylate in ethanol at 25°C reacts with phosphoro- and phosphonofluoridates replacing the fluorine atom(s) by ethoxy group(s) with the formation of sodium fluoride. All of the compounds discussed by Sass et al[387] in up to 2×10^{-3} molar concentrations, showed complete formation of sodium fluoride after a 5 min treatment with 5×10^{-2} M sodium ethylate. For more resistant fluoridates, reflux esterification could be resorted to with sodium ethylate in ethanol or with higher alkoxides in their corresponding alcohols. The determination of ionic and hydrolysable fluorine utilizes the fact that in an acid medium (pH 3) compounds such as alkyl alkylphosphonofluoridate, hydrogen alkylphosphonofluoridate, alkylphosphonofluorochloridate, and diisopropyl phosphorofluoridate show only negligible hydrolysis of fluorine over a period of at least 30 minutes. In an alkaline medium (above pH 9) the hydrolysis of fluorine is rapid. Because of the slow hydrolysis of the phosphorus monofluoridates to fluoride in acid, it is possible to determine

Table 63 - Oxygen results on organophosphorus compounds by the combustion tube method.

Compound	Recovery (%)	Compound	Recovery (%)
Dibutyl phosphate (30.5%O)	99.8	Diphenyl phosphinic acid (14.7%O)	99.7
	100.1		100.3
	99.3		100.2
	101.0		Av. 100.1
	99.8	Dioctyl phosphinic acid (11.0%O)	100.7
	Av. 100.0		101.0
Tributyl phosphate (24.0%O)	99.6		99.5
	99.8		Av. 100.4
	100.7	2-Ethylhexyl chloromethyl phosphonate (19.8%O)	100.6
	99.8		99.4
	99.7		100.0
	Av. 99.9		Av. 100.0
		Tri-n-octyl phosphine oxide (4.14%O)	100.1
			99.8
			99.1
			Av. 99.7

within 0.5% accuracy, the combined ionic fluorine (hydrogen fluoride) and hydrolysable fluorine (one fluorine from the phosphonodifluoridate) when present in these compounds, using the thorium nitrate titration.

When assaying samples of difluoridate, because one of the fluorines in the difluoridate is hydrolysable, the hydrolysable fluorine as determined should be no greater than half of the total fluorine found. Any fluorine in excess of this amount is free hydrogen fluoride.

For quantitative differentiation between ionic and hydrolysable fluorine in mixtures with ester phosphorus fluoridate compounds, a combination of acidity titrations is feasible. When titrated in iced water with 0.1 N sodium hydroxide to the end-point of methyl red bromocresol green mixed indicator (pH 5.1), all phosphonic and phosphoric acid, hydrogen fluoride, and two equivalents of acid from the hydrolysis of alkyl-phosphonodifluoridate (hydrolysis of one fluorine) are determined. When titrated in absolute ethanol with 0.1 N tertiary aliphatic amine, only the acids are titrated. The difluoridate is not titrated because it does not hydrolyse, and it esterifies only negligibly during the titration. The quantity of difluoridate can then be calculated by difference between the aqueous ice water titration and the alcoholic amine titration. The ester phosphorus fluoridates do not titrate under the above treatment. After the hydrolysable fluorine is determined, hydrogen fluoride is calculated by difference from the ionic fluorine determination. The excellent results obtained by Sass et al[387] in determinations of total fluorine and ionic and hydrolysable fluorine are shown in Tables 64 and 65 respectively. Results obtained on prepared mixtures of hydrogen fluoride and difluoridate with an ester phosphono-fluoridate using the thorium nitrate (ionic fluorine) and the acidity titration methods are shown in Table 66.

Table 64 - Total fluorine determination in phosphoro- and phosphonofluoridates by the thorium alizarin method.

Compound	Fluorine (%)		
	Calcd.	Found	Difference
Methylphosphonofluorochloridate	16.31	16.33	+0.02
		16.31	0.00
		16.31	+0.00
Methylphosphonodifluoridate	38.00	38.0	0.00
		37.99	-0.01
		37.98	-0.02
Methylphosphonofluoridic acid	19.38	19.38	0.00
		19.38	0.00
		19.40	+0.02
Ethylphosphonofluoridic acid	16.96	16.94	-0.02
		16.95	-0.01
		16.90	+0.06
Isopropyl ethylphosphonofluoridate	12.33	12.34	+0.01
		12.31	-0.02
Ethyl methylphosphonofluoridate	17.26	17.28	+0.02
		17.27	+0.01
		17.27	+0.01
Isopropyl methylphosphonofluoridate	13.56	13.58	+0.02
		13.56	+0.00
		13.54	-0.02
Diisopropyl phosphorofluoridate	10.32	10.30	-0.02
		10.33	+0.01
		10.32	0.00
N-Propyl methylphosphonofluoridate	13.56	13.53	-0.03
		13.53	-0.03
		13.55	-0.01
Pinacolyl methylphosphonofluoridate	10.43	10.42	-0.01
		10.40	-0.03
		10.39	-0.04

Orthophosphate, as mentioned, interferes with the lake methods for fluorine determination. Organophosphorus compounds of the dihydrogen alkylphosphonate (alkylphosphonic acid) type most closely resemble orthophosphate in behaviour. Alkyl dihydrogen phosphates are similar in this respect. At pH 3, these compounds show approximately 30% additive interference when calculated as fluorine. Alkyl hydrogen alkylphosphonates, dialkyl hydrogen phosphates, hydrogen alkylphosphonofluoridates (methylphosphonofluoridic acid), and alkylphosphonofluorochloridates do not interfere with the thorium-alizarin determination for fluorine. The results indicate that the presence of two groups of phosphorus, whether alkyl or alkoxy or a combination of one or the other with fluorine, precludes phosphorus interference with the titration for fluorine. In determining fluorine in mixtures of phosphonofluoridate compounds with readily hydrolys-

Table 65 - Determination of ionic and hydrolysable fluorine in ester phosphorus fluoridates

	Added impurity as F^- (%)	Ionic or hydrolysable flourine (%)	
		Calcd.	Found
Methylphosphonodifluoridate (A)		19.0	19.0 19.0 19.1
Isopropyl methylphosphonofluoridate (distilled)	0	0.0	0.2 0.2 0.2
Isopropyl methylphosphonofluoridate (distilled) +A	6.6	6.8	6.8 6.8 6.7

Table 66 - Analytical recovery of difluoridate and hydrogen fluoride in mixtures with ester phosphorus fluoridate

Compound	Hydrogen fluoride (%)			Methylphosphono-difluoridate(%)		
	Present	Added	Found	Present	Added	Found
Isopropyl methylphosphono-fluoridate (purified, 1 g)	0.2	0	0.3	0.3	4.2	4.5
	0.2	0	0.2	0.3	8.0	7.7
	0.2	1.0	1.3	0.3	1.8	2.0
	0.2	3.0	3.3	0.3	8.5	9.0
Isopropyl methylphosphono-fluoridate (crude, 1 g)	0.5	0	0.5	1.5	4.2	5.5
	0.8	1.0	1.9	2.0	1.9	3.7
	0.8	2.5	3.4	2.0	2.5	4.7
	1.0	2.0	2.9	5.3	1.5	6.7
Methylphosphonofluoridate (40 mg)	0.2	2.0	2.4	99.8	-	99.5
	0.2	0	0.3	99.8	-	99.8
	0	0	0.1	99.8	-	99.9
	0	0	0.1	99.8	-	99.6

able phosphonodichloridates, there is definite interference because the dichloridates produce alkylphosphonic acids and hydrochloric acid.

$$RP(=O)(Cl)_2 + 2H_2O \rightarrow RP(=O)(OH)_2 + 2HCl$$

The effect of organophosphorus acids and chloridates on the thorium-alizarin titration for fluorine (without sodium ethylate treatment) is summarized in Table 67.

The anhydrous sodium ethylate treatment rapidly converts the chloridates to phosphorus esters which do not interfere with the fluorine titration, while producing sodium fluoride and the ester from the fluoridate. No interference is encountered from the presence of chloride ion in the procedural range of sample concentrations. The reactions for phosphoro- and phosphonohalidates on treatment with sodium ethylate are shown in the following equation:

$$R\text{-}\overset{\overset{O}{/\!/}}{P}\text{-}X + NaOC_2H_5 \xrightarrow{C_2H_5OH} R\text{-}\overset{\overset{O}{/\!/}}{P}\text{-}OC_2H_5 + NaX$$

OR

$$(RO)_2\text{-}\overset{\overset{O}{/\!/}}{P}\text{-}X + NaOC_2H_5 \xrightarrow{C_2H_5OH} (RO)_2\text{-}\overset{\overset{O}{/\!/}}{P}\text{-}OC_2H_5 + NaX$$

X = halide

Table 67 - Effect of organophosphorus acids and chloridates on thorium-Alizarin titration of fluoride (pH 3) (Without prior treatment with sodium ethylate)

	Interference calcd. as fluoride (%)	
	Added	Found
Methylphosphonic acid	2.6	0.8
	5.5	1.7
	10.0	3.2
Methylphosphonofluoridic acid (0.3% F^-)	15.0	0.3
	20.0	0.3
	5.0	0.3
Diisopropyl phosphoric acid	10.0	0.0
	20.0	0.1
	30.5	0.0
Diisopropyl phosphorochloridate	5.5	0.0
	15.6	0.0
Ethyl methylphosphonochloridate	10.6	0.0
	20.3	0.1
Methylphosphonic dichloridate	2.5	0.6
	10.3	3.1
	20.6	6.3
Methylphosphonic dichloridate pretreated with sodium ethylate	2.5	0.0
	20.6	0.1
	30.5	0.0

In a method for the determination of the halogen in phosphonitrile halides the sample is treated with pyridine and then with water[388]. Rapid hydrolysis occurs, and the halide is titrated with silver nitrate solution, preferably potentiometrically. Chlorine in 2-chloroethyl derivatives of phospho-organic acids may be determined by dissolving the substance in ethanediol, and boiling the solution under reflux with sodium hydroxide in ethanediol[389]. In ethanolic medium, the reaction is incomplete. After addition of aqueous nitric acid, the solution is cooled and treated with excess of standard silver nitrate, the excess of which is determined by back-titration with standard ammonium thiocyanate.

G. Determination of Sulphur

Sulphur in organophosphorus compounds may be determined by fusing the sample in a bomb with sodium peroxide[390,391]. The sulphate produced is titrated with 0.02 - 0.01 N barium chloride in the presence of one drop of 0.2% aqueous nitchromazo[392]. Procedures for overcoming phosphorus interference in the determination of sulphur have been discussed[393,394] and details have been given for the determination of organically bound sulphur and phosphorus by oxygen flask combustion[395]. An oxygen flask combustion method can be used for the determination of small amounts of fluorine or phosphorus or sulphur in substances of low volatility[397]. To determine sulphur, combustion products are oxidized with a nitric acid-perchloric acid mixture, and sulphur is reduced to hydrogen sulphide and titrated with cadmium chloride solution. It was not stated whether phosphorus interferes in the determination of sulphur by these procedures although it has been pointed out that methods employing the oxygen flask combustion and subsequent titration with a barium salt usually give high results because of the slightly soluble barium phosphate formed[396]. It is necessary, therefore to eliminate the phosphate produced in the course of the combustion before an accurate measurement of sulphur can be made. Phosphate ions may be masked with iron(III) ions, since the latter chelate more readily with phosphate than with sulphate ions in an acidic solution, and the excess of iron(III) ions can be back-titrated with EDTA[397]. Sulphate is titrated by a conventional procedure using standard barium chloride[398].

Some typical results obtained by this procedure[396] are presented in Table 68.

A combustion furnace procedure for the determination of carbon, hydrogen, sulphur and phosphorus[370] and a rapid method for the determination of phosphorus - sulphur bonds in organophosphorus insecticides have been described[399].

ORGANOPLATINUM COMPOUNDS

Organoplatinum compounds may be broken down by gentle refluxing with a 50:50 mixture of concentrated hydrochloric and nitric acids followed by destruction of the remaining nitric acid by further boiling with hydrochloric acid[402]. A colorimetric analysis at 403 nm after treatment with tin(II) chloride solution enables platinum to be determined to within 1%[403].

ORGANOPOTASSIUM COMPOUNDS

Three methods for the flame photometric determination of potassium in potassium tetraphenylborate have been described[404]. In the first the potassium tetraphenylborate is precipitated in aqueous solution, and the precipitate dissolved in acetone and then examined by flame photometry. In the second method the precipitate of potassium tetraphenylborate is heated for 20 min at 350°C prior to dissolving it in water for flame photometry. In the third method the potassium tetraphenylborate is converted to potassium chloride by boiling with an aqueous solution of mercury(II) chloride.

Carbon, hydrogen, and potassium in organic samples can be determined by mixing the sample intimately with finely ground quartz in an empty tube and combusting it in a stream of oxygen[201]. Combustion in an empty tube

Table 68 - Oxygen flask method for determination of sulphur in organophosphorus compounds

Compound	Sulphur found (%)	Sulphur calcd. (%)
$C_4H_{11}O_2PS_2$ - NH_3 Salt	31.57 31.49	31.49
$C_{14}H_{19}N_2O_4PS$	9.46 9.56	9.34
$C_{13}H_{19}Cl_3NO_2PS$	8.20 8.16	8.19
$C_{13}H_{19}Cl_3NO_2PS$ (Isomer)	8.25 8.31	8.19
$C_{12}H_{17}Cl_3NO_2PS$	8.27 8.52	8.50
S-benzylthiuronium chloride	15.69 15.97 16.03 16.03	15.82

prevents absorption of carbon dioxide by the potassium residue and permits carbon and hydrogen to be determined by standard methods. The amount of potassium can then be obtained from the weight increase of the ignition tube after the combustion.

ORGANOSELENIUM COMPOUNDS

A. Determination of Selenium

Various techniques have been employed for the determination of selenium in organoselenium compounds. These include combustion techniques, oxygen flask combustion, fusion with sodium peroxide, and digestion with acids. Selenium in organic compounds containing carbon, hydrogen, oxygen, and nitrogen can be determined by tube combustion of the sample in oxygen [405]. After the ignition, the oxygen intake is replaced with a Mariotte flask serving as an aspirator, the layer of sublimed selenium dioxide is treated with water, and the selenous acid produced is determined iodimetrically. Another method[406] involves igniting the sample in a stream of oxygen and collecting the selenium dioxide produced in an absorption funnel prior to colorimetric determination with 3.3' -diaminobenzidine[407]. Sulphur or halogens do not interfere in this procedure. Combustion of organoselenium compounds in a stream of oxygen over quartz wool in a quartz tube or in an oxygen-filled flask followed by iodimetric determination yields an accuracy of ± 2%[408,409]. The combustion train is shown in Figure 19. A sample weight of 1 - 3 mg is required and this is weighed into a porcelain boat. Oxygen is passed through a preheater at 300°C then through the sample boat packed in quartz wool at 1050 - 1100°C for 10 minutes. The selenium dioxide produced is collected in an absorption tube containing water. The absorption tube is then heated to convert the selenium dioxide to selenous acid. Potassium iodide (50 mg) and 2 N sulphuric acid (10 ml) are then added and the liberated iodine titrated with 0.01 N sodium thio-

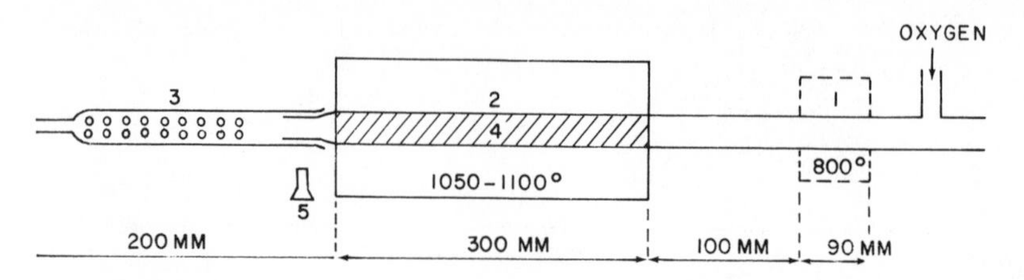

Fig. 19 - Combustion train for the determination of selenium in organosilicon compounds.

sulphate to the starch end-point. Each ml of 0.01 N sodium thiosulphate is equivalent to 0.1974 mg selenium.

$$SeO_3^{2-} + 4I^- + 6H^+ = Se + 3H_2O + 2I_2$$

Typical results obtained by the combustion tube procedure are listed in Table 69 for a range of organoselenium compounds containing carbon, hydrogen, bromine, nitrogen and oxygen.

Table 69 - Recoveries in the determination of selenium in organoselenium compounds. Tube combustion method

	mg sample	ml 0.02N $Na_2S_2O_3$	% Se Added	Found
Substance 1	1.960	1.60	32.3	32.1
	3.000	2.45	32.3	
Substance 2	3.639	3.43	37.2	37.3
	2.792	2.63	37.2	
Substance 3	2.998	3.60	47.4	47.2
	2.865	3.43	47.2	

Substance 1 contains C, H, Br, Se.
Substance 2 contains C, H, N, O, Se.
Substance 3 contains C, H, Se.

Oxygen flask combustion has been used by several workers for the determination of selenium in organic compounds[408]. The sample may be wrapped in paper and the oxygen combustion conducted in a flask containing distilled water[408]. After the combustion is completed, the selenium may be determined iodimetrically[408,410-416] or with permanganate[411,412]. The latter method has been applied to samples containing up to 62% of organically bound selenium with an absolute systematic error of less than 0.12%. In a further procedure the sample is burnt in an oxygen-filled flask and the vapours are absorbed in water[413,414]. The selenite produced is converted to selenocyanate ion by adding potassium cyanide solution. Sodium tungstate is added, the solution neutralized and a slight excess of aqueous bromine added. Unconsumed bromine is destroyed with phenol solution and the cyanogen bromide produced is reacted with potassium iodide and the liberated iodine is titrated with 0.01 N sodium thiosulphate. The results obtained by this procedure for organoselenium compounds showed a mean overall error of ±0.08%. The micro-determination of organic selenium has been carried out by fusion with sodium peroxide in a micro Parr bomb. Organic compounds containing selenium are readily decomposed by heating with sodium peroxide in a micro-bomb[467]. The product is dissolved in water, neutralized, and reacted with hydrazine, and the precipitate is filtered off in a fine-glass filter, dried at 110°C and weighed.

A colorimetric method has been described for the determination of selenium in organoselenium compounds following kjeldahl digestion[468]. Selenium is determined spectrophotometrically at 420 nm using chlor-promazine. Kjeldahl digestion with a mixture of concentrated sulphuric acid and potassium permanganate has been used as a preliminary to the iodimetric micro-determination of selenium in organic compounds[489]. Traces of selenium in organic matter can be determined using a combined spectrophotometric-isotope dilution method[470]. An earlier spectrophotometric method[407] was adapted to the micro-scale and the method improved by including an isotope-dilution procedure to compensate for the unavoidable loss of selenium. The sample (containing added ^{75}Se) is oxidized under reflux with a mixture of nitric, perchloric, and sulphuric acids and the selenium is then recovered as selenium tetrabromide by double distillation with hydrobromic acid and determined spectroscopically at 420 nm using 3,3-diaminobenzidine. X-ray emission can be used to determine down to 50 ppm of selenium in organic compounds[471]. The micro-determination of selenium in organic substances by chelatometry has been discussed[472].

A method has been described[947] for estimating selenium in selenosemi-carbazones. Selenosemicarbazone (0.01 g) is dissolved in 5 ml of ethanol and acidified with 2.5 ml of nitric acid, 0.01 N aqueous silver nitrate (10 ml) is added and the white flocculent precipitate formed is converted into the black silver selenide by being heated on a water bath for 10 to 15 min. When the suspension is cool, nitrobenzene is added to cover the precipitate and the excess of silver nitrate is titrated with 0.01 N potassium thiocyanate (about 5 ml) by the Volhard method. The average error in determination was + 1%.

B. Determination of Carbon and Hydrogen

Carbon and hydrogen in organic compounds containing selenium can be determined by combustion in oxygen in an empty tube, using finely ground quartz as a filter to retain the selenium dioxide produced[473]. An error not exceeding ±0.3% is claimed. Selenium in selenosemicarbazones can be determined by conversion to silver selenide and determination of silver by the Volhard method[474].

ORGANOSILICON COMPOUNDS

A. Determination of Silicon

The determination of silicon in organosilicon compounds has been reviewed by Belcher et al[475] and by Sykes[476].

Silicon can be determined by decomposing the organosilicon compound and then determining the silica or silicate produced. The technique generally used to bring about decomposition is wet-combustion or fusion with sodium peroxide. If a wet-combustion method is used, the resulting silica can be determined by dehydration of silicic acid by classical methods or by converting to silicofluoride ions in the presence of an excess of hydrogen ions. Silicofluoride is formed in accordance with the following equation:

$$Si(OH)_4 + 6F^- + 4H^+ = SiF_6^{2} + 4H_2O$$

Fusion with sodium peroxide leads to the formation of sodium silicate, which can be determined colorimetrically e.g. as molybdosilicic acid, volumetrically, or gravimetrically as the oxine or quinoline salt of molybdosilicic acid. A major problem associated with the analysis of organosilicon compounds is the successful decomposition of the sample. Some compounds give rise to silicon carbide when digested with acids. Generally, fusion with sodium peroxide in a Parr type bomb is the most satisfactory way of dealing with a wide variety of compounds. Alternate methods of decomposition are available, but are usually more time-consuming. Thus, when siloxanes are decomposed by heating with a mixture of sodium and potassium hydroxides in a sealed nickel crucible, quantitative decomposition may require more than three hours. The most commonly used procedure is fusion with sodium peroxide and determination of the silicate so formed by precipitation and titration of the quinoline salt of molybdosilicic acid.

Bomb Fusion Methods - Sodium Peroxide Fusion. Earlier procedures based on fusion with sodium peroxide for the determination of silicon in organosilicon compounds have been described by McHard et al[477] and by Wilson[478]. McHard et al[477] describe determinations of the silicate produced upon fusion of organosilicon compounds by colorimetry as molybdosilicic acid, volumetrically, and gravimetrically as the oxine salt of molybdosilicic acid. Wilson[478] describes a gravimetric determination of the sodium silicate as the quinoline salt of molybdosilicic acid.

Jean[479] has critically examined the procedures available for determining silicon, and Voinovitch[480] has studied methods for the determination of silicate. Ringbom et al[481] have shown that the colorimetric determination of silicon as molybdosilicic acid is subject to error because of the formation of the metastable beta form of the acid and have described a determination based on the formation of the alpha form. The colorimetric method based on the reduction of molybdosilicic acid to molybdenum blue involves use of sodium sulphite in acid solution as the reducing agent. Milton[482,483] has suggested that reduction by stannous chloride enhances the colour produced and therefore possibly gives greater sensitivity in the determination of silicon by this method.

Halzapfel and Gottschalk[484] have described a rapid procedure, based on sodium peroxide bomb fusion, for the determination of silicon and halogen in fluorine-containing organosilicon compounds and resins. The silicon is separated from the decomposition product as zinc silicate and estimated gravimetrically as silica. The filtrate is concentrated, acidified and, when necessary, reduced with sulphur dioxide. Chloride, bromide or iodide is then determined by the usual method. Fluoride can be determined in the neutral solution either gravimetrically as calcium fluoride, or volumetrically with zirconium tetrachloride or thorium nitrate. Fluoride can also be determined directly in the decomposition solution by titration with zirconium tetrachloride. The maximum error of the method is + 10%.

Christopher and Fennell[485] have described a gravimetric determination of silicon in organosilicon compounds; it applies to the milligram scale, and overcomes the disadvantages of many of the earlier procedures. The sample is fused with sodium peroxide, and silicon is determined in the extract by precipitation as quinoline silicomolybdate. This procedure is not applicable if phosphorus is present in the sample but the presence of fluorine within prescribed limits is tolerated.

Christopher and Fennell[485] made a comparison between the precisions obtained in the precipitation of the blue and yellow quinoline silicomolybdate complex from standard silicate solutions, following the fusion of organosilicon compounds with sodium peroxide in a Parr bomb. They concluded that there was no statistically significant difference in the precisions obtained and that the theoretical gravimetric factor (0.0120) could be applied to the yellow complex,

Fusion, leaching and washing a pair of Parr bombs takes about 25 minutes. By using small nickel capsules (15 mm high, 6 mm diameter, 0.5 mm walls) for fusion, this time can be cut to 10 - 15 minutes.

The same amount of peroxide was added to the samples, the flanged lid of the capsule fitted and held on by clamping the closed capsule vertically between the jaws of a small vice. The flame from a hand torch was allowed to impinge on the side of the capsule until it glowed red. The capsule was removed, cooled on an aluminium block and leached in a platinum crucible. The normal procedure was then followed.

Results obtained for the analysis of some solids by the nickel capsule (a) method are given in Table 70. The results tend to be high and are more scattered than those obtained using Parr bombs (b). However, fusion in this type of capsule, which can easily be made out of nickel rod, might be useful where high accuracy is not required.

The results in Table 70 obtained by this method on solid organosilicon compounds (0.25 ml nickel bomb), using an amount of sample containing about 0.5 mg silicon show a mean percentage recovery of 100.17% (standard deviation 1.02). Using this fusion procedure a fluorine content of 21.85% was obtained for $C_{16}H_{38}Si_3O_2F_2$, (theory 21.90).

Caustic alkali fusion. Wetters and Smith[486] have described a method employing alkali fusion in a nickel crucible for the determination of silicon in siloxane polymers and silicone-containing samples. They utilised the Technicon Auto Analyser system for the determination of silicon.

In this method about 5 g sodium hydroxide or potassium hydroxide is placed in a 75 ml nickel crucible. About 3 - 20 mg sample is then added. The lid is placed on the crucible and heat applied, gently at first, using a Meker type burner to melt the alkali. The alkali is held in the molten state for 2 - 3 min., using a hot flame to decompose the sample completely. After cooling the crucible to room temperature, the fusion mass is dissolved in 50 ml water in a Teflon-lined, stainless-steel beaker. Heat is gently applied to speed the dissolution. The crucible is removed with tongs and rinsed with water. The filtrate solution is neutralised with hydrochloric acid to a pH near 1.5, then transferred to a 500 or 1000 ml volumetric flask and diluted to volume. About 3 to 10 µg/ml of silicon solution is needed for analysis.

In an alternate alcoholic alkali evaporation and fusion procedure about 5 g alkali is added to the nickel crucible as above. With sodium hydroxide, 15 ml propan-2-ol is added. When using potassium hydroxide, 1 - 10 ml

Table 70 - Determination of silicon in pure organosilicon compounds using rapid mineralisation in (a) nickel crucible and (b) Parr bombs.

Compound	(a) Smaller nickel bombs * Sample wt. (mg)	Si found (%)	(b) Standard Parr bombs Sample wt. (mg)	Si found (%)	Si calcd. (%)
Triphenylsilanol	5.320	10.21	5.194	10.27	10.14
	4.807	10.19	5.019	10.07	
			5.288	10.12	
			5.134	10.00	
			4.963	10.15	
			5.309	9.86	
			4.958	10.02	
			5.163	10.18	
			4.863	9.96	
			4.992	10.16	
			5.115	9.98	
			5.114	10.20	
			5.042	10.48	
$C_{21}H_{24}Si_2O$	2.940	16.16	3.082	15.91	16.12
	3.299	16.04	3.250	15.07	
			3.121	15.70	
			3.060	16.01	
			3.235	16.13	
			2.996	16.19	
			2.982	16.91	
$C_{26}H_{26}Si_2O$	3.999	13.50			13.68
	3.729	13.49			
$C_{26}H_{26}Si_2O$			3.964	14.18	13.68
			3.993	13.65	
			3.757	13.61	
			3.805	13.72	
			3.427	13.62	
$C_{31}H_{28}Si_2O$	4.329	12.22	4.158	11.98	11.88
	3.997	11.93	4.202	11.87	
$C_{36}H_{30}Si_2O$	5.840	10.50			10.50
	5.334	10.48			
$C_{21}H_{24}Si_3O_3$	2.667	20.62			20.64
	2.731	20.94			
$C_{28}H_{32}Si_4O_4$	2.647	20.66	2.741	20.98	20.64
	2.460	20.53	2.772	21.98	
			2.714	19.90	
			2.480	20.29	
$(C_7H_{16}SiO)_a$	2.955	19.49			19.45
	3.224	19.45			

* 0.25 ml capacity bomb.

saturated sodium butylate solution is introduced. A sample of 3 - 20 mg is taken, and the mixture allowed to set at room temperature for 30 minutes.

The alcohol is slowly evaporated with the crucible on a hot plate at low heat setting, the final traces of alcohol being removed with high heat setting. Fusion of the alkali, sample dissolution and acidification are then carried out as described above.

Oxine precipitation (McHard et al[477]) and reduced heteropoly blue complex (Horner[487]) methods were used to estimate silicon in the digests prepared as described above.

Generally, Wetters and Smith[486] found that with linear and cyclic polydimethylsiloxanes and various other types of silicones, a higher silicon recovery was obtained by direct fusion with potassium hydroxide than was obtained using sodium hydroxide. However, recoveries of low molecular weight cyclic polydimethylsiloxanes were usually less than 50% using either fusion reagent. They found, however, that the use of an alcoholic alkali pre-treatment, using either alkali, minimised losses of such volatile silicon compounds and led to general improvement in silicon recovery. This is because, under these mild conditions, the siloxane bonds are converted to non-volatile alkali metal silanolates.

$$\equiv Si - O - Si \equiv + KOH \xrightarrow{ROH} \equiv Si - O - K + HOSi \equiv$$

$$\equiv SiOH + KOH \xrightarrow{ROH} \equiv Si - O- K + H_2O$$

Both the direct and the alcoholic alkali fusion methods gave recoveries of 90% of better with phenyl-containing silicones, the results obtained by the alcoholic alkali modification being generally higher.

Urin et al[488] studied the decomposition of siloxanes with mixtures of sodium hydroxide and potassium hydroxide in a sealed nickel crucible followed by gravimetric or colorimetric determination of the silicate produced. When fused at 210° dimethylsiloxane oil, octaphenylcyclotetra-siloxane, diphenylsilanediol, siloxanes of the varnish type, and siloxanes modified with alkyd resins are quantitatively decomposed within 1 h, 2h, 4h, 4h, and 4 h respectively.

Shanina et al[489] decomposed organosilicon compounds by fusion with potassium hydroxide in a nickel microbomb or by wet combustion with fuming sulphuric acid. Silicon is subsequently determined as molybdenum blue at 812 nm. They claim that results are accurate to within ±0.2% for silanes, siloxanes and their derivatives and also for complex organosilicon compounds also containing phosphorus, arsenic, antimony, germanium, titanium, aluminium and tin.

In an alternate method, the organosilicon compound is fused with potassium hydroxide in a nickel bomb followed by conversion of the silicate to the quinoline salt of molybdosilicic acid, which is then estimated by acidimetric titration using the procedure described by Wilson[478].

Mineralisation and Digestion. Various procedures have been described for the determination of silicon in organosilicon compounds based on digestion of the sample with a mixture of concentrated or fuming nitric and sulphuric acids. Sir and Komers[490] described the following procedure which, it is claimed, is suitable for the determination of silicon in esters of orthosilicic acid, arylsilanes, arylalkylsilanes, polysiloxanes, alkyl-silanes, and alkyl- and aryl-halogenosilanes. The determination of silicon is carried out acidimetrically. A mixture of oleum and fuming nitric acid is used for the oxidation, which is followed by ignition. Four different techniques are used according to the volatility and ease of oxidation of the compound. Non-volatile compounds are weighed directlv into a platinum crucible to which the oxidants are added. Alkylsilanes are weighed into a crucible (cooled in solid carbon dioxide) containing the oxidants. Arylhalogenosilanes are oxidised in nitric acid - sulphuric acid containing a little water; alkylhalogenosilanes are weighed into a platinum crucible (containing pyridine cooled in solid carbon dioxide) and the oxidants are added to this. The crucible is heated gently on a sand bath, then more strongly until the contents are ignited. Dissolve the residue in the minimum amount of water containing sodium (or potassium) carbonate (0.5 - 1 g), transfer to a flask, add excess of diluted hydrochloric acid (1 + 1) and boil to remove carbon dioxide. Add 10 drops of a mixture of alcoholic methyl red solution (0.1%) and aqueous bromocresol green (0.1%), neutralise exactly with sodium hydroxide, keeping the volume below 50 ml, saturate with potassium chloride or nitrate, add neutral 1% ammonium fluoride solution (10 ml) and exactly 20 ml 0.1 N hydrochloric acid. Back-titrate the excess of acid with 0.1 N sodium hydroxide until the colour changes to green.

Kreshkov et al[491] effected the decomposition of organosilicon compounds in a quartz flask using a mixture of 25% oleum and fuming nitric acid, containing 2% oxides of nitrogen. The heated mixture is treated with further amounts of fuming nitric acid until decomposition is complete, the acids are evaporated off, and the flask containing the residue of silica is weighed.

Smith[492] showed that, compared with the Kjeldahl method, the quicker crucible method of acid digestion is unsuitable for low-boiling compounds because of evaporation losses during the early stages of heating. By using a system of two crucibles, a stronger oxidising acid mixture (equal vols of 60% fuming sulphuric acid and fuming nitric acid, ice cooled) and by introducing the substance well below the acid from a sealed ampoule, losses are reduced to a minimum. Any overflow due to frothing from the inner 10 ml reaction crucible is caught in the outer 50 ml crucible. Good agreement with the theoretical silicon content was obtained, even for compounds that react violently with the acid mixture.

Mixtures of nitric acid, sulphuric acid and hydrofluoric acids have been used (Myshlyaeva et al[493]) for the determination of silicon in alkyl- or aryl-substituted organosilicon compounds such as trichlorophenylsilane, dichloromethylphenylsilane, diethoxydiphenylsilane, and triphenylsilanol with a relative error not exceeding + 15%. The sample (0.5 g) in a platinum crucible is treated with 1 - 5 ml fuming sulphuric acid (60% free SO_3) and 0.2 ml fuming nitric acid and heated slowly until sulphur trioxide evolution ceases. Then 20 ml 30% hydrofluoric acid-anhydrous acetic acid-methanol (1:1:10) is added and the mixture cooled in solid carbon dioxide-acetone. When the residue is completely dissolved the resulting hydro-flurosilicic acid solution is transferred to a polyethylene beaker and 30 ml ethanol added. The solution is then titrated conductiometrically with 0.5% benzidine in ethanol and the end-point deduced graphically.

Ammonium persulphate has been used by Myshlyaeva et a[494] as a digestion reagent for the determination of silicon in tetraalkoxysilanes and poly-meric organometallic siloxanes containing aluminium or titanium. In this method the sample containing 4 - 5 mg silicon is heated to fumes with 1 ml sulphuric acid and the cooled solution is heated with 0.2 g portions of ammonium persulphate until it is colourless. The cooled residue is diluted with 50 ml aqueous hydrogen peroxide and treated with solid potassium hydroxide (1 g more than the amount required to neutralise the solution to phenolphthalein), then heated for 5 min at 100° and neutralised with 2 M sulphuric acid.

The solution is then treated at 100°C with a boiling mixture of 10 ml 20% ammonium molybdate solution, 25 ml water, and 15 ml 2 M sulphuric acid, and heated at 100° for a further 5 minutes. The cooled solution is mixed with 35 ml 2 M sulphuric acid and molybdosilicic acid is extracted into 50 ml butan-2-ol. An aliquot (10 ml) of the extract is mixed with 50 ml acetone and titrated conductimetrically with a standard solution of tetraethylammonium hydroxide in methanol; butan-2-ol (1:1), which has been previously standardised against pure silicate solutions of known composition.

Chulkov[495] has described a procedure for the determination of silicon in polyaluminoorganosiloxane resins. In this procedure, the resin is decomposed with fuming sulphuric acid in the presence of ammonium sulphate, the solution is diluted and the aluminium complexed with tartaric acid. The finely divided resin (0.2 g) and ammonium sulphate (0.2 g) is heated in a quartz crucible with the dropwise addition of fuming sulphuric acid. Silicic acid is precipitated during this operation. The mixture is then transferred to a beaker containing 25 ml 2% tartaric acid and a few drops of methyl red indicator are added. The solution is heated to boiling, and aqueous ammonia (1:1) added until alkaline. The solution is stirred and boiled for 2 min, allowed to cool, the liquor decanted and the residue washed twice by decantation with hot water, testing the washings for absence of dissolved solids by evaporation. The silicic acid precipitate is transferred back to the quartz crucible, the filter paper ashed at 800°, and then ignited to constant weight at 1000 - 1050° and finally weighed as silica.

Terent'eva et al[496] have described a rapid method for the determination of silicon in organic silicon compounds, based on their oxidation by a mixture of chromic and sulphuric acids at 150°. The resulting silica is filtered off, dissolved in sodium hydroxide and converted into fluorosilicate ion by ammonium fluoride in the presence of an acid. The excess of acid is back-titrated with alkali. The determination takes 1.5 hours. However, when the silicon compounds is readily hydrolysable with an aqueous alkali solution with the formation of silica the determination time is reduced to 30 minutes.

Fogel'son[497] discussed a simple hydrochloric acid digestion procedure for the determination of silicon in ethyl silicate. In this method, the sample (0.1 g) is treated for a few minutes with 5 ml concentrated hydrochloric acid and the mixture is then stirred with 20 ml 20% sodium hydroxide solution to give a clear solution, which is treated with 35 ml dil. sulphuric acid (1:8) and diluted with water to 250 ml. An aliquot (25 ml) is mixed with 20 ml dil. sulphuric acid (1:8) and diluted to 250 ml and from this solution three 10 ml aliquots are taken, one of them to serve as a blank. These aliquots are diluted to 40 ml and two of them are treated with 5 ml 5% ammonium molybdate solution followed after 3 min by 10 ml 3% ammonium oxalate and then immediately by 5 ml of a 5% solution of ferrous ammonium sulphate containing 20 ml concentrated sulphuric acid per litre. The third aliquot is treated with the same reagents except the molybdate. The solutions are diluted to 100 ml and the extinctions are measured spectrophotometrically using a red filter.

Lew and Oyung[498] mention a method for the spectrophotometric determination of silicon in ethyl silicate and polyethylsilicate in which the sample is hydrolysed with acid in the presence of ammonium molybdovanadate to form the yellow molybdovonadosilicate.

Hydrolysis with Ethanolic Hydrofluoric Acid. Kreshkov et al[499,500] and Kreshkov and Kyshlyaeva[501] have studied the reaction of organosilicon compounds with ethanolic hydrofluoric acid followed by estimation of the hydrofluosilicic acid so produced by various procedures as discussed below.

1-Place 50 ml 50% ethanol, 1 - 2 ml 50% hydrofluoric acid solution (free from H_2SiF_6) and the sample in a covered polyethylene beaker, and, after 3 to 5 min, add 10% potassium hydroxide solution till the mixture is alkaline to phenolphthalein, then neutralise it with 0.2 N hydrochloric acid or sulphuric acid. Dilute to 500 - 700 ml with hot water, and titrate to the phenolphthalein end-point with sodium hydroxide. In the iodimetric method, to the hydrofluosilicic acid solution obtained as described above add 20 ml 40% calcium chloride solution, heat for 15 min, the cool to 20°. Dilute to 250 ml, and to an aliquot add 10 ml each of 10% potassium iodide solution and saturated potassium iodate solution; after 5 min. titrate the liberated iodine with standard sodium thiosulphate.

2-Kreshkov et al[500]. The sample (0.05 g) is placed in a polyethylene beaker containing 10 ml 0.2 N to 0.4 N hydrofluoric acid and 25 ml of ethanol or acetone are added. The fluosilicate is then titrated conductiometrically with a 0.5% solution of benzidine, toluidine or 2-naphthylamine in acetone, methanol or ethanol. An absolute error not exceeding ±0.2% is claimed for this method.

3-Kreshkov and Myshlyaeva[501]. In this procedure, the sample (0.03 - 0.05 g) is mixed with 10 ml 0.3 N-ethanolic hydrofluoric acid in a polyethylene beaker and treated after 2 min with 10 ml 1% ethanolic benzidine. The silicon is precipitated quantitively as the benzidine-H_2SiF_6 complex. The precipitate is collected and washed with ethanol and then decomposed with water to give hydrofluosilicic acid, which is titrated with 0.1 N potassium hydroxide in the presence of phenolphthalein. Other workers have reported procedures for the determination of organically-bound silicon, based on formation of the fluosilicate ion. Thus Popov et al[502] and Kalman and Vago[503] describe procedures in which the fluosilicate finish is accomplished by alkaline titration. Popov et al[502] also describe a gravimetric finish involving precipitation of the benzidine-fluosilicate complex, followed by ethanol dissolution of the latter and titration with standard sodium hydroxide; Myshlyaeva and Krasnoshchekov[504] describe a finish based on titration with methanol or acetone solutions of benzidine or 2-naphthylamine which, it is claimed, is applicable to the determination of silicon in alkoxysilanes such as tetramethoxysilane, tetraethoxysilane, tetrapropoxy silane, tetrabutoxysilane and in polymers and resins modified with organosilicon compounds.

Further Methods for Organically Bound Silicon. Kreshkov and Gludiva[505] dissolved the sample (0.08 mg SiO_2/ml) in 15 ml 10% potassium hydroxide solution. A portion (5 - 15 ml) of this solution was treated with 5 ml 5% ammonium molybdate in 10% acetic acid and then 5 ml of saturated solutions of sodium sulphite and of sodium sulphate were added and the solution warmed for 5 min and then cooled. This solution was treated with 20 ml of a reagent consisting of 20 g/l ammonium oxalate, 20g/l sodium carbonate and 150 ml glycerol per 1 and then diluted to 100 ml. The molybdenum blue colour was evaluated at a wavelength of 650 to 700 nm after 1 hour.

Brown and Fowles[506] analysed volatile organosilicon compounds for silicon as follows. Oxygen is bubbled through the weighed sample until evaporation is complete, then for a further 15 min using a cold bath or infrared lamp to control the rate of evaporation. The vapours are passed into a weighed silicon tube and ignited in a plug of prepared asbestos. The silica tube is removed and ignited to a constant weight at 800°.

The oxygen flask combustion technique has been used to determine silicon in organosilicon compounds (Reverchon and Legrand[507]). In this method the sample is burnt inside an oxygen-filled flask of 1l capacity constructed from pure nickel sheet. The silica produced is absorbed in N-sodium hydroxide solution and the solution adjusted to pH 1:3 - 1:7. The silicate concentration of the resulting solution is determined spectrophotometrically at a wavelength of 400 nm using the procedure described by Govett[508]. If the sample has a high silicon content, it is recommended that 60 mg sodium peroxide is included in the sodium hydroxide absorbing solution. An accuracy of within ±2% is claimed for this method.

Jenik and Juracek[508] fuzed the organosilicon compound with magnesium to produce magnesium silicide. This is then decomposed to produce gaseous silicon hydrides by addition of dil. sulphuric acid. The hydrides are absorbed in bromine water and thus hydrolysed to silicic acid. The silicic acid is converted into molybdosilicic acid and determined colorimetrically as molybdenum blue. The error is + 0.73 to - 0.54% over the range 10.53 to 37.8% of silicon. Halogens can be determined on the same sample; after removal of silicon hydrides, the elementary carbon which separates is deactivated with ferric or aluminium salts and filtered off, and halogens are determined in the filtrate by Volhard's method.

Emission spectrography has also been used for the estimation of silicon in organosilicon compounds. Kreshkov et al[509] dissolved the sample in a suitable solvent, together with a known weight of zinc chloride. The sample is placed in a fulgurator in the form of a U-shaped vessel, one limb of which consists of a hollowed graphite electrode. The test solution slowly passes to the top of this electrode and forms a film during a spark discharge between the end and a copper counter electrode. The lines silicon 634.701 nm and zinc 636.235 nm are measured.

Radell and Hunt[510] carried out the spectrograph analysis on the Todd Spectranal by the use of the emission lines at 623.7 and 636.1 nm. The sample is decomposed in acid solution by a spark discharge between platinum electrodes.

Chan[511] studied the determination of silicon in organosilicon compounds such as arylsilanes by X-ray fluorescence with a vacuum spectrograph. He used an X-ray vacuum spectrograph equipped with a power unit capable of producing 75 kV and 50 milliamps. It contained a tube having dual targets, one of tungsten and one of chromium, located inside the vacuum chamber. These targets could be switched to the operation position by remote-control mechanism. Provision was also made in this unit for dual counters, one located in front of the other, which could be operated individually or simultaneously. Two analysing crystals of different 2d spacing were placed in the vacuum spectrograph, and the remote-control device could be used for changing these crystals rapidly.

B. Determination of Phosphorus and Silicon

Fennell and Webb[512] have pointed out that silicon could not be accurately determined in the presence of phosphorus by simple wet oxidation of the organic material in a platinum crucible by weighing the silicon produced, as this method leads to high results. Fennell and Webb[512], and Fennell et al[593] found that the method of Belcher and Godbert[514] involving digestion of the sample with nitric and sulphuric acids followed by a gravimetric determination of phosphate after precipitation with Jorgensen's reagent (nitratopentamminocobaltinitrite) usually gave satisfactory results with a variety of compounds. They established that, after acid digestion of compounds containing silicon and phosphorus, the phosphate-containing acid liquid could be separated quantitatively from the silica precipitated in the glass boiling tube by use of an elongated glass filter stick (porosity 4).

The residue was washed with nitric acid (1 + 1) followed by alcohol and then dried at 100° under reduced pressure, cooled in a desiccator and weighed. The majority of this silica was then tipped into a platinum crucible, weighed, ignited to bright red heat and weighed again. The correction factor (weight after ignition/weight of silica ignited) was applied to the original weight of silica in the boiling tube and the silicon content calculated from this corrected figure.

Fennell et al[514] also described a sodium peroxide bomb fusion method for the determination of phosphorus in the presence of silicon. This method involves decomposition of the organic material (in a micro bomb) with sodium peroxide followed by precipitation of phosphate in the acidified bomb leachings as quinoline phosphomolybdate. Co-precipitation of silicomolybdate was avoided by complexing with citric acid or by removal of the silicon by volatilisation of tetrafluoride or precipitation of silica. The latter procedure was found to provide a means for the determination of silicon.

The procedure previously described (Fennell et al[514]) for the determination of phosphorus was followed except that the acidified bomb leachings were evaporated to dryness and baked at 110°. After cooling, the crystals were moistened with concentrated hydrochloric acid, dried, and baked again. The soluble salts were dissolved in hot water and filtered through a paper-pulp pad, the filtrate and washings being collected in a conical flask for precipitation of phosphate as quinoline moybdophosphate (Fennell and Webb[515]). All the insoluble residue was collected on the paper pad during the filtration and the whole was quantitatively transferred into a platinum crucible. The paper was ashed and the residue was weighed, treated with hydrofluoric acid and re-weighed. From the loss in weight, the silicon content of the organic material was calculated.

Fennell and Webb[512] state that both methods are accurate to within ±0.3% and believe that the sodium peroxide bomb fusion method is manipulatively simpler and of more universal application than the nitric acid-sulphuric acid digestion method. Results obtained in determinations of phosphorus and silicon on a range of compounds by both methods are presented in Table 71.

Phosphorus can be determined in compounds containing fluorine (Fennell et al[513]) and in compounds containing fluorine and silicon (Fennell et al[512]) Shell and Craig[516] and Harel et al[517] have described methods for the determination of silicon in compounds containing fluorine.

Fennell and Webb[518] have also described a procedure for the determination of phosphorus and silicon in fluorinated organic materials. The determination of phosphorus and silicon in the presence of fluorine presents

Table 71 - Determination of Silicon and Phosphorus by acid digestion and by bomb fusion

Compound	% Silicon (Calc.)	Found	% Phosphorus (Calc.)	Found
Nitric Acid - Sulphuric Digestion				
$C_{22}H_{32}N_2O_3Si_2$	13.09	13.22 13.18 13.14 12.94		
$C_{26}H_{60}O_9P_2Si_2$	8.84	8.50 8.52	9.75	9.75 9.76
$C_{38}H_{52}O_9P_2Si_2$	7.28	7.17 7.49 7.53 7.59	8.04	7.80 7.80 7.78
Sodium Peroxide Bomb Fusion				
$C_{14}H_{36}O_9P_2Si_2$	12.03	11.81 11.91	13.28	13.16a 13.18
$C_{26}H_{60}O_9P_2Si_2$	8.84	8.68 8.74	9.75	9.72a 9.75
$H_{38}H_{52}O_9P_2Si_2$	7.28	7.15 7.15	8.04	8.00a 7.99

(a) Using gravimetric finish for determination of phosphorus (Fennell and Web11).

some problems. Thus Schwarzkopf and Henlein[519] proposed two methods for the determination of silicon in fluorinated organic compounds. Neither method can be used if phosphorus is also to be determined, because phosphate is precipitated with lead chlorofluoride (Fennell[520]) and with oxine silicomolybdate (Brabson et al[521]). Fennell and Webb[512] in earlier work, found that interference by fluorine in the determination of silicon in organic materials could be overcome by the addition of boric acid before the precipitation of silica. They later found, however, that this method was not universally applicable, especially when larger quantities of fluorine were present.

A sodium peroxide bomb has been described by Fennell and Webb[518]; Greenfield[522]; Wilson[523]) for the determination of phosphorus and of silicon in compounds which also contain fluorine.

Silicon and phosphorus are determined, respectively, by the molybdenum blue and Vanado molybdate spectrophotometric methods.

Using these procedures Fennell and Webb[518] obtained the results shown in Table 72 on samples of triphenylsilanol and various other organic compounds containing phosphorus, silicon and fluorine. Mean recoveries, calculated from the results in Table 72 are: for silicon 101.3% (standard deviation 1.64, 18 determinations); for phosphorus 100.9% (standard deviation 3.02, 12 determinations). In spite of the possible existence of

Table 72 - Determination of Silicon, Phosphorus and Fluorine in Organosilicon Compounds

Compound	Sample wt. mg	Silicon % Calc.	Silicon % Found	Phosphorus % Calc.	Phosphorus % Found	Fluorine % Calc.
$C_{18}H_{16}O\ Si$	19.64	10.14	10.39	-	-	-
	25.68		10.40			
	28.56		10.12			
	29.68		10.28			
	30.73		10.28			
	40.05		10.34			
$C_{15}H_{37}O_5Si_2P$	13.25	14.60	15.02	8.05	8.46	-
	30.47		14.57		7.75	
$C_9H_{23}O_5Si\ P$	27.53	10.39	10.53	11.46	11.44	-
	27.77		10.51		11.45	
$C_{38}H_{52}O_9Si_2P_2$	31.27	7.29	7.36	8.04	8.06	-
	39.80		7.21		7.89	
$C_{15}H_{35}F_2O_3Si_2P$	15.54	14.46	14.54	7.97	8.43	9.8
	17.31		14.27		8.20	
$C_{20}H_{37}F_2O_3Si_2P$	24.05	12.52	12.68	6.90	7.11	8.4
	27.91		12.36		7.13	
$C_{26}H_{32}F_{28}O_2Si_2P_2$	42.71	4.93	5.13	5.44	5.34	46.7
	49.05		5.18		5.36	

a positive bias in the method, and the tendency for some results to fall outside the normally accepted limits (±0.3% absolute) for analysis of organic materials, he considers that these results are reasonable for these types of compounds.

Andrianov et al[524] determined phosphorus in phosphosilico organic compounds (POSi bonds) by dissolving the sample in water, leaving for 4 - 5 h to hydrolyse, (or for 30 - 40 min at 80°) and titrating the alkyl phosphinic acid so produced with 0.1 N alkali to the thymolphthalein end-point.

Ostrowski et al[525] developed a method for the simultaneous determination of free phosphorus, combined phosphorus, and silicon in the reaction products of tetraalkoxysilanes with phosphorus halides.

C. Determination of Aluminium and Silicon

Myshlyaeva and Shatunova[526] described a method for the determination of aluminium in organometallic compounds containing silicon, in which the sample (equivalent of 10 - 20 mg aluminium) is dissolved in 50 - 100 ml acetone. 2 N hydrochloric acid (5 ml) is added and the solution refluxed for 5 - 10 min, then cooled. Excess 0.05 M EDTA is then added and the solution again boiled for 5 to 10 min and cooled. Ammoniacal ammonium acetate pH 4.5 buffer solution (30 ml) and 0.1% dithizone in methanol or acetone (2 ml) are then added and the solution titrated with 0.05 N zinc sulphate to the end-point (grey yellow to pink colour change).

Chulkov[527] described a complexiometric method for the determination of aluminium in polyalumino-organosiloxane resins in which the sample (0.3 - 0.4 g) is decomposed by wet combustion with nitric acid, sulphuric acid and ammonium sulphate; the solution is diluted, silica is removed, and an aliquot containing not more than 20 mg aluminium is treated with excess EDTA solution and neutralised to phenolphthalein with 5% aqueous ammonia. After being boiled for 5 min, the hot solution is treated with 2 ml 2 N acetic acid and again boiled. The solution at 40° is treated with 10 ml of an acetic buffer solution (pH 6.0) and 1 ml of 0.8% haemotoxylin solution and diluted to 100 ml. The excess of EDTA is then titrated with 0.01 N aluminium sulphate solution to the colour change from yellow-green to pink. Interference is caused by iron, titanium, zinc, and manganese.

Kreshkov et al[528] describe procedures for the determination of aluminium and silicon in organic compounds. To determine aluminium, the sample, containing the equivalent of 10 - 20 mg aluminium, is dissolved in 100 ml acetone or methanol and 5 ml 2 N hydrochloric acid added and the solution boiled for 3 - 5 min. The solution is cooled and excess 0.05 M EDTA added and the solution again boiled for 2 - 3 min. The solution is then cooled, 30 ml ammoniacal buffer solution and 1 ml of 0.1% acetone or methanol solution of dithizone added, then excess EDTA is titrated with 0.05 M zinc sulphate solution to the grey-yellow to pink end-point. Kreshkov et al[528] also described a spectrographic determination of aluminium and silicon in which the sample is dissolved in dry cumene containing ferrocene as a source of iron for internal standardisation. A cooling mixture comprising ethylene glycol, water, acetone and solid carbon dioxide is placed on the outer vessel of a fulgurator, and the sample and solutions are placed in the inner vessel. Using a graphite rod as a counter electrode and the line pairs Al 309.27 and Fe 302.07 nm, and Si 252.85 nm and Fe 252.96 nm, it is possible to estimate aluminium and silicon with a coefficient of variation of 2.2% to 3.8% and a sensitivity for aluminium of 10^{-3}%.

Polarography has been used by Terent'eva[529] for the estimation of aluminium in alumino-organosiloxanes. He reported that acid Chrome blue K (C.I.Mordant Blue 31) is much more soluble and gives a more stable complex than do other dyes used by an earlier worker (Willard and Dean[530]) for the polarographic determination of aluminium. The difference between $E_{\frac{1}{2}}$ of the complex and the $E_{\frac{1}{2}}$ of the free dye is about 0.4 volts. A sample containing 3 - 7 mg aluminium is moistened with water and heated on a water bath with 5 ml of 45 - 50% hydrofluoric acid solution and 0.5 ml sulphuric acid until water is removed, then treated with 2 ml hydrofluoric acid and then evaporated to dryness. The residue is fuzed with 2 ml sodium carbonate; the melt is boiled with water, the solution filtered and the filtrate treated with 2.0 ml glacial acetic acid, 5 ml 2 N ammonium acetate and 20 ml 5% aqueous Chrome blue K, then diluted to 50 ml. After being heated at 55° to 70° for 5 min, the solution is treated with 5 drops of 1% gelatin solution in the cold, oxygen is removed by passage of nitrogen, and a polarogram is recorded over the range 0 to -0.8 V v the mercury pool.

Terent'eva et al[531] have described procedures for the determination of aluminium, carbon, and silicon in polymeric organoalumino-siloxanes. The sample is decomposed with concentrated sulphuric acid and chromic acid in a current of oxygen at 150°. The evolved gases are passed through a tube containing chromium trioxide heated to redness to ensure the complete oxidation of the products of combustion, which then pass through a tube containing an acetic acid solution of hydrazine (pH 6) to remove halogens, a drying tube and finally a weighed tube of soda lime to absorb carbon dioxide. The contents of the decomposition flask are diluted with water, silicic acid is filtered off and converted into $NaSiF_6$, and the silicon is determined volumetrically. The filtrate contains the aluminium which is separated from other elements present by electrolysis with a mercury cathode. The aluminium is then determined complexiometrically.

The same worker (Terent'eva et al[532]) also described methods for the determination in organosilicon compounds of silicon and carbon, chlorine and bromine. He used the chromium trioxide-sulphuric acid oxidation procedure described in the previous paragraph (Terent'eva et al[531]) with the modification that any incompletely ignited products are further oxidised over chromium trioxide on pumice. Carbon is determined gravimetrically as carbon dioxide, chlorine or bromine is absorbed in hydrazine hydrate solution adjusted to pH6 with acetic acid, and then determined argentimetrically. Silicic acid is precipitated during the oxidation; it is separated, and the silicon dioxide is converted into fluorosilicate and determined volumetrically as previously described (Terent'eva et al[533]). Polyorganoaluminosiloxanes are incompletely oxidised under the conditions cited and must first be decomposed by heating with concentrated sulphuric acid; the resulting siloxanes are then readily oxidised. Carbon is determined gravimetrically, silicon volumetrically, and aluminium in the filtrate after removal of silicon.

Luskina et al[534] discuss the determination of aluminium, silicon and phosphorus in complex polysiloxanes. The sample (40 - 60 mg) is dissolved in ethyl ether, the solution is added to concentrated sulphuric acid and the mixture is heated for 3 - 5 min (to break any Al-Si bonds), then potassium persulphate is added, and heating is continued for 20 - 30 mins. To the cooled solution is added water, the mixture is boiled for 10 - 15 min, silicic acid is filtered off, and silicon is determined as described previously (Terent'eva et al[533]). In the filtrate, aluminium and phosphorus are determined by complexometric and photometric (as molybdenum blue) methods respectively. The method was applied to complex polysiloxanes (also containing aluminium and phophorus) which are difficult to oxidise by other methods. For 2 - 10 mg aluminium and phosphorus the error was not greater than ±0.3 mg for each element.

D. Determination of Titanium, Silicon, Phosphorus, and Aluminium

The wet combustion procedure using potassium persulphate and concentrated sulphuric acid described in the previous paragraph (Luskina et al [534]) has also been used by the same workers (Luskina et al[535]) for the determination of titanium, phosphorus and silicon in organic compounds. Following removal of the silica by filtration, the filtrate is diluted with water and potassium hydroxide added. The solution is then passed down a Ku-2 ion exchange resin column (chloride form) which retains titanium whilst phosphorus passes into the percolate where it can be determined spectrophotometrically as molybdenum blue. Titanium is desorbed from the column with 4 N hydrochloric acid and subsequently determined spectrophotometrically in the percolate. For approximately 10%, 12% and 33% respectively of titanium, phosphorus, and silicon in the samples, the absolute errors obtained by this method of analysis did not exceed 0.5%.

Terent'eva and Smirnova[536] report a fusion method for the analysis of difficultly decomposable organic compounds containing titanium, phosphorus or silicon. The sample (15 - 50 mg) is fused with sodium fluoride (1 g), and potassium pyrosulphate (0.2 g). The cooled melt is treated with concentrated sulphuric acid (12 ml) and evaporated to fumes for 20 min in an air bath. The residue is dissolved in 70% sulphuric acid, and titanium, phosphorus and silicon are determined by the classical procedures.

Chulkov and Solov'eva[527] have described methods for the determination of titanium and aluminium in organosilicon compounds using EDTA. The sample is dissolved by heating it with fuming sulphuric acid (20% SO_3) containing a little ammonium sulphate. Concentrated nitric acid is added dropwise until the solution is decolourised and no further oxides of nitrogen are evolved. The residue is diluted, boiled for 2 min, and the solution decanted through filter paper. The precipitate is washed with 20% ammonium nitrate solution and the filtrate combined with the decanted liquid and made up to 250 ml. To an aliquot, containing approx. 20 mg aluminium, is added excess 0.05 M EDTA and then aqueous ammonia until the solution is alkaline to phenolphthalein. The solution is then boiled to decolourise, and 2 ml 2 N acetic acid is added followed by boiling for 3 min, then cooling to 40°C. Acetate buffer solution pH6 (10 ml) and 0.2% haematoxylin (1 ml) are then added followed by dilution with water to 100 ml.

Finally, the solution is heated to 60° - 70°C and titrated hot with standard potassium alum solution. The method for determining titanium is based on EDTA titration of the titanium-hydrogen peroxide complex in the presence of ferric iron, using salicyclic acid as indicator. Firstly, the ferric iron alone in the sample is titrated with EDTA at pH 1.4 - 1.6 then hydrogen peroxide is added to a further aliquot of the sample solution and titanium is titrated with standard EDTA as follows; to an aliquot of the solution is added 2 ml 0.05 M ferric iron solution (if the solution contains less than 0.8% of titanium). The solution is heated to 40 - 50°C, then 0.1 ml ethanolic salicyclic acid (10%), and 2 drops of hydrogen peroxide solution (30%) are added. The solution is then titrated with 0.05 M EDTA until the reddish brown to greenish yellow colour change occurs.

E. Determination of Tin, Silicon, and Titanium

Kohama[537] has described a simple digestion method for the determination of these elements in organic compounds. The sample (0.2 g) is weighed out in a platinum dish and dissolved by warming with 3 ml concentrated sulphuric acid. After cooling, 0.2 g crystalline ammonium nitrate is added and the mixture stirred. Ammonium nitrate additions are repeated, with warming, until a colourless solution results. This may require up to 1 g ammonium nitrate. The crucible is then heated strongly until a constant weight of metal oxide is obtained. No volatilization of tin, silicon or titanium occurs under these conditions. Determination of the individual elements is then carried out by classical methods.

F. Determination of Carbon, Hydrogen, Silicon and Halogen

Klimova et al (Klimova et al[538]; Klimova and Bereznitskaya[539,540]) have described rapid combustion methods using chromic oxide mixed with fibrous asbestos as a catalyst (to prevent the formation of silicon carbide) for the determination of carbon, hydrogen, silicon, and halogens in organosilicon compounds. To determine carbon, hydrogen, and silicon (Klimova et al[538]) the sample (3 - 12 mg) is placed at the closed end of a quartz tube 60 - 80 mm long and 7 - 8 mm diameter and the tube is three-quarters filled with a chromium oxide-asbestos catalyst, prepared by ignition of ammonium chromate or ammonium dichromate mixed with asbestos and heated in oxygen at 900° - 950°. The tube is weighed and placed towards the gas-inlet end but beyond the combustion zone of an empty combustion tube so that its open end is directed towards the furnace. The tube containing the sample and catalyst is gently heated with a burner which is moved gradually from the open to the closed end while oxygen is passed through the combustion tube and through the normal absorbers for water and carbon dioxide. The concentration of silicon in the sample is calculated from the weight of silica and catalyst. To determine carbon, hydrogen, silicon, and halogen attached to silicon, (Klimova and Bereznitskaya[539,540]) the weighed sample is placed in a small quartz vessel followed by 100 - 200 mg asbestos previously ignited to 1200°, and the whole is weighed. If the sample is volatile it is placed in a glass ampoule and the ampoule, open end down, is placed in a small beaker and the asbestos is introduced. The vessel is placed in a combustion tube connected to three absorbers. The first, of heat-resisting glass, contains silver (for halogens) and is heated to 500° - 550°. The second and third are for absorbing water and carbon dioxide respectively. If the substance contains nitrogen a further absorber is placed between the second and third. After the combustion, the beaker and the three absorbers are weighed. The method is also suitable for many other compounds containing halogen attached to silicon, but some of them give silicon carbide instead of silica. In such cases a catalyst (chromium dioxide) is introduced with the asbestos and a drop of water, the weight of which is deducted from the total weight of water found.

Standard micro dry combustion technique (Lammer[541]; Gilliam et al[542]; Balis et al[543]) are used for determining carbon and hydrogen in organo-silicon compounds. It is necessary in such procedures to ensure that silicon is completely converted to silica, and that silicon carbide formation is avoided, as this will lead to low carbon analyses. It has been suggested (Rochow and Gilliam[544]; Rochow[545]) that a two-stage combustion, in which the sample is vaporised at as low a temperature as possible, then oxidized on a platinum gauze at 850°, will overcome this difficulty. The silica formed during combustion is retained by a plug of glass-wool at the end of the tube before and after the combustion. Rochow and Gilliam[544] Rochow[545] reported that this method gave good results for a wide variety of organosiloxane polymers, but it has been suggested that the length of the combustion tube (18 cm) prevents the method from being used for volatile materials.

Aranyi and Erdey[546] discuss a modification of the method of Klimova et al[538], described above in which the use of asbestos (which is hygroscopic) is avoided. The sample is covered with a 100-fold amount of powdered chromium trioxide and heated in oxygen at 900°. The carbon dioxide and water produced are absorbed and weighed in the usual way; the content of silicon is obtained from the change in weight of the ignition tube. The packing of the catalyst is critical; loose packing allows escape of the silica formed, whereas too-dense packing can lead to explosion.

Gawargious and Macdonald[547] point out that the formation of stable silicon carbon and volatile stable silanes is the greatest source of error in combustion methods for the determination of carbon and hydrogen. They avoid this by using magnesium oxide as the catalytic filling with a furnace temperature of 900° and point out that the initial volatilization of the sample must be carried out very carefully.

Korbl and Komers[548] describe a simple semi-micro method for the determination of carbon and hydrogen in organosilicon compounds. A layer (35 cm long) of the decomposition product of $AgMnO_4$ was used as combustion catalyst. They adopted this method for the determination of compounds which have a low boiling point. The apparatus described has an arrangement by which the capillary tube containing the sample can be opened without the combustion tube being opened.

Uhle[549] used cobaltic oxide on pumice as a catalyst to avoid the formation of silicon carbide. A combustion tube containing this catalyst heated at 650° - 700° and silver wool at 450° - 500° (to absorb halogens) can be used for the analysis of compounds containing carbon, hydrogen, oxygen, silicon, chlorine, bromine and sulphur. He pointed out that if explosions are likely to occur during combustion this can be avoided by conducting the combustion in a stream of nitrogen.

Platonov[550] passed the volatiles produced by the thermal or catalytic cracking (platinum) followed by oxidation of organosilicon compounds over a wire electrode fed by a 12000 - 15000 V high frequency generator on which the finely divided silica is electrodeposited and subsequently weighed. He claims an absolute error not exceeding ±0.3% by this procedure.

Bradley[551] has described a combustion technique for the determination of carbon and hydrogen in silane-containing gas mixtures or in any gas mixture in which there is an danger of explosion. The apparatus enables successive small portions of the gas sample to be mixed with a large excess of oxygen and to be burnt without uncontrolled explosions which might rupture the combustion tube. Good results are claimed with gases containing silane and thylsilane.

Radecki[552]; and Radecki and Pikos[553] describe a combustion method for the determination of carbon and hydrogen in organo-oxyhalogenosilanes based on oxidation with mixtures of chromic, sulphuric and phosphoric acids at 150° - 180°. Carbon dioxide is absorbed in sodium hydroxide solution and determined according to the Winkler method. Chlorine or bromine is reacted with sodium hydroxide and the hypochlorite or hypobromite formed reacted with hydrogen peroxide to give the corresponding halide, which is determined by the Volhard method. The presence of sulphur does not cause interference in this method.

Kautsky et al[554] have described a method for the determination of hydrogen, oxygen and silicon in volatile or gaseous organosilicon compounds. After combustion, silicon remains as silica in the combustion tube, whilst hydrogen and oxygen are determined by standard methods. If a halogen is present, it is removed quantitatively by a boat containing powdered silver in an additional furnace placed between the combustion furnace and the tubes for absorbing water and carbon dioxide. With 30 - 40 mg sample, the method gives an accuracy of ±0.4% for silicon, ±0.3% for carbon, ±0.1% for hydrogen and ±0.2% for chlorine.

G. Determination of Halogens

Halogen in organosilicon compounds may be either directly linked to silicon or may be part of a substituent group. Total halogen is usually determined by alkali fusion methods. Halogen attached directly to silicon is easily hydrolysable by aqueous alcohols and the halogen acid so produced can be estimated by direct alkali titration or by indirect procedures.

Total Halogen. Bondarevskaya et al[555] describe a bomb fusion method for the simultaneous determination of fluorine, chlorine and silicon in organosilicon compounds. In this procedure the sample (20 - 40 mg) mixed with 4 - 5 times its weight of solid potassium hydroxide is placed in a nickel bomb. If the sample contains one fluorinated phenyl radical it is fused for 40 - 45 min at 900° - 950°; if the sample contains two fluorinated phenyl radicals it is fused for 60 min at 1000°. The melt is then allowed to cool and is dissolved in 200 ml water. Fluoride and chloride, respectively, are determined in two aliquots by titration with standard thorium nitrate and ammonium thiocyanate using standard procedures. To determine silicon a further 25 ml aliquot of the test solution is prepared by adding mixed indicator (0.1% ethanolic methyl red + 0.1% bromocresol green in 0.002% sodium hydroxide - 6:5 by volume) and then acidifying the solution with hydrochloric acid and neutralizing by controlled addition of 0.1 N sodium hydroxide. Potassium chloride (50 mg) is added, the solution again neutralized and ammonium fluoride solution and 0.1 N hydrochloric acid (10 ml) added. Finally, the excess of hydrochloric acid is titrated with standard 0.1 N sodium hydroxide to the red to yellow colour change.

Bondarevskaya et al[556] also discuss a method for the estimation of fluorine, chlorine, and silicon, based on fusion with potassium metal in a bomb. The organosilicon sample is weighed in a gelatin capsule or polyethylene ampoule. Ignition and analysis of the melt are carried out as described above (Bondarevskaya et al[555]).

Klimova and Vitalina[557] discuss the application of bomb fusion with alkali to the determination of fluorine in organofluorosilicon compounds. They recommend that the large amount of alkali present after the decomposition of the organofluoro compound is removed by the use of a cationite. The test solution (10 ml) followed by 20 ml of water, is passed through a column(80 mm x 13 mm diameter) of SDV-3 ion- exchange resin, and the combined solution is titrated with standard thorium nitrate solution, using Alizarin red S indicator.

Alcoholyzable Halogen. Magnuson and Baillargeon[558] have described a non-aqueous titration procedure for the determination of chloro-, bromo-, and iodo-silanes and siloxanes. This procedure, which can also be applied on the micro scale (Magnuson[559]) involves alcoholyzing the sample in isopropanol and titrating it with standard alcoholic potassium hydroxide using a tetrabromophenolphthalein ethyl ester(bromophthalein magenta E)end-point.

As a general rule, alkylsilanes with a boiling point above 150° may be transferred to the titration flask while open to the air. This rule is limited, of course, by the ease of hydrolysis of the specific halosilanes by atmospheric moisture. For example, although hexachlorodisilane and dichlorodiisopropoxysilane boil only 11° apart at 146 and 157° respectively, the hexachlorodisilane cannot be sampled except by sealed ampoule, whereas the dichlorodiisopropoxysilane can be sampled quantitatively in the open air by acting quickly. Bromo- and iodosilanes, whatever the boiling point, hydrolyse rapidly with atmospheric moisture and may only be sampled by means of ampoules. Analysis of iodosilanes by the above methods were not quantitative. Iodosilanes can be determined, following reaction with excess ethanolic base, by back-titration with standard acid. This method precludes the formation of ethyl iodide.

The results in Table 73 indicate that the Magnuson and Baillargeon method[558] is highly satisfactory. Standard deviations range from 0.7 parts per thousand for dichloro(2-carbomethoxyethyl)methylsilane to 2.9 parts per thousand for the volatile dichlorodimethylsilane, and to a better-than-expected 0.6 ppt for the iodotrimethylsilane by the back-titration technique.

Other workers have studied variants based on alcoholysis from the determination of alcoholyzable halogen in organosilicon compounds.

Sir and Komers[560] determined chlorine and bromine by hydrolysing the sample in a sealed tube with sodium butoxide, and the resulting chloride or bromide was determined by an argentimetric titration, with visual, or preferably potentiometric, indication of the end point.

Syavtsillo et al[561] studied the effectiveness of various media for hydrolysing alkyl- and aryl-chlorosilanes. The best medium was shown to be a mixture (1:1) of ethanol and water. After hydrolysis they determined chloride ion by a neutralisation reaction or by mercurimetry. To determine total chlorine they used the sodium in liquid ammonia method described by Watt[562].

Table 73 - Results of halogen determination in halosilanes

Halosilane	Halogen (%) Theory	Halogen (%) Found*	Number of determ-inations
Silicon tetrachloride†	83.47	83.41 ± 0.19	5
Dichlorodimethylsilane†	54.94	54.85 ± 0.16	5
Hexachlorodisliane†	79.11	78.92 ± 0.15	5
Dichlorodiisopropoxysilane	32.65	32.62 ± 0.03	8‡
Dichlorodiphenylsilane	28.00	27.96 ± 0.02	10
1,3-Dichloro-1,3-dimethyl-1,3-diphenyl-disiloxane	21.66	21.57 ± 0.05	10
Trichloro(3-chloropropyl)silane	50.17☆	50.10 ± 0.05	11¶
Dichloro(2-cyanoethyl)methylsilane	42.18	42.21 ± 0.03	10
Dichloro(2-carbomethoxyethyl)methylsilane	35.26	35.26 ± 0.02	10
Chloromethylphenylsilane	22.63	22.54 ± 0.03	9**
Bromoethyldiphenylsilane†	28.83	28.80 ± 0.04	5
Iodotrimethylsilane† ††	63.42	63.21 ± 0.04	5

* Average and standard deviation
† Sampled and weighed in glass ampoulds
‡ Total available sample
☆ Does not include chloride attached to carbon
¶ Includes a 49.95% chloride determination
** Excludes a 22.44% chloride determination
†† Reacted with aqueous-alcoholic base and back-titrated with standard acid.

Sommer et al[563] showed that halogen on β-carbon atoms attached to β silicon can be hydrolysed with sufficient ease to be titrated with dilute solutions of bases.

Hirata and Takiguchi[564] have described a volumetric method for the determination of chlorine in methyl chlorosilanes and in phenylchlorosilanes. In this method the ethyltrichlorosilane or dimethyldichlorosilane, phenyltrichlorosilane or diphenyldichlorosilane (20 - 150 mg) is hydrolysed with 50% aqueous ethanol in a sealed vessel, and chloride is titrated with silver nitrate in an aqueous suspension. Dichlorofluorescein is the best indicator. The presence of ethanol and organopolysiloxanes does not interfere in the method.

Khudyakova et al[565] described a procedure based on chronoconductiometry for the analysis of mixtures of trichlorosilane and silicon tetrachloride.

The method is based on the determination of hydrochloric acid formed during the hydrolysis of trichlorosilane and silicon tetrachloride and hydrolysis of the reaction products of trichlorosilane with mercurous chloride. One portion of the sample mixture is hydrolysed with 50% ethanol, and a second portion is treated with ethanolic mercurous chloride then hydrolysed. The hydrochloric acid formed is determined by chronoconductimetric titration with sodium acetate solution and the concentration of each substance in the sample is calculated from the total hydrochloric acid formed during the two treatments. The accuracy of the method is within ±3% for each substance

Decomposition of chlorosilanes with ammonium fluoride in glacial acetic acid has been used[566] for the determination of chlorine in organosilanes:

$$R_nSiCl_{4-n} + 6NH_4F + nCH_3COOH \rightarrow (4 - n)NH_4Cl + (NH_4)_2SiF_6 + nRH + nCH_3COONH_4$$

Chloride is determined by standard methods. For example 5 ml glacial acetic acid, 2.5 ml acetic anhydride and 4 g ammonium fluoride are added to a dry 100-ml conical flask, a sealed ampoule containing the sample (0.2 - 0.6 g) is added, and the ampoule is broken. A crystalline precipitate forms and is dissolved by adding 3 ml nitric acid (sp.gr. 1.4) and 10 - 15 ml water. The clear solution is made up to 100 ml mixed, and an aliquot is titrated to determine chloride ion. A mercurimetric titration with sodium nitroprusside an indicator is described by Chulkov[566] for the determination of chloride. Carbonates, acetates and borates do not interfere in the titration but large amounts of ammonium salts do interefere. The method has been applied to methyltrichlorosilane, dimethyldichlorosilane, diethyldichlorosilane, phenylmethyldichlorosilane and phenyltrichlorosilane.

Myshlyaeva and co-workers (Myshlyaeva[567]; Myshlyaeva et al[568]) have described methods, based on reaction with amines for the determination of chlorine directly linked to silicon in alkyl aryl chlorosilanes.

One method (Myshlyaeva[567]) is based on the formation of aniline hydrochloride according to the equation:

$$R_nSiCl_{4-n} + 2(4 - n)C_6H_5NH_2 \rightarrow R_nSi(NHC_6H_5)_{4-n} + (4 - n)C_6H_5NH_2HCl.$$

The sample (0.1 - 0.25 g) is placed in a glass ampoule, in a dried and weighed beaker with a porous bottom containing 7 - 10 ml of a mixture of anhydrous aniline with benzene or ether (1:4). The ampoule is broken and after 5 min, the liquid is sucked off and the precipitate washed with benzene or ether. The beaker containing the precipitate of aniline hydrochloride is dried to constant weight at 105°. The method is not suitable for the analysis of alkoxychlorosilanes.

Takiguchi[569] has also described a method, based on similar principles, for the determination of chlorine in organochlorosilanes of the general formula R_nSiCl_{4-n}[R = CH_3, C_2H_5 or (C_6H_5)]. These substances were shown to react with aniline in diethyl ether with 97% yield at room temperature. The sample solution in ether (0.1 - 0.5 ml) is added to anhydrous ether (water less than 0.15%) containing a few drops of ammonium thiocyanate in acetone (0.5 M) and ferric chloride in ether (0.5%). This mixture is titrated with aniline in toluene (1 or 2 M) until a red colour appears. The amount of the organosilane can then be calculated by an empirical formula.

The reaction of alkyl chlorosilanes and aromatic amines to form complex amine hydrochlorides, sometimes of low solubility, has been used for conductimetric titration of chlorine in organosilicon compounds. During titration in acetone with benzidine, aniline or other amines, with platinum electrodes, the conductivity decreases up to the end-point, then remains steady.

Myshlyaeva[570] has also described a method for the determination of chloride directly linked to a silicon in certain alkyl-and arylchlorosilanes based on the iodimetric determination of the hydrochloric acid formed on hydrolysis of the chlorosilanes. The sample (0.1 - 0.25 g) is transferred to a flask contain ng 10% potassium iodide solution (25 ml), 8% potassium iodate solution (12 ml) and water (20 - 25 ml); after 5 or 7 min the liberated iodine is titrated with 0.05 N sodium thiosulphate. This method gives low results when applied to the determination of chlorine in alkoxychlorosilanes.

Takiguchi[571] has described a volumetric method based on the use of ammonium thiocyanate for the determination of organochlorosilanes. This method is based on the observation that ammonium chloride is precipitated by double decomposition when ammonium thiocyanate is heated with dimethyldichlorosilane:

$$2NH_4CNS + (CH_3)_2Cl_2Si = 2NH_4Cl + (CH_3)_2(CNS)_2Si$$

To exactly 1 ml chlorosilane sample solution in anhydrous ether (10 - 30 ml) is added a few drops of 0.3 M ammonium thiocyanate in acetone from a 5 ml micro-burette graduated in 0.01 ml. The turbid mixture is shaken and one drop of a 1% solution of ferric chloride in ether is added. The titration is continued with vigorous shaking and cooling of the vessel to below 5° in iced water until the red colour of ferric thiocyanate persists. The concentration (C) of the chlorosilane is given by C = AV354.57/B%(w/v), where A is the molar concentration of the ammonium thiocyanate solution, V is the titre in ml and B the chlorine content in %(w/w) of the chlorosilane. The method has the advantage, over methods performed in aqueous media, that no precautions against loss of hydrochloric acid are needed.

In a further paper Takiguchi[572] showed that organochlorosilanes of general formula R_nSiCl_{4-n} ($R = CH_3$ or C_6H_5) react with potassium thiocyanate or ammonium thiocyanate in anhydrous ether to yield potassium chloride or ammonium chloride quantitatively. An ethereal solution of the silane (containing less than 100 mg chlorine (1 ml) is mixed with dry ether (10 ml) and ferric chloride in ether (0.5%, 2 drops) and titrated with potassium thiocyanate or ammonium thiocyanate (0.1 - 0.5 M) in dry acetone until the solution becomes red. The error is less than 1% (relative) for methylsilanes and less than 2% for phenylsilanes and the method is applicable to the determination of total chlorine in mixed organochlorosilanes.

Various workers have described methods for the determination of chlorine in organosilicon compounds (Swartzkopf and Henlein[519] or in organosilicon phosphorus compounds (Fennell[573]; Belcher and Macdonald[574]).

DETERMINATION OF FUNCTIONAL GROUPS

H. Determination of Alkoxy and Aryloxy Groups

The alkoxy content of alkoxysilanes is often determined, following hydrolysis, by employing the conventional analytical methods for alcohols in aqueous solution. There are several methods available for determining alcohols rapidly in this manner. However, the slow steps in such analyses are the hydrolysis and the separation of the alcohol from these silanes or siloxanes. A modified Zeisel technique (Kline[575]; Strouts et al[576]) has been used successfully for the analysis of methoxy-, ethoxy-, and propoxysilanes. The limitations of this technique are that it is time-consuming and that there is a molecular weight limitation on the alkyl iodide that can be distilled off as iodide. Sodium acetate solution is preferred as liquid washer, since anomalous results have been obtained when a solution containing cadmium acetate and sodium thiosulphate was used. The use of Ascarite soda asbestos has been suggested instead of the customary liquid washers; the advantages quoted were that the same tube could be used for several determinations without being refilled and that removal of interfering substances was complete. Accurate assays of aryloxysilanes capable of stoichiometric bromination have been reported (Smith[578]. Infra-red spectrophotometry (Brown, and Smith[578]) has been used to determine the methoxy content of siloxane polymers.

Based on the work carried out by Fritz and Schenk[579] for the determination of organic hydroxyl groups, an extremely rapid, simple and accurate method has been developed for the determination of alkoxy groups in alkoxysilanes. Both monomers and soluble high molecular weight siloxanes containing a wide variety of alkoxy and aryloxy groups can be determined quantitatively by this technique. In the method alkoxy groups bonded to silicon are acetylated by acetic anhydride with perchloric acid as catalyst in an ethyl acetate solvent. Acetylation is complete in less than 2 min at room temperature with 0.06 M perchloric acid catalyst. Following hydrolysis of excess acetic anhydride, the alkoxy content is calculated from the difference between the blank and sample titrations obtained with standard alcoholic potassium hydroxide.

The limitation of the methods appear to be the solubility of the material to be analysed, possible steric hindrance effects and those limitations expected for ordinary acetylation reactions. Substituted siloxysilyl, olefinic, nitrile, silicon hydride, γ-chlorophyl, or γ-chloroethyl groups do not interfere. Perchloric acid does not appear to attack the methyl ether groups in methoxyethoxysilanes. Water and silanol within limits, as shown by Fritz and Schenk[579] do not interfere. Compounds containing acid groups however, must first be analyzed for titratable acid and appropriate corrections made in the alkoxy content calculations.

Aminoalkyl side chains on alkoxysilanes can be quantitatively determined by acetic anhydride in the absence of acid catalysts because alkoxysilanes are inert to acetic anhydride unless certain acid catalysts are present. The nucleophilic amines require no acid catalysts for rapid acetylation by acetic anhydride.

Dostal et al[581] has also studied the determination of silicon-bound alkoxy and aryloxy groups by reaction with acetic anhydride in the presence of perchloric acid catalyst. The acetylated silicon compound is hydrolysed with water and the liberated acetic acid determined by titration with standard acetic acid. These workers then identify the alcohols present in the reaction mixture by paper chromography.

Terent'eva et al[582] used hydrolyses with aqueous hydrochloric acid followed by spectrophotometric estimation of the alcohol produced using ceric ammonium nitrate to estimate alkoxy groups between ethoxy and butoxy in various compounds including siloxanes.

Kreshkov and Bork[583] described a spectrophotometric method for the determination of phenoxy groups based on the formation of the indophenol blue colour. The sample is dissolved in 5 - 6 ml ethanol and the solution is heated under a reflux condenser on a boiling water bath with 4 ml concentrated aqueous ammonia for 3 - 4 min., is removed and washed and the filtrate, diluted to 25 ml, is heated with 30 drops of chlorine or bromine water under a reflux condenser for 5 min, and its extinction is measured after cooling.

Bomb fusion with alkali (Terent'eva and Bondarevska[584] has also been used to estimate phenoxy groups in phenoxy-phenyl and phenyl(phenyl phenoxy)-silanes. For the determination of phenoxy-groups in phenoxy-phenylsilanes, the sample (0.1 - 0.2 g) is fused with potassium hydroxide in a corrosion-resistant-steel tube heated by a flame (the reaction is complete within a few min) and an aliquot of an aqueous solution of the melt is analysed for phenoxide by titration with potassium bromide-potassium bromate solution. Phenyl-phenoxy-groups in phenyl(phenylphenoxy) silanes are determined by potassium bromide-potassium bromate titration of a solution of the sample (0.1 g) in anhydrous acetic acid.

Kreshkov and Nessanova[585] devised a procedure based on reaction with hydriodic acid for the estimation of lower alkoxy groups in organosilicon compounds. Syavtsillo and Bondarevska[586] modified this procedure so that small amounts (not less than 0.01%) of ethoxy and butoxy groups in ethylphenylpolysiloxanes could be determined with an error of + 10% of the content. For ethoxy groups, an ampoule containing the sample (about 0.02 g) is broken under 3 ml hydriodic acid (sp. gr. 1.69 - 1.70) in a reaction vessel in a stream of carbon dioxide and the gases are passed through a washing vessel containing a 10% solution of a mixture (1 + 1) of cadmium sulphate and sodium thiosulphate and into a receiver containing 4 ml 10% sodium acetate in glacial acetic acid and 5 - 6 drops bromine. The reaction mixture is boiled gently for 45 minutes. After this, the contents of the receiver are poured into a flask containing 1 g sodium acetate and treated with a few drops of formic acid to destroy free bromine and then with 2 ml dilute sulphuric acid (1 + 4) and 1 ml 10% potassium iodide solution. The liberated iodine is titrated with 0.02 N sodium thiosulphate. For butoxy groups the reaction mixture is heated at 40° for 30 min then at 60° for 30 min and then at 100° for 30 min.

Klimova et al[587] used mixtures of potassium iodide and phosphoric acid as the digestion reagent in the microdetermination of alkoxy groups in organosilicon compounds. In this procedure the sample is decomposed with potassium iodide-phosphoric acid to form the alkyl iodide, which is then distilled off and oxidized with bromine to iodic acid which is determined iodometrically. The reaction vessel is fitted with a reflux condenser which is connected via a tube containing Ascarite (to purify the alkyl iodide) to a collection vessel containing sodium acetate (20 g) dissolved in anhydrous acetic acid (200 ml) to which is added 2 ml of iodine-free bromine.

I. Determination of Acetoxy Groups

Two methods have been described in the literature for the determination of acetoxy groups bound to silicon, e.g. $R_nSi(OCOCH_3)_{4-n}$.

Drozdov and Vlasova[592] described a method based on titration of the organosilicon compound (0.04 - 0.08 g) dissolved in ethyl methyl ketone (15 ml) with methanolic sodium methoxide (0.1 N).

$$R_nSi(OCOCH_3)_4 + (4 - n)OCH_3ONa \rightarrow R_nSi(OCH_3)_{4-n} + (4 - n)CH_3COONa$$

The end-point of the titration is found visually with thymol blue as indicator, or potentiometrically with glass and saturated calomel electrodes. The error of the determination does not excess 1.4%. Alkoxyl-groups do not interfere in the procedure.

In a similar method, Kreshkov et al[593] determine acetoxy groups in alkylacetoxysilanes by dissolving the sample in ethyl methyl ketone and titrating with a solution of sodium or potassium methoxide in methanol. The end-point can be found graphically from the potentiometric curve, or visually with thymol blue as indicator. The titrating solution is standardised against recrystallized benzoic acid in ethyl methyl ketone.

J. Determination of Silicon-bound Hydroxy Groups (Silanols)

Methods proposed for the determination of silanol groups in organosilicon compounds include manometric lithium aluminium hydride and Zerevitinoff procedures, discussed later (Guenther[594]; Lees and Lobeck[595]); nuclear magnetic resonance spectroscopy (Hampton et al[596]); reaction with phenyl isocyanate (Damm and Noll[597]) and silanol condensation procedures (Haslam and Willis[598]; Kline[599]; Lucas and Martin[600], Sommer and Tyler[601]) using alkali, acid or iodine as catalyst to complete the reaction. The water produced in these condensation methods is continuously removed and recovered by azeotropic distillation with benzene or toluene using the Dean and Stark or similar apparatus, and the quantity of water collected is measured either volumetrically or by Karl Fischer titration (Damm et al[602]; Gilman and Miller[603]; Grubb[604]). A characteristic of silanol condensation procedures is that they are slow and usually incomplete and non-reproducible, particularly when applied to silicon resins.

Smith and Kellum[605] devised a catalyst system consisting of a mixture of boron trifluoride, acetic acid, and pyridine which, they claim, gives rapid and reproducible total hydroxyl analysis (silanol plus water) in a variety of silicone materials. This method avoids many of the interferences empirical calibration and miscellaneous problems such as poor solubility, incomplete reaction and interfering siloxane cleavage associated with many earlier methods.

Smith and Kellum[605] compared results obtained by the boron trifluoride catalysed condensation with results obtained by the lithium aluminium hydride procedure described by Barnes and Daughenbough[606] (Table 74).

Hydroxyl analysis of various types of polysiloxanediol fluids are similarly tabulated in Table 75. Agreement with the lithium aluminium hydride method was generally within a relative 2.5% with the monomer and fluid samples.

Total hydroxyl determinations with different types of silicone resins and representative precision of the condensation method are given in Table 76. Alkali-catalysed condensation and lithium aluminium hydride method analyses are also noted for comparison. Results obtained with the boron trifluoride, acetic acid, and pyridine catalyst system were consistently more precise, and the relative standard deviation was generally less than 3% with all types of resins tested. Precision with the miniature condensation apparatus and smaller samples was about half as good as that found using the larger system.

Table 74 - Total hydroxyl determination in monomer silanols

Sample*	% OH theory	% OH Condensation method	% OH $LiAlH_4$
$HO(CH_3)_2SiC_6H_4Si(CH_3)_2OH$	15.0	13.0	13.0
$C_6H_5Si(OH)_3$	32.7	27.0	-
$(C_6H_5)_3SiOH$	6.64	5.92	5.92
$(C_6H_5)_2CH_3SiOH$	7.95	7.87	7.68
$(C_6H_5)_2Si(OH)_2$	15.8	14.8	14.8
$(C_6H_5)_2Si(OH)_2$#	15.8	16.2	16.2

* These samples were not pure silanols. Detection of siloxane by i.r. generally accounted for the differences from theory.
\# Sample contained "free" water.
Pyridine not added in case of monomers (high results).

Table 75 - Total hydroxyl determination in silicone fluids

Sample	% OH BF_3 Condensation	% OH $LiAlH_4$
$HO[(CH_3)_2SiO]_xH$	4.13	4.19
$HO[(CH_3)_2SiO]_xH$	2.20	2.60
$HO[(C_6H_5)(CH_3)SiO]_xH$	5.76	5.85
$HO[CF_3(CH_2)_2CH_3SiO]_xH$	2.37	2.35
$HO[(CH_2=CH_2)CH_3SiO]_xH$	3.80	3.73
$HO[(CH_2=CH_2)CH_3SiO]_xH$	4.27	4.35

Table 76 - Total hydroxyl determination in silicone resins

	BF_3 condensation method				Cond. with KOH	$LiAlH_4$
Resin*	% OH	N	σ	Rel. std. dev.	% OH	% OH
1	0.786	6	0.0230	2.93	0.651	0.685
2	3.32	14	0.0760	2.28	3.51	3.58
3	0.732	8	0.0442	6.04	0.785	0.648
4	6.96	10	0.186	2.67	6.27	6.66
5	5.23	10	0.0910	1.74	5.62	4.26

* Mixtures of $RSiO_{3/2}$ and R_2SiO in which R was alkyl or phenyl substituents.

The Fritz and Schenk procedure[607] involving perchloric acid catalysed acetylation in ethyl acetate solvent as described by Dostal et al[608] and by Magnuson[609] has already been discussed in the section on the determination of alkoxy groups in organosilicon compounds. Magnuson and Cerri[610] carried out an extensive examination of the solvent used in the catalysed acetylation procedure and came to the conclusion that, for the determination of silanols, 1:2-dichloroethane is in many wavs a superior solvent to ethyl acetate as originally proposed by Fritz and Schenk[609]. In 1,2-dichloroethane, the acetylating agent can be prepared without cooling. Although the newly prepared reagent does become somewhat warm, it may be used within an hour's time. With ethyl acetate, cooling to 5° (Fritz and Schenk[609]) is necessary during one step of the acetic anhydride-perchloric acid mixing.

The acetylating agent is virtually colourless, or a very light yellow tint at most. The reagent in ethyl acetate is yellow to yellow-brown. Indicator end-points are sharp when the yellow colour is absent. The acetylating agent has a useful life, at least two months, i.e.two or three times longer than reagents in ethyl acetate.

1,2-dichloroethane is the only alternative solvent to ethyl acetate in which alkoxysilanes are quantitatively acetylated within 10 min. Table 77 shows typical results for the determination of alkoxysilanes in 1,2-dichloroethane obtained using the method described below. The relative precision (1 std. dev.) is about ±0.5% for ≡ SiOC ≡ linkages in monomeric or dimeric compounds, ±1% for alkoxy-containing poly(dimethylsiloxane) fluids of 5 - 20 dimethylsiloxy units, and ± 2 - 3% for similar fluids containing high amounts of trifunctionality ($RSiO_{1.5}$).

Table 77 - Determination of alkoxysilanes

	Per cent of theory*
Diphenyldiethoxysilane	100.1 ± 0.8
Dimethyldiisopropoxysilane	99.6 ± 0.4
7-Ocetenyltriethoxysilane	100.2 ± 0.4
2-Methoxy-2-methyl-1-thio-2-silacyclo-pentane#	100.3 ± 0.6
1,3-Di-n-propyl-1,1,3,3-tetraethoxydisiloxane	99.2 ± 0.2
Y-Chloropropylmethyldiethoxysilane	100.2 ± 0.5
Trivinyl-2-methoxyethoxysilane	99.4 ± 0.4

* Average and average deviation of triplicate determinations
# Bifunctional alkoxy- and mercaptosilane.

The gasometric procedure of Barnes and Daughenbough[606] using lithium aluminium hydride, mentioned earlier in this Chapter, gives good results for the determination of silanol groups in the presence of siloxanes, which, upon cleavage, produce volatile silicon hydrides. These workers recommend diglyme as a solvent instead of the usual butyl ether.

Kellum and Uglum[611] used lithium aluminium dibutylamide as a titrant for the determination of silanols. They titrated silanol hydroxy groups in monomeric and polymeric silanols, and in silicon resins, using N-phenyl-4-phenylazoaniline as indicator.

Guenther[612] estimated silanols gasometrically using a Grignard reagent. The gasometric apparatus and procedure used was a modification of that used by Fuchs et al[613]. A solution of the sample in di-n-butyl ether is added to a standard solution of methyl magnesium iodide (in the same solvent) in the reaction flask, through which a slow stream of methane is passed. The volume of gas collected is measured after 6 - 10 minnutes. The percentage of hydroxyl can then be calculated. The method is applicable to the hydrolysis products from alkyl and aryl trialkoxysilanes, and to commercial silicone resins. For a sample of silicone resin, most of the solvent should be removed under vacuum at room temperature before dilution with di-n-butyl ether.

Sauer[614] and Hyde[615] have also described a method based on the use of methyl magnesium iodide for the determination of silanol groups. The Zerevitinoff method works well for both large and small numbers of hydroxyl groups. Guenther[612] has analysed a variety of samples, including commercial silicone resins, by using 2 N methylmagnesium iodide in dibutyl ether as solvent and methane as inert gas. Damm and Noll[597] have also used a similar method to determine silanol groups. These procedures are subject to occasional error because of the limited solubility in the reagent of some of the partly condensed products of the reaction. None of them differentiates between Si - OH and C - OH; the latter group can be determined by reaction with acetyl chloride, and the Si - OH can then be determined by difference. Any $\equiv$ Si - $OCOCH_3$ formed by the reaction of $\equiv$ SiH groups with acetyl chloride is readily hydrolysed with water.

Noll et al[616] refluxed silicone resins, dissolved in chlorobenzene or dioxan, for 2 h with isocyanatobenzene, the excess of which is then caused to react with isobutylamine, and the residual amine is titrated with standard hydrochloric acid. This method will determine hydroxy groups present in water as well as silanol groups. To distinguish between the two types, a Karl Fischer determination is carried out; water reacts immediately but silanol groups only slowly.

K. Determination of Mercaptosilanes

Fritz and Schenk[607] reported that the catalysed acetylation procedure can be used for the direct determination of Si - S- C linkages. This acetylation method depends upon reaction of the mercaptosilane with excess acetic anhydride to give a thiolacetate and an acetoxysilane. The acetoxysilane and remaining anhydride are hydrolysed and titrated with standard base.

Berger and Magnuson[617] pointed out that the ease of hydrolysis of mercaptosilanes allows an indirect assay of the mercapto functional group by reaction with excess mercuric acetate in a methanol-toluene solvent. The excess mercuric acetate is titrated with standard hydrochloric acid in butanol to a thymol blue end-point (Kunder and Das[618]). Residual moisture in the solvents and hydrochloric acid-butanol titrant suffice for the hydrolysis of the mercaptosilane to the corresponding mercaptan.

In Table 77 are compared estimations obtained by the catalysed acetylation and by the mercuric methods on a primary, a secondary, and a tertiary alkylmercaptosilane and an aryl-mercaptosilane. The analyses may be considered quantitative except for the values by both methods for the tert-butylmercaptosilane.

Steric hindrance appears to preclude a quantitative determination of the SiSC linkage in the tert-butylmercaptosilane in reasonable lengths of time by 0.04 M perchloric acid-catalysed acetylation. Increases in the perchloric acid and acetic anhydride concentrations to 0.15 M and 2 M respectively, resulted in recoveries of only 96.0, 97.7 and 97.9% for this same silane over reaction times of 10 min., 1 h. and 2.5 h.

Table 77 - Analysis of mercaptosilanes

Mercaptosilane	Per cent of theory * Perchloric acid catalysed acetylation	Mercuric acetate
Trimethyl-n-butylmercaptosilane	98.5 ± 0.5	99.2 ± 0.1
Trimethylisopropylmercaptosilane	98.7 ± 0.2	98.2 ± 0.1
Trimethyl-tert-butylmercaptosilane	86-97‡	103.0 ± 0.1$
Trimethylphenylmercaptosilane#	99.4 ± 0.6	100.3 ± 0.3

* Average and average deviation of triplicate determinations, theory 98% pure

Not previously reported in the literature, b.p. 72°C at 8 mm. Found: C, 59.50; H, 7.80; S, 17.49; Si, 15.42. Calcd.: C, 59.27; H, 7.74; S, 17.58; Si, 15.40.

‡ Six determinations, 3 min. - 16 h. reaction time.

$ Dark ppt. formed in methanol-toluene, 400 ml. chloroform-butanol (1:1) used as solvent.
103% recovery due to occlusional mercury on mercury mercaptide precipitate.

L. Determination of Silidyne Groups (≡ SiH)

Bromination and reaction with mercuric chloride are the basis of the two principle types of methods for estimating silidyne groups. Fritz and Burd[619] showed that SiH and SiC_6H_5 groups, respectively, react quantitatively with bromine solution in acetic acid, at room temperature, and at the boiling point of the acid. Addition of excess bromine, followed by addition of excess potassium iodide and back-titration of excess iodine with 0.1 N sodium thiosulphate permit quantitative determination of either group, either seperately or in the same molecule. The accuracy of determination of the SiH group by this method (15 - 20 mg samples) is within ±5% and for the SiC_6H_5 group (30 - 100 mg samples) is within ±0.5%.

Kreshkov et al[620] describe a bromination method based on amperometric titration at 0.3 V of the silidyne groups and double-bonds in 4 M hydrochloric acid medium with 0.1 M-methanolic bromine saturated with sodium bromide with a rotating platinum micro-electrode; then analogous titration of the double-bonds after treatment with methanolic potassium hydroxide to destroy the silidyne groups. The content of silidyne groups is then found by difference.

Terenteva et al[621] treats the sample containing down to 0.0025% H as silidyne groups with bromine in acetic acid, and the unconsumed bromine is determined iodimetrically. With samples soluble in anhydrous acetic acid 0.04 - 0.05 g is dissolved in 10 - 30 ml of 0.1 N bromine in acetic acid; after 10 min. 10 ml 10% potassium iodide solution is added, and the liberated iodine is titrated with 0.1 N sodium thiosulphate.

McDougall[622] has studied the conditions for the infra-red determination of ≡ SiH in the range 0.001 to 0.03%. When radiation of 4.67 nm and the Perkin-Elmer 21 instrument are used, the standard deviation is ±0.001%.

Bork et al[623] and Kreshkov et al[624] determine small amounts of organosilicon compounds (containing SiH groups in alkyl and aryl chlorosilanes) mercurimetrically by treating a solution of the sample in acetone, ethyl methyl ketone, or mixtures of these with benzene. The SiH-containing impurities are determined from the turbidity produced by the mercurous chloride formed in accordance with the equation:

$$\equiv SiH + 2HgCl_2 \rightarrow \equiv SiCl + Hg_2Cl_2 + HCl$$

The extinction is directly proportional to the concentration of the compounds containing $\equiv$ SiH bonds. As little as 5 x 10^{-5}% of impurity (as H) can be determined by this method.

The same workers[625] in a further method determine SiH groups by amperometric titration. A suitable weight (0.07 - 0.12 g) of the organosilicon compound is dissolved in benzene and diluted to 25 ml. Benzene: methanol (1:1) containing 0.3 M lithium chloride is added to 0.5 - 2 ml of the sample solution. Using a mercury pool and dropping mercury cathode the height of the wave at - 0.8 V is measured. The solution is then titrated with 3% mercuric chloride solution in benzene: methanol (1:1) until the wave at - 0.8 V disappears.

Bork and Shvynkova[626] outline 4 methods for the determination of $\equiv$ SiH in organosilicon compounds. (i) - The phototurbidimetric method is based on measurement of the extinction of the suspension formed by the reaction of $\equiv$ SiH with mercuric chloride in a non-aqueous solvent. (ii) The gravimetric method is based on weighing the precipitate of mercurous chloride produced under the conditions described for method (i). (iii) - The photometric method is based on the oxidation of the Si - H bond with potassium permanganate solution and photometric measurement of the reduced potassium permanganate solution. (iv) - The with permanganate method is based on the same reaction.

Cermak and Dostal[627] described a polarographic determination of $\equiv$ SiH groups in the presence of silicon-silicon bonds in polyorganosiloxanes. The sample is dissolved in ethyl methyl ketone, and mercuric chloride solution is added to oxidise the Si - H bonds. The decrease in the concentration of mercuric ion is determined polarographically as the mercury-iodine complex (Schwarz[628]). The coefficient of variation is ± 0.3% for 1.5 - 7.5 mg of tetramethylcyclotetrasiloxane.

A further mercurimetric procedure for the quantitative estimation of $\equiv$ SiH in silane, dimethylsilane and trichlorosilane is based on the reaction:

$$SiH_4 + 8HgCl_2 + 4H_2O = Si(OH)_4 + 4Hg_2Cl_2 + 8HCl.$$

The quantity of mercurous chloride produced is estimated iodometrically.

Hydrogen attached directly to silicon reacts with alkali to yield a condensed silanol and hydrogen gas is evolved. The hydrogen can be collected; and its volume measured to give an estimate of the silidyne hydrogen content of the sample. When aqueous alkali does not bring about quantitative evolution of hydrogen, an alcoholic solution of alkali can be used to dissolve the sample. Kipping and Sands[629] treated polymeric organosilanes with moist piperidine. Non-quantitative evolution of hydrogen usually indicates that the sample has not been completely dissolved or that a stronger base than piperidine is required as catalyst. With polymethylhydrogen siloxane the reaction is:

$$(CH_3HSiO)_n + nH_2O \xrightarrow{\text{alkali catalyst}} (CH_3SiO_{3/2})_n + nH_2O + n/2H_2$$

The reaction is quantitative even at extremely low concentrations; it also occurs with Si - Si bonds, one such bond yielding 1 mol of hydrogen. The method described by Kipping and Sands[629] cannot, therefore, be used to determine SiH in the presence of Si - Si bonds.

M. Determination of Unsaturation

To determine unsaturation in organosilicon compounds Malyukova and Surkova[630] dissolved 0.7 - 0.2 g sample in 3 - 5 ml carbon tetrachloride. Then anhydrous acetic acid (50 ml) and concentrated hydrochloric acid (10 ml) is added and the mixture stirred for 1 minute. Electrodes are inserted and a current of - 2.5 mA established (rotating platinum anode and standard calomel cathode) and the mixture titrated with 0.5 N potassium bromide-bromate solution (50 g potassium bromide, 13.92 g potassium bromate per 1) with constant stirring until the end-point is reached. A blank determination is carried out in parallel.

Kreshkov et al[631] used amperometric titration with standard bromine solution to estimate unsaturation in organosilicon compounds. For substances not containing phenyl radicals, the sample (4 - 30 mg) is mixed with 3 ml hydrochloric acid and the solution is diluted to 25 ml with methanol (or acetic acid) saturated with sodium bromide. The solution is titrated at ± 0.2 - 0.3 V (rotating platinum anode and S.C.E. as cathode) with 0.1 N bromine in methanol (or acetic acid) saturated with sodium bromide added in 0.1 - 0.2 ml portions. For substances containing phenyl radicals, a suitable volume of the 0.1 N bromine is diluted 10 - 20 times with methanol or acetic acid), then mixed with hydrochloric acid to give a solution of 4M in hydrochloric acid and titrated with a solution of the sample in methanol (or acetic acid).

Kreshkov et al[632] also described a procedure for the estimation of unsaturated organosilicon compounds based on titration with iodine monchloride in non-aqueous medium. To determine double bonds in trisubstituted silanes, they first determined ≡ SiH bonds gravimetrically, turbidimetrically or volumetrically as described in the previous section. They then determined ≡ SiH plus - C = C - bonds by amperometric titration and calculated double bonds by difference.

Double bonds can be determined in unsaturated organosilicon compounds by i.r. spectroscopy (Mikhallenko et al[633]) using the integral absorption coefficient calculated from the Lorenz function. This coefficient is less sensitive to intermolecular interaction than is the absorption coefficient at the absorption maximum. Equations are given for calculating the integral absorption coefficient and the concentration of the substance being determined. A procedure is described for determining tetraethoxydivinyldisiloxane and triethoxyvinylsilane in carbon disulphide solution.

N. Determination of Phenyl and Alkyl Radicals

Waledziak[634] used u.v. spectroscopy for the determination of phenyl radicals in methyl phenyl-siloxane resins. The sample is dissolved in chloroform and the extract measured at 260, 266 and 272 nm in 1 mm quartz cells. The results are referred to a calibration graph prepared using pure phenyltriethyoxysilane and standard mixtures containing methyl groups. Good agreement was obtained with results obtained by i.r. spectroscopy.

Kreshkov et al[635] estimated phenyl radicals by reacting the sample (2 - 2.5 g) with 6 - 7 g anhydrous aluminium chloride per phenyl radical followed by the addition of ethyl bromide (35 - 40 g). After reflux for 2 h 30°C, 50 ml water is added, the mixture cooled, and hexaethylbenzene extracted with 5 x 50 ml portions of diethyl ether. The combined ether extract is water washed and then ether and ethyl bromide distilled off. The residue is dried over phosphorus pentoxide and weighed. The phenyl radical content can then be calculated using an empirical coefficient of ethylation of 0.91.

Voronkov and Shemyatenkova[636] have described a gasometric method for the determination of alkyl radicals in polyalkylsiloxanes and other types of organosilicon compounds. They describe an apparatus for the determination of methyl and ethyl radicals by heating the sample (0.1 - 0.15 g) with powdered potassium hydroxide (3 g) for 2 h at 250° - 270° and measuring the amount of gaseous hydrocarbon evolved. An accuracy of 1 - 2% is claimed.

O. Determination of Hydrocarbon Constituents in Organosilicon Compounds

Only unsaturated hydrocarbon groups in organosilicon compounds can conveniently be determined chemically; of these, the most important is the vinyl group, which occurs widely in certain base polymers of silicone rubbers. The use of many of the standard reagents for addition to double bonds has been tried, and the most satisfactory appears to be a solution of iodine monochloride in glacial acetic acid. The types of compound which have been analysed are $[(CH_2 = CH)CH_3SiO]_m$, $(CH_2 = CH)(CH_3)_nSi(OC_2H_5)_{3-n}$ and $CH_2 = CHSiCl_3$, in which m has a value of 12 or more, and n = 0, 1, 2 or 3.

Saturated-hydrocarbon groups are difficult to determine quantitatively, but a qualitative chemical approach has been used. Under proper conditions, substituent hydrocarbon groups can be cleaved from the polymer. Kipping and Lloyd[637] in their researches into the preparation of compounds containing an asymmetric silicon atom, found that phenyl groups were cleaved from silicon by the action of sulphuric acid. Burkhard and Norton[638] have based a scheme for the qualitative analysis of silicones on cleavage of the hydrocarbon groups by sulphuric acid. They reported a 70% yield of methane from a polydimethylsiloxane fluid and lower yields from polymers containing higher alkyl groups. They successfully identified hydrogen, methyl, ethyl, amyl and phenyl groups attached to silicon.

The proportions and kinds of organic groups present in polymeric siloxanes have been found by reaction with hydrofluoric acid and then separation of the resulting organofluorosilanes (Booth[639]; Perlson et al[640]; Booth and Freedman[641]. This method has the advantage of indicating the types and amounts of different structural groups within a silicone polymer. However, there is no simple method by which hydrocarbon groups can be determined quantitatively, and one must resort to the determination of hydrogen and carbon. The infra-red spectrum indicates qualitatively which groups are present, and, for simple molecules, results for carbon and hydrogen should allow assignment of a structure.

P. Non-aqueous Titrimetry

Kreshkov and co-workers (Kreshkov et al[642]; Kreshkov and Drozdov[643]; Kreshkov[644]; Kreshkov et al[645]) have made an extensive study of non-aqueous titrimetric methods of analysis of alkyl and aryl alkyl chlorosilanes. Kreshkov et al[646m647,648] also studied other types of organosilicon compounds containing nitrogen, carboxyl, and alkylthiocyanato functional groups.

To titrate alkylchlorosilanes (Kreshkov et al[642]) such as trimethylsilicon chloride, dimethylsilicon dichloride, methylsilicon trichloride and silicon tetrachloride, titration in methyl cyanide with 0.05 N phenazone or nitron in methyl cyanide, or with a solution of amidopyrine in benzene by visual and potentiometric methods is suggested. The indicators used are crystal violet, dimethylaminoazobenzene, bromocresol purple, methyl orange, bromophenol blue and gallo sea blue (C.I. Mordant Blue 14). The addition of benzene, toluene, chlorobenzene or carbon tetrachloride does not affect the results.

Differential conductiometric titration has been used (Kreshkov and Drozdov[649]) for the determination of trimethylsilicon chloride, dimethylsilicondichloride and methylsilicon trichloride. The method is based on the quantitative conversion of the chlorosilanes into the alkylthiocyanato-derivatives by the action of ammonium thiocyanate and conductimetric titration in methyl cyanide-ethyl ether (2:3) solvent with 0.1 M-amidopyrine in benzene. The mono-substituted compound is first titrated, then the di-substituted, and finally the tri-substituted. The smallest amounts which can be determined in the mixture are: $(CH_3)_3SiCl$, 0.0055 g: $(CH_3)_2SiCl_2$, 0.0025 g: and CH_3SiCl_3, 0.0023 g in 12 ml sample. Concentrated sample solutions must be diluted. The error in determining the individual chlorosilanes is ±2%.

Electrometric methods have been used (Kreshkov[650]) for the analysis of monomeric and polymeric silicon compounds in non-aqueous medium. The methods are applicable to alkyl and aryl chlorosilanes, aminosilanes, silylamines, weak acids containing silicon and other organic silicon compounds. In non-aqueous medium, several organic silicon compounds have strongly basic or acid properties. Solutions of tetraethylammonium hydroxide and amidopyrine in benzene or mixtures of benzene with other solvents can be used as titrants for these types of compound.

ORGANOTHALLIUM COMPOUNDS

A. Determination of Carbon, Hydrogen and Thallium

Carbon and hydrogen in organic compounds containing thallium can be determined by a combustion method in which finely ground silica is used as a filter to prevent attack by thallium on the silica combustion tube and catalyst, and to prevent the formation of thallium compounds during the determination of carbon and hydrogen in some types of compounds[425]. The combustion tube contains a silver spiral, silver turnings, cobalt(III) oxide on corundum as catalyst and a platinum spiral; this zone of the furnace is heated at 680 - 700°C. The sample is covered with finely ground silica and ignited in a stream of oxygen at 1000°C and water and carbon dioxide are determined by standard procedures. In a further procedure in which thallium is also determined the sample is placed in a silica tube and covered with a layer of powdered silica[426]. The tube is then heated in a stream of oxygen and the pyrolysis products are passed over cobalt(III) oxide at 680 - 700°C. The water and carbon dioxide formed are absorbed in Anhydrone and Ascarite, respectively, in a Pregl apparatus and determined by weighing. Thallium is determined by weighing the residue (probably thallium silicate) in the silica tube. For compounds of the types tris(ethylenediamine)thallium nitrate and bipyridylthallium chloride, the error was within ±0.3% for carbon or hydrogen and within ±0.5 for thallium. Halogens do not interfere with the determination of carbon and hydrogen but interfere with that of thallium.

ORGANOTIN COMPOUNDS

A. Determination of Tin

Numerous methods have been described for the determination of tin in organotin compounds[651-664]. These include gravimetric[655], volumetric[651, 653,655], oxygen flask combustion[656], complexometric[654,657,658], photometric [651,659], X-ray fluorescence[664], X-ray spectrophotometric[660], spectrographic [661] and polarographic[665] methods. One volumetric procedure[653] determines tin in the presence of phosphorus. Another procedure involves a volumetric determination of tin after destruction of the organotin compound with a solution of bromine in chloroform[655]. A solution of bromine in carbon tetrachloride has also been used as a preliminary treatment for more volatile types of organotin compound[666]. Titrimetric methods[651,656] for the determination of tin are usually based on a final oxidation of tin(II) to tin(IV) and suffer from the disadvantages that all traces of the oxidizing agent used in the initial combustion must be removed and that an inert atmosphere must be maintained until the final titration is complete because of the ready oxidation of tin(II) by oxygen.

Samples may be digested by heating with concentrated nitric acid, then perchloric and hydrochloric acids[654] or nitric acid and sodium sulphate [657] prior to complexometric determination of tin. Organotin compounds can be decomposed by wet oxidation with a maxture of nitric and sulphuric acids, followed by ignition to tin(IV) oxide at 900°C, which is determined gravimetrically[667,668]. A successful procedure for very volatile compounds such as stannones is to aspirate the vapour of the sample by means of a current of nitrogen into a mixture of nitric and sulphuric acids digestion mixture and continue by ignition to tin(IV) oxide as described above. A simple digestion method uses ammonium nitrate as oxidizing agent for the determination of tin, and also silicon and titanium in organometallic compounds[669]. An alternative digestion medium is sulphuric acid[670]. Oxygen flask combustion may be used for the micro-determination of tin[656]. Combustion in oxygen converts the organotin compound to a mixture of tin(IV) and tin(II) oxides. The combustion residue is warmed with freshly prepared chromium(II) sulphate solution, to dissolve the poorly soluble tin(IV) oxide. The tin(II) ions and unconsumed chromium(II) ions are oxidized by air, and the tin(IV) ions are then reduced with sodium hypophosphite and titrated with standard potassium iodate solution.

Two procedures have been described for the determination of tin in volatile organotin hydrides[671,672]. One involves bubbling oxygen through the weighed sample until evaporation is complete[671]. The vapour is passed into a weighed silica tube and ignited in a plug of prepared asbestos. The silica tube is then removed and ignited to constant weight at 800°C to obtain the weight of tin(IV) oxide produced. With very volatile samples or samples of high boiling point it is necessary to control the temperature during the evaporation with a cold bath or an i.r. lamp. A modification of this procedure is available for handling samples that readily hydrolyse to produce non-volatile products. The procedure is accurate to within ±0.5% of the determined result and requires about 4 h per analysis.

Alkyltin compounds may be oxidized with sodium peroxide in a Parr bomb[673]. The product is boiled with water in the usual way, and tin is determined spectrophotometrically using cacotheline. For the spectrographic analysis of organotin compunds the sample is dissolved in cumene containing polymethylphenylsiloxane as a silicon internal standard[661].

The solution is passed into the inner vessel of an atomizer, the outer vessel of which is cooled. Spectra are excited by a condensed spark discharge between the graphite rod in the atomizer and graphite counter electrode. The reaction of methyllithium with triphenyltin hydride has been used as the basis for the determination in the latter compound[674]. Tetrabutyl- and tetraethyltin in factory air have been analysed by oxidation with a hydrogen peroxide-sulphuric acid mixture and colorimetric determination of total tin[675].

Residues of triphenyltin compounds used in crop protection can be determined by extraction of the material with dichloromethane, phase separation of triphenyltin compounds and their water-soluble decomposition products, and determination of tin after destruction of organic matter and distillation of the tin as the tetrabromide[676]. Triphenyltin, diphenyltin, and inorganic tin compounds have been analysed by conversion to triphenyltin hydroxide, which is extracted into chloroform and determined by conventional methods[677].

The determination of triphenyltin acetate has been described by various workers[678-686]. These include its determination in ruminants[680], plants[687], celery and apply[683], potato leaves[684], sugar beet leaves and in animals feeding thereon[681] and in milk[687]. Methods have been reported for determining microgram amounts of tin in animal and vegetable matter[687], tin in foods[688], organotin fungicides used in potato blight control[689], organotin stabilizers in foods[690] and tricyclohexyltin hydroxides in fruits[691]. Work has been carried out on the determination of tin in aqueous leachates from organotin-containing antifouling paint compositions[692-695]. These include tributyltin compounds[692] and bis(tributyltin oxide)[693,694].

Tin analyses by atomic- absorption spectroscopy have been described, including the determination of butyltin compounds in textiles by graphite furnace atomic-absorption spectroscopy[696].

Vijan and Chan[696] determined tin by gas phase atomization using hydride generation and atomic-absorption spectrometry. An accuracy of 97% and a precision of 6% (relative standard deviation) is claimed. Marr and Anmar[699] carried out micro-determinations of tin in organotin compounds by flame emission and atomic-absorption spectrophotometry. Kojima[698] separated organotin compounds by using the difference in partition behaviour between hexane and methanolic buffer solution. The separated organotin compounds were determined by graphite furnace atomic-absorption spectrophotometry.

Burns et al[701] and Andreae and Byrd[700] have both applied graphite furnace atomic-absorption spectrophotometry.

In the method described by Andreae and Byrd[700] tin hydrides are detected by graphite-furnace atomic absorption, quartz cuvette atomic absorption or flame emission spectrometry, with detection limits of 50, 50 and 20 pg as tin, respectively. The effect of germanium on the determination of tin is also discussed.

B. Determination of Carbon and Hydrogen

Methods have been described for the determination of carbon and hydrogen in organic compounds containing tin[702], arsenic, antimony, bismuth, and phosphorus[703] and complex compounds containing a tin halide[704].

C. Determination of Halogens

Methods have been described for the determination of halogen in organotin compounds[652,705,706], including long-chain alkyltin compounds[707]. These include the determination of halogens in alkyltin halides by high-frequency titration[652,708]. Trimethyltriethyl- and tributyltin halides and dimethyl-diethyl- dipropyl-, and dibutyltin halides all produce halide ions upon reaction with water, which can be determined by high-frequency titration with standard silver nitrate[708]. The error is claimed to be less than 3% in the 10 - 75% halogen range.

D. Determination of Nitrogen

Nitrogen can be determined in some types of organotin compounds by Kjeldahl digestion procedures[652].

E. Determination of Sulphur

Sulphur can be determined gravimetrically[652] or by a mercurimetric procedure[709] in which the sample is dissolved in toluene, methanol and aqueous sodium hydroxide and solid thiofluorescein-sodium sulphite is added. The solution is then titrated with standard o-hydroxymercuribenzoic acid to the disappearance of the blue colour. The accuracy is claimed to be within ±0.02% of sulphur. Cleavage with standard solutions of iodine has been used for the determination of Sn - S bonds[710].

$$2R_3SnSR' + I_2 \rightarrow 2R_3SnI + R'SSR'$$

ORGANOTITANIUM COMPOUNDS

A. Determination of Elements, Titanium Phosphorus, Silicon and Chlorine

Jenik and Renger[711] described a method for the determination of iron and titanium in donor-acceptor complexes of the ferrocene bases. In this method the sample is decomposed with a mixture of nitric acid, anhydrous acetic acid and aqueous bromine. Iron is titrated complexiometrically with sulphosalicyclic acid as indicator and titanium is determined colorimetrically by extraction of its 8-hydroxyquinoline complex into chloroform.

Luskina et al[712] described a wet combustion procedure for the analysis of organic compounds containing silicon, organic titanium and phosphorus, in which the sample is heated with potassium persulphate and concentrated sulphuric acid for 3 - 4 hours. Silica is then filtered off for determination, if required (Terent'eva and Bernatskaya[713]). The filtrate is diluted with water and then excess potassium hydroxide is added. This solution is passed through an ion-exchange column (Ku-2,chloride form) which retains titanium. Phosphorus is determined photometrically as molybdate in the eluate. Titanium is eluted from the resin with 4N hydrochloric acid and subsequently determined spectrophotometrically. Absolute error did not exceed 0.5% for determinations of titanium at the 10% level in the sample. In a further method for the analysis of difficultly decomposable organic compounds containing titanium, phosphorus and silicon, Terent'eva[714] fused 10 - 50 mg of the sample with sodium fluoride (1 g) and potassium persulphate (0.2 g). The cooled melt is then treated with concentrated sulphuric acid (12 ml) and evaporated in an air bath to fuming. The residue is then dissolved in 70% sulphuric acid and titanium, phosphorus and silicon determined by standard procedures. In a variant of this digestion procedure for the polarographic determination of titanium in titanium organic compounds, the sample (containing 2 - 10 mg titanium) is added to 20% hydrofluoric acid (1.5 ml) and the solution is twice evaporated. Then concentrated sulphuric acid (3 ml) is added to the residue and the solution heated until sulphur trioxide fumes are produced. The residue is dissolved in 70% sulphuric acid to make the volume up to 100 ml and, if particles of carbon are present, is filtered. To 30 ml of this solution is added 1% gelatin solution (6 drops). After nitrogen purging the solution is polarographed with a dropping mercury cathode at $E_{\frac{1}{2}}$ = 0.37 v.s a standard calomel electrode, and the titanium content of the sample is obtained by reference to a calibration curve covering the range 10^{-3} to 10^{-4} M titanium.

Kohama[715] has described a simple digestion procedure for the determination of titanium, tin and silicon in organometallic compounds. The sample (0.2 g) is dissolved in a weighed platinum crucible by warming with sulphuric acid (3 ml). Crystalline ammonium nitrate(0.2 g) is added to the cooled solution and the mixture shaken gently. The ammonium nitrate addition is repeated, with warming, until a colourless solution is obtained (this may require up to 1 g of ammonium nitrate). The crucible contents are then heated until the weight of the metal oxide produced is constant. Titanium can then be determined by standard methods. Advantages claimed for this procedure include an absence of metal loss by volatilisation.

Kotlyar and Nazarchuk[716] have described methods for the determination of chlorine in tetra-n-butyltitanate. The method is based on the conversion of titanium into TiF_6^{2-} and the determination of chlorine by standard methods. The sample (0.8 - 1.0 g) is dissolved by stirring and heating between 110°C and 140°C with excess of a mixture of 15 ml acetic anhydride and at least 4 g ammonium fluoride (not more than 2 g ammonium fluoride per 100 ml solution). The precipitate is dissolved in the minimum of water, the mixture heated to 50°C - 55°C, cooled and made up to 250 millilitre. Chlorine is determined by titration with standard silver nitrate.

ORGANOZINC COMPOUNDS

A. Determination of Zinc

Zinc can be determined in organozinc compounds by a procedure involving complexometric titration with EDTA (disodium salt) in a suitably buffered medium[720]. The organozinc compound is diluted with an inert organic solvent and decomposed by the addition of dilute hydrochloric acid. Zinc can then be determined in the water extract by complexometric titration. Dialkylzinc preparations can be made by the reaction of an alkylaluminium compound with zinc chloride. Such preparations, even after distillation, usually contain residual amounts of organoaluminium compound. If the organozinc compound contains aluminium then this would be present in the aqueous extract and would interfere in the complexiometric determination of zinc. To overcome this the dilute hydrochloric acid extract obtained by decomposition of the organozinc sample is passed down a column of Amberlite IRA-400 ion-exchange resin. Percolation of the column with 2 N hydrochloric acid completely removes aluminium from the column, which can be determined in the eluate by complexometric titration. Subsequent percolation of the ion-exchange column with 0.2 N nitric acid then completely desorbs zinc, which can be collected seperately and determined by complexometric titration[720]. Various procedures have been described for the determination of zinc in plants[721], organic material[722], fungicides such as zineb and ziram[723] and in zinc stearate[724]. None of these procedures is particularly relevant to the determination of zinc in pyrophoric organozinc compounds.

B. Determination of Halogens

The various procedures described in detail in an earlier section for the determination of halogens in organoaluminium compounds can also be applied to organozinc compounds.

C. Determination of Zinc-Bound Alkoxide Groups

Alkoxide groups up to butoxide can be determined in organozinc compounds by hydrolysis with glacial acetic acid followed by spectrophotometric determination of the alcohol produced using cerium(IV) ammonium nitrate[725] as described in connection with the determination of alkoxide groups in organoaluminium compounds[725].

D. Determination of Zinc-Bound Lower Alkyl and Hydride Groups

Earlier is described a method devised by Crompton[726] and Crompton and Reid[727] based on alcoholysis and hydrolysis of the sample for the determination of alkyl groups up to butyl and hydride groups in organoaluminium compounds. This procedure is applicable, without modification, to the determination of the same groups in organozinc compounds. The applicability was examined of this procedure to the determination of lower alkyl and hydride groups in some diethylzinc preparations made by the triethylaluminium-zinc chloride route:

$$2Al(C_2H_5)_3 + 3ZnCl_2 = 3Zn(C_2H_5)_2 + 2AlCl_3.$$

Various weights of the diethylzinc sample were reacted at - 60°C in the gasometric apparatus with n-hexanol and then with 20% aqueous sulphuric acid reagents. The volume of gas evolved from the various sample weights together with the evolved gas compositions obtained by gas chromatography are given in Table 78. The gas evolved consists principally of ethane, a small amount of n-butane also being present. This originates from butyl impurity present in the triethylaluminium used to prepare the diethylzinc. Hydrogen was absent in the evolved gas indicating the absence of zinc-bound hydride groups and showing also that no fissioning of zinc-bound alkyl groups occurred under the reaction conditions used.

The plot of the sample weights and gas volumes shown in Table 78 passes through the origin, indicating that reaction is proceeding reproducibly, regardless of the weight of diethylzinc taken for analysis.

The procedure was applied to the determination of ethyl and butyl groups in a sample of distilled diethylzinc. This material contained some aluminium impurity originating in the triethylaluminium used in its preparation. The results obtained are presented in Table 79.

It is seen that 99% of the sample has been accounted for. The remainder of the sample is probably alkoxide. Alkoxide was not determined but undoubtedly occurs to a small extent in this reactive organometallic compound.

Philip et al [21] have discussed a rapid hydrolysis procedure for the analysis of zinc alkyls. This is described in the earlier section on organoaluminium compounds.

Table 78 - Gas yields obtained upon alcoholysis-hydrolysis of diethylzinc

Weight of diethylzinc taken for analysis (g)	Total volume of gas evolved at S.T.P. (ml)	Composition of generated gas (%v/v) ethane	n-butane	hydrogen
0.1049	36.0	98.7	1.3	nil
0.1751	62.3	98.7	1.3	nil
0.1861	70.3	98.7	1.3	nil
0.2146	75.9	98.7	1.3	nil
0.2573	88.3	98.4	1.6	nil

Table 79 - Analysis of distilled diethylzinc

Determined constituent	% wt.
ethyl	44.8
butyl	1.1
hydride	nil
zinc	47.8
aluminium	2.45
chlorine	2.75
Total	98.9

ORGANOZIRCONIUM COMPOUNDS

In one method for the determination of zirconium, the sample is fused with sodium carbonate, the cooled melt is dissolved in dilute hydrochloric acid and sulphur (if present) is oxidized to sulphate with hydrogen peroxide[729]. Zirconium is determined by direct amperometric titration at a rotating platinum electrode with 0.002 M EDTA at an applied potention of +0.9 V, with hydrochloric acid as supporting electrolyte. The error is -0.5% (absolute). In a further method zirconium is determined in organic compounds, after fusion with sodium carbonate, by amperometric titration with EDTA using a graphite electrode impregnated with paraffin wax[730]. This electrode is preferred, because of its greater accuracy and applicability over a wider range of potentials, to the rotating platinum electrode for carrying out this determination. The $E_{1/2}$ value for the electro-oxidation of EDTA in hydrochloric acid medium at pH 3 - 4 is + 1.2 V vs. SCE. The diffusion current of EDTA was measured at + 1.3 V to find the end-points in the amperometric titration of zirconium.

CHAPTER 2

TITRATION PROCEDURES

ORGANOALUMINIUM COMPOUNDS

A. Classical Titration Procedures

Aluminium in organoaluminium compounds of the type AlR_3 and AlR_2H can be determined with anhydrous ammonia[731]. A known weight of the organoaluminium sample is introduced into a nitrogen-filled reaction tube. An excess of ammonia is then passed through the sample. The apparatus is shown in Figure 20.

$$AlR_3 + NH_3 \rightarrow R_2AlnH_2 + RR$$

$$AlR_2H + NH_3 \rightarrow R_2AlNH_2 + H_2$$

Unreacted ammonia is swept away with a stream of dry nitrogen. Addition of ethanol to the reaction tube liberates ammonia, proportional in amount to the total AlR_3 plus AlR_2H content of the sample:

$$R_2AlNH_2 + C_2H_5OH \rightarrow R_2AlOC_2H_5 + NH_3$$

Finally, the liberated ammonia is swept into boric acid solution and determined by titration with standard acid. It is essential to use absolutely anhydrous nitrogen for purging. Any moisture in the nitrogen will decompose some of the dialkylaluminium amide derivative, causing loss of bound ammonia and consequent low analytical results:

$$R_2AlNH_2 + H_2O \rightarrow R_2AlOH + NH_3$$

Preliminary treatment of the nitrogen supply with a 10% solution of triisobutylaluminium dissolved in liquid paraffin is adequate for drying of nitrogen.

Compounds of the type AlR_2X where X is OR, SR, or HN_2, do not react with ammonia and therefore do not interfere in the determination of active organoaluminium compounds. Higher molecular weight dialkylaluminium halides react with ammonia. The method can be used, therefore, to determine the concentration of [AlR_2(hal)] in mixtures containing [AlR(oR)(hal)] and/or [AlR)OH)(hal)].

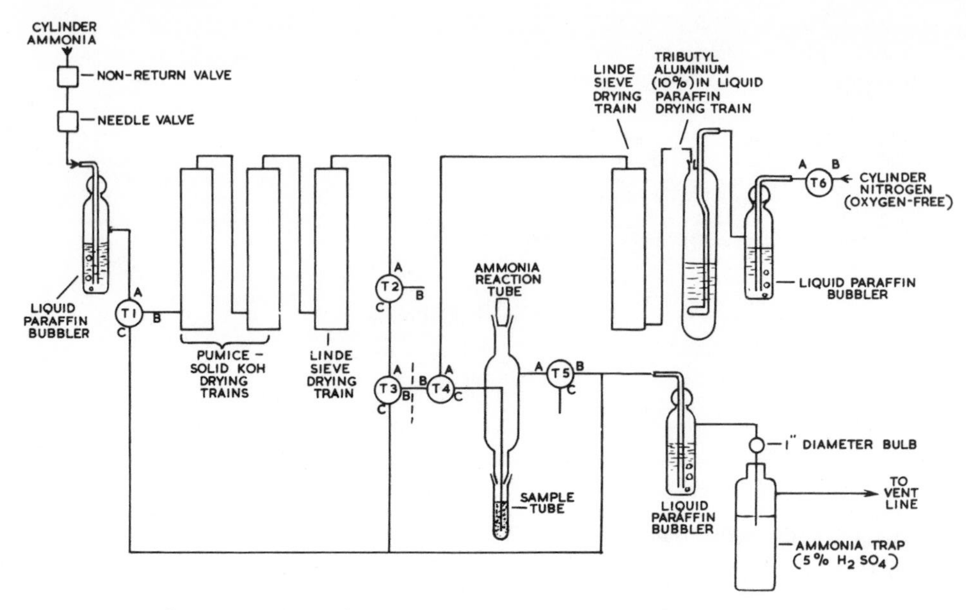

Fig. 20 - Ammonia method. General layout of apparatus.

The ammonia method does not distinguish between the two types of organoaluminium compounds, i.e. AlR_3 and AlR_2H. It is possible, however, to determine these separately when both are present in mixtures. First the dialkylaluminium hydride content of the sample is determined by either the alcoholysis-hydrolysis procedure or the N-methylaniline method described in Chapter 1 Trialkylaluminium compound is then obtained, by difference, from this hydride determination and the results obtained by the ammonia method.

The % active aluminium (AlR_3) contents of two toluene solutions of trihexadecylaluminium, obtained by the "ammonia" method, are shown in Table 80. Total aluminium which includes active aluminium (i.e. AlR_3) and "inactive aluminium" (i.e. AlR_2OR) was also determined in these samples by the EDTA method described earlier.

Due to the presence of inactive dialkylaluminium alkoxide, both samples contain a higher percentage of total aluminium than of active aluminium. The trialkylaluminium (active) and dialkylaluminium alkoxide (inactive) contents of the two samples, calculated from these figures are shown in Table 80. Application of these procedures, therefore, enables both the active and the inactive aluminium content of organoaluminium preparations beyond triisobutylaluminium to be determined.

Iodometric methods for determining organoaluminium compounds have been reported[732,733]. Triethylaluminium reacts with iodine according to the following equation[732].

$$Al(C_2H_5)_3 + 3I_2 \rightarrow AlI_3 + 3C_2H_5I$$

Dialkylaluminium chlorides and dialkylaluminium alkoxides consume, respectively, 2 and 1.25 mol of iodine per mole of organoaluminium compound[733]. This method is capable of analysing neat organoaluminium compounds or dilute

Table 80 - "Active" and "inactive" aluminium contents of toluene solutions of trihexadecylaluminium

Aluminium % wt					$Al(C_{16}H_{33})_3$ calculated from active aluminium (% wt)	$Al(C_{16}H_{33})_2(OC_{16}H_{33})$ calculated from inactive aluminium (% wt)
Active			Total	Inactive (by difference)		
1.94	1.92	1.91	2.09	0.17	42.9	3.9
1.78	1.79	1.88	1.88	0.10	39.8	2.3

Table 81 -

Organoaluminium compound analysed	Molecular weight of organoaluminium compound	Concentration of organoaluminium compound in test solution, % wt.	Sample volumes required[a] A ml	B ml
		Trialkylaluminium compounds		
$Al(CH_3)_3$	72	1.4	6	12
$Al(C_2H_5)_3$	114	2.3	6	12
$Al(C_3H_7)_3$	156	3.1	6	12
$Al(C_4H_9)_3$	198	4.0	6	12
		Dialkylaluminium chlorides		
$Al(C_2H_5)_2Cl$	120	2.4	9	18
$Al(C_4H_9)_2Cl$	176	3.5	9	18
		Dialkylaluminium alkoxides		
$Al(C_2H_5)_2(OC_2H_5)$	130	2.6	12.5	25
$Al(C_3H_7)_2(OC_3H_7)$	172	3.4	12.5	25
$Al(C_4H_9)_2(OC_4H_9)$	214	4.3	12.5	25

[a] The following relationships are used to calculate the sample volumes required:
1 Mole trialkylaluminium compound equivalent to 6 x 126.9 g iodine
1 Mole dialkylaluminium chloride equivalent to 4 x 126.9 g iodine
1 Mole dialkylaluminium alkoxide equivalent to 2.5 x 126.9 g iodine.

hydrocarbon solutions thereof, containing alkyl groups up to octadecyl at concentrations down to a few millimoles per litre with an accuracy of ± 3%. A suitable volume of the hydrocarbon solution of the organoaluminium compound is stirred with an excess of a solution of iodine dissolved in toluene (Figure 21). Alkyl groups in trialkylaluminium, dialkylaluminium chloride, and dialkylaluminium alkoxide compounds are completely iodinated within 20 min. Following the addition of dilute acetic acid, unreacted iodine is determined by titration with sodium thiosulphate solution. The concentration of organoaluminium compound is then calculated from the amount of iodine consumed in the determination. The presence of water in the iodine

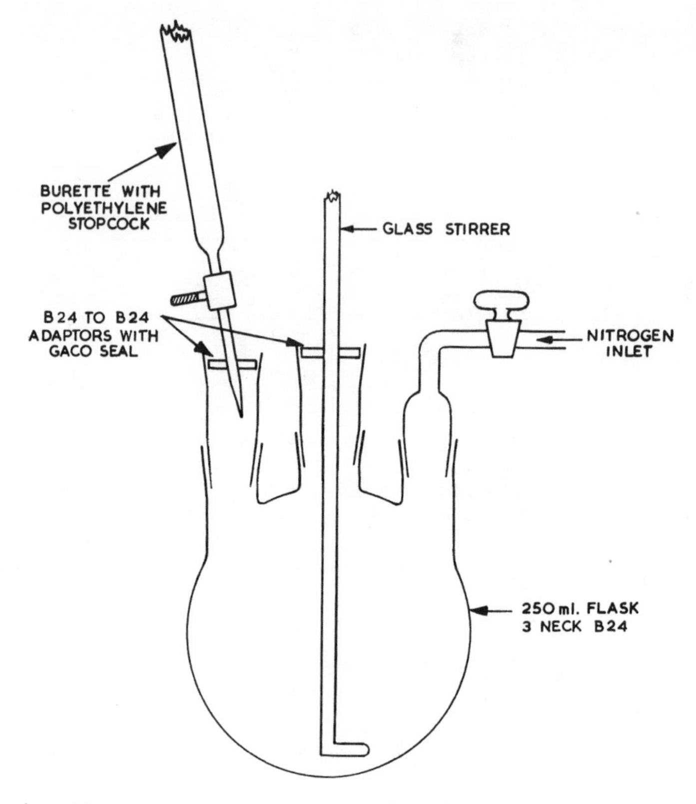

Fig. 21 - Apparatus for determination of iodine number.

reagent causes interference by reacting with some of the alkyl groups. This is corrected by a suitable 'double titration' procedure. Suitable pairs of sample volumes required for the analysis of a 200 mmol 1^{-1} solution of various types of organoaluminium compounds by the 'double titration' procedure are shown in Table 81. Correspondingly larger or smaller volumes should be taken if the concentration of organoaluminium compound in the sample differs appreciably from 200 mmol 1^{-1}.

The stoichiometry of the reactions which occur during the iodination of isooctane solutions of relatively pure specimens of various organo-aluminium compounds are shown in Table 82.

The reproducibility is shown in Table 83.

The iodine method can be used to determine the catalyst content of dilute hydrocarbon solutions of various trialkylaluminium compounds. Reasonably good agreement is obtained between the iodometric and the Bonitz [734] conductiometric isoquinoline titration methods of analysis. The iodometric method is applicable to solutions containing as little as 20 mmol 1^{-1} of catalyst.

Triisobutylaluminium has been determined by reaction with excess of standard mercaptan solution followed by amperometric titration of excess of mercaptan with standard silver nitrate in a methanolic ammonia/ammonium nitrate solution[735]. Acidic organoaluminium compounds have been titrated

Table 82 - Stoichiometry of the iodination of organoaluminium compounds

Sample description	Composition of sample of aluminium % wt/wt		Iodine consumption (moles iodine consumed per mole of organo-aluminium compound)
Triethylaluminium	$AlEt_3$	88.8	
	$AlEt_2Bu$	5.3	
	$AlEt_2H$	1.1	2.97
	$AlEt_2(OEt)$	4.3	
Tri-n-propylaluminium	$AlPr_3$	90.5	
	$Al(hexyl)_3$	1.7	3.09
	$AlPr_2H$	1.4	
	$AlPr_2(OPr)$	2.2	
Diethylaluminium chloride	$AlEt_2Cl$	93.0	
	AlEtBuCl	4.0	2.00
	AlEt(OEt)Cl	3.0	
Di-n-propylaluminium isopropoxide	$AlPr_2(OPr)$	92.7	
	$Al(hexyl)_2OPr$	5.3	1.25

Table 83 - Reproducibility of iodine number determinations

Sample description	Number of determinations	Mean iodine number (g iodine) consumed per 100 g sample)	Standard error	95% confidence limits
Triethylaluminium	6	613.8	2.6	613.8 ± 7.3
Diethylaluminium	6	158.6	1.9	158.6 ± 5.3
ethoxide	6	156.5	1.1	156.5 ± 3.1
Di-n-butylaluminium nonoxide	6	42.9	0.3	42.9 ± 0.9

with basic titrants in the presence of acid-base indicators[736]. When a dichloroethane solution of methyl violet was added to toluene, benzene, or heptane solutions of various organoaluminium compounds, a colour change occurred from violet (alkaline) to yellow or green (acidic). Addition of bases such as butyl acetate, dimethylaniline, diethyl ether, or pyridine to this solution caused a reversion of colour to violet. Organoaluminium compounds containing alkoxide groups did not affect the colour of methyl violet. Other indicators such as crystal violet and gentian violet also gave colour changes in these circumstances.

Alkyl- and arylaluminium compounds behave as strong aprotic acids, giving titration curves with organic bases which are similar to those obtained in inorganic acid-based titrations. There are considerable differences in behaviour between various trialkylaluminium compounds and alkylaluminium halides[736]. The organoaluminium sample, diluted in xylene, can be introduced by a hypodermic syringe via the septum into a flask under argon (Figure 22) to remove the last traces of oxygen and moisture and to solubilize the indicator (methyl violet, basic fuchsin, phenazin, neutral

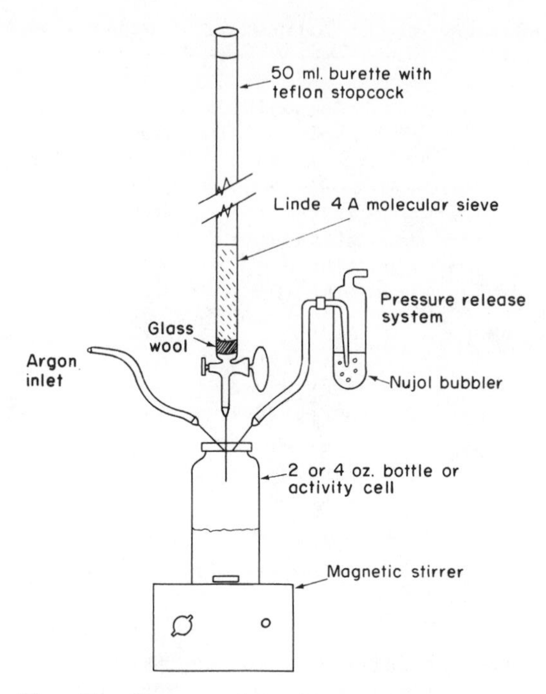

Fig. 22 - The Hagen and Leslie titration apparatus.

red, or neutral violet)[737]. The solution is then titrated to the colour change with a standard base (0.2 M pyridine or isoquinoline in xylene) in a needle-tipped burette. Excess of titrant is added to the bottle and then the organolaluminium sample injected dropwise from a syringe until the indicator changes to the excess alkyl-colour. The solution is titrated to the end-point and the syringe reweighed to obtain the weight of the organo-aluminium sample used between the two end-points.

Other types of triphenylmethane-type indicators can also function as reversible indicators[737]. For example, basic fuchsin is reversible, whereas methyl green does not display reversible complex behaviour. The triphenyl-methane-type indicators which are reversible (methyl violet and other p-rosanilines) are unstable with respect to alkylaluminium hydrides and it appears that the hydride-sensitive part of the molecule is also responsible for its ability to act as an indicator. Reversible colour changes are observed with trialkylaluminium and ketones, such as anthrone, benzil, and Michler's ketone, and aldehydes such as p-dimethylaminobenzaldehyde. Coloured complexes can be formed with trialkylaluminium compounds by the use of compounds such as pyrazine, allowing the colorimetric determination of small amounts of the alkyl. Complexes of difunctional compounds containing azomethine linkages with dialkylaluminium hydrides are much more intensely coloured than those previously reported for monofunctional compounds such as isoquinoline and pyridine. o-Phenanthroline: α,α'-dipyridyl and 6,7-dimethyl-2,3-di(2-pyridyl)quinoxaline are capable of forming complexes with dialkylaluminium hydrides which are not readily displaced with a stronger base such as pyridine, whereas m-phenanthroline forms intensely coloured AlR_2H complexes which are readily decoloured by pyridine. Neutral red and neutral violet undergo several colour changes as the end-point is approached in the titration of trialkylaluminium compounds with pyridine, isoquinoline, or alcohols. Phenazine also gives this colour transition and it appears that with excess of alkylaluminium a green complex is formed which could be attributed to a quinoid-like compound. As the trialkylaluminium compound is displaced from the indicator with pyridine, the colour changes to red and finally to yellow.

In Table 84 is shown data obtained in the titration of various C_2 to C_{30} organoaluminium compounds with standard pyridine solution and using visual indicators.

Complexes of the difunctional compounds with dialkylaluminium hydrides are much more intensely coloured than those previously reported for mono-functional compounds, such as isoquinoline and pyridine.

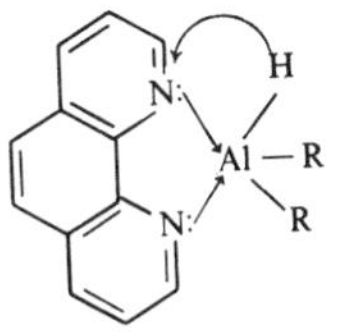

I

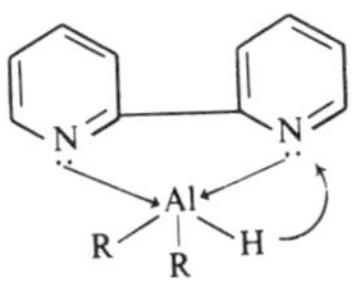

II

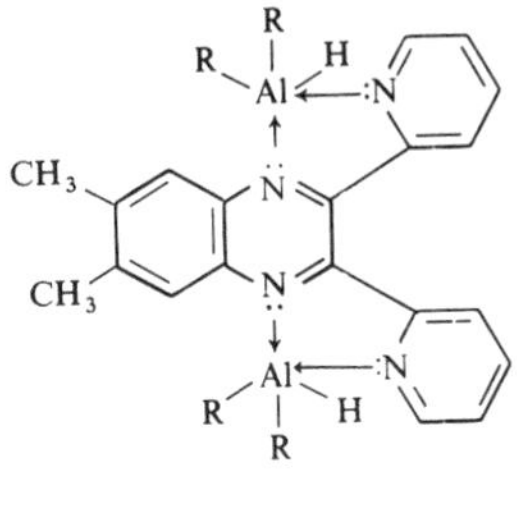

III

The stability of dialkylaluminium hydride complexes with the di-functional reagents is related to the position of the complexing sites in the electron-donating molecule. For example, o-phenanthroline(I), α,α'-dipyridyl(II) and 6,7-dimethyl-2,3-di(2-pyridyl) quinoxaline(III) are capable of forming complexes with dialkyl aluminium hydrides which are not readily displaced with a stronger base such as pyridine, while m-phenan-throline forms intensely coloured AlR_2H complexes which are readily decoloured by pyridine.

Table 84 - Titration of C_2 to C_{30} organoaluminium compounds with pyridine to visual indicator end-points.

	Per cent $AlR_3 + AlR_2H$						
Sample No.	Methyl violet	Basic fuchsin	Phenazine	Neutral red	Neutral violet	Average	Av. Dev.
1	96.3	96.3	96.6	96.8	96.6		
	96.2	96.3	96.2	96.6	96.1	96.4	0.2
2	98.8	99.3	99.2	99.4	99.6		
	98.3	99.1	98.7	98.1	98.8	99.0	0.3
3	93.1	93.1	94.0	93.3	93.0		
	93.8	92.7	93.4	92.5	93.0	93.1	0.3
4	73.6	79.1	75.1	79.9	79.5		
			75.1	77.5	78.7	77.3	2.0
5	56.9	58.6	56.6	57.2	57.4		
	57.4	58.1	56.9	56.3	56.9	57.2	0.5

Neutral red and neutral violet undergo several colour changes as the end-point is approached in the titration of trialkylaluminium compounds with pyridine, isoquinoline, or alcohols. Phenazine also gives this colour transition and it appears that with excess alkyl a green complex is formed which could be attributed to a quinoid-like compound (IV). As the trialkylaluminium compound is displaced from the indicator with pyridine, the colour changes to red and finally to yellow:

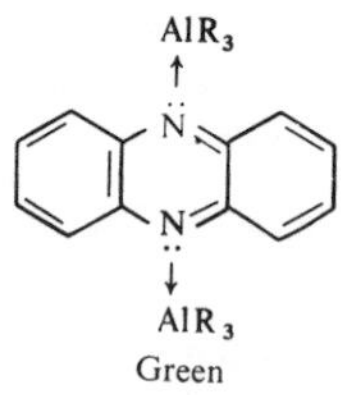

AlR_3

N

N

AlR_3

Green

IV

The titration procedure utilizing visual indicators is applicable to trialkyl samples that contain little or not hydride. Accurate determinations of dialkylaluminium hydrides are best obtained spectrophotometrically[738]. Total activity ($AlR_2H + AlR_3$) in samples containing too much dialkylaluminium hydride for the indicator titration is best obtained by a photometric titration technique.

A rapid, visual pyridine titrimetric method may be used to determine total trialkylaluminium plus diethylaluminium hydride reactivity as well as to differentiate between these moieties in mixed complex systems of aluminium alkyls[739]. Phenazine used as the indicator forms red to brown complexes with trialkylaluminium compounds and green to green-blue complexes with diethylaluminium hydride. The method is applicable to alkylaluminium compounds ranging from C_2 to at least C_{20} separately or in mixtures. The precision is ± 1% (relative standard deviation) for both high and low concentrations of trialkylaluminium and/or dialkylaluminium hydride. This indicator is not used, however, for the titration of dialkylaluminium hydride compounds because of the possible formation of an Al - N bond with the dialkylaluminium hydride[739]. Phenazine forms reversible complexes with both trialkylaluminium and dialkylaluminium hydride. Pyridine is the most suitable base for forming complexes with trialkylaluminium and dialkylaluminium hydride. Pyridine displaces phenazine quantitatively from both trialkylaluminium and dialkylaluminium hydride to form more stable 1:1 complexes. Table 85 compares the results obtained by this procedure and the isoquinoline spectrophotometric procedure[737]. The results are generally in excellent agreement. The greatest differences are observed for the trialkylaluminium values which are obtained, in each case, by difference. Of significance is the range of dialkylaluminium hydride determined by the method without alteration of the phenazine. The precision of the determination is slightly more than ± 1% relative throughout the range of dialkylaluminium hydride values. The accuracy should approach the precision[739]. Table 86 shows results obtained on typical samples. These samples vary from "pure" to partially oxidized samples. The aluminium varies from 1.31% to 29.41%. The theoretical aluminium for aluminium triethyl is 23.63%, while for aluminium diethylhydride it is 31.33%. The aluminium contents of samples A8259 and A8601 both indicate that a considerable concentration of hydride must be present. Analysis shows this to true. Additionally, Sample No. A2237 was a solvent-stripped "pure" aluminium alkyl. The aluminium concentration indicated it may or may not be triethylaluminium because various combinations of diethylaluminium hydride and triethylaluminium also yield aluminium values equal to the theoretical triethylaluminium value. A case in point is No. A1401 which was found to contain 4.4% diethylaluminium hydride. Thus, one can predict that if the aluminium content exceeds 23.63% then a molecule other than triethylaluminium (usually diethylaluminium hydride but it could also be trimethylaluminium) is present. If the aluminium is less than 23.63% aluminium triethyl, as well as other trialkylaluminium compounds and some dialkylaluminium hydride can also be present. Again, the precision in all cases is essentially ± 1% relative.

Table 85 - Comparison of results obtained by the described phenazine differential titration and the spectrophotometric isoquinoline procedures for ADAH[a] and total reactivity

	Mol-% ADAH[a]		Mol-% ATA[b]		Mol-% reactive	
Sample	Phenazine	Iso-quinoline	Phenazine	Iso-quinoline	Phenazine	Iso-quinoline
A	81.3 ± 0.85[c]	80.8	10.5	11.1	91.8 ± 0.92[c]	91.9
B	65.4	63.0	28.5	30.7	93.9	93.7
C	4.51 ± 0.05[d]	4.55	-	-	-	-
D	58.8	58.9	23.8	24.1	82.6	83.0
E	16.1	17.4	40.7	37.8	56.8	55.2
F	<0.3	-	>96.6	96.8	96.9	96.9

[a] ADAH = dialkylaluminium hydride.
[b] ATA = trialkylaluminium.
[c] Five or more replicate analyses.
[d] ATA = total reactive - ADAH.

According to Jordan[739] simple complexation is involved in the reactions of both dialkylaluminium hydride and trialkylaluminium compounds with phenazine. However, a semiquinone has been identified by E.P.R. and N.M.R. spectroscopy, which suggests that reduction as well as complexation is involved in the reaction[740]. The diethylaluminium hydride is titrated after both the excess of triethylaluminium and the phenazine-complexed triethylaluminium, indicating its relatively weak Lewis acid nature. Phenazine has also been used as an indicator in the titration of organoaluminium compounds with nitrogen bases[741].

Table 86 - Results obtained on typical samples by pyridine titration of the phenazine complexes

Sample	Al,%	ADAH	ATA[a]	Total reactive
A8131	10.98	6.1	90.6	96.7
A8132	22.86	5.0	91.7	96.9
A8136	2.01	7.9	85.9	93.8
A8259	29.41	90.1	6.7	96.8
A8601	28.60	87.2	9.5	96.7
A81151	1.31	1.5	86.2	87.7
A1361	23.03	1.0	96.5	97.5
A1401	23.75	4.4	93.2	97.6
A2152-1[b]	26.95	79.8 ± 0.08	12.6	92.4 ± 0.8
A5152-1[b]	7.56	1.4 ± 0.01	68.0	69.4 ± 0.8
A2237	23.51	<0.1	98.3	98.3
A84373	29,42	91.9	<0.1	91.9
A84474	27.54	93.6	N.D.[c]	93.6
A84487	28.00	94.5	N.D.	94.5

[a] ATA = Total reactive - ADAH
[b] Six or more determinations.
[c] None detected.

B. Conductometric Titration

Conductometric titration of various types of organolaluminium compounds with hydrocarbon solutions of either diethyl ether or isoquinoline solutions has been studied[742,743]. The sample, dissolved in anhydrous hexane, cyclohexane, or benzene, is titrated with a solution of diethylether or isoquinoline in the same solvent under dry, oxygen-free nitrogen (Figure 23). The curve obtained using diethylether as titrant shows a sharp conductivity maximum at about 2% before the molar ratio N/M = 1.00 (N = moles of diethyl ether or isoquinoline added, M = total moles of reactable compound in the sample) is obtained, i.e. at N/M = 0.98. Thus the weighed amount of triethylaluminium sample contains about 2 mol-% of a compound which is not titratable with diethyl ether. In the curve obtained by titrating triethylaluminium with isoquinoline two conductivity maxima occur. The first coincides with the single maximum obtained by titration with diethyl ether, i.e. N/M = 0.98, and corresponds to the complete titration of triethylaluminium in the sample to form a 1:1 triethylaluminium-isoquinoline complex. The minimum occurring between the two maxima obtained in the isoquinoline titration lies exactly at the molar ratio N/M = 1.00. It is at this point that the solution turns from yellow to red owing to the formation of some red 1:2 diethylaluminium hydride-isoquinoline complex. Thus, diethyl ether titrant determines only trialkylaluminium compounds and does not include the hydride.

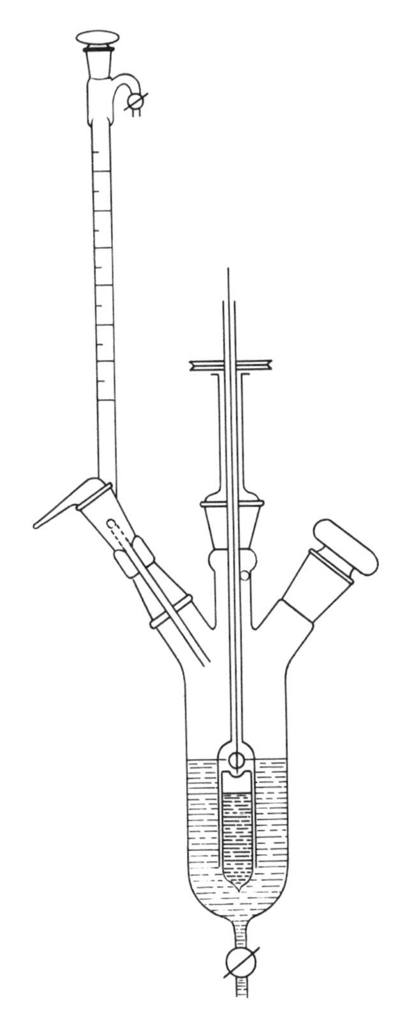

Fig. 23 - Apparatus proposed by Bonitz for conductiometric titration of organoaluminium compounds.

Conductometric titration with either diethyl ether or isoquinoline does not determine dialkylaluminium alkoxides.

Conductometric titration with isoquinoline of hydrocarbon solutions of pure dialkylaluminium hydrides such as dimethylaluminium hydride and diisobutylaluminium hydride has also been studied[742]. In both cases, no increase in conductivity occurred for values of the molar ratio N/M up to 1.00, i.e. corresponding to the formation of a 1:1 dialkylaluminium-isoquinoline complex. Beyond this point, however, up to a ratio N/M of 2.00 the conductivity increased sharply and then either flattened out or started to decrease. At a value of N/M of 1.00 the test solution starts to become red and indeed this colour change may be used as a visual indication of the end-point. The intensity of the red colour increased during the titration of dialkylaluminium hydrides and reached a maximum value when N/M reached 2.00. This corresponds to the complete conversion of dialkylaluminium hydride to the red 1:2 dialkylhydride - isoquinoline complex. Based on this work, procedures have been devised for analysing mixtures of these two types of compounds (R_3Al and R_2HAl) in the presence of each other[742]. Diethylaluminium chloride can also be determined by conductometric titration with a cyclohexane solution of isoquinoline.

A procedure for the automatic recording of conductometric titration in the titration of organoaluminium compounds with a cyclohexane solution of isoquinoline uses an automatic burette powered by a synchronous electric motor[744]. The conductivity of the test solution is measured continuously by means of an ohmmeter and registered with a compensating recording apparatus equipped with a chart recorder. The operation of the burette and the recording apparatus are synchronously linked so that a known division on the recorder chart corresponds to a known volume of titrant added to the organoaluminium sample from the burette. The titration vessel containing the sample is equipped with silver electrodes and is suitable thermostated.

As previously mentioned, the isoquinoline titration procedure provides estimates of both the trialkylaluminium and the dialkylaluminium hydride contents of organoaluminium samples. The accuracy of the isoquinoline titration procedure has been obtained by comparing results on various neat, triethylaluminium-diethylaluminium hydride mixtures, that had also been analysed by the alcoholysis-hydrolysis procedure described in Chapter 1 [745], The results in Table 87 show that the isoquinoline method gives high hydride contents and low triethylaluminium contents for samples which contain less than 10% of diethylaluminium hydride. When the samples contain more than 10% of diethylaluminium hydride the isoquinoline method gives low hydride contents (and high triethylaluminium contents). In each analysis, however, the total determined isoquinoline consumption of the trialkylaluminium and the dialkylaluminium hydride constituents agree to within 2 - 8% of the value calculated from the alcoholysis data. These results show that the isoquinoline titration procedure is unsuitable for the accurate analysis of trialkylaluminium-dialkylaluminium hydride mixtures as it neither distinguishes between different alkyl groups nor gives very accurate analyses of the individual compound types present. The more lengthy alcoholysis-hydrolysis procedures must always be used when an accurate analysis is required for these complex mixtures. The isoquinoline titration procedure does, however, give rapid and reasonably accurate estimates of the total AlR_3 plus AlR_2H organoaluminium content, even when the sample contains up to 60% of dialkylaluminium hydride. Also, the total AlR_3 plus AlR_2H organoaluminium content of hydrocarbon catalyst solutions, as dilute as 50 - 100 mmol/1, may be determined rapidly by the isoquinoline titration procedure.

A procedure for the analysis of mixtures of trialkylaluminium and dialkylaluminium compounds by conductometric titration with quinoline in light petroleum gives an error of about ± 1%[746].

C. Potentiometric Titration

The potentiometric titration of organoaluminium compounds with a standard solution of isoquinoline uses the same titration apparatus and the general procedure described for conductometric titration[742]. Potential changes may be recorded between a bare platinum wire and a bare silver wire in a pH recording amplifier. In the potentiometric titration with isoquinoline of triethylaluminium containing a small amount of diethylaluminium hydride no change in potential occurs until the ratio N/M exceeds unity (where N is moles of isoquinoline added and M is total moles of triethylaluminium plus diethylaluminium hydride in the sample). Beyond this point

Table 87 - Neat triethylaluminium-diethylaluminium hydride mixtures. Comparison of expected and determined isoquinoline consumption

Sample composition				Isoquinoline consumption of sample (g isoquinoline consumed/100 g sample)					
Based on alcoholysis-hydrolysis		Based on isoquinoline titration		Calculated from alcoholysis results			Determined by isoquinoline titration		
$AlEt_3$ + $AlEt_2Bu$ (% wt)	$AlEt_2H$ (% wt)	$AlEt_3$ + $AlEt_2Bu$[a] (% wt)	$AlEt_2H$ (% wt)	Consumed by $AlEt_3$ + $AlEt_2Bu$	Consumed by $AlEt_2H$	Total	Consumed by $AlEt_3$ + $AlEt_2Bu$	Consumed by $AlEt_2H$	Total
(1)	(2)	(3)	(4)	(5)	(6)	(7)	(8)	(9)	(10)
81.5 + 11.1 = 92.6	0.5	85.6	6.2	1.203	0.7	103.0	96.8	9.3	106.1
89.7 + 5.5 = 95.2	1.2	88.2	8.8	106.4	1.8	108.2	99.7	13.2	112.9
65.4 + 16.2 = 81.6	8.2	86.3	7.6	88.7	12.3	101.0	97.6	11.4	109.0
67.8 + 11.7 = 79.5	12.3	85.9	9.3	87.3	18.4	105.7	97.1	13.9	111.0
63.3 + 9.9 = 73.2	21.3	82.7	10.0	80.6	31.9	112.5	93.5	15.0	108.5
54.6 + 10.6 = 65.2	25.8	83.8	12.0	71.3	38.7	110.0	94.8	18.0	112.8
55.8 + 4.4 = 60.2	31.9	69.4	25.5	66.1	47.8	114.9	78.5	38.2	116.7
21.9 + 15.0 = 36.9	57.5	54.8	48.0	38.4	86.3	124.6	62/0	71.9	133.9

[a] Calculated as $AlEt_3$ as the isoquinoline procedure. does not differentiate between $AlEt_3$ and $AlEt_2Bu$.

(i.e. the formation of the 1:1 diethylaluminium hydride - isoquinoline complex) the potential starts to increase, reaching a maximum at a value of N/M of 2.00. Between values of N/M of 1.00 and 2.00 the solution becomes increasingly redder owing to the formation of the 1:2 diethylaluminium hydride - isoquinoline complex. Diethylaluminium chloride may also be titrated with isoquinoline. The automatic recording conductometric titration devise is also applicable to the automatic potentiometric titration of organo-aluminium compounds with isoquinoline[744].

A potentiometric titration procedure in which the sample is dissolved in cyclohexane and titrated potentiometrically with a standard cyclohexane solution of quinoline using platinum and silver electrodes in an inert gas atmosphere is claimed to give a more distinct end-point than the isoquinoline method[743]. The potential change occurring in the isoquinoline potentiometric titration is not always detectable or is inaccurate in the case of organo-aluminium compounds containing aluminium hydride groups[747]. Superior potentio-metric titration curves are obtained if the silver-platinum electrode system is replaced by an aluminium rod as titration electrode and an aluminium rod immersed in triethylaluminium solution as a reference electrode. The reference electrode is placed in contact with the organoaluminium sample solution by means of a porous disc (Figure 24). The establishment of the potential after addition of titrant is relatively slow. This isoquinoline titration procedure was tested against high purity samples of organoaluminium compounds dissolved in dry benzene or heptane.

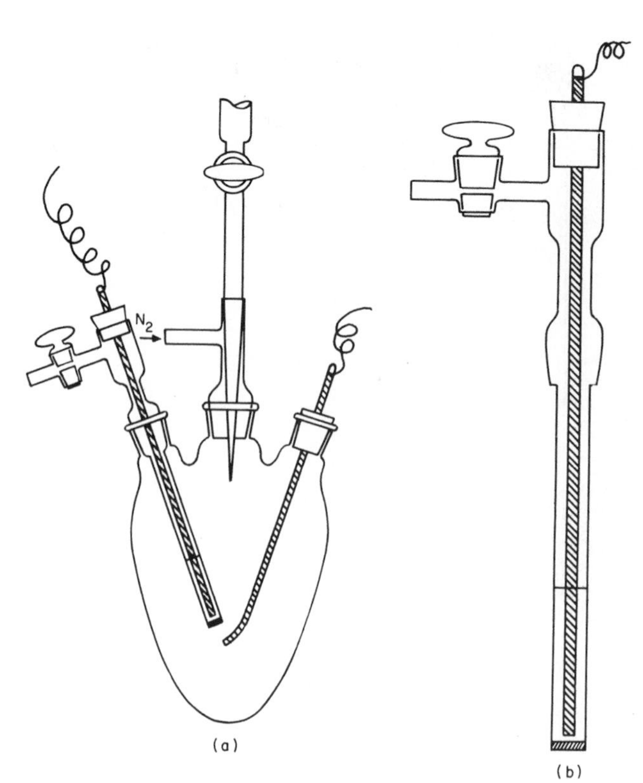

Fig. 24 - Titration cell, (b) reference electrode as proposed by Farina et al

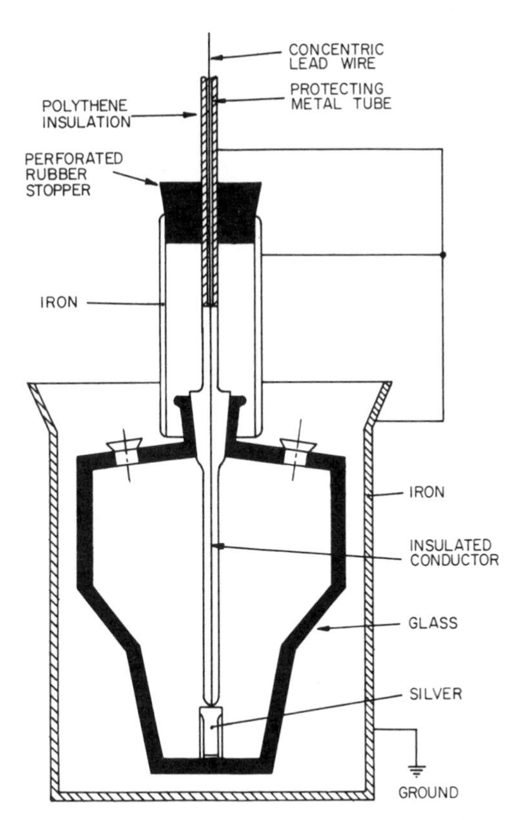

Fig. 25 - Shielded silver electrode proposed by Nebbia and Pagani for the potentiometric titration of organoaluminium compounds.

Very sharp inflections were obtained at the titration end-point during the titration of pure triethylaluminium. The total drop in potential occurring at the end-point of the titration is appreciable (300 - 600 mV cm^{-3}). Very sharp end-point breaks were obtained corresponding to purities of 99 - 99.6% with a reproducibility of ± 0.2 - ± 0.4% of the determined result.

This work emphasized the importance of ensuring that both the sample solvent and the isoquinoline titrant are completely anhydrous. Various titrations carried out with solvents having different water contents as determined by the Karl Fischer method, showed a regular decrease in isoquinoline titration as the water content of the solvent increased. It is essential, therefore, to use extremely anhydrous solvents in this analysis. Farina[747] showed that the presence of dialkylaluminium alkoxides in trialkylaluminium compounds does not affect the analysis of the latter, and that mono and dihalides of organoaluminium compounds react in a manner similar to that of trialkylaluminium compounds, exhibiting sharp breaks in the isoquinoline titration curve. Also, mixtures of dialkylaluminium monochlorides (or monoalkylaluminium dichlorides), plus trialkylaluminium compounds show a single inflection equivalent to the total concentration of these in the sample. Thus, potentiometric titration cannot determine both of these constituents in a mixture.

Farina et al[747] showed that etherates of alkylaluminium compounds can be titrated potentiometrically with isoquinoline solution. These titrations are, however, rather slow and do not have the same high precision obtained with other types of organoaluminium compounds discussed. This is due to the reduced sensitivity of the electrodes occurring in the titration of etherates (variation in potential at the break occurring in the titration curve is 80 - 150 m V cm^{-3}). It is, however, possible to analyse mixtures of trialkylaluminium compounds and trialkylaluminium etherates. Two inflections are obtained; the first one corresponding to the quantity of trialkylaluminium compound, the second to the total sum of this and the etherated trialuminium compound.

The platinum-silver cell has been re-examined using a silver electrode having a large surface area, shielded and of special construction[748]. With this electrode it is possible to obtain differences of 250 - 400 m V in titrations with isoquinoline (Figure 25). Also, a stable potential is established immediately following every addition of reagent, α β, and γ-methylpyridines, 2-ethylpyridine, 2,5-methylethylpyridine, 2,4,6-trimethylpyridine, and 2,6-lutidine are all excellent titrants, particularly the di- and tri-substituted pyridines, inasmuch as they can be easily purified by careful fractionation, and their solutions in benzene can be kept anhydrous and stable for long periods of time without change of titration. Moreover, these bases can be standardized easily and accurately by titration with perchloric acid in acetic solution using a crystal violet indicator. In these experiments Nebbia and Pagani[748] used an automatic titration apparatus. Titrations were performed in the absence of air. All the reagents and solvents used were completely anhydrous. The titrations were carried out with 10 - 30% alkylaluminium solutions in heptane or benzene. About 5 - 10 ml of sample was used and titration was carried out with 0.5 N solutions of tertiary bases. Excellent titration curves are obtained and the potential changes are appreciable. Nebbia and Pagani[748] also found that trimethylaluminium and triisobutylaluminium and monoethylaluminium dichloride were successfully titrated using their system. Diethyaluminium hydride did not, however, give a satisfactory titration curve.

Chlorodiethylaluminium and mixtures of tributylaluminium and dibutylaluminium hydride have been titrated potentiometrically with 0.2 M solution of pyridine in anhydrous benzene using aluminium and silver amalgam electrodes in an atmosphere of dry oxygen-free nitrogen[749].

D. Amperometric Titration

Amperometric titration of mixtures of triethyaluminium and diethyliodoaluminium in octane can be performed in a cell equipped with two platinum electrodes[750]. Phenetole, diethyl, ether, tetrahydrofuran, and dioxan may be used as titrants. Titrations are carried out in an atmosphere of pure, dry argon. The sequential titration of diethyliodoaluminium and triethylaluminium with dibutyl ether, dioxane or tetrahydrofuran is possible without interference from any ethoxydiethylaluminium present. The experimental error of the method is ± 1%.

E. Activity by Dielectric Constant Titration

Trialkylaluminium compounds and diethylaluminium chloride, because of their tendency to accept a lone pair of electrons, form well-defined 'donor-acceptor' compounds which have a very high dipole moment[751]. Considerable evolution of heat also accompanies the formation of these compounds. In these donor-acceptor compounds the donor (e.g. a tertiary amine) supplies both bonding electrons to the organoaluminium acceptor molecule, thus disturbing the charge symmetry (1). The donor becomes partially positively charged and the acceptor becomes partially negatively charged, and it is a consequence of this unsymmetrical charge distribution that such donor-acceptor complexes possess a considerable dipole moment, of the order of 4 - 6 Debye. Dialkylaluminium alkoxides, on the other hand do not react with these donor molecules and have a low or zero dipole moment.

$$R_3Al^{\delta-} \leftarrow N^{\delta+}R_3$$

(1)

A complexometric dielectric constant titration technique has been developed for the determination of organoaluminium compounds which form complexes of high dipole moment with suitable donor molecules[752]. The titration is carried out in an apparatus consisting of a thermostated titration vessel with magnetic stirrer, containing dry solvent and a known weight of the organoaluminium sample under pure nitrogen or argon. The dielectric measuring cell consists of a gold plated Teflon-lined immersion condenser of exactly 7 pF effective capacity used in conjunction with a Decameter. The donor titrant solution is added to the organoaluminium sample solution at a uniform rate by means of motor-driven piston burette synchronized with a suitable motor-driven chart recorder which automatically plots a curve of volume of donor solution added to the sample against dielectric constant. This procedure has been applied to the analysis of a mixture of diisobutylaluminium hydride (approx. 93 mol-%) and triisobutyl-aluminium (approx. 7 mol-%) using di-n-butyl ether as the titrant[752]. An inflection was obtained corresponding to 9.4 wt.-% of triisobutylaluminium in this sample. The slope of the curve then decreased on further addition of the ether. Figure 26 is a reproduction of a titration curve obtained during the dielectric constant titration of a mixture of diethylaluminium hydride and triethylaluminium with isoquinoline.

F. Lumometric Titration

During an investigation in which a liquid scintillation counter was being used for the determination of radioactive carbon dioxide, it was found that the absorbing solution (a toluene solution of acetyldimethylbenzyl-ammonium hydroxide) itself gave off measurable amounts of light[753]. The high background is due to luminescence associated with the quaternary ammonium compound. The phenomenon is not limited to solutioons of quaternary ammonium basis but accompanies a wide variety of reactions. It may be used as a basis for analysis of organoaluminium compounds.

The apparatus used for the lumometric titration of organoaluminium compounds with gaseous oxygen is shown in Figure 27.. The light detection system consisted of a DuMont K 1448 multiplier phototube with a preamplifier and scalar. This microburet serves the twofold purpose of metering out the gaseous titrant and serving as a circulating pump for recycling the gas through the sample.

When metal alkyls are titrated with air or oxygen in inert solvent lumo-metric curves are obtained. An example of such a curve is shown in Figure 28. The sample in this case was 2.12 mmoles of triisobutylaluminium dissolved in octane and the titrant was dry air. The peak in light emission occurs at the point where one atom of oxygen has been introduced for each atom of aluminium in the sample. A similar curve is obtained when aluminium diethylmonochloride is titrated with dry air.

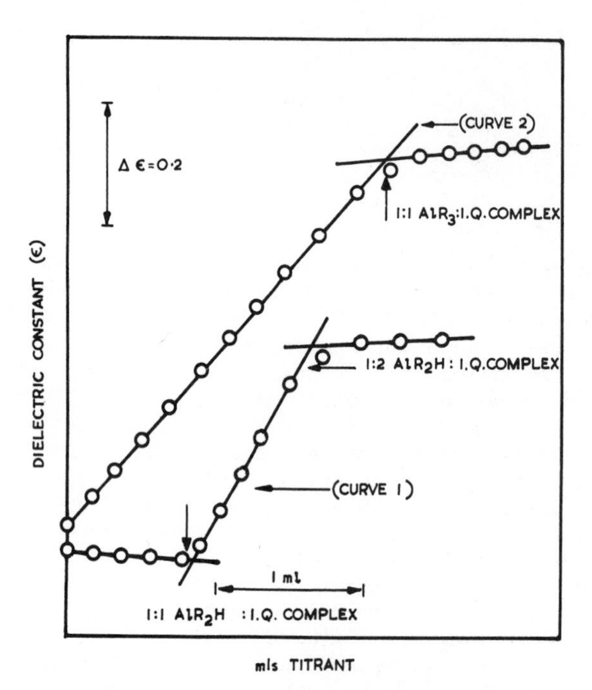

Fig. 26 - Dielectric constant titration curve of (1) 0.14 g diethylaluminium hydride, (2) 0.51 g triethylaluminium in cyclohexane with isoquinoline (29.2% w/w in cyclohexanone) at 30°C.

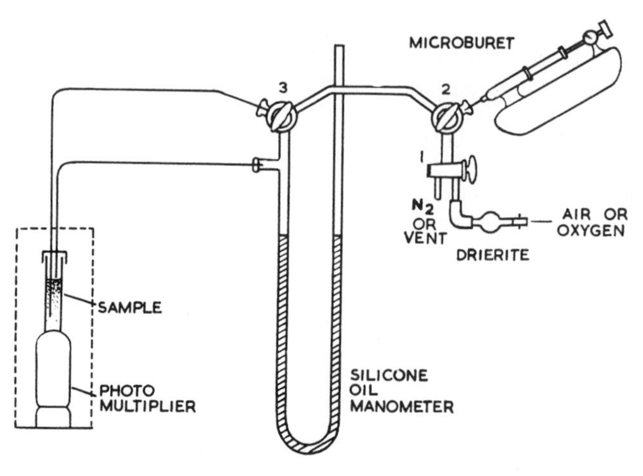

Fig 27. - Apparatus for lumometric titration of metal alkyl with oxygen or air.

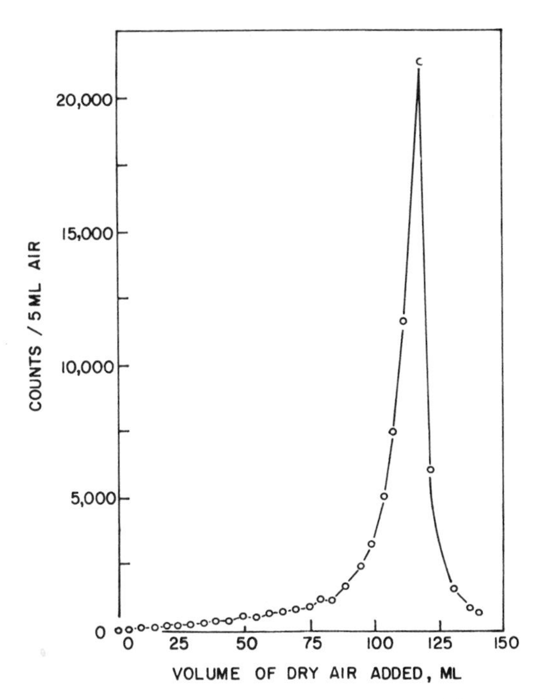

Fig. 28 - Lumometric titration of 2.12 mol of triisobutylaluminium with dry air.

G. Thermometric Titration

A suitable apparatus for carrying out the thermometric titration of oxygen- and moisture-sensitive compounds has been described[754-756]. The essential components of the apparatus are a vacuum-jacketed titration flask from which air and moisture can be excluded, a constant delivery-rate syringe burette, a thermistor connected in a Wheatstone bridge circuit to detect temperature changes, and a strip-chart recorder to indicate bridge output as a function of titrant added (Figure 29). It is essential that air and moisture are excluded from the titration vessel and that the vessel has a low heat conductivity[755]. The syringe burette is calibrated in terms of millilitres of titrant per minute.. The concentration of organoaluminium compound (mmol) is the product of this calibration factor, titrant concentration, the reciprocal of chart speed (in/min) and the distance in inches from start of titration to the intersection of lines drawn through straight segments of the curve just before and just after the inflection (see Figure 30). The various titrants used and the stoichiometries of their complexes with alkylaluminium compounds are shown in Table 88.

Triethylalmine, isoquinoline, and 2,2'-bipridyl (bipy) all give two-slope titration curves if both R_3Al- and R_2AlH-type compounds are present; typical curves obtained for the latter two titrants are shown in Figure 30. The amount of titrant consumed between the first and second inflections is a direct measure (mole for mole) of the hydride content, but the stoichiometry at the first inflection point may vary (Table 89).

Typical thermometric titration procedures are given below:

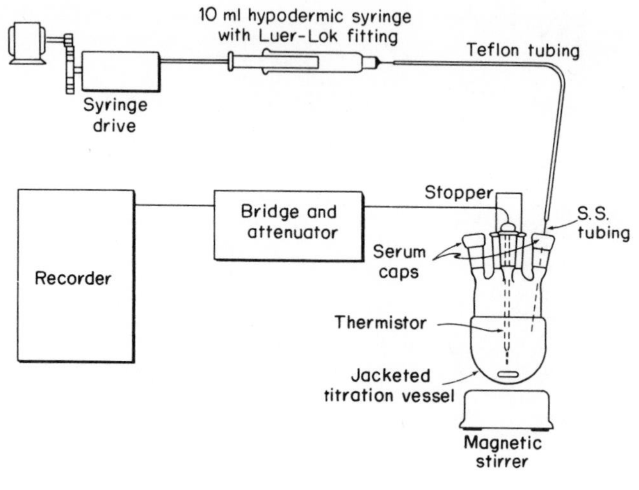

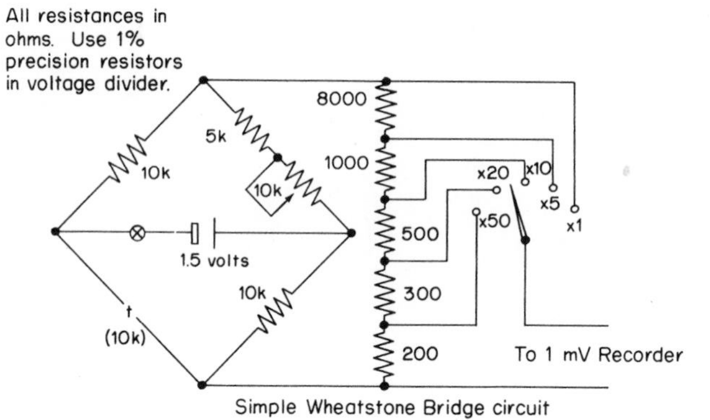

Fig. 29 - Titration assembly and bridge circuit. Thermometric titration of organometallic compounds.

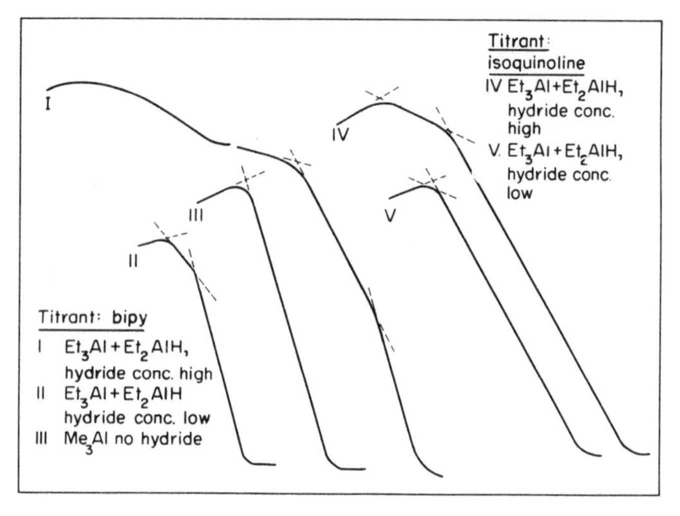

Fig. 30 - Typical thermometric titration curves obtained with 2,2'-bipyridyl (bipy) and isoquinoline.

Table 88 - Typical reactions of alkylaluminium compounds

	Reaction(s) as moles of titrant per mole of Al			
Titrant	R_3Al	R_2AlH	R_2AlCl[a]	Et_2AlOEt[a]
Triethylamine	1:1	1:1	1:1	none
Isoquinoline	1:1	1:1 and 2:1	1:1	none
2,2'Bipyridyl	1:2[b] or 1:1	1:1	1:1	none
Di-n-Butyl ether	1:1	-[c]	1:1	none
t-Butyl alcohol	1:1	1:1	1:1[d]	none
Acetone, benzophenone	none[e]	1:1[e]	none[e]	none
Oxine	1:1, 2:1, 3:1	1;1, 2:1, 3:1	1:1, 2:1, 3:1,	1:1, 2:1[f]

[a] RAlCl shows similar reactions to R_2AlCl, Et_2AlOEt was the only alkoxide studied in detail.
[b] Reaction is 1:2 with $(n\text{-}R)_3Al$ and 1:1 with $(i\text{-}R)_3Al$.
[c] Forms very weak 1:1 complex, not analytically significant.
[d] $EtAlCl_2$ reacts further to the 3:1 stage.
[e] In the presence of ether. Stoichiometry is different in the absence of ether.
[f] Slow, incomplete replacement of -OR group may occur after 2:1 reaction.

Table 89 - Titration of trialkylauminium and dialkylaluminium hydride with amines

	Moles of titrant per mole of Al, to first inflection point	
Titrant	R_3Al	R_2AlH
Triethylamine	1	0
Isoquinoline	1	1
2,2'-Bipyridyl	0.5 or 1	0

Sample titration without clean-up. The titration flask is dried at 105°C and then cooled with a stream of dry nitrogen through it. A thermistor is inserted and small side-arms are capped with rubber serum stoppers. After the addition of 40 - 50 ml of dry toluene the space above the liquid is nitrogen-flushed and the flask stoppered. A slow stream of nitrogen is continued for several min via a hypodermic needle inserted through a serum stopper; another needle, in the opposite serum stopper, serves as a vent. The solution is stirred magnetically during purging, sampling and titration.

Sample, containing preferably 2 - 3 mmoles of active aluminium, is added through a serum stopper from a hypodermic syringe. Bridge sensitivity and baseline position are adjusted. The syringe burette capillary outlet tube is inserted through a serum cap, with the tip immersed in the solution. As the recorder trace crosses a marker line the burette drive is turned on, and the chart is marked simultaneously to indicate the start of titration. Titration is continued, resetting the position of the trace if necessary, until the final inflection is obtained (some titrants may show two or more inflections).

<u>Titration with solution clean-up</u>. The above procedure is followed except that the first sample portion added (clean-up sample) is generally small (a mmole or less), and that after the addition of this sample the solution is allowed to stir for several minutes in order to eliminate the effects of reactive impurities (air, water, etc.). Also in this technique several samples are added and titrated successively, and in each titration the exact point at which titrant addition is stopped is noted on the chart as well as the point at which titration is started.

In calculating the results, the clean-up sample is generally ignored. The amount of excess titrant is added in each titration (i.e. the distance from the inflection point to the end of titrant addition) must be counted as part of the titrant consumed by the following sample and must be added to the amount shown by the titration curve (i.e. the distance from the start of titration to the inflection point). Depending somewhat on sample type and sample size, as many as five sample portions (including the clean-up sample) may be added during a series. It is generally inadvisable to exceed this number because of dilution and temperature effects.

Toluene, C.P. grade dried over calcium hydride, was used as both titration solvent and titration solvent (Oxine, not very soluble in hydrocarbons, was dissolved in anisole). Titrants were used without purification other than drying, as 1M to 2 M solutions. Ethers and amines (or their solutions) were dried over calcium hydride, and alcohols over Molecular Sieve 4A, Benzophenone was stored under dry nitrogen.

2,2'-Bipridyl is unusual in that it forms both 1:2 and 1:1 complexes with triethylaluminium and other tri-n-alkylaluminium derivatives but only a 1:1 complex with tri-alkylaluminium derivatives. This is presumably because each nitrogen atom can form a complex with an aluminium atom in $(n\text{-}R)_3Al$-type compounds, but the steric hindrance prevents a similar reaction with $(i\text{-}R)_3Al$, i.e. addition of the first molecule blocks the entry of a second one. For the 2,2'-bipryidyl-R_2AlH reaction the 1:1 ratio (rather than 2:1 as with isoquinoline and benzalaniline) is understandable on the basis that the amide formed in the initial step dimerizes rather than reacting with a second molecule of reagent.

$$C_5H_5N \cdot C_5H_5N + R_2AlH \longrightarrow$$

H H R N—Al—N R

The reaction product is deep orange-red similar to the corresponding isoquinoline compound. There is some indication (Figure 30) (curve 1) that diethylaluminium hydride reacts with a second molecule of 2,2'-bipridyl but the energy of reaction is low.

Benzalaniline (N-benzilidineaniline) reacts similarly to isoquinoline but with triethylaluminium the curve shows considerable curvature, with a suggestion of an intermediate 1:2 complex. N-Methylaniline, which reacts selectively with dialkylaluminium hydride at 30°C[756] reacts with triethylaluminium as well at room temperature and is unsuitable as a titrant.

Amine titrants generally can give, in a single titration, values of trialkylaluminium, dialkylaluminium hydride, and activity (sum of the two).

However, this advantage is more apparent than real; for hydride, in particular, the precision and accuracy leave something to be desired. At the low hydride concentrations generally found in commercial triethylaluminium, isoquinoline gives a curve (Figure 30, curve V), in which the hydride segment is too small to measure accurately or is completely obscured by the normal slight rounding of the curve near the inflection points. 2.2'-Bipridyl and triethylamine behave similarly.

At high hydride concentrations, 2.2'-bipyridyl and triethylamine give poorly defined first inflections (Figure 30, curve I), so that although the activity result is correct, the trialkylaluminium and dialkylaluminium hydride values cannot be determined very precisely. Isoquinoline gives well defined inflections with high-hydride samples but, although the result for activity (first inflection) is correct, the result for hydride tends to be low and variable. Results for trialkylaluminium are correspondingly high since, in this case, they are determined by difference. Presumably, the difficulty arises from slow reaction of the amide with a second molecule of isoquinoline; it is necessary to make potentiometric titrations with isoquinoline rather slowly in order to obtain a separate inflection for hydride[747]. These problems do not occur when a ketonic titrant is used. Alkylaluminium halides react like trialkylaluminium compounds with amine titrants forming 1:1 electron-sharing complexes.

Typical curves for oxygenated titrants are shown in Figure 31. Dialkyl ethers form electron-sharing complexes with trialkylaluminium compounds and alkylaluminium halides, giving 'normal'(type 1) curves with a single, well-defined inflection. Hydride does not interfere, although if it is present the post-inflection portion of the curve may show a slight rise, indicating the formation of weak ether - hydride complexes.

Ketones (in the absence of ethers) react readily with both R_3Al and R_2AlH compounds, although reaction with the latter is more rapid and energetic (Figure 31, curve II, a typical commercial triethylaluminium sample containing a small amount of hydride impurity). Curve III shows the reaction of acetone with a mixture containing triethylaluminium and diethylaluminium hydride in about a 5:3 molar ratio.

Alcohols such as methanol, isopropanol, t-butyl alcohol, cyclohexanol, and 2-ethylhexan-1-ol react energetically with trialkylaluminium, dialkylaluminium hydrides, and aluminium halides[751,752]. In all cases the first stage (1:1) reaction is quantitative and the inflection (ar room temperature) is sharp. With t-butyl alcohol the reaction essentially stops at this stage (but diethylaluminium chloride is an exception: see Figure 31, curve IV). Primary and secondary alcohols react further but in general the second stage reaction is slow and the third stage still slower. t-Butyl alcohol is perhaps the most useful of the alcoholic titrants because of its clear-cut, one-stage reaction with many compounds. It is the only alcohol which gives satisfactory results if the clean-up technique is used. Other alcohols may be useful in special situations, e.g. where a second-stage inflection can be obtained.

In the thermometric titration procedure complete elimination of traces of active impurities from reagents and apparatus prior to analysis is difficult and time consuming. It is considerably simpler to remove them from the reaction system just before making the analysis, and the best reagent for this purpose is the sample itself. Table 90 illustrates the effectiveness of the 'clean-up' procedure in obtaining precise results by thermometric titration.

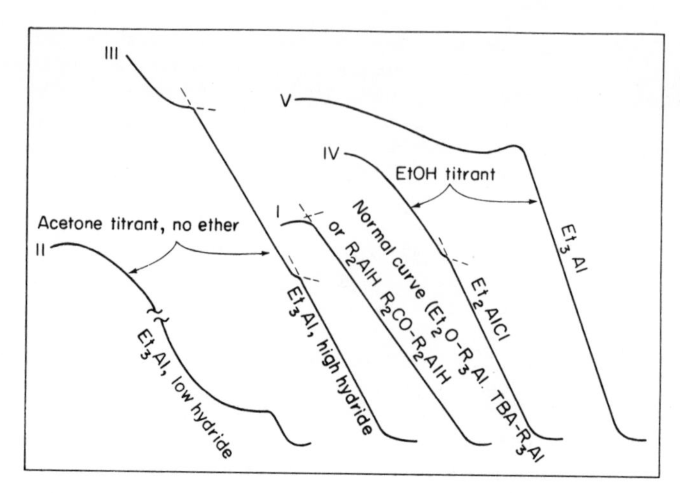

Fig 31 -Typical titration curves obtained with oxygenated titrants.

Table 90 - Typical analysis showing effect of clean-up procedure

		Determined value (mmol/g for sample portion)		
Titrant[a]	Component determined	Clean-up sample[b]	Second sample	Third sample
DNBE	$EtAlCl_2$	2.07	2.11	2.11
	R_3Al[c]	0.494	0.570	0.566
	Et_2AlCl	0.406	0.648	0.642
TBA	Me_3Al	0.609	0.665	0.679
	Activity[d]	0.645	0.680	0.683
IQ	Activity[c]	0.751	0.915	0.908
BZPH	Et_2AlH (Et_3Al present)	0.285	0.334	0.334
Bipy	Activity	0.580	0.622	0.628

[a] DNBE = di-n-butyl ether; TBA = t-butyl alcohol; IQ = isoquinoline; BZPH = benzophenone; bipy = 2,2'-bipyridly.

[b] These results are not reportable, but give some idea of the amount of reactive impurities removed by clean-up. The effect is somewhat exaggerated because the clean-up sample is usually small

[c] Commercial Et_3Al solution containing a small amount of Et_2AlH

[d] Solution of $(i\text{-}Bu)_2AlH$ and $(i\text{-}Bu)_3Al$ in about a 6:1 molar ratio.

ORGANOARSENIC COMPOUNDS

Potentiometric titration and colorimetric methods have been described for the determination of β-chlorovinyldichlorasine (Lewisite)[1006,1007]. In the latter the organic arsenic is mineralized by refluxing with aqueous sodium hydrogen carbonate and then arsenic is oxidized to the pentavalent state by the addition of aqueous iodine, followed by conversion to molybdoarsenate and reduction to molybdenum blue by boiling under reflux with a sulphuric acid solution of ammonium vanadate and hydrazine sulphate. The method is sufficiently sensitive to determine down to 3 μg of Lewisite per millilitre of sample.

The pharmaceutical sodium methyl arsinate has been determined by non-aqueous titration with mercury (II) acetate[1008].

ORGANOBORON COMPOUNDS

A. Classical Titration Procedures

A thiomercurimetric method has been described for the determination of potassium tetraphenylborate[757]. To the sample is added methanolic 0.05 N mercury (II) perchlorate and the solution is heated until dissolution is complete. Aqueous ammonia is added and the solution titrated with 0.05 N sodium mercaptoacetate in the presence of thiofluoroscein until a blue colour appears. In an alternative procedure, a known volume of 0.1 N mercury (II) acetate is added to the sample[758]. After warming and cooling and acidifying with nitric acid, the excess of mercury salt is titrated against aqueous standard ammonium thiocyanate. Another method[759] is based on reaction of the tetraphenylboron ion with the mercury (II) - EDTA complex, whereby EDTA is released in equivalent amount and is determined by titration in an acetate-buffered medium with standard zinc solution using 1-(2-pyridylazo)naphth-2-ol as indicator. Tetraphenylborate may be oxidized by chromium (VI) in concentrated sulphuric acid solution to carbon dioxide and water[760]. Excess of chromium (VI) is back-titrated with standard iron (II) solution in the presence of ferroin indicator.

The determination of dimethoxyborane in methylborate can be based on the reaction of hydridic hydrogen with iodine and subsequent titration of excess of iodine with sodium thiosulphate[761]. The method is directly applicable to dimethoxyborane in trimethyl borate solution in the dimethoxyborane concentration range 0 - 15%.

Pentaborane can be determined by its quantitative reaction with ethanol to produce hydrogen and triethyl borate, the latter being hydrolysed to boric acid, which is then converted to the mannitol complex and titrated with standard alkali[762]. Decaborane may be titrated as a monobasic acid using aqueous sodium hydroxide, although the results were consistently about 2% lower than theoretical[763]. The concentration of decaborane and diborane in air may be monitored by scrubbing the boranes out of the air into sodium hydrogen carbonate - potassium iodide electrolyte, then titrating with coulometrically generated iodine[764]. Diborane and decaborane in concentrations as low as 0.2 p.p.m. can be determined, but the various boranes cannot be differentiated. Materials that react with iodine, such as acetone and peroxides, interfere. Decaborane and tetraborane have been determined iodometrically[765]. They react instantaneously with a solution of iodine in methanol, evolving hydrogen. By back-titrating the excess of iodine with standard sodium thiosulphate solution, or measuring the volume of hydrogen evolved in a closed system, a quantitative determination of tetraborane or decaborane can be made. Since alkali titration methods for the determination of decaborane give high results owing to the presence of acidic impurities in the sample, an iodine titration has been recommended as a standard[766]. This may be based on the oxidation of decaborane with potassium iodate in glacial acetic acid, followed by an iodometric titration[767]. Although it is based on the reduction of 44 equivalents of oxygen per mole of decaborane, in actual practice the value is 3% lower than this; therefore, the reagents should be standardized with research-purity decaborane.

B. Potentiometric Titration

Various workers have described mercurimetric titration procedures for the determination of alkali metal tetraphenylborates[768-770]. Titrations may be carried out in 0.1 N sodium acetate, with standard mercury (II) nitrate or mercury (II) perchlorate solution as titrant, and the end-points detected potentiometrically with a mercury-coated platinum electrode and the standard coloured electrode joined by an agar bridge, or amperometrically with a dropping mercury electrode vs. the standard coloured electrode[768]. Potentiometric and amperometric titration procedures have been described for the determination of sodium tetraphenylborate using standard mercury (II) nitrate solution[769]. The potentiometric titration is reproducible in spite of the formation of intermediates (phenylboronic acid and diphenylmercury), the first break in the curve occurs after consumption of 3 equivalents of mercury (II) ion and the second after 8 equivalents. Boron in organic amine tetraphenylboron derivatives can be determined by potentiometric titration with standard silver nitrate solution[770]. The dry sample is dissolved in aqueous acetone (1:1) and the solution is buffered at pH 5. The resulting solution is then titrated with 0.06 N silver nitrate and a platinum indicator electrode and a shielded platinum reference electrode.

ORGANOCALCIUM COMPOUNDS

An argentimetric method has been suggested for the determination of calcium acenaphthalene and similar types of compounds. This procedure involves reaction of the sample with dialkyl or diaryl sulphides and subsequent titration of the mercaptan formed with silver ion[771].

ORGANOCOBALT COMPOUNDS

Cyanobalamin has been analysed by decomposition with sulphuric acid and hydrogen peroxide, cobalt being precipitated as cobalt (III) hydroxide by addition of sodium hydroxide and hydrogen peroxide solution, and the dissolved precipitate being titrated iodometrically. The error in this method is not greater than ± 0.5%[772].

A rapid micro-determination of cobalt carbonyl anion $[Co(CO)_4]$ in organic solvents (glacial acetic acid-toluene) can be achieved by titration with methylene blue at 0°C under nitrogen[773]. The stoichiometric reaction of 2 mol of $[Co(CO)_4]$ with 1 mol of methylene blue takes place in the pH range 1.9 - 5.2. Cyanocobalamin has been titrated amperometrically with chromium (II) ion in an EDTA medium at pH 9.5[774]. The anodic polargraphic wave at $E_{\frac{1}{2}}$ = 0.3111 V vs. SCE is probably due to the oxidation of the mercury of the electrode[775].

ORGANOCOPPER COMPOUNDS

Copper in treated fabrics may be determined by titration with 8-hydroxyquinoline[776]. The sample is extracted with hydrochloric acid at 80°C and the solution is titrated potentiometrically with 0.1 N potassium bromate. Most cations and many organic substances can be tolerated, but antimony trioxide and certain non-ionic surfactants interfere.

ORGANOGERMANIUM COMPOUNDS

Mercaptogermanes have been determined indirectly with iodine[777]. Alkoxygermanes have been determined by a modification of the Ziesel procedure[778]. Both GeOC and GeSC linkages can be determined by the perchloric acid-catalysed acetylation method[779]. The mechanism of the reaction of an alkoxy- or mercaptogermane with acetic anhydride is presumably the same as that proposed for the acetylation of SiOC[780,781] and SiSC linkages[782]. The acetylation method depends on the reaction of the alkoxy- or mercaptogermane with excess of acetic anhydride to form the corresponding acetate or thioacetate and acetoxygermane. The acetoxygermane and remaining acetic anhydride are rapidly hydrolysed to acetic acid, which is then titrated with standard base. The alkoxy or mercapto content is calculated from the difference in volume of base between the sample and blank titrations.

A further method for the assay of GeOC linkages is based on reaction with in situ generated hydrogen bromide[783]. This is based on an earlier procedure for the determination of epoxides and aziridines[784]. Under anhydrous conditions, standard perchloric acid is titrated into an acetic acid solution of tetraethylammonium bromide and the sample. The hydrogen bromide which is formed reacts with the alkoxygermane to form a bromide and the corresponding parent alcohol. The first excess of hydrogen bromide is detected by a blue BZL (Ciba 220625) indicator end-point. This method allows the quantitative determination of alkoxygermanes in the presence of alkoxysilanes and alcohols.

Procedures based on acetylation and on reaction with hydrogen bromide for the determination of alkoxy and mercapto groups in organogermanium compounds, which by elemental analysis and gas chromatography were known to have a purity between 98.7 and 99.9%, have been compared[783]. Most alkoxy- and mercaptogermanes may be determined quantitatively by either of the above methods. Acetylation, however, is preferred for mercaptogermane determinations[783].

ORGANOIRON COMPOUNDS

A. Classical Titration Procedures

In determining the standard reduction potential of haemoglobin, dialysis affects the shape of the redox titration curves, thus leading to different end-point determinations[785]. At pH 7.9, the oxidation of ferrohaemoglobin by hexacyanoferrate (III) is not a simple reversible reaction, but appears to proceed by an irreversible stepwise mechanism.

The standardization of methods for the determination of carboxyhaemoglobin has been discussed[786]. Palladium chloride is reduced in an acidic medium by means of carbon monoxide to an equivalent amount of metallic palladium. The determination of excess of palladium chloride is carried out by the use of an indirect complexometric titration after addition of potassium tetracyanonickelate (II) ($K_2[Ni(CN)_4]$), the nickel ions liberated being titrated with EDTA (disodium salt) solution using murexide as indicator.

B. Potentiometric Titration

Mason and Rosenblum[787] demonstrated that aryl ferrocenes could be titrated potentiometrically with standard potassium dichromate solution and, moreover they were able to calculate the E° values for the oxidation reaction. Peterlik and Schogl[788] carried out determinations of the molecular weight of ferrocene derivatives by potentiometric titration with potassium dichromate. The weighed sample of ferrocene derivative (10 to 15 mg) was dissolved in 2 ml of acetone, and 33 ml of acetic acid and 15 ml of water added. The solution was freed from oxygen by passage of nitrogen, then titrated with 0.01 N potassium dichromate in acetic acid (3:1) with potentiometric detection of the end-point (platinum indicator electrode and calomel reference electrode). Acyl derivatives could not be determined by this method, but could be determined if they were first reduced with lithium aluminium hydride prior to oxidation with potassium dichromate.

Nikol'skif et al[789] estimated ferrocene by addition of an acetic acid solution of the sample to a solution of ferric chloride in 2 N hydrochloric acid and measurement of the oxidation potential of the resulting clear solution:

$$Fe(C_5H_5)_2 + FeCl_3 \rightarrow Fe(C_5H_5)_2^+ + FeCl_2 + Cl^-$$

A calibration curve is used to determine the number of equivalents of ferrous chloride produced in the reaction.

Knight and Shlitt[790] devised a procedure for the determination of the per cent purity or the equivalent weight of alkyl and hydroxylalkyl ferrocenes based on potentiometric titration with standard ferric chloride solution. In addition to its rapidity, a particular advantage of this procedure is that ferrocene derivatives which have a carbonyl group adjacent to the cyclopentadiene ring do not titrate, thus it is possible to determine the amount of alkyl ferrocene in a mixture with alkyloxo ferrocene. Knight and Schlitt[790] found that potentiometric titration of alkyl and hydroxyalkylferrocenes with 0.1 N potassium dichromate in acetic acid gave erratic results and over 100% recovery, when the iron content of the sample indicated that the samples were less than 100% pure. They interpreted this as indicating that dichromate in strong acid oxidizes some of the organic material thereby causing high dichromate titrations. They corrected this by carrying out the potentiometric titration with a methanolic solution of ferric chloride. A larger potentiometric break was found using methanol as solvent than was found with acetic acid.

ORGANOLEAD COMPOUNDS

A. Classical Titration Procedures

A rapid procedure using complexone has been described for the determination of tetraethyllead in petrol[791]. The sample is pipetted into bromine solution in carbon tetrachloride until the colour persists. Methanol or ethanol is added and the solution is boiled, decolorized with a small excess of 1 N alcoholic potassium hydroxide, diluted with water, and boiled again. Lead is determined indirectly using EDTA. In a further procedure using bromine the petroleum sample is diluted with a high-boiling solvent then 30% bromine in carbon tetrachloride is added until the brown colour persists for 2 min[793]. This solution is shaken with 0.1 N nitric acid and the extract boiled to expel bromine fumes. Tartaric acid, Eriochrome Black T indicator, excess of 0.2 M magnesium chloride solution, and 0.1% potassium cyanide solution are added and the solution is buffered to pH 10 and titrated with 0.01 M EDTA.

Tetraethyllead may be extracted from petroleum into concentrated hydrochloric acid and the lead precipitated as lead sulphate[794]. The lead is determined by complexometric titration with EDTA (disodium salt) using Eriochrome Black T as indicator. The standard deviation for duplicate determinations is ± 0.002 for a range of 1.7 - 3.0 ml of tetraethyllead per gallon and results obtained were in close agreement with those obtained by the ASTM lead chromate method[795]. In a variant of this procedure, interferences caused by the presence in the petroleum of iron, dyes, and acid-extractable organic substances can be eliminated by oxidation of organic matter with sulphuric and nitric acids.

Leaded petroleum may be treated with concentrated hydrochloric acid and potassium perchlorate to extract lead into the acidic phase. This is determined by addition of excess of disodium EDTA, which is back-titrated to the Eriochrome Black T end-point with standard zinc chloride solution[802]. Alternatively, tetraethyllead may be separated from the petroleum using the ASTM hot hydrochloric acid extraction procedure[795], and the lead ion titrated with disodium EDTA[803]. Serious interferences to the end-point in this titration due to iron, petroleum dyes, and organic compounds extracted from the petroleum are reported[804]. For example, 0.3 p.p.m. iron in the petroleum renders end-point detection impossible. These may be overcome with copper-PAN indicator[804]. However, fuels containing both tetraethyllead and (methylcyclopentadienyl)manganese tricarbonyl cannot be analysed for lead content by this method. The small amount of manganese extracted by the hot hydrochloric acid frequently leads to an error of several tenths of a millilitre of tetraethyllead per gallon[804].

A tentative DIN standard[805] has been issued for the determination of tetraethyllead in petrol based on decomposition with hydrochloric acid and complexometry. The method is not applicable to samples containing multivalent metals. The sample is boiled with hydrochloric acid to decompose the tetraethyllead, diluted with water, neutralized, and buffered to a pH value of between 10 and 11. A known excess of EDTA solution is added and the excess is back-titrated with standard zinc sulphate solution. From an examination of various methods for the determination of tetraethyllead in petrol it was concluded that direct complexometric titration of the lead chloride-containing hydrochloric acid extract in the presence of tartrate was the most accurate of the methods examined[806]. Following extraction of lead by the ASTM procedure[795], tartaric acid is added to the extract, which is then made alkaline with aqueous ammonia and the lead is determined indirectly using EDTA[807]. Interferences from copper, zinc, nickel, cobalt, cadmium, and manganese, but not from calcium or magnesium, are avoided by adding solid potassium cyanide after the aqueous ammonia.

Various oxidizing agents have been used to destroy organic matter in the hydrochloric acid extracts of petroleum prior to the complexometric determination of lead using EDTA[808-813].

In a similar method Hurtado de Mendoza Riquelme[809] extracts the lead by heating the petroleum sample (100 ml) under reflux with concentrated hydrochloric acid (50 ml), the extract is evaporated, and the organic matter destroyed by heating with potassium chlorate-nitric acid. The residue is dissolved in water, aqueous sodium sulphate is added to precipitate lead sulphate, and the mixture is treated with 0.05 N sodium hydroxide until it is grey to methyl red - bromocresol green, then 0.025 N sodium hydroxide is added until a green colour appears. Nitrolotriacetic acid solution is added, which chelates with the lead and changes the colour to red; finally the mixture is titrated with 0.025 N sodium hydroxide.

Solutions of bromine in carbon tetrachloride have been used to decompose organolead compounds in petroleum. Uvarova and Vanyarkina[814] treated the sample (10 - 20 ml) with a 30% solution of bromine in carbon tetrachloride until the brown colour persists for 5 minutes. The precipitate is dissolved and a homogeneous solution is obtained by addition of methanol. The liquid is boiled and N-ethanolic potassium hydroxide is added until the brown colour is discharged and than 3 - 4 ml more. After addition of 200 ml of water the two-phase system is boiled under reflux. If the solution becomes acid to methyl red a further amount of ethanolic potassium hydroxide solution is added and the boiling continued. After this, an excess of 0.03 N EDTA (disodium salt) is added, the solution is neutralized with dilute sulphuric acid, 10 ml of a buffer solution (350 ml of 25% aqueous ammonia and 54 g of ammonium chloride dissolved in 1l), and Eriochrome Black T-sodium chloride (1:99) indicator are added, and the solution is titrated with 0.03 N zinc sulphate to a colour change from greenish blue to violet.

Gelium and Preussner[811] have described rapid and reliable methods for the analysis of tetraethyllead and tetraethyllead fluid (i.e., mixture of tetraethyllead with halogenated hydrocarbons used as anti-knock additives in petroleum). The sample (0.15 - 0.25 g of tetraethyllead) is added to 20 ml of carbon tetrachloride then shaken with 20 ml of a saturated solution of chlorine in carbon tetrachloride in a stoppered flask; 20 ml of concentrated hydrochloric acid is added, and the mixture is heated on a hot-plate until the volume is reduced to 1 - 2 ml. Water is then added, and the mixture is boiled until all the lead salt has dissolved (and the volume is 100 - 150 ml). After adding 10 ml of 50% hexamine buffer solution and xylenol-orange solution as indicator the cooled solution is titrated for lead with 0.2 N EDTA to a colour change from red-violet to yellow: with this indicator the titration is not affected by the presence of up to 0.1 mg of iron or manganese.

Dmitrievskii[812] and Saori[813] have discussed the complexiometric determination of tetraethyllead in petroleum.

A method based on titration with Karl Fischer reagent has been reported for the determination of PbOH groups[815]. An iodometric procedure for the determination of organolead compounds involves the reaction of the organolead compound with an excess of iodine. The unchanged iodine is then titrated with standard sodium thiosulphate using starch as indicator[816]. Alternatively, the sample is shaken with thiophen-free benzene, iodine solution is slowly added, and excess of iodine solution is then titrated with sodium thiosulphate[811]. In the argentiometric determination of tetraethyllead in antiknock mixtures excess of silver nitrate is added to the sample and the resulting metallic silver is filtered off, dissolved in nitric acid, and determined by the Volhard method[817]. In an organic solvent such as benzene and with a limited reaction time, tetraethyllead react quantitatively with iodine to form triethyllead iodide. This is the basis for the iodometric method from the determination of tetraethyllead[818].

With bromine or chlorine in an organic solvent, tetraethyllead reacts rapidly to form the corresponding diethyllead halide. In direct sunlight tetraethyllead is slowly converted to the triethyllead ion, then more slowly to the diethyllead ion, and finally to the simple lead ion. Under excessive heat tetraethyllead decomposes to form metallic lead and a variable mixture of hydrocarbon gases. Hexaethyllead reacts with silver nitrate to produce metallic silver, which can be separated, dissolved in nitric acid, and titrated by the Volhard method[819]. Tetraethyllead reacts similarly so that the method is not suitable for mixtures of these compounds.

Hexaethyldilead can also be determined by dissolving the sample in carbon tetrachloride, covering the solution with water containing starch indicator, and titrating in an atmosphere of nitrogen with 0.01 N iodine in potassium iodide solution. Two atoms of iodine react with one molecule of hexaethyldilead. Mixtures of tetraalkyllead, hexaethyldilead, and triethyllead chloride are analysed by titrating the hexaethyldilead with 0.01 N iodine as described above, separating the carbon tetrachloride layer, which now contains trithyllead iodide together with tetraethyllead and methyllead chloride, and treating with silver nitrate which precipitates silver from tetraethyllead, silver iodide from triethyllead iodide, and silver chloride from triethyllead chloride. The precipitate is filtered off and dissolved in nitric acid and the silver is determined by Volhard's method. The carbon tetrachloride now contains triethyllead nitrate equivalent to the hexaethyldilead plus tetraethyllead plus triethyllead chloride originally present. This is converted to lead chloride by treatment with hydrogen chloride gas.

Triethyllead ions have been determined in aqueous solution using sodium tetraphenylborate. Precipitation of triethyllead ions from acetic acid solution of pH 4 - 5 by sodium tetraphenylborate is complete within 10 min. The sample solution is filtered and excess of sodium tetraphenylborate is titrated with 0.5% benzalkonium chloride solution[820] to the Titan yellow end-point. There is no interference from inorganic lead or diethyllead ions.

Pilloni and Plazzogna[821] and Pilloni[822] and Dagnall et al[821] have used spectrophotometric titrations with EDTA to estimate diethyllead - 4-(2-pypidylazo)-resorcinol(PAR) complexes. Thus, if an excess of PAR is added to a sample of diethyllead dischloride, any lead salt impurities can be determined by a spectrophotometric titration with EDTA at 512 nm. A typical titration curve is shown in Figure 32.

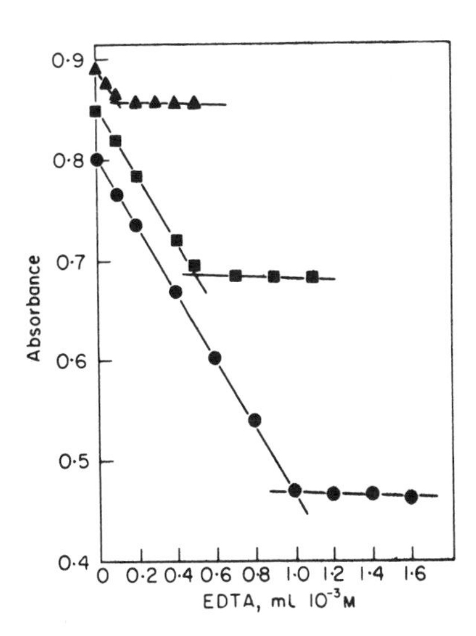

Fig. 32 - Spectrophotometric titration curves of lead chloride in $(C_2H_5)_2 PbCl_2$ with EDTA, (O) 1 ml of 1 x 10^{-3}M $(C_2H_5)_2 PbCl_2$ plus 1 ml of 1 x 10^{-3} M lead nitrate plus 2 ml 2 x 10^{-3} M PAR. ( ) 1.5 ml of 1 x 10^{-3}M $(C_2H_5)_2 PbCl_2$ + 0.5 ml 1 x 10^{-3} M lead nitrate + 2 ml 2 x 10^{-3} PAR. ( ) 1.9 ml of 1 x 10^{-3} $(C_2H_5)_2PbCl_2$ plus 0.1 ml of 1 x 10^{-3} M lead nitrate + 2 ml 2 c 10^{-3} M PAR (pH9, 1 cm cuvettes, 512 mμ V = 100 ml).

Pilloni[824] reported the stability complexes of 1 - (2-pyridyl azo-2-naphthol (PAN) with diethyllead ion spectrophotometrically in aqueous 20% v/v dioxane and showed that these had very high values when diethyllead ion is added to an aqueous dioxane solution of PAN, the yellow coloured liquid changes to a red coloured chelate which retains its colour intensity for 24 hours. The diethyllead can be directly titrated spectrophotometrically with a standard PAN solution.

Barberi et al[825] mention the spectrophotometric titration of inorganic lead as $(PbI_4)^{2-}$ in organolead compounds at a wave length of 240 nm using a standard solution of EDTA. Organolead compounds do not interfere in this titration.

B. Amperometric Titration

Plazzogna and Pilloni[826] and Pilloni[827] analysed mixtures of R_2Pb^{2+} and R_3PB^+ amperometrically at the 0.01 to 0.05 m M level with potassium ferrocyanide in 50% aqueous ethanol -0.5 M lithium nitrate base electrolyte at -0.55 V. The R_3Pb^+ is determined by the addition of an excess of sodium tetraphenylborate and then the unconsumed tetraphenylborate is titrated with thallous nitrate solution at -0.475 V v the S.C.E. in 0.01 M perchloric acid - 0.05 M sodium perchlorate as base electrolyte. A potentiometric method for determining 0.4 - 1% of lead chloride in the presence of any alkyllead chloride is also described.

Tagliavini[835] investigated the possibility of titrating hexaethyldilead by means of electrolytically generated silver ion to an amperometric end-point[832]. He also studied the conditions whereby selective titrations of organolead species in a mixture could be carried out by the utilization of quinoline as a complexing agent for silver.

Swanson and Daniels[828] have described a procedure for the determination of lead in petroleum and lubricants wherein the lead is extracted with the Schwartz reagent, (potassium chlorate and sodium chloride in nitric acid) and then, after evaporation of the acid solution, determined amperometrically by titrating with 0.01 M potassium dichromate in the presence of potassium nitrate.

C. Conductometric Titration

Jovanovic[829] determined organolead compounds in petroleum by adding to the petroleum sample a slight excess of a 1% solution of bromine in carbon tetrachloride. The precipitate of lead bromide is masked with 50% ethanol, then dissolved in and evaporated with nitric acid until a white residue remains. This residue is dissolved in 120 - 150 ml water, and the solution titrated with 0.1 N potassium chromate by a conductiometric or an oscillometric technique.

D. Coulometric Titration

Tetraalkyllead compounds have been determined coulometrically using bromine and mercury (I) ions[830,831]. The titration of the tetramethyllead is carried out to the reaction:

$$(CH_3)_4Pb + Br_2 \rightarrow (CH_3)_3PbBr + CH_3Br$$

This reaction was studied in 0.5 M methanolic ammonium bromide as basal solution.

The reaction was carried out in a cell described by Tagliavini[832]. With smaller amounts of tetramethyllead a correction was necessary for bromination of the solvent, obtained by omitting the sample from a blank run. Similar experiments in which mercuric ions were produced by constant anodic oxidation showed that these ions reacted immediately with tetramethyllead in 0.1 M methanolic lithium nitrate the presence of free mercuric ions being shown promptly by an increase in the current. The overall cell reaction is:

$$(CH_3)_4Pb + Hg_2^{2+} \rightarrow (CH_3)_2Pb^{2+} + (CH_3)_2Hg + Hg$$

Tetraethyllead behaves similarly to tetraethyllead in both the bromine and the mercuric methods.

Hexaethyldilead may be determined in the presence of tetraethyllead, by coulometric iodination at constant current, with amperometric dead-stop end-point indication[832]. The sample, in an alcoholic solution containing iodide ions, is placed in the anodic compartment of an electrolysis cell where it undergoes the following reaction:

$$R_3M\text{-}MR_3 + I_2 \rightarrow 2R_3MI$$

with the iodine electrolytically produced under constant current at the platinum anode. The end-point is observed by a rise in the indicator current, caused by excess of iodine, between a second pair of platinum electrodes sensitive to the I_3^-/I^- redox system[833]. An amperometric plot of indicator current vs. generation time can be obtained photographically in order to ensure an accurate determination. Iodine is employed in this method for the coulometric-amperometric titration of hexaethyldilead, because its rate of reaction is always greater than the rate of electrolytic generation of the iodine.

Following the discovery[819] that hexaethyldilead reacts quantitatively with silver ions:

$$(C_2H_5)_6Pb_2 + 2Ag^+ \rightarrow 2[(C_2H_5)_3Pb]^+ + 2Ag$$

the possibility of titrating hexaethyldilead by means of electrolytically generated silver ion to an amperometric end-point[833] has been investigated [834]. The conditions whereby selective titrations of organolead species in the mixtures could be carried out by the utilization of quinoline as a complexing agent for silver have also been reported[834].

E. High-frequency Titration

High frequency titration with potassium permanganate has been used[836, 837] for the determination of down to 0.01% of hexaethyldilead in tetraethyllead. The sample of tetraethyllead is diluted with acetone. Titration is carried out with a 0.05 N solution of potassium permanganate in acetone.

$$(C_2H_5)_3PbPb(C_2H_5)_3 + (O) \rightarrow Pb_2(C_2H_5)_6O$$

ORGANOLITHIUM COMPOUNDS

A. Classical Titration Procedures

A double titration procedure for the analysis of alkylithium compounds has been applied principally to n-butyllithium[838-840]. It cannot be used for the determination of methyllithium or phenyllithium owing to their low reactivity with the reagent. The total alkali (i.e. LiR + LiOR + LiOH + Li_2O) is first determined by hydrolysis of the sample solution under dry, oxygen-free nitrogen followed by titration with standard acid to the phenolphthalein end-point. To a further portion of the sample in dry diethyl-ether is added benzyl chloride dissolved in diethyl ether, which reacts with the n-butyllithium:

$$n\text{-}C_4H_9Li + C_6H_5CH_2Cl \rightarrow n\text{-}C_8H_{18} + C_6H_5CH_2C_4H_9\text{-}n + C_6H_5CH_2CH_2C_6H_5$$

After reaction the solution is hydrolysed by the addition of water and finally titrated with standard acid. This titration is equivalent to the LiOR + LiOH + Li_2O content of the sample. The n-butyllithium content of the sample is then calculated from the difference between the two titrations. A four-fold excess of benzyl chloride with respect to the alkyllithium compound, a 5-min reaction time with the benzyl chloride and the use of not less than one volume of diethyl ether per volume of sample are recommended[841].

Low results are obtained when this double titration procedure employing benzyl chloride is used to determine methyllithium and aryllithium compounds [842-845]. In addition, the purity of the diethyl ether has a marked effect when the double titration procedure is used in analysing solutions for n-butyllithium content. Reasonably accurate results are obtained in n-heptane solutions. Purification of the ether by treatment with lithium aluminium hydride leads to more satisfactory results. The low results due to the fact that benzyl chloride does not react quantitatively with alkyllithium compounds to give non-basic products can be avoided by using alternative organic halides, particularly, 1,2-dibromoethane for alkyllithium compounds and phenyllithium, and 1,1,2-tribromoethane or allyl bromide for butyllithium [842] and other alkyllithiums[847-849], as well as R_3MLi(M = Si, Ge, Sn)[850,851]. Table 91 shows some results obtained by applying the Gilman and Cartledge method[842] to methyllithium. A number of organic halides gave higher results than benzyl chloride.

Table 91 - Titrations of methyllithium. One preparation of methyllithium in diethyl ether was analysed by the double titration procedure using several different organic halides. The molarity and an average value for the yield of the preparation are given

RX compound		Deviation, %	Yield, %
None	0.880		84.5
$PhCCl_3$	0.844, 0.845	0.1	81.1
$BrCH_2CHBR_2$	0.820, 0.813	0.8	78.4
$Br_2CHCHBr_2$	0.801, 0.800	0.1	76.8
$BrCH_2CH_2Br$	0.793, 0.785	1.0	75.8
CH_2=$CHCH_2Br$	0.759, 0.755	0.5	72.8
$CHCl_3$	0,740, 0,744	0.5	71.3

Table 92 compares the results obtained for analysis of phenyllithium by the triphenyltin method and by double titration using various halides.

The halides giving the closest agreement with the yield of the tin derivative are 1,2-dibromoethane and carbon tetrachloride, 1,1,2-tribromo-ethane gives slightly higher values, andα,α,α-trichlorotoluene considerably higher.

The most probable reason for obtaining a titration value higher than the actual concentration is reaction of these organic halides with lithium alkoxide. Either 1,2-dibromoethane or 1,1,2-tribromoethane gave satisfactory analysis for a wide range of organolithium compounds using phenyllithium.

In Table 93 are shown typical titrations of n-butyllithium with both benzyl chloride and allyl bromide, being expressed as the percentage of the total basic content of the solution which is due to RLi. The first three runs are for preparations in diethyl ether, and it may be noted that the allyl bromide values are consistently 5% above those obtained with benzyl chloride. This difference corresponds to the error previously estimated for the benzyl chloride method[844]. The high percentage found with the allyl bromide titrations is taken to indicate that the n-butyllithium in the solution is more completely consumed by allyl bromide than by benzyl chloride. 1,2-Dibromoethane has also been recommended as a reagent in the double titration of cyclopropyllithium[852].

Phenyllithium may be determined by pipetting inorganic halide-free 1-bromo-2-phenylethane into a flask and adding pure di-n-butyl ether[853]. The flask is flushed with nitrogen and an aliquot of phenyllithium dissolved in pure di-n-butyl ether added. The flask is left to react, then 1.5 M nitric acid is added, and halide ion is determined by a modified Volhard procedure[854].

Table 92 - Analysis of phenyllithium. Solutions of phenyllithium in diethyl ether were analysed by double titration with several halides and by gravimetric determination of the tetraphenyltin formed on reaction with triphenyltin chloride

	None	$PhCCl_3$	$BrCH_2CHBr_2$	CCl_4	$BrCH_2CH_2Br$	C_2H_5Cl	CH_2=$CHCH_2Br$
Run 1	0.980	0.945	0.926	0.909	0.907	0.872	0.835
Run 2	1.07	1.02	0.972	0.966	0.962	0.924	0.870
Run 3	0.942	0.861	0.801	0.800	0.785	-	0.686

	Ph_3SnCl
Run 1	0.905
Run 2	0.961
Run 3	0.797

Table 93 - Double titrations of n-butyllithium. Solutions of n-butyllithium were analysed by double titration with 1,2-dibromoethane, allyl bromide, and benzyl chloride. The results are expressed as the percentage of the total basic content of the solutions which is due to C - Li

	Percentage of total base due to C-Li		
	$PhCH_2Cl$	CH_2=$CHCH_2Br$	$BrCH_2CH_2Br$
BuLi in Et_2O, run 1	88.1	93.6	
BuLi in Et_2O, run 2	90.1	95.1	
BuLi in Et_2O, run 3	76.3	81.2	
BuLi in hexane	94.7	98.6	
BuLi in hexane	97.1	99.2	
BuLi in hexane	97.0	99.1	99.1

The effect of lithium alkoxides on the determination of butyllithium (and butylsodium) compounds by the Gilman procedure was studied to establish whether any reaction between the halogen compound and any lithium alkoxide present had an effect on the accuracy of the lithium bound carbon determination[855]. Allyl bromide and 1,2-dibromoethane gave the most accurate results. When the Gilman double titration procedure is applied to tertiary lithium alkyls, then any tertiary lithium alkoxides present as an impurity in the sample react slowly with benzyl chloride or with 1,2-dibromoethane giving analyses for the alkyllithium component which are too high[856]. This may be overcome for the analysis of tertiary lithium alkyls through the observation that organic acids, including weak acids, can be titrated with sodium dimethyl sulphoxide in dimethyl sulphoxide solutions using diphenylmethane or triphenylmethane as indicator. The method involves the titration of a known amount of standard organic acid with the organolithium solution of unknown titre in dimethyl sulphoxide-monoglyme-hydrocarbon solution with triphenylmethane as indicator. Benzoic acid is used as titrant because of the relative ease of observation of the yellow to red (alkyllithium) or green to red-brown (phenyllithium) end-point and because a monoglyme solution of this acid can be standardized by an aqueous base titration. The overall reactions occurring in the system are as follows:

$$RLi + C_6H_5CO_2H \rightarrow C_6H_5CO_2Li + RH$$

and at the end-point when the standard acid is consumed:

$$RLi + (C_6H_5)_3CH \rightarrow RH + \underset{\text{(red)}}{(C_6H_5)_3C^-Li^+}$$

Analyses obtained by the Eppley and Dixon[856] and by the Gilman double titration procedures are compared in Table 94. These results show that, for tert-butyllithium, the values determined by the Gilman titration are significantly higher than those by the triphenylmethane titration. These difficulties are believed to be due to problems in the Gilman procedure resulting from analyzing hydrocarbon solutions and from the reaction of the alkoxides present with the 1,2-dibromoethane. Eppley and Dixon[856] demonstrated that lithium alkoxides did not interfere in their benzoic acid titration procedure.

The analysis of compounds of the type R_3ELi, where E = Si, Ge, or Sn, by a double titration procedure using allyl bromide and other organic halides as the reagent has been compared[850] with a method based on reaction of the R_3ELi compound with n-butyl bromide followed by titration of the released bromide ion by Volhard titration[857]. In general, better results were obtained using the allyl bromide double titration than with the n-butyl bromide Volhard analysis, triphenylgermanyllithium being an exception, (Table 95).

The present position regarding the applicability of the Gilman procedure to the assay of commercial alkyllithiums is that the ASTM has selected the Gilman benzyl chloride coupling procedure for the determination of n-butyllithium in hexane[858] and this can be taken as a measure of confidence in this procedure for this particular analysis. With experienced analysts, a reproducibility between two laboratories of about 0.2% is claimed. It has been pointed out[859-862] that, in spite of statements by earlier workers that the use of the benzyl chloride coupling reagent leads to low carbon-bound lithium values, it has frequently been possible to obtain as high as 99.2% of carbon-bound lithium (or 0.8% of non-carbon-bound lithium) on newly manufactured 15% hexane solutions of n-butyllithium. Obviously, avoidance of sample contamination by oxygen or moisture is a very important factor in obtaining these results. Thus, if benzyl chloride does give lower results for net assay (total base minus base left from Gilman coupling), the difference between 'actual' and assay values must be less than 0.8% of non-carbon-bound lithium. Results obtained by this procedure are therefore acceptably accurate. In a modification of the ASTM procedure used in Europe, a larger amount of benzyl chloride is used in the absence of ether. This method gives a result about 0.2% higher than that obtained by the ASTM assay. These comments apply strictly only to commercial organolithium preparations which contain a minimum amount of oxygenated impurities and do not necessarily apply to mixtures containing appreciable amounts of, for example, lithium alkoxides.

The ASTM method, using benzyl chloride, works well with phenyllithium provided that the coupling reaction is allowed to take place for at least 30 min. In most instances, allyl bromide can be substituted for benzyl chloride with no change in the analytical results.

Benzyl chloride does not work well with methyllithium or vinyllithium but allyl bromide does react very readily in these cases, giving consistent analyses. Owing to the very limited solubility of methyllithium in diethyl-ether, non-carbon lithium assays on solutions of 5% methyllithium in ether never exceed 0.03 - 0.05%, even if the true methoxide content of the sample is considerably higher than this.

Ethylene dibromide is recommended as a reagent for lithium alkoxides 862. It reacts with lithium t-butoxide and lithium n-butoxide in hexane, and this affects the Gilman coupling correction. Any comparison of the reactivity of a coupling reagent with lithium alkoxides should be made in the presence of an alkyllithium compound as the lithium alkoxide is actually coordinated in the alkyllithium hexamers and should be more reactive in this mixed system towards the coupling than it would be in the absence of the alkyllithium compound.

Table 94 - Titration of organolithium compounds with sodium dimethyl-sulphoxide. Comparison with Gilman method

Organolithium compound	Solvent	Alkoxide conc.	Concentration of organolithium (mol/1)	
			Eppley and Dixon (856)	Gilman (1,2-dibromo-ethane coupling reagent
Methyl	ether benzene heptane	0.581	0.451 ± 0.003	1.06 ± 0.02
sec-butyl	hexane heptane	0.365	0.292 ± 0.002	
tert-butyl	pentane benzene	0.042	0.329 ± 0.004	
tert-butyl	pentane decalin	0.345	0.966 ± 0.015	1.23 ± 0.02

Table 95 - Comparison of different methods of analysis of R_3MLi compounds. Solutions of R_3MLi compounds were analysed by the n-butyl bromide-Volhard method and by double titration with various halides. The results are expressed as percentage yield of R_3MLi compound in the preparation, and are average values of two or more determinations

R_3MLi compound	Percentage yield of R_3MLi Double titration halide used						Volhard analysis
	None	C_3H_5Br	$PhCH_2Cl$	$BrCH_2CH_2Br$	$PhCCl_3$	Br_2CHCH_2Br	
Ph_3SiLi	128.0	94.2	82.2	> 100	118.0	-	95.2
Ph_3SiLi	112.4	95.6	87.5	-	-	-	88.2
Ph_3SiLi	134.8	89.6	-	107.3	-	-	86.2
Ph_3SiLi	123.9	96.7	-	-	-	-	84.8
$Ph_2MeSiLi$	119.3	85.7	-	-	-	-	83.5
$Ph_2MeSiLi$	113.0	92.0	-	100.8	-	-	74.6
$PhMe_2SiLi$	91.2	52.7	-	70.0	-	-	-
Ph_3GeLi	117.7	80.0	-	-	-	-	84.6
Ph_3SnLi	117.2	94.2	89.5	90.3	-	-	86.1
Ph_3SnLi	118.3	94.1	75.0	81.9	110.2	86.9	74.8

A procedure has been described[863] for the determination of organolithium compounds based on cleavage of dialkyl or diarly disulphides and subsequent titration of the lithium mercaptide formed with silver ion by the silver nitrate amperometric technique[864]. Only lithium metal has been found to complicate the cleavage reaction;

$$R^1Li + RSSR \rightarrow R^1SR + LiSR$$

$$LiSR + [Ag(NH_3)_2]^+ \rightarrow RSAg + Li^+ + 2NH_3$$

Aromatic disulphides react rapidly and quantitatively with organolithium compounds in the presence or absence of ethers. A comparison of quantitative results obtained with tolyl disulphide and n-butyl disulphide in the presence and absence of ether showed good agreement. On the basis of the consistent results obtained in the presence or absence of ether with tolyl disulphide, this aromatic disulphide is recommended as the preferred reagent for the method.

Oxygen, water, and alcohols and the products of their reaction with organolithium compounds do not interfere with the disulphide cleavage procedure except for the destruction of a stoichiometric amount of the organometallic compound. The reactions with water[865] and with alcohols[866,867] are are bases for published procedures for analysing organolithium compounds. The reaction product with oxygen has been used in a procedure for analysing dilithioaromatic compounds[868]. In the analysis of organolithium compounds, substances such as lithium metal, lithium hydride, lithium hydroxide, and lithium alkoxides should be considered as possible interfering ingredients 863. Thus only lithium metal cleaved tolyl disulphide under the conditions of the analysis. Unreacted lithium metal can be readily determined[869].

Watson and Eastham[870] have described a procedure for the determination of normal, sec- and tert-butyllithium, based on titration with a standard solution of secondary butyl alcohol in xylene to the 1,10 phenanthroline or the 2,2'-biquinolyl colorimetric end-points. Addition of a few mg of 2,2'-biquinolyl and about 5 ml of 1.5 m butyllithium in hexane to 20 ml of benzene produces a yellow-green or chartreuse-coloured solution. After titration with 1 M sec-butyl alcohol in xylene, the solution is clear and colourless; the disappearance of the green colour occurs sharply after addition of one mol equivalent of titrant.

Equally good results in titrating butyllithium were obtained by forming the indicator with 1,10 phenanthroline, which gives a rust-red complex. Hydrocarbon solutions of sec-butyllithium and tert-butyllithium reagents were also satisfactorily analysed using 1,10-phenanthroline and direct titration with sec-butyl alcohol. Because they react with the reagents, ethers cannot be added to the titration mixtures with either indicator and organolithium compounds. However, there is no problem with turbidity and precipitation in the hydrocarbon titration mixture because the lithium sec-butoxide is quite soluble. If the reaction mixture is protected with an atmosphere of nitrogen, it remains bright and clear throughout titration. Reaction with n-butyl alcohol as titrant is also stoichiometric but the butoxide formed is insoluble.

Using 1,10-phenanthroline and butyllithium in benzene, Watson and Eastham[870], examined the reversibility of formation of the coloured charge-transfer complex which serves as indicator. After the end-point, addition of more lithium reagent to the titration mixture immediately regenerates the red colour. The additional reagent still requires just one mol equivalent of alcohol to come back to the end-point, which is indicated by loss of the red colour. Obviously excess alkoxide, a problem in many organometallic analyses, does not interfere with this direct titration.

A titration method utilizing sec-butyl alcohol has been developed suitable for the determination of alkyllithium compounds in ether solutions at concentrations down to 10^{-3} M[871]. This analysis causes considerable difficulty because at room temperature and even at 0°C, the presence of ethers, such as 1,2-dimethoxyethane and tetrahydrofuran, as solvents causes rapid decomposition of n-butyllithium, as indicated by titration with a standard solution of sec-butyl alcohol. If, however, the ether was cooled to -78°C prior to the addition of n-butyllithium and kept at this temperature during the titration with sec-butyl alcohol then n-butyllithium decomposition was slow. Rapid, accurate, and precise analyses were obtained in this way by titration with a standard solution of sec-butyl alcohol using 2,2'-biqüinolyl[870].

Vinyllithium can be determined by measurement of the amount of vinyltributyltin produced by reaction of the vinyllithium with tributyltin chloride[872] or the amount of tetraphenyllead obtained by reaction between tetravinyllead and phenyllithium[873,874]. Phenyllithium may be determined by measurement of the amount of tetraphenyltin produced in the transmetallation reaction between tetravinyltin and phenyllithium[874].

Organolithium compounds can be determined by iodination[875]. The organolithium compound is slowly added to an excess of a standardized diethyl ether solution of iodine and the unused iodine is back-titrated with standard sodium thiosulphate solution to the starch end-point. It is important to add the organolithium compound to an excess of the iodine solution, rather than the reverse. This ensures that interfering coupling reactions are minimized during iodination:

$$C_6H_5Li + I_2 \rightarrow C_6H_5I + LiI \text{ (iodination)}$$

$$2C_6H_5Li + I_2 \rightarrow C_6H_5 - C_6H_5 + 2LiI \text{ (coupling)}$$

This method was applied successfully to the assay of phenyllithium solutions, giving results which agreed to within 3% of the theoretical result. The method was also shown to be applicable to the assay of butyllithium.

B. Potentiometric Titration

p-Phenylenedilithium has been determined by potentiometric titration with cerium (IV) nitrate[868]. This method, which determines even small amounts of this substance in the presence of monometallo-organics, involves the oxidation and hydrolysis of the dilithium compound to form hydroquinone which is then titrated potentiometrically with standard cerium (IV) nitrate solution using a standard calomel electrode and a platinum reference electrode:

$$Li-C_6H_4-Li \xrightarrow{O_2} LiOO-C_6H_4-OOLi \xrightarrow{H_2O} LiO-C_6H_4-OLi \xrightarrow{H^+}$$

$$HO-C_6H_4-OH \xrightarrow{Ce^{4+}} O{=}C_6H_4{=}O + Ce^{3+} \quad (19)$$

As the method does not involve an acid-base titration, the presence of lithium hydroxide or other hydrolysis products does not interfere. Alkyl-lithiums and m-phenylenedilithiums give oxidation and hydrolysis products which cannot be oxidized by cerium (IV) ion; therefore, the method is selective for o- or p-phenylenedilithiums in the presence of other types of monometallo-organics.

A typical titration curve obtained by this procedure is shown in Fig. 33. As this method depends on the formation of quinone, any dimetallated aromatic which can be oxidized to a diphenol capable of being oxidized to a quinone should be assayable by the method. In this category fall either ortho or para dilithium and di-Grignard reagents as well as similar compounds of other active metals. In addition, compounds with the two metal atoms on different rings but in positions to form quinones should be analysable by this method. Compounds in this category include 2,2'-dilithio-diphenyl, 2,4-dilithiodiphenyl, etc.

As the method does not involve an acid-base titration, the presence of lithium hydroxide or other hydrolysis products does not interfere. Alkyl-lithiums and m-phenylenedilithiums give oxidation and hydrolysis products which cannot be oxidized by ceric ion; therefore, the method is selective for o- or p-phenylenedilithiums in the presence of other types of mono-metalloorganics.

n-Butyllithium in hydrocarbons has been determined by treatment with excess of vanadium pentoxide[843]. The reduced vanadium is then titrated potentiometrically with standard sulphatoaceric acid solution. A comparison or results obtained by this method with those obtained by a method for determining total alkalinity, including butyllithium, lithium butoxide, lithium hydroxide, and other basic materials, is shown in Table 96 for solutions of n- and t-butyllithium in various solvents[843]. In all cases, the concentration of butyllithium as determined by the vanadium pentoxide method is less than the concentration of total base. This is to be expected, since any air oxidation of n-butyllithium results in the formation of lithium n-butoxide, which is soluble in solutions of n-butyllithium. However, in every case the difference between the two values is 4% or less, which indicates the presence of only a small amount of soluble base other than butyllithium. The analysis of similar solutions of n-butyllithium in n-heptane by the double titration method indicated that 4 - 5% of the total base present was non-butyllithium base[844]. This blank was fairly constant and it was concluded, probably wrongly, that it was inherent in the double titration method under the experimental conditions used.

Only n-, sec-, and t-butyllithium and ethyllithium solutions were assayed by the vanadium pentoxide method, but the procedure should be generally applicable to the determination of any alkyllithium compound in a hydrocarbon solvent. It cannot be used for the determination of phenyl-lithium because, although phenyllithium rapidly reduces vanadium pentoxide, most solutions of phenyllithium contain lithium phenoxide because of air oxidation. On titration with sulphatoceric acid, the phenol is oxidized together with the reduced vanadium. This leads to high results for the phenyllithium content.

C. Thermometric Titration

n- and sec-butyllithium in hydrocarbon solution can be determined by

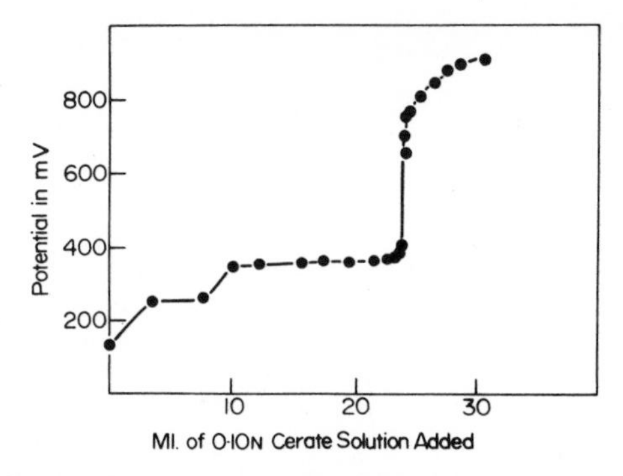

Figure 33 - Typical titration of oxidized and hydrolysed sample of p-phenylenedilithium using cerate solution.

Table 96 - Analysis of solutions of n- and t-butyllithium in various solvents

Compound	Solvent	BuLi by V_2O_5 method (M)	Total base (M)	Difference (%)
n-Butylithium	n-Heptane	1.67	1.70	1.8
	n-Heptane	2.74	2.80	2.1
	n-Heptane	1.23	1.24	0.8
	Cyclohexane	2.58	2.62	1.5
	Tolu-Sol	1.20	1.21	0.8
	Tolu-Sol	1.19	1.23	3.2
t-Butyllithium	n-Pentane	1.46	1.52	4.0

thermometric titration with a standard hydrocarbon solution of butyl alcohols (n^-, sec-, t-)[755]:

$$RLi + BuOH \rightarrow LiOBu + RH$$

The reaction is stoichiometric; lithium butoxide, normally the major impurity, does not interfere. The simplicity of the method makes it more rapid and convenient than many of the alternative methods, the method is believed to be generally applicable to compounds containing lithium - carbon bonds. The apparatus is discussed in the section on the analysis of organo-aluminium compounds[755,876]. A procedure for determining small amounts of n- or sec-butyl alcohol in the aqueous extrant from hydrolysis of butyllithium provides an independent estimate of the accuracy of the impurity correction in the double titration method (butoxide is usually the major impurity)[755]. Because of the high energy of reaction and the ease with which large samples can be handled, the sensitivity is high; the detection limit for a 50-ml sample is estimated to be well below 0.01% of butyllithium. The sensitivity of the vanadium pentoxide method appears to be comparable. Reasonable agreement is obtained between the butyl alcoholic thermometric titration method and the vanadium pentoxide method. Figure 34 shows a typical titration curve obtained in the non-aqueous titration of butyllithium. The heat of reaction, estimated from the initial slope of the curve is about 53 ± 5 kcal/mol. Because of the energetic reaction and the low heat capacity of the titration medium, temperature-rise during titration tends to be large; rises of 5° - 10°C are not uncommon. In consequence, the post-inflection

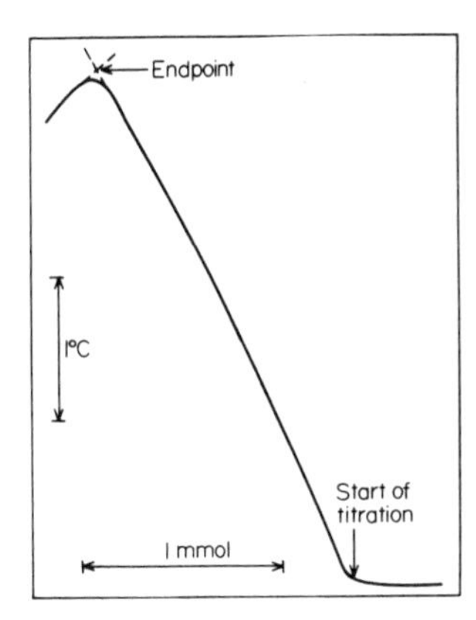

Figure 34 - Titration of sec-butyllithium with 1 M butanol.

segment of the curve generally shows a temperature decrease and the inflection is well-defined. Because of the high energy of reaction and the ease with which large samples can be handled, sensitivity is high; the detection limit for a 50 ml sample is estimated to be well below 0.01% butyllithium. The sensitivity of the vanadium pentoxide method of Collins et al [844] appears to be comparable; the double titration method[858,840] is considerably less sensitive.

Table 97 shows comparative values for samples of sec-butyllithium which were analysed by the double titration method[858,840] of the vanadium pentoxide method (Collins et al[844]) and the butanol thermometric titration method; values for the butoxide content (by dichromate oxidation) are also given. The results indicate reasonable agreement between the butanol-thermometric titration method and the vanadium pentoxide method. The double titration method gave high and erratic values for the impurity correction and correspondingly low values for butyllithium. Dichromate oxidation showed that the actual butoxide content was considerably lower than the values obtained for the benzyl chloride correction. If the lowest of these values (0.19%) be taken as most nearly correct, it can be concluded that the metallic lithium and lithium hydride content (expressed as equivalent butyllithium) is probably not over 0.1%. The difference between the butanol titration value and total lithium minus butoxide is of this magnitude (0.09%) and would also include lithium oxide if present.

Experience indicates that the butanol-thermometric titration method gives precise results, duplicate determinations seldom differing by more than 1% of their mean. Repeatability is largely a function of the operator's success in avoiding extraneous losses from oxidation and hydrolysis during sample handling prior to titration. The thermometric titration method was applied to both n- and sec-butyllithium solutions, n-butanol has been used as titrant in most cases, but sec- and tert-butanol work equally well.

D. High-frequency Titration

High-frequency titration of various alkyllithiums as well as benzyl and phenyllithium compounds with a standard solution of acetone in benzone has a sensitivity of 0.01% of organolithium compound when a 50 ml sample is used [877]. Lithium alkoxides do not interfere. Acetone is used as a titrant because it is relatively easy to obtain pure and dry, and also because under ordinary

Table 97 - Comparative values for sec-butyllithium by several methods

	sec- Butyllithium, wt. %		
Hydrolysis-acidimetric (total Li, uncorrected) (Gilman and Haubein, 1944)	Sample A 12.80 12.83	Sample B* (12.82)	Sample C* 0.845 0.850
Benzyl chloride correction (Gilman and Haubein, 1944)	0.39#		0.11
Difference (value by double titration)	12.43#		0.73
Equivalent LiOBu (dichromate oxidation)	0.04 0.06		0.020 0.021
Hydrolysis value minus Li butoxide value	12.77		0.83
Vanadium pentoxide oxidation (Collins et al)	12.37 12.50	11.74 11.88	0.80
Thermometric titration with butanol (Everson et al)	12.66 12.70	11.88	0.85 0.83 0.84 0.83

* Sample B is Sample A after standing for about a month; some deterioration is evident. Sample C is a separate dilution of Sample A.

\# Average value for five determinations of the benzyl chloride correction, ranging from 0.19 to 0.69, BuLi value using lowest value of correction, 12.63%, Manufacturer's value, 12.68%.

conditions its reaction with lithium alkyls is rapid, complete, and irreversible.

$$RLi + (CH_3)_2C = O \rightarrow (CH_3)_2\overset{\displaystyle R}{\overset{|}{C}} - O^-Li^+$$

The results obtained by high-frequency titration generally agreed within 1% with results obtained by the vanadium pentoxide method[843], and for phenyllithium compounds the high-frequency titration results were in good agreement with those obtained by the double titration procedure[840] and gave a particularly good precision (0.5% agreement) when ethylene dibromide rather than the benzyl chloride reagent was employed in the latter procedure. Gaseous products do not form in the course of the reaction, thus eliminating the necessity of venting the cell during the titration. Most important, acetone is sufficiently polar to provide very sensitive detection of its presence in excess of the end-point. A typical titration curve is shown in Figure 35.

The results for analyses by the Watson and Eastham[877] acetone titration procedure of five different organolithium reagents are shown in Table 98. The shape of a typical high frequency titration curve, shown in Figure 35 is of some interest. After the end-point, in every case a sharp continuous linear increase of instrument reading was found. This continuous linearity afterward, and the approximate linearity before the end-point, makes the

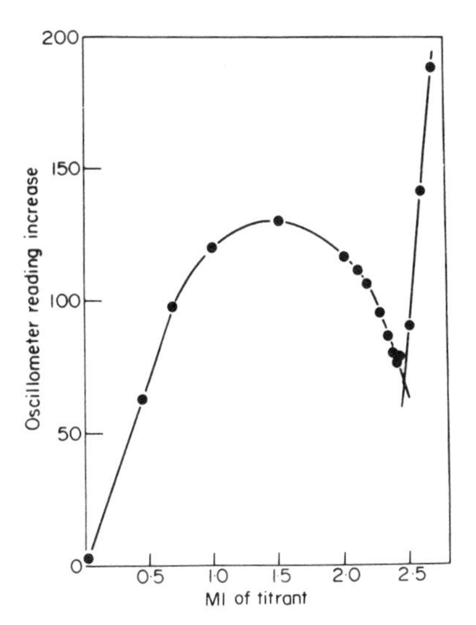

Figure 35 - High frequency titration curve of 1.412 g of sec-butyllithium-hexane solution with 0.995 M acetone.

end-point well defined in every titration. The same overall curve shape prior to the end-point shown in Figure 35 was found with all the butyllithium isomers. This curve is thought to be a result of increased alkyllithium polarity caused by its complexation with the newly formed alkoxide. The maximum in the pre-end point curve always appears around two-thirds of the way to the end-point. Complexes of bases with alkyllithium are known to be more polar than the associated alkyllithium itself. Neither lithium n-butoxide nor diethylether had any adverse effect on the determination of n-butyllithium by this procedure. Alkoxides do not precipitate out during the course of high frequency titrations with acetone. For phenyllithium compounds high frequency titration results were in good agreement with results obtained by the double titration procedure (Gilman and Haubein [840,858], Gilman and Cartledge[842], Gilman et al[839] and gave particularly good precision (0.5% agreement) when ethylene dibromide rather than the benzyl chloride reagent was employed in the latter procedure.

E. Lumometric Titration

A 2% toluene solution of n-butyllithium has been titrated lumometrically with air[878]. A very sharp increase in light intensity occurred at a point corresponding to one atom of oxygen for two atoms of lithium. Beyond this point no more oxygen was absorbed but light emission continued at a gradually decreasing rate for more than 24 h.

ORGANOMAGNESIUM COMPOUNDS

A. Classical Titration Procedures

Grignard reagents cannot be determined by titration with a standard solution of iodine in diethyl ether until a pale iodine colour persists[879].

$$RMgX + I_2 \rightarrow RI + MgXI$$

Table 98 - Results from high frequency titrations with acetone

	Wt. %	Mean	Av % Dev	Wt. % check method
n-Butyllithium in hexane	14.15			
	14.20			
	14.14	14.13	0.25	13.95*
	14.08			
	14.10			
sec-Butyllithium in hexane	11.38			
	11.40			
	11.35	11.31	0.70	11.17*
	11.22			
	11.20			
tert-Butyllithium in pentane	14.85			
	14.94			
	14.93	14.92	0.27	15.02*
	14.90			
	15.00			
Benzyllithium in c. 95:5 benzene:tetrahydrofuran	5.68			
	5.72			
	5.70	5.70	0.35	5.74*
	5.66			
	5.73			
Phenyllithium in c. 70: 30 benzene:ether	16.19			
	16.06			
	16.18	16.14	0.30	16.08#
	16.08			
	16.19			

* Average of five oxidimetric analyses
# Average of five modified double titration analyses.

owing to the occurrence of a side-reaction simultaneously with the above reaction[880,881]:

$$2RMgX + I_2 \rightarrow R-R + 2\ MgXI$$

A satisfactory analysis can be obtained by adding an aliquot of the Grignard solution to an excess of standard iodine solution, that is in the reverse order of that above, and then titrating the excess of iodine with standard sodium thiosulphate solution[882,883]. The accuracy of the iodometric method has been checked by adding known amounts of water or methanol to a Grignard reagent and checking how much the titre ran back. It is claimed to be accurate to within 1%[884]. It has been suggested[885] that methods based on acid titration[880,882,886,887] on iodine titration[879], and on Volhard titration[885] give high results. Several alternative methods for the analysis of Grignard compounds which can also be applied to diaryl magnesium compounds are described below.

Titration with sulphuric acid[880,882,886,887]. Procedure - Add 1 - 2 ml of the organomagnesium sample to water. Boil for 10 - 15 min and add, after cooling, a known excess of 0.100 N sulphuric acid. Back-titrate excess of acid with 0.1 N sodium hydroxide using phenolphthalein as indicator.

Iodine titration[879]. Procedure - Add 1 - 2 ml of organomagnesium sample to 1 N iodine solution in dry diethyl ether. Let the reaction proceed for 10 min with shaking. Determine the unreacted amount of iodine by titration with 0.1 N sodium thiosulphate using starch as indicator.

Volhard halogen titration method[888]. Procedure - Add 1 - 2 ml organomagnesium sample to water, boil until the solution is clear and adjust the pH to about 7 with sulphuric acid. Add a known excess of 0.1 N silver nitrate and heat the solution until the precipitate coagulates. Determine in the cooled solution the excess of unreacted silver nitrate with ammonium rhodanide using iron (III) sulphate as indicator.

Di-sec-butylmagnesium, i-butylmethylmagnesium, phenylmagnesium, and n-butylmagnesium chloride can be titrated directly under anhydrous conditions with sec-butyl alcohol using coloured indicators such as 1,10-phenanthroline or 2,2'-biquinolyl[889]. In a typical titration, a 1,10-phenantholine is added to an ethereal solution containing approximately 0.1 M dialkylmagnesium to obtain a violet solution. Titration of the solution with standard 1 M sec-butyl alcohol in xylene causes no significant diminution in colour until two molecules of the titrant have been added per mole of magnesium compound, when the violet colour disappears sharply. End-points are sharper in ethereal solutions of organomagnesium compounds than in hydrocarbon solutions. Also, in hydrocarbon solution, precipitation of magnesium alkoxides may cause turbidity problems. Both diethyl and di-n-butyl ether were used as solvents for direct titration of butylmagnesium chloride. An analysis of a sample of this Grignard reagent, claimed to be about 2.8 M, showed it to be 2.73 M. Phenylmagnesium chloride was also analysed satisfactorily by direct titration.

An iodometric procedure for the determination of organoalkali metal compounds such as phenyllithium may also be applied to organomagnesium compounds[890]. A procedure for the iodometric determination of arylmagnesium compounds adds a measured volume of a chlorobenzene or anisole solution of the organomagnesium sample to a solution of iodine in benzene, toluene, or diethyl ether[891]. The iodine must be in three-fold excess with respect to the arlymagnesium compound. The solution is set aside for a few minutes at room temperature, then excess of iodine is titrated.

B. Amperometric Titration

A procedure for determining organomagnesium compounds is based on their cleavage of a dialkyl or diaryl disulphide[892]. The resulting thiol is titrated amperometrically in alcoholic medium with aqueous silver nitrate solution in the absence of air.

C. Potentiometric Titration

The organomagnesium sample is added to a 20 - 30% excess of a 1 N acetone solution in dry diethyl ether and then hydrolysed with methanol[885]. A solution of hydroxylamine formate in methanol is added and the free hydroxylamine formate is titrated potentiometrically with standard perchloric acid solution in dry dioxane. A procedure has also been described for overcoming interference by basic magnesium compounds in this method[885].

ORGANOMERCURY COMPOUNDS

A. Classical Titration Procedures

Phenylmercury acetate in an aqueous acidic solution at Ph 2 - 2.5 can be titrated with a carbon tetrachloride solution of copper diethyldithiocarbamate (1.5×10^{-4} M)[897]. The end-point is reached when the solvent layer becomes pale yellow-brown in colour. Non-aqueous titration with hydrochloric acid (0.1 N) in n-butanol medium has been used for the determination of phenylmercury acetate[898]. Thymol blue or diphenyl carbazide can be used as the indicator, except in the presence of basic compounds when only diphenyl carbazide is suitable. In an alternative non-aqueous titrimetric procedure the sample is dissolved in anhydrous acetic acid to acetolyse it and the reaction product is then titrated with 0.1 N perchloric acid dissolved in acetic acid to the p-naphtholbenzein or quinaldine red indicator end-point[899]. The principle of this method is as follows. The mercurial used in pharmacy can be represented by $R\text{-}Hg\text{-}R^1$ in which one but not both of the bonds may be a carbon mercury bond. These compounds undergo solvolysis in acetic acid, as follows:

$$R\ Hg\ R^1 + HOAC \rightarrow R\ Hg\ OAC + R^1H$$

$$R\ Hg\ OAC + HOAC \rightarrow Hg(OAC) \ + RH$$

The reaction conditions and the nature of the R group determines whether the reaction is essentially stopped with the formation of the monosubstituted mercuric acetate or proceeds to mercuric acetate. If an excess of a halide salt of a base (e.g. methylaminehydrochloride) is added to the solvolysis reaction mixture, a quantity of strong base will be released which is equivalent to the acetate bound to mercury:

$$R\ Hg\ OAC + BX \rightarrow R\ HgX + BOAC$$

$$Hg(OAC)_2 + BX \rightarrow HgX_2 + 2BOAC$$

This base is titrated with starch and aqueous perchloric acid to give a measure of the original amount of organomercury compound present.

In a rapid volumetric procedure for the micro-determination of phenylmercury acetate, a weighed amount of phenylmercury acetate is dissolved in warm acetone and diluted with acetone[900]. To an aliquot is added sodium chloride and the solution is heated on a water-bath to evaporate the acetone. To the cooled solution is added nitric acid and ethanolic diphenyl carbazone, and the unconsumed sodium chloride is titrated with 0.02 N mercury (II) nitrate. The error of this method is less than ± 1%. In the case of fungicidal mixtures containing both phenylmercury acetate and organomercury halides, this method is combined with an argentimetric method[901], analysing separate sample aliquots by each method. In an alternative procedure for the determination of mercuriacetic acid in phenylmercury acetate, the sample is boiled with water and acidified with acetic acid[902]. Sodium chloride solution is added and the solution is diluted with water, then filtered. The filtrate is made alkaline with ammonia and the solution is titrated with sodium mercaptoacetate in the presence of thiofluorescein indicator.

Lanbie[903] has described a method for the identification and determination of the fungicide methoxyethylmercury chloride. This substance reacts with hydrochloric acid to produce mercuric chloride and ethylmethyl ether;

$$CH_3OC_2H_4HgCl + HCl \rightarrow HgCl_2 + CH_3OC_2H_5$$

The addition of aqueous potassium iodide to an aqueous solution of the sample produces a white precipitate which, upon the addition of concentrated hydrochloric acid, turns yellow and then forms red-coloured mercuric iodide. The subsequent addition of aqueous sodium hydroxide and aqueous ammonia produces an orange precipitate. These colour changes serve as a qualitative test for methoxyethylmercury chloride. To determine methoxyethylmercury chloride, the sample (1 g) is dissolved in aqueous hydrochloric acid (2 ml) and concentrated nitric acid (0.2 ml). The solution is boiled, cooled and diluted to 100 ml. A 20 ml aliquot is mixed with aqueous ammonia (10 ml) 0.1 N potassium cyanide (10 ml) aqueous 10% potassium iodide (1 ml) and water (70 ml). Mercury is then determined by titration with 0.1 N silver nitrate.

Phenylmercury halides in technical fungicides may be analysed by dissolving the sample in dimethylformamide at room temperature and adding sodium hydroxide to make the solution blue (alkaline) to thymolphthalein [904,905]. Water is then added and the alkali titration is continued until the blue colour returns. This sequence of titrations is continued until the blue colour persists upon addition of water. Phenylmercury halides do not precipitate out under these conditions. Chloride resulting from the hydrolysis of phenylmercury chloride is then determined by argentimetric titration (0.05 - 0.1 N silver nitrate) to the potassium chromate end-point. The procedure has been applied to pheylmercury chloride and bromide and to the determination of phenylmercury halides in formulated fungicides containing powdered talc[905].

For the determination of ethylmercury chloride in technical products and in compounded products used as fungicides, the sample is dissolved in cold dimethylformamide and neutralized to thymolphthalein with 0.1 N sodium hydroxide to the pale blue end-point[906]. The chloride ion is titrated with 0.05 N silver nitrate in the presence of 5% potassium chromate.

In a procedure for the determination of methylmercury salts in rat tissue and rat urine, the tissue is first homogenized with benzene and the extract digested with aqueous sodium sulphide and then oxidized with potassium permanganate[907]. Following decolorisation with hydroxyammonium chloride, addition of urea and EDTA, and pH adjustment to 1.5, the solution is mixed with chloroform and titrated with standard dithizone solution until the colour of the chloroform layer is intermediate between the orange of the mercury complex and the green of the dithizone solution. Concentrations down to 1 ppm can be measured. Inorganic mercury does not interfere.

Phenylmercury compounds in paints and in fungicidal preparations have been determined by an iodometric method[908]. Microbiological assaying of mercurials in pharmaceutical products such as phenylmercury compounds in amounts down to 2 - 10 ppm have been discussed[909]. A method for the determination of mercury in biological tissue is based on electrolytic decomposition of mercury from solution followed by titrimetric determination of mercury[910]. Organomercury compounds may be determined by formation of the S-organomercury derivative by reaction with excess of 2-mercaptobenzothiazide[911]. Excess of thiol is determined iodometrically.

B. Potentiometric Titration

A rapid volumetric determination of halogenated organomercury compounds is based on digestion of the sample with suitable solvents followed by argentimetric titration of the halogen, either potentiometrically or with potassium chromate as indicator[904]. Suitable solvents for use at room temperature are dimethylformamide and dioxane, and when higher temperatures are needed. methanol and ethanol may be used.

Non-aqueous titrimetry has been used for the determination of organomercury compounds such as phenylmercury nitrate, phenylmercury acetate, mercury succinamide, Thiomersal, Nitromersol, and Meralluride based on acetolysis of the sample with glacial acetic acid followed by titration (potentiometric) with a standard solution of perchloric acid dissolved in anhydrous acetic acid to the p-naphthalbenzein or the quinaldine red end-point[899].

C. Coulometric Titration

Coulometric titration has been applied to the determination of various organomercury compounds[912]. A 0.05 M solution of mercury (II) thioglycollate in acetate buffer (pH 5) can be used as generating solution with platinum and mercury electrodes. Oxygen is removed from the sample by nitrogen purging. Mercury (II) ions are generated, then the current is reversed and titration carried out with thioglycollic acid until a point of maximum potential inflection is reached (using silver amalgam and saturated calomel electrodes). The organomercury sample is added and thioglycollic acid generated to the same potential end-point as obtained previously. Samples that contain an Hg - C bond must first be heated with concentrated hydrochloric acid and brought to pH 5. The standard deviation was in the range 0.001 - 0.004 mg for 0.5 mg samples.

ORGANOPHOSPHORUS COMPOUNDS

A. Esters of Phosphoric Acid

Wilkinson and Williams[913] have described a potentiometric titration method for the estimation of di-hydrogen-n-butyl phosphate and hydrogen di-n-butyl phosphate. Solutions of tributyl phosphate and other phosphoric acid esters in kerosene and other hydrocarbon solvents have been used extensively in the atomic energy industry for the extraction of uranium from the aqueous into an organic phase. Methods for the determination of phosphoric acid esters in such hydrocarbon solutions are discussed below.

Sant and Sankar[914] transfer 2 ml of the kerosene solution to a nickel crucible, and to this is added sufficient powdered sodium hydroxide to absorb it and then further alkali to cover it. The crucible is covered and heated very slowly so that in about 1 h the mass becomes a homogeneous liquid. The cooled mass is dissolved in dil perchloric acid and diluted to 250 ml. A 100 ml portion is neutralized to litmus and 6% (v/v) perchloric acid (about 4 ml) added. The solution is then titrated with standard bismuthyl perchlorate solution using a saturated solution of di(allylthiocarbamoyl)hydrazine in chloroform as extraction indicator.

Ashbrook[915] has described a method for the determination of alkyl phosphates and amines in kerosene solutions containing uranium and sulphuric acid. The uranium is removed by extraction with dilute sulphuric acid. The excess sulphuric acid is removed by calcium acetate for the amine determination and by sodium methoxide for the alkyl phosphate determination. The sample (30 ml) is extracted with 20% v/v sulphuric acid (2 x 20 ml), and the organic phase filtered. A 5 ml aliquot of this solution is dissolved in chloroform; 0.5% (w/v) of thymol blue in ethanol is added then 0.1 N sodium methoxide until the mixture is deep blue. The excess sodium methoxide is then titrated with 0.1 N perchloric acid solution in dioxan to a blue-yellow end-point. This solution is then titrated with 0.1 N perchloric acid to a yellow-purple end-point (titre A equivalent to amine plus alkyl phosphate). To the remainder of the uranium-free filtrate is added calcium acetate (5 g) and the solution shaken for 1 min then filtered. A 5-ml aliquot in chloroform is titrated with 0.1 N perchloric acid to a yellow-purple end-point using thymol blue as indicator (titre B equivalent to amine). The difference between titres A and B is equivalent to the alkyl phosphate content of the sample. Tributyl phosphate content can be calculated from the difference between the total phosphorus in solution and the phosphorus equivalent to alkyl phosphate.

Pukhonto et al[916] have described a procedure for the simultaneous determination of butyl hydrogen phosphate, tributyl phosphate and kerosene in aqueous solution. Butyl hydrogen phosphates are determined indirectly by titration (with EDTA solution and xylenol orange as indicator) of residual zirconium after extracting a known initial amount of zirconium (in M perchloric acid medium) into butyl hydrogen phosphate solution in chloroform. The accuracy is within ± 10%, and the sensitivity for butyl hydrogen phosphate is 1 $mg l^{-1}$.

The acid-alkali titration procedure described by Bernhart and Rattenbury[917] can be applied to the determination of acid esters of phosphoric acid and of phosphoric acid but trialkyl phosphates are not determined. If esters of pyrophosphoric acid are present the results obtained are difficult to interpret. Chlorinated esters, e.g. 2-chloroethyl phosphate, lose chloride upon reaction with alkali, leading to erroneous results.

Geiger and Furer[918] have described methods for the determination of 2,2-dichlorovinyl dimethyl phosphate (DDVP). For the macro-determination this substance is hydrolysed to dichloroacetaldehyde and dimethylphosphoric acid with 0.4 N sodium hydroxide in the presence of 0.02 N iodine; the amount of iodine consumed in oxidizing dichloroacetaldehyde to dichloroacetic acid is a measure of the DDVP and is determined by titration with sodium thiosulphate. Alternatively, DDVP is hydrolysed with a measured excess of alkali at 0 - 1°C (no iodine present) and the excess is determined by back-titration with acid; interference by dichloroacetaldehyde is avoided by careful adjustment of conditions. The accuracy is within ± 0.5%. For the micro-determination, an aqueous solution of DDVP is treated with alkaline resorcinol solution and the colour is measured at 480 nm after 4 h at 20°, or 45 min at 40°C. This method is sensitive down to 5 µg DDVP, and is accurate to within ± 5%. Dichloroacetaldehyde interferes in this method.

Braundsort[919] discussed a scheme for the identification and rough estimation of tri-o-tolyl phosphate and similar plasticisers in plastics. An ether extract of the plastic is made and tested for refractive index at 20°C, phosphorus content, and saponification value. Qualitative tests are made after hydrolysis with ethanolic potassium hydroxide, acidification, and distillation to detect the presence of cresols.

Lee and Ting[920] determined mono and dilutyl phosphate in kerosene by gas chromatography of a solvent extract.

B. Esters of Phosphorous Acid

Berhart and Rattenbury[917] have described a acid-alkali titration for the estimation of dialkyl and trialkyl phosphites in the presence of each other. In this method the dialkyl phosphites are hydrolysed to sodium monoalkyl phosphites in alcoholic sodium hydroxide; 1 mol per mol is consumed within 1 min whereas trialkyl phosphites are not hydrolysed, under these conditions, within 10 min. In an acidic medium trialkyl phosphites are hydrolysed to dialkyl phosphites. The method based on these principles was applied to esters ranging from butyl to octyl phosphites. Alcohols, amine hydrochlorides and dialkane phosphonates do not interfere.

Yarden and Eger[921] have described an iodometric method for the estimation of alkyl phosphites in the presence of other organophosphorus compounds. In this method 10 ml of 0.2 N iodine solution in pyridine is added to a portion of the sample containing 50 to 150 mg of dialkylphosphite. The flask is kept stoppered for 30 min in the dark then 20 ml of peroxide-free dioxan and 5 N hydrochloric acid (24 ml in small portions) are added with cooling. The solution is then titrated with 0.1 N sodium thiosulphate until the colour changes to yellow. Water (75 ml) is added, then 1% starch (5 ml), and the titration completed. The difference between this titration and a reagent blank titration is obtained and referred to a reference curve prepared by applying the procedure to known amounts of the pure dialkyl phosphite, (this procedure overcomes the non-stoichiometry of the reaction). Thompson[922] has described a method for titrating trialkylphosphites with a methanolic solution of iodine in the presence of pyridine.

Siegfried[923] has described a procedure for the determination of acid POH groups of hydrolysis of acid susceptible groups of phosphorous acid. The procedure has an accuracy of about 1%.

C. Esters of Phosphonic, Phosphinic and Phosphorous Acids

Kreshkov et al[924] determined methylphosphonic acid and its derivatives such as isobutyl hydrogen methyl phosphonate and methyl phosphonic dichloride by non-aqueous titration with tetraethyl ammonium hydroxide or sodium methoxide in methyl cyanide, ethyl methyl ketone or benzene medium. The end-point was determined visually with 1,4-dihydroxyanthraquinone or thymolphthalein as indicator or potentiometrically. Methods are described for determining the above compounds individually or in mixtures.

Kreshkov et al[925] have also discussed the non-aqueous titrimetry of the monoisobutyl ester of methyl phosphonic acid and the diisobutyl ester of dimethylpyrophosphorous acid in mixture.

Potentiometric titration has been used by Jasinki et al[926] for the estimation of dibutyl, diphenyl, dibenzyl bis-(4-dimethylaminophenyl), and di-p-tolyl-phosphinic acids. The sample was dissolved in methanol, dimethylformamide, dimethyl sulphoxide and pyridine media and was titrated with 0.05 N-methanolic sodium methoxide (glass-SCE system). Good titration graphs were obtained. The potential change at the end-point varied between 60 and 120 m V with pyridine and from 180 to 360 m V for the other solvents. Dibutylphosphinic acid showed the smallest potential change in all the solvents. The relative error was less than 2% and the precision ranged between 0.34 and 1.1 mg for 25 mg samples.

D. Esters of Pyrophosphoric Acid and Pyrophosphonic Acid, Phosphoronahydrides, Alkyl-phosphonochloridates, Dialkylphosphorochloridates and Alkylphosphonofluoridates

Sass et al[927] developed the methods for the estimation of pyrophos-phosphonates, pyrophosphorous esters and phosphoroanhydrides and also for alkyl phosphonochloridates, dialkyl phosphorochloridates and alkyl phos-phonofluoridates, based on the formation of perphosphorous acid by reaction of the organophosphorus compound with excess sodium pyrophosphate peroxide and subsequent iodometric estimation of excess peroxide.

$$R-\overset{\overset{O}{\uparrow}}{\underset{\underset{OR'}{|}}{P}}-X + OOH \xrightarrow{PH10} R-\overset{\overset{O}{\uparrow}}{\underset{\underset{OR'}{|}}{P}}-OOH + X^-$$

$$R-\overset{\overset{O}{\uparrow}}{\underset{\underset{OR'}{|}}{P}}-OOH \xrightarrow{OOH^-} O_2 + H_2O + R-\overset{\overset{O}{\uparrow}}{\underset{\underset{OR'}{|}}{P}}O^-$$

where X = phosphorous group, as in pyrophosphate and pyrophosphonates, or chlorine or fluorine as in halidates; R = alkyl or alkoxy and R' = alkyl.

When an excess of peroxide is present in alkaline solution, the rate of peroxidation of pyrophosphonate or fluoridate is more rapid than the rate of hydrolysis. The presence of 5% alcohol further increases the stability to hydrolysis. The wet method (i.e. 50% aqueous 2-propanol) in recommended for the determination of alkyl pyrophosphonates, because this method precludes interference from chloridates, dichloridates and difluoridates when present in the alkylpyrophosphonates which hydrolyse rapidly in water to produce acids which do not interfere with the peroxide method. In the absence of chloridates and difluoridates in the sample, therefore, the original dilution of pyrophosphorous ester (or alkyl phos-phorous fluoridate) can be made with anhydrous 2-propanol instead of 50% aqueous 2-propanol as used above.

A suitable weight of sample (0.18 - 0.24 g) is weighed into a glass ampoule which is then placed in a 100 ml volumetric flask containing approximately 20 ml of 50% aqueous 2-propanol. The ampoule is broken with a glass rod, the flask swirled and the solution diluted to 100 ml with 50% aqueous 2-propanol. Peroxide reagent (8.5 - 9 g $Na_4\ P_2O_7H_2O$; 5.68 $Na_4\ B_2O_7 10H_2O$ in 500 ml water, adjusted to pH 10 with sodium hydroxide and made up to 1 litre with water) (50 ml) is pipetted into 500 ml iodine flasks, one of which is used as a blank which should be run with each series of samples. Sample solution (20 ml) is pipetted into the iodine flask with continuous swirling. The flask is stoppered and left for 2 - 4 min with occasional swirling, and 50% aqueous 2-propanol (20 ml) is added to the blank flask. To each flask is added water (30 ml), 18 N sulphuric acid (10 ml) and potassium iodide (3 g) in that order with mixing. The solutions are left in the dark for 10 min and liberated iodine titrated with 0.1 N sodium thiosulphate solution. Calculate the alkyl pyrophosphonate content of the sample from the difference between the sample and blank titration using one fourth of the alkyl pyrophosphonate molecular weight as the equivalent weight. The results in Table 99 illustrate the negligible effect of the presence of 20% of various impurities on the analysis of a sample of diiso-propyl dimethylpyrophosphonate.

Beach and Sass[928] have studied the hydrolysis characteristics of organo-phosphorus pyroester compounds and ester phosphorus fluoridates using standard acid-base reagents and have developed a simple, rapid and accurate method of assay of these types of compounds. The method is based on the large difference between the slow hydrolysis rates of these compounds in

0.1 N sodium hydroxide as opposed to the rapid hydrolysis rate of impurities present in pyroester compounds (such as other phosphorus acids) and in phosphoro and phosphonohalidates, (such as chloridates and difluoridates).

Table 99 - Effect of impurities on the analysis of diisopropyl dimethyl pyrophosphonate (bis (isopropyl methyl phosphonic) anhydride) by the sodium pyrophosphate peroxide method.

Impurity 20% (w/w) added	Sample	Sample (g)		Recovery (%)
		Added	Rec.	
Methyl phosphonic acid (dihydroxymethyl) phosphonate	Bis(isopropyl methyl) phosphonic anhydride)	0.2127	0.2116	99.5
		0.2185	0.2174	99.5
		0.2309	0.2294	99.3
		0.2017	0.2004	99.3
Bis(methyl phosphonic) anhydride	Bis(isopropyl methyl) phosphonic) anhydride	0.2064	0.2062	99.9
		0.2386	0.2372	99.4
		0.2124	0.2108	99.2
		0.2144	0.2137	99.7
Isopropyl hydrogen methyl phosphonate	Bis(isopropyl methyl) phosphonic anhydride)	0.2107	0.2098	99.5
		0.2118	0.2113	99.8

Beach and Sass[928] were able to distinguish between isopropyl methylphosphonofluoridate (GB Sarin) and bis(isopropyl methyl phosphonic)-anhydride, both of which show similar hydrolysis rates, by separating them on a silica gel column and subsequent hydrolytic titration to determine the GB Sarin[929]. The pyroester compounds (i.e. the anhydride) is estimated by difference from a total hydrolytic determination carried out on the sample both before and after separation on the silica gel column.

The hydrolytic method which has been applied to tetraethyl pyrophosphate and to its homologues and to pyrophosphonates is essentially an acid-base titration. By titrating rapidly to a methyl red end-point with 0.1 N sodium hydroxide, all free acid and acids formed on hydrolysis of impurities are neutralized. An excess of base quantitatively hydrolyses the pyroesters within 2 minutes. The quantity of base consumed in the hydrolysis is obtained by back-titration with 0.1 N hydrochloric acid to the methyl red end-point. Beach and Sass[928] tested the effect of the presence of acid on the stability of the pyroesters during the two stages of the titration. If the initial or free acid titration is accomplished within 15 s even with as high as 10% by weight of free acid, little apparent loss through acid-catalysed hydrolysis is observed for the pyroesters determined in the second or hydrolytic titration (Table 100).

Mixtures of bis(isopropyl methyl phosphonic) anhydride and its most common impurities were analysed using the hydrolytic method. The estimated purity of the test substance used as a comparison was arrived at by the macro peroxide analytical method. Table 101 shows that when the initial titration (free acid) is completed with 15 s, no real effect was observed on the percentage recovery (apparent purity). In these experiments, 2 min in the presence of base above pH 11.5 was allowed for the second (hyrolytic) phase of the titration.

Table 100 - Effect of free acid in aqueous hyrolytic titration of bis (isopropyl methyl phosphonic) anhydride ($CH_3P(=O) - CH - (CH_3)_2)_2O$

Weight (g)	Additive acid[a] (g)	Time of standing during initial titration (sec)	Purity for compound (%)[b]		Apparent loss (%)
			Original	Found	
0.15	0.001	15	96.4	96.3	0.1
0.15	0.035	15	96.4	96.4	-
0.20	0.035	30	96.4	96.3	0.1
0.20	0.035	50	96.4	95.9	0.5

[a] Equimolar mixture of isopropyl hydrogen methyl phosphonate and hydrogen chloride.
[b] Purity results calculated on an impurity-free basis.

Table 101 - Effect of phosphorus impurities on analytical recovery of diester pyrophosphates (bis(isopropyl methyl phosphonic) anhydride), hydrolytic method

Impurities added (% w/w)	Compound	Sample (g)		Recovery (%)
		Added	Recovered	
Isopropyl methyl phosphonic acid	Bis(isopropyl methyl phosphonic) anhydride			
5		0.1348	0.1345	99.8
		0.1748	0.1780	99.8
10		0.1298	0.1295	99.8
		9.1684	0.1679	99.7
Isopropyl methyl phosphonochloridate	Bis(isopropyl methyl phosphonic) anhydride			
4.2		0.1443	0.1429	99.7
		0.1779	0.1794	99.7
7.6		0.1289	0.1285	99.7
		0.1810	0.1799	99.4
Ethyl methyl phosphonic acid	Bis(isopropyl methyl phosphonic) anhydride			
3		0.1330	0.1330	99.8
		0.1798	0.1793	99.7
7		0.1350	0.1346	99.7
		0.1811	0.1802	99.5

Beach and Sass[928] applied the hydrolytic method satisfactorily to a range of pure and crude pyroesters with purities down to 30% obtaining good agreement within 1% or better with results obtained by the peroxide method (Sass et al[927]).

E. Organophosphorus Fluoridates and Chloridates

Sass et al[927] described a volumetric procedure using a peroxide reagent in an alkaline medium, for the semi-micro and macro analysis of organophosphorus fluoridates and chloridates in the presence of phosphorus esters (also ester pyrophosphates and pyrophosphonates and phosphoro anhydrides, see previous section in this Chapter). The method, which depends on the formation of perphosphorous acid by reaction of the sample with sodium pyrophosphate peroxide followed by the subsequent determination of excess peroxide, has a precision and accuracy within 1%. The most rapid, complete, and stable reaction of the peroxide with halophosphorus compounds and phosphoro anhydrides occurred at a buffered pH between 9.5 and 10.2. Reactions for all the compounds tested by Sass et al[927] were completed within 1 to 2 min with no further apparent reaction occurring over a period of one hour. The sample must be added with mixing to the peroxide buffer solution because it is necessary to maintain a strong excess of peroxide during the reaction. A reversal of the addition results in the localized depletion of peroxide with the loss of some compound to hydrolysis. Results as low as 50% of the theoretical are obtained depending on the rate of addition of peroxide to the compound. Based on these principles, Sass et al[927] described detailed procedures for the determination of phosphorofluoridates. They also described detailed procedures for the determination of alkyl phosphorochloridates and dialkylphosphorochloridates.

In general, the phosphor- and phosphonochloridates are much less stable hydrolytically than the corresponding fluoridates and pyrophosphonates. The chloridates hydrolyse rapidly. Diisopropyl phosphorochloridate ethyl methylphosphonochloridate, and isopropyl methylphosphonochloridate hydrolyse almost instantaneously. The wet method, i.e. titration in 50% aqueous 2-propanol) can be used to determine phosphoro- and phosphonofluoridates and pyrophosphonates quantitatively in the presence of chloridates that were used as intermediates in their syntheses. When determining the chloridates as intermediates, the fluorination modification yields quantitatively the more stable but highly toxic fluoridate, which gives satisfactory analytical results. When this procedure is used, all analyses should be conducted in an efficient laboratory hood. The chloridate result can be corrected for pyroester as impurity. An alternative method, although empirical and less accurate, is the dry procedure, i.e. titration in neat n-propanol. With this method, which involves less volatile and less toxic materials, consistent recoveries of 70 - 72% were obtained on the chloridates. When the dry method is used, the empirical peroxidation versus hydrolysis esterification characteristics of the particular chloridate of interest should be experimentally predetermined. Correction is then made for the difference between the analytical recovery on pure samples and that calculated for 100% recovery as found for the compound. Dichloridates interfere additively with the dry method. In this procedure, a suitable weight of sample (2 g) is weighed into a previously tared polyethylene bottle containing a weighed 15 ml volume of hydrogen fluoride solution. The bottle is swirled to mix the contents and left for 20 minutes. The container is swirled and 0.5 - 1.0 g of sample weighed into a glass ampoule. The sample is added to a 250-ml glass-stoppered Erlenmeyer flask containing 50 ml of 50% 2-propanol. A 25-ml portion of this solution is taken and analysed as described in the previous procedure.

When calculating results, it is necessary to correct the volume of the 50% 2-propanol solution (50 ml) for the additional volume contributed by the added fluorinated sample using weight directly as volume.

The % phosphorus chloridate is calculated as follows:

$$D = \frac{\text{weight of chloridate sample} \times 100}{\text{combined weights of hydrofluoric acid and chloridate sample}}$$

where D = % chloridate in hydrofluoric acid solution (15 ml).

$$F = \frac{\frac{50.0 + W}{20} \quad \frac{1}{4}(\text{molecular weight}) \times 0.1 \times (B - C) \times N}{W}$$

where F = per cent chloridate titrated
B = millilitres of sodium thiosulphate solution used in blank titrations
C = millilitres of sodium thiosulphate solution used in sample titrations
N = normality of sodium thiosulphate solution
W = weight of fluorinated aliquot (ampouled sample)
then % chloridate = F/D x 100.

To correct for pyrophosphorus ester as impurity in chloridate use the following: % chloride = (% chloridate from fluorination method) - (% pyro-ester calculated as chloridate).

Sass et al[927] applied their peroxide procedure to various dialkyl phosphonofluoridates and phosphorofluoridates (Table 102) and the fluorination-peroxide procedure to dialkylphosphorochloridates and phosphonochloridates (Table 103) of known purity. They found generally that precision and accuracy were within approx 1%.

In a further paper, Sass et al[930] discussed the determination of phosphorus fluoridates by differential fluorine methods.

A simple rapid hydrolytic method has been described by Beach and Sass[928] for the volumetric determination of ester phosphonofluoridates and ester phosphorofluoridates (which is also applicable to ester pyrophosphates). This method entails the titration of free acidic impurities in the sample followed by a rapid quantitative hydrolysis of the fluoridate. The quantity of standard sodium hydroxide used for hydrolysis is a measure of the purity of the sample. The method is based on the large difference in hydrolysis rate of ester phosphorofluoridates and ester phosphonofluoridates in 0.1 N sodium hydroxide as opposed to the rate of hydrolysis of impurities such as the rapidly hydrolysing chloridates and difluoridates on the one hand and some slowly hydrolysing or directly titratable phosphorous acids on the other.

The hydrolytic method (Beach and Sass[928]) for the determination of GB Sarin (isopropyl methylphosphonofluoridate) and its homologues such as diisopropylphosphorofluoridates and tetraethylpyrophosphate as well as their homologues, and pyrophosphonates, is essentially an acid-base titration. By titrating rapidly to a methyl red end-point with 0.1 N sodium hydroxide, all free acid and acids formed on hydrolysis are neutralized. An excess of

Table 102 - Analytical recovery and purity obtained for alkyl phosphoro and alkyl phosphono fluoridates using peroxide method.

	Sample		Recovery	Purity found
	Calculated	Recovered	(%)	(%)
Isopropyl methylphosphonofluoridate				
Calculated purity, 99.0%	0.2346	0.2337	99.6	98.6
	0.1945	0.1932	99.3	98.3
	0.1948	0.1937	99.4	98.4
	0.2266	0.2266	100.0	99.0
	0.2217	0.2204	99.4	98.9
Calculated purity, 74.8%	0.1438	0.1438	100.0	74.8
	0.1421	0.1410	99.2	74.2
	0.1350	0.1345	99.6	74.5
Diisopropyl phosphorofluoridate				
Calculated purity, 98.0%	0.2132	0.2127	99.8	97.8
	0.1887	0.1874	99.3	97.3
	0.1971	0.1961	99.5	97.5
Calculated purity, 79.2%	0.1780	0.1766	99.2	78.6
	0.1685	0.1668	99.0	78.3
	0.1565	0.1555	99.4	78.7
Isopropyl ethylphosphonofluoridate	0.2251	0.2248	99.9	93.8
Calculated purity, 93.9%	0.1984	0.1971	99.3	93.3
	0.2111	0.2091	99.1	93.0

base quantitatively hydrolyses the compound of interest within two minutes. The quantity of base consumed in the hydrolysis is obtained by back-titration with 0.1 N hydrochloric acid to the methyl red end-point.

Beach and Sass[928] used their hydrolytic purity method to determine a range of phosphorus fluoridate compounds including phosphonofluorides and phosphorofluoridates. The samples ranged in purity from pure distillation overheads to crude distillation residues and partially hydrolysed and thermally treated materials. Their results (Table 104) indicate that the method is applicable to high and low purity samples with an equal degree of precision and accuracy. Compounds having limited water-solubility were dissolved initially in ethanol and diluted subsequently with water. The quantity of n-propyl alcohol was kept below 20% by volume, and often at 10% or less. At a level of 50% alcohol the hydrolysis rate of some phosphonofluoridates is reduced by a factor of five. The solutions were titrated as soon as possible after dilution with water.

F. Phosphonium Salts and Phosphoranes

Ross and Denney[931] described a titration procedure for the determination of phosphonium salts and phosphoranes. They found that titration of the phosphoranes in acetic acid with perchloric acid was a satisfactory technique for this determination. This technique may not be satisfactory for all phosphoranes but suitable modifications as described by Fritz and Hammond[932] can be made to extend the range of applicability. Similarly, phosphonium salts can be analysed by the addition of excess mercuric acetate to the acetic acid solution, followed by titration of the liberated

Table 103 - Analytical recovery of phosphoro- and phosphonochloridates

	Fluorination followed by peroxide method					
	Sample (g)		Recovery	Purity calcd.	Purity found	Dry Method[a]
	Calcd.	Recpvered	(%)	(%)	(%)	wt(%)
Diisopropyl phosphorochloridate	0.1854	0.1850	99.8	96.0	95.8	94.1
	0.1692	0.1682	99.4		95.4	95.4
	0.1521	0.1520	99.9		95.9	93.8
	0.1439	0.1430	99.4	79.1	78.6	77.9
	0.1461	0.1448	99.1		78.3	78.4
	0.1367	0.1356	99.2		78.5	77.0
Isopropyl methylphosphonochloridate						
	0.1488	0.1488	100.0	96.6	96.6	95.8
	0.1534	0.1526	99.5		96.0	96.1
	0.1675	0.1666	99.5		95.9	94.9
	0.1607	0.1603	99.8	85.5	85.3	85.0
	0.1543	0.1543	100.0		85.5	84.9
	0.1500	0.1493	99.5		85.1	84.6
	0.1310	0.1302	99.4	65.8	65.4	64.8
	0.1263	0.1261	99.8		65.7	64.0
	0.1221	0.1211	99.2		65.3	64.3

a Empirical results corrected to a 100% basis.

Table 104 - Comparison of purity results on dialkyl phosphono and dialkyl phosphorofluoridates obtained using hydrolytic versus the macro peroxide method (Sass et al)

Compound	Purity (%)		Difference (%)
	Hydrolytic method	Macro peroxide method	
Isopropyl methyl phosphonofluoridate	99.0	98.8	0.2
	95.4	94.4	1.0
	90.8	90.0	0.8
	88.1	88.0	0.1
	88.6	87.5	1.1
	77.3	77.0	0.3
	35.2	34.5	0.7
	10.5	10.2	0.3
Diisopropyl phosphorofluoridate	98.0	97.5	0.5
	98.0	97.6	0.4
	78.5	78.2	0.3
	78.5	78.3	0.2
Isopropylethyl phosphonofluoridate	93.9	93.8	0.1
	75.7	75.7	0.0

acetate ion with perchloric acid. These techniques have been used in the past for the determination of ammonium salts and weakly basic amines, and have been described in considerable detail by Fritz and Hammond[932].

G. Substituted Phosphines

Streuli[933,934] has determined the relative basicity of various substituted phosphines by non-aqueous titrimetry with N-hydrochloric acid in nitromethane. Approximately 0.5 - 1 mmol of the phosphine in 100 ml of solvent was used for the titration. Some titrations were conducted under a nitrogen blanket to avoid aerial oxidation to the corresponding phosphine oxide. Diphenylguanidine was used as a reference standard in non-aqueous titrations and a pH 7 buffer used to standardize the electrodes for mixed solvent experiments (methanol - water). Streuli claims that non-aqueous titration along these lines gives a simple analysis for the less easily oxidized substituted phosphines.

Fritz and Pappenburg[935] have described a method for the determination of the phosphine group in hydrolysable element-phosphorus compounds in which phosphorus is linked to a less electronegative element such as diethyl (trimethylsilyl)phosphine.

In this method the sample is weighed in nitrogen in a gelatin capsule, and the capsule is dropped into a 100-ml flask containing 10% aqueous sodium hydroxide (20 ml) under nitrogen and fitted with an inlet tube for nitrogen and a reflux condenser. The outlet of the condenser is connected with two absorption tubes in series, each containing 20 ml of 5% aqueous mercuric chloride. The condenser is cooled by water maintained at a suitable temperature, e.g. 65°C for diethylphosphine, and the mixture is boiled for 30 min under a slow stream of nitrogen. The precipitate formed in the first absorption tube is a compound of the type (chloromercuri)diethylphosphine;

it is determined by adding potassium iodide and iodine solution and titrating the residual iodine

H. Trisubstituted Phosphine Oxides and Sulphides and Bis(disubstituted-phosphinyl) alkanes

Wimer[936,937] reported the potentiometric titration of trimethylphosphine oxide, trioctylphosphine oxide and triphenylphosphine oxide using perchloric acid in dioxane as the titrant and acetic anhydride as the solvent.

Henderson et al[938] have developed a procedure for the titration of triarylphosphine oxides in nitromethane medium with perchloric acid. The titrations were carried out using a pH meter with calomel and glass electrodes. The aqueous bridge in a sleeve-type calomel electrode was replaced with a saturated solution of anhydrous lithium perchlorate in acetic anhydride. A fibre-type calomel electrode with an aqueous salt bridge can be used, but the potential readings are somewhat unstable.

Results obtained by Henderson et al[938] in the titration of three basic types of phosphoryl compounds are shown in Table 105. Trisubstituted phosphine oxides (R_3PO), bis(disubstituted phosphinyl)alkanes, ($R_2P(O)-(CH_2)_xP(O)R_2$) and β-ketophosphine oxides, ($R_2P(O)CH_2C(O)R$).

The titration of compounds in which the phosphoryl group was surrounded by either highly electronegative groups (e.g. triphenylphosphine oxide) or by very bulky groups (e.g. tris(2-ethylhexyl) phosphine oxide) gave poorly defined end-points. Very weakly basic compounds such as di-n-octylphosphine oxide and di-n-butyl n-butylphosphonate could not be titrated. The potential breaks obtained at the end-points are given in Table 105. As can be seen, the magnitude of the potential break is dependent upon the volume of titrant used. In the acetic anhydride solution the acid which acts as the titrant is CH_3CO^+ (as "acetyl perchlorate"). Increasing the acetic acid concentration causes the formation of species less acidic than "acetyl perchlorate" and results in poorer end-points. Perchloric acid in acetic anhydride gradually decomposes giving rise to a yellow solution. However, this decomposition is slow enough to permit accurate determinations to be made. Thus, at least in certain cases, the phosphoryl group is basic enough to be titrated. This affords a rapid and convenient method for quantitative determinations to be made.

I. Dialkyl Phosphorodithioates, Dialkyldithiophosphates, Phosphorothioates, Dialkylphosphonodithioates and other P - S Compounds

O,O-dialkyl hydrogen phosphorodithioates (Batora and Vesela[939]) can be determined by conversion into the nickel salt, which is determined iodimetrically without preliminary isolation. Interference by hydrogen sulphide or thiols is eliminated by the formation of insoluble nickel sulphide or mercaptide, respectively. In this method, the sample (0.5 g) is added to a saturated solution of nickel sulphate (3 ml), mixed, then water (40 ml) added, then a known excess of 0.1 N iodine added until the solution becomes yellow. Iodine is back-titrated with standard sodium thiosulphate to the starch end-point. When determining O,O-dialkyl sodium phosphorodithioate in the presence of the free acid, a preliminary neutralization of the sample to methyl red must be carried out.

Sul'man et al[940] established that the nickel salt of diethyl phosphorodithioate (dissolved in 0.01 M perchloric acid) could be titrated potentiometrically with 0.01 M iodine in potassium iodide solution or with 0.01 M

Table 105 - Potentiometric titration of phosphoryl compounds[a]

Compound	% Purity		Potential break[a] (mv/ml)[d] at the end-point for a given volume of titrant		
			5 ml	10 ml	20 ml
1. Tri-n-octylphosphine oxide	99.8	100.0	255	165	-
2. Tris(2-ethylhexyl)phosphine oxide	b				
3. Di-2-ethylhexylmethylphosphine oxide	83.3				
4. Triphenylphosphine oxide	b				
5. Di-n-octylphosphine oxide	c				
6. Di-n-butyl n-butylphosphonate	c				
7. Di-n-heptylphosphinic acid	c				
8. Bis(di-n-hexylphosphinyl)methane	100.6	100.5		100	
9. Bis(dicylohexylphosphinyl) methane	97.6	98.2	250		
10. Bis (di-2-ethylbutylphosphinyl) methane	99.3		190		
11. Bis (di-2-ethylhexylphosphinyl) methane	b				
12. [(Di-n-octylphosphinyl)(diphenylphosphinyl)] methane	102.4	102.7	170		
13. Bis (di-n-hexylphosphinyl) ethane	100.5	101.1	-	65	
14. Bis (di-n-hexylphosphinyl) propane	99.6	100.4		120	65
15. Bis (di-n-hexylphosphinyl) butane	100.2	99.8	150	100	
16. Acetonyldi-n-hexylphosphine oxide	102.2	101.8	205		
17. Phenacyldi-n-hexylphosphine oxide	97.6	97.7			
18. Phenacyldiphenylphosphine oxide	b				

[a] Compounds 1, 2, 4 and 6 were obtained commercially and were analysed without purification; all other compounds were synthesized and were purified before being titrated. Titrant was perchloric acid in a glacial acetic acid-acetic anhydride mixture.
[b] Poorly defined end-point
[c] Not titratable
[d] Defined as the potential at 0.5 ml after the end-point minus the potential 0.5 ml before the end-point.

silver perchlorate. Platinum immersed in a solution saturated with an organic disulphide and containing diethyl phosphorodithioate was used as indicator-electrode. The relative error was ± 0.3% with either titrant. Similarly, the titration can be carried out with mercuric perchlorate solution.

Bode and Arnswald[941] discussed the determination of sodium diethylphosphorodithioate by iodimetry and also photometrically as the bismuth complex. Lewkowitsch[942,943] has described a volumetric procedure for the determination of the molecular weight of such salts via the copper salt.

O,O,S-Trimethyl phosphorodithioate has been determined in amounts from 5% to 20% in dimethylphosphorodithioate (Yamaleev and Pal'yanova[944]) by a saponification procedure in which the sample (1.5 - 2 g) is dissolved in 2 - 3 ml of methanol, and the acid is neutralized to phenolphthalein with 2.5 N methanolic potassium hydroxide, the final adjustment being with 0.5 N hydrochloric acid and 0.5 N potassium hydroxide. The mixture is heated under reflux for 10 min with 10 ml of N-methanolic potassium hydroxide and the residual potassium hydroxide in the cooled solution is titrated with 0.5 N hydrochloric acid.

J. Alkyl Phosphonic Chlorides

Gefter[945] has pointed out that many acid chlorides such as ethylphosphonic dichloride, 2-chlorethylphosphonic dichloride, methyl phenylphosphinic chloride, and ethyl phosphorodichloridate react so rapidly with potassium hydroxide that they can be determined by addition of excess potassium hydroxide and back-titration of unused alkali.

K. Alkyl Silylphosphates

Kreshkov et al[946] have described a non-aqueous titration procedure for the analysis of alkylsilylphosphates. Trimethylsilyl dihydrogen phosphate ($(CH_3)_3SiOPO(OH)_2$) and its analogues can be titrated in non-aqueous medium with solutions of lithium, sodium, or potassium methoxides and tetraethylammonium hydroxide. The end-point is determined potentiometrically with glass and saturated mercurous chloride electrodes, or visually in the presence of quinizarin, bromophenol blue, brilliant green, alkaline blue or methyl red (the most accurate results are obtained with the latter). Methyl cyanide, acetone, methyl ethyl ketone, methanol, ethanol, isopropyl alcohol, butanol and benzyl alcohol can be used as titration media. In the titration of trimethylsilyl dihydrogen phosphate with lithium methoxide solution in methanolic medium, the reaction proceeds stepwise, with the use of first one and then two equivalents of the titrant. In the other alcohols (except benzyl alcohol), the ketones and methyl cyanide, the reaction proceeds until three equivalents of lithium methoxide have been used per mole of trimethylsilyl dihydrogen phosphate. The titration of trimethylsilyl dihydrogen phosphate with potassium methoxide or tetraethylammonium hydroxide in all the media mentioned above gave a single potential jump, corresponding to the reaction:

$$(CH_3)_3SiOPO(OH)_2 + CH_3OK \rightarrow (CH_3)_3SiOCH_3 + KH_2PO_4$$

The titration of trimethylsilyl dihydrogen phosphate with sodium methoxide in a medium of isopropyl alcohol, butanol or benzyl alcohol proceeds similarly. In methanol and ethanol, two jumps were obtained, corresponding to two equivalents of titrant, as in the titration of trimethylsilyl dihydrogen phosphate with sodium methoxide in a methanolic medium.

ORGANOPOTASSIUM COMPOUNDS

Phenylisopropylpotassium has been determined by reaction with excess of p-ditolyl sulphide to produce a mercaptide which is determined by titration with standard silver nitrate solution[948].

ORGANOSELENIUM COMPOUNDS

In a method for determining nitrobenzene selenyl bromides, thiocyanates alkoxides and amides, and 2,4-dinitrobenzenselenyl, the substance is dissolved in ethyl acetate[949]. After addition of 96% ethanol and glacial acetic acid, standard sodium thiosulphate is added and the excess of sodium thiosulphate is back-titrated with standard iodine solution. When organoselenium compounds are dissolved in sulphuric acid, selenium is present as elemental selenium, 'dissolved' selenium selenous acid, and organoselenium compounds[950,951]. A method for the separation and determination of elemental and 'dissolved' selenium and selenous acid is based on titration with sodium thiosulphate. An error of less than 1% is claimed for this method for the determination of total selenium in organic selenium compounds, and less than ± 0.3% for the determination of 'dissolved selenium' and selenous acid.

ORGANOSODIUM COMPOUNDS

An iodometric method has been suggested for the assay of pentane solutions of amylsodium, although the absolute accuracy of the method is unknown[165]. A procedure for the high-frequency titration of cyclopentadienylsodium is based on titration of combined sodium with 0.5 N hydrochloric acid [952]. Indicators are excluded in this titration owing to the intense colour of the analysis solution. Also, potentiometric and conductometric methods are ruled out owing to electrode fouling.

The Gilman titration procedure can be used to determine butylsodium compounds containing various alkali metal alkoxides[953]. Allyl bromide is used to react with the organometallic compound for 1 min. Accurate assays can be obtained on organometallics in the presence of large amounts of alkoxides. Even in mixtures containing potassium t-butoxide at twice the concentration of butylsodium, accurate determination of the carbon-bound sodium was obtained.

ORGANOTHALLIUM COMPOUNDS

Pepe et al[954] have described a procedure for the estimation of phenylthallium (III) dichloride. This compound forms a complex with xylenol orange which can be titrated with EDTA in aqueous ethanol medium at pH 4.75 with 9.1% aqueous xylenol orange as indicator. The end-point is shown by a sudden colour change from red to lemon yellow and corresponds to the formation of a 1:1 complex of phenylthallium dichloride with EDTA.

ORGANOSILICON COMPOUNDS

A. Non-Aqueous Titration

Kreshkov and co-workers (Kreshkov et al[955]; Kreshkov and Drozdov[956];

Kreshkov[957]; Kreshkov et al[958] have made an extensive study of non-aqueous titrimetric methods of analysis of alkyl and aryl alkyl chlorosilanes. Kreshkov et al[959,960,961] also studied other types of organosilicon compounds containing nitrogen, carboxyl, and alkylthiocyanato functional groups.

To titrate alkylchlorosilanes (Kreshkov et al[955] such as trimethylsilicon chloride, dimethylsilicon dichloride, methylsilicon trichloride and silicon tetrachloride, titration in methyl cyanide with 0.05 N phenazone or nitron in methyl cyanide, or with a solution of amidopyrine in benzene by visual and potentiometric methods is suggested. The indicators used are crystal violet, dimethylaminoazobenzene, bromocresol purple, methyl orange, bromophenol blue and gallo sea blue (C.I. Mordant Blue 14); the addition of benzene, toluene, chlorobenzene or carbon tetrachloride does not affect the results.

B. Non-Aqueous Conductiometric Titration

Non-aqueous conductiometric titration has been used (Kreshkov and Drozdov[956]) for the determination of trimethylsilicon chloride, dimethylsilicon dichloride and methylsilicon trichloride. The method is based on the quantitative conversion of the chlorosilanes into alkylthiocyanato-derivatives by the action of ammonium thiocyanate and conductiometric titration in methyl cyanide-ethyl ether (2:3) solvent with 0.1 M-amido pyrine in benzene. The mono-substituted compound is first titrated, then the di-substituted, and finally the tri-substituted. The smallest amounts which can be determined in the mixture are: $(CH_3)_3SiCl$, 0.0055 g; $(CH_3)_2SiCl_2$, 0.0025 g; and CH_3SiCl_3, 0.0023 g in 12 ml sample. The error in determining the individual chlorosilanes is + 2%.

Non-aqueous electrometric methods have been used (Kreshkov[957]) for the analysis of monomeric and polymeric silicon compounds in non-aqueous medium. The methods are applicable to alkyl and aryl chlorosilanes, aminosilanes, silyamines, weak acids containing silicon and other organic silicon compounds. In non-aqueous medium, several organic silicon compounds have strongly basic or acid properties. Solutions of tetraethylammonium hydroxide and amidopyrine in benzene or mixtures of benzene with other solvents can be used as titrants for these types of compound.

C. Non-Aqueous Amperometric Titration

Kreshkov et al[958] has also used amperometric titration for silane and trimethylchlorosilane. Two methods are described:

In methanol benzene - medium. About 20 - 30 mg alkylchlorosilane is dissolved in 15 ml 0.3 M lithium nitrate in methanol - benzene (1:2) and titrated amperometrically (mercury anode; dropping mercury cathode, potential - 1.2 V) with 0.2 N lead acetate dissolved in methanol or dissolved in methanol benzene (1:2). The error is less than ± 1.5% and sensititivity 0.01% of chloride in the sample.

In anhydrous acetic acid medium. About 10 - 25 mg sample is dissolved in 15 ml saturated lithium nitrate solution in anhydrous acetic acid. The solution is titrated amperometrically (- 1.4 V) with 0.15 N cadmium nitrate dissolved in anhydrous acetic acid.

To determine nitrogen-containing organosilicon compounds (Kreshkov et al[959]) the sample (0.04 - 0.1 g) is dissolved in 6 ml methyl cyanide or nitromethane, an equal volume of benzene or dioxan is added, and the solution is titrated potentiometrically with 0.1 N perchloric acid in glacial acetic acid. Alternatively, the end-point may be found visually by use of one of a number of suitable indicators such as crystal violet or bromocresol purple. With carboxyl-containing compounds, the solution in the mixture of organic solvents is titrated with a solution of tetraethyl-ammonium hydroxide in benzene-methanol. In an alternate non-aqueous titration method for nitrogen-containing compounds (Kreshkov et al[960]) it is suggested that the sample be titrated with perchloric acid solution in methyl cyanide, methyl nitrile, or their mixtures with benzene or dioxan, by visual or potentiometric methods. The indicators used were crystal violet, bromocresol purple, bromophenol blue, cresol red, dimethylaminoazobenzene, thymol blue, methyl red and methyl orange. Compounds containing nitrogen attached directly to silicon and compounds with nitrogen in an organic radical were titrated. The potential jumps for titration in methyl nitrite are 10 m V larger than those in methyl cyanide.

The non-aqueous titration of alkyl thiocyanatosilanes (Kreshkov et al [961] is carried out with 0.1 N sodium methoxide in methyl cyanide or methanol, either visually with a hydroxyanthraquinone as indicator, or potentiometrically with the sample dissolved in the same solvent.

Tarasyants[962] has described the titration of alkylsilyl sulphates in anhydrous solvents. Alkylsilyl sulphates were determined by potentiometric and conductiometric titration in anhydrous alcohols, ketones, methyl cyanide or nitromethane. Titration curves were the same for all solvents. In potentiometry, titration with sodium, potassium, or lithium methoxide gave curves with one step; titration with quaternary ammonium compounds gave curves with two steps. In conductimetry, titration with quaternary ammonium compounds gave curves with no inflection, whereas titration with alkali metal methoxides gave curves with a sharp break at the end-point.

Drozdov et al[963] described the non-aqueous conductiometric titration of tri-alkylsilyl phosphates and bis (trialkylsilyl) sulphates. These workers used methanol, ethanol, butanol, acetone, ethylmethylketone, nitromethane, nitrobenzene, methyl cyanide, dimethylformamide, tetra-hydrofuran, butyl acetate and ethyl formate as solvents, and methanolic sodium, potassium and lithium methoxides and tetraethylammonium hydroxide as titrants. For the phosphates, characteristic curves were obtained only in nitromethane and methyl cyanide. Titrating in nitromethane with 0.15 N methanolic potassium methoxide gave the best results. Alcohols, ketones, nitromethane and methyl cyanide were suitable solvents for the sulphates, but use of tetraethylammonium hydroxide as titrant did not give good results.

Gilman et al[964] have described a method for the titration of organo-silymetallic compounds including those of lithium. They showed that these compounds react with n-butyl bromide to yield an equivalent amount of metallic bromide, which is titrated argentimetrically by the Volhard procedure. To a flask containing 5 ml n-butyl bromide, under an atmosphere of nitrogen, is added 5 ml sample (0.25 - 0.45 N) with gentle swirling. The solution is left for 10 min, then about 5 ml 0.1 N sodium hydroxide added and the excess alkali titrated with 0.1 N sulphuric acid to the colourless phenolphthalein end-point. An excess of standard silver nitrate solution is then added and the excess silver back-titrated with standard ammonium thiocyanate solution to the ferric alum end-point. Gilman et al[964] describe an alternate procedure in which after the neutralization, the organic phase is separated and washed with water, the combined aqueous phase and washings made up to 200 ml with water, and a 50 ml aliquot titrated as described above. This method gives lower results but is claimed to be more accurate.

ORGANOTIN COMPOUNDS

A. Classical Titration Procedures

A method for the determination of monobutyltin trichloride in technical dibutyltin dichloride and in dibutyltin oxide depends on the fact that both monobutyltin trichloride and dibutyltin dichloride form blue complexes with catechol violet; between pH 1.2 and 2.3, however, only the complex of monobutyltin trichloride is decomposed by EDTA, and this facilitates the determination of this substance in dibutyltindichloride[965]. The sample is dissolved in methanol, ethanolic catechol violet is added, and the resulting blue or green solution is titrated with 0.05 M EDTA to a reddish violet to red end-point. To determine butanestannonic acid in technical dibutyltin oxide, the sample is dissolved in warm methanolic hydrochloric acid, the solution is diluted with methanol, and catechol violet solution is added. Methanolic potassium hydroxide is added dropwise until the colour changes from red or green, and the monobutyltin trichloride produced is then determined as described above.

Salts of the di- and tri-basic di- and monoalkyl (or-aryl) compounds of tin can be detected by the deep blue complex formed with catechol violet; the tetraalkyl (or arlyl) tin compounds and the salts of the monobasic compounds do not give the reaction[966]. The blue complexes are quantitatively destroyed by EDTA. Diphenyltin diacetate can be accurately determined in the presence of triphenyltin acetate by adding a solution of catechol violet to the test solution in methanol and titrating to a yellow colour with EDTA solution (disodium salt). Tin in PVC can be determined by complexometric titration with EDTA[967]. The sample is dissolved in hot tetrahydrofuran and treated with 50 ml of ethanol. The precipitated PVC is filtered off and washed with ethanol, and 0.1% catechol violet solution is added dropwise to the combined filtrate until the solution becomes blue. This blue solution is then titrated to a green end-point with 0.001 M EDTA. Complexometric titration using EDTA and back-titration with standard zinc acetate has been used for the determination of dibutyltin oxide and dibutyltin dichloride[968]. Analysis of the triphenyltin hydroxide - bis(triphenyltin) oxide system has been reported[969].

A Karl Fischer titration procedure can be used for the determination of and differentiation between trialkyl (aryl) organotin hydroxides and the corresponding oxides[970,971]. This method is based on the observation that in the determination of water by Karl Fischer reagent in silanols and silanediols consistently high water contents are obtained[972]. Investigation led to the conclusion that not only was the water content being determined, but also that the silanol was reacting quantitatively with the reagent. The Karl Fischer reagent is not only effective in the quantitative determination of triaryl- and trialkyltin hydroxides, but is also applicable to bis(triaryltin) oxides[970]. The following reactions are postulated:

$$(R_3Sn)_2O + I_2 + SO_2 + CH_3OH \rightarrow 2R_3SnI + HSO_4CH_3$$

$$R_3SnOH + I_2 + SO_2 + CH_3OH \rightarrow R_3SnI + HSO_4CH_3 + HI$$

The R_3SnOH class of compounds consumes 1 mol of iodine for each tin atom, whereas with the $(R_3Sn)_2O$ type of compound the ratio is 0.5. Table 106 illustrates the usefulness of this procedure to differentiate between a triorganotin hydroxide and its bis(triorganotin) oxide equivalent where both exist. As shown, the difference in carbon analysis between these two compounds is only 1.47%; the difference in tin is 0.79% and the difference in hydrogen 0.17%. However, the difference in magnitude of the value determined by Karl Fischer titration is of the order of 100%. The method, therefore, is also useful in determining the composition of a mixture of the two compounds. Typical analytical data for other trialkyltin hydroxides and oxides are shown in Table 107.

Table 106 - Comparative analytical data for triphenyltin hydroxide and bis (triphenyltin) oxide

	Triphenyltin hydroxide		Bis(triphenyltin)oxide	
	Calcd.	Found	Calcd.	Found
Sn.%	32.36	32.37	33.15	33.10,33.20
C,%	58.90	58.69,58.74	60.37	60.22,60.17
H,%	4.39	4.32, 4.45	4.22	4.40, 4.33
mp		118.5 - 120 (decomp.)	-	122 - 123.5
H_2O (apparent), %	4.90	4.83	2.52	2.60, 2.62
I_2/Sn atom, mol	1.00	0.984	0.500	0.506

Table 107 - Analytical data for trialkyltin hydroxides and bis(trialkyltin) oxides

	Trimethyltin hydroxide	Bis(tri-n-propyl tin) oxide	Bis(tri-n-butyl tin) oxide*
Sn(calc.).%	65.60	46.38	39.83
Sn(found).%	65.35	46.29	39.73-40.09 (range)
H_2O apparent (calcd.).%	9.95	3.51	3.02
H_2O apparent (found).%	9.88	3.68	3.12,3.08,3.14,3.07
MoI_2/Sn(Calc.)	1.00	0.500	0.500
MoI_2/Sn(found)	0.992	0.509	0.523,0.510,0.520, 0.508

* Various commercial samples, minimum purity 96%.

Organotin carboxylates have been determined by distillation with phosphoric acid and by titration in non-aqueous medium[973]. Allytin compounds in the presence of propenyltin compounds can be determined by titration with a benzene solution of iodine[974]. Trace studies by the radioactivation method have been used for the semi-quantitative determination of residual tin compounds in laundered sanitized nylon cloth[975]. Non-aqueous titrimetry and atomic-absorption spectrometry have been used for the determination of organotin biocides in insect-proofed textile materials[976]. Triphenyltin-lithium has been determined by a double titration procedure using allyl bromide[977]. Triphenyltinlithium has been estimated by Gilman[989] by a double titration procedure using allyl bromide.

B. Potentiometric Titration

Organotin compounds of the general type $R_{(4-n)}SnCl_n$(R = methyl, ethyl, propyl, butyl, or phenyl; n = 2 or 3) have been titration by potentiometric procedures using tetraphenylarsonium chloride or tetraethylammonium chloride or bromide in acetonitrile as titrants[978,979] Potentiometric titrations of organotin compounds with halogen bases, as well as the reverse titrations, were performed at 25° in a cell (Figure 36).

A known volume of the halogen base was transferred to vessel A from burette B_1 (10 ml capacity) containing a molybdenum wire M_1; the tip of the burette was immersed in the sample solution so that wire M_1 formed one electrode of a concentration cell, the other one being the molybdenum wire M_2. The titration was performed by adding the titrant from burette B_2. Figure 37 shows a typical titration curve of tetraphenylarsonium chloride with phenyltin chloride and Table 108 shows results obtained for a range of organotin halides.

Table 108 - Titration of tetraphenylarsonium chloride with $R_{4-n}SnCl_n$ compounds (n= 4,3,2,1) in acetonitrile at 25°

Conc. Ph_4AsCl ($M.10^2$)	Amount Ph_4AsCl titrated (ml)	Titrant	Conc. $R_{4-n}SnCl_n$ ($M.10^2$)	Titrant added ml at end-point Calc.	Found	x = $[R_{4-n}SnCl_n]$ / $[Ph_4AsCl]$ at e.p.
1.344	10	$SnCl_4$	7.07	1.90	0.93-1.90	0.49-1.000
1.344	10	$MeSnCl_3$	5.00	2.69	2.70	1.002
1.344	12	Me_2SnCl_2	5.00	3.23	3.15	0.975
1.344	10	$EtSnCl_3$	5.00	2.69	2.70	1.002
1.344	12	Et_2SnCl_2	5.40	2.99	2.90	0.970
1.344	12	Pr_2SnCl_2	5.40	2.99	2.95	0.986
1.344	9	$BuSnCl_3$	4.48	2.70	2.70	1.000
1.344	14	Bu_2SnCl_2	5.00	3.77	3.80	1.007
1.344	10	$PhSnCl_3$	4.80	2.80	2.70	0.964
1.344	12	Ph_2SnCl_2	5.00	3.23	3.15	0.975
1.344	10	Ph_3SnCl	2.80	4.80	4.95	1.031

Mixtures of R_2Sn^{2+} and R_3Sn^+ compounds where R is methyl or ethyl have been determined in the 0.01 - 0.05 mM range by potentiometric titration with standard potassium hydroxide to determine total organotin ions, followed by amperometric titration of R_2Sn^{2+} on a second aliquot using 2 nM 8-hydroxyquinoline as titrant at pH 9.2 (aqueous ammonia-ammonium nitrate buffer)[980]. Amperometric titration was applied to the determination of dialkyltin perchlorates using as titrant either 8-hydroxyquinoline (R = CH_3, C_2H_5) or with hexacyanoferrate(II) ions (R = C_4H_9, C_6H_5). Amperometric titration is carried out to - 1.4 V against a standard calomel electrode. The amperometric and potentiometric titration of R_3Sn^+ and R_2Sn^{2+} compounds have also been studied[981], as well as the potentiometric titration of methyltin chlorides and bromides[982].

C. Amperometric and Coulometric Titration

Amperometric titration with electrolytically generated iodine, bromine, or silver ion has been applied to the titration of hexaorganoditin compounds such as hexamethylditin, hexaethylditin, hexapropylditin, hexabutylditin,

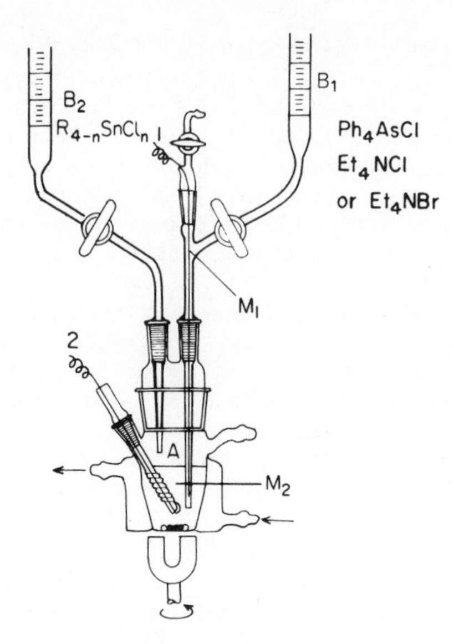

Figure 36 - Cell assembly, potentiometric titration of organtin compounds.

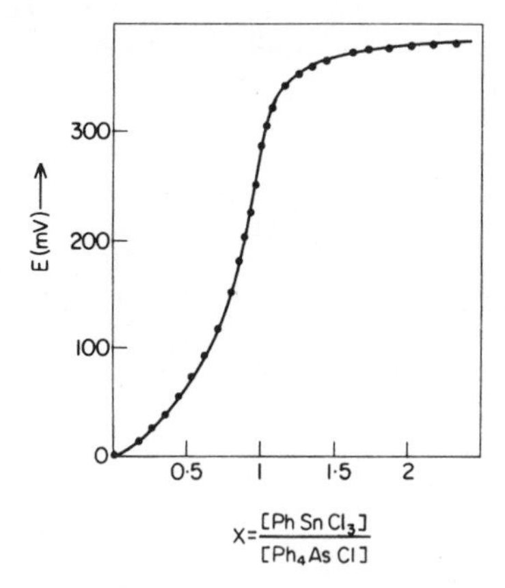

Figure 37 - Potentiometric titration plot of tetraphenylarsonium chloride with phenyltin trichloride in acetonitrile, 25°C.

hexaphenylditin, and trimethyl-triphenylditin[983]. In early work the hexaalkylditin compound in an alcoholic solution containing bromide ions, X^-, was placed in the anodic compartment of an electrolysis cell, where it underwent the following reaction:

$$R_3M - MR_3 + X_2 \rightarrow 2R_3MX$$

owing to the X_2 (bromine) electrolytically produced under constant current at the platinum anode. The end-point was observed by a rise in the indicator current, caused by excess of halogen, between a second pair of platinum electrodes sensitive to the X_3^-/X^- redox system. An amperometric plot of indicator current against generation time was obtained photographically in order to ensure an accurate determination. In the determination of hexamethylditin, bromine had to be used because, unlike iodine, its rate of reaction with hexamethylditin was greater than the rate of electrolytic generation of bromine. For a definite rate of generation of bromine, the titration depends on an appropriate rate of reaction of hexamethylditin with the halogen. The rate of reaction can be modified not only by changing the temperature, but also by choosing a proper concentration of halide ion (X^-) according to the equilibrium.

$$X_2 + X^- \rightleftarrows X_3^-$$

An attempt has been made to verify which of hexaethyl-, hexapropyl- hexabutyl-, hexaphenylditins, and trimethyltriphenylditin could be titrated with bromide and which could be iodinated[983]. In addition, since the quantitative reaction

$$Et_6Pb_2 + 2Ag^+ \rightarrow 2Et_3Pb^+ + 2Ag$$

have been verified for hexadiethyllead[984], the possibility of titrating each of the above organotin compounds by means of electrolytically generated silver ion to an amperometric end-point[985] was explored[983]. Conditions were also studied whereby a selective titration of one species in the presence of another could be carried out by the utilization of a complexing agent (quinoline) for silver. In the iodination of hexaalkylditin compounds, even when the iodide concentration is greatly reduced, analytically correct results are obtained only for hexaethylditin. All of the ditin compounds could be determined with bromine except for hexaphenylditin. Even at elevated temperatures and with a bromide concentration of 0.01 M, the hexaphenyl compound gave unsatisfactory results.As shown in Figure 38, the end-points are readily obtained and can be obtained by extrapolating the linear part of the indicator current caused by excess reagent, to the zero current line.

Tagliavini[983] found that in the present of the quinoline-silver ion in the anodic solution, the reaction rate of hexaphenylditin was appreciably slower than it was in the solution containing uncomplexed silver ion. This made it possible to estimate very accurately low concentrations of trimethyl triphenylditin in a large excess of hexaphenylditin. Because of the extreme accuracy in reading the end-point, satisfactory results were also obtained in titrations of small quantities of hexaphenylditin in an excess of trimethyltriphenylditin (Table 109).

Amperometric plots of these titrations are shown in Figure 39. The slopes of the titration curve (a) and (b), varies from the slope of the blank titration curve (c), because of the subsequent reaction of hexaphenylditin. The indicator reaches a constant value, indicating that the reaction rate of hexaphenylditin is equal to the generation rate of silver ion. In spite of this, the extra-polated values are in good agreement with the calculated values.

The titration of hexaalkylditin compounds with silver ion has been verified for hexaphenylditin, hexamethylditin, and trimethyltriphenylditin [983].

$$R_3Sn - SnR_3 + 2Ag^+ \rightarrow 2R_3Sn^+ + 2Ag$$

Known amounts of these substances dissolved in ethanol or alcohol-benzene

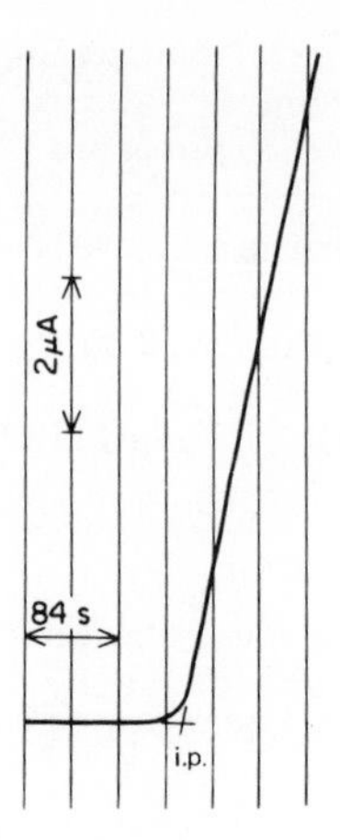

Figure 38 - Silver ion titration of hexamethylditin.

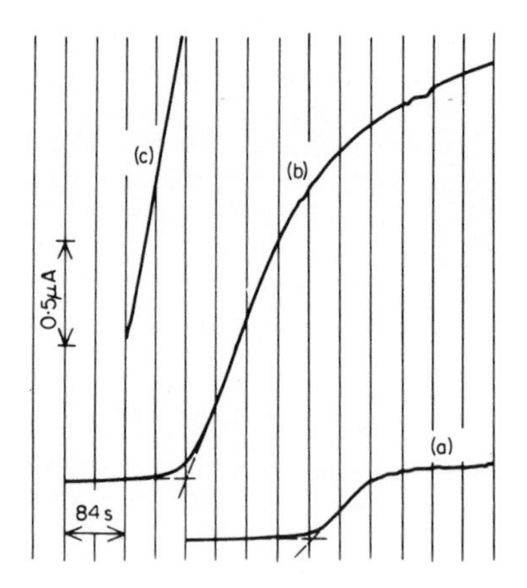

Figure 39 - Silver ion titration of hexaphenylditin-trimethyl triphenylditin mixture.

mixtures were added to alcoholic silver nitrate. The precipitated metallic silver was then separated and titrated by the Volhard method. Sodium fluoride or sodium tetraphenylborate was added to the filtrate to precipitate the triphenyl and trimethyl ions. Coulometric-amperometric titrations with silver ions can be carried out for all the hexaalkylditin compounds.

Dibutyltin dichloride and ioctyltin dichloride can be determined by amperometric titration in weakly acidic medium with standard oxalic acid solution[986]. Coulometric titration has been used for the titration of dialkyltin perchlorates[987]. A coulometric method for the determination of hexamethylditin in tetramethyltin with amperometric indication of the end-point employs 0.5 M methanolic ammonium bromide solution as basal electroyte [988].

Table 109 - Silver ion titration of $Me_3SnSnPh_3$-Ph_6Sn_2 mixtures in ethanol: benzene (80:20) (lithium nitrate, 0.1 M at 20°C).

Run no.	$(CH_3)_3Sn$-$Sn(C_6H_5)_3$ Coulombs calc.	$(C_6H_5)_3Sn$-$Sn(C_6H_5)_3$ Coulombs calc.	(Quinoline) (M)	Generator current (A.10^3)
1	0.718	-	-	1.95
2	0.718	0.025	0.03	1.85
3	0.359	0.276	0.03	1.86
4	0.150	1.69	0.03	0.9
5	0.150	6.76	0.03	0.9
6	0.150	6.76	0.03	0.9

ORGANOZINC COMPOUNDS

A. Classical Titration Procedures

Iodometric methods as described for the determination of organo-aluminium compounds are also applicable to the determination of organozinc compounds[990]. The procedure is capable of an accuracy of ± 3%. The reactions involved when iodine reacts with various types of organoaluminium and organozinc compounds are as follows:

$$AlR_3 + 3I_2 \rightarrow AlI_3 + 3RI$$

$$AlR_2OR + I_2 \rightarrow AlI_2OR + R - R$$

$$AlR_2Cl + 2I_2 \rightarrow AlI_2Cl + 2RI$$

$$ZnR_2 + 2I_2 \rightarrow ZnI_2 + 2RI$$

In this method a suitable volume of the hydrocarbon solution of the sample is stirred with an excess of a solution of iodine dissolved in toluene. The alkyl groups are completely iodinated within 20 min. Following the addition of dilute acetic acid, unreacted iodine is determined by titration with sodium thiosulphate solution under conditions of vigorous stirring. The concentration of dialkylzinc compound is then calculated from the amount of iodine consumed in the determination. The presence of water in the iodine reagent causes interferences by reacting with some of the alkyl groups. This is corrected by a suitable 'double titration' procedure.

For the iodometric determination of diethylzinc the titration vessel is first dried at 150°C and cooled[991]. Solid potassium iodide is introduced and the vessel put under a dry oxygen-free nitrogen purge. A measured volume of 0.4 N iodine is then introduced through the stopper, followed by the heptane-diluted diethylzinc sample. After 5 min, glacial acetic acid is added and excess of iodine is titrated with 0.2 N sodium solution to the starch indicator end-point. A reagent blank determination is run in parallel.

An unsuccessful attempt has been made to apply the amperometric silver nitrate titration procedure to the determination of diethylzinc[948]. In this procedure the sample is treated with dialkyl or diarly disulphide to produce a thiol, which is then titrated amperometrically with standard silver nitrate. Although the method was applied successfully to organolithium, diethylmagnesium, triisobutylaluminium and isopropylphenylpotassium, it was found that diethylzinc did not cleave the disulphide at a high enough rate to be of any practical use.

In a volumetric method for the determination of zinc in zinc dialkylphosphorodithioate the sample is ashed at 800 - 900°C and the ash is dissolved in hydrochloric acid containing concentrated nitric acid, then neutralized to methyl orange[992]. Zinc is determined complexometrically using standard EDTA. An alternative analysis of this substance is based on its precipitation as a silver salt or by oxidation to disulphate by means of iodine solution[993]. Zinc benzothiazolyl mercaptide can be determined iodometrically, either directly or after decomposition by acid.

B. Thermometric Titration

The thermometric titration of organozinc compounds[994] can be carried out in the same apparatus as used for organoaluminium compounds[754,755]. The following titrants were examined in detail[994].

Compound	Concentration (M)	Solvent
o-Phenanthroline (phen)	0.8	Anisole
2,2'-Bipridyl (bipy)	1.0 - 1.3	Toluene
3-Hydroxyquinoline (oxine)	1.6 - 1.7	Anisole
Ethanol (absolute)	2.0	Toluene
Water	2.2	Dioxane

For titration, a weighed amount of 0.2 - 2.5 M diethylzinc in toluene was added by hypodermic syringe to 40 - 50 ml of toluene in a dry, nitrogen-purged titration flask. Figure 40 shows typical titration curves obtained by titrating diethylzinc with phen and bipy. Both compounds react exothermically, and the heats of reaction are about the same (10 ± 2 kal/mol). However, bipy gives curves which are somewhat rounded, possibly indicating an unfavourable equilibrium; phen is a much better titrant, giving sharp, well defined breaks and considerably better precision.

Figure 41 shows typical curves obtained in the titration of diethylzinc with oxine. The reaction is more exothermic than either the phen or the bipy reactions (33 ± 4 kcal/mol). Curve 1 in Figure 41 shows the type of curve obtained with relatively pure samples of diethylzinc. It shows three breaks. The first is fairly well defined and corresponds to the reaction of 1 mol of oxine per mole of diethylzinc. The second is poorly defined and sometimes is not observed; it is thought to be related to reaction of the Et - Zn bond in compounds of the Et - ZnO type, but agreement with this assumption is not good. In any event, this break is not useful analytically. The third break is sharply defined, usually with a slight characteristic peak at the inflection point; it represents the reaction of 2 mol of oxine per mole of Et_2Zn,EtZnOEt or $Zn(OEt)_2$. Reaction with EtZnOH proceeds past the 1:1 stage but is not stoichiometric. If the amount of oxidation products present in the diethylzinc sample is large, or if oxine is used to titrate a sample which was previously titrated with bipy or phen, then the curve obtained is similar to II in Figure 41. In this case only the final break is significant, and it is sharply defined. If much ethylzinc hydroxide is present in the sample, then the final break is considerably less sharp and may take the form of a smoothly rounded dome where no definite end-point can be located. The steep rise at the start of curve II (Figure 41) is abnormal; it is observed only when the oxine

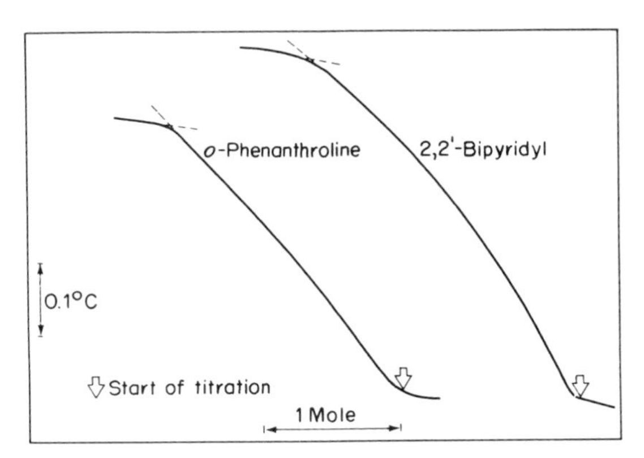

Figure 40 - Thermometric titration of diethylzinc with o-phenanthroline and 2,2'-bipyridyl.

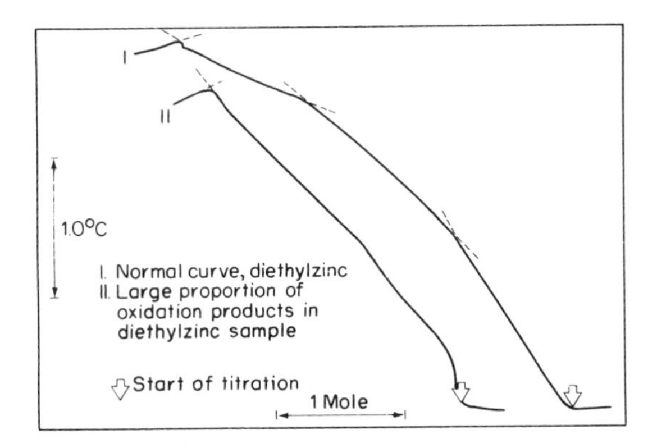

Figure 41 - Thermometric titration of diethylzinc with oxine.

titration is made shortly after addition of ethanol or water to the diethylzinc sample. Similar segments of abnormally high slope are observed when such badly contaminated samples are titrated with phen or bipy.

There is little difference between results obtained by simple titration and those obtained using the 'clean-up' technique as described in the section on the thermometric titration of organoaluminium compounds. In general, it is considered that this technique is preferable to the simple titration because it is less vulnerable to contamination errors, and because comparison of results for the 'clean-up' sample and subsequent samples give some idea of the level of impurities present in the titration system and the effectiveness of the solvent purification and sample-vessel preparation techniques in use.

The thermometric titration of diethylzinc with gaseous oxygen from a motor-driven syringe burette has also been investigated to see whether quantitative indication of the reaction could be obtained using gas as titrant[994]. The oxidized solutions were subsquently titrated with oxine.

Table 110 summarizes the results of these experiments.

These tests gave some indication that the reaction of diethylzinc with oxygen or reaction with ethanol gives equivalent products. Oxygen is not considered to be a practical titrant because of the large volume required, the need for slow addition, and the fact that the titration curve was not very well defined. The initial portion was normal, but a break was obtained

Table 110 - Titration of diethylzinc with air or oxygen[a], then oxine

		Calculated composition			Found by oxine titration		
Test No.	Oxygen added as	Et_2Zn	EtZnOEt	$Zn(OEt)_2$	Et_2Zn	EtZnOEt	Total Zinc
1	-	-	-	-	0.435	0.040	0.478
2	Air,0.020	0.418	0.060		0.436	0.044	0.480
3	O_2, 0.099	0.339	0.139	-	0.329	0.147	0.476
4	O_2, 491[c]	-	-	0.478	-	-	0.458

[a] All results in mmol/g
[b] Et_2Zn from first oxine break, total Zn from final break, EtZnOEt by difference.
[c] Sample titrated with oxygen.

Table 111 - Titration of diethylzinc with phen, bipy and Oxine

Titration technique	Et_2Zn (mmol g^{-1}) by titration with Oxine			
	Phen	Bipy	Oxine To first break	Oxine To final break
Simple titration[a]	1.134		1.137	1.142
	1.137[b]		1.115	1.138
			1.112[b]	1.140[b]
"Clean-up" titration[c]	(1.009)[d]	(1.130)[d]		
	1.142	1.223		
	1.144	1.130		
"Clean-up" titration[c] repeated	(1.143)[d]			
	1.135			
	1.136			
Average, excluding "clean-up" samples	1.138	1.176	1.121	1.140

[a] 1.5 - 2.1 g of sample added to 47 ml of toluene and titrated. Rate of titrant addition, 0 5 ml min^{-1} except as noted.
[b] Rate of titrant addition 0.2 ml min^{-1}.
[c] Small "clean-up" sample added and allowed to react with reactive impurities, if any, in titration system before main samples were added. Excess reagent from each titration counted as part of the following titration.
[d] "Clean-up" sample. Variation of this result from those following is a rough indication of the amount of reactive impurities, if any, present before starting the analysis. These results are not considered valid.

at an O:Zn ratio of 0.65 rather than the expected 1.0. The temperature then remained approximately constant (i.e. heat production was balanced by heat loss) up to a second break at an O:Zn ratio of 2.06, after which the temperature decreased. The end product was assumed to be $Zn(OEt)_2$ rather than the monoperoxide, EtZnOOEt. There was no evidence of formation of the diperoxide.

Thermometric titration with phen provides a simple, rapid and precise method for determining the net diethylzinc content of a solution which may also contain its oxidation or hydrolysis products. Titration with oxine gives a precise measure of the total zinc content in diethylzinc solutions which contain its oxidation products, but is not reliable if hydrolysis products are present. Under favourable circumstances oxine gives, in the same titration, a measure of both total zinc and diethylzinc.

In Table 111 are shown results obtained by titration with phen, bipy, and oxine of a 1 M toluene solution of diethylzinc.

The results in Table 111 indicate that both phen and oxine give precise results. For comparison, the total zinc was determined by hydrolysis and titration with EDTA; the values obtained were 1.124 and 1.131 mmol g^{-1}, in good agreement with the phen and oxine values. The rate of titrant addition causes no significant variation in results. The first oxine break is less sharply defined than the final break and gives results that are somewhat lower and of poorer precision. The bipy results lack precision.

CHAPTER 3

SPECTROSCOPIC TECHNIQUES: VISIBLE AND ULTRAVIOLET SPECTROSCOPY

ORGANOALUMINIUM COMPOUNDS

A. Visible Spectroscopy

The ultraviolet and visible spectra of isoquinoline alone (curve 1) and of mixtures of isoquinoline and diethylaluminium hydride (curves 2 and 3) are shown in Figure 42. In Figure 43 are shown the spectra of isoquinoline alone (curve 1), mixtures of isoquinoline and diethylaluminium ethoxide (curve 2), and isoquinoline and triethylaluminium (curve 3). Comparison of curves 1 and 2 shows that the spectrum of diethylaluminium ethoxide remains unchanged in the presence of isoquinoline, suggesting that no reaction occurs between these two substances. The addition of triethylaluminium to isoquinoline, however, produces a distinct change in its spectrum, suggesting that complex formation occurs. Analysis of the difference between curves 1 and 3 (Figure 43) shows that triethylaluminium and isoquinoline form a colourless 1:1 complex. Diethylaluminium hydride forms two complexes with isoquinoline. The 1:1 diethylaluminium hydride isoquinoline complex is colourless and absorbs only in the ultraviolet region of the spectrum (Figure 42, curve 2). The 1:2 diethylaluminium hydride - isoquinoline complex has an intense red coloration and this is shown in the absorption spectrum (see the strong absorption occurring above 400 nm in curve 3, Figure 42). Procedures have been devised based upon these observations for determining trialkylaluminium compounds and dialkylaluminium hydrides either singly or in the presence of each other[995].

The absorption of the triethylaluminium - isoquinoline complex occurring at about 328 nm (see arrow on curve 3, Figure 43) can be used for the colorimetric determination of trialkylaluminium compounds (isoquinoline itself does not absorb at this wavelength). The dialkylaluminium hydride - isoquinoline comple also absorbs at 328 nm and would, of course, interfere in this method of determination of trialkylaluminium compounds. Also, the absorption of the red 1:2 diethylaluminium hydride - isoquinoline complex occurring at about 500 nm (see curve 3, Figure 42) can be used to determine dialkylaluminium hydrides without interference from any trialkylaluminium compounds present in the sample, as the colourless 1:1 trialkylaluminium - isoquinoline complex does not absorb at 500 nm.

Dialkylaluminium hydrides can be used as visual indicators in the titration of organoaluminium compounds that form a 1:1 complex with iso-

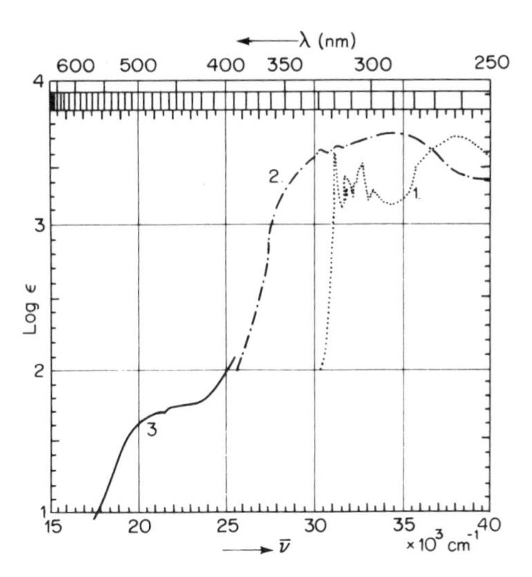

Figure 42 - Absorption spectrum in cyclohexane (at 20 °C) of isoquinoline alone (1); isoquinoline aluminium hydride, i.e. colourless 1:1 complex (2); and excess of isoquinoline plus diethylaluminium hydride, i.e. red 1:2 diethylaluminium hydride - isoquinoline complex (3)

1. N

2. N → $Al(C_2H_5)_2$ | H

3. Red 1:2 diethylaluminium hydride - isoquinoline complex obtained with diethylaluminium hydride in the presence of 10 M excess of isoquinoline.

quinoline. Thus, to determine the concentration of a trialkylaluminium compound in a solution, a small volume of diethylaluminium hydride is added (often the sample will contain a small amount of this as an impurity left in from the manufacture) and the solution is titrated with isoquinoline. When the trialkylaluminium compound and the dialkylaluminium hydride have both formed 1:1 complexes with isoquinoline then the solution suddenly becomes red owing to the formation of some 1:2 diethylaluminium hydride - isoquinoline complex. The volume of isoquinoline corresponding to the first appearance of the red colour can be taken as the end-point of the titration and is equivalent to the total trialkylaluminium plus dialkylaluminium hydride content of the sample. Every precaution must be taken to avoid contact of substances in the cell with air and moisture which would have a serious influence on the spectra obtained[995].

Neumann[996,997,998] continued the work started by Bonitz[995] on the reaction occurring between one mole of dialkylaluminium hydrides and two moles of isoquinoline to form strongly coloured 1:2 complexes. He believed that reactions of this type might form the basis for a spectro-

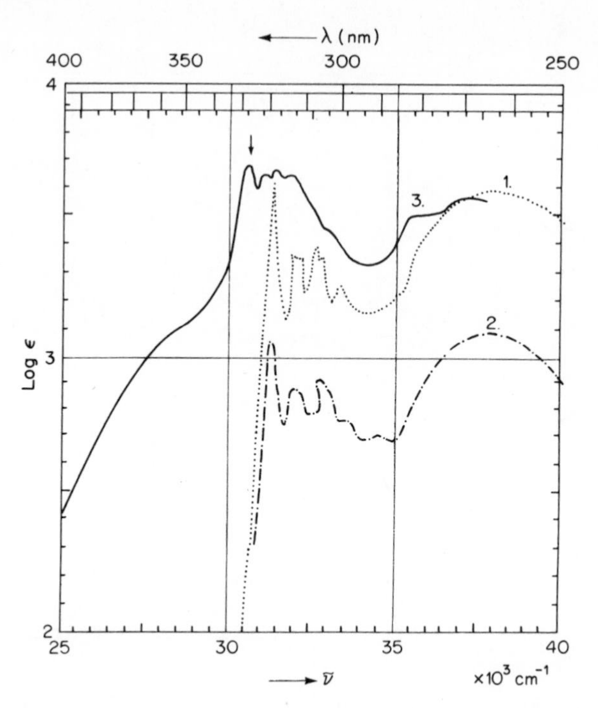

Figure 43 - Absorption spectra in cyclohexane (at 20°C) of isoquinoline alone (1), isoquinoline plus diethylaluminium ethoxide (2), and isoquinoline plus triethylaluminium (3).

photometric method for estimating low concentrations of dialkylaluminium hydrides in trialkylaluminium compounds. He studied the reaction of dialkylaluminium hydrides with isoquinoline and with various other azomethines which form coloured 1:2 complexes (Neumann[996]) and concluded that isoquinoline and benzalaniline were the two most interesting azomethines to study further. He postulated the reactions occurring as follows:

1:1 AIR_2H - azomethine complex

R_2AlH + RCH = NR = $RCH_2N(AlR_2)R$(colourless)

1:2 AlR_2H - azomethine complex

$RCH_2N(AlR_2)R$ + RCH = NR = $RCH_2N(AlR_2)R$
RCH = NR
(coloured)

The azomethine used in this reaction may be either open-chain, (e.g. benzalaniline) or cyclic (e.g. isoquinoline). Trialkylaluminium compounds produce 1:1 molecular compounds upon reaction with azomethines. These, however, absorb at shorter wavelengths than the 1:2 products obtained with dialkylaluminium hydrides. Compounds of the type $AlR_2(OR)$ do not react with azomethines to produce coloured derivatives.

In Figure 44 (curves C and D) are shown respectively, the absorption curves in the range 350 - 550 nm obtained for a mixture of dialkylaluminium

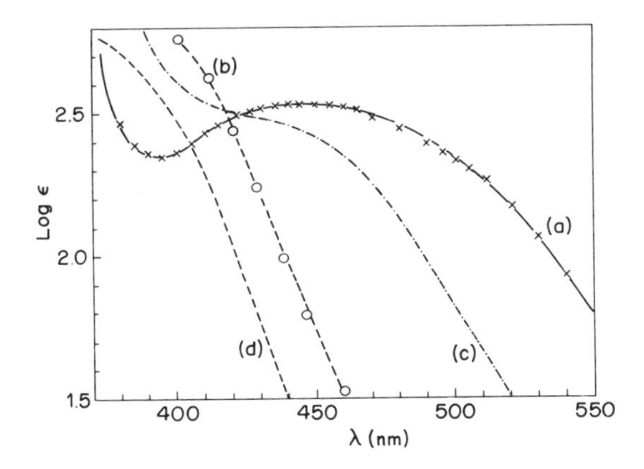

Figure 44 - Absorption spectra in the region 350 - 550 nm. (a) Dialkylaluminium hydride - benzalaniline complex; (b) 1:1 trialkylaluminium - benzalaniline complex; (c) 1:2 dialkylaluminium hydride - isoquinoline complex; (d) 1:1 trialkylaluminium hydride - isoquinoline.

hydride with isoquinoline and a trialkylaluminium compound with isoquinoline. It is seen that only the dialkylaluminium hydride - isoquinoline reaction product absorbs and this can be conveniently measured at a wavelength of 460 nm (log = 2.39). At this wavelength the trialkylaluminium-isoquinoline reaction product exhibits no absorption whatsoever. However, occassionally, due to the presence of colloidal metals, organoaluminium preparations are dark coloured and these colours sometimes interfere in measurements carried out at 460 nm. This necessitates troublesome corrections when carrying out measurements of optical density at 460 nm of the dialkylaluminium hydride - isoquinoline complex. Furthermore, the isoquinoline colour does not correspond to any definite maximum in the spectrum. The extinction curve at 460 nm shows only a flat shoulder which rises sharply in the direction of the shorter wavelengths (Figure 44, curve C). A considerable improvement was obtained by the use of benzalaniline as reagent. It causes the formation of a colour with a broad absorption band at 450 nm (log = 2.54) (Figure 44 curve A). The evaluation can also be made at 500 nm (log = 2.31) and this has the advantage that any colour due to colloidal metals in the samples does not interfere in the analysis. The absorption at 500 nm obtained with benzalaniline obeys the Lambert-Beer Law. In Table 112 it is seen that errors are usually within ± 2% (relative). The coloured molecular compound formed between dialkylaluminium hydride and benzalaniline still contains Al-C linkages and is, therefore, a very sensitive metal alkyl compound. If any of the Al-C linkages in this product are destroyed, for example by hydrolysis or oxidation, then the molecular compound decomposes at once and the colour disappears (i.e. low hydride contents will be obtained). Hence, it is essential to avoid completely contamination of the cell solution with moisture or oxygen during the spectrophotometric analysis. Neumann achieved this by using very dry benzene as a solvent. Also, the benzalaniline reagent itself contained some alkylaluminium which completely dried this reagent. Flow-through glass spectrophotometer cells were employed in order to reduce sample contamination to nil by atmospheric water and oxygen.

Mitchen[999] also studied the spectrophotometric titration of isoquinoline at 460 nm with dialkylaluminium hydrides and trialkylaluminium compounds. He used a modified sample cell compartment. The sample cell was fitted with

Table 112 - Accuracy of spectrophotometric determination of diisobutylaluminium hydride at 500 nm. Diisobutylaluminium hydride - benzalaniline complex

	Weight of diisobutylaluminium hydride (mg)		
Sample Number	Added	Found	Error rel (%)
1	34.8	34.0	-2.2
2	58.0	57.8	-0.3
3	70.0	71.4	+2.0
4	101.5	100.2	-1.2
5	125.8	124.2	-1.3
6	141.0	142.5	+1.2
7	155.7	149.4	-4.1
8	216.8		+1.4
9	231.0	231.2	+0.1
10	283.5	281.6	-0.8
11	300.5	300.3	+1.9

a rubber serum cap and sample transfers were made using a hypodermic syringe which was weighed before and after transfer in order to obtain the weight of the sample added to the spectrophotometer cell. The benzene solvent is thoroughly dried by distillation from phosphorus pentoxide. Oxygen and moisture are rigorously excluded from the sample solution during the analysis.

In the analysis a 25 ml portion of a benzene solution of isoquinoline is transferred to the nitrogen filled spectrophotometer cell. A solution of diethylaluminium hydride in dry benzene is then added until a stable red colour is obtained having an absorbance of about 0.3 at 460 nm (i.e. the 1:2 diethylaluminium - isoquinoline complex is formed). To determine dialkylaluminium hydride in an unknown sample a weighed portion of benzene diluted sample is added to the cell contents and the absorbance recorded. The increase in absorbance is due to dialkylaluminium hydride in the sample. According to Mitchen this procedure is most conveniently calibrated by determining the absorbance factors obtained for various known weights of a pure sample of diethylaluminium hydride.

To determine the trialkylaluminium content of the same sample the addition of sample is continued until the absorbance passes through a maximum (see Figure 45). At maximum absorbance all the isoquinoline is bound as the red 1:2 dialkylaluminium hydride - isoquinoline complex and as the colourless 1:1 trialkylaluminium - isoquinoline complex. Thereafter, further sample addition destroys the 1:2 red complex in preference to the 1:1 complexes, decreasing the absorbance at 460 nm. Sample addition is continued until a convenient absorbance for measurement by the spectrometer is obtained (e.g. 1:5). This absorbance and the additional volume of sample added since the maximum absorbance are both recorded. A second weighed portion of sample solution is then added to the cell and the volume and absorbance again recorded. All absorbance volumes should be corrected to the absorbance equivalent at the original volume (25 ml) of benzene solution of isoquinoline put into the spectrophotometer cell at the beginning of the analysis.

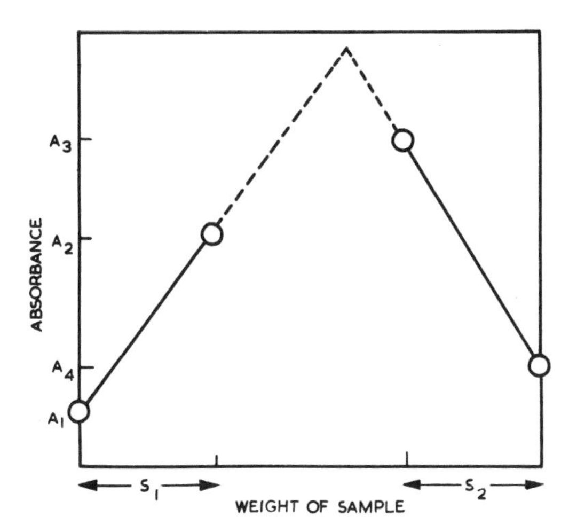

Figure 45 - Absorbance changes in the Mitchen spectrophotometric method for estimation of dialkylaluminium hydrides and trialkylaluminium compounds.

The decrease in absorbance is due to both the dialkylaluminium hydride and trialkylaluminium. Because the curve is calibrated as dialkylaluminium hydride, the decrease in colour observed is that which would be produced by the dialkylaluminium hydride (already determined) plus the dialkylaluminium hydride equivalent of the trialkylaluminium present in the sample. It is assumed in Figure 45 that the sample size (S_1) used to determine the dialkylaluminium hydride (original colour increase produced) is identical with the sample size (S_2) used to determine trialkylaluminium. Let x represent mg AlR_2H and y represent mg AlR_3 in the weighed sample. Then (A_2 - A_1) represents colour due to x mg AlR_2H. The dotted line from A_2 to A_3 represents additional unweighed sample which complexes all the isoquinoline at the maximum absorbance and then decreases the colour as the stronger trialkylaluminium isoquinoline and dialkylhydride - isoquinoline 1:1 complexes are formed at the expense of the 1:2 dialkylaluminium hydride-isoquinoline complex (A_3 - A_4) represents the further absorbance decrease due to the second weighed sample (S_2) containing x mg dialkylaluminium hydride and y mg trialkylaluminium compound. The difference (A_3 - A_4) - (A_2 - A_1) is the dialkylaluminium hydride equivalent of y mg trialkylaluminium compound. A standard curve was prepared by a reversal of the working procedure. Weighed increments of isoquinoline were added to a solution of the 1:1 complex in 25 ml benzene. Since accurate addition of small amounts of isoquinoline necessitated the use of a dilute solution of isoquinoline in benzene, the volumes added were of insignificant magnitude. Therefore, the absorbance value for each addition was corrected for the dilution effect.

In a further study[1000] of the isoquinoline spectrophotometric method for the determination of dialkylaluminium hydrides and trialkylaluminium, the dialkylaluminium hydride reacts with isoquinoline to form red coloured 1:2 complexes which have an absorption maximum at 460 nm. Trialkylaluminium compounds form only a colourless 1:1 complex which has no absorption at 460 nm. Samples were analysed by this method and by the ammonia method, (Table 113).

In a further method for the spectrophotometric titration of isoquinoline at 460 nm with dialkylaluminium hydrides and trialkylaluminium compounds a sample cell fitted with a rubber serum cap is used[999]. Sample transfers are made using a hypodermic syringe which is weighed before and after transfer in order to obtain the weight of sample added to the spectrophotometer cell. In the analysis a portion of a benzene solution of isoquinoline is transferred to the nitrogen-filled spectrophotometer cell and a solution of diethylaluminium hydride in dry benzene is then added until a stable red colour is obtained having an absorbance of about 0.3 at 470 nm (i.e. the 1:2 diethylaluminium - isoquinoline complex is formed). To determine dialkylaluminium hydride in an unknown sample a weighed portion of benzene-diluted sample is added to the cell contents and the absorbance recorded. The increase in absorbance is due to dialkylaluminium hydride in the sample.

To determine the trialkylaluminium content of the same sample the addition of sample is continued until the absorbance passes through a maximum, when all the 1:1 complex with isoquinoline is estimated. Samples were analysed by this method and by the ammonia method[1001]. Good agreement is obtained by the two methods of analysis.

In a further study of the isoquinoline spectrophotometric method for the determination of dialkylaluminium hydrides and trialkylaluminium, decomposition by air and moisture of organoaluminium samples in syringes was considerably reduced by using a syringe with a smooth-bore barrel and a machined Teflon plunger[1002]. The syringes require no lubrication and it is claimed that organoaluminium samples can be stored in the syringe for several days without severe decomposition.

Table 113 - Analysis of trialkylaluminium compounds. Comparison of the Wadelin[1000] isoquinoline spectrophotometric titration method and the Ziegler and Gellert[1001] ammonia method

Compound	Activity (mmol/g)[a]	
	Isoquinoline method	Ammonia method
Tri-n-propylaluminium	6.00	6.02
	5.96	5.92
	6.58	6.57
	6.41	6.37
	6.30	6.34
	5.67	5.81
	5.47	5.81
Tri-iso-butylaluminium	4.85	4.80
Triisohexylaluminium	3.52	3.44

[a] Each result is the average of duplicate determinations.

ORGANOARSENIC COMPOUNDS

Phenarsazine derivatives can be determined spectrophotometrically as the disodium salt of dinitrophenarsazinic acid at 520 nm[1003]. The sample is dissolved in glacial acetic acid and oxidized and nitrated with an excess of nitric acid to form dinitrophenarazinic acid. Addition of excess of sodium hydroxide yields a violet disodium salt suitable for photometric evaluation. From 1 to 8 μg/ml of phenarsazine can be determined by this method with an error of ± 4%. A spectrophotometric method has been described for the determination of 4 - hydroxy - 3- nitrophenylarsenic acid in animal feeds[1004].

Two digestion methods have been compared for their effectiveness in releasing arsenic from three organoarsenicals introduced into wastewater samples[1005]. The digestive methods utilized included a wet method employing hydrogen peroxide - sulphuric acid and ultraviolet photodecomposition. The organoarsenicals investigated were disodium methanearsonate, dimethylarsinic acid, and triphenylarsine oxide. All the digestive methods gave quantitative arsenic recoveries when applied to wastewater samples. The ultraviolet photodecomposition proved to be an effective digestive technique, requiring a 4 h irridation to decompose a primary settled raw wastewater sample containing spiked amounts of the three organoarsenicals. Arsenic was determined in the digests by the silver diethyldithiocarbonate spectrophotometric method.

ORGANOBORON COMPOUNDS

Colorimetric methods based on the use of triphenyltetrazolium chloride have been described for the determination of pentaborane and for monitoring concentrations of pentaborane in air[1010,1011]. Instruments have been described for monitoring atmospheres: a portable field model and an automatic differential reflectance photometric analyser. Both methods depend on non-specific, highly sensitive reductions of the reagent by boron hydrides to give a red coloration. The field model can detect 0.1 ppm of decaborane and 0.5 ppm of pentaborane and the automatic instrument is capable of detecting 0.1 ppm of either compound.

An alternative excellent procedure for the determination of pentaborane in air samples is based on the formation of a coloured pyridine complex[1012]. Using toluene as the solvent, Beer's law is followed through the concentration range 2 - 12 μg/ml at 400 nm. A trapping system is described which permits dynamic air sampling at rates as high as 15 1/min with over 95% efficiency. Using a 30-1 air sample, the method is suitable for pentaborane concentrations as low as 0.1 ppm.

Non-specific colorimetric methods have been described for the determination of decaborane based on colour formation with triphenyltetrazolium chloride[1010,1011], as well as colorimetric methods based on reaction with quinoline[1013,1014] and β-naphthoquinoline in which red addition products with an absorption maxima at 490 nm are produced by addition of solutions of quinoline or β-naphthoquinoline in xylene medium to the sample. These methods are very efficient and accurate for dynamic air analysis, only one absorption bulb (containing 2% quinoline in xylene) being needed. Interference by diborane and pentaborane is negligible. A colorimetric procedure based on the use of 1,2-di(4-pyridyl)ethylene is claimed to have greater sensitivity, more rapid colour development and lower reagent blanks than the quinoline method[1015]. This produces with decaborane a pink to red colour with an absorption maximum at 515 nm. The coloured complex formed between decaborane and benzo(f)quinoline is claimed to be more stable than that formed with quinoline and is more suited to spectrophotometric determination at 486 nm[1016]. The absorption maximum of the decaborane - reagent complex occurs at 506 nm and of the diethylborane - reagent complex at 525 nm.

A colorimetric method using indigo carmine for the determination of decaborane, dimethylaminoborane, trimethylaminoborane, and pyridineborane is sensitive enough to detect 1 - 10 μg of boron[1017]. The method does not distinguish between boron present as boron hydrides, boric acid, or boron oxides.

A method in which decaborane in water or cyclohexane is treated with an excess of N-diethylnicotinamide in water is claimed to be free from interference by boric acid, boron salts, diborane, and pentaborane[1018]. After 90 min the orange - red colour intensity is measured at 435 nm and the decaborane concentration is derived from standards. In the colorimetric determination of boron hydrogen compounds with molybdophosphoric acid, addition of molybdophosphoric acid to decaborane, dimethylamineborane, or sodium borohydride produces a blue solution[1019]. The colour intensity is directly related to the amount of boron present. A direct micro method for the determination of decaborane involves measurement of the strong u.v. absorption occurring at 265 - 270 nm of the solution in triethanolamine[1014]. Beer's law is valid and from 1 - 25 < g/ml of boron can be determined accurately. The method is unaffected by the very slow hydrolysis of decaborane in the solvent and can be applied to the dynamic and static analysis of air and gases containing decaborane. The air or vapour is bubbled slowly through two glass bulbs containing aqueous triethanolamine the contents are afterwards combined and diluted and the u.v. absorption is determined; the decaborane recovery is 98%. Static air samples are taken with a gas pipette (250 ml) into which 5 ml of the triethanolamine solution are introduced.

The photometric determination at 590 nm of diphenylborinic acid and its esters using diphenylcarbazide as reagent has been discussed[1020]. In a colorimetric method for the determination of borinic acids in biological materials the frozen tissue sample is mixed with calcium hydroxide and concentrated sulphuric acid and rendered colourless by heating with hydrogen peroxide[1021]. After complete destruction of hydrogen peroxide the determination is completed by reaction with 1,1-dianthrimide in concentrated sulphuric acid at 90°C and spectrophotometric evaluation at 620 nm. The error does not exceed 15%. An integrating monitor has been designed for determining low concentrations of gaseous boron hydrides in air[1022]. The boron hydride vapours are quantitatively converted by burning into boron oxide, which is then determined colorimetrically with carmine at 585 nm. Tetraphenylborate ions have been determined spectrophotometrically, based on reaction with excess of standard rosaniline solution at pH 4.6 The precipitate is filtered off and the absorbance of the coloured filtrate is measured at 546 nm against water; the decrease in molar absorptivity of the filtrate is proportional to the concentration of tetraphenylborate ion.

ORGANOCHROMIUM COMPOUNDS

Malik and Ahmed[1046] have devised rapid spectrophotometric methods for the estimation of chromium stearate used as additives in lubricating oils. They found that when chromium stearate is heated with o- or m-toluidine at 180° to 200°C for about 20 min, complexes are formed having maximum extinctions at 480 and 540 nm respectively. Beer's law is obeyed for the concentrations of 1.2 to 10.7 mg per ml, so that the metal ion content of the soap can be determined even in dilute solution. These workers[1047,1048] also studied the spectrophotometric behaviour at 425 nm of chromium (III) and stearates, in amounts between 1 and 14 nmole 1^{-1}, in non-aqueous solution.

ORGANOCOBALT COMPOUNDS

Cyanocobalamin

A general theory of partition has been developed and applied to the spectroscopic determination of cyanocobalamin and hydroxocobalamin[1023], In this method, the determination of the apparent partition coefficient is based on the spectrophotometric determination of total cobalamin in each phase of a benzyl alcohol and water partition at a wavelength (356 nm) at which both components have the same molar absorptivity. The total cobalamin concentration in each phase is given by:

$$\frac{356 \text{ nm}}{0.0174} \ \mu g/ml$$

and the apparent distribution coefficient for any particular mixture is given experimentally by:

$$K = \frac{356 \text{ nm (solvent}}{356 \text{ nm (water)}}$$

Good agreement is reported between determined and known concentrations of mixtures of cyanocobalamin and hydroxocobalamin.

Newark and Leff[1024] proposed a test for determining the purity of cyanocobalamin injection U.S.P. The sample (containing 5 - 10 mg of cyanocobalamin) is extracted with 3 x 5 ml portions of a mixture of cresol-carbon tetrachloride (1:1). The combined extract is washed with 0.1% aqueous disodium EDTA solution (10 ml) and water (2 x 10 ml). Acetone (100 ml) is then added with constant stirring, then light petroleum (100 ml) added and the mixture left to stand for 1 h. It is then filtered through sintered glass, the precipitate washed with acetone and dried at 105° - 110°C in vacuo (5 mm Hg) for 2 h prior to weighing. Finally, the precipitate is dissolved in water, diluted to 200 ml and the cyanocobalamin determined spectrophotometrically at 361 nm. Bruening et al[1025] devised a method for measuring the relative purity of cyanocobalamin in pharmaceutical products. They showed that pure cyanocobalamin has values of $E^{1\%}_{1cm}$ at 341 nm and 376 nm of 80.4 and 80.9 respectively; the average ratio E(341 nm)/E(376 nm) being 0.990. They suggest the following maximum values to limit the amount of "red pigments" in injections made from cyanocobalamin - E(341 nm/E(376 nm) > 1.020, $E^{1\%}_{1cm}$ (341 nm) > 83.0, $E^{1\%}_{1cm}$ (376 nm) > 82.5. The validity of this procedure was established by comparing results with the purity index found by a combination of tracer and spectrophotometric (361 nm) methods. The average difference between duplicate determinations of the extinction ratio was 0.6%. In a later paper Bruening and Kline[1026] state that results by the extinction-ratio test agreed with those by the U.S.P. XV1 limit test for cyanocobalamin solids for 20 samples of injections, of which 11 were authentic. The validity of both methods was supported by inspectional evidence obtained by the Food and Drug Administration. Bayer[1027] pointed out that when exposed to light, cyanocobalamin in aqueous solution is converted into hydroxocobalamin, which has a lower extinction coefficient at 361 nm. Errors in the spectrophotometric assay of cyanocobalamin may be avoided by avoiding exposure to light; or by measuring the extinction at 356 nm, which is isosbestic for cyanocobalamin and hydroxocobalamin or by using abs ethanol as solvent. Covello and Schettino[1028] described a method for the chromatographic separation and spectrophotometric determination of cyano and hydroxocobalamins in association with other pharmaceutical products. The two cobalamins are converted into the dicyano derivatives by treatment with potassium cyanide, and concentrated by extraction with butanol of an aqueous solution saturated with ammonium sulphate. Impurities are removed by paper chromatography, and the two purified cobalamins are eluted with water from the spots. The extinctions are then measured spectrophotometrically at 367 nm (cyanocobalamin) and at 580 nm (hydroxocobalamin) and the concentrations of both cobalamins calculated using standard calibration curves. A method has been described for the determination of cyanocobalamin in injection liquids and in purified liver extracts[1029].

A differential spectrophotometric method was developed for the determination of cyanocobalamin and hydroxocobalamin[1030,1031]. This is claimed to be much more precise than the direct spectrophotometry at 361 nm. To determine hydroxocobalamin in the presence of cyanocobalamin, the difference is measured between the absorbances at 349 nm in 0.01 N sodium hydroxide and in 0.01 N hydrochloric acid solution. To determine cyanocobalamin in the presence of hydroxocobalamin the ratio of the absorbances at 351 and 361 is measured in 0.01 N hydrochloric acid solution. A spectrophotometric method has been used for the determination of cyanocobalamin in concentrates and in supplements for compound feeding stuffs[1032]. Cyanocobalamin is determined from absorbance measurements made at 360 and 535 nm[1033]. A chromatographic separation and spectrophotometric determination can be used to determine cyanocobalamin in orange juice[1034]. The cyanocobalamin is determined by measuring the absorbance at 530 nm.

A method suitable for the determination of cyanocobalamin in injections, tablets, and liver extracts depends on the colour produced with nitroso-R salt at controlled pH after oxidation of the sample with hydrogen peroxide [1035]. The absorbance of the final solution is measured at 420 nm. Beer's law is obeyed for the range 100 - 600 μg of cyanocobalamin and the recovery is 100 - 103.8%.

An automated procedure has been developed for the determination of cyanocobalamin in pharmaceutical dosage forms[1036]. Diluted samples are mixed with sulphuric acid and passed to a photolysis device. The hydrogen cyanide produced is distilled as it is formed, and trapped in sodium hydroxide, and this solution is mixed with sodium dihydrogen phosphate solution and chloramine T solution in an ice-bath, reacted with saturated aqueous 3-methyl-1-phenyl-2-pyrazolin-5-one, and the absorbance is measured at 620 nm. A spectrophotometric method for the determination of down to 0.25% of cyanocobalamin in hydroxocobalamin depends on photolysing the cyanocobalamin to hydrogen cyanide, which is then collected by diffusion and measured colorimetrically at 506 nm[1037].

Dominguez et al[1038] modified the original Rudkin and Taylor[1039] spectrophotometric method for the determination of cyanocobalamin in liver extracts and polyvitamin mixtures. Liver extracts were first extracted with benzyl alcohol, then treated with butanol to remove impurities that interfere in the spectrophotometric determination of cyanocobalamin. Results obtained agreed with those of the U.S.P. microbiological method; the Dominguez et al method has the advantages of simplicity and speed. Polyvitamin mixtures were added to columns of Amberlite IRA-400 or strongly basic Zerolit FF resin, which retained interfering riboflavine, and then cyanocobalamin was determined spectrophotometrically in the percolate.

In a further method for the determination of cyanocobalamin in pharmaceutical products in the presence of liver extract and salts of cobalt, iron, and copper, the absorbance of a sample extract is measured at 570 nm[1040]. The 3:4 complex formed by cyanocobalamin with Fast Navy 2R (CI Mordant Blue 9) has a maximum at 550 nm and an inflection at 620 nm; the latter point is used for the spectrophotometric determination of cobalt[1041]. The presence of calcium, magnesium, manganese, lead, and zinc does not cause interference if EDTA is used as masking agent; interference from lead can also be suppressed by the use of tartaric acid, from manganese by the use of ascorbic acid, and from nickel by treatment with EDTA solution at pH 8.5 for 15 min. Further spectrophotometric methods have been described for the indirect determination of cyanocobalamin in cyanocobalamin preparations, liver extracts, and multivitamin preparations[1042].

The determination of cobalt in aqueous solutions of cyanocobalamin can be based on the catalytic effect of cobalt on the fluorescence produced by hydrogen peroxide with luminol[1043-1045].

Mader and John[1049] developed a counter-current method of analysis of cyanocobalamin which can be applied to the determination of the purity of crystalline cyanocobalamin; oral grade solids and simple solutions. The presence or absence of pseudocobalamin and other non-cobalamin fractions may also be determined. Results agree well with those obtained by radio-active tracer methods. To a sample containing about 50 μg of cyanocobalamin are added 50 ml of water, 5 ml of potassium cyanide solution (10%) and 5 ml of sodium nitrite (2%) and the pH of the resulting solution is adjusted to 4 with 6 N acetic acid. The solution is boiled, then cooled, and 1 ml of formaldehyde solution (37%) is added before filtration. Half of the combined filtrate and washings is extracted with cresol-carbon tetrachloride (1:1) and the extract is washed first with 5N sulphuric acid and then with sodium bicarbonate (2%) - potassium cyanide (0.4%) solution. To the washed extract are added 20 ml of chloroform, 5 ml of butanol and 5 ml of water. The aqueous layer is separated centrifugally and the extraction is repeated with two further 5 ml portions of water; the volume of the extract is made up to 25 ml. Aliquots of this solution are used for the counter-current distribution in 8 tubes; the solvent system employed is benzyl alcohol shaken with an equal volume of water till both phases are mutually saturated. The cyanocobalamin content of the phases is determined micro-biologically or spectrophotometrically.

Hydroxocobalmin, Cyanocobalamin and other Cobalamins

Bayer[1030,1031] has described a spectrophotometric method for the determination of hydroxocobalamin and cyanocobalamin. This worker points out that a differential spectrophotometric method for the determination of cyanocobalamin shows a mean deviation of ± 0.15% and is much more precise than direct spectrophotometry at 361 nm. To determine hydroxocobalamin in the presence of cyanocobalamin Bayer[1031] measured the difference between the extinction at 349 nm in 0.01 N sodium hydroxide and in 0.01 N hydrochloric acid solutions. To determine cyanocobalamin in the presence of hydroxo-cobalamin, Bayer[1031] measured the ratio of the extinctions at 351 and 361 nm in 0.01 N hydrochloric acid solution.

For the standardization of hydroxocobalamin the absorbance can be measured near 351 nm against the buffer solution ($E^{1\%}_{1cm}$ for the pure dry-substance is 190)[1050]. For identification, the absorption spectrum from 250 to 600 nm is compared with that of a reference sample. The absorption spectra of cyano-, aquo-, and sulphitocobalamins in acidic, basic, and neutral solution, together with changes in the spectra resulting from ageing of the solution, have been studied[1051]. The dissociation of cobalamins in acidic solution and the stability of the coordinate bond between cobalt and the nitrogen at position 3 of benzimidole are discussed together with the behaviour of sulphitocobalamin and its transformation to hydroxocobalamin.

Miscellaneous. The absorbance maxima of cobalt myristate and palmitate have been measured. In pyridine, these were found to be 550 nm. A linear relationship existed between the absorbance and the concentration of the soap. This method is more rapid and precise than the gravimetric techniques normally used[1052].

ORGANOCOPPER COMPOUNDS

The spectrophotometric determination of copper 8-hydroxyquinolinate in fabrics is subject to interference from some of the acid-soluble dyes used, but this can be overcome by examining a blank containing no 8-hydroxyquinolinate but otherwise treated and dyed in a similar manner[1053]. A chromatographic method has also been described, but the spectrophotometric method is considered to be the most suitable.

For the determination of copper 8-hydroxyquinolinate in paint the copper complex is extracted by boiling the paint film with 0.5 M sulphuric acid and 8-hydroxyquinoline is then determined spectrophotometrically at 307, 317, or 355 nm[1054,1055]. To determine copper the film sample is treated with sodium sulphide solution and acidified. Barium sulphate is added to collect the precipitated copper sulphide, which is dissolved in nitric acid and the copper is determined photometrically with benzoin - oxime or diphenylcarbazone. The determination of copper naphthenate is dealt with in both a British Standard[1056] and in a paper[1057].

ORGANOGERMANIUM COMPOUNDS

Kreingol'd and Bozhevov'nov[1058] have described a method for the determination of iron, copper, cobalt and manganese in high purity germanium tetrachloride. The germanium tetrachloride is slowly distilled and the residue used for analysis. Iron is determined spectrophotometrically with stilbexion (4,41-bis-bis(carboxymethyl)amino-stilbene -2'2' disulphuric acid) and hydrogen peroxide in acidic medium. Cobalt is determined with salicyl-fluorane and hydrogen peroxide in alkaline medium. Copper is determined by dimerization of lumo cupferron (((4-dimethylaminobenzylidene hippuric acid) in weakly alkaline medium and manganese is determined by reaction with lumomagnesion (5-5 chloro-2-hydroxy-3 sulpho-phenylazo) barbituric acid) and hydrogen peroxide.

ORGANOIRON COMPOUNDS

Haemoglobin in Blood

Many of the earlier methods for determination of haemoglobin in blood were based on direct visible spectrophotometry of the strongly coloured solution in a suitable medium. Gladyshevskaya et al[1059] showed with dilutions of blood from 1:500 to 1:500, that haemoglobin could be determined by measurements of the absorption at 417 nm (Sorets method). They compared results obtained by this procedure with those obtained by a gasometric method. Popov and Sobchuk[1060] reported measurements of the extinction at

430 nm of haemin chloride, obtained by dissolving blood in dilute hydrochloric acid solution.

Scaife[1062] critically examined the alkaline haematin method (Clegg[1063]) for determining haemoglobin and myoglobin in blood and has revealed some of the conditions affecting the behaviour of alkaline haematin. Scaife[1062] modified this method to make it suitable as a comparative method of analysis for either blood haemoglobin or tissue myoglobin. The determination is made spectrophotometrically at 380 nm. He reported that standard solutions of haemin in alkali are not stable but show progressive fading, which is accelerated by exposure of the solution to light or heat, by agitation with air or by the presence of traces of copper.

The cyanomethaemoglobin (hemiglobincyanide) method for the determination of haemoglobin in blood has been extensively discussed in the literature. In an earlier description of this method van Kampen and Kijlstra 1063 add homogenized blood (20 µl) to 5 ml of a solution containing 200 mg of potassium ferricyanide, 50 mg of potassium cyanide, 140 mg of potassium dihydrogen phosphate and 0.5 ml of detergent, Sterox SE, in 1 litre of water. The solution is mixed and the extinction at 540 nm read off in a 1 cm cell vs. distilled water in the reference cell. A conversion factor of 37.7 is used to convert the extinction to the concentration of haemoglobin in g/100 ml. Ressler et al[1065] described an artificial standard for haemoglobin determinations by the cyanomethaemoglobin method. The proposed standard is a stable solution containing nickelic chloride hexahydrate (43,3 g) and cobaltic chloride hexahydrate (10.3 g) dissolved in 100 ml of 10% acetic acid. The extinction is determined at 542 nm in a spectrophotometer having a half-intensity band-width of not greater than 5 nm, and is equated with a concentration of 60.1 mg of cyanomethaemoglobin per 100 ml. Results on samples are then calculated by application of Beer's law. Over the range 540 to 545 nm the absorption curve of the standard is almost identical with that of cyanomethaemoglobin. Zijlstra and van Kampen[1066] described the preparation of a stable cyanomethaemoglobin standard for use in conjunction with their method of analysis (van Kampen and Zijlstra[1064]). A concentrated oxyhaemoglobin solution is prepared by collecting about 90 ml of fresh blood in a 200-ml flask containing 20 ml of 3.2% sodium citrate (5.5 H_2O)solution. The red cells are separated and washed twice with 0.9% sodium chloride solution and then mixed with an equal volume of water and 40% of their volume of toluene. The mixture is stored overnight at 4°, then centrifuged, and the top layers are discarded. The filtrate from the remaining oxyhaemoglobin is stored in the refrigerator. The standard solution contains per litre, oxyhaemoglobin solution (5 ml), potassium ferricyanide (240 mg), potassium cyanide (50 mg) and sodium bicarbonate (1 g). When stored in the dark at 0°C these solutions remain stable for at least 3 h, but when stored at room temperature they deteriorate about after 90 min (as measured by periodic extinction measurements at 540 and 504 nm).

Zijlstra and van Kampen[1067] using the cyanomethaemoglobin method introduced by Stadie[1068] as a basis for the standardization obtained an extinction coefficient at 540 nm of cyanomethaemoglobin of 11.0. Tentori et al[1069] using samples of haemoglobin prepared by chromatography on carboxymethylcellulose converted it to cyanomethaemoglobin (nitrogen content checked by Dumas method) and obtained a calculated mean value of the 0.25 millimolar extinction coefficient at 540 nm of 10.90.

The cyanomethaemoglobin (hemiglobincyanide) method for the determination of haemoglobin in blood has been extensively discussed[1063-1071]. The haemoglobin is reacted under various conditions with potassium cyanide and the resulting cyanomethaemoglobin evaluated spectrophotometrically at a wavelength in the vicinity of 540 nm. A portion of blood or haemolysate is added by pipette to cyanide solution buffered at pH 7. The absorbance of this solution is measured at 540 nm between 1 and 4 h after mixing the reactants. The mean error ranges from 0.001 to 0.002 absorbance unit[1070, 1071]. A wedge-shaped adjustable photometer cell that permits calibration of the instrument with a single cyanomethaemoglobin solution has been described[1064]. A method in which total blood haemoglobin is converted into an azide complex of methaemoglobin is presented as an alternative to the cyanomethaemoglobin kethod[1072]. Owing to the similarity of the absorption spectra of both coloured species, both procedures yield identical results. For the proposed method, a single reagent, containing potassium hexacyanoferrate(III) and sodium azide is required. A spectrophotometric iron (III) thiocyanate method is available for determining haemoglobin as iron[1073]. A comparison of three methods for the assay of haemoglobin, determined as cyanomethaemoglobin, oxyhaemoglobin, and pyridine haemoglobin, indicated that the latter method gives the correct values, but the differences between the results are small enough to be negligible in routine work[1074]. A comparison of results obtained in haemoglobin in blood determinations using seven commercial instruments, concluded that for routine haemoglobinimetry in skilled and practised hands an absorptiometric method is suitable[1075]. For survey work and in general practice, the American Optical Spenser haemoglobin meter provided an adequate and rapid determination. The British Standards Institution[1076] specifies sealed glass cells containing suitable solutions for the photometric determination of haemoglobin. Haemoglobin in trout and carp has been determined using the acid haematin method and the cyanohaemoglobin method[1077].

Haemoglobin in Plasma

An ultraviolet spectrophotometric method for the determination of haemoglobin in plasma is based on the absorption of a solution of oxyhaemoglobin at 415 nm (i.e. in the Soret band) and also at 380 and 450 nm (to eliminate background absorption)[1078]. This method is simple, each analysis taking only 5 min, and it can be applied to other media containing haemoglobin but not to icteric plasma or sera. The benzidine-hydrogen peroxide method[1079,1080] involves reaction of the blood sample with these reagents in a suitable medium and photometric evaluation of the colour produced. The method is capable of determining 0.16 - 2.6 mg of haemoglobin per 100 ml of plasma and has been applied to the determination of haemoglobin in plasma and urine[1081] and for the determination of small amounts of haemoglobin by the haemoglobin-haptoglobin-peroxidase reaction[1082]. In the latter method a completely haemolysis-free serum, which gives only a slight benzidine- or guaiacol-positive reaction, is used as the source of haptoglobin. The sample (0.5 ml) is mixed with this seru (0.5 ml) and incubated for 10 min at 37°C. Benzidine solution 0.1% in 10% acetic acid) or guaiacol solution (550 mg in 150 ml of 0.1 M-acetate buffer of pH 0.5) (5 ml) is added and the solution placed in a water bath at 30°C. Two aliquots (2.9 ml) are placed in cells and the photometer is set with one (blank). To the other is added 2% aqueous hydrogen peroxide (0.1 ml), the solution stirred rapidly and the extinction measured at 470 nm every 20 s. The increase in extinction per s is used to obtain the haemoglobin concentration. Solutions of known haemoglobin concentration are used to prepare the standard graph. Down to 5 μg of haemoglobin per ml can be determined.

The cyanomethaemoglobin method has been used for the determination of haemoglobin in plasma[1083,1084]. Haemoglobin is converted to methaemoglobin by reaction with potassium cyanide. Both methaemoglobin and cyanomethaemoglobin obey Beer's law over the range 540 - 554 nm[1083]. The molar absorptivity of cyanomethaemoglobin is 74.21. mol^{-1} cm^{-1} at 420 nm and 7.31. mol^{-1} cm^{-1} at 540 nm. The absorptions of the methaemoglobin and cyanomethaemoglobin solutions are measured at a specific wavelength and the difference gives a value which is proportional to the haemoglobin content of the plasma sample. The method can be used for the determination of low concentrations of haemoglobin, in the range 1 - 12 mg per 100 ml in plasma and has a relative standard deviation of 1.09%. In applying it to plasma the turbidity of the plasma must be taken into account by making a correction or by adding a protein solubiliser[1084].

Haemoglobin in Serum and Urine

Haemoglobin in serum has been determined by the benzidine-hydrogen peroxide photometric method[1081]. The o-toluidine-hydrogen peroxide method has been used for both serum and urine[1085,1086]. Various spectrophotometric methods[1087] have been applied to the determination of haemoglobin and bilrubin in serum and compared with a chemical method[1087]. In general, good agreement was obtained. Direct and indirect bilrubin have essentially the same absorption spectra. Haemoglobin calculated from absorption equations showed little correlation at low levels with results obtained by the benzidine method.

Oxyhaemoglobin in Blood

A simple spectrophotometric method for the analysis of uncontaminated blood for oxyhaemoglobin involves measurement of the absorbance at 660 nm against a saturated blood sample as blank[1089]. A simply construced cuvette can be used for the spectrophotometric determination of haemoglobin and oxyhaemoglobin by absorbance measurements at 660 and 850 nm[1090]. Oxyhaemoglobin, haemoglobin, carboxyhaemoglobin, and methaemoglobin have been quantitatively determined spectrophotometrically in small blood samples[1091]. Procedures have been described for the following determinations: (1) total haemoglobin, by conversion with cyanomethaemoglobin by treatment of the blood with potassium hexacyanoferrate (III) and sodium cyanide and measuring the absorbance at 540, 545, and 551 nm; (2) oxyhaemoglobin mixed with reduced haemoglobin by measuring the absorbance at 560 and 506 nm; (3) carboxyhaemoglobin mixed with oxyhaemoglobin by measuring the absorbance at 562 and 540 nm; (4) methaemoglobin mixed with oxyhaemoglobin in 0.1% aqueous ammonia solution by measuring the absorbance at 540 and 524 nm. The absorbance of oxyhaemoglobin at both the maximum (576 nm) and the minimum absorption (560 nm) has been measured[1092]. Measurement of the absorbance of oxyhaemoglobin at the Soret band (412 - 415 nm) and how interference effects due to bilirubin at this wavelength may be overcome has been described[1093].

Carboxyhaemoglobin in Blood

Carboxyhaemoglobin mixed with oxyhaemoglobin can be determined by measuring the absorbance at 562 and 540 nm[1091]. An earlier spectrophotometric method[1094] for the determination of carboxyhaemoglobin at 555 nm has been modified[1095]. A spectrophotometric method for the determination of carboxyhaemoglobin in blood in which the absorption (A) of a 1:2000 dilution of blood in 0.04% aqueous ammonia is read at 576, 560 and 541 nm, and the carboxyhaemoglobin calculated from the ratios of A_{541} to A_{560} and A_{576} to A_{560} with the aid of calibration graphs, sometimes gives erroneously high values[1096]. This may be due to the formation of other haemoglobin derivatives (e.g. methaemoglobin or haematin), or to turbidity. The error due to turbidity can be corrected for by reading the absorbance at 700 and 660 nm (at which wavelengths the absorptions of oxyhaemoglobin and carboxyhaemoglobin are minimal), extrapolating the results to obtain blank values for 576, 560, and 541 nm, and subtracting these from the test readings.

For the determination of carboxyhaemoglobin in the presence of oxyhaemoglobin the absorbance is determined at 576, 560 and 541 nm[1097-1100]. The ratios A_{541}/A_{560} and A_{576}/A_{560} are calculated and the percentage of carboxyhaemoglobin is read off from a calibration graph.

The method of Commins and Lawther[1101] is also based on a spectrophotometric principle, but uses the difference in the Soret bands (absorption band due to porphyrin nucleus) for oxyhaemoglobin and carboxyhaemoglobin; that for oxyhaemoglobin has a maximum at 415 nm and that for carboxyhaemoglobin has a maximum at 420 nm, the latter being the greater[1102]. The spectrum of oxyhaemoglobin from 360 to 610 nm is shown in Figure 46. In addition to Soret bank there are two bands with maxima at 544 and 578 nm and a maximum at 564 nm. These double peaks were used by Commins and Lawther[1101] for the evaluation of the absolute haemoglobin concentration in the test solution. The spectrum of carboxyhaemoglobin from 360 to 610 nm is shown in Figure 47. This is similar to that of oxyhaemoglobin except that the peaks are in different positions. The difference spectrum is shown in Figure 48. This shows a sharp peak at 422 nm, which represents the absorption of carboxyhaemoglobin. Commins and Lawther[1101] used as a baseline the absorption at 414 and 426 nm, these values corresponding to zero % carboxyhaemoglobin.

Based on this procedure, Knight et al[1103] have developed a visible spectrophotometric method utilizing a Unicam SP8000A automatic UV spectrophotometer, for the estimation of carboxyhaemoglobin in blood.

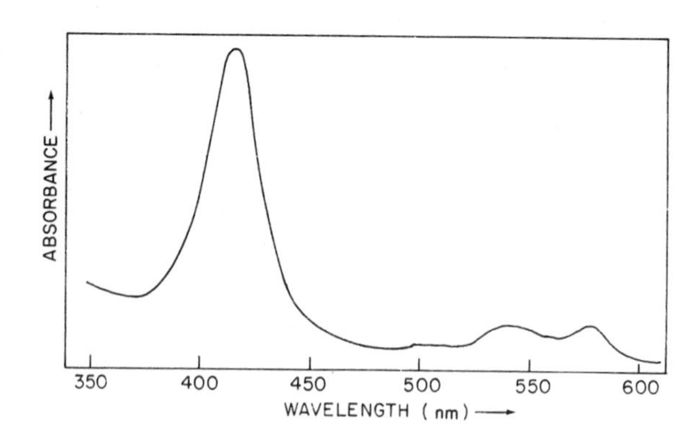

Figure 46 - Visible spectrum of oxyhaemoglobin.

Knight et al[1103] evaluated this method on a series of fourteen volunteers. All were healthy laboratory workers; nine were non-smokers and the remaining five were cigarette smokers. A sample of venous blood was obtained from each by venepuncture without stasis which was subsequently anticoagulated with sequestrene. These samples were immediately analysed using a Coulter Counter Model S. A layer of liquid paraffin was added to

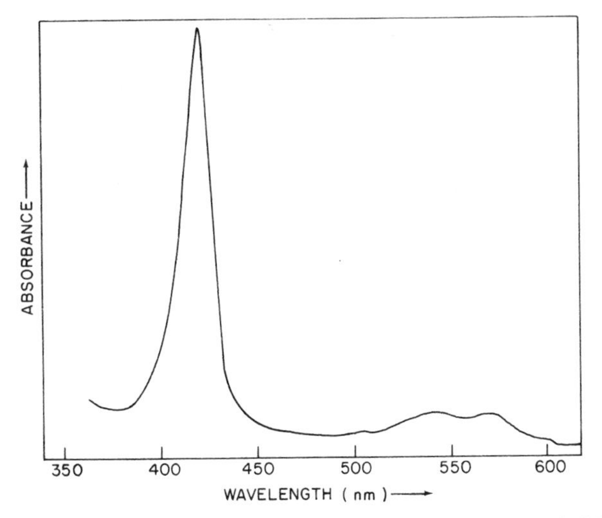

Figure 47 - Visible spectrum of carboxyhaemoglobin.

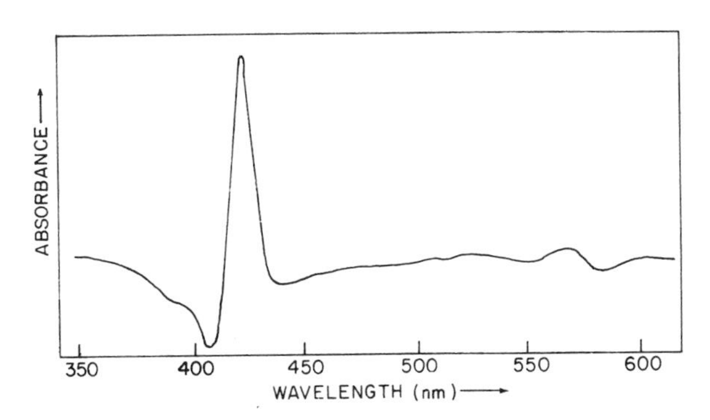

Figure 48 - Visible spectrum of carboxyhaemoglobin with oxyhaemoglobin as reference solution.

the sample to prevent loss of carboxyhaemoglobin to the atmosphere. The carboxyhaemoglobin content was then estimated and the results are shown in Table 114.

It was found that in non-smokers the mean carbon monoxide level was 0.52% with a range of 0 - 1.8%, and in smokers, whose consumption of filger-tipped cigarettes ranged between 20 and 40 per day, was 4.4%, with a range of 2.3 - 7.7%. These levels were lower than those found by Whitehead and Worthington[1095]. This may be due to the difference in the principles of the two methods. The method of Whitehead and Worthington would appear to be technically more difficult due to the precipitation procedures involved. Knight et al[1103] were unable to find any significant correlation between the biochemical and haematological parameters.

A further method for determining carboxyhaemoglobin is based on the reduction of palladium(II) chloride to metallic palladium by the abstraction of carbon monoxide from the carboxyhaemoglobin:

$$PdCl_2 + CO + H_2O \rightarrow CO_2 + 2HCl + Pd$$

Unchanged palladium(II) chloride can be determined colorimetrically after the addition of diethyl-p-nitrosaniline[1104]. Alternatively, the blood sample may be treated with dilute sulphuric acid and the volume of carbon monoxide measured[1105-1107] or the palladium complex may be determined by ultraviolet spectroscopy at 278 nm[1108]. Determination of carboxyhaemoglobin in the blood by the Van Slyke gasometric, photometric, and spectrophotometric methods have been compared[1109]. All three methods give satisfactory and comparable results. The photometric method is recommended for convenience and simplicity. An account of the spectrophotometric determination at four wavelengths of carboxyhaemoglobin, and methaemoglobin makes special reference to the simultaneous determination of carboxyhaemoglobin and methaemoglobin in human blood[1110].

Table 114 - Spectrophotometric estimation of carboxyhaemoglobin. Biochemical parameters obtained during investigation

Status *	No.	Sex	Number of cigarettes per day	$A(A_{421} - A_{408})$ Blank height	Test Height	50% COHb height	Concentration COHb (%)
S	1	F	20	1.4	4.0	29.0	4.7
S	2	M	30	0	4.0	26.0	7.7
NS	3	F	0	1.0	1.0	32.0	0
S	4	M	40	1.0	3.0	44.0	2.3
NS	5	F	0	1.0	1.0	25.0	0
NS	6	F	0	0.8	0.8	30.0	0
NS	7	M	0	0.6	0.9	26.0	0.6
S	8	F	20	0	2.8	30.0	4.7
S	9	M	20	0.8	3.0	40.5	2.8
NS	10	F	0	0	1.0	29.0	1.7
NS	11	F	0	1.0	1.0	26.0	0
NS	12	F	0	1.0	1.0	42.0	0
NS	13	F	0	1.0	2.0	33.0	0.6
NS	14	F	0	0	1.0	28.0	1.8

* S = smokers; NS = non-smokers.

Methaemoglobin in Blood

Several groups have studied the determination of methaemoglobin in blood[1091,1110-1115]. Methaemoglobin in mixtures with oxyhaemoglobin may be determined in a 1.0% aqueous ammonia medium by measuring the absorbance at 540 and 524 nm[1091]. Methaemoglobin, which has a characteristic absorption peak at 630 nm, can be determined by first measuring the absorbance at 630 nm before and after addition of sodium cyanide. Another fraction of buffered haemolysed red cell solution is centrifuged, and the supernatent liquid treated with potassium hexacyanoferrate(III). Finally, the total haemoglobin content is determined by Crosby's method[1116]. The concentration of methaemoglobin in the sample can then be calculated. In another method the absorbance of the whole blook haemolysate is measured at 578 and 525 nm[1112]. The ratio of the absorbance at 578 nm to that at 525 nm corresponds to the content of methaemoglobin. The results compare well with those obtained by established methods, even at low concentrations of methaemoglobin. When a blook haemolysate is treated with sodium cyanide methaemoglobin is concerted into cyanomethaemoglobin, and the reduction in the absorption of the solution at 630 nm is proportional to the amount of methaemoglobin in solution[1113].

Iron Carbonyl

Low concentrations of iron pentacarbonyl in commercial carbon monoxide can be determined by passing a gas sample through a train in which the iron pentacarbonyl is either condensed in a trap of solid carbon dioxide or absorbed in pure methanol[1117]. After the system has been flushed with nitrogen, the concentration of iron pentacarbonyl is determined spectrophotometrically at 235 nm in a methanol solution. The error is about ± 1.3% on samples containing not less than 0.04 mg of iron pentacarbonyl. To determine iron pentacarbonyl in air the sample (50 1) is drawn at 2 - 31/min through bubblers containing acidic iodine in potassium iodide, which traps 99% of the iron pentacarbonyl[1118]. Iron is reduced to the iron(II) state with hydroxylammonium chloride solution and determined at 508 nm using 1,10-phenanthroline. For the determination of iron pentacarbonyl in amounts down to 0.01 ppm in town gas, the carbonyl is trapped in iodine monochloride solution and, after reduction, iron is determined by the ammonium thicyanate procedure[1118]. Iron pentacarbonyl in carbon monoxide and carbon dioxide has been determined by passing the gas sample through a silica tube containing a plug of silica-wool kept at 350°C[1120]. The tube is allowed to cool while argon is flowing through it. The deposited iron is dissolved in hydrochloric acid and determined by conventional absorptiometric procedures.

ORGANOLEAD COMPOUNDS

Determination of Mono, Di and Trisubstituted Organolead Compounds

Earlier work in the early 1960s on the analytical application of dithizone to the spectrophotometric determination of these compounds was reported by Irving and Cox[1121], Barbieri et al[1122], Henderson and Syder [1123], Cremer[1124], and Aldridge and Cremer[1125].

Cremer[1124] in studies of the toxicological properties of tetraethyllead and its decomposition products used a dithizone colorimetric method in which the colour of the dithizone freed from the triethyllead-dithizone complex by acidification was used as a measure of the triethyllead ion. Interference from inorganic lead was eliminated by complexation with EDTA.

Barbieri et al[1122] converted mixtures of di- and tri-alkyllead chlorides in aqueous solution into dithizonates, which were then extracted with chloroform and determined spectrophotometrically. The absorption maxima are at 498 nm - 430 nm for dialkyllead dichlorides and tri-alkyllead chlorides (methyl or ethyl derivitives) and the proportions are obtained by differential analysis. If lead ions were present they were complexed with EDTA to prevent interference.

Henderson and Snyder[1123,1126] also Griffing et al[1127] described a simple and rapid spectrophotometric method using dithizone for the determination of triethyllead, diethyllead and inorganic lead ions.

On the basis of their experimental work, these workers concluded that with the exception of the monoethyllead ion which for all practical considerations is nonexistent, there were in their ethyl organolead preparations four components requiring analytical consideration: tetraethyllead, triethyllead ion, diethyllead ion and inorganic lead ion. Since dithizone does not react with tetraethyllead the field of analysis is narrowed to three components. Each of these three forms of lead has a distinctive chloroform-soluble dithizonate. Triethyllead ion forms a canary yellow compex, diethyllead ion an orange compex and inorganic lead ion a red complex. Absorption maxima are at wavelengths of 435, 487 and 520 nm respectively. By extracting the combined dithizonates and measuring the absorbances of this solution at three wavelengths they concluded that it should be possible to calculate each component provided the colour system of the dithizonates follows Beer's law and the absorption curves of the individual dithizonates have sufficent differentiation.

The results in Table 115 show analyses obtained by this method in mixtures of triethyllead chloride, diethyllead dichloride, and lead nitrate.

Henderson and Snyder[1123] point out that for the accurate determination of the individual diorganolead ions other than the ethyl or methyl homologs, the calibration data reported in their paper may not be applicable. The absorption curve of diphenyllead dithizonate differed significantly from that of the diethyllead dithizonate. New calibration data would definitely be needed for the accurate analysis of a system containing diphenyllead ions. For the analysis of systems containing other diorganolead ions, further confirmation of the calibration data would be advisable. They also emphasize that organic compounds of zinc, mercury, thallium, tin and bismuth may interfere where the simple metal ions would not. Unlike inorganic mercury, the ethylmercury ion is not complexed by gross amounts of cyanide. Organomanganese compounds interfere, but manganous compounds do not. If the system contains organometallic compounds of metals forming coloured dithizonates, the possible interference should be investigated. Interference from decomposed or partially decomposed dithizone, although negligible in the determination of inorganic lead, is more serious when trioganolead dithizonates are being measured at 424 nm. When the decomposed dithizone reaches a concentration sufficient to give a blank absorbance of greater than 0.20, it is advisable to prepare a fresh solution of dithizone.

Some workers prefer 4-(2-pyridylazo)resorcinol (PAR) to the less selective dithizone[1128,1129]. PAR reacts with compounds of the type R_2PbCl_2 (R = Me,Et,Ph) to form stable coloured complexes and forms no coloured complexes with R_3PbCl. The wavelengths of maximum absorption of $(C_2H_5)_2PbCl_2$ are unchanged at 514 nm in the pH range 5 - 10. The maximum colour intensity of the $(C_2H_5)_2Pb$ - PAR complex develops immediately and is stable for at least 24 h. A plot of the absorbance at constant wavelength against pH indicated that the optimum pH is 9 for $(C_2H_5)_2PbCl_2$. Application of the method of continuous variations and the slope-ratio method show that only a 1:1 complex of PAR is formed with diethyllead dichloride. Beer's law is obeyed between 2 x 10^{-6} and 10^{-4} M $(C_2H_5)_2Pb$ - PAR and the complex has a molar absorptivity coefficient of 41,000 l/mol. cm. For the diethyllead carbon, spectrophotometric determination is possible provided that the molar ratio of reagent to cation exceeds 4. The minimum detectable amount of cation is about 0.7 ppm for 0.1 absorbance in a 19 mm cell. Because of the great solubility of PAR (monosodium salt) in water, the dilution factor may be a low as 2 and hence the sensitivity ($\varepsilon \times 10^{-3}$ divided by dilution factor) may reach 20. In the case of diethyllead dichloride, the complex is so stable that spectrophotometric titrations with standard PAR solution can be carried out. Compounds of the type R_2PbCl_2 readily decompose to lead chloride both in the solid state and in solution, and hence samples of the former often contain some of the latter compound. The absorbance of the $(C_2H_5)_2Pb$ - PAR complex does not change on addition of EDTA[1128,1129], whereas the Pb - PAR complex[1130] is quantitatively destroyed by a stoichiometric

Table 115 - Analysis* of samples containing known amounts of triethyllead chloride, diethyllead chloride and lead nitrate

Added				Found				Error			
Tri	Di	Pb^{+2}	Total	Tri	Di	Pb^{+2}	Total	Tri	Di	Pb^{+2}	Total
0.0	57.0	50.0	107.0	0.6	54.4	49.2	104.2	+0.6	-2.6	-0.8	-2.8
0.0	95.0	0.0	95.0	0.0	95.2	0.0	95.2	0.0	+0.2	0.0	+0.2
0.0	0.0	107.0	107.0	0.1	-0.3	107.3	107.1	+0.1	-0.3	+0.3	+0.1
45.5	57.0	50.0	152.5	46.3	52.8	52.8	151.9	+0.8	-4.2	+2.8	-0.6
91.0	57.0	0.0	148.0	91.7	55.8	0.6	148.1	+0.7	-1.2	+0.6	+0.1
91.0	0.0	50.0	141.0	92.9	1.0	49.5	143.5	+1.9	+1.0	-0.5	+2.4
91.0	19.0	50.0	160.0	92.7	17.3	48.7	158.7	+1.7	-1.7	-1.3	-1.3
91.0	57.0	25.0	173.0	92.5	56.6	27.4	176.5	+1.5	-0.4	+2.4	+3.5
182.0	0.0	0.0	182.0	182.0	0.0	0.0	182.0	0.0	0.0	0.0	0.0
								Mean error	-1.0	+0.3	+0.1
								Std. dev. of error	±0.7	±1.4	±1.9

* All data expressed in micrograms of lead.

amount of EDTA. Accordingly, if an excess of PAR is added to a sample of diethyllead dichloride, any lead salt impurity can be determined by spectrophotometric titration with EDTA at 513 nm.

The stability constants of 1-(2-pyridylazo)-2-naphthol (PAN) with the diethyllead ion in aqueous 20% v/v dioxane are very high[1131]. When dialkyllead ion is added to an aqueous dioxane solution of PAN, the yellow liquid changes to a red chelate which retains its colour intensity for at least 24 h. The diethyllead ion can be directly titrated spectrophotometrically with a standard PAN solution or determined as the $(C_2H_5)_2$ - Pb(PAN)OH complex photometrically be measuring the absorbance of a chloroform extract at 555 nm which corresponds to the maximum of the uncharged complex[1132]. Alkyllead ions form stable coordination compounds with 8-hydroxyquinoline [1132,1133], 2,2'-bipyridyl[1133,1134], o-phenanthroline[1134-1136], acetylacetone[1136,1137] and picolonic acid[1136].

Triethyllead ions can be determined in blood and urine by extracting the triethyllead selectively into benzene from urine or deproteinized blood that has been almost saturated with sodium chloride[1138]. After re-extraction into dilute nitric acid, the triethyllead is decomposed with sulphuric and nitric acid and the lead is determined by the dithizone method. The sensitivity of the method is 2 µg of lead and the precision is ± 6% but the extraction from blood samples is only 90% efficient (100% from urine).

Tetrasubstituted Organolead Compounds

Tetraethyl and other tetraorganolead compounds may be determined by converting them to the ionic form by reaction with iodine and measuring the increase in ionic lead[1123,1126,1127]. The precision of the method thus applied to tetraorganolead compounds is ± 1.3% of the amount present. The method was used successfully in the determination of PbR_4 where R = Me,Et, n-Pr,i-Bu,i-Am,n-Am,Ch_2 = CH and C_6H_5 individually, or in the presence of appreciable amounts of ionized lead.

Tetraethyllead does not react with a chloroform solution of dithizone to form a complex. Analysis of solutions by the dithizone method[1123,1126] will, therefore, give values for only the ionized lead present in the sample. If two equal portions of the sample are analysed, one by the dithizone method and the other one with an added oxidation step by which the tetraethyllead is converted to ionic lead, the difference in the two analyses will be equal to the tetraethyllead originally present.

Results obtained in analyses of various tetraorganolead compounds are presented in Table 116. Results obtained by the dithizone procedure are in good agreement with those obtained by the classical iodimetric titration procedure described by Newman et al[1139].

In a spectrophotometric method capable of determining down to 10 ppm alkyllead in petroleum, the lead is extracted from the petroleum with a solution of potassium chlorate and sodium chloride in dilute nitric acid (Schwartz reagent) and then determined colorimetrically, as lead sulphide, by comparing the colour developed with that obtained by adding lead nitrate solution to sodium sulphide solution[1127,1140,1141]. In an alternative method using dithizone, bromine is added to the sample to convert the tetraethyllead to lead bromide[1127]. In a further method diisobutylene and bromine are added to produce hydrogen bromide gas, which then completes the decomposi-

Table 116 - Analyses of high purity tetraorganolead compounds. Weight per cent lead

	Theoretical	Iodimetric	Gravimetric sulphate	Dithizone	Error (%)
Tetramethyllead	77.50	76.40	76.20	76.40	+0.26
Tetraethyllead	64.06	63.61	63.44	63.16	-0.44
Tetra-n-propyllead	54.59	54.70	54.43	54.88	+0.83
Tetra-i-butyllead	47.58	57.28	47.41	47.35	-0.13
Tetra-i-amyllead	42.14	41.95	41.83	42.14	+0.74
Tetra-n-amyllead	42.14	41.06	41.42	40.95	-1.13
Tetravinyllead	65.70	*	65.24	65.20	-0.06
Tetraphenyllead	40.19	*	40.01	40.01	0.0
				Mean error	+0.01
				Std. dev.or error	+0.63%

* Reaction not stoichiometric to the trioranolead stage as described by Newman et al[1139].

tion of the organolead compound to inorganic lead ions. The lead is measured spectrophotometrically in the first procedure and by visual colour comparison in the second. The concentration range is 0.5 - 10 µg of lead and the accuracy and precision of both methods are within ± 0.01 ppm. Spectrophotometric determination as lead sulphide has been used for the determination of tetraethyllead in petroleum[1142].

Aldridge and Street[1252] studied spectrofluorometric and fluorometric methods for the determination of organolead compounds using dithizone and 3-hydroxfluorine.

Hexaethyl dilead has been determined by spectrophtometric methods[1143].

ORGANOMAGNESIUM COMPOUNDS

Spectrophotometric methods for the determination of alkyl-magnesium compounds based on reaction with Michlers ketone and with benzophenone and with hydroxylamine have been described[1144].

When a Grignard reagent is allowed to react with 4,4'-bis-(dimethyl amino) benzophenone (Michler's ketone) a typical colouration occurs. To a series of samples of the organomagnesium solution is added a known amount of water or methanol in dry benzene followed by a known excess of Michler's ketone solution. The quantity of unreacted Michler's ketone is estimated photometrically. By addition of different quantities of water or methanol to each sample, it is possible to estimate (graphically) the amount necessary to destroy exactly the whole quantity of Grignard reagent. This amount gives the concentration of reagent, assuming that the following reactions occur quantitatively, ($-C_6H_4N(CH_3)_2$ denoted by Ar).

$$\text{"RMgX"} + CH_3OH \rightarrow RH + \text{"}CH_3OMgX\text{"}$$

$$\text{"RMgX"} + ArCOAr \rightarrow Ar - \underset{\displaystyle R}{\underset{|}{C}} = \langle \bigcirc \rangle = \overset{+}{N}(CH_3)_2$$

The second method is based on the fact that benzophenone has a much greater molar extraction at 333 nm than most of the products formed by the reaction with an organomagnesium halide. The extinctions of the reaction products at thiw wavelength are negligible except for tert and some sec-alkylmagnesium halides. In that case, 1,4-addition (II) as well as dehydration (III) still cause high absorption. It is stated that this method only gives good results when applied to aliphatic non-tertiary types of Grignard reagents. With phenylmagnesium halides, the oxidation product (phenol) of the reagent causes appreciable errors.

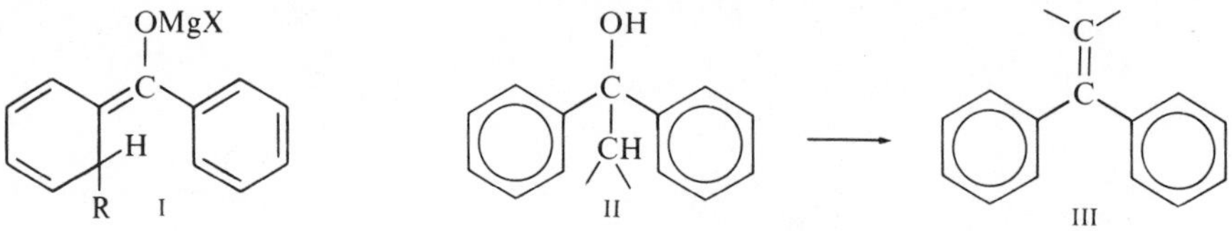

In the hydroxylamine method the Grignard reagent is reacted with an excess of acetone and the unreacted ketone is determined using hydroxylamineformate.

Errors may arise because the ketone influenced by the Grignard salt can enolize as follows:

$$RMgX + CH_3COCH_3 \rightarrow CH_2 = \underset{\displaystyle OMgX}{\underset{|}{C}} - CH_3 + RH$$

Procedure - Preparation of the hydroxylamineformate in methanol. Dissolve 5.6 g potassium hydroxide in 100 ml methanol. Add to this 4.6 g pure formic acid and 80 ml methanol in which 7.0 g hydroxylamine chlorhydrate is dissolved. Dilute to 200 ml, shake well, and filter the precipitate.

Add 1 - 2 ml of the organomagnesium sample to 5 ml 1 N acetone solution in dry ether, so as to provide a 20% - 30% excess of acetone. Hydrolyse with 10 ml methanol. Add to this 10 ml of a hydroxylamineformate solution in methanol. Leave to react for 15 min and titrate the free hydroxylamineformate potentiometrically with 0.2 N perchloric acid solution in dry dioxane. Two different blanks should be carried out.

Blank 1 - Estimate the titre of hydroxylamineformate by titrating with 0.2 N perchloric acid in dioxane using thymol blue as indicator.

Blank 11 - Basic magnesium compounds in the sample interfere in the titration with perchloric acid. To correct for this, take the same quantity of sample solution as used in the original analysis. Add it to 5 ml 1 N acetone solution in dry ether. Hydrolyse with methanol and add a known excess of 0.2 N perchloric acid in dioxane. After approximately 30 min back-titrate with 0.1 N sodium hydroxide solution using p-nitrophenol as indicator.

Table 117 tabulates results obtained by D'Hollander and Anteunis[1144] in a comparison of results obtained by the classical methods and their hydroxylamine and benzophenone spectrophotometric methods, for a range or organomagnesium compounds of the type RMgX where R ranges from methyl to phenyl.

Table 117 - Estimation of concentration of some Grignard reagents RMgX by different methods

Grignard Reagent R	Acidimetry H_2SO_4	Acidimetry $HClO_4$*	Volhard	I_2	Hydroxylamine	Spectrophotometry benzophenone
CH_3	3.072 N	3.339 N	3.339 N	2.864 N	2.841 N	2.852 N
Et	1.682 N	1.745 N	1.745 N	1.375 N	1.236 N	1.212 N
n-propyl	1.483 N	1.622 N	1.565 N	1.346 N	1.169 N	1.184 N
n-butyl	1.073 N	1.168 N	1.113 N	1.037 N	0.763 N	0.757 N
isobutyl	1.160 N	1.341 N	1.343 N	0.893 N	0.674 N	0.905 N
phenyl	1.984 N	2.116 N	1.918 N	1.539 N	1.295 N	0.976 N

* Obtained as blank II during hydroxylamine method.

The results obtained by the hydroxylamine and by the spectrophotometric methods are lower than those obtained either by acidimetry or by iodine titration. The agreement between the spectrophotometric and hydroxylamine methods is excellent and D'Hollander and Anteunis claimed that these are the correct results. In the case of isobutylmagnesium bromide, the low titre obtained with the hydroxylamine method is due to the enolization. In the case of phenylmagnesium bromide the spectrophotometric method gives low results owing to the presence of benzene formed by oxidation of the reagent.

Furthermore it is noteworthy that in all the cases, except for phenylmagnesium bromide, the results obtained with the Volhard bromine titration and with the perchloric acid titration agree well. D'Hollander and Anteunis [1144] claim that both of these methods are suspect and point out that, in the acidimetric titration, the result obtained depends on the nature of the acid employed. Results considerably below theoretical were obtained by the spectrophotometric method using Michler's ketone.

Blomberg et al[1145] on the other hand, showed that estimations of ethylmagnesium bromide obtained by the D'Hollander and Anteunis benzophenone spectrophotometric method, by an acid-base titration method, and by a gas evolution method all agreed to within 1% - 3%.

These workers point out that in the Gilman et al[1146,1147] acid-base titration technique, in addition to alkylmagnesium halides, oxidized or hydrolysed Grignard compounds ("R - O - Mg - X" or "H - O - Mg - X") also yield "basic" magnesium compounds. Therefore, the acid-base titration method does not give the exact amount of active alkyl or aryl groups. However, when working under exclusion of oxygen and moisture, Gilman et al [1147] demonstrated that the number of equivalents of "basic" magnesium agrees to within 1% - 2% with the number of moles of alkane formed on hydrolysis of alkylmagnesium halides.

Blomberg et al[1145] carried out analyses with a sample of ethylmagnesium bromide prepared with rigorous exclusion of oxygen and water[1148]. Ampoules containing approximately 2 ml of 1 N Grignard solution in diethyl ether were broken in a sealed glass tube containing the required amount of purified benzophenone, dissolved in the same ether. After shaking for 40 min, the sealed glass tube was opened and the reaction mixture treated with methanol and acetic acid, followed by the determination of the amount of unreacted benzophenone as by D'Hollander and Anteunis[1144]. At the same time, in an aliquot part of the methanol solution, the amount of total magnesium was determined titrimetrically using EDTA (Flaschka[1149]) (methanol was removed by evaporation before titration). The ratio "total" magnesium/"basic" magnesium was determined on the same ethylmagnesium bromide solution in another ampoule, and so a comparison of the results obtained by the two methods under discussion was possible. The difference was not more than 1%. This value was obtained on the assumption that the reaction products do not absorb at 333 nm. However, when the Grignard compound was in excess (1.5 - 3 fold) the final solution showed a small absorption even after thorough purification of the benzophenone. Making a correction for this absorption, the amount of the Grignard compound, found by the benzophenone spectrophotometric method, differed by not more than 2.8% of the amount obtained titrimetrically. Also they compared the gas evolution method with the acid-base titration method using an ampoule of ether-free ethylmagnesium bromide. On reaction with an excess of water (about 100 ml for approximately 6 mg equivalents of "basic" magnesium) in a sealed glass tube, the gas obtained was transported to a vessel connected with a manometer, by cooling this vessel with liquid nitrogen. The gas was dried with phosphorus pentoxide and its amount determined: this corresponded to within 1% with the number of mg-equivalents of "basic" magnesium, determined titrimetrically in the aqueous layer.

Blomberg et al[1145] concluded that, when applied to ethylmagnesium bromide uncontaminated by water or by oxygen, the benzophenone spectrophotometric method and the acid-base titration and the gasometric method all agreed within 1% - 3%. Moreover, they state that the large discrepancies obtained by D'Hollander and Anteunis[1144] in the determination of the concentration of ethylmagnesium bromide using their benzophenone photometric method and the acid-base titration method can only be ascribed to the way these workers treated their Grignard solution (i.e. they did not completely exclude oxygen and moisture. By taking a sample of 1 - 2 ml of 1 N Grignard solution with the aid of a pipette, contamination with oxygen and moisture was inevitable and so the aliquot, which is used for further reaction with benzophenone was not a representative part of the original sample. In the case where the amount of Grignard compound in this sample is determined by acid-base titration, their results were not influenced by oxidized and hydrolysed products.

Blomberg and his co-workers[1145] conclude therefore that the photometric method proposed by D'Hollander and Anteunis[1144] can be a useful one, when working under rigorous conditions, for the determination of active Grignard compounds. For ordinary laboratory use, the acid-base titration is preferable because it involves a relatively easy technique.

ORGANOMANGANESE COMPOUNDS

Cyclopentadienylmanganese tricarbonyl and methylcyclopentadienylmanganese tricarbonyl anti-knock compounds have been determined in petroleum[1150] by a spectrophotometric procedure in which the petroleum (50 ml) is extracted first with bromine (3 to 8 ml), then with sulphuric acid - nitric acid-water (1:1:2 ml), and the extracted manganese is oxidised with potassium periodate (0.4 g). The resulting permanganate is determined spectrophotometrically, at 520 nm. A determination in the range of 0.03 to 0.3 g of manganese per U.S. gallon has an average deviation of 0.002 to 0.004 g per gallon, and can be completed in about 2 h.

ORGANOMERCURY COMPOUNDS

A method for the determination of diphyenylmercury, either alone or in the presence of other phenylmercury compounds, has been described[1151], For the determination of diphenylmercury alone, it is extracted from a chloroform solution with 9 N hydrochloric acid. Phenylmercury chloride is formed quantitatively and is determined by reaction with dithizone and measurement of the colour produced at 629 nm. Diethyl-mercury is determined in the same way except that 12 N hydrochloric acid is used for the hydrolysis. For the determination of diphenyl- or diethylmercury in the presence of phenyl- or ethylmercury compounds, the latter are removed by extraction into acidified sodium thiosulphate, and the diphenyl- or diethylmercury is determined as above. Beer's law is obeyed over the ranges 1 - 30 and 90 - 120 μg with an error of less than 5%.

Phenylmercury compounds react with diphenylcarbazone in the same way as do inorganic mercury compounds but are resistant to reduction by zinc dust. In a method for the determination of microgram amounts of some phenylmercury compounds and their separation from inorganic mercury salts, the sample solution is added to ethanolic diphenylcarbazone solution and the colour is measured against standards[1152], which gives total mercury. The analysis is repeated after shaking the sample solution with zinc powder for 40 min, giving the phenylmercury content of the sample. Inorganic mercury is then obtained by difference. In a spectrophotometric method for the differential determination of phenylmercury acetate and inorganic mercury the mercury solution is shaken with hydrochloric acid and 2×10^{-4} M dithizone in carbon tetrachloride and the extract placed on an alumina column[1153]. Successive elution with carbon tetrachloride, carbon tetrachloride-chloroform (19:1), and carbon tetrachloride-chloroform (1:1) gives a yellow eluate containing phenylmercury dithizonate and then an orange eluate containing mercury dithizonate. These eluates are stabilized with anhydrous acetic acid, and the absorptions are measured at 480 and 490 nm for phenylmercury and mercury dithizonates respectively. Beer's law is obeyed for each compound up to 30 μg of mercury. The error in this procedure does not exceed 5%. Phenylmercury acetate forms a stable complex with diphenylcarbazone in alkaline media (maximum absorption in chloroform at 580 nm) that may be used for colorimetric analysis[1154].

Various workers have described methods for the determination of mercury in tissues. Mercury in urine and kidney can be determined by forming a complex in aqueous trichloroacetic acid between potassium bromide and the mercury in certain organic mercurials[1155]. After adjustment to pH 5.0 with formate buffer solution, the mercury in this form is extracted with dithizone solution in chloroform for spectrophotometric determination at 475 nm. The procedure can be used to detect as little as 1 μg of mercury. Methods based on the use of dithizone can be used to determine phenylmercury acetate in urine, kidney, liver, muscle, spleen, and brain[1156].

For the determination of phenylmercury compounds and total mercury in paints, phenylmercury compounds are converted into phenylmercury acetate by boiling with 1 N acetic acid; the solution is neutralized to pH 6 with sodium bicarbonate and phenylmercury acetate is extracted with benzene[1157]. The colour obtained on the addition of 0.5% ethanolic diphenylcarbazone solution is compared with that given by the gradual addition of standard phenylmercury acetate to a reagent blank. The results are low because of absorption losses. Total mercury is determined by the Schoniger combustion technique, followed by reaction with ethanolic diphenylcarbazone; the sensitivity of this reaction can be reduced to a convenient level (about 5 μg of mercury) by the addition of potassium cyanide. Methods based on the use of dithizone have been described for the determination of mercury as methylmercury, ethylmercury, and phenylmercury in soil, turf, and grain samples[1158,1159]. This is preceded by wet oxidation of the organic material with dilute sulphuric acid and nitric acids in an apparatus in which the vapour from the digestion is condensed into a reservoir from which it can be collected or returned to the digestion flask as required[1160]. The combined oxidized residue and condensate is diluted until the acid concentration is 1 N and nitrate is removed by addition of hydroxylammonium chloride with boiling. Any fat can be removed from the cooled solution with carbon tetrachloride and the liquid is then extracted with a solution of dithizone in carbon tetrachloride and mercury is determined spectrophotometrically at 485 nm using dithizone. Procedures for the determination of micro amounts of mercury in biological materials involving destruction of organic matter and the use of dithizone for mercury extraction have been reviewed[1161].

Dithizone procedures for the determination of phenylmercury chloride in fungicidal dust preparations and phenylmercury acetate in emulsifiable concentrates have been described[1162,1163]. For the determination of ethylmercury phosphate in emulsifiable concentrates the sample is dissolved in 4% acetic acid in aqueous methanol and 30% sodium chloride solution added[1162,1163]. An aliquot containing about 50 μg of mercury is shaken with chloroform and mercury is determined using dithizone at 478 nm. Calcium and magnesium stearates do not interfere in the procedure. Ethylmercury compounds may be separated from phenyl- and tolymercury compounds by heating mixtures of these compounds with 1:3 hydrochloric acid and methanol[1164,1165]. This decomposes phenyl- and tolymercury compounds in 30 - 60 min. Ethylmercury compounds remain unchanged and can be determined in a chloroform extract by the dithizone method described above.

Stary et al[1166] studied in detail the distribution of labelled methylmercury, ethylmercury and phenylmercury species between aqueous phases and pure carbon tetrachloride or dithizone solutions in carbon tetrachloride. The direct determination of the extraction constants is rather difficult because of the high values of these constants and because indirect determinations, e.g. by exchange reactions with silver dithizonate, are complicated by the formation of secondary dithizonates. The aim of their study was to determine the extraction constants of methylmercury, ethylmercury and phenylmercury dithizonates and to investigate the influence of various masking agents in order to compare the distribution data found experimentally with those calculated from the stability constants.

To determine stability and stability constant carbon tetrachloride (5.0 ml) was shaken with an equal volume of an aqueous phase containing labelled methylmercury chloride (or ethylmercury or phenylmercury chloride), perchloric acid, hydrochloric acid or silver nitrate. The total concentration of organomercurials was 10^{-6} M, and the total ionic strength I = 1.0. After shaking for 2 min, aliquots of the organic and aqueous phases were measured radiometrically to determine the distribution ratio. The shaking was then continued for another 3 min, and again aliquots were measured to verify that the distribution equilibrium had been reached.

The distrubution ratio of organomercurials between dithizone solutions to carbon tetrachloride and 1 M hydrochloric acid was determined in a similar way as described above. In these experiments, however, aliquots of the organic phase were also measured spectrophotometrically at 620 nm for the determination of the equilibrium concentration of dithizone[1167]. The values of the distribution ratio obtained after 2 and 5 min of shaking were the same, which suggests that the kinetics of the extraction of dithizonates is very fast.

The distribution of methylmercury (or ethylmercury or phenylmercury) dithizonate in the absence and in the presence of complexing agents (total concentration 0.01 M) was investigated at 25°C; 2 x 10^{-5} M dithizone solution in carbon tetrachloride was used in all experiments. The composition of the aqueous phase was changed from 5 M hydrochloric acid to 1 M sodium hydroxide. After shaking for 2 min, the equilibrium pH of the aqueous phase was measured simultaneously with the measurements of the radio-activity of both phases.

The distribution constants, K_o, were determined for aqueous phases containing 0.999 M perchloric acid and 0.001 M hydrochloric acid (or hydrobromic or hydroiodic acid). The results obtained are summarized in Table 118.

Table 118 - Survey of the logarithmic equilibrium constants found (25°C, 1 = 1.0)

Reaction	CH_3Hg^+	$C_3H_5Hg^+$	$C_6H_5Hg^+$
$RHg^+ + Cl^- = RHgCl$	5.51 ± 0.04[a]	5.32± 0.02	5.77 ± 0.03
$RH_gCl = (RHgCL)_{CCl_4}$	-0.05 ± 0.01	0.50 ± 0.01	1.19 ± 0.01
$RH_g{}^+ + Cl^- = (RHgCl)_{CCl_4}$	5.46 ± 0.05	5.82 ± 0.03	6.98 ± 0.04
$RH_gBr = (RH_gBr)_{CCl_4}$	0.81 ± 0.02	1.35 ± 0.02	2.25 ± 0.03
$RH_gl = (RH_gl)_{CCl_4}$	1.60 ± 0.03	1.41 ± 0.03	3.17 ± 0.03
$RH_g{}^+ + (H_2Dz)_{CCl_4} = (RH_gHDz)_{CCl_4} + H^+$	10.03 ± 0.03	10.11 ± 0.03	11.96 ± 0.06
$RH^+ + HDz^- = (RH_gHDz)_{CCl_4}$	19.1 ± 0.1	19.2 ± 0.1	21.0 ± 0.1

[a] Standard deviation, n > 10, is quoted in all cases.

For the determination of the stability constants of the chloride complexes, the distribution of methylmercury, ethylmercury and phenylmercury between pure carbon tetrachloride and perchloric acid was studied in the presence of an excess of silver ions (I = 1.0). The equilibrium concentration of chloride ions was calculated from the solubility product of silver chloride (pK_s = 9.59 at 25°C and I = 1.0). The results (see Figure 49) fulfill the equation from which the stability constants were calculated. The stability constant of methylmercury chloride determined here (log β Cl = 5.51) is similar to the value log β Cl = 5.45 (24 - 25°C, I not given) reported by Walton and co-workers[1168] but is higher than that determined at 20°C in 0.1 M KNO_3 (log β Cl 0 5.25)[1169]. Ethylmercury chloride (log β Cl = 5.32) is less stable whereas phenylmercury chloride (log β Cl = 5.77) is more stable than methylmercury chloride.

$$\beta\, Cl = \frac{D^0}{(KD - D^0)\,[Cl^-]}$$

where:

$$K_D = \frac{[R\ HgCl]}{[R\ HgCl]}\ org$$

D^0 = Distribution ratio in absence of dithizone

$[Cl^-)$ = Concentration of chloride ions (mg/litre)

βCl = Stability constant.

The values of the distribution ratios obtained in the presence of dithizone are summarized in Table 119. Because of the small difference between the K_D value and the D^o value found when 1 M hydrochloric acid was used as the aqueous phase, the formation of complexes of the type $RHgCl_2{}^-$ can be neglected for the present conditions.

Stary et al[1166] also studied the influence of various complexing agents on the extraction of methylmercury dithizonate into carbon tetrachloride, (Table 120). Chlorides have no effect on the extraction within the pH range 1 - 13. In the presence of 0.01 M bromide, the methylmercury is extracted

Table 119 - Determination of extraction constants, K_{ex}, for methylmercury, ethylmercury and phenylmercury dithizonates (25°C, 1.0 M HCl, CCl_4)

CH_3HgHDz (D^o = 0.88)			C_2H_5HgHDz (D^o = 3.11)			C_6H_5HgHDz (D^o = 15.3		
$-\log (H_2Dz)_{org}$	D	$\log K_{ex}$	$-\log (H_2DZ)_{org}$	D	$\log K_{ex}$	$-\log (H_2Dz)_{org}$	D	$\log K_{ez}$
5.25	1.05	9.99	4.86	3.88	10.07	5.43	19.6	11.83
5.08	1.13	9.99	4.69	4.24	10.08	5.37	22.0	11.97
4.95	1.22	9.99	4.57	4.74	10.10	5.14	25.3	11.91
4.84	1.35	10.02	4.47	5.15	10.10	5.04	28.7	11.94
4.76	1.44	10.02	4.40	5.63	10.12	4.96	30.3	11.91
4.67	1.62	10.05	4.28	6.45	10.12	4.92	35.8	12.00
4.55	1.80	10.02	4.11	8.34	10 15	4.75	48.0	12.03
4.38	2.29	10.04	3.97	9.57	10.10	4.61	60.5	12.03
4.25	2.87	10.06	3.87	11.80	10.13	4.51	66.8	11.99
4.07	3.74	10.04	3.79	13.10	10.11	4.42	85.9	12.04
3.95	5.13	10.09						
Mean value		10.03 ±0.03			10.11 ±0.03			11.96 ±0.06

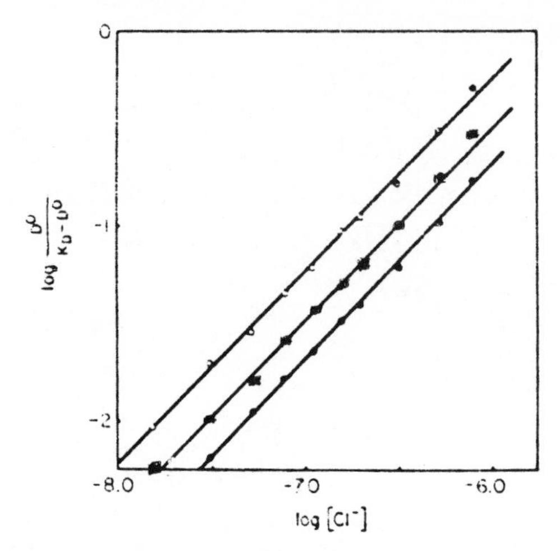

Figure 49 - The dependence of log $D^*/(K_D - D^o)$ vs. log (Cl^-) for ethylmercury (•), methylmercury (■) and phenylmercury (◦).

at pH 1 as a mixture of methylmercury bromide and methylmercury dithizonate whereas above pH 2 more than 99% of the methylmercury is extracted as the dithizonate. Methylmercury is extracted predominantly as an iodide complex at pH = 1.0; only above pH 5 is more than 99% of the methylmercury extracted as the dithizonate.

Cyanide ions form a very stable complex with methylmercury cation. The influence of sulphide ions is rather complicated because, under the present experimental conditions, the formation of CH_3HgS^- and of $(CH_3Hg)_2S$ must be considered; the distribution depends therefore on the total concentration of methylmercury.

In a rapid colorimetric method for the analysis of ethylmercury chloride the sample is dissolved in dimethylformamide and then mercury is determined spectrophotometrically using diphenylcarbazone[1170]. Phenylmercury acetate has been determined spectrophotometrically in amounts down to 2 μg by evaluation at 550 nm of the coloured complex produced with Richemann's purple, [2-(1,3-dioxoindan-2-yl)iminoindane-1,3-dione][1171].

Eldridge and Sweet[1172] have demonstrated that acidified aqueous solutions of phenylmercury acetate exhibit absorption maxima at 250, 256 and 262 nm; based on measurements at 256 and 262 nm they devised a method suitable for the determination of 0.01 - 0.1 g phenylmercury acetate with an accuracy of ± 1%. They carried out a detailed study of interference effects on this determination and concluded that the addition of perchloric acid to the sample solution considerably reduces the error caused by impurities.

Diphenylmercury and phenylmercury chloride do not interefere in the determination of phenylmercury acetate (Table 121).

Eldridge and Sweet[1172] compared analyses of phenylmercury acetate fungicidal and herbicidal preparations obtained by their ultraviolet spectroscopic methods and by the thiocyanate method reported previously by Kolthoff and Sandell[1173]. Table 120 shows that for pure phenylmercuric

Table 120 - The influence of complexing agents on the extraction of methylmercury dithizonate into carbon tetrachloride [X] = 0.01 M $[H_2Dz] = 2 \times 10^{-5}$ M)

Complexing agent, X	log βx(5 - 8)	pH	log D_e	
			Calculated	Found
Cl^-	5.5	1.2 - 13	>3	>2
Br^-	6.7	1	1.8	1.9
		>2	>2	>2
I^-	8.7	1	1.6	1.6
		>5	>2	>2
SCN^-	6.1	2 - 13	>3	>2
CN^-	14.1	1	2.3	2.2
		13	2.3	1.7
HPO_4^{2-}	5.0	1 - 13	>4	>2
S^{2-}	21.2	2	1.8	1.6
CH_3HgS^-	16.3	13	-2.0	-1.7
SO_3^{2-}	8.1	1 - 13	>4	>2
$S_2O_3^{2-}$	10.9	2	-1.6	-1.3
		>6	>2	>2
$Fe(CN)_6^{4-}$		1 - 13		>2
NH_3	7.6	1 - 13	>4	>2
CH_3COO^-	3.6	1 - 13	>4	>2
$Oxal^{2-}$		1 - 13		>2
$Tart^{2-}$		1 - 13		>2
Cit^{3-}		1 - 13		>2
$HEDTA^{3-}$	6.2	1 - 13	>4	>2

Table 121 - Interference of organic mercurials

Impurity	Impurity Taken, g	C_6H_5HgOAc Taken, g	% Error		
			250 nm	256 nm	262 nm
$(C_6H_5)_2Hg$	0.0050	0.0455	0.2	0.7	0.7
	0.0050	0.0455	0.4	0.4	0.2
C_2H_5HgCl	0.0028	0.0484	1.8	0.7	0.1
	0.0045	0.0494	-0.6	-0.4	-0.4
	0.0130	0.0370	0.6	-0.2	0.0
Ceresan M	0.0045	0.0459	12.9	8.1	7.8
	0.0578	0.0476	67.6	41.2	42.2
Semesan	0.0088	0.0487	18.9	21.1	16.6
	0.0138	0.0440	25.5	18.2	23.6

Ceresan M: 7.7% N(ethylmercuri)p-toluenesulphonamide.

Semesan; 30% 2-chloro-4(hydroxymercuriphenol).

acetate the thiocyanate method is as accurate as the spectrophotometric method. It is also rapid and convenient. However mercuric salts, such as mercuric acetate, interfere causing errors of up to 120% as shown in Table 122.

Eldridge and Sweet[1172] also found good agreement between their spectrophotometric method and the iodide-thiosulphate method described by Gran[1174] (Table 123). The spectrophotometric method is more rapid and convenient.

Table 122 - Determination of phenylmercuric acetate by titration with thiocyanate

Known weight of C_6H_5HgOAc (g)	$Hg(OAc)_2$ added (mol)	C_6H_5HgOAc Found (g)	% Error
0.0527	-	0.0527	0.0
0.0554	-	0.0550	-0.7
0.0429	4.74×10^{-5}	0.0736	72
0.0425	4.72×10^{-5}	0.0935	120

Table 123 - Comparison of iodide-thiosulphate titrimetric and spectrophotometric methods

		Taken, g	Found, g	Error, %
Titrimetric		0.0505	0.0490	-1.2
		0.0621	0.0618	-0.5
		0.0550	0.0547	-0.6
Spectrophotometric	250 nm	0.0534	0.0539	0.9
		0.0477	0.0481	0.8
		0.0524	0.0522	-0.4
"	256 nm	0.0534	0.0539	0.9
		0.0477	0.0477	0.0
		0.0524	0.0528	0.8
"	262 nm	0.0534	0.0538	0.8
		0.0477	0.0475	-0.4
		0.0524	0.0522	-0.4

A colorimetric method using dithizone has been used for the direct determination of 1 - 100 μg of methylmercury dicyanamide or methylmercury (II) chloride in fungicidal preparations[1175]. Copper, cobalt, cadmium, iron, lead, nickel, silver, zinc, bismuth, and mercury (II) ions in amounts less than 1 mg do not interfere in this procedure. For the analysis of seed disinfectants based on phenylmercuricatechol the sample is shaken with 10% sodium hydroxide solution to produce catechol[1176]. After preliminary work-up the catechol is determined spectrophotometrically at 560 nm using aqueous 4-aminophenazone and aqueous potassium hexacyanoferrate (III). The determination of methylmercury dicyanamide has been discussed[1158].

ORGANONICKEL COMPOUNDS

McCarley et al[1177] have described a sensitive, continuously recording instrument for detecting nickel carbonyl in the air. A stream of air (500 ml per min) flows through a nozzle and impinges on a hot borosilicate-glass disc, on which nickel carbonyl, if present, is deposited. A collimated beam of plane-polarised light falls on the disc at the Brewsterian angle for borosilicate glass, its plane of polarisation being perpendicular to the plane of incidence. This arrangement results in extinction, so that no light is reflected from the disc until a deposit is formed. The intensity of the reflected light is then measured by a recording photomultiplier photometer calibrated in p.p.m. of nickel carbonyl. Concentrations in the range 0.05 to 4 p.p.m. by volume can be measured and at a concentration of 1 p.p.m. the accuracy is ± 0.2 p.p.m. The detector is also sensitive to iron carbonyl. Fikhtengol'ts and Kozlova[1178] in their method for determining 0.002 mg or more of nickel tetracarbonyl in 1 litre of air, pass a 1 litre air sample through 2 ml of a 1.5% solution of iodine in carbon tetrachloride. After treatment with sodium sulphite the solution is analysed colorimetrically for nickel. Modifications to the method enable it to be extended below the 0.002 mg nickel tetracarbonyl per litre range.

In a further method for determining nickel tetracarbonyl in air (Suzuki and Oishi[1179]) the air (500 litres) is dried by passage over calcium chloride, then passed at 2 litres per min through 20 ml of 0.05% ethanolic iodine in a special absorber cooled to - 30°C in a bath of trichloroethylene - solid carbon dioxide. The absorbent is then evaporated to dryness, the residue is dissolved in water, and nickel in the resulting solution is determined photometrically with dimethylglyoxime. The calibration graph is rectilinear for 16 to 64 μg of nickel per 50 ml of the aqueous solution.

Brief et al[1180] describe a method for determining nickel tetracarbonyl in air or in process gas samples in which, after passage through a filter to remove any solids, the sample is bubbled into 3% aqueous hydrochloric (10 - 15 ml) for 20 - 30 min at 0.1 ft^3 per min. To the solution are added phenolphthalein indicator (2 drops) 20 to 30% aqueous ammonia (4 drops), 20% aqueous sodium hydroxide (to the phenolphthalein end-point, then 3 drops in excess), 1% furildioxime solution in 50% aqueous alcohol (3 ml) and chloroform (5 ml). The mixture is shaken, the chloroform layer is separated and its colour is compared either visually or spectrophotometrically at 435 nm with standards.

Densham et al[1181] have described a method for the determination of nickel and iron carbonyls in town gas. In the spectrophotometric method the gas sample (0.1 - 2.0 standard ft) is passed through a sintered bubbler containing M iodine monochloride in anhydrous acetic acid (10 ml) at 1 standard ft per hour. The bubbler contents are then poured into a beaker and the bubbler washed with 5% sulphuric acid (10 ml). Deposited sulphur is dissolved in nitric acid and this solution transferred to the beaker. The solution is evaporated to dryness, then treated with a few drops of nitric acid, the sulphuric acid is boiled off, and the organic matter is then removed by heating over a flame. The residue is dissolved in hydrochloric acid (1:1) and the nickel determined with dimethylglyoxime. The detection limit of this method is 0.006 p.p.m. nickel tetracarbonyl.

In a method for the determination of nickel tetracarbonyl in carbon monoxide and carbon dioxide, developed by workers at the United Kingdom Atomic Energy Authority[1182] the gas sample is passed through a silica tube containing a plug of silica wool kept at 350°C. The tube is allowed to cool while argon is flowing through it. The deposited nickel is dissolved in hydrochloric acid and determined by conventional absorptiometric procedures.

ORGANOPHOSPHORUS COMPOUNDS

Trialkyl Phosphates

Fudge and Hutton[1186] describe a method in which a kerosine solution containing about 40 mg of tributyl phosphate is hydrolysed by heating under reflux with 2 ml of 55% hydriodic acid for 15 minutes. After cooling, 10 ml of water and 3.5 ml of concentrated nitric acid are added through the condenser to decompose the excess hydriodic acid; when the reaction has subsided the condenser is removed, and the liquid is heated until colourless. After being cooled, the solution is made up to 50 ml, and about 10 ml is filtered through a dry 541 filter paper. A 5 ml aliquot of the filtrate is treated with a little water, 5 ml of nitric acid (1:2), 5 ml of 0.25% ammonium vanadate and 5 ml of 5% ammonium molybdate, and made up to 50 ml. The extinction of the molybdovanadophosphate colour is read against a blank, in a Spekker photoelectric absorptiometer, with a mercury-vapour lamp and H 503 and Ilford 601 filters. The results are referred to a standard curve prepared from solution of di-potassium hydrogen phosphate containing from 0.1 - 1.0 mg of phosphorus. Recovery is 99 - 100%. The final solution should not contain more than 1.0 mg of phosphorus in 5 ml; above this, Beer's law is not obeyed.

Dialkyl Hydrogen Phosphites, Trialkyl and Triaryl Phosphites

Sass and Cassidy[1187] have described a method for the detection and determination of dialkyl hydrogen phosphites in the presence of other phosphatics such as monoalkyl and trialkyl phosphites based on the reaction of phosphites with sym. 1:3:5 trinitrobenzene to form a red coloured complex with an absorption maximum at 465 nm. This method is sensitive to 0.2 to 0.5 µg of dialkylphosphite per ml of solution. This method is based on the evidence of Sass and Cassidy[1187] that dialkyl hydrogen phosphite exists as an equilibrium between the hydrogen phosphite and phosphonate, as is indicated by its hydrolytic characteristics:

$$(CH_3O)(HO)P(\rightarrow O)H + HOH \xrightarrow{HCl} P(OH)_3 + CH_3OH$$

$$\underset{(90\%)}{(CH_3O)_2P(\rightarrow O){-}H} \rightleftarrows \underset{(10\%)}{(CH_3O)_2P{-}OH}$$

$$(CH_3O)_2P(\rightarrow O){-}H + HOH \xrightarrow{HCl} (CH_3O)(HO)P(\rightarrow O)H + CH_3OH$$

In this method trinitrobenzene is treated with the dialkylhydrogen phosphites in the presence of alcoholic sodium hydroxide. The mechanism postulated by Sass and Cassidy[1187] for the formation of a coloured product, extended from the work of Baerstein[1188] and Whitmore[1189] with ketones, is shown below:

NO_2; O_2N; NO_2 + NaOR (R = H, Et) ⟶ $\ominus$ $-O-N-ONa^+$; O_2N; NO_2; RO H $\ominus$ + $R'O$, $R'O$ $\rangle$P—OH ⟶ NO_2Na; O_2N; NO_2; H $OP(OR')_2$

One difference in the method for the determination of dialkyl hydrogen phosphites is that the keto product, while it is coloured in the basic solution, loses the colour upon acidification. The colour formed with the dialkyl hydrogen phosphites remains after acidification. In fact, the method described below is based on acidification after prior reaction and colour formation when basic. This treatment was required because the trinitrobenzene itself is strongly coloured in basic solution.

Practically no interference in the method was encountered in the presence of the compounds listed in Table 124.

Saunders and Stark[1190] studied colour formation between dialkyl phosphites and alkaline solutions of m-dinitrobenzene and 3,5 dinitrobenzoic acid. They obtained fairly stable colours if the reaction was carried out in a saturated solution of sodium bicarbonate, and they concluded that quantitative methods of analysis could be based on these reactions.

Bass[1191] used t-butyl hydroperoxide for the quantitative determination of tributyl and triphenyl phosphites. In this method excess t-butyl hydroperoxide dissolved in 1 ml of octane is added to the test solution (0.1 - 10 mg phosphite per ml of octane) and butanol (5 ml) and glacial acetic acid (10 ml) added. The cold solution is then saturated with carbon dioxide and 50% potassium iodide added (0.5 ml saturated with carbon dioxide). The solution is diluted with glacial acetic acid (saturated with carbon dioxide) to 25 ml. The flask is then closed with a ground glass stopper and set aside for 1 h at room temp in the dark. The extinction of the coloured solution is then measured at 363 nm against a blank solution in which t-butyl hydroperoxide is absent. The method permits the determination of phosphite at concentrations down to 10^{-6} mol l^{-1}. Alcohols, carbonyl compounds; ethers, weak organic acids, and organic phosphates do not interfere in this procedure.

Various methods have been described for the determination of Polygard antioxidant (tris(nonylphenyl)phosphite) in polymeric materials. Serafin et al[1192] determine Polygard in styrene-butadiene resin by first extracting the antioxidant from the resin under reflux with isopropyl alcohol for 30 minutes. The extinction of neutral and alkaline potassium hydroxide portions of the extract are measured at 296 nm. The antioxidant content is proportional to the difference between the two extinctions. The standard deviation of the method is ± 0.028% and the relative error does not exceed ± 0.9%.

Brandt[1192] noted that a solution of Polygard in 2,2,4-trimethylpentane had a strong u.v. spectrum peak at 273 nm. Upon adding a strong base the

Table 124 - Specificity of method for dialkyl hydrogen phosphites

Sample	Dimethyl hydrogen phosphite	
	Added	Recovered
Trimethyl phosphite (1000 µg)	0	<1
	30	30
Triisopropyl phosphite (1000 µg)	0	1
	150	152
Bis (hydroxymethylphosphonic acid) anhydride (150 µg)	0	0
	167	166
Methylphosphonic acid (120 µg)	0	0
	25	26
Triethyl phosphate (1000 µg)	0	0
	180	182
Dimethyl methylphosphonate (200 µg)	0	0
	30	30
Phosphorous acid (100 µg)	0	1
	90	89
Monomethyl hydrogen phosphonate (1000 µg)	0[a]	0
	200	201

[a] Prepared in aqueous solution by treating dimethyl hydrogen phosphite with hydrochloric acid.

Polygard is hydrolysed and the peak shifts to 296 nm. The difference between the extinctions at 299 nm of the neutral and alkaline solution is directly proportional to the quantity of Polygard present in the styrene-butadiene resin. Recoveries of Polygard added to styrene-butadiene latex were in excess of 98%.

0,0-Dialkyl-1-Hydroxylphosphonates

Giang et al[1194] have described a colorimetric method for the determination of 0,0-dialkyl-1-hydroxyphosphonates such as 0,0-dimethyl-2:2:2-trichloro-1-hydroxyethyl phosphonate derived from the reaction of chloral with dialkylphosphites. In this procedure the sample (25 - 300 µg) is heated for 1 h at about 550°C in a micro-furnace and the pyrolysis products are drawn steadily through 6 ml of a pyridine-water mixture (50 ml of water and 400 ml of redistilled colourless pyridine) in an adsorption tube that is cooled in an ice-salt mixture (- 15° - 20°C). A further 3 ml of the pyridine-water mixture are added to the absorption tube followed by 1 ml of 0.25 N sodium hydroxide. The tube is loosely stoppered, heated for 3 min in boiling water and cooled immediately in running water. The solution is filtered through paper, and the colour, which is stable for 3 min is measured in a colorimeter with a green filter (500 - 570 nm). The amount of ester is derived by reference to a standard curve prepared from the pure ester (recrystallized from light petroleum containing a little benzene).

Tetraalkyl Pyrophosphate

Fournier[1195] described a colorimetric method for the micro determination of tetraethylpyrophosphate insecticide in the atmosphere. The method is based on saponification and oxidation of the pyrophosphate with aqueous sodium persulphate (10%) followed by treatment with 10 N sulphuric acid, aqueous ammonium molybdate (7.5%) and aqueous stannous chloride (0.2%). The resulting colour is evaluated at 4 min after addition of reagents at 640 nm using a red filter.

Gehauf et al[1196] described a sensitive colorimetric method for the estimation of tetraethylpyrophosphate based on measurement of the yellow colour produced by the oxidation of benzidine. This method has also been applied to the determination of traces of phosphoroanhydrides.

Sass et al[1197] have discussed a colorimetric method, described below, involving the use of diisonitrosoacetone or its monosodium salt for the determination of tetraalkyl pyrophosphates in amounts down to 1 μg.

A measured volume (1 ml) of the sample containing 5 - 60 μg of pyrophosphate ester is pipetted into a colorimeter tube. Diisonitrosoacetone reagent (1 ml) is added and the solution mixed prior to the addition of 3 ml of buffer solution. After 7 min the optical density of the solution is measured at 486 nm or 580 nm.

Sass et al[1197] demonstrated that, in addition to ester pyrophosphates, the following compounds produce a magenta colour with diinitroacetone and would interfere consequently in the analysis, methylphosphonochloridate, diisopropyl phosphorofluoridate. Tabun (ethyl dimethyl phosphoroamido cyanidate). Saran, phosphorus oxychloride, Paraxon, Parathion, Methyl phosphonic acid, diisopropyl hydrogen phosphite, diisopropyl methyl phosphonate, isopropyl methylphosphonic acid and diethyl phosphoric acid do not produce a colour with the reagent. They attained very close to 100% recovery in the analysis of 75 - 132 μg quantities of tetraethyl pyrophosphate dissolved in benzene without interference by 200 μg quantities of its chief hydrolysis products (diethyl hydrogen ortho phosphate and triethylphosphate).

Further studies on the application of peroxide-amine reactions to the colorimetric or fluorimetric determination of organophosphorus pyrophosphates have been carried out by Marsh and Neale[1198] and by Gehauf and Goldenson[1199].

Phosphorophenoxides

Gehauf et al[1196] reported an extremely sensitive method for the colorimetric estimation of phosphorophenoxides based on the measurement of the yellow colour produced by the oxidation of amine bases, such as benzidine.

Phosphonium Salts

Kolmerten and Epstein[1200] described a procedure for the colorimetric determination of tetramethylphosphonium ion involving conversion of the tetramethylphosphonium chloride into phosphate by oxidation with ammonium persulphate in dilute alkaline solution. Phosphate is then determined colorimetrically by the molybdenum blue procedure of Dickman and Bray[1201]. They showed that although tetramethylphosphonium chloride in distilled water can be quantitatively oxidized to the orthophosphate ion, the conversion is incomplete when the solution contains an excess of chloride ion (Table 125).

The less than theoretical recovery was shown not to be due to the interference of chloride ion in the formation of the molybdenum blue colour from orthophosphate. From 98 to 99% recovery of orthophosphate ion (0.0735 mg) was achieved even when as much as 700 mg of sodium chloride was added. It is believed that chloride ion interferes in the initial oxidation of tetramethylphosphonium chloride, and the interference may be overcome by performing the oxidation in a strongly alkaline solution (Table 126).

The initial concentration of sodium hydroxide needed for complete oxidation is dependent upon the quantity of chloride ion in the sample. For example, theoretical conversion to orthophosphate by ammonium persulphate in this procedure is obtained when the initial sample contains 0.1 mg of tetramethylphosphonium chloride and 6 mg of sodium chloride, and is 0.47 M with respect to sodium hydroxide. On the other hand, theoretical recoveries (within the limits of experimental error) are obtained even with solutions containing as much as 90 mg of sodium chloride in a sample of 0.1 mg of tetramethylphosphonium chloride, if the initial concentration of sodium hydroxide is 1.0 M.

Kolmerten and Epstein[1200] showed that the interference of chloride ion in the oxidation of tetramethylphosphonium chloride by ammonium persulphate is due to a simultaneous, competing reaction of chloride ion with ammonium persulphate in acid and neutral media, and that the role of hydroxyl ion is to prevent this reaction. They demonstrated that, in neutral and acid solution, a volatile product was formed (most probably, chlorine) which was capable of oxidizing iodide ion to iodine. Under the same experimental conditions, no such material was formed in the absence of chloride ion. Probably there is either no reaction between chloride and ammonium persulphate in alkaline solution, or a very slow one. If there were a reaction, it might be expected that sodium hypochlorite would be one of the products formed. Sodium hypochlorite does not convert tetramethylphosphonium chloride in alkaline solution to the orthophosphate ion. As this reaction does take place in alkaline solution with quantities of sodium chloride sufficient to use all the available oxygen in the ammonium persulphate, it would appear reasonable to conclude that sodium hypochlorite is formed, if at all, in negligible quantities.

Boiling an alkaline tetramethylphosphonium chloride solution does not result in the formation of a compound more easily oxidizable to the orthophosphate ion than tetramethylphosphonium chloride itself. It thus appears that the role of hydroxyl ion in this procedure is to prevent the oxidation of chloride ion by ammonium persulphate.

Other phosphorus compounds in which the phosphorus atom is contained in the anion moiety of the compound, or neutral phosphorus compounds, can be separated from tetramethylphosphonium chloride by using cationic exchange resins (Anderson and Keeler[1202]). The tetramethylphosphonium ion, which is exchanged and affixed to the resin, can then be eluted with hydrochloric acid. Some typical results obtained in experiments using an ion-exchanger (Dowex 50,hydrogen form) are shown in Table 127.

Substituted Phosphine Oxides

O'Laughlin et al[1203] determined the solubilities in water of four bis (dialkylphosphinyl) alkanes and trioctylphosphine oxide by extraction of the yellow adduct of the phosphine oxide with Ti^{4+} and potassium thiocyanate from acid solution into chloroform. The absorption maxima were at 390 mm,

Table 125 - Recoveries of tetramethylphosphonium chloride (TMPC) as orthophosphate ion. (After treatment with ammonium persulphate solutions)

NaCl (mg)	TMPC (mg)	
	Added	Recovered
0	0.1030	0.1045 ± 0.0056[a]
0	0.2760	0.2590
1.40	0.2760	0.1904
2.82	0.2760	0.1716
4.68	0.2760	0.0469

[a] Average of seven individual determinations.

Table 126 - Recovery of tetramethylphosphonium chloride (TMPC) as orthophosphate ion in the presence of sodium chloride[a]. (After treatment in alkaline solution with ammonium persulphate)

NaOH concn.[b] ($mol\ l^{-1}$)	TMPC (mg)	
	Added	Recovered
0.036	0.1032	0.0108
0.30	0.1035	0.0730
0.36	0.1032	0.0708
0.40	0.1035	0.0788
0.50	0 1035	0.0932
0.60	0.1035	0.0954
0.70	0.1035	0.0966
0.90	0.1035	0.1010
1.00	0.2070	0.1932
1.40	0.1035	0.1088
1.80	0.1035	0.1056

[a] NaCl, 24 mg
[b] Initial concentration of sodium hydroxide in reaction mixture.

and Beer's law was obeyed up to 27 μg of bis(dihexylphosphinyl) methane per ml in the final solution. Replicate analyses gave coefficients of variation of about 4.8%. The molecular extinction coefficients were in the region 13,500 to 14,500 and suggested a combining ratio of phosphine oxide to titanium of three. Tributyl phosphate appeared not to form the adduct. The effect of other organic substances was not investigated, although the reaction is not unique. These workers also described a sensitive reagent for the detection on paper of phosphorus compounds. The reagent was prepared by adding potassium titanyl oxalate (0.5 g), potassium thiocyanate (5 g) and hydrochloric acid (5 ml) to 100 ml of water. Yellow to orange zones were obtained when paper chromatograms containing phosphine oxides, organic phosphates, phosphonates, phosphinates and acidic organic phosphorus compounds were sprayed with the reagent.

O'Laughlin and Banks[1204] described a spectrophotometric method for the estimation of phosphine oxides.

Table 127 - Recoveries of tetra methyl phosphonium chloride (TMPC) after resin treatment[a]

Run	Solution contents[a]	Conc. p.p.m.	Added (mg)	Found[c] (mg)	
				Effluent	Eluent
1	TMPC	231	23.1	-	22.2
2	TMPC	486	24.3	-	22.2
3	TMPC	241	3.61	-	3.42
4	TMPC	241	3.61	-	3.34
5	TMPC	2.5	4.88	-	4.18
	NaCl	250			
6	TMPC	206	3.10	-	3.30
	DEEP	260			
	DAP	423			
7	TMPC	60	2.38	-	2.31
	DEEP	54)			
	EMPA	77)	4.70 as PO_4^{3-}	4.08	
	TEP	50)			-
	DAP	56	1.23	1.27	-
8	DEEP	68)			
	EMPA	68)	5.56 as PO_4^{3-}	4.63	0.24 as PO_4^{3-}
	TEP	80)			
	DAP	56	1.23	1.50	

[a] Resin column 6 to 12 inches in a 25- or 50-ml buret. Flow rate varied from 0.10 to 3.7 ml min^{-1}.

[b] TMPC tetramethylphosphonium chloride; DEEP, ethyl ethanephosphonate; EMPA, ethyl ethanoephosphonic acid; TEP, ethyl phosphite; DAP, di-potassium hydrogen phosphate.

[c] Analyses on effluent and on eluent were for orthophosphate ion content. All solutions were oxidized with ammonium persulphate prior to development of the phosphomolybdate colour except in runs 6, 7 and 8, where analyses were also made for orthophosphate ion without oxidation to determine DAP content. The column, after water washing, was eluted with 50 to 100 ml of 5 N HCl. The eluent was evaporated to dryness and tested for TMPC according to the procedure given herein.

Alkyl Chloridates and Fluoridates and Alkyl Phosphoroamidocyanidate

Alkyl chloridates are toxic compounds and should be handled with great care in an efficient fume chamber.

Sass et al[1197] have described a colorimetric method, involving the use of diisonitrosoacetone reagent or its monosodium salt for the determination of small quantities down to 1 µg of organophosphorus halides and ester pyrophosphates. Absorbance measurements were made on a spectrophotometer over a range of 400 - 700 nm. Measurements can also be made at 580 nm. The spectral characteristics are shown in Figure 50. Some typical sensitivity data obtained by this procedure are quoted in Table 128.

Sass et al[1197] found from studies made at pH values from 5.0 - 12.0 that, for phosphorus compounds such as tetraethyl pyrophosphate, GB Sarin (isopropylmethylphosphonofluoridate) and diisopropylphosphorofluoridate (Table 128), the optimum pH for colour development - i.e. for best

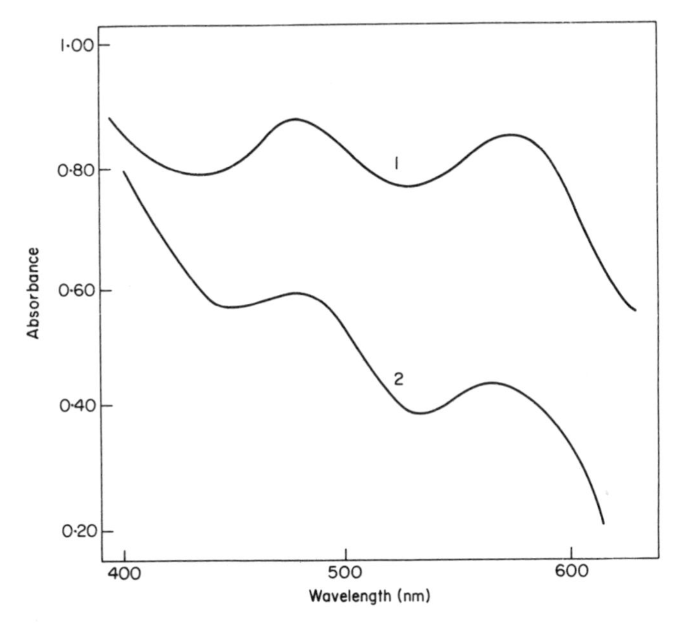

Figure 50 - Spectral characteristics of: (1) Sarin and diisonitroso-acetone; (2) monosodium salt of diisonitrosoacetone irradiated at 253.7 μm.

Table 128 - Dialkyl phosphono and phosphoro fluoridates and other compounds tested with diisonitrosoacetone[f]

	Colour	Sensitivity, μg per 4 ml
GA Tabun[b]	Magenta	1
GB Sarin[c]	Magenta	1
DFP[d]	Magenta	1
TEPP[e]	Magenta	2
Methyl phosphonodichloridate	Magenta	2
Phosphorus oxychloride	Magenta	2
Paraoxon	Magenta[a]	15
Parathion	Magenta[a]	25
Methylphosphonic acid	No colour	-
Diisopropyl hydrogen phosphite	No colour	-
Diisopropyl methylphosphonate	No colour	-
Isopropyl methylphosphonic acid	No colour	-
Diethylphosphoric acid	No colour	-

[a] Paraoxon and parathion required a longer period for colour development; parathion was the slower of the two.
[b] ethyl dimethyl phosphoroamidocyanidate.
[c] isopropyl methylphosphono fluoridate.
[d] diisopropyl phosphorofluoridate.
[e] tetraethyl pyrophosphate.
[f] approx. 0.02 g of substance added to 5 ml of a solution containing 2×10^{-3} mol 1^{-1} of diisonitrosoacetone and 2% sodium bicarbonate.

sensitivity and stability - was approximately 8.5. At a pH higher than 8.5 the colour formed more rapidly but persisted for a shorter period of time. Below pH 8.3 the colour developed more slowly, never reaching the maximum sensitivity obtained at the optimum pH. Between pH 8.3 and 8.5 the maximum sensitivity of colour was reached within 7 min and this colour at peak sensitivity persisted for at least 10 minutes.

The optimum concentration of diisonitrosoacetone to compound to be determined was found to be in the range of 95 mol to 1 mol at pH 8.5. Above or below this range of relative reagent to compound concentration, a definite gradual decrease in sensitivity was noted.

At pH 8.3 - 8.6, final total volumes (reagent, solvent and compound being tested) were prepared from 10 ml to 1 ml. It was found that the sensitivities vs. absorbance follows a linear pattern - i.e. 10 ml of solution containing 10 μg of GB Sarin showed the same approximate sensitivity as 1 ml of solution containing 1 μg of GB Sarin.

The results in Table 129 show that the method gives good recoveries in the determination of GB Sarin dissolved in 2-propanol and is not subject to interference by the chief hydrolysis products of this substance.

Table 129 - Analytical recovery of GB Sarin in presence of common impurities

Compound	Impurity (Hydrolysis products of Sarin)	Compound	
		Added (μg)	Found (μg)
GB Sarin	Methylphosphonofluoridic acid, 150 μg	35	34
		160	162
	Isopropyl methylphosphonic acid, 200 μg	76	76
	Diisopropyl methylphosphonate, 100 μg	45	46

Epstein and Koblin[1205] mentioned that ammonium persulphate is a useful reagent for converting GB Sarin (isopropyl methylphosphonofluoridate) and GA Tabun (ethyl dimethylphosphoramidocyanidate) to the orthophosphate ion which may then be estimated colorimetrically by conventional methods.

Golderson[1199] pointed out that because of their very reactive P-F or P-CN linkages, GB Sarin and GA Tabun can be detected by the chemiluminescence produced from a solution of the gas with 3-aminophthalhydrazide (luminol) and sodium perborate. A plot of maximum luminosity values versus the concentration of nerve gas gives an approx linear relationship. As little as 0.5 μg of the nerve gas can be readily detected by this method.

Gehauf et al[1196] have described a rapid colorimetric method for the detection of as little as 1 μ of nerve gases such as GA Tabun and GB Sarin. The method is based on the measurement of the yellow colour produced by the oxidation of amine bases, such as benzidine by these compounds. The method is applicable also to other types of organophosphorus compounds including phosphono and phosphorohalides, phosphoroanhydrides and phosphorophenolates. If the concentration of GB Sarin in the sample solution is as low as 0.1 part/10^6 then, after reagent addition and colour development, the colour may be extracted from a large volume of sample into a relatively small volume of xylene in order to achieve a concentration factor.

Ions commonly found in water (such as sodium, potassium, calcium, magnesium, chloride, nitrate, sulphate, and phosphate) certainly in amounts up to 1 mg in the sample, interfere in the determination of GB Sarin in water. Materials such as copper, molybdenum, iron and manganese, which are capable of reacting with hydrogen peroxide to form peracids, will interfere but this may be overcome by the addition to the sample solution of a small amount of sodium hexametaphosphate.

To estimate nerve gases in air, the air sample is drawn through a bubbler containing distilled diethyl phthalate (completely free from phthalic anhydride). Sufficient acetone is then added to effect a single phase when the benzidine and aqueous perborate reagents are added and the analysis continued as described above.

Kobin and Epstein[1207] have described a rapid method of sampling and analysing nerve gases in air in the field. The method for absorbing active undergraded GB Sarin or GA Tabun from air prior to analysis involved absorption in hexylene glycol and colorimetric determination by a peroxide-base reaction.

$$RO - \overset{\overset{O}{\|}}{\underset{\underset{R}{|}}{P}} - X + H_2O_2 \rightarrow RO - \overset{\overset{O}{\|}}{\underset{\underset{R}{|}}{P}} - O - OH$$

$$RO - \overset{\overset{O}{\|}}{\underset{\underset{R}{|}}{P}} - O - OH + \text{Dianisidine} \rightarrow \text{yellow-orange product}$$

Using a sample (1l) of air and 10 ml of hexylene glycol as absorbing solution and a total of 7 ml subsequently added reagents, the method is sufficiently sensitive to detect down to 4 parts/10^6 of nerve gas in air.

An alternative but less specific method, which determines total active plus degraded nerve gas, involves passage of the air sample through 2% aqueous sodium hydroxide solution at a flow rate of 1 litre per min.

Total GA Tabun can be estimated by determination of cyanide in the alkali absorbing solution. Total GB Sarin can be estimated similarly by determination of fluoride or by the subsequent addition of ammonium persulphate, which converts the organophosphonate to orthophosphate, which is then determined by classical molybdenum blue techniques.

Aqueous 2% hydrogen peroxide is an alternate absorbent. Ten ml of a 2% solution is quite satisfactory for absorbing 95% GB Sarin. In the case of GB Sarin, phosphate and fluoride analyses can be performed. In the case of GA Tabun, phosphate and cyanide analyses can be determined.

Neale and Perry[1207] have described a method for the sampling and colorimetric analysis of GA Tabun and GB Sarin in air, based on their reaction with hydrogen peroxide and an aromatic base of the benzidine type in alkaline solution. A measured amount of air is drawn through cyclohexanol in a spined bubbler. The sample is then adjusted to 40 ml with a phosphate buffer (pH 5.5) and a suitable aliquot is treated with hydrogen peroxide and o-dianisidine solution. The pH is then raised to 11.0 by the addition of another buffer solution, the volume adjusted to 25 ml with an acetone-water mixture and the colour measured in an absorptiometer. In a modification of the method, all operations can be effected without removal of the sample from the specially designed bubbler, the normal range being covered by the use of 0.25 - 3 cm cells in the absorptiometer.

G A Tabun

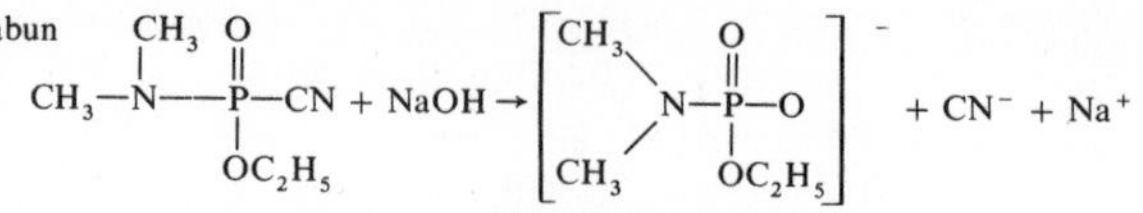

$$CH_3-N(CH_3)-P(=O)(OC_2H_5)-CN + NaOH \rightarrow \left[(CH_3)_2N-P(=O)(OC_2H_5)-O\right]^- + CN^- + Na^+$$

Dimethylamido phosphorate

GB Sarin

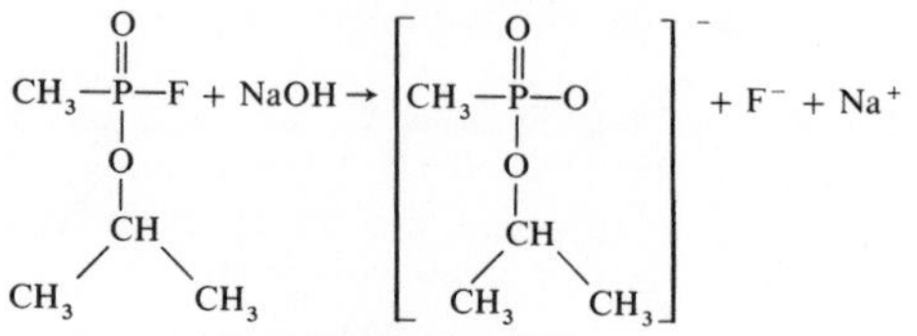

$$CH_3-P(=O)(F)-O-CH(CH_3)_2 + NaOH \rightarrow \left[CH_3-P(=O)(O)-O-CH(CH_3)_2\right]^- + F^- + Na^+$$

Isopropyl methylphosphonate

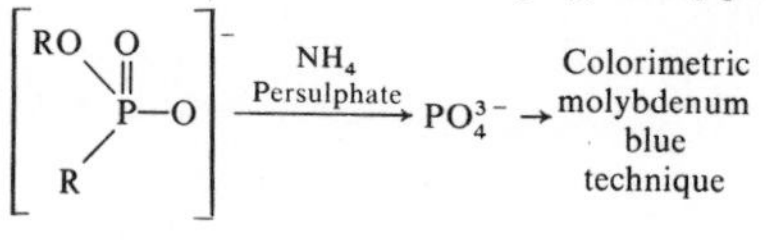

$$\left[(RO)(R)P(=O)-O\right]^- \xrightarrow[\text{Persulphate}]{NH_4} PO_4^{3-} \rightarrow \text{Colorimetric molybdenum blue technique}$$

Dialkyldithiophosphoric Acids

Masoero and Perini[1208] studied the absorption spectra in the visible region for nickel salts of some dialkyldithiophosphoric acids in benzene solutions and established conditions for colorimetric qualitative analysis of these compounds. The method is applicable to the titration of crude industrial acids, especially for dimethyl- and diethyl-dithiophosphoric acids. The technique has an accuracy within 1%.

In a method for the determination of sodium hydrogen S-(2-aminoethyl) phosphorothioate (sodium hydrogen cysteamine S-phosphate) the sample is added to a solution containing mercury (II) acetate in aqueous acetic acid, metol, and molybdate[1209,1210]. The absorbance is determined at 660 nm.

Petrov and Lur'e[1211] determined phosphorodithioates in industrial effluents as follows:

The neutralized sample, containing not more than 0.5 mg of the phosphorodithioate, is treated with 5 ml of borate buffer solution (pH 8.2) and 0.5 ml of 0.05% cupric sulphate pentahydrate solution, and the mixture is extracted with 5-ml portions of carbon tetrachloride. The combined extracts are filtered, the filtrate is diluted to 25 ml and its extinction is measured at 420 nm. Interference from xanthates can be prevented by preliminary acidification to a pH of not more than 2.7, and that from cyanides by the addition of formaldehyde.

Kovac and Paulinyova[1212] have described the determination of S-(p-chlorophenylthiomethyl) O,O-diethyl phosphorodithioate in commercial products. In this procedure the sample (60 mg) is dissolved in benzene, diluted to 100 ml, mixed with sodium bicarbonate solution (2%) (1 ml) in a special separating funnel by passage of 20 ml of air. The aqueous layer is removed, water (1 ml) added to the organic phase, which is mixed again by passage of 2 x 20 ml portions of air. The aqueous layer is removed, the benzene layer removed into a test-tube and the separating funnel rinsed with acetone (2 x 0.5 ml). The washings are added to the benzene layer. The mixed solvents are evaporated by applying a stream of inert gas to the liquid surface in the tube. To the residue is added perchloric acid (70%) (1 ml) the test-tube covered with aluminium foil, and placed in a heating block at 260°C for 30 min, then cooled. Molybdovanadate reagent (5 ml) is added (dissolve 40 g of $(NH_4)_6Mo_7O_{24}$ in 400 ml of warm water, dissolve 2 g of NV_4VO_3 in 250 ml of warm water and add 450 ml of 70% perchloric acid, add the cooled molybdate solution with stirring, to the vanadate solution and dilute to 2 litres), and the solutions mixed and diluted to 25 ml. The solution is evaluated at 400 nm against standard solutions.

Akerfeldt[1214] described a method for the determination of sodium hydrogen S-(2 aminoethyl) phosphorothioate (sodium hydrogen cysteamine S-phosphate). The determination is effected by adding to 0.5 ml of sample solution (containing up to 1 µmol) 0.1 ml of a solution containing 70 mg of mercuric acetate in 100 ml of 5% aqueous acetic acid, 2.9 ml of water, 0.50 ml of metol solution and 1 ml of molybdate solution. After 1 h the extinction is determined at 660 nm.

Garnier and Wakli[1214] have described a procedure for the identification of nitrothiophosphoric esters in the toxicological examination of tissue or viscera. A diethylether extract of the sample is evaporated to dryness and ethanol (10 ml) water (20 ml), 6 N hydrochloric acid (2 ml), liquid paraffin (1 g) and zinc (0.5 g) are added and the solution heated under reflux in a steam bath for 10 minutes. The slightly cooled solution is filtered on cotton wool, washed with 3 ml of water, diazotized with 1 ml of 0.25% sodium nitrite solution and left to stand for 10 minutes. Then 2.5% ammonium sulphate solution (1 ml) is added and the mixture left for 10 minutes. N-1-naphthylethylenediamine dihydrochloride solution (1 ml) is then added and after 10 min, 6 N hydrochloric acid (2 ml) and ethanol (50 ml) are added and the solution made up to 100 ml. The pink-peach to rose-violet colour produced attains a maximum intensity after 10 min; it is very stable and can be compared with similarly prepared standard solutions.

Glycerophosphates

The Morton[1215] spectrophotometric procedure, using specific dehydrogenases and diphosphopyridine nucleotide has also been applied to 1-glycerophosphate and D-glyceraldehyde phosphate. A similar procedure has been described for the determination of L-α-glycerophosphate in animal tissues. The reduction by L-α-glycerophosphate of diphosphopyridine nucleotide to dehydrodiphonsphopyridine nucleotide is catalysed by L-α-glycerophosphate dehydrogenase, and the spectrophotometric determination of the dehydrodiphosphopyridine nucleotide produced permits L-α-glycerophosphate to be determined. The rate of reduction of diphosphopyridine nucleotide during the first 20 s of reaction is measured by the change in extinction at 340 nm, preferably with a recording spectrophotometer. Values agree with those obtained by the non-enzymatic colorimetric method of Leva and Rappoport[1216]. The method is sensitive down to 2 $\mu mol\ ml^{-1}$.

Boltralik and Noll[1217] modified the two-stage enzymic assay of glycerol described by Bublitz and Kennedy[1217] so that glycerokinase and glycerophosphate dehydrogenase can be assayed in the presence of each other with adenosine triphosphate and diphosphopyridine nucleotide. The product dehydrodiphosphopyridine nucleotide, is determined spectrophotometrically in 50% aqueous ethanol at 340 nm. Recoveries of 0.2 - 1 µmol of glycerol of 96.5 ± 2% are claimed.

Methods have also been described for the determination of phospholipids, total lipids and cholesterol in cerebrospinal fluid (Roboz et al[1183]) and phospholipids, lecithin, cephalin, and sphingomyelin in butter (McDowell[1184]). The method described by Roboz like that of Dawson, described above, includes a preliminary extraction of the sample with chloroform-methanol, followed by determination of phospholipids by wet oxidation and colorimetric estimation of phosphate with ascorbic acid-molybdate reagent in amounts down to 0.15 µg.

Richardson and Tolbert[1185] have described various methods for the determination of phosphoglycollic acid. Each method depends on the cleavage of the acid into phosphoric acid and glycollic acid by phosphoglycollic acid phosphatase. In the first method, the oxygen consumed when the liberated glycollic acid is oxidized by glycollic acid oxidase is determined manometrically. In the second method, the free glycollic acid is determined colorimetrically by the standard method with 2,7-dihydroxynaphthalene. In the third method, the inorganic phosphorus is determined by standard molybdate.

ORGANOSELENIUM COMPOUNDS

In a method for the determination of small amounts of hydrogen selenide in air the sample is passed through an absorber containing hydrogen bromide solution with 18% of free bromine[1219]. Unchanged bromine is destroyed with hydroxylammonium chloride. The colour is determined photometrically by the 3,3'-diaminobenzidine method[1220]. Analytical aspects of organoselenium compounds have been reviewed[1221].

ORGANOSILICON COMPOUNDS

Borisov et al[1123] have discussed a spectrophotometric procedure, based on the formation of a coloured complex with antimony pentachloride in carbon tetrachloride, for the determination of 0.01 - 1.0% biphenyl in phenyltrichlorosilane. They also described a spectrophotometric procedure for the determination of 0.01 - 10% biphenyl in phenyltrichlorosilane based on measurement of the difference in extinction of solutions of the sample and a standard in chloroform - ethanol at 251 nm. Papov and Lulchuch[1224] described a procedure for the determination of biphenyl in phenyl trichlorosilane and its hydrolysis products. The extinctions of pure trichlorotrichlorosilane and the sample are measured at 251 nm in chloroform:ethanol 1:1; the difference is related to the content of biphenyl, which absorbs 60 times as strongly as phenyltrichlorosilane at that wavelength; 0.05 - 10% biphenyl may be determined.

Spectrophotometric methods have been described for the determination of copper, iron, cobalt and manganese in trichlorosilane and silicon tetrachloride[1225], boron in silicontetrachloride[1226] and in hexachlorosilane[1227] and iron and aluminium in silicone polymers[1228].

ORGANOTHALLIUM COMPOUNDS

Pepe et al[1222] have described a procedure for the estimation of phenylthallium III dichloride. This compound forms with xylenol orange a complex having a maximum absorption at 576 nm, with a mean mol. extinction sufficient of 3.69×10^4 in the presence of 1% - 2% of methanol; The optimum concentration range for determination of phenylthallium dichloride 6 - 30 μM. It can also be titrated with EDTA in aqueous method methanol medium at pH 4.75 with 0.1% aqueous xylenol orange as indicator. The end-point is shown by a sudden colour change from red to lemon-yellow and corresponds to the formation of a 1:1 complex of phenylthallium dichloride and EDTA. The coefficient of variation for the titrimetric method is 0.30%.

ORGANOTIN COMPOUNDS

A colorimetric method, based on the formation of a dithizone complex, can be used for the determination of diethyltin and triethyltin chloride and sulphate, either singly or as mixtures[1229-1231]. The absorption maximum for the diethyltin - dithizone complex, after being shaken with 10% trichloroacetic acid, is at 510 nm, whereas triethyltin does not react under these conditions. In the presence of borate buffer of pH 8.4, diethyltin and triethyltin compounds, with dithizone, give absorption maxima at 485 and 440 bn, respectively. At 510 nm the triethyltin-dithizone complex and dithizone have the same absorption. The separation and determination of diethyltin and triethyltin compounds can be based upon these findings and their distribution between chloroform and aqueous media. Interference from other metals is avoided by the use of EDTA.

In the Aldridge and Cremer procedure[1230] 1 ml of aqueous solution containing up to 24 μg of organotin compounds is taken and 3 ml of borate-EDTA buffer and 5 ml of chloroform added. The solution is mixed well by moving a mushroom-ended glass rod up and down for 1 minute. Diethyltin remains in the aqueous phase. The layers are left to separate, and 3 ml of this phase taken. 5 ml of chloroform and 1 ml of dithizone reagent are added and the solution again mixed for 1 minute. The aqueous layer is removed by suction. The difference in absorption between a control and the unknown is read at 510 mμ. The reading so obtained is referred to a calibration curve prepared in an identical manner (see Table 130) from standard solutions of diethyltin.

The chloroform phase from the original distribution contains 90% - 91% of the triethyltin. The remaining borate-EDTA buffer is removed and 4 ml of the chloroform layer taken. 1 ml of chloroform, 1 ml of dithizone reagent and 3 ml of borate-EDTA buffer are added. The solution is again mixed thoroughly in subdued light with a glass rod for 1 min, and the aqueous layer removed by suction. The difference in absorption between the control and unknown at 610 nm is referred to a calibration curve prepared in an identical manner (see Table 130) from standard solutions of triethyltin. It may be readily shown that at 510 nm there is no diethyltin present. If the triethyltin-dithizone complex in chloroform is exposed to bright light absorption at 510 nm becomes apparent.

Results for the calibration curves are shown in Table 130. The relation between triethyltin and optical density at 610 nm is linear up to 30 μg, whereas that for diethyltin is not, and in practice under these conditions not more than 20 μg of diethyltin should be taken. The range of the determination of diethyltin may be increased by using more dithizone.

This procedure should also be applicable to other types of organotin compounds, although these were not investigated as thoroughly as diethyltinchloride and triethyltinsulphate.

Barbieri et al[1232] and Irving and Cox[1233] have also studied the applicability of dithizone to the determination of organotin ions.

Table 130 - Calibration results for the separation and determination of triethyltin and diethyltin

Results for triethyltin					
Triethyltin present, μg	5.9	11.8	17.7	23.6	29.5
Optical density at 610 nm	0.115	0.235	0.360	0.460	0.555
Optical density at 510 mμ	0.01	0.01	0.01	0.004	0
Results for diethyltin					
Diethyltin present, μg	5.1	10.3	15.4	20.6	25.7
Optical density at 510 nm	0.115	0.22	0.305	0.35	0.38

The use of dithizone for the determination of trialkyltin compounds has the serious disadvantage that exposure to bright light causes a rapid change in the colour of the trialkytin dithizone complex. This colour change is consistent with the conversion of the trialkyltin complex to the dialkyltin complex. Measurements of trialkyltin dithizone complexes must therefore be made in subdued light.

A spectrophotometric method[1234] for the determination of down to 3 mg of dibutyltin dichloride in the presence of mono-, di-, and tetrabutyltin chlorides and several inorganic ions including zinc, manganese (II), iron (III), iron (II), lead, copper (II), copper (I), cadmium, tin (II) and tin (IV) depends on the fact that diphenylcarbazone produces a red colour with dibutyltin dichloride at pH 8.4[1235]. Dibutyltin dichloride in the sample reacts with diphenylcarbazone at pH 1.8 to form a 3:1 complex with an absorption maximum at 510 nm. Butyltin trichloride, the only one of the butyltin compounds to interfere at this pH, is removed by extraction with EDTA. The success of this procedure in being able to determine only dibutyltin dichloride in the presence of a mixture of the various butyltin chlorides is due to the selective extraction of butyltin trichloride with EDTA. Dibutyltin dichloride together with butyltin trichloride is extracted in varying amounts at higher pH. Conceivably the tributyl tinchloride would not be extracted within this range and it could then be determined with dithizone[1230].

Table 131 shows the results obtained by Skeel and Bricker[1234] in determinations of low levels of dibutyltin chloride in mixtures of monobutyltin trichloride, tributyltinchloride and tetrabutyltin.

4-(2-Pyridylazo) resorcinol (PAR)[1236,1237] and 1-(2-pyridylazo)-2-naphthol (PAN)[1237,1238] have been studied as reagents for the spectrophotometric determination of organotin ions, PAR reacts with compounds of the type $R_2SnCl_2 (R = CH_3, C_2H_5, C_6H_5)$ to form coloured complexes, whereas compounds of the type R_3SnCl do not produce a colour using a 2×10^{-3} M solution of PAR as reagent[1236]. The wavelength of maximum absorption (514 nm) of the 1:1 diethyltin dichloride - PAR complex is independent of pH in the pH range 3 - 8, with an optimum at pH 6. The colour develops immediately and is stable for 24 h. Reagent absorption is very small at this wavelength. Beer's law is obeyed in the range $2 \times 10^{-6} - 10^{-4}$ M of organotin ion, the molar absorptivity of the diethyltin dichloride - PAR complex being 42,500 1/mol. cm. A four-fold molar excess of reagent over organotin ion is necessary.

Dimethyl-, diethyl-, and di-n-butyltin dichlorides were examined spectrophotometrically using PAN as reagent[1237]. In solutions of R_2Sn^{2+} there is considerable uncertainty regarding the ionic species which may be present. Potentiometric investigations of dimethyltin ion hydrolysis have shown the existence of the $(CH_3)_2Sn^{2+}$ ion alone up to a limiting pH of about 2[1239-1242]. These studies have been extended to other organotin ions in order to establish the pH values within which the ions do not undergo hydrolytic equilibria. Addition of dialkyltin ions to PAN solution in dioxane produces a colour change from yellow to red which is stable for at least 24 h. The wavelengths for maximum absorption and the corresponding molar absorptivities are listed in Table 132. The stability of the complexes of R_2Sn^{2+} with PAN decreases in the order $C_6H_5 > C_4H_9 > C_2H_5 > CH_3$. The high stability of these chelates agrees well with the established action of PAN as a tridenate ligand chelate system with five membered rings being formed[1242-1244].

Table 131 - Determination of dibutyltin dichloride (DBTD) in presence of butyltin trichloride (BTT), tributyltinchloride (TBTC) and tetrabutyltin (TBT)

Sample Weight, mg					
DBTD	BTT	TBTC	TBT	DBTD Found	Error, mg
0.134	1.55	0.935	-	0.136	+0.002
0.134	1.08	-	0.78	0.126	-0.008
0.134	-	0.670	0.312	0.126	-0.008
0.134	0.775	0.286	1.56	0.129	-0.005
0.096	0.310	1.34	-	0.096	±0.000
0.096	1.55	-	1.09	0.108	+0.012
0.096	-	0.935	0.780	0.090	-0.006
0.096	1.08	0.670	0.312	0.100	+0.004
0.041	0.775	0.268	-	0.041	±0.000
0.041	0.310	-	1.56	0.032	-0.009
0.041	-	1.34	1.09	0.035	-0.006
0.041	1.55	0.935	0.780	0.044	±0.003
0.014	1.08	0.670	-	0.011	-0.003
0.014	0.775	-	0.312	0.008	-0.006
0.014	-	0.268	1.56	0.005	-0.009
0.014	0.310	1.34	1.09	0.012	-0.002

Table 132 - Absorption data for PAN chelates, aqueous 20% dioxane

Compound	max (nm)	x 10^{-3} (1/mol.cm)
$(CH_3)_2Sn^{2+}$	532	21.1
$(C_2H_5)_2Sn^{2+}$	538	22.6
$(C_4H_9)_2Sn^{2+}$	540	22.5
$(C_6H_5)_2Sn^{2+}$	542	22.0
$(C_2H_5)_2Pb^{2+}$	540	22.3

Other compounds that form stable co-ordination compounds with alkyltin ions include 8-hydroxyquinoline[1245-1247], 2,2'-bipyridyl[1245,1247,1248], phenanthroline[1245,1248,1249], acetylacetone[1249,1250], picolinic acid[1249], and Alizarin Red S[1251]. The analytical chemistry of these complexes has not been extensively studied. Sodium dimethyldithiocarbamate can be used instead of dithizone for the determination of both types of organotin compounds. The alkyltin complexes with this compound show maximum absoption in the u.v. region, at about 280 nm, but unlike the dithizone derivative are not decomposed photochemically in daylight.

Aldridge and Street[1252] have studied spectrometric and fluorometric methods for the determination of tri- and di-organotin compounds using dithizone and 3-hydroxylfl

Arakawa et al[1253] studied the extraction and fluorometric determination of organotin compounds using Morin. Detection limits are 1×10^{-9} M for dialkyltins, 1×10^{-7} M for monoalkyltin, 5×10^{-7} M for trialkyltins, and 1×10^{-7} M for triphenyltin. Recoveries of organotins added to tissues at the 1,0 - 100-nmol level range from 91.0 to 99.7% depending on the organotin species.

Frankel et al[1254] have described two colorimetric methods for the determination of organotin hydrides. In one of these methods the organotin hydride reduces isatin in alcoholic medium in the presence of azobis (isobutylronitrile) to colourless dioxindole (hydroxyindolin-2-one). In the second method ninhydrin in alcoholic medium is reduced by organotin hydrides but not by other organotin compounds, to 2-hydroxyindan-1,3-dione, and a blue-violet colour is formed that is stable in the absence of air. The change in extinction produced with either reagent can be used to determine organotin hydrides in concentrations down to 10^{-4}M. Other reducing agents must be absent.

Isatin is reduced by organotin hydrides as follows to produce the colourless dioxindole:

isatin $\xrightarrow[\text{azobisbutyronitrile}]{\text{Reduction}}$ dioxindole

The formation of colourless dioxindole from the coloured isatin was utilized for the quantitative determination of organotin hydrides. The change in the optical density of isatin solution is proportional to the concentration of added hydride.

Frankel et al[1254] observed that organotin hydrides, such as tri-n-butyltin hydride, triphenyltin hydride, and di-isobutyltin dihydride give a blue-violet colour with ninhydrin. The reaction can be carried out in various solvents in which both components are soluble, such as in alcohols, pyridine, acetone, dioxane and chlorobenzene. The colour fades on exposure to air, but on careful exclusion of oxygen, and by use of alcohol, distilled under argon, as solvent, the optical density does not change for at least one hour. This colour reaction was not given by other types of organotin compounds such as oxides, hydroxides or halides and is specific for organotin hydrides. The fact that the colour develops also in the presence of relatively large amounts of free radical scavengers such as ditertiary butyl hydroquinone, implies that the reaction does not involve a free radical mechanism.

The visible spectrum of the coloured reaction mixture in ethanol with tributyl or triphenyltin hydride shows a maximum absorption at 490 nm, and with di-isobutyltin dihydride the maximum was 525 nm. The absorption maxima are independent of the relative concentration of the reactants. The same maxima were obtained with a ten-fold excess of tributyltin hydride or a five-fold excess of ninhydrin.

The optical density in alcohol was found to be proportional to the concentration of the hydride, thus enabling the quantitative determination of organotin hydrides.

These workers suggest that their experimental evidence indicates that the coloured product obtained upon reaction with organotinhydrides originates by a reduction of ninhydrin to 2-hydroxyindanedione tin salt (I) in the case of R_3SnH compounds, and to compound II in the case of R_2SnH_2 compounds.

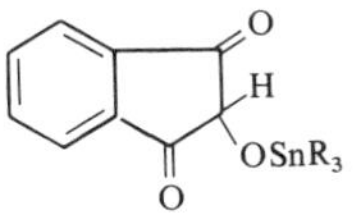

I

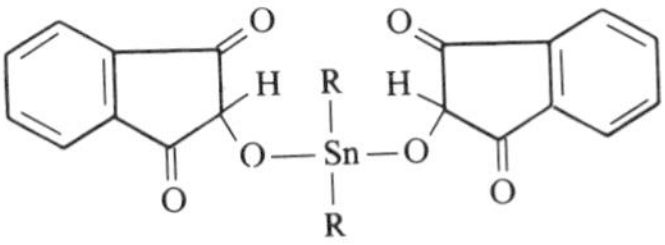

II

Dibutyltin compounds have been used for a number of years as stabilizers for poly(vinyl chloride). They are generally present to the extent of 1 - 2% in the finished polymer. Other dialkyltin compounds, and in particular those of dioctylin, are equally effective as stabilizers whilst having no demonstrable mammalian toxicity. The organotin compound present in PVC can usually be determined by wet ashing a 5 g sample with sulphuric nitric acids, reduction with aluminium or nickel, and titration of the tin (II) so formed with standard iodate solution. Precautions are necessary to prevent loss of tin by volatilization during the rapid evolution of hydrochloric acid in the early stages of the decomposition of the PVC. An alternative method for determining dialkyltin compounds involves extraction from the polymer with 1,2-dichloethane and spectrophotometric determination with dithizone at 490 nm[1255]. In a method for differentiating between dibutyltin and ioctyltin compounds in PVC, the dibutyltin compounds are extracted from chloroform solution by 1 N sodium hydroxide, but those of dioctyltin are only slightly soluble, and can be readily detected in the chloroform solution with dithizone after extraction. An aliquot of the dichloroethane solution from the tin determination is transferred to a separating funnel and tetrahydrofuran and 1 N-sodium hydroxide are added. The funnel is shaken and the two layers are allowed to separate. The organic layer is transferred to another separating funnel and shaken with trichloroacetic acid and dithizone solution. In the presence of a dibutyltin compound the dithizone remains unchanged while a dioctyltin compound produces a red coloration.

Dithizone forms coloured complexes with certain tin compounds and these complexes can be used for the determination of several dialkyltin and trialkyltin compounds[1230]. When a solution of, for example, diethyltin dichloride is shaken with a chloroform solution of dithizone in the presence of 10% trichloracetic acid, a red colour is produced, but under the same conditions neither the triethyltin nor the tetraethyltin compound reacts. In the presence of a borate buffer solution of pH 8.4, however, both tetraethyltin and triethyltin compounds gave a yellow colour, while the diethyltin derivative produces an orange colour.

A method for the estimation of dialkyltin stabilizers in aqueous extracts of PVC used in foodstuffs and drug packaging applications used 4-(2-pyridylazo) resorcinol - EDTA reagent[1256]. The absorbance of the separated chloroform layer is measured at 518 nm and referred to a calibration graph prepared with 1 - 10 µm dibutyltin maleate. Various other aspects of the migratory tendencies of organotin stabilizers from PVC which have been studied include the analysis of aqueous and non-aqueous extracts of rigid and of plasticized PVC[1257], and of tin in injection fluids packaged in PVC ampoules[1258].

A colorimetric determination of dialkyltin compounds in fats and olive oil containing 2% of added oleic acid is suitable for testing the migration of substances from plastic packaging and has been used for determining dialkyltin stabilizers added to PVC film and extractable in small amounts by fats and olive oil under the conditions of accelerated tests[1259]. To the sample of the used olive oil extractant is added alcoholic catechol violet solution. The absorbance of the clear upper layer is measured at 550 nm against a blank solution of similarly treated olive oil and is referred to a calibration graph perpared with standards. Triphenyltin residues in plant material can be determined by a spectrophotometric dithizone procedure[1260, 1261]. A fluorometric method has been used for the determination of triphenyltin compounds in water[1262].

The reagent diphenylthiocarbazone (dithizone) forms coloured complexes with certain tin compounds and these complexes can be used for the determination of several dialkyltin and trialkyltin compounds (Aldridge and Cremer[1230]).

When a solution of, for example, diethyltin dichloride is shaken with a chloroform solution of dithizone in the presence of 10% trichloracetic acid, a red colour is produced, but under the same conditions neither the triethyltin nor the tetraethyltin compound reacts. In the presence of a borate buffer solution of pH 8.4, however, both tetraethyltin and triethyltin compounds gave a yellow colour, while the diethyltin derivative produces an orange colour.

ORGANOZINC COMPOUNDS

For the spectrophotometric determination of zinc diethyldithiocarbamate in rubbers as the copper complex, the finely divided rubber vulcanizate is extracted with benzene and the zinc compound separated from interfering substances on acetylated paper prior to spectrophotometric evaluation at 430 nm[1263]. Zinc diethyldithiocarbamate cannot be determined in the presence of thiuram disulphide, which forms diethyl-dithiocarbamate during vulcanization and extraction. Other dialkyldithiocarbamates may also be determined by this method.

A procedure for the determination of zinc ethylenebisdithiocarbamate (zineb) residues on crops depends on the measurement of the absorbance at 434 nm of copper diethyldithiocarbamate in the presence of alkali[1264].

ULTRAVIOLET SPECTROSCOPY

Organoboron Compounds

Decaborane can be determined using the absorption maximum at 272 nm in cyclohexane solution (molar absorptivity = 3000 1/mol. cm)[1265]. This is a very useful method, but the samples analysed must be free from other materials which absorb at 272 nm.

Organolithium Compounds

The absorbance of diluted solutions of organolithium compounds calibrated by n.m.r. spectroscopy can directly measure the carbon-bound lithium content[1266]. Suitable glass apparatus which connects the two physical methods such that adventitious contamination can be eliminated by pre-purging with sample within a closed system is desirable. The absorbance over the range 275 - 305 nm has been measured for a series of butyllithium concentrations from 6 x 10^{-3} to 3.1 x 10^{-2}mol/1. The maximum absorption shifted from 278 nm at the lowest concentration to 282 nm at the highest and changed monotonically with concentration. Beer's law was obeyed at 285 nm and a molar absorptivity of 91 1/mol. cm was calculated at this wavelength. Concentrations of approximately 10^{-3} mol/1 in butyllithium can be satisfactorily determined to within 5% using the combined n.m.r.-u.v. technique. A check was made on the accuracy of the n.m.r. result by hydrolysing a sample of butyllithium - benzene with water and titrating against standard hydrochloric acid. The titration results were consistently 2 - 3% higher than the n.m.r. results. This is to be expected since titration gives the total lithium content, consisting of butyllithium, butoxide, and hydroxide.

Organomagnesium Compounds

Grignard reagents may be determined by breaking ampoules containing the Grignard solution in diethyl ether in a sealed glass tube containing the required amount of purified benzophenone, dissolved in the same ether[1267, 1268]. After shaking, the sealed glass tube was opened and the reaction mixture treated with methanol and acetic acid, followed by the determination of the amount of unreacted benzophenone at 333 nm[1169]. At the same time, in an aliquot of the methanol solution, the amount of total magnesium was determined titrimetrically using EDTA[1270]. When applied to ethylmagnesium bromide uncontaminated by water or by oxygen, the benzophenone spectrophotometric method, the acid-base titration, and the gasometric method all agree within 1 - 3%[1267]. Moreover, the large discrepancies obtained by others[1269] in the determination of the concentration of ethylmagnesium bromide using the benzophenone photometric method and the acid-base titration method can only be ascribed to the way these others treated their Grignard solution[1267].

When a Grignard reagent is allowed to react with 4,4'-bis(dimethylamino) benzophenone (Michler's ketone) a typical coloration occurs, providing a method for determining Grignard compounds[1271]. After mixing with a 10% solution of Michler's ketone in dry chloroform the sample is hydrolysed with methanol - acetic acid - water (70:20:10). Then a solution of iodine in acetic acid is added and the concentration of unreacted Michler's ketone is determined at 615 nm.

Organomercury Compounds

The marked difference in absorption between diphenylmercury and phenylmercury compounds in the 226 nm region makes possible a determination of diphenylmercury in the presence of phenylmercury compounds[1272,1273]. Methods have been described for the determination of organomercury dusts and vapours in air. The sample of air is drawn through a furnace at 800°C in which the organomercury compounds are decomposed to metallic mercury, and finally through a u.v. spectrophotometer in which total mercury (original organo-, and metallic mercury) is determined[1274]. Acidified aqueous solutions of phenylmercury acetate exhibit absorption maxima at 250, 256, and 262 nm; a method suitable for the determination of 0.01 - 0.1 g of phyenylmercury acetate with an accuracy of ± 1% is based on measurements at 256 and 262 nm.[1272]. From detailed study of interference effects in this determination it was concluded that the addition of perchloric acid to the sample solution considerably reduces the error caused by impurities. Diphenylmercury and phenylmercury chloride do not interfere in the determination of phenylmercury acetate by this procedure.

Organosilicon Compounds

Smith and McHand[1415] have reviewed the applicability of ultra-violet spectroscopy to organosilicon compounds, also mass, Raman, and other spectroscopic techniques as discussed in the three following sections. Alkyl-substituted silanes and siloxanes are not u.v. absorbers. Buckhard and Winslow[1417] found several absorption bands for methylsiloxanes. These compounds do not, however, absorb above 220 nm and the bands observed by Buckhard must arise from aromatic impurities. This conclusion is confirmed by Eaborn[1418] who found no absorption for trimethylsilanol, hexamethyldisiloxane, or trimethylmethoxysilane. Phenyl substitution on silicon, however, gives rise to a series of absorption peaks located at 248, 254, 260, 264, 266 and 271 mμ. These bands are common to triphenylsilane, triphenylethoxysilane and triphenylsilanol (Gilman and Dunn[1419]), and trimethylphenylsilane (Bowden and Braude[1420]). Para-substituted phenyl groups produce an intense absorption at about 270 μm (Gilman and Dunn[1419]). It is obvious that, at its present stage of development, u.v. spectroscopy is not a very good tool for identifying silicones. It is, however, very useful in detecting trace amounts of aromatic contaminants in dimethylsiloxane fluids or determining traces of aromatic-substituted silicones in the presence of non-absorbing materials, (Pozefsky and Grenoble[1421]).

Organotin Compounds

Some u.v. absorption bands for organotin comounds have been discussed [1273]. It is now generally accepted that, in organodistannanes, there is intense absorption associated with the Sn - Sn bond which is not dependent upon the presence of aromatic groups joined to tin[1274-1276], although no observation of maxima in hexabutyldistannane was possible[1276]. Similar absorption bands are observed with compounds in which tin is joined to other Group IVB metals. The intense absorptions recorded for some simple butyltin compounds are remarkable[1273] and possible origins of these bands have been discussed[1277]. Detailed analyses have been made of the u.v. absorption spectra of the phenyltin chlorides[1276], of some vinyltin compounds[1275], of compounds of the type $Me(SnMe_2)_nMe$[1276] and also the corresponding ethyl compounds[1277].

CHAPTER 4

OTHER SPECTROSCOPIC TECHNIQUES

ORGANOALUMINIUM COMPOUNDS

A. Infrared Spectroscopy

Triethylaluminium Compounds. Pitzer and Gutowski[1278] have measured the infrared spectrum of trimethylaluminium (Table 133). From this data one can conclude little because it does not extend below about 700 cm^{-1}. However, the lack of any marked deviations from the normal carbon-hydrogen stretching frequencies favours polar binding rather than protonated double bond character in the bridge. Also, the very high absorption coefficient for the aluminium-carbon bands near 700 cm^{-1} indicates a highly polar bond.

Table 133 - Wave lengths, frequencies and intensities of infrared absorption maxima of aluminium trimethyl

(μ)	cm^{-1}	Intensity
3.380	2958	7
3.425	2919	0[a]
3.49	2865	1[a]
3.735	2677	0(?)
6.685	1496	0(?)
6.925	1444	1[b]
7.67	1303	1
7.985	1252	7
8.295	1205	9
11.495	869)	
11.56	866)	2
12.875	779	10+
14.005	715)	
14.35	696)	10+[b]

[a] Indicates a shoulder. [b] A broad band

The high molar polarization found by Wiswall and Smyth[1279] for aluminium trimethyl seems almost certainly to be due to large atomic polarization rather than a dipole moment. No likely structure gives a permanent dipole. The large atomic polarisation is consistent with the high infrared absorption intensities. The diamagnetic character and lack of colour in aluminium methyl are consistent with this and other structures.

Pitzer and Gutowsky[1278] conclude that aluminium trimethyl is completely dimeric while the ethyl and n-propyl compounds show measureable dissociation of the dimer. On the other hand, aluminium isopropyl is completely monomeric. Also, a mixed methyl-isopropyl compound was found to be more highly dimerised than the ethyl aluminium compounds. Work by Schomburg[1280] and by Hoffman[1281] is consistent with a dimeric formulation of trimethyl aluminium. The generally accepted model of this dimer is that of Lewis and Rundle[1282] whose X-ray data showed the skeletal symmetry of the molecule to be D_{2h}.

After making corrections for deviations from Raoult's Law, association constants were calculated for triethyl and tri-n-propylaluminium.

Consideration of these data together with the Raman spectral data leads to the conclusion that the binding is primarily a polar attraction of the positive aluminium atoms for the negative carbon atoms. This agrees with the fact that the alkyls of metals more electro-positive than aluminium are all polymerized, usually into solids, while the alkyls of more electronegative metals are monomeric.

In a later paper, Pitzer and Sheline[1283] report the infrared spectrum of trimethylaluminium in a different wavelength range to that reported previously (Pitzer and Gutowsky[1278]) (see Table 133). They determined the spectrum of the same sample of trimethylaluminium used previously in the range 500 - 800 cm^{-1} (see Figure 51). This spectrum shows that in the range expected for Al-C vibrations (500 - 800 cm^{-1}) there are four very intense bands, 776, 694, 604 and 563 cm^{-1}. These are all very markedly stronger than C - H bending vibrations which might also be as low as 800 cm^{-1}. Thus their interpretation seems unequivocal. Also, the intensity of these bands indicates very considerable polarity (or ionic character) in the Al-C bond.

Gray[1284] and Gray et al[1285] discuss the vapour phase infrared spectra at low pressure of trimethylaluminium, dimethylaluminium chloride methyl-aluminium dichloride and some of their deuterated derivatives. He determined infrared spectra of the compounds $(CH_3)_6Al_2$, $(CD_3)_6Al_2$,$(CH_3)_4Al_2Cl_2$,(CD_3) Al_2Cl_2 and $(CH_3)_2Al_2Cl_4$ in the region 300 - 4000 cm^{-1}.

The observed absorption bands of the various compounds under discussion are shown in Table 134.

These workers have listed the normal modes in trimethylaluminium which are expected to give rise to infrared absorption or Raman scattering in the frequency range 800 - 300 cm^{-1}. Their approximate descriptions are given in Table 135, listed according to the symmetry species of the D_{2h} point group. Methyl deformations taking place perpendicular to the planes containing the carbon atoms are referred to as "wags", while those parallel to these planes are "rocks". The A_g and B_g species are Raman active, the B_u species are infrared active, and A_u is inactive in both infrared and Raman. The assignments given in Table 135 for the infrared active species of trimethylaluminium have been arrived at by application of the usual principles of vibrational analysis and by some arguments based on analogous systems.

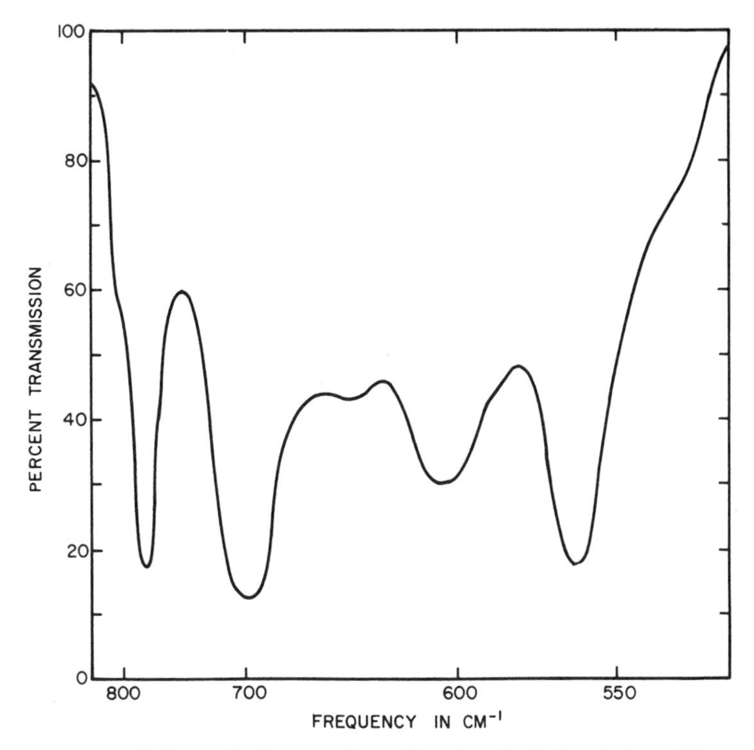

Figure 51 - The infrared absorption of trimethylaluminium dimer; cell length 9 cm; pressure, 11.5 cm.

Table 134 - Infrared spectra of various organoaluminium compounds

$Al_2(CH_3)_6$	$Al_2(CH_3)_4Cl_2$	$Al_2(CH_3)_2Cl_4$	$Al_2(CD_3)_6$	$Al_2(CD_3)_4Cl_2$
2941 s			2200 m	
2899 m			2170 w	
2837 w			2110 w	
1445 uw	1440 uw		1140 uw	
1255 m			1036 m	1020 w
1208 s	1212 s		955 s	957 s
1055 uw	1050 uw			
722 s			677 s	664 s
700 us	720 us	700 m	579 us	583 us
652 w				
616 w	588 R)m	675 m	472 us	530 w
572 s	581 P)m	587 m	435 m	483?w
480 m	487?w	483 s	315 m	430?w
369 s	440?w	423?m		331 m
	343 m	310 m		307 s
	310 s			

Dimethyl Aluminium Chloride. The dimeric nature of dimethylaluminium chloride has been established by Brockway and Davidson[1286] by electron diffraction and by Davidson and Brown[1287] by vapour pressure measurements. Infrared and Raman data have been reported by Groenewege[1288] and by

Hoffmann[1281]. Raman data on this compound and on a variety of other halogen derivatives, were also reported by Van der Kelen and Herman[1289]. It now seems well established that the bridging for this and analogous alkylaluminium halides is through the halogen atoms.

Gray[1284] and Gray et al[1285] determined the infrared spectra for this compound and its deuterium derivative, and considered certain inconsistencies which appear between their work and previously published spectra. The infrared spectrum of the vapour of dimethylaluminium chloride (see Table 134) shows two weak bands at 487 and 400 cm^{-1}. In solution, Hoffmann [1281] similarly finds two weak bands; one near 500 and the other at 443 cm^{-1} the relative intensities are, however, reversed.

Groenewege's solution spectra, on the other hand, only barely indicate absorption in this area, although a double-beam instrument was used and the intensity of the other bands indicates that it should be observed, if real. Groenewege[1288] does not find the weak polarized lines at 480 and 410 cm^{-1} found by Hoffmann[1281] although they are in agreement on other lines of comparable intensity. Groenewege[1292] has pointed out that the discrepancies between his Raman work and that of Van der Kelen and Hermann[1289] who found six additional lines at 94.5, 343, 480, 531, 685 and 727 cm^{-1}, all polarized. Groenewege[1288] considered the 480 and 531 lines as originating from 588 shifts of the f and g mercury lines, although Hoffmann[1281] finds the 480 line also.

All these observations are accounted for on the assumption that Van der Kelen and Hermann's[1289] material was quite heavily contaminated with aluminium sesquichloride, $(CH_3)_3Al_2Cl_2$, which evidently exists as a mixed dimer, and that the spectra reported by Gray[1284] and Gray et al[1285] and Hoffmann[1281] are similarly but slightly impure. The sesquichloride is the initial product obtained in the usual preparation of dimethylaluminium chloride from the reaction of methyl chloride and aluminium (Cross and Mavity[1292]). The 487 cm^{-1} band in the infrared is probably an Al - Cl outer stretch and the 400 cm^{-1} band, a bridge mode of the sesquichloride corresponding to either ν_6 or ν_{13} in aluminium chloride, which occur at 438 and 420 cm^{-1} respectively. Corresponding moles are found in the deuterated spectrum (Table 134) at 483 and 430 cm^{-1}. The sesquichloride has no symmetry and all Raman lines are polarized in agreement with those of the impurity.

Having disposed of these bands, the spectrum of dimethylaluminium chloride dimer is seen to be quite simple, and the assignments follow from the discussion previously given for trimethylaluminium. It is observed that the band centred at 585 cm^{-1} appears to be a doublet with maxima at 588 and 581 cm^{-1}, the lowest frequency being somewhat stronger. However, a calculation of the expected band contour for a B_{3u} mode (Badger and Zumwalt[1293]) predicts a separation of P and R branches of 6.8 cm^{-1}, in good agreement with that observed. Evidently the Q branch is diffuse and shifted to lower frequency for this particular mode due to a difference in upper- and lower-state inertial moments and the population of many K levels (Herzberg[1294]). This behaviour is common for heavy molecules.

The assignment of the Raman modes in this region is essentially the same as that of Groenwege[1288].

Methylaluminium Dichloride. The infrared spectrum of this compound has been reported by Groenewege[1292,1296] and the Raman by Van der Kelen and Hermann[1289]. A priori there would seem to be little doubt by anology with

Table 135 - Infrared assignments of organoaluminium compounds

D_{2h}	Approximate description	$Al_2(CH_3)_6$	ν_H/ν_D	$Al_2(CD_3)_6$	$Al_2Cl_2(CH_3)_4$	ν_H/ν_D	$Al_2Cl_2(CD_3)_4$	$Al_2Cl_4(CH_3)_2$		Al_2Cl_6
A_g	Outer CH_3 rock									
	Al - X stretch ν_1	592[a]			588[a]			582	492[c]	506[d]
	Bridge stretch ν_2	453			330				342	340
	Al $\langle^{X}_{X}$ bend ν_3	313			282					217
A_u	Outer CH_3 wag									
	Outer CH_3 wag									
B_{1g}	Bridge CH_3 rock	632[b]?								
	Bridge stretch ν_6	495?								438
B_{2g}	Outer CH_3 rock	725			713			689		606
	Al - X stretch ν_{11}	683			620					
B_{3g}	Outer CH_3 wag									
	Bridge CH_3 wag									
B_{1u}	Al - X stretch ν_8	772	1.14	677	720	1.08	664			625
	Outer CH_3 rock	700	1.21	579	700	1.20	583			
	Bridge CH_3 wag	700	1.21	579						
B_{2u}	Outer CH_3 wag							675		
	Bridge stretch ν_{13}	652?							423?	420
	Outer CH_3 rock	700	1.21	579	700	1.20	583	700		
	Bridge CH_3 rock	572	1.21	472						
B_{3u}	Al - X stretch ν_{16}	616	1.08	570	585	1.10	530	587	483	484
	Bridge stretch ν_{17}	480	1.10	435	310	1.01	307		310	301
	Al $\langle^{X}_{X}$ bend ν_{18}	369	1.17	315	343	1.04	331			

a Raman data, Ref Hoffmann[1281]. b Questionable assignments. c Raman data, Ref Van der Kelen and Herman[1289]. d Infrared and Raman data, observed and calculated Ref Klemperer[1290]. Skeletal normal modes unmbered according to the scheme of Bell and Longuet Higgins[1291].

its closest relatives that the molecule is dimeric and bridged through chlorine atoms but one must consider three possible structures for the dimer, corresponding to those of dichloroethylene; the trans-1,2-dichloro (C_{2h}) cis-1,2-dichloro (C_{2v}) and 1,1-dichloro (C_{2v}) isomers.

The major features of the vapour phase spectrum of this compound is shown in Table 134. In order to obtain these bands with reasonable intensity, it was necessary to heat the gas cell to approximately 65°C. The spectrum was first scanned in the NaCl region and was then scanned in the CsBr region. The dashed bands in the CsBr spectrum are considered questionable due to the possibility that the sample had reacted with Kel-F stopcock grease. Groenewege's[1288] spectrum is to be preferred.

A consideration of the selection rules for the various possible forms utilising the assignments which have been proposed for the compounds previously discussed, indicates that the observed spectra are those of the trans form. The cis isomer is predicted to have 13 infrared-active modes falling in the range 800 - 300 cm^{-1}, and the "symmetrical" isomer should have 11. Moreover all of these should be Raman active. On the other hand, for the C_{2h} trans model one expects in the infrared, one methyl rock, one methyl wag, one Al - C stretch, one Al - Cl stretch and two bridge modes. The observed bands fit this scheme extremely well, the corresponding modes being assigned to 700, 675, 587, 483, 402 (or 423) and 310 cm^{-1}, respectively. The Raman spectrum should show a similar distribution of lines, one methyl rock, one Al - C stretch, one Al - Cl stretch and one bridge stretch (ν_2) all totally symmetric and, therefore, polarized. Depolarized lines should consist of one methyl wag and one bridge stretch (ν_6). Again, as seen in Table 135, the Raman data of van der Kelen and Hermann[1289] are in agreement with this scheme. The polarized methyl rock corresponding to that in A_g of the D_{2h} point group is not found, nor is the depolarized bridge stretch corresponding to ν_6. However, neither of these were found in any of the spectra of related molecules previously considered.

The assignments are collected in Table 135, those associated with Cl modes and comparable to $AlCl_3$ to the right and those involving methyl modes and comparable to methyl-containing compounds to the left of the column.

Alkyl Group Exchange and Association in Alkylaluminium Compounds. Hoffman[1281] discusses in detail the phenomenon of alkyl group exchange and association in organoaluminium compounds. It is known that trialkylaluminium compounds are associated to dimers unless their alkyl chains are branched in the α or βposition with respect to the aluminium atom, and that the infrared (Figure 52) and Raman spectra of the associated and non-associated trialkylaluminium types of compounds show characteristic differences; originating from differences in their molecular state. Hoffmann illustrates this in the case of dimer trimethylaluminium and the monomer trisobutylaluminium. It was proved that after mixing these compounds an alkyl group exchange occurs leading to the formation of mixed alkylaluminium compounds and these mixed alkylaluminium compounds can now participate in dimerisation. This can be seen from the vibration spectrum of the mixture (Figure 53) and can also be detected by means molecular weight measurements. The formation of new association bonds is also accompanied by an additional heat of reaction, the measurement of which should constitute a means of estimating of heat of association across alkyl bridges.

Proton magnetic resonance measurements show that alkyl group exchange is a very fast reaction. The proton magnetic resonance spectrum of the mixed alkylaluminium compounds differ remarkably from the individual spectra of the pure trialkylaluminium compounds before mixing.

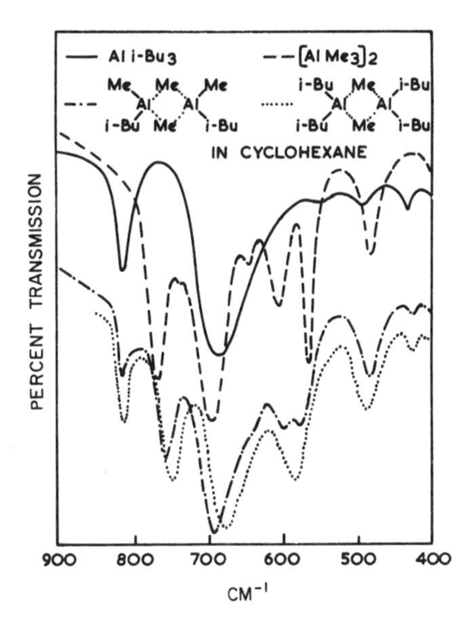

Figure 52 - Infrared spectra of cyclohexane diluted $Al(CH_2CH(CH_3)_2)_3$, $(Al(CH_3)_3)_2$ and their 2:1 and 1:2 mixtures demonstrating alkyl group exchange (concentration w.r.t. dimers ca. 0.4 mol/l in a 0.05 mm thick sample cell).

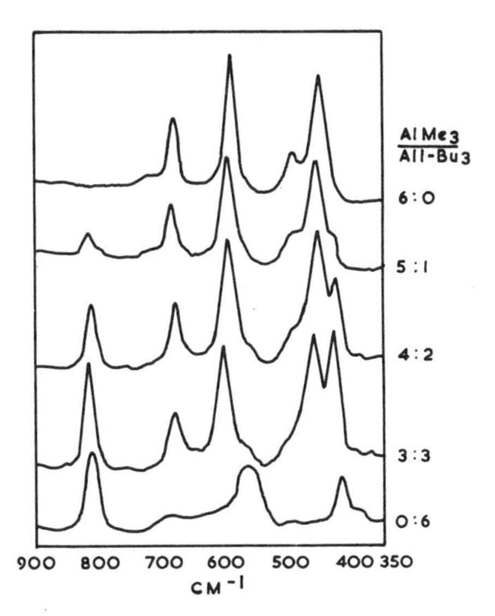

Figure 53 - Raman spectra of undiluted mixtures of $Al(iso\ Bu)_3$ with $Al(Me)_3$ in various proportions scanned with a model 81 Cary spectrophotometer in a 4.5 cm^3 sample tube.

According to the ratio of the amounts of the two mixed trailkylaluminium compounds the methyl group signal changes its position through a considerable range of the spectrum obeying a linear relation between methyl contents and signal position. At small methyl concentrations in mixtures when all the methyl groups are bridge bonded, the position of the signal becomes constant as would be expected. The signal position of the non-bridged outer methyl group can be extrapolated easily. At - 70°C the alkyl group exchange is prevented; the spectrum of trimethylaluminium shows two signals, one each from the pure methyl bridge and the outer methyl groups.

Dialkylaluminium Hydrides. Hoffman and Schomburg[1296] studied the infrared absorption and association of various dialkylaluminium hydrides. Based on cryoscopic measurements they showed that these compounds exist as trimers either when in the pure state or as solutions in neat solvents. At about 1800 cm^{-1} the infrared spectra of dialkylaluminium hydrides has an Al-N valence vibration which is considerably extended when association occurs between AlR_2H molecules. In Figure 54 (spectra (a) and (b) are compared the infrared spectrum of diethylaluminium hydride (neat and diluted) with that of neat triethylaluminium.

The infrared spectrum of the diethylaluminium hydride differs from that of the corresponding triethyl compound in that intensive absorptions occur at 1800 cm^{-1} and at 750 cm^{-1} due to Al - H valence vibration. An Al - H deformation vibration may be held responsible for the latter (by comparison with the spectra of a large number of other hydride compounds). AlH_4 also shows considerable absorptions at 760 cm^{-1}.

The absorption band at 1800 cm^{-1} shows a marked interaction with the association condition of the compound. If the hydride is added to α-olefins or ethylene to produce a trialkylaluminium compound then the absorption band at 1800 cm^{-1} disappears from the spectrum. If the hydride hydrogen is replaced by deuterium then the band moves from 1777 to 1280 ~ 1777 $\sqrt{2}$ cm^{-1} (see Figure 54, spectrum (c)). There is thus no question of its belonging to an Al - H valence vibration.

Hoffman and Schomburg[1296] also showed that dialkylaluminium hydrides do not form stable etherates. Ether can be almost completely distilled off again from a hydride-ether mixture. It is, nevertheless, possible to give clear spectroscopic (and dielectric) proof of an interaction. In Figure 55 is shown that Al - H valence vibration band of $(i - C_4H_9)_2$ AlH in mixtures of hexane and dibutyl ether at the same hydride concentration with an increasing mol ratio of diisobutylaluminium hydride to ether. As can be seen, it is only in the case of fairly considerable excess of ether that equilibrium is completely on the side of the donor-acceptor complex.

In the reaction between diisobutylaluminium hydride and various other donor molecules (e.g. triethylamine, tetrahydrofuran, etc.) the equilibrium is displaced with increasing donor activity in favour of a donor-acceptor complex. Using triethylamine to form the complex the equilibrium lies practically wholly on the right side. Even in the case of a stoichiometric mole ratio all the diisobutyl aluminium hydride is converted to amine complex.

The fact that the Al - H absorption band becomes sharply distinguished in the complex bonding of aluminium hydrides to tertiary amines can be put to analytical use. It is possible to establish, for a series of variously substituted dialkylaluminium hydrides, that this sharp absorption band has a molar extinction coefficient of 2.10 x 10^5 cm^2 mol^{-1} within the limits of error (± 4%) independent of the molecular weight. As absorption is extremely intensive, this is a highly sensitive method for estimating the hydride content of aluminium trialkyls. Care must be taken to ensure that all organoaluminium compounds present, e.g. trialkylaluminium compounds, are complexed with amine. The sensitivity of the method for estimating

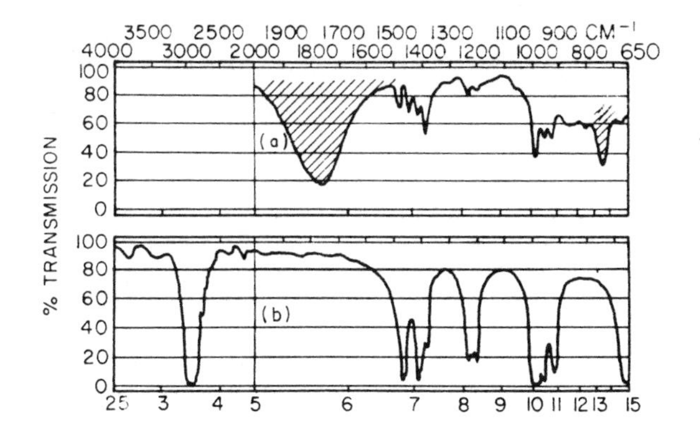

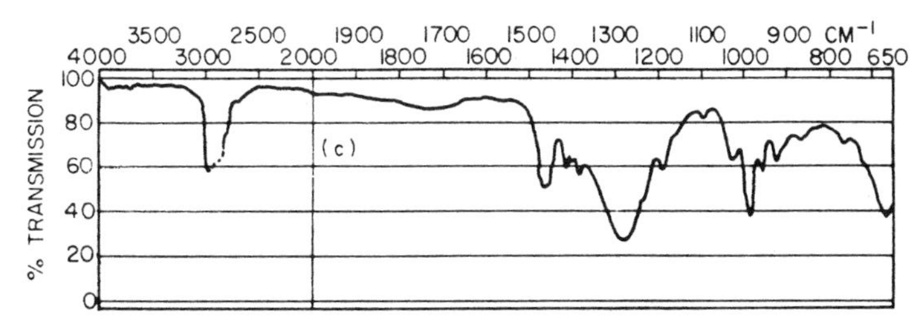

Figure 54 - Infrared spectra in the range 650 - 4,000 cm^{-1}. (a) $(C_2H_5)_2$ AlH in cyclohexane, (b) $(C_2H_5)_2$Al neat, (c) $(C_2H_5)_2$Al D in cyclohexane.

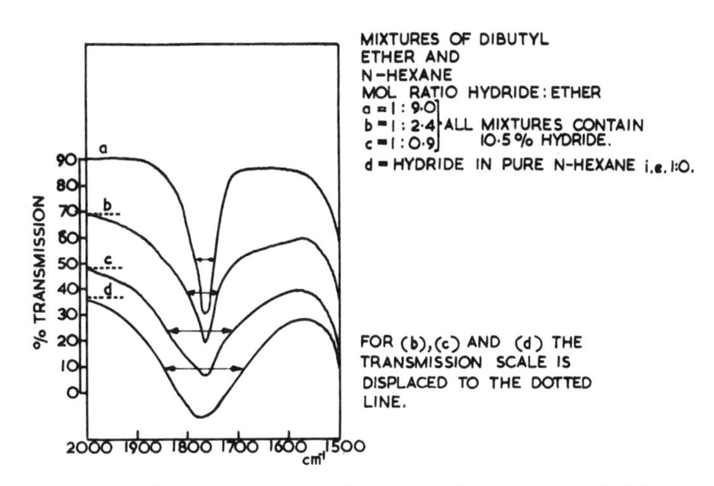

Figure 55 - Infrared spectra of diisobutylaluminium hydride mixed with varying ratios of dibutyl ether.

dialkylaluminium hydrides then depends on the degree of success in compensating for slight background absorptions of the aluminium trialkylaminates in the region of 1800 cm^{-1}.

Hoffmann and Schomburg[1297] have also determined the infrared spectra of a wide range of compounds containing aluminium hydride linkages; oxygen and water were rigorously excluded from test solutions during the preparation of spectra. Also, only solvents (e.g. cyclohexane) which were completely inert towards the organoaluminium compounds, examined were used to prepare solutions for spectroscopy. Hoffmann and Schomburg[1297] obtained distinct evidence that electron donor solvents such as carbon tetrachloride and carbon disulphide and also solvents containing active hydrogen atoms are unsuitable spectroscopic solvents as they react with the organoaluminium compound and have appreciable influence on the recorded spectrum.

In Figure 56 is given a survey of the Al - H valence frequencies obtained by Hoffmann and Schomburg[1297] for various hydride containing organoaluminium compounds. These results show that the Al - H valence frequencies cover a surprisingly wide range from 1700 - 1900 cm^{-1}. The lowest frequency is obtained with a donor - acceptor complex with two moles donor per aluminium atom, whilst the highest frequency is obtained with a dibromoaluminium hydride-ether complex. Unusually large half-width values were obtained for associated dialkylaluminium hydrides and for tetrahydrofuran complexes, or solutions of aluminium hydride compounds in tetrahydrofuran.

In the course of their work, Hoffmann and Schomburg[1297] measured the infrared spectra of a range of co-ordination complexes of aluminium hydride compounds with trimethylamine and triethylamine and also the infrared spectra of various aluminium hydride derivatives containing negatively charged substituents.

Hudson[1298] has described an infrared spectroscopic method for the determination of diethylaluminium hydride and diethylaluminium ethoxide in amounts down to 2% in triethylaluminium.

Alkylaluminium Alkoxides. Bell et al[1299] studied the infrared spectra of aluminium triisopropoxides. They showed that the isopropoxide group had characteristic absorption bands at 8.5, 8.8 and 9.0 μ (see Figure 57). Guertin et al[1300] prepared and infrared spectra of aluminium isopropoxide aluminium secondary butoxide and aluminium 2-pentoxide and also of the corresponding alcohols (namely, isopropanol, secondary butanol and pentanol-2). Interpretation of these spectra (Figure 58) lead Guertin to assign the frequencies between 1033 and 1070 cm^{-1} to the Al - OC linkage and to assign the band at 990 cm^{-1} to an Al - O - Al type linkage.

Barraclough et al[1301] in their extensive survey of the infrared spectra of various metal alkoxides determined the spectrum of aluminium triisopropoxide. They find the following bands:

Strong (cm^{-1}) 1185, 1174, 1138, 1124, 1036, 951, 835, 678, 566
Medium (cm^{-1}) 861, 699, 610, 535.

Barraclough[1301] claims that the bands occurring at 1170 and 1150 cm^{-1} are both split into doublets, probably by intramolecular coupling and claim that their conclusions are in good agreement with those of previous investigators (Bell et al[1299]; Zeitler and Brown[1302]).

Barraclough et al[1301] claim that the band at 1000 cm^{-1} claimed by Guertin et al[1300] to be due to the Al - O stretching mode are, in fact, due predominantly to C - O stretching modes of the Al - O- C groups and that the Al - O stretching modes occur at lower frequencies, not covered by the

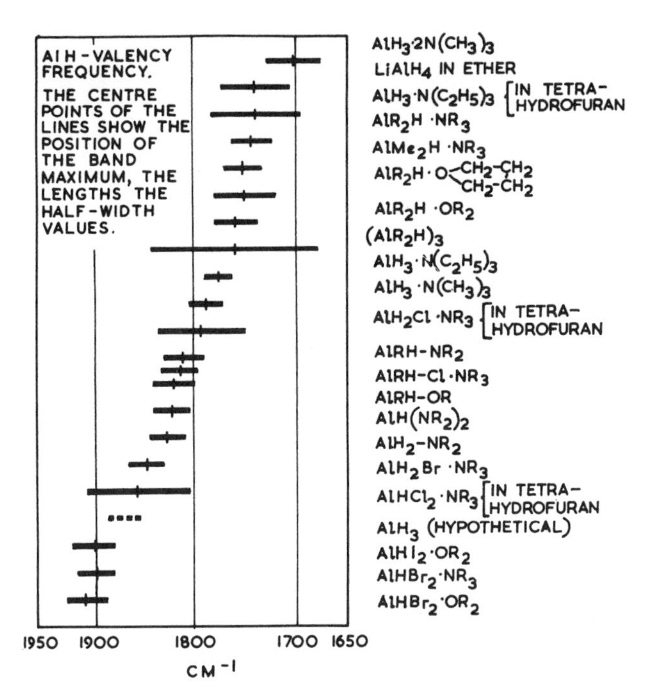

Figure 56 - Survey of Al - H frequencies obtained for various compounds by infrared spectroscopy.

investigations of Guertin et al[1300]. Comparison of the infrared spectrum of aluminium triisopropoxide with that of isopropanol permitted Barraclough et al[1301] to make the following tentative assignments:

C - O 1036 cm^{-1}
Al - O 699, 678, 610, 566, 535 cm^{-1}

He claims that the Al - O region of the aluminium isopropoxide spectrum provides more evidence for intermolecular bonding through oxygen atoms. In monomeric aluminium isopropoxide there could be a maximum of only three stretching vibrations, whatever the stereochemistry of the molecule. Actually five bonds are observed in this region and the presence of the bridging oxygen atoms seems to be the most reasonable explanation of this. Unfortunately, only one C - O stretching frequency was definitely identified whereas at least two would be expected. This may arise from the difficulty in assigning frequencies in this part of the spectrum and the two bonds tentatively assigned to coupling of the isopropyl skeletal vibrations may in fact be C - O stretching vibrations.

Wilhoit et al[1303] studied the infrared spectrum of samples of aluminium triethoxides made by several different synthetic routes. They determined the spectrum using sodium chloride optics in the 1 - 14 μ region and potassium bromide optics in the 10 - 25 μ region. In Table 136 are enumerated assignments obtained by Wilhoit et al[1303] on the Nujol mulls prepared from the purest available specimen of aluminium triethoxide.

These results include only those absorption peaks which can definitely be ascribed to aluminium ethoxide. Other weak bands may be present which

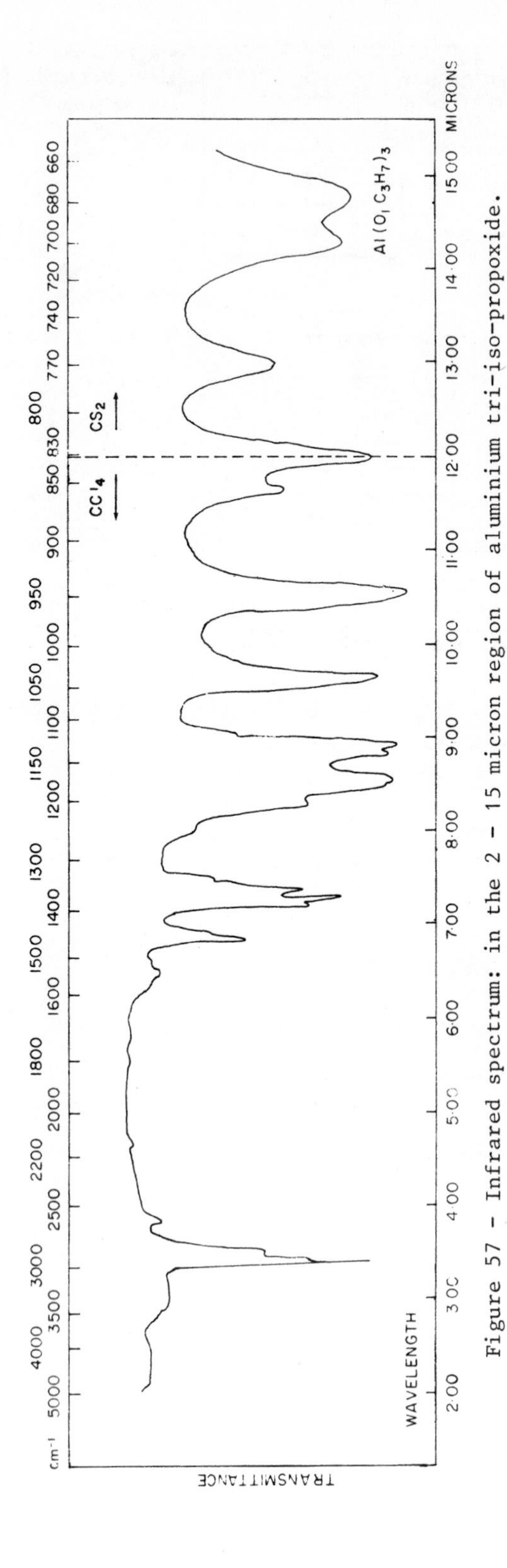

Figure 57 - Infrared spectrum: in the 2 - 15 micron region of aluminium tri-iso-propoxide.

are masked by the absorption of the Nujol. If the assignment made for the 3340 and 1350 cm^{-1} bands are correct, they must be the result of a small amount of hydrolysis of the sample during handling. Most of the bands were confirmed by the examination of the slurries suspended in xylene but the spectra obtained from the K Br pellet technique showed significant variation from this data. Because of the possibility of exchange of bromides and ethoxide ions and of the increased possibility of hydrolysis of samples it was felt that the spectra of the K Br pellets is not typical of pure aluminium ethoxide.

Table 136 - Absorption spectra of "pure" aluminium ethoxide, 1 - 25 μ

Observed frequency (cm^{-1})	Intensity	Assignment
6250	w	
4250	m	
3340	w	O-H stretching vibration in (-Al-OH-Al-) polymer chains
2740 - 1520 (21 equally spaced bands)	w	
1460	s	$-CH_2$ deformation
1382	s	$-CH_3$ deformation
1350	w	-OH bending
1178	m	ethoxide skeletal vibration
1105	m	ethoxide skeletal vibration
1059	s	Al-O-C stretching vibration
935	w	Al-O-Al stretching vibration
896	s	
707	s	
650	s	
515	s	
465	w	

s = strong, m = intermediate, w = weak.

The spectra of partially hydrolysed samples of aluminium ethoxide depends primarily upon the proportion of aluminium present. As the degree of hydrolysis increases the following changes have been found to take place:

1. The intensity of all absorption bands except those at 3340 and 935 cm^{-1} decreases.
2. The 3340 cm^{-1} peak becomes sharper, more intense, and shifts to slightly higher frequencies.
3. The 935 cm^{-1} peak becomes broader and shifts to slightly lower frequencies.

Wilhoit et al[1303] confirmed the absorption observed by Guertin et al [1300] at 1033 cm^{-1} in aluminium isopropoxide and observed corresponding peaks in aluminium tri-n-propoxide and aluminium tri-n-butoxide. They enumerate all the presently available data on the Al-O-C absorption frequencies (Table 137) and on the Al-OAl absorption frequencies (Table 138).

The infrared spectroscopy of aluminium isopropoxides has been discussed by Brown and Mazdiyasni[1304] and Lynch et al[1305].

Table 137 - Absorption frequencies for the Al-O-C vibration in aluminium alkoxides

Alkoxide group	Frequency (cm^{-1})
n-propoxide	1015
iso-propoxide	1033
n-butoxide	1048
s-butoxide	1058
ethoxide	1059
2-pentoxide	1070

Table 138 - Absorption frequencies for the Al-O-Al vibration in aluminium alkoxides and soaps[a]

Alkoxide group	Frequenccy (cm^{-1})
n-propoxide	898
ethoxide	935
iso-propoxide	950
Al soaps	983
n-butoxide	995

[a] Compiled from data of Bell et al[1299], Scott et al[1306] and Wilhoit et al [1303].

Trialkylaluminium - Trialkylboron Compounds. Sutton and Schneider[1307] have applied infrared spectroscopy to the determination of mixtures of triethylaluminium (15%) and triethylboron (85%), which are used as hypergolic ignitors in the fuel lines of rockets to initiate ignition. The absolute concentrations of these substances could not be determined as Beer's law was not obeyed. However, the infrared results could be correlated with a critical factor in rocket operation, namely, ignition delay.

B. Raman Spectroscopy

Van der Kelen and Hermann[1289] have determined the Raman spectra in the liquid state of various dimethylaluminium mono-halide $((CH_3)_2AlX)_2$ where X is Cl, Br or I, and of monomethylaluminium dihalides $(CH_3AlX_2)_2$ where X is Cl or Br. For $((CH_3)_2AlCl)_2$ and $(CH_3AlCl_2)_2$ depolarization factor measurements were reported. Based on Raman spectroscopy they found that for both series of compounds the results they obtained agreed best with a dimeric structure of symmetry C_{2c} and D_{2h}.

Van der Kelen and Herman[1289] state that dimerization occurs by formation of a ring comprising two aluminium atoms and two methyl groups.

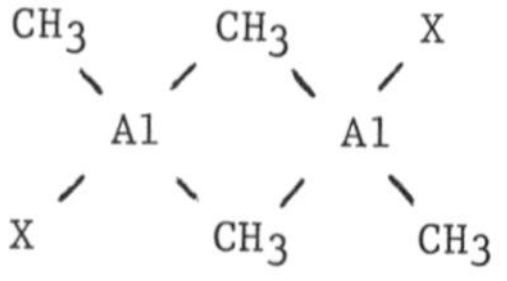

```
CH3     CH3     X
   \   /   \   /
    Al      Al
   /  \   /   \
X      CH3     CH3
```

X = halogen

Yamamoto[1308] investigated the Raman spectra of diethylaluminium monohalides. He states that it has been established that some lower members of the trialkylaluminium series have the molecular structure of the dimeric bridged form represented by

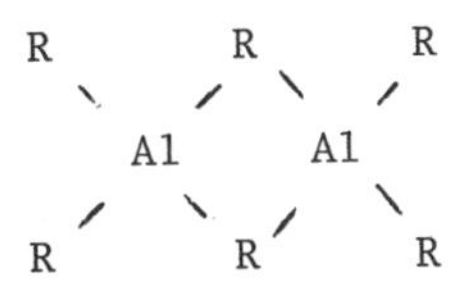

in which the plane containing two terminal alkyl groups and aluminium atom is perpendicular to the bridge plane (Kohlrausch and Wagner[1309], Lewis and Rundle[1282]). These compounds have been defined as "electron deficient compounds", and according to Rundle[1310] the bondings between the aluminium and bridging alkyl groups are formed by two "half bonds". Monohalogen derivatives of trialkylaluminium also exist as dimers, and the following structural formulae can be considered for them.

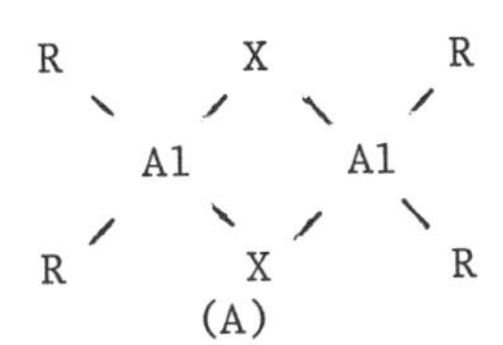

(A)

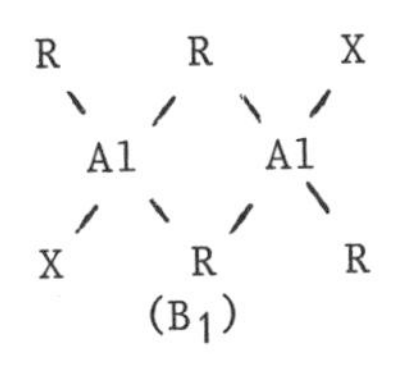

(B_1)

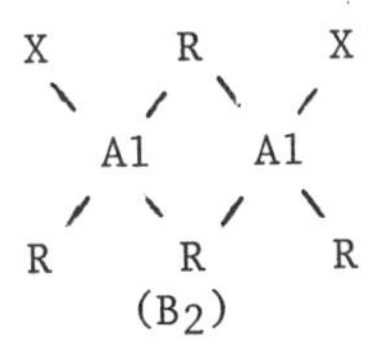

(B_2)

Formula A with symmetry D_{2h} has a halogen bridge which would be formed by the coordination of lone pair electrons in halogen atoms to aluminium Formula B_1 with symmetry C_{2h} corresponds to the "trans" form, in which two halogen atoms are present on opposite sides of the bridge plane, while formula B_2 with symmetry C_2 corresponds to the "cis" form where the halogen atoms are on the same side of the bridge plane. In the latter two, the alkyl bridge would be formed by the half bonds, as in the case of trimethyl-aluminium.

Raman spectral data obtained by Yamamato[1308] is taubulated in Table 139. This table shows that the Raman lines are divided into three groups: 0 - 700 cm^{-1}, 900 - 1500 cm^{-1}, and 2700 - 3000 cm^{-1}, which represent skeletal vibrations, CH deformation vibrations and CH stretching vibrations respectively. Of these three kinds of vibrations, the CH deformations and stretching vibrations are not so important as far as the molecular structure is concerned. Furthermore, the CH deformations have been discussed in some detail by Hoffmann[1281]. Yamamoto[1308] was interested mainly in the skeletal vibrations and did not explore the CH stretching vibrations.

Van der Kelen and Hermann[1289] suggested, after a study of the Raman spectra, that aluminium alkylhalides would have the structure B_2, (see above) while Groenewege et al[1295] and Groenewege[1288] proposed structure A for dimethylaluminium chloride as a result of Raman and proton nuclear magnetic resonance spectroscopy. Electron diffraction studies by Brockway and Davidson[1286] also indicated that the compound would have the structure A. Hoffmann[1281] reported Raman and infrared spectral data on some members of trialkylaluminium and dimethyl and diethylaluminium chlorides, but his

Table 139 - Raman spectra of triethylaluminium and diethylaluminium aluminium monohalides

$(Al(C_2H_5)_3)_2$		$(Al(C_2H_5)_2Cl)_2$		$(Al(C_2H_5)_2Br)_2$		$(Al(C_2H_5)_2I))_2$	
$\Delta\nu$, cm^{-1}	I	$\Delta\nu$, cm^{-1}	I	$\Delta\nu$, cm^{-1}	I	$\Delta\nu$, cm^{-1}	I
119	30 dp	114	44 dp	95	66 dp	81	100 dp
196	sh dp	164	3 dp	183	sh dp	170	85 p
269	91 dp	239	27 dp	197	100 p	202	25 p
362	18 p	263	19 p	263	11 dp	268	20 dp
430	100 p	339	56 p	266	26 p	321	25 p
560	66 p	480	5 p	319	23 p	544	100 p
640	30 dp	556	100 p	547	100 p	664	34 dp
		670	27 dp	664	27 dp		
922	8 dp	922	5 dp	921	3 dp	921	3 p
955	12 p	957	10 p	954	12 p	956	8 p
979	41 dp	990	45 dp	988	24 dp	987	19 dp
1190	62 p	1107	1 p	1074	3 p	1098	3 p
1384	1 dp	1197	74 p	1132	3 p	1132	3 p
1404	19 dp	1379	1 p	1190	84 p	1189	154 p
1460	36 dp	1406	23 dp	1377	2 dp	1381	5 dp
		1463	46 dp	1402	15 dp	1404	11 dp
				1459	25 dp	1462	22 dp
2724	9 p	2732	17 p	2733	8 p	2731	6 p
2792	18 p	2786	14 p	2789	22 p	2790	24 p
2827	54 p	2870	420 p	2866	300 p	2867	360 p
2864	565 p	2904	115 dp	2900	73 dp	2901	33 dp
2896	153 dp	2943	230 dp	2939	160 dp	2940	78 dp
2938	260 dp						

work was mainly concerned with CH deformations and little consideration was given to the structure of the diethylaluminium monohalogenides.

Yamamoto[1308] made tentative assignments for the skeletal vibrations in dialkylaluminium halides by comparison with the known spectra of aluminium trihalides (Bell and Longuet-Higgins[1291]: Klemperer[1290]; trimethyl aluminium (Kohlrausch and Wagner[1309]) and dimethylaluminium chloride (Groenewege[1311]) for which the bridged structure has been firmly established by many workers. The skeletal vibration lines obtained for the compounds under discussion are schematically drawn in Figure 59 in which the notation for the frequencies of Bell and Longuet-Higgins[1291] is used.

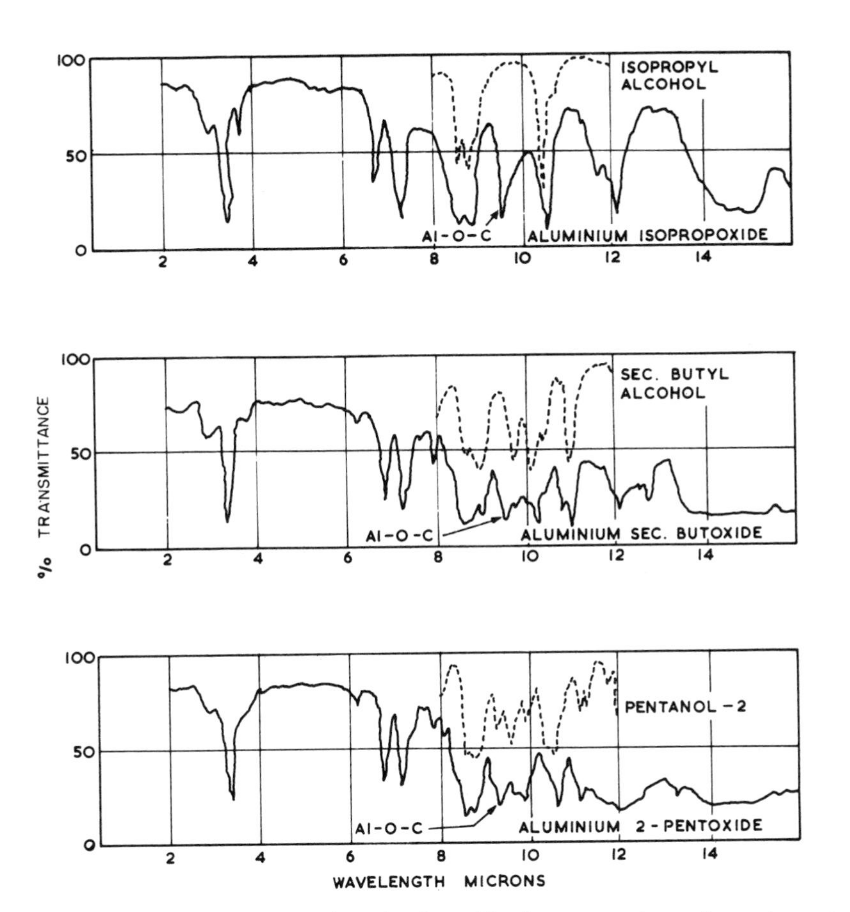

Figure 58 - Infrared spectra in the 2 - 15 micron region of various trialkoxyaluminium compounds.

In the Raman spectra of triethylaluminium and diethylaluminium monohalogenides the following two series of strong lines are observed:

X of $Al(C_2H_5)_2X$	Series 1 (cm^{-1})	Series II (cm^{-1})
Et	560	430
Cl	556	339
Br	547	197
I	544	170

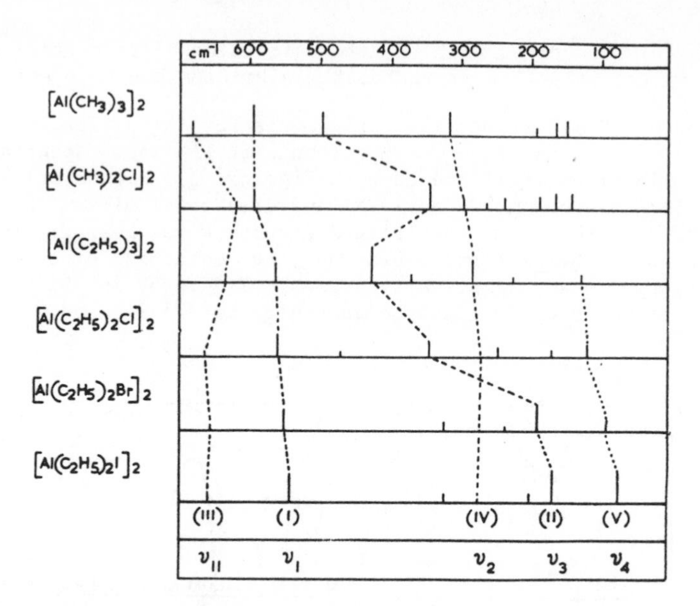

Figure 59 - Schematic representation of Raman spectra on 0 - 700 cm^{-1} region of dimethylaluminium halides and related compounds.

Since both series of lines are polarized, these must be either ν_1, ν_2, ν_3, or ν_4 for D_{2h} symmetry. Furthermore, it is reasonable to assume that series I and II are assigned to either ν_1 or ν_2 by comparison with the assignment of the spectrum of trimethylaluminium.

If the halogen bridge model A with symmetry D_{2h} is acceptable, series II must be assigned to ν_2, the bridge stretching vibration, because it varies in frequency with the increase in the atomic weight of substituents while series I does not. The frequencies of the bridge stretching vibration, ν_2, of aluminium trihalides are 340, 204 and 146 cm^{-1} for chloride, bromide, and iodide respectively, as is shown in Table 140 where the Raman lines in the skeletal regions of trimethylaluminium, dimethylaluminium chloride and aluminium trihalogenides are listed with the normal vibrations to which they are assigned, Thus, it can be found that there is a close relationship in ν_2 between the trihalides and diethylaluminium monohalides. When series II is assigned to ν_2, series I is naturally assigned to ν_1, the outer stretching vibration.

If the halogen bridge model A is assumed, the remaining outer vibrations ν_2, ν_7, ν_{11} and ν_{12} must not vary in frequency with the increase in the atomic weight of the substituents. Thus, the series III, 640(Et), 670(Cl), 664(Br) and 664(I)cm^{-1} is assigned to the outer stretching vibration ν_{11} by comparison with methyl compounds and by a consideration of polarization factors. Series IV, 269(Et), 263(Cl), 266(Br) and 268(I) cm^{-1} may be assigned to the outer bending vibration ν_3. The depolarised series V, 119(Et), 114(Cl), 95(Br) and 81(I)cm^{-1} might be ν_4 by intensity consideration, although ν_4 has to be a polarized vibration.

These assignments have been performed with the halogen bridge model. If the alkyl bridge model, B_1 with symmetry C_{2h} (trans form), is adopted, the series which does not vary in frequency is the bridge variation, so that polarized series I should be a bridge vibration. But even in this case, triethylaluminium must belong to the same symmetry class D_{2h} as trimethylaluminium, so that, for example, 560 cm^{-1} of triethylaluminium in series I should be assigned to ν_1, the outer vibration, by comparison with ν_1 and ν_2 of trimethylaluminium. With the model B, all of the monohalides have the alkyl bridge. Then strong lines at approximately 430 cm^{-1} should be also observed as the bridge stretching vibration in the spectra of the monohalides as in the case of triethylaluminium. Actually it is not the case. Furthermore, it is impossible to interpret the marked correlation of frequencies between the triethylaluminium and the diethylaluminium monohalogenides obtained in Table 140.

Yamamoto[1308] concluded that the Raman spectra of diethylaluminium chloride, bromide and iodide are all consistent with a dimeric bridge structure over the halogen atoms with symmetry D_{2h}.

C. Proton Magnetic Resonance and Nuclear Magnetic Resonance Spectroscopy

An excellent general discussion on the application of proton magnetic resonance spectroscopy (P.M.R.) to the elucidation of the structure of organic compounds has been given by Hoffmann[1312] who includes in his discussion a description of the proton resonance spectrum of diethylaluminium hydride at 40 and 60 MHz.

Brownstein et al[1313] discuss the proton resonance spectrum of trimethylaluminium dimer and dimethylaluminium chloride dimer. From the observation of van der Kelen and Hermann[1289] that in halogenated methylaluminium compounds the methyl groups are bridged between the two aluminium atoms in the dimer; Brownstein et al[1313] concludes that in trimethylaluminium and in dimethylaluminium chloride dimers there should be two different kinds of methyl groups. These two kinds of methyl groups had previously been observed by Muller and Pritchard[1314] using proton magnetic resonance spectroscopy at - 75°C. Muller and Pritchard[1314] found that at higher temperatures than - 75°C the two peaks coalesce to a single one due to rapid chemical exchange of the methyl groups.

Brownstein et al[1313] carried out proton resonance spectroscopy on cyclohexane solutions of trimethylaluminium, dimethylaluminium chloride and methylaluminium dichloride and in each case did not observe line splitting, even when the spectra were measured at 0°C. To explain this observation they suggest that in dimethylaluminium chloride dimer a methyl group exchange readily occurs between bridging and non-bridging groups within the dimer.

Smidt et al[1315] have applied proton magnetic resonance spectroscopy to the study of exchange of alkyl groups occurring in triethylaluminium, diethylaluminium chloride and mixtures of these and also to a mixture of methylaluminium dichloride and dimethylaluminium chloride. The claim that proton magnetic resonance showed that there is an exchange between the outer and bridge alkyl group in dimeric triethylaluminium at room temperature whilst at lower temperature the rate of exchange diminishes and at - 60°C is hardly noticeable (Figure 60).

No exchange of alkyl groups was found in the case of diethylaluminium chloride dimer between room temperature and - 60°C and this observation seems to fit well the concept of chloride bridges in this dimer. In a 1:1 mixture of triethylaluminium and diethylaluminium chloride exchange of alkyl groups was observed at room temperature. Upon lowering of the temperature of the mixture first the exchange occurring between the alkyl groups in the two different compounds stops and then the exchange between the alkyl groups in triethylaluminium also stops. Similar findings were obtained in the case of 1:1 mixtures of methylaluminium dichloride and dimethylaluminium chloride.

Table 140 - Raman lines in the region of 0 - 700 cm^{-1} of trimethylaluminium, dimethylaluminium chloride and aluminium trihalogenides

Reference	$(Al(CH_3)_3)_2$ (Wiswall and Smythe[1279])	$(Al(CH_3)_2Cl)_2$ (Badger and Zumwalt[1293])	Al_2Cl_6 (Schomberg[1280])	Al_2Cl_6 (Brockway and Davidson[1286])	Al_2Br_6 (Schomberg[1280])	Al_2I_6 (Schomberg[1280])
1	590 (10p)	588 (10)	606 (2½p)	506	491 (3)	406 (3)
v_2	441 (8p)	332 (7)	340 (10p)	340	204 (10)	146 (10)
v_3	314 (9p)	286 (4)	217 (5p)	217	140 (5)	94 (6)
v_4	148 (5)	131 (6)	112 (6dp?)	112	73 (6)	53 (6)
v_{11}	681 (5dp)	620 (2)	506 (3)	606	407 (2)	344 (2½)
v_{12}	188 (1)	173 (5)	164 (3dp)	164	112 (3)	-
v_6	563 (00)	247 (3)	438 (½dp?)	438	(291?)	(195?)
v_7	164 (3)	152 (5)	284 (2dp)	164	176 (2)	-
v_{15}	-	214 (4)	-	(160)	-	-

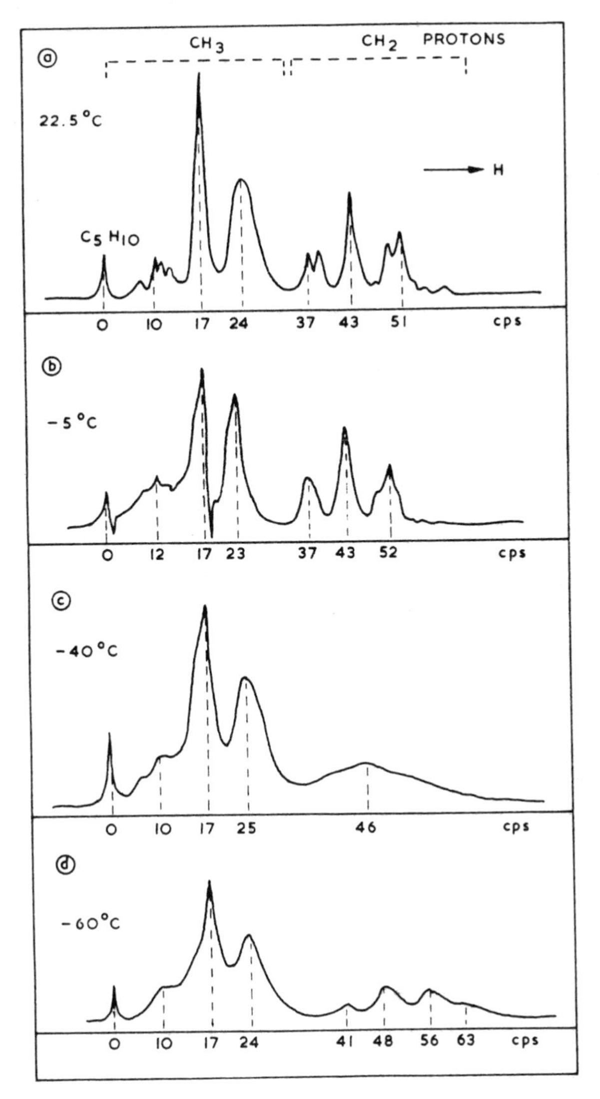

Figure 60 - PMR spectra of triethylaluminium at four temperatures. The spectrum at 22.5°C was taken without the variable temperature insert, therefore, the homogeneity of the magnetic field is better than in the other cases.

Hoffmann[1281] has applied proton magnetic resonance to a study of alkyl group exchange occurring in mixtures of trimethylaluminium dimer and triisobutylaluminium, which due to the branching of the alkyl groups, can exist only in the monomeric form. He showed that the alkyl group exchange is an extremely fast reaction and that due to this exchange the PMR spectra of mixtures were appreciably different from that of either single compound, supporting conclusions reached by infrared and Raman spectroscopy. According to the ratio of trimethylaluminium dimer to triisobutylaluminium monomer in the mixture the methyl group signal changes the position through a considable range of the spectrum obeying a linear relation between methyl contents and signal position. At small methyl concentrations when all the methyl groups are bridge bounded, the position of the signal becomes constant as would be expected. The signal position of the non-bridged outer methyl group can be extrapolated easily. At lower temperatures (- 70°C) the exchange is prevented: the spectrum of trimethylaluminium shows two methyl signals, one each for pure methyl bridge and outer non-bridged methyl groups.

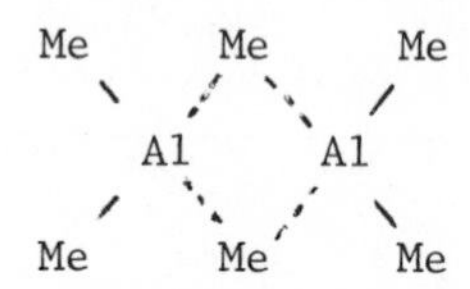

Hoffmann[1281] claims that the NMR spectrum of trimethylaluminium dimer show two signals corresponding to non-bridged and bridged methyls with relative intensities 2:1. Thus in Figure 61 (a) and (b) are shown the NMR spectra of a pentane solution of trimethylaluminium at -67°C and +28°C. At -67°C the two signals are seen in a ratio 2:1. At 28°C the two signals collapse into one sharp signal nearer to the position of the signal of the non-bridged groups.

Groenewege et al[1295] have studied the NMR spectra at +22°C and -75°C of trimethylaluminium, dimethylaluminium chloride and methylaluminium dichloride dimers at various concentrations in cyclohexane solution and measured chemical shifts in ppm relative to cyclopentane. They compare their results with those obtained by Brownstein et al[1313] and by Muller and Pritchard[1314]. Brownstein's[1314] chemical shifts based on cyclohexane were reduced to cyclopentane by adding - 0.07 p.p.m., this being the difference in chemical shift between cyclopentane and cyclohexane.

It can be seen in Table 141 that the effect of dilution in cyclopentane solution is within the experimental error. The room temperature peak of $Al_2(CH_3)_6$ occurs at the weighted average of the two low temperature peaks. The effect of temperature is negligible in the case of $Al_2(CH_3)_4Cl_2$. The low temperature spectrum of $Al_2(CH_3)_2Cl_4$ could not be observed because of the low solubility in cyclopentane at temperatures below 0°C.

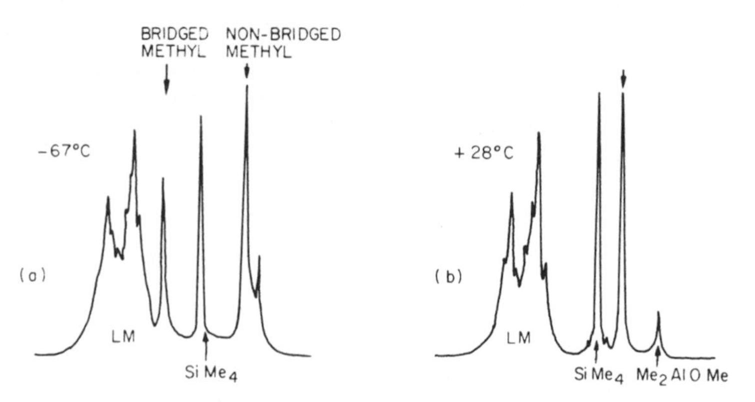

Figure 61 - NMR methyl group signals of $Al(Me)_3$ (40% in pentane at (a) -67°C (b) +28°C. $(Al(Me_2(OMe))_3$ and Si Me_4 as internal standards. LM denotes solvent signal.

Groenewege[1295] reach the following conclusions:

$Al_2(CH_3)_6$: that their own observations and conclusions are in good agreement with those of Muller and Pritchard[1314] and Brownstein et al[1313]. The bridge CH_3 group protons are less shielded than those of the outer CH_3 groups. At room temperature a rapid exchange of bridge and outer CH_3 groups makes the two resonances coalesce into a single peak.

$Al_2(CH_3)_4Cl_2$: The unsplit low temperature resonance can be explained readily in terms of the chlorine bridge model exhibiting four equivalent outer CH_3 groups. A shift over about 0.35 p.p.m. to a lower field with respect to the outer CH_3 group protons in $Al_2(CH_3)_6$ is seen as a consequence of the substitution of CH_3 by Cl.

The low temperature spectrum observed does not support the methyl bridge model proposed by Brownstein et al[1313] and by van der Kelen and Hermann[1289]. The bridge and outer CH_3 groups of this model would have to exchange rapidly even at - 75°. Moreover, the substitution of two CH_3 groups by Cl atoms in $Al_2(CH_3)_6$ is expected to shift the weight average resonance to a lower field.

$Al_2(CH_3)_2Cl_4$: The low temperature spectrum is expected to be almost identical with the room temperature spectrum. There is little reason to expect any splitting by unequivalent positions in this molecule. The effect of temperature on the resonance of $Al_2(CH_3)_4Cl_2$ was seen to be very small. The shift to a lower field over about 0·25 p.p.m. with respect to $Al_2(CH_3)_4Cl_2$ and over about 0.60 p.p.m. with respect to the outer CH_3 groups of $Al_2(CH_3)_6$ suggests that there are outer CH_3 groups in this molecule as well. Finally it is improbable that the bridge CH_3 group protons of $Al_2(CH_3)_6$ would shift over about 0.5 p.p.m. to a higher field if four CH_3 groups should be replaced by Cl atoms. Consequently, Groenewege et al[1295] propose that their NMR data support the chlorine bridge hypothesis in both $Al_2(CH_3)_4Cl_2$ and $Al_2(CH_3)_2Cl_4$.

Shiner et al[1316] measured the molecular weight and NMR spectra of aluminium t-butoxide over a range of temperatures between - 14°C and + 74°C and conclude that in solution, this substance exists as a cyclic dimer:

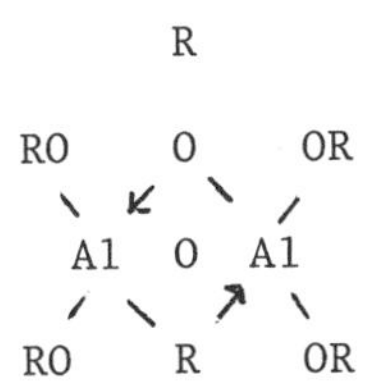

They found no evidence that this substance dissociates at higher temperatures or when dissolved in such basic solvents as dioxane or t-butyl alcohol. They also found that aluminium isopropoxide in organic solvents has the tetrameric structure proposed by Bradley[1317]. Shiner showed that at higher temperatures, either on melting or in solution, this material is converted into a cyclic trimer which only slowly reverts to the tetramer at lower temperatures in the super cooled melt or in solution:

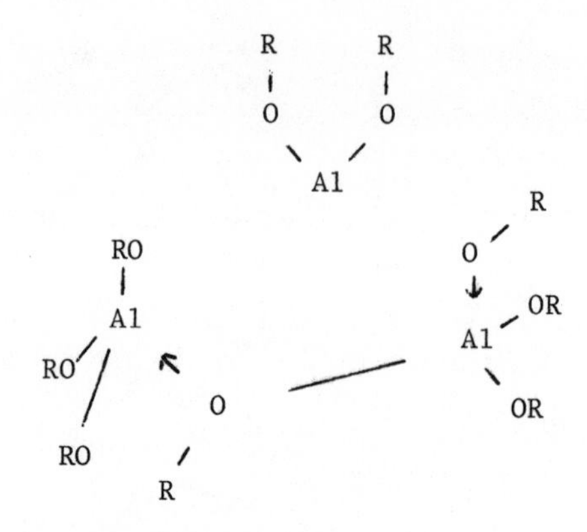

aluminium triisopropoxide cyclic trimer in melt.

Table 141 - N.M.R. chemical shifts relative to C_5H_{10} in ppm + 0.03

Compound	Concentration[a]	22°	-75°
			- 10[b]
	1.80	- 1.81	- 2.20[c]
	0.52	- 1.85	
$Al_2(CH_3)_6$	0.21	- 1.83	1.05[b]
			- 2.20[e]
			- 1.04[d]
		- 1.79[d]	- 21.7[d]
		- 1.82[c]	
	1.20	- 1.83	- 1.85
$Al_2(CH_3)_4Cl_2$	0.60	- 1.87	
	0.24	- 1.85	- 1.80
		- 1.82[e]	
	0.93	- 1.60	
	0.46	- 1.60	
$Al_2(CH_3)_2Cl_4$	0.19	- 1.62	
		- 1.52[e]	

[a] Concentration in mol/1^{-1} of monomer.
[b] Bridge CH_3 groups.
[c] Outer CH_3 groups.
[d] Ref(Brownstein et al[1313].
[e] Ref (Muller and Pritchard[1314].
[f] Solubility too low.

O'Reilly[1318] discusses in some detail the NMR chemical shifts for Al^{27} for aluminium alky halides, halides, alkyls, alkoxides, AlH_4 and $Al(OH)_4$. He claims that the magnitude of the resonance line width of aluminium compounds as liquids or in solution is indicative of the cubic or non-cubic molecular symmetry around the Al^{27} nucleus. O'Reilly discusses the calculation of the chemical shift of Al^{27} in AlH_4 by a variational procedure using both valence bond and molecular orbital type wave functions.

Bortnikov et al[1319] have reported the chemical shifts of aluminium isopropoxides (Table 142). The solid aluminium isopropoxide is tetrameric (Shiner et al[1316]).

Table 142 - Chemical shifts for isopropyl alcohol and some metal isopropoxides

```
              CH3
             /
M - O - C
        | \
        H   CH3
```

Element M	CH proton peaks (Hz)	CH rel. to isopropanol (Hz)	CH_3 proton shifts (Hz)	CH_3 rel. to isopropanol (Hz)	Coupling constants (J)
H[a]	237.8 septet	0	69.8	0	6.3
Al[b]	264.3 (295.5 - 246.5:9 pks)	26.5	76.5	6.7	6.0
Al[c]	261.5 (293.0 - 244: 14 pks)	23.7	78.6	8.8	6.0

Internal Standard: TMS.

a Neat
b Satd soln of $M(OC_3H_7)_n \cdot C_3H_7OH$ in benzene n = 3 or 4.
c Satd soln of distilled product (unsolvated) in benzene.

A saturated solution of the solid aluminium isopropoxide in benzene shows a major CH_3-proton doublet and two minor doublets in agreement with the report of Shiner et al[1316]. In the CH proton region, only 9 peaks can be detected. Liquid Al-isopropoxide, a supercooled melt obtained by distillation from the solid, is known to be essentially trimeric which slowly reverts to the tetrameric form at lower temperatures (Shiner et al[1316]); Mehrotra.

A saturated benzene solution of the liquid aluminium isopropoxide shows the major CH_3 doublet and three minor doublets. In the CH proton region, 14 peaks can be detected. The important difference between the liquid and solid aluminium isopropoxide is a change in the degree of polymerization (from 3 to 4), and this change is reflected somewhat in the lower field chemical shifts.

NMR spectroscopy has been applied to the elucidation of the structure of the reaction products of alkylaluminium compounds with isoquinoline[1321].

D. Mass Spectrometry

Table 143 lists a few of the m/e values obtained by Brown and Mazdiyasni 1322 in the mass spectrum of freshly distilled aluminium isopropoxide. Weak molecular ion peaks were found for the pentamer and tetramer. However, the presence of a very strong peak at 757 m/e, which corresponds to the tetramer minus an isopropoxy group, indicates that the compound is predominantly tetrameric in agreement with the structure recently reported by Fieggan and coworkers[1323]. Weak peaks were also observed at m/e values corresponding to molecular ions for trimeric, dimeric, and monomeric species but the fragmentation data tend to show that these peaks result from fragmentation of the larger polymers.

Table 143 - Mass spectrometric analysis of aluminium isopropoxide. Model 21 - 110B mass spectrometer, ionizing energy 12 and 70 at 140 - 250°C. Perfluorokerosine reference

Measured m/e	Identity
1020 w	$Al_5O_{15}C_{45}H_{105}$
961 m	$Al_5O_{14}C_{42}H_{98}$
801 m	$Al_4O_{12}C_{35}H_{81}$
757 vs	$Al_4O_{11}C_{33}H_{77}$
655 s	$Al_4O_{10}C_{27}H_{63}$
613 w	$Al_4O_{10}C_{24}H_{57}$
553 vs	$Al_3O_8C_{24}H_{56}$
451 vs	$Al_3O_7C_{18}H_{42}$
437 s	$Al_3O_7C_{17}H_{40}$
409 s	$Al_3O_7C_{15}H_{36}$
393 s	$Al_2O_6C_{17}H_{41}$
349 s	$Al_2O_5C_{15}H_{35}$
335 s	$Al_2O_5C_{14}H_{33}$
307 m	$Al_2O_5C_{12}H_{29}$
291 m	$Al_2O_4C_{12}H_{29}$
58 s	OC_3H_6
57 s	OC_3H_5
44 m	OC_2H_4
43 m	C_3H_7
42 m	C_3H_6
40 m	C_3H_4
38 m	C_2H_2

Approximate intensities designate as follows: vs = very strong, s = strong, m = medium, w = weak.

ORGANOBORON COMPOUNDS

A. Infrared Spectroscopy

Nadeau and Oakes[1324] have described an infrared method for the determination of diborane, dichloroborane, and trichloroborane in mixtures. Myers and Putnam[1325] point out, however, that this method has its limitations in that it does not include the determination of monochlorodiborane, hydrogen chloride or hydrogen. The Nadeau and Oakes[1324] method is useful since it is rapid and relatively more sensitive than wet chemical analysis because of the strong absorption in the i.r. region shown by boron compounds. The individual components are determined in mixtures directly from their characteristic absorption peaks - diborane at 5.35 μ, dichloroborane at 8.98 μ and trichloroborane at 10.24 μ. All three components have mutual absorption bands in the regions most likely to be used for analysis and so this method is applied by using Vierordt's method with simultaneous equations to correct for overlapping absorption. The procedure has an overall precision on synthetic samples of ± 2% of the three compounds in the concentration ranges studied. General interference with the analysis can be expected in the presence of compounds containing B-N and/or B-Cl linkages.

McCarty et al[1326] have described an infrared spectroscopic procedure for the determination of diborane (4.05 μ), tetraborane (4.65 μ), pentaborane (5.54 μ) and decaborane (6.15 μ) in samples which contain pentaborane as the major constituent. Agreement between the amounts of these substances present in the sample and those determined is claimed to be within 2%.

Kuhns[1327] determined decaborane in isooctane cells with sodium chloride windows and a path length of 0.2 mm were used. The instrument was calibrated with decaborane of research purity at 9.92 μm, where the boron hydride polymer does not interfere, and this calibration was used to calculate the decaborane concentrations of the samples.

Le Sachs and Garrigues[1328] studied the infrared spectroscopy of decaborane. The spectrum was compared with that of pentaborane and diborane. The solid sample of decaborane was analysed by dispersing it in liquid paraffin and compressing it between potassium bromide plates. Le Sachs and Garrigues discuss the structure of decaborane and indicate that it has five characteristic bands at 718, 763, 858, 1008 and 1551 cm^{-1}. They suggest that quantitative analysis is made by dissolving the sample in carbon disulphide and examining it at 718 cm^{-1}.

Infrared spectroscopy has been used by Mackey et al[1329] for the determination of hydroxylic impurities such as methanol and boric acid in methylborate. The concentration of hydroxyl groups can be estimated with a precision of 0.05 mequiv. in the range of 0 - 0.5 mequiv./ml by i.r. analysis. Mackey et al[1329] describe methods for establishing working curves using the base-line method and the area method.

Kaufman et al[1330] have studied the infrared spectra of ^{11}B labelled diboranes. They showed that the spectra of $^{11}B_2H_6$ and $^{11}B_2D_6$ are similar to those of normal diborane; however, there are pronounced differences in band structure, in shape of rotation envelopes, and positions of maxima of bands. With sodium chloride prism resolution, peaks which should show three separate bands, $^{11}B_2H_6$, $^{11}B^{10}BH_6$, and $^{10}B_2H_6$, show only one with slight shifts in positions of maxima. These differences are directly related to percent ^{10}B content.

By calibrating an instrument with samples of diborane of known ^{10}B content, Kaufman et al[1330] were able to determine the boron isotopic composition of diborane to within 1% on an ordinary single or double-beam spectrometer. No vibrations, which should become i.r. active in $^{11}B^{10}BH_6$ because of lowering of symmetry from v_h to $C_{2}\nu$, were observed. Studies employing a grating i.r. spectrometer permitted complete separation of $^{11}B_2$, $^{11}B^{10}B$, and $^{10}B_2$ peaks of ν_{14} and ν_{18} vibrations for diboranes containing 1.3, 18.83, 56.4, 73.5, and 94% ^{10}B (Figure 62). The resolution was good enough to permit the estimation of the fractions of each of the species $^{11}B_2$, $^{11}B^{10}B$, and $^{10}B_2$ present in the samples.

Kaufman and Koski[1331] also carried out an i.r. study of the exchange of deuterium between decaborane and diborane. They work on partially dueterated decaborane samples prepared by exchaning deuterium between diborane and decaborane at 100°C and compared the spectra obtained with those of partially deuterated decaborane in which the deuterium is randomly distributed. Their results demonstrated that the exchange proceeds on the terminal hydrogen atoms in decaborane and that the bridge hydrogen atoms are not involved at all in the reaction under the experimental conditions used.

These workers Kaufmann and Koski[1332] have described further infrared studies on the deutero diborane-pentaborane exchange reaction. They

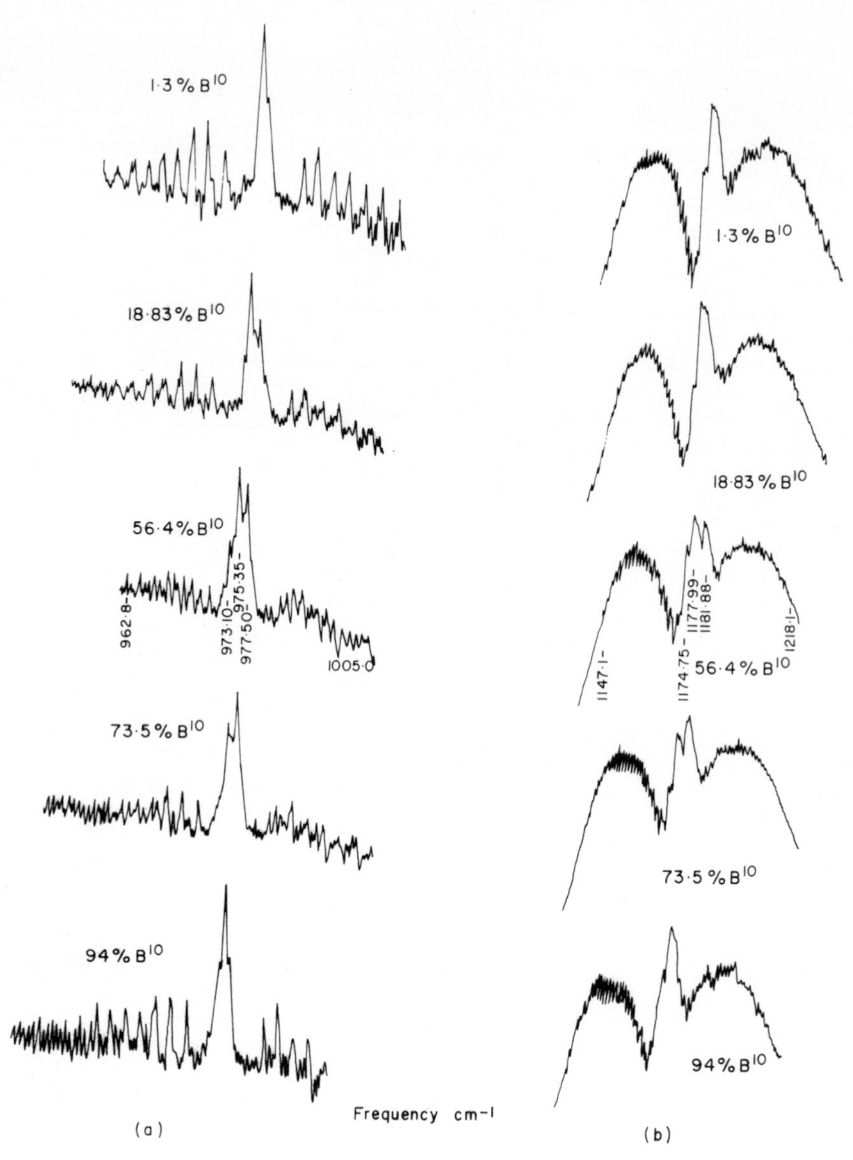

Figure 62 - (a) i.r. spectrum of ν_{14} in diborane, (b) i.r. spectrum of ν_{18} in diborane.

prepared spectra of the partially deuterated pentaborane produced in this reaction and showed by comparison of this data with that obtained by mass spectrometry that when the exchange reaction takes place at 80°C only the non bridged hydrogen positions in the pentaborane are involved. They showed that one of the most favourable regions in the pentaborane spectrum for observation of the entrance of deuterium into the nonbridge positions is the 5.14 μm absorption band which is due to a B-D stretch vibration (Hrostowski and Pimental[1333]). It was shown that pentaborane having a total of 28.8% deuterium statistically distributed absorbs appreciably less than an exchange sample having 26.5% deuterium. Likewise, a statistical sample of 47.3%

deuterium absorbs considerably less than a 37% exchange sample. These observations indicate that, qualitatively, the exchange is proceeding preferentially in the non-bridge positions. This and other data are expressed in quantitative form in Table 144. Here it is seen that to within experimental error all of the deuterium that has entered the pentaborane molecule by exchange can be accounted for by the amount of deuterium in the non-bridge positions, indicating that the exchange reaction proceeds preferentially into these positions.

Other less favourable regions of the pentaborane spectrum also reflect this behaviour. Kaufman and Koski[1331,1332] point out that p.n.m.r. measurements made on partially dueterated pentaborane, prepared by exchange of deuterium between di- and pentaborane, also support these conclusions.

Sutton and Schneider[1334] have described an i.r. spectroscopic method for the determination of mixtures of triethylaluminium and triethylboron based on measurements at 10.1 to 10.2 μ (triethylaluminium) and at 9.7 to 9.8 μ (triethylboron).

Table 144 - Deuterium analysis of partially deuterated pentaborane obtained from the B_2D_6-B_5H_9 exchange

Percent deuterium		
In exchangeable positions (mass spectra)	In nonbridge positions (I.r. spectra)	
	5.14 μM	13.4 μM
2.6	2.5	
10.9	11.8	10.9
40.5	35.4	35.0
56.5	51.5	54.5

B. Nuclear Magnetic Resonance Spectroscopy

Koski et al[1335] studied, by n.m.r. spectroscopy, the exchange of deuterium between diborane and pentaborane. They found that the exchange reaction proceeded preferentially in the terminal hydrogen positions in pentaborane. The rate of exchange of the apex hydrogen appeared to be within ± 10% of the exchange rate of the base terminal hydrogen. Under the conditions studied, the bridge hydrogen atoms in pentaborane did not participate

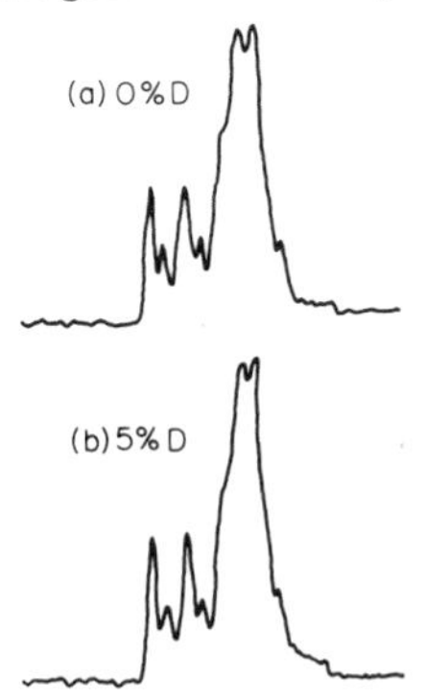

Figure 63 - High resolution p.m.r. spectra of pentaborane (a) normal isotopic content, (b) 5% deuterium,

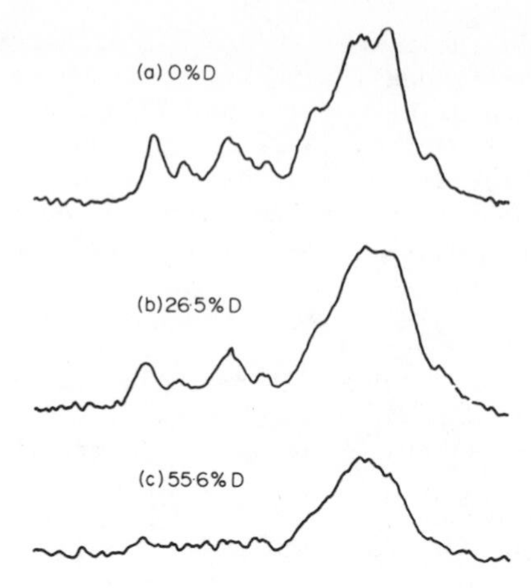

Figure 64 - High resolution p.m.r. spectra of pentaborane containing (a) normal isotopic abundance, (b) 26.5% deuterium and (c) 55.6% deuterium.

in the exchange. Some n.m.r. spectra of pentaborane and deuterated pentaborane obtained by these workers are shown in Figures 63 and 64.

C. Mass Spectrometry

Koski and Kaufman[1336] carried out a mass spectrometric study of the dideuterborane-pentaborane exchange reaction. They studied the distribution of partially deuterated pentaborane species resulting from the reaction between B_2D_6-B_5H_9 and B_2H_6-B_5D_9. It was found that the deuterium enters the pentaborane by two main processes; namely exchange and synthesis by pyrolysis of diborane. The number of readily exchangeable positions in the pentaborane under their experimental conditions was five. If the remaining four hydrogens exchange, they do so at a very much slower rate. Examination of the mass spectrum indicated that the deuterium enters the molecule singly during exchange. Tracer studies with ^{10}B indicated that boron atoms are not involved in the exchange.

Koski et al[1337] also carried out a mass spectroscopic appearance potential study of boron trichloride, tribromide and triiodide. Their measurements permitted a determination of the average B - I bond dissociation energy in boron triiodide. The value obtained was 2.77 ± 0.2 eV.AB-I bond length of 2.03 A was estimated independently. A plot of the bond energy vs bond length in the boron trihalides gave a linear relationship. A set of apparently self-consistent ionization potentials for the boron trihalides and fragments formed from these molecules were estimated and compared with ionization potentials of other BY_3 molecules as well as for BY_2 and BY fragments (where Y can be H, halogen or alkyl substituents). Comparison of the ionization potentials of BY_2 radicals indicated a correlation between the ionization potentials and the nature of the group attached to the boron. CY_3 radicals exhibit a similar behaviour. The mass spectra of the boron trihalides were also run at 70 e V ionizing voltage to obtain a set of fragmentation patterns under constant conditions.

Kaufman et al[1338] measured the appearance and ionization potentials of some fragments ($B_{10}H_{14}^+$, $B_{10}H_{12}^+$, $B_{10}H_{10}^+$, $B_{10}H_8^+$, $B_{10}H_7^+$, $B_{10}H_6^+$) from labelled decaborane, $^{11}B_{10}\ ^{1}H_{14}$. Kaufman et al measured the appearance and ionization potentials of selected fragments ($B_5H_9^+$, $B_5H_8^+$, $B_5H_7^+$, $B_4H_6^+$, $B_5D_9^+$) from isotopically labelled pentaborane. These quantities were determined by mass spectrometric electron impact measurements. Using these data, monosiotopic fragmentation patterns for normal and deuterated pentaboranes were calculated from the mass spectra of the compounds at 70 e V.

Wilson and Shapiro[1339] analysed the mass spectra of normal and isotopically labelled 1:1- and 1:2-dimethyldiboranes and their ethyl analogues and recorded some alkyl rearrangement with the 1:2-dialkyldiboranes.

ORGANOCHROMIUM COMPOUNDS

Mass Spectrometry

Boaker and Isenhour[1440] have reported on the mass spectrum of chromium (III) hexafluoroacetyl-acetonate.

ORGANOCOBALT COMPOUNDS

Infrared Spectroscopy

Ruzicka[1341] carried out a spectrophotometric investigation of dicobalt octacarbonyl. He showed that ultraviolet radiation causes decomposition of $Co_2(CO)_8$ to $Co_4(CO)_{12}$ plus carbon monoxide. The infrared spectra of the carbonyl in the 1800 - 1900 cm^{-1} range, containing the absorption bands of the bridging carbonyl groups, are sufficiently improved by the use of a double monochromator (sodium chloride prism and diffraction grating) to permit the determination of the two carbonyls when present together by means of the extinction ratio of the 1857 cm^{-1} and 1866 cm^{-1} bands. The error rises from ± 1% for up to 30% to ± 2% for 50% and to ± 5% for greater than 60% of $Co_4(CO)_{12}$ in the sample. The resolution in the 2000 to 2150 cm^{-1} range, where the absorption bands of the terminal carbonyl groups are located is not improved by the double monochromator technique.

ORGANOCOPPER COMPOUNDS

Mass Spectrometry

Benyon et al[1342] obtained a mass spectrum of copper tetrachlorophthalocyanine. They achieved this by mounting the sample in a thin Pyrex-glass tube surrounded by a heater and projecting into the ionization chamber. The envelope of the mass-spectrometer tube was also heated to prevent sample condensation on the insulators and electrodes. Features of the observed spectrum were discussed briefly.

ORGANOGALLIUM COMPOUNDS

Electron Probe Microanalysis

Nakaming and Oric[1343] have shown that ultrasonically treated gallium oxinate-epoxy resin mixtures are suitable standards for quantitative electron probe microanalysis of biological thin sections.

ORGANOGERMANIUM COMPOUNDS

A. Infrared Spectroscopy

Bayer [1344] and Brown et al[1345] have reported on the i.r. spectroscopy of organogermanium compounds. Brown et al discuss the spectra of the methyl derivatives of germanium between 400 and 1500 cm^{-1}, also of dimethylgermanium oxide (trimer, tetramer and high polymer), dimethylgermanium sulphide (trimer) and bis(trimethylgermanium) oxide.

B. Mass Spectrometry

Mass spectrometry has been used extensively for the determination of traces of impurities in these compounds. Shafran and Kuraeva[1346] and Vescernyes and Hangos[1347] have described procedures for the submicro-determination of boron in germanium tetrachloride. Vescernyes and Hangos add the germanium tetrachloride sample slowly to 100 mg of chlorotriphenylmethane, then set the sample aside for 30 min and evaporate to dryness under an infrared lamp under a stream of dry air or nitrogen. The ground residue is transferred to a graphite electrode and boron is determined spectrophotographically. The ranged determination is from 10^{-9} to 10^{-5} g boron per g of sample.

Larin et al[1348] have described a method of quantitative mass spectrometric analysis of germane for 1 parts/10^6 to 10% of phosphine, ethyl ether and ethylene. An equation is derived for the relationship between the relative sensitivity of a mass spectrometer to an impurity and the concentration of that impurity. The relative sensitivity varies with the gas pressure in the ion source, the applied potential and the emission current of the ionising electrons. The minimum determinable concentration ranges from 0.01 - 10 parts/10^6 for germane. At least two mixtures of known composition must be analysed in order to establish a calibration graph; this avoids any effects due to sensitivity changes. Devyatykh et al[1349] described mass spectrometric procedures for the determination of traces of phosphine in germane using a mass spectrometer fitted with a secondary electron multiplier. The relative sensitivity depends on the concentration of impurity, the gas pressure in the ion source, the potential of the ejector electrode, and the cathodic emission current. The recommended conditions are - potential of the ejector electrode, 2 V; cathodic emission current, 1.5 mA; and intensity of the characteristic peak of the main substance in the mass spectrum, greater than 20 volts. Under these conditions, the sensitivity is 2 x 10^{-6}% of phosphine in germane.

ORGANOIRON COMPOUNDS

A. Infrared Spectroscopy

Winter et al[1350] have discussed the infrared spectrum and structure of crystalline ferrocene. They examined the theoretical aspects of the ferrocene molecular and crystal structure with respect to point group, site group and factor group symmetry considerations. The growth and orientation of single crystals of ferrocene are described, together with practical details for recording the i.r. spectra of a single crystal in the region 450 - 6000 cm^{-1} using plane-polarised exiting radiation. The observed i.r. frequencies, together with tentative assignments and species, are tabulated and support the assignment of Lippencott and Nelson[1351]. The polarisation data indicate that the molecular symmetry axis lies nearly in the t_1t_2 plane of the unit cell and the angle between it and the t_2 axis is approx $\pi/4$.

Reid et al[1352] have discussed the characterisation of solid ferrocene and of various cyclopentadienyl metal compounds by their near-infrared reflection spectra. A simple technique for obtaining reflection spectra is described, and the near-i.r. combination spectra of several cyclopentadienyl compounds have been measured. The method makes possible the spectral identification of the cyclopentadienyl ring structure in reactive solid compounds, and its use is complementary to identification by nuclear magnetic resonance and i.r. absorption measurements.

They recorded the near-infrared solution spectra of ferrocene and of $C_5H_5TiCl_3$ and the reflection spectra of a range of solid cyclopentadienyl metal compounds with one to four rings per molecule, or with various halogens or other groups also attached to the metal. They showed that the near-infrared spectra of these compounds which are due to combinations of the fundamental frequencies all have a striking similarity above 3800 cm^{-1} which is characteristic of a C_5H_5 ring bonded with five-fold symmetry to a metal atom.

Reid et al[1352] comment that the fundamental infrared spectra of cyclopentadienyl metal compounds also show marked similarities, a point which has been discussed by a number of other authors (Lewis and Wilkins [1353]; Wilkinson and Cotton[1354]; Fischer[1355]; Wilkinson et al[1356]; Lippencott and Nelson[1351]; Nakamoto[1357]. The fundamental spectra are largely derived from the vibrations of the C_5H_5 moiety, but ring-metal vibrational frequencies in the 100 - 800 cm^{-1} region (Winter et al[1350], Lippencott and Nelson[1351]) also contribute.

Reid et al[1352] recorded the spectra of solid compounds from 3600 - 30 000 cm^{-1} as solid diffuse spectra, using a spectrophotometer fitted with a reflectance attachment and using a blank plate covered with magnesium oxide as a reference.

In Figure 65 are shown the near-infrared solid and solution spectra of ferrocene. The spectra above 3800 cm^{-1} of ferrocene and of a wide range of other metal cyclopentadienyl compounds are extremely similar. The equivalence of the five hydrogen atom positions in the C_5H_5 ring in ferrocene and in the other metal cyclopentadienyls examined results in a single sharp CH stretching overtone at 6100 to 6200 cm^{-1}. This is to be contrasted with the spectra of gaseous or liquid cyclopentadiene or with that of molecules containing rings sigma bonded through a single carbon atom (thereby retaining the diene structure of the ring), all of which have a complex CH overtone bond structure. Apart from the CH stretching frequencies the highest fundamental frequency in the spectrum of ferrocene is approximately 1580 cm^{-1}.

The vibrational assignment of the ferrocene fundamental spectrum given by Lippencott and Nelson[1351] shows that, assuming a D_{5d} symmetry (the staggered ring molecular structure) there are eight vibrational symmetry species, of which only two are infrared active and three Raman active. However, combination of the symmetry species by the rules given by Wilson et al[1358] show that ten combinations are infrared active and that all assigned fundamental frequencies whether active or inactive enter at least once into the combination frequencies. A similar result is obtained if D_{5h} symmetry is considered, i.e. that for an isolated C_5H_5 planar ring or an "eclipsed sandwich" di-cyclopentadienyl metal structure. The values of the frequencies to be assigned are almost identical for both cases, owing to the low degree of coupling of the vibrational modes of the two rings. Similar considerations apply for most other overall molecular symmetries to which a C_5H_5 ring can be placed.

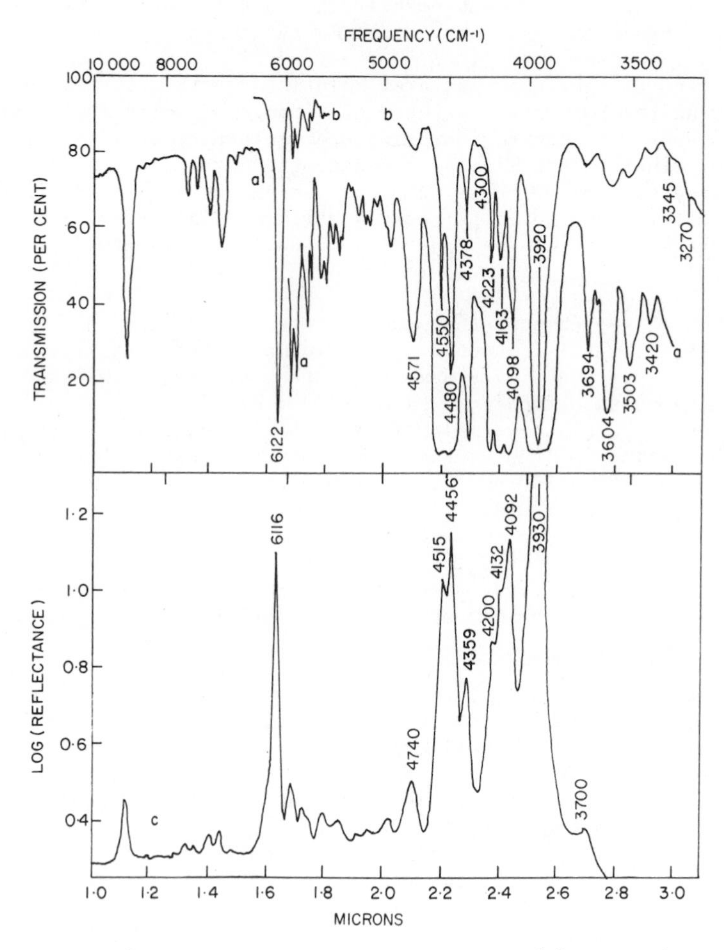

Figure 65 - Near infrared spectrum of ferrocene. (a) 0.3 M in carbon tetrachloride, 100 mm; (b) 0.3 M in carbon tetrachloride, 10 mm; (c) solid reflection spectrum.

The fundamental frequencies of absorption for ferrocene in solution or as a vapour and for the crystal have been collated by Wilkinson[1356] and Winter et al[1350] together with an approximate estimate of relative intensities for the spectra of the solid, derived from the spectra given by Winter et al[1350] and by Wilkinson et al[1351] for a single crystal in transmitted non-polarized light.

Pellegrini et al[1359] has listed infrared spectral bands of various ferrocene derivatives as shown below:

triphenylstannylferrocene, 3040, 1475, 1425, 1135, 1102, 1072, 996, 835, 815, 730 and 696 cm^{-1} (in Nujol and Fluorolube).
triphenylplumbylferrocene, 3040, 1570, 1465, 1425, 1135, 1105, 1065, 1020, 996, 836, 815, 725 and 695 cm^{-1} (in Nujol and Fluorolube).
tributylstannylferrocene, 3040, 2960-2890, 1470, 1465, 1375, 1135, 1102, 1072, 1022, 998, 958, 875, 816, 770 and 690 cm^{-1}.

Spilners and Pellegrini[1360] have reported that the infrared spectrum of polyferrocenylene ($H-(C_5H_4FeC_5H_4)_nH$) in Nujol has a strong doublet at 1110 cm^{-1}, strong singlets at 1030 and 1000 cm^{-1}, and a strong absorption at 815 to 820 cm^{-1}. Weaker absorption bands occur at 3120, 1700 - 1600, 1410, 878, 860, 845, 720 - 722, 700 and 675 cm^{-1}. This compound has absorption maxima in the ultraviolet at 220 nm ($E^{1\%}_{1cm}$ 1.12 x 10^3) and at 298 nm ($E^{1\%}_{1cm}$ 2.18 x 10^2). They also discuss the infrared spectrum of bis (1-(1'-ethylferrocenyl)) and report strong bands at 1110, 1030, 1020, 855 and 810 cm^{-1} and weak bands at 1225, 1000, 910, 840 and 725 cm^{-1}. The infrared spectrum of this substance has also been reported by Korshak et al[1361].Spilners[1362] has reported the infrared spectra in Nujol and Fluorolube and the ultraviolet spectra of various ferricinium salts.

B. Nuclear Magnetic Resonance, Proton Magnetic Resonance and Electron Spin Resonance Spectroscopy

Spilners and Pellegrini[1360] have discussed the NMR spectrum of polyferrocenylene in carbon disulphide which they showed to have one intense singlet at 6.17 (unsubstituted ring protons) and weaker doublets at 5.77 and 5.98 (substituted ring protons) and apparent singlets at 6.03 and 6.21. They also determined the NMR spectrum of bis(1-(1'-ethylferrocene)) and reported a singlet at 6.26 (probably protons on ethylcyclopentadienyl ring), a multiple at 5.93 (cyclopentadienyl protons), a quadruplet at 7.97 (methylene protons) and a triplet at 8.99 (methyl protons).

Spilners[1362] reported that the NMR spectrum of ferricinium tetrachloroferrate (in deuterium oxide) and of 1,1'-bis(triphenylsilyl) ferricinium tetrachloroferrate (in deuterated acetone) did not show any lines. He also reported that the ESR spectrum of the ferricinium salt reaction product of ferrocene and mercuric chloride has a line with a g value of 2.087 and a line width of 691 gauss.

Spilners and Pellegrini[1360] have applied NMR (and mass spectrometry) to the elucidation of the structure of reaction products formed by the cleavage of ferrocene by aluminium chloride, 1,1'-(1,3 cyclopentenyl) ferrocene was identified as a major reaction product and its NMR spectrum is reported. The NMR spectra of substituted ferrocenes has been discussed by various other workers including Gold'berg[1363], Rosenblum et al[1364], Dunitz et al[1365], Leto et al[1366], Laing and Trueblood[1367], Rinehart et al[1368], Paul[1369], Lippencott and Nelson[1351], Fritz[1370], Pauson et al[1371].

The PMB spectra of hetero annular disubstituted ferrocenes has been discussed by Dvoryantseva et al[1372]. These compounds have different substituents in both five membered rings. On the basis of these results and of those obtained in their earlier work on monosubstituted ferrocenes (Dvoryantseva et al[1373]), these workers were able to obtain information regarding the structure of homo-annular isomeric amides of methyl and ethyl derivatives, and of nitriles of ethyl and phenyl-ferrocenecarboxylic acids.

C. Mass Spectrometry

Several workers have presented and interpreted the mass spectra of ferrocene and several substituted ferrocenes (Friedman et al[1374]; Reed and Tabizi[1375]; Schumacher and Taubenesi[1376]; Moaz et al[1377]; Cordes and Rinehard[1378]; Slocum[1379]; Egger[1380]; Muller and D'or[1381]; Roberts et al[1382,1383,1384] and of 1,1'-biferrocenylene (Hedberg and Rosenberg[1385] and indenyl iron derivatives (King[1386]). Also the utility of mass spectrometry in analysing ferrocenes has been demonstrated (Clancy and Spilners[1387]), (Spilners and Pellegrini[1388,1389]).

Clancy and Spilners[1387] have reported a detailed study of the mass spectra of 14 different alkylferrocenes, 2 biferrocenyls and several other substituted ferrocenes at 70 eV.

Clancy and Spilners[1387] obtained low voltage (8 eV) mass spectra of several substituted ferrocenes at an inlet temperature of 350°C. For all the compounds listed in Table 145 they obtained very intense molecule ion peaks with little or no fragmentation, and this permitted the determination of the molecular weights of these compounds. Compounds containing acid substituents such as carboxylic, sulphonic and $B(OH)_2$ decomposed in the mass spectrometer under these conditions and their spectra could not be obtained.

Spilners and Pellegrini[1390] have applied low voltage mass spectroscopy at 8 eV and mass spectrometry at 70 eV (and NMR) to the elucidation of the structure of reaction products formed by the cleavage of ferrocene by aluminium chloride and showed that 1,1'-(1,3 cyclopentenyl) ferrocene is a major reaction product. Rinehart et al[1391] reported the mass spectrum of diferrocenylethane. Spilners and Pellegrini[1388,1390] found that the low voltage (8 eV) mass spectrometry of polyferrocenylene gave molecular ion peaks for ferrocene (m/e 186), cyclopentadienylferrocene (m/e 250, biferrocenyl (m/e 368, 370, 371, 372) and terferrocenyl (m/e 554). They also determined the low voltage (8 eV) mass spectra of various other compounds including ethylbiferrocenyl, diethylbiferrocenyl and triethylbiferrocenyl and the low voltage (8 eV) and high voltage (70 eV) mass spectra of poly-(ferrocenyl ketone).

Table 145 - Intensities of ferrocene molecule ions

	m/e or molecule ion	Intensity of molecule ion, chart divisions
Ferrocene	186	811.0
Biferrocenyl	370	222.0
1,1'-Dichloroferrocene	254	307.0
1,1'-Diethylferrocene	242	708.0
1,1'-Di-n-butylferrocene	298	646.0
1,1'-Dibutyrylferrocene	326	594.0
Trimethylsilylferrocene	258	399.0
1,1'-Bis(trimethylsilyl)-ferrocene	330	2244.0
Triphenylsilylferrocene	444	439.0
Dimethylethoxysilylferrocene	288	3740.0
1,1'-Bis(dimethylethoxysilyl)ferrocene	390	3210.0
Vinyldimethylsilylferrocene	270	1226.0
1.1'-Bis(vinyldimethylsilyl)ferrocene	354	436.0

Schulten et al[1392] used laser assisted field desorption mass spectrometry of haemoglobin for elemental analysis and for molecular weight determination and isotope determination.

D. X-Ray Diffraction and X-Ray Emission Spectrography

Baun[1393] has used X-ray diffraction for the identification of pure polycrystalline ferrocenes and found that these substances had individualistic diffraction patterns, and claimed that this technique is valuable where chemical similarities make detection and quantitative analysis difficult by infrared absorption techniques. The following ferrocenes were studied, ferrocene, hexadecanoylferrocene, dioctanoylferrocene, acetylferrocene, dihexadecanoylferrocene, diacetylferrocene, didodecanoylferrocene, N-phenylferrocene, bisferrocenyl mercury, chloromercuriferrocene, ditridecanoylferrocene, hexadecylferrocene, 2-diphenylferrocene, 3-methylbenzoylferrocene, 3-tolyferrocene, 1,1'-dibenylferrocene, benzoylferrocene, benzylferrocene, 1,1'dibenzoylferrocene, and 2-methylbenzylferrocene.

Eiland and Pepinsky[1394] and Kealy and Pauson[1395] have also studied the crystallography of ferrocene and confirmed its pentagonal antiprismatic structure, they did not agree, however, on the size of the unit cell. De Bruyne[1396] used dispersive (lithium fluoride crystal) and non-dispersive X-ray fluorescence analysis for the determination of arsenic, antimony and tin in iron pentacarbonyl. For samples of iron carbonyl containing added arsenious and antimonious oxides and tin powder, the pulse rates measured with a Geiger or scintillation counter are given as a function of channel height. It is reported that accuracy is not improved by the use of aluminium -titanium (20:1) filter. Calibration graphs of pulse rate vs impurity concentration, showed that the best results are obtained by using the dispersive technique (analysing angles - As, 34.1°; Sb, 13.65°; and Sn, 13.9°), together with pulse-height discrimination.

Morningstar et al[1397] measured the millimolar extinction coefficient of cyanomethaemoglobin from direct measurements of haemoglobin iron by x-ray emission spectrography.

ORGANOLEAD COMPOUNDS

Infrared Spectroscopy

Bajer[1398] has discussed the i.r. spectra of organolead compounds. Curry [1399] has described an i.r. procedure for the determination of naphthalene and ethyl chloride in tetraethyllead.

Howard et al[1400] have shown that it is possible to determine by mass spectrometry tetraethyllead/tetramethyllead ratios in mixtures. This procedure does not distinguish between trimethylethyllead, dimethyldiethyllead, methyltriethyllead in mixtures. Bate et al[1401] and Ulrych and Russell [1402] both discuss the application of gas source spectrometry to tetramethyllead as a means of conducting isotopic analysis for ^{204}Pb, ^{206}Pb, ^{207}Pb and ^{208}Pb of a lead sample. Ulrych and Russell[1402] describe a method for the preliminary conversion of the metallic lead sample to tetramethyllead prior to carrying out mass spectrometry.

ORGANOLITHIUM COMPOUNDS

Nuclear Magnetic Resonance Spectroscopy

Urwin and Reed[1403] described methods based on NMR spectroscopy for the analysis of n-butyl lithium solutions. The NMR spectrum of n-butyllithium in benzene (Figure 66) showed a triplet 48 cps upfield from tetramethylsilane (TMS) due to the methylene protons adjacent to the lithium. This triplet can be integrated and compared with the integrated spectrum of a suitable reference liquid. All that is required is to mix a carefully measured volume of the reference liquid with a given volume of the lithium alkyl solution. In this way NMR gives a direct measure of the carbon-bound lithium content. The reference substance should have a large and preferably single peak which does not overlap with the spectrum of butyllithium. A low vapour pressure is preferable since the integrated peak is a measure of the reference in the liquid phase. Mesitylene was chosen as the reference since its NMR spectrum consists of a nine proton singlet 132 cps downfield from TMS (Figure 66) giving therefore a large separation of reference and n-butyllithium peaks. Furthermore its vapour pressure at 20° (the temperature at which all experimental measurements were recorded) is negligible. The NMR method is clearly suitable for analysis down to concentrations where the integrated signal can be estimated with reasonable accuracy, that is in the range above 10^{-1}M.

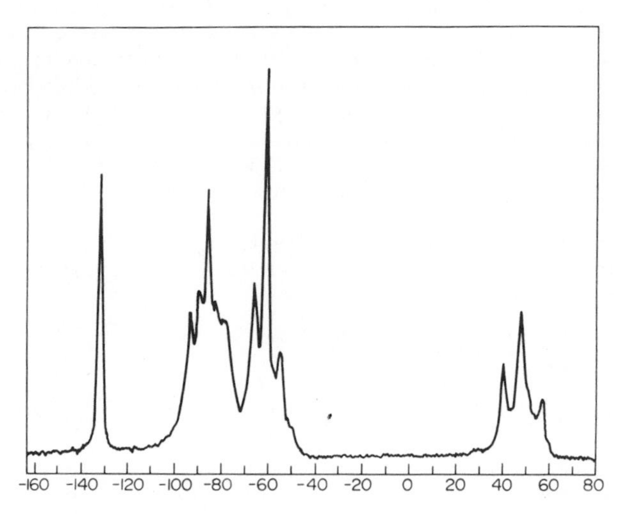

Figure 66 - Resonance spectrum at 20° of butyllithium in benzene with mesitylene as reference. Chemical shifts (in cps) are referred to the resonance position of tetramethylsilane.

ORGANOMANGANESE COMPOUNDS

Infrared Spectroscopy

Susi and Rector[1404] have developed a quantitative infrared analysis procedure for analysis by the potassium bromide technique of mixtures of four pesticides, namely manganese ethylenebis (dithiocarbamate) hydrate fungicide, pp'-dichlorobenzil and mixtures of 1:1-bis(4-chlorophenyl)-trichloroethanol and 1:1-bis(4-chlorophenyl)trichloroethylene. Their results suggest that quantitative analysis is possible by this technique provided that samples of known composition are available as standards and uniform grinding, and mixing procedures are adopted.

Figure 67 gives the infrared spectra, from 2000 to 650 cm^{-1}, of the four substances studied. In Figure 68 the effective concentration of three disks containing manganese ethylenebis-(dithiocarbamate) hydrate is plotted against base line absorbances at three different frequencies. Straight lines going through the origin were observed for each band.

ORGANOMERCURY COMPOUNDS

Infrared Spectroscopy

Gauthier et al[1405] have determined the i.r. spectra of phenylmercury bromide, phenylmercury iodide and phenylmercury acetate using the potassium bromide disc technique. The spectra of the halides are similar, but the spectrum of phenylmercury acetate shows an intense band of 7.6 μ and two others at 7.2 and 7.3 μ. The spectrum of a freshly prepared solution of phenylmercury acetate in chloroform also shows these bands. Gauthier suggests that the spectrum of phenylmercury acetate reported previously is that of a double-decomposition product between this compound and potassium bromide.

ORGANONICKEL COMPOUNDS

Mass Spectrometry

Campana and Risby[1406] have carried out a mass spectrometric examination of nickel carbonyl in connection with their studies of automobile catalysts. They obtained a minimum detection limit of the tetracarbonyl of 10 ppb.

ORGANOPHOSPHORUS COMPOUNDS

A. Infrared Spectroscopy

Phosphate esters. Shesterikov et al[1407] and Pentin et al[1408] describe the infrared determination of tributyl phosphate and diisopentylmethylphosphate in kerosene and other saturated hydrocarbons. The determination is carried out using the bands at 1025, 870 on 835 cm^{-1} for tributylphosphate, and 1005 cm^{-1} for diisopentylmethylphosphate. The extinction coefficient of these bands is proportional to the extractant content which is found from a calibration curve. Determinations can be carried out in several ranges between 0.01% and 5% of the phosphates. These workers also describe a method for the determination of diisopentylmethylphosphonate in kerosene using the absorption band at 1005 cm^{-1}. Other workers[1409] have determined tributylphosphate in kerosene by a procedure involving washing the sample with sodium carbonate, then water and measuring the extinction due to P = 0 at 1275 cm^{-1}.

Phenoxyphosphorous compounds. Lorenz and Kaiser[1410] determined the characteristic infrared absorptions of phenoxyphosphorus compounds. They showed that the infrared spectra of eight compounds containing the C_6H_5,OP group had two characteristic absorption regions (850 - 970 cm^{-1}) and the hitherto unknown absorption at 715 - 790 cm^{-1}. Tervalent phosphorus

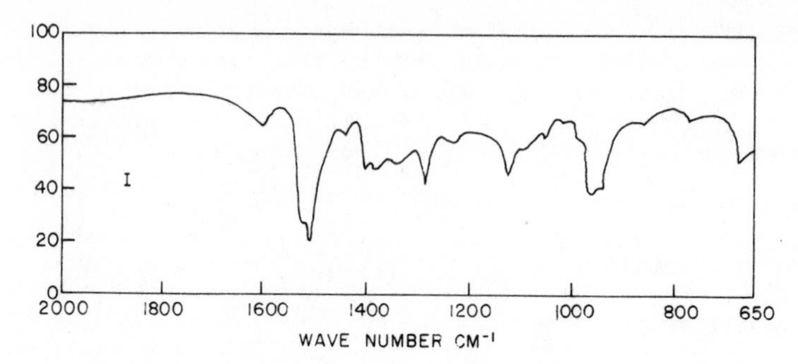

Figure 67 - Infrared spectrum in potassium bromide of manganese ethylene-bis(dithiocarbamate).

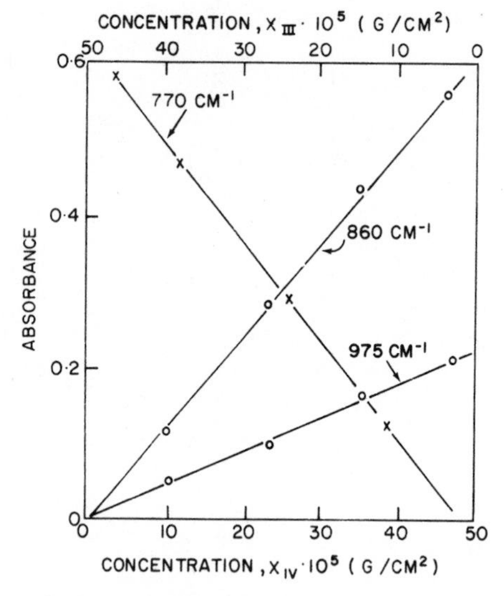

Figure 68 - Observed infrared absorbances of the 1,1-bis(4 chlorophenyl) trichloroethanol (compound iii). - 1,1-bis (4 chlorophenyl)-trichloroethylene (compound iv) system in potassium bromide. X = Absorption band of compound iii. O = Absorption band of compound iv.

compounds tend to absorb at lower wave numbers than do the corresponding quinquevalent phosphorus compounds.

Ferraro[1411] has discussed solvent effects on the infrared spectra of organophosphorus compounds.

Infrared spectroscopy has been used by Steger and Stahlberg[2732] to detect the PNP frequency at 1300 cm^{-1} in phosphonitrilic halides ($P_3N_3X_6$).

Infrared spectroscopy has been applied to the determination of organophosphorus insecticides and pesticides including Gusathion and Phosdrun[2733], Parathion[2734,2736], Ethion[2737] and dimethyl and diethyl parathion[2738].

B. Mass Spectrometry

Occolouitz and White[1412] demonstrated that the mass spectra of trialkylphosphates show ions produced by migration of hydrogen as well as ions formed by simple bond fusion. Trimethyl and triphenyl phosphites give fewer hydrogen rearrangement ions than do triethylphosphite and higher homologues. Comparison of the spectra with those of other organophosphorus esters indicates that the migration of hydrogen in the trialkylphosphites takes place both in the phosphorus and the oxygen atoms to give ions containing pentavalent phosphorus. Spectra of some dialkylphosphates were recorded by these workers and they showed that the m/e value for the base peak differentiates these compounds from the isomeric trialkylphosphites.

C, Nuclear Magnetic Resonance Spectroscopy

O'Neill[1413] has applied phosphorus 31 Fourier transform NMR to the determination of 2-aminoethyldehydrogen phosphate. This technique has also been applied to the analysis of organophosphorus insecticides[1414].

ORGANOSILICON COMPOUNDS

Smith and McHard[1415] have discussed the applications of infrared, ultraviolet, mass and NMR spectroscopy to organosilicon compounds.

A. Infrared Spectroscopy

Infrared spectroscopy has certain limitations when applied to organosilicon compounds. First, non-silicon substituents present in small amounts may be overlooked. Secondly, infrared is not very sensitive to differences in molecular weight or molecular weight distribution. Organosilicon monomers hydrolyse very easily to form siloxanes and hydrogen chloride and must be handled with great care.

Monomers. Chlorosilanes are most conveniently run as dilute solutions in carbon tetrachloride (2 - 7.5 μ) and carbon disulphide (7.5 - 25 μ). Chloro, bromo and iodosilanes hydrolyse very easily and must be handled in a dry glovebox. Silanes with a high vapour pressure can be run in the gas on vapour state. Adsorptions characteristic of the Si-Cl bonds appear in the potassium bromide prism region of 16 - 25 μ. If more than one chloride atom is present, two bands appear, corresponding to the symmetrical and asymmetrical frequencies. A general asymmetrical stretch appears at a shorter wavelength and is the more intense of the two frequencies. The Si-Cl bands are the basis of an analytical method for the determination of other chlorosilanes appearing as impurities in trimethylchlorosilane (Brown and Smith[1422]).

Frequency correlations as applied to silicon-containing polymers and silane monomers also apply to the chlorosilanes. However, electronegative substituents such as chlorine may shift some of these vibrational bands to shorter wavelengths. The siloxane band in hexamethyl disiloxane, for example, is located at 9.50 mincrons. As the methyl groups are replaced by chlorine, the wavelength of the absorption gradually shifts until it is found at 8.99 μ for hexachlorosiloxane. The shift of the Si- H stretching frequency in substituted silanes can be predicted precisely from the structure of the molecule (Smith and Angelotti[1423]).

Analysis of polymers. The classification of silicone polymers into resins, fluids, etc., has been discussed (Kline[1424]). The following remarks apply principally to pure silicone polymers. Some polymers are soluble in carbon disulfide and carbon tetrachloride, and solutions in these materials is a convenient method for obtaining the spectra for such polymers. However, high viscosity dimethylpolysiloxanes have a limited solubility in carbon disulfide and it may be necessary, if a solution spectrum is desired, to use carbon tetrachloride throughout the entire spectral region or to use some solvent such as 2,2,4-trimethylpentane in the 11- to 16- μ region. Frequently, sufficiently good spectra may be obtained by simply casting a film of the polymer on a polished salt plate from a low boiling solvent such as chloroform. Hard, brittle materials such as cured resins can be run by the mineral oil mull (Bradley and Potts[1425]) or potassium bromide pressed plate (Ingebrigtson and Smith[1426]; Scheidt[1423]) techniques. In cases where a filler or a reinforcing agent is present, the pyrolysis technique of Harms[1428] is very useful.

Smith and McHard[1415] discussed the interpretation of infrared spectra of organosilicon compounds in some detail. They point out that the infrared spectra of silicones are notable for two things; the characteristic vibrations of substituents vary only slightly in wavelength, regardless of the compound in which they are found; and the intensity of the absorption bands lying beyond 7 μ is 5 - 10 times greater than is normal with most organic compounds (Wright and Hunter[1429]). These characteristics make silicons relatively easy to identify by their i.r. spectra.

Silicone polymers are distinguished by a strong absorption in the 9- to 10- μ region arising from an Si - O stretching vibration in the siloxane backbond of the polymer. Linear and cyclic polymers containing eight or fewer units can be characterized by the shape and position of the siloxane band (Wright and Hunter[1429]).

In commercial polymers the silicon atoms are usually substituted by some combination of methyl, ethyl, phenyl, hydrogen, alkoxy, or hydroxyl groups. Dimethyl substituted polysiloxanes have strong bands at 4.6 and 11.2 μ which indicate the presence of Si - H, and the Si - Me group has a characteristic band at 7.95 μ.

An all methyl silicone resin with a high (Me Si $O3/_2$) content shows Si $O3/_3$ units which can be recognised by a strong band at 9.0 microns. An ethyl substituted resin is essentially a silsesquinoxane. The Et - Si group is characterized by absorptions at 8.0, 9.9, and 10.3 microns.

Silicons used to modify organic resins usually contain either the dimethylsiloxane (Me_2SiO) group which absorbs at 7.9 and 12.5 μ, or the phenylsilicon group, which shows a needle-sharp band at 7.0 μ or both. A broad absorption in the 9- to 10-μ region indicates the presence of a siloxane. The spectra of other silicone polymers have been discussed by Wright and Hunter[1429] and by Young et al[1430]. The dimethylpolysiloxane fluids have been studied by Smith et al[1431] and the identification of silicones in cosmetics was reported by Pozefsky and Grenoble[1421].

Silicone emulsions, resins and elastomers. This subject has been extensively reviewed by workers at Midland Silicones Ltd.[1432].

Emulsions. Oil-in-water emulsions of silicone fluids are used in polishes and as release agents and water-repellent treatments. Such emulsions are normally examined with a view to determining the amount and nature of the silicone fluid they contain and sometimes the emulsifying agents. For a "non-active" fluid, i.e., one containing no active groups such as SiH and SiOH, the silicone fluid content can be determined by adding ethanol and acid or alkali to break the emulsion and centrifuging the mixture, the silicone fluid separates as a clear upper layer. For emulsions containing "active" fluids, the water can be removed by freeze-drying, and the fluids can then be characterized by infrared spectroscopy.

Resins. Silicone resins are used as water-repellents, electrical varnishes, protective coatings etc., and vary in structure much more than do the fluids. A wide variety of substituent groups is used to achieve certain desired properties, and the ratio of difunctional (R_2SiO) to trifunctional ($RSiO_{3/2}$) units can be varied widely. As with silicone fluids, the infrared spectrum of a silicone resin can be of value to the analyst. Substituent groups can be identified and once a reference file of spectra has been compiled, it is possible to assess the ratio of difunctional to trifunctional material. The spectra are best run by evaporating, at room temperature, a small amount of the sample on to a sodium chloride plate.

Infrared spectroscopy has been used for the identification of silicone wire enamels (Hummel[1433]) and silicone resins (Fishl and Oyoung [1434]; Hummel [1435]; Kagarise and Weinberger[1436]).

Elastomers. The chemical examination of a silicone rubber is usually aimed at identifying the polymer, the filler, and any other additives present. If the rubber is in the raw unvulvanized state, the polymer can be separated from the filler and other inorganic material by solvent extraction using solvents such as toluene, ethyl acetate, chloroform, and diethylamine. The separated polymer can then be identified by its i.r. spectrum and silicon content. Any organic peroxide vulcanizing agent in the rubber will also be removed with the organic solvent. In order to avoid cross-linking, therefore, it is essential to remove solvent at a low temperature and reduced pressure.

If the sample of rubber has already been cured, the polymer cannot be isolated by solvent extraction, and one must resort to depolymerisation in order to characterise the polymer. Depolymerisation can be carried out by heating the rubber in vacuo and collecting the pyrolysate, which is then examined by infrared spectroscopy; its spectrum closely resembles that of the original polymer. Harms[1428] had devised a simple dry-distillation method for complex organic materials and has described several examples of the pyrolysate - i.r. approach.

In a silicone rubber based on a methylphenyl polymer the absorption of the C_6H_5-Si bond can be clearly seen at 8.9, 13.5 and 14.6 μ. This technique has also been studied by Smith et al[1437] and by Yonemoto and Senzoch[1438].

Davidson and Bates[1439] have discussed the i.r. spectroscopic analysis of elastomers. They describe techniques for preparing samples of cured and uncured materials for spectroscopic examination. The same workers also discussed the application of infrared spectroscopy to the identification of silicone elastomers.

Ecknig et al[1440] have studied the use of i.r. spectroscopy in conjunction with gas chromatography for the examination of organosilicon isomers.

Infrared spectra - structure correlations. Information on this subject has been tabulated by various workers including Bellamy[1441]; Gilman[1442]; Kreshkov et al[1443]; Smith and McHard[1444]; and Smith[1445]. Kreshkov et al[1443] gave diagrams of the i.r. absorption spectra of eleven silicon - organic compounds between 2 and 10 microns. Wavelengths (microns) representative of various groupings are assigned, viz., CH_3, 3.45 and 6.8 - 7.1; C_6H_5, 3.25, 6.25, 6.7 and 9.7; ≡ Si - C, 7.95 - 8.0; ≡ Si - O, 9.5 - 9.65; O - R, 8.6 - 8.7; and CH ≡ CH_2, 6.15.

Smith[1445] in his very detailed review pointed out that the concept of group frequencies is particularly useful in the interpretation of the i.r. spectrum of organosilicon compounds. This is because of the insulating effect of the silicon atom which causes the spectra of these compounds to be approximately the same as the sum of independent vibrations of the substituent groups. Small shifts from these characteristic frequencies are to some extent predictable, so the assignment of molecular structure is in many cases comparatively simple. Such perturbations of group frequencies as do occur usually arise from:

(1) The inductive effect of neighbouring groups or atoms.
(2) Coupling or interaction of vibrations.
(3) Steric effects such as bond strain.

The vibrational insulation provided by the silicon atom can probably be attributed, at least in part, to the fact that its size and mass are greater than those of carbon.

Another characteristic of the spectra of silicones is the large absorption intensities of bands falling at wavelengths greater than 7.5 μ. This intensity may result at least in part from the electronegativity difference between carbon and silicon which results in a greater ionic character of the Si - C bonds. The C - H stretching and bending bands in the 2 - 7.5 μ region are correspondingly weak compared with those in conventional organic compounds, probably because of the smaller C - H dipole resulting from withdrawal of electrons from the C - H bond by the strong Si^+C^- dipole.

Smith[1445] has published a chart which summarizes structure correlations. Wavelength assignments shown were determined from solution spectra and except where indicated as tentative, were based on at least five different compounds containing the group in question.

Peents et al[1446] have determined impurities in silicon tetrafluoride by infrared spectroscopy and Fourier transform mass spectrometry. The detection limit for impurities was about 300 ppm.

Infrared spectroscopy of specific functional groups. Alkyl and aryl groups. Methyl - In addition to the weakness of their absorption, the CH stretching vibrations of methyl groups on silicon may be displaced from their normal position in aliphatic compounds. Both the symmetric and antisymmetric vibrations are influenced by the presence of adjacent electron-withdrawing groups such as chlorine or oxygen. The effect of electronegative elements on frequencies in organosilicon compounds has been discussed by Kriegsmann[1473]. There is also a surprisingly large difference between the intensities of the two modes, with the antisymmetric vibration absorption being the more intense. The reverse is true with the bending modes, where the symmetric deformation at 7.9 μ is much more intense than the anti-symmetric deformation absorption at 7.1 μ.

The band most characteristic of methyl on silicon is the symmetric deformation at 7.8 - 8.0 μ. This band is intense, sharp, and not easily misidentified. Its exact position depends on the nature of the other substituents on the silicon (Wright and Hunter[1474]; Smith et al[1466]). If three methyl groups are present on the same silicon, vibrational interactions usually split this band into two components of unequal intensity.

The 8 μ band is always accompanied by one or more euqally intense bands in the 11.8 - 13.1 μ region, which arise from methyl rocking and Si - C stretch vibrations (Young et al[1475]; Smith [1476]). This region also gives useful clues about the structure of the molecule. A single methyl group on silicon absorbs at 13.1 μ, two methyls at 11.7 and 12.5 μ, and three methyls at 11.9 and 13.1 μ. Tetramethylsilane has an intense band at 11.6 μ arising from a degenerate methyl rocking vibration (Smith[1476]; Young et al [1471]). If an alkyl group as well as n methyl groups are attached to a silicon, the spectral pattern in the 11 - 13 μ range usually resembles that obtained from (n + 1) methyl groups.

Ethyl. In ethyl substituted silicon compounds, both the CH_3 and the CH_2 modes are influenced by the silicon. An exact interpretation of the band envelope is difficult because a calcium fluoride prism does not resolve it fully, and overtone and combination bands may confuse the pattern. Assignments are therefore tentative.

An ethyl group on silicon is characterized by a band of medium intensity at 8 - 8.2 μ, arising from a symmetrical CH_2 deformation. A pair of bands found at 9.8 - 9.9 and 10.3 - 10.6 μ further serve to identify the ethyl-silicon grouping. A strong absorption arising from the SiC stretching mode occurs in the 12 - 15 μ region.

Other alkyl groups. It might be expected that with longer chain alkyl substituents, the silicon would have progressively less influence on the hydrogen motions, and this is indeed the case. It is somewhat surprising, however, that the CH stretch bands, even in butyl trichlorosilane, are considerably displaced from their normal positions in organics. As one proceeds through propyl, butyl, amyl and higher alkyl substituents, the 3.4, 6.8 and 7.2 μ bands increase in intensity (Kaye and Tannbaum[1478]; Harvey et al[1479]; Westermark[1480]). The lower members of the series show characteristic absorption patters which are remarkably constant in going from one molecule to another and which are useful for identification. The $SiCH_2$- deformation absorption becomes progressively weaker, appearing at 8.2 - 8.38 μ in propyl-Si, 8.35 - 8.4 μ in butyl-Si, and at about 8.4 - 8.5 μ in longer chain alkyl substituents (Kaye and Tannbaum[1478]; Harvey et al[1479]).

The straight chain alkyl compounds also show absorption in the 13 - 15 μ region, due at least in part to CH_2 rocking in the methylene chains (Bellamy[1481]). Shorter linear and branched chain alkyl silicon compounds have a number of characteristic absorptions in the 7.5 - 13 μ range, but none of them are very intense.

Heterocyclic rings which include silicon show some characteristic absorptions. The spectra of dialkyl cyclopentamethylene silanes have been discussed by Oshesky and Bentley[1482]. These compounds show characteristic bands at 10.93 - 11.00 μ.

Vinyl substituted silanes always show a strong CH in-plane bending absorption at 7.1 - 7.2 μ and a C = C stretching frequency at 6.2 μ, as well as a distinctive doublet at 9.8 - 10 and 10.2 - 10.5 μ (Curry[1483]; Scott and Frisch[1484], Kanazashi[1485]; Frisch et al[1486]). These latter bands arise out-of-plane CH bending vibrations (Schull et al[1487]). A weak but distinctive overtone at 5.2 μ is also characteristic.

Alkyl groups on silicon, which absorb at 6.10 μ and give several bands in the 8 - 11 μ range (Frisch et al[1486]; Scott and Frisch[1484]; Bailey and Pines[1488]; Nasiak and Post[1489]) are easily distinguished from β -methyl vinyl substituents, which have a band at 10.3 μ as well as absorption in the CH stretching and bending regions characteristic of a methyl group (Bailey and Pines[1488]). Allyl groups also absorb in the 12.4 - 13 μ region (Scott and Frisch[1484]). Unsaturation farther removed from the silicon gives a conventional absorption pattern corresponding to its substitution and symmetry (Bellamy[1490]).

Phenyl. The phenyl group when attached to silicon usually shows as least three sharp bands in the aromatic CH stretching region. Further investigations are needed to arrive at an exact assignment of these absorptions, but their frequencies remain quite constant from one compound to another, except where they are influenced by strongly electronegative groups attached to the silicon.

Bands characteristic of phenyl on silicon include a sharp, narrow band at 6.98, a broader band at 8.9, weak absorptions at 9.7 and 10.0, and two or three bands at 13.5 - 14.4 μ (Young et al[1475]; Richards and Thompson[1491]; Harvey et al[1492]; Bellamy [1481]). All absorptions except the latter are essentially invariant in wavelength regardless of the other substituents on the silicon. The 8.9 μ band. however, often splits into two components when two phenyl groups are present on the same silicon (Young et al[1475]), but reverts to a single band for the Ph_3Si group. Taken all together, these bands identify the phenyl-silicon linkage. Only the 8.9 μ band is truly characteristic, however, since compounds containing the groups Sn - Ph, Pb - Ph and Ge - Ph show a pattern essentially identical to that of Si - Ph except for a shift in this one band (Noltes et al[1443]).

Spectra of a number of phenyl-containing silanes and siloxanes have been studied by Kriegsmann and Schowtka[1494]), who have tabulated their i.r. and Raman bands and assigned their skeletal frequencies. A partial frequency assignment of the inner vibrations of the phenyl ring has been made by Spialter et al[1495]. By analogy with organic aromatic compounds, these authors attribute the 13.5 and 14.3 μ bands to the out-of-plane hydrogen deformation modes. They describe the 6.25, 6.73, 7.66, 7.97, 8.95 and 10.02 μ absorptions as characteristic of CH and C = C vibrations and ring deformations.

Assignment of the ring vibrations is not easy, because not only are the frequencies different from those of analogous organic molecules, but the intensity distribution of the bands is also different. It has been shown that the 8.9 μ band is affected by the electronegativity of the substituent (Kross and Fassel[1496]) and this band may arise from an "X-sensitive" planar ring vibration which has some Si - C stretch character (Whitten[1497]). The three bands found at 6.27, 6.73 and 6.99 μ must arise from C - C ring stretching vibrations (Whitten[1497]). The usual intensity relationship of the 6.27 and 6.73 μ bands is reversed in silicon-substituted phenyl compounds, with the former band being the stronger.

In spite of the great similarity of the spectra of compounds containing the phenyl-silicon group, specific compounds can often be distinguished by slight differences in the 13 - 15 μ absorption bands (Beck et al[1498]).

Other aromatics. A number of substituted aryl trimethyl silanes have been characterized by Clark et al[1499]. They point out that the 5 - 6 μ range, useful in characterizing many substituted benzenes (Young et al[1500]) is applicable to most aryl silane derivatives. In many such compounds the 12 - 13.5 μ region, often used to distinguish isomers (Colthup[1501]) is confused by the presence of both the out-of-plane bending modes of aromatic hydrogen and the methyl rocking and SiC stretch absorptions. In some cases isomers of a given substituent such as chlorine will show characteristic absorption patterns. The p-biphenylyl groups absorbs characteristically at 6.48, 7.40, 9.91 and 12.0 μ (Spialter et al[1495]). Spectra have also been published for disilyl xylenes (Wilson et al[1502]). The spectra of several naphthyl silanes have also been discussed (Gilkey and Tyler[1503]; West and Rochow[1504]).

Procedures for the determination of alkyl and aryl groups. Several workers have studied the applicability of i.r. spectroscopy to the determination of alkyl and aryl compounds in organosilicon compounds and polymers. Snowacka et al[1505] investigated the determination of methyl and phenyl groups in silixane resins at 5% w/v solutions in carbon disulphide using the band at 8 μ to determine methyl groups, and that at 14.36 μ for phenyl groups. The monomers used in preparing the resins were employed for preparing the calibration graphs. As the extinction coefficient for the phenyl groups occurring in the compounds of the types $(C_6H_5)_2SiO$ and C_6H_5SiO varied, two calibration graphs were used when determining phenyl groups, and results were calculated as mean values from both graphs. For the determination of methyl groups, the coefficient of variation was 2.8% (26 determinations): for phenyl groups it was 2.0% (31 determinations).

Burnacka and Wokroj[1506] continued this work and modified the above method to differentiate between the methyl groups attached to the benzene ring and those attached to the siloxane chain. The peak of the band at 1265 cm^{-1} shifts from 1257 cm^{-1} for diethoxydimethylsilane to 1260 cm^{-1} for diethoxymethylphenylsilane, and to 1266 cm^{-1} for triethoxymethylsilane; hence the compensation procedure gives the content of the methyl groups in the two positions. The phenyl groups are determined by measuring the extinction at 1430 and 1595 cm^{-1}. The absorption at 1430 cm^{-1} is due to the (C_6H_5) - Si linkages, but compensation is required for distortion caused by the nearby ethoxy-group absorption band. Although well-shaped the 1595 cm^{-1} band is weak and the extinction coefficient is seriously affected by the direct neighbours of the phenyl group. The identification of the $Si(C_6H_5)$ and the $Si(C_6H_5)_2$ groups from the absorption spectrum between 400 and 600 cm^{-1} was also studied by these workers.

Lady et al[1507] determined the ratio of methyl to phenyl groups in silicone polymers by measuring the intensity of the methylsilicon and phenyl-silicon bands at 7.92 and 6.97 μ respectively. Reasonable agreement was reached with theoretical values.

Grant and Smith[1508] have also discussed the measurement of methyl to phenyl ratio in silicones. Fishl and Young[1509] applied the absorbance ratio method to the determination of the relative amounts of phenyl and alkyl groups in silicone resins. Owing to the lack of a suitable standard, no determination of absolute concentration could be made. To obtain suitable samples, the resins were spread on flat salt plates and air-dried for a few minutes; the procedure was repeated until a check on the bands used (3030 cm^{-1} for phenyl and 2960 cm^{-1} for alkyl) showed absorptions of suitable intensity. The samples were then cured. The spectra were recorded on a Perkin-Elmer Model 12C spectrometer with a sodium chloride prism. The precision estimated for the phenyl to alkyl ratio is no better than ± 10%.

Kubota and Takamura[1510] determined methyl and phenyl groups in methylphenylsiloxane polyer by n.m.r. spectroscopy. Proton spin resonance spectra were recorded for the three binary mixtures benzene - dioxan, benzene - hexamethylsiloxane, and dioxan - cyclooctaphenyltetrasiloxane. In all the systems the peak area due to the resonating proton of the methyl or phenyl group was found to be directly proportional to the number of protons giving rise to the peaks of the respective groups, over a wide range of recording conditions. The determination of methyl and phenyl groups in methylphenylsiloxane polymer was carried out by the use of dioxan as an internal standard. The polymer was mixed with dioxan and the spectrum of the mixture recorded, the number of each group being then calculated from the appropriate peak area. The results are in good agreement with those obtained by i.r. spectral analysis.

Urin and Hakamada[1511] applied i.r. and u.v. spectroscopy to the determination of methyl and phenyl groups in silicones. They showed that while siloxanes containing only methyl groups have no absorption in the region above 240 mμ, those containing phenyl groups adjacent to silicon have marked absorption at 270, 264, 259 and 253 mμ. The molar extinction coefficient corresponding to one phenyl group is practically constant (335) at 264 mμ for high molecular weight siloxanes. This is useful for the determination of the phenyl group. Compounds in which two phenyl groups are attached to silicon have much greater extinction above 240 mμ than those in which one phenyl group is attached to silicon. This method could be applied to the analysis of silicone lubricants silicone varnish and silicone rubber. For the identification of methylsilicones such as $CH_3SiO_{1.5}$, $(CH_3)_2SiO$ and $CH_3SiO_{0.5}$ by i.r. spectroscopy (Urin et al[1512]) characteristic absorptions at 11.7, 12.5 and 13.0 μ respectively, have been used. Since almost all commercial silicones are binary condensation products of $CH_3SiO_{1.5}$ and $(CH_3)_2SiO$, their quantitative analysis can be effected by an empirical method.

The sample is analysed as a 2% solution in carbon disulphide using a 0.1 mm cell. Urin et al[1513] also showed that, in the i.r. determination of the ratio of the methyl to the phenyl radical in silicones the following relationship holds between the amounts of methyl and phenyl groups bound to silicon - $\log[Me]/[Ph] = 1.507 \log(E_{Me}/E_{Ph}) + 0.202$ where Me/Ph is the ratio of methyl to phenyl, E_{Me} is the extinction at 3.4 μ, and E_{Ph} is that at 3.3 μ. This method compares favourably with that of Kuratani[1514] which makes use of the C - H stretching at 3.38 μ and the Si - CH_3 vibration at 7.97 μ for the methyl group, and the corresponding vibrations at 3.28 and 6 97 μ for the phenyl group.

Margoshes and Fassel[1515] have described an i.r. method for the determination of the concentration ratio of phenyl and p-tolyl substituent groups in tetraarylsilanes and hexaaryldisilanes. The absorption maxima for the p-tolyl group bands fall within the range 12.47 to 12.50 μ and the phenyl group bands within the ranges 13.40 to 13,45 μ and 14.30 to 14.33 μ. The molar absorptivities of the functional groups are not constant, but accurate results are possible. Certain substituent groups on the silicon atom can interfere; aliphatic groups absorb strongly in the 12 - 15 μ region silanols absorb at 12.4 μ in solution and the Si - H group gives an absorption band at about 12.5 μ.

Efremova and Popkov[1516] used u.v. spectroscopy for the determination of a number of phenyl radicals.

Shull[1517] has described infrared procedures for the determination of methyl and phenyl also hydroxyl groups in dimethyldiphenyl silicone resins. These procedures are described in detail in the following section on the determination of hydroxyl groups.

Hydroxyl. Silanols exhibit the usual OH stretching bands between 2.7 and 3 μ as found in organic alcohols, corresponding to monomers and hydrogen-bonded polymers. Although these bands are not unique, the broad, asymmetric SiO stretching absorption at 11 and 12 μ which is also present in the spectra of silanols serves to characterize the SiOH group.

Several investigators have studied hydrogen bonding in silanols. Kakudo et al[1518,1519] concluded that in crystalline silanols, two hydroxyl groups interact in such a way as to form a four-membered ring. Studies of association in solutions and in silicone resins (Damm and Noll[1520]; Kastochkin et al[1521]; Kriegsmann[1522]; Ryskin and Voronkov[1523]) have been carried out and assignment of the stretching and bending vibrations discussed (Kriegsmann 1473; Kriegsmann[1522]; Kantor[1524]; Tyskin[1525]). There have also been a number of studies on the interaction of SiOH groups on silica with water vapour (Benesi and Jones[1526]; Young[1527]) and other molecules (Folman and Yates[1528]; McDonald[1529]) and their behaviour on thermal dehydration (Lygin[1530]; McDonald 1531).

Shull[1517] described infrared methods for the estimation of free hydroxyl (2.8 μ), hydrogen-bonded hydroxyl (2.9 μ), also phenyl (3.3 μ) and methyl (3.4 μ) groups in dimethyldiphenyl silicone resins.

Alkoxy Groups. All compounds containing the group ROSi where R is a saturated aliphatic radical, have at least one strong band between 9 and 10 μ which arises from the C - O stretching vibration (Simon and McMahon[1532]; Kreschkov et al[1533]; Okaware[1534]). Although this band is often obscured by the intense SiOSi absorption which falls in the same range, other bands appear which usually permit identification of the alkoxy group.

Methoxy compounds have a sharp, distinctive band at 3.55 μ which varies only slightly in wavelength with a change in the other substituents on the silicon (see Table 147). Also characteristic is an absorption at 8.4 μ, attributed to $SiOCH_3$ rocking (Forneris and Funck[1535]), which has been used for quantitative estimation of CH_3OSi (Brown and Smith[1536]). The C - O stretch falls at 9.1 μ, and an SiO mode falls between 11.8 and 12.5 μ (Forneris and Funck[1535]; Kriegsmann and Light[1537]; Stuart et al[1538]). The vibrational spectra of methoxy substituted silanes have been the subject of many investigations (Forneris and Funck[1537]; Kriegsmann and Light[1537]; Stuart et al[1538]; Kreshkov et al[1539]; Hayashi[1540]).

Ethoxy substituents show a doubling of the SiOC band at 9 - 9.3 μ. In addition, analytically useful bands occur at 8.5 - 8.6 and 10.35 - 10.65 μ. Vibrational assignments for ethoxy-substituted silanes are not at all firm in spite of the large amount of work which has been done on molecules containing this group (Simon and McMahon[1532]; Okaware[1534]; Stuart et al[1538]; Kreshkov et al[1539]; Hayashi[1540]; Bulanin et al[1541]; Lazarev and Voronkov 1542).

Longer chain alkoxy silanes show analogous absorptions at 8.5 - 8.8,

9.1 - 9.3 and 10.1 - 10.6 μ. Alkoxy groups with branching on the α-carbon, such as iPrO and tBuO, do not absorb between 10 and 11 μ (Kriegsmann and Light[1537]; Bulanin et al[1541]; Lazarev[1543]; Lazarev and Voronkov[1542]; Breedervelt and Watermann[1544]).

Table 146 - Hydroxyl determinations by i.r. of dimethyldiphenyl silicone resins

Sample	i.r.*			Chemical		
	CO - H	C_{OH} ... O	C_{OH}	C_{OH}	Diff.#	(%)Diff.#
A	0.10	0.10	0.20	0.20	0.00	0.0
B	0.16	0.43	0.59	0.65	-0.06	-9.2
C	0.24	0.76	1.00	1.11	-0.11	-9.9
D	0.26	0.85	1.11	1.28	-0.17	-13.3
E	0.28	1.34	1.62	1.45	+0.18	+12.4
F	0.30	1.32	1.62	1.52	+0.10	+ 6.6
G	0.33	1.66	1.99	1.98	-0.01	+ 0.5
H	0.30	1.55	1.85	2.08	-0.23	-11.0
I	0.36	1.65	2.01	2.12	-0.11	- 5.1
J	0.39	2.38	2.77	2.66	+0.11	+ 4.1

* Symbols are defined below.
C_x content (in weight %) of the desired functional group, where x is O - H, the free hydroxyl; OH ... O, the hydrogen-bonded hydroxyl; OH, the total hydroxyl;
Difference relative to chemical values.

Table 147 - Frequencies of CH stretching bands for methoxysilanes in CCl_4 solution

Compound	Absorption frequencies (cm^{-1})				
$MeSiCl_3$		2913		2975	
$MeSi(OMe)Cl_2$	2846	2916 (sh)	2948	2979	3054 (sh)
$MeSi(OMe)_2Cl$	2842	2918 (sh)	2946	2976	3044 (sh)
$MeSi(OMe)_3$	2837	2916 (sh)	2941	2967	-
Assignment	(s)OCH_3	(s)CH_3	(a)OCH_3	(a)CH_3	Combination bands

sh = shoulder; a = antisymmetric; s = symmetric.

Brown and Smith[1536] reported the following infrared quantitative analysis data from the determination of methoxy groups in siloxane polymers.

No.	Component Name	Component Formula	Range %	Accuracy %	B.L. Pts.	Slit (mm)	Conc. mg/ml Length mm
1	Methoxy	$\equiv SiOCH_3$	0.40	± 0.5	8.40 μ 8.1 - 8.6 μ	0.31 0.028 μ	20 0.2

Instrument: Baird Model AB-1, normal split program
Sample Phase: Solution in carbon disulfide

Cell Windows: NaCl
Absorbance Measurement by Base line
Calculation: Graphical
Relative Absorbances - Analytical Matrix:

Component	8.40 μ
1	0.900

Material Purity: reference compounds 95 - 99% pure.

Tanaka[1545] compared the i.r. spectra of the lower members of methoxy-, methylmethoxy: and methoxy-end-blocked dimethylpolysiloxanes with those of dimethyl-ether, dimethylpolysiloxanes, etc. The characteristic absorption band of the $SiOCH_3$ group appears near 1190 cm^{-1}. Except in the case of the first members of these alkoxysiloxanes, a spectral sequence corresponding to the structural unit sequence of SiO_4, SiO_3C and SiO_2C_2 is observed in the region of 1000 - 1100 cm^{-1}.

Okawara [1546] showed that in ethoxy- and methylethoxy-polysiloxanes characteristic bands due to the $(Si)OC_2H_5$ group are located near 960 and 1160 cm^{-1} and near 840 cm^{-1} in ethoxy end-blocked dimethylpolysiloxanes, and near 790 cm^{-1} in ethoxypolysiloxanes.

Acetoxy groups. Compounds containing the grouping SiOOCR show, besides the expected C = O absorption at 5.7 - 5.9 μ, bands analogous to those of the alkoxy silanes. Where R is CH_3, characteristic bands appear at 8.2 and 10.6 μ. The SiOC stretch, which absorbs strongly at 9.2 μ in aliphatic silicon ethers, is probably responsible for the 8.1 μ absorption in acetoxy compounds, by analogy with aliphatic acetates (Bellamy[1547]). Other acetoxy compounds absorb in the region 8.4 - 8.8 and 10.3 - 10.8 μ. (Lanning and Moore[1548]).

Silicon-containing vinyl acetylenes. The mass spectra of the following compounds have been investigated (Khmel'nitskil et al[1549]): trimethyl-(vinylacetylenyl)silane, trimethyl(isopropenylacetylenyl)silane, trimethyl (propenylacetylenyl)silane, triethyl(vinylacetylenyl)silane, dimethyl (vinylacetylenyl)silane, and methyl(vinyl)-di(vinylacetylenyl)silane. The mechanism of formation of the most intense ions in the mass spectra has been proposed. Dissociative ionisation of the similarly constructed compounds trimethyl(vinylacetylenyl)silane, trimethyl(isopropenylacetylenyl)silane, and trimethyl(propylacetylenyl)silane proceeds by the same mechanism, with predominant formation of ions produced by the loss of a methyl group from the molecular ion. The formation of ions by the loss of methyl, ethyl and vinyl radicals is typical of the more unsaturated silane derivatives. These suggested mechanisms are confined by the presence of the corresponding metastable ions in the spectrum. The identification of substituted silicon-containing vinylacetylenes is also discussed by Khmel'nitskil et al[1549].

Silanic hydrogen. The SiH group is usually easy to identify, since its stretching absorption is quite intense and falls in the 4.4 - 4.8 μ range, where there is very little interference from other bands. Its position depends on the inductive power of the other substituents on the silicon, and in many cases can be predicted exactly (Smith and Angelotti[1550]). The bending modes of a single hydrogen on silicon fall between 10.9 and 13.1 μ (Kaplan[1551]). The SiH_2 scissors mode has been assigned to the range 10.2 - 10.8, the SiH_2 wagging between 10.5 and 11.8, the SiH_2 twist between 13.5 and 16.0, and the SiH_2 rock between 19 and 24 μ (Ebsworth et al[1552]), although in most cases these limits are somewhat narrower (Kniseley et al 1553; Kriegsmann[1554]). The SiH_3 group absorbs strongly at about 10.8 μ.

Both the symmetric and antisymmetric SiH_3 deformation vibrations absorb in the 10.5 - 11 μ region, whereas the SiH_3 rocking mode absorbs between 15 and 17 μ (Kriegsmann and Schowtka[1555]; Kniseley et al[1553]; Dixon and Sheppard[1556]; Newmann et al[1559]). Again, the wavelengths of these absorption bands depend on the nature of the other silicon substitents (Westermark[1558]; West and Rochow[1559]; Ebsworth et al[1552]). Because of interferences, the longer wavelength SiH bands are not as useful for characterization as the stretching frequency, although the 10 - 11 μ region is often useful. The only interferences from organosilicon compounds in this region are from SiOH, SiF, SiNSi and SiOM, where M is metal.

The following conditions are suitable for the determination of silanic hydrogen in silicone polymers.

Component			Range	Accuracy	μ	Slit	Conc.
No.	Name	Formula	%	%		(mm)	Length (mm)
1	Silanic Hydrogen	≡Si - H	0.001 - 0.03	± 0.001	4.67	0.032	undiluted 0.100

Instrument: Perkin-Elmer Model 21, NaCl Prism
Sample Phase: Undiluted Liquid
Cell Windows: NaCl
Absorbance Measurement: Base line
Calculations: Graphical
Relative Absorbances - Analytical Matrix:

Component/	4.67 μ
1	1.418

Material Purity: Reference compound 99 + % pure 1,1,1,3,5,5,5 heptamethyltrisiloxane
Comments; (1) Accuracy expressed as standard deviation.
(2) Absorbance expressed as slope of Beer's law curve in absorbance per 100% of constituent (hydrogen).

Di- and trimethylsilanes. Hershenson[1560] has reviewed i.r. spectra of organosilanes. Ball et al[1561] studied the i.r. and Raman spectra of these compounds and some of their deuterated derivatives. Their results suggest that coupling between bending motions of SiH and SiH_2 groups and other motions prevents the assignment of SiH bending modes. Otherwise complete assignments were made by these workers.

$Si(CH_2)_nSi$. Vibrations involving the disilyl-methylene linkage also give rise to a strong absorption in the 9 - 10 μ range, but this band is rather narrow and thus can usually be distinguished from siloxane absorption. This band has been attributed to the CH_2 twisting frequency (Kriegsmann[1562]) but assignment to the CH_2 wag may be more reasonable. The wavelength of this band is affected by mass and inductive effects of the other silicon substituents but to a lesser degree than is the siloxane absorption. A weak but distinctive band appears at 7.4 μ which is characteristic of $SiCH_2Si$. The asymmetric SiCSi stretch falls at 13.0 μ (Kriegsmann[1562]).

Disilylethylene groups conveniently show two hands in the 8 - 10 μ region which are quite characteristic if not hidden by other absorptions.

Disilylpropylene and longer chain bridges give absorption patterns in the 8 - 10 μ interval which are somewhat reminiscent of propylsilicon, butylsilicon, etc. groups. This observation is in accord with expectations since at least some of the characteristic absorptions in this region arise from motions of hydrogens on the carbon adjacent to the silicon.

Siloxanes (SiOSi). Almost all siloxane compounds show at least one strong band between 9 and 10 μ arising from an asymmetric SiOSi stretch vibration (Wright and Hunter[1563]; Lord et al[1564]) and corresponding to the COC stretching absorption in aliphatic ethers. In simple disiloxanes, the position of the band varies from 8.9 - 9.8 μ depending on the mass and inductive effect of the substituents on the silicon. With progressively longer chain compounds (trisiloxanes, tetrasiloxanes, etc.), the absorption splits into two or more overlapping components, until with an essentially infinite siloxane chain, the absorption occupies almost the entire interval between 9 and 10 μ with points of maximum absorption at 9.2 and 9.8 μ (Wright and Hunter[1563]; Richards and Thompson[1565]).

Cyclic polysiloxanes also absorb in this region, with a characteristic cyclotrisiloxane band at 9.8 μ (Wright and Hunter[1563]; Young et al[1566]). Cyclic tetramers and pentamers have a single band at about 9.2 μ which gradually widens and splits with larger rings (Young et al[1566]). The positions of these bands are essentially independent of the other silicon substituents. The shift of the siloxane band to longer wavelengths in the case of the trimer is due to ring strain, and the siloxane absorption may move to as far as 11 μ in highly strained ring systems. Such extreme shifts are rare, however, and compounds showing them are unstable. The cyclotrisiloxane ring is planar (Kriegsmann[1567]) but the cyclotetrasiloxane ring has a non-planar (and probably unstrained) structure (Kriegsmann[1568]).

The siloxane band is occasionally shifted to shorter wavelengths by the inductive effect of other substituents. Thus, in $Cl_3SiOSiCl_3$, the siloxane absorption occurs at 8.9 μ. Silsesquioxane ($RSiO_{3/2}$) structures usually exhibit their siloxane bands at about 9.0 μ (Barry et al[1569]). The symmetric SiOSi stretching vibration lies at much longer wavelengths. In the case of $Me_3SiOSiMe_3$, it falls at about 520 cm^{-1} (Kriegsmann[1562]) for disiloxane, at about 600 cm^{-1} (Lord et al[1564]; McKean[1570]). Symmetrical vibrations such as this usually show a large intensity in the Raman effect, and may be missing or only weakly active in the infrared.

Launer and Grenoble[1571] reported the determination by i.r. of poly (dimethylsiloxane) fluids.

Phenylsiloxy monomer units. Yamamoto[1572] has shown that phenylsiloxy monomers in the form $C_6H_5SiO_{1.5}$, $(C_6H_5)SiO$ and $C_6H_5CH_3SiO$ can be identified by i.r. absorption at 20.6 μ. The ratio of the units $C_6H_5SiO_{1.5}$ and $(C_6H_5)_2SiO$ is determined by measuring the extinction (D) at 19.5 and 20.6 μ with the aid of the empirical formula -

$$\text{molar fraction of } (C_6H_5)_2SiO = \frac{1.19D_{19.5}}{D_{19.5} + D_{20.6}} - 0.19$$

Silicon-halogen bonds. As expected, the Si-halogen stretching frequencies fall in the same order as the atomic masses of the halogens. Because of inductive effects, however, the shifts in frequency upon going from one member to another are larger than would be predicted by a simple Hooke's law calculation. The stretching vibrations of the SiF group fall in the 10 - 12 μ region (Ebsworth et al[1553]; Kriegsmann[1567]; Newman et al[1554]; Anderson and Bok[1573]; Collins and Nielsen[1574]; Hayashi[1575]; Jones et al[1575]; Okawara[1577]). For SiF_2 and SiF_3, as is usually the case when two or more identical single atoms are attached to silicon, the asymmetric stretching mode falls at a higher frequency and is more intense than the symmetric mode.

SiCl bonds absorb strongly between 16 and 24 μ (Smith[1570]; Gibian and McKinney[1579]). Earlier work (Richards and Thompson[1580]) mis-identified an SiH bending mode at 12.5 μ in $HSiCl_3$ as being due to SiCl, and this error has been repeated by many later workers. The fact is that the wavelength extremes of the SiCl stretching vibrations are represented by $SiCl_4$ with bands at 621 and 424 cm^{-1}(16.3 μ and 23.6 μ; the latter band is infrared inactive). Highly electronegative groups or atoms such as fluorine on the same silicon could, however, displace the SiCl stretch to shorter wavelengths.

Infrared studies of methyl chlorosilanes (Smith[1581]; Burnelle and Duchesne[1582]; Shimizu and Murata[1583]; Tobin[1584]) and other substituted chlorosilanes (Forneris and Funck[1585]; Kriegsmann[1586]; Grenoble and Launer [1587]) show that the wavelength of the SiCl bands varies in a regular manner. These absorptions are useful for quantitative (Grenoble and Launer[1587]; Brown and Smith[1588]) as well as qualitative analysis.

SiBr and SiI fundamentals may fall beyond the range of the potassium bromide prism, but can usually be detected with caesium bromide optics (Ebsworth et al[1552]; Kriegsmann[1554]; Dixon and Sheppard[1556]; Newman et al[1557]; Linton and Nixon[1589]; Mayo et al[1590]; Opitz et al[1591]).

The tendency of the halosilanes to react with traces of moisture to form siloxanes must not be overlooked when working with these materials (Smith and McHard[1592]).

Elfremova and Popkov[1593] showed that mixtures of chlorosilanes can be identified and determined by Raman spectroscopy. Genoble and Launer[1587] showed that the phenyltrichlorosilane in methylphenyldichlorosilane can be determined in the range 0 - 10% by weight by adding 0.5 ml of the sample to 10 ml carbon disulphide and taking measurements at 17.1 μ with a Perkin-Elmer model 321 instrument. A matched cell containing 4.76% by volume of pure methylphenyldichlorosilane is placed in the reference cell.

These workers (Grenoble and Launer[1594]) have also carried out infrared spectroscopy of phenylchlorosilanes in the caesium bromide region of the spectrum. They measured the infrared spectrum of seven phenylchlorosilanes over the range 2 - 30 μ. The range 2 - 15 μ, in which the spectra are very similar, was recorded with a Perkin-Elmer model 21 spectrophotometer with a sodium chloride prism, and the region 15 - 30 μ with a Perkin-Elmer model 321 double-beam double-pass spectrometer and a caesium bromide prism. The latter instrument was flushed with nitrogen or dry air to eliminate interference by water-vapour absorption, and in this region marked differences between the compounds were recorded and attributed to Si - Cl stretching and aromatic-ring deformation vibrations. The region was also recorded successfully on a Perkin-Elmer KBr Infracord. Quantitative analyses of mixtures of diphenyldichlorosilane, phenyltrichlorosilane, and chlorobenzene were made and small amounts (0.5%) of phenyltrichlorosilane in diphenyldichlorosilane were determined.

Brown and Smith[1588] applied infrared absorption spectroscopy to the determination of chlorosilanes impurities in trimethylchlorosilane.

Silicon-nitrogen compounds. Primary silyl amines show two hydrogen stretching frequencies between 2.8 and 2.95 μ, as well as a strong NH_2 deformation falling at 6.5 μ. The SiNHSi linkage is characterized by a group of bands including the NH stretch at 2.95 μ, an NH bending at 8.5 μ, and the asymmetric SiNSi stretch at 10.7 μ (Kriegsmann[1586]; Cerato et al [1597]). Cyclic silazanes show behaviour analogous to the cyclic siloxanes, with the SiNSi vibrations absorbing at 10.85 μ in hexamethyl cyclotrisilazone (Kriegsmann[1567]) and 10.65 μ in the octamethyl cyclotetrasilazane. There is also a shift of the NH deformation from 8.7 μ in the trimer to 8.5 μ in the tetramer.

Trisilyl amine has a planar structure with the asymmetric SiN stretch falling about 10 μ (Ebsworth et al[1598]; Kriegsmann and Foster[1599]; Robinson[1600]). In di(trimethylsilyl)methylamine, however, the SiN stretch is found at 11 μ (Kriegsmann[1562]).

Because of their hydrolytic instability, silazane compounds are seldom encountered commercially.

Although the preparation and infrared spectra of siliconisocyanide compounds have been reported (Ebsworth[1552]; McBridge and Beachell[1601]; Prober[1602]), isotopic studies (Linton and Nixon[1603,1604]) as well as chemical evidence indicate that only the normal cyanide is formed. The CN stretching vibration falls between 4.5 and 4.6 μ, which is at slightly longer wavelengths than in the corresponding organic compounds. Cyanosilanes hydrolyze readily in moist air to form hydrogen cyanide.

Linton and Nixon[1604] have reported on the infrared spectra of gaseous SiH_3CN and SiD_3CN at 28°, also the vibrational assignments and structural interpretations. For SiH_3CN the chief absorption bands are at about 600, 920, 1360 and 2200 cm^{-1}, whilst for SiD_3CN they are at about 600, 700, 1080, 1600 and 2210 cm^{-1}. They showed that there is very strong evidence for a normal cyanide rather than isocyanide structure for SiD_3CN. Launer and Crouse[1605] developed a method for the determination of trichloro-e-cyanopropylsilane in dichloro-(3-cyanopropyl)-methylsilane.

Silicon-sulphur compounds. Infrared data on silicon-sulphur compounds is rather meagre. The SiSH stretch in Me_3SiSH falls at 3.9 μ (Okawara[1546]), but there are apparently no characteristic SiS bands in the NaCl prism range. In $MeSiSSiMe_3$, the SiS stretch vibrations are found at 20 μ (Kriegsmann[1586]) in $H_3SiSSiH_3$ at 19.5 μ (Ebsworth et al[1606]; Emeleus et al[1607]; Linton et al[1608]) and in Me_3SiSH at 22 μ (Jones et al[1609]). Disilyl selenide has also been studied (Ebsworth et al[1606]; Emeleus et al[1607]).

Several silyl isothiocyanates have been prepared and their spectra recorded (Goubeau and Rushing[1610]; Macdiarmid and Maddock[1611]). The Si-Si vibration which is usually inactive in the infrared, falls at about 400 cm^{-1} in the Raman spectrum of hexamethyl-disilane (Nurata and Shimazu[1612]).

Miscellaneous. Several metal derivatives of trimethylsilanol, triethylsilanol and diphenylsilanediol were prepared by Tatlock and Rochow[1613]. The curves show that SiONa, SiOK and SiOSn absorb strongly between 10 and 11 μ. SiOHg shows a strong band at about 11 μ. The SiO - Tl vibration is found at 10.9 μ (Zeitler and Brown[1614,1615]). Work on silicon-phosphorus compounds shows a broad Si-OP absorption to be at 9.7 μ (Keeber and Post[1616]) and the SiP band at 22 μ (Linton and Nixon[1617]).

C. Mass Spectrometry

Aulinger and Reerink[1595] applied mass spectrometry to the investigation of the pyrolysis of methylchlorosilanes. They included the series $H_3Si\ CH_2\ SiH_3$ to $(CH_3)_3\ Si\ CH_2Si(CH_3)_3$ and the cyclic compounds $Si(CH_2)_3H_6$ to $Si(CH_2)_3(CH_3)_6$. Hirt[1596] has described a mass spectrometric method for the determination of methyl and phenyl-chlorosilanes in mixtures. The peaks chosen for analysis were those of mass numbers 126, 148, 210 and 217. Calibrations were based on known mixtures, and the coefficient of variation was less than 0.5 mol % for all the components present. Garzo et al[2335] also studied the thermal decomposition of branched chain methylpolysiloxane resins and used mass spectrometry to identify up to 17 low molecular weight methyl siloxane reaction products. These workers list retention indices on three different stationary phases, which enable conclusions to be drawn regarding the structure of the individual pyrolysis products and hence of the resin degredation process.

The mass spectrometric method is generally applicable to the identification as well as the quantitative analysis of silicones, because of the distinctive fingerprint formed by the natural abundance of isotopes silicon-28, silicon-29, and silicon-30. An additional advantage of mass spectrometry for identification purposes is that it gives directly the molecular weight of a compound.

Sampling is straightforward; the only restriction is that the sample must have a vapour pressure of at least 1 mm mercury at 225°. This condition, of course, excludes most siloxane polymers, but the method is applicable to monomers, to low molecular weight polymers, and to decomposition products of polymers.

Published mass spectra for organosilicon compounds include trimethylsilyl derivatives of aliphatic alcohols (Sharkess et al[1447]), tetramethylsilane (Dibeler[1448]), hexamethyldisiloxane, octamethyltrisiloxane and symtetramethyldiphenyldisiloxane (Dibeler et al[1449]) and several alkyl and aryl chlorosilanes (Sokolov et al[1450]).

D. Raman Spectroscopy

Raman spectra can be used for identification of silicones. In practice, however, restrictions on the type of sample which can be run severely limit the applicability of the method. It is not practical to examine solids, for example, by this technique. Nevertheless, Raman spectroscopy has some unique advantage which can be used advantageously in the analysis of certain types of silicones.

The first advantage lies in the fact that vibrations of low frequency are easily observed. Consequently, the presence of Si - Br and Si - I, whose stretching vibrations fall in the range 150 - 450 cm^{-1} is easily determined. A second advantage is that Si - Si, which because of symmetry considerations gives very weak or no absorption in the infrared, can be detected (Bethke and Wilson[1451]). In addition, double or triple bonds at a center of symmetry give no characteristic infrared bands, but are very strong in the Raman spectrum (Batnev et al[1452]).

In general, group frequency correlations hold for Raman spectra (Ulbrich[1431], Yegorov and Bazhulin[1454]) and because intensities are additive, quantitative group analyses can be carried out (Batnev et al[1457], Bazhulin et al[1456]). Some structures for which spectra are given in the literature include methylchlorosilanes (Goubeau et al[1456]), methylbromosilanes (Murata and Hayashi[1457]), chlorosilanes (Goubeau and Warneke[1458]), ethylchlorosilanes (Batnev et al[1459], Murata et al[1460], Sanidan[1461]), other alkylhalosilanes (Mataka[1462]) and siloxanes (Cerato et al[1463], Murata[1464]; Slobodin et al [1465]; Smith et al[1466]).

E. N.M.R. Spectroscopy

Nuclear magnetic resonance promises to be a useful aid in the elucidation of the structure of silicon-containing molecules (Currie and Harrison [1467], Goodman et al[1468]). Chemical shifts have been given for proton (Schnell and Rochow[1469]); fluorine-19 (Schnell and Rochow[1469,1470]); and silicon-29 (Halzman et al[1471]) resonances in organosilicon compounds. Proton shifts have also been used to obtain information about molecular motions in liquid and solid methylchlorosilanes, polysiloxanes, and silicone rubber (Rochow and Le Clair[1472]).

F. P.M.R. Spectroscopy

Munro et al[1416] have applied proton magnetic resonance spectroscopy to the determination of the purity of bis(trimethylsilyl)acetamide. Low concentrations of hexamethyldisiloxane and monosilylacetanide were found.

ORGANOTIN COMPOUNDS

A. Infrared and Raman Spectroscopy

The measurement of the specific absorption of Sn-H bands at about 1800 - 1850 cm^{-1} in the infrared has been suggested for the quantitative determination of organotin hydrides (Newmann et al[1618,1619]). Friebe and Kelker[1620] worked out a procedure for the analysis of mixtures of triphenyltin hydroxide and bis(triphenyltin) oxide. These substances can be codetermined by triphenyltin analysis, but only in a solid phase (suspension of liquid paraffin). Triphenyltin has an absorption band at 896 cm^{-1}. Standard curves for the determination of the components of such systems are given. Solutions of the two compounds in carbon disulphide, for example, show identical spectra.

Domange and Guy[1621] attempted to analyse mixtures of diethyltindiiodide and triethyltiniodide quantitatively. They prepared i.r. spectra of 0.01 mm thick sections of the two pure compounds and observed that in the 10 micron region there is a double peak for the di-iodo(solid) compound and a single peak for the iodo (liquid) compound. The additional peak observed with the di-iodo compound was attributed to the fact that this substance was examined in the solid state. Comparisons of the spectra of these compounds in carbon disulphide at different concentrations enables a plot of log concentration vs. transmitting (%) to be made. At 8.12 and 9.80 μ, the curve for the di-iodo compound lies above and parallel to that for the iodo compound and at 8.48 μ the order is reversed. The transmission of a mixture of the two lies between the curves and is distinguishable at 10% of the iodo compound, more particularly at 8.12 μ.

Reference is made below to i.r. spectra of organotin compounds which may be of analytical interest. Whilst i.r. spectra in the sodium chloride region can yield much useful information it must be remembered that, save for the ν-(Sn - H) mode, almost all fundamental stretching vibrations involving the tin atom occur at frequencies below 650 cm^{-1}. Reference is made here only to absorptions due to molecular vibrations which directly involve the tin atom and secondary effects, for example, the peturbation of the ν (V ≡ C) mode which occurs when an acetylenic carbon atom is attached to tin.

Tin-carbon stretching frequencies. The aliphatic carbon tin stretching frequencies are the most commonly quoted i.r. bands[1626-1654].

Tin halogen stretching frequencies. Poller[1658] has also summarized tin-halogen stretching frequency assignments for alkyl- and phenyltin halides, (Gastilovick et al[1659], Clark et al[1641], Tomsalu and Wood[1645], Lohmann[1646], Butcher et al[1648], Clark and Williams[1655], Kriegsmann and Pauly[1656], Brown et al[1660], Poller [1661], Srivastava[1662], Kriegsmann and Geissler[1663], Geissler and Kriegsmann[1664].

Overall ranges are ν(Sn-Cl) 385 - 318 cm^{-1} ν(Sn-Br) 264 - 222 cm^{1} and v(Sn-I) 207 - 170 cm^{-1}. Less information is available on organotin fluorides; these compounds are associated with fluorine bridges and hence are structurally dissimilar to the other organotin halides. Poller[1658] points out that the highly tentative range of ν(Sn-F) 372 - 328 cm^{-1} is lower than would be expected from simple atomic mass considerations and this is probably a consequence of the highly associated structure.

The tin-halogen stretching frequencies are very sensitive to changes in the co-ordination of the tin atom (Poller[1658], Clark et al[1641] Tanaka [1651], Clark and Wilkins[1652], Poller and Toley[1665], McWhinnie et al[1654]). Conversion of a 4-coordinate dichloride to a 6-coordinate adduct with a Lewis base causes a reduction in the frequency of the ν(Sn-Cl) bands by some 100 cm^{-1}. Similar reductions in absorption frequencies are observed for the other halides which form 6-coordinate complexes although the effect is less marked with the iodides.

Tin-oxygen stretching frequencies. According to Poller[1658] the interpretation of the spectra containing SnO bands in rather difficult though several problems of assignment have now been solved (Mendelsohn et al[1626], Cummins[1627], Kawakami and Okawana[1634], Tanaka et al[1666], Schumann et al [1667], Okawana and Yasuda[1642], Lohmann[1646], Haire and Onaki[1668], Tobias and Friedline[1653], Kriegsmann and Geissler[1663], Kriegsmann et al[1669], Friebe and Kelken[1670], Kushlefsky et al[1671], Poller[1672], Vyshiniskii and Rudnerskii [1673], Cummins and Evons[1674], Marchard et al[1675], Yasuda et al[1676], Yasuda and Okawara [1677], Kawasaki et al[1678], Schmidtbauer and Hussek[1679], Clark et al[1680], Kawasaki et al[1681].

The trialkytin hydroxides show a medium to strong intensity band at about 900 cm^{-1} due to the Sn - OH deformation mode. Why triphenyltin hydroxides show a doublet in that region (replaced on deuteration by a single band) is not clear. It has been suggested (Cummins[1638]) that there is a Fermi resonance with an out-of-plane CH vibration.

Other element-tin stretching frequencies. Absorption associated with the tin-hydrogen stretching frequency is usually intense and Kuivila[1628] lists absorption bands in the range 1880 - 1790 cm^{-1} for various organotin hydrides. The position of the tin-hydrogen stretching frequency depends upon the electro-negativities of the groups attached to tin (Egarov et al[1682]) the ν(Sn-H) band in dialkytin halides, R_2SnXH, occurs in the range 1874 - 1820 cm^{-1}, (Kawakami et al[1630]; Sawyer and Brown[1683]). The tin-sulphur stretching vibration occurs at approximately 350 cm^{-1} (Lohmann[1646]; Schumann and Schmidt[1684]; Finch et al[1685]; Poller and Spillman[1686]), some examples are given by Poller[1658]. In symetrically substituted distannanes, R_3SnSnR_3 the ν(Sn-Sn) mode is i.r. inactive but it can be observed in the Raman spectrum or in the i.r. spectrum of an unsymmetrically substituted compound (Gager et al[1687]; Carey and Clark[1688]). Values for some tin-tin and other tin-metal stretching frequencies have been reviewed by Poller[1658].

The ν(Sn-N) absorption bands, like the ν(Sn-O) bands, occur over a wide range of frequencies and appear to be very sensitive to changes in the molecular environment of the Sn-N group. The tin-nitrogen stretching frequency in N-trimethylstannylaniline has been shown (Randall et al[1689]) to occur at 843 cm^{-1}. There is disagreement concerning the bands in $(Me_3Sn)_3N$, the most recent assignments being ν as (NSn_3) 672 cm^{-1} and ν/s (NSn_3) 514 cm^{-1} whereas a band at 728 cm^{-1} has also been assigned to the antisymmetric stretching mode in this compound (Sisido and Kozima[1690]). Absorption bands of much lower frequency have been assigned to the stretching vibrations of N - Sn co-ordination bands in compounds in which the nitrogen atom is incorporated into an aromatic ring. Thus in the organotin oxinates, bands in the region 406 - 387 cm^{-1} have been assigned to ν(Sn-N modes) (Kawakami and Okawara[1634]; Tanaka et al[1666]) and the tin-nitrogen stretching vibrations in the complexes Me_2SnX_2. 2pyridine are thought to occur at about 200 cm^{-1} (Tanaka et al[1636]).

Recent assignments made for compounds in which tin is attached to other Group V elements are as follows: ν as $(PhSn_3)$ 351 - 347 cm^{-1} and ν $(PhSn_3)$ 296 - 284 cm^{-1} in $(R_3Sn)_3P$ (R = methyl (Hester and Jones[1691]), phenyl (Engelhardt et al[1692]; ν(Sn-P) 351 cm^{-1} in Ph_3SnPPh_2 (Schumann et al[1693]; $\nu s(AsSn_3)$ 233 cm^{-1} and $\nu s(AsSn_3)$209 cm^{-1} in $(Me_3Sn)_3As$ (Hester and Jones[1691]).

B. Nuclear Magnetic Resonance Spectroscopy

A review article published in 1965 (Maddox et al[1694]) dealt with the application of n.m.r. spectroscopy to the study of organometallic compounds and includes a section on organotin compounds.

Of the isotopes of tin there are three in which the nuclear spin quantum number 1 = ½, of these the abundance of the ^{115}Sn isotope (0.34%) is so low that n.m.r. measurements involving this nucleus are rarely possible. Hence the nuclei which are significant in magnetic resonance measurements are those of the ^{117}Sn(7.54% abundance) and ^{119}Sn(8.62% abundance) isotopes. A few direct measurements of ^{119}Sn resonances have been made but there has, so far, been very limited application of this technique to structure investigations. By observing the ^{119}Sn -C - H coupling and using a heteronuclear double resonance technique it is possible to obtain, indirectly, ^{119}Sn chemical shift values. The greater part of the published work, however, is concerned with p.m.r. measurements involving the hydrogen atoms of the organic groups attached to tin. The magnitudes of the indirect tin-proton coupling constants J(^{117}Sn - C - H) and J(^{119}Sn - C - H) are often informative in structural studies. In proton magnetic resonance measurements on the organotin hydrides direct spin-spin coupling between tin and hydrogen nuclei occurs and the parameters J(^{117}Sn - H) and J(^{119}Sn - H) can be evaluated. The results of proton magnetic resonance measurements on a number of methyltin compounds have been reviewed by Poller[1658] and by various workers[1695-1715].

C. Proton Magnetic Resonance Spectroscopy

Proton magnetic resonance measurements have been carried out predominantly on methyltin compounds with substantially less work reported for aryl- and higher alkyl-tin derivatives. The protons of a phenyl group attached to tin show chemical shifts of $\tau = 2.7$ (Kula et al[1716]) and o-bis (trimethylstannly) benzene showed a complex multiplet of $\tau = 2.47 - 3.1$ parts/10^6 due to the protons of the aromatic ring (Evnin and Seyforth[1703]). Although separated by two carbon atoms. spin-spin coupling between the ortho-protons and the ^{119}Sn and ^{117}Sn nuclei can be observed in the spectra of phenyltin compounds. Coupling between tin nuclei and the meta protons has also been observed in phenyltin compounds (Kula et al[1716]).

Results of p.m.r. measurements on ethyltin compounds (Lorberth and Vakrenkamp[1700]; Narasimhan and Rogers[1717]; Tanaka et al[1718]; Verdonck and Van der Kelen[1719]), have been reviewed by Poller[1720]. The longer range coupling constants, J(Sn - C - C - H) are always greater than the J(Sn - C - H) values and the chemical shifts of the methylene protons are more sensitive than those of the methyl protons to changes in the extent of halogenation of the tin atoms.

Poller[1720] reviewed p.m.r. measurements on other types of organotin compounds (Verdonck and Van der Kelen[1721]; Verdonck et al[1722]; Verdonck and Van der Kelen[1723]; Fritz and Kreiter[1724]; Simonnin[1725]). Leusink et al[1726] have published a comprehensive list of p.m.r. parameters for olefinic organotin compounds.

Proton magnetic resonance spectroscopy is a valuable method for assigning configurations to alkenyltin compounds (Delmas et al[1727]; Seyferth and Vaughan[1728]; Vaughan and Seyferth[1720]) for example the tin-proton coupling constants in the unit SnC = CH are approximately doubled when the relative position of tin and hydrogen changes from cis to trans (Leusink et al[1726]).

D. Mass Spectrometry

Hordon et al[1730] have described an accurate mass spectrometric technique which permits the determination of tetramethyltin, tetraethyltin or mixtures of these two compounds. A specific advantage of this technique is that it is capable of rapidly determining mixtures of these two substances. The concentration of tetramethyltin and tetraethyltin is determined by comparing the mass spectrum of the sample with mass spectra of two synthetic blends, one having a known concentration of tetramethyltin and the other a known concentration of tetraethyltin.

ORGANOZINC COMPOUNDS

Infrared and Raman Spectroscopy

Infrared spectroscopy has been applied to the study of dimethyl zinc (Gutowsky[1731]; Boyd et al[1732]), vinyl and ethyl zinc compounds (Kaesz and Stone[1733]) and diethyl zinc-alcohol epoxide polymerization catalyst systems (Ishimori and Tsurura[1734]). Raman spectroscopy has been applied to dimethylzinc (Gutowsky[1731]). Proton magnetic resonance spectroscopy has been applied to diethylzinc, and Gordy and McCormick[1735] (Nasimhan and Rogers[1736]) have carried out microwave investigations of radiation effects in solid dimethyl and diethylzinc. Lyttle and Rexroad[1737] have applied electron resonance spectroscopy to gamma irradiated di-n-butylzinc.

Everson and Ramirez[1738] observed that the 2,2'-bipyridyl(BIPY) and o-phenanthroline (PHEN) complexes of diethylzinc have rather similar spectral curves, showing a pronounced shoulder with maximum absorbance at about 425 nm (See Chapter 3).

CHAPTER 5

POLAROGRAPHIC TECHNIQUES

ORGANOARSENIC COMPOUNDS

Substituted diarsines ($R_2As-AsR_2$) can be determined polarographically, [1739], the $E_{1/2}$ of the anodic wave being independent of the nature of the substituent. The analysis must be carried out in the absence of oxygen, which oxidizes the As - As bond. Concentrations of diarsines down to about 10^{-4}M can be determined by this procedure

Various other workers have reported on the polarographic reduction of organoarsenic compounds[1747-1751]. Various methods have been described for the determination of arsine in air and other gases. Various absorbing solutions have been used by different workers, including mixtures of potassium permanganate, concentrated sulphuric acid and bromine[1740], a mixture of silver nitrate and silver diethyldithiocarbamate[1741], and a solution of 1 N ammonium nitrate (9:1 v/v) in 95% ethanol[1742]. The last solution is also a suitable medium for the polarographic determination of arsine. The method is sensitive enough to determine down to 5 x 10^{-4} M of arsine in gas mixtures. Phosphine interferes in this determination.

Arsenic (III) and arsenic (V), monomethylarsonate, and dimethylarsinate have been determined by differential pulse polarography after separation by ion-exchange chromatography[1743], detection limits for the latter two are 18 and 8 ppb, respectively. Diphenylarsenic acid has been studied polarographically [2745].

Anodic stripping volammetry has been used to determine total arsenic species[1746]. Pulse polarographic methods have been applied to aqueous and non-aqueous solutions of methane arsenic acid and dimethylarsenic acids at concentration levels down to 0.1 μg/ml[1744]. These arsenicals are electr-active in aqueous buffers and in non-aqueous media in which the acidic supporting electrolyte, guanidinium perchlorate, is employed. A direct method of analysis, based on differential pulse polarography is reported. Detection limits of roughly 0.1 μg/mL (for methane arsonic acid) and 0.3 μg/mL (for dimethylarsinic acid) are achieved with non-aqueous electroyltes, and working curves are linear over at least 3 orders of magnitude change in concentration. A procedure for separately analysing methane arsonic acid and dimethylarsinic acid in solutions containing both acids is given, based on prior separation by ion-exchange chromatography. The mechanism of the reduction of dimethylarsinic acid in pH 4 buffer was studied, and dimethylarsine was the major product identified.

The determination of inorganic arsenic by differential pulse polarography and differential pulse anodic stripping voltammetry has been reported[1752,1753]. The polarographic reduction of dimethylarsinic acid and methylarsonic acid has also been reported[1749]. Bess et al[1750] studied the differential pulse polarography of a series of alkylarsonic and dialkylarsinic acids below pH 2. They found the peak potentials were pH dependent, shifting to more anodic values at lower pH values. Recently, Bess et al [1751] reported on the differential pulse polarography of aromatic arsonic and arsinic acids. They found peak potentials as well as peak currents were pH dependent below pH 2.

Phenylarsonic acid and phenyl arsonous acid were studied extensively by Watson and Svehla[1747,1748] using dc polarography at a dropping mercury electrode. They suggested that phenylarsine oxide existing in aqueous solution as phenyl arsonous acid and the diffusion current was pH independent. Unfortunately, above pH 2 their waves were poorly formed and exhibited broad maxima. No data were presented above this pH, and most of the work was carried out in 0.1 M hydrochloric acid at concentrations below 1×10^{-4} M, where phenylarsine oxide showed two main reduction processes. The half-wave potentials were shown to be pH dependent as the reduction processes became increasingly irreversible with increasing pH. A reaction scheme for phenylarsine oxide in 0.1 M hydrochloric acid was proposed, where the reduction product reacts with the electroactive species with the formation of an insoluble polymeric product. The first wave was attributed to the reduction of phenylarsine oxide to phenylarsine, where each mole of phenylarsine combines with an additional two moles of phenylarsine oxide to form the insoluble polymeric product. The second wave was due to an increase in the fraction of phenylarsine oxide molecules that undergoes reduction to phenylarsine and a decrease in the fraction of phenylarsine oxide molecules that reacts with the phenylarsine. At this wave there is a net increase in the average number of electrons consumed per molecule of phenylarsine oxide.

Phenylarsine oxide is used as a titrant for the direct and indirect determination of residual chlorine and ozone in water and wastewater[1754]. Preliminary investigations on the direct measurement of phenylarsine oxide by differential pulse polarography indicate that this technique is a promising method for lowering the detection limits in the indirect measurement of these oxidants[1755]. Since the control of pH is a necessary consideration in free and combined chlorine analysis with phenylarsine oxide[1754,1756] as well as the stability and measurement of ozone[1754,1757], an investigation of the effect of pH on the differential pulse polarography of phenylarsine oxide was mandatory.

Such a study has been conducted by Lewry et al[1758]. pH was shown to exert a strong influence on the differential pulse polarographic behaviour of phenyl arsine oxide. In general, phenylarsine oxide exhibited three reduction peaks, A, B, and C, with varying pH dependencies, as shown in Figure 69. Two unresolved peaks, B and C, occurred at the more cathodic potential. At low pH, peak B predominates while at high pH value, peak C predominates. At about pH 4.7, both peaks B and C approach equal height. Better resolution could not be obtained with decreases in scan rate (0.5 mV/s) or smaller modulation amplitudes (5, 10, 25 mV). The pH dependence of the current is inconsistent with the results reported by previous investigators. It is apparent that the pH dependence is not an artifact of differential pulse polarography. To determine if this behaviour is characteristic of a dropping mercury electrode only, an investigation of the mechanism

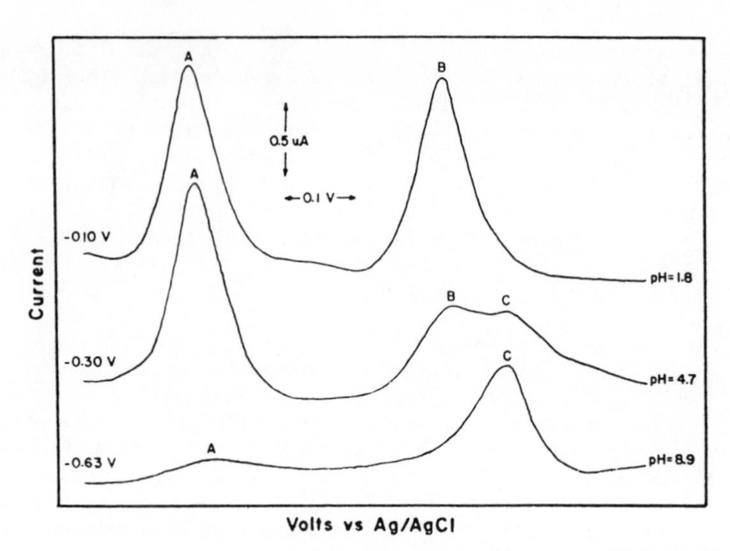

Figure 69 - DPP of PAO at various pH values. (PAO) = 3.08 x 10^{-6} M; scan rate = 2 mV/s; drop time = 2 s; pulse amplitude = 199 mV; ionic strength (μ) = 0.025.

of reduction of phenylarsine oxide by stationary electrode voltammetry and coulometry was carried out. The optimum sensitivity is obtained by using peak A in the lower pH range, over a phenyl arsine oxide range of 1.23 x 10^{-6} M to 1.23 x 10^{-5}.

ORGANOBORON COMPOUNDS

A polarographic wave of $E_{1/2}$ = 1.55 V vs. the mercury-pool electrode is obtained from a solution of potassium tetraphenylboron in dimethylformamide with tetrabutylammonium iodide as the supporting electrolyte[1759]. The wave heights are proportional to concentration over the range 0.0002 - 0.0075 M corresponding to 0.08 - 8.00 mg of potassium in the cell. Interference is caused by ammonium, rubidium, and caesium. The precision is within 3.0%.

ORGANOCHROMIUM COMPOUNDS

Holloway et al[1760] applied a vacuum electrochemical cell to the oxidation of bis(cyclopentadienyl) chromium.

ORGANOCOBALT COMPOUNDS

Geyer and Gliem[1761] have studied the polarographic determination of cobalt carbonyl in the products of the Oxo process. They showed that dicobalt octacarbonyl gives a cathodic wave at -0.45 V versus the standard calomal electrode in M-lithium chloride in isobutyl alcohol, which can be used for the determination of dicobalt octacarbonyl if the polarogram is recorded immediately after mixing the sample and the basal solution. Dicobalt octacarbonyl disproportionates to $Co(CO)_4^-$ and solvated cobaltous ions in the basal medium. $Co(CO)_4^-$ gives an anodic wave, also at -0.45 V, but the solvated cobaltous ion gives no wave. In this way, the concentration of dicobalt octacarbonyl (4×10^{-4}M to 1.4×10^{-3} M) can be determined from the height of the anodic wave when disproportionation is complete. There is no interference from nickel tetracarbonyl, iron pentacarbonyl or aldehydes and their oxidation products.

In a study of the polarographic behaviour of cobalt palmitate soaps in lithium chloride, methyl hydrogen sulphate, and potassium chloride base electrolytes polarograms were obtained for the pyridine complexes of the soaps in benzene-methanol (1:1) for lithium chloride and methyl hydrogen sulphate and ethanediol for potassium chloride base electrolytes[1781]. Well defined waves were found, except for cobalt soaps in methyl hydrogen sulphate base electrolyte. The cobalt soaps were reducible at the dropping mercury electrode in the presence of these electrolytes. The diffusion current was a rectilinear function of concentration so that this method can be used for the determination of the metal content of soaps.

ORGANOIRON COMPOUNDS

Various workers have described polarographic methods for the determination of ferrocene[1762] and nitroferrocene[1763]. A voltammetric method with a rotating platinum electrode was used for the determination of ferrocene in dimethylformamide medium[1762], using 0.1 M sodium perchlorate in dimethylformamide as the base electrolyte, $E_{1/2}$ = +0.88 V (reversible wave) relative to the silver-silver chloride saturated tetraethylammonium chloride electrode. Polarographic reduction of nitroferrocene in neutral of alkaline buffer solution is a diffusion-controlled, concentration-dependent, six-electron process[1763]. The half-wave potential of the reduction moves 58 m V per pH unit. In situ electrochemical reduction of nitroferrocene within an electron spin resonance spectrometer produced an unstable radical. In aqueous buffer solution, nitroferrocene undergoes photochemical decomposition.

In a polarographic method for the determination of iron (III) dimethyldithiocarbamate (Ferbam), a freshly prepared acetone solution of Ferbam is diluted with a solution of disodium hydrogen phosphate and trisodium citrate and the solution is analysed by conventional polarography at 0.8 - 0/2 V vs. the saturated calomel electrode (SCE)[1764]. Alternatively, a solution of sodium acetate and trisodium citrate is used to dilute the acetone solution of the sample and the mixture is analysed by cathode-ray polarography with a start potential of 0.5 V vs. the mercury pool. The limits of detection of Ferbam by the conventional and cathode-ray polarographic procedures are reported to be 2 and 0.02 µg/ml respectively.

Price and Baldwin[1765] determined ferrocenecarboxaldehyde at a chemically modified (alkylamine) platinum electrode in amounts down to 10^{-7} M. One hundredfold excess of ferrocene did not interfere. Differential pulse voltammetry was used to carry out this determination.

ORGANOLEAD COMPOUNDS

In a procedure for the determination of lead in petroleum and lubricants, the lead is extracted with Schwartz reagent (potassium chlorate and sodium chloride in nitric acid) and then, after evaporation of the acidic solution, determined either polarographically in the presence of ammonium acetate and magenta or amperometrically by titrating with 0.01 M potassium dichromate in the presence of potassium nitrate[1766]. By these methods increased accuracy and reduced time of operation are secured. With petroleum, agreement is obtained with standard methods within the specified limits of ± 0.04 ml per imperial gallon. In a rapid polarographic method for the determination of tetraethyllead in petroleum the sample is dissolved in 2-ethoxy-ethanol (Cellulose)-hydrogen chloride solution, which simultaneously decomposes the tetraethyllead to lead chloride and extracts the latter[1767]. If the 2-ethoxyethanol is cooled in an ice-bath during acidification with anhydrous hydrogen chloride, the residual current can be measured with great reliability.

In an alternative rapid polarographic method for the determination of tetraethyllead in petroleum at concentrations between 80 and 200 mg/l, the tetraethyllead is decomposed to lead chloride by the direct addition of concentrated hydrochloric acid to the petroleum[1768]. The mixture is shaken, refluxed, then extracted with water, and 0.5% gelatin is added to the combined aqueous extracts. A portion of this solution is deoxygenated by nitrogen purging in the polariographic cell and the wave recorded between -0.2 and 0.7 V. A 0.02% solution of dithizone in chloroform at pH 8.5 - 9.0 in the presence of citrate to mask iron has been used to extract lead from petroleum prior to its determination by differential oscillopolarography[1769]. The chloroform in the extract is volatilized, the brown residue is heated to fumes with 65% nitric acid - 70% perchloric acid (1:1), and the resulting white residue is dissolved rapidly and completely in the basal electrolyte (0.5 M ammonium tartrate - 0.1 M tartaric acid). The oscillopolarogram is recorded between - 0.2 and 0.7 V. When thallium is present the basal electrolyte used to dissolve the white residue must be changed to EDTA - acetate buffer solution (1:1) so that the waves of lead and thallium are clearly separated. In a further procedure a slight excess of a 10% solution of bromine in carbon tetrachloride is added to the petroleum sample[1770]. The precipitate of lead bromide obtained is washed with 50% ethanol, then dissolved in and evaporated with concentrated nitric acid until a white residue remains. This residue is dissolved in water, and the solution is titrated with 0.1 N potassium chromate by a conductometric or oscillometric technique. Other polarographic procedures for the determination of lead in petroleum have been described[1771-1778].

A procedure for the analysis of tetraethyllead involves decomposition by bromination and solution in dilute nitric acid[1774]. The lead is determined by anodic decomposition and an empirical correction factor of 0.86 is applied for the conversion of lead to lead dioxide. The solution is compared with a standard by placing the two solutions in two cells connected in parallel to an adjustable potential divider via calomel electrodes and twin electrodes. A null galvanometer between the two standard electrodes indicates any e.m.f. caused by a difference in concentration of lead ions between the two solutions. This method of comparison can be applied directly to a solution of the petroleum in 2-ethoxyethanol containing hydrogen chloride [1767]. The method is rapid and accurate for normal concentrations of tetraethyllead in petroleums that are not rich in unsaturated hydrocarbons. An anodic stripping technique for the determination of lead in petroleum uses a solution of bromine in chloroform to decompose the tetraethyllead and then extraction of the lead ions into 0.1 M nitric acid[1775]. After suitable working up, the solution is do-oxygenated with argon and submitted to anodic-stripping voltammetry in a modified cell with an SCE as reference electrode. Pre-electrolysis is carried out at -0.8 V for 1 - 5 min and lead is determined at - 0.39 V. Capillary tubes in the cell are made water-repellent with paraffin wax. The detection limit is 8 parts of lead per 10[1766], with a coefficient of variation on 7%.

Hexaethyldilead in tetraethyllead and in triethyllead chloride has also been determined polarographically. Because of the ease of hydrolysis of hexaethyldilead, it is necessary to conduct the titration in anhydrous ethanol[1776]. Tetreathylammonium hydroxide is used as the base electrolyte and analysis is conducted in the absence of oxygen. Under these conditions hexaethyldilead has an $E_{1/2}$ value between 1.8 and 2.0 V. Between 0.5 and 10% of this substance could be determined in mixtures of tetraethyllead and triethyllead chloride with a mean error of 7%. An alternative polarographic method for the determination of hexaethyldilead and triethyllead chloride in tetraethyllead involves a direct polarographic measurement at $E_{1/2}$ = - 0.24 V vs. SCE of the concentration of hexaethyldilead, using 1:1 v/v benzene-methanol solvent with lithium chloride as supporting electrolyte medium[1777] (Figure 70). There is no interference by tetraethyllead or other lead compounds. Concentrations of hexaethyldilead equivalent to the 0.1% level in tetraethyllead samples can be determined rapidly and accurately. The concentration of triethyllead chloride can be determined from the same polarogram, as it exhibits a separate polarographic wave at 0.98 V vs. SCE. Samples containing hexaethyldilead can be analysed simultaneously for triethyllead chloride[1777]. Oxygen is a common contaminant and is reduced in the voltage range corresponding to the $E_{1/2}$ of triethyllead chloride. Care must be exercised to remove oxygen completely for this determination. The second wave for triethyllead chloride exhibited a maximum which could be suppressed by Triton X-100. No use was made of a suppressor in the measurements for triethyllead chloride as the pre-wave at - 0.98 V could be measured without interference from the maximum.

Polarographic techniques have been used for the analysis of leachates of antifouling paints containing triphenyllead acetate[1778]. Polarography has been applied to the determination of compounds of the type A_3PX[1779,1780].

ORGANOMERCURY COMPOUNDS

In a polarographic method for determining phenylmercury halides in fungicidal preparations, a dimethylformamide extract of the sample is prepared and to the extract is added a solution of lithium monohydrate and 0.5% gelatin[1787]. The mixture is diluted with lithium hydroxide and nitrogen is passed through for 25 min. The polarogram is then recorded at 25°C from - 0.1 to - 0.8 V; the error is ± 1.5%. Polarography has also been used for the determination of down to 4 μg/ml of phenylmercury chloride[1783]. The phenylmercury chloride is extracted into chloroform from acidified aqueous solution, the chloroform is removed, and the residue, dissolved in ethanol, is treated with 0.1 M potassium chloride - boric acid - sodium hydroxide buffer of pH 10 containing a small amount of Triton X-100. The polarogram is recorded between - 0.4 and - 1.6 V vs. SCE. Mercury (II) chloride is not extracted with chloroform but can be identified by re-extracting the aqueous sample solution with diethylether, evaporating the ether, dissolving the residue in dilute acid, and testing for mercury with thioacetamide.

In a polarographic procedure for the determination of ethylmercury chloride fungicide in mixtures with talc and mineral oil, fungicide is digested with dimethylformamide, a portion of the filtrate and a buffer solution (0.1 N boric acid - 0.1 N sodium hydroxide, 1:3) are mixed, and nitrogen is passed through to sweep out dissolved oxygen[1784]. The polarogram is recorded between - 0.2 and - 0.7 V at a sensitivity of 0.05. Ethylmercury

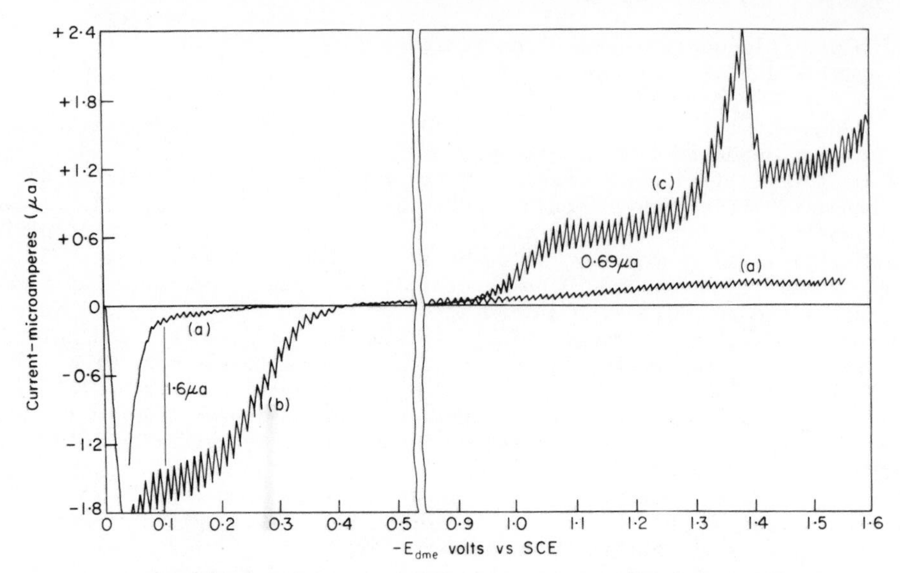

Figure 70 - Polarograms of hexaethyldilead and triethyllead chloride. (a) Solvent-electrolyte blank. (b) Hexaethyldilead (approx. 3 x 10^{-4} M) in solvent-electrolyte. (c) Triethyllead chloride (approx. 1.5 x 10^{-4} M) in solvent-electrolyte. Sargent Model XXI polarograph. Sensitivity: 0.015 μa/mm.

$$\text{Average id} = \frac{\text{mm wave height} \times 0.015 \times 6}{7}$$

chloride has an $E_{1/2}$ value of - 0.435 V vs. SCE. Calibration graphs are prepared under identical conditions using recrystallized ethyl-mercury chloride as a reference standard.

Ethylmercury chloride and methoxyethylmercury chloride in mixtures have been determined polarographically in 0.1 M potassium nitrate containing Britton - Robinson buffer and 0.01% of gelatin[1785]. Ethylmercury chloride gives a two-step wave ($E_{1/2}$ = 0.49 V and - 1.6 V vs. SCE) at pH 1.9 - 11.8. The wave height is proportional to concentration up to 3 x 10^{-4} M, and is independent of pH. In the same solution methoxyethylmercury chloride gives one wave ($E_{1/2}$ = 0.49 V) at pH < 2.9, the height being proportional to concentration up to 2.5 x 10^{-4} M. At pH > 8, the methoxyethylmercury chloride gives a two-step wave ($E_{1/2}$ = - 0.49 V and - 1.2 V), the height of the first wave being proportional to concentration, but almost half of that occurring at pH 2.9. Since the law of addition holds for their wave heights at - 0.49 V, irrespective of the pH, a binary mixture can be analysed by measurement of the wave height at pH 2.9 and 10.0 using an empirical formula. The deviation is ca. 1.5%.

Polarography has been used for the determination of phenylmercury acetate. In neutral or slighly alkaline base electrolytes, phenylmercury acetate gives two waves, the $E_{1/2}$ value of the first wave being constant over a wide pH range whilst the $E_{1/2}$ of the second wave decreases with increase in both the pH and the concentration of phenylmercury acetate[1786, 1787].

Polarographic determination of (a) ethylmercury chloride plus phenylmercury acetate in 0.2 M potassium nitrate base electrolyte at pH 10 and (b) ethylmercury chloride in 0.1 M tetramethylammonium bromide base electrolyte enables the phenylmercury acetate content of the sample to be obtained by difference[1788]. In 0.1 M tetramethylammonium bromide solution of pH < 9 phenylmercury acetate exhibits a very low wave ($E_{1/2}$ = - 0.27 V vs. SCE), with an almost constant wave height and this does not cause interference in the determination of ethylmercury chloride ($E_{1/2}$ = - 0.40 V, 0.4 - 1.6 mg per 10 ml). In 0.2 M potassium nitrate of pH about 10, both organomercury compounds give a reduction wave.

Polarographic and classical methods have been compared for the analysis of organomercury drugs[1789,1790]. A detailed study of the cathode-ray polarographic determination of merbromin showed that in Britton - Robinson buffer solution (pH 7.5) - potassium chloride solution merbromin exhibits reduction peaks at -0.28 V and - 1.07 V vs. the silver-silver chloride anode, and the current at - 1.07 V is directly proportional to concentration from 10 to 100 μg/ml of merbromin[51].

ORGANONICKEL COMPOUNDS

A dropping mercury electrode polarographic procedure has been applied to the determination of nickel content of the pyridine complexes of nickel myristate and palmitate[1791]. Lithium chloride or methyl hydrogen sulphate, each dissolved in benzene-methanol (1:1) or in potassium chloride in ethanediol, were used as supporting electrolytes, and polarography was carried out in an atmosphere of purified nitrogen. The diffusion current varied directly with the concentration of the nickel soap in pyridine.

Bontempelli et al[1792] have studied the cathodic behaviour of trans dicyano (diethylphenylphosphine) nickel complex in acetonitrile solution using cyclic voltammetry, controlled potential analysis and spectrophotometry. Cyclic voltammetry revealed three cathodic peaks located at - 1.44, - 2.00 and - 2.60 V.

ORGANOPHOSPHORUS COMPOUNDS

Alkyldithiophosphates have been determined by polarographic procedures. For compounds up to sodium dibutyldithiophosphates they may be polarographed versus the standard calomel electrode with 0.1 N perchloric acid as base electrolyte, and a dropping mercury or a flowing junction platinum anode 1793. Electrolysis with micro-electrodes serves to isolate mercuric dialkyldithiophosphates as products from the mercury electrode and di (OO-dialkylthiophosphoryl)disulphides from the platinum electrode, suggesting that the mercury takes part chemically in the reaction. The current versus concentration plot for the micro-electrode is linear up to 10^{-3} M. $E_{1/2}$ becomes increasingly more negative as the molecular weight of the sample increases. The electrocatalytic oxidation of dihydronicotinamide adenosine diphosphate with quinones and modified quinone electrodes has been reported[1794] and a polarographic method for the determination of glyphosphate residues as their N-nitroso derivatives in natural waters has been described[1795].

Nangniot[1796] has described a cathode-ray polarographic method for the determination of traces of phosphoric acid ester pesticides. Esters that contain the groups P = S or P - S give rise to adsorption peaks in cathode-ray polarography. Measurement of such peaks provides a means of determining the esters in concentrations of not less than 0.5 μg/ml^{-1}.

Kovac[2739] has described a polarographic determination of 0,0-dialkyl 1-hydroxyethylphosphonates derived from chloral. He showed that 0,0-dimethyl 2:2:2-trichloro-1-hydroxyethylphosphate (Dipterex) yields a polarographic wave ($E_{1/2}$ = - 1.08 V versus the standard calamel electrode in a supporting electrolyte containing 30% by volume of ethanol, 70% of phosphate buffer solution (pH 6.5) and 0.01% of gelatin. In the procedure 0.05 g of the sample is dissolved in ethanol (96%) and made up to 50 ml. A 2 ml aliquot of this solution and 5 ml of the phosphate buffer solution are introduced into the polarographic cell and the wave height at - 0.4 V recorded.

Sohr[1797] has discussed the oscillopolarographic behaviour of organophosphorus compounds. Giang and Caswell[2740] have determined dipertex organophosphorus insecticide in technical materials using polarography.

Tse and Kuwana[2336] studied the electrocatalysis of dihydronicotinamide adenosine diphosphate with quinones and modified quinone electrodes.

ORGANOSELENIUM COMPOUNDS

The polarography of 2-aminoethaneselenosulphuric acid and 2-aminoethanethiosulphuric acid has been studied under various conditions of pH, buffer composition, ionic strength, and temperature[1798]. The former compound is reduced in two steps and the latter gives a single wave, all waves being irreversible at the dropping mercury electrode. Mechanisms for the electro-reduction of both compounds were given. The second wave of aminoethaneselenosulphuric acid is eliminated by the addition of a surface-active agent such as gelatine or polyacrylamide, which produces dithionate, which is inactive at the mercury electrode.

A further application of polarography to organoselenium compounds includes the determination of piazselenol (benzo-2,1,3-selenadiazole) and piazthiol (benzo-2,1,3-thiadiazole) in aqueous solutions[1799]. The reduction of these compounds at the dropping mercury electrode in 0.1 M aqueous lithium perchlorate involves six electrons and yields o-phenylenediamine and hydrogen selenide and hydrogen sulphide, respectively.

ORGANOTIN COMPOUNDS

Both a.c. and d.c. polarography have been used for the analysis of alkyltin chlorides. Tributyltin chloride in dibutyltin dichloride have been determined by a.c. polarography[1800,1801], in various base electrolytes, and also by d.c. polarography[1802]. A commonly used base electrolyte in these methods is Britton - Robinson buffer at pH values between 9.3 and 10.3 containing also potassium chloride (0.5 M) and isopropanol (30%). Voltametric methods for the determination of tributyltin chloride compounds at concentrations down to 0.5% (5×10^{-6} M) in dibutyltin dichloride have an average error of ± 5%[1803]. The accuracy and sensitivity of a.c. polarography are claimed to be better than those obtained by d.c. polarography. By rectifying the alternating current using a phase-selective rectifier it is possible to suppress the much higher capacitive part in the a.c. polarography because of its different phase angle, leading to higher sensitivities in the reduction process. Further increases in sensitivity result from the use of square wave polarography (Mehner[1803]). This technique records the current at the end of one drop life. During this interval the increase in the surface area per unit time is small. Further, the measurement is carried out in a small part at the end of each square wave. With this method it was possible to determine 5 x 10^{-7} tributyltin chloride very accurately. Within the range 0.5 - 12 x 10^{-6} M the height of the first reduction wave is a linear function of the concentration (see Figure 71). For the second reduction step, the wave grows up in this range in a progressive way.

A much lower detection limit for organotin compounds can be obtained with the time-consuming anodic-stripping technique. Before the real electrochemical determination, a concentration process on a stationary mercury electrode takes place (pre-electrolysis). For a definite time the organotin compound is reduced at a potential beyond the second step (Mehner[1804]). Then the polarogram is recorded in the anodic direction and a sharp characteristic peak occurs. Down to 5 x 10^{-8} of tributyltin can be analysed by this method with an average relative error of ± 20%.

Mehner et al[1803] found, in their fundamental investigations on the surface-active properties of organotin compounds, that tributyltin derivatives have the strongest surface activity of the butyl series. They took advantage of this behaviour to simplify the analytical method. The organotin compound is enriched currentless at the potential of maximum adsorption. Then the adsorbed amount is determined by AC polarography in the normal cathodic direction. With this method it was possible to carry out quantitative analyses of down to 10^{-8} M tributyltinchloride in dibutyltindichloride. Inorganic impurities do not interfere strongly in this technique, but detergents do. To overcome this all organic substances were separated with the aid of multiply-activated charcoal.

A detailed study of the a.c. polarographic capacity effects of organotin halides in alcoholic base electrolytes showed a capacity decrease caused by even low concentrations of the organotin compound[1805].

Geyer and Seiditz[1806] described a method for the polarographic determination of dialkyltin compounds such as dibutyltindichloride, dibutyltinthioglycollate and dioctyltin bis thioglycollic acid acetylester. The behaviour of such cations is polarographically closely similar, so that a single calibration graph serves for all the compounds. In this procedure the sample (10 - 50 mg tin) is dissolved in ethanol, if necessary with the addition of a little chloroform and diluted to 25 ml with ethanol. This solution is then diluted 1:25 with supporting electrolyte (M-hydrochloric acid-ethanol (1:1), containing 0.5% of gelatin). Samples containing sulphur liberate hydrogen sulphide and give a cloudy solution, but this does not interfere. The polarogram is recorded from - 0.4 to - 0.9 V v a mercury-pool anode. The $E_{\frac{1}{2}}$ value varies from - 0.57 to - 0.66 V. Dibutyltin dichloride or dioctyltin oxide, dissolved in acetic acid, are used for calibration.

A study of the polarographic behaviour of organotin compounds in strongly polar solvents and the electrode processes involved in the reduction of butyltin chlorides at the dropping mercury electrode established that butyltin trichloride, dibutyltin dichloride and tributyltin chloride are all completely hydrolysed at concentrations up to 1 x 10^{-4} M in aqueous solution [1807,1808]. Also, the reduction potentials of these organotin halides move towards more negative values in many electrodes corresponding to decreasing polarity with an increasing degree of substitution of the tin. Parallel to an increase in polarizability, the potentials are shifted to more positive values when the chain lengths of the alkytin derivatives increase, corres-

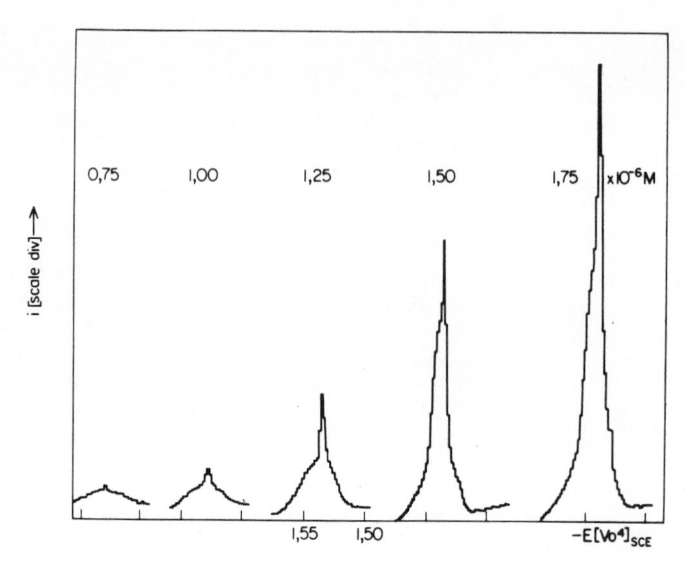

Figure 71 - Square wave polarogram of tributyltin chloride. Solution 0.5 M potassium chloride, 30% in isopropyl alcohol.

ponding to a more facile reaction. The reduction of organotin compounds at the mercury electrode proceeds via two steps[1808,1809]. The first step involves an electron transfer of n = 3 for butyltin trichloride, n = 2 for dibutyltin dichloride, and n = 1 for tributyltin chloride. The second reduction wave is kinetic and strongly irreversible. Almost identical polarographic waves were obtained for compounds as different as tributyltin chloride and hexabutyldistannoxane.

Oscillographic polarography has been applied to butyltin trichloride, dibutyltin dichloride, dibutyltin diacetate, tributyltin chloride, tributyltin acetate, triphenyltin acetate and tetrabutyltin[1810,1811]. These substances could be determined at concentrations down to 0.005 mol^{-96}.

The polarographic behaviour of some trialkytin compounds of the type Et_3SnX, where X = F, Cl, Br, I, and R_3SnCl, where R = Pr, Bu, has been described[1812,1813]. The polarogram of, for example, triethyltin chloride shows three distinct waves, but the position and size of these waves depend on a number of variables, particularly the pH of the solution. Thus, in acidic solution up to pH 7 only the first wave is visible, whereas on increasing the pH above 7 the second and third waves appear. Further, the reduction potential of the first wave becomes progressively more electronegative with increasing pH, at the same time becoming smaller, finally disappearing at pH 12. In addition, a plot of the height of the first wave against the concentration of the trialkytin compound is not linear, although the second and third waves give linear graphs. Unfortunately, the first wave is the best defined and it is not possible to make accurate measurements of the other waves owing to irregularities in their shape.

This work was extended to the determination of mixtures of dialkyltin and trialkyltin compounds by the examination of the polarographic behaviour of some dialkytin compounds and by the use of a derivative circuit. With triethyltin hydroxide, a neutral solution containing isopropyl alcohol, potassium chloride, and gelatin gives three poorly defined waves which cannot be measured with any degree of accuracy. With a derivative circuit, however, the second and third waves give well defined peaks, the heights of which are directly proportional to the concentration of the organotin compound. This direct proportionality does not extend over a wide concentration range, but at concentrations between 0.2 and 0.6 mg/ml accurate and reproducible results can be obtained. With tributyltin compounds in hydrochloric acid solution a single wave is obtained which, again over a limited concentration range (0.1 - 0.4 mg/ml) is directly proportional to the concentration of the compound[1812-1814]. This wave is well defined using both the direct and the derivative circuit. Experiments with dibutyltin dichloride in the same medium show an irregular wave which is not proportional to the concentration, followed by a well defined wave similar to that produced by the tributyltin compound and separated from it by about 0.5 V. The height of the second wave is directly proportional to concentration over a limited range. It is thus possible to make a quantitative determination of a mixture of the dibutyltin and tributyltin compounds by either direct or derivative polarography, using the second wave of the dibutyltin compound and the single wave of the tributyltin compound.

Further work has been described on the application of polarography to various types of organotin chlorides[1800,1815-1820], and to the determination of various particular compounds such as butyltin trichloride in dibutyltin dichloride and tributyltin chloride[1821], trichlorethyltin[1822], diethyldichlorotin[1824], triethyltin halides[1825], diethylchlorostannane[1826], methyl-, ethyl-, and phenyltin trichlorides[1827], triphenyltin fluoride[1828], dialkyltin compounds[1821], organotin (IV) halides[1827], and dialkyldichlorotin compounds in water-methanol solutions[1830]. The voltammetry of the aquodiethyltin (IV) cation-poly[diethyltin (II)] system has been discussed[1831].

Detection limits of 1 - 100 μg are claimed for the polarography of trialkyl-substituted organotin compounds[1832]. A polarographic method for determining triphenyltin acetate residues in vegetables can determine as little as 25 μg of triphenyltin acetate with a precision of ± 5%[1833].

The oscillographic properties of various organotin acetates, including dibutyltin diacetate, tributyltin acetate and triphenyltin acetate, have been investigated[1811]. Chronopotentiometry has also been applied to the determination of triphenyltin acetate at very low concentrations in plant material. In this method a hanging-drop electrode is used, at which the ions are reduced in a pre-electrolysis step at - 0.7 V vs. the silver-silver chloride saturated potassium chloride electrode for 5 min; the potential is then increased gradually to - 0.1 V, and the anodic diffusion current is registered at about - 0.45 V. A peak height of about 0.3 A is obtained for a concentration of about 0.8 g/ml of tin.

Triphenyltin acetate has been determined polarographically in fresh leaves[1834]. Tetraphenyltin has been determined in PVC by treating the sample with hydrogen peroxide solution and concentrated sulphuric acid[1835]. When reaction ceases, the mixture is heated until it darkens, then concentrated hydrochloric acid is added and the mixture is boiled until polymer decomposition is complete. The cooled solution is diluted with 4 N ammonium chloride-10% hydrochloric acid (1:1). The solution is de-aerated and the tin (IV) polarogram starting at - 0.2 V is recorded. Polarography has been applied to the determination of triethyltin hydroxide, triethyltin oxide and $R_3SnOCOC(Me) = CH_2$ where R = Et or Bu[1836], and amperometric and polarographic titration has been used to determine dialkytin oxides in weakly acidic solution with standard oxalic acid solution.

Geyar and Sudlitz[2837] discuss the polarographic determination of tetraphenyltin. Polarographic titration has been used to estimate dialkyltin oxides in weakly acidic solutions with standard oxalic acid solution[1538]. Svehla and Glockling[1839] carried out a voltammetric study of some tris-(trimethylsilyl) methyltin derivatives. They used classical polarography, supplemented by other voltammetric techniques, to elucidate the mechanism of reduction of di- and tri-halide derivatives of tris(trimethylsilyl) methyltin, despite the difficulties encountered in work at very negative potentials in ethanolic solutions. The dihalides are reduced in two irreversible one-electron steps; the trihalides are reduced in three one-electron steps, the first of which is quasi-reversible, the other two being irreversible. The first step in each case is adsorption-controlled while the others are diffusion-controlled. D.C. polarography is suitable for the determination of these compounds down to 5 x 10^{-5} M.

While all other compounds studied, including the mono-halides, showed little or no electroactivity, the di- and tri-halides gave fairly well-defined cathodic waves (Figures 72 and 73). The polarograms of the dihalides were practically identical (except for small differences in $E_{\frac{1}{2}}$ values), as were those of the trihalides. The diiodide exhibited a small wave ($E_{\frac{1}{2}}$ - 0.2 V vs the standard calomel electrode which was not observed with the other dihalides but it was not studied further. Each electroactive halide also gave a strong maximum of the first kind between - 1.6 V and - 1.8 V vs. the standard calomel electrode. The polarographic characteristics of the di- and tri-halides are summarized in Table 148.

This investigation showed that classical d.c. polarography, with a simple mercury pool in contact with tetramethyl ammonium chloride as reference electrode, is suitable for the determination of these compounds at concentrations from 10^{-1} M down to about 3 x 10^{-5} M. Lower concentrations could undoubtedly be determined by differential pulse polarography or by utilizing the adsorption of the compounds on the mercury drop through anodic stripping voltammetry.

Fleet and Fouzder[1840] and Fouzder and Fleet[1841] have reviewed the polarography of organotin compounds.

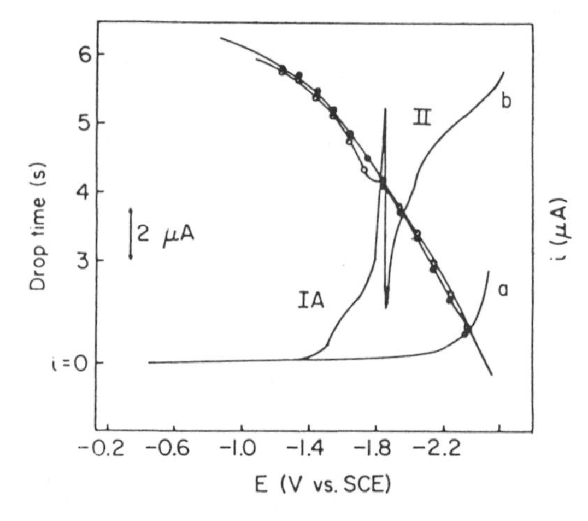

Figure 72 - Polarography of organotin dihalides.

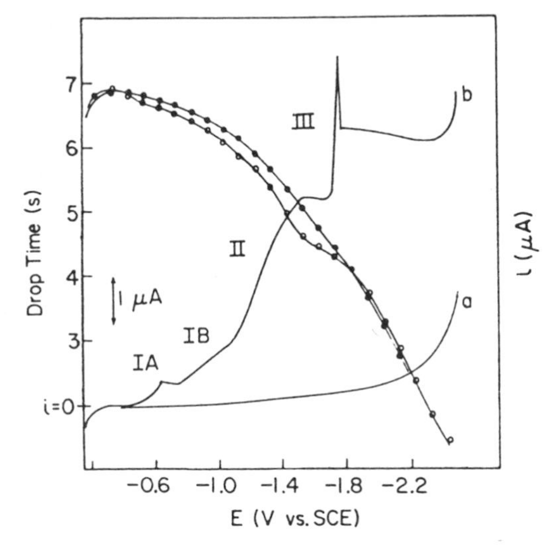

Figure 73 - Polarography of organotin trihalides.

Table 148 - Polarographic characteristics of the di- and tri-halide derivatives. (Solution concentrations, 5.0×10^{-4} M for dihalides and 1.8×10^{-4} M for trihalides. $E_{1/2}$ in V vs the standard calomal electrode; α is the transfer coefficient and n the number of electrons involved in the electrode process)

Compound	Wave 1			Wave 11			Wave 111		
	$E_{1/2}$	αn	n	$E_{1/2}$	αn	n	$E_{1/2}$	αn	n
$RSnMeCl_2$	-1.63[a]	0.54	1.05	-1.98[a]	0.62	N.A.	-	-	-
$RSnMeBr_2$	-1.61[a]	0.53	0.91	-1.97[a]	0.82	N.A.	-	-	-
$RSnMeI_2$	-1.54[a]	0.56	1.13	-1.90[a]	0.57	N.A.	-	-	-
$RSnCl_3$	-0.72[b]	0.45	1.10	-1.34[a]	0.46	0.99	-1.66[a]	(0.84)[c]	(0.98)[c]
$RSnBr_3$	-0.70[b]	0.44	1.09	-1.34[a]	0.52	1.00	-1.69[a]	(0.80)[c]	(1.46)[c]

[a] Irreversible process. [b] Quasi-reversible process. [c] Uncertainty in the value.

ORGANOZINC COMPOUNDS

The polarographic determination of zinc dialkyldithiophosphate in lubricating oils has been discussed[1842].

CHAPTER 6

GAS CHROMATOGRAPHY

ORGANOALUMINIUM COMPOUNDS

Longi and Mazzochi[1843] have reported a direct gas chromatographic analysis of organoaluminium compounds. To avoid sample decomposition they used a special sample injector. They purge the sample with helium carrier gas on a 1 metre column of Chromosorb W containing 7.5% of paraffin wax mixed with triphenylamine (17:3) at a column temperature of 73 - 165°C. A thermistor detector was used by these workers.

Bortnikov et al[1844] separated organoaluminium and organogallium compounds on a column containing silicone elastomer E-301 on diatomaceous brick operated at 110°C. They used helium at 100 ml per minute as a carrier gas and a dual kathorometer as a detector.

Brown and Mazdiyasni[1845] separated aluminium, germanium, silicon and titanium alkoxides on a 1 ft column of 1% Apiezon L on Chromosorb W packed in a Teflon tube. Operating temperatures were programmed in the range 60 - 150°C at 15°C per minute. The injection port and detector temperatures were respectively 203 and 270°C. To minimise moisture contamination, a glove bag was attached to the injection port and continuously purged with dry helium during operation of the chromatograph. These workers tried several solid supports and Chromosorb W and Gas Pack F gave equally good results. Best results were obtained with lightly loaded columns (1% liquid phase) using Apiezon L, Silicon Gum rubber SE-30 and Silicone Oil DC-200. Under these conditions, the group IV isoproxides eluded in the order of their volability, silicon in 45 seconds, germanium in 105 seconds, titanium in 255 seconds and aluminium isopropoxide in 405 seconds (Figure 74).

Lioznova and Genusov[1846] and Hagen et al[1847] have discussed gasometric methods utilizing gas chromatography for the estimation of organoaluminium compounds.

The microwave emission detector has been demonstrated to be useful for the detection of organoaluminium compounds[1848,1849]. Feldman and Batistoni [1850] have described a design of a glow discharge as a detector for the gas chromatography of organoaluminium compounds. This technique is described in detail in the section on arsenic.

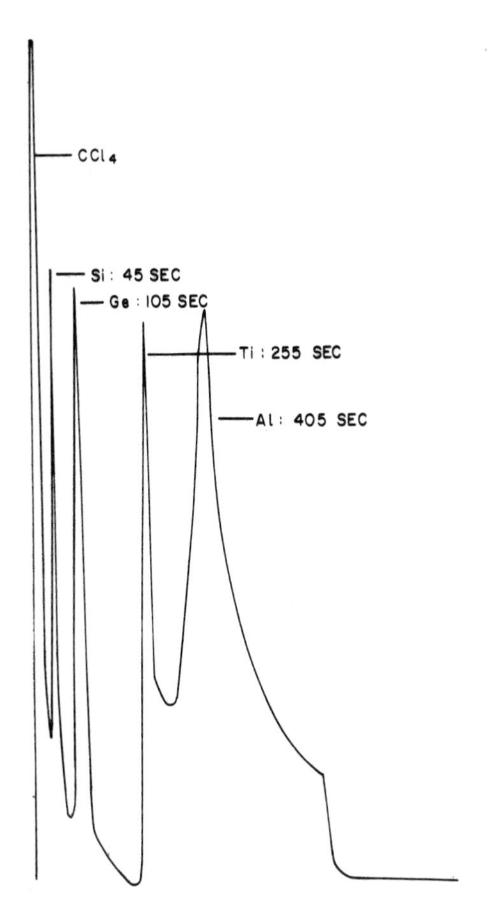

Figure 74 - Gas chromatography of mixed silicon, germanium, titanium and aluminium isopropoxides. Sample 2.5 µl mixed isopropoxides 1:2:4:4 in carbon tetrachloride. Teflon tubing 1 ft x 1/4 in. (o.d.) 1% Apiezon L on Chromosorb W (60 - 70 mesh). Injection port 203°C. Temperature programme: 60 - 150°C at 15°C per min. Upper limit interval 4 min.

ORGANOANTIMONY COMPOUNDS

Longi and Mazzocchi[1843] have described a gas chromatographic method for the analysis of organoantimony compounds. A special sample injector is used to avoid oxidation of the sample. Separation was achieved on a 1-m column of Chromosorb W containing 7.5% of paraffin wax (m.p. 63 - 64°C)-triphenylamine (17:3), and using dry purified helium as a carrier gas and a thermistor detector. Column temperatures ranged from 73° - 175°C, depending on the type of compound being determined.

ORGANOARSENIC COMPOUNDS

A. Alkyl and Aryl Arsines

Gudzinowicz and Martin[1851] applied gas chromatography to the separation of eight substituted organoarsines and substituted organobromarsines of the type RAsR'R" where R is an alkyl or aryl group, R' is an alkyl group, CF_3 or C_3F_7 and R" is CF_3 or C_3F_7, ranging in molecular weight from 156 to 306. They found that an almost linear relationship existed between log retention time and either the boiling point or the molecular weights of each component of a homologous series. Chromatography was carried out on a column (6 ft x 0.25 in.) of 5% of SE-30 silicone gum rubber on 80 - 100 mesh Chromosorb W; the column was operated at 290°C with argon at 40 ml per minute as carrier gas; an argon ionization detector was used.

Gudzinowicz and Martin[1852] used a modified Barker Coleman Model 10 argon ionization detector chromatograph (Gudzinowicz and Smith) for quantitative studies of organoarsenic and organo-bromo-arsenic compound mixtures. Using isothermal column operating conditions, the chromatograms for the separation of the arsenic derivitives investigated (Table 149) were obtained with a 6 ft by 1/4 inch o.d. stainless steel column packed with 5% by weight dimethylsilicone polymer (General Electric SE-30 silicone gum rubber) as liquid stationary phase on 80 to 100 mesh Chromosorb W.

Table 149 - Organo- and Organobromo-Arsines

Compound	Molecular weight	Boiling Point, °C
Dimethylbromoarsine	185	128 - 130/720mm
Methylethylbromoarsine	199	152 - 155
Methylbutylbromarsine	227	172 - 178/720mm
Tributylarsine	246	114/10mm
Trivinylarsine	156	130
Triphenylarsine	306	...
Methyldibromoarsine	250	179 - 181/720mm
Vinyldibromoarsine	262	74 - 76/14mm

Triphenylarsine (b.p. 360°C) was eluted in 4.2. minutes using the following higher temperature operating conditions:
column temperature, 290°C; flash heater, 340°C; detector temperature, 365°C; argon pressure, 30 p.s.i.; flow rate, 40 cc per minute.

Table 150 shows the relative retention time data (corrected for air time) for the separation of a five-component mixture containing dissimilar structures.

Table 150 - Retention Times of specific organo- and organobromo-arsines relative to methylbutylbromarsine

Compound	Relative Retention Time
Dimethylbromarsine	0.43
Methylethylbromoarsine	0.70
Methylbutylbromoarsine	1.00
Vinyldibromoarsine	1.53
Tributylarsine	2.13

Operating conditions: column temperature, 95°C; flash heater, 165°C; detector temperature, 195°C; cell voltage, 1250 v.; argon flow rate, 40 cc/min.

Figure 75 shows the effect of the alkyl chain length on the retention of organobromoarsines. A plot of the logarithm of the net retention time of these compounds vs. the boiling point of each component gives a nearly linear relationship, Figure 76. However, when the same operating conditions were used, methyldibromoarsine could not be resolved from methylbutylbromoarsine.

Parris et al[1853] have described a procedure utilizing a commercial atomic absorption spectrophotometer with a heated graphite tube furnace atomizer linked to a gas chromatograph for the determination of trimethylarsine in respirant gases produced in microbiological reactions. The overall GC-AA system used by these workers is illustrated in Figure 77. The apparatus consists of a dual column gas chromatograph with flame ionization detectors fitted with glass columns 1/8 inch o.d. by 6 ft long. Columns were packed with 5% SP-2100 (methyl silicone) and 3% SP-2401 (fluoropropyl silicone) on 80/100 mesh Supelcon AWDMCS support. The oven was programmed for isothermal operation at 40°C. Argon, flowing at 20 ml per minute, was used as carrier gas. Where required, hydrogen was added to the carrier gas with mixing accomplished by the hydrogen jet of the cold detector.

Parris et al[1853] investigated the efficiency of atomization (i.e. observed response per unit analyte) for trimethylarsine on two different tubes (bare graphite and silica-lined graphite) using inert (pure argon) and reactive (argon with 10% hydrogen) carrier gases. The results are shown in Figure 78. With inert carrier gas, very little atomization occurs at dial settings less than 1400°C on either surface. At the upper operating range on silica (1600°C) atomization is observed but the maximum sensitivity achievable is less than 2 x 10^5 µv.s/µg As and the steepness of the curve (Figure 78) in this temperature range suggests that variance in reproducing the furnace temperature of + 10°C would induce a 9% limit on precision. On bare graphite, not only can higher temperatures be routinely used, but also it is observed that, even with inert carrier gas, atomization begins at lower temperature than on silica.

Odanaka et al[1854] determined inorganic arsenic and methylarsenic compounds by gas chromatography and multiple ion detection mass spectrometry after conversion of the arsenic compounds to hydride using a heptane cold trap. The detection limit using 5.0 ml samples was 0.2 - 0.4 µg/ml. Relative standard deviations ranged from 2 to 5% for aqueous solutions spiked at the 10 ppb level.

B. Perfluoroorganoarsenic Compounds

The separation of perfluorated organoarsenic compounds has also been studied in detail by Gudzinowicz and Driscoll[1852]. In this work they used a Perkin-Elmer Model 143C Vapour Fractometer with helium as the carrier gas.

The chromatographic data for the separation of the arsenic compounds investigated (Table 151) were obtained using isothermal column operating conditions with two columns containing different stationary liquid phases and solid supports: (a) at 5½ foot by 1/4 inch o.d. copper column packed with 15% by weight dimethyl silicone polymer. (General Electric SE-30 silicone gum rubber) on Fluoropak 80 (Fluorocarbon Co.) and (b) a 25 foot by 1/4 inch o.d. copper column with 33% by weight Kel-F Wax 400 (Minesota Mining and Manufacturing Co.) on 30 / 60 mesh Chromosorb W (Johns-Manville Co.).

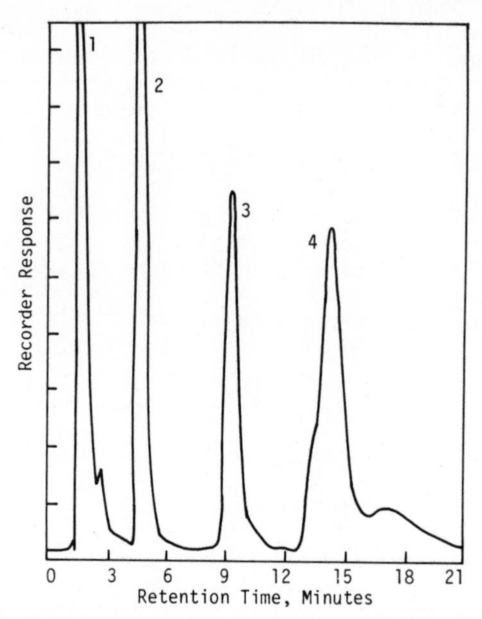

Figure 75 - Gas chromatogram of (1) methylene chloride, (2) dimethylbromoarsine, (3) methylethylbromoarsine, (4) methylbutylbromoarsine.

Operating conditions; column temperature, 68°C; flash heater, 132°C; detector temperature, 200°C; cell voltage 1250 v.; argon flow rate 44 cc/min.

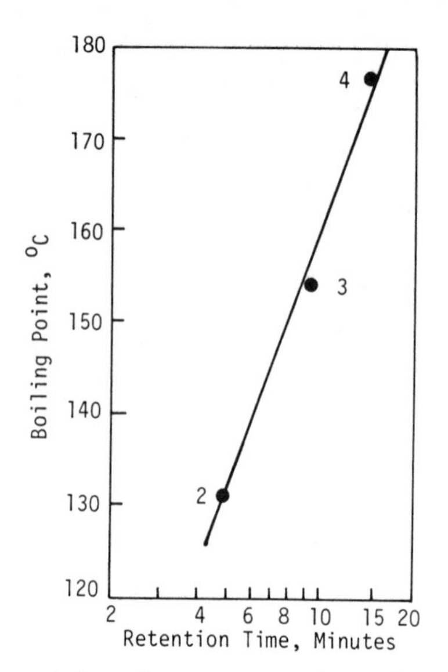

Figure 76 - Plot of logarithm of net retention time vs. boiling point. (2) dimethylbromoarsine, (3) methylethylbromoarsine, (4) methylbutylbromoarsine.

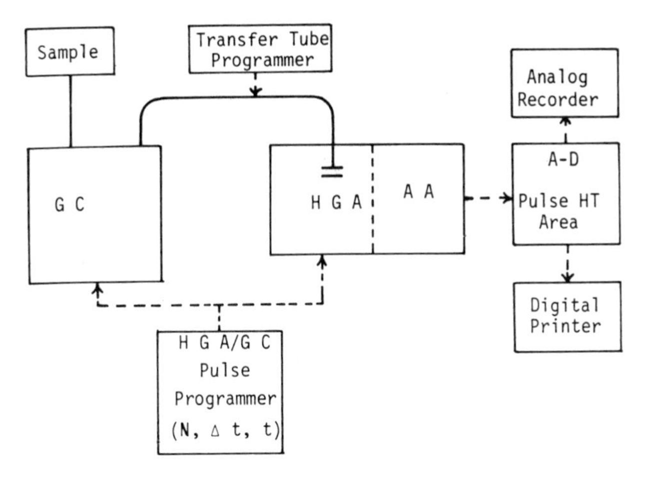

Figure 77 - Schematic diagram of the GC-AA system.

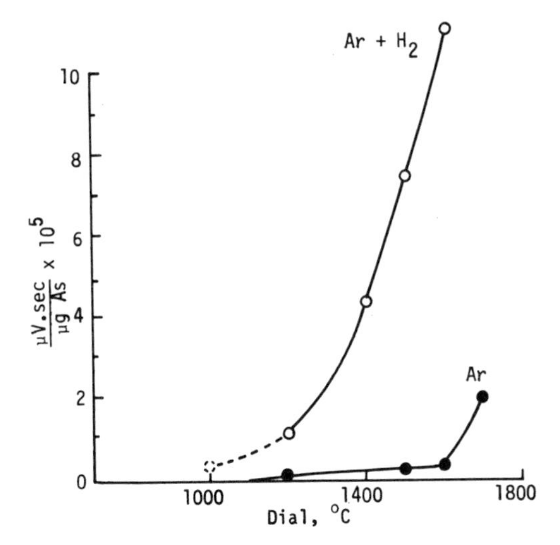

Figure 78 - Comparison of atomization efficiency curves for trimethylarsine in a silica-lined graphite furnace with and without hydrogen added to the carrier gas.

The results in Table 151 show relative retention volumes, corrected for void volume and pressure drop obtained for various arsines using different chromatographic columns and operating conditions. A comparison of retention volumes for several aryl arsenic derivatives at 200°C on both columns (conditions "C" and "D" in Table 151) indicates that these compounds can be separated at markedly decreased retention times with the shorter SE-30 column, provided, however, that diphenyl-perfluoromethyl- and diphenyl-fluoropropyl arsines are not present in the same mixture (Figure 79).

Table 151 - Relative retention volumes for several alkyl/aryl and perfluorinated organoarsines at various operating conditions (a)

Compound	Mol Wt.	B.P. °C	A	B	C	D
			Rel ret.vol.	Rel ret.vol.	Rel ret.vol.	Rel ret.vol.
$C_2H_5As(CF_3)_2$	242	77	0.51			
$(C_2H_5)\ AsCF_3$	201	112	1.06			0.33
$C_4H_9As(CF_3)_2$	270	118	1.00(b)			0.23
$(C_4H_9)_2AsCF_3$	258	186	1.64			
$C_6H_5{-}As(C_3F_7)_2$	490	128/68mm		0.65	0.66	0.84
$C_6H_5{-}As(CH_3)(C_3F_7)$	336	123/69mm		1.00(c)	1.00(d)	1.00(e)
$(C_6H_5{-})_2AsCF_3$	298				4.91	4.58
$(C_6H_5{-})_2AsC_3F_7$	398				5.16	

(a) Operating conditions:

A. 5½ ft column with 15% by weight SE-30 on Fluoropak 80, 100°C. Column temperature, 35 cc/min helium flow rate at Iatm and 25°C.
B. 5½ ft column with 15% by weight SE-30 on Fluoropak 80, 150°C. Column temperature, 25.4 cc/min helium flow rate at Iatm and 25°C.
C. 5½ ft column with 15% by weight SE-30 on Fluoropak 80, 200°C. Column temperature, 52 cc/min helium flow rate at Iatm and 25°C.
D. 25 ft column with 33% by weight Kel-F Wax 400 on Chromosorb W, 200°C. Column temperature, 88 cc/min helium flow rate at Iatm and 25°C.

(b) Retention time = 9.75 minutes

(c) Retention time = 22.40 minutes

(d) Retention time = 4.15 minutes

(e) Retention time = 14.42 minutes.

C. Detectors for organoarsenic compounds

The gas chromatography-microwave plasma detector technique has been applied to the analysis of alkyl arsenic acids in environmental samples [1856,1857].

Bramen et al[1858] investigated methods involving reduction to produce hydrides followed by separation and detection by an emission type detector for the analysis of organoarsenic compound.

Feldman and Batistoni[1850] modified the design of a glow discharge tube proposed earlier as an element specific detector for gas chromatography by Bramen and Dynako[1859] to overcome its principal drawback, namely the fact that it appeared to be subject to coating of the tube walls by decomposition products of the sample, thus attenuating the light signal as chromatographic peaks passed through the discharge.

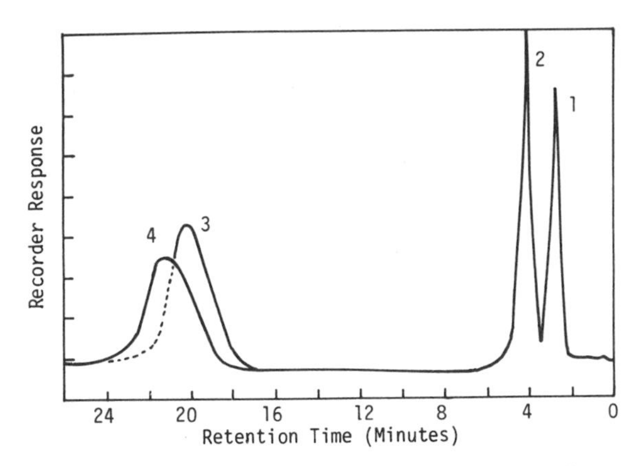

Figure 79 - Gas chromatogram of (1) $C_6H_5As(C_3F_7)_2$; (2) $C_6H_5AsCH_3C_3F_7$; (3) $(C_6H_5)_2AsCF_3$ and (4) $(C_6H_5)_2AsC_3F_7$.

Operating conditions: 5½ ft column packed with 15% w/w SE-30 on Fluoropak 80; column temperature, 200°C; helium gas carrier pressure 20 psi.; helium flow rate 52 cc/m.

Feldman and Batistoni[1850] also comment that several of the earlier workers who have developed optical emission detectors have remarked that spectral background correction would be beneficial, but did not perform such a correction. Even when used, the possible advantages of background correction have not always been fully enjoyed, because most of the devices proposed for the purpose have had a fixed wavelength interval between the line of interest and the position at which background intensity was measured. Inevitably occasions arise upon which these positions are occupied by other lines or bands, so that the simple background correction hoped for cannot be performed. They comment that this difficulty can often be circumvented by making the separation of the two observation points variable, as suggested by J. D. Defreese and H. V. Malmstadt[1860]. To prevent mechanical interference between the two photomultipliers, the spectrally resolved beam is split inside the monochromator, with the normal fixed slit at one exit and the moveable slit at the other. The splitting can be accomplished with a conventional beam splitting mirror[1860]. but less ultraviolet energy is lost if instead one replaces the existing mirror by a quartz-substrate mirror with a latticework coating pattern, or if one cuts away the upper half of the existing deflecting mirror, as was done in the present case. If the latter course is chosen, the aperture of the monochromator should be filled as completely as possible, in order to minimize the effects of the stirgmaticity of the optical system on the relative intensity of the beams.

Based on these principals, Feldman and Batistoni[1850] developed a simple helium glow discharge detector with a stable but inexpensive power supply to detect various metals (Al, As, Cr, Cu) also Pi, Si, C and S in gas chromatographic effluents. Improved glow chamber design prevents degradation products from coating the observation window. (Figure 80).

The optical system used by Feldman and Batistoni[1850] is depicted in Figure 81. A plano-convex fused quartz lens (75 mm focal length, 25 mm diameter) was placed at the slit of the Jarrell-Ash Ebert Mark V Monochromator (see Figure 81). An image of the glow discharge was focused on the centre of the entrance aperture of the collimating mirror. Interchange-

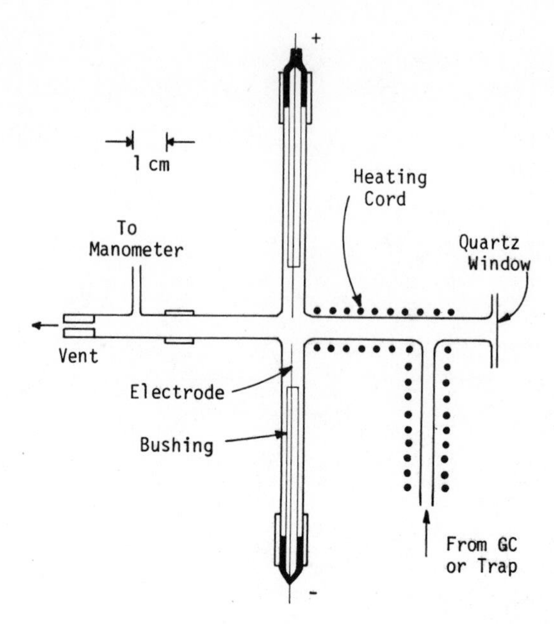

Figure 80 - Glow discharge detector (schematic).

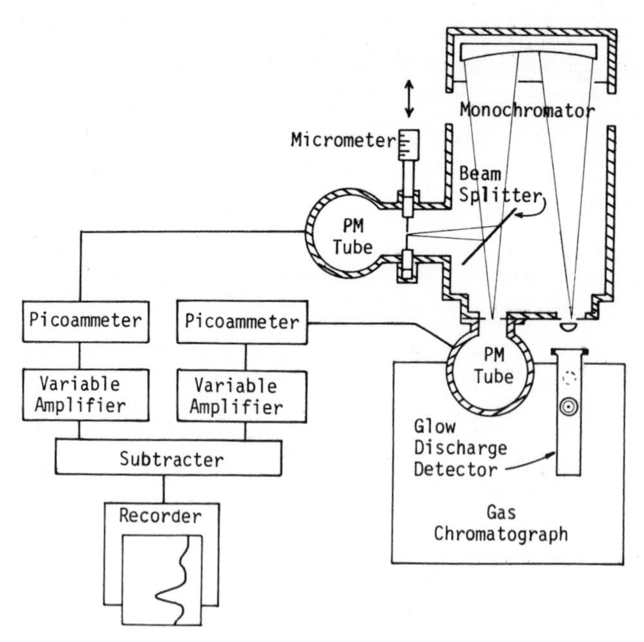

Figure 81 - Experimental arrangement (schematic).

able gratings blazed at 1900 Å (500 nm) were used for the ultraviolet and visible ranges of the spectrum, respectively. During measurements in the visible region, a yellow filter was placed at the slit in order to eliminate interference from second-order ultraviolet radiation. Entrance and exit slit widths were usually 50 or 70 μm (equivalent to 0.08 or 0.12 nm bandpass) respectively. The upper half of the internal deflecting mirror of a Jarrell-Ash Ebert Mark V 0.5 meter scanning monochromator was removed, so that the upper half of the diffracted and partially refocused exit beam went to the normal exit slit and the lower half to the side exit port (see Figure 81). The side exit port slit housing (Catalogue No. 82 - 018) was modified as follows. The dovetailed end section which normally holds the bayonet-type slit was milled away and replaced by a light-tight slit holder which could be moved 15 mm horizontally by means of a micrometer screw. Dovetailed ways inside the new holder located the slit in the original focal plane of the deflected beam. The horizontal (i.e. wavelength) limits of travel of this slit holder were adjusted so that a narrow wavelength band (e.g. 0.4 - 1 nm) could be observed through it at any wavelength within 20 nm of the line of interest on either side. The spectral displacement of this slit relative to the normal exit slit was calibrated with the aid of a low pressure mercury discharge lamp. Each exit channel was provided with an RCA IP28 or 4840 photomultiplier tube, a John Fluke Model 415B power supply (John Fluke Co., Mountlake Terrace, Wash. 98043), and a Gencom Model 1012 picoammeter (EMI-Gencom., Inc., Plainview, N.Y. 11803) to which a 7-stage output damping control was added. These signals were fed to a two-channel variable amplifier which amplified each signal 0- to 2.5-fold. The degree of amplification was controlled by a 10-turn potentiometer in each channel. The main channel potentiometer was usually set for maximum gain; the side exit channel potentiometer was set somewhere lower. The side-exit signal was then subtracted from the normal exit signal, and the difference signal fed to a Varian A-25 recorder. The side-exit signal was usually the background or reference signal, but gain-control potentiometers were provided in both channels of the amplifier in case it was ever desirable to reverse these roles.

Table 152 also tabulates limits of detection for arsenic and various other elements studied by Feldman and Batistoni[1850].

Feldman and Batistoni[1850] point out that the detection limit for a given element achieved by their system is governed by a number of factors. These included the characteristics of the glow discharge and its chamber, characteristics of the optical system outside and inside the monochromator, and those of the electrical detection/amplification system.

The selectivity of the technique was measured by Feldman and Batistoni 1850 by chromatography of a silylized mixture of phenylarsonic acid, nonanoic acid, undeconic acid, and three aliphatic hydrocarbons. The effluent was split; one stream was directed through a flame ionization detector and the other through the glow discharge detector. In one set of runs, the pass band of the monochromator was 228.12 - 228.20 nm (to detect Si 288.16 nm). In the other, the pass band was 228.77 - 288.85 nm (to detect As 228.81 nm). In each case a flame ionization detection trace was obtained simultaneously with the glow discharge detector trace. The selectivity of a given Si or As line was defined as the ratio of the peak height obtained per gram atom of carbon in the form of the interfering compound tested.

Figure 82 shows results obtained in the gas chromatography of a silylated mixture of aliphatic acids, phenylarsenic acid and hydrocarbons using a flame ionization detector and a glow discharge detector set at the silicon and arsenic wave lengths.

Table 152 - Limits of detection for various elements using the helium glow discharge detector (preliminary values)

Atom or molecule obsd	Compound used	Wavelength of line or bandhead, nm (4)	Detection limit of element[b] without background ng	correction ng/cm^3
P	Tributyl phosphite	213.62	6.4	0.13
P	Tributyl phosphate	213.62	18	0.4
Si	BSTFA ester or undecanoic acid	251.61	10	0.2
Al	Al trifluoroacetylacetonate	396.15	12	0.2
As	AsH_3	228.81	0.3[a]	0.001
Cr	Cr acetylacetonate	425.43	1.4	0.028
Cu	Cu hexafluoroacetylacetonate	324.75	110	2.2

[a] Observed by release of AsH_3, from a liquid nitrogen-cooled trap, not by gas chromatograph. Carrier gas velocity 600 cm^3/min.
[b] Quantity required to give a peak twice as high as the peak to peak base line noise.

D. Miscellaneous Organoarsenic Compounds

Ives and Guiffrida[1861] investigated the applicability of the potassium chloride thermionic detector and the flame ionization detector to the determination of organoarsenic compounds in the presence of organophosphorus and nitrogen-containing compounds.

Weston et al[1862] investigated the determination of arsanilic acid and carbarsone in animal feeding stuffs. The additives were extracted from the food with water; any carbarsone present converted to arsanilic acid which was then reduced to aniline which can be separated by steam distillation of the reaction mixture. The aniline was determined on a column of Versamid 900 containing 2% potassium hydroxide on Chromosorb W at 120°C using nitrogen as carrier gas and a flame ionization detector.

E. Arsine

Gas chromatography has been used (Iguchi et al[1863]) to determine arsine in hydrogen rich mixtures resulting from the reaction of arsenious oxide with zinc and dilute sulphuric acid. The arsine was detected on a column comprising dioctylphthalate on polyoxyethylene glycol as absorbant and using hydrogen as carrier gas. The limit of detection as arsenious oxide was 0.001 mg.

Zorin et al[1864] have described a method for the determination of down to 1.5×10^{-3} g/l. arsine in silane using a column (8mm x 5 mm) of alumina moistened with VK-Zh-94B silicone oil operated at 0° or down to 4.2×10^{-4} g/l. arsine in silane using a column (4M x 4 mm) of diatomic brick treated with PFMS-4F silicone oil operated at 30°C. Both procedures utilized dry nitrogen as carrier gas and a katharometer detector after passage of the dry gas issuing from the column through a furnace at 1000°C to decompose the arsine to hydrogen.

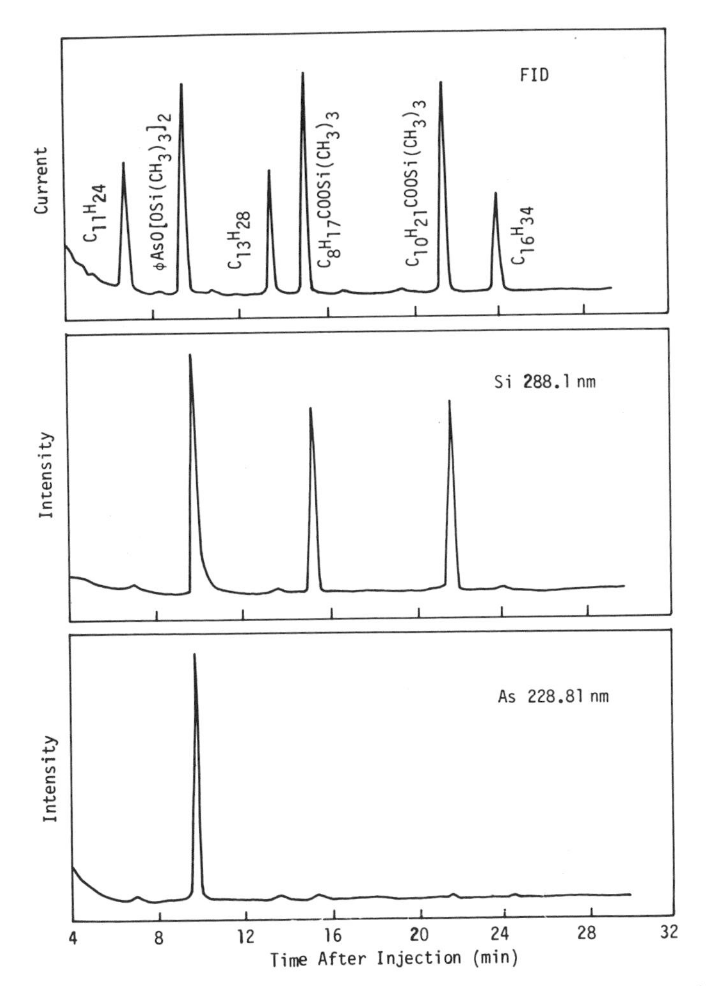

Figure 82 - Gas chromatograms of a silylated mixture of aliphatic acids, phenylarsonic acid, and hydrocarbons. Top curve: flame ionization detector (FID). Centre and bottom curves: glow discharge detector (GDD), with monochromator set as indicated.

Covello et al[1865] have described a procedure in which arsenic in toxiological samples is reduced to arsine which is then swept out of the reaction flask with a stream of helium carrier gas on to a column of silica gel at 50 to 90°C. Detection is achieved by thermal conductivity at the level of down to 2 μg arsenic.

Molodyk et al[1866] have described a gas chromatographic procedure for the determination of parts per million of methane, ethylene, acetylene and ethane in arsine.

Gifford and Bruckenstein[1867] separated and determined arsine, stannane and stibene by gas chromatography with a gold-gas-porous electrode detector. Detection limits for 5 ml samples were AsIII 0.2 ppb, SnII 0.8 ppb and SbIII 0.2 ppb.

ORGANOBERYLLIUM COMPOUNDS

Longi and Mazzochi[1843] describe the gas chromatography of the organometallic compounds of beryllium, zinc, boron, aluminium and antimony.

These workers claim that organoberyllium compounds can be chromatographed, using a special sample injection method and a 1 metre column of Chromosorb W containing 7 5% of paraffin wax (M.63 - 64°C) - triphenylamine (17:3), employing dry purified helium as a carrier gas. They used column temperatures between 73 and 165°C according to the compound being determined. A thermistor detector was used. Using this procedure separation was achieved of diethyl beryllium and triethyaluminium.

The microwave emission detector has been demonstrated to be useful for the detection of organoberyllium compounds[1848].

ORGANOBORON COMPOUNDS

A. Boron Alkyls

Although they are not classified as organoboron compounds, boron hydrides, chloroboranes, boron halides and boric acid esters are discussed in the first part of this chapter as also is the determination of hydridic and active hydrogen in these types of compounds. Any worker dealing with reactions mixtures containing organoboron compounds would be likely to be interested in methods for determining these substances which are intimately tied up with either the starting materials used in the synthesis of organoboron compounds or could occur as side products in the reaction mixture.

One of the early papers on the gas chromatographic analysis of alkyl boron compounds was that of Schomberg, Koster and Henneberg[1868]. They describe the analysis of mixtures of boron alkyls ranging from triethylboron to tri-n-propylboron. A 1 metre column packed with silicone oil on Celite at a carrier gas flow rate of 100 ml/min. at 80°C separated seven compounds in 13 minutes. A very similar analysis of boron alkyl compounds by Koster and Bruno[1868] is shown in Figure 83. This figure illustrates the usefulness of gas chromatography in separating the complex mixtures encountered, in this case, in the product obtained in a catalytic exchange reaction between trialkyl boron and trialkyl aluminium.

Longi and Mazzochi[1843] describe a method for chromatography using a thermistor detector of organoboron compounds on a 1 metre column of Chromosorb W containing 7.5% paraffin wax (melting point 63 - 64°C) - triphenyl amine 17:3 using dry pure hydrogen as the carrier gas and a column temperature between 73°C and 165°C, depending on the type of compound.

Kuhns et al[1870] compared methods for the determination of decaborane in commercial samples in the 90 - 97% purity range. Gas chromatographic, infra-red, ultra-violet, iodine titration and iodometric procedures gave comparable results at the 95% confidence level. Decaborane was determined by these methods without significant influence by the impurities present.

The gas chromatographic analyses were performed on a 3-metre 3/8 inch column packed with 60 - 80 mesh Celite impregnated with 20 weight % Apiezon L. It had a column efficiency of 12.00 theoretical plates; the retention time for decaborane relative to n-decane was 2.65 and to naphthalene, 0.730.

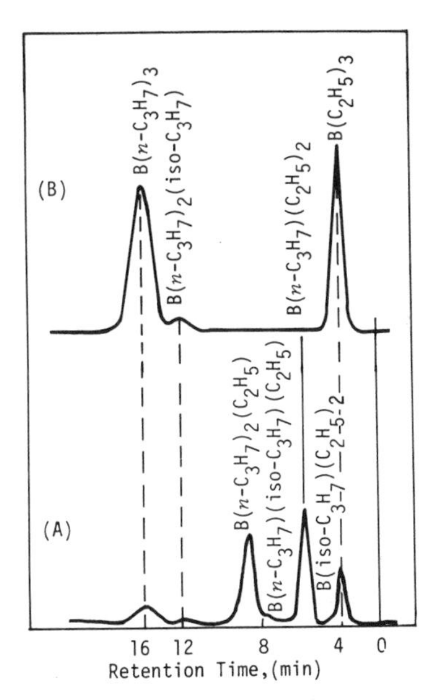

Figure 83 - Gas chromatograms demonstrating catalytic alkyl exchange between boron trialkyl and aluminium trialkyl. (1) Initial mixture of boron triethyl-boron tri-n-propyl. (2) Mixture after standing with approximately 5% triethylaluminium and immediately hydrolysed.

The helium flow rate was 340 ml per minute and the column and detector temperature was 150°C. Cyclohexane was used as a solvent for the decaborane. Alternate possible conditions include temperatures ranging from 90° - 220°C. Squalane, Apiezon L or M, silicone grease and Flurolube are suitable partitioning liquids. For decaborane samples the recommended conditions are: a 0.5-metre column (1/4 inch diameter) packed with 100 - 120 mesh Celite impregnated with 20 weight % of squalane. The operating temperature should be about 140°C and the helium flow rate 50 cc per minute. Naphthalene is recommended as an internal standard, and this method is recommended when impurities are present which interfere with other methods.

Seely et al[1871] and Parsons et al[1872] determined the specific retention volumes (Vg), as a function of temperature for diborane, trimethylborane, and the methyldiboranes in columns packed with mineral oil on pulverized firebrick, and operated at 0°C. The chromatographic apparatus used by these workers consisted of two columns containing a packing made of mineral oil on 32 - 65 mesh Johns-Manville firebrick in the weight proportions 38 - 100 respectively. The short column contained 8.69 grams of packing in a volume 55 cm long x 0.6 cm in inner diameter; the long column contained 8.90 g of packing in a volume 125 cm long x 0.4 cm in inner diameter. A typical chromatogram obtained with the long column is shown in Figure 84.

In Figure 85 are shown the specific retention volumes (Vg), at several temperatures for trimethylborane and borane and the five methyldiboranes. Not shown are the values for diborane which appear invariant with the temperature from the average 0.22 ± 0.02 ml per gram. Values for trimethyldiborane and tetramethyldiborane were obtained on the short column with dimethyldiborane included as a standard; the others were measured on the

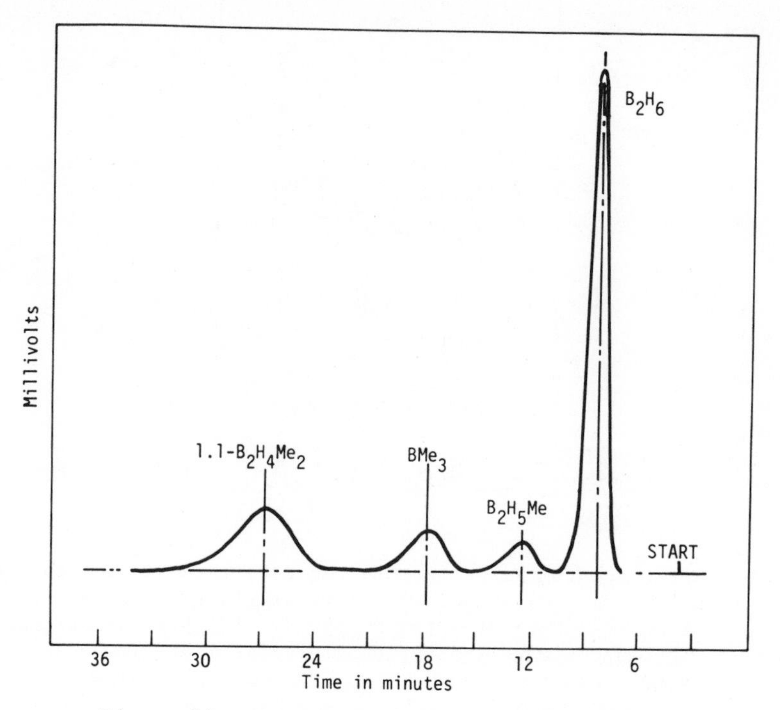

Figure 84 - Gas chromatogram of boron alkyls.

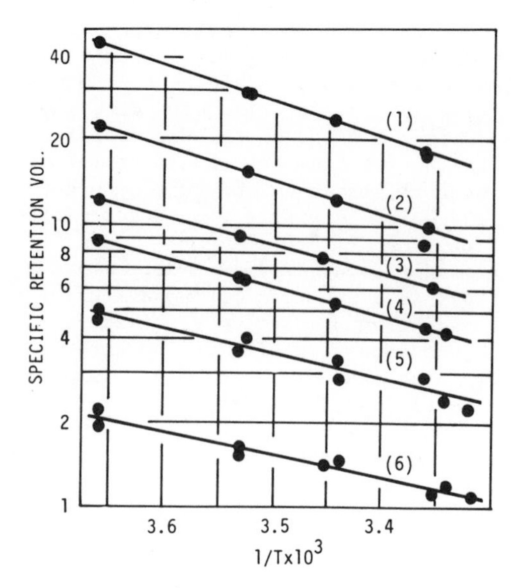

Figure 85 - Specific retention volumes, (1) $B_2H_2Me_4$ (2) $B_2H_3Me_3$ (3) 1,2-$B_2H_4Me_2$ (4) 1,1-$B_2H_4Me_2$ (5) B Me_3 (6) B_2H_5Me.

long column. The linear relations are described by the equation:

$\log Vg = \frac{A}{T} - B$ the constants for which are found in Table 153.

Table 153 - Constants for equations in log Vg and I/T.

Substance	A	B
Monomethyldiborane	0.825	2.698
Trimethylborane	0.898	2.595
1,1-Dimethyldiborane	1.070	2.964
1,2-Dimethyldiborane	1.057	2.771
Trimethyldiborane	1.213	3.085
Tetramethyldiborane	1.302	3.112

B. Borazoles

Semlyen and Phillips[1873] used an empirical method to calculate the retention times at 1,3,5,-trialkylborazoles. Retention parameters for individual alkyl groups were calculated from the log t_R (logarithm of the individual retention times relative to mesitylene log t_R = 2) values of the symetrical 1,3,5,-derivatives by subtracting the log t_R of borazole (0.54) and dividing by 3. The results of applying these individual retention time values to the calculation of log t_R of some mixed 1,3,5,-borazoles are shown in Table 154 and compared to experimental values. The agreement is good. It is better when the difference in size of the alkyl groups in the mixed derivative is small. The columns used contained 13% Carbowax 400 at 100°C using a flame ionization detector and oxygen-free hydrogen as carrier gas. The solid support was Embacel (May and Baker, 60 - 100 mesh, acid washed) made inactive by treatment with hexamethyldizilazane. This technique permits a good estimation for retention times of mixed alkyls when standards are not available.

Table 154 - Logarithm of retention times for boron alkyls. Difference between observed and calculated values

	Difference		Difference
c = 1*			
B-Et_2-(n-Pr)	- 0.02	B-Me-Et_2	+ 0.01
B-Et-$(n\text{-}Pr)_2$	- 0.01	B-Et_2-(i-Pr)	+ 0.01
B-H-Me_2	+ 0.01	B-Et-$(i\text{-}Pr)_2$	+ 0.01
B-Me_2-Et	+ 0.01		
c = 2*			
B-Me_2-(n-Pr)	+ 0.03	B-H_2-Et	+ 0.07
B-Me-$(n\text{-}Pr)_2$	+ 0.03	B-H-Et_2	+ 0.07

* c is the difference in carbon number between the radicals.

The corresponding values for alkyl groups attached to the benzene ring were found by subtracting the log t_R of benzene (0.85) and dividing by 3, (Table 155).

In general, the retention parameters of B-alkylborazoles are closer to those of alkylbenzenes than to N-alkylborazole values. The difference

Table 155 - Retention parameters for alkyl groups.

	H	Me	Et	Pr^n	Pr^i	Bu^t	Bu^s	Bu^i
Tri-N-alkylborazoles	0.00	0.33	0.50	0.77	0.65	0.85	0.88	0.94
Tri-B-alkylborazoles	0.00	0.32	0.65	0.90	0.77	-	-	-
Trialkylbenzens	0.00	0.38	0.63	(0.87)*	0.72	(0.95)*	(0.98)*	(1.04)*

* The trialkylbenzene Pr^n, Bu^t, Bu^s and Bu^i parameters were estimated from the tri-N-alkylborazole values.

between values for B- and N-alkylborazoles may be related to the free N - H groups (of polar character) in the former compounds so that N-substituted values are about 0.13 less than B-substituted values or 0.10 less than the aromatic values. On this basis the methyl group parameters are anomalous, low for B-methyl- and high for N-methyl-borazoles. The high N-methyl value may be related to an effect observed by James and Martin[1874], the unusual retardation of the tertiary amine, trimethylamine, in the ethereal column liquid "Lubrol MO". They attributed this retardation to "active", methyl groups of the amine forming hydrogen bridges with oxygen in the ether. This conclusion is supported by the unusually high boiling point of trimethylamine relative to other tertiary amines and alkyl analogues of both boron and carbon. The alkyl group retention parameters may be used to calculate the log t_R values of mixed 1,3,5,-trialkyl derivatives.

Semlyen and Phillips[1873] calculated the log t_R values of methyl- and ethyl-borazoles, substituted on both boron and nitrogen atoms of the ring from retention parameters in the same way as the unsymmetrical trialkyl derivatives. Table 156 lists the numerical difference in the observed and calculated log t_R values.

It can be seen that alkyl groups on adjacent atoms make retention times of these borazoles very different from those calculated from the simple alkyl group parameters. Thus hexaethylborazole has a retention time less than half that calculated, and hexamethylborazole a retention time twice that calculated. However, an underlying pattern is again apparent, indicating perhaps once again that alkyl groups attached to the borazole ring are affecting the retention times in a regular manner.

Table 156 - Differences between observed and calculated log t_R values of methyl- and ethyl-borazoles.

	$B-Me_3$	$B-Me_2$	B-Me	B-Et	$B-Et_2$	$B-Et_3$
$N-Me_3$	+ 0.30	+ 0.21	+ 0.11	+ 0.05	+ 0.05	+ 0.02
$N-Me_2$	+ 0.18	-	-	-	+ 0.03	+ 0.04
N-Me	+ 0.08	-	-	-	-	+ 0.01
N-Et	+ 0.07	-	-	+ 0.10	+ 0.05	- 0.03
$N-Et_2$	+ 0.10	+ 0.05	+ 0.07	+ 0.08	- 0.04	- 0.12
$N-Et_3$	+ 0.13	+ 0.07	+ 0.03	+ 0.07	- 0.19	- 0.35

Similar calculations of the log t_R values of methylbenzenes measured by Chang and Karr[1875,1876] give positive deviations of the same order of magnitude as found for methylborazoles (Table 157), these deviations are again reflected in boiling point values.

Table 157 - Differences between observed and calculated log t_R values and boiling points of methyl-benzenes

	1,2,3,5-Tetramethyl-benzene	Pentamethyl-benzene	Hexamethyl-benzene
log t_R, observed - calculated value	+ 0.12	+ 0.25	-
Boiling Point, observed - calculated value (°C)	+ 3.1	+ 9.8	+ 14.7

Semlyen and Phillips[1873] studied the effect of alkyl groups on the retention times of trialkylborazoles on Carbowax 4000. The specific retardations of borazoles with free N - H groups on Carbowax 4000 have been discussed by other workers in the case of N-alkylborazoles[1877], methyl-borazoles[1878] and ethylborazoles[1879] suggested that hydrogen bonding occurred between N - H groups and oxygen atoms in the column liquid and that "steric effects" of B-alkyl groups might influence the retardations of borazoles with free N - H groups.

The retention behaviour of tri-B-alkylborazoles on Carbowax 4000 is of most interest in this respect, for with three N - H groups "sterically hindered" by a range of adjacent alkyl groups, they form a very convenient system for such a study. A useful method of comparing the behaviour of compounds in two different liquid phases is to plot the log t_R values against each other, for when this is done, the vertical distance between a compound and the paraffin line on the graph gives a measure of its specific retard-ation in the polar phase. Figure 86 shows a plot for sixteen tri-B-alkyl-borazoles. The distance of the symmetrical tri-B-alkylborazoles from the paraffin line were divided by 3, to give four distances representing (if the theory is correct) the effect of each alkyl group on the retardations due to the free N - H groups. In log t_R units these are: Me 0.29; Et 0.22= Pr^n 0.19; Pr^i 0.14.

As expected the "steric effects" are in the order Pr^i>Pr^n>Et>Me. When the distance of mixed tri-B-alkylborazoles from the paraffin line are compared with those calculated by simply adding the individuals alkyl group values, very good agreement is obtained (Table 158). Thus, if the log t_R values of tri-B-alkylborazoles are known in squalane, their retention times in Carbowax 4000 can be quite accurately predicted.

Table 158 - Retardations of tri-B-alkylborazoles on Carbowax 4000

Tri-B-alkylborazole	Distance (log t_R units)		Tri-B-alkylborazole	Distance (log t_R units)	
	(obs)	(calc)		(obs)	(calc)
$EtPr^iPr^i$	0.49	0.50	$MePr^mPr^n$	0.68	0.67
$MePr^iPr^i$	0.56	0.57	$MeEtPr^n$	0.71	0.70
$EtEtPr^i$	0.57	0.58	$MeMrPr^i$	0.71	0.72
$EtPr^nPr^n$	0.61	0.60	MeEtEt	0.74	0.73
$EtEtPr^n$	0.65	0.63	$MeMePr^n$	0.78	0.77
$MeEtPr^i$	0.66	0.65	MeMeEt	0.80	0.80

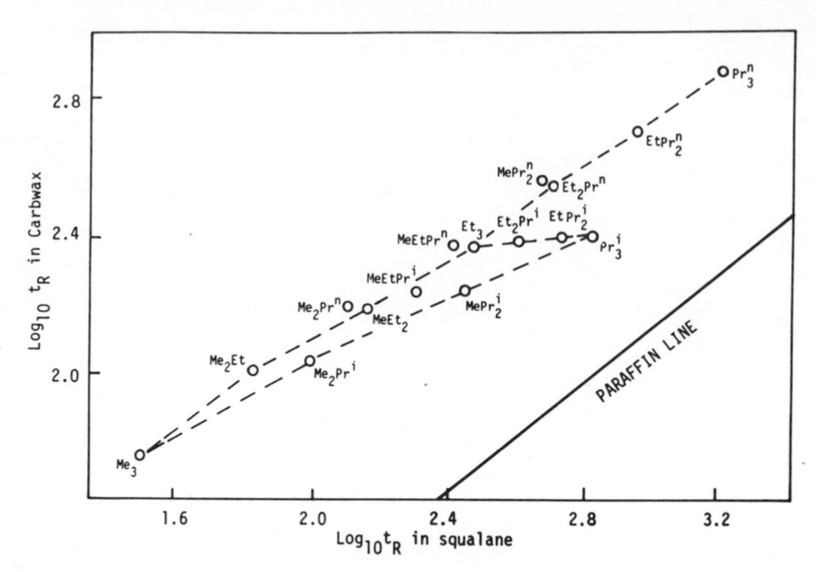

Figure 86 - Comparison of tri-boran alkylborazole retention times on Carbowax 4000 and on squalane.

C. Boron Hydrides

Kaufman et al[1880] discuss the application of gas chromatography to the analysis of boron hydrides. Using as a column packing Celite coated with paraffin oil (Octoils) or with tricresyl phosphate, they resolved diborane, tetraborane and pentaborane without decomposition on the column. Extensive decomposition occurred on the column, however, in the case of dihydropentaborane.

Kaufman et al[1880] incorporated a vacuum system in the gas chromatographic apparatus (Figure 87), because boron hydrides must be handled in the absence of air. Samples were introduced into the apparatus through a removable, calibrated sample bulb, filled to the desired pressure with a standard gas or a mixture. A vacuum connection was provided to provide for the removal of air before the sample was introduced. The sample was distilled through the inlet into one arm of a bypass system situated at the inlet of the gas chromatographic column. Helium was used as carrier gas. Helium from the tank (Figure 87) flows through a T-joint both into the reference side of a thermal conductivity detector cell and past a flowmeter through the chromatographic column to the sensing side of the detector cell. A series of traps after the sensing side of the cell are used independently or together to collect the fractions from the chromatographic seperation. During a run, liquid nitrogen is kept around the collecting traps. Helium escapes through a valve consisting of a glass frit covered with mercury. This valve minimizes pressure and flow fluctuations. A thermal conductivity cell employing thermistors was used for detection.

The chromatography columns are made of copper tubing 1/4 inch in outside diameter and 10 feet long. They are packed with keiseiguhr (Celite 545 (between 60 and 100 mesh). The particles are coated with 30% of various high boiling liquids as stationary phase solvents: paraffin oil, Octoil S, and tricresyl phosphate. This affords a good range of nonpolar to polar solvents.

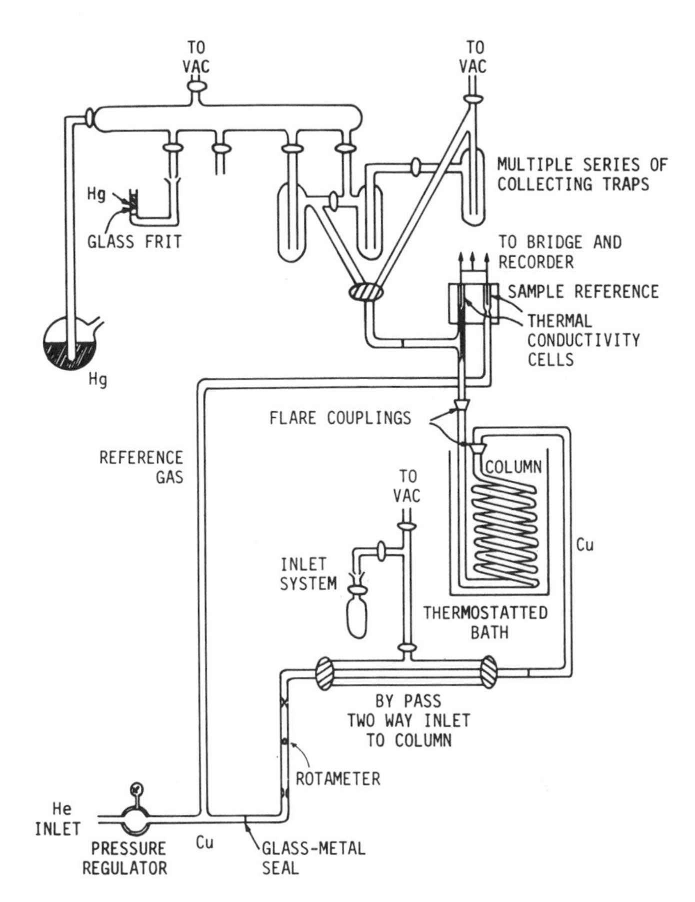

Figure 87 - Schematic diagram of gas chromatography apparatus and vacuum system.

After a chromatographic column is packed, and connected in the apparatus, the column is evacuated and degassed for several days under vacuum at a temperature higher than that at which the runs are made. Helium is swept through the system and the entire system again evacuated. The chromatographic apparatus is then isolated from the vacuum system and permitted to stand overnight. The pressure is checked after 15 hours with a McLeod gage; if it is better than 10^{-1} mm, the apparatus is considered fit for use with the boron hydrides.

To calibrate the procedure, a pure sample of each boron hydride is individually run through every column tested, in order to obtain its retention time under the operating conditions of the column.

Figure 88 shows a gas chromatogram obtained by Kaufman et al[1880] of a mixture of diborane, tetraborane, pentaborane and dihydropentaborane. Whilst the first three compounds were resolved on the Celite-paraffin oil column with almost no decomposition, a different but interesting behaviour was exhibited by dihydropentaborane (B_5H_{11}). When a small pure sample of this compound was put through the gas column, virtually complete conversion into tetraborane and diborane resulted. These compounds were trapped out after

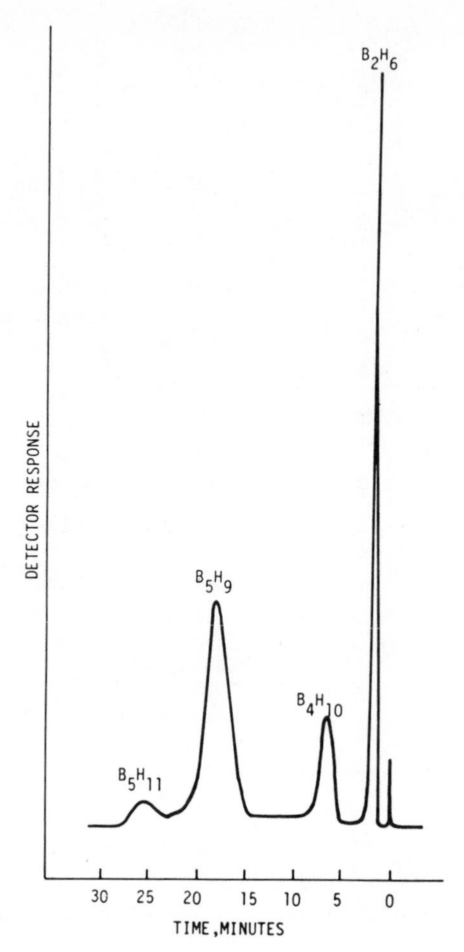

Figure 88 - Gas chromatogram of boron hydride mixture. Celite-paraffin oil column at 27°C. Helium flow rate 111 ml per minute.

emerging from the column and identified by means of their infra-red spectra.

Of the three stationary phase liquids used by Kaufman et al[1880] in preparation of the chromatographic columns, paraffin oil proved best for separation of the boron hydrides themselves. The retention times for the boron hydrides were longest on the Celite-tricresyl phosphate column, and shortest on the Celite-paraffin oil column. The peaks were well resolved on all columns in all cases, except where dihydropentaborane decomposition occurred.

Seely et al[1871] describe a gas chromatographic method for the analysis of mixtures containing methyldiboranes. They showed that mixtures of methyldiboranes can be almost completely resolved and determined on chromatographic columns of mineral oil on crushed firebrick, operating at 0°C. A quantitative determination can be carried out by area measurement and is accurate to between 1 and 2% of the components present in a mixture.

Schomberg et al[1868] studied the combination of gas chromatography with mass spectrography for the analysis of mixtures of alkyl boranes. They report on experimental conditions and on results obtained on samples of trialkylboranes.

Koski et al[1881] have studied the use of pyrolysis followed by the use of thermal conductivity cells for the determination of deuterium in deuterated boron hydrides, their organic derivatives and nitrogen compounds. The volatile boron compounds were first passed over with hot uranium metal (500 - 800°C) which pyrolysed various compounds with a recovery between 95% and 100% of hydrogen and deuterium. This gas mixture was then analysed using a thermal conductivity cell which had been previously calibrated against standard mixtures of hydrogen and deuterium.

Sawinski and Suffet[1882] applied a Melpar flame photometric system to the determination of boron hydrides, the detector was fitted with an optical filter having a maximum transmittance of 546 nm. The column packing was OV-1 on Chromosorb W. Down to 0.7 ng decaborane could be detected with this equipment.

Zorin et al[1883] applied gas chromatography to the determination of organic impurities in diborane. The bulk of the sample was removed on a precolumn and the impurities were then determined on a squalane or liquid paraffin column. Nitrogen was used as carrier gas and a flame ionization detector was used.

D. Chloroboranes

Myers and Putnam[1884] describe a gas chromatographic separation of diborane, chloroboranes (B_2H_5Cl, $BHCl_2$ and Bcl_3) and hydrogen and hydrogen chloride on low temperature (- 78°C) columns containing powdered Teflon, silicone oil 703, Flurolube GR 362, Kel F oil, liquid paraffin or hexadecane, It was apparent to these workers that to find a single partition column to resolve all of the components would be unrealistic. Thus, analytical requirements were successfully met through the development of several different columns with specific applications. Their procedure is described below.

Because of the volatility, reactivity and toxicity of these materials, the gas chromatography unit was constructed as an integral part of a borosilicate glass high-vacuum system of conventional design. With this arrangement, purified samples from fractionation traps were introduced directly to the chromatography sample loop, unknown gas sample mixtures were admitted to the system through evacuated lines, and emerging chromatograph peaks were frozen selectively in vacuum-line fractionating traps for later identification. A section in the glass system was provided for quantitative aqueous hydrolysis and measurement of the resulting products: hydrogen, boric oxide and hydrogen chloride. With stopcocks in the vacuum system lubricated by the low vapour pressure fluorocarbon greases, the entire apparatus could be routinely evacuated to 10^{-5} mm of Hg.

Relative retention volumes obtained for the components on various columns are given in Table 159.

Myers and Putnam[1884] applied their procedure to samples of diborane containing hydrogen, hydrogen chloride and monochlorodiborane. In Figure 89 is shown a gas chromatograph obtained from this mixture using a column comprising 60 - 80 mesh Chromosorb coated with 20% wt silicone oil at 0°C with a helium flow rate of 100 ml per minute. Although the carrier gas used was helium, positive hydrogen peaks were invariably recorded and precise

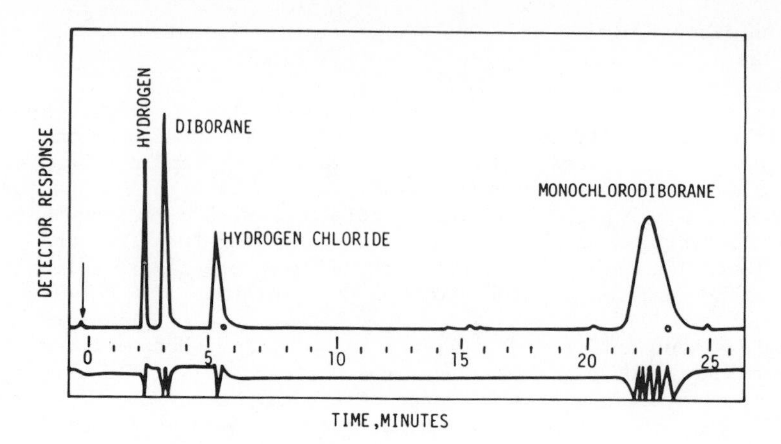

Figure 89 - Gas chromatographic separation of diborane and monochlorodiborane on a 0°C silicone oil column.

calibrations were obtained. Apparently the hydrogen segment leaving the 0°C column becomes warmer than the helium stream before it reaches the ambient temperature detector. Subsequently, the temperature of the hydrogen becomes more important than thermal conductivity properties. At the thermistor, then, less heat is conducted to the hydrogen segment than is lost to the cold helium flow, and a positive peak results. Once a constant helium flow is established, the temperature gradient between column exit and detector remains constant. A chromatogram illustrating the use of a longer 0°C silicone oil column (40 feet) for trace analysis of hydrogen chloride in diborane is given in Figure 90.

Table 159 - Relative retention volumes

Component	Column			
	Teflon - 78°C	Silicone Oil 0°C	Flurolube 25°C	n-Hexadecane 40°C
H_2	-	0.46	0.95	-
B_2H_6	-	0.62	-	0.97
HCl	1.00	1.00	1.00	1.00
B_2H_5Cl	-	4.05	-	2.65
BHCl	1.81	-	-	4.70
BCl_3	4.54	-	2.72	6.30

The quantitative results obtained by Myers and Putnam [1884] for a representative number of experimental mixtures are given in Table 160.

E. Boron Halides

Pappas and Milliou[1885] have described a gas chromatographic technique for the examination of mixtures of corrosive fluorides such as boron-, tri- and pentafluoride, free chlorine, chloryl fluoride, dichlorotetrafluoroethane, hydrofluoric acid and uranium hexafluoride. These workers used columns constructed of Teflon packed with a Teflon support and using either Kel-F or fluorocarbon oil as liquid phase. The technique was applied to halocarbon coolant analysis and in uranium fuel recovery.

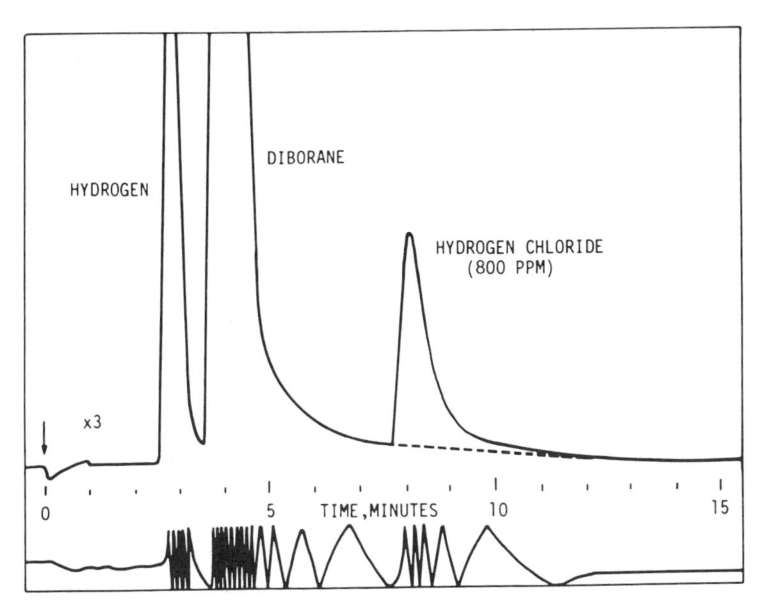

Figure 90 - Gas chromatographic separation of diborane and hydrogen for a low hydrogen chloride concentration: 0°C silicone oil column.

Table 160 - Quantitative determinations for experimental mixtures

	Known, Mole %						Found, Mole %					
	H_2	B_2H_6	Hcl	B_2H_5Cl	$BHCl_2$	BCl_3	H_2	B_2H_6	HCl	B_2H_5Cl	$BHCl_2$	BCl_3
Teflon	-	-	1.2	-	5.0	93.8	-	-	1.4	-	4.7	92.0
	-	-	4.0	-	2.9	93.1	-	-	3.7	-	3.0	95.6
	-	-	0.50	-	9.6	89.9	-	-	0.54	-	9.1	88.0
Silicone Oil	89.8	-	-	10.2	-	-	91.1	-	-	9.8	-	-
	7.2	11.0	-	81.2	-	-	7.3	11.7	-	80.7	-	-
	9.8	86.7	3.5	-	-	-	10.3	86.1	3.5	-	-	-
Fluoro-lube	-	-	75.1	-	-	24.9	-	-	74.2	-	-	25.2
	-	-	2.6	-	-	97.4	-	-	2.7	-	-	95.7

Brazhnikov and Sakodynskii[1886] determined the retention time of boron trichloride on various supports and stationary phases and compared its behaviour to that of boronalkyls. Zelyaev et al[1887] used a molybdenum glass column packed with poly (methylsiloxane) on Spherchrom 1 to separate in amounts down to 0.1 ppm of various trace impurities in boron trichloride, such as chlorine, phosgene, silicon tetrachloride, chloroform, carbon tetrachloride and dichloroethane. They used a column temperature of 60°C with nitrogen as carrier gas and employed thermionic and flame ionization detectors.

Dazard and Mongeot[1888] separated boron trifluoride from difluoroborane by gas chromatography at low temperature. The column used was 50% Kel-F of a Chromosorb W at - 196°C. Hydrogen was used as carrier gas.

Li Lorenzo[1889] carried out a quantitative gas chromatographic analysis of mixtures of boron trichloride, boron and boron nitride i.e. the gas products of partial decomposition of boron trichloride in a nitrogen plasma jet. The gaseous products of partial decomposition of boron trichloride in the plasma jet were analysed on a PTFE column (2 metres x 5 mm) packed with 10% of Kel-Fe on Teflon 6 and operated at 40°C, with helium (containing approximately 1% of boron trichloride to prevent irreversible absorption of the small amount of boron trichloride to be detected) as carrier gas (40 ml per minute) and thermal conductivity detection. The solid products (boron and boron nitride) were exposed to chlorine at 500°C for 1 minute in a specially designed reaction-injection system, and the reaction products were swept on to the column through a gas-sampling valve.

F. Methyl Borate

Toshiyuki and Matsuda[1890] discuss the gas chromatography of methyl borate. The methyl borate sample is dissolved in benzene dilutent containing cyclohexane internal standard. Using a column (1 metre x 5 mm) of Silicone DC - 550 on Celite 545 (60 - 100 mesh) and nitrogen (200 ml per minute) as carrier gas at 65°C methyl borate gives a peak that is separated from those of the internal standard and the solvent.

G. Boronates

Wood and Siddiqui[1891] studied the gas chromatography of the butane boronates of carbohydrates and their trimethylsilyl ethers. The carbohydrate (15 mg) was dissolved in pyridine (1 ml) at 60°, butaneboronic acid (25 mg) was added, and the mixture was set aside for periods of up to 4 hr. Hexamethyldisilazane (0.2 ml) and chlorotrimethylsilane (0.1 ml) were then added. The derivatives were separated on glass columns (9 ft x 0.25 in.) packed with 3% of ECNSS-M on Gas-Chrom Q temp.-programmed from 100° at 2° per min, and operated with nitrogen as carrier gas (60 ml per min) and flame ionisation detection. Mixtures of various pentoses and hexoses gave single peaks for each sugar, but D-ribose, t-rhamnose, erythritol, D-arabinitol, xylitol and D-flucitol did not.

Brooks and Harvey[1892] carried out comparative gas chromatographic studies of corticosteroid boronates. Cyclic boronates were prepared by treating the steroid (10 μmoles) with, e.g., methane-, butane-, cyclohexane- or benzene-boronic acid (10 μmoles) in 1 ml of ethyl acetate at room temperature for 5 min., 17 ,20- and 20,21 -diols and 17 ,20,21-triols were fully esterified under these conditions. The gas chromatography was carried out on glass columns (6 ft x 4 mm) packed with 1% of OV-17 on Gas-Chrom Q (100 - 120 mesh). Retention indices and mass-spectral results are reported for 21 steroids esterified with each of the boronic acids. When hydroxy-groups other than those in the side chain are present, the use of an excess of the boronic acid leads to a peak with a much lower retention time and with severe tailing. Sometimes, the acyclic esters causing this effect can be displaced by trimethylsilyl groups and the resulting derivatives can be separated well, but the treatment is unsatisfactory for boronates of oxo- or ketol-steroids which decompose under these conditions.

Poole and Morgan[1893] used a nitrogen thermal conductivity detector to determine steroid boronates. Rubidinium chloride and caesium salt tips were investigated. Gas chromatographic conditions for this analysis are listed in detail in Table 161.

Table 161

Variable	RbCl tip	CsBr tip	Comments
Salt tip height	0.02 mm	0.000	Response with RbCl varies markedly with height above the flame, having a maximum at the position indicated. With CsBr height is less critical and maximum response is observed at 0.000 mm.
Hydrogen flow rate	37 ml min^{-1}	37 ml min^{-1}	Some tolerance in setting for RbCl but it is critical for CsBr.
Carrier gas flow rate	40 ml min^{-1}	40 ml min^{-1}	Convenient for both resolution and detector response.
Air flow rate	375 ml min^{-1}	500 ml min^{-1}	The flow rate is optimum for RbCl and values either side of this show a poorer response. A maximum response with CsBr is reached at 400 ml min^{-1} and increases beyond this have little effect.
Detector oven temperature	300°C	300°X	Sufficiently high to prevent condensation of sample material.
Least detectable amount	50 x 10^{-9}g	10 x 10^{-9}g	Expressed as g of 2 ,3 dihydroxy-5a-cholestane. With RbCl tip, base line stability was very good and an attenuation setting of x 20 could be used. Base line stability of CsBr was good but the maximum attenuation factor which could be used was x 100.
G.L.C. Conditions			0.45 m column of 1% OV-101 on CQ, 250°C phenylboronic anhydride Rt = 0.75 min (used to establish the original detector profile) 2 , 3 dihydroxy-5a-cholestane n-butyl boronate Rt = 3.90 mins.

H. Determination of Boron-hydrogen Bonding

Putnam and Myers[1894] have developed a method for the determination of boron-hydrogen bonding using chemically active gas chromatography. Boranes and chlorinated boranes may be determined by passing them with argon as carrier gas, through a column of molecular sieve 5A moistened with water, then through a column of dry molecular sieve 5A to estimate the quantity of liberated hydrogen produced by hydrolysis of the B-H bonds in the sample. If the original sample contains hydrogen as impurity it is frozen out in a cold trap at - 78°C, then gaseous hydrogen is swept out of the trap with a current of argon, and then the residual boranes in the trap are allowed to warm up to room temperature and are collected in a gas burette. This sample is now ready for analysis by chemically active gas chromatography. Putnam and Myers[1894] based their method on earlier work by Greene and Pust [1895].

Samples containing a chloroborane or a boron hydride are introduced into the system. The hydrogen peak resulting from complete hydrolysis is recorded and the millilitres hydrogen (STP) are calculated. In the case of dichloroborane which hydrolyzes according to the following equation:

$$BHCl_2 + 3HOH \rightarrow B(OH)_3 + 2HCl + H_2$$

the dichloroborane present in the sample is exactly equal to the hydrogen produced. With the other materials tested, monochlorodiborane, diborane and pentaborane, hydrolysis at 60°C yields stoichiometric quantities of hydrogen that are 5, 6 and 9.5 times the original material volume, respectively. As a result, the precision of these determinations is greater than that of dichloroborane.

This method was applied by Putnam and Myers[1894] to a wide range of mixtures containing monochlorodiborane, dichlorodiborane, diborane, pentaborane and hydrogen. Typical results are presented in Table 162.

Table 162 - Accuracy of the chemically active gas chromatography method in analysis of boron-hydrogen bonding

Standard samples introduced (ml of Gas, STP)					Analytical results (ml of Gas, STP)				
$BCHCl_2$	B_2H_5Cl	B_2H_6	B_5H_9	H_2	$BHCl_2$	B_2H_5Cl	B_2H_6	B_5H_9	H_2
0.182	-	-	-	-	0.177	-	-	-	-
0.182	-	-	-	-	0.181	-	-	-	-
0.181	-	-	-	-	0.175	-	-	-	-
-	-	0.126	-	0.095	-	-	0.125	-	0.095
-	-	0.125	-	0.095	-	-	0.119	-	0.095
-	-	0.066	-	-	-	-	0.066	-	-
-	-	0.331	-	-	-	-	0.323	-	-
-	0.062	-	-	-	-	0.060	-	-	-
-	-	-	0.191	-	-	-	-	0.188	-
-	-	-	0.195	-	-	-	-	0.185	-
-	-	-	1.196	-	-	-	-	1.162	-
-	-	-	1.120	-	-	-	-	1.124	-

Zhigach et al[1896] have studied the quantitative reaction occurring between pentaborane and ethyl alcohol to produce hydrogen and triethyl borate. He describes an apparatus for carrying out this reaction under nitrogen and for measurement of the volume of hydrogen liberated.

A very detailed study of gas chromatographic technique for the determination of hydridic and active hydrogen in borane compounds has been reported by Lysyj and Greenough[1897]. This technique is based on the chemical liberation of active or hydridic hydrogen in a micro-reaction cell incorporated in a gas chromatographic flow system and measurement of the hydrogen gas band in a nitrogen gas carrier by thermoconductivity detection. The differential in thermal conductivity between hydrogen and nitrogen is detected with a Teflon coated hot wire detector. The hydrogen gas is formed from the borane compounds by acid hydrolysis. The main advantage of this method is that it can be applied for the analysis of a variety of borane compounds using conventional commercially available equipment, i.e. gas chromatograph of any make with a gas sampling valve of any design.

The generated hydrogen gas was determined in the nitrogen carrier gas stream by thermal conductivity detection. The technique used provides an excellent means for the quantitative measurement of minute quantities of hydrogen gas. The high sensitivity of the readout system, combined with extreme eimplicity of procedure, permits fast micro- and submicro-analyses of a variety of organic functional groups.

Lysyj and Greenough[1897] applied their procedure to the determination of boron-hydrogen bonds in trimethylaminomonoborane, ($(CH_3)_3N\ BH_3$) and borazole ($B_3N_3H_6$).

ORGANOCHROMIUM COMPOUNDS

Veening et al[1898-1900] have studied the gas chromatographic separation and quantitative determination of arene tricarbonyl chromium complexes such as benzenetricarbonyl-chromium, toluene tricarbonyl-chromium, mesitylene tricarbonyl-chromium, durene tricarbonyl-chromium, hexamethylbenzene-tricarbonyl-chromium, fluorobenzene tricarbonyl-chromium and methylbenzoate tricarbonyl-chromium.

A Hewlett-Packard (F and M) Model 700 gas chromatograph equipped with flame ionization and tritium foil electron capture detectors was used to separate and detect the chromium complexes. The column used was a four foot length of coiled borosilicate glass tubing, 4 mm i.e., packed with 100/120 mesh Gas-Chrom-Q coated with 3.6% SE-30. For the flame ionization work, the carrier gas was nitrogen whereas for the electron capture measurements, a mixture of 95% argon and 5% methane was used.

The following instrumental conditions were employed for measurements using flame ionization detector: nitrogen carrier gas flow rate, 60 ml/min; hydrogen flow, 60 ml/min; air flow, 480 ml/min; column temperature, programmed from 100 to 200°C at 7.5°C per min; injection port temperature, 200°C; detector temperature, 200°C; electrometer range setting, 10^2.

For the electron capture work, the conditions were: argon-methane carrier gas flow, 60 ml/min to 75 ml/min; purge flow, 60 ml/min; column temperature, 135 to 145°C (isothermal); injection port temperature, 200°C; detector temperature, 200°C; electrometer range setting, 10^2; pulse rate, 15.

It was found that the benzene, toluene, mesitylene, durene and hexamethylbenzene tricarbonyl chromium could best be separated and quantitatively determined in mixtures by use of temperature programming with the flame ionization detector. Figure 91 illustrates the separation achieved. It can be seen that the resolution is excellent, and that the retention times of the early components are far enough removed from the tail of the solvent peak that measurements from the original base line are possible in each case. Under isothermal conditions (135°C) using flame ionization, the benzene tricarbonylchromium and toluene tricarbonylchormium peaks eluted on the tail of the benzene peaks and the retention time of the hexamethylbenzene tricarbonylchromium was 32.6 minutes. This peak was also considerably broader and flatter when eluted isothermally. For this determination therefore, programmed temperature operation is the most efficient means of analysis; not only is the resolution improved, but also the total time of analysis is reduced to 15 minutes. Retention data for isothermal and temperature programmed operation are given in Table 163.

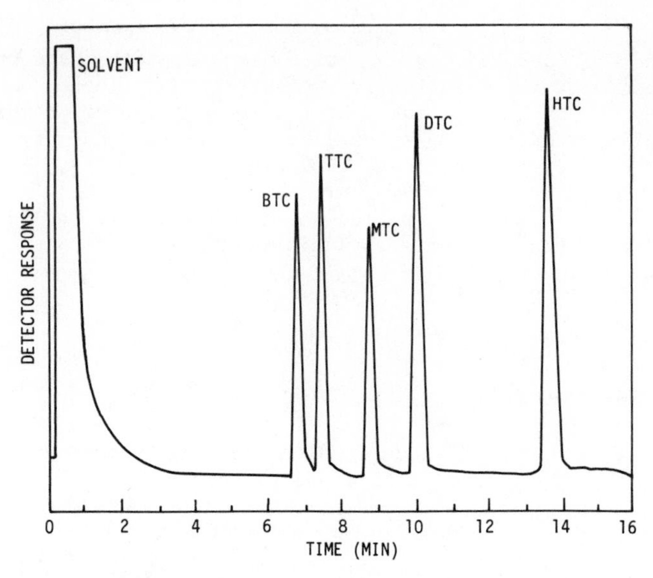

Figure 91 - Gas chromatographic separation of arene tricarbonyl chromium complexes
8.8 g. BTC (benzene tricarbonyl chromium)
8.6 g. TTC (toluene tricarbonyl chromium)
5.4 g. MTC (mesitylene tricarbonyl chromium)
8.7 g. DTC (durene tricarbonyl)
8.7 g. HTC (hexamethylbenzene tricarbonyl chromium)
Column temperature: programmed from 100 - 200°C at 7.5°/min
Carrier gas: nitrogen 60 ml/m.
Detector: Flame ionization
Attenuation: x5

Table 163 - Retention data flame detector

	Temperature programmed 100-200°C at 7.5 degrees/min		Isothermal (135°C)	
	Rel. Retention (MTC = 1.00)	Retention time (min)	Rel. Retention (MTC = 1.00)	Retention time (min)
benzene TC	0.77 ± 0.01	6.8	0.56 ± 0.05	3.1
toluene TC	0.84 ± 0.01	7.4	0.66 ± 0.06	3.7
mesitylene TC (std)	1.00	8.8	1.00	5.6
durene TC	1.25 ± 0.01	10.1	1.54 ± 0.04	8.6
hexamethylbenzene TC	1.56 ± 0.02	13.6	5.82 ± 0.07	32.6

The flame ionization response for these compounds increases as the number of methyl groups susbtituted on the benzene ring increases. This effect is predictable. For an equimolar (1.0×10^{-2}M) mixture of the five components separated isothermally at 135°C, the ratios of the peak heights of each component to benzene tricarbonylchromium increase roughly 0.13 unit in relative peak height for each additional methyl group on the ligand. When subsequent dilutions of this equimolar mixture were made and the resulting solutions were injected, it was found that the limit of detectability by flame ionization for these compounds was approximately 1×10^{-4} M (corresponding to absolute weights of complex ranging from 0.1 μg for benzene tricarbonylchromium to 0.2 μg for hexamethylbenzene tricarbonylchromium.

Arene tricarbonylchromium complexes can also be monitored with an electron capture detector, and indeed, much of the exploratory work of Veening et al[1900] was carried out by the use of electron capture. One disadvantage of using this detector is that the relative responses for the complexes decrease drastically as the number of substitute methyls on the aromatic ring increases. While electron capture and flame detectors compare favourably in their sensitivity toward the benzene,toluene and mesitylene tricarbonylchromium complexes the electron capture detector was less useful for the higher molecular weight complexes. It was also found that the use of temperature programming in combination with electron capture detection was very difficult due to severe base line changes as the column temperature increased above 150°C. Figure 92 illustrates the isothermal separation and electron capture detection of a four-component mixture of benzene, toluene, mesitylene and durene tricarbonylchromium dissolved in benzene. It is seen that approximately four time. as much mesitylene tricarbonylchromium and ten times as much durene tricarbonylchromium in comparison to benzene and toluene tricarbonylchromium are required to attain roughly the same signal for each component. The advantage of the electron capture detector in this analysis is the fact that the solvent (benzene) is a hydrocarbon and yields a very narrow peak in comparison to the flame detector.

Using the flame ionization detector and temperature programming Veening et al[1900] obtained results for a mixture of five arene tricarbonylchromium compounds. Calibration curves were linear between 1.0×10^{-4} and 1.0×10^{-3} g/ml. The accuracy of the method is within ± 3.2% relative error.

More recently Veening and coworkers[1901] have reported the gas chromatographic separation and determination of ring isometric methylbenzene tricarbonylchromium complexes such as di, tri, and tetra-methylbenzene tricarbonylchromium. A gas chromatograph equipped with a flame ionization detector was used to separate and detect in benzene or carbon tetrachloride solution the complexes. Helium was used as a carrier gas, and prepurified hydrogen and air were used to operate the flame ionization detector.

Two types of columns were employed by Veening et al[1901]. One was a 6 ft (or 12 ft) length of borosilicate glass tubing, 2 mm i.d., packed with 100 - 120 mesh Gas Chrom-Q coated with 3.6% SE-30; the second column consisted of a 100 ft x 0.5 mm i.d. stainless steel, support coated open tubular (SCOT) column coated with m-bis(m-phenoxyphenoxy) benzene and Apiezon L (Perkin-Elmer). The injection block for the SCOT column was equipped with a split ratio restrictor of 1:4. The hydrogen flow rate was 24 ml/min while that of air was 300 ml/min.

The isometric trimethylbenzene complexes, 1,2,3-TBTC, 1,2,4-TBTC, and 1,3,5-TBTC could be separated and determined on the packed column as shown in Figure 93, (see key).1,2,3,4-TMTC and p-XTC were included in this mixture and served as the internal standards. Identification of the peaks was accomplished by trapping the eluted components in hexane and recording an

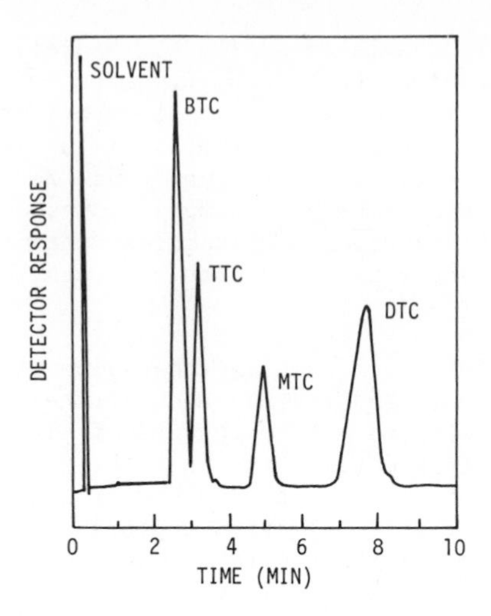

Figure 92 - Gas chromatographic seperation of arene tricarbonylchromium complexes.
3.7 g. BTC (benzenetricarbonyl chromium)
3.7 g. TTC (toluene tricarbonyl chromium)
16 g. MTC (mesitylene tricarbonyl chromium)
37 g. DTC (durene tricarbonyl chromium)
Column temperature 140°C. Carrier gas 95% argon, 5% methane, 75 ml/min.
Detector, electron capture
Attenuation x 5

ultraviolet spectrum of the resulting solution. In each case the spectrum of the eluted sample was indentical with that of the known compound. While arene tricarbonylchromium complexes survive gas chromatographic analysis intact on a packed glass column, elution of these compounds on the SCOT column results in complete decomposition.

Van der Heuvel et al[1902] have studied the gas chromatography - mass spectrometry of the three xylene tricarbonylchromium complexes, also of benzene, toluene, mesitylene, p-methoxyanisole and 2,3-dimethylnapthalene tricarbonylchromium. These workers used as a column a 12 ft x 4 mm ID glass W-tube; 2% F-60 (DC - 560) coated over 1.5% SE-30 on 80 - 100 mesh acid-washed and silanized Gas-Chrom P; the carrier gas was argon (45 ml/min) and detection was achieved by flame ionization. A variety of column temperatures in the range 120 to 180°C were employed. Chromatographic conditions: 5 ft x 3 mm ID glass spiral column, packing same as above except 4% F-60; column temperature programmed from 135 to 170°C and held at the upper limit; 30 ml/min helium. The arene tricarbonylchromium complexes were injected in benzene/hexane solution (1 - 3 μg of each compound).

The gas chromatographic separation of a mixture of substituted benzene tricarbonylchromium complexes is presented in Figure 94; the influence of molecular weight upon retention behaviour with the non-selective stationary phase is clear to see and as expected. Complete resolution of the three isomeric xylene complexes (o-, m- and p-), as well as of the isomeric complexes of the three trimethylbenzenes and the three tetramethylbenzenes, was also achieved by these workers[1902] using a capillary column.

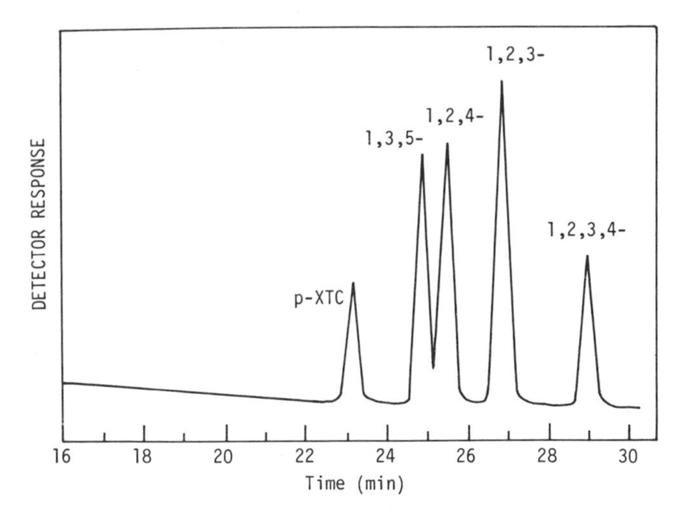

Figure 93 - Gas chromatographic separation of 1,3,5,TBTC, 1,2,4,TBTC and 1,2,3,TBTC on a packed column.
Internal standards: p XTC and 1,2,3,4, TMTC
Column temperature: 80 - 200°C programmed at 4°/m.
Injection port temperature: 145°C
Column flow: 11 ml/m.
Amounts injected: 1.49 μg. 1,3,5, TBTC
1.68 μg. 1,2,4, TBTC
2.09 μg. 1,2,3, TBTC

Key:
1,2 - dimethylbenzenecarbonyl chromium (Ø-XTC)
1,3 - dimethylbenzenetricarbonyl chromium (m-XTC)
1,4 - dimethylbenzenetricarbonyl chromium (p-XTC)
1,2,3 - trimethylbenzenetricarbonyl chromium (1,2,3-TBTC)
1,2,4 - trimethylbenzenetricarbonyl chromium (1,2,4-TBTC)
1,3,5 - trimethylbenzenetricarbonyl chromium (1,3,5-TBTC)
1,2,3,4 - tetramethylbenzenetricarbonyl chromium (1,2,3,4-TMTC)
1,2,3,5 - tetramethylbenzenetricarbonyl chromium (1,2,3,5-TMTC)
1,2,4,5 - tetramethylbenzenetricarbonyl chromium (1,2,4,5-TMTC)

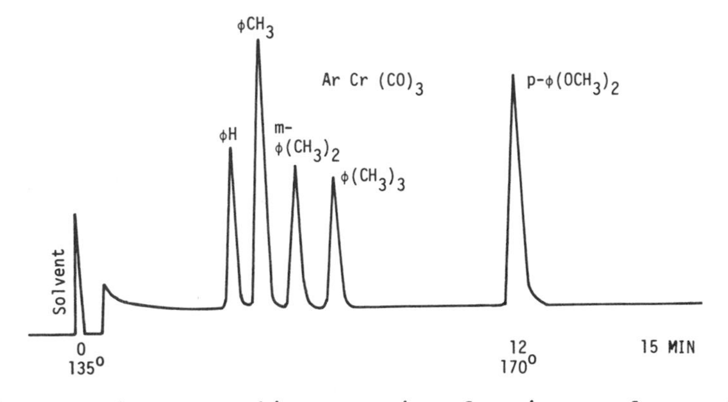

Figure 94 - Gas chromatographic separation of a mixture of arene tricarbonyl complexes.
The compounds are in order of increasing retention time, the complexes of benzene (ØH), toluene ($ØCH_3$), m-xylene (m-Ø $(CH_3)_2$), mesitylene ($Ø(CH_3)_3$) and p-methoxyanisole (p-$Ø(OCH_3)_2$).
Column conditions (LKB) given in the text.

Krueger and McClosky[1903] have reported that thermal effects with the metal jet-orifice separator are precluded by the very high velocity and resulting short residence time of the sample components in the separator. The tricarbonylchromium complex of 2,3-dimethylnapthalene appears to be an exceptionally labile compound and might serve as a sensitive test substance for investigating the possible tendency of structural alteration occurring to compounds during their passage between a gas chromatograph and a mass spectrometer.

A combination of gas chromatography and nuclear magnetic resonance spectroscopy has also been used in studies of arenetricarbonylchromium complexes, (Segard et al[1904]). These workers studied thirteen benzenetricarbonylchromium complexes on columns operated between 80 and 180°C and showed by mass spectroscopy of the eluted compounds that no decomposition occurred on the column; gas chromatography was carried out on a Girdel chromatograph (Giravious Dorand) equipped with a flame ionization detector. The two columns used (2 mm ID x 1.5 M) were filled with Chromosorb W (80 - 100 mesh, acid washed) coated with 10% SE-30 or Apiezon L grease. (0.29 and 0.30 g SE-30 and Apiezon L, respectively, charge weight of columns). The samples were injected either neat with a solid sampler or as a solution in benzene.

Segard et al[1904] found when studying the chromatographic behaviour of each pure complex that injections of solids lead to the appearance of memory effects on both columns used; when injecting a pure compound A, a peak corresponding to the compound B injected in the previous analysis appears. This effect was not observed with injections of the complexes dissolved in benzene. This procedure thus seems better, although a very slow decomposition of these solutions occurs.

Table 164 gives the specific retention volumes of the arenetricarbonylchromium derivatives on SE-30 and Apiezon L, and their relative retentions to benzenetricarbonylchromium measured at 145°C on both phases. At the same temperature, the retentions on Apiezon L are about 3 to 4 times greater than the retentions on SE-30 silicone gum.

At constant temperature on the SE-30 column, the separation of para- and meta-xylenetricarbonylchromiums is poor; the complexes of orthoxylene and ethylbenzene are eluted in the same time and are not completely resolved from the mesitylene complex. The selectivity of the Apiezon L column is greater for the xylene derivatives and the separation of o-xylene and ethylbenzene complexes is improved but the resolution of mesitylene- and pseudocumeme-tricarbonylchromiums is bad.

Improved separations are obtained by programming the column temperature. In these conditions, all the compounds studied can be separated from each other on one of the two columns except the o-xylene and ethylbenzene complexes. Analysis is time-consuming and this phase is more specially interesting for a better resolution of the first compounds of the series. However, it is noteworthy that no decomposition seems to affect the peak shape of the more retained solutes.

Mironova et al[1905] examined the applicability of five non-polar stationary phases (Apiezon L, SE-30, FS-303, E-301 and PMS-100) for the separation of bis(ethylbenzene)-chromium from the mixtures of TF complexes of chromium and aromatic hydrocarbons. A 100 cm x 3 mm column was used with helium (40 ml per min) as carrier gas and a katharometer detector. A

Table 164 - Specific retention volumes (V_g) and relative retentions (γ) of arenetricarbonylchromium complexes on SE-30 and Apiezon L greases

Ligand	Peak No.	SE-30		Apiezon L	
		$V_g$ml		$V_g$ml	
Benzene	1	385	1	1150	1
Toluene	2	485	1.25	1450	1.25
p-xylene	3	510	1.35	1750	1.50
m-xylene	4	530	1.40	1900	1.65
o-xylene	5	575	1.50	2160	1.90
Ethylbenzene	6	585	1.52	2270	2.00
Mesitylene (1,3,5-trimethylbenzene)	7	605	1.57	2490	2.15
Pseudocumene (1,2,4-trimethylbenzene)	8	680	1.75	2640	2.30
Cumene (isopropyl-benzene)	9	720	1.85	2860	2.50
n-propylbenzene	10	845	2.20	3380	2.95
tert-butylbenzene	11	960	2.50	3880	3.35
Isobutylbenzene	12	1010	2.65	4280	3.70
n-butylbenzene	13	1290	3.35	5650	4.80

comparison of SE-30 and PMS-100 with respect to the temperature for the separation of bis(ethylbenzene)-chromium from its homologues indicated that the optimum temperatures are 200 and 220°C, respectively. The selectivity coefficient and the separation coefficient were determined as functions of the column temperature, and it is shown that the PMS-100 phase is superior to SE-30 in its selectivity. Mironova et al[1905] used this technique to examine products from the Friedal Crafts reaction. Jackson and Jennings[1906] have also studied the isomer distribution for the Friedel Crafts acetylation of alkylbenzene tricarbonylchromium complexes.

Devyatykh et al[1907] have determined impurities in bis(ethylbenzene)-chromium obtained by a Friedel Crafts reaction. Chromatography was carried out on a 1 metre column of crushed brick or Chromosorb G supporting 10% of SE-30 at 225°, (with helium (50 ml per min) as carrier gas) and detection by katharometer. The impurities determined are benzene, toluene, ethylbenzene o-ethyltoluene, o-diethylbenzene, triethylbenzene, bis(benzene)chromium, bis(toluene)chromium, (benzene) (ethylbenzene) chromium, (ethylbenzene) (o-diethylbenzene) chromium, bis (diethylbenzene)chromium and bis(triethyl-benzene) chromium, contents of which vary with the conditions of preparation of the bis(ethylbenzene)chromium.

Gracey et al[1908,1909] have used gas chromatography to study cis-trans isomer ratios of alkindane tricarbonylchromium complexes.

Segard et al[1910] have studied the gas chromatography of arene tricar-bonylchromium complexes of thiophenes of the types indicated below (Table 165). They used silicone SE-30 as stationary phase (10%) on Chromosorb W at 140°C (injection block temperature 220°C) using helium as carrier gas. Under these conditions they separated benzene, toluene, p-xylene, m-xylene, o-xylene and ethylbenzene tricarbonyl compounds and at a lower injection block and detector temperature of 160°C they separated various arene tricarbonylchromium complexes. Segard et al[1910] also discuss the thermal decomposition on gas chromatographic columns of benzene and thiophen complexes of tricarbonylchromium and conclude that the thiophen complexes are the least stable of the two.

Feldman and Batistoni[1850] have described a design of a glow discharge tube as a detector for gas chromatography of organochromium compounds. This technique is described in detail in the section on arsenic. The microwave emission detector has been demonstrated to be useful for the detection of organochromium compounds[1843,1911].

Table 165 - Thiophene derivatives (Segard[1910])

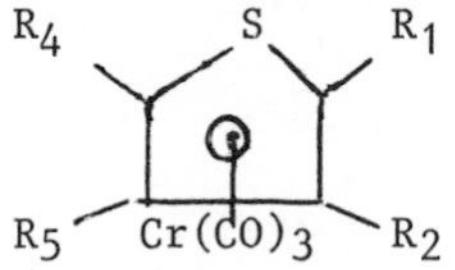

R4, S, R1, R5, $Cr(CO)_3$, R2

	R1	R2	R3	R4
IC	H	H	H	H
IIC	CH_3	H	H	H
IIC	C_2H_5	H	H	H
	$n\text{-}C_3H_7$	H	H	H
	$t\text{-}C_4H_9$	H	H	H
	H	CH_3	H	H
	H	$t\text{-}C_4H_9$	H	H
	CH_3	CH_3	H	H
	CH_3	H	CH_3	H
XC	CH_3	H	H	CH_3
XIC	H	CH_3	CH_3	H
	$t\text{-}C_4H_9$	H	H	$t\text{-}C_4H_9$
	CH_3	H	H	OCH_3
	C_6H_5	H	H	H
	$C_6H_5CH_2$	H	H	H

ORGANOCOBALT COMPOUNDS

Yanotovski et al[1912] have discussed the gas chromatography of the vitamen K series and ubiquinones. They showed that the retention times of these vitamins is dependent on the number of isoprene units present. They suggest that it may be possible to identify such compounds using logarithmic plots.

ORGANOCOPPER COMPOUNDS

Feldman and Batistoni[1850] have described a design of a glow discharge tube as a detector for organocopper compounds. This technique is described in detail in the section on arsenic. The microwave emission detector has been demonstrated to be useful for the detection of organocopper compounds [1913].

ORGANOGALLIUM COMPOUNDS

Fukin et al[1914] determined various components in dimethylgallium using a column (2 metres) of Chromasorb supporting 15% of E-301 elastomer and operated at 70°C with helium (20 ml per minute) as carrier gas and detection by katharometer. Peaks due to methane plus hydrogen, trimethylgallium, dimethylgallium chloride, methylgallium dichloride and dimethylaluminium chloride were obtained. The compounds were identified by condensing the various fractions at the outlet of the column followed by analysis by suitable means.

The microwave emission detector has been demonstrated to be useful for the detection of organogallium compounds[1913].

ORGANOGERMANIUM COMPOUNDS

A. Alkylgermanium Compounds

Alkylgermanium compounds exhibit very similar behaviour on a gas chromatographic column to those of silicon and may be separated on the same types of columns as reported by a number of authors: Abel et al[1915], Phillips and Timms[1916], Pollard et al[1917], Semlyen et al[1918], Semlyen and Phillips[1919], Timms et al[1920], Snegova et al[1921], Garzo et al[1922]. Garzo et al[1922] used columns containing Apiezon L, SE-30, QF-1, XF 112 and o-nitrotoluene as stationary phase with flame ionization and thermal conductivity detectors. Snegova et al[1921] analysed mixtures of organo-germanium and organosilicon compounds using fluorosilicone oil as a stationary phase on a column operated at 150°C with helium as carrier gas.

In Table 166 are shown determined retention times (Semlyen and Phillips 1919) for a number of alkylgermanium compounds compared with empirically calculated values. Retention data are expressed as logarithms of retention times relative to mesitylene = 100. Estimates of the retention values of mixed alkyl germanes are made from observations on symmetrical tetraalkyl germanium compounds. The log t_R values of the latter were divided by four which gave constants representing the effect of single alkyl groups on the retention time of mixed alkyls. These are: methyl, 0.14; ethyl, 0.54, n-propyl, 0.69; n-butyl, 0.93. A constant of 0.14 was added to calculate the germane series. Figure 95 is a plot by Phillips and Timms[1916] of the logarithm of the relative retention time of n-isomers of germanium, silicon, and silicogermanium alkanes as a function of the number of combined silicon and germanium atoms in the chain. The n-isomers fall on straight lines. In addition, the mixed germanium-silicon compounds form straight lines, indicated as dashed lines in the figure. These relationships permit prediction of retention times of more complex isomers.

The alkyl germanes were analysed on columns of 2 - 13% squalane at 100°C using a flame ionization detector and oxygen-free hydrogen as carrier gas. The solid support was Embacel (60 - 100) acid washed which was made inactive by treatment with hexamethyldisilazane.

Semlyen and Phillips[1919] and other workers[1918,1920] claim that the use of retention parameters for the calculation of retention time works well for silanes, germanes, silicogermanes and germanium tetraalkyls. They obtained the methyl group parameter of 0.10 from log t_R values of tetra-methylgermane found by graphical extrapolation of GeR_3Me, and GeR_2Me results. These parameters can be used to calculate the log t_R values of mixed tetra-alkylsilanes direct; the same parameters can be used for tetraalkylgermanes when a constant of 0.14 is added to the results.

Pollard et al[1917] used a flame ionization gauge to determine retention data for germanium tetramethyl. Tetramethylgermanium had the following

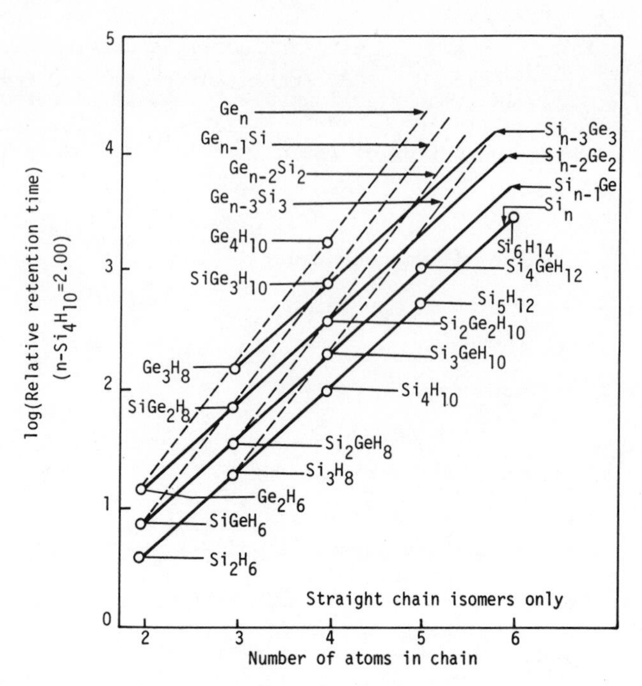

Figure 95 - Silanes, germanes and silico-germanes. Log retention times versus number of silicon and/or germanium atoms.

specific retention volumes (Vgml) on a 15% E 301 silicone oil on Celite column;

80°C	20.1 ± 0.05
100°C	12.37± 0.12
110.5°C	9 29± 0.1

Table 166 - Logarithm of retention times for alkyl germanes, Squalane at 100°C

Alkyl Groups	log t_R Obs.	Calc.	Alkyl Groups	log t_R Obs.	Calc.
Me_3(nBu)	1.42	1.49	Me(nBu)$(nPr)_2$	2.61	2.58
$MeEt_3$	1.57	1.63	MeEt$(nBu)_2$	2.64	2.59
$Me_2(nPr)_2$	1.77	1.79	Et$(nPr)_3$	2.66	2.66
Me_2Et(nBu)	1.77	1.80	$(nPr)_4$	2.89	2.88
$MeEt_2$(nPr)	1.84	1.87	Et$(nPr)_2$(nBu)	2.91	2.89
Et_4	1.94	1.94	$Et_2(nBu)_2$	2.95	2.90
Me_2(nPr)nBu	2.04	2.04	Me$(nBu)_3$	3.14	3.07
MeEt$(nPr)_2$	2.09	2.10	Pr_3(nBu)	3.13	3.13
Et_3(nPr)	2.18	2.18	Et(nPr)$(nBu)_2$	3.16	3.14
$Me_2(nBu)_2$	2.31	2.28	$(nPr)_2(nBu)_2$	3.38	3.37
Me(nPr_3)	2.25	2.24	Et$(nBu)_3$	3.40	3.38
$Et_2(nPr)_2$	2.42	2.41	(nPr)$(nBu)_3$	3.61	3.62
Et_3(nBu)	2.45	2.42	$(nBu)_4$	3.85	3.86

Bortnikov et al[1923] have discussed the gas chromatography of alkyl germanium compounds. They determined retention volumes by gas chromatography and heats of dissolution or absorption for the tri- and tetraethyl derivatives of germane on a stainless steel column (1 metre x 3 mm) packed with silanised Chromosorb W supporting 20% of Apiezon L or 15% of Carbowax 20M and operated at 120° or 90°C respectively, with helium as carrier gas and thermal conductivity of flame ionization detection - and by gas-solid chromatography on a column of Carbochrom-1 (graphitised thermal carbon black with 0.01% of Apiezon L) operated at 190°C or Silichrom C-80 (modified with chlorotrimethylsilane, hydroxylated or dehydroxylated) operated at 100°C, 147°C or 203°C respectively.

B. Miscellaneous Organogermanium Compounds

Snegova et al[1921] have developed a method for the gas chromatography of halogen-containing organogermanium compounds such as $Cl_2GeH(Cl)CH_3$, $Cl_2GeCH_2CH_2Cl$, $Cl_3GeCH(Cl)CH_2CH_3$, $Cl_3GeCH_2CH_2CH_2Cl$ boiling up to 280°C.

Vyazankin et al[1924] separated the cis and trans isomers of 1-trimethylgermanyl ethylene and of bis (trimethylgermanyl)ethylene on a metre column of Apiezon L on Chromasorb W or of graphitized carbon operated at 100 - 200°C using helium as carrier gas and a thermal conductivity or a flame ionization detector.

Retention volumes for compounds of the type $Et_3M-M^1Et_3$ where M and M^1 are germanium or tin, have been determined by Pappas and Milliou[3793] on a column of Apiezon L or Carbowax 20 M on silanized Chromasorb W at 222 - 234°C with helium as carrier gas. A column of graphitized carbon black also gave good separations at 226°C. Bortnikov et al[3794] separated compounds containing bonds between 2 or 3 like or unlike heteratoms (silicon, germanium tin or sulphur) on columns of silanized Chromasorb W coated with 20% Apiezon L, 15% of polyoxyethylene glycol 20 M or Reoplex 400, also by gas-solid chromatography on graphitized thermal carbon black with helium as carrier gas and a flame ionization or thermal conductivity detector.

Bortnikov[1925] also gas chromatographed bis(triethyl germanyl) sulphide separating it from its silicon, selenium, sulphur, tellurium and tin analogues. Separation was achieved at 254°C on a stainless steel column (100 cm x 0.4 cm) packed with Chromasorb W supporting 20% Apiezon L with helium as carrier gas and thermal conductivity detection.

Brazhnikov and Sakodynskii[1926] have discussed the separation of germanium tetraiodide by preparative gas chromatography on a carbon black filled column.

Frolov[1927] has applied gas chromatography to the determination in germane of down to 10^{-4} to 10^{-5}% of methane, ethane and ethylene. Gas chromatography is carried out at 40°C on a 5 metre column containing porous glass previously treated with 3-N hydrochloric acid at 50°C and ignited for 5 hours at 700°C with nitrogen as carrier gas, a flame ionization detector was used. Silica furnaces heated to 400° are placed before and after the column, to decompose the germane and prevent decomposition of germanium dioxide in the detector. The error of the determination is claimed to be less than ± 20%.

Gorbackev and Tret'yakov[1928] applied gas chromatography to the determination of dissolved oxygen and nitrogen in germanium tetrachloride. The separation was carried out at 60 to 90°C on a 2.5 m x 6 mm column containing 20% fluorosilicone oil 169 on firebrick (approx. 0.2 mm) using hydrogen as a carrier gas at a flow rate of 125/150 ml per minute.

Various workers[1929-1931] have discussed the gas chromatographic determination of germanium tetrachloride. Several attempts toward a quantitative gas chromatographic determination of germanium as the hydride have been made by Devyatykh et al[1932], Fiser[1933], and Kadeg and Christian [1934].

Iatridis and Parissakis[1935] have developed a method for the determination of germanium in oxides, ores and alloys involving the chlorination of the sample and gas chromatographic determination of the germanium tetrachloride produced. The method has relative accuracy and precision better than - 4.6% and 0.88%, respectively. It is applicable to a germanium level of 1 to 99%. It is also rapid, since two samples per hour can be analyzed, and sensitive, since an amount of 10^{-5} g of germanium can be detected.

In this method a weighed amount of the sample is sealed into a borosilicate tube together with dry carbon tetrachloride. Chlorination of the ore proceeds at 575°C. The capsule is then broken by a crushing device and the liberated volables swept into a gas chromatograph by the carrier gas.

As column packing material, silicon oil DC 550 20% w/w on Celite 545 was used because of its inertness toward germanium tetrachloride, and its very good resolution for a number of chlorine compounds. A gas chromatograph equipped with a thermal conductivity detector was used, modified in order to keep the oven temperature constant within ± 0.1°C. Nitrogen was used as carrier gas. This was dried by passing through activated molecular sieve and P_2O_5 traps at flow rates of 10 - 100 ml/min. The detector used showed a satisfactory response toward inorganic chlorides. From time to time the detector was washed with acetone-hydrochloric acid solution in order to prevent alteration of response owing to deposit of hydrolysis or reaction products upon the filaments. To avoid reaction with the highly corrosive chlorides, glass columns were used (4 mm i.e., 61 mm o.d., 183 cm long).

Figure 96 represents a chromatogram of the reaction products of carbon tetrachloride with a germanium ore in the glass capsule. Peaks were observed due to chlorine carbon dioxide, phosgene and carbon disulphide.

ORGANOIRON COMPOUNDS

A. Ferrocene and Derivatives

Tanikawa and Arakawa[1937] determined ferrocene derivatives by gas chromatography on an Apiezon L (2.5%) on a Chromosorb W column (1.4 M x 4 mm) at 200°C. The chromatograph was equipped with a thermal conductivity detector and helium was used as carrier gas. These workers claimed complete separation of ferrocene derivatives. Nesmeyanov et al[1938] seperated four 1-ferrocene and 2 ruthenocene derivatives on 2,2-dimethylpropane -1,3-dioadipate; polyoxyethylene glycol and Apiezon L (1 - 5%) on Celite 545, (80 - 100 mesh) at 100 to 200°C. They used packed columns (1 to 1.2 M x 0.4 cm) in glass and steel tubes, and a capillary column (45 M x 0.25 mm) and aβ-ray detector. Best separations were achieved on 2,2-dimethylpropane-1, 3 diol adipate and polyoxyethylene glycol M20 columns. Nesmeyanov et al[1938] report relative retention volume data and discuss the separation of mixed isomers of ferrocene derivatives. Nitro-, dicyano-, diphenyl- and diacyl-

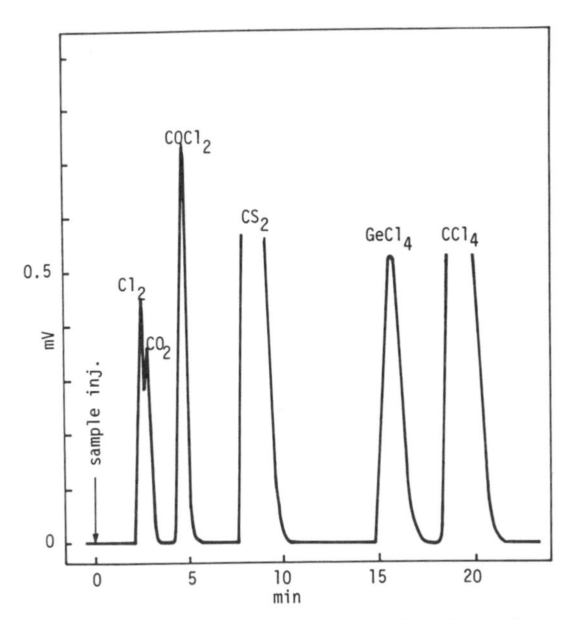

Figure 96 - Gas-liquid chromatogram of chlorination of germanite ore with CCl_4(6 min, 575°C). Chromatographic conditions: glass column length = 183 cm, o.d. = 6 mm, i.d. = 4 mm.
Packing material: silicon oil DC 550 20% w/w on Celite 545.
Carrier gas flow rate: 14 ml/min N_2.
Detector: TCD (Gow-Mac 4 tungsten filaments), bridge current 150 mA.
Temperatures °C: capsule chamber, 120; inj. port, 110; column, 80; detector, 100.

ferrocenes could not be separated owing to their poor thermal stability at the column operating temperature.

Ayers et al[1939] have described gas chromatographic procedures for the separation of benzene, chloroform and carbon tetrachloride solutions of ferrocene and for butyl-, ethyl-, vinyl-, 1,1'-dibutyl-, acetyl-, 1,1-diacetyl-, hydroxymethyl- and 1,1'-bis(hydroxymethyl)- ferrocenes. These workers used a gas chromatograph equipped with thermal conductivity and flame ionization detectors. The column composed: a stainless steel column (5 ft x 1/8 inch o.d.) packed with 5% by weight of General Electric SE-30 (methyl silicone gum rubber) on 60/80 mesh Chromosorb W. This column did not cause significant decomposition of ferrocene derivatives. The column was preconditioned at 250°C for 72 hours prior to use. Column temperatures of 125, 150, 175 and 200°C were used, and the injector and collector temperatures were maintained at 20 to 40°C above the column temperatures in each case. Ferrocene itself can be conveniently chromatographed at 125°C. The flow rate of helium carrier gas was 30 ml/min.

A typical separation obtained by Ayers et al[1939] of a mixture of ferrocene, acetylferrocene and 1,1'-diacetylferrocene is shown in Figure 97. At a temperature of 175°C, the first peak after the solvent has a retention time of 1.15 minutes from the point of injection and is due to ferrocene. The acetylferrocene and 1,1'-diacetylferrocenes, respectively, are eluted after 3.2 and 9.7 minutes.

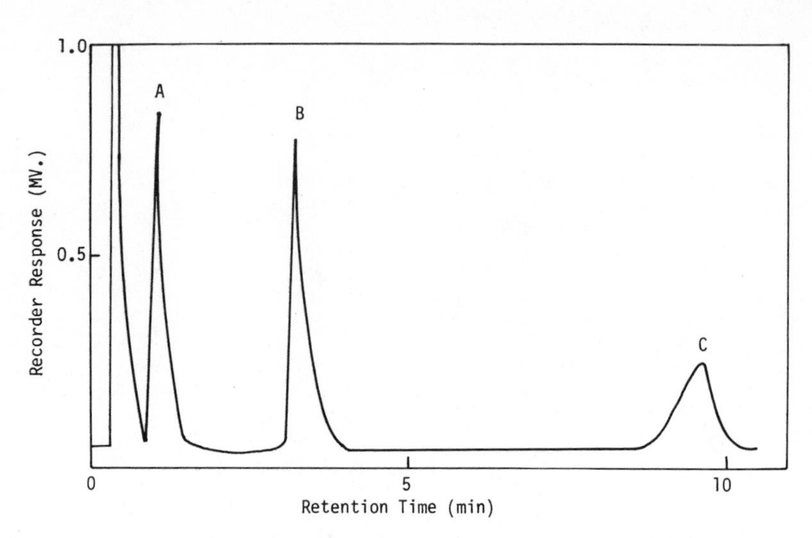

Figure 97 - Gas chromatography of mixture of ferrocene (A) acetylferrocene, (B), and 1,1'-diacetylferrocene (C).

The retention times of various other ferrocene derivatives at several temperatures are given in Table 167. Although all of the analyses were conducted under isothermal conditions, it is apparent that temperature programming would be desirable in the separation and analysis of mixtures containing both volatile and relatively nonvolatile ferrocene derivatives.

Pommier and Guichon[1940] have studied the retention of ferrocene derivatives on various stationary phases. The retention was shown to be made up of contributions from the ferrocene moity and from each of the substituents. The logarithms of these contributions are additive, provided that there is no interaction between the substituents or with the cyclopentadienyl rings, e.g. with methyl- and 1,3-dialkylferrocenes. These workers tabulate retention data for separations on SE-30, Apiezon L and polyoxyethylene glycol.

Yamakawa et al[1941,1942] have described methods for the gas chromatography of mixtures of mono substituted (1 position) and disbustituted ferrocenes containing small amounts of 1,2- and 1,3-disubstituted derivatives, such as ferrocene, 1,1'-diacetylferrocene, dimethyl 1,1'-ferrocenedicarboxylate, acetylferrocene, ferrocene methanol, 1,1'-ferrocenedimethanol, ferrocene mercury chloride, iodoferrocene and phenylferrocene. They used 140 cm x 4 mm i.d. columns in which Chromosorb W or glass microbeads coated with 0.5 to 2.5% Apiezon L or silicone rubber was packed. Detection was by thermal conductivity. A stainless column 140 cm (70 cm U-shape x 2) x 4 mm i.d. was used, but 210 cm x 4 mm i.d. stainless column was used in the column C. The conditions of the column were as follows: temperature, 170 to 200°C; sample heater and the detector temperature, 30° and 40° higher than the column temperature, respectively. Carrier gas helium is introduced into a column at 30 to 100 ml/min. The samples were generally dissolved in chloroform (3 to 4%). The most suitable temperature to separate ferrocene derivatives was 170 to 200°C depending on the column packing used.

Benkeser and coworkers and Forbes et al[1943-1947] have reported the separation and analyses of ferrocene derivatives by gas chromatography. These workers separated various acetylferrocenes and the isomers of acetylalkyl-

Table 167 - Retention times for ferrocene derivatives

Compound	Melting point (°C)	Molecular weight	Retention time (minutes) 125°C	150°C	175°C	200°C
Ferrocene	173 - 174	185.95	3.7	1.75	1.15	0.65
n-Butylferrocene	b.p. 84 - 86/0.2 mm	242.15	16.0	6.0	2.80	1.31
Ethylferrocene	b.p. 74 - 76/0.2 mm	213.97	6.7	2.64	1.48	0.83
Vinylferrocene	48 - 49	211.97	6.8	3.05	1.52	0.90
1,1'-di-n-Butylferrocene		298.22	73.0	20.2	7.7	2.90
Acetylferrocene	85 - 86	227.97	18.5	7.55	3.2	1.49
1,1'-Diacetyl-ferrocene	122 - 124	269.99	77.0	26.2	9.7	3.70
Hydroxymethyl-ferrocene	76 - 78	215.96	15.0	5.9	2.3	1.20
1,1'-Dihydroxy-methylferrocene	85 - 86	245.97	76.0	20.5	9.2	3.40

ferrocenes (1,1'-, 1,2-, 1,3-disubstituted ferrocenes). The analyses were carried out with a diethylene glycol sebacate packed (15 ft x 0.25 in) column at 215°C. Separation of triethylsilylisopropylferrocene, 1-t-butyl-1-trimethylsilylferrocene and 1-methyl 1-1'-triethylsilylferrocene were also carried using an Apiezon L column, 16 ft x 0.25 in., at 280, 290 and 290°C.

Uden et al[1948] have discussed the gas chromatography of ferrocene and other iron group metallocenes. The microwave emission detector has been demonstrated to be useful for the detection of organoiron compounds[1849].

B. Miscellaneous Organoiron Compounds

Forbes et al[3795] gas chromatographed diene tricarbonyliron complexes and reported excellent separations of the tricarbonyl-iron^{III} complexes of cyclopenta-, cyclohexa-1,3-, cyclohepta-1,3-, cycloocta-1,3-, and cycloocta-1,5-diene, bicyclo(2,2,1) heptadiene, cycloheptadiene, tropone, eucarvone, tropyl alcohol and eucarvol on a PTFE column (3 ft x 3/16 in o.d.) packed with Universal B supporting 15% of E301 methylsilicone. The column was operated isothermally at between 50 and 80°C, the carrier gas was nitrogen (100 ml per min) and detection was by flame ionization. Samples were injected as solution in xylene or hexane, a 'solvent-flush' technique being used to ensure rapid injection. Symmetrical peaks were obtained for the conjugated-diene complexes; non-conjugaged-diene complexes gave slightly broader peaks, with some 'tailing'. There was no evidence of decomposition of the complexes and these workers conclude that gas chromatography is a suitable technique for the separation and determination at the microgram level of such complexes.

Hill and Aue[1949] used a spectrophotometric detector incorporating a conventional flame ionization detector to determine volatile organoiron compounds. They found that a hydrogen rich flame was preferable giving sensititivities at the nanogram level and a strong discrimination against hydrocarbon sample solvents.

ORGANOLEAD COMPOUNDS

Almost all of the published work on the gas chromatography of organolead compounds is concerned with their analysis in petroleum solutions, as, of course, certain types of organolead compounds tetramethyllead, trimethyllead, dimethyllead, methyltriethyllead and tetramethyllead, are used as antiknock additives in petroleum. The analytical problem is complicated by the fact that the lead alkyls must be separated and analyzed in the presence of a complex mixture of hydrocarbons. The volatility of the lead compounds is such that their peaks are superimposed upon the hydrocarbon peaks. Selective detectors are required.

An early approach to the problem was that of Parker et al[1950] who used gas-liquid chromatographic columns to fractionate the lead alkyls and then determined the amount of lead in the fraction containing both lead and hydrocarbon compounds by a spectrophotometric method. In these procedures, the lead alkyls are separated by a chromatographic column, individually collected in iodine scrubbers, and measured by a dithizone spectrophotometric lead analysis procedure.

Parker and Hudson[1951] modified the above technique employing a much simpler chromatographic unit. Basically, this unit consists of a thick-walled aluminium tube which serves as the column. A uniform temperature is maintained by means of electrical heating tape wrapped around the column, with control effected by a variable transformer. Carrier gas flow is controlled by the pressure regulator at the supply cylinder and is measured by a bubble flow meter. No detector elements are necessary since the retention times are determined by calibration and remain unchanged under the fixed conditions of use.

The principal changes from the original procedure (Parker et al[1950]) are the column packing and the method of establishing lead alkyl retention times. Apiezon M on Chromosorb W is recommended because it provides more reproducible retention times than the originally recommended Nujol on Chromosorb P. Calibration is based on the fact that the retention time for each lead alkyl bears a fixed relationship to that of tetraethyllead under the recommended operating conditions. Although changes in gasoline base stocks may slightly alter the retention times, the relationship of each alkyl to tetraethyllead remains constant. Since the tetraethyllead retention time may be determined by analysis of a series of effluent samples, it is not necessary to use a conventional recording chromatograph at any point in this procedure.

Apart from innovations such as coupling spectrophotometric detection (Parker et al[1950]), Parker and Hudson[1951] or titrimetry (Parker et al[1950]) at the outlet end of the gas chromatographic column, conventional thermal conductivity or flame ionization methods of gas chromatographic analysis are not effective in the chromatography of the trace tetraalkyl leads due to the extreme complexity of the gasoline base stock.

Lovelock and Gregory[1952] have discussed the use of an ionization detector for the chromatography of tetraethyllead in petroleum.

Several other methods have been reported for the analysis of tetraalkylleads in gasoline using gas chromatography. The methods of detection included: an electron capture detector[1953,1954], a flame-emission detector [1956], and a hydrogen-rich flame ionization detector[1955]. An atomic-absorption spectrometer has been used as a detector for gas chromatography[957].

Coker[1958] reported a gas chromatography/atomic absorption spectrometry technique for alkylleads in which he could analyze a sample in 5 min with a detection limit of 0.2 ppm lead, a value, he points out suitable for determining trace lead in unleaded gasoline. Alkyllead has recently been determined in the atmosphere with a gas chromatographic microwave plasma detector by Reamer et al[1959].

Other gas chromatographic detection methods that have been investigated for alkyllead analysis include several electron capture procedures[1960,1961], a flame ionization method which requires a complicated separation and derivation step[1962], and a dithizone spectrophotometric technique[1963,1964].

Laveskag[1965] has described a gas chromatographic mass-spectrometric method for the determination of tetramethyllead and tetramethyllead in air.

Lovelock and Zlatkis[1966] used a selective detector, in this case electron capture, to detect tetraethyllead with essentially no interference from hydrocarbons which have a much lower response factor and a photoionization detector to measure the total amount of hydrocarbons. They did not attempt to determine the various lead alkyls separately.

A schematic diagram of their apparatus is illustrated in Figure 98. The electron absorption detector consisted of a plane parallel ionization chamber with an electrode separation of 1 cm. The chamber was swept continuously by a stream of dry clean argon containing 1% hydrogen at a rate of 100 cc per minute, at atmospheric pressure. Electrons were released in the chamber by soft radiation from a sheet of metal coated with a thin layer of titanium tritride; this radiation source also served as the cathode of the chamber. The density of free electrons in the chamber was measured by applying to the anode every 20 microseconds, a rectangular pulse 50 volts in amplitude and 1 microsecond in duration. The duration of this sampling pluse is just sufficient to collect all the electrons set free in the chamber, i.e. capable of with-drawing a saturation current from the chamber. The vapours of the test sample are introduced into the gas stream flowing to the chamber after their passage through the chromatographic column. The concentration of vapour in the gas stream is measured by a photoionization detector (Lovelock[1967]) connected in parallel with the chamber for observing electron capture. The current flow from the two ionization detectors is amplified by an electrometer amplifier and recorded using two separate potentiometric recorders.

The chromatographic column was 85 ft x 0.02 inch i.d. stainless steel coated with Apiezon L and operated at 90°C and 10 p.s.i. Conventional packed columns are also satisfactory. The flash heater is kept below 100°C to prevent thermal decomposition of lead alkyls.

In Figure 99 is shown the chromatogram obtained from the two detectors in parallel using pure tetraethyllead containing a chlorobenzene marker. The relative electron absorption coefficient of tetraethyllead was 115, indicating it to be a strongly absorbing compound. The reference standard, chlorobenzene, was assigned the value of unity. The photoionization detector served a dual purpose in computing the co-efficient and in determining the elution time of the tetraethyllead.

Figure 100 shows typical chromatograms with samples of leaded and unleaded gasolines. Lead scavengers, ethylene dibromide and ethylene dichloride, are present in the gasoline. They are strong electron absorbers

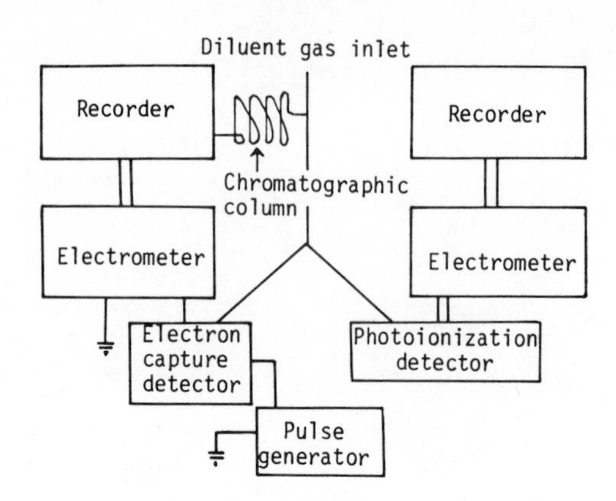

Figure 98 - Gas chromatography of lead alkyls. Schematic diagram of chromographic and detection systems.

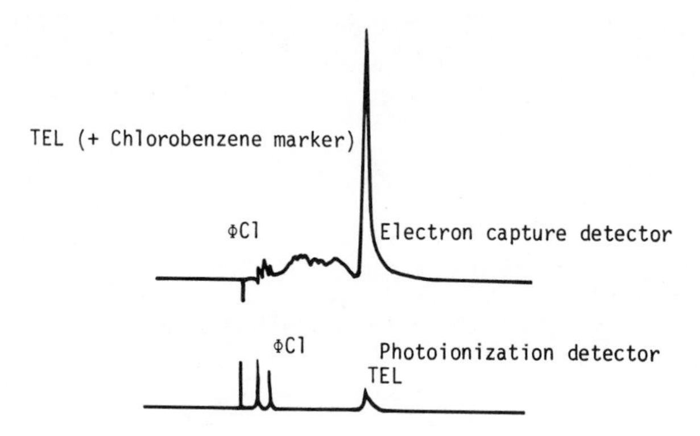

Figure 99 - Gas chromatogram of pure tetraethyllead from parallel detection system.

and could be analysed by this same technique. At the retention time of tetraethyllead in the leaded gasoline, a well resolved peak is absent in the unleaded gas. The photoionization detector records unresolved components and is complementary to the electron capture detector. The time for complete analysis of tetraethyllead in gasoline can be shortened to less than 2 minutes, simply by increasing the linear gas velocity through the column.

In addition to the methyl-ethyl lead alkyls, gasolines frequently contain ethylene dichloride and dibromide as scavengers. These latter compounds also give a high response in the electron capture detector and frequently elute at the same time as one of the lead alkyls. Dawson[1968] overcame this difficulty by using a chemically active stationary phase, silver nitrate in Carbowax 400, as a precolumn before the detector to remove the scavengers. The analyzing column contained a silicone rubber on Chromosorb W. Good separation of the five methyl-ethyl lead alkyls was

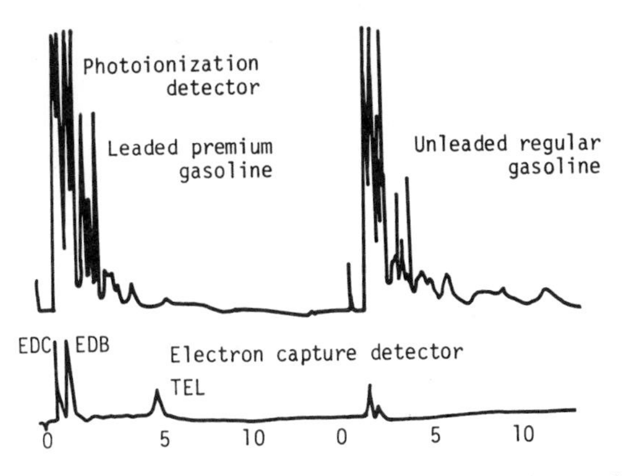

Figure 100 - Gas chromatogram of leaded and unleaded gasoline from parallel detection systems.

obtained. However, the sensitivity of the electron capture detector varied markedly with the applied voltage which required careful control of the operating conditions for quantitative analysis. Dawson pointed out that interchange of methyl and ethyl radicals between tetramethyllead and tetraethyllead occurs on a column of 5% SE-30 silicone rubber on acid-washed Chromosorb W.

Coating the Chromosorb with sodium hydroxide before the stationary liquid phase is applied reduces interchange of radicals to a non-detectable level. Slight interchange takes place when the silver nitrate packing is located at the column inlet. When this packing is located at the column exit, the lead alkyls are separated before contact with the silver nitrate and interchange is avoided.

Typical results obtained by this procedure are shown in Figures 101 and 102.

The method was applied to the determination of individual lead alkyls in commercial petroleumns. Results for total lead agreed well with those obtained by x-ray fluorescence; Table 168 compares data on four petroleums. Samples A and B contained only tetraethyllead. Samples C and D contained the five lead alkyls. There was no indication of interference by additives.

Table 168 - Total lead in gasolines by gas chromatography and X-ray fluorescence

	Grams of lead per gallon	
	Gas chromatography	X-ray fluorescence
Gasoline A	2.98	3.12
Gasoline B	0.45	0.50
Gasoline C	2.21	2.09
Gasoline D	2.91	2.76

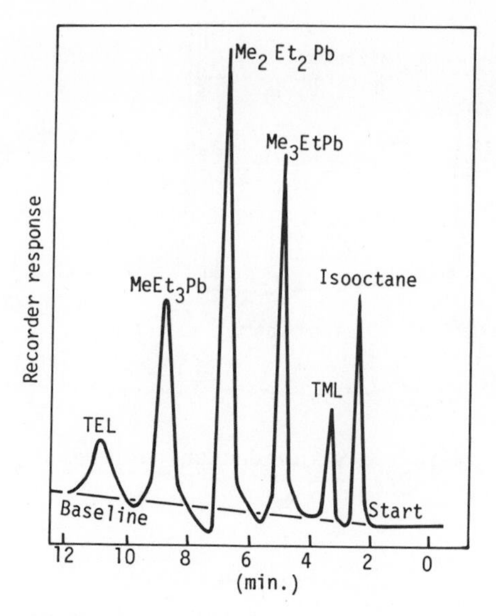

Figure 101 - Gas chromatographic separation of lead alkyls in petroleum with 28 volts across the detector.

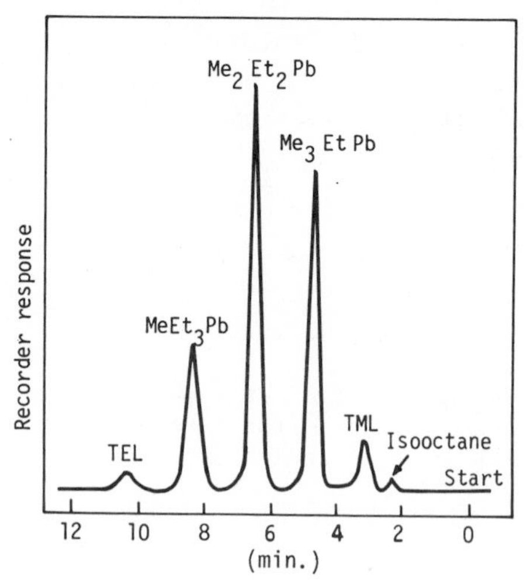

Figure 102 - Gas chromatographic separation of lead alkyls in petroleum with 37 volts across the detector.

Barrall and Ballinger[1969] found that 10% tris-1,2,3-(2-cyanoethoxy) propane, a very polar liquid, when used as a packing material gave good resolution of the methyl ethyl lead alkyls and retained the halogenated scavengers beyond the elution times for the lead alkyls so that these did not interfere. They studied a number of variables that affected the performance of the electron capture detector including cell geometry, carrier gas flow rate, electrometer voltage etc., and concluded that frequent calibration was necessary for maximum accuracy. Relative response factors for the alkyl leads were calculated for the flow conditions of this analysis.

Detector response was found to decrease with increasing molecular weight of the lead alkyl. The retention characteristics of the tetraalkyl leads on Apiezon L, Silicone SF 96, and 1,2,3-tris (2-cyanoethoxy) propane columns at two temperatures are reported. Several scavenger columns were developed by these workers for the purification of the carrier gas and the removal of interfering peaks. An absolute accuracy of ± 3% was obtained on standard samples made from weighed amounts of tetraethyl and tetramethyllead.

In Figure 103 is shown a chromatogram of a mixture of five lead alkyls together with the chloroform internal standard peak obtained using a 1,2,3 tris (2 cyanoethoxy) propane on Chromosorb W column and a parallel plate detector in the Jarrell-Ash instrument. Due to drift in the detectors, partial pre-calibration is required daily. For maximum accuracy and precision it is necessary to inject each sample in triplicate and bracket these either side with two standard solutions. Using these precautions it was found that for synthetic solutions of tetramethyllead and tetraethyllead, the determined results were usually within ± 3% of the expected results.

Barrall and Ballinger[1969] conclude that the electron affinity detector furnishes a simple and direct means for the analysis of alkyllead isomers normally found in petroleum.

Bonelli and Hartmann[1970] combined the use of 1,2,3-tris (2-cyanoethoxy) propane for the analyzing column as proposed by Barrall and Ballinger[1969] with a short scrubber column of silver nitrate on Carbowax 400 before the detector to remove the lead scavengers. These workers claim an analysis time for the five methylethyl lead alkyls of 10 minutes with an overall standard deviation of about 4% for each compound. An electron capture detector was used.

The method consists of separating the lead alkyls on a 10 foot x 1/8 inch stainless steel column of 10% 1,2,3-tris (2-cyanoethoxy) propane (TCEP) on 80/100 Chromosorb W, hexamethyldisilazane (HMDS) treated. The scrubber section, a 6 inch x 1/8 inch stainless steel column composed of 20% Carbowax 400 (saturated with silver nitrate) on 30/60 Chromosorb W pre-coated with 8% potassium hydroxide is attached between the analytical column and the detector.

The separating efficiency of this technique is shown in Figure 104. The five lead alkyls are separated according to increasing molecular weight. The trace represents a standard mixture containing 6.25 mole % of tetramethyllead, 25 mole % trimethylethyllead, 37.5 mole % dimethyldiethyllead, 25 mole % methyltriethyllead, and 6.25 mole % tetraethyllead. This mixture was diluted with spectrograde hexane to 4 ml/gal tetraethyllead. The resulting solution was then diluted 1:200 with the same solvent so that the sample would not exceed the linearity range. Operating conditions are: column, 72°C; injector, 95°C; detector, 150°C; carrier, nitrogen at flow rate, 27 ml/min; cell potential, 22 volts; attenuation, 10X-4X.

Basic work on the gas chromatography, respectively, of tetraethyllead and tetramethyllead has been reported by Pollard et al[1917] and by Abel et al[1971]. They are of the opinion that electron capture detection, whilst sensitive, is not specific enough nor is it a very easy method of detection of organolead compounds. It requires extreme care, cleanliness, and rigid adherence to microchemical techniques. An alternative method consists of running the gasoline sample through the gas chromatograph to separate the components, which are then introduced, one by one, directly to the atomic

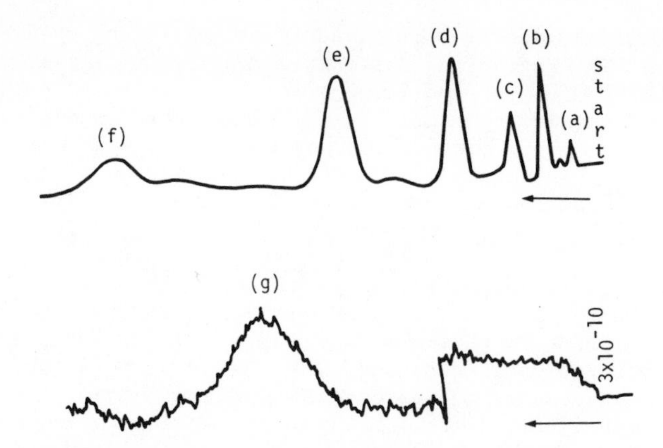

Figure 103 - Gas chromatograms of methylethyl tetraalkyl leads. 10 ft., 10% 1,2,3 tris (2 cyanoethoxy) propane on Chromosorb W 175 ml. per min, carrier gas flow through the parallel plate detector at 18 volts,
(a) air, (b) chloroform, (c) tetramethyllead, (d) trimethyllead, (e) dimethyldiethyllead, (f) methyltriethyllead, (g) tetraethyllead.

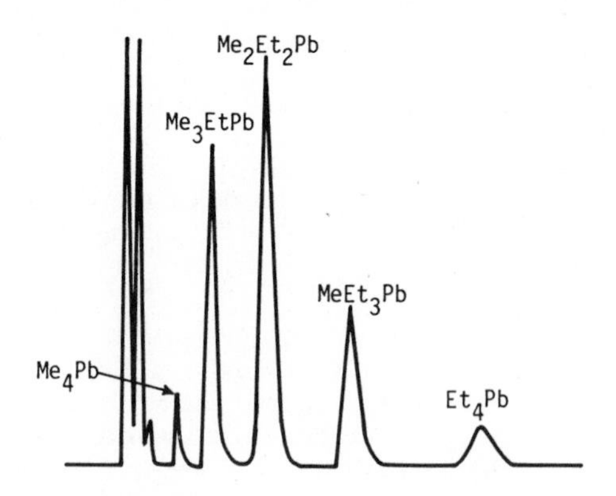

Figure 104 - Gas chromatography of tetraethyllead. Sample 4.0 cc per gallon as tetraethyllead diluted 1:200 0.4 µl. injection.

absorption burner. The atomic absorption spectrophotometer, which is set up for lead determination, records a peak absorption for each lead compound as it passes from the chromatograph. The method is standardized by using mixtures of known composition. Gasoline sample sizes are typically 1 µl; as little as 20 nanograms of lead as lead alkyl can be detected.

Dupuis and Hill[1972] have demonstrated the use of a hydrogen atmosphere flame ionization detector (HAFID)[1973-1977] modified from a commercial flame ionization detector for the determination of organolead compounds in gasoline. By a simple one to ten dilution of a leaded gasoline, alkyllead compounds were detected with no interference from overlapping chromatographic peaks of hydrocarbons. Detection limits were calculated to be 7.3 x 10^{-12} g/s of lead.

The hydrogen atmosphere flame ionization detector introduced in 1972 1973-1977 and subsequently developed, is a sensitive and selective gas chromatographic detector for organometallic compounds. Hill and Aue showed that this detector had a selectivity for tetraethyllead over dodecane of 10^{-4} with a minimum detectable amount of 51 pg tetraethyllead injected and were able to detect tetraethyllead in a gasoline diluted 1 to 100 with gasoline[1976,1977]. Minimum detectable amounts for certain metal containing compounds extend to the low picogram and sub-picogram range with selectivities of 10^4 and 10^5 when compared to n-hydrocarbon responses.

The minimum detectable amount of 30 pg obtained for tetraethyllead compares favourably with that of 51 pg reported earlier by Hill and Aue [1976]. The detector's minimum detectable limit for tetraethyllead was calculated to be 1.1 x 10^{-11} or 7.2 x 10^{-12} g/s for lead.

Various workers (Reamer et al[1978], and Quimby et al[1979]) have examined the applicability of helium in microwave glow discharge detectors for the detection of organolead compounds leaving a gas chromatographic column. A microwave emission detector was first described by McCormack et al[1980].

Quimby et al[1979] examined the applicability of this type of detector to a range of elements including lead, silicon, phosphorus, mercury and manganese. The results they obtained for organolead compounds are discussed below and those for the other elements are discussed elsewhere. Quimby et al[1979] used a microwave emission detection system for gas chromatography which utilizes the TM_{010} resonant cavity to sustain a plasma in helium at atmospheric pressure. The effluent from the gas chromatograph is split between a flame ionization detector and a heated transfer line directing it to a small auxillary oven containing a high temperature valve. The valve allows the effluent to be directed either to a vent or to the plasma. Atomic emission from the lead entering the discharge is observed axially with an echelle grating spectrometer. The system allows for highly selective and sensitive detection of lead by monitoring an appropriate wavelength.

The type of detector devised by McCormack et al[1980] involved eluting the compounds from the gas chromatograph and directing them into a microwave discharge which is sustained in either argon or helium. Observation of the optical emission spectrum resulting from the fragmentation and excitation of compounds entering the plasm affords sensitive, element selective detection. This type of detector has been used by various workers 1981-1991 to detect a wide variety of metallic and non-metallic elements. With the types of resonant cavities employed in the above studies, discharges can be sustained in argon at either atmospheric or reduced pressure (usually 5-50 Torr) and in helium at reduced pressure only. Line emission is observed for all elements in the helium plasma. Since the measurement of band spectra has presented previous workers[1980,1992] with problems when applied for selective detection of elements, the helium plasma, although somewhat less convenient to operate, is preferred by Quimby et al[1979] because of its wider range of applicability. To overcome the necessity of operating the helium plasma at reduced pressure, Beenakker[1993] introduced the TM_{010} cylindrical resonant cavity[1993,1994]. Owing to its increased efficiency for transfer of microwave power to the discharge, an atmospheric pressure helium (or argon) plasma can be sustained at the same lower power levels as used with previous cavities. A further advantage offered by this design is the ability to view light emitted from the plasma axially. With earlier cavities in which the helium plasma is operated at reduced pressure, the plasma is usually viewed transversely through the walls of the quartz discharge tube. Deposition of materials on the discharge tube walls and devitrification of the quartz when using the helium plasma result in gradual attenuation of sample response with time. The addition of small amounts of oxygen or nitrogen to the helium to act as a scavenger gas[1991] reduces carbonaceous deposits, but deposition of metals and devitrification still present limitations. Since the plasma can be viewed directly when the TM_{010} cavity is employed, the enhanced sensitivity of the atmospheric pressure helium plasma and the high optical resolution of the echelle grating monochromator are combined to afford low detection limits and high selectivity for lead.

Quimby et al[1979] used a Varian 2440 gas chromatograph equipped with a flame ionization detector. Stainless steel chromatographic columns were used. These were a 3 ft x 1/8 inch o.d. column of 5% OV-17 on 100/120 mesh Chromosorb 750 (adapted for on-column injection) which was used for the mercury evaluation, a 3 ft x 1/8 inch o.d. column of 3% QF-1 on 100/120 mesh Varaport 30 (adapted for on-column injection) for the phosphorus evaluation, a 6 ft x 1/8 inch o.d. column of 6% Carbowax 20M on 100/120 mesh Chromosorb P for the silicon determination, and a 6 ft x 1/8 inch o.d. column of 2.5% Dexsil 300 on 100/120 mesh Chromosorb 750 for the remainder of the elements investigated. Helium carrier gas flow rates were 50 ml/min. The injector, detector, transfer line and interface oven were maintained at 170°C for lead, column temperature 130°C.

Estes et al[1995] have applied high resolution gas chromatography with an inert solvent venting interface for microwave excited helium plasma detection to the determination of trialkyllead chlorides. The system incorporates a chemically deactivated, low-volume, valveless fluidic logic gas switching interface for venting large quantities of eluent solvent that would disrupt the helium discharge as sustained by the TM_{010} cylindrical resonance cavity. The same workers[1996] determined n-butylated trialkyllead compounds by gas chromatography with microwave plasma emission detection. Trialkyllead chlorides were converted quantitatively into n-butyltrialkyllead derivatives with a Grignard reagent after extraction. Precolumn trap enrichment allows for determination at the low parts-per-bliion level.

Bonelli and Hartmann[1953], like Barrall and Ballinger[1969], point out that the sensitivities of compounds which have electron affinity properties vary with the conditions of analysis such as column temperature, detector temperature, flow rate, voltage applied across the cell, and the cleanliness of the source and that for this reason, it is advisable to calibrate the instrument frequently with known standards. They also emphasise the importance of the part played by the scrubber section. It adsorbs the column material and thereby maintains the full sensitivity of the detector. It also removes the halogenated lead scavengers by reacting with silver nitrate in the packing. Since these scavengers elute at approximately twice the retention time of tetraethyllead, the time of analysis may be considerably shortened by their removal. If the analysis of these lead scavengers is important, it may be included with the lead analysis by simply using a scrubber without silver nitrate.

Soulages[1997] described an ingenious but complicated analysis scheme. The whole petroleum sample, including the alkyl leads and scavengers, was chromatographed at 60 to 70°C on a column (30 cm x 0.4 cm) of 10% of polyoxypropylene glycol 400 on 30 to 60 mesh Chromosorb P using hydrogen as carrier gas (40 ml/min). The eluant from this column was passed through a hydrogenator where the alkyl leads were catalytically converted over nickel at 140°C to ethane and methane. 1.2-dichlorethane and 1.2-dibromethane scavengers are similarly converted to ethane. The petroleum was almost unaffected. The hydrogenated eluant was then passed through a short column comprising 3% liquid paraffin supported on charcoal which operated at 60 to 70°C retained almost permanently all materials above propane. A flame ionization detector placed at the end of the charcoal column detected the methane and/or ethane from the lead alkyl. In this way the petroleum "background" was separated from chemically produced ethane and methane by a specific chemical reaction followed by sorption. This method avoids the need for a specific detector. Figure 105 shows the three stages of the process.

Soulages[1997] analysed commercial gasolines containing tetramethyl and/or tetraethyllead and various halogenated scavengers. A typical chromatogram is shown in Figure 106.

In later work Soulages[1998] modified the above method in respect of the stationary phase to permit the simultaneous determination of the lead alkyls and the scavengers. The copper column (150 cm x 0.4 cm) used was packed with 20% of 1,2,3-tris-(2-cyanoethoxy)-propane on Chromosorb P (30 to 60 mesh) precoated with 1% of potassium hydroxide and operated at 80°C with hydrogen (40 ml per min) as carrier gas. Recoveries were between 98 and 102% for each tetraalkyllead component in the range equivalent to 0.026 to 0.2 g of lead per litre. Similar results were obtained for the scavengers (dibromo- and dichloro-ethane).

A major advance in lead alkyl analysis was made possible by the development of an electron capture detector capable of operating at high temperatures. The methods described earlier in this section using electron capture detection utilized tritium detectors with a safe upper operating temperature of 225°C. At this temperature high boiling components of the sample and column substrate can condense on the detector giving erratic response thus requiring frequent cleaning and calibration. Green[1999] used a Ni^{63} electron capture detector operating at 300°C and found excellent long-term stability. A 6-foot analyzing column containing 20% 1,2,3-tris (2-cyanoethoxy) propane operated at 90°C separated the lead alkyls and scavengers in about 25 minutes. The lead alkyls could be determined in either petroleum or fuel oil with a sensitivity of 0.15 ppm.

Kramer[2000] used an electron capture detector for the determination of lead alkyls in petroleum. He claims that a complete analysis for tetramethyllead and tetraethyllead in petrol can be achieved on a column (3 metres) of 10% of Apiezon L on Chromosorb W at 120°C with bromobenzene as internal standard. To separate the mixed alkyls (ethyltrimethyllead, diethyldimethyllead and triethylmethyllead, tetramethyllead and tetraethyllead) it is necessary to combine the above analysis with use of a pre-column (5 cm) of 20% of Carbowax 400 saturated with silver nitrate on Chromosorb W impregnated with 6% of potassium hydroxide. The pre-column retains halogen-containing scavengers in the petrol, which would otherwise mask the ethyltrimethyllead peak. As little as 0.002 g of tetramethyllead and tetraethyllead can be detected in 1 litre of petroleum in a 45 minute analysis.

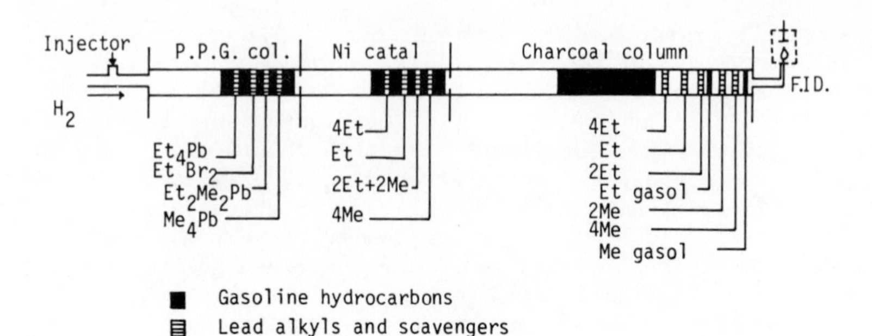

Figure 105 - Catalytic chromatographic process for separation of lead alkyls and scavengers in gasoline.

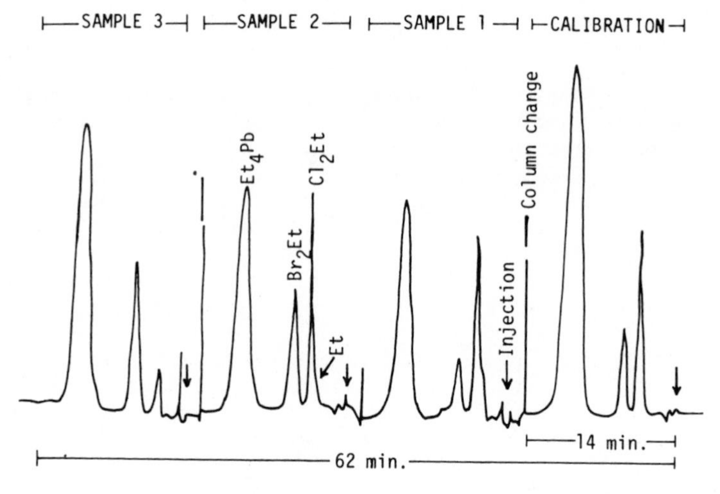

Figure 106 - Continuous analysis of gasolines with ethyl fluid.

Svob[2001] used a combination of gas chromatography and mass spectrometry to analyse lead alkyls. Mutsaars and Van Steen[2002] used a combination of gas chromatography with a flame photometric detector to determine lead alkyls in petroleum. An aluminium column (12 ft x 0.25 in i.d.), packed with 10% of 1,2,3-tris-(2-cyanoethoxy)propane on Chromosorb W HMDS (80 to 100 mesh) was followed by a scrubber (stainless steel, 6 in x 0.125 in) packed with a 1:4 mixture of a saturated solution of silver nitrate in Carbowax 400 and Chromosorb W (30 to 60 mesh) pre-coated with 8% of potassium hydroxide. The column was operated at 90°C with nitrogen as carrier gas (50 ml per min). The spectrophotometer with an oxygen-hydrogen flame and photomultiplier unit were operated at 405.8 nm. The response was rectilinear for samples containing 10 to 1000 p.p.m. of alkyllead compounds.

Pollard et al[2003] have determined retention data for tetraethyllead at various temperatures on a column containing 15% E 301 Silicone oil, (Table 169).

Table 169 - Specific retention volumes V_g (ml) on 15% E 301 silicone oil

Compound	100°	110-5°	140°
$PbEt_4$	1171.5 ± 9.0	713.3 ± 3.3	293.1 ± 5.0

Longi and Mazzochhi[1843] have gas chromatographed mixtures of lead, tin and antimony tetraethyl on columns comprising 7.5% paraffin wax on Chromosorb W at 143°C with an injection temperature of 196°C and using helium as a carrier gas.

Dressler et al[2004] have studied the response of the alkali flame detector to lead alkyls.

Hill and Aue[2005] have described a spectrophotometric detector incorporating a conventional flame ionization detector to determine volatile organolead compounds. They preferred the hydrogen rich flame as this gave sensitivities in the lower nanogram region and had a strong discrimination against hydrocarbon sample solvents, this made the technique eminently suitable for the determination of organolead compounds in gasoline.

A very elegant analytical technique for the lead alkyls is that of Ballinger and Whittemore[2006]. They combined pressure programming with use of an atomic absorption spectrophotometer as a specific detector to produce a rapid, precise, and sensitive analytical technique. A 10 foot column packed with 20% 1,2,3,-tris-(cyanoethoxy)-propane on 60/80 mesh Chromosorb P coated with 1% potassium hydroxide was operated at 85°C. Flow rates were programmed from 10 - 1.00 ml/min. Analysis of the five lead alkyls was completed in less than one and a half minutes. The amount of lead was determined by the absorption of the lead 2833 Å emission line. The method could detect as little as 2- nanograms of lead as lead alkyl. The application of atomic absorption spectroscopy to the determination of lead alkyls separated by gas chromatography has also been discussed by other workers [2006-2011].

Kolb and others[2011] used a process whereby a column effluent from the gas chromatograph was introduced by means of a heated transfer tube to the nebulizer of a flame atomic absorption spectrometer. The same procedure was adopted by Katou and Nakagawa[2012] and Chau et al[2013], Coker[2014] utilized a slightly different method of interfacing. Instead of introducing the column effluent to the nebulizer, it was passed directly to the burner head itself. In this way a detection limit of about 20 ng expressed as lead could be obtained for each alkyllead compound.

Some authors have also described the use of non-flame atomization systems. Chau et al[2008] utilized a heated silica furnace in the a.a.s. unit, and in this way obtained a detection limit of about 0.1 ng. Segar[2015] interfaced the gas chromatograph directly to a graphite furnace; a tungsten transfer line introduced the column effluent through an enlarged sample entry hole of the graphite tube, so that the effluent impinged directly on the heated graphite surface. A sensitivity of approximately 10 ng was thus obtained for the determination of trimethylethyllead. A similar method was described by Radzuik et al[2010]. The effluent from the g.c., however, was lead to the entry hole of the graphite tube through a tantalum connecting piece. Robinson et al[2016] designed a "hollow-T" carbon atomizer. The column effluent was transferred to the bottom of this furnace by means of a heated stainless steel transfer line. For tetramethyllead the detection limit was about 0.04 ng.

A comparison of the various atomizing systems leads to the conclusion that the heated graphite furnace atomizer gives the highest sensitivity for the determination of tetraalkylleads separated by gas-liquid chromatography[2017,2018]. Nevertheless, there are some demerits to the methods of interfacing proposed to date. Segar[2015] reported that his gas chromatographic furnace atomic absorption spectrometric system was only crudely achieved and that he could not control the temperature of the tungsten transfer line. Moreover, he reported the inconvenience caused by solvent peaks, even with deuterium correction. Radzuik et al[2010] reported that the "solvent peak" absorption, thought earlier to result from molecular absorption of radiation at the lead wavelength, was not caused by the solvent itself but by deposition and remobilization of lead. The tantalum transfer tube held at high temperature was held responsible for this deposition. In the same apparatus also, the quartz windows were removed from the furnace assembly to allow an optimal gas flow from the gas chromatograph, a modification which could easily lead to a decrease in sensitivity. The system of Robinson et al[2016] did not suffer from these problems since its special design seems to have allowed total destruction of the solvent molecules before they reached the light path, reducing the molecular absorption considerably. Further, no part of the transfer line was exposed to very high temperatures, thus avoiding any possible deposition and remobilization of lead. The adaption of the atomic absorption spectrometric unit to function as a gas chromatographic detector is, however, rather difficult and requires a specially designed atomizing system.

De Jonghe et al[2019] have described an alternative way of interfacing a gas-liquid chromatograph with a graphite-furnace atomic absorption spectrometer for the determination of tetraalkyllead compounds, in which the difficulties confronted by Segar[2015] and Radziuk et al[2010] are removed. Moreover, commonly used gas chromatographic and atomic absorption units are utilized without any drastic modifications, so that the spectrometer can be made operational as a gas chromatographic detector in less than one minute. These workers reported a systematical study of the behaviour or organolead compounds in this system, as well as the optimization of various instrumental parameters, and applied the procedure to gasoline samples. Absolute detection limits range between 40 pg for tetramethyllead and 90 pg for tetraethyllead. For all species the reproducibility is better than 2% at the nanogram level.

In this method a Varian 3700 gas-liquid chromatograph is equipped with two identical glass columns (6.2 mm o.d., 2 mm i.d., 180 cm long) packed with 3% OV-101 on 100 - 120 mesh Gaschrom Q operated at 75°C or programmed from 50 to 175°C at 20°C per minute. One column is connected to a flame ionization detector, maintained at 200°C; the other is connected to a Perkin-Elmer HGA-74 graphite furnace atomizer in the following way: in the cavity of the g.c. provided for a second detector, a closely fitting copper cylindrical block is mounted with a concentric hole of about 6.2 mm diameter. Through this hole, a glass transfer tube (6.2 mm o.d. by 0.5 mm i.d.) is inserted and connected to the column by means of Swagelok fittings. On the other end of the transfer tube is mounted a silica T-piece which is connected to both the inner gas flow entrances of the graphite furnace. In this way the column effluent is forced to follow the same flow pattern and in fact replaces the normal gas flow inside the graphite tube. It follows a symmetric path from both ends of the tube and exits through the normal sample introduction hole. During the gas chromatographic elution of the tetraalkylleads, the graphite furnace is maintained continuously at the required atomization temperature. It is necessary to heat the transfer line to about 200°C. This is accomplished by wrapping a heating wire around it and shielding it with asbestos tape. No part of the transfer tube is exposed to temperatures above 200°C. Further, as the column effluent is introduced symmetrically from both sides of the graphite tube, effective decomposition of the largest part of the solvent molecules occurs and remobilization effects of lead are prevented. In this way, deuterium background correction is sufficient to correct for the background absorption.

The atomic absorption spectrometer was a Perkin-Elmer 503 model equipped with deuterium background correction. The working conditions of the flame ionization detector are: oxygen flow rate 300 ml min^{-1} and hydrogen flow rate 30 ml min^{-1}.

For accurate analysis of the tetraalkyllead compounds present in high concentrations, dilutions of the gasoline with pure diisopropyl ether are necessary. Some results are summarized in Table 170, which also shows the total lead concentration obtained by graphite-furnace atomic absorption spectrometry, after wet digestion of the samples. For the decomposition, 5 µg of gasoline is shaken vigorously for 5 min with 1 ml of concentrated nitric acid, as proposed by Chau et al[2013]. The results for total lead are somewhat higher than the combined results for the alkyllead compounds. The reason for this is probably the presence of inorganic lead and other lead species of the type PbR, X, $PbR_2^{2+}X_2^{2-}$ which would not be determined in the gas chromatography atomic absorption spectrometric procedure.

Table 170 - Organolead concentrations in gasoline (ppm)

Manufacturer	Grade	TML	TMEL	DMDEL	MTEL	TEL	Total	Total by digestion
A	High octane	112	187	90	19	7.9	416	426
	Normal	118	200	93	21	8.9	439	447
B	High octane	130	246	121	25	5.7	528	569
	Normal	139	253	117	21	9.4	539	564
C	High octane	406	0.7	0.5	0.5	2.1	410	414
	Normal	395	0.8	0.5	0.3	2.7	399	410
D	High octane	2.9	0.3	0.4	0.7	474	478	500
	Normal	2.8	0.2	0.2	0.7	351	355	367
E	High octane	346	0.4	0.4	0.3	72	419	441
	Normal	130	0.5	0.3	2.1	353	486	455
F	High octane	3.1	0.6	0.5	0.9	445	450	461
	Normal	4.5	0.4	0.2	0.6	343	349	357

Robinson et al[2009] have studied in detail the atomization processes occurring in furnace analyses during the atomic absorption spectrometry of organolead compounds in gasoline previously separated on a gas chromatograph.

Forsythe and Marshall[2020] have evaluated the performance of an automated gas chromatograph-silica furnace atomic absorption spectrometer for the determination of alkyllead compounds. These workers ascertained that the decomposition-atomization of alkyllead compounds is a hydrogen radical mediated process.

ORGANOLITHIUM COMPOUNDS

A.Vinyl Lithium

Earlier workers[2021-2024] have studied the hydrolysis or alcohoysis of vinyl lithium to ethylene as a basis for estimating this compound. Hydrolysis coupled with identification of the gas produced by mass spectrometry[2024-2026] and by infra-red analysis[2024,2025] have also been used. Leonhardt et al[2027] based their gas chromatographic method for estimating vinyl lithium in tetrahydrofuran and in diethyl ether on a similar reaction. Their method involves the use of two gas chromatographic units connected in series. An aliquot of a solution of vinyl lithium is hydrolysed in a short precolumn containing a solid support coated with a high boiling polyol:

$$CH_2 = CH - Li + H_2O \rightarrow CH_2 = CH_2 + Li\ OH$$

The vinyl lithium concentration of the solution is determined quantitatively by comparing the amount of ethylene formed on hydrolysis with known ethylene standards. Impurities such as acetylene and 1 - 3 butadiene are separated from ethylene on the second gas chromatograph.

Leonhardt et al[2027] developed a gas chromatographic method of assay for vinyl lithium containing lithium acetylide as impurity. These workers connected two gas chromatographs in series. This scheme permitted two quantitative separations requiring different columns and different gas chromatographic conditions via a single injection. The column specifications and instrument conditions employed are given in Table 171.

Table 171 - Two-column gas chromatographs conditions for vinyl lithium assay

Chromatographs	Wilkins A-90-P	Wilkins A-700
Column	5' x 1/4" (SS) X 20% didecyl phthalate on 60/80 firebrick (C-22)	20' x 1/4" (SS) X 20% dimethylsulfolane on 60/80 firebrick (C-22)
Column temperature	75°C (isothermal)	35°C (isothermal)
Injector temperature	ambient	ambient
Detector temperature	270°C	270°C
Gas sampling valve temp.	31°C	
Filament current	175 ma	175 ma
Carrier	helium	helium
Carrier gas flow rate	90 psi (in)	30 cc/min

Figure 107 displays a chromatogram obtained from a typical vinyl lithium analysis run on the two instrument hook-up. This Figure also shows the ethylene peak from the pre-column hydrolysis. This peak includes residual ethylene as well as acetylene. The large broad peak is the solvent tetrahydrofuran.

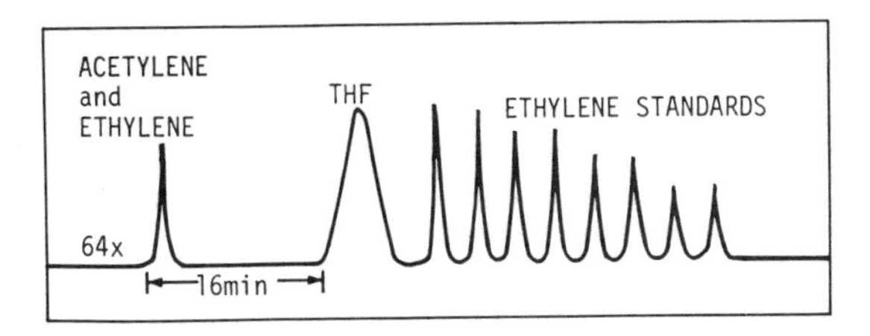

Figure 107 - Analysis of vinyl lithium by hydrolosis - gas chromatography. Ethylene plus acetylene unresolved on precolumn.

Leonhardt et al[2027] obtained good agreement between their gas chromatographic method and the vanadium pentoxide method[2028]. They believe the method is applicable to the assay of most organolithium compounds which quantitatively react with active hydrogen containing compounds to replace the lithium in the organolithium with active hydrogen:

$$RLi + H^+ \rightarrow RH + Li^+$$

B. Phenyl Lithium

Bernstein[2030] has also described a technique based on gas chromatography for the determination of organolithium compounds. In this procedure, a measured volume of phenyllithium solution is slowly transferred under nitrogen into an excess of a solution of iodine in diethyl ether, with stirring. The excess of iodine is removed from the ethereal phase by shaking with dilute sodium hydroxide solution and the concentration of iodobenzene in the ether phase is then determined by gas chromatography. Other organic compounds of lithium do not interfere. The accuracy was within 2%.

C. Methyl Lithium

To determine methyllithium, House and Respers[2030] reacted the sample with dimethylphenylchlorosilane (Dow Corning) in ether solution to form phenyltrimethylsilane:

$$CH_3 - Li + C_6H_5 - \underset{CH_3}{\overset{CH_3}{\underset{|}{\overset{|}{Si}}}} - Cl \rightarrow C_6H_5 - \underset{CH_3}{\overset{CH_3}{\underset{|}{\overset{|}{Si}}}} - CH_3 + C_6H_5Li$$

Reaction is complete in less than 1 hour at room temperature with an excess of chlorosilane. The ether solution is washed with aqueous ammonium chloride and the quantity of trimethylphenylsilane produced is estimated by gas chromatography using cumene as an internal standard and Silicone Fluid No. 710 on Chromosorb P as column packing.

ORGANOMAGNESIUM COMPOUNDS

A. Alkyl and Aryl Magnesium Halides

D'Hollander and Anteunis[2031] mention the use of gas chromatography in the analysis of n-propyl magnesium bromide. They reacted the organomagnesium compound with an excess of diisopropyl ketone to form a compound which at elevated temperatures produces propylene in almost quantitative yield. The propylene was determined gas chromatographically.

$$n - C_3H_7MgBr + (nC_3H_7)_2\ C = O \rightarrow (n\ C_3H_7)_2\ C\begin{matrix} OMg\ Br \\ H \end{matrix} + C_3H_6$$

It was found that enolization reactions occur to an extent of less than 1% and that condensation reactions occur to a negligible extent.

House and Respess[2032] have studied the gas chromatography of organomagnesium compounds much more exhaustively and have described a method for the determination of methyl magnesium bromide and magnesium based on reaction of the sample with dimethylphenylchlorosilane to produce phenyltrimethylsilane which is then estimated by gas chromatography.

$$CH_3MgBr + C_6H_5 - \underset{\underset{(I)}{CH_3}}{\overset{CH_3}{Si}} - Cl \rightarrow C_6H_5 - \underset{\underset{(II)}{CH_3}}{\overset{CH_3}{Si}} - CH_3 + Mg\ Br\ Cl$$

The reaction is effected in ether solution at room temperature with an excess of the chlorosilane and is complete in less than 1 hour. After the ether solution has been washed with aqueous ammonium chloride, the quantity of the phenyltrimethylsilane produced can be determined. In practice, an aliquot of an ethereal solution containing excess chlorosilane and a known weight of the internal standard, cumene, is added to an aliquot of an ether solution of methyllithium or the methylmagnesium derivative. After the reaction is complete and the reaction mixture has been washed, the ether solution is separated and analyzed by gas chromatography.

Using a column packed with Silicone Fluid 710 suspended on Chromosorb P, magnesium methoxide present in samples of methyl magnesium did not interfere in the determination of these substances.

Molinari et al[2033] have described a procedure for the determination of organomagnesium derivatives by gas chromatography.

In this method, solutions of organomagnesium derivatives are analysed by hydrolysis with concentrated phosphoric acid in a micro-reactor and passing the hydrocarbons evolved through a bypass injector into a gas-chromatographic apparatus (with an activated silica-gel column, for C_1 - C_3 hydrocarbons). Concentrations of alkyl groups are then calculated from the peak areas in the usual way.

Guild et al[2034] applied gas chromatography to the analysis of arylmagnesium compounds. In particular, they discuss the analysis of para-tert butylphenylmagnesium bromide. This method distinguishes between active Grignard reagent and hydrolysed Grignard reagent. The separation was carried out on a column comprising polypropyleneglycol supported on C-LL firebrick. The analysis was performed with a sample tube constructed as shown in Figure 108. This tube facilitated the transfer of samples with a minimum of exposure to the atmosphere. The total volume of the tube was approximately 10 ml. The end of the tube were fitted with silicone seals as shown, and the gas chromatographic column was also fitted with a silicone seal.

Several milliliters of the arylmagnesium compound are added to the

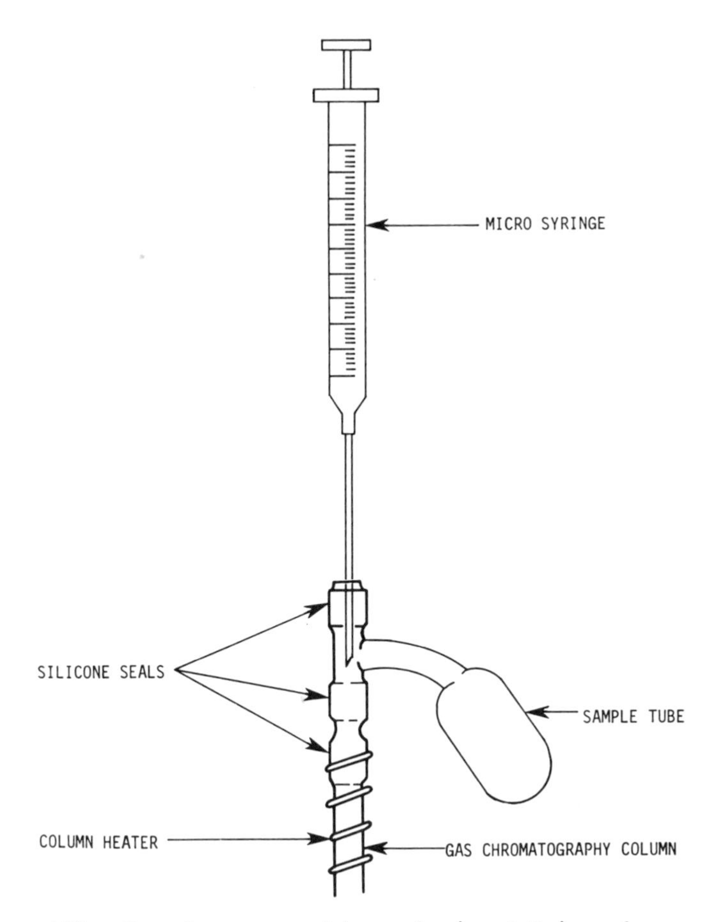

Figure 108 - Gas chromatographic analysis of Grignard compounds.

tared sample tube. A weighed amount, approximately 1.0 ml, of an internal standard n-nonane, is added by a syringe, and the tube again weighed. The internal standard and the sample reagent are mixed by shaking. A micro-syringe is then inserted through one of the silicone seals in the sample tube, and after the tube is tilted to trap liquid above the seal approximately 0.005 ml of the sample is drawn into the syringe care being taken to free the sample of gas bubbles. With the syringe still in place the tube is positioned as shown in Figure 108 and pressed tightly against the top of the silicone seal in the gas chromatographic column. The needle is then forced down through the double seal compressing the seal on the sample tube and the seal in the top of the chromatographic column and the sample injected. These data are used to determine the amount of organo-magnesium component of the sample that had undergone hydrolysis prior to analysis. After the chromatogram is obtained the unreacted organomagnesium compound in the column is destroyed by the addition of methanol and water to the column (0.05 ml) which themselves are removed by elution before proceeding. Using a syringe, excess methanol is added to the sample tube, and 5 μl of sample is again injected into the chromatographic column by the technique already described. From the chromatogram the total arly magnesium content originally present in the sample could be obtained.

B. Vinyl Magnesium Halides

Wowk and Di Giovanni[2035] have described a gas chromatographic method for the analysis of vinyl Grignard reagents. The sample is treated with a large excess of tributyltinchloride in tetrahydrofuran and the magnesium salts formed (or their tetrahydrofuran derivatives) are precipitated with hexane, and filtered off. The filtrate is heated to evaporate the solvent, and the residue, containing unconsumed butylin compounds, is analysed chromatographically. This method distinguishes the vinyl magnesium from other compounds resulting from hydrolysis, oxidation or decomposition of the Grignard compound.

Wowk and Di Giovanni[2035] found that after a vinyl Grignard compound had been stored for some time, then hydrolyzed with dilute acid, the gases produced, in addition to the expected ethylene, also contained considerable amounts of ethane, hydrogen and several C_4 and C_5 alkanes and alkenes, making a gas-volumetric method unsuitable for assay purposes.

Their proposed method is essentially a test in which a vinyl Grignard is used to vinylate an organometallic halide. An aliquot of the sample being tested is reacted with a large excess of tributyltinchloride. The inorganic magnesium salts formed (or their THF complexes) are then precipitated with an alkane and filtered. The solvents are evaporated and the residue containing only butyltin compounds is analyzed by gas-liquid chromatography. The organotin mixtures contained as their main components tributyl-vinyltin and the large excess of tributyltinchloride. If it is assumed that the vinylmagnesium chloride couples quantitatively with the large excess of tributyltinchloride, the amount of tributylvinyltin in the coupling product is a measure of the reactive $CH_2 = CH\ Mg$ moiety in the Grignard reagent being assayed.

A representative chromatogram is shown in Figure 109.

For most samples of freshly prepared vinyl Grignard reagent, the total product is volatile and determinable to within ± 2% of the sample weight. However, in the case of an aged vinyl Grignard the volatiles total only 70% of the sample weight indicating the presence of 30% non-volatile material in the coupling product. Where the sole objective is to determine the concentration of the vinyl Grignard reagent, only the amounts of tributylvinyltin and dibutyldivinyltin need be determined.

C. Etioporphyrin Magnesium Chelates.

Karayannis and Corwin[2036] have applied hyperpressure gas chromatography to the etioporphyrin magnesium chelates. Chromatography was carried out at 145°C and 1000 - 1700 psi using dischorodifluoromethane as carrier gas and 10% Epon 1001 Chromosorb W as column packing. These conditions separated the etioporphyrin chelates of magnesium II, copper II, nickel II, cobalt III, vanadyl, titanyl, manganese II, zinc II, platinum II and palladium II.

ORGANOMANGANESE COMPOUNDS

Turkel'taub et al[2037] have described a gas chromatographic procedure for the determination of tricarbonylcyclopentadienyl-manganese, tritolyl phosphate and various solvents in anti-detonator preparations. These are analysed on a column of Apiezon L on Celite 545, temperature programmed from 60 to 240°C using helium as a carrier gas.

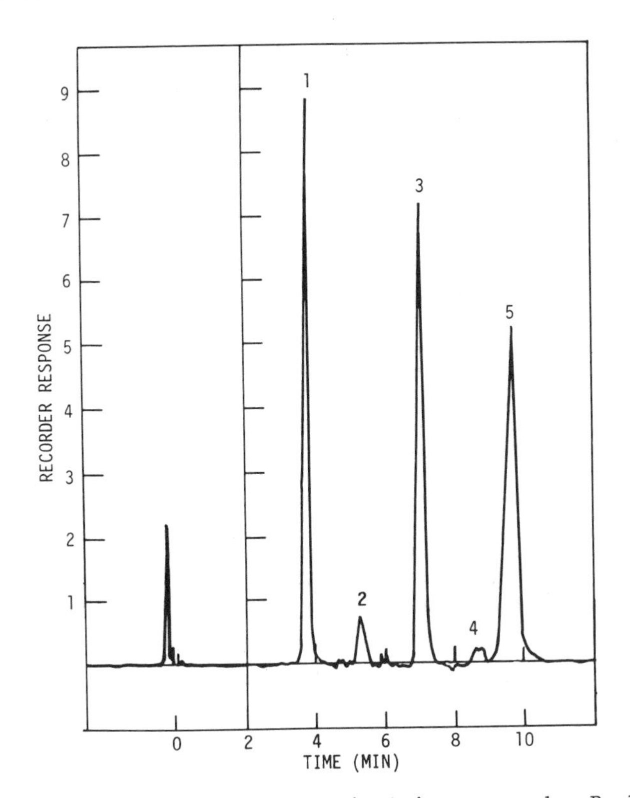

Figure 109 - Gas chromatogram of butlyvinyltin compounds. Peak 1 dodecane, 2 dibutylvinyltin, 3 tributylvinyltin, 4 tetrabutyltin, tributyltinchloride.

Uden et al[2038] have described a method for the determination of methylcyclopentadienylmanganesetricarbonyl ($CH_3C_5H_4Mn(CO)_3$) in gasoline utilizing gas chromatographic separation with interfaced specific manganese detection by means of dc argon plasma emission spectroscopy. The procedure is rapid, free of interference, specific and requires little sample preparation. The use of cyclopentadienylmanganesetricarbonyl (cymantrene) as an internal reference yield a precision of ± 0.8 - 3.4% relative standard deviation. The limit of detection is approximately 3 ng of manganese metal as the complex.

These workers used a prototype spectraspan III dc plasma echelle spectrometer, 510 - 512, (Spectrametrics Inc., Andover, Mais.). They adapted a Varian 1200 gas chromatograph for on-column injection onto a 6 ft x 1/8 in o.d. stainless steel column packed with 2% Dexsil 300 GC on Chromosorb 750, 100;120 mesh (Johns-Manville Corp., Denver, Col.). Column effluent was split by an approximately 1:1 ratio between the flame ionization detector of the gas chromatograph and a heated, thermal, and electrically insulated 1/16 in p.d. stainless steel transfer line to the dc plasma. Preheated argon sheath gas was required in addition to the argon supplied to sustain the plasma, in order to optimize spectral sensitivity. The column and injection port temperature were set at 130 and 160°C, respectively, and the interface temperature was 1700°C. Helium carrier gas flow rate was 25 ml/min.

The overall method for gasoline requires no sample preparation other than the addition of internal references. The method is fast since samples may be run at 3-min intervals, interference-free and specific for methylcyclopentadienylmanganesetricarbonyl. Analysis time comparison is favourable with respect to atomic absorption analysis which typically required about 15 min per sample when run in batches. Further, the procedure is readily adapted for use with open tubular columns. Porous layer open tubular columns are best suited to accomodate the direct injection of gasoline samples. The improvement in analysis time for methylcyclopentadienylmanganesetricarbonyl determination by the latter method may outweigh the somewhat poorer limit of detection resulting from limitations on the sample size.

The typical dc plasma emission spectrometer response from a 5 µl injection of a standard iso-octane solution containing 10 µg/ml (i.e. 50 µg) methylcyclopentadienylmanganesetricarbonyl and 20 µg/ml (i.e. 100 ng) cyamantrene is shown in Figure 110. The small negative response directly upon injection corresponds to the passage of the solvent through the plasma. The dc plasma tolerates large volumes of eluted solvent. This is an important advantage of this system. Baseline resolution was obtained under isothermal conditions at 130°C, and no peak distortion was noted by comparison with flame ionization detection (although in this case the peaks were superimposed on the iso-octane solvent tail). No degradation of gas chromatographic resolution for this system imparted by the heated transfer line to the plasma is apparent.

Uden et al[2038] studied the long-term stability of the response of the dc argon plasma over a period of some hours. The results for the ratio of methylcyclopentadienylmanganesetricarbonyl to cymantrene indicated that no notable variation was evident after 4 h, and the precision obtained was ± 3%. Experience has indicated that, for the same standard solutions, this reproducibility can be maintained throughout a working day. During long term operation, the plasma electrodes are gradually consumed, and this slightly affects the spacial relationship of the discharge to the spectrometer entrance slit. However, the methylcyclopentadienylmanganesetricarbonyl to cymantrene ratio was found constant for the concentration range of interest over a number of days.

A typical response curve for methylcyclopentadienylmanganesetricarbonyl using the Mn 1 279.83 nm line is given in Figure 111. This resonance line is among the more sensitive lines for dc argon emission of manganese. Linearity is observed over the full range of analytical interest, the upper limit being taken as the injection of 2 µl of a 170 µg/ml solution of methylcyclopentadienylmanganesetricarbonyl (i.e. 340 ng methylcyclopentadienylmanganese tricarbonyl). All commercial gasoline samples examined fell well below this limit for methylcyclopentadienylmanganesetricarbonyl. The flame ionization detection response for methylcyclopentadienylmanganesetricarbonyl standards is also linear over the same sample size ranges.

Typical relative standard deviation based on the repetitive injections for standards and samples ranged between 0.8% and ± 3.4% depending on the degree of the plasma stability. To obtain good precision, all parameters which affect the plasma stability, notably the various gas flow rates, were carefully controlled.

Methylcyclopentadienylmanganesetricarbonyl values obtained for a range of commercial gasolines ranged from 15 to 120 µg per ml.

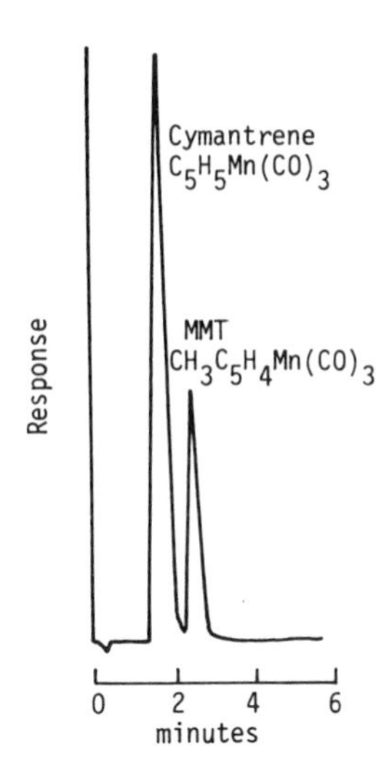

Figure 110 - DC argon plasma emission chromatogram of cyclopentadienylmanganesetricarbonyl (cymantrene) (100 ng) and methylcyclopentadienylmanganesetricarbonyl (MMT) (50 ng) in 5 μl isooctane solution. Column 2 meter, stainless steel, 1/8 in. o.d. 2% Dexsil 300 GC on 100 - 120 mesh Chromosorb 750, 130°C.

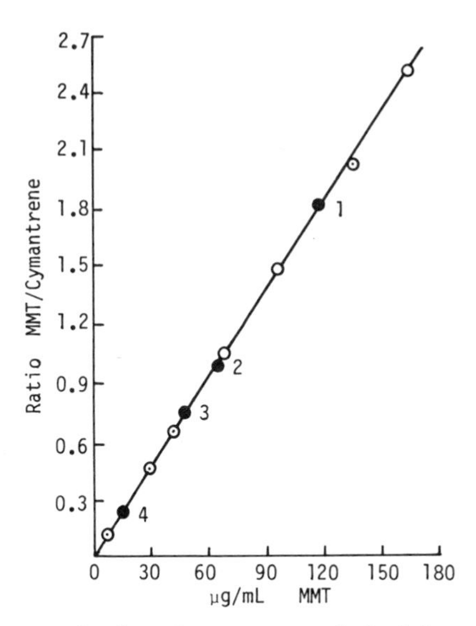

Figure 111 - DC plasma emission detector peak height response for MMT as a function of MMT/cymantrene ratio. Detection at 279.83 nm (each calibration point ⊙ represents three replicate 2 μl injections of isooctane solution). Response ratios for commercial gasolines represented as 0.

DePuis and Hill[1972] have described a method for the determination of methylcyclopentadienylmanganese in amounts down to 1.7 x 10^{-14} g/s of manganese. In this work they used a hydrogen atmosphere flame ionization detector (HAFID) modified from a commercial flame ionization detector. The apparatus is described in detail in the section on organolead compounds. In the HAFID detector, the detector temperature was maintained at 250°C and the total hydrogen flow rate was held at 1600 ml/minute and for optimal response was doped with 34 ppm of silane by mixing pure hydrogen and hydrogen doped with 100 ppm of silane. Air, 120 ml/min was enriched with 150 ml/min of oxygen before entering the jet tip. On the flame ionization detector, the detector temperature was maintained at 250°C. Flow rates used were 30 ml per minute for hydrogen and 240 ml per minute for air. DePuis and Hill[1972] used a Hewlett-Packard 5830A gas chromatograph with a dual flame ionization detector (one modified to HAFID). The column 6 ft x 1/4 in p.d., 2 mm i.d.) borosilicate packed with 80/100 mesh Ultra Bond 20 M (RFR Corp. Hope, R.I.) was interchanged between detectors. The injection port temperature 225°C and the carrier gas flow was 20 ml/min of helium. A column temperature of 140°C was used. Standard solutions of methylcyclopentadienylmanganese in isoctane were prepared fresh each day in the 10 µg/µ 1 to 10 pg/µ 1 concentration range in brown bottles to avoid light induced decomposition of the organomanganese compound. Gasoline samples for analysis were analyzed by injecting 0.25 µ 1 of the gasoline directly into the column.

Methylcyclopentadienylmanganese had a minimum detectable amount of 200µg and a detectable limit of 6.6 x 10^{-14} g/s for methylcyclopentadienylmanganese or 1.7 x 10^{-14} g/s when calculated as manganese. The amount of methylcyclopentadienylmanganese found in unleaded gasoline was 24 mg/gal (6 mg/gal for Mn). Since methylcyclopentadienylmanganese decomposes readily, the manganese value obtained with this method accurately reflects the level of active antiknock agent but does not necessarily represent the total manganese concentration in the sample.

Quimby et al[1979] used a helium carrier gas flow rate of 50 ml per minute and an injection point temperature of 200°C. The wavelength setting of the monochromator was optimized for manganese mercury (257.6 nm) using a hollow cathode lamp and a small mirror placed between the lens and the cavity. The wavelength setting was optimized by introducing small amounts of methylcyclopentadienylmanganese vapour into the plasma by connecting with a tee to a hydrocarbon solution of this compound. Quimby et al[1979] investigated the effect of the total flow rate of helium through the discharge tube on response to organomanganese compounds by repeatedly injecting a standard solution while varying the "helium plasma" flow with the carrier gas flow rate and column temperature maintained constant. The response increases with increasing flow rate reaching a maximum at 220 ml/min decreasing sharply thereafter.

A gas chromatogram demonstrating the sensitivity of the microwave emission detector to methylcyclopentadienylmanganese is shown in Figure 112. The quantity of methylcyclopentadienylmanganese referred to is the amount entering the plasma (i.e. amount injected corrected for the split ratio). The very large selectivity ratio wavelength of analysis (ratio of peak response per gram - atom of manganese to the peak response per gram - atom of carbon as n-decane) obtained for manganese results from a combination of two factors, (i) the high sensitivity observed for this element, and (ii) the favourable wavelength region employed with respect both to optical resolution of the monochromator and the minimal interference by molecular band emission from hydrocarbons.

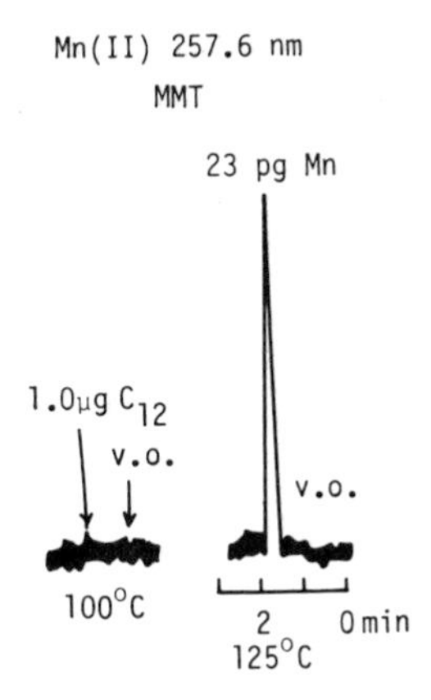

Figure 112 - Element selective gas chromatogram with methylcyclopentadienylmanganese GC-MED system. Column temperature indicated on chromatogram.

The detection limit, defined as the main flow rate of element entering the plasma required to produce a signal to noise ratio of two as listed below, together with the selectivity. Manganese 257.6 nm, detection limit 0.25 pg/s, selectivity 1.9×10^6.

Coe et al[2039] used a combination of gas chromatography and atomic absorption spectrometry to determine methylcyclopentadienylmanganestricarbonyl.

ORGANOMERCURY COMPOUNDS

Broderson and Schlenker[2040] have applied gas chromatography to the separation of mixtures of compounds of the type RHg Br where R is methyl, ethyl, n-propyl or n-butyl. Hydrogen was used as carrier gas and the column was packed with Dow Corning silicone 550 and maintained at 190 - 220°C. The sample (50 μl) was introduced into the column as a 10% solution in tetrahydrofuran.

In Table 172 are shown retention data for the four alkyl mercury bromides at column temperatures of 190 and 200°C.

Table 172

	Column temperature (°C) at 60 psi H_2 - carrier	Retention time min.
CH_3-Hg-Br	190	3.0
	220	1.7
C_2H_5-Hg-Br	190	5.6
	220	3.0
nC_3H_7-Hg-Br	190	6.2
	220	3.5
nC_4H_9-Hg-Br	190	9.4
	220	4.7

Nishi and Horimoto[2041] identified organomercury II compounds by gas chromatography using subtractive techniques. This technique described can be used to identify methylmercury, ethylmercury and phenylmercury compounds. The peaks obtained on a column (1 metre x 4 mm) of 2 or 3% of poly (dimethylene glycol) on Chromosorb W (60 - 80 mesh) containing 5% of sodium chloride at 130°C (for methyl and ethyl mercury compounds) or 180°C, with nitrogen (60 ml per min.) as carrier gas, disappear when the sample solution (e.g. 1 ml of benzene solution containing 1 part per 10^9 of methyl mercury) is shaken with 1 ml of an aqueous solution of an inorganic sulphide or thiosulphate for 5 min., or with an aqueous suspension (0.5 ml) of aluminium, iron, nickel, zinc or Devarda alloy (100 mg for 60 min.). The peaks also disappear if a column (20 cm x 4 mm) containing the metal powder is incorporated between the separation column and the detector.

Dressman[2042] showed that some phenyl-mercury salts upon injection into the gas chromatographic column are converted to diphenylmercury and phenylmercury chloride. He injected solutions of phenylmercury salts into a glass column (93 cm x 4 mm) packed with 5% of DC-200 plus 3% of QF-1 on Gas-Chrom Q. A dual flame ionization detector was used. The injection port was fitted with a silica insert held at 150° - 250°C. The column was temperature-programmed from 140° - 200°C at 10° per minute and the carrier gas was nitrogen (50 ml per minute). Fractions were collected in capilliary tubes and analysed by re-injection into the chromatograph, by infrared spectrometry and spectrometry on a time-of-flight instrument. The chief thermolytic product from the salts was found to be diphenylmercury. Phenylmercury chloride shows very little thermal decomposition and Dressman[2043] recommends that organomercury salts should be converted into their respective chlorides before gas chromatographic analysis.

Baughman et al[2044] studied the gas chromatographic behaviour of methylmercury compounds on a glass column (6 ft x 0.25 in) packed with 5% of DEGS on Chromosorb W and operated at 160°C, and of phenylmercury compounds on a similar column packed with 3% of OV-1 on Chromosorb W and operated at 150°C. Flame ionization and ^{63}Ni electron-capture detectors were used. Dimethyl- and diphenylmercury were stable under these conditions, but combined glc-ms confirmed that methyl- and phenylmercury salts decompose during gas chromatography. Reliable determination of methylmercury salts were achieved only on columns specially treated so as to make the decomposition reaction reproducible. Phenylmercury salts, which decompose extensively, could not be determined by gas chromatography.

Teramoto et al[2045] used a column comprising 25% diethyleneglycol succinate on Chromosorb for the separation of methylmercury compounds.

Hey[2046] has used atomic absorption spectrophotometry as a mercury specific detector system for the gas chromatography of organomercury compounds. Organomercury compounds in the effluent from the column are burnt in the flame ionization detector, and the resulting gases are passed through a 10% solution of stannous chloride in 20% aqueous sulphuric acid to reduce all compounds to the metal. The gases containing the mercury vapour are passed to an atomic absorption spectrophotometer and the mercury is determined at 253.7 nm. The limit of detection is 10 μg. Hey discusses precautions necessary to avoid the loss of mercury from the system. The determination of alkyl mercury compounds as halides by gas chromatography has been discussed by Rodriguez-Vasquez[2047].

Bache and Lisk[2049] determined dimethylmercury and methylmercurychloride in fish by emission spectrometry in a helium plasma. Dimethylmercury was chromatographed on a 2 ft glass column of Chromosorb 101. Methyl mercury salts were separated on a 6 ft column of 1:1 OV-17/QF-1. The separated compounds were detected by their emission spectra at the 2537A° atomic mercury line which gave a linear response for 0.1 - 100 µg of injected methylmercurychloride.

More recent work on the determination of alkylmercury compounds has been centred on the applicability of the helium microwave glow discharge detector as a gas chromatographic detector. This has been discussed by Bache and Lisk[2049.2050], Grossman[2051], Talmi[2052] and Quimby et al[1979], who examined the applicability of helium microwave glow discharge detectors to the detection of diphenylmercury. Details of the microwave emission detector and the experimental set-up are described in detail in the section dealing with organolead compounds. These workers used an atmospheric pressure helium (or argon) plasma as this leads to enhanced sensitivity and high optical resolution and selectivity.

Quimby et al[1979] used a helium carrier gas flow rate of 70 ml per minute and an injection port temperature of 200°C. The wavelength setting of the monochromation was optimized for mercury using a hollow cathode lamp and a small mirror placed between the lens and the cavity. The wavelength setting was optimized by introducing small amounts of dimethylmercury vapour into the plasma by connecting with a hydrocarbon solution of this compound. Quimby et al[1979] investigated the effect of the total flow rate of helium through the discharge tube on response to organomercury compounds by repeatedly injecting a standard solution while varying the "helium plasma" flow with the carrier gas flow rate and column temperature maintained constant. The response for organomercury compounds was found to be significantly effected by the total flow rate of helium through the discharge tube. The response remains constant over the range 42 - 50 ml per min., then decreases sharply with increasing flow rate.

A gas chromatograph demonstrating the sensitivity of the microwave emission detector to diphenylmercury is shown in Figure 113. The quantity of diphenylmercury referred to is the amount entering the plasma. (i.e. amount injected corrected for the split ratio).

The selectivity for mercury at the wavelength of analysis is the ratio of peak response per gram - atom of mercury to the peak response per gram - atom of carbon as n-decane. The very large selectivity ratio obtained for mercury results from a combination of two factors (i) the high sensitivity observed for this element, and (ii) the favourable wavelength region employed with respect both to optical resolution of the mono-chromator and the minimal interference by molecular band emission from hydrocarbons.

The detection limit, defined as the mean flow rate of element entering the plasma required to produce a signal to noise ratio of two is listed below together with the selectivity: Mercury 253.7 nm, detection limit; 1 pg/s, selectivity 9.1×10^4.

ORGANOMOLYBDENUM COMPOUNDS

A reference to the gas chromatography of organomolybdenum compounds found is to that of the work of Umilin and Tyutyaev[2053] on impurities in bis(ethylbenzene)molybdenum obtained in the Friedel Crafts reaction. In this method the sample (1 to 2 µl) is applied to a column (1 metre) of Chromosorb W or diatomaceous firebrick TND-TS-M supporting 15% of siloxane

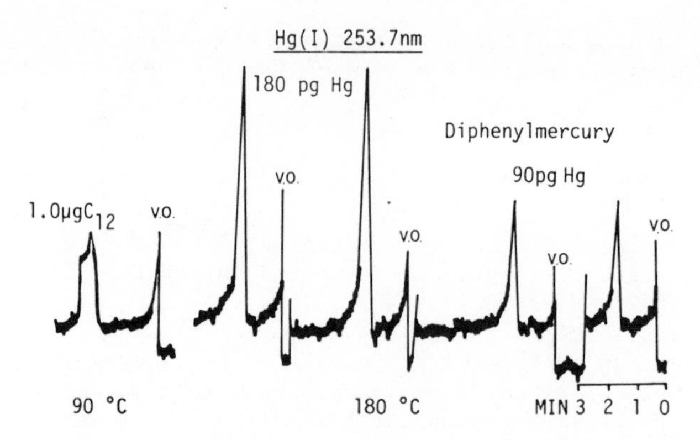

Figure 113 - Element selective gas chromatograms for disphenylmercury obtained with GC-MED system. Column temperature indicated on chromatogram.

elastomer E-301 or SE130 that has been modified by the passage of a mixture of similar bis(arene) chromium -compounds to deactivate certain sites on the elastomer. The column was operated at 190 to 210°C with helium (containing less than 10 p.p.m. of oxygen) (50 ml per min.) as carrier gas and a katharometer detector. Separate peaks for total hydrocarbons, (benzene) ethylbenzenemolybdenum, bis(ethylbenzene)molybdenum (diethylbenzene) ethylbenzenemolybdenum and bis(diethylbenzene)molybdenum are thus obtained.

Veening et al[2054] have discussed the gas chromatography of molybdenum tricarbonyls.

ORGANOPHOSPHORUS COMPOUNDS

Davis et al[2055] reported a limited review of the literature on the gas chromatographic analysis of organophosphorus compounds, including phosphorus chloride, phosphorus oxychloride, $PSCl_3$, phosphonitrilic halides, cyclic phosphonitrilic chloride trimer and tetramer, triethyl and triaryl phosphates, trialkyl phosphines and their oxidation products, alkyl and allyl phosphites, trialkyl phosphonyl propionate, and tetramethyl phosphonylsuccinate.

Davis et al[2055] emphasized several points of importance in the gas chromatographic analysis of organophosphorus compounds. They recommended that water should be excluded from the analysis system, with the carrier gas dried very carefully to avoid sample hydrolysis, for example by passing the gas through a trap containing type 5A molecular-sieve. This trap can be dried periodically at 300 - 400°C. They found that the use of glass, as opposed to metal columns, and direct injection of the sample on to the column both serve to reduce the sample decomposition and tailing of peaks. Davis et al[2055] obtained much greater reproducibility in a variety of analyses of different types of substances when glass inlets and columns were used instead of metal. They recommended silanizing the column support (in their case acid and base washed Chromasorb W) and pointed out that they have not found any case where silanizing the support produces a deterioration in the separation achieved and that in many cases it reduced tailing. They recommended temperature-programming for handling wide boiling-range mixtures in order to facilitate resolution and minimize sample decomposition, particularly of the lower boiling components.

A. Organic Phosphates and Thiophosphates

The micropyrolytic gas chromatographic technique has been applied by Legate and Burnham[2056] to the identification of organic radicals in organic phosphates and metal dialkylthiophosphates. The latter are used commercially as antioxidants and load-carrying agents in commercial motor oils. The compound is pyrolysed in the inlet system of a gas chromatograph and the volatile pyrolysis products (generally olefins) are fractionated and collected individually for identification by mass or infrared spectrometry. The olefins are formed generally by the breaking of a carbon-oxygen bond and abstraction of hydrogen from a beta carbon atom with no skeletal isomerization. Thus, the structures of the olefins are directly related to the structure of the alkyl groups initially present. Only when hydrogen is not available on a beta carbon atom - e.g. neopentyl radicals - are olefins formed by carbon-skeletal rearrangements. Legate and Burnham[2056] determined the exact configuration of the alkyl radicals in several model organic phosphates and metal dialkyl thionothiophosphates and described a gas chromatographic inlet system for pyrolysis or for conventional vaporization.

Flash pyrolysis of dithiophosphates produced olefins in good yield. If the pyrolysis is conducted in the inlet system of a gas liquid chromatographic column, the olefins separated can be identified by their retention volumes, infrared, and mass spectra. An example of a reaction which occurs during pyrolysis is:

$$\mathrm{Pb}\left[-\mathrm{S}-\overset{\overset{\displaystyle S}{\uparrow}}{\mathrm{P}}-\mathrm{O}-\overset{\overset{\displaystyle C}{|}}{\mathrm{C}}-\mathrm{C}-\overset{\overset{\displaystyle C}{|}}{\mathrm{C}}-\mathrm{C}\right]_{2\,2} \xrightarrow[275]{A} \mathrm{Pb}\left[-\mathrm{S}-\overset{\overset{\displaystyle S}{\uparrow}}{\mathrm{P}}(\mathrm{OH})_2\right]_2 +$$

(hypothetical non-volatile product)

$$\overset{\overset{\displaystyle C}{|}}{\mathrm{C}}=\overset{\overset{\displaystyle C}{|}}{\mathrm{C}}-\mathrm{C}-\mathrm{C}+\overset{\overset{\displaystyle C}{|}}{\mathrm{C}}-\mathrm{C}-\overset{\overset{\displaystyle C}{|}}{\mathrm{C}}-\mathrm{C}$$

cis + trans

The structure of the olefins are directly related to those of the alkyl groups initially present in the dithiophosphate and thus can be used to determine the structure of this type of compound.

Legate and Burnham[2056] proved the validity of their technique by identifying various types of alkyl, and one type of aryl side-chains in organic phosphates and thiophosphates. Table 173 lists the purity of the compounds, and Table 174 summarizes the pyrolyses. As shown in Table 174 olefins are formed generally by cleavage of the C - O bond, followed by simple abstraction of hydrogen from a beta carbon atom with no skeletal isomerization. Thus the structure of the olefins are, generally, related directly to the structure of the alkyl groups present initially.

Hardy[2057] studied the gas chromatography of the alkyl esters of phosphoric acid. He converted these esters into their more volatile methyl esters by reaction with diazomethane and applied the technique to the analysis of impurities in tributyl phosphate and also to the analysis of alkyl phosphoric acids in general.

Table 173 - Purity of organic phosphates and metal dialkylthiophosphates for GC-pyrolysis

Compound	Analysis					
	Theory			Found		
	%M	%P	%S	%M	%P	%S
Lead salts of 0,0'-di-n-amyl thionothiophosphate	27.8	8.3	17.2	24.3	6.0	13.1
Zinc salt of 0,0'-di-n-dodecyl thionothiophosphate	6.6	6.2	12.9	6.5	5.8	10.8
Zinc salt of 0,0'-dineopentyl thionothiophosphate	10.8	10.3	21.2	10.8	10.5	20.3
Potassium salt of 0,0'-diisopropyl thionothiophosphate	Commercial grade					
Lead salt of 0,0'-di-(4-methyl-2-pentyl) thionothiophosphate	25.8	7.7	16.0	23.8	7.7	14.7
Zinc salt of 0,0'-di-cyclohexyl thionothiophosphate	11.9	11.2	23.2	8.9	8.4	14.7
2-Ethyl-1-hexyl diphenyl phosphate	Commercial grade					

Table 174 - GC pyrolysis of model phosphate and thiophosphate compounds

Model compound pyrolysed	Temperature (°C)		Principal volatile products
	Pyroliser	GC column	
Primary Alkyl Structure			
Lead salt of 0,0'-di-n-amyl thionothiophosphate	315	31.0	1-Pentene
Zinc salt of 0,0'-di-n-dodecyl thionothiophosphate	290	150.0	1-Dodecene
Zinc salt of 0,0'-dineopentyl thionothiophosphate	460	30.0	2-Methyl-1-butene (42% w.w)[a] 2-Methyl-2-butene (56w/w) C_4 and lighter olefins (2%w/w)
Secondary Alkyl Structure			
Potassium salts of 0,0'-diisopropyl thionothiophosphate	300	30.0	Propene
Lead salt of 0,0'-di(4-methyl-2-pentyl) thionothiophosphate	275	30.0	cis- and trans-4-Methyl-2-pentene (59% w/w) 4-Methyl-1-pentene (35% w/w) 2-Methyl-2-pentene (6% w/w)
Zinc salt of 0,0'-dicyclohexyl thionothiophosphate	450	30.0	Cyclohexene

(continued)

Table 174 (Continued)

Alkyl Aryl Structure			
2-Ethyl-1-hexyl diphenyl phosphate	230	108.0	2-Ethyl-1-hexene (63%w/w) Phenol (37%w/w)

a Basis total volatile pyrolysis products.

Table 175 shows the retention volumes obtained on silicone columns for 2 µl samples of methyl esters of various methyl butyl phosphates dissolved in tributyl phosphate.

Table 175 - Retention volumes (relative to tributyl phosphate = 10) of alkyl phosphates

Compound	Liquid phase: silicone	
	DC 200, 188°	E 301, 197°
Monomethyl di-n-butyl phosphate	4.1	4.4
Dimethyl mono-n-butyl phosphate	1.6	1.9
Trimethyl phosphate (2 peaks, see text)	0.4 and 0.55	0.4 and 0.6
Ether	0.01	0.01

Table 176 shows the results obtained by this procedure in the analysis of a synthetic mixture of dihydrogen butyl phosphate and hydrogen dibutyl phosphate in tributyl phosphate on a silicone E301 silicone column at 197°C. Individual constituents can be determined in amounts down to 1% with an accuracy of about ± 10%.

Table 176 - Gas chromatography of synthetic mixture of alkyl esters of ortho phosphoric acid

Components	% (w/w) present	% (w/w) found
Tributyl phosphate	89.8	90.0
Hydrogen dibutyl phosphate	4.9	4.9
Dihydrogen butyl phosphate	5.3	4.5
Phosphoric acid[a]		0.6

a Impurity in dihydrogen butyl phosphate.

Hardy[2057] successfully chromatographed mono-n-octyl phosphoric acid, di-n-octyl phosphoric acid and tributyl phosphate, as their methyl esters, on a silicone DC 200 column at 204°C. A 2 µl sample of impure mono-n-octyl phosphoric acid gave a main peak due to dimethyl mono-n-octyl phosphate and five small peaks at lower retention volumes due to unidentified impurities, one of which was probably trimethyl phosphate from phosphoric acid. A 2 µl sample of di-n-octyl phosphoric acid gave one broad peak due to methyl di-n-octyl phosphate and a small sharp peak which could have been due to trimethyl phosphate. A 2 µl sample of equal parts of tri-n-propyl phosphate and tri-butyl phosphate was well-resolved at a column temperature of 204°C (Table 177).

Table 177 - Retention volumes (relative to tributyl phosphate = 10) of alkyl phosphates

Tri-n-propyl phosphate	3.7
Dimethyl mono-n-octyl phosphate	7.9
Methyl di-n-octyl phosphate	73

Apelblat and Hornik[2058-2060] and Apelblat[2061,2062] and Strache[2063] carried out an extensive series of investigations on the determination of tributyl phosphate and related compounds in kerosene-water mixtures used in industrial processes for the extraction of uranyl nitrate (atomic energy industry) and other metallic nitrates.

In an earlier paper Apelblat and Hornik[2058] described the analysis of synthetic mixtures of tributyl phosphate, kerosene (or dodecane) and butyl alcohol (Figure 114) carried out on a 2 in x 1/4 in. o.d. column filled with 20% (w/w) Apiezon-L on 60 - 80 mesh Diaport-S with a helium flow-rate of 50 ml per min (conditions, column temperature programmed 115°C (1 min) to 270°C (4 min) at a rate of 20°C per min, injector port temp 250°C, detector temp 265°C). They obtained the following retention times:

	Seconds
Butyl alcohol and water	22
Tributyl phosphate	650
Dodecane	460

In further papers on the technological applications of tributyl phosphate to metal extraction, Apelblat and Hornik[2060] described their work on the determination of the activity coefficients and excess thermodynamic functions at infinite dilution in the binary systems diluent plus solvate of uranyl nitrate, diluent plus tributyl phosphate, diluent plus dodecane and in the ternary system diluent plus dodecane plus tributyl phosphate using the following as diluents: benzene derivatives, paraffins, halogen-substituted hydrocarbons, cyclohexane, acetone, dioxane and carbon disulphide. They used gas chromatography extensively in this work.

Some specific retention volume data obtained by these workers are shown in Figure 115. This specific retention volume data is corrected for the pressure drop across the column and extrapolated to the zero sample.

Apelblat[2061] also applied the same gas chromatographic procedure as described above (Apelblat and Hornik[2061]) to the solvent extraction system di-n-butyl phosphoric acid using the following diluent, benzene, toluene, hexane, cyclohexane, carbon tetrachloride, chloroform, acetone, methanol, carbon disulphide, and water. The specific retention volume (V_g) data obtained as a function of temperature for the diluent di-n-butyl phosphate and diluent mono-n-butyl phosphate systems are presented in Figure 116. The value of V_g is extrapolated to zero sample. Except for acetone, methanol and water, the peaks are always symmetrical or with negligible trailing (chloroform-mono-n-butyl phosphate system).

Apelblat and Hornik[2059,2060] and Apelblat[2061] applied essentially the same gas chromatographic system as described above to the analysis of the system tributyl phosphate-nitric acid-water-diluent (diluents benzene, carbon tetrachloride, chloroform).

Using gas chromatographic techniques similar to those described above,

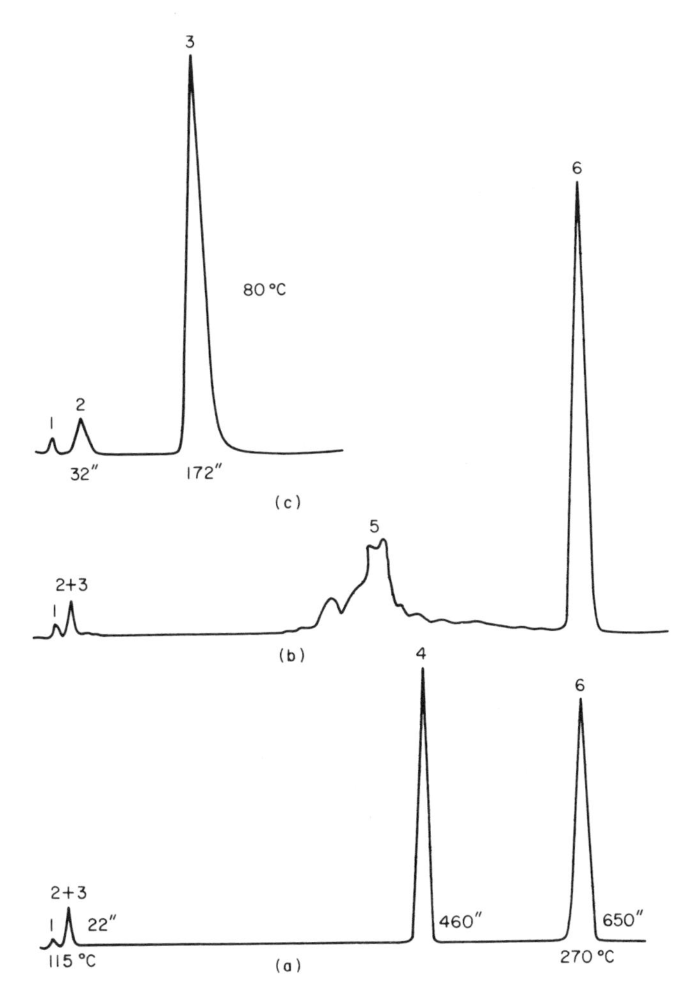

Figure 114 - Gas chromatographic analysis of tributyl phosphate-diluent mixtures. (a) 60% (v/v) tributyl phosphate + 40% (v/v) dodecane, saturated with water, 1.5 µl; (b) tributyl phosphate + kerosene, saturated with water, 3 µl; (c) Butyl alcohol + water, 1 µl; 1 = air, 2 = water, 3 = butanol, 4 = dodecane, 5 = kerosene, 6 = tri-n-butyl phosphate.

Apelblat and Hornik[2059,2060] and Apelblat[2061,2062] also evaluated activity coefficient and excess thermodynamic functions at infinite dilution for a number of binary diphenyl methyl phosphate diluent systems using the following as solvents: cyclohexane, benzene, toluene, carbon tetrachloride, chloroform, acetone, dioxane, methanol, ethanol and water.

Campbell[2064] has discussed the gas chromatography of tributyl phosphate, bis-(2 ethylhexyl) hydrogen phosphate, dibutyl butylphosphonate, trioctyl-phosphine oxide and other solvents used in reactor fuel reprocessing and fission product recovery. He used a column (5 ft x 0.25 in) of 2.0% of Apiezon N on Chromosorb W (60 - 80 mesh) operated isothermally at 150° to 325°C depending on the solvent, with helium as carrier gas and hydrogen-flame detection. Quantitative calibrations for each of the organophosphorus

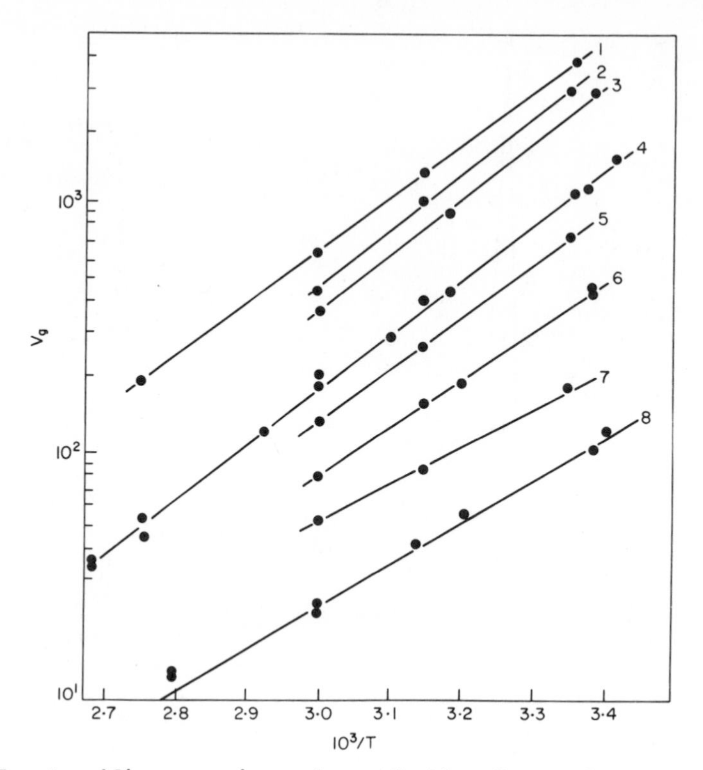

Figure 115 - Specific retention volume of chloroform and water as a function of temperature: (a) Chloroform: 1, tributyl phosphate; 2, 70% tributyl phosphate + 30% dodecane; 3, 40% tributyl phosphate + 60% dodecane; (b) Water: 4, tributylphosphate, experimental points from two different columns of tributyl phosphate; 5, 70% tributylphosphate + 30% dodecane; 6, 40% tributylphosphate + 60% dodecane; 7, 20% tributylphosphate + 80% dodecane; 8, $UO_2(NO_3)_2 2TPB$.

compounds examined in hydrocarbon and carbon tetrachloride diluents gave standard deviations of ± 0.2 to ± 0.5%.

Workers at the United Kingdom Atomic Energy Authority [2065] have described a gas chromatographic procedure for the determination of down to 0.005% tri-isobutyl phosphate in tri-n-butyl phosphate. In this procedure tridecane (0.05 ml) is added to the sample (49.95 ml) and 0.5 μl of the mixture is analysed on a glass column (5 ft x 4 mm (packed with 10% of Apiezon L on silanized Celite (40 - 60 mesh) at 200°C; the carrier gas is nitrogen (50 ml min^{-1}) and a hydrogen-flame ionization detector is used. The apparatus is calibrated with standard solutions and the result is derived from the ratio of the peak heights for tridecane and tri-isobutyl phosphate.

Methods have been described by Moffat and Thompson[2066] for the determination of butyl nitrate and water in tributyl phosphate.

Berlin et al[2067] have discussed the gas chromatography of tributyl phosphate and other types of organophosphorus compounds, using Silicone SE 30 columns.

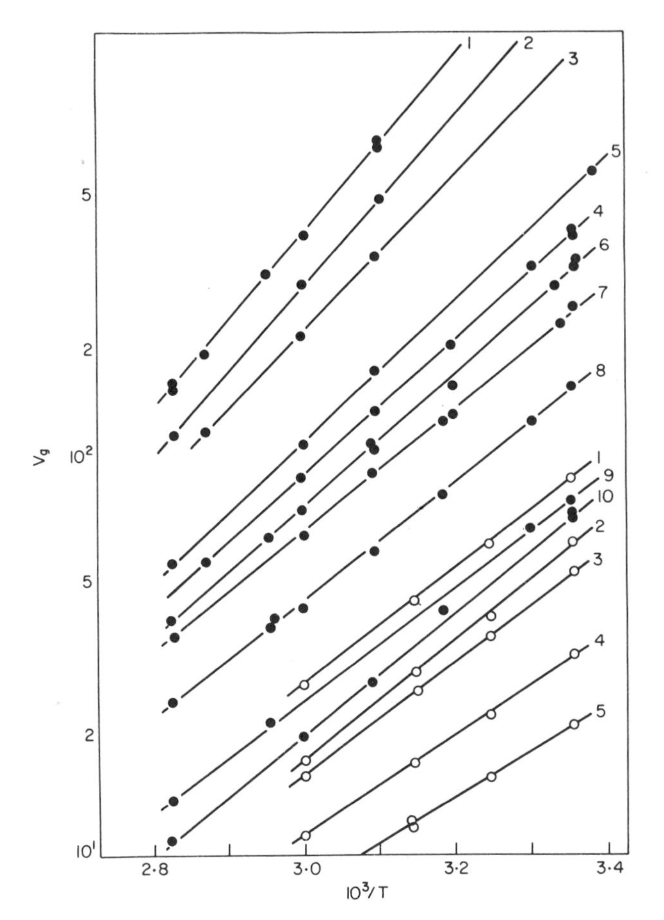

Figure 116 - Specific retention volume as a function of temperature. (a) Dibutyl phosphate: 1, water; 2, methanol; 3, toluene; 4, benzene; 5, acetone; 6, chloroform; 7, carbon tetrachloride; 8, cyclohexane; 9, hexane; 10, carbon disulphide. (b) Monobutylphosphate; 1, benzene; 2, chloroform; 3, carbon tetrachloride; 4, cyclohexane; 5, carbon disulphide.

Lewis and Patton[2068] described conditions for the gas chromatographic separation of triethyl phosphate, triphenyl phosphate and tris(m-tolyl) phosphate on a column 60 cm of Apiezon oil K on Celite 545 at 283°C.

B. Trialkyl and Triaryl Phosphites

Shipotofsky[2069] separated dimethyl phosphite from diethylphosphite on a di-n-butyl phthalate on Fluoropak 80 at 107°C. Davis et al[2055] have described conditions for the separation of crude trimethyl phosphite (Table 178) on a column of SE 20 on Chromasorb W 250. The same column, with slightly different temperature programs, also served for analysis of triallyl phosphite and for the separation of dimethyl fumarate, trimethyl phosphonylpropionate, and tetramethyl phosphonylsuccinate. For triethyl phosphite, in order to achieve separation of a particular impurity it is necessary to use 10% diethyl glycol adipate in place of 20% SE 30. In all cases there were no difficulties with tailing.

Table 178 - Instrument conditions, gas chromatography of trimethyl phosphite

Instrument	Burrell K-2
Column	20% SE 20 on Chromasorb W 250 glass U-tube
Carrier	Helium
Initial temperature	90°C
Programme rate	Isothermal for 30 min, then raise temp as rapidly as possible to 250°C
Final temperature	250°C
Flow rate	40 ml min^{-1}
Sample size	5 μl
Detector	Thermal conductivity

Feinland et al[2070] have discussed the gas chromatographic separation of mixtures of various organophosphorus compounds including tributyl phosphite and various substituted phosphines, phosphine oxides, phosphinites, phosphonates, phosphinates and phosphates. They used a Reoplex 400 or a silicone grease on a Chromasorb W column. Berlin et al[2067] have studied the gas chromatography of organic phosphites, phosphine oxides, phosphine sulphides, phosphinites, phosphonates, phosphates, phosphinic acids, phosphorylated amidines, phosphines, phosphinites and phosphonites. They used SE 30 on Chromasorb W, Apiezon L on Chromasorb W, Apiezon K on Fluoropak, silicon rubber on Chromasorb W or Fluoropak as columns and a hydrogen flame ionization detector.

Lee and Ting[2071] have described a method for the determination of di- and mono-butyl phosphates in tributyl phosphate kerosene mixtures, by solvent extraction and gas chromatography.

C. Trialkyl Phosphines, Phosphine Oxides and other Oxygenated Organophosphorous Compounds

The analysis of these types of compounds has been discussed by various workers including De Rose et al[2072] whose used columns of Apiezon or Squalane on Celite to determine the composition of the reaction mixtures from a study of the ester interchange among dialkyl hydrogen phosphonates and to analyse organic phosphites, Gudzinowicz and Campbell[2073,2074] who analysed tertiary phosphines and phosphine oxides using 5% SE 30 on Chromasorb W and Apiezon L on firebrick and Plumb and Griffin[2075] who used SE 30 and Carbowax 20 M on Carbosorb P for the analysis of photoinitiated oxidation products of phosphites.

Due to the very low vapour pressure of many organophosphorus compounds, several workers have avoided the problem by the use of pyrolysis, followed by gas chromatographic analysis of the resulting volatiles. Legate and Burnham[2056] used infrared and mass spectroscopy, after gas chromatographic separation to identify the organic pyrolysis products of complex lead, zinc or potassium phosphates or thiophosphates. Dulon et al[2076] used gas chromatography to identify pyrolysis products of organophosphorus compounds. Williams et al[2077] irradiated tributyl phosphate saturated with water with a Van de Graaf generator and studied the resulting products by gas chromatography.

Grob and McCrea[2078] pointed out that the separation of phosphine oxides from reaction products presented difficulties with standard types of gas chromatographic columns, due mainly to their high melting point. They worked out a separation system using Porapak Q as column material and collected products emerging from the column for identification by infrared spectroscopy.

In Figure 117 is shown the chromatogram obtained in the seperation of a mixture of four phosphine oxides using the uncoated (column A) and the SE 30 coated (column B) Porapak Q columns, respectively. The coating of the Porapak Q beads has produced sharper peaks but a decrease in retention time, with the consequence that phosphine oxides of lower boiling point do not separate from the methanol solvent peak. The uncoated Porapak A at higher temperature allows the phosphine oxides to be separated from the methanol solvent peak.

Gudzinowicz and Campbell[2073] achieved separations of the following types of compounds with molecular weights between 262 and 368 on a 5% SE 30 silicone gum rubber column programmed from 240 - 350°C, triphenylphosphine, p-hydroxyphenyl diphenylphosphine, p-methoxyphenyldiphenylphosphine, (p-methoxyphenyl) phenyl phosphine, tri(p-methoxylphenyl) phosphine, triphenyl phosphine oxide, p-methoxyphenyldiphenylphosphine oxide and tri(p-methoxyphenyl) phosphine oxide.

Buckler[2079] used gas chromatography to separate trialkylphosphines and their oxidation products.

Feinland et al[2070] developed a gas chromatographic method for the determination of tributylphosphine, tricyclohexylphosphine and a wide range of their oxidation products, including the oxides. Separation was achieved in a Reoplex 400 or silicone grease on Chromasorb W column. The relative retention distances of tributylphosphine and its oxidation products versus hexadecane on both Silicone Grease and Reoplex 400 columns at 200°C and 206°C, respectively, are given in Table 179.

The compounds tailing the most on Silicone Grease are tributylphosphine, tributylphosphine oxide, and dibutylphosphine oxide. On Reoplex, these compounds give symmetrical peaks. Reoplex is also more effective than Silicone in separating tributylphosphine from butyl dibutylphosphinite, tributyl phosphite from dibutylphosphine oxide, and tributyl phosphate from butyl dibutylphosphinate.

All of the cyclohexylphosphines and their oxidation products were resolved adequately by Silicone Grease at 240°C, thus eliminating the need for the Reoplex column. The relative retention distances of these compounds versus dibutyl sebacate are given in Table 180. No peak was obtained for tricyclohexyl phosphate on Silicone Grease at 240°C or at column temperatures up to 350°C. The only peaks observed arose from various low-boiling decomposition products which could not be duplicated with repeated injection. A 4 ft Reoplex 400 column at 220°C was also tried but without success.

Feinland et al[2070] carried out qualitative analyses of known mixtures of tributylphosphine, dibutyl butylphosphonate, and tributylphosphine oxide in benzene solvent. Dibenzyl was added to each as an internal standard and each mixture was chromatographed three or four times on the Silicone column at 200°C. The dibutyl butylphosphonate and tributylphosphine oxide gave linear results upon plotting weight ratio versus area ratio of each with respect to dibenzyl. Relative standard deviations of 1.0 and 1.9% were

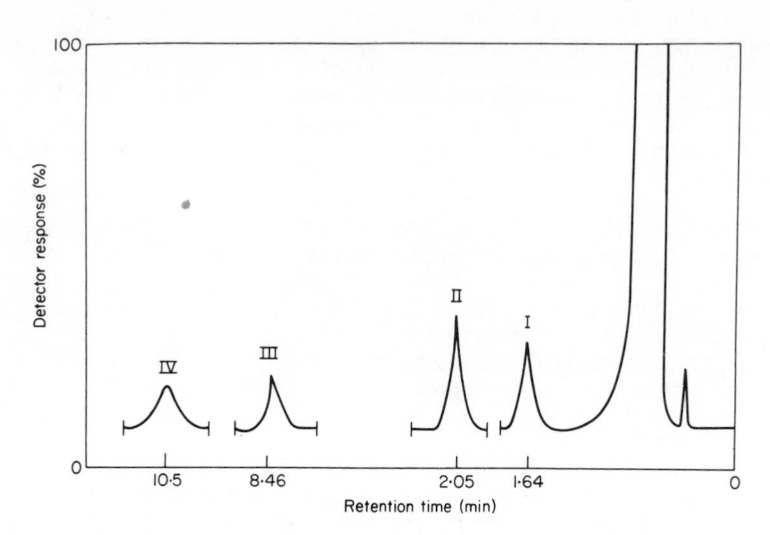

Figure 117 - Gas chromatography of phosphine oxides. Column 6 ft x 0.25 in. o.d. of 80/100 mesh Porapak Q at 245°C. Helium flow-rate 61.5 ml min^{-1}, I tris-2-furylphosphine oxide, II tris-s-thienylphosphine oxide, III tris-(p-chlorophenyl) phosphine oxide, IV tris-(p-hydroxyphenyl) phosphine oxide.

Table 179 - Relative retention distances of tributylphosphine and oxidation products versus hexadecane[a]

Compound	Formula	Relative retention of	
		Silicone at 200°C	Reoplex at 206°C
n-Hexadecane		1.00 (5.7 min)	1.00 (2.0min)
Tributylphosphine	Bu_3P	0.45	0.78
Butyl dibutylphosphinite	Bu_2POBu	0.45	0.27
Tributyl phosphite	$(BuO)_3P$	0.56	0.17
Dibutylphosphine oxide	$Bu_2P(O)H$	0.57	7.74
Dibutyl butylphosphonate	$BuP(O)(OBu)_2$	0.96	4.94
Tributyl phosphate	$(BuO)_3PO$	1.07	5.99
Butyl dibutylphosphinate	$Bu_2P(O)OBu$	1.11	6.76
Tributylphosphine oxide	Bu_3PO	1.48	13.8

[a] Relative retention distances were measured from point of injections.

obtained for the area ratios of phosphonate and oxide, respectively, versus dibenzyl. The tailing of tributylphosphine on Silicone resulted in non-linear response and a relative standard deviation of 5.3% for the area ratio relative to dibenzyl. Standard solutions of tributylphosphine and dibenzyl eluted on Reoplex column at 206°C, gave a symmetrical phosphine peak. The response was linear and the relative standard deviation improved considerably to a value of 1.9%.

Davis et al[2055] made recommendations for the analysis of organophosphorus compounds which included the use of glass columns and direct on-column injection to avoid contact of the materials with metal portions of the apparatus, also the use of pre-dried carrier gas to avoid decomposition of products.

Berlin et al[2080], Berlin and Nagabhushanam[2081], Berlin et al[2067] and Berlin and Austin[2067], have carried out an extensive study of the application of the hydrogen flame ionization detector to the gas chromatography of a wide range of organophosphorus compounds, covering a wide boiling range, including phosphine oxides, phosphine sulphides, phosphinates, phosphonates, phosphates, phosphinic acids, phosphorylated amidines, phosphines, phosphinites, phosphonites and phosphites. These workers found that commercially available equipment is sufficient without modification, to give good resolution and rapid qualitative and quantitative analysis on these types of organophosphorus compounds. The polar nature of many of these materials, especially the tetrasubstituted phosphorus derivatives, resulted in asymmetrical peaks and tailing in many instances and this problem precluded the accurate measurement of chromatograms when quantitative results were desired. Several columns were studied before these problems were allieviated. A gas chromatograph with a hydrogen flame ionization detector was chosen to analyse the mixtures since in general this detector is more sensitive to trace quantities of compounds than a conventional thermal conductivity cell and small quantities of water present was not a critical problem in the analyses.

Table 180 - Relative retention distances of cyclohexylphosphines and oxidation products versus dibutyl sebacate[a]

Compound	Formula	Relative retention of silicone at 240°C
Dibutyl sebacate		1.00 (6.7 min)
Cyclohexyl dibutylphosphinate	$Bu_2P(O)OC_6H_{11}$	0.52
Butyl dicyclohexylphosphinate	$(C_6H_{11})_2P(O)OBu$	1.14
Tricyclohexylphosphine	$(C_6H_{11})_3P$	1.26
Dicyclohexyl cyclohexylphosphonate	$(C_6H_{11})P(O)(OC_6H_{11})_2$	1.84
Cyclohexyl dicyclohexylphosphinate	$(C_6H_{11})_2P(O)OC_6H_{11}$	2.19
Tricyclohexylphosphine oxide	$(C_6H_{11})_3PO$	3.11
Tricyclohexyl phosphate	$(C_6H_{11}O)_3PO$	did not elute

The types of columns used by Berlin et al[2067] are shown in Table 181. Several available standard supports and substrates give adequate results with travilent organophosphorus materials. For example, Figure 118 shows the chromatogram of a mixture of triphenylphosphine, methyl diphenylphosphinite, and dimethyl phenylphosphonite, on a column packed with silicone rubber on Chromasorb W. Table 182 lists several trivalent organophosphorus compounds which were separated on this type of column. When this same mixture was injected on either SE 30 or Apiezon columns, tailing of the peaks became acute, and the retention times were long unless high temperatures were employed.

Although the silicone rubber on Chromasorb W gave very satisfactory results with all the trivalent organophosphorus compounds under study, the resulting chromatograms from the analysis of tetrasubstituted phos-

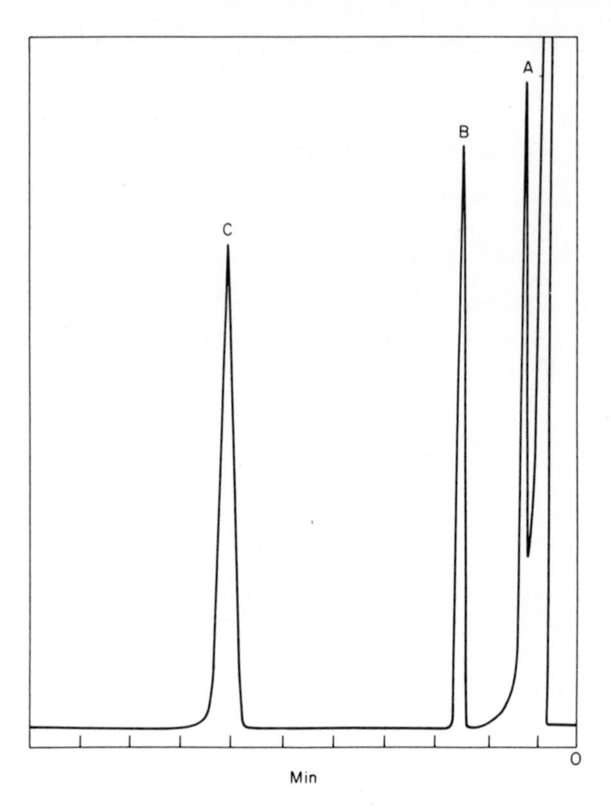

Figure 118 - Gas chromatogram of a mixture of dimethyl phenylphosphonite (A), methyl diphenylphosphinite(B), and triphenylphosphine (C) in benzene. Column A see Table 181 and 182. T = 268°C, Flow Rates N_2/H_2 = 31/32 ml min^{-1}.

phorus derivatives exhibited unsymmetrical peaks due to excessive tailing. Consequently, for phosphates, phosphonates, phosphinates, and phosphine oxides, it was necessary to use Chromasorb G, a high density support, which had been acid-washed and pretreated with dimethyldichlorosilane (DMCS) and coated with 5% Silicone 30.

The use of silanized column supports (Davis et al[2055]) generally improves the elution of the polar phosphorus compounds. Berlin et al[2067] stated that silanized Chromasorb G as a column support had not previously been reported in the literature for the analysis of organophosphorus compounds.

D. Phosphoronitril Halides and Phosphorus Chlorides and Oxychlorides

Several workers have published methods for the gas chromatographic separation of mixtures of phosphorus chlorides with other phosphorus compounds. Shipotofsky and Moser[2082] used Kel F-90 on Fluoropack 80 to separate $PSCl_3$ - phosphorus trichloride and phosphorus oxychloride - phosphorus trichloride mixtures. They did not achieve a separation of $PSCl_3$ and phosphorus oxychloride. A method was also developed for the analysis of mixtures of dimethyl and diethyl phosphite, phosphorus trichloride and $PSCl_3$ or phosphorus trichloride and phosphorus oxychloride. Pyrex glass was

Table 181 - Columns[a] used for gas chromatography of organophosphorus compounds

Substrate	Support[b]	Mesh size	Concentration (w/w %)	Length (ft)
SE30	Chromasorb W	60/80	20	4
SE30	Chromasorb W	80/100	10	6
SE30	Chromasorb W	80/100	10	5
SE30	Chromasorb W	60/80	5	5
Apiezon L	Chromasorb W	60/80	20	4
Apiezon K	Fluoropak		10	8
Silicone Rubber (A)	Chromasorb W	80/100	10	8
Silicone Rubber	Chromasorb W	80/100	10	6
Silicone Rubber	Fluoropak		10	8
Silicone 30 (B)	Chromasorb G/DMCS[c]	60/80	5	6

[a] All of the columns were obtained commercially and were of 1/8 in. stainless steel.
[b] The Chromasorb W was acid-washed in all cases.
[c] The Chromasorb G was acid-washed and pretreated with dimethyldichlorisilane (DMCS) by the supplier.

used for the column and for the body of the thermal-conductivity cell, and tantalum wire was employed for the sensing elements.

Stanford[2083] also separated phosphorus trichloride from phosphorus oxychloride. He used Silicone Elastomer E301 (SE30) on Celite 545 to do the separation (column 6 ft x 0.25 in. containing 23% (2/2) of silicone E301 on 80 - 120 mesh Celite 545; temperature 63°C; carrier gas 1.75 litres of nitrogen per h; inlet pressure 63 cm, outlet 24 cm; bridge current 100 mA).

Abe[2084] used DC-703 with good results to separate phosphorus trichloride, hydrogen chloride, trichlorsilane, silicon tetrachloride and boron trichloride. He also investigated several other substances, some of which caused hydrolysis and degradation on the column.

Table 183 gives the conditions used by Davis et al[2055] for the separation of mixtures of phosphorus trichloride, phosphorus oxychloride, and $PSCl_3$.

Several authors have published methods for analysing cyclic phosphonitrilic halides. Mou et al[2085] and Chapman et al[2086] used silicone elastomer E301 (SE30) on Celite 545. Gimblett[2087] used the same substrate and support to separate the trimer and tetramer of cyclic phosphonitrilic chloride.

Table 184 gives the conditions used by Davis et al[2055] for the analysis of cyclic phosphonitric trimer to heptamer mixtures. They found that the addition of a small amount of Carbowax 20 M markedly reduced tailing. Figure 119 is a chromatogram of a typical crude mixture of phosphonitrilic chlorides.

Rotzsche et al[2088] have described a gas chromatographic method for separating phosphonitrilic chlorides/bromides. They used a column (200 x 0.6 cm) of 5% poly(dimethylsiloxane) (Gi 7100) on Sterchamol operated at 205°C

Table 182 - Chromatographic analysis of organophosphorus compounds

Class	Formula	Retention time (min)	Temp. (°C)	N_2/H_2 ml min^{-1}	Column
Phosphine Oxides					
$(C_4H_9)_3P{\to}O$	$C_{12}H_{27}OP$	2.2	247	30/24	B
$(C_6H_5)_2P(O)CH_3$	$C_{13}H_{13}OP$	3.6	246	29/25	B
$(C_6H_5)_2P(O)C_3H_7$	$C_{15}H_{17}OP$	7.2	240	30/25	B
$(C_6H_5)_2P(O)CH{=}CHCH_3$	$C_{15}H_{15}OP$	7.5	244	30/25	B
$(C_6H_5)_2P(O)CH_2CH{=}CH_2$	$C_{15}H_{15}OP$	6.9	245	30/25	B
$(C_6H_5)_2P(O)C_4H_9$	$C_{16}H_{19}OP$	2.4	300	28/26	B
$(C_6H_5)_2P(O)CH(CH_3)CH_2CH_3$	$C_{16}H_{19}OP$	2.0	300	28/26	B
$(C_6H_5)_3P{\to}O$	$C_{18}H_{15}OP$	9.3	263	30/28	B
$(C_6H_5)_2P(O)CH_2C_6H_5$	$C_{19}H_{17}OP$	2.8	286	29/25	B
Phosphine Sulphides					
$(C_6H_5)_3P{\to}S$	$C_{18}H_{15}PS$	8.7	286	28/22	B
Phosphinates					
$(C_6H_5)_2P(O)OCH_3$	$C_{13}H_{13}O_2P$	2.2	263	30/28	B
$(C_6H_5)_2P(O)OC_6H_4$	$C_{18}H_{15}O_2P$	7.1	263	30/28	B
Phosphonates					
$C_6H_5P(O)(OCH_3)_2$	$C_8H_{11}O_3P$	4.2	202	31/23	A
$C_6H_5P(O)(OC_6H_5)_2$	$C_{18}H_{15}O_3P$	10.2	246	30/30	B
Phosphates					
$(CH_3O)_3P{\to}O$	$C_3H_9O_4P$	2.4	134	31/23	A
$(C_4H_9O)P{\to}O$	$C_{12}H_{27}O_4P$	3.5	204	30/25	B
$(C_6H_5O)_3P{\to}O$	$C_{18}H_{15}O_4P$	10.0	252	30/30	B
Phosphinic Acids					
$(C_6H_5)_2P(O)OH$	$C_{12}H_{11}O_2P$	1.8	297	28/22	B
Phosphorylated Amidines					
$(C_2H_5O)_2P(\to O){-}N{=}C(N(CH_3)_2){-}CH(CH_3)_2$	$C_{10}H_{23}N_2O_3P$	2.8	250	28/22	B
$(C_4H_9O)_2P(\to O){-}N{=}C(N(CH_3)_2){-}CH(CH_3)_2$	$C_{14}H_{31}N_2O_3P$	6.4	250	28/22	B
$(C_6H_5)_2P(\to O){-}N{=}C(N(CH_3)_2){=}CH(CH_3)_2$	$C_{18}H_{23}N_2OP$	20.7	250	28/22	B
Phosphines					
$(C_4H_9)_3P$	$C_{12}H_{27}P$	1.1	247	30/24	B
$(C_6H_5)_3P$	$C_{18}H_{15}P$	5.2	281	31/23	A
Phosphinites					
$(C_6H_5)_2POCH_3$	$C_{13}H_{13}OP$	7.8	202	31/23	A
$C_6H_5P(OCH_3)_2$	$C_8H_{11}O_2P$	4.9	202	31/23	A

(continued)

Table 182 (Continued)

Phosphites					
$(CH_3O)_3P$	$C_3H_9O_3P$	1.0	134	31/23	A
$(C_2H_5O)_3P$	$C_6H_{15}O_3P$	1.0	132	30/30	B
$(C_4H_9O)_3P$	$C_{12}H_{27}O_3P$	3.9	172	30/30	B

[a] Letters contained in this column refer to the columns listed in Table 181.

Table 183 - Gas chromatographic separation of phosphorus trichloride phosphorus oxychloride and $PSCl_3$

Instrument	Burrell K-2
Column	Silicone grease (Burrell cat. no. 341 -136) on glass
Carrier gas	Helium
Initial temp	60°C
Programme rate	3 V min^{-1}
Final temp	250°C
Flow rate	87 ml at 4 lb in^{-2}
Sample size	5 µl
Detector	Thermal conductivity

Table 184 - Conditions for gas chromatographic separation of phosphonitrilic chlorides

Instrument	Burrell K-2
Column	3% Dow 11 plus 0.15% Carbowax 20 M on Columnpak 80 - 100 mesh 100 cm glass U-tube
Carrier gas	Helium
Initial temp	30°C
Programme rate	2 V min^{-1}
Final temp	230°C
Flow rate	50 ml min^{-1}
Sample size	20 µl
Detector	thermal conductivity
Solvent	CH_2Cl_2

with hydrogen as carrier gas, and a thermal conductivity detector (internal standard, trichlorsilane and $(CH_3)_3SiCl$). Relative retention times for all the compounds having the general formula $P_3N_3Cl_nBr_{6-n}$ (n = 0 to 6) are reported in Table 185. A plot of log (retention time) versus n in the equation $P_3N_3ClnBr_{6-n}$ gives a rectilinear graph.

Infrared spectroscopy has been used (Steger and Stahlberg[2089]) for the detection of the PNP frequency occurring at 1300 cm^{-1} in phosphonitrilic halides and this technique could, no doubt, be coupled with gas chromatography for the examination of separated fractions.

The gas chromatography of phosphonitrilic chlorides, fluoridates and bromides has been studied respectively by Gimblett[2087] and Chapman et al [2086,2088].

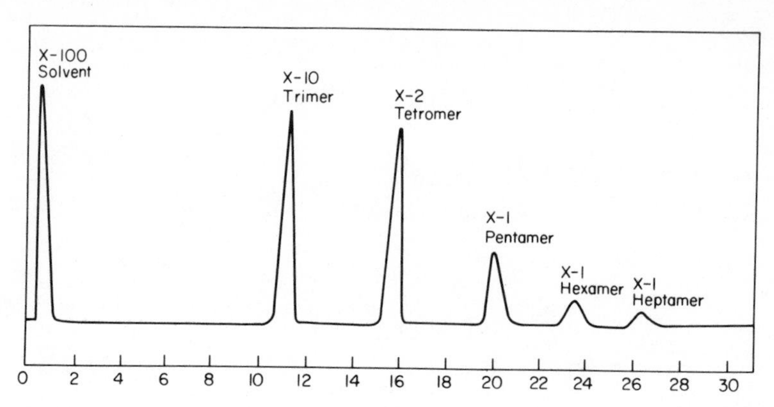

Figure 119 - Gas chromatogram of phosphonitrilic chlorides $(PNCl_2)_x$.

Schulte and Shive[2090] determined phosphoryl chloride and phosphorus trichloride in electronic grade trichlorosilane by gas chromatography with thermionic detection.

Table 185 - Relative retention times of $P_3N_3Cl_nBr_{6-n}$

Compound	Relative retention time ($P_3N_3Cl_6$ - 1.00
$P_3N_3Cl_6$	1.00
$P_3N_3Cl_5Br$	1.38
$P_3N_3Cl_4Br_2$	2.00
$P_3N_3Cl_3Br_3$	2.72
$P_3N_3Cl_2Br_4$	3.94
$P_3N_3ClBr_5$	5.75
$P_3N_3Br_6$	8.13

E. Insecticides and Pesticides

Electron-capture gas chromatography (Lovelock and Lipsky[2091]; Lovelock 2092) has been studied extensively in connection with the separation of organophosphorus insecticides and pesticides. Egan et al[2093] found that the Lovelock electron-capture ionization detector, which had been previously successfully used by Goodwin et al[2094] for detecting organochlorine insecticide residues, was unsuitable for detecting organophosphorus compounds, although Parathion, which has a highly electrophilic nitro group, was detected with reasonable sensitivity. Egan et al[2039] described practical conditions for the general application of electron-capture gas chromatography to organophosphorus residue analysis. Egan et al[2039] applied this method to the extraction and clean-up of organophosphorus pesticide residues for a number of plant tissues. The method has been applied to Chlorthion, Ethion, Fenchlorphos, Malathion, Parathion, Phenkapton and Trithion. Table 186 gives the recoveries obtained. They experienced some difficulty in eluting Malathion from an alumina column, and a column containing 30 g of magnesium oxide was used in place of the alumina column.

Kawahara et al[2095] described a procedure for the determination of phosphorothioate pesticides including Parathion and Parathion methyl. It consists of solvent extraction, clean-up by thin-layer chromatography on Silica gel (0.25 mm layer), and identification by gas chromatography in an aluminium column (120 x 0.6 cm o.d.) packed with equal portions of acid-washed Chromosorb P supporting 5% of DC 200 silicone oil, and unwashed Chromosorb W supporting 5% of Dow-11 silicone; the column is operated at 180°C with argon-methan (9:1) as carrier gas (120 ml min^{-1}) and electron-capture detection. If the sample volume is sufficient identification can be confirmed by infrared spectrophotometry. This method was used to follow accidental contamination of river water by pesticides.

Crasso et al[2096] determined Parathion and its oxygen analogue Paraoxon in nanogram amounts in biological material by gas chromatography with a Sr^{90} electron-capture detector. Following the cleanup procedure, the benzene extract (1 - 10 µl) is chromatographed on a column (80 cm x 2.5 mm) comprising of 5% of Apiezon L on Chromosorb W at 202°C using nitrogen (100 ml min^{-1}) as carrier gas. The amounts of Parathion and Paraoxon were determined from standards prepared by adding known amounts of these pure substances to control tissue extracts. Recoveries of both insecticides at the 3 - 10 pg level were 85 - 90% and limits of detection of the two substances were 0.1 and 10 pg respectively.

Table 186 - Extraction of organophosphorus pesticides with methyl ethyl ketone-hexane and estimation by gas-liquid chromatography

Organo-phosphorus pesticide	Source	Added pesticide part/10^6	Column clean-up	Recovery (%)
Chlorthion	Lettuce	0.08	Magnesia	84
Ethion	Onions	0.15	Alumina	86
	Sugar beet	0.10	Alumina	91
	Apples	0.20	Alumina	80
	Lettuce	0.50	Magnesia	80
Fenchlorphos	Lettuce	0.08	Magnesia	73
Malathion	Barley	0.20	a	80
Parathion	Apples	0.20	Alumina	86
	Sugarbeet	0.10	Alumina	91
	Brussels sprouts	0.20	Alumina	75
	Lettuce	0.30	Magnesia	80
Phenkapton	Apples	0.10	Alumina	82
Trithion	Apples	0.20	Magnesia	84

a No clean-up required.

Hrivnak and Pastorek[2097] studied the gas chromatographic separation of O,O-dialkyl O-(4-nitrophenyl) phosphorothioates (Parathions) and found that separation was best achieved on Apiezon L on Chromosorb W (60 - 80 mesh). More polar stationary phases than Apiezon L led to higher retention values for compounds containing a methyl group. They used a flame ionization detector and nitrogen as carrier gas using a separation column (80 cm x 3.5 mm).

Horiguchi et al[2098] applied gas-liquid partition chromatography to the separation of nine insecticides on three stationary phases, 1% SE30, 1% FC 1265 (QF-1) and 1% poly (diethylene glycol succinate), each supported on

Chromasorb W. They applied the method to commercial 1.5% Malathion dusts.

Hrivnak et al[2099] separated mixtures of Thiometon, Disulfoton and O-ethyl S-2 (ethylthio) ethyl O-methyl phosphorodithioate in technical mixtures. The compounds were separated at 140°C in a glass tube (80 cm x 3.5 mm) packed with 5% of silicone E301 on silanized Chromosorb W, with nitrogen as carrier gas and flame ionization detection. O-(4-Chlorophenyl)-O,O-dimethyl phosphorothioate was used as internal standard. The relative error was less than ± 1.5%.

Thornton and Anderson[2100] applied gas chromatography to the determination of S,S,S-tributyl phosphorotrithioate in light petroleum extracts of cotton seed and cotton seed oil. Separation was achieved on a column of SE30 at 200°C using an electron-capture detector. Down to 0.1 part/10^6 of insecticide can be determined in the sample by this procedure.

Hartmann[2101] describes a specific, highly sensitive detector for phosphorus-containing compounds for use with single and dual-channel gas chromatographic columns. It is a modified electron-capture detector in which the flame burns above a compressed caesium bromide tip. The detector is simple in construction, rapidly interchangeable, easy to operate, has high reliability, and shows good rectilinearity of response. Hartmann gives examples of the use of this method in the identification of phosphorothioate pesticides in mixtures with chlorinated hydrocarbon pesticides. The limit of detection is 3×10^{-12} g for Phorate and 12×10^{-12} g for Malathion.

Using a combination of six sets of operating conditions with three columns containing, respectively, high vacuum silicone grease, silicone compound DC-11 and fluorosilicone FS-1265, Kanazawa et al[2102] separated mixtures of up to 19 organophosphorus pesticides into their components. In all experiments, 32 - 48 mesh Celite 545 was used as support, the carrier gas was helium, and a thermal conductivity detector was used.

McCaulley[2103] has described an approach to the separation, identification and determination of at least ten organophosphorus insecticides in agricultural products such as apples, tomatoes, spinach, lettuce, cauliflower, cabbage, turnips and potatoes. Extraction of the sample with acetonitrile is followed by vacuum sublimation, gas-liquid chromatography, and infrared analysis. For chromatography, a glass tube (21 in x 0.25 in) is packed with 2% of 2,2-dimethylpropane01,3-diol adipate on Anachrom ABS (50 - 60 mesh). This column is operated at 160°C (or temperature programmed) with argon (100 ml min^{-1}) as carrier gas, and a radium ionisation (or thermistor or flame) detector.

Abbott et al[2104] have described a procedure involving semi-preparative gas-liquid chromatography for the detection, determination and identification of organophosphorus pesticide residues in vegetable material.

Kanazawa and Kawahara[2105] applied electron-capture gas chromatography to the determination of pesticides. They investigated the retention time, peak-area sensitivity, and rectilinearity of the calibration graph for various pesticides under two sets of operating conditions. The columns used were 5% of Dow silicone 11 at 170°C and 2% of polyoxyethylene glycol adipate at 180°C. For organophosphorus compounds, the polyoxyethylene glycol adipate is the more satisfactory stationary phase, peak-area sensitivities on this column being higher than those on the silicone column. The concentration ranges for rectilinear calibration graphs are fairly varied.

Guiffrida et al[2106] studied the effects of varying the operating conditions for electron-capture and flame ionization detectors. Use of a hydrogen flame burning in an atmosphere or alkali-metal salt (e.g. potassium chloride) results in increased sensitivity and complete specificity in the detection of organophosphorus pesticides. Typical conditions used by these workers include a column (6 ft x 4 mm) of 10% of DC 200 silicone fluid on 80 - 100 mesh Gas-Chromb Q, operated at 190° - 230°C, with nitrogen as carrier gas. A technique involving the parallel use of an electron-capture detector and a modified flame ionization detector is described. A method for the specific detection of sub-nanogram quantities of organophosphorus compounds by flame ionization has been described by Kamen[2107]. This detection is specific for phosphorus and halogens.

The flame photometric detector, which is specific for phosphorus and sulphur has been used (Brody and Chaney[2108]) for the detection of subnanogram quantities of pesticides containing these elements. The detector is constructed by modifying existing flame ionization equipment; narrow band-pass filters are used to isolate the phosphorus and sulphur emission bands at 526 nm and 394 nm respectively. The detector is sensitive to parts/10^9 of phosphorus and to less than 1 part/10^6 of sulphur, but is insensitive to hydrocarbons and to compounds containing chlorine, nitrogen and oxygen which are burned inside the burner tip, which is shielded from the photomultiplier tube. Response to phosphorus is rectilinear for up to 63 parts/10^6; the response to sulphur varies exponentially up to 100 parts/10^6. Nitrogen is used as carrier gas; the flow of hydrogen plus air is kept at 200 ml min^{-1} and the ratio of nitrogen to oxygen is maintained at 4:1.

The application of the microcoulometric gas chromatographic detector to the selective detection of phosphorus, sulphur, and halogen in pesticides has been studied by Burchfield et al[2109]. After the separation of phosphorus- sulphur- or chlorine-containing pesticides on a column (1 m x 3 mm) of Anakrom ABS supporting 10% of DC 200 silicone oil, with hydrogen (15 ml min^{-1}) as carrier gas and temperature programming from 175° - 230°C in 10 min, the column effluent is passed through a small furnace at 950°C, in which the compounds are converted into phosphine, hydrogen sulphide, or hydrogen chloride. These gases are measured with a micro-coulometric titration cell equipped with silver electrodes. The insertion of a short tube containing alumina between the furnace and the titration cell removes hydrogen sulphide and hydrogen chloride, whereas a short tube containing silica gell similarly placed retains hydrogen chloride and separates phosphine and hydrogen sulphide.

In a further paper Burchfield et al[2110] further studied the selective detection of phosphorus-, sulphur-, and halogen-containing pesticides. They discuss ionization detectors and the possibility of increasing their specificity. They also describe microcoulometric methods for the analysis of pesticides containing the above three elements. These methods are briefly reviewed below:

Halogens and sulphur. The sample solution in an organic solvent is carried through the chromatography column by means of nitrogen and oxygen is added to the effluent before combustion at 800°C. To determine halogens, the combustion products are passed through a titration cell in which halides (except flouride) cause precipitation of silver chloride and the current required to generate sufficient silver ions to replace those consumed is recorded. Down to 0.001 μg of chloride ion can be detected. To determine sulphur the combustion products are passed through a cell in which the sulphur dioxide is automatically titrated with iodine, the current required to generate sufficient iodine being recorded also.

Phosphorus. The sample organophosphorus compounds are first separated on a chromatographic column, and are then reduced at 950°C to phosphine by means of hydrogen. The reduction products pass through a titration cell in which silver phosphides are precipitated and the current necessary to regenerate silver ions equivalent to those precipitated is recorded. To prevent interference by hydrogen chloride and hydrogen sulphide a "subtraction" column of alumina or silica gel is inserted between the reduction tube and the titration cell. The determination of chlorine plus sulphur plus halogen in a sample mixture is possible by omitting the subtraction column. The methods have been applied to the determination of insecticides such as Mevinphos, Diazinon, and Carbophenothion.

Ives and Giuffrida[2111] have investigated thermionic detector response in the gas chromatography of compounds containing phosphorus, nitrogen, arsenic and chlorine. They found that the response to Group 5 elements depends on the salt cation used in the detector, and is similar for different salts of the same cation. Phosphorus can be distinguished from nitrogen and arsenic by comparing the responses of the thermionic and a flame ionization detector. Use of potassium chloride as source is preferred for phosphorus compounds, and rubidium chloride is preferred for nitrogen compounds. Therionic detection is especially useful for compounds containing several nitrogen atoms in the molecule. The thermionic response to various pesticides was also examined by these workers.

Rubin and Bayne[2112] optimized the design of the nitrogen-phosphorus specific gas chromatographic detector.

ORGANORUTHENIUM COMPOUNDS

Nesmeyanov et al[2113] have described methods for the separation of ruthenocene (and ferrocene) derivatives by gas chromatography. The separation of 41 ferrocenes and 2 ruthenocene derivatives on 2,2-dimethylpropane-1,3-diol adipate, polyoxyethylene glycol adipate, polyoxyethylene glycol M-20, polyoxypropylene glycol and Apiezon L (1 to 5% on Celite 545, 80 to 100 mesh) at 100° to 200°C is discussed. Packed columns (1 to 1.2 metres x 0.4 cm) in glass and steel tubes, and a capillary column (45 metres x 0.25 mm), were used, together with a β ray ionization detector. Best results were obtained with the use of 2,2-dimethylpropane-1,3-diol and polyoxyethylene glycol M-20 columns.

Ballschmitter[2114] has described a procedure for converting ruthenium to its volatile 4(hexachlorobicycloheptane-2-methylene) thiosemicarbazide and its determination by gas chromatography using an electron capture detector.

ORGANOSELENIUM COMPOUNDS

Evans and Johnson[2115] have studied the gas chromatography of a range of organoselenium compounds including dialkylselenides and ethylselenocyanate. They used a Wilkins Instrument and Research Inc. Hi-Fi 6000C gas chromatograph with either a hydrogen flame ionization detector or an electron capture detector.

The also carried out preparative scale gas chromatograph (Wilkins Instrument and Research Inc. Model, Aerograph A-90-P) for the purification of synthesized organoselenium compounds. The column used for the preparative gas chromatography was 5 ft x 1/4 in, Silicone Fluid (methyl) SF-96 on 60/80 fire brick. Helium was used as carrier gas with a flow rate of 35 - 40 ml/minute.

Three columns were used in the Hi-Fi 6000C instrument, (i) 5 ft x 1/8 in 20% polymetaphenyl ether on 60/80 Chromosorb W coated with hexamethyldisilazane. (ii) 10 ft x 1/8 in, 20% Carbowax 20M on 60/80 Chromosorb W treated with hexamethyldisilazane. (iii) 10 ft x 1/8 in 20% silicone oil DC 550 on 60/80 Chromosorb W coated with dimethyldichlorosilane. Nitrogen was used as carrier gas at flow rate between 20 and 30 ml/min. Using the hydrogen flame ionization detector, the retention times of the alkyl selenium compounds were determined on each of the three columns at column temperature, within the range of 35 - 175°C. The injector temperature was set at 50 to 100°C higher than the column temperature. One per cent solutions of each selenium compound in carbon disulphide were used for the determinations, as carbon disulphide gives very little response with this detector system.

Tables 187 - 189 give the retention times for each compound on the three columns used. A comparison of the degree of resolution of the organoselenium compounds which was obtained from each of the three columns, showed that the best resolution was achieved on the polymetaphenylether column, but both Carbowax 20 M and Silicone Oil DC 550 columns gave satisfactory resolution of all the alkyl selenium compounds studied. However, the DC 550 column was unsatisfactory for use above 150°C as a large amount of continuous column bleeding occurred above this temperature.

Table 187 - Retention times of alkyl selenium compounds on the polymetaphenylether column.
Nitrogen carrier gas flow rate 25 cc/minute

	Retention time (minutes)					
Column Temperature °C	50	75	100	125	150	175
Injector Temperature °C	100	180	180	180	225	220
Dimethyl selenide	2.3	1.4	0.8	0.6	0.5	
Diethyl selenide	12.0	5.8	2.4	1.8	1.0	
Dipropyl selenide	35.0	20.0	8.4	4.3	2.2	
Dimethyl diselenide		22.5	10.2	5.2	2.7	2.0
Diethyl diselenide			28.5	12.6	6.0	4.2
Dipropyl diselenide				35.5	12.4	9.0
Ethyl selenocyanate		30.5	11.5	7.0	3.2	2.7

Figure 120 shows the complete resolution of the seven alkyl selenium compounds on the polymetaphenylether column, at column temperature of 150°C. Two additional peaks occurred at retention times 4.0 min, and 8.2 min, which were tentatively identified as methylethyldiselenide, and ethylpropyl diselenide on the basis of their correspoondence with the theoretical retention times of these compounds as determined from the logarithmic plot of the retention time vs. number of carbon atoms in the alkyldiselenides. Similarly the logarithmic plot of retention time vs. boiling point suggested that the

Table 186 - Retention times of alkyl selenium compounds on the Carbowax 20M column

	Retention time (minutes)					
Column temperature °C	35	55	70	100	120	160
Injector temperature °C	100	100	180	180	180	220
Dimethyl selenide	7.0	4.0	2.8	1.7	1.3	0.8
Diethyl selenide		16.0	6.7	3.8	2.5	1.4
Dipropyl selenide			16.2	9.0	5.2	2.3
Dimethyl diselenide			36.5	16.0	8.9	3.8
Diethyl diselenide				32.0	14.7	6.4
Dipropyl diselenide				72.0	30.0	11.0
Ethyl selenocyanate				48.0	19.5	6.5

Table 189 - Retention times of alkyl selenium compounds on the Silicone Oil DC550 column,
Nitrogen carrier gas flow rate: 25 cc/minute

	Retention time (minutes)		
Column Temperature °C	100	125	150
Injector Temperature °C	180	180	225
Dimethyl selenide	5.2	3.7	2.5
Diethyl selenide	15.5	9.2	5.8
Dipropyl selenide		28.0	14.0
Dimethyl diselenide		28.0	13.0
Diethyl diselenide			29.0
Ethyl selenocyanate		27.5	14.0

additional peaks corresponded to methylethyldiselenide, and ethylpropyldiselenide. It appears that these mixed alkyl diselenides could have been formed in the mixture by interaction between dimethyl, diethyl and dipropylselenides.

Parris et al[2116] have described a procedure utilizing a commercial atomic absorption spectrophotometer with a heated graphite tube furnace atomizer linked to a gas chromatograph for the determination of dimethylselenium, (also trimethylarsenic and tetramethyltin) in respirent gases produced in microbiological reactions. Chau et al[2117] have applied a combination of gas chromatography with atomic absorption spectrometry to the determination of dimethylselenide.

The gas chromatography-microwave plasma detector technique has been applied to the analysis of organoselenium compounds[2118].

Bortnikov et al[2119] separated bistriethylsilyl sulphide and related compounds containing silicon, selenium, sulphur, germanium, tellurium and/or tin as the central linking atom at 254°C on a stainless steel column (100 cm x 0.4 cm) packed with Chromosorb W, supporting 20% of Apiezon L, with helium as carrier gas and a thermal conductivity detector. Specific retention volume data are reported.

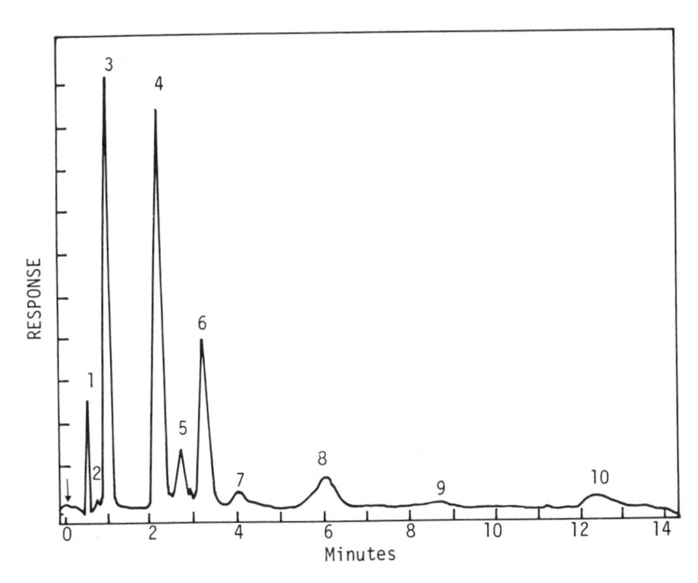

Figure 120 - The separation of alkyl selenium compounds on a polymetaphenylether column with hydrogen flame ionization detector. Column temperature 150°C., injector temperature 225°C., nitrogen carrier gas; flow rate 25 ml/min., (->) injection point at time 0 min. 1% solution of each compound in carbon disulphide. (1) dimethyl selenide, (2) carbon disulphide, (3) diethyl selenide, (4) dipropyl selenide, (5) dimethyl diselenide, (6) ethylselenocyanate, (7) methyl ethyl diselenide, (8) diethyldiselenide, (9) ethyl propyl diselenide, (10) dipropyldiselenide.

Vlazakova et al[2120] applied gas chromatography to analysis for trace amounts of volatile metabilites such as dimethyl selenide and dimethyl diselenide in expired air. Exhaled air from rats injected with $Na_2{}^{75}SeO_3$ solution was drawn through a column of Alusil, and this column was then heated at 230°C. The desorbed gases were swept by nitrogen (50 ml/min) into the column (1.2 metres x 4 mm) packed with 15% of SE-30 or 10% of Apiezon L) or Gas-Chrom P (80 to 100 mesh). The column temperature was kept at 20°C for 20 min., then programmed at 6°C per min. to 240°C; detection was by flame ionization. The eluted compounds were counted for ^{75}Se with a NaI (TI) detector. The main products identified were dimethyl selenide and dimethyl diselenide.

Nakashima and Toei[2121] showed that inorganic selenium in the form of selenium acid reacts with 4-chlorololphenylenediamine hydrochloride to produce 5-chlorobenzene-2,13 selenadiazole which is then extracted into toluene preparatory to determination on a column consisting of SE-30 on Chromosorb W operated at 200°C. Nitrogen is used as a carrier gas and detection was achieved by an electron capture detector.

Shimoishi and Toei[2122] determined ultramicroamounts of selenium in pure sulphuric acid by oxidizing it to selenous acid with bromine-bromide redox buffer solution followed by quantitative conversion to 5-nitropiaselenol, as reaction with 4.nitro-o-phenyldiamine, and extraction into toluene and estimation of the 5-nitropiaselenol under the same conditions as described above.

Narquant and Belford[2337] determined impurities in carbondiselenide. These included methylene dichloride, chloroform, 1,1,2,2,-tetrachloroethane and tetrachloroethylene, tetrahydrofuran, toluene, OCSe and SCSe. Separation was achieved on a column comprising 0.4% Carbowax 1400 on 60/80 mesh Carbopack A, programmed between 50 and 175°C at 10°C per minute and using helium as carrier gas. A flame ionization detector was used.

ORGANOSILICON COMPOUNDS

A. Silicon Hydrides

Phillips and Timms[2123] demonstrated many of the advantages of using gas chromatography in the analysis of the organic compounds of silicon. Figure 121 shows the rapid separation they obtained of trisilane and the mixture resulting after the passage of an electrical discharge through trisilane. For mixed silicon-germanium hydrides, it was found that compounds with similar retention volumes gave an approximately molar response so that a gas density balance could be used to measure molecular weights.

In Table 190 are shown observed retention volumes reported by Timms et al[2124] for a range of silicon hydrides. Observed retention volumes are calculated according to the method described by Semlyen et al[2125].

Feher and Strack[2126] separated silanes up to $Si_{18}H_{18}$ on a column of 15% squalane on kieselguhr at 110°C using helium as carrier gas (flow rate 119 ml per minute).

B. Alkyl and Arylsilanes

Phillips and Timms[2123] demonstrated that for alkyl silicon compounds log retention time versus the number of silicon atoms gave straight lines in a manner similar to that reported for the hydrocarbons (Figure 122). In this figure the column packing was Silicone 702 operating at 100°C. The solid circles and line represent the straight chain compounds. The open circles are the branched isomers. The latter lie below the line and are analogous in behaviour to the hydrocarbons.

A somewhat different approach was taken by Pollard et al[2127] who showed that approximately straight line plots were obtained by plotting the log of corrected retention volume versus the total number of carbon atoms in the alkyl groups (Figure 123). This relationship provides a method for identifying additional compounds. Pollard et al[2127,2128] have carried out an extensive tabulation of retention data on Silicone Oil E301. Other column packings have been discussed by Garzo et al[2129] and by Timms et al [2124].

Semlyen et al[2125] have listed retention volumes for a range of alkyl-silanes.

Pollard et al[2128] have described an ionization guage detector which they used for the detection of organosilicon compounds emerging from a column comprising 15% E 301 Silicone Oil on 30 - 60 mesh treated Celite operated at various temperatures between 80 and 140°C. Nitrogen was used as carrier gas. In Table 191 are shown the specific retention volume data obtained by Pollard et al[2128] for a range of organosilicon compounds, It is seen that the specific retention volume increases with the molecular weight of the alkyl silane.

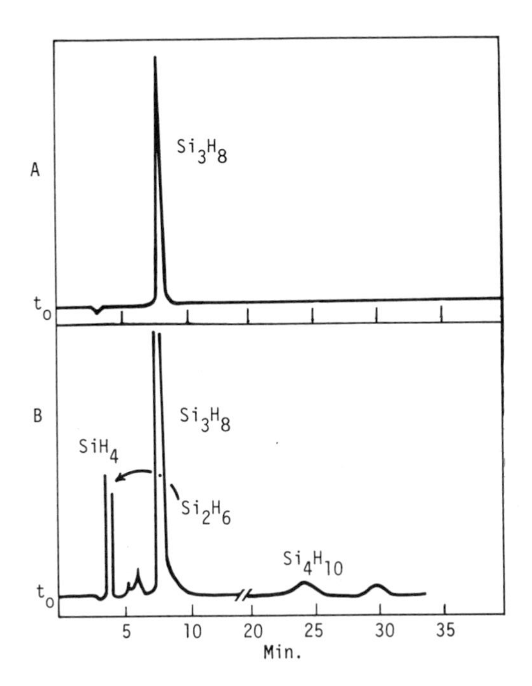

Figure 121 - Gas chromatograms of (a) trisilane and (b) products obtained after passing an electric discharge through disilane.

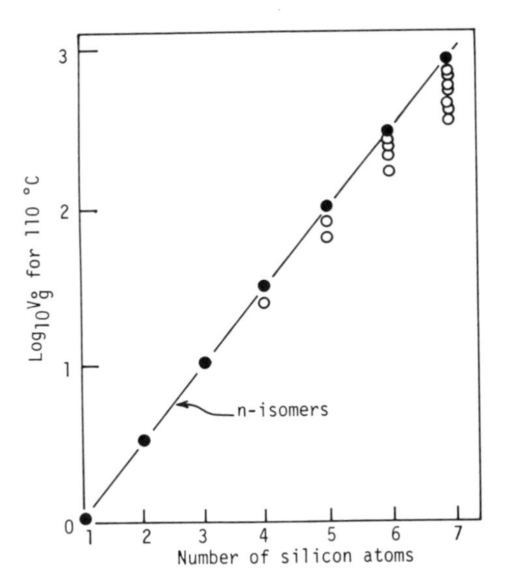

Figure 122 - Gas chromatography of silanes; log (retention times) versus number of silicon atoms.

Table 190 - Logarithm of retention times for silicon hydrides - comparison of calculated and observed values

Compound	Silicone 702				Tritolyl Phosphate			
	20°C		64.5°C		20°C		64.5°C	
	Obs.	Calc.	Obs.	Calc.	Obs.	Calc.	Obs.	Calc.
Si_3H_8	0.66	0.65	0.98	0.98	0.79	0.78	1.08	1.06
iso-Si_4H_{10}	1.25	1.24	1.41	1.42	1.32	1.31	1.47	1.47
n-Si_4H_{10}	1.41	1.40	1.54	1.55	1.48	1.48	1.59	1.60
iso-Si_5H_{12}	2.00	2,00	2.00	2.00	2.00	2.01	2.00	2.01
n-Si_5H_{12}	2.17	2.16	2.13	2.13	2.18	2.18	2.13	2.14
iso-Si_6H_{14}			2.58	2.57	2.68	2.70	2.54	2.55
n-Si_6H_{14}			2.72	2.70	2.87	2.87	2.69	2.68
iso-Si_7H_{16}			3.14	3.13			3.07	3.10
n-Si_7H_{16}			3.28	3.26			3.22	3.23

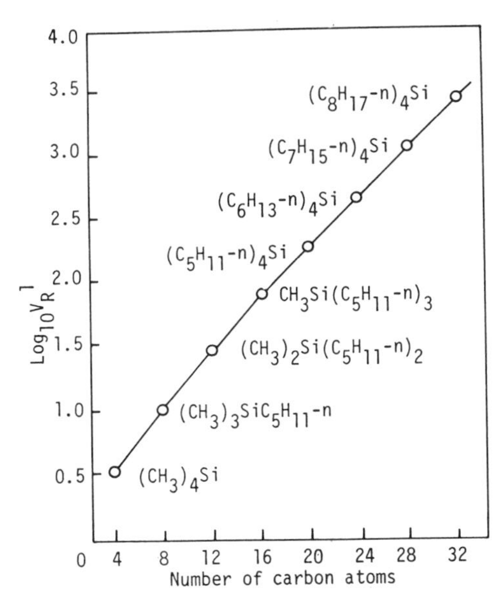

Figure 123 - Gas chromatography of organosilicon compounds. Mixed alkyls containing CH_3 and n-C_5H_{11} to nC_8H_{17} groups in various proportions.

Table 191 - Specific retention volumes V_g (ml) of organosilicon compounds on 15% E301 Silicone Oil

Compound	80°C	100°C	110.5°C	140°C
$SiMe_4$	11.82±0.15	7.65±0.07	6.31±0.07	3.09±0.03
Me_3SiEt	31.4±0.2	18.45±0.06	13.77±0.2	7.15±0.05
Me_2SiEt_2	79.2±0.5	43.2±0.2	32.0±0.7	16.4±0.2
$MeSiEt_2$	190.9±0.9	98.7±0.7	68.3±1.8	34.1±0.1
$SiEt_4$	446.8±2.7	216.8±3.0	147.7±2.9	69.3±0.4
Me_3SiPr-n	61.6±0.6	34.3±0.4	24.9±0.1	13.33±0.2
Me_3SiBu-n	129.3±2.0	68.0±0.2	47.7±0.2	24.3±0.1
Et_3SiH	118.8±0.4	64.6±1.1	44.8±0.6	23.2±0.2
$Me_3SiOSiMe_3$	74.4±0.5	40.1±0.1	28.4±0.4	14.7±0.05

Pollard et al[2130] have also studied qualitatively the thermal decomposition of some aryltrimethylsilanes in the presence of hydrogen in a continuous flow reactor, using gas chromatography for the analysis of the reaction products.

Semlyen and Phillips[2131] have published gas chromatographic retention data on a range of alkylsilanes and show the patterns between log t_R values which enable retention times to be predicted. They also carried out a comparison between the alkyl silanes and the analogous hydrocarbon systems. The alkylsilanes were analysed on columns of 2 - 13% squalane at 100°C using a flame ionization detector and oxygen-free hydrogen as carrier gas. The solid support was Embacel (60 - 100 mesh acid-washed) which had been made inactive by treatment with hexamethyldisilanzane[2132]. Semlyen and Phillips[2131] express retention data as the logarithm of the retention time relative to that of mesitylene (log t_R = 2). Semlyen and Phillips[2131] and

and other workers[2124,2133] claim that the use of retention parameters for the calculation of retention times works well for silanes and tetraalkylsilanes.

Peetre and Smith[2134] and Peetra et al[2135] have developed equations for the calculation of retention indices for mixed tetraalkylsilanes using the same methods as they applied earlier for the tetraalkoxysilanes[2132,2134-2137]. They discuss the significance of the sign of the charge on the terminal methyl carbon atom of the alkyl groups bonded to silicon in determining the magnitude of the retention index and I value. They also give rules for the estimation of the temperature dependence of retention indices for the tetraalkylsilanes[2134] and show that linear relationships exist between retention indices for homologous series of mixed tetraalkylsilanes on the one hand and retention indices of symmetrical tetraalkylsilanes or of tetraalkoxysilanes on the other. They show the the two-phase plot of I^{APL}_{160} versus I^{XE}_{160} for the tetraalkylsilanes is completely different from that for the tetraalkoxysilanes. Peetre and Smith[2134] used a Varian Model 1400 gas chromatograph with a flame ionization detector, and equipped with an auxiliary thermocouple and temperature gauge for the retention index measurements. The temperature could be read to ± 0.1° and was constant within this range during a measurement. Any significant deviation of the observed temperature from the measurement temperature in the tables was corrected for.

Perkin-Elmer capillary steel columns (50 m x 0.25 mm) containing Apiezon L and cyanosilicone XE-60, respectively, as stationary phases were used throughout. For the determination of accurate retention indices they used a computer method[2134].

In investigations of tetraalkoxysilanes[2135,2136], their change in retention index with temperature was found to be always negative and to increase with molecular size. For the tetraalkylsilanes, the temperature increments are considerably lower and alternate between positive and negative values in an apparently erratic manner. However, there is also a tendancy towards increasing negative values for larger molecules.

Pollard et al[2127] have reported a linear relationship between the logarithm of the retention volume and the carbon number for tetraalkylsilanes belonging to the series R_4Si-R_3-SiR'-$R_2SiR'_2$-RSiR'$_3$-SiR'$_4$. However, this is true only for certain series. Thus for the seven series Me_nSiR_{4-n} (R = ethyl to pentyl) and Et_nSiR_{4-n} (R = propyl to pentyl) they studied the above statement is approximately correct only for the two series $Me_nSi Et_{4-n}$ and Et_nSiR_{4-n}. For the other five series, there will be an increased deviation from linearity with increase in the difference in size between the two alkyl groups in the tetraalkylsilane, as a consequence of the increased departure from additivity of the group retention indices.

Hailey and Nickless[2138] have reported specific retention volumes for some tetraalkylsilanes mainly containing branched alkyl groups, Wurst and Churuacek[2139] measured a quantity called the silicon index (ISi) for tetramethyl, tetraethyl and tetrapropyl silanes.

Nametkin et al[2140] reported a study of the behaviour of compounds of the types:

$$(CH_2)_3Si(CH_2)_nSi(CH_3)_3,\quad (CH_3)_2\overline{Si-CH_2-CH_3)_n}^{CH_2} \quad \text{and} \quad (CH_3)_3Si(CH_2)_nCH_3$$

in comparison with normal paraffins during separation at 50°C - 150°C using helium carrier gas on columns (200 cm x 0.6 cm) containing 20% of polymethylphenylsiloxane or polyoxyethylene glycol 1540 on firebrick. The relative retention volumes (nonane as standard) graphs of the relationship between the log retention times of the substances investigated on the two stationary phases, and values of the structural increments for calculating the relative retention time are presented by these workers for several compounds[2141].

Bortnikov et al[2156] measured the retention volume for triethyl and tetraethyl silanes on a stainless steel column (1 metre x 3 mm) packed with silanized Chromosorb W supporting 20% of Apiezon L or 15% of Carbowax 20M and operated at 120°C and 90°C respectively using helium as carrier gas and thermal conductivity or flame ionization detection. They also carried out measurement by gas-solid chromatography on a column of Carbochrom-1 (graphitized thermal carbon black with 0.01% of Apiezon L) operated at 190°C or Silichrom C-80 (modified by chlorotrimethylsilane or dehydroxylated) operated at 100, 147 or 203°C, respectively.

Bersadchi et al[2157] separated trichlorosilane and silicon tetrachloride mixtures in carbon tetrachloride using transformer oil activated by glycerol as the liquid phase.

C. Chlorosilanes

Palamarchuk et al[2158] studied the separation of mixtures of di and trichlorisilane, silicon tetrachloride, methylchloride and also various methylchlorosilanes on columns (270 - 350 mm x 0.4 cm) operated at 30°C and containing 10% of benzyl benzoate, dibutyl phthalate or diethylphthalate supported on Celite or on diatomite brick previously treated with methyl silicon dichloride vapour. Helium was used as a carrier gas with a katharometer detector.

Abe[2159] used dibutyl phthalate, also tritotyl phosphate or Silicone Oils KF96 and DC703 (Dow Corning) as stationary phases with helium as the carrier gas for the separation of silicon tetrachloride and phosphorus trichloride in trichlorosilane[2160]. He found that the esters vitiated results due to the production of hydrogen chloride by hydrolysis of the sample. He recommends a column (4 mm x 750 mm) containing Silicone Oil DC703.

Fritz and Ksinsik[2161] showed that chlorosilanes could not be separated by gas absorption chromatography on any of the usual solid absorbents. By gas-liquid partition chromatography, quantitative separation of the lower-boiling silicon compounds was achieved on a column containing Silicone Oil DC200/350 and dibutyl phthalate on kieselguhr (3:1:5, by wt) at 30°C and 60°C with hydrogen as carrier gas, and of the higher-boiling compounds on a column containing silicon oil on kieselguhr (3:10) at 150°C.

Rotzsche[2162] separated mixtures of mono, di and trichlorosilane, silicon tetrachloride and silane (SiH_4). He used two separation columns, both operated at room temperature. This mixture was fully resolved on a stationary phase of silicon oil with the exception that any nitrogen in the sample elutes with silane (Figure 124). He separated nitrogen from silane on a column of silica gel[2163].

Vronti-Piscou and Parissakis[2164] have reported on the elution characteristics of various silicon chlorides of the type $Si_nCl_{2n} + 2$ obtained by the pyrolysis of $(SiCl_2)_n$. They successfully chromatographed mixtures of these compounds in benzene solution on a column (1 metre x 6 mm) comprising

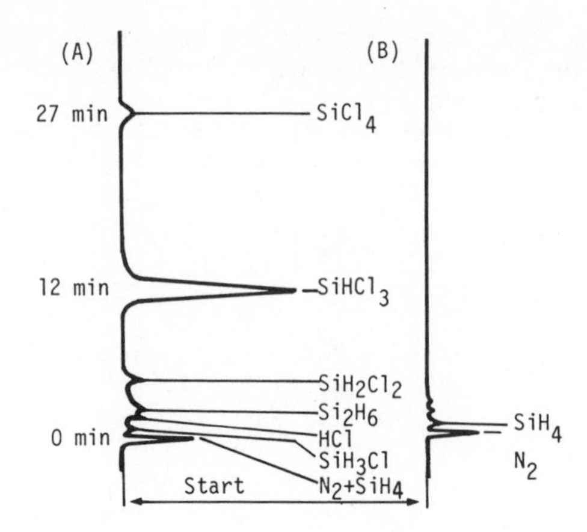

Figure 124 - Gas chromatogram of chlorosilanes.

Silicone oil 550 (20%) on Celite (30 - 60 mesh) using nitrogen as the carrier gas and a thermal conductivity detector with temperature programming between 150 and 290°C at 140°C/min., (detector temperature 150°C).

Basu et al[2165] separated mixtures of silicon tetrachloride, chlorine, phosgene, argon, nitrogen, carbon monoxide and carbon dioxide obtained in the reductive chlorination of metal ores on a three column, three detector gas chromatograph. The three gas chromatograms were equipped with the following columns in series: Column 1 (20 foot x 0.25 inch) was filled with 15% of Kel-F 10 on Chromosorb T (30 - 60 mesh) and operated at 60°C. Column 2 (5 foot x 0.25 inch) was filled with silica gel (30 - 60 mesh) and operated at 139°C. Column 3 (5 foot x 0.25 inch) was filled with molecular sieve 5A (30 - 60 mesh) and operated at 90°C. Thermal-conductivity detectors were used in all instances and the carrier gas was helium (80 ml per minute). Silicon tetrachloride, chlorine and phosgene were determined on column 1, carbon dioxide on column 2 and carbon monoxide, argon and nitrogen on column 3. It was possible to detect hydrogen chloride on column 1 by operation at a lower column temperature.

Impurities such as methane, ethane, propane, isobutane, butane, chloromethane, chloroethane, hydrogen chloride, silicon tetrachloride and methyltrichlorosilane can be determined in trichlorosilane[2166] by separation on a glass column (4.5 metres x 4 mm) packed with silanized silica gel operated at 50°C using nitrogen as carrier gas and flame ionization or thermal conductivity detectors.

D. Alkyl, aryl and vinylchlorosilanes

Friedrich[2161] separated six silanes including trimethylchlorosilane and silicon tetrachloride using nitrobenzene on infusorial earth as the stationary phase. The gas stream was passed into 0.02 N potassium chloride and changes in acidity are determined by conductivity measurements.

Joklik[2168] separated ethylchlorosilanes on a chromatographic column packed with methylsilicone oil MSO 150 (10% on ground unglazed tile). The column is maintained at 90°C and nitrogen is used as carrier gas. He gives elution rates for nine compounds.

Kawazumi et al[2169] showed that for the separation of mono, di and trimethylchlorosilane CH_3SiHCl_2 and silicon tetrachloride, liquid paraffin and insulating oil H-132 are more satisfactory than Silicone oil TS 953 M, squalane or Electron oil A on firebrick (30 - 60 mesh) as stationary phases. By the use of two 4 metre columns containing liquid paraffin and H-132, respectively, the five compounds are satisfactorily separated at 70°C. Helium was used as carrier gas at a flow rate of 30 - 70 ml per minute.

Turel'taub et al[2170] studied the effect of the nature of the stationary phase, and the solid support, the amount of stationary phase, the rate of flow and moisture content of the carrier gas, and the column temperature on the gas-liquid chromatographic separation of the components of a mixture obtained in the preparation of methylchlorosilanes. They developed three methods for the analysis of such mixtures. The mixtures of hydrogen chloride, methylchloride, silicon tetrachloride, trimethylsilicon tetrachloride, dimethyl silicon dichloride and methyl silicon trichloride can be analysed on a column (200 cm x 0.4 cm) containing 20% of nitrobenzene on firebrick of particle size 0.25 - 0.5 mm. To maintain a constant amount of this stationary phase in the chromatographic column, another column (200 cm x 0.5 cm) with a similar packing, is placed before it. A mixture containing the above substances plus tetramethyl silicon, trichlorislane ($HSiCl_3$) and CH_3HSiCl_2 is analysed on two successive columns (100 cm x 0.4 cm and 300 cm x 0.4 cm) containing 15% of polymethylphenylsiloxane liquid and 15% of petroleum jelly respectively, on Celite. A mixture of all the above substances except tetramethylsilicon can be analysed on a column (400 x 0.4 cm) containing 20% of polymethylphenylsiloxane liquid on firebrick. In all cases, the column temperature is 40°C; the carrier gas is helium passed at 80 ml per minute; and the detector is katharometer. The analysis takes 20 - 30 minutes, and the error of the determinations is 1 - 2% absolute.

Garzo et al[2171] developed an apparatus for the analysis of a mixture containing trichlorosilane, CH_3SiHCl_2, silicon tetrachloride, trimethylchlorosilane, dimethyldichlorosilane, methyltrichlorosilane by gas-liquid chromatography on a 160 cm column of nitrobenzene supported on Celite 545. The column is eluted with nitrogen (50 - 55 ml per minute), and the emergent gas is absorbed in flowing 0.01 N potassium chloride. Hydrolysis of the silanes yields hydrochloric acid which alters the electrical resistance of the solution during elution thus permits quantitative analysis of the silane mixture.

Oiwa et al[2172] separated methylchlorosilanes on two 3 metre columns (in series) containing, respectively, 20% of tritolyl phosphate and dioctyl phthalate supported on Kieselguhr. The columns are operated at 58°C with helium as carrier gas (75 ml per minute). Isopropyl ether is used as internal standard.

Lengyel et al[2173] studied the separation of methylchlorosilanes on different supports and with different liquids. Celite and Sterchamol were both satisfactory supports but Termalite (firebrick) had a strong selective-absorption effect, resulting in asymmetric elution curves. Of the stationary phases tested (liquid paraffin, silicone oil, 1-chloronaphthalene, dibutylphthalate and nitrobenzene), those with the highest dipole moments were the most effective. Celite impregnated with 23% of nitrobenzene gave complete separation of a six-component mixture.

Rotzsche[2174] using a nitrile silicone oil and m-nitroluene stationary phase with a katharometer detector separated a mixture of various methylchlorosilanes and chlorosilanes obtained in reactions between ferrosilicon and methylchloride. Typical chromatograms are shown in Figure 125. In general good agreement is obtained between the determined and the expected results.

Avdonin et al[2175] separated mixtures of methylchlorosilanes produced in the reaction of methylchloride with silicon copper alloys on a column of FS-16 (Dow DC methyl p-chloro phenyl siloxane polymer) on firebrick at 70°C using nitrogen carrier gas.

Cermak et al[2176] also developed gas chromatographic methods for the analysis of the methylchlorosilane products mixture (boiling range 70 - 220°C) produced by the reaction of silicon with methyl chloride. The 70/220°C fraction was distilled in vacuo, the distillate was fractionated on a high-efficiency column, and each fraction was further rectified. The fractions were then purified by preparative gas chromatography. The column used for this purpose (5 metres x 30 mm) consisted of Rysorb impregnated with 20% of a polymethylphenylsiloxane oil and was operated at a temperature 30 - 50°C below the boiling point of the particular fraction. The carrier gas was nitrogen (60 - 70 litres per hour). The purified fractions so obtained were analysed for elements and were also examined by infrared spectrophotometry. Their elution times relative to dimethyldichlorosilane were determined on a similar column (180 cm long) operated at 130°C with a nitrogen flow rate of 3 litres per hour.

Sivitsova et al[2177] examined the effect of 31 stationary phases on the relative volatility of silicon tetrachloride and methylchlorosilanes. Chromatography was carried out at 25°C on columns (2.5 metres x 4 mm) of firebrick containing 20% of stationary phase. Various stationary phases (2-chloroethyl ether, ethyl chloroacetate and 1,1,3; trichloropropane) were suitable for separating multi-component mixtures of silicon tetrachloride and methylchlorosilanes.

Popov et al[2178] measured the relative retention volumes and retention indices for silicon tetrachloride, methyltrichloro- and dimethyldichlorosilane, phosphorus trichloride and phosphorus oxychloride at 50°C on the following stationary phases (each in a concentration of 20% on the support): Vaseline oil, polysiloxane liquids VKZh-94, PFMS-2 and DS-701 fluorosilicone oil 169, dinonylphthalate and dibutyl phthalate. The retention index for phosphorus oxychloride increased with increase in the polarity of the stationary phase; that for silicon tetrachloride remained practically constant. The maximum difference in retention volume was obtained on a non-polar stationary phase for phosphorus trichloride and the chlorosilanes and on a polar stationary phase for phosphorus oxychloride and the chlorosilanes.

Turkel'taub and Palamarchuk[2179] have investigated the determination of low concentrations of methyl chlorosilanes using a flame ionization detector. They investigated detector sensitivity using a 2 metre column of 15% silicon FS-16 on silanized INZ-600 brickdust at 40°C using nitrogen (40 ml/min) as carrier gas. They found that contamination of the detector by silica could be avoided by venting most of the column effluent, before it reached the detector. The sensitivity of the detector was found to increase with increasing number of carbon atoms in the organosilicon compound, but the response was less than half that for compounds with the same number of carbon atoms but no silicon atoms.

Gas chromatography has been applied to the analysis of the chlorination

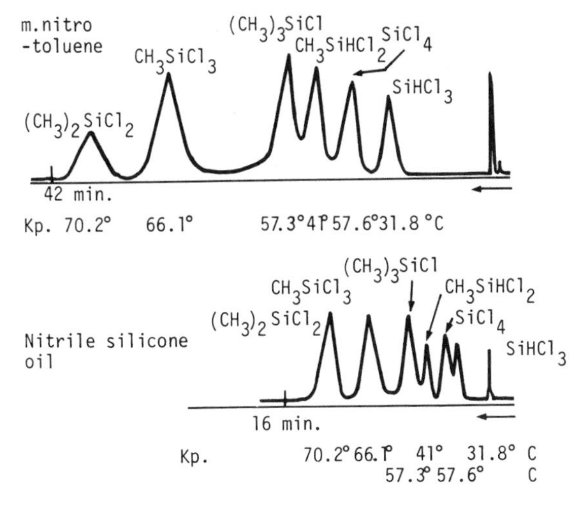

Figure 125 - Gas chromatography of chlorosilanes on nitrotoluene and nitrile silicone columns.

products of alkylchlorosilanes (Kreshkov et al[2180]). The separation and determination of the chlorination products of trichloroethylsilane or dichlorodimethylsilane are effected at 90°C or 80°C, respectively, on a column (1 metre x 4 mm) packed with 15% of PFMS-4 poly (methylphenylsiloxane) on Çelite 545, with helium (43 or 50 ml per minute) as carrier gas and katharometer detection. The chlorination products of trichloromethylsilane are analysed at 75°C on a column (1 metre x 4 mm) packed with 15% of SKTFT-50-siloxane rubber on TND-TS-M with a helium flow rate of 60 ml per minute.

With the exception of the earlier work of Turkel'taub and Palamarchuk 2179 all the work on the gas chromatography of alkylchlorosilanes discussed so far in this section utilises a katharometer detector. Turkel'taub and Palamarchuk[2179] used a flame ionization detector and commented on the fact that this type of detector had to be protected from silica produced by thermal decomposition of the organosilicon compound. Garzo and Till[2181] have investigated further the applicability of the flame ionization detector to gas chromatography of methylchlorosilane and report that this detector is unsuitable due to adverse dissociation processes occuring in the ionisation space in the detector. The results they obtained however, did permit some conclusions to be drawn on the dissociation processes occuring in the ionization space and they established relative retention data of some organosilicon compounds on a column packed with Apiezon M and silicone elastomer using a Pye argon chromatograph and showed that quantitative analysis of these compounds was possible.

Garzo and Till[2181] analysed mixtures of the following compounds: tetramethylchlorisilane, dimethyldichlorosilane, triethyltrichlorosilane, $SiHCH_3$, $MeSiHCl_2$, and carbon tetrachloride. The experiments were carried out on a Pye Argon chromatograph using specially purified argon (30/40 ml/min) as carrier gas and a glass column packed with 10% of dioctylphthlate on 80 - 100 mesh Celite, as stationary phase. On account of the high hydrolysis rate of the introduced samples the mixture was sealed in glass capillaries, according to the method of Joklik[2182], and the capillaries were broken while constantly flushing the breaking device with argon. Less than 1 mg of silane could be weighed this way.

Garzo and Till[2181] also investigated the utilizibility of the Pye argon chromatograph for the analysis of other organosilicon compounds. Thus they assayed linear and cyclic polymers of low molecular weight and methylethoxysilanes. Apiezon M(10%) and a devolatilised linear siloxane polymer (the Hungarian product HV-Au-120, mol. wt 59000) were used as stationary liquids supported on material on a glass column. Samples were introduced by 0.025 - 0.1μl pipette, using the following operating operating conditions: cell voltage 1000V; argon flow rate: 20 - 25 ml/min.

Ainshtein and Shulyat'eva[2183] have tabulated retention indices for alkylchlorosilanes and propylchlorosilanes on a polymethylsiloxane liquid PMS-20,000 and on a fluorosiloxane liquid FS-303 mobile phases.

Burson and Kenner[2184] have studied the gas chromatographic separation of chlorosilanes, methylchlorosilanes and their associated siloxanes and report retention times for these on 11 different stationary phases. These workers commented that many of the previous investigations dealt primarily with the separation of the chlorosilanes and methylchlorosilanes and ignored the presence of siloxanes which may be present in the sample, produced by hydrolysis, but which were not eluted in a reasonable length of time under isothermal conditions. Temperature programing is necessary for the elution of the siloxanes but many of the stationary phases reported in the literature cannot be used under these conditions because of their high volatility at elevated temperatures. Burson and Kenner[2184] specifically studied the problem of separating and quantitatively determining mixtures of chlorosilanes, methylchlorosilanes and their associated siloxanes.

Burson and Kenner[2184] used an F & M model 700 dual column, programmed-temperature instrument equipped with a thermal conductivity detector with tungsten-rhenium filaments. The detector was held at 230°C and the injection port at 180°C. Helium was used as the carrier in all cases. The following liquids or gums were used as stationary phases with acid-washed Chromosorb P as the inert support: dimethylphthalate; dibutylphthalate; dinonylphthalate; diethylphthalate; dipropyltetrachlorophthalate; QF-1 Silicone oil and XE-60 Silicone gum (Applied Science); LSX-3-0295 Silicone gum (Analabs); DC-710 and DC-704 Silicone oils (Dow Corning) and SF-96 Silicone oil (General Electric). Retention times for the components in the chlorosilane mixtures are listed in Table 192. Burson and Kenner[2184] identified the siloxane peaks by trapping the column effluent for analysis by mass spectrometry.

Burson and Kenner[2184] found that the dimethyl-, diethyl- and dibutylphthalates were the most effective of the phthalate ester stationary phases in making complete separations of chlorosilanes and methylchlorosilanes. The maximum temperature for these three phases is less than 100°C and, consequently, considerable column bleed is evident when the program nears this temperature. Dinonylphthalate has a higher temperature limit, but is not so effective in resolving the methyltrichlorosilane and dimethyldichlorosilanes. Dipropyltetrachlorphthalate is even less effective.

Franc and Wurst[2185] suggested a method for the analysis of phenylchlorosilanes (a mixture of phenyltrichlorosilane, diphenyldichlorosilane, benzene, chlorobenzene and biphenyl) by gas-liquid chromatography at 240°C on a column of length 150 cm containing a silicone elastomer (20%) on Chromosorb. The carrier gas is nitrogen (1.5 litres per hour) and the weight of sample 10 mg. After the separation the gases are passed into an ignition tube containing cupric oxide wire, pumice coated with reduced silver, and finally with iron filings. The organic substances are ignited at 480° -

600°C, the chlorine formed is absorbed, and the water formed is decomposed with the liberation of hydrogen. To the tube are attached a drying tube containing anhydrone and soda asbestos, and a katharometer to measure the hydrogen evolved from the organic material. The relative error is 3% and the analysis takes 15 minutes.

The same workers (Franc and Wurst[2186]) separated technical mixtures of benzene, chlorobenzene, phenyltrichlorosilane and diphenyl and diphenyldichlorosilane produced during the manufacture of silicones. These compounds were separated at 240°C on a column of Chromosorb impregnated with 20% of Silicone-elastomer E 301 (Griffin and George). The issuing gases are burnt to carbon dioxide and water in a combustion tube, the carbon dioxide is absorbed and the water is reduced to hydrogen with heated iron turnings and detected by thermal conductivity.

Wurst[2187] has described a gas chromatographic method using a gas density balance detector for the determination of chlorinated phenyl chlorosilanes. Seperation is achieved on a column (2 metres x 4 mm) packed with Celite 545 supporting 10% of Lukopren G 1000 (a Silicone elastomer), with nitrogen (60 ml per minute) as carrier gas. The column was operated isothermally at various temperatures from 140° - 210°C or temperature programed at 3° per minute over the range 100° - 200°C. Identification of the components was achieved by determination of retention values.

Wurst[2187] has tabulated the retention values for the components of commercial silane which include hydrocarbons and chlorinated hydrocarbons.

Wurst and Churacek[2139] have listed retention data for 44 alkyl-chloro and chlorophenylsilanes, 34 linear and branched dimethyl and methyl-vinylsiloxanes, 14 cyclic dimethyl and methylvinylsiloxanes and 6 linear methylphenyl and ethoxyvinyl-siloxanes. They propose a modification of the retention index which is calculated from the retention times of the organosilicon compounds and the retention times of the most closely related dimethylsiloxanes. Furthermore, they claim to validify the new retention index by analysis of mixtures of low molecular weight organosilicon compounds.

Wurst[2188,2189] studied the separation and determination of dichloromethylvinylsilanes and the chlorovinylsilane and impurities therein. Compounds in trichlorovinylsilane were separated on a column (1.6 metres x 5 mm) of 10% of silicone elastomer supported on Celite (or 10% of dibutylphthalate supported on Rysorb BLK) and operated at 40°C with nitrogen carrier gas (33 ml per minute). For impurities in trichloromethylvinylsilane best results were obtained with a column (1.8 metres x 4 mm) of 30% of nitrobenzene supported on kieselguhr and operated at 25°C, the carrier gas again being nitrogen (50 ml per minute). The mean error was generally less than ± 5% for the determination of individual compounds. Figure 126 is a typical chromatogram obtained for a 60° - 90° distillation cut of a sample of methylvinyldichlorosilane.

In the high temperature condensation method for the preparation of dichloromethylvinylsilane from dichloromethylsilane and vinylchloride a product is obtained that is contaminated with trichlorovinylsilane, which cannot be separated by distillation. For the quantitative analysis of the reaction product, a gas-chromatographic method was developed by Knauz et al[2190]. A 160 cm column, packed with 23% of dibutylphthalate on Celite (80 - 130 mesh), is operated at 27°C with nitrogen as carrier gas. The chromatogram shows eight characteristic peaks, including those of trichlorovinylsilane and dichloromethylvinylsilane which can be determined quantitatively from the peak areas.

Table 192 - Retention times for components of silane mixtures

Stationary phase	Retention time (minutes)				
	Tri-chlorosilane	Silicon tetra-chloride	Trimethyl-chlorosilane	Methyltril-chlorosilane	Dimethyldi-chlorosilane
Dimethylphthalate	8.5	10.2	15.0	20.1	23.3
Diethylphthalate	7.9	9.5	13.8	18.8	20.9
Dibutylphthalate	5.7	7.1	14.1	17.9	20.8
Dinonylphthalate	5.5	9.2	13.6	16.4	17.3
Dipropyltetrachlorophthalate	3.0	3.9	4.8	6.7	6.8
SF-96 Silicone oil	3.6	7.8	6.8	10.8	10.8
DC-704 Silicone oil	3.6	5.5	4.6	9.7	9.7
OF-1 Trifluoropropyl Silicone oil	2.2	3.1	4.3	5.2	6.0
XE-60 Cyanoethylmethyl Silicone oil	7.3	8.8	12.3	15.0	17.0
DC-710 Phenyl Silicone oil	2.8	4.3	4.5	6.7	6.7
DC-LSX-3-0295 Trifluoropropyl silicone gum	5.8	7.8	10.2	11.5	13.0

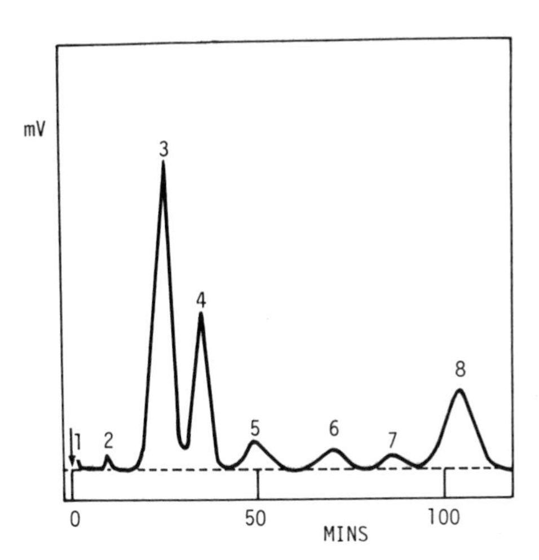

Figure 126 - Gas chromatogram of methylvinyl-dichlorosilane, 68° - 90°C distillation cut. Column 10:3 kieselguhr: nitrobenzene at 25°C. Carrier gas, nitrogen, 50 ml/min. (1) Hydrogen chloride; (2) Methyl-dichlorosilane; (3) Methyltrichlorosilane; (4) Dimethyldichlorosilane; (5) Dimethylvinylchlorosilane; (6) Vinyltrichlorosilane; (7) Unknown; (8) Methylvinyldichlorosilane.

Snegova et al[2141] have described a gas chromatographic procedure for the determination of fluoro and chloro-fluoro organosilicon compounds of the type $(CF_3CH_2CH_2)_3$ SiH_2, $CF3(CH_2CH_2)_3$ SiCl and $(CF_3CH_2CH_2)_4Si$.

E. Alkoxy Silanes

Wurst and Dusek[2160] showed that the components formed during the Grignard synthesis of the methylphenyldiethoxysilanes could be separated by gas chromatography. Four different chromatographic columns were used and the substances were identified. They discuss application of the method to the determination of methylphenylethoxysilane in reaction mixtures. These workers also show that the components of the mixtures produced in the reaction between vinyltrichlorosilanes and ethanol, 2-ethoxyethanol or 2-methoxy-ethanol could be separated by gas chromatography on a column of Silicone E-301 supported on Chromosorb, with nitrogen as carrier gas (1.7 - 3 litres per hour). Wurst and Dusek[2160] also discuss the suitability of the method for the determination of vinylethoxysilanes.

Volkov and Sakodynskii[2191] separated a mixture of deuterated ethoxymethylsilanes, obtained by the Grignard reaction, from tetraethoxysilane deuterated, hydrogen bromide and containing ethoxytrimethylsilane, diethoxydimethylsilane, triethoxymethylsilane, unchanged tetraethoxysilane and ethylether, on a column (7 metres x 33 mm) of 20% of dinonylphthalate on firebrick, operated at 95°C with nitrogen as carrier gas at 1.8 litres per minute. The liquid sample (10 ml) is injected through a rubber cap into an evaporator at 150°C. The separated components are detected by thermal conductivity and the fractions are automatically directed into separate liquid-nitrogen traps.

Gabor and Takacs[2192] carried out an investigation of the products formed during the preparation of ethoxyphenylsilanes by the Grignard reaction. The sample (5 µl) was injected on to an aluminium spiral column (2 metres x 4 mm) packed with 10% of SE-30 on Chromosorb W (60 - 80 mesh), with hydrogen (25 ml per minute) as carrier gas and a thermal-conductivity detector. The column temperature was programmed from 160° - 310°C at 0.44°C per minute. Tetraethoxy-, triethoxy-, phenyl, diethoxydiphenyl- and ethoxytriphenyl-silanes were separated and identified by these workers.

Compounds having the general formula $Si(OC_2H_5)_n(OC_3H_7)_{4-n}$, where n = 0 - 4, have been separated on a column (4 foot x 0.125 inch o.d.) of 20% of Triton X-100 on Chromosorb P (60 - 80 mesh) temperature programmed from 80° - 160°C at 5° per minute with helium as carrier gas (70 ml per minute) and flame ionization detection[2193]. The fractions emerging from the column were trapped in glass U-tubes cooled in solid carbon dioxide and analysed by infrared spectrophotometry.

Kirichenko and Markov[2194] separated and identified methoxy- and ethoxy-chlorosilanes. They studied the behaviour of these compounds on columns (1 or 2 metres x 3 mm) packed with Rysorb or Chemasorb coated with SE-30 FS-60 Carbowax 6000, Apiezon L or DC-550 with nitrogen (40 - 50 ml per minute) as carrier gas. Separation was best on the polar stationary phase.

Heunish[2195] analysed the thermal decomposition products of tetraethoxy-silane (ethylsilicate) on columns comprising polypropylene glycol on silanized Chromosorb P at 40°C using a thermal conductivity detector. The main impurities in the sample were ethyl alcohol, acetaldehyde, ethylene and ethane.

Thrash[2196] measured the retention of tetraethyoxysilane relative to toluene on three different stationary phases. Wurst[2197] has reported the specific retention volume of tetraethyoxysilane on a column containing silicone oil as the stationary phase. Taylor[2198] studied the gas chromatographic separation of a number of symmetrical and mixed tetraalkoxysilanes present in a heat transfer liquid. Garzo et al[2129] reported retention indices for a considerable number of organosilicon compounds, but only two of the compounds were tetraalkoxysilanes, namely tetramethoxy- and tetra-ethoxy-silane.

Ellren et al[2199] and Peetre[2136] have carried out extensive detailed work on the determination and calculation of Kovats retention indices for tetraalkoxysilanes with non-branched alkyl groups, while Peetre[2200] has extended this study to tetraalkoxysilanes with branched alkyl groups. Ellren et al[2201] have devised a procedure involving the use of refractive index in conjunction with Kovats retention indices for the identification of tetra-alkoxy silanes with either straight chain or branched alkoxy groups.

Ellren et al[2199] used a slope detector that permitted the measurement of retention indices to within ± 0.1 second. Most of the work was performed using a Varian Model 1400 gas chromatograph with a flame ionization detector. As the temperature scale on this instrument could not be read to better than 1°C an auxiliary thermocouple and temperature gauge were used, which enabled the temperature to be measured to ± 0.1°C (Mettler TM15). Steel columns (0.3 - 6.0 m 1/8 inch o.d.) were packed with either Apiezon M or cyanosilicone GE XE-60 (Applied Science Labs. State College, Pa., USA) 4% and 5% (w/w) respectively, on acid-washed and DMCS-treated Chromosorb G, 80 - 100 mesh. The column length was chosen to give a maximal retention time of about 1 hour. The columns were conditioned for 12 hours at 260°C (Apiezon M) and 220°C (XE-60) before use. The carrier gas was nitrogen at a flow rate of about 30 ml/min.

The work of Ellren et al[2199] showed that relationships could be readily established for the temperature dependence of retention indices of mixed tetraalkoxysilanes.

Ellren et al[2201] in a more recent paper have developed further their work on the use of the Kovats retention index of tetraalkoxysilanes in conjunction with their refractive index for the identification of these compounds. They convert the refraction index into a quantity known as the refractive number by linear interpolation between the refractive indices of n-alkanes. As this quantity is a linear function of the retention index for homologous series of compounds the refractive number - retention index plot can be utilised for the identification of tetraalkoxysilanes.

Ettre and Billeb[2202] have pointed out the usefulness of retention index versus temperature plots for the evaluation of the optimum column temperature when analyzing mixtures of compounds with similar boiling points. For example, plots for the tetraalkoxysilanes $(MeO)_nSi(OEt)_{4-n}$ (n = 0 - 4) give straight lines that converge towards higher temperatures. Hence, it appears that the best separation of the compounds in question will be obtained at the lowest possible column temperature.

Pollard et al[2127] have stated that a linear relationship exists between the logarithm of the retention volume and the carbon number for tetraalkylsilanes belonging to the series R_4Si-R_3SiR' $_2RSiR'_3$ - SiR'_4. For the corresponding series of tetraalkoxysilanes, an approximately linear relationship between retention index and carbon number will exist only for the series $(PrO)_nSi(OBu)_{4-n}$ (n = 4 - 0), but not for the series $(MeO)_n$ Si $(OBu)_{4-n}$ (n = 4 - 0).

Peetre[2136] has developed a modified equation for the calculation of retention indices of mixed tetraalkoxysilanes from the retention indices of the symmetrical counterparts and compared it with earlier equations. The new equation is more accurate.

An important conclusion of the work of Peetre[2136] is that although the retention indices obtained may differ from one column to another, the equation previously derived for the calculation of retention indices can always be applied. Only the set of group retention indices to be used changes. It therefore appears that the set of group retention indices deduced from the experimental data, obtained in a certain column, is primarily useful for calculating and comparing retention indices obtained on the same or a closely similar column.

Peetre[2136,2200] has applied the methods for the calculation of retention indices of mixed tetraalkoxysilanes with normal alkoxy groups to some counterparts which also contained branched alkoxy groups (isopropoxy, isobutoxy and sec-butoxy). Peetre[2200] found that best agreement between experimental and calculated values for retention index was obtained with the Apiezon M stationary phase, the average difference being about 3 under units. For the cyanosilicone XE-60 stationary phases retention indices, good agreement was obtained only for tetraalkoxysilanes that contain isobutoxy groups.

F. Siloxanes

Moore and Dewhurst[2203] separated and identified the cis- and trans-isomers of 2,4,6-trimethyl 2,4,6-triphenylcyclotrisiloxane on a 2 foot column of silicone rubber or Apiezol L grease on Chromosorb at less than 290°C with helium as carrier gas. Partial separation of the tetramer isomers was also achieved.

Wurst[2204,2205] separated and determined linear and cyclic polydimethylsiloxanes by chromatography on a column of 20% of silicone elastomer (Lukopren M) on Rysorb BLK at a temperature of 150°, 165°, 180° or 195°C according to the components to be determined using nitrogen as carrier gas. The relative error of measurement of a single component of a mixture was less than ± 7.5%.

Luskina et al[2206] studied the determination of methyl phenylpolysiloxanes of boiling point up to 500°C. In this work the theoretically derived equation, $T_B/T = 0.216$ (log r - log w + log u + constant), in which T_B is the boiling point of the substance; T is the column temperature; u is the volumetric rate of flow of carrier gas; r is the retention time, and w is the concentration of the stationary phase, was verified under the condition of constant retention time by the use of a 1 metre column of silanised crushed firebrick containing various amounts of polymethylsiloxane liquid. A katharometer detector was used. Column temperature was reduced by increasing the carrier gas rate and/or reducing the concentration of stationary phase. With a mixture of ten methylphenylpolysiloxanes (of b.p. from 100° - 500°C) satisfactory separation without decomposition were obtained by Luskina et al[2206] at 246°C on a 1 metre column containing 5% of stationary phase with helium (75 ml per minute) as carrier gas.

Otto and Doubek[2207] developed an all glass gas chromatographic apparatus which they used for the analysis of mixtures of siloxanes and of other mixtures. Rotzsche and Rosler[2208] studied the gas chromatography of methylhydrocyclosiloxanes.

Carmichael et al[2209] reported the gas chromatographic analysis of cyclic and linear products up to $[(CH_3)_2SiO]_{40}$ of molecular weight up to 2530 and 2960 formed upon heating $(CH_3)_3Si[CH_3)_2SiO]_8$ $O\text{-}Si(CH_3)_3$ with acid clay catalyst at 25°C. Prior to this work only cyclic methylsiloxanes of molecular weight up to 1120 and cyclic siloxanes in high molecular weight equilibriated siloxane polymer had been successfully analysed with any degree of success.

Carmichael et al[2209] used a F & M Model 810 linear programmed temperature gas chromatograph with a thermal conductivity detector. The column used were 2 foot x 1/4 inch stainless steel packed with 9% diphenylsiloxane-dimethylsiloxane copolymer gum (Dow Corning proprietary gum of undisclosed composition) on 60 - 80 mesh Chromosorb W, non-acid washed. Column temperatures were programmed form 50 - 400°C as follows: detector temperature, 350°C; injection point temperature, 350 - 360°C; column and reference helium flow rates, 100 ml/min; bridge current, 140 ma.

The calculation of weight per cents utilising component areas and response factors, was accomplished on an IBM 1401 computer utilising a Fortran program. A second computer program was prepared to statistically analyse the molecular weight distribution. Figure 127 shows a chromatogram obtained in the rearrangement of M D_8M.

Carmichael et al[2209] reported in detail on the analysis of a cyclic extract from a high molecular weight dimethylsiloxane polymer. The cyclic portion was separated from five high molecular weight dimethylsiloxane polymers (M = 7.5 - 16 x 10) by dissolving the polymer to 2 - 10% by weight

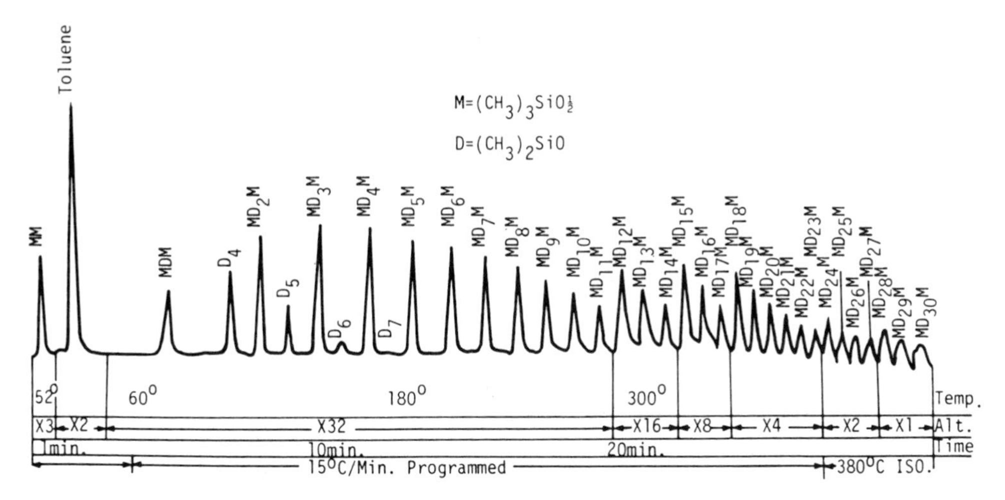

Figure 127 - Gas chromatogram of equilibrium siloxane distribution from MD_8M on 0.25 in silicone gum packed column.

in benzene and adding methanol in a ratio of 1/1.6 to benzene to precipitate the high molecular weight portion of the polymer. The filtrate was then separated and evaporated. No linear molecules are evident in the chromatogram.

Franc et al[2210] carried out a structural analysis of polysiloxanes by gas chromatography and by pyrolysis. They reported on the behaviour of 33 methyl phenylsiloxanes on three stationary phases, (i) 20% of the 3,5,-dinitrobenzoyl ester of 2(2-butoxy ethoxy) ethanol on Rysorb BLK at 180°C and (ii) 10% of Apiezon on Rysorb BLK at 260°C and (iii) 20% of silicone elastomer on Chromosorb P at 260°C. The columns were 150 cm long and 5 mm in diameter. Nitrogen was used as carrier gas at a flow rate of 3 litres per hour. They also discussed a technique involving reaction of the sample with boron trifluoride followed by analysis of the gaseous reaction products on a column (3 metres x 6 mm) containing 20% of fluorinated silicone oil F-16 on kieselguhr. For the pyrolysis, a sample (2 - 5 mg) in a platinum boat was heated in an oven temperature programed at 10° - 12°C per minute with nitrogen as carrier gas. The volatile products passed to a combustion oven, and thence through a carbon dioxide absorber to a thermal-conductivity detector. A 2-channel recorder indicated both the temperature and the detector response. The methods permitted determination of the molecular weight and of the number of phenyl groups and of some structural features of the compounds.

Heylmun and Pikula[2211] also reacted methylpolysiloxanes with boron trifluoride diethyletherate to produce methylfluorosilanes prior to gas chromatography. This method determines the relative amounts of Me_3Si-, $Me_2Si =$ and $MeSi \equiv$ groups in methylsiloxane polymers. Polymers which cannot be analysed in their original condition are depolymerized to methylfluorosilane monomers with boron trifluoride etherate in an enclosed system:

$$2\ MeSiO_{3/2} + 2\ BF_3.Et_2O \rightarrow 2\ MeSiF_3 + B_2O_3 + Et_2O$$

$$3\ Me_2SiO + 2\ BF_3.Et_2O \rightarrow 3\ Me_2SiF_2 + B_2O_3 + 2\ Et_2O$$

$$6\ Me_2SiO_{1/2} + 2\ BF_3.Et_2O \rightarrow 6\ Me_3SiF + B_2O_3 + 2\ Et_2O$$

These workers also describe a modified procedure for siloxanes containing silanic hydrogen.

The gas chromatography of siloxanes associated with chlorosilanes and methyl chlorosilanes has been discussed by Burson and Kenner[2184]. This work is discussed in the earlier section on methylchlorosilanes. Burson and Kenner[2212] have also described a simple trap arrangement for collecting gas chromatographic fractions for mass spectometric analysis. They applied this technique to the examination of siloxanes.

Ivanova et al[2213] have identified eight linear components of polymethylphenylsiloxanes by the use of columns (0.5 or 1 metre x 8 mm) of Aerosilogel A-380 or Silochrom C-80 supporting 3% of 3,3'-oxydipropionitrile and operated at 30° or 50°C with rates of flow of heptane of 0.37 or 0.49 ml per minute. The separated components were detected by ultra-violet spectroscopy.

Alexander and Garzo[2214,2215] have carried out structural identifications of polycyclic methylpolysiloxanes of the general formula $[(CH_3SiO_{1.5})]_x$ $[(CH_3)_2SiO]_y$, x and y being between 2 and 5 based on retention index sensitivity. Retention data for a series of siloxanes were determined on the phases squalane (at 110°C), Apiezon L (125°C) QF-1 (95°C) and polysorbate 60 (125°C) The value of the retention index is related to the structural characteristics of the molecule, which are regarded as additive and to a stationary phase factor.

Branched low molecular weight methyl and methylhydrosiloxanes have been separated and identified by gas chromatography - mass spectrometry (Wurst et al[2216,2217]. Separations were carried out on stainless-steel columns (2 metres x 2 mm) packed with 5% Apiezon L or 5% of Silicone oil AK 30,000 on Aw-DMCS Chromosorb G. Helium was used as carrier gas (15 ml per minute) and the columns were temperature programed form 100°C - 200°C at 4° per minute. Connection to the mass spectrometer was through a Watson-Beimann separator heated to 200°C. Mass spectra were recorded at 70 or 12 eV. All compounds exhibited significant peaks at m/e 73 and 147. Molecular ions were generally absent, but the molecular weight of the organosilicon compound could be determined from the presence of $(M - 15)^+$ ions.

Wurst[2218] analysed technical hexamethyldisiloxane on 1 - 4 µl samples, on a column (2 metres x 4 mm) packed with Chromosorb W.AW (0.12 - 0.15 mm mesh), impregnated with 0.2% of dibutyl phthalate. For the higher molecular weight fractions the carrier gas was impregnated with 0.1% of silicone elastomer Lukopren G 1000. Nitrogen (60 m per minute) was used as the carrier gas. The first separation was carried out at 35°C; the later one at 120°C. The separated compounds were identified from their retention times, densities and boiling points. A combined gas chromatographic mass spectrometric procedure is also described in which a column (2 metres x 2 mm) of Chromosorb G-AW DMCS impregnated with 0.05% of silicone oil is temperature programed from 100°C - 200°C at 4° per minute. Helium (15 ml per minute) was used as the carrier gas in this procedure.

High boiling organopolycyclosiloxanes have been analysed. (Turkel'taub and Luskina[2219]) on a composite column of two stationary phases 5% Apiezol L (2 metres x 4 mm) and 5% of PMS-100 (1 metre x 4 mm) with temperature programing from 155°C - 310°C at 15° per minute using a thermal conductivity detector.

Sutter and Fuchs[2220] have applied gas chromatography to the determination of the low molecular weight portion in poly (dimethylsiloxane). Polymers of molecular weight up to 700 in poly (dimethylsiloxane) can be determined on a steel column (1.2 metres x 4 mm) packed with 3% of OV - 1 on Chromosorb W AW-HP (80 - 100 mesh) with helium as carrier gas (45 ml per minute) and a flame ionization detector, The column temperature is programmed from 50°C to 300°C at 10° per minute.

Tochner et al[2221] have discussed the application of time sharing to data reduction in gas chromatographic analysis of polydimethylsiloxane fluids.

G. Silyl Ethers and Silyl Thioethers

Pollard et al[2222] measured retention data on a homologous series of trimethyl silyl ethers and trimethyl silylthioethers. The particular problems encountered in the gas chromatography of these compounds include the tendency to hydrolyse and breakdown on some stationary phases and under some conditions. To overcome these difficulties, high column temperatures are needed to produce reasonable retention volumes for the thioethers.

The stationary phases used were present to 20% w/w extent on acid-washed Silicel C-22 firebrick 60 - 85 mesh; the liquids used were (i) polyethylene glycol 400, (ii) dinonyl phthalate, (iii) tritolyl phosphate, (iv) squalane and (v) Apiezol L grease. Great care was taken to ensure that no moisture was present on the liquid phase or support.

The two stationary phases used by Pollard et al[2222] for systematic measurements of the oxygen ethers were Apiezon L and tritolyl phosphate, neither of which caused any appreciable amount of hydrolysis of the samples injected on to the column. Pollard et al[2222] showed that an amount of hydrolysis of the oxygen ethers to hexamethyldisiloxane occured when squalane (2.6, 10, 15, 19, 23-hexamethyltetracosane) was used as a stationary phase as the column temperature was raised from 75 to 110°C, breakdown being almost complete above 120°C. Similar results were obtained with polyethylene glycol 400 except the hydrolysis was greater at low temperatures; this was due to the presence of hydroxyl groups present on the phase which become active at these temperatures. The only phase found suitable for the gas chromatography of the silyl thioethers was Apiezon L. Complete breakdown of the ethers was noted on tritolyl phosphate even at 80°C.

The values of V_g given in Tables 193 - 195 are the mean values at each temperature of determination at inlet/outlet pressure ratios of 1.2, 1.4 and 1.6. Errors given are on the mean of several determinations and are in almost every case less than 2%.

H. Miscellaneous Organosilicon Compounds

Silyl Sulphides. Bortnikov et al[2223] have discussed the gas chromatography of bis (triethylsilyl) sulphide and its separation from its sulphur, germanium, tellurium and tin analogues on a stainless steel column (100 cm x 0.4 cm) at 254°C packed with Chromosorb W supporting 20% of Apiezon L. Helium was used as the carrier gas.

Silicon - Nitrogen Compounds. Smith[2224] analysed bis (trimethylsilyl) acetamide on a column of unsilanized Chromosorb W coated with 5% potassium hydroxide from methanol solution, (solvent was evaporated at 60°C in air). The support was heated at 500°C overnight, then coated with 20% of liquid paraffin from toluene. By using this column, commercial samples of bis-

Table 193 - Gas chromatography of silyl ethers and silyl thioethers* Specific retention volumes on Apiezon L

Solvent	Apiezon L				
Temperature °C	58°	80°	100°	120°	140°
$(CH_3)_3Si-O-R$					
R = CH_3	16.30±0.5	6.73±0.03	4.28±0.08	2.57±0.06	1.50±0.01
C_2H_5	26.2±1.1	10.07±0.26	6.22±0.07	3.72±0.07	1.03±0.06
iso-C_3H_7	33.9±0.4	12.8±0.54	7.93±0.16	4.71±0.09	2.69±0.03
n-C_3H_7	53.7±1.3	19.1±0.7	11.50±0.13	6.96±0.06	3.83±0.08
n-C_4H_9	116.0±0.8	38.8_1.0	22.1±0.7	12.42±0.12	6.64±0.14
$Si(CH_3)_3$	33.7±1.5	12.7±0.5	7.74±0.12	4.65±0.06	2.59±0.04

* The specific retention volume V_g is defined as V_R0, the retention volume fully corrected for dead volume, pressure drop across the column and measured at the column temperature, divided by the weight of the solvent.

Table 194 - Gas chromatography of silyl ethers. Specific retention volumes on tritolyl phosphate

Solvent	Tritolyl phosphate		
Temperature	80°	90°	100°
$(CH_3)_3Si-O-R$			
R = CH_3	5.65±0.03	3.70±0.03	2.69±0.04
C_2H_5	7.88±0.11	5.14±0.08	3.64±0.04
iso-C_3H_7	8.76±0.06	6.18±0.04	4.30±0.05
n-C_3H_7	13.70±0.27	9.10±0.13	6.40±0.09
tert-C_4H_9	12.02±0.07	7.99±0.04	5.65±0.01
n-C_4H_9	26.52±0.53	16.75±0.20	11.±910.23
$Si(CH_3)_3$	6.57±0.80	4.36±0.08	3.03±0.04

Table 195 - Gas chromatography of silyl thioethers. Specific retention volumes on Apiezon L

Solvent	Apiezon L		
Temperature °C	160°	±80°	200°
$(CH_3)_3Si-S-R$			
R = C_2H_5	4.22±-0.08	2.89±0.03	2.10±0.04
iso-CH_3H_7	5.31±0.08	3.43±0.11	2.48±0.07
n-C_3H_7	6.53±0.09	4.25±0.07	3.07±0.05
tert-C_3H_7	7.65±0.03	5.18±0.14	3.70±0.02
n-C_4H_9	9.95±0.08	6.52±0.08	4.49±0.03
$Si(CH_3)_3$	7.16±0.09	4.80±0.11	3.31±0.04

(trimethylsilyl) acetamide were found to be 90 to 97% pure, the main impurities being hexamethyldisiloxane and trimethylsilylacetamide. Molar response date for these compounds were determined relative to nonane as an internal standard.

Silicon-containing Phenols. Bortnikov et al[2225] separated silicon containing stericallyhindered phenols and their oxidation products including 4 tert butyl 2,6-bis (trimethylsilyl) phenol and 2,4-ditert butyl-6 trimethyl silyl phenol on columns (1 metre x 3 mm) of 20% Apiezon L or 3% of OV-17 on Chromosorb W, operated at 200°C with helium as carrier gas (30 ml per min) and detection by katharometer (Apiezon L column) or flame ionization.

Silicon Heteroelement Compounds. Bortnikov et al[2226] measured retention volumes for compounds of the type $(C_2H_5)_3M.M'(C_2H_5)_3$, where M and M' are silicon, germanium or tin on columns (100 cm x 0.4 cm) of silanised Chromosorb W supporting Apiezon L or Carbowax 20 M and operated at 222°C to 234°C with helium as carrier gas and thermal-conductivity or flame ionization detection, and they also carried out gas-solid chromatography on columns of graphitised carbon black operated at 266°C under similar conditions. Mixtures of the compounds were separated satisfactorily.

Vyazankin et al[2227] gas chromatographed 1,2-bis (trimethylsilyl) - substituted ethylenes and the germanium analogues. They showed that separation of the cis- and the trans- isomers of 1,2-bis (trimethylsilyl)-, 1-trimethylgermanyl-2-trimethylsilyl- and bis (trimethylgermanyl) ethylene is possible by gas-solid chromatography on carbon black graphitised at 2700°C to 3000°C and by gas-liquid chromatography on Chromosorb W (80 to 100 mesh) supporting 20% of Apiezon L. The columns (1 metre x 4 mm) were operated at 100°C to 200°C with helium as carrier gas and a thermal-conductivity of flame ionisation detector.

Bortnikov et al[2228] studied the separation of compounds containing bonds between two or three unlike heteroatoms (silicon, germanium, tin or sulphur) also heterocyclic compounds with silicon atoms in the ring by gas liquid chromatography on columns of silanized Chromosorb W coated with 20% of Apiezon L, 15% polyoxyethylene glycol 20M or Reoplex 400, also by gas-solid chromatography on graphitised thermal carbon black, with helium as carrier gas and flame ionisation or thermal conductivity detection. Gas-solid chromatography provided greater selectivity than gas-liquid chromatography, especially for the o- and p-isomers of sterically hindered silicon containing phenol derivatives.

Silyl Arenes. High boiling-point bis (dimethylsilyl) arenes have been separated by Sivtosova[2229]. With 1,4-bis (dimethylsilyl) benzene the components of the product were separated on a column (6 metres x 4 mm) of INZ-600 fire-brick supporting 5% of poly(4-dimethylsiloxyphenyl 4-dimethylsilylphenyl ether) operated at 140°C with helium (4 to 6 litres per hour) as carrier gas and katharometer detection. With bis-(4-dimethylsilylphenyl) ether the products were separated on a column (3 metres x 5 mm) of TND-TSM supporting 10% of polymetacarboranesiloxane operated at 210°C with helium (9 litres per hour) as carrier gas and katharometer detection. Mixtures of bis (4-dimethylsilylphenyl) ether and dimethyl (phenoxyphenyl) silane can be analysed by the use of dibutyl phthalate as internal standard.

Cyclosilazanes. Hailey and Nickless[2230] used gas chromatography to study the interchanges involving different cyclosilazanes occuring during redistribution reactions of tris (N-alkyl hexamethyl) cyclosilazanes.

Impurities in Organosilicon Compounds. Franc[2231] has used gas chromatography for the determination in organosilicon monomers of down to 0.001% of benzene, chlorobenzene, chlorodiphenylsilane, chlorophenylsilane, biphenyl, chlorobiphenyl, dichlorophenylsilane and o- and p- terphenyls.

The column comprised 135 cm x 5mm of Rysorb BLK supporting 10% of polymethyl phenylsiloxane and was operated at 210°C with a 3.1 litre per hour nitrogen carrier gas flow.

Turkel'taub et al[2232] have described gas chromatographic methods for the determination of ethanol in triethoxysilane. They used a column containing 20% of silicone PFMS-3 supported on firebrick with hydrogen or helium as carrier gas and a katharometer detector. Turkel'taub et al have also described a method for the determination of ethanol in triethoxysilane based on chromatography of 25°C on a 200 cm column containing 2% of petroleum jelly on firebrick (particle size 0.025 to 0.050 cm previously dried at 150 to 200°C). A flame ionization detector was used and hydrogen was employed as carrier gas (25 ml per minute).

Turkel'taub et al[2233] described a method for determining benzene in trichlorosilane by separation on a column containing 20% of petroleum oil supported on brick with a katharometer or flame ionization detector; the thermal enrichment method is used, by introducing the sample into the columns at - 70°C and carrying out the chromatography at 70°C. The sensitivity of the method is 0.05%.

In a further method for the determination of down to 0.0001% benzene in silicon tetrachloride Turkel'taub et al[2232] chromatographs the sample at 25°C on a 200 cm column containing 2% of petroleum jelly on firebrick (particle size 0.025 to 0.050 cm) previously dried at 150°C to 200°C. A flame ionization detector is used, and hydrogen (25 ml per min) is used as the carrier gas. The minimum determinable amount of benzene is 10^{-5} mg, with a relative error of 7.2%. The method was also applied to the determination of benzene in trichlorosilane.

I. Gas Chromatographic Detectors for Organosilicon Compounds

Detailed studies have been performed on various types of detectors in the gas chromatography of organosilicon compounds. These include argon ionization[2234], flame ionization[2129,2245,2235,2236], alkali flame[2237] glow discharge tubes[1850], microwave emission (MED)[1979,2238-2241,2243,2244], hydrogen atmosphere flame ionization (HAFID)[2243-2246], atomic emission/ absorption and atmospheric pressure microwave induced helium plasma[2247] detectors.

Garzo et al[2234] studied the applicability of the argon ionization detector to the gas liquid chromatography of organosilicon compounds. They found that a Pye argon chromatograph was unsuitable for the analysis of methylchlorosilanes, because the sensitivity of the instrument to these compounds was low, and varied widely between individual substances. This was attributed to ion recombination caused by the high affinity of these chloro-compounds for free electrons. The instrument gave reproducible results in the assay of polymers of low molecular weight and of methylethoxysilanes, with Apiezol M or a devolatilised linear siloxane polymer as stationary phase. The addition of a pyrolysis chamber permitted the analysis of both linear and cross-linked methylsiloxane polymers of high molecular weight. Reproducible chromatograms were obtained. No attempt was made by these workers to identify the peaks.

Fritz et al[2235] and Garzo et al[2236] studied the anomalous response of the flame ionization detector in the analysis of organosilicon compounds. The behaviour of the compounds was studied on columns of silicone elastomer on Chromosorb or Celite. Above a critical sample size, the top of the peak base became inverted, the maximum height remaining the same, but the depth of the inverted portion increasing with increasing sample size. When the carbon mass flow rate was plotted against ion current, the graph for hexane increased continuously, whereas that for the organosilicon compounds showed a maximum value. The position of this maximum depended on the ratio of C to Si and could be used to determine this ratio. The effects of gas flow-rate and composition and detector voltage were also studied by Fritz et al[2235]. When the fuel gas was a mixture of hydrogen and acetylene many organosilicon compounds gave a negative peak, even when they gave no response in a pure hydrogen flow.

Lengyel et al[2248] stated that the inversion of peaks of organosilicon compounds observed by Fritz et al[2235] has no relation to the slight deposition of silica which occurs on the electrodes of the flame ionization detector. Instead, these workers state that the inversion of the peak, i.e. the decrease of the ion-current at increasing mass flow rates takes place when the silicon mass flow rate exceeds a critical value in the flame. The critical values of the silicon mass flow rate, as well as the maximum ion current, depend on the following factors; detector geometry, detector voltage, the flowrates of the carrier, fuel, and scavenging gases and the actual carbon mass flow rate in the flame, the extent of which is determined by the composition of the organosilicon compound tested. Except this last factor, these are the same as those which influence the sensitivity and the linearity range of a flame ionization detector when measuring organic compounds. All experimental conditions being constant, the maximum height of the inverted peak depends only on the carbon/silicon molar ratio of the test substance.

Lengyel et al[2248] measured the response curves of the detector (containing two platinum electrodes with a variable gap) relative to another detector (of which the jet is the anode and a platinum cylinder the cathode). The workers describe a method by means of which the response curves of a flame ionization detector may be recorded on the basis of a single chromatogram. The column effluent is divided into two equal streams which simultaneously enter a reference detector and the detector being examined. The relation between the two detector signals is recorded with the aid of an X-Y recorder. Providing that the performance of the reference detector is linear, and by determining its response factor, the abscissa of the X-Y diagram represents a scale of mass flow rates, and the diagram itself the response curve of the detector tested. They used the method for the investigation of flame ionization detector performance when detecting organosilicon compounds. Inversion of the gas chromatographic peaks takes place and the response curves display a maximum under certain operating conditions. The shape of the response curves, the height and situation of the maximum depend on the detector voltage, the electrode gap, the flow rate of gases and the carbon/silicon ratio of the compound detected. The estimation of the carbon/silicon ratio on the basis of the maximum ion current is suggested by Lengyel et al[2248].

Garzo et al[2129] determined the retention indexes of 68 organosilicon compounds on columns containing Apiezon L, SE 30, QF-1, XF-112 and o-nitrotoluene as stationary phases and with flame ionization and thermal conductivity detectors. Some generalizations and rules are reported of character relationships based on ΔI values and retention index increments ($\Delta\delta I$), for successive homologous compounds.

Dressler et al[2237] have carried out a detailed study of the application of the alkali flame detector to organosilicon, organotin and organolead compounds.

Feldman and Batistoni[1850] have described a design of a glow discharge tube as a detector for the gas chromatography of organosilicon compounds. This technique is described in detail in the section on arsenic.

The microwave emission detector has been demonstrated to be useful for the detection of organosilicon compounds[2238]. Quimby et al[1979] have also examined the applicability of helium microwave glow discharge detectors to the detection of tetravinylsilane. He gives details of the microwave emission detector and the experimental set-up. These workers used an atmosphere pressure helium (or argon) plasma as this led to enhanced sensitivity and high optical resolution and selectivity.

Quimby et al[1979] used helium carrier gas flow rate of 70 ml per minute and an injection point temperature of 200°C. The wavelength setting of the monochromator was optimized for silicon using a hollow cathode lamp and a small mirror placed between the lens and the cavity. The wavelength setting was optimized by introducing small amounts of tetravinyl silane vapour into the plasma by correcting with a hydrocarbon solution of this compound. Quimby et al[1979] investigated the effect of the total flow rate of helium through the discharge tube on response to organosilicon compounds by repeatedly injecting a standard solution while varying the "helium plasma" flow with the carrier gas flow rate and column temperature maintained constant.

Because it is a constituent of the discharge tube. silicon exhibits unique behaviour with respect to flow rate. Note only does its response contibually increase with increasing flow rate up to the value at which the plasma becomes unstable (ca. 475 ml/min), but the selectivity increases in a similar fashion.

With the gas chromatographic microwave emission detector, for many elements the linear range extends from the detection limit up to a concentration at which the deposition of carbon and/or quenching significantly alter the plasma characteristics (ca. 1 μg/s of organic material entering the plasma). In the case of certain elements (e.g. lead, silicon), however, the linear range is only 10^3, and the upper limit as element sample levels 2 - 3 orders of magnitude below the upper limits for the halogens. This behaviour is due to the deposition of the element in question onto the walls of the discharge tube when large amounts of the element containing species enter the plasma. This effect with respect to selectivity is observed in the case of silicon. The element is present in the discharge tube wall and does not necessarily derive from a sample overload.

Hill and Aue[2243] reported that when silane is added to an atmosphere of hydrogen in which an oxygen-fed flame burns, organometallic compounds efficiently ionize in the flame, providing a sensitive and selective gas chromatographic detection method for metal-containing compounds. In a subsequent study[2244] the effect on organosilane response of doping the hydrogen atmosphere with organometallics was investigated. As expected, they found that the metal doped flame system enhanced silicon ionization, but it did not produce a linear calibration curve. However, a non-doped hydrogen atmosphere flame system was found to exhibit a linear response curve while still providing good selectivity for silicon-containing compounds. The severe peak tailing reported in their work for silicon-containing compounds was attributed to chromatographic conditions rather than detector design.

Osman et al[2245] carried out more extensive characterization of the non-doped silicon selective detector system in which they modified a commercial flame ionization detector to a detector which is specific for silicon-containing compounds by simply interchanging the oxygen and hydrogen inlets so that the flame burned in a hydrogen atmosphere (hydrogen atmospheric flame ionization detector - HAFID). The optimal configuration of various detector geometries was found to be a narrow cylindrical tube with the collector electrode positioned more than 10 cm from the flame). This configuration eliminated peak tailing which is prevelent in other designs. Helium was shown to be the carrier gas of choice since nitrogen reduced the sensitivity of the detector to silicon-containing compounds, but not to hydrocarbon compounds. The calibration curve for tetraethylsilane proved to be linear for about three orders of magnitude with a minimum detectable amount of 4 ng and a selectivity over decane of 2600. This corresponds to a silicon to carbon selectivity better than 1×10^4.

Osman et al[2245] investigated three different designs of hydrogen atmosphere flame ionization detector. Figure 128 depicts the finally selected detector design. In this the internal volume of the detector channel between the flame and collector electrode is kept low. This design also features a stainless steel tube (8 mm i.d.) to extend into the detector well, terminating below the jet tip and just above the base of the detector. Hydrogen travelled down the outside of this tube to mix with the oxygen, gas chromatographic carrier gas, and eluting compounds at the exit of the jet. All gases were then exhausted through the narrow chimney.

These workers claim that the following conditions are suitable for achieving best selectivity in the hydrogen atmospheric flame ionization detector. Highest selectivity occurred at parameter settings where reasonably good sensitivity was also obtained. A hydrogen flow of 1600 ml/min, an oxygen flow of 130 ml/min, a nitrogen purge flow of 0 ml/min and an electrode height of 103 mm above the jet tip. A linear response for silane from its minimal detectable amount of 4 ng to the low microgram region where the compound begins to saturate the detector. This is a linear range of three orders of magnitude.

Osman et al[2245] reported the analysis of a mixture of n-hydrocarbons and n-alcohols. The alcohols were converted to their trimethylsilyl derivatives with Tri-Sil Z and the mixture was chromatographed on two different instruments, one equipped with a standard flame ionization detector, and the other with the third design of the silicon sensitive hydrogen atmosphere flame ionization detector.

Figure 129 is the flame ionization detector response tracing of this mixture in which each component is detected with approximately equal sensitivity. Hydrocarbons were chosen that could be easily separated from the silylated derivative but, in a real sample, where such separation is not assured, the presence of hydrocarbons may interfere with quantitative results. Figure 130 is the hydrogen atmosphere silicon selective detector tracing of the same test mixture under similar chromatographic conditions. With the exception of the solvent, only compounds containing silicon are detected. In this case, overlapping hydrocarbon peaks may not interfere with quantitative results.

J. Micropyrolysis - Gas Chromatography of Organosilicon Compounds

Reaction-gas chromatography has been used[2349,2250] to detect alkyl, chloralkyl, phenyl, chlorphenyl, alkoxy and vinyl groups also $\equiv$ Si-Si $\equiv$ and

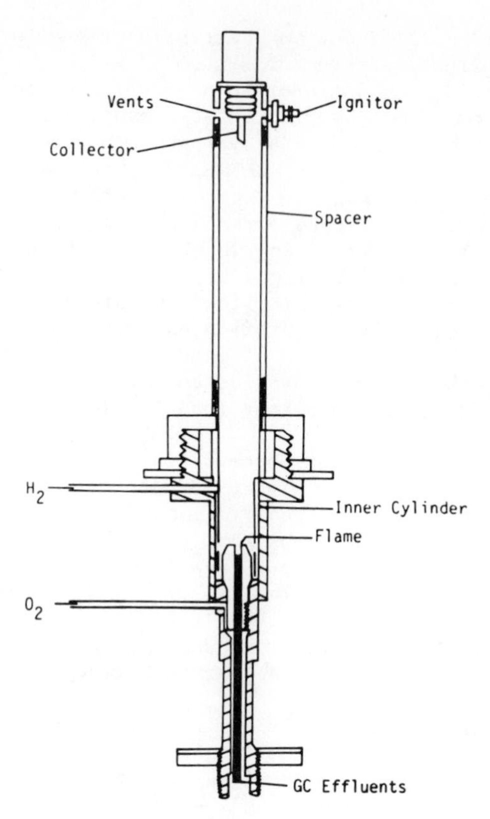

Figure 128 - Third design of silicon detector.

≡ Si-H in organosilicon compounds. Frank and Dvoracek[2249] showed that treatment of organosilicon compounds with boron tribromide at 130°C converts silicon bond alkoxy and phenyl groups to the corresponding alcohols and benzene respectively. Acetaldehyde and propionaldehyde are produced from ethylene oxide and propylene oxide chains in organosilicon compounds on treatment with potassium bisulphate or persulphate at 260°C.

Garzo and Till[2234] have applied the micropyrolytic gas chromatography technique to the identification of siloxane polymers and as an aid to the control of the reproducibility of cross linked siloxane resins. The pyrolysis chamber they used is illustrated in Figure 131. This apparatus may be dismounted at plane grinding A assembled with a clamping spring. The platinum wire coil B can be heated to about 600 - 700°C. The sample, generally dissolved in an appropriate solvent, such as butyl acetate, may be dripped directly on the coil or alternatively it may be slid into the coil in a small, tin walled silica tube. The bulk of the solvent is removed by gently heating the coil prior to the fitting of the chamber on the column. The residual traces of the solvent are removed from the sample after the assembling of head and column by similar heating in an argon stream. The complete removal of the solvent is indicated by the detector. The coil was heated for a substantial period in an argon stream to remove traces of the preceding sample.

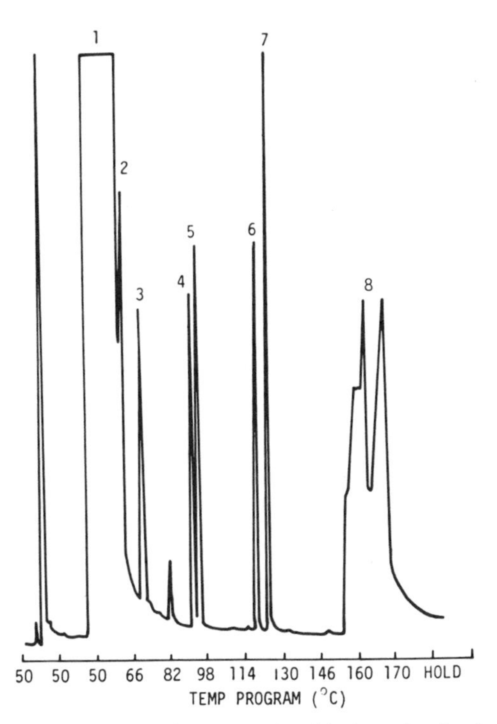

Figure 129 - FID tracing of a mixture of silylated alcohols and n-hydrocarbons. (1) Pyridine (solvent) (2) Decane (3) $CH_3(CH_2)_5OSi(CH_3)_3$ (4) Dodecane (5) $CH_3(CH_2)_7OSi(CH_3)_3$ (6) Tetradecane (7) $CH_3(CH_2)_9OSi(CH_3)_3$ (8) Tri-Sil Z.

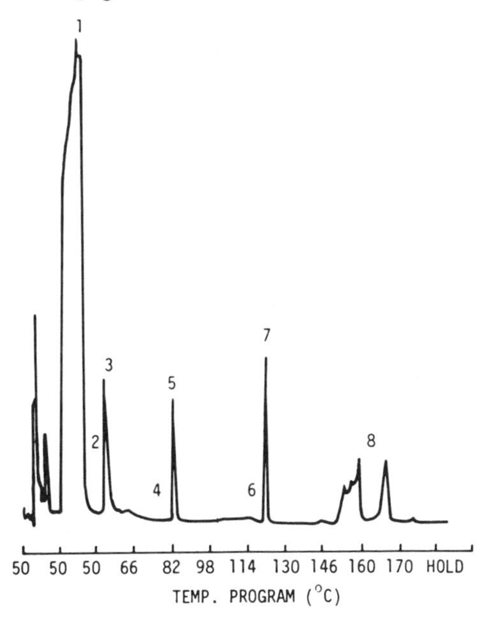

Figure 130 - HAFID tracing of a mixture of silylated alcohols and n-hydrocarbons. (1) Pyridine (solvent) (2) Decane (3) $CH_3(CH_2)_5OSi(CH_3)_3$ (4) Dodecane (5) $CH_3(CH_2)_7OSi(CH_3)_3$ (6) Tetradecane (7) $CH_3(CH_2)_9OSi(CH_3)_3$ (8) Tri-Sil Z.

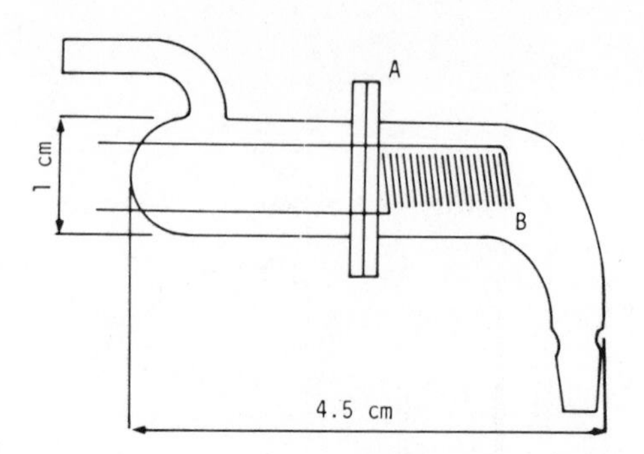

Figure 131 - Pyrolysis Chamber.

Operating conditions in the case in linear polymers were as follows - the column packing was 10% Apiezon M and a devolatilized linear siloxane polymer (the Hungarian product Hu-Au-120, mo. wt 59000) supported on Celite (80 - 120 mesh), detector voltage: 1000 V; column temperature: 100°C argon flow rate: 20 to 30 ml/min. The wire coil was heated for varying periods with a current of 3.1 A. The sample was applied in these experiments in a solvent-free state on the coil.

When comparing the pyrolysis chromatograms of three linear methyl-siloxane polymers (mol. wt. 217,000 (D3), 425, 400 (D4) and 755,200 (D5) respectively), results similar to that shown in Figure 132 were obtained. No splitting of C - H and C - Si bonds was observed. The characteristic peaks occurring on the chromatogram correspond, on the basis of the retention volumes, to D_3, D_4 and D_5. Independently of the quantity and the molecular weight of the sample D_4 is represented by about 45% in the pyrolysis products. With the decrease in height of the D_4 peak, which is to be considered as a measure of the quantity of the sample, the quantity of D_3 is increased at the expense of D_5. The quantity of the sample may be substantially reduced.

Garzo and Till[2234] investigated cross-linked siloxane polymers as solutions in butyl acetate. The chromatogram of the resin shown in Figure 133 is obtained under the following conditions; column and solid support as described below containing 10% linear siloxane polymer as stationary phase; detector voltage; 1750 V, argon flow rate: 30 ml/min. column temperature; 50°C the same coil as in the pyrolysis of the linear polymers, heated with a current of 3.1 A for 2 sec. The pyrolysis chromatograms of the 20% solution of cross-linked methylsiloxane polymers in butyl acetate is quantitatively identical with that shown in Figure 133 independently of the CH_3/Si ratio of the sample.

Franc and Dvoracek[2240] showed that some functional groups and bond types may be identified by degradation of the sample in a micro-reactor, and gas chromatographic separation of the volatile reaction products. Details of the method and apparatus are given, and results are presented for the degradation of alkyl- aryl- and chloroalkyl- silanes and siloxanes (with concentrated sulphuric acid saturated with vanadium pentoxide) of alkoxysilanes (with hydriodic acid at 75°C) and of vinylsilanes (by saturation with chlorine on a water bath followed by decomposition with concentrated sulphuric acid at 175°C). Alkyl, chloroalkyl, phenyl, chlorophenyl, alkoxyl and vinyl groups, and ≡ Si-Si ≡ and ≡ Si - H bonds were all detected by these procedures.

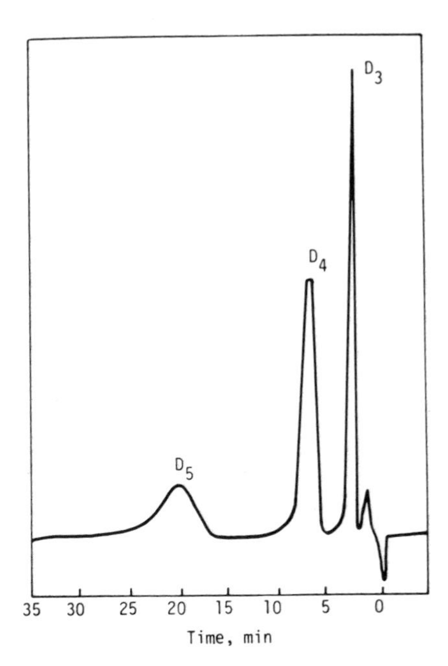

Figure 132 - Pyrolysis chromatogram of a linear siloxane polymer.

Thrash et al[2251] studied the analysis of phenylmethyldichlorosilane by reaction chromatography with sodium fluosilicate involving an on column conversion in a pre-column to the corresponding fluorosilanes which are then separated on the chromatographic column. Due to the similarity of their boiling points, phenyl methyl silicon dichloride and phenyl silicon trichloride cannot be separated by normal chromatographic procedures.

The reactions occuring on the precolumn are as follows:-

$$\phi CH_3SiCl_2 + Na_2SiF_6 \rightarrow \phi CH_3SiF_2 + SiF_4 + 2NaCl$$

(142°C) (201°C)

$$2\phi SiCl_3 + Na_2SiF_6 \rightarrow 2\phi SiF_3 + 3SiF_4 + 6NaCl$$

(101°C)

In Figure 134 is shown a chromatogram of a mixture of silicon tetrafluoride, phenylsilicon trifluoride and phenyl methyl silicon difluoride obtained by fluorination of a mixture of phenyl methyl silicon dichloride and phenyl silicon trichloride. Using a 10 μl sample and a thermal conductivity detector, phenyl silicon trifluoride can be accurately determined in phenyl methyl silicon dichloride down to 0.01%. In addition, this procedure is applicable to the separation of other high boiling chlorosilanes which are difficult to separate by distillation or normal chromatographic procedures. Thrash et al[2252] also achieved the separation of δ cyanopropylmethyldichlorosilane δ cyanopropyltrichlorosilane, and δ cyanopropyldimethylchlorosilane by this procedure.

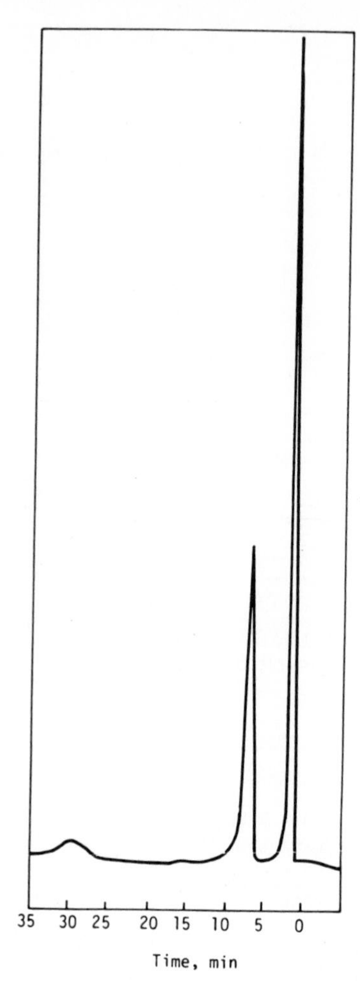

Figure 133 - Pyrolysis chromatogram of a linear siloxane polymer.

Various workers[2252-2257] have investigated the thermal pyrolysis of tetraalkylsilanes using gas chromatography combined with mass, infra-red and nuclear magnetic resonance spectroscopic techniques. The chemical cleavage of silicon to carbon bonds especially where at least one of the carbon atoms is attached to a silicon atom has been examined.

Pollard et al[2130] have investigated the thermal decomposition of arylsilanes. They identified the pyrolysis products using retention time techniques. A graph of log retention time against carbon number indicated that the assignments were correct. Figure 135 shows a plot of boiling point versus log corrected retention time (t'R). The compounds studied were phenyltrimethylsilane, benzyltrimethylsilane, p-methylbenzyltrimethylsilane, o-tolyltrimethylsilane and p-tolyltrimethylsilane. A specimen pyrogram is shown in Figure 136.

Various workers[2324-2327] investigated the application reaction gas chromatography to the determination of alkyl and phenyl groups in organosilicon compounds.

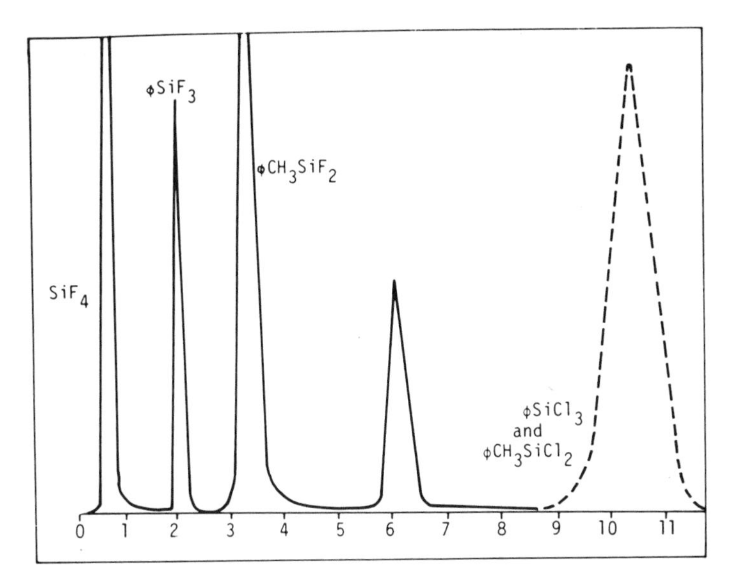

Figure 134 - Reaction chromatography of phenylmethyldichlorosilane. Chromatogram showing separation of reaction products with the broken line indicating the position of the unreacted chlorosilane.

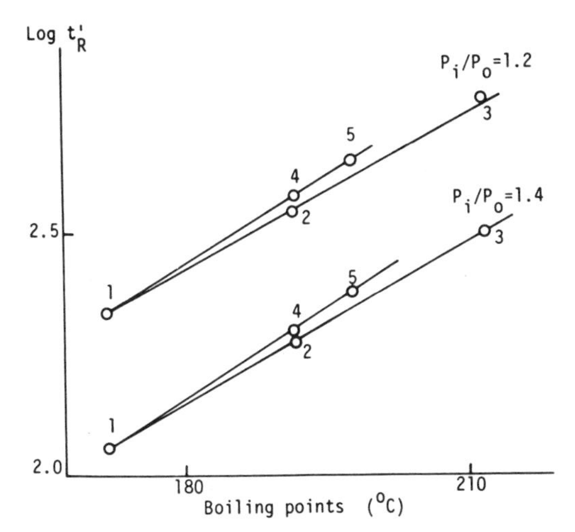

Figure 135 - Graph of $\log_{10}(t'R)$ against boiling points for a series of aryltrimethylsilanes at various Pi/Po ratios: 1 = Phenyltrimethylsilane, 2 = benzyl trimethylsilane, 3 = methylbenzyltrimethylsilane, 4 = o-tolyltrimethylsilane, 5 = p-tolyltrimethylsilane.

Krasikova et al[2325] determined ethyl and phenyl groups in siloxane polymers. These groups were split off with phosphorus pentoxide and water and the liberated ethane and benzene determined by gas-solid chromatography

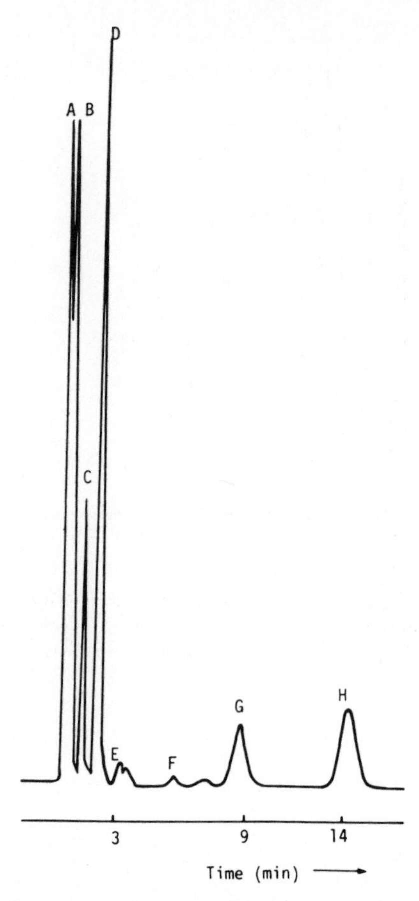

Figure 136 - Pyrolysis pattern of benzyltrimethylsilane. A = CH_4, B = $(CH_4)_3SiH$, C = C_6H_6, D = $C_6H_6CH_3$, E = xylenes, F = $C_6H_5Si(CH_3)_3$ G = $C_6H_5CH_2Si(CH_3)_3$, H = o- or p-$CH_3C_6H_4CH_2Si(CH_3)_3$.

on a column (170 cm x 6 mm) packed with aluminium oxide de-activated with 2% sodium hydroxide solution and operated at 30° or 120°C for ethane or benzene respectively with argon as carrier gas (60 or 180 ml per minute respectively) and flame ionization detection. The error in this procedure is claimed to be less than 10%.

Treatment with aqueous potassium hydroxide (60%) for 2 hours at 120°C has been used to convert phenyl groups to benzene which was determined gas chromatographically.

Schlueter and Siggia[2326-2327] have applied the technique of alkali fusion reaction gas chromatography to the determination of alkyl and aryl groups in polysiloxanes. The method involves the quantitative cleavage of all organic substituents bonded to silicon by fusing with powdered potassium hydroxide producing the corresponding hydrocarbons. Alkyl and aryl groups were converted to the corresponding aliphatic acid and aliphatic hydrocarbons. Reactions are driven to completion with no apparent decomposition, in less that 10 minutes by fusing the sample with potassium hydroxide in inert atmosphere. After concentration of the volatile products, they are separated and determined by gas chromatography. Sample losses are minimized by performing the total analysis in a single piece of apparatus. (Figure 137).

Fluids, gum rubbers, and resins are handled with equal ease. The percent relative standard deviation of the method is 1%; the average deviation between experimental and theoretical of check method results is 0.5% absolute.

Table 196 summarizes the results obtained in applying this method to a series of polydimethylsiloxanes. Methane gas was used to prepare the calibration curve for GE Viscasil 60,000. This polymer was then used as a secondary standard for preparing subsequent curves. The precision of these determinations varied from 0.2 to 0.8% relative. Theoretical weight percent values were calculated using manufacturers' average molecular weight data and the assumption that the polymers were linear and chain-terminated with trimethylsilyl groups. With the exception of the lowest molecular weight sample, the recovery of methane was within 0.5% absolute of theory. The low value obtained from the GE SF96 - 20 samples may be due to vapourization of low molecular weight oligomers before reaction.

Table 196 - Analysis of polydimethylsiloxanes by alkali fusion reaction gas chromatography

$$CH_3-\underset{CH_3}{\overset{CH_3}{|}}\!\!\!\!Si-\left[O-\underset{CH_3}{\overset{CH_3}{|}}\!\!\!\!Si-\right]_X-O-\underset{CH_3}{\overset{CH_3}{|}}\!\!\!\!Si-CH_3$$

Sample	a	Methyl content wt % Alkali fusion[b]	Theory	% recovery of theoretical methane
GE SF96-20	4	39.74 ± 0.30	45.86	86.7
GE SF96-50	45	40.74 ± 0.16	41.26	98.7
GE SF96-100	85	40.75 ± 0.06	40.93	99.6
DC 200-100cs	85	41.15 ± 0.35	40.93	100.5
GE SF96-1000	340	40.50 ± 0.28	40.65	99.6
DC 200-12500cs	850	40.74 ± 0.49	40.49	100.4
GE Viscasil 60000	1240	40.55 ± 0.24	40.58	99.9
GE SE-30	....	40.47 ± 0.24	40.55	99.8
GE SE-33	...	39.93 ± 0.42	c	...
DC Silastic 430	...	39.75 ± 0.30	c	...
GE SE-31	...	38.14 ± 0.33	c	...

[a] Average number of dimethylsiloxane repeat units based upon manufacturers' data.
[b] The standard deviation is based upon five or more determinations.
[c] Sample also contained a low percentage of vinyl groups.

A gas chromatogram of the reaction products obtained from DC 704, a polymethylphenylsiloxane, is shown in Figure 138. Methane and benzene are the only volatile reaction products observed; the water is liberated from the reagent.

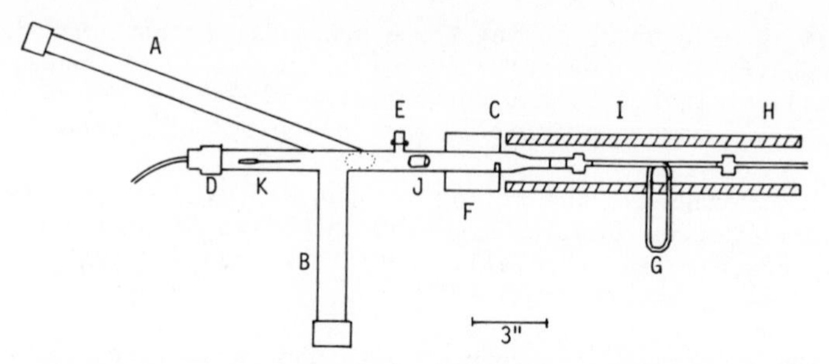

Figure 137 - Diagram of the reaction and trapping unit used for alkali fusion reaction GC. The mounting rack and electrical control units are not shown. A, storage area for unreacted samples; C, quartz reaction zone; D, carrier gas linlet; E, side arm with rubber septum; F, variable temperature furnace; G. trap loop; H, injector assembly into gas chromatograph; I, combustion furnace surrounding transfer tubing; J, metal cylinder behind platinum sample boat; K, magnetic retriever.

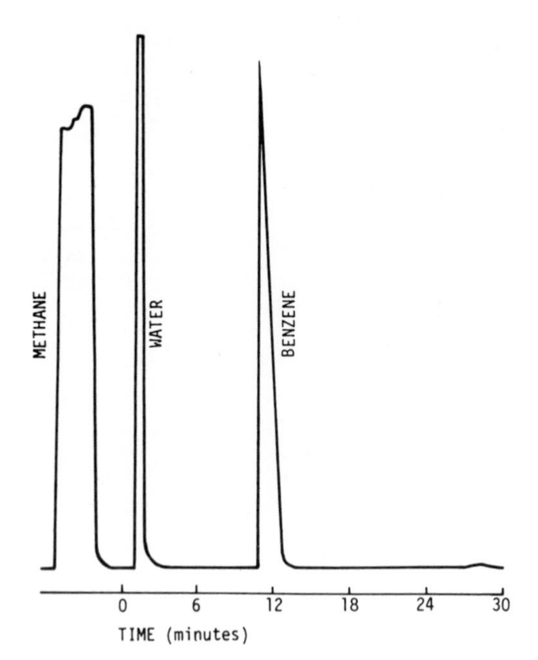

Figure 138 - Gas chromatogram of the methane and benzene produced from the alkali fusion of DC 704 polymethylphenylsiloxane.

Vinyl Groups. Reaction gas chromatography has been used to determine vinyl groups in organosilicon compounds, particularly siloxane polymers.

Generally, these methods involve reaction of the sample with hot phosphorus pentoxide[2328-2330], syrup phosphoric acid[2330], 90% sulphuric acid[2331] or potassium hydroxide[2331,2332] or by pyrolysis[2333] followed by estimation of the ethylene so produced by flame ionization gas chromatography. Heylmun et al[2330] reported a method for determining vinyl groups involving cleavage with phosphorus pentoxide to liberate vinyl groups as ethylene which was estimated by gas chromatography.

Krasikova and Kaganova[2325,2329] developed a method based on the chromatographic determination of the ethylene formed from the vinyl groups during interaction between the polymers, phosphorus pentoxide and water. An alumina column (170 cm x 6 mm) is used and detection is by flame ionization. The relative errors in determining 0.05% and 20% of vinyl groups are less than 10% and 4% respectively.

Bissell and Fields[2331] have carried out a thorough investigation of methods for determining vinyl groups in siloxane polymers and conclude that in the range 0.1 to 2% of vinyl groups generally found in such polymers, these methods all give precisions no better than ± 5 to 10% of the amount present. Bissell and Fields[2331] found that reaction conditions were critical and that temperatures above 400°C were required for the complete cleavage of the sample. To facilitate achieving this temperature they adopted sulphuric acid rather than phosphoric acid for the cleavage of alkyl groups in siloxanes as this is reactive at lower temperatures than phosphoric acid.

The column they used was a 6 ft x 3/16 in o.d., stainless steel tubing packed with 80/100 mesh Poropak S. Samples were introduced with a 8-port Carl sampling valve using 4 ml loops. The detector was of hot wire, thermal conductivity type operated at 175 mA. The sampling valve and column were operated at ambient temperature and the detector at 190°C. The carrier gas was helium at a flow rate of 71 ml/min. These conditions are not critical as long as they can be reproduced accurately from run to run. The retention time for ethylene was approximately 3.5 min, that for ethane about 5 min.

The results of analyses for a number of siloxanes containing only vinyl, phenyl, methyl, and hydrogen groups are given in Table 197. Molecular weights calculated from the vinyl contents (assuming difunctionality) agree well with molecular weights calculated from gel permation chromatography and vapour pressure osmometry.

Evdokimova et al[2333] determined vinyl groups in alkylcyclosiloxanes and (polymethyl) vinylsiloxanes by pulse pyrolysis under gas chromatographic conditions. The organosilicon compounds were analyzed by pulse pyrolysis of μg amounts (deposited on a tungsten filament) in the carrier gas (argon) stream. The vinyl-group content of alkylcyclosiloxanes and polysiloxanes is obtained from the ethylene-to-methane ratio of the pyrolysis gases, determined by gas-solid chromatography on silica gel.

Alkoxy Groups. Hanson and Smith[2332] have described a procedure for the determination of alkoxy and vinyl substituents in amounts between 0.001 and 1% in siloxane materials using alkali fusion and gas chromatography. In this method the sample is fused with potassium hydroxide and the alcohol and ethylene produced by cleavage of the vinyl and alkoxy groups up to butoxy are determined in the measured volume of gas evolved. A relative standard deviation of less than 20% is claimed in the alkoxy analyses and less than 10% in the vinyl analysis.

Table 197 - Vinyl content of silicon resin (wt %)

Silicon Type	Method A[b]		Method B[b]		Titration	
	Ave		Ave		Ave	
Dimethylpolysiloxane A	0.142	0.003	0.148	0.005	0.17	0.03
Dimethylpolysiloxane B	0.070	0.005	0.092	0.005	0.14	0.02
Dimethylpolysiloxane C	0.369	0.008	-	-	0.31	0.05
Dimethylpolysiloxane D	0.203	0.007	0.192	0.005	0.13	0.02
Gelled dimethylpolysiloxane	0.076	0.003	0.079	0.004	Not titratable	
Methyl hydrogen polysiloxane	0.129	0.003	0.137	0.004	Not titratable	
Methylphenylpolysiloxane	-	-	0.025	0.003	0.18a	-
Methylphenylpolysiloxane	-	-	0.075	0.004	0.070a	-
Methylphenylpolysiloxane	-	-	0.162	0.008	0.149a	-

a Nominal based on ratio of monomers used in preparation
b Minor variants of the Bissell and Fields method[2331].

In general the alkoxy method is most applicable to samples containing 0.001 to 1% ≡ SiOR and to materials with high boiling points above 300°C. Lower boiling materials usually will vaporize before the alkali melt reaction becomes effective at about 350°C. The qualitative distinction of alkoxy types was readily accomplished, since the C_1 to C_4 alcohols were resolved by gas chromatography. (= COR) will interfere with ≡ SiOR analysis because of significant OR cleavage from carbon to give alcohols. This latter reaction was not quantitative so total alkoxy content would not be obtained if both ≡ COR and ≡ SiOR were present. The method is capable of detecting 0.001% vinyl in siloxane fluids and gums. The upper use range was generally limited to 1.0%. The relative standard deviation usually was within 10%. The relative error vs. expected values was generally less than 10% and a maximum of 15%.

Silicon-Hydride Groups. Franc and Mikes[2334] described the microdetermination of silicon-hydrogen bonds in compounds such as dichlorodimethylsilane by destruction of the SiH bond in the presence of potassium hydroxide to produce an equivalent volume of hydrogen which is determined by gas chromatography. The sample is injected on to a porous mass (e.g. Rysorb P) saturated with aqueous potassium hydroxide. The reaction products are led into a column (180 cm long) of activated charcoal operated at 25°C with nitrogen (1 litre per hr.) as carrier gas and thermal-conductivity detection. The error is ± 6.5% for a content of 0.1% of dichloroethylsilane, and the sensitivity is 0.005%.

ORGANOTELLURIUM COMPOUNDS

Bortnikov et al[2258] have separated bis(triethylsilyl) sulphide and a series of related compounds containing sulphur, selenium, germanium, tellurium and/or tin as the central or linking atom. Separation was achieved on a 100 cm x 0.4 cm stainless steel column packed with 20% Apiezon L on Chromosorb W at 254°C using helium as carrier gas and thermal conductivity detection. Specific retention volume data are presented for this range of compounds.

ORGANOTIN COMPOUNDS

A. Alkyltin and Vinyl Tin Compounds

The gas chromatography of organotin compounds has been extensively studied. The subject has been reviewed by Ingham et al up to 1960[2259] and by Luijten and Van der Kerk[2260] up to 1966. Pollard et al published the first of their series of papers on the gas chromatography of alkyltin compounds (Pollard et al[2261]) and of reaction products of trimethylstannane (Pollard et al[2262]). They point out that two factors which render the gas chromatography of σ- and π-bonded organotin compounds difficult are their instability towards oxygen and moisture and their thermal instability.

They further point out that if a solid column support is insufficiently covered by the stationary liquid phase, e.g. 2 - 5%, absorption on the exposed siliceous sites becomes significant with polar solutes, and trailing occurs. As a consequence, retention volumes are no longer just directly proportional to the weight of solvent, and hence specific retention volumes can only be measured with columns containing a high proportion of stationary phase. Where organometallic solutes are involved, this adsorption becomes very important, the band spreading is so bad that squalane columns can hardly be used for analysis of mixtures of such materials.

Chemical instability gives rise to chemical change as the compound passes through the chromatographic column. This usually occurs through formation of bonds between the compound and reactive groups either on the column support (e.g. acid sites), or the stationary phase (e.g. hydroxy groups as in polyethylene glycol). This phenomenon is termed transesterification (Bohemen et al[2263]) and has been observed in the organotin hydrides chlorosilanes, and amino-compounds, e.g. hexamethyldisilazane.

Earlier approaches to pre-treatment of support to remove this activity was to add small amounts of highly polar and involatile liquids to the support (Bohemen et al[2263], Eggersten and Groennings[2264]) or to acid and then alkali wash the support (James and Martin[2265], Liberti[2266], Orr and Callen 2267. More recently there have been attempts to deposit solids such as silver on the support surface (Ormerod and Scott[2268]) but unfortunately, this method cannot be used in the presence of thio-compounds, e.g. silyl thioethers. The alternative method is to treat the active sites of the support, (which are presumed to be hydroxyl groups (-Si-O-H), and replace these be groups which should yield at least a weekly adsorbing site. Both trimethylchlorosilane (Kiselev et al[2271], Kohlschutter et al[2272]) and dimethyldichlorosilane (Howard and Martin[2269]), Kwantes and Rynders[2270]) have been used successfully to reduce the activity, the surface reaction is presumed to be of the type:

$$(CH_3)_3SiCl + -\underset{|}{\overset{OH}{\overset{|}{S}}} \longrightarrow \underset{|}{\overset{O\,-\,Si(CH_3)_3}{\overset{|}{Si}}}- \quad + HCl$$

or:

$$-\underset{OH}{\underset{|}{\overset{|}{Si}}}-O-\underset{OH}{\underset{|}{\overset{|}{Si}}} + (CH_3)_2SiCl_2 \longrightarrow -\underset{O}{\underset{|}{\overset{|}{Si}}}-O-\underset{O}{\underset{|}{\overset{|}{Si}}}- \text{ (both O bridged by } Si(Me)_2\text{)} + 2HCl$$

When the hydroxyl groups are not adjacent, then a chlorosilyl ether > SiOSi $(CH_3)_2Cl$ may be left, which is not beneficial since it will be as reactive as the grouping replaced because of the chlorine grouping.

As an alternative to this treatment, hexamethyldisilazane has been used since it reacts quantitatively with hydroxyl groups (Langer et al [2273, 2274]) and was used by Bohemen et al[2263] and has now been used to treat all the common solid supports (Perrett and Purnell[2275]). Many advantages have been claimed for hexamethyldisilazane, but it is expensive and gives a surface similar to the trimethylchlorosilane.

Pollard et al[2261] point out that in the gas chromatography organotin compounds, careful consideration must be given to the detector and its design since often when a compound is eluted from a column, decomposition occurs in the detector, invalidating the elution process. When such decomposition occurs, the metal is deposited on the wires or filaments of the katharometer, or on the collector plates of a flame ionization gauge possibly causing the formation of tarry, and finally carbonaceous deposits which foul the katharometer filaments (Ackman et al[2276]). Recommended treatment in such cases is regular flushing of the detector block, with both polar and non-polar solvents. Although such treatment is beneficial in the course of time, however, the tarry deposits carbonize, leading to permanent changes in the katharometer resistance. The partial contacts of carbon deposits between the helices of the coiled filament, presumably are responsible for the increase in recorder base line noise. When finely-divided powder metal is deposited in the katharometer, especially on the filaments, a similar situation arises, but the bridge becomes permanently out of balance since unfortunately no cleaning procedure can be used. A similar situation is found with the flame ionization detector, especially the conventional types where the collector electrode plate is vertically above the flame. A modified detector is needed, and even when detection can occur, attention must be paid to saturation limits, since non-linearity of signal response, and the inversion effect as reported by Novak and Janak[2227] has been observed.

Pollard et al[2261] showed that trans-esterification of organotin compounds can be overcome by treatment of the supporting phase (Bohemen et al [2263]) (Celite 545, 36 - 60 mesh B.S.S.) by baking at 300°C for 5 hours, acid- and alkali-washing, drying at 50°C and treatment with trimethylchlorosilane. After such treatment it is possible to chromatograph and separate the methylchlorosilanes and organotin hydrides. However, as indicated later, the choice of stationary phase is important for this type of compounds.

He carried out the gas-liquid chromatography of thermally unstable organometallic compounds followed by combustion in a conventional microanalytical furnace, absorption of water, and detection of the carbon dioxide with a Stuve katharometer (Stuve[2278]). The metal oxide deposited in the furnace gradually poisons the copper oxide furnace packing, and has to be replaced frequently. The column was 3 ft of 25% w/v di-2-ethylhexyl sebacate on Celite (36 - 60 mesh) at 56°. Carrier gas was oxygen free nitrogen at 50 ml/min.

Tin tetraalkyls and related compounds could be detected by a thermal conductivity cell, a modified flame ionization gauge and a commercial gas-density balance unit, (Martin and James[2279]). The latter unit has many advantages for such work not the least of these being that the sample is not subjected to a temperature greater than the column temperature;

presumably a temperature at which the compound is stable. Pollard et al [2261] compared results obtained with these three detectors. These workers showed that the gas density balance was excellent for the gas chromatography of alkyltin mono and dihydrides without decomposition on the column. A separation of trimethyltin hydride, tetramethyltin and dipropyltin dihydride is given in Figure 139. The conditions were as follows:

Column 6 ft 15% w/w silicone oil E301 on Celite 545 (36 - 60 mesh).
Column temperature 80°C.
Balance temperature 100°C
Flow rate 30.0 ml of nitrogen per ml.

It was possible to elute trimethyltin hydride through Silicone E301, Apiezon L, and dinonylphthalate phases, but when attempts to elute down a squalane (hexamethyltetracosane) column were made decomposition occurred and an elution pattern of the type shown in Figure 140 was obtained. The shape of the peak suggests that the negative flat portion is interpreted as saturation of the gas-density balance, and the small positive peak is identified as tetramethyltin, showing that some re-arrangement reaction must take place in or on the stationary phase. The reaction of the hydride with the phase was not unexpected, since the hydride readily adds on to triple bonds (Van der Kerk and Noltes[2280]) and less easily to a double bond.

$$Me_3SnH + R - C \equiv CH \rightarrow R - CH = CSnMe_3 + H$$

The hydrogen atom formed may then abstract hydrogen from further trimethyltin hydride forming hydrogen molecules and the radical Me_3Sn. Such a radical could again react with trimethyltin hydride abstracting a methyl group and so forming the tetramethyltin. Compounds of the type R - CH = $CHSnMe_3$ are perfectly stable and can be chromatographed without decomposition, when the molecular weight is low i.e. up to R = C_7 (alkyl).

In a later paper, Pollard et al[2281] discuss the use of the flame ionization detector in the gas chromatography of tetramethyltin, trimethyltin, dimethyldiethyltin, methyltriethyltin and tetraethyltin. They discuss the unusual sensitivity characteristic of the chromatography of these compounds and present and correlate specific dynamic properties obtained and various thermodynamic properties obtained on columns comprising 15% Silicone oil E301 (mol. wt. 700,000) on Celite (James and Martin[2265]) (treated by dry sieving to mesh 36 - 60, washing with concentrated hydrochloric acid, methanol and distilled water, followed by drying at 300°C).

The specific retention volumes of some organotin compounds are listed in Table 198.

Table 198 - Specific retention volumes (Vg ml) of organotin compounds on 15% E301 Silicone oil on celite

$SnMe_4$	50.1 ± 0.3	28.4 ± 0.1	21.0 ± 0.4	11.2 ± 0.1
Me_3SnEt	116.6 ± 1.0	62.3 ± 0.3	43.9 ± 0.3	22.5 ± 0.3
Me_2SnEt_2	264.3 ± 3.2	131.2 ± 1.0	91.3 ± 1.2	44.0 ± 0.4
$MeSnEt_3$	577.0 ± 9.1	270.5 ± 1.8	181.9 ± 0.9	82.7 ± 1.0
$SnEt_3$	122.4 ± 10	546.4 ± 7.0	355.0 ± 3.3	133.9 ± 2.4

Putnam and Pu[2282,2283] determined the retention indices of 14 organotin compounds on columns of (i) 40% of Apiezol on Celite, (ii) 20% of Carbowax

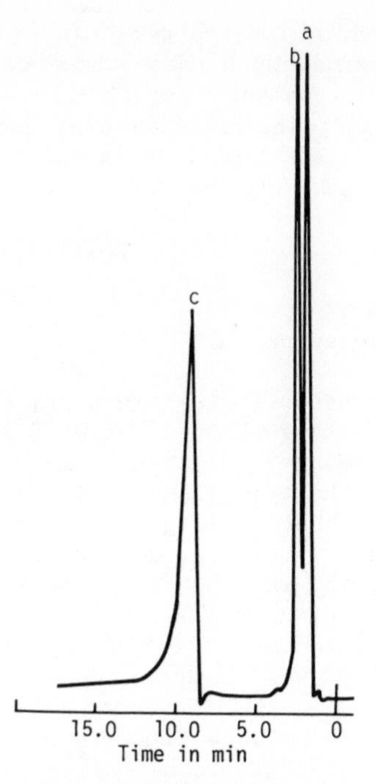

Figure 139 - Elution pattern of alkyltin hydrides. (a) Trimethyltin hydride, (b) Tetramethyltin, (c) Dipropyltin hydride.

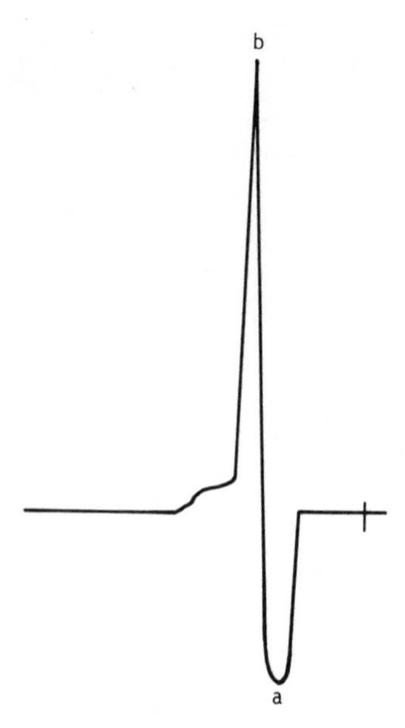

Figure 140 - Elution pattern of trimethyltin hydrides on squalane (a) Hydrogen (b) Tetramethyltin.

1500 on diatomaceous earth and (iii) 20% of polyaceous earth. The columns were all operated at 190°C and before use were conditioned at 200°C overnight and pre-treated with hexamethyldisilazane and dichlorodimethylsilane at room temperature. The nature of the compounds had little effect on the retention times of alkanes except on column (i) where the number of carbon atoms was less than 10. A correlation between retention index and boiling point existed for column (i), so that the retention index could be predicted from the boiling point with an average error of 1%.

Distinct differences between saturated and unsaturated compounds were observed on the polar columns (ii) and (iii), and average values of retention index divided by molecular weight increased with two exceptions, as the refractive index increased. A correlation of increasing retention index with increasing calculated molar refraction applied without exception to (i) and with a few exceptions to (ii) and (iii). The retention indices of the 14 compounds mentioned by these workers are tabulated in Table 199.

Table 199

	Retention Indices				
	Apiezon L on Celite Column	Carbowax 15 on diatomaceous earth Column	Polyethylene glycol succinate on diatomaceous earth Column	B.P.°C	M.W.
Me_4Sn	630	676	692	77.4	178.8
Et_4Sn	1049	1074	1097	179.5	234.9
Pr_4Sn	1327	1347	1352	223.5	291.0
iPr_4Sn	1355	1364	1371		291.0
Bu_4Sn	1599	1606	1642	267.5	347.1
iBu_4Sn	1466	1515	1489		347.1
Me_3EtSn	728	745	744	108.2	192.7
Me_3PrSn	833	874	836	131.7	206.9
Me_3iPrSn	794	849	818	123	206.9
Me_3t-BuSn	820	856	848	134	220.8
Me_3CyhexSn	1208	1243	1300		247.0
Me_3ViSn	703	800	800	99.5	190.7
Me_3PhSn	1172	1498	1531	203	240.9
Vi_4Sn	911	1153	1188	160	226.8
		Av. 1114	Av. 1122		

A graph (Figure 141) of the retention index on Apiezon versus the total number of carbon atoms shows that the homologous series, tetraethyl, tetrapropyl, tetrabutyl increases almost linearly, but that those compounds not in the series deviate irregularly due to boiling point and polarity differences. A plot of the boiling points versus the total number of carbon atoms also shows some degree of regularity.

An extensive collection of retention times as a function of temperature for the methylethyltin compounds has been given by Proesch and Zoepfl[2284] who used a packing consisting of Sterchamal brick (impregnated with 15% of high vacuum stopcock grease, (Zeiss No. 20) operated isothermally at 70 - 95°C and using argon as carrier gas. Pollard et al[2127] demonstrated that plots of log retention volume versus the number of chain carbon atoms in the alkyl group gave straight lines for tetra-n-propyltin through tetra-n-butyltin compounds.

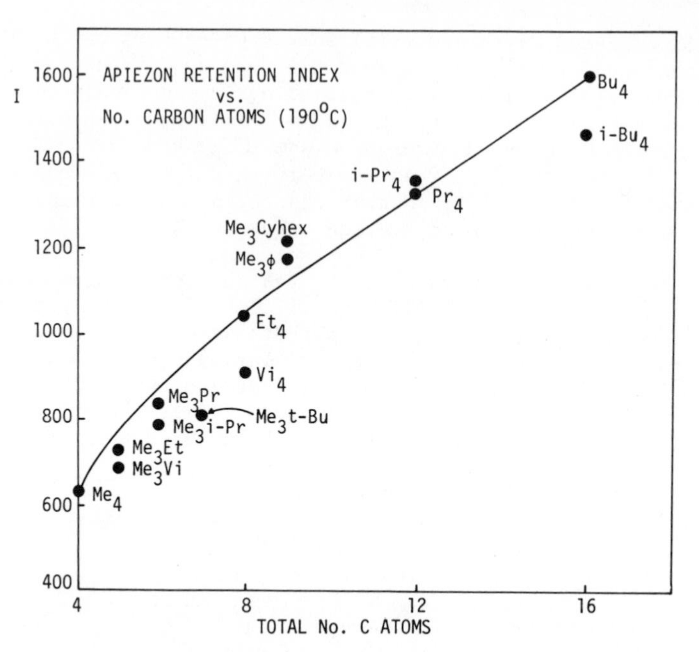

Figure 141 - Gas chromatography of alkyl, aryl and vinyl tin compounds. Apiezon retention index versus number of carbon atoms (190°C)

Pollard et al[2261] have correlated specific retention data and various thermodynamic properties for tetramethyltin, trimethylethyltin, dimethyldiethyltin, methyltriethyltin and tetraethyltin using columns comprising 15% silicone oil E301 (mo. wt. 700,00) on Celite, (treated by dry sieving to mesh 36 - 60, washing with concentrated hydrochloric acid, methanol and distilled water, followed by drying at 300°C).

Abel et al[2285] have published details of separation of the tetramethyl compounds of tin and various other elements, using Apiezon L columns. Matsuda and Matsuda[3796] have discussed the separation of tetraalkyltin compounds and alkyl-tinhalides. They encountered difficulties in the analysis of these compounds due to redistribution effects.

Hoppner et al[2286] purified tetraalkyltin compounds by preparative gas chromatography. Diethyldimethyltin was separated from impurities (mainly ethyltrimethyltin and triethylmethyltin) by passing a 6-ml sample through a column (13 metres x 25 mm) of kieselguhr (particle size 0.4 to 0.5 mm) containing 18% of polyoxyethylene glycol 400. The column was operated at 105°C and hydrogen was used as carrier gas (64.5 litres per hr.). The main fraction (retention time 29.5 min) contained only 10^{-3}% of impurity.

Hoppner and Zoepfl[2287] and Wowk and Dogiovanni[2288] described gas chromagrographic procedures for the separation of mixtures of tetraalkyltin and other organotin compounds. Umilin and Tainvoi[2289] used gas chromatography and mass spectrometry to investigate impurities such as tributyltinchloride in tetrabutyltin.

Jitsu et al[2290] determined tetrabutyl, tetraactyl and tetraphenyl tin compounds and their mono, di and tri halogen derivatives by reacting them with a Grignard reagent in tetrahydrofuran and analyzing the products on columns of either DC 550, DC high vacuum grease or DC high vacuum grease on Celite 545.

Vyakhirev and Chereshnya[2291] studied the gas chromatographic behaviour of tetramethyl-, ethyl-, and butyl-tin with the use of eight stationary phases and five supports. The dependence of the HETP value on the flow rate of carrier gas was investigated, and the rectilinear dependence of the log of the retention volume on the inverse of the column temperature and on the number of carbon atoms in the molecule was established. The best separation of the above substances was attained on a column (1.2 metres x 4 mm) packed with silicone E-301 or Apiezon L on INZ-600 brick (calcined at 1000°C) or on Spherochrom-1 and operated at 135°C with argon (18 ml per min.) as carrier gas and a β-ionisation detector.

Chernoplekova et al[2292] measured retention volumes for 60 organotin compounds at 180°C to 220°C on columns (2 metres x 3 mm) packed with 5% of e.g., SE-30 polyoxyethylene glycol 40,000 or Apiezon L on silanised Chromosorb G. In conflict with the findings of Vyakhirev et al[2291] these workers found that the log of the retention was not rectilinearly related to the total number of carbon atoms in compounds of the type R_4Sn.

Selikhova and Umilin[2293] studied the thermal decomposition products of tetrabutyltin. They analyzed these products by gas-solid chromatography at 50°C and 130°C on a column (5 metres x 4 mm) of silica gel (particle size 0.25 to 0.5 mm and calcifed at 150°C to 200°C before use). They used nitrogen as carrier gas (60 ml per min.) and a flame ionization or katharometer detector. Hydrogen and both saturated and unsaturated C_1 to C_8 hydrocarbons were detected in the decomposition products.

Cooke et al[2294] cited work on the separation of trimethyltin hydride, tetramethyltin, and trimethylpentyltin. Few details of the gas chromatography column are given.

Devyatayk et al[2295] discuss the determination in tetrabutylstannane of impurities of the type:

$R_nBu_{4-n}Sn$, where R = Et, Pr and n + 0 - 4, and $Bu_{4-n}SnCl_n$

(where n = 1 - 3)

Nelson[2296] has used gas chromatography to determine the solubility of tin tetraalkyls in water. Longi and Mazzochi[1543] developed an apparatus for use in conjunction with gas chromatography which permits sampling and sample injection into the column in the absence of air and water contamination of the sample.

Jergen and Figge[2297] and Brazhnikov and Sakodynskii[2298] have analyzed volatile methyltin compounds by gas chromatography in conjunction with flame ionization detection. Brazhnikov and Sakodynskii[2298] have carried out the same analysis using an electron capture detector to detect tetramethyltin in respirant gases. The detection limits of these methods are in the 1 - 100 ng range.

A procedure was developed by Gauer et al[2299] for the determination of various cyclohexyltin bromides. It was not, however, applicable to the phenyltin bromides due to their low volatility.

Vickrey et al[2338] coupled a graphite furnace atomic absorption detector to a gas chromatograph for the analysis of methyltin compounds. These workers reported that zirconium treatment of the graphite cuvettes in the atomic absorption spectrophotometer improved the response to organotin compounds.

B. Organotin Halides

Geissler and Kriegsmann[2300] reported on the gas chromatography of mixtures of the various butyltin chlorides. They showed that at a column temperature of 160°C to 180°C butyltin trichloride reacted with tetrabutyltin, but that mixture containing tetrabutyltin, tributyltin chloride and dibutyltin dichloride were unaltered. Figure 142). The 2-metre column used was packed with 18% of silicone oil OE 4011 on Sterchamol, and the carrier gas was hydrogen (3.5 litres per hour). A thermal-conductivity detector was used.

These workers[2300] also studied the gas chromatography of the four butyltin chlorides on a variety of stationary phases. Of the 12 stationary phases tested, separation was achieved only with 20% of GI 7100 FF or 18% of OE 4007 D or OE 4011, supported on Sterchamol (column diameter = 0.25 mm) and operated at 175°C, 178°C or 194°C, respectively. A thermal-conductivity detector was used, with hydrigen as carrier gas.

Tonge[2301] studied the gas chromatographic analysis of butyl, octyl and phenyl tin halides, compounds used as intermediates for the manufacture of stabilisers for plastics, fungicides for paints, and certain other biologica and agricultural chemicals. They used a gas chromatograph equipped with a hydrogen flame ionization detector.

Column temperatures of 110°C for butyltin halides, 180°C for phenyltin halides and 210°C for octyltin halides were found to be satisfactory. Small variations were made from these temperatures as the columns were progressively exhausted after continuous use. This was particularly noticeable at the high temperatures required for the analysis of mixtures of octytin bromides. They used a column comprising a 17 cm U-shaped stainless steel section of 4 mm internal diameter. The stationary phase was 5% w/w silicone oil (Midland Silicone MS 200) supported on Celite 545 alkali treated.

Figures 143 and 144 illustrate the resolution for typical butyltin and octyltin mixtures. A quantitative calibration is also made for the corresponding alkyl halide present in the mixture usually at less than 5% w/w.

Steinmeyer et al[2303] supplied gas chromatography using a thermal conductivity detector to the analysis of butyltin bromides. They showed that the butyltin bromides may be quantitatively converted to their more volatile butylmethyltin analog with methyl Grignard reagent and then determined by gas liquid chromatography with an estimated accuracy of about ± 2%. Direct chromatography of the alkyltin bromides appeared to produce severe sample decomposition. The method is stated to be applicable to a number of tetralkyl and/or tetraaryltins and alkyl or aryltin halides.

Jitsu et al[2290] have described a method for the gas chromatographic analysis of tetrabutyl, tetraoctyl and tetraphenyl-tin and their mono-, di- and tri-chloro and bromo derivatives in which these compounds are converted into tetraalkyl tin compounds with Grignard reagents in tetrahydrofuran (propyl-mercuric bromide for the butyl derivatives and pentylmagnesium

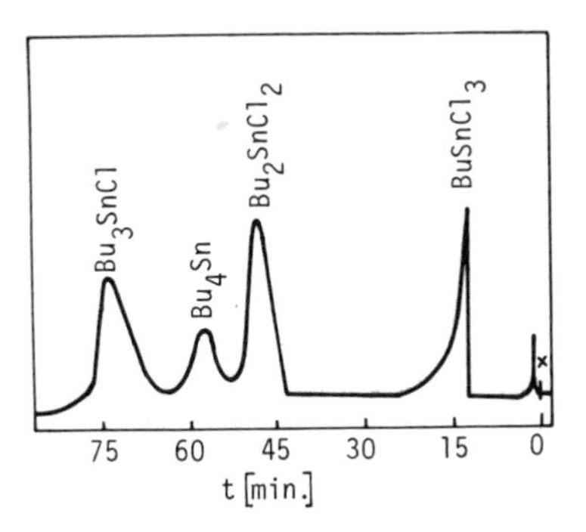

Figure 142 - Gas chromatogram of mixture of chlorobutyltins at 160°C. 3.5 l hydrogen per h. x = injection point.

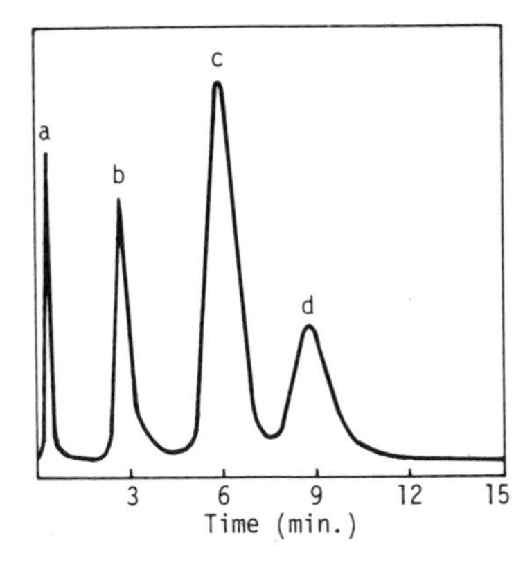

Figure 143 - Gas chromatogram of butyltin bromides, (a) butyl bromide, (b) butyltin bromide, (c) dibutyltindibromide, (d) tributyltin bromide.

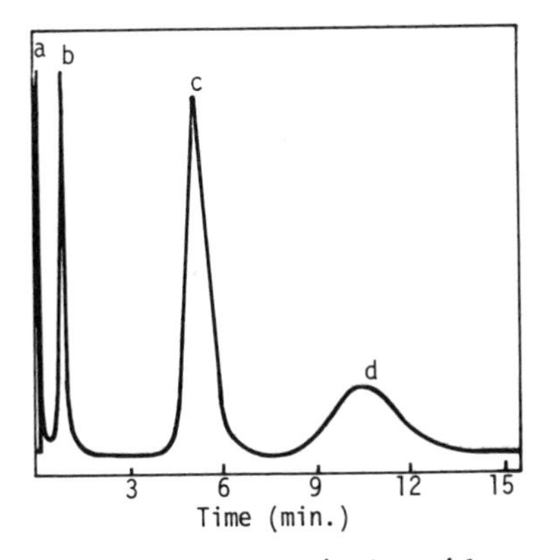

Figure 144 - Gas chromatogram of octyltin bromides, (a) octyl bromide, (b) octyltintribromide, (c) dioctyltindibromide, (d) trioctyltin-bromide.

bromide for the octyl and phenyl derivatives) and submitted to gas chromatography (column 3 mm x 75 cm), with thermal-conductivity detection. The stationary phases are 25% of silicone DC 550, DC HV grease and DC HV grease on Celite 545 (80 to 100 mesh). The helium flow rates are 50, 90 and 50 ml per min. Column temperatures 190°C, 280°C and 260°C. Octadecane, trioctylpropyltin and butyltriphenyltin are used as internal standards for the butyl, octyl and phenyl derivatives, respectively. The retention time increases with increase in total number of carbon atoms in the tetraalkyl compounds and their peaks are satisfactorily separated for each of the butyl, octyl and phenyl series. The mono-, di-, and trichloro derivatives of octyltin and tetrahydrofuran solution of the tetraalkyl compounds are treated with saturated ammonium chloride solution (7 ml), and the organic layer is evaporated to 3 ml, diluted with benzene (30 ml), washed with water (2 x 20 ml) and evaporated to = 2 ml and a 1μ 1 portion is used for gas chromatography.

Geissler and Kriegsmann[2304] investigated the gas chromatographic behaviour of n-butyltintrichloride, di-n-butyltindichloride, tri-n-butyltinchloride and tetra-n-butyltin. They measured the relative retention volumes of the butyltin compounds on ten different liquid phases using a katharometer and hydrogen as carrier gas. Their results in Table 200 clearly show the effect of liquid phase polarity on the relative retention volumes. On non-polar liquid phases the compounds are separated according to their molecular weights, whereas in polar liquid phases the relative retention volumes of compounds containing the Sn-Cl bond increase with increasing polarity of the liquid phase. In the case of Carbowax, tetra-n-butyltin is eluted as the first compound.

Table 200 - Relative retention volumes of butyltin compounds (a) methylsilicone rubber, (b) methyl-phenylsilicone oil, (c) methylcyanopropylsilicone oil, (d) 1.2.3.-trimethyl-pentaphenyltrisiloxane, (e) 1.3-dimethyl-hexaphenyltrisiloxane

Liquid Phase	T °C	V_{rel}			
		$BuSnCl_3$	Bu_2SnCl_2	Bu_3SnCl	Bu_4Sn
GI 7100 FF[a)]	176	0.16	0.51	0.88	1
SE 30[a)]	162	0.17	0.55	0.95	1
OE 4011[b)]	178	0.22	0.84	1.26	1
OE 4178[c)]	176	0.25	1.00	1.42	1
OE 4006D+ OE 4178(2:1)	180	0.37	1.16	1.42	1
OE 4007[d)]	194	0.39	1.19	1.46	1
OE 4006[e)]	180	0.45	1.58	1.58	1
Benzyldiphenyl	176		1.79	1.97	1
Reoplex 400	163	1.75	3.19	2.07	1
Carbowax 20M	165	2.92	3.32	2.32	1

It is known that various alkyltin compounds can interact in the following way:

$$R_3SnCl + RSnCl_3 \text{ --- } 2\ R_2SnCl_2 \quad (1)$$

$$R_4Sn + RSnCl_3 \text{ --- } R_3SnCl + R_2SnCl_2 \quad (2)$$

$$R_4Sn + R_2SnCl_2 \text{ --- } 2\ R_3SnCl \quad (3)$$

Geissler and Kriegsmann[2304] examined whether these redistribution reactions which would make quantiative analysis impossible in the gas chromatograph. They showed that tributyltinchloride does not react with butyltintrichloride in any of the used liquid phases. On the other hand, tetrabutyltin reacts with butyltintrichloride according to equation (2) in all the liquid phases used. Reaction (3) proceeds in all of the liquid phases applied except Carbowax 20M.

They concluded that it was not possible to analyse mixtures of the four n-butyltin-halides since tetra-n-butyl consistently reacts with n-butyltin chloride to form tri-n-butyltinchloride. Mixtures of n-butyltin-trichloride, di-n-butyltintrichloride and tri-n-butyltinchloride can be successfully analysed in the liquid phase OE 4011 or GI 7100FF. The analysis of mixtures of di-n-butyltindichloride, tri-n-butyltinchloride and tetra-n-butyltin is possible only on Carbowax 20 M column. In Table 201 are shown results obtained on this column for a range of synthetic mixtures. The agreement between theoretical and observed analysis is within ± 2.5%.

Table 201 - Quantitative analysis of di-n-butyltindichloride, tri-n-butyltinchloride and tetrabutyltin mixtures.
Conditions: 5% Carbowax 20 M on brickdust, 3 metre column at 165°c, carrier gas 3.5 litres per hour, hydrogen tetradecane internal standard

	Composition % w/w	
Compound	Theoretical	Found
Bu_4Sn	3.73	3.73
Bu_3SnCl	3.84	3.85
Bu_2SnCl_2	91.43	92.4
Bu_4Sn	15.2	14.4
Bu_3SnCl	16.4	16.0
Bu_2SnCl_2	68.4	69.6
Bu_4Sn	26.1	26.4
Bu_3SnCl	31.8	31.7
Bu_2SnCl_2	42.1	42.9
Bu_4Sn	0.47	0.47
Bu_3SnCl	0.71	0.67
Bu_2SnCl_2	98.82	98.9

Franc et al[2305] studied the gas chromatography of a series of organotin compounds prepared by reaction of stannic chloride by the Grignard reaction and determined various physical properties, including R_f values. They used Chromosorb and Celite columns impregnated with 20% to 30% of silicone with nitrogen carrier gas and a thermal conductivity detector and reported that separations were not very successful. Gas chromatographic separations of alkyltin halides have been reported by other workers. (Gerrard and Mooney 2306, Chromy et al[2307], Newmann and Burkhardt[2308], Bladk[2309]).

Jitsu et al[2290] reported on the separation of organotin halides after total acetylation.

Keller[2310] reviewed the literature on the gas chromatography of volatile metal compounds and in this discusses the chromatography of stannic chloride using squalane, n-octadecane, silicone oil, paraffin wax and Apiezol grease as partitioning agents supported on Chromosorb at temperatures between 100 and 200°C.

Kaesz et al[2311,2312] have discussed the possibility of eluting perfluoroalkyl and perfluorovinyltin compounds on a polyethylene glycol column.

Barrell and Platt[2313] have studied the photolysis of tetramethylstannane, trimethyltinhydride and hexamethylditin vapours and examined the alkane reaction products on Poropak at 50°C or Apiezon at 100°C.

Bortnikov et al[2314] determined the retention volume of hexaethylditin on a column of Apiezon L or Carbowax 20M on silanized Chromosorb W at 222 - 234°C using helium as carrier gas. These workers also used a column of graphitized carbon black at 266°C to separate mixtures of compounds of the type Et_3M-MEt_3 (where M is tin, germanium or silicon). Bortnikov et al also separated organometallic compounds containing bonds between two or three like or unlike heteroatoms (tin, silicon, germanium, sulphur) on columns of silanized Chromosorb W coated with 20% of Apiezon, 15% of polyoxyethyleneglycol 20M or Reoplex 400. They also carried out separations of graphitised thermal carbon black using helium as carrier gas and flame ionization or thermal conductivity detection. Bortnikov et al[2315] also gas chromatographed a series of compounds containing silicon, selenium, sulphur, germanium, tellurium and/or tin as the central or linking atoms at 254°C in a stainless steel column (100 cm x 0.4 cm) packed with Chromosorb W supporting 20% of Apiezon L, with helium as carrier gas and thermal-conductivity detection. Specific retention volume and differential heats of solution were measured for Apiezon L as solvent. The heats of solution were directly proportional to the molecular weight of the organometallic compounds.

Dressler er al[2316] have studied the response of alkali flame detectors to organotin compounds.

Hill and Aue[2317,2318] have described a spectrophotometric detector incorporating a dual flame ionization detector to determine volatile organotin compounds. They preferred the hydrogen rich flame giving sensitivities in the lower nanogram level for organotin compounds and a strong discrimination against hydrocarbon sample solvents.

C. Organotin Oxides and Hydroxides

Hanzen et al[2319] compared a flame ionization detector and a hydrogen atmosphere flame ionization detector (HAFID), to identify similarities and differences in their responses to organotin compounds. They found in studies on various organotin oxides (bis(tributyltin)oxide) and organotin hydroxides (triphenyltin hydroxide and tricyclohexyltin hydroxide) that the HAFID detector is not significantly more sensitive than the flame ionization detector. Its principal advantage over the flame ionization detector is that it has a tremendous selectivity against hydrocarbons, which permits the determination of small quantities of organotins in the presence of large quantities of organic material. Selectivity of the hydrogen atmosphere flame ionization detector was found to be a function of the temperature of the precombustion zone of the flame where an organic compound is apparently oxidized prior to entering the flame, reducing the ionization efficiency in the flame. Prior to gas chromatography the triphenyltin hydroxides were converted to methyltriphenyl tin by the use of a Grignard reagent. Various types of separation columns were used in this work.

ORGANOZINC COMPOUNDS

A. Alkylzinc Compounds

The alcoholysis-hydrolysis procedure, described in detail in Section 3A can be used for the determination of lower alkyl and hydride groups in organozinc compounds. In this procedure, a weighed amount of the organozinc compound is reacted at - 60°C in the gasometric apparatus with n-hexanol and then with 20% aqueous sulphuric acid reagents. The volume of gas evolved from the sample is measured and the evolved gas compositions obtained by gas chromatography. In Table 202 are given results obtained by applying this procedure to various weights of diethylzinc. The gas evolved consists principally of ethane, a small amount of n-butane also being present. This originates from butyl impurity present in the triethyl-aluminium used to prepare the diethylzinc. Hydrogen was absent in the evolved gas indicating the absence of zinc bound hydride groups and showing also, that no fissioning of zinc bound alkyl groups occurred under the reaction conditions used. The plot of the sample weights and gas volumes in Table 202 passes through the origin, indicating that reaction is proceeding reproducibly, regardless of the weight of diethylzinc taken for analysis.

The procedure was applied to the determination of ethyl and butyl groups in a sample of distilled diethylzinc. This material contained some aluminium impurity originating in the triethyl-aluminium used in its preparation. The results obtained are presented in Table 203. It is seen that 99 per cent of the sample has been accounted for. The remainder of the sample is probably alkoxide. Alkoxide was not determined but undoubtedly occurs to a small extent in this reactive organometallic compound.

Table 202 - Gas yields obtained upon alcoholysis/hydrolysis of diethylzinc

Weight of diethylzinc taken for analysis g	Total volume of gas evolved at STP ml	Composition of generated gas % v/v		
		ethane	n-butane	hydrogen
0.1049	36.0	98.7	1.3	nil
0.1751	62.3	98.7	1.3	nil
0.1961	70.3	98.7	1.3	nil
0.2146	75.9	98.7	1.3	nil
0.2573	88.3	98.4	1.6	nil

Table 203 - Analysis of distilled diethylzinc

Determined Constituent	% wt
ethyl	44.8
butyl	1.1
hydride	nil
zinc	47.8
aluminium	2.45
chloride	2.75
Total	98.9

Only one reference has been found in the literature to the direct gas chromatography of organozinc compounds. Longi and Mazzochi[2320] claim to have gas chromatographed this type of compound by gas chromatography on a 1 metre column of Chromosorb W containing 7.5% of paraffin wax (mp 63 - 64°C) - triphenylamine '17:3) using dry purified helium as a carrier gas and a thermistor detector. They used a column temperature of 73 - 165°C, according to the type of organozinc compound being examined.

Karayannis and Corwin[2321] have applied hyperpressure gas chromatography to the analysis of various etioporphyrin II metal chelates including that of zinc. Gas chromatography is carried out at 145°C and 1000 - 1700 psi pressure using dichlorodifluoroethylene as carrier gas and 10% Epon 1001 on Chromosorb or 10% XE-60 (cyanoethylmethyl silicone polymer) on Chromosorb as stationary phase. Up to fifteen different etioprorphyrins were separated by this procedure.

B. Zinc Dialkyldithiophosphates

The pyrolysis-gas chromatographic technique has been used to identify microgram amounts of zinc dialkyldithiophosphates, isolated from mineral oils. (Perry[2322]). The zinc dithiophosphates are non-volatile and cannot therefore be eluted from a gas chromatographic column into an ionization detector. However, they are thermally degraded to volatile olefins and a non-volatile inorganic residue.

Earlier work by Legate and Burnham[2323] leads to a method for the characterization of zinc diethyldithiophosphates using pyrolysis - gas chromatography. This method, however, utilized a katharometer detector and consequently was not as sensitive as that of Perry which utilized a hydrogen flame ionization detector.

Perry[2322] developed a technique for the recovery of zinc dialkyldithiophosphates from a thin-layer chromatogram and for the pyrolysis of the resulting separated compounds and for the gas chromatographic separation and detection of the olefinic fractions thus formed. The presence of zinc-containing substances on a thin-layer chromatogram can be revealed by spraying the thin absorbent layer of silica with dithizone, which reacts with zinc to give a pink coloration. In this way one locates the area of the absorbent layer in which the zinc dialkyldithiophosphates is to be found after developing the chromatogram under standardised conditions. To identify the alkyl groups of the zinc dialkyl dithiophosphates both dithizone and the chromatographic eluant (in this case a mixture of pyridine and acetic acid) must be absent from the material recovered for pyrolysis since, being organic in nature, they too will give fragments on pyrolysis. Accordingly, it is necessary to recover the zinc dialkyl dithiophosphates from the thin-layer chromatogram in a pure state. This is done by extraction from the appropriate area of the chromatogram with methylene chloride and then evaporating the methylene chloride, and any pyridine and acetic acid, on a water bath in a stream of nitrogen.

The pyrolysis-gas chromatographic apparatus is shown schematically in Figure 145. The pyrolysis vessel is shown in Figure 146. The platinum dish on which the sample is deposited and then pyrolysed, is about 5 mm x 4 mm and 22 gauge in thickness. A gas chromatograph with hydrogen flame ionization detector and 4-way gas sampling valve was used. Gas chromatographic separation of the pyrolysis products was achieved with a 6' x 1/4" diameter copper column packed with 10% silicone oil on 60 - 100 mesh "Embacel" (both from May and Baker Ltd., Dagenham, Essex) operated at 23°C (room temperature)

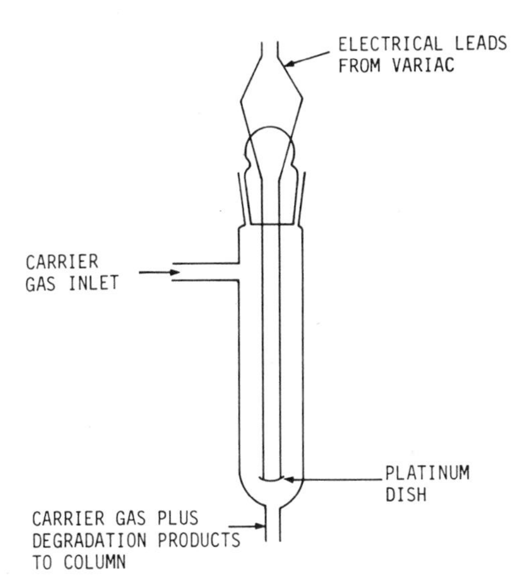

Figure 145 - Block diagram of apparatus for identification by pyrolysis-gas chromatography.

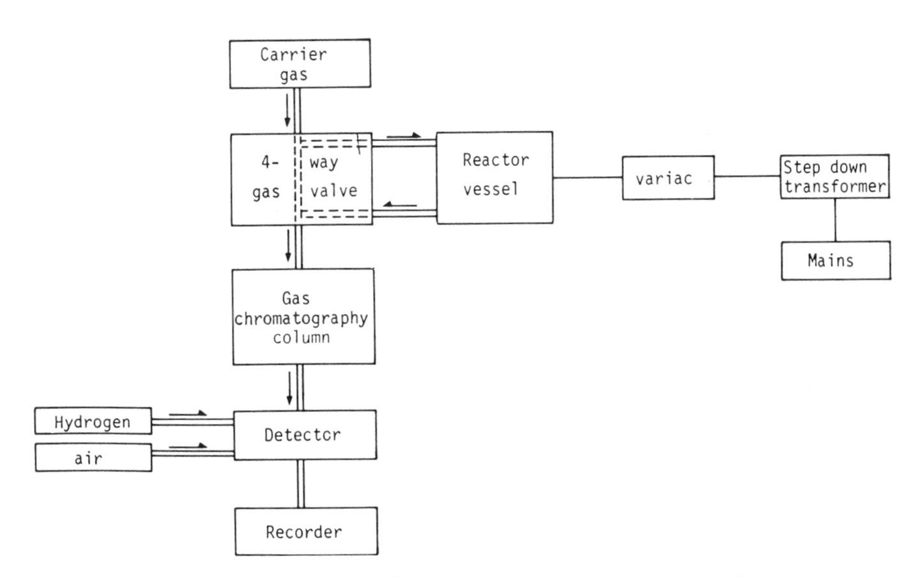

Figure 146 - Reaction vessel used for pyrolysis.

The nitrogen carrier gas flow rate was 30 ml min^{-1} respectively. In Table 204 are shown the analysis of pyrolysis fragments obtained from zinc dialkyldithiophosphates containing known alkyl groups.

Table 204 - Pyrolysis fragments of zinc dialkyl dithiophosphates (weights per cent of volatile products).

olefinic fragment	ZDDP alkyl group				
	isoC$_3$	isoC$_4$	1-pentyl	3-pentyl	4-methyl 2-pentyl
C$_2$"	100	3	2	-	-
C$_4$"-1/isoC$_4$"	-	85	1	-	-
C$_4$"-2	-	12	-	8	-
C$_5$"-1	-	-	60	92	-
C$_5$"-2	-	-	36	-	-
3MeC$_4$"-1	-	-	1	-	3
2MeC$_4$"-1	-	-	-	-	1
2MeC"$_5$-2	-	-	-	-	34
4MeC"$_5$-1	-	-	-	-	62
4MeC$_5$"-2	-	-	-	-	

The only significant decomposition products observed arose from simple-cleavage of the carbon-oxygen bond. Only very small amounts (5%) of methane or C$_2$ hydrocarbons were observed in any of the cracking patterns, suggesting that the carbon-carbon bonds in the alkyl structure were much more stable than the C - O bond and that further degradation of the olefinic fragments was negligible.

Pyrolysis of zinc di-isopropyl dithiophosphate leads to formation of propylene alone. It would be impossible to distinguish between dialkyldithiophosphates made from normal or isopropanol by this technique, propylene being the only possible product from either. The iso-butyl compound similarly gives almost entirely iso-butene. The cracking pattern given by zinc di-(4-methyl 2-pentyl) dithiophosphate is more informative, and the results obtained may be compared with those obtained by Legate and Burnham[2323]. Having two beta carbon atoms from which hydrogen may be removed, two olefins are likely to be formed, 4-methyl pentene-1 and 4-methyl pentene-2.

```
          CH3        CH2        CH3
          |          |          |
P - O -   CH         CH         CH
          |          |          ||
          CH2        CH         CH
          |          |          |
          CH-CH3 --> CH-CH3  +  CH-CH3
          |          |          |
          CH3        CH3        CH3

          (1)        (11)       (111)
```

In Table 205 the product proportions observed are compared with those of Legate and Burnham[2323] for the lead salt. Product distribution is almost identical in the two cases.

The preponderance of 111 over 11 in the products, shows that the preferred direction of hydrogen abstraction is to give the olefin having the more centrally placed double bond. The same orientation occurs in pyrolysis of hydrocarbons, such as 2-methyl pentane-1 which gives a pentene-2 : pentene-1 ratio of 6:1. This is in accord with the known relative stabilities of isometric olefins.

Pyrolysis of the 1-pentyl and 3-pentyl structures shows that in the former case although the 1-ene is the main product, a considerable amount of double bond migration to the 2 position occurs. With the 3-pentyl compound, over 90% of the product appears as pentene-2, only 7.7% pentene-1 being observed.

Table 205 - Pyrolysis of zinc and lead di-methyl pentyl dithiophosphates (weights per cent)

Product	Zinc salt	Lead salt
$4MeC_5$''-2	62	59
$4MeC_5$''-1	34	35
$2MeC_5$''-2	1	6
$2MeC_5$''-1	3	-

Perry[2322] also separated five zinc dialkyldithiophosphates with unknown alkyl groups from commercial preparations by thin-layer chromatography and then examined by the pyrolysis-gas chromatography procedure.

Legate and Burnham[2323] have also studied the micropyrolytic-gas chromatographic technique for the analysis of substances such as the zinc salts of 0,0^1 di-n-dodecyl thionothiophosphate, 0,0^1-dineophentyl thionothiophosphate and 0,0^1- dicyclohexyl thionothiophosphate. The compound is pyrolysed in the inlet system of a gas chromatograph, and the volatile pyrolysis products, generally olefins are separated and collected individually for identification by mass or infrared spectrometry. The olefins are formed generally by the breaking of a carbon-oxygen bond and abstraction of hydrogen from a beta carbon atom with no skeletal isomerization. The structure of the olefins produced is thus directly related to the structure of the alkyl groups in the original zinc compound. Only when hydrogen is not available on a beta carbon atom - e.g. neopenyl radicals - are olefins formed by carbon skeletal rearrangement. Legate and Burnham[2323] give examples of the determination of the exact configuration of the alkyl radicals in several model zinc dialkylthionothiophosphates.

CHAPTER 7

CHROMATOGRAPHY OF METAL CHELATES

A. Introduction

It has long been an ambition of chemists to be able to analyse mixtures of metals by converting them to volatile derivatives which are amenable to chromatographic separations. For gas chromatographic separation it would be necessary for the derivatives to be sufficiently volatile to chromatograph and yet be stable enough not to undergo thermal decomposition during the separation. For liquid chromatographic methods derivative volatility would not be a problem although stability at operating temperatures is still a requirement.

As well as their analytical implications such methods would offer the potential of preparing small quantities of metals in a very high state of purity and would provide much useful information on physical problems such as metal to carbon bonding.

The earliest attempts to carry out gas chromatography of metals involved analysis of the volatile metal halides, usually fluorides or chlorides. This work started about 1960 (see Table 206). However, despite limited success, the technique has not been very extensively used for various reasons. Due to the corrosive nature of many metals halides, severe experimental difficulties were encountered. Also, many metals either do not form halides or do not form halides which are volatile enough for gas chromatography.

Anvaer and Drugov[2339] reviewed the literature on the gas chromatography of inorganic materials for the period 1963 to 1970 with special reference to water, metals and isotopic analyses (300 references).

This led to a search for derivatives of metals which had the required properties without the disadvantages of metal halides. The search was rapidly narrowed down to metal chelates which, in addition to volatility, had in many, but not all, instances the additional advantage of thermal stability under gas chromatographic conditions.

A chelate is a cyclic compound in which the ring is closed by co-ordination bonds with an acceptor atom. For the compounds of interest, this acceptor is a metal cation, and the chelating agent is an organic anion.

Table 206 - Gas chromatography of metal chlorides and fluorides

Fluorides

Year	Elements	Reference
1960 - 63	U	2340, 2341
1966	Te U Mo	2342
	M W Al Cr Rh	2343
	Miscellaneous	2344
1967	Miscellaneous	2345
1968	U	2346
1969	U	2347
1970	Mo U Sb	2348
1971	As Be Ge Mo P Re Sb Se Si W Te	2349
	W Mo Re S P Si S	2350
1972	U	2351
1973	S Se Te	2352

Chlorides

Year	Elements	Reference
1959 - 1960	Sb Ti Su	2353, 2354
1961 - 1964	Ge As Te Hg Su	2355, 2356
1966	Sn	2343
1968	Nb Zr Ta Hf Al	2357
	U	2346
1969	Sn	2358
	Rare earths	2359
	Mo U Sb	2360
1970	transuranic elements	2361
	Sn Sc Ge V Ti P As Sb	2362
	Si P Ge Su Ti P U As Sb	2363
1971	Ti V	2364
	Ti Si Ge Sn P As	2365
	Si P Ge Sn Ti B P U As Sb	2366
1972	Bi Be In Sn Zn Te Pb Cd	2367
	As	2368, 2369
1977	Sb	2370

For reasons to be examined in this chapter, certain chelates lend themselves very well to the analysis of metals by gas chromatography. The first studies were published in 1959. In recent years, efforts have been directed primarily toward an improvement of the sensitivity of analyses of trace elements with the use of more sophisticated detectors and the application of the method to a larger number of metals by means of new chelating agents.

The metal compounds which can be analyzed by gas chromatography are quite small in number; hydrides, halides, alkyl derivatives, metal carbonyls and metallocenes. Each of these groups has its own particular shortcomings; low thermal stability, low volatility, corrosive properties, or impossibility of applying the method to a large number of metals. The chelates, particularly those of the β-diketones and their fluorinated derivatives, have properties making them more suited than other organometallic compounds for the separation, purification or quantitative analyses of the metals by gas chromatography.

The metal chelates can be formed quantitatively, rapidly and reproducibly. The reactions can be carried out under different pH conditions, in different solvents, and, with a suitable choice of chelating agent, almost all metal elements can react with the formation of compounds of sufficient volatility and thermal stability to permit their elution from a gas chromatographic column. Thus, for example, by replacing acetylacetone by its fluorinated derivatives, the chelates obtained have a shorter retention time the higher their degree of fluorination. This phenomenon, which is also observed in other series of organic compounds (acetates and trifluoroacetates, etc.), seems to be the result of a reduction of van der Waals forces since the polarizability of the fluorine atom is lower than that of the hydrogen atom, and perhaps of a decrease of intermolecular hydrogen bonds in the fluorine-containing compounds[2371]. Another advantage of these fluorinated chelates is the possibility of detecting them at trace level with an electron capture detector.

The solubility of chelates in various organic solvents and their insolubility in water greatly facilitate their separation from the aqueous medium which generally contains the metal ions and contribute toward realization of quantitative yields in their preparation. The chromatographic analysis is carried out with a fraction of the organic phase. Numerous details on the preparation and extraction of the complexes have been given by Moshier and Sievers[2372].

Uden and Henderson[2373] have reviewed methods for the determination of metals by gas chromatography of metal chelates and Minear and Palesh[2374] have discussed the application of this technique to the determination of trace metals in water.

Almost all studies on the gas chromatography of metal chelates make use of β-diketones containing oxygen donor atoms. Below are successively examined the analysis of acetylacetones, trifluoroacetylacetonates, and hexafluoroacetylacetonates, and their various other chelates. Studies on the thermodynamic dissolution parameters of some of these complexes in liquid stationary phases have been discussed[2375,2376].

Also discussed have been the gas chromatography of certain metal porphyrins performed under special experimental conditions at very high pressures.

More recent work has discussed the applications of liquid chromatography to the analysis of metal chelates.

B. Acetylacetonates

In 1955, in a paper on the analysis of rare earths by paper chromatography, M. Lederer for the first time considered the possibility of separating volatile metal complexes such as the acetylacetonates by gas chromatography[2377]. The first studies appeared only three years later. Dustwald observed that it was possible to elute the beryllium, zinc, and scandium acetylacetonates on columns containing a silicone oil or ethylene glycol polyadipate, but the retention times of the three complexes were identical[2378]. Floutz repudiated these results and assumed that the peaks attributed to zinc and scandium are due to traces of beryllium present in the samples[2379]. However, the occurrence of thermal decomposition could not be ruled out[2380].

More conclusive results were obtained in the next few years by Brandt

2381-2383, Janak[2384,2385], Sievers et al[2386,2387], Hill et al[2388,2389], Melcher[2390] Heneran[2391], Biermann and Gesser[2392], Baldwin[2393], Duswalt[2394], Janak[2385] and Brandt[2383]. Each briefly described work on metal acetylacetonate analysis by gas chromatography. Biermann and Gesser[2392] reported in the same year their work on the gas chromatographic analysis of aluminium and chromium acetlyacetonates, Floutz[2379] investigated eleven metal acetylacetonates by gas chromatography but successfully eluted only the aluminium, beryllium and chromium chelates. Copper (II), vanadium (II) and iron (III) acetylacetonates were also chromatographed successfully using column temperatures between 150 and 200°C (Brandt[2363], Janek[2385], Biermann and Gesser[2392], Melcher[2390], Heveran[2391], Hill[2386], Sievers et al[2386,2387], Hill and Gesser[2389]). Attempts to chromatograph the acetylacetonates of zirconium (IV), hafnium (IV) cobalt (III), thorium (IV) and titanium (IV) at column temperatures between 150 and 250°C were all unsuccessful (Sievers et al[2386]). At lower temperatures only one peak, attributable to the sample solvent, is produced. At higher temperatures the chelates generally decompose rapidly to yield a volatile product which may appear as a well-defined peak on the chromatogram or may be obscured by the sample solvent peak. With few exceptions, the metal acetylacetonates have proven to be unsatisfactory for quantitative metal analysis owing to thermal and solvolytic instability and, with a few exceptions such as beryllium, chromium, aluminium and scandium to low volability during gas chromatography as, in general, have other organometallic compounds such as carbonyls, alkyls, cyclopentdienyls, fluorides, chlorides, bromides and iodides.

The beryllium, aluminium, chromium, and even copper (II), iron (III), and vanadium (in the form of vanadyl ion) acetylacetonates were analyzed from their solutions in carbon tetrachloride or disulfide, with the latter having the advantage of a low response factor in the flame ionization, detector[2389]. The most symmetrical peaks were obtained on columns having a small liquid-phase loading ratio and with Apiezon grease or silicone oil deposited on glass microbeads. The compounds are eluted between 100 and 150°C in the order of their increasing boiling points of 270, 314, and 340°C for beryllium, aluminium and chromium acetylacetonates, respectively[2392].

Except for the copper and iron complexes, no decomposition was observed in the injector. The eluted fractions could be collected and identified by their melting point and UV-spectrum. Quantitative determinations of chromium[2391] and beryllium in Be-Cr alloys were obtained with relative errors in the order of 3 - 5%.

Figure 147 shows the separation of beryllium, aluminium and chromium acetylacetonates on a silicone oil SE-30 column at 165°C[2389].

Scandium acetylacetonate can be eluted without apparent decomposition and separated from the beryllium derivative on Apiezon L at about 210°C, but its resolution with the aluminium chelate is incomplete[2395]. The response of the katharometer used in this work is linear between 1 and 6 mg of $Sc(acac)_3$ (acetylacetonate anion is designated by acac). The samples are injected in benzene solution[2395] or in solid form[2396].

Many acetylacetonates do not have a sufficient thermal stability to be analysed quantitatively or even to be eluted from a gas chromatographic column. It is impossible to analyse the complexes of cobalt, zirconium, hafnium, and thorium between 150 and 230°C[2371,2397]; depending on the column temperature, the compounds are either completely retained or decomposed.

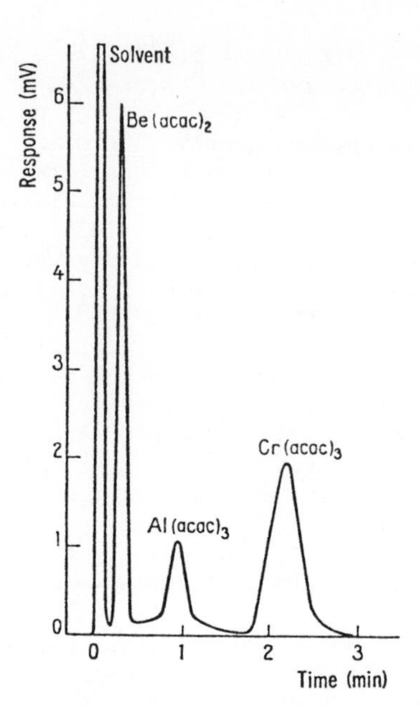

Figure 147 - Separation of beryllium, aluminium, and chromium acetylacetonates. Conditions: 30 cm long column, 7.5% SE.30 on 40 - 60 mesh fire brick; temperature, 165°C; carrier gas: nitrogen; chelates in CCl_4 solution.

While Yamakawa et al[2398] mention the elution of a large number of acetylacetonates, including those of Co(III), Ni, Cu, Fe (III), Mg, Mn (II), Ba, Ca, Cd (II), Mo (II), Ti (III), Th, Zn, and Zr on SE-30 between 100 and 190°C, no verification of the nature of the component detected at the column outlet was made, and the authors acknowledge that the peaks relative to these derivatives are broad, while those obtained with the Be, Al, Cr and V (IV) complexes are narrow. This leads to the suspicion that thermal decomposition of the first series of chelates occurred in the chromatographic system.

The use of acetylacetonates for gas chromatographic analysis thus remains limited to a few elements such as beryllium, aluminium, chromium, scandium and vanadium. It then became necessary to find chelates of higher volatility and thermal stability which would still be simple to prepare. The fluoroacetylacetonates satisfy these requirements. The most commonly used complexes of 1,1,1-trifluoro-2,4-pentanedione and 1,1,1,5,5,5-hexafluoro-2,4-pentanedione for which the anions are commonly designated by tfa and hfa:

$$CF_3 - \overset{\overset{O}{\vdots}}{C} \overset{(-)}{\cdots} CH \cdots \overset{\overset{O}{\vdots}}{C} - CH_3 \quad (tfa)$$

$$CF_3 - \overset{\overset{O}{\vdots}}{C} \cdots \overset{(-)}{CH} \cdots \overset{\overset{O}{\vdots}}{C} - CF_3 \quad (hfa)$$

C. Fluorinated Acetylacetonates

Many of the fluorinated β-diketone chelates of metals are considerably more volatile requiring for example a column temperature of approximately 100°C as compared to 170 - 200°C required for the unfluorinated chelate of the same metal.

Table 207 tabulates the vapour pressure of chromium acetylacetonate $Cr(acac)_3$ and a range of fluorinated chromium diketonates and shows the marked reduction in vapour pressure of the chelate that accompanies the introduction of fluorine into the chelate molecule.

Table 207 - Vapour pressures of chromium acetylacetonate and of fluorinated chelates

	R^1-C(=O)-CH =	C(O^-)-R^2		Vapour pressure(mm) of Cr chelate at 125°C	Temp. 0°C of which Cr chelate exhibits a vapour pressure of 1 mm
acetyl acetone	CH_3-	$-CH_3$	acac	0.01	179
1,1,1 trifluoro-2,4 pentanedione (acetyl-acetonate)	CF_3-	$-CH_3$	tfa	0.7	128
1,1,1,5,5,5, hexafluoro-2, 4 pentane dione (acetylacetonate)	CF_3-	$-CF_3$	hfa	50.0	67
2,2,6,6 tetramethyl-3, 5-heptanedione	$C(CH_3)_3$-	$-C(CH_3)_3$	thd	-	-
1,1,1,2,2,3,3 hepta-fluoro-7,7 dimethyl-4, 6-octanedione	$CF_3CF_2CF_2$-	$-C(CH_3)_3$	fod	0.6	135
1,1,1,2,2,3,3,7,7,7, -decafluoro-4, 6-heptanedione	$CF_3CF_2CF_2$-	$-CF_3$	dfhd	-	-

With nonpolar liquid stationary phases the ease with which the complexes can be eluted roughly parallels their saturation vapor pressures. While column temperatures between 150 and 200° are required for the acetylacetonates (often with extensive sample decomposition), much lower column temperatures (100 to 150°) can be used for the trifluoroacetylacetonates and still lower (30 to 80°) for the more extensively fluorine-substituted hexafluoroacetylacetonates (Sievers et al[2399,2400]). Some of the hexafluoroacetylacetonates, e.g. Cr (III), Rh (III), Fe (III) are so volatile that they can be easily steam distilled (Sievers et al[2399]. Collman et al[2401]). This discovery greatly reduced problems with sample decomposition and broadened greatly the applicability of gas chromatography to trace metal analysis. It has also led to the development of an empirical rule that enables one to predict the identity of unknown compounds from a knowledge of their chromatographic retention behaviour. For complexes of a given metal ion, increasing substitution of fluorine for hydrogen reduces the retention time.

Early work on these derivatives was carried out by Sievers and co-workers (Sievers et al[2386,2387,2402], Sievers[2403], Sievers[2404], Schwarberg [2405], Schwarberg et al[2406], Moshier et al[2407], Moshier and Schwarberg[2408], and Moshier and Sievers[2409] recommended the use of metal trifluoro-acetyl-acetonates (i.e. metal chelates of 1,1,1 trifluoro-2, 4 pentanedione, tfa) and hexafluoroacetylacetonates (i.e. metal chelates of 1,1,1,5,5,5, hexa-fluoro-2, 4 pentanedione, hfa) because of their greater volatilty as compared to the acetylacetonates and consequently their reduced thermal decomposition at the lower column and injection port temperature required for separation on a gas chromatographic column.

The more commonly used fluorinated acetylacetonates are now discussed individually in more detail:

Analysis of trifluoroacetylacetonates. The analysis of trifluoro-acetylacetonates, more volatile than their unfluorinated homologs, can be carried out at lower temperatures with less risk of thermal decomposition. The chelates of Be, Al, Cr, In, Cu (II), Rh, Zr, and Hf are eluted between 100 and 150°C[2371,2397,2410]. Sievers has studied the behaviour of a number of trifluoroacetylacetonates in benzene solution injected into a silicone oil column at 135°C[2411-2413]. He divides these compounds into three groups (Table 208).

Table 208 - Gas chromatography of trifluoroacetylacetone complexes - column Dow Corning Silicone 710 oil on silanized glass beads (60 - 20 mesh) in borosilicate glass tube (4 ft x 4 mm d). Column temp. 100 - 150°C, injection port temp. 155°C. Sample solvent, benzene, instrument F. & M. Model 500 with thermal conductivity detector all with W-2 filaments. Helium flow rate 83 ml/min.

Class I	Class II	Class III
(decreasing ease of elution)		
beryllium	iron (III)	neodimium (iii)
aluminium	manganese (III)	
gallium (III) Schwarberg (1964)	zirconium	
scandium (III)	hafnium	
copper (II)	zinc	
chromium (III)		
vanadium		
indium (III)		
rhodium (III)		

The complexes of Group I are analyzed without apparent decomposition. They are listed in the order of increasing retention times. The retention volumes relative to the chromium chelate vary from 0.35 for beryllium to 5.9 for rhodium. Among the nine metals listed in this group, we find those which have been successfully analyzed in the form of the acetylacetonates.

Chromatography of the Group II complexes must be conducted carefully while verifying that the compound eluted at the column outlet has not been modified. However, a slight residue in the injector indicates a minor degree of decomposition in the majority of cases. The neodymium complex is the only one which could not be volatilized without significant decomposition.

Furthermore, according to Schwarberg et al[2414,2415], the trifluoroacetylacetonates of TI(I), Ga (III), and Th(IV) are analyzed without apparent decomposition, but the elution product of the indium chelate is decolored and has a 1 - 2°C lower boiling point than that of the injected compound. It is therefore appropriate to add Tl (I) and Th (IV) to Group I and to place In (III) into Group II.

Figures 148 and 149 show the type of separation obtained. For these compounds, the peak tails are always long if the solid support is Chromosorb whether silanized or not. More narrow peaks are obtained on glass microbeads but the use of this support only permits low degrees of impregnation, less than 1%. Apparent contradictions found among authors concerning the possibility of eluting certain complexes with or without decomposition thus can often be explained by the influence of the surface of the solid support.

Veening et al prepared and analyzed cobalt and ruthenium trifluoroacetylacetonates on SE-30 at 110°C without observing decomposition[2416]. The two complexes are perfectly separated, but the retention times of $Co(tfa)_3$ and $Cr(tfa)_3$ are too similar to permit complete resolution of the two peaks on the column used.

Vanadyl trifluoroacetylacetonate ($VO(tfa)_2$) has been eluted from a column packed with glass beads coated with silicon oil DC-710, between 90 and 130°C[2417]. The response of the flame ionization detector is linear, but the calibration curve (peak area vs sample size) does not go through the origin; the chelate is probably either partly decomposed or irreversibly adsorbed in the column. The same authors have shown that Co(II) and Ni can be analyzed in aqueous solutions by gas chromatography of the complexes made with trifluoroacetylacetone and dimethylformamide (DMF). The two metallic chelates, $M(tfa)_2$ and $M(DMF)_2$, could not be separated on the column used, but both are quantitatively eluted so the determination of each metal is possible[2418]. Dimethylformamide replaces the hydration water molecules of the hydrated chelate, which allows a better vaporization of the complex.

The use of temperature programming often improves the separations. Figure 150 shows a chromatogram obtained by programming the column temperatures between 100 and 135°C. It is interesting to note that, under these conditions, beryllium, aluminium and scandium are perfectly separated. Thus the method permits their analysis in rare earth ores where they are often present together.

It seems difficult to separate the copper and iron trifluoroacetylacetonates completely on Se-30 even with temperature programming[2389]. In contrast, very good resolution of the two complexes is obtained at 92°C on polyethylene deposited on glass microbeads[2420] or on Chromosorb W impregnated with XE-60 or Daifloil 200[2419]. The iron and copper complexes are eluted in this order and separated from the gallium complex[2419].

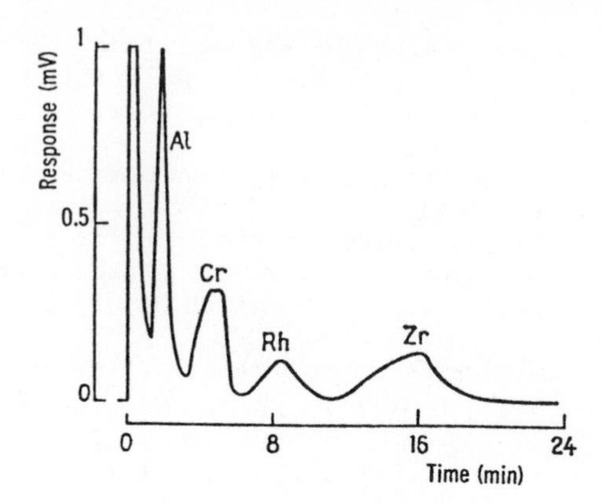

Figure 148 - Separation of trifluoroacetylacetonates. Conditions: 1.20 m long column; 0.5% silicone grease DC on glass microbeads of 60 - 80 mesh; temperature, 135°C; helium flow rate, 130 ml/min: chelates in C_6H_6 solution.

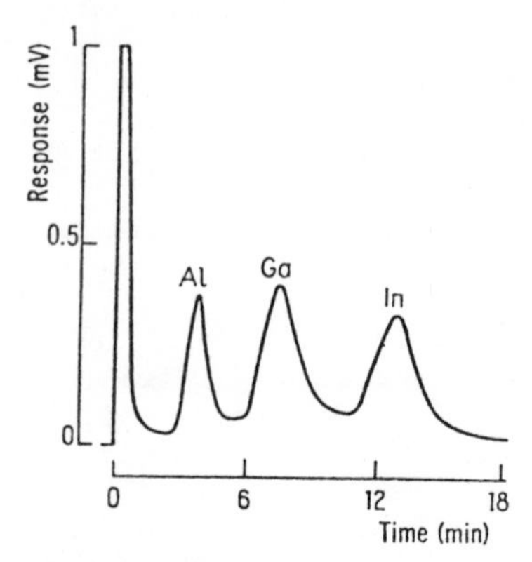

Figure 149 - Seperation of trifluoroacetylacetonates. Conditions: 1.20 m long column; 0.5% silicone oil DC-710 on glass microbeads of 60 - 80 mesh; temperature, 120°C; helium flow rate, 79 ml/min; chelates in C_6H_6 solution.

The retention of trifluoroacetylacetonates depends highly on the nature of the surface of the solid support used and its degree of impregnation; Figure 151 shows the chromatograms of the chromium complex obtained on bare Chromosorb W with different degrees of silanization. Adsorption on the solid decreases when the amount of silanizing reagent used increases, although it can be demonstrated that adsorption cannot be neglected even with 10 - 15% of impregnation by Apiezon L or squalane[2421]. It is probably less significant when polar phases such as polyesters are used. It seems, however, that the best packing for the analysis of these compounds consists of glass microbeads with low impregnation.

Quantitative analysis of trifluoracetylacetonates. By performing the operations under well-defined conditions, a certain selectivity in the

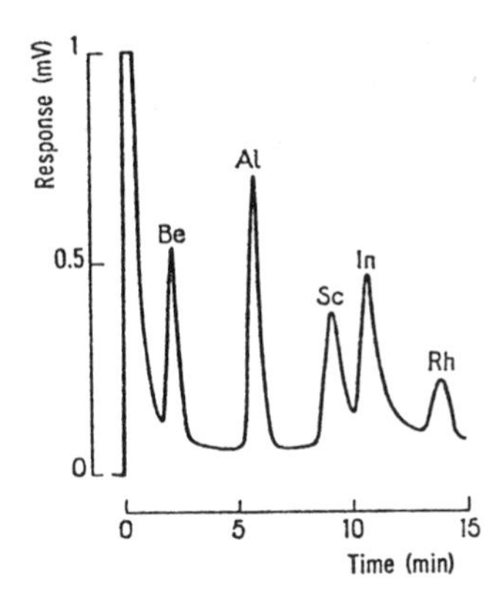

Figure 150 - Separation of trifluoroacetylacetonates. Conditions: 1.20 m long column; 0.5% silicone oil DC-710 on glass microbeads of 60 - 70 mesh; temperature programming, between 100 and 135°C at 3°C/min; helium flow rate, 83 ml/min; chelates in benzene solution.

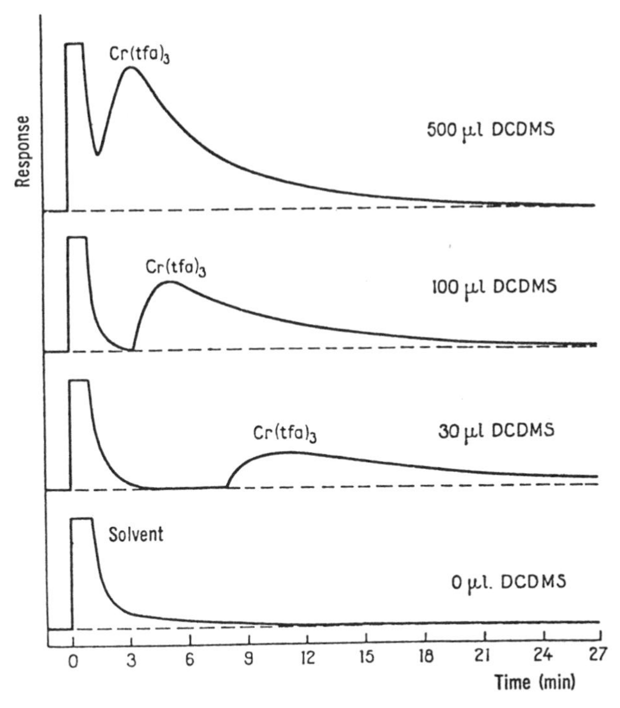

Figure 151 - Chromatograms of Cr(tfa) obtained on unimpregnated Chromosorb W with different degrees of silanization. Conditions: 1 m long glass column; temperature, 120°C; carrier gas velocity, 4.0 cm/sec; injections of 30 μg $Cr(tra)_3$ in solution of 3 μl CCl_4.

metal ion chelation and extraction of the formed complexes permits the elimination of the metals which may interfere in the gas chromatographic analysis.

The aluminium, copper and iron chelates can be extracted quantitatively from an aqueous solution with chloroform at about pH 4 [2421,2423]. Under these conditions, the manganese, nickel zinc and magnesium complexes are not extracted. Solvent extraction and separation by gas chromatography thus are combined to furnish a very interesting analytical method. The mean relative error here is 1.39% for aluminium 0.19% for iron and 0.89% for copper[2423].

Furthermore, the aluminium, gallium, and indium trifluoroacetylacetonates extracted with benzene can be analyzed by gas-liquid chromatography on silicone oil DC-550[2424]. The relative error in the determination of the three metals in the mixture is 1 - 4%. Aluminium and iron are separated and determined under similar conditions with an accuracy in the same order of magnitude[2425,2426]. Interference from copper and nickel can be avoided by complexing these metals with picolinic acid.

The scandium, beryllium, and aluminium trifluoroacetylacetonates are separated and analyzed quantitatively on silicone oil at 170°C[2395]. The response of the katharometer used as a detector in this study is linear for between 1 and 10 µl of $Sc(tfa)_3$ solution, whether the chromatographic peak heights or areas are determined.

Trace analyses of trifluoroacetylacetonates. Gas chromatography of trifluoroacetylacetonates with the use of an electron capture detector, which is very sensitive for halogenated compounnds, lends itself particularly well for the analysis of metals of the trace level. Table 209 compares the detection limits obtained in the analysis of beryllium, chromium, rhodium, aluminium, cobalt, palladium and nickel by gas-liquid chromatography combined with solvent extraction and by other physical methods[2427-2430]. The numbers shown do not represent the ultimate limit either for gas chromatography or the other techniques because improvements are constantly being made. They are nonetheless useful for comparative purposes. It can be seen from these data that the high sensitivity of the gas chromatographic approach is one of several important advantages of the technique. For some elements, gas chromatography is more sensitive than any other methods. The detection limits listed for the gas chromatographic method have all been experimentally established with no extrapolations or assumptions. They are conservative values that will doubtless improve with advances in instrumentation and technique, just as the sensitivity of the other methods continues to improve.

With such concentrations, extreme precautions must be taken in the extraction of the complex with the organic solvent. An excess of chelating agent avoids adsorption of the metal compound on the vessel surface. However, the electron capture detector has a high response factor for trifluoroacetylacetone; consequently in order to avoid interference between its peak and the peaks of the chelates, the excess of complexing agent is eliminated by washing the organic phase with dilute sodium hydroxide solution. The use of tracers such as ^{9}Be shows that about 1 - 2% of the beryllium is retained on the vessel walls. The calibration curves can be determined either on the basis of the peak heights or with an internal standard, for example, hexachloroethane. The direct method is simpler but requires a more precise injection system and more careful control of the equipment parameters (tempatures, pressures, flow rates).

Table 209 - Gas chromatography of metal chelates

Comparison of absolute detection limits (in g) with other methods of analysis.

Metal	Gas Chromatography	Atomic Absorption	Neutron Activation	Emission Spectrography	Spark-source mass spectroscopy
Be	4×10^{-13}	1×10^{-8}	-	2×10^{-10}	8×10^{-12}
Cr	2×10^{-14}	2×10^{-9}	1×10^{-6}	1×10^{-9}	5 c 10^{-11}
Rh	2×10^{-12}	2×10^{-8}	5×10^{-11}	2×10^{-6}	9×10^{-11}
Al	7×10^{-11}	1×10^{-6}	1×10^{-9}	3×10^{-9}	2×10^{-11}
Co	1×10^{-11}	2×10^{-9}	5×10^{-10}	1×10^{-6}	5×10^{-11}
Pd	1×10^{-10}	3×10^{-8}	5×10^{-11}	1×10^{-6}	5×10^{-10}
Ni	1×10^{-11}	2×10^{-9}	5×10^{-9}	8×10^{-8}	7×10^{-11}

The procedure for analyzing a metal in aqueous solution thus consists of the following steps:

1. add an excess of organic solution of the chelating agent;
2. adjust the pH;
3. perform the extraction;
4. wash the organic phase with dilute sodium hydroxide solution;
5. inject an aliquot.

This method makes it possible to analyze the chromium present in serum after decomposition of the latter by nitric acid[2431].

Direct reaction of the chelating agent with the alloy or metal compound, when possible, simplifies this process. Trifluoroacetylacetone can react with chromium and iron at an acceptable rate in the presence of catalytic quantities of an inorganic acid. A method for analysing chromium in iron alloys by this technique has been described[2432] in which the direct reaction of the chelating agent is accelerated by radio-frequency waves. Furthermore, beryllium trifluoroacetylacetonate is formed quantitatively in a fused tube by the direct reaction of trifluoroacetylacetone with biological liquids containing Be^{2+} [2433].

Traces of aluminium in uranium (a few ppm to 0.1 ppm) have been analyzed by gas chromatography after benzene extraction of the complexe[2434]. The extraction yield from the uranyl nitrate solution is 100% at a pH of about 6. The excess of trifluoroacetylacetone is removed by washing with 10^{-2} M ammonia solution. The only elements which might interfere with aluminium are beryllium and gallium, but with a sufficiently long column the chelates are completely separated.

Work has been done to improve the techniques of microanalysis, especially for the chelates extraction step. Analysis of picogram amounts of beryllium and chromium in biological samples (blood, plasma)[2435-2437] or mineral samples (lunar rocks and dusts)[2438] have been achieved.

The sensitivity of the electron capture detector has sometimes been matched or even exceeded by the mass spectrometer[2436].

Gosiuk[2439] has described a gas chromatographic method for the rapid simultaneous determination of picogram quantities of alumium and chromium in water as their trifluoroacetylacetonates. The trifluoroacetylacetonates of aluminium and chromium are extracted into benzene from warm buffered aqueous solution. Two μl portions of the benzene extract are then analysed by gas chromatography using an electron capture detector which is very sensitive for the fluorine-containing metal chelates. Results are quoted for standard solutions but no results are given for actual water samples.

Analyses of geometric isomers and trifluoroacetylacetonates. The trifluoroacetylacetonates of trivalent metals exist in two forms of geometric isomers (Figure 152). Gas chromatography, like chromatography on an alumina column[2440] permits the resolution of these isomers. Thus, cis- and trans-trifluoroacetylacetonates of chromium[2371,2410,2441] and rhodium[2427] in benzene solution are separated at 115 and 105°C respectively, on a silicone grease column; the generally more stable trans- form is eluted first. A partial separation of the Ga(tfa)$_3$ isomers can be obtained on XE-60 at 140°C[2419], but the cis- isomer shows the least retention.

Thus, the trifluoroacetylacetonates offer many more possibilities for the analysis of metals by gas chromatography than their nonfluorinated homologs. However, it must be pointed out that, while the volatilities of the former are higher, the separations obtained by fractional sublimation are better with the acetylacetonates[2442].

D. Hexafluoroacetylacetonates

The hexafluoroacetylacetonates are eluted much faster than their trifluorinated homologs and the analyses can be often performed at slightly above ambient temperatures. This is not because they are more volatile, as is often claimed, for their vapour pressure is comparable to that of the corresponding tfa complexes, but because they are much less soluble in most liquid phases[2443].

Thus, chromium hexafluoroacetylacetonate, melting at 84°C and sublimable at room temperature at 0.05 mm Hg, is rapidly eluted at 30°C. To obtain the elution of the acetylacetonate in the same period of time, the column temperature must be increased by 100°C[2371]. At 30°C, Cr(hfa)$_3$ and Ru(hfa)$_3$ are separated with retention times of 8 and 16 min on a 1.20 m column packed with silicone grease coated glass microbeads. The aluminium and chromium complexes are separated at 65°C on silicone grease DC-710 and are eluted before the solvents used - toluene or chloroform[2444] while on SE-30 at 50°C, carbon tetrachloride is eluted before the beryllium, aluminium and chromium hexafluoroacetylacetonates which are well seperated[2389].

These few data show the extraordinary volatility of these metal derivatives. According to Moshier and Sievers[2407] it is sufficient for a compound to sublime rapidly at 0.05 mm Hg at the column temperature in order to be elutable under acceptable conditions. This comes from the fact that these fluoro compounds have a very small solubility in many organic phases, with very large activity coefficients. It should then be possible to analyse the Fe (II), Ni (II), Co (II), Mn (II), Cu (II) and Th (IV) hexafluoroacetylacetonates by gas chromatography. Unfortunately, it seems that the majority of these complexes form hydrates, and the effect of water molecules, although not well known, may interfere with elution[2445].

In contrast, the solvates of rare earth hexafluoroacetylacetonates formed with tributylphosphate are volatile and more stable than the unsolved

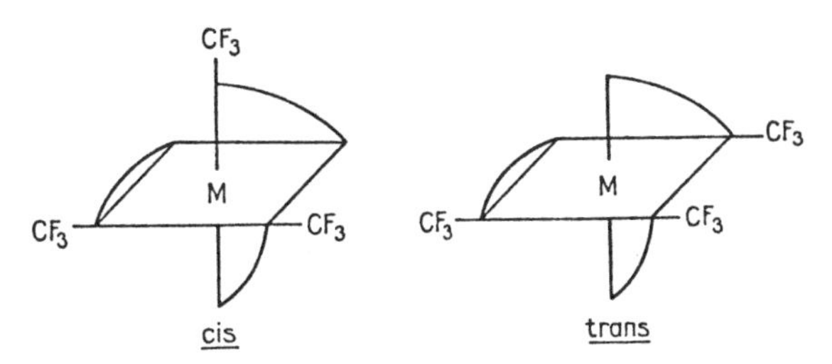

Figure 152 - Geometric isomers of octahedral trifluoroacetylacetonates.

chelates since it has been possible to elute them on silicone SE-30 between 150 and 200°C[2446]. The addition of tributylphosphate to the aqueous hexafluoroacetylacetonate solution furthermore results in better extraction of the complex by hexane.

The beryllium, aluminium, copper and chromium hexafluoroacetylacetonates have been separated at 60°C on QF-1, and the aluminium, beryllium, chromium and iron complexes on Fluorolube HG-1200[2447]. It may be noted that the normal elution order of the beryllium and aluminium chelates is reversed on the latter phase. The same study reports very good separation of iron (III) and copper (II) complexes on SE-52 at 30°C with retention times of about 2 and 15 min respectively.

Quantitative determinations of chromium[2389,2448], beryllium, aluminium [2389,2388] cobalt, and ruthenium[2416] from the hexafluoroacetylacetonates have been obtained by gas chromatography. Hexafluoroacetylacetone can directly react with certain metal chlorides such as aluminium trichloride[2449], thus simplifying the sample preparation procedure. With other chlorides such as titanium tetrachloride, chelation is incomplete:

$$TiCl_4 + 2H(hfa) \rightarrow TiCl_2(hfa)_2 + 2HCl$$

The mixed compounds formed are volatile and can be easily eluted in gas chromatography[2450] but, like the chlorides, they are very sensitive to hydrolysis.

The hexafluoroacetylacetonates of trivalent cations have two optical isomers (Figure 153) which should not be confused with the geometrical cis- and trans- isomers of trivalent metal trifluoroacetylacetonates (see Figure 152). The seperation of these optical isomers by chromatography requires an optically active retention phase so that a preferential adsorption or dissolution of one of the species can be obtained. Gas-solid chromatography on a 3.6 m column packed with dextrorotatory quartz permits a partial resolution of the two forms of $Cr(hfa)_3$ at 55°C[2410,2441,2451], but gas-liquid chromatographic experiments with d-dibutyltartrate have been fruitless since the complex is solvated too highly by the stationary phase and is not eluted[2371].

The hexafluoroacetylacetonates do not have geometrical isomers, but mixed chelates can form with mixtures of trifluoro- and hexafluoroacetylacetone. There are seven chromium compounds with the formula $Cr(tfa)_x(hfa)_{3-x}$ two isomers for x = 3, three cis-cis, cis-trans, and trans-cis forms for x = 2; and a single geometrical form for x = 1 and x = 0. It seems that gas-liquid chromatography can resolve these mixtures in the following elution order: $Cr(hfa)_3$, $Cr(hfa)_2(tfa)$, three peaks for $Cr(tfa)_2(hfa)$, then trans- and cis-$Cr(tfa)_3$[2452,2453]. However, the authors do not show a chromatogram or give any details on this seperation.

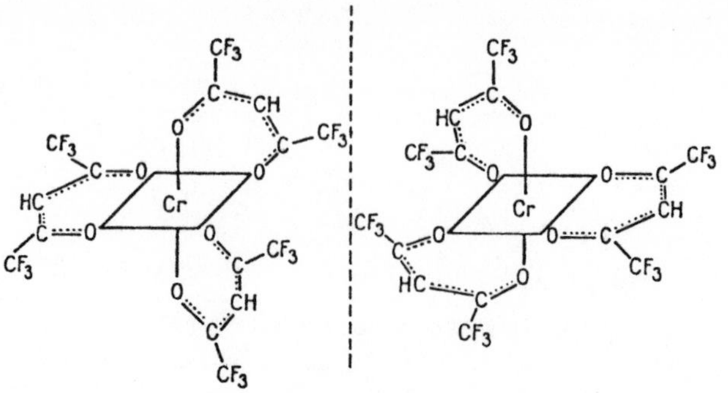

Figure 153 - Optical isomers of chromium hexafluoroacetylacetonate.

Gas chromatography has also been used to measure the kinetics and equilibrium constant of the reactions of formation of these mixed complexes. The most widely studied system has been:

$$Al(acac)_3 + Al(hfa)_2(acac) \rightleftarrows 2Al(acac)_2(hfa).$$

These compounds, all volatile, are separated on silicone grease with temperature programming between 65 and 135°C and are eluted in the order of decreasing degree of fluorination[2454].

Studies devoted to the chromatography of hexafluoroacetylacetonates have remained less numerous than those published on trifluoroacetylacetonates. The extraordinary volatility of the former is a great advantage, but hexafluoroacetylacetone reacts with water to form 1,1,1,5,5,5-hexafluoro-2,2,4,4-tetrahydroxypentane. This property may be undesirable during solvent extraction of the metal ions in aqueous solution and the method described above with trifluoroacetylacetone is often preferable.

E. β-Diketone Derivatives

Acetylacetone and its fluorinated derivatives permit a number of metals to be converted into volatile compounds, allowing their analysis by gas chromatography; however, this method is powerless with alkali, alkaline-earth metals, and especially with the lanthanides, which often present difficult analytical problems. Various other chelating agents have been used in the attempt to fill this gap.

Interesting results have been obtained by replacing the methyl groups in acetylacetone by t-butyl or perfluoroalkyl radicals. A thermogravimetric study of a number of β-diketonates of metals shows that the volatility of the fluorinated chelates is higher than that of unfluorinated species - a phenomenon already pointed out for the acetylacetonates. On the other hand, substitution of the methyl groups of acetylacetone by phenyl radicals or of a hydrogen atom in position 3 by bromine decreases the volatility of the complex[2455].

Dipivaloylmethane. Fluoro derivatives of dipivaloylmethane $(CH_3)_3C.CO.CH_2CO.C(CH_3)_3$ have been used in order to form fairly volatile chelates with alkali metals[2456,2457].

Sievers et al prepared and investigated complexes of scandium, yttrium, and thirteen lanthanides formed with 2,2,6,6-tetramethyl-3,5-heptanedione (dipivaloylmethane)[2458,2459]. These compounds $M(thd)_3$ sublime rapidly between 100 and 200°C in vacuum without apparent decomposition; their retention volumes on Apiezon greases vary regularly with the ionic radius of the metal (Figure 154). Figure 155 shows a separation obtained at 157°C on Apiezon M. The chelates prepared in ethanol solution from the nitrates $M(NO_3)_3$, $6H_2O$, are injected in benzene solution. Their vapor pressures as well as their vaporization and sublimation enthalpies have been determined[2460].

The physicochemical behaviour of the ytterbium complex $Yb(thd)_3$ has been studied on Apiezon L and QF-1[2461]. A knowledge of the excess molar enthalpy and entropy, of interest from the standpoint of the thermodynamics of the solutions, may also be useful in the choice of the stationary phases and chelating agents likely to furnish the best results in a given analysis.

Trifluoroacetylpivaloylmethane. The chelates of scandium, yttrium, and eleven rare earths with 1,1,1-trifluoro-5,5-dimethyl-2,4-hexanedione (trifluoroacetylpivaloylmethane H[tpm]) have been synthesized and then eluted on silicon grease SE-30[2462-2464]. According to the thermogravimetric curves presented by the two groups of researchers, all complexes except that of neodymium are volatilized above 200°C without decomposition. The scandium lutecium erbium, dysprosium and samerium derivatives in n-hexane solution are separated, and an essentially linear variation of the retention volumes with the metal cation radii is obtained[2463,2464].

A very good separation of aluminium, chromium, and iron complexes $M(tpm)_3$ in chloroform solution has been obtained on Apiezon L at 152°C[2465]. A quantitative analysis of Cu (II), Fe (III), and Al (III) chelates is possible on a column containing silicone oil SE-30[2466].

The lead chelates, as well as the similar complexes derived from 1,1,1,2,2,3,3-heptafluoro-7,7-dimethyloctanedione-4,6, have been eluted from a column packed with Univeral-B impregnated with Apiezon[2467]. However, the peaks are large, tailing badly, and there is no detector signal with samples smaller than 50 - 100 μg which establish that strong adsorption occurs in the column.

In general, the chromatographic peaks are less symmetrical and the column efficiency is not so good with all of these metal compounds than with organic compounds of comparable volatility. An attempt to extrapolate the analytical studies to a preparative scale has shown that the injection of large amounts of metal chelate samples on highly loaded stationary phases (10% or more) results in abnormal phenomena[2465]. In addition to tails, which essentially are due to adsorption on the support and have been reported earlier[2421], a displacement effect is observed; the injection of the iron complex several hours after normal elution of the analogous aluminium complex produces a peak of the latter[2465]. Contrary to the tails, this phenomenon does not diminish when a silanized support or Teflon powder is used, but increases with the polarity of the liquid phase. It would thus seem that very strong and practically irreversible interactions occur between the chelates and the retention phases. Several types of β-diketonates and several liquid phases have been studied from this point of view. The interactions are weakest between triflouroacetylpivaloylmethane complexes and Apiezon L. Moreover, the peaks of iron chelates show a shoulder even when a column is used into which no other compound had been injected previously, as if a small part of the sample underwent only incomplete interaction with the liquid phase and were eluted before the rest, according to the authors

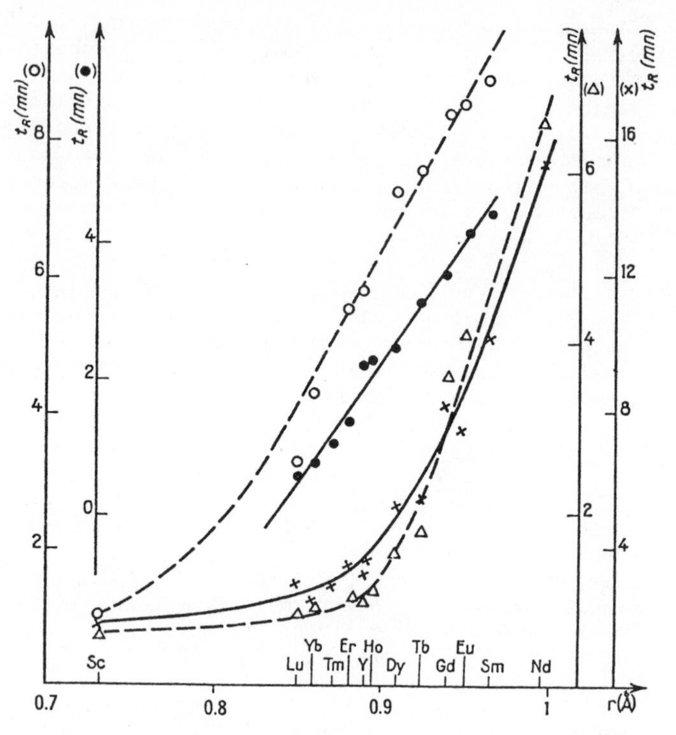

Figure 154 - Relations between the retention times of rare earth chelates and the metal cation radius.
Chelates formed with:
x dipivaloylmethane (according to [66])
o trifluoroacetylpivaloylmethane (according to [71])
O trifluoroacetylpivaloylmethane (according to [72])
1,1,1,2,2,3,3-heptafluoro-7,7-dimethyl-4,6-octanedione (according to [79]).

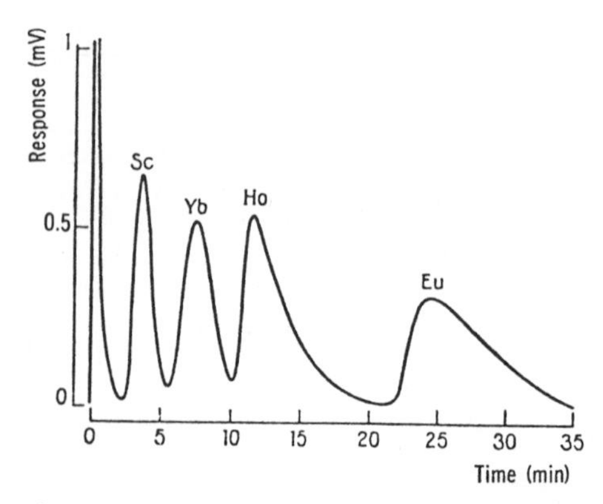

Figure 155 - Separation of rare earth chelates $M(thd)_3$. Conditions: 15 cm long Teflon column; 2% Apiezon H on Gas-Pack (60 - 80 mesh); temperature, 157°C; helium flow rate, 100 ml/min; chelates in C_6H_6 solution.

2465. A more thorough study of these phenomena should be made, however, before such an explanation is accepted.

Heptafluorodimethyloctane dione. 1,1,1,2,2,3,3, heptafluoro-7,7 dimethyl -4,6 as 1,1,1-trifluoro-2 octane dione (tfa) has proved to exhibit extraordinary properties which aid in the analysis of ultra-trace amounts of metal. These compounds can be used readily to extract metals from aqueous solution by solvent extraction methods (Morie and Sweet[2424,2425], Moshier and Sievers [2409], Savory et al[2468,2469], Scribner et al[2470], Scribner et al[2471], Gentry et al[2472]). These ligands will react directly (without another solvent) with most metals and with many of the metal salts (Ross and Sievers[2432], Sievers et al[2473] and Taylor et al[2433]).

Sievers et al[2473] studied the reaction of H fod (1,1,1,2,2,3,3 heptafluoro 7,7-dimethyl 4,6 octanedione) with various metals, alloys and ores. The metal and excess reagent were heated at 170°C in a sealed tube under dry air, i.e. a microreactor technique. Reaction times extended between 2 and 8 hours. The following substances reacted completely producing the colours indicated: iron and iron ores (deep red), chromium (violet), zirconium and hafnium (orange), manganese (dark brown), beryllium and beryllium oxide, titanium alloys (dark blue green), zinc (colourless, gassing) and lanthanum (amber). The following substances reacted if 2 mg water was included in the ampoule: titanium (blue to amber), aluminium (light yellow), gallium and indium (metallic globules), scandium (yellow) and vanadium (dark red). The following substances either underwent incomplete reaction (copper, molybdenum, silicon, tin ores, nickel, bauxite, bismuth and cobalt) or did not react at all (brass, tungsten, magnesium, niobium, tantalum, rhenium and cadmium).

The reaction of iron with H fod is very rapid and is accompanied by the production of hydrogen suggesting the following reaction.

$$Fe + 3\,H(fod) \rightarrow Fe(fod)_3 + {}^3/_2\,H_2$$

This is in sharp contrast to the reaction with acetylacetone for which very little reaction occured unless oxygen is present.

The need for water to be present in the reaction of scandium suggests that these reactions are water catalysed. Unlike the cases considered above, the reaction of metal oxides with 1,1,1,2,2,3,3 heptafluoro -7,7 dimethyl -4,6 octanedione (H(fod)) does not ordinarily involve an oxidation-reduction process. The reaction of beryllium oxide with this chelating reagent is illustrative of what might be considered a simple neutralization reaction:

$$BeO + 2H(fod) \rightarrow Be(fod)_2 + H_2O$$

Application of the solvent extraction/gas chromatographic combination has been largely limited to those metal ions which form co-ordination saturated chelates with the ligand. Divalent metals such as zinc and nickel extracted very poorly into chloroform-trifluoroacetylacetone mixtures presumably because of hydrate

1,1,1,2,2,3,3-heptafluoro-7,7-dimethyl-4,6-octanedione, H(fod), can react directly with a large number of metals, alloys, and metal compounds. The preparation of samples for gas chromatographic analysis is possible without going through an aqueous organic solution, and a very small quantity of sample is needed. The latter is weighed into a glass capillary tube, closed at one end, and an excess of chelating agent is added. The capillery is then fused at its other end, heated, and broken in the injection port of a chromatograph.

Be (II), Al (III), Pd (II), Fe (III), Cu (II), Cr (III), and Y (III) complexes are vaporized at a temperature below 200°C without decomposition but $Ni(fod)_2,2H_2O$ which sublimes at about 220°C, leaves a small residue[2473]. The retention volumes relative to n-hexadecane at 170°C on SE-30 are 0.36 (Be), 1.17 (Al), 1.17 (Cu), 1.29 (Cr), 1.11 (Fe), 1.50 (Y), and 1.90 (Pd). The extreme volatility of these compounds which are eluted shortly after hexadecane, will be noted while the trivalent metal complexes contain 27 carbon atoms and numerous heavy atoms (molecular weight = 870 for the aluminium complex and 226 for hexadecane). On Apiezon L, elution of the nickel complex is more reproducible, and the cis- and trans-isomers of $Cr(fod)_3$ can be resolved.

The cis- and trans- isomers of $Co(fod)_3$ have been separated by gas chromatography, and the detection limit of cobalt is in the order of 10^{-11} g with an electron capture detector[2474]; it has been possible to analyse this metal in vitamin B-12.

Thorium and uranium chelates have been prepared from the tetrachlorides of these metals; their composition as well as that of the products eluted from the column corresponds to the formula $M(fod)_4$, and they can be separated at 170°C on glass microbeads coated with silicone oil QF-1[2475].

The effect of pH on the extraction yield of the two metals as fod chelates has been studied, and a procedure for the quantitative analysis of uranium and thorium in aqueous solutions described. The quantitative analysis of uranium is possible, however, with $U(fod)_6$ and not with $U(fod)_4$, which makes it necessary to oxidize U (IV) into U (VI) before preparation and extraction of the complex. Analysis of transuranium elements by chromatography of their complexes can now be expected.

The $M(fod)_3$ complexes of scandium, yttrium, and thirteen rare earths have been synthesized[2476]. All except the scandium complex are isolated in the form of monohydrates and are easily dehydrated on phosphorus pentoxide in vacuum. Thermogravimetry shows that they are volatilized between 150 and 220°C without decomposition. On silicone oil SE-30, their retention volume varies with the ionic radius of the metal in a manner similar to that observed in the $M(thd)_3$ complexes (see Figure 154).

Figure 154 summarizes the data of [2458,2463,2464,2476] on the variation of the retention times t_R of rare earth complexes with the ionic radius of the metal. Even when the scandium and neodymium complexes which were not studied by the Japanese authors are neglected, it is difficult to assume the curves relative to the chelates analysed by Sievers et al to be straight lines. For these complexes, the variation of log t_R is no longer a straight line; the increase of the retention time with the radius of the cation M^{3+} is more rapid than an exponential increase.

Both 1,1,1 trifluoro-2,4 pentanedione and 1,1,1,2,2,3,3, heptafluoro-7,7, dimethyl-4,6 octanedione form relatively stable complexes, and the presence of fluorocarbon groups in the molecule results in compounds which respond extremely well to electron capture detection on the gas chromatographic column. Further advantages of the direct reaction technique are reduction in analysis time and reduction of reagent blank values because of the need for fewer reagents to dissolve metals and convert them to their metal chelates. Detection limits of considerably less than 1 picrogram have been established for beryllium and chromium trifluoroacetylacetonates, Ross and Sievers[2426,2428,2429]. Detection limits for the 1,1,1,2,2,3,3, heptafluoro-7,7, dimethyl-4,6 octanedione chelates are about an order of magnitude less, but these compounds appear, in some cases, to be more stable in the chromatographic column. 1,1,1,5,5,5, hexafluoro-2,4 pentanedione (hfa) chelates have even lower detection limits than the trifluoroacetylacetonates but quantitative solvent extractions are not effected as readily.

Bis(acetylpivaloylmethane) ethylene diamine. Belcher et al[2477] have described a method for the simultaneous determination of copper and nickel by gas chromatography of a solvent extract containing these metal chelates.

Other β-diketones. The chelates of diisobutylmethane (2,6-dimethyl-3, 5-heptanedione) and of the divalent (copper, beryllium), trivalent (aluminium chromium, scandium, and iron), and tetravalent (thorium) cations have been synthesized and analyzed by gas chromatography on silicone grease (Figure 156)[2478]. In general, the volatility of these complexes is intermediate between that of the acetylacetonates and that of the dipivaloylmethane complexes and thus increases with the steric hindrance of the chelating group. The derivatives obtained by the reaction of an aqueous rare earth nitrate solution and an alcoholic solution of diisobutylmethane [H(dibm)] have a composition corresponding to the formula $M(dibm)_2OH$. They cannot be volatilized without decomposition, and only the lutetium and erbium chelates give a well-defined chromatographic peak. The yttrium, scandium and rare earth chelates of 2,2,6-trimethylheptanedione 3,5 have been analysed on silicone oil DC between 210 and 280°C[2488].

The cobalt, nickel, and palladium monothioacetylacetonates have been prepared in chloroform solution and separated on Apiezon grease with temperature programming between 170 and 220°C (Figure 157)[2479]. The peaks are symmetrical and no decomposition is observed even at 240°C. The stability and volatility of these chelates are well above those of the acetylacetonates and the study of other metal complexes might therefore be of certain interest. Traces of nickel have been analysed as monothioacetylacetonate: the detection limit with an electron capture detector was 5×10^{-11} g[2480]. The chelates of monothio- β-diketones with a wide range of bivalent metals, including nickel, cobalt, zinc, palladium and platinum, are stable and volatile and have been chromatographed successfully[2481,2482].

In general though, those chelates are difficult to prepare. A notable separation is nickel which can be determined down to 10^{-11} g as its monothiotrifluoroacetylacetonate[2483].

Some chelates can be analyzed only in special conditions under very high pressures; thus, the nickel, copper (II) and zinc salicyladoximates have been eluted on Kel-F oil at 130°C under a pressure of 1000 psi, as has been done for the thenoyltrifluoroacetonates of the same metals and of trivalent (Sc, Y, Eu, Al, Fe, Co) and tetravalent metals (Zr, Th)[2484].

Aluminium was determined in the form of 2-thenoyltrifluoroacetonate on Apiezon L at 250°C with a katharometer as the detector[2485].

Among the other chelating agents used for gas chromatography of metal chelates, we should still mention decafluoro-2,4-heptanedione, octafluoro-2,4-hexanedione, and hexachlorobicycloheptene-2-methylenethiosemicarbazide[2486]. The Cu (II) and Ni (II) complexes of β-ketoimine derivatives of 2,4-pentanedione and of salicylic aldebyde have been analyzed on Apiezon L and SE-30 at between 150 and 190°C[2487].

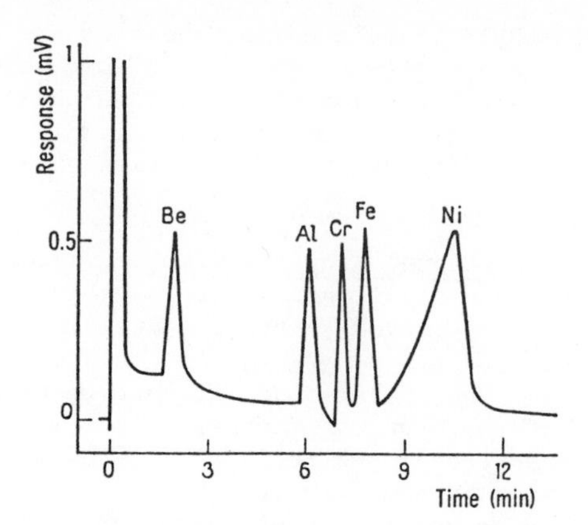

Figure 156 - Separation of diisobutyrylmethane chelates.
Conditions: 75 cm long column packed with Chromosorb W (60 - 80 mesh) impregnated with 5% silicone grease DC; column temperature programmed from 140 to 250°C at 8°C/min; injector temperature 300°C; helium flow rate 45 ml/min.

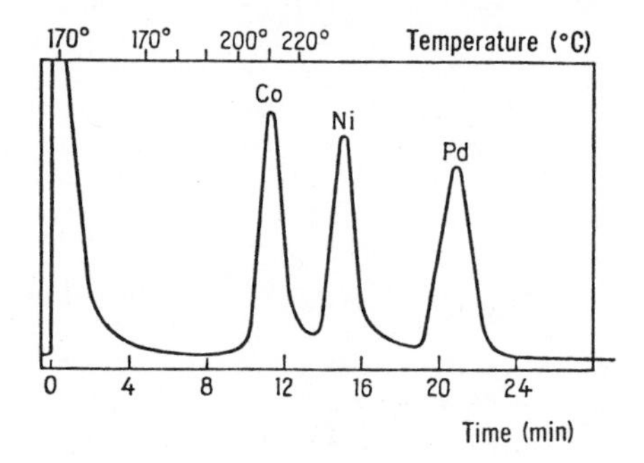

Figure 157 - Separation of nickel (II), cobalt (II) and palladium monothioacetylacetonates.
Conditions: Teflon 90 cm long column, 5% Apiezon on Univeral B (60 - 80 mesh); column temperature programmed from 170 to 220°C; injector temperature, 240°C; nitrogen flow rate, 120 ml/min; complexes in $CHCl_3$ solution.

The thenoyltrifluoroacetylacetonates of divalent (beryllium, cobalt, nickel and copper), trivalent (aluminium, scandium and chromium), and tetravalent (vanadium) metals have been eluted through columns packed with glass beads coated with silicone oil DC 710 ([2489,2490]). Some separations have been achieved (beryllium, aluminium and chromium). The chelates of nickel and cobalt give tailing peaks. This situation is improved by recrystallizing the hydrated complexes in diethylamine: the mixed chelates give more symmetrical peaks.

Chelates of many divalent metals (copper, zinc, iron, nickel, palladium, platinum, cadmium and lead) with monothiotrifluoroacetylacetone are volatile enough to be analysed by gas chromatography[2491]. The chelates of nickel, palladium and platinum have been resolved on a glass column packed with Chromosorb W impregnated with 3% OV-17.

Bis-acetylacetone ethylenediamine has been used to prepare chelates of transition metals[2492].

Uranium VI and thorium IV have been separated as mixed complexes $UO_2(hfa)_2DBSO$ and $Th(hfa)_4DBSO$, with hexafluoroacetylacetone and di-n-butylsulfoxide[2493]. Similar complexes have also been prepared with trifluoroacetylacetone and 1,1,1,2,2,6,6,7,7,7-decafluoroheptanedione-3,5. These complexes have been eluted at 200°C on columns packed with Chromosorb W impregnated with SE-30 and QF-1.

The lithium and sodium derivatives of hexafluoroacetylacetone or of 1,1,1-trifluoro-5,5-dimethyl 2,4-hexanedione can be rapidly sublimed above 185°C without apparant decomposition. Elution of these compounds on SE-30 at about 200°C however, has not been successful[2494] perhaps because of too high a reactivity of these chelates for the liquid phase.

The complexes of 1,1,1,2,2-pentafluoro-6,6-dimethyl-3,5-heptanedione and 1,1,1,2,2,3,3-heptafluoro-7,7-dimethyl-4,6-octanedione prepared from aqueous hydroxide solutions and extracted with ether or ethyl acetate have been sublimed at 0.1 torr between 140 and 190°C[2495]. Their mass spectra were obtained at about 200°C. They were eluted at the same temperature from a silicone rubber E-30± column. The retention times of the two derivatives of the same metal are practically identical, while those of the chelates of the three investigated alkali metals (lithium, sodium and potassium) differ noticeably. However, no peak resolution could be observed with mixtures.

The mixed complexes with the formula $M'ML_4$ (M = rare earth, M' = alkali metal) in which the complexing agent L is the anion of trifluoroacetylacetone, hexafluoroacetylacetone, thenoyltrifluoroacetone or 1,1,1-trifluoro-5,5-dimethyl-2,4-hexanedione have been synthesized, but their volatility appears to be insufficient[2496].

The chelates formed between alkaline earth cations and tetramethylheptanedione have been eluted on Apiezon L at about 250°C, but mixtures of these complexes cannot be resolved because of the formation of heteronuclear polymerized species[2497]. The chromatogram of a mixture of calcium and strontium chelates thus has only a single peak corresponding to a compound formed by the reaction:

$$[Ca(thd)_2]_2 + [Sr(thd)_2]_2 \rightarrow 2CaSr(thd)_4.$$

F. Amino Substituted β-Diketones.

Amino-substituted β-diketones also show potential as chelating agents [2498]. Dilli and Maitra[2500] developed a simultaneous gas chromatographic method for the analysis of copper, nickel and vanadium at ultra trace levels as chelates of fluorinated Schiffs reagent; (1,1,1,1',1',1',-hexafluoro-4,4'-(1-methylethane-1,2-diyldiimino)bis(pent-3-en-2-one). The reagent reacts readily with copper and nickel and gives chelates whose detection limits by gas chromatography are about 5 pg. The reagent does not chelate readily with vanadium so that the method cannot be applied to that metal. The appropriate column components and chromatographic conditions are discussed. The presence of excess ligand reagent caused problems. The copper chelates tended to be less stable than those of nickel.

Bidentate and tetradentate β-aminoketonates have also found favour for some elements (copper, nickel, platinum and palladium)[2501,2502]; the tetradenates show excellent stability and chromatographic behaviour, and have been investigated[2503] for the determination of copper and nickel at the picogram level, with fluorinated derivatives and electron capture detection.

It is now well established that in terms of thermal stability and gas chromatographic applicability, tetradentate β-ketoamine chelates have many superior features for the analysis of divalent transition metals e.g. copper, nickel, palladium[2508-2512] and, recently, vanadium[2513].

Khalique[2514] have described the preparation of the Schiff's bases derived from 1,2-diaminoethane and a series of 5-alkyl substituted-1,1,1-trifluoropentan-2:4-diones, including the elusive compound obtained from 1,1,1-trifluoro-5,5,5-trimethylpentan-2:4-dione, and trivially denoted as $H_2[en(TPM)_2]$. Gas chromatography of the copper complexes of these ligands shows increasing retention times as the size of the alkyl substituent is increased, with the exception of $Cu[en(TPM)_2]$ which is unusually volatile and has a much reduced retention time. It is most probable that this great difference in properties occurs because of the considerable steric hindrance of the tert-butyl groups in the original β-diketone, which prevents the normal condensation of reaction with 1,2-diaminoethane taking place. The properties of the metal complexes of $H_2[en(TPM)_2]$ make them ideally suited to quantitative gas chromatography.

Khalique et al[2514] prepared the copper chelates of ethylene-bis (trifluoroacetylpivalylmethaneimine, $H_2[en(TPM)_2]$ and other chelating agents, (Table 210).

Table 210 - Structures of ligands prepared and retention times of copper chelates at 220°C

R_1	R_2	Abbreviation for ligand	Retention time(s)
CF_3	CH_3	$H_2[en(TFA)_2]$	567
CF_3	C_2H_5	$H_2[en(TPrM)_2]$	708
CF_3	C_3H_7-iso	$H_2[en(TiBM)_2]$	795
CF_3	C_4H_8-iso	$H_2[en(TPnM)_2]$	1068
C_4H_9-tert	CF_3	$H_2[en(TPM)_2]$	116 [a]
CF_3	C_5H_{11}-iso	$H_2[en(TiHxM)_2]$	1835

[a] At 170°C

Gas chromatography on a 6 ft x 1/8 in o.d. stainless steel column packed with 3% AF.1 on Varaport 30, in order to assess trends in elution order as the substituent groups were increased in size. A temperature of 220°C was chosen as the most suitable for all complexes with the notable exception of $Cu[en(TPM)_2]$ which had an exceptionally short retention time, even at 170°C. Retention times are given in Table 210. Increased size of the substituent groups results in increased chelate retention times, in marked contrast to the behaviour on bridge amyl substitution. This effect persists for both normal and iso-substituent groups.

A typical chromatographic separation and the peak characteristics of both the free ligand and its copper chelate are shown in Figure 158 for $H_2[en(TiBM)_2]$.

It can be seen that resolution of $Cu[en(TFA)_2]$ and $Ni[en(TFA)_2]$ is almost identical with than between $Cu[en(TFA)_2]$ and $Ni[en(TFA)_2]$ on the QF.1 column which is optimal for such separations. Retention times for these four complexes are 795 s, 711 s, 567 s and 510 s respectively, and thus there is no advantage for copper - nickel resolutions in using the heavier ligand. Similar or less resolution is noted for other copper - nickel pairs. All chelates appear to be eluted unchanged, even those complexes which are retained on the column for prolonged times.

G. Dialkyldithiophosphates

These chelates have been used for the separation of zinc, nickel, palladium and platinum[2504].

H. Salicylaldimes

This chelating agent has been used for the separation of copper, nickel, and zinc[2505].

I. Diethyldithiocarbamates

Cardwell et al[2515] have reported on the gas chromatography of some volatile metal diethyldithiocarbamates.

D'Ascenzo and Wendlandt[2506] have established that some metal dithiocarbamates are volatile. It was therefore considered of interest to study such chelates thoroughly to determine their suitability for gas chromatographic dithiocarbamate for the gas chromatographic determination of arsenic.

J. Gas Chromatography of Non-chelated Metal Complexes

Although the majority of work on the gas chromatography of metals has been performed using various chelates, a limited amount of work has been carried out on non-chelated metal complexes. A good example of this approach is the determination of selenium in natural waters using 1,2-diamino-3,5-dibromobenzene[2516].

The total selenium and selenium (IV) contents in sea water and river waver can be determined directly without preconcentration. The reagent reacts only with selenium (IV) to form a 4,6-dibromopiazselenol; other oxidation states of selenium must therefore be converted to the tetravalent state for total selenium determinations. The piazselenol formed can be extracted quantitatively into 1 ml of toluene from 500 ml of sample water. A method is proposed for the determination of selenium (IV) and total selenium in natural waters at levels as low as 2 ng l^{-1}. Coastal sea water and river water in Japan was found to contain 8 - 30 ng of Se (IV) and 20 - 50 ng of total Se per liter.

Table 211 gives the chemical structure of the principal chelating agents other than acetylacetonates and their fluorinated derivatives and also lists some of the metals that have been examined with each chelating agent.

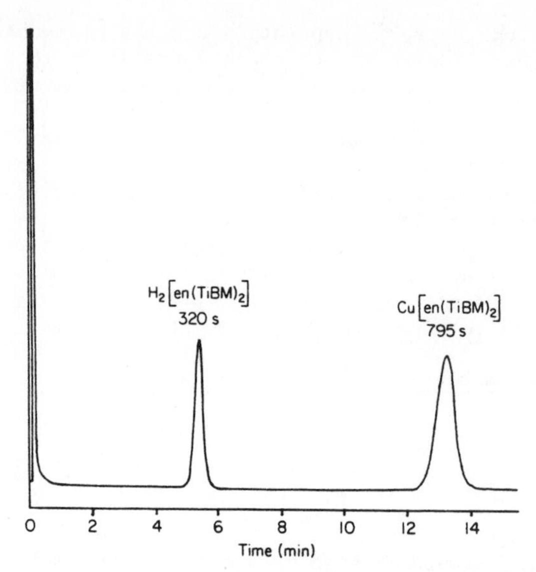

Figure 158 - Gas chromatographic separation of $H_2[en(TiBM)_2]$. Column 6 ft x 1/8 in o.d. packed with 3% QF.1 fluorosilicone oil on Varaport 30, 80 - 100 mesh. Conditions: column, 220°C; injector and detector, 240°C. Helium carrier gas flow, 40 cm^3 min^{-1}.

K. Metal Porphyrins.

Certain metal derivatives cannot be eluted under the usual conditions of gas chromatography because of their insufficient volatility and stability at temperatures where their vapor pressure becomes significant. A technique which is used for the analysis of metallic porphyrins, will undoubtedly permit a solution of these problems, at least for a certain number of families. This is chromatography under very high pressure, using light, more or less polar or polarizable organic compounds as a carrier gas at a temperature above their critical point.

In 1962, Klesper, Corwin, and Turner mentioned the analysis of metallopolyphyrins under supercritical conditions[2543]. Since then, the equipment has been improved[2484] and theoretical studies have been developed[2544,2545]. Dichlorodifluoromethane with a critical temperature of 111.5°C, which is not flammable, is of low corrosiveness, and is a good solvent of porphyrins, serves as the carrier gas. The majority of the conventional detection methods cannot be applied because of the high density of the carrier fluid; consequently, a spectrophotometer with circulation cells is employed.

In addition to difficulties of manipulation at pressures of about 200 atm, the main problem arises in the choice of stationary phases. Chromosorb P and W as well as potassium chloride crystals are perfectly suited as solid supports, but the majority of the usual liquid phases are inapplicable. Thus, Apiezon M, Ucon 50-HB-2000, silicone rubber SE-52, Carbowax 20 M, and Harflex 370 are decomposed more or less completely or dissolved by the mobile phase; however, Kel F wax and particularly the Epon 1001 and 1009 resins can be used for the analysis of low-volatility acetylacetonates and metalloporphyrins[2484].

Table 211 - Principal chelates other than acetylacetonates and their fluorinated derivatives analyzed by gas chromatography

Chelating Agent	Metals	References
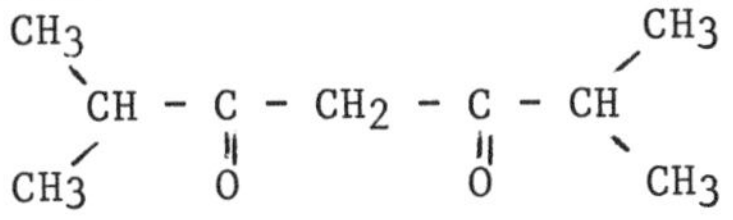 $(CH_3)_2CH-C(=O)-CH_2-C(=O)-CH(CH_3)_2$ [H(dibm)]	Be, Al, Cr, Fe, Ni	2478
	Al, Cr, Be	2517, 2518
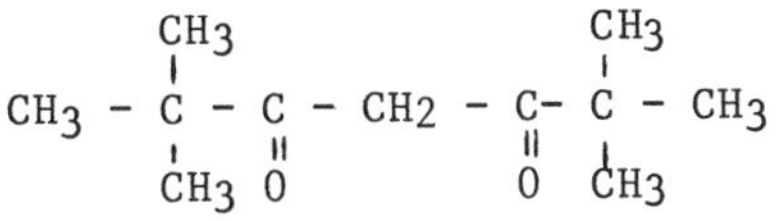 $CH_3-C(CH_3)_2-C(=O)-CH_2-C(=O)-C(CH_3)_2-CH_3$ [H(thd)]	Sc, Y, rare earths	2458, 2459, 2496
	Cr, Fe, Al	2519
	Alkali metals	2520
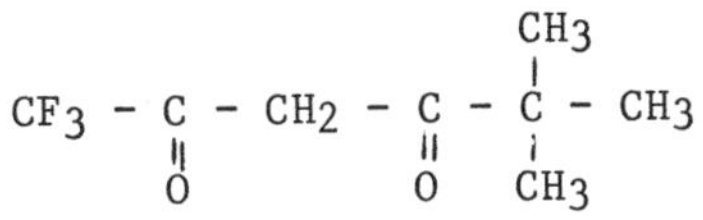 $CF_3-C(=O)-CH_2-C(=O)-C(CH_3)_2-CH_3$ [H(tpm)]	Sc, rare earths	2462, 2464
	Al, Cr, Fe	2465, 2519
	Al, Fe, Co, Cu, Ga	2466
	Be, Al, Fe, In, Ga	2521
	Pb	2522
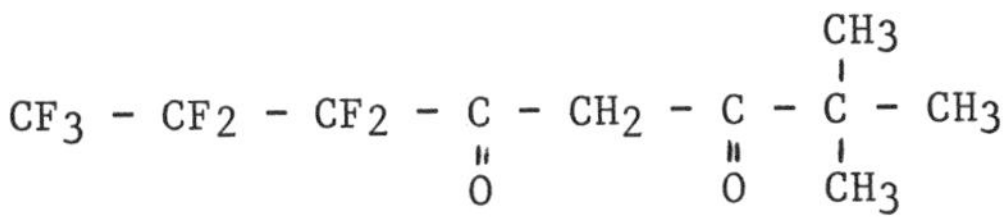 $CF_3-CF_2-CF_2-C(=O)-CH_2-C(=O)-C(CH_3)_2-CH_3$ [H(fod)]	Be, Al, Pd, Fe, Cu, Cr, Ni	2473
	Sc, Y, rare earths	2476
	Al	2473, 2485
	Al, Cr, Rh, Cu, Fe, Pd, Cr, Fe, Al	2523
	Fr, Al, Be, Cu, Fe, Pd, Y, Pd	2524
	Pd, Cr, Rh, Cu, Fe	2523
	u	2528
	Th	2475, 2528
	Co	2474, 2529
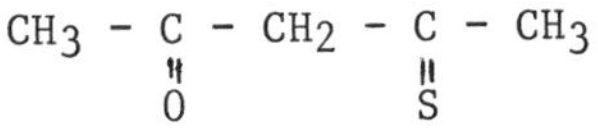 $CH_3-C(=O)-CH_2-C(=S)-CH_3$ [H(tacac)] (monothioacetylacetonate)	Co. Ni, Pd	2479
	Al, Be, Cr, Sc, V, Co, Ni, Cu	2530
	rare earths	2531
1,1,1 trifluoro-4-mercapto pent-3 cm-one	Ni, Co. Pt, Mg, Ca, Ba, Hg, Cd, Zn, Mn, Cu	2532
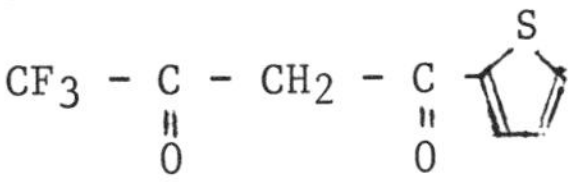 $CF_3-C(=O)-CH_2-C(=O)-C_4H_3S$ (thienyl)	Al	2485

cont.

Table 211 (cont.)

β - ketoimines of acetylacetone and salicylic aldehyde	Cu, Ni	2487
3 methyl (or 3 ethyl or 3 propyl) derivatives of pentane 2.4 dione	Al, Be, Cr, Cu	2533
1,1,1,2,2 pentaphoro-6,6-dimethyl-heptane-3,5-dione	Al, Cr, Fe	2524
	Pb	2522
Perfluoropivaloylmethanes	rare earths	2534
Benzoyltrifluoroacetone	rare earths	2531
Isobutyrylpivalalylmethane	rare earths	2535
1,1,1,2,2,6,6,7,7,7 decafluoro-heptane-3,5-dione-dibutylsulphoxide	rare earths	2536
Oxyacetates and oxypropionates	Zn, Be	2537
Etioporphyrin	Pd, Ti, V, Cu, Ni, Mg, Co, Mn, Zn, Pt	2538 - 2540
hfa-d-n-butyl sulphoxide mixed ligand couples	Th, U	2541
1,2,2-trifluoroacetylacetone-dimethylformamide	Ni	2542

The elution of magnesium, copper, nickel, and vanadyl etioporphyrin II is obtained on Epon 1001 resin[2546]. The first two derivatives are eluted practically at the same time and are separated only partially from the nickel complex. Table 212 shows the relative retentions on three stationary phases. After trapping at the column outlet and dissolving in dichloro-ethylene or pyridine, the eluted compounds have an absorption spectrum identical to that of the injected samples. No decomposition occurs during the chromatographic analysis.

This analysis can be quantitative; the peak area relative to the copper derivative varies linearly with the injected quantity between 10 and 80 μg. The same complexes of other elements, such as Co(II), Co(III), Mn(II), Zn, Pt(II), Pd(II), Ag(II), Fe(III), and TiO have been studied and separations have been obtained on Epon 1001 or XE-60[2547].

Chromatography under hypercritical pressures thus seems to be an interesting method for the analysis of compounds which have low volatility or low thermal stability and cannot be eluted under conventional conditions.

L. Gas Chromatographic Detection of Metal Chelates

Metal chelates are easily detected by conventional flame ionization or thermal conductivity detectors. However, one of the reasons for the extraordinary development of the analysis of these compounds by gas chromatography is the possibility of their detection in extremely small concentrations by means of the electron capture detector.

Table 212 - Relative retentions of porphyrins (according to ref. 2546).

Compound	Epon 1001 at 140°C	Versamid 900 at 153°C	XE-60 at 140°C
Cu etioporphyrin II	1	1	1
Mg etioporphyrin II	0.83	0.96	1.10
Ni etioporphyrin II	1.13	1.17	1.32
VO etioporphyrin II	2.15	2.47	2.87
Etioporphyrin II	0.88	1.03	1.34

However, no matter how sensitive the detector used the results obtained will be of little or no value if thermal decomposition of the metal chelates have occurred in the separation column, as this will lead to misshapen peaks or impurity peaks due to the breakdown products. Table 213 provides a summary of some information on the thermal stability of the chelates of various metals and indicates those elements which might exhibit instability[2548,2557]. In these instances, the analyst should proceed with caution.

The complexes with three asterisks can be eluted without apparent evidence of decomposition. The complexes denoted with two asterisks produce chromatographic peaks, but with those compounds there is evidence of decomposition. A single asterisk marks complexes known to be volatile at temperatures at which their chromatography should be possible, but their properties have not been thoroughly examined. In the experiments of Genty et al[2472] radioactive tracers were employed to determine whether any part of the samples was irreversibly adsorbed or decomposed in the injection port or chromatographic column. This is experimentally more difficult to perform, but it is more definitive than visual examination of the injection port liner for sample decomposition. Sample adsorption and instability become most important when gas chromatography is used for ultratrace metal analysis. In other applications some sample loss can occasionally be tolerated.

In order to provide an acceptable method for the analysis of mixtures of metals, the gas chromatographic determination of metal chelates requires good resolution of components, minimum interferences and, in particular, sensitivities at least as high as those of other methods in common use, e.g. atomic-absorption and emission spectroscopy. Hence the choice of detector is of major importance.

The katharometer is non-selective and provides sensitivities in the range 10^{-4} to 10^{-8}. In addition, some decomposition of chelates occurs on the hot metal filaments. Flame ionisation detectors have also been widely used but their response to metal chelates is less sensitive than that to organic compounds containing no metals. The presence of fluorine atoms and the metal atom itself in the chelate decreases the response of the detector and only moderate sensitivities have been obtained[2558].

The electron-capture detector has been widely used for the determination of fluorinated chelates[2556,2559] and is highly sensitive[2558-2561]. The electron affinity of the chelates is a function of both the metal and any halogen atoms present: however, the detector is much more sensitive to halogenated chelates. Under comparable conditions, the limit of detection for chromium (III) acetylacetonate is 2.5×10^{-10} mol, for the trifluoroacetylacetonate 1.8×10^{-13} mol and for the hexafluoroacetylacetonate 4.9×10^{-14} mol[2559].

Table 213 - Gas chromatography of metal chelates

Li thd *	Be tfa *** hfa *** acac *** fod ***											
Na thd *	Mg thd ** dfhd **											Al tfa *** hta *** acac *** fod ***
K thd *	Ca dfhd ** thd **	Sc fod *** thd *** tfa *** acac ***	Ti(IV) hfa ***	V(IV) tfa *	Cr(III) tfa *** hfa *** acac *** fod ***	Mn(III) tfa **	Fe(III) tfa ** hfa *** fod ***	Co(III) hfa *** thd * tfa ***	Ni(II) hfa * thd * fod ***	Cu(11) tfa *** hfa ** acac ** fod ***	Zn tfa ** hfa * thd *	Ga tfa ***
Rb	Sr dfhd ** thd **	Y fod *** thd ***	Zr tta ** dfhd * thd *	Nb hfa **	Mo	Tc	Ru(III) tfa *** hfa ***	Rh(III) tfa *** hfa ***	Pd fod ***	Ag	Cd hfg *	In tfa ***
Cs	Ba dfhd ** thd **	57-71	Hf tfa ** dfhd *	Ta hfa **	W	Re	Os	Ir	Pt	Au	Hg	Tl
Fr	Ra	89-103 Th-tfa *										
											Yb fod *** thd ***	Lu fod *** thd ***
La thd ***	Ce thd *	Pr fod *** thd ***	Nd fod *** thd ***	Pm	Sm fod *** thd ***	Eu fod *** thd ***	Gd fod *** thd ***	Tb fod *** thd ***	Dy fod *** thd ***	Ho fod *** thd ***	Er fod *** thd ***	Tm fod *** thd ***

Often the choice of chromatographic conditions is limited by the volatility and stability of the chelates used and hence complete resolution of mixtures is not always possible. Selective detection of individual metals would, therefore, be of considerable interest, but very little work has been reported on the application of selective detectors.

The electron affinity and detection limit of complexes depend on the nature of the metal, the degree of fluorination of the chelating radical and, to a lesser extent, the nature of the carrier gas. Table 214 shows the values cited by Ross for the fluorinated and unfluorinated chromium and aluminium acetylacetonates with nitrogen as carrier gas.

Albert reports even lower detection limits and compares them to those obtained with a flame ionization detector (Table 215)[2562].

A certain correlation seems to exist between the response of an electron capture detector and the ultraviolet wavelength of the absorption maximum of a chelate[2563]. Figure 159 shows the variation of the electron capture coefficient K relative to the scandium complex with this wavelength λ_{max} for several trifluoroacetylacetonates. Coefficient K is the ratio of the chromatographic peak areas multiplied by 3/2 for copper and beryllium and by 3/4 for thorium so as to equalize the number of tfa radicals. According to the authors, this phenomenon can be explained as follows. The shift in the metal complexes of the absorption band of trifluoroacetylacetone at 282 mμ is related to the energy level of the lowest unoccupied molecular orbital which also determines the electron affinity of the chelate if electron capture does not result in dissociation.

The flame photometry detector is recommended by Juvet and Durbin[2564, 2565]. It has the advantage of a certain selectivity, with each compound being detected at the wavelength of the maximum of its emission peak. Even if the chromatographic separations are not complete, quantitative analysis remains possible. On the contrary, if the column is sufficiently efficient, the inlet slit width of the spectrometric system can be increased, resulting in better sensitivity. In general, the detector is much more sensitive for the investigated metal compounds than the thermal conductivity detector. The response expressed in peak area is linear as a function of the quantity of injected metal and the detection limits for $Rh(hfa)_3$ and $Cr(hfa)_3$ are 10^{-11} and 10^{-10} mole, respectively, making the sensitivity of this detector comparable to that of the flame ionization detector.

Durbin viewed the emissions from oxy-hydrogen flame by using a Beckmann DU spectrophotometer. Although the sensitivity of the detector response was not high, good selectivity and a range of linearity of 10^4 were obtained. A similar system was used by Juvet and Zado[2566] but with less sensitive results.

Mass spectrometry has been applied in the determination of chromium and beryllium chelates at the 10^{-12} g level[2567]. The negative ion chemical ionization mass spectrometry of volable metal tris-β-diketonates using a gas chromatograph chemical ionization mass spectrometer-computer system has been reported. Sensitivity measurements are reported for several chromium tris-β-diketonates.

Microwave-excited emissive detectors have proved highly sensitive and fairly selective in organic analysis[2568,2569]. The potential usefulness of this type of emission for the determination of metals has been shown by the work of Bache and Lisk[2570] who used emission at the 253.7 nm atomic mercury line from a low-pressure helium plasma in order to measure the

Table 214 - Detection of metal chelates by electron capture (according to ref. 2444)

Chelate	Signal/Noise Ratio	No. of Moles detected
$Cr(acac)_3$	10	2.5×10^{-16}
$Cr(tfa)_3$	4	1.8×10^{-13}
$Cr(hfa)_3$	7	4.9×10^{-14}
$Al(acac)_3$	17	1.6×10^{-7}
$Al(tfa)_3$	2	5.9×10^{-12}
$Al(hfa)_3$	2	7.4×10^{-13}

Table 215 - Comparison of the detection limits by electron capture and flame ionization in miles (carrier gas, nitrogen; signal/noise ratio = 2) (according to reference

Compound	Electron capture	Flame ionization
$Cr(acac)_3$	4.1×10^{-13}	1.2×10^{-10}
$Cr(tfa)_3$	3.1×10^{-15}	3.7×10^{-10}
$Al(tfa)_3$	1.2×10^{-13}	1.7×10^{-10}
$Cu (tfa)_3$	2.0×10^{-10}	1.6×10^{-9}

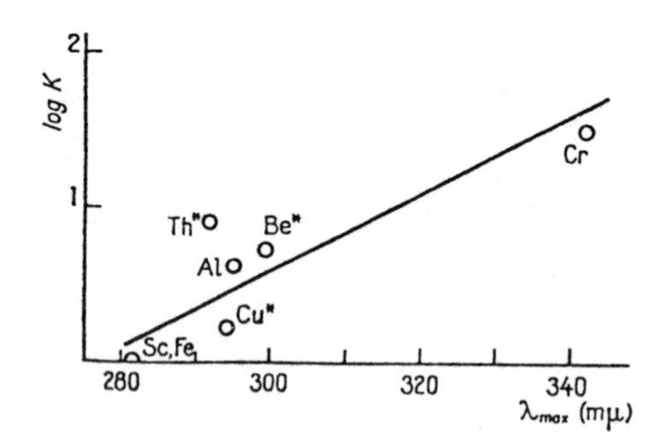

Figure 159 - Relation between the electron capture detector response to trifluoroacetylacetonates and the wavelength of their maximum UV absorption.
K is multiplied by 3/2 for Cu and Be and by 3/4 for Th.

amounts of organomercury compounds present in fish, and by the work of Runnels and Gibson[2571] and Dagnall et al[2572,2573]. These two groups of workers used microwave plasma in order to excite metal emissions from compounds that had been flash evaporated into the plasma from a platinum loop. This technique is highly sensitive, but is limited by the small volume (about 1 μl) of solution that can be retained on the loop and to those compounds which can be volatilised under these conditions.

Black and Sievers[2574] determined chromium in human blood by gas chromatography of a β-ketonate chelate using a microwave emission detector. They managed to eliminate the matrix problems associated with the determination of points per billion levels of chromium in blood serum. Kawazuchi et al[2575] and Serravallo and Risky[2576] have also applied microwave plasma emission detectors to the analysis of metal chelates.

It was Dagnall et al[2577] who carried out a thorough investigation of the applicability of the microwave excited emissive detector operated at atmospheric pressure to the detection of metal chelates of acetylacetone and trifluoroacetylacetone. These workers investigated the acetylacetonate of aluminium (III), chromium (III), copper (II), and iron (III) and scandium (II), and the trifluoroacetylacetonates of aluminium (III), chromium (III), copper (II), gallium (III), iron (III), scandium (III), and vanadium (IV).

In order to avoid decomposition or condensation of the metal chelates the silica tubing containing the plasmas was connected directly to the column exit and the monochromator was positions so as to view the plasma just above the roof of the oven. In order to reduce further the risk of condensation, the silica tubing between the column and the plasma was heated by means of a short coil or wire and a Variac variable transformer. The samples were injected directly on to a glass-wool pad in the top of the column, before the packing, and the top of the column was heated by means of a small electrical furnace. The temperature of the top of the column could be maintained at about 30°C above the oven temperature by applying a potential of 40 V across the furnace.

To investigate the parameters of the system Dagnall et al[2577] made 1 µl injections of a 950 p.p.m. (m/V) solution of gallium trifluoroacetyl-acetone in benzene. Gallium trifluoroacetylacetonate was used because it was found to be conveniently eluted at 125°C after 3.1 minutes with a fairly good peak shape and the intense gallium atomic line at 294.4 nm lies in a region of very low background. Any variation of detector response at this wavelength could, therefore, be associated directly with changes in the metal emission. Emissions characteristic of the metallic elements were obtained from the first 10 mm of the plasma. In practice, in order to obtain a linear calibration graph that passes through the origin, it was found necessary to position the plasma so that emissions from the first 3 mm of the plasma could be monitored, and this position was used for all measurements of chelate emissions.

The variation of the peak height at 294.4 nm with microwave power was measured by repeated injections of the gallium trifluoroacetylacetonate solution. For each new power setting, the length of the plasma altered and the position of the plasma had to be re-optimised. The maximum peak height was obtained at a power of 70 W. The variation of noise with power was found to be very small the maximum signal to noise ratio also occurred at 70 W.

Table 216 tabulates the emission characteristics found for the chelates of aluminium, chromium, copper, gallium, iron, scandium and vanadyl.

The area response of the detector was found to be proportional to the reciprocal of the flow-rate, and the slowest convenient flow-rate (3.3 $l\ h^{-1}$) was used.

Dagnall et al[2577] determined the signal to noise ratios for the major emissions characteristic of the metallic elements (Table 216) for solutions of each of the trifluoroacetylacetonates. In addition, the detector response at each wavelength to a 1000 p.p.m. solution of benzyl alcohol in benzene was measured in order to determine the selectivities of the metal emissions over carbon compounds. Benzyl alcohol was used because it has convenient retention times (1.0 to 3.5 minutes) on the columns used. The molar selectivity ratio was calculated in each instance as the ratio of the number of gram atoms of carbon injected to the number of gram atoms of the metal required to produce the same response. Benzene was used as the solvent for all the chelates. It was eluted after only 10 to 20 s and the plasma was not initiated until about 30 s after injection.

The optimum experimental conditions used are shown in Table 217. All the trifluroacetylacetonates were chromatographed successfully, as were the acetylacetonates of aluminium, chromium and scandium. However, no peaks could be obtained for the copper and iron acetylacetonates. The most sensitive wavelength found for each metal, with the limit of detection and selectivity are shown in Table 218.

Table 216 - Emission characteristics of metal chelates

Metal species	Principal emissions/nm	Metal species	Principal emissions/nm
Al	396.2	Fe	344.1 357.0 373.5 249.0 (248.8, 249.0, 249.1)
Cr	357.9 425.4 520.5 (520.8, 520.6, 520.4)	Sc	361.4 363.1 364.4 (364.5, 364.3) 357.5 (357.3, 357.6, 358.0) 424.7
Cu	324.7 327.4		
Ga	287.4 294.4 (294.3, 294.4) 417.2 403.3	VO	318.4 (318.3, 318.4, 318.5) 292.4 (292.4, 292.5) 438.1

Table 217 - Optimised operating parameter

Chelate	Column	Temp °C	Retention time/minutes
$Al(acac)_3$	II	175	1.67
$Al(tfa)_3$	I	100	4.83
$Cr(acac)_3$	I	190	4.80
$Cr(tfa)_3$	I	130	3.30
$Cu(tfa)_3$	II	140	2.00
$Ga(tfa)_3$	I	125	3.10
$Fe(tfa)_3$	I	135	2.90
$Sc(acac)_3$	I	135	2.90
$Sc(tfa)_3$	I	160	4.00
$VO(tfa)_3$	II	160	1.45

Column 1 0.6 m x 4.8 mm i.d. borosilicate glass packed with universal B precoated with 10% Apiezon L

Column 2 as above packed with 0.5% Apiezon L on glass micro beads (0.2 mm diameter).

Columns conditioned for 36 h at 200°C in a stream of argon.

Table 218 - Limits of detection and selectivities for some metal chelates

Chelate	Wavelength/nm	Limit of detection g s-1 of metal	Selectivity
$Al(acac)_3$	396.2	2.0×10^{-11}	-
$Al(tfa)_3$	396.2	1.9×10^{-11}	900
$Cr(acac)_3$	357.9	2.9×10^{-12}	-
$Cr(tfa)_3$	357.9	3.6×10^{-12}	3930
$Cu(tfa)_3$	324.7	8.0×10^{-12}	2250
$Ga(tfa)_3$	294.4	2.7×10^{-12}	1170
$Fe(tfa)_3$	344.1	1.3×10^{-11}	1610
$Sc(acac)_3$	361.4	2.1×10^{-12}	-
$Sc(tfa)_3$	361.4	3.0×10^{-12}	1620
$VO(tfa)_2$	318.4	8.5×10^{-12}	1400

Dagnall et al[2577] concluded that the microwave excited emissive detector operated at atmospheric pressure responded to all the chelates they investigated, both non-selectively by monitoring atomic carbon and selectively by using metal emissions.

The detector response was linear and highly sensitive to the metals studied with limits of detection between 2×10^{-12} and 2×10^{-11} g s-1 (Table 218). The limits of detection of the acetylacetonates and trifluoro-acetylacetonates of the same metals were approximately equal, indicating that the detector response may well prove to be independent of the chelating agent used. The sensitivity of this detector is considerably higher than those of other selective detectors used in the analysis of metal chelates, with the exception of mass spectrometry, and is of the same order as that of the electron-capture detector. Further, the dependence of the detector response on the metal, rather than on the chelate as a whole, avoids the limitations of having a halogen atom or other electron-capturing species present in order to achieve high sensitivity.

It is highly selective and, further, has the advantage that the pattern of atomic metal lines can be rapidly scanned in order to confirm the presence of a compound of a particular metal.

The limitations of this detector for the analysis of mixtures of metals are similar to those for organic analysis. The principal problem is that of overloading, which imposes an upper limit on the working range of the detector. In addition, the formation of metallic deposits requires the tubing to be changed frequently when concentrated samples are used.

Black and Braumer[2578] converted metals to volatile β-diketonates carried the vapours over by argon into a stabilized conductively coupled argon plasma. Black et al[2579] determined metal chelates in biological materials by continuose nebulization inductivly coupled plasma atomic emission spectrometry. They obtained a reproducibility of 2 to 3% relative standard deviation and detection limits in the ug/ml range.

M. Gas Chromatography, Conclusions

Among the metal compounds which can be analyzed by gas chromatography, the chelates have the advantage of being easily prepared with a large number of elements. By varying the nature of the chelating agent, it is possible to obtain compounds of sufficient volatility and thermal stability in the majority of cases. Thus, a general method is available to the chemist which, when combined with solvent extraction, permits the qualitative and quantitative determination of most metal ions in aqueous solution. Table 219 summarizes some of the studies published in this field and the reasonable prospects of new discoveries.

The liquid stationary phases in general use are the silicone and Apiezon greases. In order to prevent adsorptions or secondary reactions of the injected compounds either with the solid supports or the inside column surface, the liquid phase is deposited on glass microbeads and Pyrex glass columns are preferred to metal tubes. The good results obtained with glass bead supports has certainly to do with the usually very low solubility of the chelates with most liquid phases, the main contribution to the retention being the adsorption at the gas-liquid interface.

The application of gas solid chromatography to metal chelates has received limited attention. The adsorbent used was a fluorinated resin, Fluroplast-4 [2580].

Gas chromatography of fluorinated chelates combined with electron capture detection is one of the most sensitive methods known for the determination of metallic elements. Traces in the order of 10^{-13} mole are easily analyzed.

Because of the large energy of vaporization of the chelates from the stationary phase, the retention times vary much faster than usual with temperature, doubling for a temperature decrease of 10 - 15°C. Consequently, temperature programming will often be necessary to analyze mixtures, and changes in column temperature and in liquid phase loading will help much in separating compounds which interfere in first experiments.

It is now realised that much of the early work on the gas chromatography of metal chelates is of limited value or is described in insufficient detail to be of value to the analyst. This is understandable in what was a brand new technique which, it was believed, would solve many or all problems in metal analysis. With the progress of continued work in the field we now hold a more mature view and see the technique as one which has great potential for many problems whilst being of limited value in other instances.

In the final sections are given some good examples of the application of the technique which illustrate its extreme value in analysing metal mixtures with high sensitivity.

N. Liquid chromatography of metal chelates

Work on the application of this technique started later than the gas chromatographic technique and is, as yet, limited. Applications of liquid chromatography and high pressure liquid chromatography are seen as having great potention, will undoubtedly be the subject of much further study, and, in some areas, will in the future undoubtedly surplant gas chromatography of chelates.

The recent technique of chromatography under hypercritical conditions of the carrier fluid undoubtedly will permit a study of certain low-volatility metal derivatives. However, numerous technological problems make the application of this method difficult; liquid-liquid high-pressure chromatography would then be of certain interest for the analysis of these compounds. Some work has already been published in this field: Separation

Table 219 - Chelates analyzed or analyzable by gas chromatography

1. acetylacetonate; 2. trifluoroacetylacetonate; 3. hexafluoroacetylacetonate; 4. other chelates.

a. Complex assumed to be sufficiently volatile for chromatography
b. Complex with confirmed elution but slight decomposition
c. Complex which can be eluted without decomposition.

H																	He
Li 4a	Be 1r 2c 3c 4c											B	C	N	O	P	Ne
Na 4a	Mg 4c											Al 1c 2c 3c 4c	Si	P	S	Cl	Ar
K 4a	Ca 4c	Sc 1c 2c	Ti 3c	V 1c 2c	Cr 1c 2c 3c 4c	Mn 2b 3a	Fe 1b 2b 3c 4c	Co 2c 3c 4c	Ni 3a 4c	Cu 1b 2c 3a 4c	Zn 2b 3a 4a	Ga 2c	Ge	As	Sc	Br	Kr
Rb	Sr 4b	Y 4c	Zr 2b 4a	Nb 3b	Mo	Tc	Ru 2c 3c	Rh 2c 3c	Pd 4c	Ag	Cd	In 2c	Sn	Sn	Te	I	Xe
Cs	Ba 4b	Lanth. 3a 4c	Hf 2b 4a	Ta 3b	W	Re	Os	Ir	Pt	Au	Hg	Tl 2c	Pb	Bi	Po	At	Rn
Fr	Ra	Actin. Th, U 2c 3a 4c															

of $Co(fod)_2$ and $Co(fod)_3$ on a polyurethane column with benzene as the carrier liquid[2581]; Be, Al, Cr(III), Fe(III), Co(II), Co(III), Ni(II), Cu(II), Zr, Rh(III) acetylacetonates and trifluoroacetylacetonates, using the two-liquid phases at equilibrium in a ternary system (water, ethanol, and isooctane)[2582 2583].

Numerous complexes of Group I-VI metals have been eluted at about 115°C by chromatography at pressures of 800 - 1000 psi[2584]. Under these conditions, thermal decomposition does not seem to take place, but the retentions of the different chelates are very similar and the resolution is generally poor.

Aden and Bigley[2585] applied high pressure liquid chromatography to mixtures of metal diethyldithiocarbamates using ultra-violet and d.c. argon plasma with emission spectroscopic detection. Uden et al[2585] separated geometric isomers and mixed ligand forms of cobalt II and chromium (III) β-diketonates by high pressure liquid chromatography.

Fay and Piper[2725,2726] separated fac and mer isomers of the octahedral benzoylacetonates and trifluoroacetylacetonates of Cr(III), Co(III) and Rh(III) by column chromatography on alumina, and their data suggest the application of adsorption high-pressure liquid chromatography for separation of fac and mer isomers and mixed ligand species. Huber et al[2727] were the first to report the high-pressure liquid chromatography separation of metal β-diketonates, using a ternary liquid-liquid partition system of water, ethanol and iso-octane. Other metal chelate systems studied have included tetradentate Schiff base salicylaldiminato and β-ketoaminato complexes[2728-2730] and diethyldithiocarbamates[2731].

Gaetani et al[2586] separated cobalt II, nickel II and copper II as their tetradentate β-ketoamine chelates by high-pressure liquid-liquid partition chromatography. An ultraviolet detector was used. Hanjo et al [2587] have studied the solvent effect on the liquid-liquid distribution of monothiotheonylfluoroacetone and its copper(II), cobalt(II), and zinc(II) chelates.

High performance liquid chromatography using the partition liquid-liquid technique with direct or reversed phase, has been used for the chromatography of organochromium compounds[2588,2589], and organoiron compounds [2590] and for the chromatography of the diketonates of several divalent and trivalent metals[2591].

Gactani et al[2592] have briefly discussed the separation of copper and nickel using the ligand $H_2(en)AA$ (N, N'ethylene bis(acetylacetoneimine).

Uden and Walters[2593] also reported on this separation using liquid-solid chromatography with microparticulate silica as a stationary phase.

Metal dithizonates have been separated on a column[2594]. Liquid chromatography has been used as a powerful tool for separation in the field of porphyrin chemistry. Columns packed with silica gel[2595], alumina[2596], calcium carbonate[2597], or magnesium silicate-cellulose mixture[2598], paper[2599], and thin layers of silica gel or alumina[2600,2601] have been effectively used for the purification of synthetic porphyrin compounds or separation of naturally occurring metalloporphyrins. However, systematic investigation of chromatography for these compounds has hardly been carried out from the viewpoints of the effects of stationary phase substance and mobile phase solvent. In addition, most of the published work has dealt with porphyrin chelates of limited metal ions, for example, chelates of magnesium or iron, which are known as chlorophyll or hemin.

Q. Size Exclusion Chromatography of Metal Chelates

Size-exclusion chromatography is a technique for separating and characterizing solute compounds primarily on the basis of their sizes in solution, although other secondary effects related to the mutual interactions between the solute, solvent, and gel are also frequently involved. This technique has been applied extensively to investigations involving organic polymers or biological molecules, while the application to low molecular weight compounds or inorganic compounds has been limited partly because the behaviour of such compounds is much more subject to the secondary effects.

In recent years, however, the application of size-exclusion chromatography to metal complexes has been reported. The size-exclusion chromatographic behaviour of metal β-diketonato complexes has been systematically investigated in various organic solvent systems[2602,2610,2613,2614]. Saitch and Suzuki[2608] have proposed a new theoretical treatment of the solute distribution in practical size-exclusion chromatography, taking metal acetylacetonates and normal alkanes as model compounds. They have applied the concept of a regular solution to the estimation of the contribution of the secondary effects to the separation mechanism in size-exclusion chromatography. Their view is such that p-dioxane is the nearest to the ideal eluent solvent for the poly(vinyl acetate) gel column; that is, the poly(vinyl acetate) gel-p-dioxane system permits the solute compounds to be differentiated only by the size exclusion effect[2608]. This means that the effective volumes of the solutes in p-dioxane can be estimated from their distribution coefficients obtained in size-exclusion chromatography. On the basis of these observations, they have explained the chromatographic behaviour of metal thenoyltrifluoroacetonates in the poly(vinyl acetate) gel-p-dioxane system in terms of the difference in effective volumes of these complexes reflected by the adduct formation with p-dioxane[2610].

The theoretical results presented by Saitch and Suzuki, however, have been supported only by the chromatographic data for Be(II) and Cr(III) complexes with acetylacetone and normal alkanes[2608]. Therefore, the validity of the theoretical approach for the distribution of other metal complexes in size-exclusion chromatography has not been evaluated.

Saito et al[2615] investigated the correlation between the distribution coefficient and the molar volume for six β-diketones and their metal complexes, based on the theory proposed by Ogston[2611] and Laurent and Killander [2612] in order to examine the utility of the poly(vinyl acetate) gel-p-dioxane system for the size estimation of unknown species by size-exclusion chromatography. They showed that the relationship between the distribution coefficient and the molar volume depends on the type of substituent in the β -diketone and on the eluant used.

Saitah and Suzuki[2610] studied the gel chromatographic behaviour of metal(II III) theonyltrifluoroacetonates in the polyvinyl acetate gel-p-dioxane system. The elution volume of the chelates was in the order of central metal ions such as Ni(II)<Cu(II) = Zn(II<Cr(III) = Co(III) = Al(III) = Fe(III)<Be(II) = Pd(II) (< chelating agent).

P. High Performance Thin-Layer Chromatography of Metal Chelates

Saitoh et al[2616] separated seven tetraphenylporphyrin (TPP) chelates on C_{18} bonded silica using a two dimentional developing method. The elements separated were magnesium (II), nickel (II), copper (II), zinc (II), cadmium (II), manganese (III) and iron (III). Separation was achieved on cellulose, silica gel, and C_8-bonded and C_{18}-bonded silica thin layers with various organic developing solvents. The seperation between the magnesium (II) and cadmium (II) chelates is not successful with thin-layer systems other than cellulose. With this exception, every chelate pair can be resolved on C_{18}-bonded silica layer. Mutual separation of seven metal tetraphenylporphyrin chelates is demonstrated on C_{18}-bonded silica layer by a two-dimensional developing method in which the 20:80 (v/v) acetone-propylene carbonate mixture and acetone are used as the primary and the secondary developers, respectively.

Other work on the separation of metal tetraphenylporphyrin chelates includes that of Sato and Kwan[2617] and Hui et al[2618]. In this work the stationary phase materials were limited to silica gel and alumina, and a few kinds of solvent mixture were used as the mobile phases.

Q. Index of Metal Analyses by Gas Chromatography of their Volatile Chelates

In the following are compiled the principal information on the separation of mixtures of metal elements by gas chromatography of their chelates for the convenience of the analytical chemist. The chelating agents are noted by the following abbreviations:

acac = acetylacetonate
dibm = 2,6-dimethyl-3,5-heptanedionate
fod = 1,1,1,2,2,3,3,-heptafluoro-7,7-dimethyl-4,6-octanedionate
hfa = hexafluoroacetylacetonate
tacac = monothioacetylacetonate
tfa = trifluoroacetylacetonate
thd = 2,2,6,6-tetramethyl-3,5-heptanedionate
tpm = 1,1,1-trifluoro-5,5-dimethyl-2,4-hexanedionate

Aluminium

In mixture with	Complex	Phase	References
Be, Sc	acac	Apiezon L	2395
Be	acac	Apiezon L	2392
Cr	acac	Apiezon L	2342
Be, Cr	acac	SE.30	2389, 2398, 2666 - 2672
Cr, Fe, Cu	acac	PTFE	2660
Be, Cr	acac	Apiezon L	2664, 2665
Cr	acac	Dow Silicone 710 R	2677, 2678
Cr, Fe	acac	SE.30, Apiezon L, Diethyl-glycol adipate	2519
Be, Sc	tfa	Silicone oil	2395, 2635
Cr, Rh	tfa	SE.30	2427
Cr, Ru	tfa	Tergipol NPX	2371
Cu, Fe	tfa	Polyethylene	2420, 2422, 2423, 2687, 2688, 2713
Ga, In	tfa	DC-550	2619, 2659
Fe,	tfa	DC-550	2689
Cr, Rh, Zr	tfa	Silicone grease	2371
Be, Ga, In, Tl	tfa	Silicone 710	2414, 2681

Be, Cu	tfa	Dow Silicone 710	2711
Cu, Fe, Be, Cr, Fe	tfa	SE-30	2690
Cr, Fe	tfa	SE-30, Apiezon L, Diethyl-glycoladipate	2524
Be, Sc, In, Rh	tfa	Dow Silicone 710.R	2682-2684
Be, Cr, U, Th	tfa	DC-559	2685, 2686
Be, Ce	tfa	SE-30	2518
Be	tfa	Dow high vacuum silicone	2679
Sc	tfa	Dow Silicone 710.R	2680
Traces in biolog-ical materials	tfa	OV-17	2650
Traces in poly-ethylene	tfa	Chromator NAW-DMCS	2668, 2669
Cr	tfa	PTFE	2660
Cr	tfa	Dow Silicone 710.R	2667, 2668
Traces in U	tfa	-	2434
Traces	tfa	-	2427 - 2430, 2444
Traces	hfa	-	2427 - 2430, 2449, 2629
Cr	hfa	Dow Silicone 710.R	2667, 2668
Be, Cr	hfa	DC-710, SE-30	2389
Be, Cr, Cu	hfa	QF-1	2453
Be, Cr, Fe	hfa	Fluorolube	2453
Be, Fe	hfa	Dow Silicone 710.R	2694
Be, Ga, Fe, Cr, Cu, Rh	hfa	QFI and SE.52	2625, 2626
Cr, Rh, Cu, Fe, Pd	hfa and fod	Squalane or Apiezon L or QF.1	2523
Be, Cu, Cr, Fe, Pd, Y	fod	SE-30	2473
Cr, Rh, Cu, Fe, Pd	fod	Squalane, Apiezon L, QF1	2523
Cr, Fe	fod	SE-30, Apiezon L, Diethyl-glycol adipate	2524
Cr, Fe	tpm	Apiezon L	2465
Cr, Fe, Cu	tpm	SE-30	2466
Be, Fe, In, Cu	tpm	SE-60	2521
Cr, Fe	tpm	Se-30, Apiezon L, Diethyl-eneglycol adipate	2524
Be, Cr, Fe, Ni	dibm	Silicone grease	2478
Be, Cr	dibm	High vacuum silicone grease	2517
Cr, Fe	thd	SE-30, Apiezon L, Diethyl-ene glycol adipate	2519
Be, Cr, Cu	3-methyl 3-ethyl 3-propyl 3-butyl derivs of pentane 2.4 dione	Apiezon L or SE.30	2533
Cr, Mn, Fe, Co, Ni Cu, Ga, Zn, In	trifluoro-acetylpiv-aloylmethane	Se-30	2695
Cr, Fe	acac 1,1,1,2,2-pentafluoro-6,6-dimethyl-heptane-3,5-dione	SE-30, Apiezon L, Diethy-lene glycal adipate	2524

Be, Cr, Sc, V, Co, Ni, Cu	2-theonyl trifluoro-acetylacitonates	DC-710	2530
Cr, Fe	various	-	2659
Ga, In	various	silicone grease	2720
		Alkaline Earths	
Traces Ca, Se,	thd	Apiezon L	2709
Mg in presence of Ti, V, Cu, Ni, Co, Mn, Zn, Pt, Pd	otioporphyrin	Epon 1001, XE-60	2538 - 2540
		Alkali Metals	
Traces	thd	-	2520
		Beryllium	
Al, Sc	acac	Apiezon L	2392
Cu	acac	Silicone oil	2393
Al, Cr	acac	SE-30	2389, 2398, 2672, 2676
Cr, Al	acac	Apiezon L	2664, 2665
Al, Sc	tfa	Silicone oil	2395
Al, Ga, Tl, In	tfa	Silicone 710	2414, 2681
Traces in blood, urine, tissue	tfa	-	2427 - 2430, 2630 2690 - 2692
Cr biological materials	tfa	UC-W98	2710
Al, Cu	tfa	Dow Silicone 710	2711
Al, Cu, Fe	tfa	SE-30	2690
Traces in lunar soil	tfa	-	2714
Traces in standard rocks	tfa	-	2715, 2716
Al, Fe	tfa	Dow Silicone 710 R	2694
Al, Sc, In, Rh	tfa	Dow Silicone 710 R	2682 - 2684
Al, Cr, U, Th	tfa	DC 559	2685, 2686
Al, Cr	tfa	SE 30	2578
Al	tfa	Dow high vacuum silicone	2679
Traces in blood	tfa	SE 52	2656
Fe, Traces in urine	tfa	SE 52	2657
Al, Co, Cu, Fe, Mg, Mn, Zn, Ca Na, K, traces in blood, urine organs	tfa	SE 52	2661,
Ca, K, Na, Al, Co, Cu, Fe, Mg, Mn, Zn	tfa	SE 52	2662, 2663
Traces in air	tfa	W-98 silicone (Union Carbide)	2653
Fe, Al, Traces in biological material and air	tfa	SE 52	2654
Traces in biological materials, blood,		SE 52	2655

Al, Cr	tfa	SE-30	2389
Al, Cr, Cu	hfa	QF-1	2453
Al, Cr, Fe	hfa	Fluorolube	2453
Al, Ga, Fe, Cr, Cu, Rh	hfa	QF1 and SE-52	2625, 2626
Al, Cu, Cr, Fe, Pd, Y	fod	SE-30	2473, 2631, 2632
Al, Fe, In, Cu	tpm	SE-60	2521
Al, Cr, Fe, Ni	dibm	Silicone grease	2478
Al, Cr	dibm	High vacuum silicone grease	2517
Al, Cr, Sc, V, Co, Ni, Cu	trifluoroacetyl acetonate	DC-710	2530
Zn	Oxyacetates and oxypropionates	Apiezon L	2537
Al, Cr, Cu	3-methyl, 3-ethyl, 3-propyl and 3-butyl derivs of pentane 2.4 dione	Apiezon L or SE-30	2533

Chromium

Al, Be	acac	SE-30	2389, 2398
Al	acac	Apiezon L	2392
Al, Be	acac	Apiezon L	2664, 2665
Be, Fe	acac	-	2673, 2674
Cu	acac	Apiezon L	2675
Al, Fe	acac	SE-30, Apiezon L, Diethylene glycol adipate	2538
Al	acac	Dow Silicone 710.R	2677, 2678
Al, Rh	tfa	SE-30	2427
Al, Ru, Zr	tfa	Tergipol NPX	2371
Al	tfa	Dow Silicone 710.R	2677, 2678
Be,	tfa	-	2676
Al, Be, U, Th	tfa	DC-559	2685, 2686
Be, Al	tfa	SE-30	2518
Be, biological materials and air	tfa	UC - W98	2717
Traces	tfa	Dextro	2670
Ru	tfa	Apiezon L	2706
Traces in tissue, linen	tfa	OV-225	2642
Traces in serum	tfa	QF-1	2643, 2644
Traces in serum	tfa	SE-52	2645
Traces in serum	tfa	-	2649
Traces in blood	tfa	LSX-3-0295 silicone grease	2648
Traces in KCRO4	tfa and hfa	Se-30	2641
Traces in Fe	tfa	-	2421, 2432, 2621 2622, 2449, 2454
Traces	tfa	Dow Silicone 710.R	2427 - 2430, 2474, 2717, 2718
Al, Fe	tfa	SE-30 Apiezon L, Diethylene glycol adipate	2524
Traces (geometrical isomers)	tfa	-	2637 - 2639

Traces	hfa	Dow Silicone 710.R	2427 - 2430, 2434, 2717, 2718
Al	hfa	Dow Silicone 710.R	2677, 2678
Rh	hfa	Dow high vacuum silicone grease	2707
Rh, Fe	hfa	DC-200	2708
Al, Be	hfa	SE-30, silicone 710	2389, 2398, 2444, 2448, 2666 - 2672
Al, Be, Cu	hfa	QF-1	2447
Al, Be, Fe	hfa	Fluorolube	2447
Rh	hfa	Silicone grease	2371
Be, Al, Ga, Fe, Cu, Rh	hfa	QF.1 and SE.52	2625, 2626
Al, Rh, Cu, Fe, Pd	hfa	Squalane, Apiezon L, QF.1	2523
Traces (geometrical isomers)	hfa	-	2637 - 2639
Al, Be, Cu, Fe, Pd, Y	fod	SE-30	2473, 2631, 2632
Al, Fe	fod	SE-30, Apiezon L, Diethylene glycol adipate	2519
Al, Rh, Cu, Fe, Pd	fod	Squalane or Apiezon or QF.1	2523
Al, Fe	tpm	Apiezon L	2465
Al, Fe, Cu	tpm	SE-30	2466
Al, Fe	tpm	SE-30, Apiezon L Diethylene glycol adipate	2519
Be, Al	dibm	High vacuum silicone grease	2517
Al, Be, Fe, Ni	dibm	Silicone grease	2478
Al, Fe	1,1,1,2,2-pentafluoro-6,6-dimethylheptane 3,5-dione	SE-30, Apiezon L, Diethylene glycol adipate	2519
Traces (geometrical isomers	1,1,1 trifluoro--2,4-pentanedione	polyurethane	2640
Al, Mn, Fe, Co, Ni, Cu, Ga, Zn, In	Trifluoro-acetylpivaloylmethane	SE-30	2695
Al, Be, Cu	3-methyl, 3-ethyl 3-propyl and 3-butyl derivs of pentane 2.4 dione	Apiezon L or SE-30	2533
Al, Be, Sc, V, Co, Ni, Cu	2-theonyl trifluoroacetylacetonate	DC-710	2530
Traces in serum	-	-	2623
Al, Fe	Various	-	2659

		Cobalt	
Ru	tra, hfa	SE-30	2416
Ni, Pd	tacac	Apiezon	2479
Traces and organocobalamin in liver	fod	LSX-3-0295	2474, 2529
Al, Be, Cr, Sc, V, Ni, Cu	2-theonyl trifluoro-acetylacetonate	DC-710	2530
Al, Cr, Mn, Fe, Ni, Cu, Zn, Ga, In	trifluoroacetyl-pivalyl methane	SE-30	2695
Ni	1,2,2 trifluoro-acetylacetone-dimethylformamide	XE-60	2542
Ti, V, Cu, Ni, Mg, Mn, Zn, Pt, Pd	etioporphyrin	Epon 1001, XE-60	2338 - 2540
		Copper	
Be	acac	Silicone oil	2393
Al, Fe	acac	PTFE	2660
Al, Fe	tfa	PTFE	2660
Be, Al	tfa	Dow Silicone 710	2711
Be, Al, Fe	tfa	SE-30	2690
Al, Fe	tfa	-	2426
Al, Fe	tfa	Polyethylene	2420, 2422, 2423 2687, 2688, 2713
Fe	tfa	Daifl-Oil 200	2419
Al, Be, Cr	hfa	QF-1	2447
Fe	hfa	SE-52	2447
Be, Al, Ga, Fe, Cr, Rh	hfa	QF1 and SE.52	2625, 2626
Al, Be, Cr, Fe, Pd, Y	fod	SE-30	2631, 2632, 2473
Al, Be, Fe, In	tpm	SE-60	252
Al, Cr, Fe	tpm	SE-30	2466
Al, Be, Cu, Sc, V, Co, Ni	2-theonyl trifluoro-acetylacetonate	DC-710	2530
Be, Al, Cr	3-methyl, 3-ethyl, 3-propyl and 3-butyl derivs of pentane 2,4-dione	Apiezon L or SE-30	2533
		Gallium	
Al, In	tfa	DC-550	2414, 2619, 2689
Al, Be, In, Tl	tfa	Silicone 710	2414, 2681
Al, Be, Fe, Cr, Cu, Rh	tfa	QF1 and SE-52	2625, 2626
Al, Cr, Mn, Fe, Co, Ni, Cu, Zn, Fn	trifluoroacetyl-pivaloyl-methane	SE-30	2695
		Indium	
Al, Ga	tfa	DC-550	2424, 2619, 2689
Al, Be, Ga, Tl	tfa	Silicone 710	2414, 2681
Al, Be, Sc, Rh	tfa	Dow Silicone 710R	2682, 2683
Al, Be, Fe, Cu	tpm	SE-60	2521
Al, Cr, Mn, Fe, Ni, Cu, Zn, Ga	trifluroacetyl-pivaloyl-methane	SE-30	2695

		Iron	
Cr, Cu	acac	PTFE	2660
Cr, Al	acac	SE-30, Apiezon L, Diethylene glycol-adipate	2519
Cr, Cu	tfa	PTFE	2660
Cr, Al	tfa	SE-30, Apiezon L, Diethylene glycol-adipate	2519
Al, Be	tfa	Dow Silicone 710.R	2694
Be, Al, Cu	tfa	SE-30	2690
Al Cu	tfa	Polyethylene	2420, 2422, 2423, 2687, 2688, 2713
Al Cu	tfa	DC-550	2426, 2689
Cr	tfa	-	2452, 2622
Cu	tfa	Daifl-Oil 200	2419
Cr Rh	hfa	DC-200	2708
Al Be, Cr	hfa	Fluorolube	2447
Cu	hfa	SE-52	2447
Be Al, Ga, Cr, Cu, Rh	hfa	QF1 and SE-52	2625, 2626
Al, Cr, Cu, Rh, Pd	hfa	Squalane or Apiezon L on QF1	2523
Al, Cr, Cu, Rh, Fe, Pd	fod	Squalane, Apiezon L, QF1	2523
Cr, Al	fod	SE-30, Apiezon L Diethylene glycol adipate	2519
Cr, Al	tpm	SE-30, Apiezon, Diethylene glycol adipate	2519
Al, Cr	tpm	Apiezon L	2465
Al, Cr, Cu	tpm	SE-30	2466
Al, Be, In, Cu	tpm	SE-10	2521
Al, Be, Cr, Ni	dibm	Silicone grease	2478
Cr, Al,	thd	SE-30, Apiezon L diethylene glycol adipate	2519
Cr, Al	1,1,1,2,2-penta-fluoro-6,6-dimethyl-heptane-3,5-dione	SE-30, Apiezon L, diethyleneglycol adipate	2519
Al, Cr, Mn, Co, Ni, Cu, Zn, Ga, In	trifluoroacetyl-pivoloyl methane	SE-30	2695
In ore	-	-	2621
Al, Cr	various	-	2659
Pt, Mu, Ti, V	Etioporphyrin	Epon 1001	2538 - 2540
		Lead	
Traces	tpm, 1,1,1,2,2, penta-fluoro-6,6-di-methylbutane, fod	Apiezon L	2522

	Manganese		
Al, Cr, Fe, Co, Zn, Ga, In	Trifluoroacetyl pivaloyl methane	SE-30	2695
Ti, V, Cu, Mg, Co, Zn, Pt, Fe, Ti, V	Etioporphyrin	Epon 1001, XE-60	2538 - 2540
	Molybdenum		
U, W			2624
	Nickel		
Co, Pd	tacac	Apiezon	2479
Co, Pt, Mg, Ca, Ba, Hg, Cd, Zn, Mn, Cu	1,1,1-trifluoro-4-mercapto-pent-3 en-ore	Silicone Gum E350	2532
Al, Be, Cr, Fe	dibm	Silicone grease	2478
Al, Be, Cr, Cu, Fe, Pd, Y	fod	SE-30	2473
Al, Be, Cn, Sc, V, Co, Cu	2-theonyl trifluoroacetly-acetonate	DC-710	2530
Al, Cr, Mn, Fe, Co, Cu, Zn, Ga, In	Trifluoroacetylpivoloyl methane	GE-30	2695
Ti, V, Mg, Co, Mn, Zn, Cu, Pt, Pd	Etioporphyrin	Epon 1001, XE-60	2538 2540
Co	1:2:2 trifluoroacetone-dimethyl formamide	XE-60	2542
	Palladium		
Al, Be, Cr, Cu, Fe, Y	fod	SE-30	2473
Traces	tfa and fod	-	2479
Co, Ni	tacac	Apiezon	2479
Al, Cr, Rh, Cu, Fe	hfa and fod	Chromosorb-W-HP (DHCS treated with squalane or QF1	159
Ti, V, Cu, Ni, Mg, Co, Mn, Zn, Pt	Etioporhyrim	Epon 1001	2538 - 2540
	Platinum		
Ti, V, Cu, Ni, Mg, Co, Mn, Zn, Pt	Etioporphyrin	Epon 1001 XE-60	2538 - 2540
	Rare Earths		
Sc	thd	Apiezon H	2458, 2459
Sc, Y	tpm	SE-30	2462
Sc, Y	fod	SE-30	2476
Traces	thd	-	2696 - 2698
Traces	thd and fod	-	2699
Traces	Perfluoro-pivaloylmethane	-	2534
Traces	Benzoyltrifluoroacetone and complexes with hta and tri-n-butylphosphate	-	2531
Traces	hfa - tri-n-butyl phosphate mixed chelates	-	2700

Traces	-	-	2701
Traces	Mixed ligand	-	2705
Traces	tpm	Dow high vacuum Silicone grease	2702
Traces	Isobutyl pivol-aloyl methane	Dow high vacuum silicone grease	2535
Traces	2 different β-diketones	-	2703
Traces	tpm - tri-n-butyl phosphate, tpm-triactylphosphine		
Traces	1,1,1,2,2,6,6,7,7 decafluoroheptane -3,5 dione - dibutyl sulphoxide	Dexsil 300 Gc	2536
		Rhodium	
Al, Cr	tfa	SE-30	2427
Al, Cr	tfa	Tergipol NPX	2371
Al, Cr, Zr	tfa	Silicone oil	2371
Traces	tfa	SE-30	2427 - 2430
Traces	tfa	-	2627, 2628
Al, Be, Sc, In,	tfa	Dow Silicone 710R	2682 - 2684
Traces	tfa	-	2517
Cr	hfa	Dow high vacuum silicone grease	2707
Cr, Fe	hfa	DC-200	2708
Be, Al, Ga, Fe, Cr, Cu	hfa	QF1, SE-52	2625, 2626
Al, Cr, Cu, Fe, Pd	hfa	Squalane or Apiezon L or QF1	2523
Al, Cr, Cu, Fe, Pd	fod	Squalane, Apiezon L, QF1	2523
		Ruthenium	
Cr	tfa	Apiezon L	2706
Co	tfa	SE-30	2416
Cr	hfa	Silicone oil	2371
Co	hfa	SE-30	2416
		Scandium	
Al, Be	acac	Apiezon L	2395
Traces	acac	Apiezon L	2636
Al, Be	tfa	Silicone Oil	2395, 2635
Be, Al, In, Rh	tfa	Dow Silicone R.710	2682 - 2684
Al	tfa	Dow Silicone R.710	2680
Rare earths, Y	fod	SE-30	2466
Rare earths, Y	tpm	SE-30	2495, 2476
Rare earths	thd	Apiezon H	2487
Be, Al, Cr, V, Co, Ni, Cu	2-theonyl tri-fluoro-acetyl-acetonate	DC-710	2530
		Titanium	
In bauxite		-	2620
V, Cu, Ni, Mg, Co, Mn, Zn, Pd, Pt	Etioporphyrin	Epon 1001, XE-60	2538 - 2540

		Tungsten	
U, Mo	-	-	2624
		Thorium	
U	fod	QF.1	2475
U	hfa di-n-butyl sulphoxide mixed ligand complex	QF.1	2541
U	fod	-	2528
Be, Al, Cr, Th	tfa	DC.559	2685, 2886
		Uranium	
Al, Be, Cr, Th	tfa	DC.559	2685, 2686
Th	fod	QF.1	2475
Al	-	-	2421
W, Mo	-	-	2624
Th	fod	-	2528
Th	hfa di-n-butyl-sulphoxide mixed ligand complex	QF.1	2541
		Vanadium	
Traces	tfa	Various	2633
Al, Be, Cr, Sc, Co, Ni, Cu	2-theonyl tri-fluoroacetyl-acetonate	DC-710	2530
Ti, Cu, Ni, Mg, Co, Mn, Zn, Pt, Pd	Etioporphyrin	Epon 1001, XE-60	2538 - 2540
		Yttrium	
Al, Be, Cr, Cu, Fe, Pd	fod	SE-30	2473, 2631, 2632
Sc, rare earths	tpm	SE-30	2462 - 2464
Sc, rare earths	fod	SE-30	2476
		Zinc	
Traces	Pivaloyltrifluoro-acetonate	OV-10L on Chromosorb B	2634
Be	Oxyacetate and oxypropionate	Apiezon L	2537
Al, Cr, Mn, Fe, Co, Ni, Cu, Ga, In	Trifluoroacetyl pivoloyl methane	SE-30	2695
Ti, V, Cu, Ni, Mg, Co, Mn, Pt, Pd	Etioporphyrin	Epon 1001 XE-60	2535 - 2540
		Zirconium	
Al, Cr, Rh	tfa	Silicone grease	2371

ILLUSTRATIVE EXAMPLES OF THE APPLICATIONS OF CHROMATOGRAPHY TO METAL CHELATES

R. Gas Chromatography

Cobalt in water and cyanocobalamin (Vitamin B_{12}) in liver extracts. Cobalt has been determined in amounts down to 4×10^{-11} by a method involving chelation with 1,1,1,2,2,3,3 heptafluoro-7,7-dimethyl-4,6, octanedione (H fod) followed by gas chromatography using an electron capture detector (Ross et al[2529]).

Cobalt in aqueous solution. An aliquot (2 ml) of the aqueous sample containing approximately 1×10^{-4} g cobalt is transferred to a 10 ml Pyrex tube fitted with a screw cap. A small Teflon-coated stirring bar is placed in the culture tube which is positioned over a stirrer hot plate. The stirring mechanism is turned on before the addition of the reagents, A 0.1 ml aliquot of 0.5 N sodium hydroxide is added, followed by 1.0 ml of 0.1 M, 1,1,1,2,2,3,3 heptafluoro-7,7 dimethyl 1-4,6-octane (H fod)dione in benzene (present in excess to ensure complete reaction of cobalt), and 0.2 ml of 35% hydrogen peroxide to ensure cobalt is completely converted to the trivalent state). A small piece of Teflon tape is placed over the mouth of the tube and a cap is screwed on tightly to ensure no leakage of vapour during the reaction. The reagents are placed in a water bath heated to 84 - 86°C and allowed to react for 30 minutes with continuous stirring. The temperature should not be increased much in excess of 75°C to avoid the formation of 1,1,1,2,2,3,3 heptafluoro-7,7 dimethyl-4,6 octanedione dihydrate which does not readily gas chromatograph. The reagents in the tube are allowed to separate into two layers and the organic layer is decanted and put into a screw top vial. A 1 ml aliquot of 0.1 N sodium hydroxide is added to the benzene layer for back-washing. A thick, white precipitate forms immediately. This precipitate is the sodium salt of 1,1,1,2,2,3,3 heptafluoro-7,7 dimethyl-4,6 octanedione (Na fod), which is not sufficiently soluble to be entirely dissolved in this volume of aqueous layer. Four millititers of distilled water are added and the contents shaken vigorously. The precipitate usually disappears, however, if some remains, washing with distilled water will remove it. The benzene layer can now be decanted and is now ready for chromatography.

Cobalt in Vitamin B_{12}. The procedure for this analysis is very similar to the one described above for cobalt in aqueous solution. The Vitamin B_{12} is dissolved in distilled or deionized water. At higher concentrations the B_{12} solutions are bright red, and after the reagents are added and heating begins, the red colour in the aqueous layer quickly disappears and the green colour of cobalt 1,1,1,2,2,3,3 heptafluoro-7,7 dimethyl 4,6 octanedionate Co $(fod)_3$ soon appears in the organic layer, due to the conversion of cobalt in the cyanocobalamin complex to the cobalt β-diketonate.

This method depends upon the efficiency of benzene extraction and the measurement of the formed chelate in the organic layer of the reaction system. The sensitivity of the electron capture detector is essential for analysis of minute amounts of metals present in biological fluids. The major problem, however, with this detector is its sensitivity, for the free ligand 1,1,1,2,2,3,3 heptafluoro-7,7 dimethyl-4,6 octanedione (H(fod)). This free ligand which must be present in excess for quantitative reaction of the metal, desensitizes the detector making the analysis impossible. The procedure must, therefore, include a crucial back-washing with sodium hydroxide as described above to remove the excess ligand from the organic layer without removing the cobalt chelate.

The sensitivity of the method can be enhanced two or three orders of magnitude at this point by evaporating the benzene while in contact with the aqueous layer and redissolving it in from 10 ml to 0.01 ml of benzene, depending on the desired sensitivity. An alternate method of concentration is to leave the tube open during the reaction to permit evaporation of the benzene.

The gas chromatograph instrument conditions for analysis are listed below:
Instrument, Hewlett Packard Model 402.
Column, 2 ft x 1/4 in o.d. Pyrex tube packed with 5% Dow Corning LSX-3-0295 on 60 - 80 mesh Gas Chrom P.
Column temperature, 135°C.
Helium eluant, 60 ml/minute
Auxiliary gas, 95% argon and 5% methane, 100 ml/minute
Injection port temperature 135°C
Detector temperature 180°C
Known aliquots of the benzene solutions are injected into the chromatograph. Peak heights are obtained and compared with the calibration curve produced from the analyses of a standard cobalt 1,1,1,2,2,3,3 heptafluoro-7,7 dimethyl-4,6 octanedionate solutions.
When these particular conditions are used, the cis and the trans isomers of cobalt 1,1,1,2,2,3,3 heptafluoro-7,7 dimethyl-4,6 octanedionate are both eluted as one peak, thereby simplifying the analysis.

Ross et al[2529] determined recoveries by radioactivation techniques on cobalt solutions containing known additions of Co^{60}. They checked recoveries of cobalt 1,1,1,2,2,3,3 heptafluoro-7,7 dimethyl-4,6 octanedionate (Co(fod)). Using 2 ml aqueous cobalt sample solution, 4 ml 0.1 M, 1,1,1,2,2,3,3, hepta-fluoro-7,7 dimethyl-4,6 octanedione (Hfod) in benzene, 0.2 ml 0.5 N sodium hydroxide and 0.2 ml 30% hydrogen peroxide they obtained cobalt recoveries of 96% or higher. Cobalt recoveries fell to less than 10% when 4 ml of a 0.01 M or a 0.001 M benzene solution of 1,1,1,2,2,3,3, heptafluoro-7,7-dimethyl-4,6 octanedione was substituted for the 0.1 M reagent in the above recovery experiments. The following conditions lead to a cobalt recovery in excess of 97%: To 2.00 ml aqueous Co II solution (1 x 10^{-4} g Co/ml) were added 0.20 ml of 0.54 sodium hydroxide, 4.00 ml of a 0.1 M benzene solution of 1,1,1,2,2,3,3, heptafluoro-7,7 dimethyl-4,6 octanedione H(fod), and 0.20 ml of 30% hydrogen peroxide, with shaking after each addition. The mixture is heated, capped or open to the atmosphere, for 30 minutes in a water bath at 84 - 86°C. After heating, the open systems are treated with 4.00 ml of benzene, 0.50 ml of 2 M perchloric acid if desired, shaken 5 minutes and centrifuged 5 minutes. Closed systems can be treated with 0.5 ml of 2 M perchloric acid if desired, shaken 5 minutes and centrifuged 5 minutes. Table 220 shows data obtained in the analysis of cobalt in aqueous solutions of Vitamin B_{12}. Because the results are consistently about 15% low, a correction factor could be employed in this method. The method is rapid (less than one hour) and sensitive enough to be useful for analysis of Vitamin B_{12} in pharmaceutical preparations, liver extracts and in B_{12} feed concentrates. Ross et al[2529] report that other biological compounds in liver extracts do not interfere in this method of analysis.

Beryllium in particulate matter collected from air filters. A routine method of analysis for ultratrace concentrations of beryllium in particulate matter collected on air filters and in water has been described by Ross and Sievers[2653]. The particulate matter on the filters is ashed, the residue is digested with acid, and the resulting solution is buffered to the optimum pH for solvent extraction. Potential interfering metals are eliminated by masking with EDTA. A benzene solution of trifluoroacetylacetone is allowed to react with the beryllium in the aqueous solution to form beryllium tri-fluoroacetylacetonate. The organic extract is injected into a gas chromatograph equipped with an electron capture detector. The chromatographic peak resulting from beryllium trifluoroacetylacetonate formed from the unknown solution is compared with standard beryllium trifluoroacetylacetonate solutions for quantitation. This gas chromatographic method is more sensitive and reproducible than other methods used for the determination of beryllium in environmental samples, and has been applied to the determination of beryllium at concentrations from 0.0001 to 0.0005 microgram of beryllium per cubic meter of air. The method described below is therefore suitable for determining beryllium in air with good precision at levels two orders of magnitude below the A.E.C. recommended maximum allowable concentration of 0.01 μg beryllium/m^3 air. The technique is rapid, requiring about 30 minutes for the extraction procedure and 10 minutes for the gas chromatographic analysis.

Table 220 - Analysis for cobalt in vitamin B_{12} by gas chromatography

Cobalt in vitamin B_{12} added - g	Cobalt in vitamin B_{12} found by gas chromatography - g	Mean g	Mean recovery %	Mean Error	Relative Error %
4.35×10^{-19}	3.54×10^{-10}	3.67×10^{-10}	84.4	0.68×10^{-10}	15.6
	3.56×10^{-10}				
	4.11×10^{-10}				
	3.54×10^{-10}				
	3.60×10^{-10}				

Reagents. 0.164 M trifluoroacetylacetone in Nanograde Benzene (Mallinckrodt). 2 ml of freshly distilled trifluoroacetylacetone (Pierce Chemical Co.) diluted in 100 ml of benzene. The solution was stored in a silanised 100 ml borosilicate volumetric flask. Benzene solutions of trifluoroacetylacetone and also neat trifluoroacetylacetone are subject to slow decomposition on standing which produces spurious gas chromatographic peaks.

3N sodium hydroxide solution for neutralising reagent and 0.18 NaOH back-washing reagent. Prepared from Matheson, Coleman and Bell (MCB) Analyzed Reagent Grade sodium hydroxide and deionized water.

Ethylenediaminetetraacetic Acid (EDTA) - Buffer Solution. Weigh out 5.15 g of $Na_2EDTA.H_2O$ (G.F. Smith Chemical Co.) 85 g of $NaOAc.3H_2O$ (MCB, reagent) and 6.25 ml of glacial acetic acid (Baker, Ultrex) and dissolve these in 500 ml of deionized water.

Standard Beryllium Solutions. Beryllium trifluoroacetylacetonate solutions prepared by dissolving beryllium in trifluoroacetylacetonate, purified by sublimation, in Nanograde (Mallinckrodt) benzene. Further dilutions made until the resulting concentration was 1.0×10^{-11} g of beryllium / μl. This solution should be prepared every two weeks. The standard aqueous beryllium solutions are prepared by dissolving a beryllium metal chip in concentrated hydrochloric acid and diluting with deionized water to an ultimate concentration of 9.2×10^{-12} g of beryllium / μl.

Silanization of Glassware. The standard samples and trifluoroacetylacetone reagent should be stored in borosilicate volumetric flasks which had been cleaned in chromic acid cleaning solution and then silanized to reduce any reactive sites. Silanization can be achieved by filling the flasks with a 20% solution of hexamethyldisilazane (HMDS) in benzene and letting them stand overnight. The reaction vessels (5 ml culture tubes) are treated by a similar method.

Preparation of filter digests. Air filters were obtained by drawing known volumes of air through porous filters. Glass fiber filters which retain particles as small as 0.3 micron are recommended in conjunction with high volume samplers. The filter size was 7 inches by 9 inches of exposed surface. These filters permitted passage of between 50 and 60 feet of air per minute. A total air sample of 2200 cubic meters collected over a 24 h period was used. One-inch strips were cut from these filters and were digested as follows. The strips are placed in borosilicate glass boats and placed in a Tracerlabe Low-temperature Asher at ca. 150°C for 1 h at 1 mm chamber pressure with an oxygen flow of 50 cc per minutes.

The ashed filter is placed in a glass thimble, and the thimble placed in an extraction tube. A 125 ml Erlenmeyer flask is charged with 8 ml of constant boiling hydrochloric acid (19%) and 32 ml of nitric acid (40% redistilled ACS grade). The flask is attached to the extraction tube and the extraction tube fitted with an Ahlin condenser. The acid is refluxed over the sample for 3 h. The extraction tube and the condenser are removed and the flask fitted with a thermometer adapter. The extracted liquid is concentrated to 1 - 2 ml on a hot plate, cooled, and allowed to stand overnight. This material was quantitatively transferred to a graduated 15 ml centrifuge tube with three washings of 5 to 10 drops of diluted acid. The samples are diluted to 4.4 ml with distilled water. The tubes are then centrifuged at 2000 rpm for 30 minutes, and the supernatant liquid transferred to polypropylene tubes ready for analysis.

Procedure for gas chromatographic analysis. One milliliter of the aqua-regia filter digest is accurately pipetted into a 5 ml borosilicate glass culture tube fitted with a screw cap and 1 ml EDTA-sodium acetate buffer solution added to the tube. Adequate amounts of a 3 N sodium hydroxide solution are added to bring the pH to between 5.5 and 6.0. The requirement is between 0.8 and 1.3 ml, depending upon the original pH of the digest. The tube containing a small Teflon coated magnetic stirrer is capped and inserted into an oil bath maintained at 93°C on a stirrer hot plate. The sample is heated and stirred vigorously for 10 minutes.

The tube is cooled, and 1 ml of 0.164 N trifluoroacetylacetone in benzene added. The tube is capped and placed above the stirring mechanism of the hot plate and stirred for another 15 minutes at ambient temperature.

The tube is permitted to stand until organic and aqueous layers are separated, and then the organic layer transferred with a medicine dropper to a 2-dram vial and 2 ml of 0.1 N sodium hydroxide added. This mixture is quickly shaken by hand for 5 s to remove excess trifluoroacetylacetone from the organic layer. The two phases are immediately separated by withdrawing the organic layer with a medicine dropper and transferring it to another 2-dram vial which is then capped. This step must be performed rapidly and reproducible because intolerably large amounts of beryllium will otherwise be lost and both precision and accuracy will suffer.

Five 1 μl aliquots of organic layer are repetitively injected into the gas chromatograph. The mean of peak heights from the beryllium trifluoroacetylacetonate peaks in the unknown solutions are compared with the average from five analysis of a standard solution of beryllium trifluoroacetylacetonate. Over certain beryllium concentration range the response is essentially linear so quantitation is simplified and within this range peak height ratio can be used for the calculation of the quantity of beryllium in the sample.

Instrument Conditions.

Gas Chromatograph: Hewlett Packard Model 402 high efficiency gas chromatograph equipped with a tritium source electron capture detector.
Column: A 2 meter x 3 millimeter ID borosilicate glass column packed with 2.8% W-98 Silicone (Union Carbide) on Diataport S (Hewlett Packard).
Carrier Gas: Methane 10% argon 90%
Column Temperature: 110°C
Detector Temperature: 200°C
On-Column injection (no additional heat at site of injection)

Ross and Sievers[2653] found in beryllium recovery determinataions carried out using the above method that this element could be determined at the $8-9 \times 10^{-9}$ g/ml level in air filter digests with a relative error of 4.3% (Table 221).

Table 221 - Determination of beryllium in air by gas chromatography - Beryllium recovery from known aqueous solutions

	Conc. in known aqueous soln. (g/ml)	Beryllium found (g/ml)
	9.2×10^{-9}	9.1×10^{-9}
		8.8×10^{-9}
		7.9×10^{-9}
		8.8×10^{-9}
		9.4×10^{-9}
	Mean	8.8×10^{-9}

Mean error = 0.4×10^{-9}
Relative error = 4.3%

In a precision study of the measurement of beryllium in a typical authentic filter in which the aqueous digest of one filter was carried through the analytical procedure five times, they obtained the data shown in Table 222.

Table 222 - Beryllium found in air filter digest precision study

	Amount detected μ Be/ml aqueous digest	Amount detected (μg Be/m³ Air)
	2.3×10^{-8}	4.1×10^{-4}
	2.4×10^{-8}	4.5×10^{-4}
	2.3×10^{-8}	4.1×10^{-4}
	2.4×10^{-8}	4.3×10^{-4}
	2.4×10^{-8}	4.3×10^{-4}
Mean	2.4×10^{-8}	Relative standard deviation = 3.0%.

* Mean of five GC injections compared with five GC analyses of standard.

Ross and Sievers[2653] showed that concentrations in the air sample of iron (3.19 µg/m^3), lead 6.0 µg/m^3), nickel (0.01 µg/m^3), copper (o.61 µg/m^3) and magnesium (0.13 µg/m^3), all present at much higher concentrations than beryllium, did not interfere in this gas chromatographic procedure. In the first step of the analytical procedure EDTA is used to remove most of several of the metals present as major constituents. Most of the metals which remained are removed rapidly by the sodium hydroxide backwash step. Any remaining metals are removed or were separated by the chromatographic column so that they did not interfere. Any aluminium and chromium that remain after the above treatment are indicated by two peaks eluting as aluminium trifluoroacetylacetonate and chromium trifluoroacetylacetonate which have higher retention times than beryllium trifluoroacetylacetonate.

Beryllium in human and rat urine. Foreman et al[2657] have reported the following method for determining beryllium in amounts down to 1 ug/ml in urine.

Foreman et al[2657] used a chromatograph fitted with an electron capture detector or with a flame-ionisation detector. Operating conditions were chosen so that the beryllium complex was eluted in a minute without decomposition, and was resolved from the bulk of the solvent and from solvent and complexing agent impurities. The most suitable stationary phase was found to be the methyl phenyl silicone gum Gas Chrom Z with 5% SE52 stationary phase packed in PTFE.

Adsorption effects can be minimised by using column plugs of PTFE yarn rather than glass yarn. Total adsorption occurs on the porous polymer bead column. The treatment of a PTFE support with Cetrimide gave a significant improvement over a similar, untreated support. The use of PTFE columns in conjunction with an electron capture detector does not cause contamination of the detector, provided that, both prior to and after packing, the column is conditioned for several days at 200°C.

Calibration. Calibration of the electron capture detector was carried out with standard solutions of bis-(trifluoroacetylacetonate)-beryllium (II) in benzene by successive dilution of a 100 µg ml-1 primary standard. The detector is readily overloaded and this can give rise to some curious peak shapes making quantitative analysis difficult. An upper limit of detection of 10 µg beryllium/ml was achieved with this apparatus whilst the lower limit was 0.5 mg beryllium ml. It is recommended that all standard solutions are stored in borosilicate glass to minimize adsorption effects.

Method. EDTA is added to a known volume of the aqueous sample which is then adjusted to pH 6 with ammonia solution and buffered with M sodium acetate. The same volume of 0.05 M trifluoroacetylacetone in thiophen-free benzene is added, and extraction carried out for 30 minutes. In a typical analysis of 100 mg of EDTA disodium salt, 5 ml of sample and of complexing solution and 1 ml of sodium acetate solution are used. The excess of complexing agent in the organic extract is destroyed by shaking it with 1 ml of 0.1 M sodium hydroxide for several seconds. Microlitre aliquots of the extract can then be analysed. Failure to destroy the excess of trifluoroacetylacetone results in gross interference in the gas-chromatographic analysis and the electron capture detector may remain unusable for many minutes.

Foreman et al[2657] applied the above procedure to standard aqueous solutions of beryllium sulphate, covering the range 0.05 to 0.5 µg/ml^{-1}. Nearly complete recovery of the beryllium was observed in all cases.

The analysis of samples containing beryllium in the presence of large amounts of other metals is open to error when these metals also form - diketonates.

It is necessary to ensure that sufficient β-diketone is added to comple all the metals or, if convenient, the interfering metals can be complexed by the addition of EDTA prior to adding the β-diketone. To demonstrate this the analysis of an aqueous solution containing 0.1 μg ml^{-1} of beryllium and 50 μg ml^{-1} of iron and 1.8 μg/l phosphoric acid was carried out. In the absence of EDTA, most of the β-diketone was used in complexing the excess of iron, and only 0.02 μg/ml beryllium was detected. In the presence of excess of EDTA, however, a 0.10 μg/ml recovery of beryllium was obtained.

Foreman et al also applied the method to human urine, to which known amounts of beryllium sulphate had been added, under the following operating conditions: the injection, column and detector temperature were 100°, 100° and 125°C, respectively; column and purge gas flow rates were 67 ml minutes- (nitrogen gas);the column was of PTFE, 4 ft. long by 1/8 inch o.d., containi 5% SE 52 on 72 to 85 mesh Gas Chrom Z; and the electron capture detector had a pulse period of 150 μs, width 0.75 μs, amplitude 47 to 60 V and a source of 10 - mCi nickel-63.

Table 223 - Spiked Urine Analysis

Extraction after wet combustion beryllium concentration μg ml^{-1}			Direct extraction beryllium concentration μg ml^{-1}		
Sample No.	Found	Added	Sample No.	Found	Added
1	2.7	3.4	10	0.82	1.0
2	1.6	1.3	11	0.80	1.0
3	1.3	1.3	12	0.11	0.10
4	1.0	1.0	13	0.11	0.10
5	0.40	0.37	14	0.050	0.050
6	0.25	0.37	15	0.050	0.050
7	0.10	0.10	16	0.009	0.010
8	0.036	0.050	17	0.005	0.005
9	Nil	Nil	18	0.001	0.001
			19	Nil	Nil

Beryllium was recovered both by wet oxidation and by direct extraction by shaking buffered urine with the complexing agent in benzene, in the manner described above. Background interference can be minimised by successively shaking the organic extract with water. The results are summarised in Table 223, from which it can be seen that both procedures yield satisfactory results which are the mean values of triplicate analyses. Chromatograms of urine extracts are shown in Figure 160.

Preparative scale separation of mixtures of chromium, aluminium and iron 1,1,1 trifluoro-5,5-dimethyl hexane, 2,4-dionates ($M(tpm)_3$). Belcher et al[2519] showed that these complexes could be eluted with reasonable retention times at temperatures as low as 160°C on a 6.1 m x 9.5 m aluminium preparative column containing 15% Apiezon L on Universal B. Separation was good when samples are containing 33mg of each complex were injected (dotted line in Figure 161).

An excellent separation was achieved when all three elements were present at similar concentrations. The relative retention times for the aluminium,

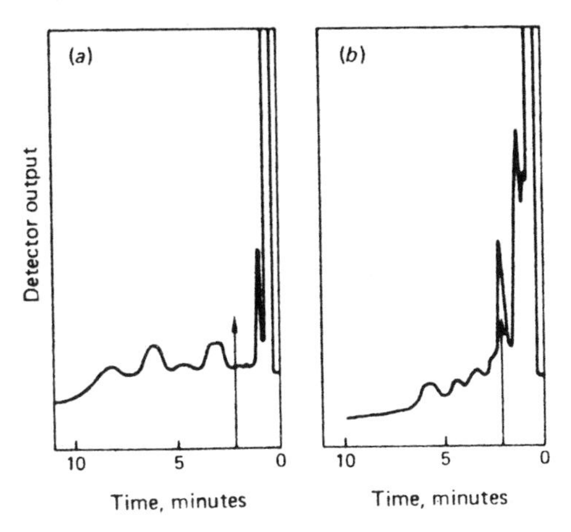

Figure 160 - Gas chromatographic determination of beryllium in urine.

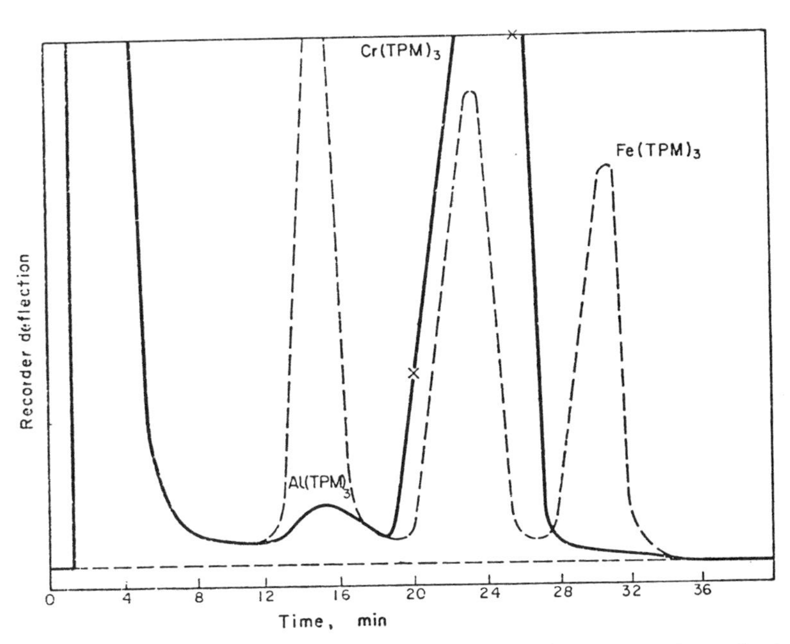

Figure 161 - Preparative scale gas chromatographic separation of chromium, aluminium and iron tpm chelates.

chromium and iron complexes were 9, 14.5 and 20.5 minutes. In Figure 161 (continuose line) is shown the result of an attempt to seperate the chromium TPM complex from 2% w/w each of the aluminium and iron TPM complexes. A total 0.1 g of mixed complexes was injected automatically in 0.5 ml portions with a syringe operated by an air piston. A small peak corresponding to the aluminium complex was observed in addition to the main chromium -TPM peak but the katharometer did not detect any iron complex. On comparing the chromatogram with that obtained in the earlier separation (dotted line), it was assumed that either the iron had been eluted after the chromium peak but not detected, or it had been retained or decomposed on the column. Since the analysis of the collected chromium chelate showed very low levels of iron impurity, the assumption that the iron chelate was not emerging from the oolumn in any considerable quantity during the collection period seems justified. Furthermore these low levels suggest that any iron possibly present as an impurity in the support material was not being leached out to any measurable extent. No detectable quantity of iron complex could be trapped out between the end of the predicted position of the iron peak and the beginning of the next collection period. Nevertheless, as a general rule, a time lapse of some 5 min was allowed between the end of the chromium peaks and the reactivation of the injection cycle, thus giving a total time lapse of about 25 min. between chromium collection periods.

Separation of aluminium, chromium, rhodium, iron, copper and palladium complexes. Wolf et al[2523] determined retention data over a range of temperatur es for aluminium, chromium, rhodium, iron, copper and palladium trifluoroacetylacetonates (tfa), hexafluoroacetylacetonates (hfa) and 1,1,1,2,2,3,3, hepta fluoro-7,7 dimethyl-4,6 octanedionates (fod). They used an F & M Model 810 research chromatograph equipped with a thermal conductivity detector. Columns made of Teflon tubing and glass liners in the injection port were used to minimise the amount of metal surface in the system. Peak retention times from the air peak were measured to ± 0.005 min. This accuracy necessitated that the column temperature was controlled to within 0.2° at 50°C and 2°C at 200°C over the whole length.

Saturated solutions of the chelates in toluene, n-heptane, or n-pentane were used and sample volume was used so that a symmetrical peak of 30 - 50% full-scale deflection was observed at the lowest attenuation available. This corresponded to approximately 0.01 - 0.1 μmol of chelate. Sample sizes 10 times as great as this showed no change in peak retention times. A slight tailing of the peaks was noticed for several systems at larger sample sizes. This was attributed to incomplete volatilization in the injection port.

At flow rates of carrier gas above 60 cm^3/min the error in reproducibility of peak retention times increased to about 5%. This may be due to nonequilibrium conditions arising from the low solubility of these chelates in the liquid phase studied.

Three types of column were used by Wolf et al[2523], acid washed Chromosorb W-HP (acid washed) DMCS treated (Johns Manville) was used for all three columns. Liquid phase; squalane (2,6,10,15,19,23-hexamethyltetracosane) was used since it is a nonpolar pure hydrocarbon that should have no specific interactions with the metal chelates. This phase is useful only to about 110°C before excessive-bleeding makes the amount of phase in the column uncertain. The chelates containing the hexafluoroacetylacetone ligand can be successfully chromatographed below 100°C but the chelates containing ether ligands require somewhat higher temperatures (ca 150°C) for reasonably rapid elution. Liquid phase; Apiezon L used at column temperatures of about 150°C. Liquid phase; QF-1.

The data in Table 224 demonstrates successful separations under various experimental conditions. Thus copper and palladium and copper are separated as their 1,1,1,2,2,3,3 heptafluoro 7,7-dimethyl-4,6 heptane dionates (fod) on a column comprising 23.4% Apiezon L on 60/80 mesh Chromosorb W-HP at a column temperature of 162 - 192°C.

Table 224 - Constants for specific retention volume equation (A) and heats of solution

Chelate	Column	A	H, R cal/mol	Temp. range °C
			Squalane	
$Al(hfa)_3$	a	10.34	- 9.6 ± 0.1	69 - 98
$Cr(hfa)_3$	a	11.74	-11.2 ± 0.1	69 - 98
$Rh(hfa)_3$	a	12.28	-12.0 ± 0.1	69 - 98
			Apiezon L	
$Al(hfa)_3$	f	14.31	-11.4 ± 0.2	52 - 72
$Cr(hfa)_3$	f	14.43	-12.0 ± 0.4	52 - 77
Rh(hfa)	f	15.45	-13.1 ± 0.1	127 -152
$Al(tfa)_3$	c	11.52	-13.3 ± 0.1	167 -202
$Cr(tfa)_3$	b	11.54	-14.1 ± 0.1	167 -202
$Al(fod)_3$	c	13.36	-15.7 ± 0.2	132 -162
$Cr(fod)_3$	b	12.14	-14.9 ± 0.1	167 -202
$Fe(fod)_3$	c	13.14	-15.8 ± 0.2	132 -192
$Cu(fod)_3$	c	12.34	-16.0 ± 0.3	162 -192
$Pd(fod)_3$	c	13.64	-17.6 ± 0.2	162 -192
			QF-1	
$Al(hfa)_3$	d	14.34	-14.0 ± 0.2	67 - 92
$Cr(hfa)_3$	d	15.11	-15.1 ± 0.1	67 -107
$Rh(hfa)_3$	d	15.44	-15.6 ± 0.1	147 -167
$Al(tfa)_3$	d	13.71	-16.1 ± 0.3	147 -172
$Cr(tfa)_3$	d	12.41	-15.5 ± 0.1	147 -172
$Al(fod)_3$	d	14.02	-17.3 ± 0.1	147 -172
$Cr(fod)_3$	d	13.31	-16.8 ± 0.1	147 -164
$Cu(fod)_2$	e	13.90	-17.3 ± 0.1	147 -164
$Pd(fod)_2$	e	13.93	-17.5 ± 0.2	147 -164

Column (a) 24.9% squalane (1.353 g), length 33 in; (b) 23.4% Apiezon L (0.793 g) 46 in; (c) 23.4% Apiezon L (0.387 g) 24 in; (d) 19.1% QF-1 (0.301 g) 24 in; (e) 19.1% QF-1 (0.291 g) 24 in; (f) 16.6% Apiezon L (0.665 g 60 in;
Columns (a) and (f) were 3/16 in i.d. Teflon tubing; the others were 1/8 in;
Column (a) was 80/100 mesh Chromosorb W-HP; the others were 60/80 mesh Chromosorb W-HP.
Specific retention volume (Vg°) is obtained from the following equation:

$$\ln Vg^\circ = - \frac{Hn}{R\,t} + A$$

Gas chromatography of volatile metal diethyldithiocarbamates. Cardwell et al[2719] prepared and separated the diethyldithiocarbamates of nickel, copper II, palladium II, zinc, cadmium, mercury II, lead, platinum II, silver I, iron III and cobalt III.

Gas chromatographic studies were carried out on two instruments, a Perkin-Elmer F30 and a Varian 2440, both with flame ionization detectors. U-shaped pyrex glass columns (6.5 mm o.d., 2.8 mm i.d. length 35 cm) were fitted to the Perkin-Elmer F30, and coiled stainless steel columns (3.2 mm o.d. 2.16 mm i.d. length 1 mm) were used with the Varian 2440. The packing material was acid-washed Chromosorb W (60 - 80 mesh) which was exhaustively silanized by refluxing with a 5% dimethyldichlorosilane (DMCS) solution in ether, before it was coated with 5% OV-101. Eluted samples were collected and identified by mass spectrometry.

They found the most satisfactory chromatographic results were obtained with 5% OV-101 on a highly deactivated diatomaceous earth support, acid-washed Chromosorb W pretreated with DMCS. It is imperative that the support be exhaustively silanized before being coated with the liquid phase.

All the complexes, except for those of Fe(III) and Ag(I) could be eluted successfully at column temperatures in the range 220 - 245°C, with injector and detector temperatures of 250°C. Glass and stainless steel were equally effective for column construction and well defined peaks were obtained with no chromatographic evidence of decomposition taking place. Eluted samples were analysed by mass spectrometry, which confirmed that elution was effected without decomposition, e.g. when a sample was collected after injection of the nickel chelate, the major ions observed in the mass spectrum occurred at m/e 355, 323, 229 and 207, identical to the spectrum of the pure chelate. Column efficiencies of up to 1000 theoretical plates were achieved with the narrower bore stainless steel columns.

Figure 162 shows an example of the separation of zinc, cadmium and lead at 220°C. The order of elution is in close agreement with the volatilities of the complexes observed in thermal analysis.

Figure 163 shows a separation of zinc from nickel, palladium and platinum; as in the previous example, the order of elution agrees with the order of volatility observed in thermal analysis. Cadmium and mercury were also separated.

Table 225 summarizes the chromatographic data for the metal chelates examined. Although the survey is by no means complete, the metal dithiocarbamates clearly show considerable promise for metal analysis by gas chromatography. There is considerable scope for variation of alkyl substituents which may lead to improved resolution of metals which have similar retention times for the diethyl derivatives. The range of metals could also be extended to include some non-transition metals such as arsenic, antimony and bismuth. Other advantages of the dithiocarbamate system are the ready availability of chelating agents, the simplicity of chelate preparation, and the wealth of information available on extraction conditions and physical properties of the chelats.

Table 225 - Gas chromatographic data (Injector/detector, 250°C; carrier gas flow, 60 cm^3 min^{-1}: glass column, 35 cm x 2.8 mm i.d.)

Chelate	Ni	Pd	Pt	Zn	Cd	Cu	Pb	Hg	Cu
Oven temp (°C)	230	235	245	205	225	230	220	220	245
Retention time (min)	2.81	4.63	4.17	4.00	3.00	2.36	3.01	5.80	7.64

Separation of copper and nickel by eluent extraction and gas chromatography using bis(acetylprivalyl-methane) ethylenediimine as reagent. In common with the metal complexes of other ligands of this type, the copper(II) and nickel(II) complexes of bis(acetylpivalyl-methane)ethylenediimine (H_2[en (APM)$_2$] showed sufficient volatility and thermal stability to allow for

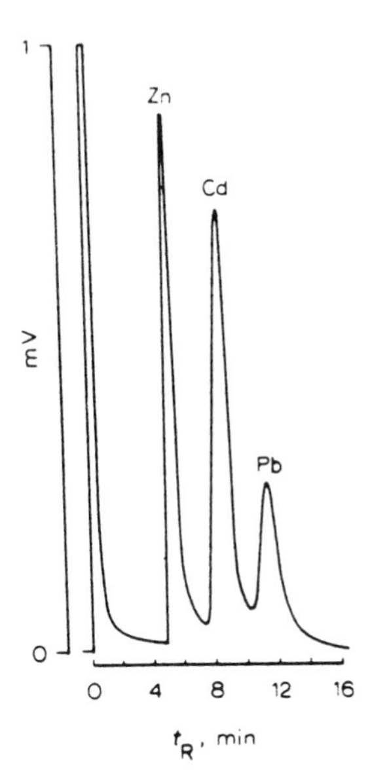

Figure 162 - Gas chromatographic separation of zinc, cadmium and lead diethyl-dithiocarbamates.

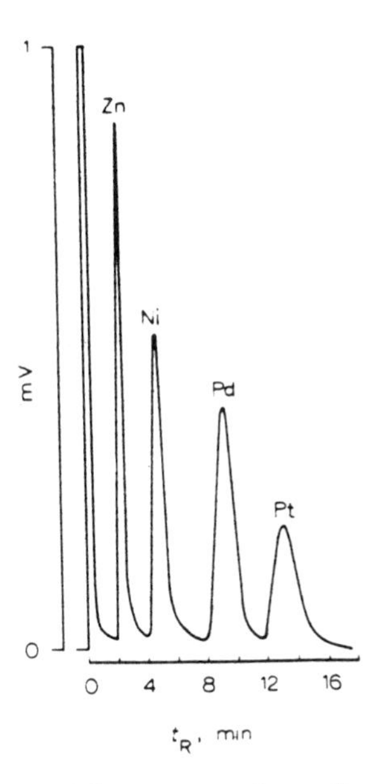

Figure 163 - Gas chromatographic separation of zinc, nickel, palladium and platinum diethyl-dithiocarbamates.

their successful gas chromatography. Although somewhat less volatile than the complexes of the tetradentate β-ketoamines derived from fluorinated β-diketones, such as 1,1,1-trifluoroacetylacetonate, the chelates of $H_2[en(APM)_2]$ possessed a remarkable thermal stability and could be eluted undecomposed from a gas chromatograph at temperatures as high as 300°C. The relative ease with which the copper (II) and nickel chelates of $H_2[en(APM)_2]$ were formed in aqueous solution and their facile extraction into a wide range of water-immiscible organic solvents indicated to Belcher et al[2721] the possible usefulness of this ligand in quantitative analysis for copper and nickel. Only with this ligand was the separation of the copper (II) and nickel chelates by gas chromatography consistently reproducible and effective enough for quantitative purposes.

The investigation of Belcher et al[2721] also included a study of the chelates of $H_2[en(TPM)_2]$ and $H_2[pn(TPM)_2]$ derived from the β-diketone, trifluoroacetylpivalylmethane, HTPM. The copper (II) and nickel chelates of this tetradentate ligand are remarkable in possessing exceptionally high volatilities, the highest recorded for this kind of β-ketoamine complex. Unfortunately, the preparation of $H_2[en(TPM)_2]$ in amounts suitable for analytical use is not an attractive proposition, because of unusually low yields and the exploitation of the extraordinary properties of the chelates of $H_2[en(TPM)_2]$ in analytical procedures could not easily be realized.

In the development of $H_2[en(APM)_2]$ as an analytical reagent, the thermogravimetric spectrophotometric and gas chromatographic characteristics of its copper (II), nickel (II) and palladium (II) chelates were studied together with the solvent extraction of these chelates from aqueous solution. The bulky t-butyl groups in the metal complexes were responsible for enhanced solubility of the chelates in organic solvents, particularly non-polar solvents, and provided efficient extraction of the chelates for subsequent examination, either by spectrophotometry or, preferably, by gas chromatography.

The resolution of a mixture of the copper (II) and nickel (II) complexes of $H_2[en(APM)_2]$ by gas chromatography was successfully achieved over a sufficiently wide range of chromatographic conditions to encourage the development of a simple gas chromatograph method for the determination of both elements in a variety of samples.

Gas chromatography was carried out on a gas chromatograph equipped with flame ionization detection. Hydrogen and air were used for the detector system. The glass columns (1.5 m x 4 mm i.d.) used were packed with 5% silicone gum rubber E-350 on Universal B support (60 - 85 mesh). All columns were preconditioned for 16 h at the maximum operating temperature. For the chromatography of the individual metal complexes, injection port and detector oven temperatures at 260°C and 270°C respectively, were used. Carrier gas flow as usually 60 $cm^3\ min^{-1}$. The shapes of the eluted peaks were strongly temperature-dependent. The broad peak obtained at a column temperature of 230°C improved in shape as the temperature was raised. At 290°C, sharp, symmetrical peaks with no evidence of decomposition were obtained, but unfortunately, were not too well resolved from the solvent peak. For most practical purposes, a temperature of 250°C was optimal; not only was the peak shape quite satisfactory, but the resolution of complex and solvent peaks was good. The other chelates $Ni[en(APM)_2]$ and $Pd[en(APD)_2]$ were also satisfactorily eluted under these conditions. Good Gaussian-shaped elution peaks were obtained for all three complexes, with no indication of decomposition or sample loss on the column, and with good base-line return. At 250°C, the retention times for the three chelates were 260 s (copper), 297 s (nickel) and 495 s (palladium).

To determine traces of copper in water, aliquots of water (100 cm^3) were treated with acetate buffer (1.0 M, pH 7.0, 10 cm^3). A solution of $H_2[en(APM)_2]$ in n-hexane (0.02 M, 5 cm^3) was added to each container, which was tightly closed and shaken on a mechanical shaker for 1 h. The organic phase was removed and suitable volumes (1 - 5 μl) were examined chromatographically under the conditions described above.

Simultaneous determination of copper and nickel[2721]. Solutions (1 cm^3) containing copper and nickel (0.02 - 0.12 mg cm^{-3}) were taken in 2-dram vials, and 1 M sodium acetate solution (1 cm^3) was added to each to give pH 8.0. A solution of $H_2[en(APM)_2]$ in 2:1 ethanol-water (1 cm^3) was added and the mixture was heated on a boiling water bath for 15 min. Cyclohexane (2 cm^3) was added to the cooled mixture and the vials were sealed. The copper and nickel chelates were extracted by shaking the vials for 1 h on the mechanical shaker. Portions (1 - 5 μl) of the organic phase were subjected to gas chromatography with the standard column, maintained at 260°C and a carrier gas flow rate of 10 cm^3 min^{-1}.

For the analysis of alloys, the sample (0.5 g) was dissolved in aqua regia, with warming to expel free chlorine and nitrous fumes. The solution was diluted to 250 cm^3 and 25 cm^3 of this solution were further diluted to 250 cm^3, after adjusting the pH to 6.0 - 7.0 with 0.1 M sodium hydroxide. An aliquot (1.0 cm^3) of this solution was taken for analysis as above.

Linear calibration curves were obtained over the range 10^{-7} to 10^{-8} g of copper and nickel, the practical limit of detection for each chelate being about 5 x 10^{-9} g, corresponding to approximately 1 μg of metal.

The analysis of tap water for copper was a particularly tacile operation involving equilibration of the buffered (pH 7.0) water sample with a solution of the ligand in n-hexane, separation of the phases, and injection of a suitable aliquot of the organic solution into the chromatograph. Calibration curves were prepared by extracting the copper complex from 100 cm^3 portions of aqueous solutions containing known amounts of copper (II) in the range 10 - 100 μg. The straight-line graphs obtained were used to determine the amount of copper in the test samples. Results for domestic water samples containing about 0.25 μg cm^{-3} of copper (S_r = 6% for 4 determinations) agreed well with those found by the well-known photometric procedure based on sodium diethyldithiocarbamate.

At the other extreme of the concentration range, two samples of copper-nickel alloy were analyzed. Despite the necessity for extensive dilution of the test sample solutions before the solvent extraction procedure could be performed, the results were in surprisingly good agreement with the actual metal content of these alloys (Table 226).

Table 226 - Analysis of metal alloys

Material	Copper			Nickel		
	Expected (%)	Found(%)	S_r(%)	Expected(%)	Found(%)	S_r(%)
Alloy	67.3	69.5	0.6[a]	30.8	29.4	1.8[a]
Coin	75	72.8	5.4[b]	25	23.8	3.9[b]

a 3 analyses. b 4 analyses.

Gas chromatographic determination of selenium IV and total selenium in natural waters with 1,2-diamino 3,5-dibromobenzene. Shimoishi and Toei [2616] have described a gas chromatographic determination of selenium in natural waters based on 1,2-diamino-3,5-dibromobenzene with an extraction procedure that is specific for selenium (IV). Total selenium is determined by treatment of natural water with titanium trichloride and with a bromine-bromide redox buffer to convert selenide, elemental selenium and selenate to selenious acid. After reaction, the 4,6-dibromopiazselenol formed from as little as 1 ng of selenium can be extracted quantitatively into 1 ml of toluene from 500 ml of natural water; up to 2 ng 1^{-1} of selenium (IV) and total selenium can be determined. The percentage of selenium (IV) in the total selenium in sea water and river water varies from 35 to 70%.

The Shimadzu Model GC-3AE gas chromatograph used was equipped with an electron-capture detector. A glass column (1 m long, 4 mm i.d.) was packed with 15% SE-30 on 60 - 80 mesh Chromosorb W. The column and the detector were maintained at 200°C. The nitrogen flow-rate was 28 ml min^{-1}. A Shimadzu Model 250A recorder was used at a chart speed of 5 mm min^{-1}.

Procedures for total selenium content. Immediately after sampling of the sea water or river water, acidify with concentrated hydrochloric acid (1 ml 1^{-1}). Filter the sample through a membrane filter (pore size 0.45 µm) as soon as possible. Transfer 500 ml of the sample to a 500 ml beaker; add 5 ml of selenium-free concentrated sulphuric acid and 0.5 ml of titanium trichloride solution (20% minimum), and heat the mixture on a water bath (at about 90°C)for 10 min. Selenite and selenate are reduced to elemental selenium. Then add the bromine - bromide redox buffer solution until the light violet colour becomes pale yellow, heating again at the same temperature for 10 min. Elemental selenium and selenide are thus oxidized to selenious acid.

After cooling, transfer the solution to a 500 ml separating funnel. Add 25 ml of toluene, and shake vigorously for 5 min to saturate the solution with toluene and extract the toluene-soluble materials. This minimizes interferences in the gas chromatographic determination.

After separation of the phases, transfer the aqueous phase to another 500 ml separating funnel; add 10 ml of the 0.12% 1,2-diamino-3,5-dibromobenzene solution and allow to stand for 2 h. Then extract the 4,6-dibromopiazselenol formed by shaking for 5 min with 1 ml of toluene on a mechanical shaker; was the toluene extract twice with 3 ml of perchloric acid solution (2 + 1) by shaking for 1 min each time. Inject 2 µl of the toluene extract into the gas chromatograph and measure the peak height.

For selenium (IV) content. Only the tetravalent state of selenium can react with the reagent to form a 4,6-dibromopiazselenol. Therefore, selenium (IV) can be determined directly without either reduction or oxidation procedure.

Transfer the filtered 500 ml sample to a 500 ml separating funnel and add 20 ml of concentrated hydrochloric acid for pH control. Saturate the solution with 25 ml of toluene, and remove the toluene-soluble materials as before. Then follow the procedure given above.

No interference occurred in this method due to sodium, potassium, magnesium and calcium present at the concentrations normally found in sea water. No interference was encountered from the following ions, even when a 10.000 excess of foreign ion was present; aluminium III, chromium III, iron III, manganese II, cobalt II, nickel II, zinc, barium and uranyl ion. A relative standard deviation of 6.9% was obtained for total selenium at the 33.1 ug Sc C^{-1} level.

A gas chromatogram obtained in the determination of total selenium and selenium IV by this method is shown in Figure 164.

S. High Performance Liquid Chromatography

Tetradendate β-Ketoamines. Gaetani et al[2586] applied high pressure liquid-liquid partition chromatography to some metal chelates of the tetradebtate β-ketoamines; N,N' ethylenebis (acetylacetoneimine), (H_2(en)AA); N,N'-trimethylenebis (acetylacetoneimine), (H_2(tm)AA) and N,N'-ethylenebis (benzoylacetoneimine) (H_2(en)BA).

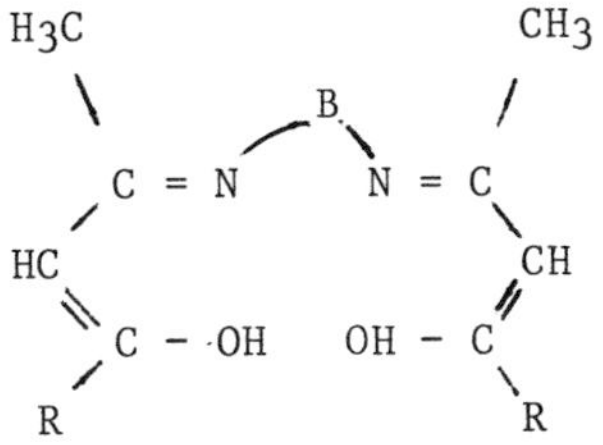

```
 H3C                    CH3
   \          B         /
    \      /    \      /
     C = N        N = C
    /                  \
 HC                     CH
   \\                  //
     C - OH      OH - C
    /                  \
   R                    R
```

with B = $(CH_2)_2$ R = CH_3 (H_2(en)AA)
B = $(CH_2)_3$ R = CH_3 (H_2(tm)AA)
B = $(CH_2)_2$ R = C_6H_5 (H_2(en)BA)

The complexes Co^{11}, Ni^{11}, Cu^{11}, Pd^{11}(en)AA, Ni^{11}, Cu^{22}(en)BA, Cu^{11}(tm)AA have been considered. Nickel and copper were separated by using H_2(en)AA and H_2(en)BA on two different columns. Palladium was successfully separated from copper but not from nickel. The dependence of the response of the uv detector (254 nm) on the amount of metal in aqueous solutions is reported for Ni and Cu(en)AA. The detection limits are about 0.2 and 0.5 ng of metal injected for nickel and copper respectively.

Separations were achieved on a Micro Pak CH column using various mobile phases and using blends of methanol and aqueous buffers of different pH as the eluting phase.

Separation of a methanol solution of H_2(en)AA, Co, Ni, Cu(en)AA was achieved by using a methanol/buffer volume ratio 65/35 and pH 7.8 (Figure 165a). The resolution factors are 1.44 for Cu/Ni(en)AA, 2.85 for Ni/Co(en) AA, 4.25 for Cu/Co(en)AA. The peaks were identified by injecting the single chelates. The cobalt chelate decomposes in solution, even in nitrogen atmosphere; the solution turns brown, and successive injections show a second peak immediately after that attributed to Co(en)AA.

Separation of Ni(en)AA and Pd(en)AA was not successful, Pd(en)AA being only slightly more retained (Figure 165b) for every volume ratio between methanol and buffer solution. The separation of Pd(en)AA and Cu(en)AA in methanolic solution is shown in Figure 165c. The resolution factor for Cu/ Pd(en)AA under the conditions quoted in the figure is 1.94.

The behaviour of Cu(en)AA is not considerably influenced by the pH of the buffer on the range 7.0 - 11.0 (phosphate buffers up to 8.0 and borate buffers between 8.0 and 11.0 were used); in this pH range, the uncorrected retention time lengthens from 1.7 to 1.9 min for a methanol/buffer volume ratio 70/30: the area of the peak is constant for the same quantity of complex. The copper chelates Cu(en)AA, Cu(tm)AA, Cu(en)BA in the THF solution are separated using a 50/50 methanol/buffer pH 7.8 mixture. The presence of the phenyl ring in Cu(en)BA determine a major affinity of the

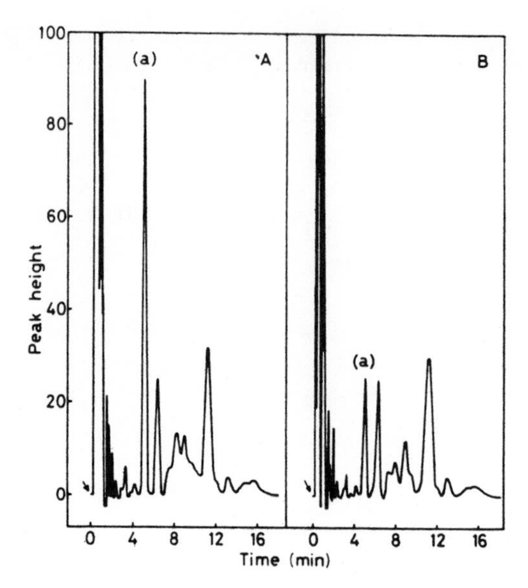

Figure 164 - Gas chromatogram of 4,6 dibromo pia 3 selenol complex.

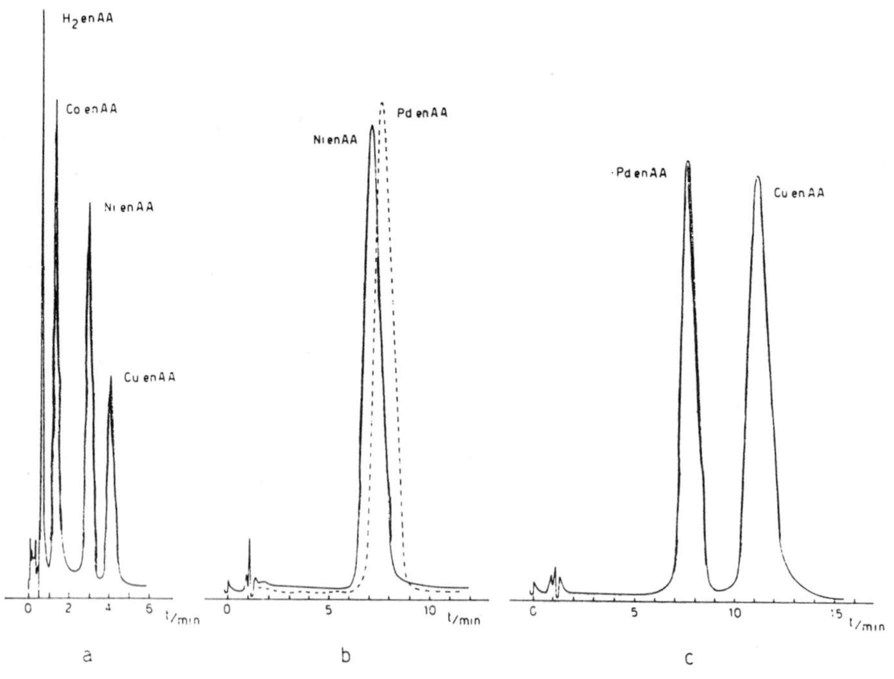

Figure 165 - (a) Separation of H_2(en)AA, Co^{II}(en)AA, Ni^{II}(en)AA, Cu^{II}(en)AA. Column: MicroPak CH 25 cm x 0.2 cm; moving phase: methanol/ phosphate buffer pH 7.8, 65/35 (v/v): flow rate: 1 cm^3 min^{-1} Theoretical plate number for Cu(e)AA: 480. (b) Superimposed chromatograms of Ni(en)AA and Pd(en)AA. Moving phase: methanol/ phosphate buffer pH 7.8, 45/55(v/v); flow rate: 1 cm^3 min^{-1}. (c) Separation: Pd(en)AA and Cu(en)AA at the same condition of (b). Theoretical plate number for Cu(en)AA: 410.

complex for the stationary phase and the retention time increases sensibly. With this ligand, nickel and copper are not seperated: on the contrary, Ni(en)BA and Cu(en)BA are separated on the $-NH_2$ column, using a mobile phase with a volume ratio methanol/buffer pH 7.8, 40/60.

Both these chelates are stable and have high adsorption in the uv region. Their molar absorptivities at 254 nm (e ~ 20,300 and 26,000 mol^{-1} $l.cm^{-1}$ for Cu(en)BA and Ni(en)BA, respectively), could allow a good sensitivity in the chromatographic determination of traces of nickel and copper.

To estimate the possibility of an analytical application of the liquid-liquid partition chromatography of tetradentate ketoamines, Gaetoni et al [2586] verified the dependence of the detector response on the amount of the metal for Ni(en)AA and Cu(en)AA. For this purpose, the MicroPak CH column was used; the mobile phase was a methanol/borate buffer pH 10, 70/30 mixture. Calibration curves were set up both with solutions of the pure chelates and starting from the metals in aqueous solution. In the latter procedure, methanolic solutions of an excess of ligand (500 - 50 times) were added to equal volumes of aqueous solutions of the metal acetates in the range of concentration 0.2 - 2 and 10 - 100 ppm of metal, buffered at pH 10. The reaction was carried out three times for each sample; the mixed solutions were directly injected into the column. The injections were replicated five times and the peak area was determined by means of an integrator. 2.5 ng of metal could be detected by this procedure. The calibration curves obtained starting from the aqueous solutions of the metals are linear in both ranges of concentration, corresponding to 0.5 - 5 and 25 - 250 ng of metal injected; the relative standard deviation of each mean value of the peak area is about 2% in the higher range of concentration and between 3 - 6% in the lower one. The plots obtained by using methanolic solutions of the two chelates coincide with those obtained starting from the metals, within the experimental errors. The detection limits, corresponding to a signal-to-noise ratio 2:1, for Ni and Cu are about 0.2 and 0.5 ng. The lower detection limit for Ni is attributable to the higher molar absortivity at 254 nm of the nickel chelate and especially to its lower retention time. Moreover, it is noticed that the metals can be extracted by means of a chloroform solution of the ligand and the extracts can be concentrated; in this way, it is possible to start from more dilute aqueous metal solutions.

Separation of geometrical isomers and mixed ligand forms of cobalt III and chromium III β-diketonates. Uden et al[2722] have separated these compounds by high performance liquid chromatography on silica. For isocratic chromatography, a Tracor model 3100 instrument was employed, with single-wavelength u.v. detection at 254 nm, a full-scale sensitivity of 0.01 absorbance units and an 8 µl flow cell. The system used for gradient elution consisted of an Altex Scientific Inc. Model 312 pumping system with a Model 410 solvent programmer. The detector used was a Spectromonitor II variable wavelength unit (Laboratory Data Control, Inc.) with a full-scale sensitivity of 0.01 absorbance and an 8 µl flow cell. The columns used were a 10 µm Partisil silica column, 30 cm x 4 mm i.d. (Whatman Inc.) and a 16 cm x 4 mm i.d. column, slurry-packed in the laboratory with a 5 µm Hypersil silica (Shandon Scientific Inc.).

Adsorption high performance liquid chromatography permitted ready resolution of isomer pairs of non-volatile complexes and the use of a solvent system of a medium polarity (6% acetonitrile in methylene chloride) gives complete separation of chelate pairs such as $Co(BAA)_3$ and $Co(PAM)_3$.

The resolution factor for the $Co(BAA)_3$ isomers was 2.1 and for the $Co(PAM)_3$ isomers was 6.5. Column efficiencies for all peaks were approximately 600 theoretical plates for the irregular 10 µm silica employed. To confirm the identify of the peaks eluted, the eluted fractions were collected and after solvent had been evaporated the solid residues were inserted into the direct probe of the mass spectrometer.

Table 227 summarizes the high performance liquid chromatographic behaviour of representative fac and mer isomer pairs for some cobalt and chromium complexes. In all cases, the mer isomer elutes before the fac isomer, and, for a given ligand, chromium complex isomers elute before the analogous cobalt species. There is also a noteworthy difference in the polarities of the mobile phase required to elute the isomers and achieve resolution for fluorinated complexes as compared with their non-fluorinated analogues. Trifluoroacetylacetonate isomers were eluted and resolved by a relatively non-polar mobile phase (8% methylene chloride in hexane); in contrast the non-fluorinated complexes required a more polar eluent (6% acetonitrile in methylene chloride) to achieve similar isomer resolution. This difference in elution behaviour reflects the stronger adsorption on silica of the non-fluorinated complexes.

Table 227 - Retention volumes for octahedral cobalt and chromium β-diketonate complexes on a column (30 cm x 4.0 mm i.d.) containing 10 µm Partisil (silica)

Complex	Solvent	Retention volume (ml)	
		mer	fac
$Co(TFA)_3$	8% Ch_2Cl_2 in C_6H_{14}	23.7	51.0
$Cr(TFA)_3$	8% CH_2Cl_2 in C_6H_{14}	21.8	44.2
$Co(BAA)_3$	6% CH_3CN in CH_2Cl_2	3.6	7.1
$Cr(BAA)_3$	6% CH_3CN in CH_2Cl_2	2.9	4.7
$Co(PAM)_3$	6% CH_3CN in CH_2Cl_2	3.9	14.6

This study served to illustrate several important features regarding the application of high performance liquid chromatography to inorganic metal complex chemistry. First, there is the applicability of adsorption high-resolution separation of chelate isomers with efficiencies in excess of those achieved preparatively by classical methods. Further high performance liquid chromatography is available for systems where gas chromatographic methods are unsuitable for reasons of thermal instability of involatility. High performance liquid chromatography appears to be attractive for following metal chelate reactions and preparative procedures and potentially for extension to larger scale applications. Finally, the scope of the method may be greatly extended by gradient elution procedures which enable complexes with a wide range of polarities to be resolved in a single chromatogram.

Metal chelates of dialkyldithiophosphoric acids. Cardwell et al[2723,2724] showed that mixtures of cobalt (III) dialkyldithiophosphates undergo rapid ligand exchange in solution at 40°C and the components may be resolved by adsorption high performance liquid chromatography. Under optimum conditions, additional components are observed and these have been identified as disulfide dimers arising from oxidation of the ligand during syntheses of the metal complexes. Exchange reactions in the metal chelates are accompanied by exchange in the organic disulfides.

These workers have reported that excellent resolution of tris(dialkyl-dithiophosphato)chromium (III) chelates and their mixed-ligand derivatives can be achieved on a silica column using eluents containing 0.1% or less alcohol in a hydrocarbon[2723]. After equilibration of the column with ethanol-hexane (0.08:99.92), the chromatogram of a mixture of diethyl- and diisopropyl-dithiophosphates of cobalt (III) showed four peaks as expected (Figure 166).

$$CoA_3 + CoB_3 \rightleftarrows CoA_2B + CoAB_2$$

where A = $(i\text{-}C_3H_7O)_2PS_2^-$ and B = $(C_2H_5O)_2PS_2^-$. The peaks with retention volumes of 4.4 and 5.4 cm^3 were identified as CoA_3 and CoB_3, respectively. Assuming that the additional peaks are due to the formation of the mixed ligand species according to the redistribution reaction above, the order of elution is $CoA_3 < CoA_2B < CoAB_2 < CoB_3$ by analogy with the chromium(III) chelates [2723].

A decrease in the ethanol content of the eluent to 0.06% or less produced a surprising result with more than four peaks being observed in the chromatogram of a mixture of cobalt (III) diethyl- and diisopropyldithio-phosphates which had been maintained at 40°C for about 2 h. The use of alcohols other than ethanol in the eluent had the same effect although the number of peaks observed depended on the composition of the eluent. The most striking examples of the chromatography of these reaction mixtures are shown in Figure 167 using isobutanol-hexane (0.05:99.95) as the eluent. Seven peaks were resolved for the redistribution reaction of equimolar amounts of cobalt (III) diethyl- and diisopropyldithiophosphates which had reached equilibrium (Figure 167a). For identification of the individual components in the mixture, chromatograms of the initial reactants, CoA_3 and CoB_3, were obtained (Figures 167b and 167c). It is concluded that peaks 1 and 3 are impurities, in the synthesised samples of CoA_3 and CoB_3, the latter eluting as peaks 4 and 7, respectively, in Figure 167a.

T. Size Exclusion Chromatography of Beryllium and Chromium Chelates of -diketones

The compounds studied by Saito et al[2615,2616] are listed in Table 228.

Table 228 - β-diketones and their metal complexes studied[a]

Compound	Abbrev-iation	R_1	R_2
acetylacetone	HAA	CH_3	CH_3
benzoylacetone	HBA	C_6H_5	CH_3
dibenzoylmethane	HDBM	C_6H_5	C_6H_5
trifluoroacetylacetone	HTFA	CH_3	CF_3
benzoyltrifluoroacetone	HBFA	C_6H_5	CF_3
thenoyltrifluoroacetone	HTTA	C_4H_3S	CF_3
bis(acetylacetonato)beryllium(II)	$Be(AA)_2$		
bis(benzoylacetonato)beryllium(II)	$Be(BA)_2$		
bis(dibenzoylmethanato)beryllium(II)	$Be(DBM)_2$		
bis(trifluoroacetylacetonato)beryllium(II)	$Be(TFA)_2$		
bis(benzoyltrifluoroacetonato)beryllium(II)	$Be(BFA)_2$		
bis(thenoyltrifluoroacetonato)beryllium(II)	$Be(TTA)_2$		
tris(acetylacetonato)chromium(III)	$Cr(AA)_3$		
tris(benzoylacetonato)chromium(III)	$Cr(BA)_3$		
tris(dibenzoylmethanato)chromium(III)	$Cr(DBM)_2$		
tris(trifluoroacetylacetonato)chromium(III)	$Cr(TFA)_3$		
tris(benzoyltrifluoroacetonato)chromium(III)	$Cr(BFA)_3$		
tris(thenoyltrifluoroacetonato)chromium(III)	$Cr(TTA)_3$		

[a] The β-diketones have the following general formula (keto form): R_1-$COCH_2$ CO-R_2

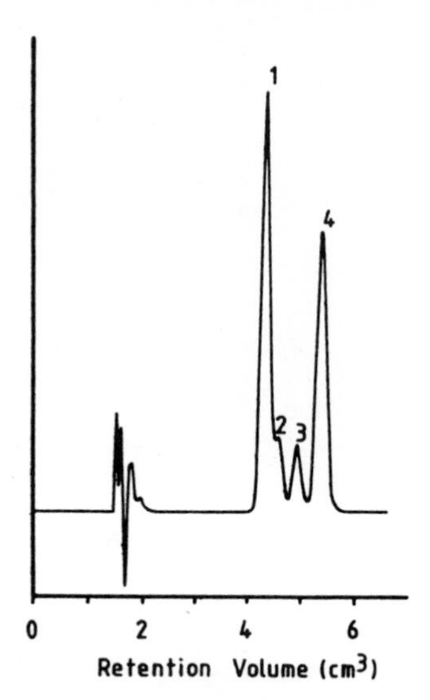

Figure 166 - Liquid chromatogram of a mixture of diethyl- and diisopropyl-dithiophosphates of cobalt(III) sampled shortly after mixing. Eluent: ethanol-hexane (0.08:99.92) at 1.0 cm^3 min^{-1}. Column: 150 x 4.6 mm i.d. packed with 7 µm Zorbax silica. Wavelength: 254 nm. Peak identifies: 1 = CoA_3; 2 = CoA_2B; 3 = $CoAB_2$; 4 = CoB_3, where A = $(i\text{-}C_3H_7O)_2PS_2^-$ and B = $(C_2H_5O)_2PS_2^-$; abbreviations as in the text.

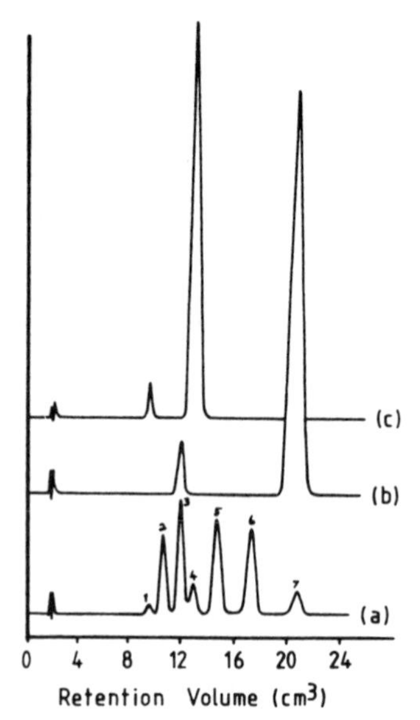

Figure 167 - (a) Liquid chromatogram of an equilibrated exchange reaction mixture of CoA_3 and CoB_3 containing traces of disulfide impurities. Eluent: isobutanol-hexane (0.05:99.95) at 1.0 cm^3 min^{-1}. Other conditions as in Figure 166. Peak identities 1 = $[(i\text{-}C_3H_7O)_2PS_2]_2$; 2 = $(i\text{-}C_3H_7O)_2PS_2S_2(OC_2H_5)_2$; 3 = $[(C_2H_5O)_2PS_2]_2$; 4 = CoA_3; 5 = CoA_2B; 6 = $CoAb_2$; 7 = CoB_3; abbreviations as in the text. (b) Chromatogram of CoB_3 containing the disulfide impurity, $[(C_2H_5O)_2PS_2]_2$. (c) Chromatogram of CoA_3 containing the disulfide impurity, $[(i\text{-}C_3H_7O)_2PS_2]_2$.

Purified benzene, acetone, p-dioxane, chloroform and tetrahydrofuran were used to prepare solutions of these compounds in 0.1 - 8 ug/l range. The separation was carried out on tractogel PVA 2000 (cross linked polyvinly acetate). The dry powder particle size 32 - 63 µm, E. Merck, Darmstadt, was successively washed with acetone, water, and methanol, in this order. The gel beads were then dried overnight at 60°C.

Most of the parts of the equipment coming into contact with the solvent solutions were made from Teflon or Pyrex as the β-diketonates have a high reactivity with metals. A 5 mm x 550 mm Pyrex column, of which the inner wall had been treated with dimethyldichlorosilane, was packed with Fractogel PVA 2000 swollen by the eluant solvent to be used. The solvent reservoir was a commercially available glass syringe with the 200 cm^3 capacity. A Hitachi Model 139 spectrophotometer modified to accomodate a flow-through cell (path length of 20 mm and volume of 0.063 cm^3) or a Model R1-2 differential refractometer (Japan Analytical Industry Co.) was used as a detector. The elution of HAA, HTFA, $Be(AA)_2$, $Be(TFA)_2$, and standard polystyrene in benzene system was monitored with a differential refractometer, while for the other compounds in these systems the UV-VIS spectrometric detection was carried out at an optimum wavelength for each compound. In other solvent systems, the spectrophotometer operating at 260 nm was used for the monitoring. The detection signal was fed into a Shimadzu Chromatopac C-RIA data processor.

The molar volumes (Vm, moles) obtained by Saito et al[2615,2616] for the β -diketones and their beryllium II and chromium chelates in p-dioxane are reported in Table 229.

It is thus considered that the molar volumes of the β-diketones and their Be(II) and Cr(III) chelates used can be ragarded as substantially constant regardless of the type of the organic solvents. Although β-diketones exist as a mixture of keto and enol tautomers, the molar volumes of both species may be treated as being approximately identical.

The distribution coefficient denoted by K_{av} is frequently adopted for the characterization of a solute molecule in size-exclusion chromatographic elution not only for practical purposes but also in theoretical work. It is derived from the following equation and corresponds to the distribution coefficient of the solute between the interstitial solution phase and the swollen gel phase.

$$V_e = V_o + K_{av}V_x$$

where V_e, V_o and V_x are the elution volume of a sample compound, the column void volume, and the volume of the swollen gel phase, respectively[2612]. If the total volume of the gel bed, V_t ($V_t = V_o + V_x$), is introduced, the above equation can be rearranged to:

$$K_{av} = \frac{V_e - V_o}{V_x} = \frac{V_e - V_o}{V_t - V_o}$$

The K_{av} values of the sample compounds were calculated according to this equation (see Table 230. The V_t value of each column used in this work was 10.82 cm^3. The V_o value was determined from measurements of the elution volume of standard polystyrene (molecular weight 9000 or 233,000), which can be regarded as excluded completely from the network of Fractogel PVA 2000. The dead volume relating to the spaces between the sample injector and the column inlet and between the column outlet and the detector cell was corrected for in the calculation of the K_{av} value.

No compounds gave an elution profile with excessive distortion, and the reproducibility of the K_{av} values obtained was satisfactory (relative standard deviation less than 1.0%). In Table 230 are the mean values from triplicate determinations.

Table 229 - Molar volumes for β-diketones and their beryllium(II) and chromium(III) chelates in p-dioxane

Compd	molar vol eq $V_{m,M}$	highest mole fraction of compd studied	$V_{m,2}$ cm^3/mol[c]	$V_{m,M}$[a]
HAA	$85.718 + 17.14X_2$	0.005 191	102.9	±0.006
HBA	$85.700 + 63.48X_2$	0.003 290	149.2	0.004
HDBM	$85.712 + 106.2X_2$	0.002 627	191.9	0.005
HBFA	$85.699 + 74.41X_2$	0.001 692	160.1	0.004
$Be(AA)_2$	$85.710 + 106.9X_2$	0.002 969	192.6	0.004
$Be(BA)_2$	$85.712 + 202.3X_2$	0.001 336	288.0	0.002
$Be(DBM)_2$[b]	$89.422 + 273.2X_2$	0.000 770 2	362.6	0.002
$Be(BFA)_2$	$85.717 + 232.4X_2$	0.000 710 9	318.1	0.002
$Be(TTA)_2$	$85.694 + 228.1X_2$	0.000 663 7	313.8	0.003
$Cr(AA)_2$	$85.694 + 183.9X_2$	0.000 997 0	269.6	0.002
$Cr(BA)_2$	$85.701 + 338.1X_2$	0.000 448 7	423.8	0.005
$Cr(DBM)_3$[b]	$89.419 + 407.8X_2$	0.000 437 4	560.2	0.001
$Cr(TFA)_3$	$85.695 + 261.9X_2$	0.000 838 1	347.6	0.003
$Cr(BFA)_3$	$85.703 + 406.0X_2$	0.000 667 4	491.7	0.004
$Cr(TTA)_3$	$85.700 + 373.7X_2$	0.000 606 8	459.4	0.001

[a] Standard deviation with respect to $V_{m,M}$ of experimental points from the best straight line derived from least-square analysis.
[b] Result on the benzene solution.
[c] Determined from slopes and intercepts of plots of $V_{m,M}$ aagainst X_2 where X_2 = mole fraction of chelate in test solution.

Figure 168 shows a typical relationship between molar value (V_m, moles) and log distribution coefficient (log Kav) for p-dioxane solutions of the β-diketonates studied and their various chromium and beryllium chelates. The plots for each β-diketone and its metal salts give straight lines which do not coincide with each other. If the chromatographic elutions of solute compounds are based only on the size exclusion mechanism, all the plots should form a straight line. However the plots for each β-diketone and its metal II III chelates give a straight line which do not coincide with each other. The result shown in Figure 168 implies that some secondary effects other than the size exclusion effect are involved in the elution process even in the poly(vinyl acetate) gel-p-dioxane system. The slope and the intercept of the plot depend strongly on the type of the substituents in the β-diketone. For example, the slope of the plot for each fluorinated β -diketone and its metal chelates in smaller than that for the nonfluorinated β-diketone and its metal chelates is smaller than that for the non-fluorinated compounds.

Similar plots in other solvent systems show that not in all cases is there the same linear relationship, see for example, the results obtained in chloroform (Figure 169).

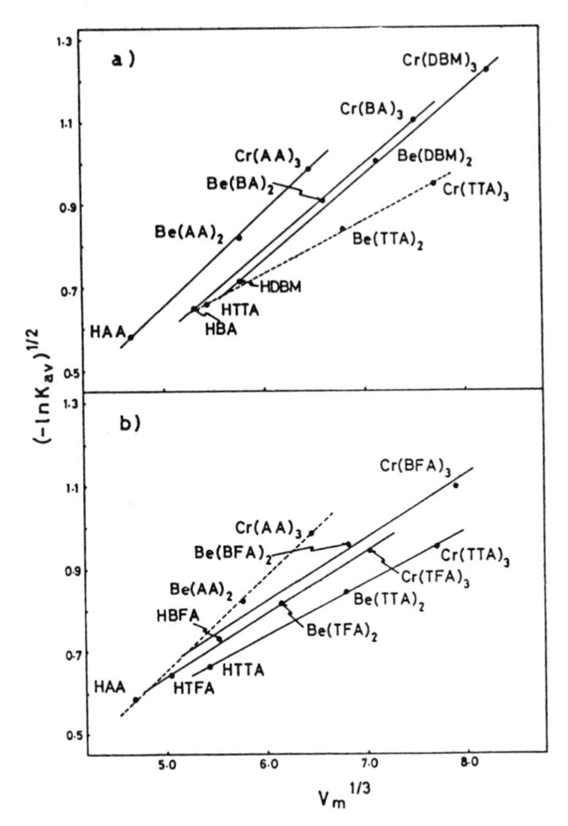

Figure 168 - Relationship between K_{av} and V_m for β-diketones and their metal (II, III) complexes; (a) nonfluorinated compounds; (b) fluorinated compounds. Solvent was p-dioxane.

In all instances, K_{av} values of a β-diketone and its Be(II) and Cr(III) chelates decrease in that order, which indicates that the size exclusion effect is a dominant factor in the separation mechanism, although the behaviour of these compounds cannot be explained only by the exclusion effect. From these observations, it is obvious that a plot for a given set of standard compounds, such as normal alkanes, cannot be used as a universal calibration curve for the size estimation. However, the linear relationships between $(-\ln K_{av})^{1/2}$ and $V_m{}^{1/3}$ in the Fractogel PVA 2000-p-dioxane system suggest that the effective size of a given metal chelate dissolved in p-dioxane can be estimated from the K_{av} value by using the plot for the free ligand and a series of its metal chelates which do not form the solvated complexes.

U. High Performance Thin-Layer Chromatography of Metal Tetraphenylporphyrin Chelates

Saitoh et al[2616] have described the high performance thin-layer chromatography of metal chelates of 5,10,,15,20-tetraphenylporphyrin (TPP) which is one of the typical synthetic porphyrins. Seven metals were studied; magnesium II, nickel II, copper II, zinc II, cadmium II, manganese III and iron III. The four high performance thin-layer stationary phases included cellulose and silica gel for normal phase distribution and two kinds

Table 230 - K_{av} for β-diketones and their beryllium(II) and chromium(II) chelates on Fractogel PVA 2000 with organic solvent systems

	K_{av}				
Compd	p-dioxane	benzene	chloroform	acetone	tetrahydrofuran
HAA	0.712	0.762	0.496	0.669	0.649
HBA	0.654	0.746	0.455	0.726	0.587
HDBM	0.596	0.727	0.415	0.769	0.516
HTFA	0.663	0.760	0.586	0.567	0.559
HBFA	0.586	0.689	0.504	0.610	0.485
HTTA	0.646	0.864	0.567	0.674	0.520
$Be(AA)_2$	0.509	0.503	0.297	0.589	0.564
$Be(BA)_2$	0.439	0.469	0.258	0.653	0.445
$Be(DBM)_2$	0.363	0.412	0.219	0.668	0.330
$Be(TFA)_2$	0.512	0.531	0.423	0.406	0.402
$Be(BFA)_2$	0.396	0.427	0.319	0.414	0.298
$Be(TTA)_2$	0.491	0.603	0.409	0.463	0.354
$Cr(AA)_3$	0.378	0.304	0.243	0.569	0.466
$Cr(BA)_3$	0.295	0.276	0.188	0.532	0.321
$Cr(DBM)_3$	0.222	0.227	0.135	0.435	0.210
$Cr(TFA)_3$	0.412	0.398	0.372	0.340	0.316
$Cr(BFA)_3$	0.305	0.314	0.241	0.303	0.216
$Cr(TTA)_3$	0.405	0.501	0.345	0.360	0.280

of alkylated silicas for reversed phase distribution; a wide range of mobile phases were studied.

Chromatography was carried out in a thermostatically controlled room at 25°C. The sample solution of H_2TPP or its metal chelate was prepared in chloroform at a concentration of about 1 mg/mL. In the case of NiTPP, the concentration of the sample solution was lower than 0.5 mg/mL because of its lower solubility. A 0.5 μL portion of each sample solution was put on the start line placed at 15 mm from an edge of the chromatographic plate. The chromatogram was developed until the solvent front was 75 mm from the origin by means of a sandwich method with a covering glass plate placed 1.6 mm apart from the plate. Every spot on the chromatogram was easily detected owing to the characteristic intense color of each TPP chelate in day light. Under ultraviolet light (254 and 356 nm), MgTPP, ZnTPP, CdTPP, and H_2TPP were fluorescent.

The best separations of the seven metals were achieved on the C_8 and C_{18} bonded alkylated silicas, the C_{18} giving the best separation factor under comparable conditions of solvent (Tables 231 and 232).

On the plates of the C_8-bonded and C_{18}-bonded silicas, most of the metal TPP chelates migrated nearly to the solvent front with such developing solvents as carbon tetrachloride, dichloromethane, and pyridine. The R_f value for a compound is larger on a C_8-bonded plate than on a C_{18}-bonded plate. This tendency was not always consistent in the case of FeTPP and MnTPP. The R_t sequence observed with the 20 - 80 (v/v) acetone-propylene carbonate mixture is MnTPP > ZnTPP > FeTPP > MgTPP H_2TPP CdTPP > NiTPP > CuTPP   MgTPP > NiTPP   MnTPP > CuTPP. These facts show that the resolution among these compounds can be regulated by changing the composition of the above solvent mixture.

Table 231 - R_f values of the metal chelates of tetraphenylporphyrin on C_8-bonded silica[a] with various single-component developers at 25°C

Developer solvent	R_f							
	MgTPP	NiTPP	CuTPP	ZnTPP	CdTPP	MnTPP	FeTPP	H_2TPP
methanol	0.44	0.34	0.33	0.63	0.44	0.15	0.12	0.44
ethanol	0.59	0.53	0.51	0.70	0.58	0.93	0.73	0.58
1-propanol	0.97	0.85	0.84	0.97	0.96	1.0	0.99	0.95
2-propanol	0.79	0.77	0.73	0.89	0.80	0.95	0.83	0.79
2-butanol	0.89	0.89	0.87	0.95	0.89	0.96	0.91	0.87
acetonitrile	0.24	0.20	0.19	0.45	0.23	0.34	0.33	0.23
propylene carbonate	0.25	0.31	0.24	0.57	0.25	0.36	0.39	0.24
acetone	0.81	0.80	0.77	0.87	0.81	0.61	0.85	0.81
N,N-dimethylformamide	0.87	0.87	0.85	0.89	0.85	0.24	0.71	0.85
carbon disulfide	0.96	0.97	0.97	0.95	0.97	0.24	0.96	0.97
dichloromethane	0.91	0.96	0.95	0.95	0.93	0.93	0.95	0.93
pyridine	0.95	0.94	0.94	0.96	0.94	1.0	0.95	0.94

[a] HPTLC plate RP-8 F

Table 232 - R_d values of the metal chelates of tetraphenylporphyrin in C_{18}-bonded silica[a] with various single-component developers at 25°C

Developer solvent	R_f							
	MgTPP	NiTPP	CuTPP	ZnTPP	CdTPP	MnTPP	FeTPP	H_2TPP
methanol	0.11	0.05	0.04	0.35	0.11	0.21	0.13	0.11
ethanol	0.23	0.15	0.13	0.48	0.24	0.52	0.27	0.24
1-propanol	0.58	0.41	0.35	0.78	0.57	0.99	0.93	0.57
2-propanol	0.44	0.37	0.31	0.72	0.45	0.91	0.60	0.45
2-butanol	0.71	0.63	0.56	0.89	0.71	0.96	0.80	0.69
acetonitrile	0.12	0.04	0.04	0.21	0.11	0.20	0.13	0.11
propylene carbonate	0.13	0.13	0.07	0.36	0.13	0.27	0.21	0.13
acetone	0.70	0.67	0.61	0.82	0.73	0.64	0.78	0.73
N,N-dimethylformamide	0.71	0.70	0.65	0.80	0.73	0.13	0.67	0.73
carbon disulfide	0.95	0.98	0.97	0.95	0.94	0.13	0.92	0.91
1,4-dioxane	0.93	0.93	0.95	0.97	0.94	0.96	0,96	0.93
pyridine	0.94	0.95	0.94	0.95	0.93	0.96	0.95	0.92

[a] HPTLC plate RP-18F

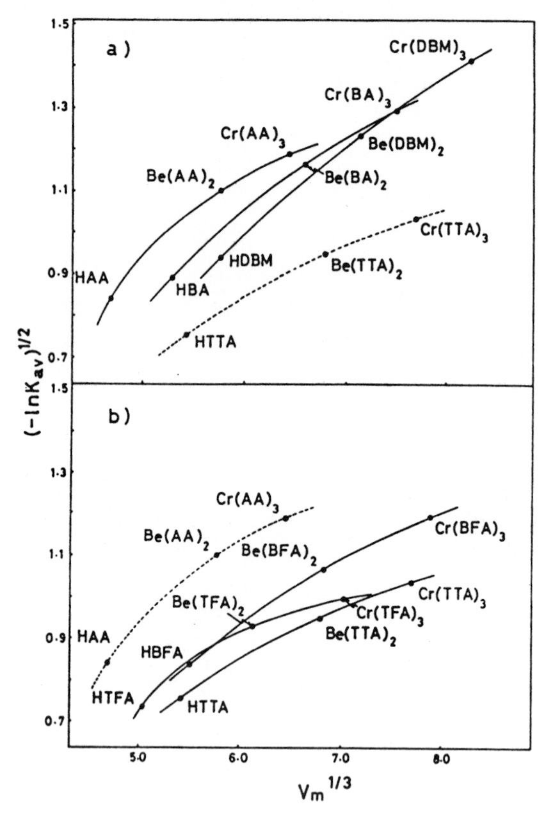

Figure 169 - Relationship between K_{av} and V_m for β-diketones and their metal (II, III) complexes. Solvent was chloroform. For other details see Figure 168.

CHAPTER 8

OTHER CHROMATOGRAPHIC TECHNIQUES

A. THIN LAYER CHROMATOGRAPHY

Organoantimony Compounds

Thin-layer chromatography has been used to separate triphenylantimony from its phosphorus, arsenic, and bismuth analogues. R_F values have been reported for these compounds on alumina plates, using light petroleum as developing solvent[2742].

Organoarsenic Compounds

Thin-layer chromatography on alumina using light petroleum as the solvent has been used to separate $(C_6H_5)_3M$, where M is phosphorus, arsenic, antimony, or bismuth[2742]. R_F values have been determined on silica gel G and neutral alumina for mixtures of labelled triphenylarsine, triphenyl phosphate, triphenyl phosphite, triphenylphosphine oxide, triphenylphosphine and tritolylphosphate using acetone-light petroleum, benzene-acetone (1:1 and 9:1) and chloroform-acetone (3:2 and 9:1) as solvents[2743]. The chromatograms were automatically examined radiometrically. A range of aromatic compounds containing arsenic or phosphorus have been separated on silica gel, aluminium oxide, and magnesium silicate[2744]. Twenty-four solvents were studied and R_D values and their standard deviations tabulated.

Organobismuth Compounds

Thin-layer chromatography on alumina has been applied to the separation of compounds of the type $(C_6H_5)M$, where M is bismuth, phosphorus, arsenic or antimony[2742]. Light petroleum is the recommended elution solvent.

Organoboron Compounds

Thin-layer chromatography has been used to separate some βββ-trichloroborazines[2745]. 2,4,6-trichloroborazine and its 1,3,5-trimethyl and -triphenyl derivatives are separated from each other on 0.5 mm layers of microcrystalline cellulose with ethyl acetate or pyridine-ethyl acetate (1:1000) as solvent. For detection, the plates are sprayed with 6 N hydrochloric acid and then with a tincture of curcuma and heated on a hot-plate at 80°C. A red-brown spot on a yellow background indicates the presence of boron compounds. Pentaborane, decaborane, and some chlorinated boron compounds have been separated on silica gel G or binder-free alumina[2746]. Between 3 and 50 µg of sample were applied, and the spots were detected with 1% potassium permanganate solution or iodine vapour, followed by spraying with a solution containing 5% each of silver nitrate and ethylenediamine. In many cases, silver nitrate alone was sufficient. For pentaborane and decaborane and some halogenated compounds, a 10% solution of ethyl isonicotinate in hexane was a suitable reagent. R_F values in 10 solvents were listed for several boranes, and the chromatographic behaviour of the products of their reaction with some Lewis bases such as amines, hydrazine, trialkyphosphines and dialkylsulphides was discussed. Difficulties were observed with low-boiling compounds or with those that were oxidized or hydrolysed readily. Hydrolysis occurred to some extent even in aprotic solvents, although in many instances the speed of the separation minimized the effect.

Organocobalt Compounds

For the thin-layer chromatographic separation of mixtures of cyano-cobalamin and cobinamide the sample is applied to a thin-layer of silica gel impregnated with thymol, and the chromatogram is developed with an aqueous solution containing 0.4% of pyridine, 3% of phenol, 0.01% of sodium cyanide, and 10% of acetic acid, saturated with thymol2747. To determine cyanocobalamin the chromatogram is eluted with a 0.1% solution of sodium cyanide in 75% methanol, and the absorbance is measured.

Cyanocobalamin and hydroxocobalamin have been separated on thin films of silica gel[2748]. Two methods of development were used, the first involving Merck Kieselgel G and elution with butanol-acetic acid -0.066 M potassium dihydrogen phosphate-methanol (4:2:4:1). In the second method the silica gel is suspended in 0.066 M potassium dihydrogen phosphate and development is effected with butanol-acetic acid-water-methanol (4:2:4:1). Identical results were obtained by both methods, but the latter method is more rapid, requiring only 5 - 6 h for development. If the spots are visible they are separately cut from the plate and extracted with 1% aqueous Tween 80 at 40 - 45°C, the extraction being repeated if necessary after centrifuging. The absorbance of the extract is measured, for cyanocobalamin at 361 nm and 548 nm. For hydroxocobalamin, the corresponding figures are 351 nm (165) and 527 nm (56). When the spots are not visible, duplicate plates are made, one of which is used for locating the cobalamin spots bioautographically, by incubation in contact with a plate of 3% agar in USP Difco assay medium, containing 0.02% of triphenyltetrazolium chloride, and innoculation with Lactobacillus leichmannii ATCC 7830. The corresponding zones are then removed from the second plate and eluted. The displacement of hydroxo-cobalamin relative to cyanocobalamin is 0.5 ± 0.05.

By using alumina G as adsorbent, and anhydrous acetic acid-water-methanol-chloroform-butanol (2:9:10:20:50) as solvent, cyanocobalamin was completely separated, but not hydroxocobalamin[2749]. By using silica gel G as adsorbent, and anhydrous acetic acid-water-methanol-chloroform-butanol (9:11:5:10:25) as solvent, vitamin derivates could be successfully separated, viz., hydroxocobalamin (R_F 0.05), cyanocobalamin (R_F 0.23), unknown substance (R_D 0.33), and thymidine (R_F 0.7). Spots were detected by bio-autography on the plate with vitamin B_{12} assay agar medium and Lactobacillus ;eocj,ammoo ATCC 7830 as test organism. The minimum amounts of cyano-cobalamin detected on silica gel G and alumina G were 0.005 and 0.025 x 10^{-9} g, respectively. A thin-layer chromatographic-spectroscopic procedure for the determination of hydroxo- and cyanocobalamins associated with other medicaments in various pharmaceutical forms is based on extraction of the vitamins, purification by thin-layer chromatography, and spectro-photometric determination as dicyanocobalamin[2750].

Methods for extracting and determining cobalamins have been reviewed under the headings, biological, chemical, radiometric, and spectrophotometric[2751]. A method particularly recommended that is applicable to all forms of cobalamin present in any type of extract or liver hydrolysate involves three double extractions with p-chlorophenol solution with intermediate washings with water, followed by three single extractions with p-chlorophenol solution. The extracts are combined and concentrated by several alternate extractions with decreasing volumes of water and p-chlorophenol solution, the alternate aqueous and p-chlorophenol extracts being discarded. The organic concentrate so obtained is shaken with diethylether, then with added water, and then with added ethyl oleate. After centrifuging, the organic phase is discarded, and the aqueous phase containing the cobalamins is washed with ether, then submitted to thin-layer chromatography on silica gel G, previously dried at 38°C by development with 50% aqueous ethanol for 3 h. Cyanocobalamin and coenzyme B_{12} are extracted individually with 50% ethanol. The hydroxocobalamin zone is treated with potassium cyanide solution, adjusted to pH 6 with acetic acid, and the resulting cyanocobalamin is extracted into 50% ethanol. Finally, cyanocobalamin and coenzyme B_{12} are determined spectrophotometrically at 550 and 525 nm, respectively.

In a further method for separating hydroxocobalamin and cyanocobalamin by thin-layer chromatography, spots of aqueous solution adjusted to pH 8.5 with dilute aqueous ammonia were applied to layers of dry neutral alumina and developed with isobutyl alcohol-isopropyl alcohol-water (6:4:5) (with addition of aqueous ammonia to pH 8.5)[2752]. In this system, the R_F values of hydroxo- and cyanocobalamins were 0.30 and 0.46, respectively; 0.5 µg of each substance could be detected. For thin-layer chromatography of mixtures of cyanocobalamin (vitamin B_{12}), factor B_{12}, pseudo-vitamin B_{12}, factor A, factor B, and factor VnB, a 5% solution of sodium cyanide was added to an aqueous solution of the sample[2753]. Spots of the solution were applied to a layer (1 mm thick) of alumina and developed in isobutyl alcohol-isopropyl alcohol-water (1:1:1). All the above substances were well separated. For the analysis of a phenol solution of three B_{12} coenzymes, removal of inorganic ions was necessary for good results[2754]. This was achieved by washing the phenol phase with water. Plates were coated with cellulose MN 300 or MN 300-CM and were developed with the lower layer of butan-2-ol-0.1 M acetate buffer (pH 3.5)-methanol (4:12:1).

Organocopper Compounds

Copper chlorophyllins in preserves have been determined by thin-layer chromatography[2755]. For the extraction of chromopigments the sample is homogenized and acidified with concentrated formic acid, then extracted by stirring with isopropyl alcohol-acetone-diethyl ether (5:3:2). The concentrates are developed on thin-layers of silica gel. For identifying copper chlorophyllins the chromatograms are developed with benzene-ethyl acetate-methanol (17:1:2); copper chlorophyllins remain at the origin as dark brown spots.

Organoiron Compounds

Thin-layer chromatography on Merek Kieseigel G has been applied to the separation of ferrocene and its derivatives[2756]. Ferrocene compounds have been separated by thin-layer chromatography on methylene chloride solutions on a plate coated with a 0.25 mm layer of Mallinckrodt Silic AR TLC-7G, a silicic acid gypsum absorbent, that has been activated for 1 h at 92°C[2757]. The elution solvent was benzene-acetone (30:1). Detection of ferrocene compounds was done visually and detection of benzenoid compounds by means of iodine vapour treatment. In most cases the ferrocenyl analogue has a lower R_F than the phenyl compound. To ensure that no reaction had taken place on the adsorbent, a methylene chloride solution of ferrocenylphenylcarbinol was allowed to stand overnight in contact with Silic AR TLC-7G, then filtered and evaporated to dryness. The infrared spectrum proved to be identical with the original, indicating that no reaction had taken place.

Bosak and Fukuda[2757] found that all but two of the ferrocenyl analogues had a lower R_F than the phenyl compound (Table 233).

Table 233 - Thin-layer chromatographic data of ferrocene compounds and their benzenoid analogues

No,	Compound*	R_f value	No.	Compound	R_f value
1	ϕCHO	0.75	9	ϕ$CHOHCH_3$	0.29
2	ØCHO	0.30	10	Ø$CHOHCH_3$	0.25
3	ϕCOϕ	0.69	11	ϕ$CH=CH-COCH_3$	0.32
4	ØCOϕ	0.30	12	Ø$CH=CH-COCH_3$	0.28
5	ϕ$COCH_3$	0.45	13	ØCH=CHCOϕ	0.61
6	Ø$COCH_3$	0.25	14	ϕCH=CHCOϕ	0.41
7	$ϕ_2$CHOH	0.30	15	ØCH=CHCOØ	0.26
8	ØCHoOH	0.32	16	ϕCHoCH_2COϕ	0.73
			17	ØCHoCH_2COϕ	0.77

* The symbol Ø stands for $C_{10}H_9Fe$.
The symbol ϕ stands for benzenoid group.

Organolead compounds

The dithizonates of trialkyllead halide (yellow) and dialkyllead dihalide (red) compounds can be separated by thin-layer chromatography[2758]. Mixtures containing much tributyllead acetate and a small proportion of dibutyllead diacetate are best chromatographed in acetone-water (1:1), and other mixtures can be separated in benzene. Following chromatography, the thin-layer plates are sprayed with a dithizone solution in chloroform (0.1%), after which yellow and/or red spots appear on a light green-blue background. Inorganic lead does not migrate and causes red spots at the baseline on the chromatogram.

Organophosphorus Compounds

Donner and Lohs[2759] separated various organic phosphates and phosphites on Kieselgel D using hexane-benzene-methanol (2:1:1) or hexane-methanol-diethyl ether (6:1:1) as the solvent. Satisfactory colour reactions were obtained by spraying with a solution (1%) of cobalt chloride in acetone, either in the cold or on warming.

Reuter and Hanke[2760] separated mixtures of phosphoric acid and ethyl dihydrogen phosphate and diethyl hydrogen phosphate obtained by the hydrolysis of esters of polyphosphoric acids on a thin layer of Kieselgel G or D, using methanol-dichlormethane-25% aqueous ammonia (7:10:2) as solvent. To reveal the spots the chromatogram was sprayed with ammonium molybdate solution, heated to 120°C for 15 min, and then sprayed with 20% sodium bisulphite solution.

Neubert[2761] separated alkyl and aryl phosphites and phosphates on

Silica gel G (20 cm x 20 cm x 0.25 mm) activated by heating for 1 h at 110°C. The chromatogram was developed with chloroform-acetone (19:1) and the spots revealed by spraying with a 2% solution of iodine in chloroform, then, after a short interval, with 0.1 N aqueous iodine-dil sulphuric acid (1:1). Three groups of compounds were separated, basic (triaryl phosphites), neutral trialkylphosphites; trialkyl or aryl phosphates, dialkyl or aryl phosphonates) and acidic (alkyl or aryl phosphonates and phosphates, dialkyl or diaryl phosphates, and the parent acids).

Robinowicz et al[2762] used thin-layer chromatography for some monoesters of phosphoric and sulphuric acids. To separate phosphoric acid from methyl dihydrogen and ethyl dihydrogen phosphates, they chromatographed the sample on a 0.25 mm layer of cellulose MN 300 using butanol-acetic acid-water (5:2:3) as solvent. They located the separated compounds using the perchlorate-molybdate reagent described by Hanes and Isherwood[2766].

Berei[2763] carried out thin-layer radio-chromatography of labelled triphenyl phosphate, triphenyl phosphite, triphenyl phosphine oxide, triphenyl phosphine, triphenylarsine and tritolyl phosphate. Solvent systems tested included acetone, light petroleum, benzene, benzene-acetone (1:1) and (9:1) and chloroform-acetone (3:2 and 9:1). The radio chromatograms were scanned automatically.

Braun[2764] separated organic phosphate plasticizers from other types of plasticizers present in plastics by thin-layer chromatography using dichloromethane as the mobile phase. Suitable spray reagents for this separation were discussed.

Narney[2765] has pointed out that the Hanes and Isherwood[2761] molybdate reagent for the detection of organic phosphates, and compounds that produce phosphates upon hydrolysis, suffers from several limitations. Many organophosphorus compounds are not readily hydrolysed to orthophosphate by the reagent. Bunyan[2767] mentioned that the Hanes reagent is unstable (requires fresh solution at frequent intervals). The contrast between the spots and the background is poor and changes rapidly with time, and the sensitivity is poor, probably because the hydrolysis reaction is incomplete. Barney[2765] developed two new spray tests for detecting many types of organophosphorus compounds, including organophosphates, organophosphonates, organophosphoric acids, organophosphonic acids, and the related thio-compounds in thin-layer plates. These reagents are simple to use, are stable, and will detect down to 1 µg of most organophosphorus compounds. To detect organic phosphates, phosphorothioates and the corresponding acids, the plate is sprayed with hydriodic acid reagent (11.2 ml of 57% hydroidic acid - free of inhibitors - and 50 ml glacial acetic acid diluted to 100 ml with water), and the plate heated on a hot plate at 250°C for 15 minutes. The plate is then cooled, sprayed with 2% ammonium molybdate solution in 1 M hydrochloric acid, and after 3 min treated with benzidine reagent (0.05 g benzidine; 10 ml acetic acid and a minimum quantity of water mixed until the benzidine dissolves. Sodium acetate trihydrate (22.5 g) is then added to the solution which is then diluted to 100 ml with water). A blue colour is produced in the presence of organophosphorus compounds.

An alternate spray reagent described by Barney[2765] for the detection of organic phosphates and phosphonates is used as follows. The plate is sprayed with the hydriodic acid reagent described above, heating for 5 min at 250°C, then sprayed with fresh 5.7% ammonium persulphate solution. The plate is heated again, cooled, then treated with molybdate and benzidine reagents as described above. A blue colour is produced in the presence of organophosphates and organophosphonates.

Barney[2765] detected phosphate and phosphorothioate esters and the corresponding acids on thin-layer chromatograms by spraying with hydriodic acid (11.2 ml of 57% hydriodic acid and 50 ml anhydrous acetic acid, diluted to 100 ml with water). The plate was then heated at 250°C, for 15 min, cooled, sprayed with 2% ammonium molybdate solution in M hydrochloric acid and, after 3 min, treated with benzidine reagent to produce a blue colour in the presence of phosphate and phosphorothioate esters (0.05 g of sodium acetate trihydrate and the mixture diluted to 100 ml with water). To detect phosphates and phosphonates, this worker sprayed the plate with the hydriodic acid reagent described above, heated it for 5 min at 250°C, sprayed with fresh 5.7% aqueous ammonium persulphate solution, heated again, cooled, and then treated with the molybdate and benzidine reagents, as described above, to produce a blue spot in the presence of phosphate or phorphonate esters.

Alkyl and Aryl Phosphites

Thin-layer chromatography has been used for the separation of alkyl and aryl phosphites from other types of organophosphorous compounds.

Berei[2763] used silica gel G and neutral alumina to achieve separations of labelled triphenyl phosphate, triphenyl phosphite, triphenyl phosphine oxide and triphenylphosphine. Mixtures of acetone, light petroleum and benzene and of chloroform-acetone (3:2 and 9:1) were used as development solvents.

Neubert[2761] separated organophosphites, organophosphates, and organophosphonates by thin-layer chromatography on Silica gel G. In this procedure aryl and alkyl phosphites and phosphates are separated on gel activated by heating for 1 h at 110°C. Spots of test solution (0.01 ml) containing 10 μg of phosphorus were applied 1.5 cm apart. The chromatogram was developed with chloroform-acetone (19:1), and the spots are revealed (both P^{III} and P^{V}) by spraying with 2% iodine in chloroform and, after a short interval, with 0.1 N iodine-dil sulphuric acid (1:1). Three groups were separated, the so-called basic (triaryl phosphites), neutral (trialkyl phosphites, trialkyl (or aryl) phosphates and dialkyl (or aryl) phosphonates) and acid (alkyl (or aryl) phosphonates and phosphates, dialkyl (or aryl) phosphates and the parent acids) groups.

Donner and Lohs[2759] separated organic phosphites and phosphates on Kieselgel D and Kieselgel G using hexane-benzene-methanol (2:1:1) or hexane-methanol-ethyl ether (6:1:1) as the solvent. Colour reactions are given by spraying with a 1% solution of cobalt chloride in acetone, either in the cold or on warming. The sensitivity depends on the nature of the adsorbent as well as the phosphorus compound, and is typically in the range from 10 - 250 μg.

Phosphonic Acids

Taulli[2768] has described a chromogenic reagent for the detection of organophosphonic acids separated on thin-layer chromatographic plates. The phosphonic acid is converted to a phosphate by spraying the plate with sodium hypochlorite solution and the phosphate then detected by successive spraying with ammonium molybdate and a reducing agent (Hanes and Isherwood) 2766.

Barney[2765] has pointed out the limitations of the Hanes and Isherwood molybdate reagent. He detected phosphonate and phosphate esters on thin-layer chromatographic plates coated with silica Gel G by spraying the plate with hydriodic acid reagent (11:2 ml of 57% hydriodic acid, and 50 ml of anhydrous acetic acid, diluted to 100 ml with water) then heating the plate for 5 min at 250°C, spraying with fresh 5.7% aqueous ammonium persulphate solution, heating again, cooling and then treating with 2% ammonium molybdate in 1 M hydrochloric acid then with benzidine reagent (0.05 g benzidine dissolved in 10 ml dilute acetic acid and 22.5 g anhydrous sodium acetate added and the mixture diluted to 100 ml with water. The formation of a blue spot indicates the presence of esters of phosphonic and phosphoric acid. Compounds separated included O,O-diethylchloromethylphosphonate, O,O-dipropylchloromethylphosphonate, O,O-dibutylchloromethylphosphonate, O,O-diethylethoxycarbonylmethylphosphonate, O,O-diethylcyanomethylphosphonate, methyl phosphonic acid, O-methyl methylphosphonic acid, ethylphosphonic acid, O-ethyl ethylphosphonic acid, chloromethylphosphonic acid and mono-ethyl acid orthophosphate.

Phosphinic Acids

Hecker and Hein[2769] showed that dialkylphosphinic acids (and phosphine oxides) gives deep blue colorations when heated with anhydrous nickel bromide in toluene. Some phosphine sulphides and tertiary phosphines produce green colours. Acetone and methyl ethyl ketone interfere.

Separations of carbohydrate phosphites and phosphonites have been achieved by Nifant'ev[2770] using thin-layer chromatography on 0.5 mm thick layers of alumina and chloroform, carbon tetrachloride, nitromethane and dimethyl formamide as solvents. The separated substances were revealed with sulphuric acid and subsequent warming under an i.r. lamp, or by other specific reactions. Acid compounds were identified by spraying the chromatogram with 1% congo red solution. Reagents that revealed tervalent, but not quinquevalent phosphorous compounds were 2% potassium permanganate (brown spots) and 10% potassium dichromate solution in 20% sulphuric acid (blue spots). Dialkyl phosphites were revealed as violet spots by using a slightly alkaline solution of sodium, 3,5-dinitrobenzoate followed by gentle heating with an i.r. lamp. The separating of various neutral and acidic organophosphorus compounds was studied.

Substituted Phosphines

Thin-layer chromatography combined with infrared spectrophotometry have been used (Schindbauer and Mitterhofer[2771]) for the identification of the products of thermal decomposition of triphenylphosphine.

The thin-layer separation on alumina, silica and magnesium silicate of aryl derivatives of phosphine and phosphine oxide has been discussed by Berei and Vasana[2772].

Substituted Phosphine Oxides

The thin-layer separation on alumina, silica and magnesium silicate of aryl derivatives of phosphine oxide and of phosphine has been described by Berei and Vasana[2772].

Phosphorothioc Acid Esters

Petschik and Steger[2773] have separated ethyl esters of phosphorothioc acids on a layer of aluminium oxide bound with starch, with a mixture of heptane and acetone (10:1) as the mobile phase. The spots were made visible by spraying the chromatogram with a solution of periodic acid in 70% perchloric acid containing about 0.002% of vanadium pentoxide. These workers (Petschik and Steger[2774]) have described an alternate reagent for the detection and determination of alkyl phosphorothioates on thin-layer chromatograms. The spots (detected by spraying with perchloric acid-periodic acid) are marked with pencil. The marked areas are scraped off into separate hard glass test-tubes, and the contents of each tube are oxidised with perchloric acid-periodic acid and centrifuged. The phosphorous in the supernatant solution is determined spectrophotometrically (at 407 nm) with ammonium metavanadate (NH_4VO_3) and $(NH_4)_6MoH_{24},4H_2O$. The amount of phosphorus from each spot is calculated as a fraction of the total. In spite of losses (15 - 50%), the fraction for each spot can be determined with a coefficient of variation of 1.2%. This method has been successfully applied to mixtures containing 7 - 70% of each of three esters.

Barney[2765] detected phosphorothionates and phosphoric acid esters on thin-layer chromatograms by a method in which the plate is sprayed with hydriodic acid reagent (11.2 ml of 57% hydriodic acid and 50 ml of anhydrous acetic acid, diluted to 100 ml with water) heated on a hot-plate at 250°C for 15 min, cooled, sprayed with 2% ammonium molybdate solution in M-hydrochloric acid and after 3 min treated with benzidine reagent (0.05 g of benzidine is dissolved in 10 ml of dilute acetic acid, 22.5 g of hydrated sodium acetate is added, and the mixture is diluted to 100 ml). A blue spot indicates the presence of a phosphorothioate or a phosphate ester.

Simple alkyl ester phosphorothioates have been separated (Hankiewicz and Studniarski[2775]) from tetraalkyl dithiobisphosphonates, tetraalkyl and tetra-allyl pyrophosphates and dialkyl(alkoxy)(alkyl)pyrophosphonates by chromatography on silica gel plates using various mobile phases and detection solutions. Detailed procedures and tables of R_f values obtained under optimum conditions are given.

Mastryukova et al[2776] separated phosphorothionic esters, phosphorothiolic esters and some thiopyrophosphoric acids, phosphorodithionic and phosphorothionic and phosphorothiolic isomers by thin-layer chromatography on alumina (100 mesh) or on KSK silica gel (~ 150 mesh) containing 6% of water. Separation takes 18 - 20 min with hexane-acetone (4:1 or 10:1) as solvent. The chromatograms are treated with potassium permangante in an acid medium. The optimum amounts of phosphorothionic and phosphorothiolic esters were found to be 20 - 40 μg and 90 - 100 μg respectively.

O,O-dimethyl O-(3-methyl-4-nitrophenyl) phosphorothioate has been determined in technical products after preliminary removal of associated compounds by thin-layer chromatography (Kovac[2777]). O,O-dimethyl O-(3-methyl-4-nitrophenyl) phosphorothioate can be separated from O-methyl O,O-di-(3-methyl-4-nitrophenyl) phosphorothioate, dimethyl 3-methyl-4-nitrophenyl phosphate and 3-methyl-4-nitrophenol by chromatography on thin layers of silica gel with a mixture of light petroleum (boiling range 60° - 80°C) and 1.4% (v/v) of acetone. The spots of the unwanted compounds are removed by sunction into a column of adsorbent, from which they are eluted with 4 ml of methanol, and the eluate is examined polarographically. The amount of O,O-dimethyl O-(3-methyl nitrophenyl) phosphorothioate present is then calculated. The relative error is ± 2%.

Zinc O,O-dialkyl phosphorodithioates have been identified in mineral oil products by thin-layer chromatography (Geldern[2778]). The lubricant or additive concentrate is diluted to an additive concentration of 0.5% with a paraffinic solvent. A portion of the solution (2 vol) is shaken with diethylamine (1 vol) to convert the additive into the ammonium salt, and a 5-μl aliquot is applied to a plate of aluminium oxide G (activated at 120°C). The oil is removed by development with hexane, and the chromatogram is then developed with hexane-acetone-ethanol-triethylamine (20:30:1:1). Increasing the concentration of triethylamine increases the R_f values. The spots are made visible by spraying with iodine azide (IN_3) then with starch solution. Spots due to other compounds, particularly sulphur compounds, are visible before spraying. The R_f values increase with increasing size of the alkyl groups. In instances where the R_f values are nearly equal, the extracted spots can be identified by infra-red spectrophotometry.

Miscellaneous Organophosphorous Compounds

Amormino and Cingolani[2779] have shown that determination by thin-layer chromatography of mixtures containing adenosine and uridine phosphates, pyridoxal 5'-phosphate, thiamine pyrophosphate, thiamine phosphate and vitamin B_{12}, can be effected on layers (0.5 m thick) of Ecteola cellulose MN 300 and cellulose MN 300. The test solution (25 - 30 μl containing less than 50 μg of each compound) is applied to the plate, dried in warm air and ascending development carried out in the dark for 40 min with 0.05 M ammonium chloride. The plate is dried at 60°C. Some compounds form coloured spots, others are detected by ultraviolet light, R_f values are: adenosine triphosphate 0.04, adenosine pyrophosphate 0.12, adenosine 5'-phosphate 0.39.

Wagner[2780] has described a thin-layer chromatographic procedure for the determination of phosphatidylethanolamine (colaminecephalin) and lecithin in pharmaceutical preparations, soya bean oil, egg yolk, heart muscle and brain tissue.

Organophosphorous Insecticides and Pesticides

Dorathion diethyl ester isomers have been analysed by thin-layer chromatography. Bunyan[2767] has discussed the thin-layer chromatography and detection of pesticides including Azinphosmethyl, Demeton O, Demeton S, Diazinon, Dimethoate, Ethion, Fenchlorphos, Malathion, Mevinphos, Morphothion, Parathion-methyl, Phenkapton, Phorate, Phosphamidon, Schradan, Vamidothion and Vamidothion sulphone. Barney[2765] has used thin-layer chromatography in the analysis of the following insecticides and pesticides, Co-Ral, Vapona (Dichlorovos), Delnav, Dipertex, Di-Syston, EPN, Guthion, Meta-Systox R, Meta-Systox 1, Systox, Malathion, Methyl Parathion, Paraoxon, Parathion, Phosfon and Ruelene. Melchoirri et al[2781] have used thin-layer chromatography to determine Diazinon, Malathion, Parathion, Parathion methyl, Chlorthion, EPN, Dicaptan, Trichlorphon, Dimethoate, Phosdrin and Schradan in vegetable oils and Smark and Hill[2782] used the same technique to estimate Parathion ethyl, Parathion ethyl, Dimethoate, Menazon, Vamidothion, Demeton-S-,methyl oxide in fruit and vegetables. Braithwaite[2783] determined Parathion methyl and Diazinon, Parathion ethyl and phosphorothioate insecticide in crops and foodstuffs by thin-layer chromatography and this technique has also been applied to the determination of Fenthion, Guthion, Malathion and Diazinon in potatoes, wheat, maize (Pejkovic-Tadic and Vitorvic,[2784], Malathion, Parathion, Dimethoate in plant residues (Pantovic[2785], Steller and Curry[2786]) and Cidial in honey (Bazzi et al[2787]). Thin-layer chromatography has been used to separate prior to its colorimetric determination, Potosan (0,0-diethyl 0-(4-methyl-7-coumarinyl) phosphorothioate) from Coumaphos (0-(3-chloro-4-methyl-7-coumarinyl) 0,0-diethyl phosphorothioate) (Wasleski[2788]), and for the estimation of Dimethoate in emulsion concentrates (Grimmer et al[2789]) and for the assay of Phorate (Thimet) (Blinn[2790]).

Methods for detecting separated organophosphorous insecticides and pesticides on thin-layer chromatograms have been discussed by various workers (Table 234).

Organosilicon Compounds

Thin-layer chromatography has been applied (Uhie[2806]) to the analysis of silanols, siloxanes, and silanes. Mixtures of these substances can be separated into their components on 0.5 mm layers of Kieselgel D with benzene (or benzene-acetone), light petroleum or light petroleum-benzene (25:1), respectively, as solvent. For the separation of the three classes of compounds, light petroleum-benzene (10:1) was the best solvent; silanols had the lowest R_F values and silanes had the highest.

Organotellurium Compounds

Analytical aspects of organotellurium compounds have been discovered [2807]. Mixtures of the o-, m-, and p-isomers of ditolytellurium dissolved in light petroleum have been separated by thin-layer chromatography on alumina[2808]. In the separation of organotellurium compounds containing phenyl, tolyl or naphthyl radicals on alumina (activity III-IV, particle size 0.06 - 0.09 mm), the substances are revealed as yellow to red-brown spots by treating the chromatogram with iodine vapour[2809]. Light petroleum is used as solvent to separate substances of the type R_2Te, where R is one of the organic radicals mentioned above, and ethyl acetate-methanol (1:1) and benzene-ethanol (9:1 or 4:1) are used to separate derivatives of the types R_2TeCl_2 and R_3TeCl.

Organotin Compounds

Thin-layer chromatography has been used to determine lubricants and organotin stabilizers in carbon tetrachloride extracts of PVC[2810]. Radiometric methods have been used to study the migration of bis(2-ethylhexyl) [di(1-^{14}C]octyl)stannylene)dithiodiacetate from rigid PVC into edible oils [2811]. The extractability of dibutyl- and dioctyltin stabilizers from PVC into foodstuffs simulation liquids such as water, 3% acetic acid, 10% aqueous ethanol, and diethyl ether can be measured by extracting the organotin compound with chloroform from the simulation liquid and then applying the chloroform solution to a glass plate coated with a suspension of silica gel in water (1:23) that has been dried at 20°C and activated at 105°C just before use. The thin-layer plate is then developed with diisopropyl ether containing 1.5% of anhydrous acetic acid. The chromatogram is dried at 105°C and the separated organotin compounds are revealed by spraying with a 0.02% solution of dithizone in chloroform. The limit of detection is 10 µg of organotin compound. The procedure permits distinction between dibutyl- and dioctyltin compounds. Organotin stabilizers in PVC were identified using thin-layer chromatography[2812]. The organotin compound is isolated from the polymer by shaking the diethyl ether extract with aqueous EDTA (disodium salt) followed by extraction with a chloroform solution of dithizone. Alternatively, the polymer film may be soaked in an oil (e.g. olive oil) and the organotin compounds absorbed from the oil on to activated silica gel. The mobile phase is either butanol-acetic acid (50 - 100:1) saturated with water or water-butanol-ethanol-acetic acid (20:10:10:1), and the spots are revealed with dithizone or diphenylcarbazone. Dibutyl-, dioctyl- and

Table 234 - Detection of organophosphorus insecticides and pesticides on thin-layer chromatography.

Compounds mentioned	Coating	Migration solvent	Detection method	Sensitivity	Reference
(a) Thin-layer chromatography					
Potosan and coumaphos	silica gel	solvent	u.v. light	-	Wasleski[2788]
Dimethoate	alumina	acetone-heptone (1:2)	0.5% $PdCl_2$ spray	-	Grimmer et al 2789
Organothion-phosphorus insecticides (47 compounds mentioned)	silica gel	ethyl acetate-hexane (1:3)	bromine vapour exposure then $FeCl_3$-2(2-hydroxyphenyl) benzooxazole spray (fluorescent blue spots in u.v. light). Spraying with Congo red then destroyed fluorescence giving blue spots on a red background.	0.2 - 5 µg	Ragab[2791]
Organophosphorus pesticides (31 compounds mentioned)	silica gel G or H, alumina G, Kieselguhr G	cyclohexane benzene acetone ethylacetate isopropyl alcohol	iodine (not specific)	-	Stanley[2792]
Organophosphorus insecticides (10 insecticides mentioned	silica gel	16 solvents investigated	-	-	Salame[2793]
Organophosphorus insecticides including phosphorothioc acids and chlorinated pesticides	-		phosphorothioic acids-$PdCl_2$ spray chlorinated insecticides. p-amino N,N dimethylaniline spray	-	Baumier and Rippstein[2794]

Table 234 (cont.)

Pesticides	-	-	Acidic ($HCl/HClO_4$) aqueous acetone solution of ammonium molybdate spray followed by exposure consecutively to infrared and ultraviolet radiation to produce a blue colour	-	Hanes and Isherwood[2766] Bandurski and Axelrod[2795] Hais and Macek[2796] Doman and Kagen[3797] Bunyan[2767] Otter[2798]
Organophosphorus insecticides and pesticides (16 compounds mentioned)	silica gel	-	Aqueous acetic acid - hydriodic acid spray then heated to 250°C and sprayed with HCl-ammonium molybdate solution then with acetic acid solution of benzidine (blue colour produced)	2 µg	Barney[2765]
Organophosphorus pesticides (17 compounds mentioned)	silica gel neutral alumina	10% acetone in chloroform, benzene or hexane	(i)bromophenol blue spray, heated to 80° sprayed with aqueous acetic acid (coloured spots on blue background)	-	Bunyan[2767] Cook[2799] Wood[2800]
			(ii)exposure to bromine vapour then bromo-phenol blue spray (iii)exposure to bromine vapour water spray, then contact with filter paper treated with blood plasma, bromophenol blue and sodium hydroxide. Finally sprayed with acetylcholine (blue spots)		
Thiophosphate pesticides	silica gel	-	2,6-dichloro-p-benzoquinone-4-chlorimine (non-specific)	0.1 µg	Menn et al[2801] Braithwaite[2783]
Organophosphorus insecticides and pesticides	silica gel	-	ferric chloride-sulphosalicyclic acid spray (non-specific)		MacRae and McKinley[2802] Wade and Morgan 2803

Organophosphorus pesticides (P or PS type)	-	-	4-(p-nitrobenzyl) pyridine in acetone spray used in conjunction with tetraethylenepentamine (red or blue spots)	-	Watts[2804]
Organophosphorus insecticides (11 compounds mentioned)	silica gel G	hexane-acetone	(i)Palladium chloride spray (ii)Ammoniacal silver nitrate spray (not specific)	-	Takeshi[2805]
Dimethoate	silica gel	acetone-chloroform (3:1)	molybdenum blue method	-	Steller and Curry[2786]
Dimethoate Menazon, Vamidothion Demeton-S-methyl S-oxide	silica gel HF 254	1:2 dichloroethane-benzene (2:1)(one direction)cis 1:2-dichloroethylene (other direction)	-	0.1 part $/10^6$	Smart and Hill 2782
Parathion methyl Diazanon, etc.	silica gel, plaster of Paris pH4 (with hydrochloric acid)	-	0.5% 2:6 dichloro- (or dibromo) p-benzoquinone-4-chlorimine (red spot) heat to 120°C (specific for P-S bond)	1 µg	Braithwaite[2783]
Malathion, Parathion, Dimethoate	-	acetone-hexane (1:9)	molybdenum blue method	1 µg	Pantovic[2785]
Diazinon, Malathion, Parathion, Parathion methyl, Chlorthion, Dicaptan, Trichlorphon, Dimethoate, Phosdrin, Schradan	silica gel GF 254	various	ultraviolet light or aqueous palladium chloride or Dragendorf reagent sprays	-	Melchoirri et al 2781
Cidial	silica gel (ascending)	-	0.5% cyclohexane solution of 2,6-dibromo-o-benzoquinone 4-chlorimine spray (yellow spot, becomes red on heating to 70°/75°)	0.3 part $/10^6$	Bazzi et al[2787]

dibenzyltin salts were separated and identified and mixtures of dialkyl- and trialkyltin compounds could be separated.

Traces of mono-, di- and tervalent and uncharged tin compounds can be separated using silica gel as adsorbent and various solvents as the mobile phase[2812]. The R_F values of the individual compounds are such that cations can be identified; the anionic part of the molecule has no effect. The recommended mobile phase for general use is a mixture (2:1) of isopropyl alcohol and ammonium carbonate solution (2 parts of 10% aqueous ammonium carbonate and 1 part of aqueous ammonia). For the separation of bivalent organotin compounds, isopropyl alcohol-1 N sodium acetate can be used. For monovalent compounds, butanol - 2.5% aqueous ammonia is the most suitable. Spots are located by spraying the chromatogram with catechol violet solution and examining it in u.v. light. Phenyltin compounds dissolved in 96% ethanol are decomposed by u.v. light to give phenol, which can be detected with 4-aminophenazone solution. The method has been applied to the determination of the extractability of organotin stabilizers from poly(vinyl chloride) foil and the solubility of such stabilizers in fats and oils[2813].

Thin-layer chromatography of organotin compounds on silica gel can be carried out with 1.5% acetic acid in diisopropyl ether as developing solvent [2814]. Sulphur-containing compounds are identified by spraying the chromatogram with 2% ethanolic molybdophosphoric acid, and other tin compounds are located by examination under u.v. light and by treatment with 0.5% ethanolic catechol violet. In a procedure for the identification of organotin compounds commonly used as stabilizers in poly(vinyl chloride) packaging materials, the stabilizers are extracted from the material with diethyl ether and separated on layers of silica gel G by radial development with 2,2,4-trimethylpentane-diisopropyl ether-acetic acid (80:3:8)[2815]. The developed plates are exposed to bromine vapour then sprayed with a solution of Rhodamine B and catechol violet in acetone and examined under daylight and under u.v. irradiation (254 and 366 nm). Migration values (relative to dioctyltin diacetate) have been reported for 10 compounds. The detection limit is 1 μg.

Thin-layer chromatography has been applied to a wide variety of other types of organotin compounds, including bis(2-ethylhexyl)tin compounds[2816], sulphur-containing organotin compounds[2817], impurities in di-n-octyltin stabilizers[2818,2819], triphenyltin compounds[2820], mixed allylphenyl and isobutylphenyltin compounds[2821], mixed phenyltin chloride[2822], and alkyltin compounds as their quercetin chelates[2823]. Thus, dialkytin compounds in amounts of 5 - 100 μg can be separated from other organotin species by thin-layer chromatography. They are determined by treatment with dithizone, elution, and photometric determination[2817]. It is not usually easy to detect, for example on thin-layer chromatographic plates, tetraalkyltin compounds or trialkyl- or triaryltin halide compounds with the usual colorimetric reagents since complexes are either not formed or are too unstable. This difficulty can be overcome by exposing the plates to bromine vapour[2824], so that tin - carbon bond cleavage occurs and the product has sufficient Lewis acidity to form a stable complex with the reagent.

Organozinc Compounds

A thin-layer chromatographic procedure for the determination of zineb in tobacco plants has a sensitivity of 10 μg of zineb per 100 g of tobacco [2825].

B. PAPER CHROMATOGRAPHY

Organoarsenic Compounds

Six organoarsenic compounds (arsanilic acid, arsenosobenzene, arsphenamine, 4-hydroxy-3-nitrophenylarsonic acid, 4-nitrophenylarsonic acid, and p-ureidophenylarsonic acid) were separated by two-dimensional chromatography on sheets of Whatman No. 1 paper, with a solvent consisting of water and nitric acid diluted with methyl cyanide[2826]. The compounds were located on the chromatogram and identified by their quenched or fluorescent areas in u.v. light. Final identification was made by spraying with ethanolic ammoniacal silver nitrate or ethanolic pyrogallol, with air drying between the sprayings. The identification limit is 1 μg for each compound.

Organobismuth Compounds

Basic bismuth gallate and free tribromophenol can be detected in small samples of bismuth tribromophenoxide by the simultaneous ascending chromatography of a neutral ethanolic suspension of one portion of the sample and an acidified ethanolic solution of another portion[2827]. The two spots are placed side by side and developed with butanol saturated with water. After 2 h the chromatogram is treated with aqueous iron (II) ammonium sulphate solution at about halfway between the origin and the front and silver nitrate solution is applied close to the front. The presence of tribromophenol is shown by an orange fleck with silver nitrate on the neutral chromatogram and basic bismuth gallate by a blue-black spot with iron (II) ammonium sulphate solution on the acid chromatogram. An orange fleck on the acid chromatogram is due to the acid decomposition of the bismuth tribromophenoxide.

Organocobalt Compounds

For the determination of cobalamins, total cobalamins are determined spectroscopically in a pH 5.5 acetate buffer containing 25 - 30 μg (dry matter) of sample per millilitre[2828]. Individual cobalamins are determined by applying a portion of the sample solution in a band to Whatman 3MM paper. Development is carried out for 18 h by the descending technique with water-saturated sec-butyl alcohol to which is added 1% of acetic acid and water to incipient turbidity. The individual cobalamin bands are eluted with measured volumes of 0.1% potassium cyanide solution (pH 6.0). Each individual substance is determined spectrophotometrically as cyanocobalamin. Hydroxo- and cyanocobalamins can be determined by converting them into the dicyano derivatives by treatment with potassium cyanide prior to paper chromatography[2829]. The dicyano derivatives are concentrated by extraction with butanol of an aqueous solution of the cyanide-treated sample saturated with ammonium sulphate. Impurities are removed by paper chromatography, and the purified compounds are eluted from the spots with water. The absorbances are then measured spectrophotometrically at 367 and 570 nm, and the concentrations of the cobalamins are calculated with refeerence to calibration graphs.

Cyanocobalamin and its derivatives can be separated on paper with chloroform-phenol-butanol-water (12:2:5:20)[2830,283±]. Several R_F values were tabulated[2830]. Cyanocobalamin can be extracted from the paper with water and determined absorptiometrically; the error was ± 2%[2831]. Hydroxycobalamin has been determined in parenteral injection solution by paper chromatography followed by spectroscopy[2832]. Photodensitometric determinations of cyanocobalamin and hydroxocobalamin after separation on Whatman No. 1 paper, on thin layers of alumina or silica gel, or on Amberlite WA-2 paper show that the sensitivity decreases from 5 μg on Whatman paper to 15 μg on Amberlite paper[2833]. Hydroxo- and cyano cobalamins give identical calibration graphs. The chromatogram is developed in a dark chamber at 20 ± 2°C in an atmosphere saturated with the solvent. The results for hydroxocobalamin are more reproducible than those for cyanocobalamin owing to losses of cyanide ion from cyanocobalamin during chromatography. The errors range from -5 to -8% for up to 30 μg of cyanocobalamin on paper, alumina, or silica gel, and from -1.8 to -2.4% on paper for 50 - 200 μg of cyanocobalamin. Results for the determination of cyanocobalamin by a paper chromatographic[2834] and by a cup plate agar diffusion method[2835] have been shown to be equally accurate.

Organocopper Compounds

In a paper chromatographic method for the quantitative determination of copper and zinc 8-hydroxyquinolinates, the 8-hydroxyquinolate, after extraction with 10% sulphuric acid and then with chloroform at pH 5 - 7, is chromatographed with butanol-hydrochloric acid - water (8:1:1) on Whatman No. 3 paper[2836]. Two bands, one due to 8-hydroxyquinoline and Cu^{2+} alone (R_F = 0.4) are formed. Both bands are cut out, and the copper is extracted with dilute acetic acid and titrated with EDTA, with 1-(2-pyridylazo)napth-2-ol as indicator.

Organoiron Compounds

Paper chromatography of ferrocenes has been discussed[2837,2838]. In the paper chromatography of iron complexes of porphyrins the iron of uro-, copro-, haemato-, deutero-, meso, and protoporphyrins in the form of corresponding haemin, haematin, or haematin acetate are separated by reversed-phase paper chromatography with water-propanol-pyridine as the solvent system and silicone (Dow Corning No. 550) as the stationary phase[2839]. A simplified lutidine - water system has been described for the determination of the number of carboxyl groups in haemlins and free phosphyrins. Methods for the determination in plant extracts of iron chelates such as Fe - EDTA and Fe - 1,2-diaminocyclohexanetetraacetic acid depend upon the removal of unwanted plant constituents and concentration of the chelates by passage through a column of cation-exchange resin or of activated charcoal, followed by evaporation and paper chromatography with phenol - water (4:1 w/v) in the presence of aqueous ammonia and potassium cyanide[2840]. For the quantitative determination of free EDTA and 1,1-diaminocyclohexanetetraacetic acid, colorimetric methods depending on either the iron-(III)-salicylate complex or the copper - biscyclohexanoneoxalyldihydrazone complex are described. The iron chelates of free EDTA and 1,2-diaminocyclohexanetetraacetic acid may be determined by a modified 1,10-phenanthroline method; up to 200 μg of each can be determined.

Organolead Compounds

Mixtures of R_3PbCl, R_2PbCl_2, and R_4Pb (R = methyl, ethyl, or phenyl) can be separated by paper chromatography followed by conversion to the $[PbCl_4]^{2-}$ ion and spectrophotometric determination of the latter at 357 nm[2841]. When applying this method to samples containing a single organolead compound, a suitable sized portion of the sample is mixed with 3 volumes of ethanol saturated with iodine. After 5 min the mixture is evaporated to dryness on a water-bath, the iodine is removed by heating on a hot-plate, and the residue is dissolved in 4 M potassium iodide and examined spectrophotometrically. Recoveries of lead range from 98 to 102.2% of theoretical. The solvents used for the di- and trimethyl- and di- and triethyllead chlorides are water and methanol, for the di-, tri-, and tetraphenyl compounds chloroform, acetone, or methanol, and for the tetramethyl and tetraethyl compounds n-heptane. Mixtures of di- and trimethyl- and di- and triethyllead chlorides with lead chloride containing 25 - 40 µg/ml of total lead can be separated chromatographically on paper[2842]. The spots located by means of a reference strip, are cut out and extracted, and the extracts are treated as described above. The errors range from - 2.0 to + 1.9%. Treatment of the spots on the paper without previous elution was more rapid but less precise, the errors amounting to as much as 12.5%. Ionic lead present as an impurity can be determined by dissolving the sample in water, adding potassium iodide to a concentration of 4 M, and measuring the absorption at 357 nm, the organolead compounds having no absorption at this wavelength; lead ions are converted into $[PbI_4]^{2-}$.

Schafer[2843] has described a procedure for the paper chromatographic separation of trialkyllead chlorides and bromides. He used two solvents - (i) benzene (6 vol) and cyclohexane (3 vol) saturated with water, the organic phase is filtered through a paper wetted with benzene and mixed with acetic acid (1 vol); (ii) cyclohexane saturated with water. The bottom of the chromatography tank is covered with water. Ascending development on 2043b paper (Schleicher and Schull) is used and 5 µl of sample solution (containing ~ 20 µg lead) is applied. The solvent front is allowed to travel 23 - 25 cm. After a 10 min irratiation with u.v. light, the compounds are detected by spraying with ammonium sulphide solution. Schafer lists R_f values for several compounds in the two solvents. The water content of the solvents is critical; a lower water content leads to an increase in R_f, especially for slowly moving components. Better separations are obtained with trialkylead chlorides than with trialkyllead bromides.

Paper chromatography[2844] has been successfully applied to the separation of mixtures of alkyllead compounds in petroleum. Tetramethyl- and tetraethyllead are detected by chromatography of their bromo-derivatives, (R_3PbBr). To the sample (10 - 25 ml of petrol) is added aqueous bromine (5 ml) and the mixture shaken for 30 seconds. If the colour of the bromine is not appreciably changed, repeat with a larger volume of sample. Then concentrated aqueous ammonia is added in excess. The phases are left to separate then a few drops of the aqueous phase are placed on Whatman No. 1 paper, and developed by descending chromatography for 12 h with the organic phase of butanol-concentrated aqueous ammonia (1:1). The air-dried paper is sprayed with 0.1% dithizone in chloroform. The lead compounds appear as yellow to orange spots on an evanescent green background. The original compounds are identified from the R_f values of the derivatives, viz., tetramethyllead 0.6; ethyltrimethyllead, 0.7; diethyldimethyllead, 0.7 or 0.81; triethylmethyllead, 0.81; and tetraethyllead, 0.9. The mixed-alkyl compounds result from metathetical reactions that occur on storage when both tetramethyl- and tetraethyllead are present. For the pure monohalogen derivatives of both single- and mixed-alkyl compounds, the following equations have been derived - for R_3PbCl, R_f = 3160 - 3528 x; for R_3PbBr, R_f = 3190 - 4143 x; and for R_3Pbl, R_f = 3200 - 4686 x, where x is the lead content (%) divided by 100 and R is - CH_3 or - C_2H_5.

Organomercury Compounds

Bartlett and Curtis[2845] have developed a procedure for the paper chromatography of organomercury compounds. They found that several developing systems could be used, if aqueous ammonia is one of the components. The most suitable combination is 1 - butanol - 95% ethanol - 28% ammonia (8:1:3). (Table 361). A chloroform solution of dithizone is the preferred spray reagent (Miller et al[2846]), alternatively an aqueous sodium stannite solution can be used. The former is more sensitive but less selective than the latter. Both the ascending and the descending techniques were used. This procedure, described below, is useful for detecting relatively small amounts of organo-mercurials in the presence of large amounts of inorganic mercurials.

Table 235 - R_f values of some organic mercury compounds (n-butanol, ethanol ammonia)

Compound	Small scale (ascending)	Large scale (descending)
Phenylmercuric chloride	0.38	0.39
Tolylmercuric chloride	0.48	0.49
o-Chloromercuri phenyl	0.28	0.22
p-Chloromercuriphenol	0.07	0.07
Di-p-tolylmercury		0.94
Mercaptomerin sodium*	0.16	0.22
Mercuhydrin #	0.17	0.15
Mercurital ‡	0.43	0.41
Merthiolate*	0.70	0.75
Salyrgan#	0.44	0.39

* N-y-carboxymethylmercaptomercuri- -methoxy -propylcamphoramic acid disodium salt,

\# N-(β-hydroxymercuri-y-methoxy)-propyl)-N'-succinylurea.

‡ Sodium salt of o-(3-hydroxymercuri-2-methoxypropyl)carbamyl)phenoxyacetic acid.

\# Sodium ethylmercurithiosalicylate.

Bartlett and Curtis[2845] state that the R_f values for the ethyl alcohol-ethyl ether system are similar to those found in Table 235 but the spots were not as sharply defined.

In one procedure for the paper chromatography of organomercury compounds an aqueous solution is spotted on to a filter-paper and chromatographed with n-butanol saturated with 1 N aqueous ammonia solution[2847,2848]. The spots are detected with 1% diphenylcarbazone in ethanol. The R_F value increases with increasing number of carbon atoms in the organic radical (phenylmercury acetate 0.39), phenylmercury chloride 0.40, ethylmercury chloride 0.27, methoxyethylmercury chloride 0.18, methylmercury chloride 0.17; at 30°C); Hg_2^{2+} and Hg^{2+} remain at the baseline. Another method for the paper chromatographic separation of alkyl- and arylmercury compounds used as agricultural chemicals uses n-butanol-pyridine-1 N aqueous ammonia solution (35:34:31) [2848]. The R_F values are as follows: p-tolymercury acetate, 0.77; phenylmercury acetate, 0.68; di(ethylmercury phosphate, 0.62); 2-methoxyethylmercury chloride, 0.47; methylmercury chloride 0.41; and mercury (II) chloride, 0.06. The spots are revealed with a solution of 1% diphenylcarbazone in ethanol. Various methods based on paper chromatography, color reactions, and u.v. spectrophotometry have been used for the detection of organomercury compounds[2849]. In paper chromatography, butanol-glacial acetic acid - water (4:1:5) is the most useful solvent, and dithizone is a sensitive detection reagent. Both the Reinsch sublimation method and colour reaction with 2,2'-bipridyl are suitable for qualitative analysis, but colour reactions with di-2-naphthylthiocarbazone and dithizone are unsuitable.

Radio paper chromatography has been used to separate mixtures of mercury (II) bromide, phenylmercury bromide and p-bromophenylmercury bromide[2852]. Various other aryl- and alkylmercury bromides were also separated. Various migration solvents were used, including n-butanol-ammonia, tetrahydrofuran-n-butanol, I N ammonia (15:35:20), methyl acetate-n-butanol-2 N ammonia (47:40:13), and $PO[N(CH_3)_2]_3$-n-butanol-2 N ammonia (1:1:1) (Table 236).

Table 236 - R_f of alkyl- and arylmercury bromides at 20°C

Solvent	Compound	R_f Range	R_f Mean
n-BuOH-1 N NH_4OH	BrHgBr	0.00	0.00
	CH_3HgBr	0.10 - 0.20	0.15
	C_2H_5HgBr	0.20 - 0.30	0.25
	n-C_3H_7HgBr	0.22 - 0.41	0.31
	n-C_4H_9HgBr	0.74 - 0.97	0.86
	p-$CH_3OC_6H_4HgBr$	0.15 - 0.35	0.25
	C_6H_5HgBr	0.26 - 0.43	0.35
	p-BrC_6H_4HgBr	0.84 - 1.00	0.92
THF-n-BuOH	C_6H_5HgBr	0.00 - 0.03	0.01
- 1 N NH_4OH (15:35:20)	p-$CH_3OC_6H_4HgBr$	0.17 - 0.36	0.27
CH_3COOCH_3 - n-BuOH- 2N NH_4OH (47:40:13)	p-$CH_3OC_6H_4HgBr$	0.19 - 0.32	0.26
$PO(Ni(CH_3)_2)_3$ -	C_6H_5HgBr	0.66 - 0.94	0.82
n-BuOH-2 N NH_4OH (1:1:1)	p-$CH_3OC_6H_4HgBr$	0.91 - 1.00	0.96

Organophosphorus Compounds

Alkyl and aryl phosphates and phosphites. In a technique for the separation of dialkyl and monoalkyl phosphates in trialkyl phosphates, Shvedov and Rosyanov[2853] impregnated the paper with 0.1 N oxalic acid. They studied several solvent sysmtes; the best was butanol - ethanol - water (4:1:2) which gave R_f = 1 for trialkyl phosphates. To detect the chromatographic zones, the chromatogram was treated with a reagent comprising a mixture of 5 ml of 60% perchloric acid, 10 ml of N-hydrochloric acid and 25 ml of 4% ammonium molybdate solution diluted to 100 ml, then dried for 7 min at 85°. The chromatogram was set aside in the air to become saturated with moisture, then placed in an atmosphere of hydrogen sulphide for 5 - 10 min. The monoalkyl and dialkyl phosphates gave blue zones, trialkyl phosphates gave yellow-brown zones. The R_M values for the monoalkyl and dialkyl phosphates are linear functions of the number of methylene groups present.

Cvjeticanin and Cvoric[2854] have described a separation of monobutyl dihydrogen ortho-phosphate, dibutyl hydrogen ortho-phosphate and phosphoric acid and their separate determination in a tributyl phosphate-kerosene mixture by paper chromatography. The separation was performed with the following solvents; methanol-dioxan-N-aqueous ammonia (1:1:1); pentanol-pyridine-25% aqueous ammonia (3:7:10); pentanol-pyridine-2N-hydrochloric acid (2:2:1). The separated components were detected by dipping the strip in an acetone-molybdate solution and exposing the dried strip to u.v. light.

Shvedov and Rosyanov[2853] reported a separation of phosphoric acid and its mono- and dibutyl esters. They used paper strips 2.5 cm wide; 0.2 ml of a solution of the acids was placed 2 cm from the edge and dried. The following mixtures of solvents were used - (i) butanol-methanol-water (4:1:2 (ii) isoamyl alcohol-pyridine - 25% aqueous ammonia (2:7:10); (iii) chloroform-methanol-water (4:5:1); (iv) propanol - 25% aqueous ammonia-water(6:1:3 and (v) methanol - 25% aqueous ammonia-water (6:1:3). The chromatograms were sprayed with a mixture of 60% perchloric acid (5 ml), hydrochloric acid (10 ml) and 4% ammonium molybdate solution (25 ml), made up to 100 ml with water, dried in warm air, heated at 85°C for 7 min, set aside in the air to become saturated with moisture, and finally suspended in a vessel containing hydrogen sulphide for 5 or 10 minutes. The developed spots were cut out, ashed with concentrated sulphuric acid + concentrated nitric acid (1+ 1) and the orthophosphate was determined. If chloroform saturated with water is used as chromatographic solvent, the dibutyl ester is not displaced from its point of introduction; this may be used to separate it from tributy phosphate.

Gabov and Shafiev[2854] have described procedures for the separation of C_1 to C_4 alkyl phosphites and alkyl phosphates using 70% formic acid-pentanol (1:1); propanol-concentrated ammonia (4:1) or isoamyl alcohol-pyridine-concentrated ammonia (3:2:1) as solvent. They describe a scheme based on the use of both acid and alkaline solvents for the complete analysi of a mixture of three alkyl phosphates and three alkyl phosphites.

Cerrai et al[2855] used n-butanol-formic acid-water as solvent to separat phosphoric acid, monobutyl phosphate and dibutyl phosphate. The separated spots were located by the procedure of Hanes and Isherwood[2856] which involves spraying the paper with acid molybdate solution then exposing it to hydrogen sulphide. Alternatively, to locate the separated compounds Cerrai et al[2855] sprayed the paper with bromophenol blue solution. Hardy and Scargill[2857] separated phosphoric acid and mono- and di-butyl phosphate with n-butanol-acetone-water-ammonia as solvent; spraying with a ferric solution, followed by spraying with sulphosalicyclic acid solution was used to locate spots, as recommended by Wade and Morgan[2858]. This spray system produces colourless spots on a mauve background.

Plapp and Casida[2859] have described the separation of certain mono- and dialkyl phosphates derived from the breakdown of insecticides. They used isopropanol-water-ammonia mobile phases; to reveal the spots they sprayed the papers with an acidic molybdate solution and then exposed them to ultraviolet light as recommended by Crowther[2860].

Tyszkiewicz[2861] separated phosphate esters from plant extracts by descending two dimensional paper chromatography at room temp on Whatman No. 4 paper washed in aqueous EDTA (Eggleston and Hems[2862]). The solvents recommended are: (i) isobutyric acid-N-aqueous ammonia-0.1 M EDTA (100:60: 1.6) and (ii) butanol-propionic acid-water (75:36:49).

Hardy and Scargill[2857] developed a qualitative rapid paper chromatographic method for the analysis of mixtures of phosphoric acid, dihydrogen monobutyl phosphate, hydrogen dibutyl phosphate and tributyl phosphate. Drummond and Blair[2863] and Towle and Farrand[2864] developed this method into a quantitative method for the determination of the above impurities in solutions of tributyl phosphate in hydrocarbon diluents. Moule and Greenfield[2865] have also discussed the separation of dihydrogen butyl phosphate and hydrogen di-n-butyl phosphate by paper chromatography.

Runeckles and Krotov[2866] developed a system for the paper ionophoresis and chromatography of phosphate esters and organic acids. Electrophoresis was conducted in 2.5% pyridine in 0.25% aqueous acetic acid by applying 800 v d.c. (20 v per cm) across the paper. Conventional paper chromatography was then used in the second dimension using 2-methoxyethanol-pyridine-glacial acetic acid-water (8:4:1:1), containing 0.15% (w/v) of 8-hydroxyquinoline, as migration solvent.

Moule and Greenfield[2867] studied the separation by paper chromatography of esters of phosphoric and phosphorous acids. Samples studied included tris-2-chlorethyl phosphite, octyl diphenyl phosphite, triallyl phosphite and trioctyl phosphite, and also mixtures containing mono-, di- or trialkyl phosphates, free phosphoric acid, and pyrophosphoric acid and its esters.

They showed that the separated components could be determined quantitatively by wet ashing the spots and determining the phosphorus using the method suggested by Hanes and Isherwood[2856].

Typical separations achieved by these workers are shown in Tables 237 and 238.

Table 237 - R_f values for esters of phosphorous acid. (Sodium carbonate stationary phase, 90 / 10 petrol-n-butanol moble phase)

Tris-2-chloroethyl phosphite	0.95
Bis-2-chloroethyl phosphite	0.35
Tris-2-chloroethyl phosphite	1.00
Bis-2-chloroethyl phosphite	0.73

Table 238 - R_f values for esters of phosphorous acid and associated impurities. (Polyethylene glycol 400 stationary phase, 80 / 20 petrol-chloroform mobile phase)

	Ester	R_f Value
Triallyl phosphite	$(CH_2 = CH,CH_2O)_3P$	0.97
Diallyl phosphite	$(CH_2 = CH,CH_2O)_2POH$	0.43
Triallyl phosphate	$(CH_2 = CH,CH_2O)_3PO$	0.67
Diallyl phosphate	$(CH_2 = CH,CH_2O)_2POOH$	0.31
Tris-2-chloroethyl phosphite	$(ClCH_2,CH_2O)_3P$	0.56
Bis-2-chloroethyl phosphite	$(ClCH_2,CH_2O)_2POH$	0.09
Tris-2-chloropropyl phosphite	$(CH_2,CH(Cl),CH_2O)_3P$	0.93
Bis-2-chloropropyl phosphite	$(CH_2,CH(Cl),CH_2O)_2POH$	0.25

Phosphites are revealed as yellow spots on a pink background by the use of alkaline permanganate; the background colour faces rapidly, and the spots must be outlined as soon as possible. For acid phosphates, alkaline Universal Indicator shows the spots distinctly and is quite sensitive. The traces are yellow on a blue-green background, fading slowly as the paper dries to red on a yellow background.

If trialkyl phosphates are suspected, the paper should be sprayed with the ferric thiocyanate solution. Trialkyl phosphates from C_3 upwards give bright red spots; acid phosphate esters give colourless spots on a brown-pink background. This reagent is not as sensitive as Univeral Indicator to the acidic components.

Esters of phosphorous acid. Samples of tris-2-chloroethyl phosphite or tris-2-chloropropyl phosphite may be separated on sodium carbonate papers with 90 / 10 petrol-n-butanol mixture as mobile phase. The sodium carbonate serves only to maintain a mildly alkaline environment for the sample, which is sensitive to acids. As might be expected the moisture content of the paper affects the R_f' values and Table 238 gives typical values obtained with papers treated as described above.

Gerlach et al[2867] have discussed solvents for the paper chromatography of phosphoric acid esters in various materials including tissue extracts. They describe three solvent mixtures composed of various combinations of formic acid, trichloroacetic acid and other constituents such as ammonia, 8-hydroxyquinoline and EDTA. The mixtures are all suitable for the simultaneous separation of numerous phosphorus compounds, even from tissue extracts rich in ions; they can also be used for one- and two-dimensional ascending or descending chromatography and a saturated vapour atmosphere need not be maintained during their use. Details of the properties and methods of application of these mixtures are given by these workers together with the advantages of their use with different types of compound. Three further mixtures of organic solvents containing ammonia are given; the sharpness of the separation of nucleotides and phytic acid is greatly increased in the ammoniacal solvent by the addition of 8-hydroxyquinoline with or without EDTA (disodium salt).

Various workers have described spray reagents suitable for the detection of phosphate esters on paper chromatograms. Haney and Loughheed[2868] locate phosphate esters by spraying with molybdate reagent and irradiating with u.v. lamps to produce coloured spots. The colour intensity is increased if both sides of the paper are irradiated, and these workers describe a simple irradiation box containing four u.v. lamps, holding the chromatogram suspended on nylon monofilament between pairs of lamps.

To locate organic phosphates and phosphites Donner and Lohs[2869] spray the paper with a 1% solution of cobalt chloride in acetone, either in the cold or on warming. Between 10 and 250 µg of phosphate ester can be detected by this procedure.

Macho[2870] dries the chromatogram at 80°C, moistens it with an acidic solution of molybdate and heats for 7 min at 85°C, to hydrolyse the phosphate esters. Addition of a freshly prepared 0.2% solution of stannous chloride in ether followed by warming for 10 to 30 s reveals the organic phosphates as blue spots.

In a further detection method (Rorem[2871]) the dried paper is immersed in a solution (0.5%) of quinine sulphate dihydrate in abs ethanol. The paper is dried at room temp for a few mins and examined under u.v. light in darkness to reveal the phosphate esters as light spots on an intense grey-blue fluorescent background. As little as 0.007 µg phosphorus can be detected by this method. Purines, pyrimidines, nucleosides and free sugars and amino acids also give a positive test with this reagent.

Otter[2872] described a general method for locating tertiary alkyl phosphates, thiolphosphates and thionophosphates on paper chromatograms in amounts down to 7 µg. The spots are first treated with N-bromo-succinamide dissolved in acetone; this facilitates subsequent hydrolysis of the organophosphorous compound. Treatment with ammonium molybdate (Hanes and Isherwood [2856]) then gives blue spots after 10 - 20 min exposure to u.v. light.

Gordon et al[2873] used 2-(o-hydroxylphenyl)benzoxazole to detect phosphate esters on paper chromatograms. This fluorescence-forming reaction is sensitive to about 10^{-9} mol of phosphorus. The paper chromatogram is saturated with a solution (0.54%) of ferric chloride hexahydrate in 80% ethanol diluted 1:50 with acetone. After allowing the paper to dry, a solution of 0.0024 M 2-(o-hydroxyphenyl)benzoxazole and 2% dimethyl sulphoxide in methylethyl ketone is applied and the chromatogram examined under long wavelength u.v. light. Ferric-complexing compounds fluoresce yellow on acidic paper and blue on alkaline paper.

Sokolowska[2874] has used paper chromatography for the determination of free phenols in tritolylphosphate. The ethanolic sample solution (1 µl) is applied to Whatman No. 1 paper previously impregnated with 20% ethanolic acetamide, and the chromatogram developed with cyclohexane-chloroform-ethanol (45:3:3) by the ascending solvent technique for 3 h. The chromatogram is dried at 20°C, and sprayed with diazotized-4-nitroaniline reagent (saturated solution of 4-nitroanline in 0.33 N hydrochloric acid (2 ml) mixed with 1% aqueous sodium nitrite (2 ml) and 5% aqueous urea (2 ml)); after 5 min the mixture is diluted with water (14 ml) this reagent is stable for 2 h), then with 10% aqueous sodium carbonate. The phenols present in the sample can be identified by comparing their colours and R_f values with those of standard pure compounds.

Gabov and Shafiev[2875] separated alkyl phosphites and phosphates by paper chromatography using 70% formic acid-pentanol (1:1), propanol-conc ammonia (4:1) or isoamyl alcohol-pyridine-conc ammonia (3:2:1) as solvent.

Moule and Greenfield[2865] also separated organic phosphites and phosphates on paper.

Phosphonous, Phosphinic and Phosphonic Acid Esters

Weil[2876] separated mixtures of phosphonous, phosphinic and phosphonic acids and their isomers by ascending paper chromatography on Watman No. 1 sheets (rolled into cylinders); 50 µg of the substances (0.005 ml of a 1% alcoholic solution) were applied. The solvents used were collidine-water or butanol:2 N aqueous ammonia. The spots were developed either by spraying with a 0.1% solution of phenol red in alcohol, adjusted to alkaline (purple) with 2 N sodium hydroxide and subsequent stabilization in ammonia vapour (yellow spots on violet background); or by spraying with a saturated alcoholic solution of silver nitrate and subsequent exposure to sunlight (white spots on a brown background), Weil[2876] tabulated the R_f values of 15 organophosphorous compounds [$R,P(OH)_2$, $RPO(OH)_2$ and $R_2,PO.OH$] separated by both methods of elution.

Halmann and Kugel[2877] have discussed the paper chromatography and paper electrophoresis of methylphosphonic, methylphosphinic and dimethylphosphinic acids and of trimethylphosphine oxide from each other and from the common phosphorus oxy-acids. Chromatography was carried out by the ascending technique on Whatman No. 1 paper (washed with 2% EDTA solution) using propanol-conc ammonia-water (6:3:1) as developing solvent. Electrophoresis was carried out on strips (2.5 cm x 90 cm) of Whatman No. 3 MM paper, soaked in 0.1 M lactic acid, pressed between plates of silicone-coated glass, the best results being obtained at 750 V and 3 mA, no cooling being necessary. Phosphorous-containing spots were detected either by spraying with ammonium molybdate solution or by neutron activation, in which the dried strips are exposed for 30 sec to a neutron flux of 2.5×10^{12} neutrons cm^{-2} s^{-1}. After 1 week, to allow ^{24}Na to decay, the ^{32}P activity of each strip was counted with an end-window proportional counter.

Gabov[2878] carried out ascending paper chromatography of chlorinated methyl phosphinic and chlorinated methyl phosphonic acids in the presence of methyl phosphonic acid and phosphate and phosphite ions using acetone-pyridine-25% aqueous ammonia (8:1:5) as the elution solvent. The spots were located by means of the molybdate reagent (Hanes and Isherwood[2856]).

Siuda[2879] discussed the separation of inorganic phosphorus and phenyl phosphorus compounds by paper chromatography and paper electrophoresis with sulphuric acid and subsequent warming under an i.r. lamp, or of other substances, phenylphosphoric acid, triphenylphosphine, triphenylphosphine oxide $(NH_4)_3PO_4$, $Na_4P_2O_6$, $K_4P_2O_7$, NaH_2PO_2 and Na_2HPO_3. The compounds were separated by paper chromatography on Whatman No. 3 MM paper previously treated with 2% EDTA solution. Development for 33 cm was carried out with propanol - 25% aqueous ammonia (2:1) by the ascending technique. In the electrophoresis, paper strips were soaked in ammonium oxalate solution of pH 5.5 and pressed between two silicone-treated glass plates. The separation took 70 min with a potential gradient of 45 V cm^{-1}. The spots were located by neutron activation or by the Hanes and Isherwood reagent[2856] followed by hydrogen sulphide. The R_f and R_p values (ratio of distances of peak maxima to those of orthophosphate) obtained by paper chromatography and electrophoresis, respectively, are reported.

Trialkyl and Triaryl Phosphines and Phosphine Oxides

Siuda[2879] discussed the separation of ionorganic and phenylphosphorus compounds by paper chromatography and paper electrophoresis. The compounds studied included triphenylphosphine, triphenylphosphine oxide, diphenylphosphinic acid, phenyl phosphonous acid, phenylphosphoric acid, $(NH_4)_3PO_4$, $Na_4P_2O_6$, $K_4P_2O_7$, NaH_2PO_2 and Na_2HPO_3.

Siuda[2879] has discussed the separation of inorganic and phenylphosphorus compounds by paper chromatography and paper electrophoresis. The compounds studied included triphenylphosphine oxide, triphenylphosphine, diphenylphosphinic acid, phenyl phosphonous acid, phenylphosphoric acid $(NH_4)_3PO_4$, $Na_4P_2O_6$, $K_4P_2O_7$, NaH_2PO_2, and Na_2HPO_3. This method is discussed more fully in Section F of this chapter.

Halmann and Kugel[2877] discussed the paper chromatography and paper electrophoresis of mixtures of trimethylphosphine oxide with methylphosphonic, methylphosphinic, dimethylphosphinic acids and the common phosphorous oxy-acids. Paper chromatography was carried out on Whatman No. 1 paper (washed with 2% EDTA solution) by the ascending technique using propanol-conc ammonia-water (6:3:1) as solvent. Electrophoresis was carried out on strips of Whatman No. 3 MM paper, soaked in 0.1 M lactic acid, pressed between sheets of silicone-coated glass. The best results were obtained at 750 V and 3 mA, no cooling being necessary. Phosphorus-containing spots were detected either by use of the ammonium molybdate reagent or by neutron activation of the strips.

Pyrophosphoric Acid Esters

Goedde et al[2880] described a method for the determination of synthetic (±)- α-hydroxyethyl-2-thiamine pyrophosphate. Thiamine pyrophosphate and (±)- α-hydroxyethyl-2-thiamine pyrophosphate are distinguished by differences in their ultra-violet spectra at 248 nm in phosphate buffer solution at pH 8.0. The isosbestic point of (±)-α-hydroxyethyl-2-thiamine pyrophosphate in the pH range 5 to 8 is at 270 nm. The concentration of thiamine pyrophosphate in (±)- α-hydroxyethyl-2-thiamine pyrophosphate can be determined by enzymatic dephosphorylation, separation of thiamine, and hydroxyethylthiamine by development on paper with butanol-ethanol-water (4:1:1), elution, and determination of thiamine by the thiochrome method. Equimolar amounts of the two substances give equal responses to the thiochrome reaction.

Phenylphosphorothioates and Phenylphosphonthioates

Heavy paper chromatography has been used for the determination of o-ethyl O-p-nitrophenyl-phenylphosphonothioate in industrial products of purity 60 - 85% (Sato[2881]). The sample (< 500 mg) is separated from impurities on thick paper with a mixture of 2-chloroethanol and methanol (1:1) as stationary phase and hexane as developer. The sample solution is applied as a line on a thick (0.7 mm) paper (12 cm x 20 cm, cut to a point at the lower end), treated with the stationary phase, and hexane is run downwards until the o-ethyl O-p-nitrophenyl phenylphosphonothioate front reaches the lower end of the paper. The eluate is then collected in a weighing bottle, and the o-ethyl O-p-nitrophenyl phenylphosphonothioate weighed after evaporating off the solvent.

Paper inpregnated with silicone grease has been used for the inverse phase chromatography of diethyl 4-methyl-2-isopropyl-pyrimid-6-yl phosphorothionate and diethyl 4-methyl-2-isopropylpyrimid-6-yl phosphorothionate and diethyl 4-methyl-2-isopropylpyrimid-6-yl phosphate. (Vigne et al[2882]). Conc ammonia-ethanol-water (1:3:5) is used as the mobile phase. The spots were detected with iodine or with a solution of bismuthyl nitrate in trichloracetic acid and potassium iodide solution. The same workers in a more recent paper (Vigne et al[2883]) pointed out that the separated compounds mentioned above can be located by treatment with potassium permanganate (intensitied by benzidine acetate) or by mercuric perchlorate followed by ammonium sulphide or by ultraviolet absorption at 253.7 nm.

Vigne et al[2884] still discussing the two compounds mentioned above, have described a new method of separation, depending on the simultaneous operation of two chromatographic systems, one inverse-phase with a substance migrating in the oil, one direct-phase with a substance migrating in the water imbibed by the cellulose. If only the lower half of the paper is oil-impregnated, diethyl 4-methyl-2-isopropylpyrimid-6-yl phosphorothionate forms an almost stationary spot, but an unidentified impurity can be eluted pure.

Carbohydrate Esters of Phosphoric Acid

Various workers have described methods for the estimation of sugar esters of phosphoric acid on paper chromatograms. Nagai and Kimura[2885] devised a sensitive colorimetric test for detecting inositol and inositol diphosphate spots by dipping the chromatogram in a solution of 10 ml of conc nitric acid in 90 ml of ethanol and heated at 95° - 100°C for 10 mins. The paper is then exposed to ammonia vapour for 1 - 2 s to remove excess of acid and then dipped in a mixture of n-butanol (40 ml), glacial acetic acid (10 ml), aqueous 10% (v/v) barium acetate solution (10 ml), and calcium chloride (0.5 g). After heating the paper for 4 - 5 min at 90° - 100°C, an orange colour due to barium rhodizonate develops. The colour is sufficient to detect inositol in quantities above 8 μg, but below this concentration the colour must be intensitified by spraying with 5% (w/v) calcium chloride solution in 50% ethanol and heating for 2 min at 90° - 100°C, when the orange colour changes through cherry red to rose pink; 1 μg of inositol can then be detected. When the test is applied to the detection of inositol diphosphate, the ester is first hydrolysed by spraying the chromatogram with N-hydrochloric acid and heating at 95°C for 4 minutes. This may be repeated twice to give sufficient hydrolysis. The procedure is then similar to that described above except that the treatment with nitric acid-ethanol is repeated twice. The colouration is weaker than that given by free inositol but detection is easy. R_f values of inositol and inositol diphosphate in a solvent system of 62% (v/v) tert-butyl alcohol containing 10% (w/c) trichloroacetic acid were 0.35 to 0.36 and 0.38 to 0.39 respectively, and in isopropyl alcohol-acetic acid-water (3:1:1, by volume) were 0.19 - 0.20 and 0.13 - 0.15, respectively.

Harrap[2886] detected phosphate esters such as fructose 1,6-diphosphate, glucose 1-phosphate, glucose 6-phosphate and phytic acid on paper by hydrolysis to phosphoric acid, which is treated with an acid molybdate spray or dip; the complex formed is then reduced to a blue compound when the paper is treated with hydrogen sulphide or exposed to u.v. light. After either treatment the paper gradually becomes blue, and later identification of faintly coloured spots becomes difficult. Also, certain naturally occurring acids (e.g. citric, tartaric, and ascorbic acids) form a blue colour slowly. To overcome these disadvantages the filter-paper is dried in air and dipped in a mixture of 5 ml each of 60% perchloric acid and 20% molybdate solution, 10 ml of N hydrochloric acid and 80 ml of acetone. The paper, dried in air, is exposed to u.v. radiation then immediately dipped in a 2.5% (w/v) solution of benzoin α-oxime in methanol and dried.This provides the paper with a background of a white precipitate stable for some weeks.

Harrap[2887] has pointed out that the addition of sufficient boric acid to saturate 2-methoxyethanol-ethyl methyl ketone - 3 N aqueous ammonia (7:2:3) gives a sharp separation on paper of glucose 1-phosphate from glucose 6-phosphate, fructose 1:6:diphosphate and inorganic phosphate.

Miscellaneous Organophosphorus Compounds

2-Phospho- and 3-phospho-glyceric acids can be separated on paper impregnated with molybdate salts (Cowgill,[2888]), Whatman No. 52 paper is passed through 0.5% (w/v) $Na_2MoO_4.2H_2O$ solution and dried in air. Descending chromatograms are run with 88 - 90% formic acid-water-95% ethanol (1:29:70) for 20 h at room temp. The spots are developed by a modification of the acid molybdate spray and u.v. light technique of Axelrod and Bandurski[2889].

Heald[2890] has discussed a system of analysis of acid soluble radioactive phosphates such as adenosine triphosphate, inorganic phosphate and creatine phosphate in extracts of cerebral tissues, based on a paper electrophoretic method. This system can be used in the study of the metabolism of ^{32}P. Trichloroacetic acid extracts of the tissue are fractionated into groups by precipitation as barium salts, which are then decomposed by a cation-exchange resin. The resulting solution is freeze-dried and the phosphates are then separated by paper electrophoresis. After elution of the various fractions, the phosphorus is determined by standard procedures. The method is useful for the fractionation of small amounts of phosphates (120 - 130 μg of total P).

The ultra-violet absorption spectra of adenosine-5'-triphosphate and related 5'-ribonucleotides have been reported (Bock et al[2891]). Extinction values for purified adenosine 5'-triphosphate in 0.01 N hydrochloric acid (pH 2), 0.02 M disodium hydrogen phosphate (pH 7) and 0.002 N sodium hydroxide (pH 11) are 14.7, 15.4 and 15.4 (x 10^3) at λ_{max} 257, 259 and 259 nm, respectively. Extinction values and ratios under similar conditions are also presented for adenosine, adenosine 5'-di- and -mono-phosphates, for cytidine, uridine, and guanosine and their 5'-tri-,-di- and-mono-phosphates.

Cerletti and Siliprandi[2892] have described a paper chromatographic method for the separation of adenosine triphosphate and its organic and inorganic derivatives using propanol-water-trichloroacetic acid-22° Be-aqueous ammonia (5:20:5:0:3) as the mobile phase. They detected the separated adenosine phosphate and related compounds by ultraviolet light, and inorganic phosphates by a molybdic acid reagent. Quoted R_r values include adenosine triphosphate (R_f 0.03), inosine triphosphate (0.02), inosine diphosphate (0.07), inorganic pyrophosphate (0.24), adenylic acid (0.40 - 0.37), adenosine, (0.55), adenine (0.67 - 0.64), and inorganic orthophosphate (0.65 - 0.64). Adenine, adenosine, adenosine phosphoric acid and pyridine nucleotides have been separated by paper chromatography (Gerlach and Doring[2893]). The chromatograms are sprayed with aqueous 0.07% potassium permanganate and hung immediately in an atmosphere of chlorine for 15 s, then dried for 5 min at 100°C. This causes the wine colour to fade and all adenine- or adenosine-containing areas to appear as yellow to orange spots. Spraying with 3 N potassium hydroxide then changes the colour to deep red; the background of the paper is white when dry. Moist ammonia vapour should replace the potassium hydroxide spray if the phosphate components are to be determined on the same sheet with molybdate reagent. The reaction is sensitive to 0.2 µg ml^{-1} and is specific for free nucleoside and nucleotide-bound adenosine. As mentioned in Section N, Hanes and Isherwood[2856] have described a paper chromatographic method for separating adenosine phosphate from hexose monophosphate.

Dawson and Eichberg[2894] have described a procedure for the determination of diphosphoinositide and triphosphoinositide and their changes post mortem. The polyphosphoinositides are extracted quantitatively into chloroform: methanol (2:1) containing 0.25% of conc hydrochloric acid after a preliminary extraction with chloroform-methanol (1:1) to remove most of the other phospholipids. When the extract is washed with M hydrochloric acid the polyphosphoinositides pass completely into the lower chloroform-rich phase. Their concentration is determined by chromatography on formaldehyde-treated paper, or chromatography and electrophoresis of the products of acid hydrolysis.

Witter et al[2895] have described a procedure for the circular paper chromatography of phospholipids. Octanol-lutidine-acetic acid (18:1:2), chloroform-lutidine-acetic acid (4:6:1), 3-methylbutan-2-one-acetic acid (10:1= and especially 2:6-dimethylheptan-4-one-acetic acid (6:1) are good solvents for the circular chromatography of phospholipids on Whatman No. 1 paper. The sample containing 3 - 30 µg of phospholipid is applied to the paper as a solution in benzene-isoamyl alcohol (1:1) (4 - 20 µl), and the chromatogram developed at 21°C to 23°C for 1 - 2 h (or for 5 h with first solvent system). The chromatogram is allowed to dry, washed with water, and re-dried.

<u>Organophosphorus Insecticides and Pesticides</u>

Paper chromatography has been applied to the determination of Parathion ethyl ester ionomers and derivatives[2896,2897].

Paper chromatography has also been applied (Cook[2898]) to the separation of Systox, Isosystox isomer, EPN, Parathion, Diazinon, Chlorothion, Methyl Parathion and Malthion. Cortes and Gilmore[2899] showed that Diazinon and Sulphotep and their breakdown products (tetraethylpyrophosphate, tetraethylthiopyrophosphate and O,O-diethyl O-(2-isopropyl-6-methyl-4-pyrimidinyl) phosphate) can be separated on paper and the separated spots detected by cholinesterase-inhibition. Waggon et al[2900] have discussed the paper and thin-layer chromatography and detection of microgram quantities of phosphorothioate insecticides of the Demeton-S-methyl type, particularly Tinox (dimethyl S-(methylthio) ethyl phosphorothiolate).

Paper chromatography has been used for the separation of Gusathion (Guthion)(O,O-dimethyl S-(4-OXO-3H-1 1:2:3 benzotriazine-3-methyl) phosphorodithioate) from Phosdrin (Mevinphos) (Sato[2901]), the estimation of Schradan (Pestox) (octamethylpyrophosphoramide) in the presence of related phosphoramide esters (Marsh et al[2902]), Demeton (O,O diethyl O-2-ethylthioethylthionophosphate) in the presence of plant metabolites (Marsh et al[2902]) and in insect tissue extracts (Menn et al[2903]), Dimethoate in technical products and formulations (Bazzi et al[2904]), Dipertex (O,O-dimethyl 2,2,2-trichloro-1-hydroxyethyl phosphonate) and its vinyl derivative (Yamashita [2905]), Denlav in neat preparations and spray washes (Casapieri and Keppler [2906]), Phorate (Thimet) (O,O diethyl S-(ethylthio) methyl phosphorodithioate) in insect tissue extracts (Menn et al[2903]), Parathion, Chlorthion, Diazinon, Malathion in fruit and vegetables (Muller et al[2907]), Thiometon in pears (Jucker[2908]), Trithion, Diazinon, Parathion in fruit and vegetables (Getz[2909]), phosphorothioate insecticides in vegetables and animal tissue (Drudayasamy and Natarjan[2910]), various organophosphorus insecticides (Zadroginska[2911]) and phosphamidon (Anliker and Menzer[2912]).

Various workers have discussed the use of reagents for the location of separated organophosphorus insecticides and pesticides on paper chromatograms (Table 239).

I. ORGANOSELENIUM COMPOUNDS

Selenomethionine has been separated from methionine by paper chromatography[2924]. Layers (0.25 mm thick) of silica gel G were used with isopropyl alcohol-butanol-water (1:3:1) as the solvent system. The spots were located with ninhydrin solution in butanol. The limit of detection was 0.2 μg of either compound. In a method for the identification of selenoamino acids by paper chromatography the sample containing selenomethionine and selenocystine is applied to the paper, and then exposed to the vapour of 15% hydrogen peroxide solution for 45 - 60 s; this oxidizes the selenium to selenoxide, but does not affect the sulphur of sulphur-containing amino acids[2925]. After solvent development, the chromatogram is allowed to dry and is then lightly sprayed with starch-sodium iodide-hydrochloric acid solution. The selenium-containing zones give a purple colour in 45 s, and this colour changes to brown when the chromatogram is dried at 50°C. Sulphur-containing amino acids may be detected either by the ninhydrin reaction,after decolorization of the paper by exposure to ammonia vapour, or by a process similar to that described above, with hydrogen peroxide oxidation for 30 min to form the sulphoxides.

J. ORGANOSILICON COMPOUNDS

Fritz and Wick[2925] studied the paper chromatographic separation of pyrolysis products of tetramethylsilane. Following preliminary separation on alumina columns, one of these fractions was subjected to benzene extraction and components of the extract separated by paper chromatography.

Table 239 - Methods of separating and identifying organophosphorus insecticides and pesticides on paper chromatograms

Compounds mentioned	Coating	Migration solvent	Detection method	Sensitivity	Reference
(b) Paper Chromatography					
Gusathion, and Phosdrin	ascending	butanol-glacial acetic acid-water (4:1:5) or chloroform-ethanol-water (5:5:3)	2-phenyl-1-naphthylamine in acetic acid spray, then 18% hydrochloric acid spray, Blue spot obtained with Gusathion. Subsequent spraying with ammonium molybdate reveals Phosdrin as a blue spot.	500 μg	Sato[2901]
Demeton (Demeton separated from its thio isomer O,O-diethyl S-2-ethylthioethyl thiophosphate and plant metabolites)	silicone impregnated paper or paper impregnated with propylene glycol	upper phase of system chloroform-ethanol-water (10:10:6 by vol) petrol-toluene (4:1 by vol) mixture saturated with propylene glycol	-	-	Marsh et al 2902
Schradan and related phosphoramide esters	paper treated with propylene glycol	benzotrichloride-carbon tetrachloride toluene (1:1:1 by vol) saturated with propylene glycol	spray with 72% perchloric acid - 4% ammonium molybdate-N-hydrochloric acid-water (5:25:10:60 by vol) heat to 80°C then expose to u.v. (365 nm) to produce blue spots, or measurement of anticholinesterase activity of segments of chromatogram	-	Marsh et al 2902

Table 239 (Continued)

Dimethoate (Roger) in technical products	Schleicher and Schull paper 2045 b	carbon tetrachloride methyl alcohol-nitro methane (18:1:1)	70% perchloric acid-N-hydrochloric acid - 4% ammonium molybdate-water (5:10:25:60). Blue spot with Dimethoate phosphorus determined in eluate	-	Bazzi et al 2904
Sulphotep, Diazinon and their breakdown products (tetraethyl pyrophosphate, tetra-ethylthiopyrophosphate and O,O-diethyl O-(2 isopropyl-6-methyl-4 pyrimidinyl)phosphate	Whatman 3MM paper impreg-nated with light mineral oil and Epon 828 resin	35% aqueous acetonitrile	cholinesterase inhibition technique	-	Cortes and Gilmore[2899]
Organophosphorus pesticides	paper	-	cholinesterase inhibition technique	-	McKinley and Johan[2913] Cook[2914]
Tinox and phosphoro-thioates of the Demeton-S-methyl type	paper	toluene-isopropyl alcohol-methanol-methyl cyanide-water (40:16:16:20:9)	potassium chloroplatinate or palladium chloride sprays	1 μg	Waggon et al 2900
Organophosphorus insecticides	reversed phase paper chromato-graphy	ethanol-light petroleum	palladium chloride or aqueous ammoniacal silver nitrate sprays	-	Takeshi[2915]
Sulphur containing organophosphorus insecticides	paper	-	silver nitrate spray	0.2 - 1 μg	Wood[2916]
Systox, Parathion, Chlorthion, Methyl-parathion, Malathion, EPN, Diazinon	paper	-	N-bromo-succinamide spray then fluorescein spray to produce fluorescent spot	-	Cook[2898]

Estimation of principal biologically active constituent in Denlav spray washes	ascending on Whatman No. 2 paper impregnated with liquid paraffin	upper phase of the mixture chloroform-ethanol-water (5:5:3)	0.2 M mercuric chloride then 0.08 M 2:4-dinitrophenylhydrazine in 32% perchloric acid sprays (brown spot) and subsequent estimation of phosphorus in spot	-	Casapieri and Keppler[2906]
Organophosphorus insecticides including Diazinon	paper	-	2% 4-(p-nitrobenzyl) pyridine in acetone/tetraethylene pentamine spray (Diazinon-red spot all other organophosphate pesticides, thio or non-thio, give blue spots)	-	Watts[2917]
Thiophosphate insecticides	paper	-	0.5% 2:6-dibromo-p-benzoquinone-4-chlorimine in cyclohexane spray then heated to 110°C (yellow to red spots)	1 μg	Menn et al 2918
Pre-chromatographic purification of insecticides from insect tissue extracts (Demeton, Phorate mentioned)	Whatman paper No. 4	methyl cyanide	0.5% N,N-dimethyl-p-1-naphthyl-azoaniline	-	Menn et al 2918,2903
Parathion, Diazinon, Chlorthion, Malathion Meta-Systox, Thiometon and their saponification products	Test tube chromatography	acetone-water-1% aqueous ammonia (19:1:1)	sodium azide-iodine dip. sulphur compounds colourless on brown-black background	1-2 μg	Fishcher and Otterbeck[2919]
Phosphamidon	paper	light petroleum-toluene-methanol-water (5:5:7:3)	(a) dip paper in 0.1% aqueous tetrazolium blue-2 N sodium hydroxide (1:9) (b) molybdenum blue method	0.2 - 0.4 part/10^6	Anliker and Menzer[2912]
Organophosphorus insecticides	paper or acetylated paper	-	rabbit liver homogenate, then RR(C.I.azoic diazo component 24) sprays, grey brown colour in area free from insecticides)	-	McKinley and Read[2920]

(continued)

Table 239 (Continued)

Dimethoate, Menazon, Vamidthion, Demeton-S-methyl-S-oxide	formamide impregnaged paper	benzene-chloroform (3:2 and 9:1, two dimensional)	-	0.1 part/10^6	Smart and Hill[2921]
Thiometon	Dow silicone No. 550 impregnated paper	-	bromine vapour and fluorescein spray	4 μg	Jucker[2908]
Trithion, Diazinon, Parathion, Demeton, Guthion	paper	-	bromophenol blue-silver nitrate in aqueous acetone spray	-	Getz[2909]
Phosphorothioate insecticides	liquid paraffin impregnated paper (ascending)	dimethyl formamide-butanol-water	bromine vapour then Congo red in aqueous ethanol (1:1) spray (blue spot on red background)	0.5 μg	Irudayasamy and Natarjan[2910]
Parathion isomers	-	-	isolated spots solvent extracted and isomer determined by polarography	-	Kovac[2896]
Organophosphorus insecticides	paper	-	exposure of paper to bromine vapour, liver homogenate spray (cholinesterase inhibition) then 1-naphthyl acetate/Fast Blue RR (Cl Azoic Diazo Component 24) spray, yellow spots with pesticide on purple brown background	-	McKinley[2922]
Phosphorothioates (Tinox mentioned)	kieselgel	toluene-isopropanol-methanol-methyl cyanide-water (40:16:16:20:9)	potassium chloroplatinate and palladium chloride sprays	1 μg	Waggon et al[2900]

K. ORGANOTIN COMPOUNDS

Williams and Price[2926,2927] have described procedures for paper partition chromatography of the compounds of the type $RnSnX_{4-n}$ using butanol-pyridine-water, butanol-ethanol-water, butanol-ammonia-water and aqueous pyridine. They detected the organotin compounds by spraying with catechol violet. Compounds of the type R_4Sn and R_3SnX are initially oxidized by u.v. irradiation before spraying with catechol violet.

It is seen in Table 240 that the nature of the acid radical in $RnSnX_{4-n}$ does not affect the R_f value obtained, indicating that the same solvents must cause hydrolysis of these compounds to the corresponding hydroxides of these hydroxides, those of trialkyl and triaryltin are stable, but those of the dicompounds tend to lose water to form polymeric oxides $(R_2SnO)_n$. These are insoluble in most solvents and would lead to zero R_f values, as was found in 60% pyridine for the dibutyl compounds upwards. Because of the stability of the tricompounds, and their general insolubility in water, coupled with their solubility in organic solvents, high R_f values may be expected.

Compounds of the type R_4Sn and R_3SnX are initially oxidized by u.v. irradiation before spraying with catechol violet. For the paper chromatography of diaryltin compounds the paper is impregnated with a 10% solution of olive oil in light petroleum[2928]. Solutions of various concentrations of methanol in 1 N hydrochloric acid are used for development. The diaryltin spots are detected with a saturated solution of diphenylcarbazone in 50% aqueous methanol (purple spots). The minimum detectable amount of diphenyltin dichloride is 0.5 µg.

Tanaka and Morikawa[2929] separated and identified butyltin compounds (Bu_nSnX_{4-n}) by reversed phase chromatography using liquid paraffin as the stationary phase. Tetrabutyltin (R_f 0.0), tributyltinchloride, dibutyltin dichloride, (0.50), (0.82) and butyltintrichloride (tailing) were separated by development with acetone-water acetic acid (20:10:1); dichlorodioctyltin (0.69) and dibutyltindichloride (0.95) with propanol-water-acetic acid (20:20:1); and butyltintrichloride and stannic chloride with butanol saturated with 2 N hydrochloric acid in the absence of liquid paraffin.

Gasparic[2930] have reported the reversed phase paper chromatography of compounds of the type $(alkyl)_nSnX_{4-n}$ (where X = halogen or residue of monocarboxylic acid or dicarboxylic acid half ester), using 1-bromonaphthalene as stationary phase with 50% to 70% acetic acid saturated with 1-bromonaphthalene as the development solvent. Spots were revealed by spraying with dithizone or catechol violet solution.

Franc et al[2931] carried out paper chromatography of the range of alkyltin chlorides using Whatman No. 2 paper and a 30% solution of petroleum (boiling point 190 - 275°C) in benzene as stationary phase and ethanol-acetic acid-water 20:1:14 as mobile phase. Spots were revealed by dithizone after 16 - 18 h at 20°C. Barbieri et al[2932] have also discussed the paper chromatographic separation of organotin compounds.

Cassol et al[2933] have discussed the application of paper chromatography and paper electrophoresis to the separation of organotin chloride, bromide and iodide systems (Cassol and Barbieri[2934]) and of methyltin trifluoride, dimethyltin difluoride and trimethyltin fluoride.

A method for the chromatographic identification of dibutyltin compounds

Table 240 - R_f values

	Developing solvent			
	Pyridine, 60%	Butanol-pyridine-water	Butanol-ammonia-water	Butanol-ethanol-water
Me_2SnCl_2	0.36[c]	0.55[b]	0.03	0.67[b]
Me_3SnCl	0.35	0.25	-	0.32
$EtSnCl_3$	Streaks to 0.8	Streaks length of paper	Streaks length of paper	Streaks length of paper
Et_2SnCl_2	0.36[c]	0.80[b]	0.16[c]	0.98[b]
Et_2SnOAc	0.40	0.85[b]	-	0.95[b]
Et_3SnOH	0.88	0.94	0.94	0.95
Pr_2SnCl_2	0.00[b]	0.9[b]	Streaks to 0.52	0.98
$(Pr_3Sn)_2O$	0.85	0.94	-	-
$(isoPr_3Sn)_2O$	0.87	0.94	-	-
$Pr_3SnCH_2CH_2COONa$	0.82	0.90	0.83	-
$BuSnCl_3$	Streaks to 0.8	0.99[b]	Streaks length of paper	Streaks length of paper
Bu_2SnCl_2	0.00[a]	0.93	0.00[b]	0.98[a]
Bu_2Sn dilaurate	0 00[a]	0.95	-	0.96
Bu_3SnCl	0.85	0.92	0.95	0.94
Bu_3Sn laurate	0.82	0.92	-	0.94
$(Bu_3Sn)_2O$	0.82	0.97	-	0.96
Bu_3Sn abietate	0.84	0.97	0.94	-
Hexabutyl distannane	-	0.94(0.8,0.3[d])	0.95(0.8[d],0.3[d])	-
Bu_4Sn	0.00	0.93	0.94	0.92
n-$Hexyl_2Sn$ dilaurate	0.00	0.95	-	0.95
n-$Octyl_2Sn$ dilaurate	0.00	-	-	0.94
n-$Octyl_3SnCl$	0.00	0.94	0.93	0.95
n-$Octyl_4Sn$	0.00	0.90	0.94	0.83
$PhSnCl_3$	Streaks to 0.85	0.99[b]	Streaks length of paper	Streaks length of paper
Ph_2SnCl_2	0.00	0.00	0.00	0.95[b]
Ph_3SnCl	0.85	0.95	0.97	0.97
$Ph_3SnO\overline{Ac}$	0.85	0.95	0.97	0.97
Ph_4Sn	0.00	0.00	0.00	0.00
$SnCl_4$	0.00	0.00	0.00	0.00[a]

[a] Slight tail
[b] Severe tailing
[c] Elongated spot
[d] Probably impurity

in the presence of dioctyltin compounds in PVC subjects an acetone solution to chromatography on Whatman No. 4 paper (descending technique) with butanol-acetic acid-methanol-water (5:1:20:24)[2935]. The air-dried chromatogram is sprayed with 0.1% butanolic diphenylcarbazone to produce red-violet spots. All of the dibutyltin compounds have an R_F value of 0.75 and the dioctyltin compounds remain at the origin. This procedure was also applied to PVC extracts prepared by extracting PVC with carbon tetrachloride, then converting the organotin compound into a chloride with concentrated hydrochloric acid[2936]. Upon chromatographing as described above the dioctyltin stabilizers remained at the baseline, whereas the dibutyltin compounds migrated.

In a procedure for the identification of sulphur-containing organotin stabilizers such as dibutyltindithioacetate in the presence of sulphur-free organotin compounds, chromatograms of the stabilizers extracted with diethyl ether from the plastics are developed with hexane-acetic acid (12:1) and sprayed with a solution of dithizone in chloroform or ethanolic catechol violet. The dioctyl, dibutyl, and trioctyl derivatives are clearly separated. The dialkyl compounds are revealed as red spots with dithizone reagent or as blue spots with catechol violet, and the trioctyl compounds as yellow spots with either reagent. The sulphur-containing and the sulphur-free compounds, e.g. dioctyltin di(butyllthio)acetate and dioctyltin di(butyl maleate), have the same R_F values but can be distinguished from one another, as sulphur-containing compounds form additional spots (blue with dithizone and yellow with catechol violet). Trioctyltin compounds can best be detected by conversion into the corresponding dioctyl compounds by exposing duplicate plates to bromine vapour for 10 min; the dioctyl compounds are then detected as indicated above. The minimum detectable amount by this method is 10 µg.

COLUMN CHROMATOGRAPHY

Organoarsenic Compounds

The arsenic species, arsenite (AsO_3^{3-}), arsenate (AsO_4^{3-}), monomethyl arsonate and dimethyl arsinate have been separated by liquid column chromatography and detected by use of differential pulse polarography[2937,2938] or flameless atomic absorption[2939]. The separations are done by using a gravity column chromatograph and by collecting fractions as they elute. Such procedures are very time consuming.

In other recent work, a Dionex ion chromatograph has been used to separate (AsO_4^{3-}) from other common anions and to quantify it by using conductivity detection[2940].

Morita et al[2941] determined methyl arsonic acid, dimethylarsinic, arsenobetamine, arsenite and arsenate in biological samples by liquid chromatography with inductively coupled argon plasma emission spectrophotometric detection. These substances were separated on anion and cation exchange columns using a phosphate buffer. Arsenic emission was observed at 193.6 nm. The detection limit is 2.6 ng arsenic.

Ricci et al[2942] have described a highly sensitive, automated technique for the determination of monomethyl arsonate, dimethyl arsinate, p-aminophenyl arsonate, arsenite acid arsonate. This procedure is based on ion-chromatography on a Dionex 3 x 150 mm column. When this column is used with 0.0024 M $NaHCO_3$/0.0019 M Na_2CO_3/0.001 M $Na_2B_4O_7$ eluent, all the compounds except arsenite and dimethyl arsinate are separated effectively. For separation of dimethyl arsinate from arsenite, a lower ionic strength eluent (0.005 M $Na_2B_4O_7$) can be used in a separate analysis. The detection system utilizes a continuous arsine generation system followed by heated quartz furnace atomization and atomic absorption spectrometry. Detection limits of less than 10 ng/ml were obtained for each species.

A typical gradient elution of the five components is shown in Figure

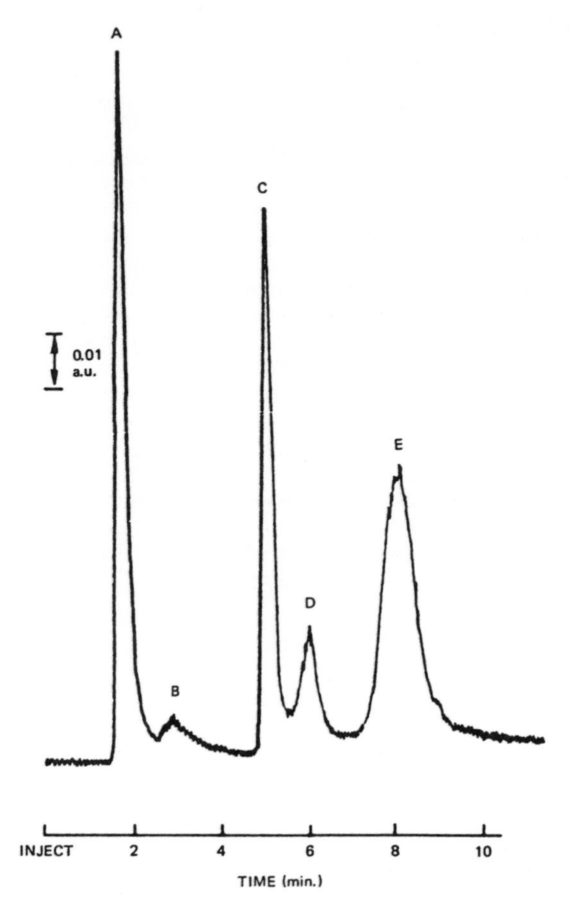

Figure 170 - IC-AAS chromatogram of trace inorganic and organic arsenic species separated by gradient elution. Peak identification is as follows: peak A, DMA (20 ng/mL); peak B, AsO_3^{3-} (20 ng/mL); peak C, MMA (20 ng/mL); peak D, p-APA (20 ng/mL); peak E, AsO_4^{3-} (60 ng/mL).

170. Precision was approximately 11% at the 20 ng/mL concentration. Detection limits (ng/ml) were as follows: dimethyl arsinate (6:5); arsenite (4.9); monomethylarsonate (3.2); p-aminophenyl arsonate (9.3) and arsenate (20.0).

Organocobalt Compounds

In an ion-exchange chromatographic method for the determination of traces of cyanocobalamin, the solution is passed through a column of Amberlite CG-50 (100 - 150 mesh)[2943]. The column is washed with 0.1 N hydrochloric acid and then the cyanocobalamin is eluted from the column with dioxane - 0.25 N hydrochloric acid (3:2). To the elute is added 2% sodium chloride and concentrated nitric acid and the solution is evaporated to dryness, then redissolved in perchloric acid and evaporated to dryness again. The residue is dissolved in acetate buffer solution (pH 6) and nitroso R salt solution, and then acidified. The absorption of this solution is measured at 520 nm against a reagent blank, and cyanocobalamin is determined by reference to a calibration graph or by means of an internal standard. Other vitamin B_{12} homologues interfere in this procedure.

In a procedure for the determination of hydroxo, aquo, and other cobalamins in injection solutions, the cobalamins are converted into cyanocobalamin by treatment with potassium cyanide at pH 7.5 (citrate buffer solution)[2944]. The solution is adjusted to pH 4 with citric acid and applied to a column of Amberlite XE-97, which is then washed successively with water, citrate buffer solution, 0.1 N hydrochloric acid, 85% acetone, and 0.1 N hydrochloric acid to remove impurities. The red cyanocobalamin fraction is then eluted with dioxane into 0.1 N hydrochloric acid and converted into the dicyano compound by treatment with potassium cyanide. The absorption is read within 20 - 30 min at 578 nm, and compared with that of control and standard solutions.

Cyanocobalamin in a solution of the vitamin B complex has been separated and determined using ion-exchange chromatography on acidic Wafatit CP 300, sodium form[2945]. The sample is placed on the column and then the various strengths of the resin are washed successively with water. This removes all the B vitamins other than cyanocobalamin. Cyanocobalamin is then eluted with acetone-water (1:1) containing 7.5% sodium chloride solution until a suitable volume has been collected. Finally, the absorption of this solution is measured at 361 nm against that of the solvent in the reference beam. Ion-exchange chromatography has been applied to the determination of cyanocobalamin in pharmaceutical syrups[2946]. The sample at pH 4.5 - 5.1 is passed through a column of Amberlite XE-97 ion-exchange resin. The column is washed with 0.1 N hydrochloric acid, 80% acetone, and 0.1 N hydrochloric acid. Cyanocobalamin is then eluted from the column with dioxane - 0.1 N hydrochloric acid (3:2) and the eluate collected and its absorption measured at 361 nm. Recoveries are claimed to be between 98 and 99.4% of the amount of cyanocobalamin added.

Neutral and basic cobalamins such as hydroxo- and cyanocobalamin and cobinamide have been separated on various ion-exchange papers under a variety of conditions[2947]. R_D values have been listed for the vitamins of the B_{12} group on cellulose ion exchangers, alginic acid, and alginic acid-cellulose preparations using water, mixtures of water with various alcohols, and water with hydrogen cyanide as solvent[2948].

Organoiron Compounds

Ferrocenes have been separated on columns[2949]. Separation on a carboxymethylcellulose column was used in a simple and rapid method for the quantitative determination of haemoglobin A_2[2950]. Packed red cells are lysed with 0.01 M phosphate buffer (pH 6.2) and diluted with buffer solution. A 5% potassium hexacyanoferrate (III) solution and 2% potassium cyanide are added. An aliquot is added to a column of carboxymethylcellulose and suspended in phosphate buffer solution (pH 6.2). The haemoglobin components other than A_2, are eluted first with 0.01 M phosphate buffer (pH 7.1) and then, when the red band reaches the bottom of the column, with buffer solution of pH 7.3. Haemoglobin A_2 is eluted with buffer solution of pH 7.7. The absorptions of the eluates are read at 540 nm. The behaviour of haemoglobins during chromatography is influenced by the pH and ionic concentration of the chromatographic developers, the state of equilibrium of the Amberlite IRC-50 resin with the developer, the temperature during equilibration and chromatography, the amount of haemoglobin on the chromatogram, and the oxidation state of the haem in the haemoglobin[2951].

Two chromatographic procedures have been described for the determination of human haemoglobins A, B, C, and F on Amberlite IRC 50 (EX64) resin [2952]. The results compare favourably with those of older methods. An investigation of the methods for determining foetal (Hb_f) and sickle-cell (Hb_b) haemoglobins in the presence of adult haemoglobin (Hb_a) led to the conclusion that for the determination of Hb_f the alkali denaturation method is not sufficient in the presence of Hb_a and Hb_b[2953]. Amino acid composition provides a more definite answer to the presence of this protein. Only electrophoretic measurements can be used for the detection and determination of Hb_b. By the use of the four different methods it appeared that the alkali resistant fraction, sometimes found in the blood of patients with sickle-cell anaemia, is not foetal haemoglobin[2953].

The influence of gel structure and pore size of synthetic gels such as acrylamide gel on the resolution of various haemoglobins has been studied[2954]. Human, horse, and dog haemoglobins have been separated on cation-exchange cellulose[2955]. Chromatography of haemoglobins in the carboxymethylcellulose eluted with 0.01 M sodium phosphate buffers to give a gradient of increasing pH. The eluted fractions are examined spectroscopically and, after concentration, by paper electrophoresis and sedimentation methods. The heterogeneous nature of adult, human, and horse carboxyhaemoglobins and the effects of temperature and of concentration of the urea solution on the chromatography were discussed. The chromatographic behaviour of normal haemoglobin and seven different abnormal human haemoglobins on the cation exchanger Amberlite IRC-50 has been studied[2956,2957], and the separation by electrophoresis, column chromatography, solubility, and alkaline denaturation of normal and abnormal human haemoglobins reviewed [2958].

Separation on Sephadex gels has been used for the determination of haemoglobin and myoglobin[2958]. Separation was effected on Sephadex G-50 or G-75 (200 - 400 mesh). The columns were equilibrated at 4°C overnight with 0.05 M phosphate buffer (pH 7.4), which was also 0.05 M in sodium chloride. A mixture of the two proteins in buffer solution containing 25% of sucrose was applied to the column. The proteins were eluted with the buffer solution containing sodium chloride but no sucrose. The bands were located and determined by absorption measurements at 280, 410 and 577 nm. Recoveries exceeding 95% were usually obtained.

Organolead Compounds

High-performance liquid chromatography enables concentrations as low as 0.01 g of tetraethyllead as lead per imperial gallon in gasoline to be determined[2960].

Liquid column chromatography has been employed for the analysis of alkyllead compounds in petroleum[2961,2962]. Ruo et al[2961] used high performance liquid chromatography. The method is based on the separation of tetraethyllead from other ultra violet absorbing material on silica gel and quantification of the ultra violet detector response. Tetraethyllead concentrations corresponding to as little as 0.03 μg lead in the sample (corresponding to 0.01 g Pb/imp. gal) could be determined quantitatively. The response of other alkylleads (tetramethyllead) and mixed alkylleads differs appreciably from that of tetraethyllead, precluding the use of this method for unknown samples. However, the analysis can be done in 5 min using commercially available equipment and can be used in all cases where the type of alkyllead present in the gasoline is known. High performance liquid chromatography of gasolines on microparticulate silica has shown that tetraethyllead or other alkylleads appear together with the saturate peak and can be detecated over the saturate background using an ultra violet detector at a fixed wavelength of 254 nm. This technique can be used for the determination of tetraethyllead (and other alkylleads) in gasolines. The sample is injected onto the column without any pretreatment, the alkyllead is separated from other ultra violet absorbing material (aromatics and olefins) and the ultra violet response is quantified.

Liquid chromatography has been applied to the analysis of gasoline for a single tetraalkyllead compound using flame atomic absorption spectrometric detection[2963], and conventional molecular spectrophotometric detection at 254 nm[2964]. The liquid chromatographic technique has also been used in conjunction with Zeeman-effect atomic absorption spectrometry for the speciation of tetramethyllead and tetraethyllead in a gasoline standard reference material[2965]. A standard mixture of five tetraalkyllead compounds has been separated by liquid chromatography using Zeeman-effect atomic absorption spectrometric detection[2966].

Mossman and Rains[2962] also investigated the applicability of liquid chromatography coupled with atomic absorption spectrometry to the rapid determination of the tetraalkyllead compounds in petroleum. Separation of the individual tetraalkyllead species is achieved by reversed-phase liquid chromatography using an acetonitrile/water mobile phase. The effluent from the liquid chromatograph is introduced directly into the aspiration uptake capillary of the nebulizer of an air/acetylene flame atomic absorption spectrometer. Special interferences due to coeluting hydrocarbon matrix constituents were not observed at the 282.3 nm resonance line as lead used for analysis. Detection limits of this technique, based on a 20 µl injection, are approximately 10 ng Pb for each tetraalkyllead compound.

The potential advantages of the liquid chromatography - atomic absorption spectrometric hybrid analytical system over liquid chromatography with conventional ultra violet absorbance detection were investigated in the determination of tetraalkyllead compounds in a gasoline antiknock additive and in commercial gasoline. The effluent from the liquid chromatographic system was introduced directly into the nebulizer of the flame atomic absorption spectrometer detection system to allow direct comparison of their selectivity and sensitivity. Retention time and peak area reproducibilities were also compared for both systems. Figure 171 shows a comparison of chromatograms for the determination of tetraalkyllead compounds in gasoline antiknock additive MLA 500 Dilute which has a theoretical tetraalkyllead composition of 16% by weight.

The tetraalkyllead distribution (in mole percent) as provided by the manufacturer is nominally 6% tetramethyllead, 25% trimethylethyllead, 38% dimethyldiethyllead, 25% methyltriethyllead and 6% tetraethyllead. Even in such a relatively simple matrix, the metal selectivity of the liquid chromatographic atinuc absorption spectrometric system is demonstrated. Whereas only tetraalkllead compounds are monitored at the 283.3 nm analysis line with the liquid chromatographic atomic absorption spectrometric system, an additional component of MLA 500 Dilute (undoubtedly xylene which is the predominant diluent) absorbs at the 254 nm wavelength with the liquid chromatographic ultra violet system. With ultra violet detection the diluent peak is shown to elute just prior to tetramethyllead and could possibly preclude the accurate determination of this tetraalkyllead species.

An attractive feature of the liquid chromatographic atomic absorption spectrometric system compared to liquid chromatography ultra violet spectro-

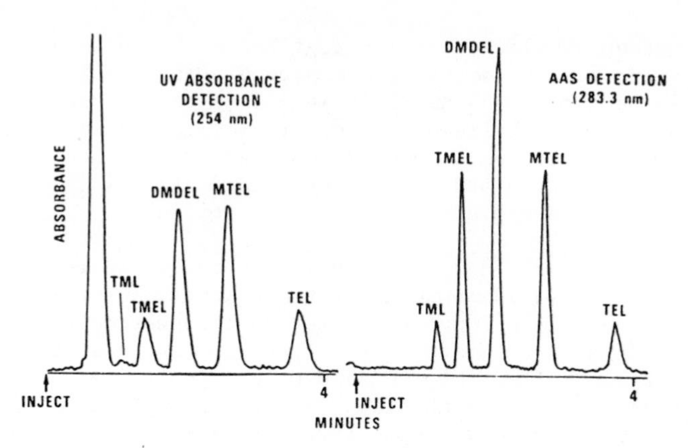

Figure 171 - Separation of tetraalkyllead compounds in gasoline antiknock additive MLA 500 Dilute by reversed-phase LC: 0.18 µg of TML (0.14 µg of Pb), 0.77 µg of TMEL (0.56 µg of Pb), 1.20 µg of DMDEL (0.84 Mg of Pb), 0.84 µg of MTEL (0.56 µg of Pb), and 0.22 µg of TEL (0.14 µg of Pb). Analysis conditions are described in the text.

scopy for this particular analysis is the relatively greater convenience in calibration for quantitative purposes. A well-documented problem in the flame atomic absorption spectrometric determination of total lead in gasoline concerns appropriate calibration when more than one tetraalkyllead compound is present in the sample. This dependence of the atomic absorption signal on the chemical form of lead is not observed in the system described by Messman and Rains[2962].

Vickray et al[2967] have investigated post column digestion methods for liquid chromatography - graphite furnace atomic absorption speciation of organolead compounds. The addition of iodine improves absolute response and inter sample precision for the determination of alkyllead compounds by this technique.

Vickray et al[2968] have investigated post column digestion methods for liquid chromatography - graphite furnace atomic absorption speciation of organolead compounds. The chemical digestion of organolead compounds induced in the graphite cuvette of an atomic absorption spectrometer. The metal-containing components are digested and analyzed after the molecular species are separated by reversed phase liquid chromatography. The five lead species, tetramethyl-, trimethylethyl-, dimethyldiethyl-, methyltriethyl-, and tetraethyllead, are analyzed by using the peak storage sampling method and are digested with methanolic iodine which improves the lead atomization efficiency and improves the precision of the lead analysis. The application of these procedures to reduce species-dependent sensitivity in atomic absorption analysis of liquid chromatographic eluents is discussed by these workers.

Table 241 shows some typical analyses obtained by this procedure.

Table 241 - LCGFAA analysis of a standard alkyllead mixture[a,b]

Component	% total Pb (as Pb)		% component in solution	
	actual	found	actual[c]	found
tetramethyllead	4.1	3.2 ± 0.2	2.1	1.6 1 0.1
trimethylethyllead	23.2	25 ± 1.7	12.3	13.2 ± 0.9
dimethyldiethyllead	40.7	40 ± 1.5	22.7	22.6 ± 0.8
triethylmethyllead	25.9	25 ± 1.9	15.3	15 ± 1
tetraethyllead	6.±	6.2 ± 0.3	3.8	3.9 ± 0.1
total alkyllead	100	99.4 ± 5.6	56.1	56 ± 3

a Average values found with post column 1_2 digestion on four diluted solutions containing 2.6 - 12.1 ppm total lead.
b Liquid chromatography - graphite furnance atomic absorption spectrometry recovery is 89.8% based on direct graphite furnace atomic absorption spectrometry analysis. "Found" values were corrected for recovery and are based on chromatographic leap analysis. The individual peak areas were compared with peak areas obtained from standard solutions of each alkllead component.
c As alkyllead (wt %).

Organomercury Compounds

High-performance liquid chromatography has enabled dibenzo-18-crown-6-complexes of divalent mercury to be separated[2969]. Methyl-, ethyl-, and phenylmercury cations have been determined in amounts down to 2 µg/g using liquid chromatography with differential pulse electrochemical detection[2970, 2971].

Koizumi et al[2972,2973] have combined a Zeeman atomic absorption detector with liquid chromatography. High-performance liquid chromatography has been used for the determination of methyl mercury compounds in fish[2974].

Mercuric compounds - inorganic and alkyl mercury - have been separated with high-performance liquid chromatography and detected by use of flame atomic absorption[2975], ultra violet[2976,2977], graphite furnace atomic absorption[2978,2979], electrochemical detector[2980,2981] or ultra violet absorption [2982-2985]. The sensitivities of these detectors are, however, not satisfactory for the determination of trace mercuric compounds in column effluent from high-performance liquid chromatography and the atomic absorption detector could not afford a continuous chromatogram. Cold vapour atomic absorption had not been applied to monitoring mercuric compounds in column effluent from high performance liquid chromatography because of difficulty in coupling the cold vapour atomic absorption with high performance liquid chromatography. The difficulty mainly arises from the fact that cold vapour atomic absorption is usually operated in a stepwise fashion, while the effluent from high performance liquid chromatography is a continuous flow of liquid. Fujita and Takebatake[2986] have described a simple apparatus combining high performance liquid chromatography and cold vapour atomic absorption spectrometry permitting the continuose and sensitive monitoring of mercury compounds separated by high performance liquid chromatography. This method uses a reducing vessel which couples liquid chromatography with cold vapour atomic absorption spectrometry. The reducing vessel continuously reduces mercuric compounds, which are separated by liquid chromatography, and the volatillized mercury is swept into the cold vapour atomic absorption spectrometer with a continuous flow of air, where the atomic absorption is measured at 253.7 nm. Thus, cold vapour atomic absorption works effectively as a continuous detector for mercury in the effluent from liquid chromatography. When stannous chloride is used as reductant, mercury alkane thiolates are monitored. For the detection of inorganic, methyl, and ethyl mercuric compounds, sodium borohydride is used. The system detects mercury with high sensitivity and specificity partly due to less matrix interference than in ultra violet absorption detection. At a low concentration such as 20 ng each of mercury thiolate, the peaks of thiolates are indistinguishable from the background signals on the chromatogram of ultra violet detector. However, the cold vapour atomic absorption detector offers a chromatogram with larger peaks of thiolates and with less background signals. This is because cold vapour atomic absorption detector selectively picks up mercury compounds and has little matrix interference, resulting in much better sensitivity than ultra violet detector. Although stannous chloride does not reduce methylmercury compounds, cadmium chloride mixed with stannous chloride does as also does sodium borohydride at low pH ranges.

Mac Creham[2987] has studied the application of differential pulse detection to the liquid chromatography of methyl mercury species. He examined the advantages of differential pulse over amperometry electrochemical detection and found that the former technique had added selectivity, higher sensitivity and lower base-line drift and that electrode fouling problems were overcome.

Organonickel Compounds

High performance liquid chromatography above critical temperatures has been used (Klesper et al[2988]) to separate nickel actioporphyrin II and nickel hesoporphyrin 1 x dimethylesters. Dichlorodifluoromethane at a starting pressure of 1830 psi was used to separate the two nickel porphyrins in 1 mg quantities on a column of polyoxyethylene glycol (33%) on Chromosorb W at 150 - 170°C. The porphyrins do not move on the column at a gas pressure of less than 600 psi.

Organophosphorous Compounds

Phosphoric Acid Esters. Schnitzerling and Schunter[2989] have described a method for the column chromatography in non-aqueous medium of acid esters of phosphoric acid, such as diethyl hydrogen phosphate. They used silicic acid washed with ethanol and ethanol-hexane (3:37) to prepare a column of diam 3 cm. The impure acid esters (0.75 g) dissolved in a small amount of ethanol-hexane (3:37) are applied to the column, followed by the same solvent mixture at 3 ml min^{-1}. The eluate is monitored by infrared spectrophotmetry at 950 cm^{-1} in a 0.1 mm sodium chloride flow-through cell, and fractions are collected according to the extinction.

Various workers have studied the solvent extraction of organic phosphates (Blimmer and Bunch[2990]; Kumler and Eiler[2991]; Peppard et al[2992,2993]; Stewart and Crandell[2994]).

Plapp and Casida[2995] achieved a partial separation of phosphoric acid and mono and dimethyl phosphates by gradient elution from a column of Dowex 1 x 8.

Higgins and Baldwin[2996] separated phosphoric acid and mono-n-butyl dihydrogen and di-n-butyl hydrogen phosphates on a column of Dowex 2 x 8 in the chloride form.

Schmitz and Walpurger[2997] separated phosphoric acid esters by ion exchange chromatography on Dowex 2 x 10. Chromatography at pH 6 - 6.5 permits the quantitative separation of complex mixtures of nucleotides and other phosphate esters (10 - 20 μmol of each component) derived from tissue extracts. Elution with dilute formic acid is practically quantitative, but recovery depends in part on the temperature since certain of these esters decompose at 20°C.

Jakob et al[2998] have described salting out chromatographic procedures for the analysis of mixtures of dialkyl phosphoric acids, alkane phosphonic acids containing alkyl groups (methyl to n-butyl) and orthophosphoric acid. Varon et al[3004] described a procedure for the separation of mixtures of the monoalkyl esters of alkane-phosphonic acids and dialkyl esters of phosphoric acid.

Jakob et al[2998] separated the compounds on a low capacity (less than 5 meq g^{-1}) Dowex 50 x 4 cation-exchange resin, using 4 M lithium chloride in M hydrochloric acid as eluant. Analysis of the fractions was carried out by conversion of the organic acids into phosphate which was determined by a molybdovanadate method, with recoveries of 94 - 108%. Brief details of the method are discussed below.

The preparation of columns for ion-exchange separation has been described by Rieman and Sargent[2999]. The resin (capacity 3.8 meq g^{-1}) is equilibrated with the eluent, 4.0 M in lithium chloride and 1.0 M in hydrochloric acid, by passing 1 litre of this solution through the column. The eluent is then drained to the level of the top of the resin bed. The height of the resin column at this point is 61.8 cm. A sample (1.00 ml) is placed by pipette on top of the resin bed and allowed to drain into the resin. At this point, collection of the effluent is initiated. The sample is then completely washed into the resin with the aid of three small portions of the eluent (3 - 4 ml). After each addition the solution is permitted to drain into the resin. Any eluent changes are made after draining the previous eluent completely into the resin bed. The column is maintained at 50 ± 2°C. The hydrostatic head of the eluent is adjusted (1.5 m) so as to give an initial flow rate of 0.11 cm min^{-1}. Small fractions (2.76 or 3.10 ml) are collected with the aid of a siphon pipette and a fraction collector.

The procedure used by Jakob et al[2998] for the determination of phosphorus in the eluate involved the conversion of the organophosphorus acids to orthophosphate, and the determination of the resulting orthphosphate by the molybdovanadate method (Barton[3000]; Quinlan and De Sesa[3001]; Misson[3002]).

The investigation of Kolmerten and Epstein[3003] indicated the vital role played by the ratio of the chloride ion to sodium hydroxide concentrations in the persulphate oxidation of organophosphorus compounds. They used sodium sulphite to destroy the excess persulphate that remained after heating for 1 hour. This step was eliminated in the work of Jakob et al[2998] by extending the period of heating to 2 hours. The phosphovanadomolybdate method, originally proposed by Misson[3002] was used to determine the orthophosphate concentrations using absorbance at 400 nm.

Each fraction, containing 2.76 or 3.10 ml of a solution 4.0 M in lithium. Figure 172 shows the separation achieved for the mixture of nine compounds.

In the separation of monoalkyl esters of alkanephosphoric acids and

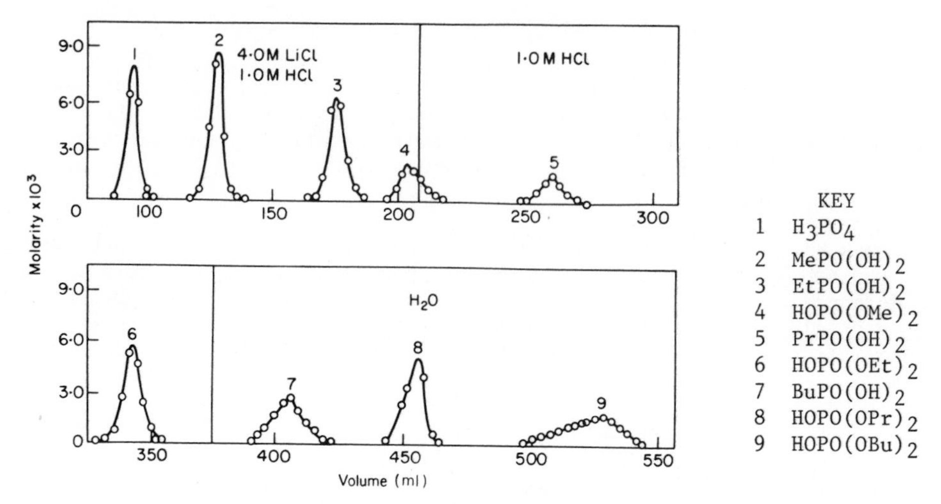

Figure 172 - Separation of nine component mixture of phosphoric acid esters. 61.8 cm x 2.76 cm^2 Dowex 50W-X4, 325 mesh, 3.82 meq g. For key to compounds see Table

dialkyl esters of phosphoric acid by salting-out chromatography, Varon et al[3004] found that chromatography with low-capacity Dowex cation-exchange resin was more successful than anion-exchange chromatography because of the small differences in the ionization constants of these compounds and in the selectivity coefficients of their conjugate bases. Elution was carried out at 50°C with various concentrations of lithium chloride and hydrochloric acid. After separation on the column, phosphorus was determined spectrophotometrically in the various fractions.

Lew et al[3005] have discussed the separation by salting-out chromatography of mixtures of orthophosphoric acid and its mono and dialkyl esters on columns of low capacity Dowex 50-X4 resin using 2 M - 4 M lithium chloride or N hydrochloric acid, or both, as eluants. Recoveries of the methyl, ethyl and butyl esters ranged from 06 - 99%. The column also separated dimers of some of the monomethyl esters of phosphoric acid. The method used by these workers is discussed below. They used low capacity resins, identical to Dowex 50X4, except that their capacities were lower than the usual value of 5.2 meq g^{-1}. Each low-capacity resin consisted of spheres that passed through a 325-mesh screen and were cross-linked with 4% nominal divinylbenzene.

The preparation of columns for ion-exchange and salting-out separations has been described by Rieman and Sargent[2999]. The columns were maintained at a temperature of 50 ± 2°C. The organophosphorus compounds in the extracted fractions were oxidized with ammonium persulphate in strongly alkaline solution as described previously (Jakob et al[2998]; Kolmerton and Epstein[3003]) The oxidations were carried out in polypropylene test tubes in order to avoid interference from silica (Lew and Jakob[3006]). The resulting orthophosphate was determined by the phosphovanadomolybdate method (Barton[3000]). All absorbance measurements were made with a Beckman DU spectrophotometer (1 cm Corex cells).

Figure 173 shows a chromatogram obtained by Lew et al[3005] for a freshly

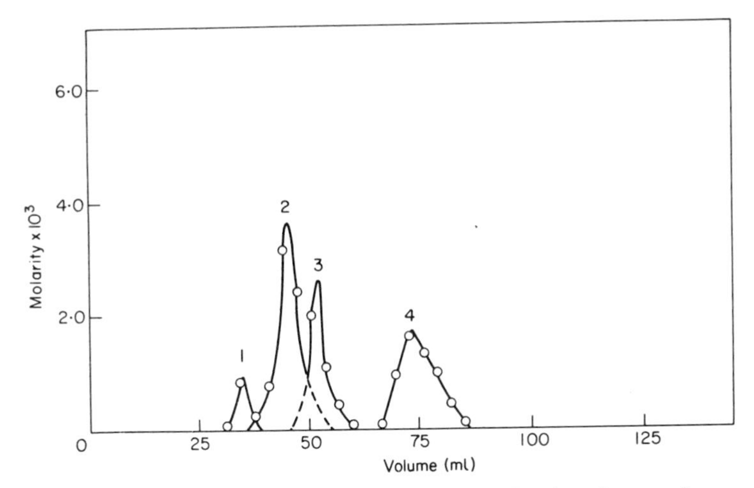

Figure 173 - Elution of mono- and di-methyl orthophosphate mixture with 4.0 M lithium chloride and 1.0 M hydrochloric acid. (16.7 cm x 2.76 cm^2 Dowex 50W-X4, 325 mesh, 3.82 meq g. linear flow rate 0.17 cm min^{-1}). (1) Orthophosphoric acid; (2) monoethyl ester of orthophosphoric acid; (3) dimer of monomethylester of orthophosphoric acid; (4) dimethyl ester of orthophosphoric acid.

prepared solution of the monoethyl and dimethyl esters of orthophosphoric acid. They identified the four peaks in order of elution from the column as being due to orthophosphoric acid, monomethyl dihydrogen orthophosphate, the dimer of monomethyl dihydrogen phosphate and dimethyl hydrogen orthophosphate.

Various Italian workers (Cerrai and Gadda[3007]; Cesaranol and Lepseky 3008) have used ion exclusion chromatography (Wheaton and Bauman[3009,3010]) to achieve a partial separation of phosphoric acid, mono-n-butyl phosphate and di-n-butyl phosphate. A 180 cm column of Dowex 50X8 was required. A small quantity of a dimer of mono-n-butyl phosphate was incompletely separated from the phosphoric acid. Cesaranol and Lepseky[3008] suggested that this technique might be applicable to the separation of other alkyl phosphates.

Alkyl Chloridates and Fluoridates

Beach and Sass[3011] carried out a particular study of two compounds, isopropylmethylphosphonofluoridate (GB Sarin) and bis(isopropylmethylphosphonic) anhydride. They developed simple separation procedures for these substances involving silica gel column chromatography and liquid-liquid partitioning and subsequently determined GB Sarin by hydrolytic titration and obtained the pyroester (anhydride) content by difference from a total hydrolytic determination of the original sample.

By saturating a silica gel column with water and then washing with isopropyl ether, all of the pyroester in a sample was retained in the column while the GB Sarin in that sample could be almost quantitatively eluted (> 99%) with isopropyl ether.

Carbohydrate Phosphates

Caldwell and Black[3012] have described a procedure for the determination of inositol hexaphosphate in soils and manures. Separation is achieved on a De-acidite anion exchange column.

Mehta [3013] has discussed the factors affecting the estimation in soil of phosphate esters such as inositol hexaphosphate, glycerophosphate, glucose 1-phosphate and nucleic acids. He critically reviews existing methods for the determination of organic phosphate in soil and presents results obtained by various methods on a calcareous soil and an acid clay loam and points out that the type of soil can influence the hydrolysis of some esters during extraction by standard procedures and that several types of esters are not recovered after acid pre-treatment. He describes a modification of this procedure in which the soil is first extracted with 0.3 N sodium hydroxide; this recovers esters of the above types which would otherwise be completely or partially hydrolysed during the acid extraction stage.

In a method for the separation of phosphate esters in deproteinized tissue extracts Rappoport and Chen[3014] treat the sample with Dowex-50 ion-exchange resin (H + form) and then apply to Whatman No. 41 H paper for ascending development with ethyl acetate-acetic acid-water (3:3:1) for 15 hours. After drying, the chromatograms are developed by the descending technique in the second dimension with ethyl methyl ketone-2-methoxy-ethanol (7:2) purified by passage through Dowex-50 resin (H + form) and then diluted with 0.7 part of conc ammonia and 2.3 parts of water. Development is for 22 h, after which the carbohydrate phosphates are detected by two spray reagents or (for ^{32}P derivatives) by an autoradiographic technique.

Morton[3015] has described a method for the determination of D-hexose phosphates using substrate specific enzymes (specific dehydrogenases) coupled with reduction of diphosphopyridine nucleotide, which is determined spectroscopically.

Miscellaneous Organophosphoric Compounds

Fluorothionopyrophosphoric acid alkylamides have been determined by ascending chromatography on mineral oil impregnated paper (Stolzer and Simon[3016]) using 50% aqueous glacial acetic acid as the mobile phase. The chromatogram is dried at 160°C for 15 min, sprayed with molybdate reagent and heated for 7 min at 85 - 85°C. After 4 to 7 d the chromatogram is sprayed with very dilute stannous chloride solution to reveal the fluoro-thionopyrophosphoric acid alkylamide.

Hirotoshi et al[3017] have isolated and identified 2-aminoethyl phosphonic acid from bovine brain using ion-exchange chromatography followed by paper chromatography followed by elemental analysis and infrared spectroscopy.

Organophosphorus Pesticides and Insecticides

Parathion diethylester isomers and derivatives have been analysed by column chromatography (Paulus and Mallack[3018]) and ion-exchange chromatography (Paulus and Mallack[3019]).

Fischer and Otterbeck[3020] have used chromatography to separate Parathion, Diazinon, Chlorthion, Malathion, Meta Systox (Demeton methyl), Thiometon and their saponification products in amounts down to 1 μg. Sandi

3021 applied column chromatography and subsequent polarography to the detection and determination of the following analogues of Parathion methyl, Chlorthion (O-(3-chloro-4-nitrophenyl) O,O-dimethyl phosphorothioate), isochlorthion (O-(4-chloro-3-nitrophenyl) O,O-dimethyl phosphorothioate), isomeric chlorothion (O-(2-chloro-4-nitrophenyl) O,O-dimethyl phosphorothioate), Bromthion (O-(2-bromo-4-nitrophenyl) O,O-dimethyl phosphorothioate), and Para-oxon (diethyl p-nitrophenyl phosphate).

Column chromatography has been used to determine Malathion (S-(1:2-dicarbethylethyl)-O,O-dimethyl dithiophosphate) in dusts and fruit fly bait (Double[3022]), Dimethoate or Rogor (O,O-dimethyl S-(N-methylcarbomoyl methyl) phosphorodithioate) in technical products and formulations (Bazzi et al[3023]), Delnav (2:3-p-dioxandithion S,S-bis-(O,O-diethyl)phosphorodithioate) in organic materials, oils and fats (Dunn[3024]) and Co-Ral (O-(3-chloro-4-methylumbelliferone) O,O-diethyl phosphorothioate) in animal tissues (Anderson et al3025). Column chromatography has also been applied to the determination of organophosphorus insecticides and pesticides in technical products and formulations (Dupee et al[3024]).

Tittarelli and Mascherpa[3027] have described a method based on liquid chromatography with graphite furnace atomic absorption spectrometric detection for the determination of organophosphorus additives in lubricating oils in amounts down to 2 μg (as phosphorus). The plasma chromatography of organophosphorus esters has been investigated[3028].

Ethyl- and methylparathion at the parts per 10^9 level have been determined in run-off waters using high performance chromatography[3029].

Organoselenium Compounds

Selenocystine and selenomethionine may be separated from their sulphur analogues and from leucine by chromatography on a column of sulphonated styrene-divinylbenzene copolymer resin; elution is carried out at 33°C with 0.2 N sodium citrate buffer, first at pH 3.28, changing to pH 4.25 after the passage of 300 ml through the column[3030]. Selenocystine leaves the column in the eluate fraction between 405 and 415 ml, and selenomethionine between 450 and 500 ml.

Blatcky et al[2741] determined trimethylselnnium ion in urine by ion-exchange chromatography and molecular neutron activation analysis.

Organosilicon Compounds

Fritz and Wick[3031] studied the column chromatographic separation of pyrolysis products of tetramethylsilane. The high boiling silicon compounds formed by pyrolysis of tetramethylsilane could not be separated unchanged by distillation, but were separated on alumina columns into three fractions. The first fraction was not adsorbed from benzene solution and could be further fractionated on an alumina column with pentane. The second fraction was eluted with benzene and contained strongly fluorescent substances which were characterised and separated by paper chromatography, and had characteristic absorption and fluorescence spectra. The third fraction was eluted with benzene - methanol (10:1) and was further fractionated on alumina columns with other mixtures of benzene and methanol.

Oligomers of methylphenylsiloxane, dimethylsiloxane, methylhydrosiloxane and methylhydrocyclosiloxane have been separated (Sarishvili et al 3032) into fractions of different molecular weight on a column (127 cm x

2 cm) containing Silica gel MSM, activated carbon BAU or quartz sand. The sample was applied to the column in light petroleum solution and elution was effected with light petroleum - benzene and with benzene. The fractions were evaporated in vacuo, and the molecular weights were determined either cryoscopically or ebullioscopically. Quartz sand (particle size 0.25 - 0.1 mm) was the most effective column packing. The greatest molecular weight distribution was observed for dimethylsiloxane, which had a molecular weight range of 600 - 3200 and an average molecular weight of approximately 1850. The smallest distribution was found for methylhydrocyclosiloxane with a molecular weight range of 600 - 1200 and an average molecular weight of about 600.

Nagy and Brandt-Petrik[3033] studied the chromatographic adsorption analysis of alkyl and alkyl-aryl siloxanes. The sample in methanolic solution is applied to a column of activated carbon and the straight-chain components i.e. $(CH_3)_3SiO[Si(CH_3)(C_6H_5)O]_nSi(CH_3)_3$, are eluted (as a mixture) with methanol, followed by the cyclic species, e.g., $[Si(CH_3)(C_6H_5)O]_n$, which are eluted with benzene. Some separation of the straight chain components was achieved by successive elution with methanol, ethanol, and butanol, and of the cyclic components by successive elution with acetone and benzene.

Organotin Compounds

Burns et al[3034] have investigated the determination of tin tetraalkyls and alkyltin tin chlorides by high performance liquid chromatography.

Vickray et al[2968] have investigated post column digestion methods for liquid chromatography - graphite furnace atomic absorption spectrum of organotin compounds. The chemical digestion of organotin compounds is induced in the graphite cuvette of an atomic absorption spectrometer. The metal containing species are separated by reverse phase liquid chromatography. The organotin compounds analysed included tetrapropyl, tetrabutyl, tetraphenyl and dibutyldichlorotin. These compounds are decomposed in the graphite furnace by the zirconium treated surface of the cuvette. The application of this procedure to reduce species - dependent sensitivity in atomic absorption analysis is discussed by these workers.

From the data in Table 242 the species-dependent sensitivity (in terms of AU/ng) for the several tin compounds can be seen. The graphite cuvettes treated with zirconium show a uniform response to these compounds. That is the species-dependent sensitivity is removed to a greater extent than in the alkyllead plus iodine case.

The untreated tubes yield responses from 0.004 AU/ng '$(CH_3)_3SnCl$) to 0.113 AU/ng $(Bu_3Sn)_2SO_4$). The average signal per nanogram is 0.06 ± 0.04 AU/ng or a 60% relative standard deviation of observed response with different tin species. The average signal obtained for this series of compounds on a zirconium-treated cuvette is 0.204 ± 0.020 AU/ng of zinc. This is the average signal per nanogram of Sn, and the precision indicates the range of response values for the different species (the observed precision of individual analysis is ca. 6%). The signal obtained for an unknown compound can be related directly to the aqueous standard. The 95% confidence level estimate of the sensitivity of an unknown will be 0.204 ± 0.039 AU/ng. This twofold range of observed responses is much more uniform than the untreated tubes.

MacCrehan has studied the application of differential pulse detection to the liquid chromatography of methyl, ethyl and butyl tin species. He examined the advantages of differential pulse over amperometric electrochemical detection and found that the former technique had added selectivity, higher sensitivity and lower base-line drift. Differential pulse detection also eliminated the electrode fouling problem.

Table 242 - Effect of Zirconium-modified graphite cuvette surface on the atomic absorption signal obtained for several organotin species

Compound	mass (ng) as Sn	signal (absorption units) untreated	treated	signal per ng (treated)[a]	signal (treated)/signal (untreated)
$SnCl_4$	1.00	0.074	0.229	0.229	3.1
$SnPh_4$	1.86	0.110	0.359	0.193	3.3
$Sn(Bu)_2$	1.40	0.110	0.269	0.192	2.5
$Sn(Bu)_2Cl_2$	1.44	0.074	0.309	0.215	4.2
$SnMe_3Cl$	1.46	0.007	0.257	0.176	36.7
$SnMe_2Cl_2$	1.23	0.050	0.248	0.202	6.0
$(SnBu_3)_2SO_4$	1.94	0.220	0.385	0.198	1.8
$SnPr_2Cl_2$	1.24	0.068	0.267	0.215	3.9
$Sn(C_6H_{11})_2Br_2$	1.94	0.130	0.352	0.181	2.7
$SnPh_3Cl$	1.02	0.076	0.247	0.241	3.3

a For the treated tubes the average signal/ng Sn was 0.204 ± 0.020 absorption units/ng Sn for the compounds listed.

Organovanadium Compounds

Fish and Koulenic[3035] have applied liquid chromatography with graphite-furnace atomic absorption detection to the molecular charaterization and profile identification of vanadyl compounds (porphyrin and non-porphyrin types) in heavy crude petroleums. They attempted to speciate the vanadyl porphyrin and non-porphyrin compounds by comparison of their size - exclusion and polar aminocyano separated vanadium histograms to authentic compounds.

The heavy crude oils, the extracted oils, and the extracts were all analyzed with the 50/100 size exclusion chromatographic column. The 50/100/1000 A combination was also used with the four heavy crude petroleums. All size exclusion chromatographic runs were performed with tetrahydrofuran as the mobile phase at a flow rate of 0.5 mL/min. Further, the four pyridine/water extracts were also analyzed by using gradient elution chromatography with the polar aminocyano column. The polar aminocyano separations were obtained with a solvent gradient consisting of an initial linear ramp from 100% n-hexane to 25% methylene chloride - tetrahydrofuran (1:1 v/v) from 0 to 3 min and a second linear ramp from 25% to 100% methylene chloride - tetrahydrofuran (1:1 v/v) from 27 to 30 min at a flow rate of 2.0 mL/min. Recalibration of the polar aminocyano column after each run was accomplished by ramping to 100% n-hexane over 3 min and holding until a minimum of 10 column volumes of n-hexane had eluted.

D. ELECTROPHORESIS

Organoiron Compounds

Several electrophoretic methods have been used for separating hamoglobin A from haemoglobin F. One uses starch hydrolysed in hydrochloric acid-acetone (1:100, v/v) at 38.5°C[3036]. Electrophoresis is carried out at 20°C for 3 h under a potential gradient of 12 V/cm. A second method uses a layer of 2% agar in buffer solution (pH 5.7 - 5.9) supported on a glass plate covered with filter-paper[3037]. Light green SF(CI Acid Green 5) is used for staining, followed by washing with 2% acetic acid until almost decolorized. The haemoglobins are well separated, haemoglobin A remaining near the point of application. Another method uses gel phase ion-exchange electrophoresis [3038]. Carboxymethylcellulose gel is the supporting medium and electrophoresis is carried out in a sodium phosphate buffer (0.17 - 0.04 M sodium ion). The best overall separation is achieved with a sodium ion concentration of 0.07 M Asensitive two-dimensional paper - agar electrophoretic method detects small amounts of haemoglobin A in the presence of haemoglobin F, and detects minor components in a haemoglobin solution[3039]. Electrophoresis is carried out at pH 8.2 on paper. A strip cut along the centre of the paper in the direction of migration is inserted in a slit in the agar gel (pH 6.2), prepared in a plastic tray. Electrophoresis is then carried out at right-angl to the length of the paper strip. Haemoglobins A and F have been separated by the cation-exchange dextran gels SE-Sephadex C-50 and CM-Sephadex C-50 [3040,3041]. The samples were converted into carboxyhaemoglobin and the haemoglobin was determined, after conversion, into cyanomethaemoglobin, by measurement of the absorption at 420 or 540 nm. Both foetal and adult haemoglobin were eluted from each gel with 40 MN phosphate buffer (pH 6.0) containing an increasing concentration of sodium chloride.

In a rapid method for haemoglobin fractionation on cellulose acetate, the buffer solution (pH 8.6) is prepared from Tris, EDTA, and anhydrous boric acid[3042,3043]. A constant current of 0.4 - 0.5 mA per centimetre of strip width is applied for about 40 min, a clear separation of haemoglobins A, C, S, and A_2 being achieved. The spots are eluted into the buffer solution and the absorption of the eluate is measured at 400 - 410 nm. The factors influencing the chromatography and differentiation of similar haemoglobins have been discussed[2951].

Various methods have been described for the determination of haptoglobin complexes. For zone electrophoresis o-dianisidine is the most satisfactory detection reagent[3044]. Zones possessing peroxidase activity are stained brown-pink. To determine the haptoglobin group in sera containing low concentration of haptoglobin, haemoglobin is added to the sera in excess of the haemoglobin-binding capacity, and an electrophoretic separation is performed with a gel prepared by boiling starch with a phosphate buffer[3045]. The free haemoglobin migrates towards the cathode, producing a clear separation of the haptoglobin groups, which migrate towards the anode. A solution of benzidine, acetic acid, and hydrogen peroxide is used to develop the electrophoretic pattern. Electrophoresis on acrylamide gel has been used to determine haptoglobin types. The three types of haptoglobin observed after separation by electrophoresis in starch or acrylamide gel can be distinguished readily by adding 10% normal adult haemoglobin solution to fresh, clear, non-haemolysed serum to form the complexes, and staining the electropherogram with a benzidine reagent[3046]. After electrophoresis, the gel is placed in the reagent and then washed with 15% acetic acid solution. In a specific separation of serum haptoglobin as a haemoglobin complex the serum is diluted five-fold with phosphate or acetate buffer solution and applied to a diethylaminoethylcellulose column, which has been previously washed with 0.2 M acetic acid, then with 0.2 M sodium hydroxide, and then with the buffer solution[3047,3048]. The column is washed with the buffer solution to elute the albumin and the globulins, then washed again with the buffer solution containing a small amount of sodium chloride to elute other proteins. A calculated amount of horse haemoglobin dissolved in the buffer solution is added to saturate the haptoglobin, then the complexed haptoglobin -haemoglobin is eluted with buffer solution containing a higher concentration of sodium chloride than before. The complex is eluted rapidly and specifically.

In a discontinuous buffer system for human haemoglobins the electrode vessels are filled with barbitone buffer solution (pH 8.6) and the paper is immersed in a buffer solution (pH 9.1), prepared by dissolving tris (hydroxymethyl)aminomethane (Tris), disodium EDTA dihydrate, and boric acid in water[3049]. The paper is placed in the electrophoresis chamber, buffered haemoglobin solution is applied, and the process is carried out in the usual way. The resolution of abnormal haemoglobins is superior to that in either buffer solution alone, and enables haemoglobin A_2 to be detected in small samples of haemolysate. The staining with Amido Black 10B (CI Acid Black 1) of patterns that on visual inspection show only haemoglobin A reveals the A_2 component if present. Agar gel[3050], starch gel[3051-3056], paper[3057,3058], cellulose acetate[3057-3061] and acrylamide gel electrophoresis [3062] have been applied to the determination of haemoglobin A_2. One group subjected the blood sample to electrophoresis on agar gel at 90 - 100 V and 50 - 50 mA[3050]. After drying, the electropherograms are stained with Amido Black 10B (CI Acid Black 1) solution and measured in a denitometer at 500 nm. A good separation of haemoglobins A_1 and A_2 was achieved and up to three unidentified non-haemoglobin fractions were observed. Excellent resolution has been obtained in the starch-gel electrophoresis of normal and abnormal haemoglobins with relatively little trailing of components[3054]. Separation occurs of haemoglobin A_2, two other pigmented fractions, and two non-pigmented protein fractions in normal red cell haemolysates, and the method can be used to distinguish between foetal haemoglobin and haemoglobin A when both are present.

Starch-gel electrophoresis of haemoglobin A_2 has been carried out by the vertical technique[3063], with a gel made up in 0.025 M borate buffer (pH 8.6)[3055,3056]. After separation, the haemoglobin fractions are cut out and placed in 0.06 N aqueous ammonia and stored in a refrigerator for 3 days. The absorption is measured at 540 nm, and the haemoglobin A_2 content calculated.

Haemoglobin A_2 in blood can be determined quantitatively by paper electrophoresis[3057]. The blood samples are collected with EDTA (dipotassium salt) as anticoagulant, haemolysed with water and toluene, and converted into carboxyhaemoglobin. Electrophoresis is carried out in vertical tanks with strips of Whatman No. 3 MM paper and tris(hydroxymethyl)aminomethane (Tris) buffer adjusted to pH 8.6 with boric acid. The treated samples are applied to the strips, which are then moistened with buffer on both sides to within 0.25 in of the point of application. After passing a current of 225 V and 2 mA per strip for 16 h, the strips are dried at 90 - 100°C, fixed in fresh 10% mercury(II) chloride solution and 10% glacial acetic acid in ethanol, redried, washed in water for 10 min, and then dried again. They are then stained in 1% bromophenol and 1% acetic acid in ethanol. The stained bands are cut out and eluted with 1.5% sodium carbonate solution in 50% methanol, and the absorbances of the eluates are read at 595 nm in a spectrophotometer. Another method uses electrophoresis of the sample prepared as a solution in 50% glycerol on paper for 3 h in barbitone buffer (pH 9.1) at 350 V[3058]. After drying, the paper is stained with bromophenol blue and then examined visually or with a densitometer and automatic integrator to evaluate the separated haemoglobin A_2. A quantitative determination of haemoglobin A_2 by electrophoresis on cellulose acetate uses an electrophoretic cell solution of 0.05 M sodium carbonate, and a membrane buffer solution of Tris-EDTA solution adjusted to pH 8.8 with aqueous boric acid[3061]. The strips are stained with Ponceau S (CI Acid Red 9), washed with 5% acetic acid, and the bands are eluted with 0.268 M aqueous ammonia. Acrylamide gel is superior to starch as it is easier to work with, gives more reproducible layers and faster separations, and affords fractions that can be evaluated by direct densitometry[3062]. Haemolysates are diluted with an equal volume of Tris-borate-EDTA buffer solution and electrophoresis is carried out for 1 h at 300 V. The unstained gel patterns are scanned with a recording densitometer with a 500 nm interference filter. The normal range for haemoglobin A_2 found by this method was 1.4 - 4.4%.

A rapid method for the determination of haemoglobins uses microelectrophoresis on cellulose acetate[3064]. The red-cell haemolysate is applied to a cellulose acetate membrane and subjected to electrophoresis for 1 - 1.5 h in a microelectrophoresis apparatus. The strips are then stained with 0.2% Ponceau S (CI Acid Red 9) solution in 3% trichloroacetic acid containing 5% of sulphosalicylic acid. After being washed with 5% acetic acid, the strips are cleared by immersion in methanol-acetic acid '7:3) and scanned in a photodensitometer. Cellulose acetate electrophoresis has also been used for haemoglobin fractionation[3065]. The sample is subjected to electrophoresis on cellulose acetate with Tris-EDTA A-boric acid buffer solution (pH 8.8). The electropherograms are stained with Ponceau S (CI Acid Red 9) rinsed in 5% acetic acid, cleared in acetic acid -95% ethanol (3:17), and then examined with a densitometer equipped with a blue filter.

Several groups have studied the application of starch gel electrophoresis to the separation of haemoglobins[3063,3065-3068]. A simple inexpensive electrophoresis apparatus with water cooling is available[3066]. Errors, resulting from the instability of the haemoglobin solution can be reduced by adding a small amount of potassium cyanide. Reagents for the detection of haemoglobins separated by starch gel electrophoresis have been described. After completion of electrophoresis, the gel is cut into two portions, one of which is stained with benzidine and hydrogen peroxide to identify and locate the fractions. Zones containing the individual fractions are then cut out of the second portion of gel and dissolved in 10% sodium hydroxide solution. Each resulting solution is mixed with 1% benzidine solution in acetic acid and 1% (v/v) hydrogen peroxide solution and set aside, then 10% acetic acid is added. The mixture is again set aside and then absorption is measured at 515 nm.

A study of combined agar gel-paper electrophoresis for the resolution of haemoglobins showed that by carrying out the electrophoresis in highly purified agar gel with a buffer of decreased ionic strength and a high potential gradient, separation can be attained in 15 min[3069,3070]. For agar gel electrophoresis for the identification of haemoglobin types[3071], blood samples may be prepared by centrifuging and addition of oxalic acid or heparinizing and addition of citric acid then removing the plasma and adding an equal volume of water[3072]. The solution is shaken to haemolyse and the solution added to lead acetate barbitone solution at pH 8.6. After 30 min the solution is centrifuged at 2000 rev/min. Half saturated phosphate solution is then added and the solution is again centrifuged. It is claimed that this solution upon electrophoresis gives a sharp resolution of the constituent haemoglobins.

In an electrophoretic method for the separation of haemoglobin and cytochrome C, the sample components are separated on cellulose acetate at pH 8.6, and zones are revealed by staining with ethanolic Amido Black 10G (CI Acid Black 1)[3073]. Similar procedures were used to separate haemoglobin from myoglobin.

Organolead Compounds

To separate 50 μg amounts of lead chloride, diethylleaddichloride and triethylleadchloride by paper electrophoresis, 2 - 3 M lithium chloride is used as the supporting electrolyte on a 5.5 x 40 cm strip of Whatman No. 1 paper, applying a potential of 135 V for 2 h[3074].

Organophosphorus Compounds

Vander Leiden[3075] has described a method for the high voltage electrophoresis separation, identification and determination of ^{32}P-labelled phosphate esters from erythrocytes. The electrophoresis (2000 V for 2 h) is carried out on acid-washed Whatman No. 3MM filter-paper in a pyridine - acetic acid - water (5:16:229) buffer system (pH 3.9). Samples (10 μl) are applied, and, to enable a correction for electro-osmosis to be made when calculating the relative electrophoretic migration of the different compounds, 10 μl of 50% glucose solution is also applied. The phosphate spots are located by observation in u.v. light or treatment with a molybdate reagent, or detected by autoradiography, followed by elution and quantitative determination, or radioactive counting. Hanes and Isherwood[3076] have described methods for separating on paper such compounds as hexose monophosphate and adenosine phosphate using an alcohol - water - ammonia mobile phase and acid molybdate followed by exposure to hydrogen sulphide for visualization of the separated compounds on the paper.

Runeckles and Krotov[3077] have shown that complex mixtures of inorganic phosphates and organic phosphates and other metabolites, sugars and amino acids can be separated by electrophoresis on paper followed by chromatography in the second dimension.

Electrophoresis has also been applied to the determination of adenosine phosphates (Heald[3078]), esters of phosphonic, phosphinic and phosphonous acids and free acids (Siuda[3079]) and trisubstituted phosphine oxides (Helman and Kugel[3080]).

Organothallium Compounds

Paper electrophoresis has been carried out on phenyl- and methylthallium compounds, using 10 V/cm for 1 h and aqueous sodium chloride containing 0.1 M hydrochloric and as the supporting electrolyte[3081]. Unlabelled compounds were detected by spraying the paper with potassium rhodizonate and Thallium labelled compounds by autoradiography. Good separations were obtained between organothallium(III) chlorides and thallium(I) and thallium(III) chlorides.

CHAPTER 9

ORGANOMETALLIC COMPOUNDS IN THE ENVIRONMENT

A. INTRODUCTION

A surprisingly large number of organometallic compounds occur in the environment. In this context the environment means inland waterways, potable water supplies, the oceans and also sedimentary matter, vegetation and animal life in inland waterways and the oceans. Some of the more volatile organometallic compounds are found in air. Organometallic compounds are also found in plant material, crops and biological materials such as fish, animal and human tissues and body fluids. Thus organometallic compounds occur fairly widely throughout the environment principally as compounds of mercury, tin, lead, arsenic and magnesium and, to a lesser extent of compounds of germanium, antimony, copper, silicon, manganese and nickel. The distribution of these compounds throughout the environment - air, water, food - has toxiological complications and as such, is of concern from the point of view of the health of humans, animals, fish, insects, birdlife who, in one way or another, are all subject to contamination by organometallic compounds and who are all part of a food chain.

Organometallic compounds enter the environment by three main routes. The first are the organometallic compounds produced by man such as those of lead and mercury. Lead enters chiefly as alklyllead compounds, used as an additive in gasoline which enters the air via gasoline spillages and possibly to some extent, from automobile exhausts and then contaminates water-ways and consequently river sediments and fish and plant life and also enters crops and animals in the fields. Similar comments apply to organomanganese compounds which to some extent is displacing lead as a petroleum additive, and to organotin compounds which enter the environment as for example, antimolluscticide paints used on the hulls of ships and in harbour works. The second source of organometallics are those which enter the environment as inorganic substances and are subsequently converted in vivo to organometallic compounds. Classic cases of this are the organomercury compounds and organoarsenic compounds which in the environment can be converted by bacteria or in animal tissue to methylmercury and methylarsenic compounds. Probably the chief source of organomercury contamination is inorganic mercury entering rivers as an effluent from industries such as chloroalkali works and industries which use alkalies in large quantities such as papermaking. It is believed that tin entering the environment via industrial effluents or mining operations as inorganic tin can be similarly converted in nature to organotin compounds. The third group of organometallic compounds occuring in the environment are those that are produced in waterways from naturally occuring concentrations of metals, as opposed to man-made sources of metal pollution. These include methylarsenic compounds produced by methylation of inorganic arsenic in fish or on sediment deposits and probably include other elements such as antimony and selenium.

The list of organometallic compounds found in trace amounts in the environment has increased dramatically over the past few years and has necessitated the development of analytical methodology both for the purposes of first identifying new compounds and secondly, of monitoring the concentrations of such compounds so that trends can be followed and working hypotheses developed. The occurence of organometallic compounds in the environment is now systematically discussed under the following headings.

Organomercury Compounds

Organomercury compounds are more toxic than metallic mercury[3082,3083,3090,3091] and inorganic mercury forms and when present in the environment, may cause serious illness in extremely polluted areas. Methylmercury has been stated to be neurotoxic[3092].

A growing public interest in environmental quality has led to the development of analytical techniques for the monitoring of environmental pollutants. Due to its acute toxicity and its tendency to bioaccumulate, mercury is of prime interest. Being extremely volatile in the organic and elemental forms, mercury is well dispersed in the atmosphere. The activity of certain bacteria, molds, and enzymes in the soil or sediment can produce methylated mercury from elemental or inorganic mercury[3084-3089]. The organic mercury compounds produced, primarily dimethylmercury and methylmercury halides, are potentially more toxic than inorganic mercury forms. Therefore, recent studies of environmental mercury have been concerned with its chemical speciation to determine not only the amounts of mercury present but the chemical forms as well. More extensive data in this area will assist in determining the role of of organic mercury in the global cycling of the element. It has been reported that organomercury compounds are significantly concentrated in fish[3093-3099], predominantly as methylmercury compounds[3094-3100]. The syntheses of methylmercury compounds by microorganisms in freshwater sediments have been investigated by some workers[3100,3101].

Although methylmercury has been found in aquatic organisms, its origin is not clearly known. It is generally assumed that methylmercury exists in natural waters and that the organisms concentrate it, because it has been detected in many aquatic organisms; also the methylation of inorganic mercury in sediments has been reported by several investigators[3102-3104].

Organotin Compounds

Organotin compounds have been widely used not only as plastic stabilizers or catalytic agents in industry but also as biocidal compounds, i.e. bactericides, fungicides, anthelminthics, and insecticides, in agriculture and medicine and antifouling agents for ships. Recently, deep concern has been expressed concerning the safety of these compounds in the environmental cycle.

There is demand for analytical techniques capable of speciating organotin compounds in environmental studies for two obvious reasons. First is the increasing use of inorganic and organotin compounds in many industrial, chemical, and agricultural areas. Very little being known about their environmental fate; secondly there is a great difference in toxicity of the various organotin compounds according to the variation of the organic moiety in the molecules.

There is a special interest in the biotic and abiotic methylation of tin compounds[3105-3107] and the fate of some industrial organotins in the aquatic ecosystems. One possible route as discussed by Brinckman[3108] is the dealkylation of the trialkyltin species eventually to Sn(IV), and the microbial methylation of Sn(IV) to the various methyltin species. Increasing methyltin concentrations with increasing anthropogenic tin influxes has been noted in the Chesapeake Bay.

Organolead Compounds

The use of tetraalklylleads as antiknock additives/octane enhancers for automotive gasolines has been reduced due to environmental considerations in the United States, although not elsewhere. However, the complete elimination of tetraalkyllead additives is unlikely.

The high toxicity of tetraalkylleads is attributed to their ability to undergo the following decomposition in the environment[3109-3112]:

$$R_4Pb \rightarrow R_3Pb^+ \rightarrow R_2Pb^{2+} \rightarrow Pb^{2+}$$

The formation of alkyllead salts, probably associated with proteins, arising in tissues from rapid metabolic dealkylation of tetraalkyllead compounds is of toxicological importance in evaluating exposure to tetraalkylleads[3113]. The toxic effect of tetraalkylleads to mammals has been attributed to the formation of trialkyllead compounds in body fluids and tissues[3114,3115].

The possibility of biomethylation of lead or organolead ionic species by microorganisms[3116,3117] reversing the decomposition mechanism given above, may add to the problem of lead toxicity already faced by man, although the area is presently much disputed[3118,3119].

Organically bound lead is a minor but important contribution to total lead intake by humans and animals. It has been shown that alkyl lead salts such as trialkyllead carbonates, nitrites and/or sulfates arising in tissues from rapid metabolic dealkylation of tetraalkyllead compounds are important in low toxicity[3120].

Recently a renewed interest in the speciation of lead in environmental samples has resulted from several diverse lines of investigation. Organolead compounds have been detected in cod, lobster, mackerel, and flounder meal (10 to 90% of the total lead burden)[3121], freshwater fish[3122], air[3123-3126], street dust[3127], and human brains[3128]. A steady input of organoleads into the environment results from the continued use of tetraalkylleads as antiknock additives. In addition evidence for the chemical[3129-3132] and biological[3133-3136] alkylation of organolead salts or of lead(II) salts have been obtained.

Although organoleads may make only a small contribution to the total lead intake of an organism, it has been demonstrated that trialkyllead salts arising in tissues from the degradation of tetraalkylleads are important in lead toxicity. The conversion of R_4Pb to R_3Pb^+ occurs rapidly in liver homogenates from rats and rabbits[3137,3138]. Acute toxicities of tetraalkylleads and of trialkyllead salts are similar[3137,3139] and are at least an order of magnitude greater than dialkyllead salts or inorganic lead salts. Relatively little is known either of the effect of chronic exposure to small amounts of such compounds or the levels of organic lead compounds, such as the tetraalkylleads, in biological and food material. Dialkyllead salts cause symptoms of toxicity similar to those produced by inorganic lead salts and they exhibit an affinity for thiol compounds[3137]. Triorganolead salts inhibit oxidative phosphorylation and have been reported to bind to proteins[3140,3141].

Specitiation of alkyllead compounds including molecular and ionic, volatile, and solvated forms has become immensely important and in urgent demand in studies related to toxicity and environmental consequences. The highly polar dialkyl and trialkyl forms in particular are more important species because of their high toxicity to mammals[3142] and the consequence of the formation as a result of degradation of tetraalkyllead in aqueous medium[3143].

Organoarsenic Compounds

Organoarsenic species are known to vary considerably in their toxicity to humans and animals[3145]. Large fluxes of inorganic arsenic into the aquatic environment can be traced to geothermal systems[3146], base metal smelter emissions[3147] and localized arsenite treatments for aquatic weed control. The methylated arsenicals have entered the environment either directly as pesticides or by the biological transformation of the inorganic species[3148-3151].

Organoarsenical pesticides such as sodium methanearsonate and dimethylarsinic acid are used in agriculture as herbicides and fungicides. It is possible that these arsenicals enter soil, plant, and consequently humans. On the other hand, arsenic is a ubiquitous element on the earth, and the presence of inorganic arsenic and several methylated forms of arsenic as monomethyl-, dimethyl-, and trimethylarsenic compounds in the environment has been well documented[3152-3155]. The occurrence of biomethylation of arsenic in microorganisms[3156,3157], soil[3159-3161], animals[3162-3165] and humans[3166,3167] has been also demonstrated. Therefore, further investigation of the fate of arsenicals in the physical environment and living organisms requires analytical methods for the complete speciation of these arsenicals.

It has been shown that arsenic is incorporated into both marine and freshwater organisms in the form of water-soluble and lipid-soluble arsenic compounds[3168-3170]. Recent studies to identify the chemical forms of these arsenic compounds have shown the presence of arsenite (As(III), arsenate (As(V)), methylarsonic acid, dimethylarsinic acid, and arsenobetaine (AB) [3152,3171,3172]. Methylated arsenicals also appear in the urine and plasma of mammals, including man, by biotransformation of inorganic arsenic compounds[3173,3174,3166,3175,3163]. Several methods have been devised to characterize these arsenicals.

B. ORGANOMETALLIC COMPOUNDS IN WATER

Organomercury Compounds

Stability of Samples. May et al[3176] have carried out radiochemical studies using 203 Hg labelled compounds on the behaviour under different storage and working conditions, of mercuric chloride and methylmercury chloride in inland and sea water. In this investigation they used cold vapour atomic absorption spectroscopy. The application of ^{203}Hg unambiguously revealed that the loss of mercury observed upon storage of unacidified sea water samples in polyethylene bottles was due to adsorption and to the diffusion of metallic Hg(Hg^o) through the container wall.

For the chemical speciation of mercury compounds the storage time and the kind of storage are of paramount importance. After a three-day storage in brown glass bottles 47% of $^{203}HgCl$, added to the sea water became reduced to Hg^o, but a complete reduction of mercury (II) in the samples by stannous chloride was not achievable. The Hg (II) species not reduced by stannous chloride may be to a distinct extent iodides and sulfides. Within 35 days of storage $CH_3{}^{203}HgCl$ decomposed into Hg^o and Hg(II), 35% could be identified as Hg^o and 27% as reactive mercury. Strong solar radiation does not influence the transformation of ^{203}Hg(II) into Hg^o, but after a strong 3 day solar radiation of a sample containing $CH_3{}^{203}HgCl$ besides Hg^o and reactive Hg(II) also the formation of dimethylmercury could be observed. Thus, cooling and darkness during storage are important prerequisites for the subsequent differentiation of Hg-species. Experiments with sodium borohydride, as an alternative reducing agent in comparison to stannous chloride, showed a quantitative reduction of all Hg(II) species in river and sea water and even with $CH_3{}^{203}Hg$ gave a yield of nearly 80%.

Olsen[3177] has studied the disappearance of methylmercury and the lability of the methylmercury bond by volatilisation during incubation for 96 hours in deionized water, tap water, fish tank waters and various synthetic media including sodium chloride solution and synthetic sea water. These studies were performed on sub micromolar amounts of $^{14}CH_3HgCl$ and $CH_3{}^{203}HgCl$ and the concentrations variations of these occurring during incubation were evaluated by liquid scintillation counting.

The most significant finding of this study was the failure of deionized water to lose substantial quantities of methylmercury during the 96 h experimental period. The percent of ^{203}Hg lost from deionized water is at most attributable primarily to the relatively large loss at the onset of the experiment as no more than 2 to 3% of the isotope is lost thereafter.

Loss of mercury from tap water and aquarium (fish) water was quantitatively and chronologically similar to values reported by Burrows and Krenkel[3178,3179] and confirm their findings. Less than 5% of either was lost from New Brunswick tap water and, with the exception of an initial 10% decrease in ^{203}Hg levels in one experiment, neither mercurial was lost from fish water during the 96 h experimental period.

Inorganic mercury appeared to be lost from the basic medium (1% NaOH) as 10 to 14% of the mercury-203 was lost without any corresponding loss of carbon-14. The short time course for this loss (within 8 h) might reflect, in part, loss of inorganic contaminants. Quantitatively similar losses of mercury-203 (6 to 14%) from the acid media were also observed.In experiments where ethanol had been added to the water, there was a concomitant 14% loss of carbon-14 and the concentration of both mercurials decreased slowly during the entire experiment suggesting some CH_3-Hg cleavage. The solution chemistry for methylmercuric salts at various pH's can involve formation of $(CH_3Hg)_2OH^+$, $(CH_3Hg)_3O^+$,, CH_3HgOH, $CH_3HgOH_2{}^+$ and possibly others; thus, ethanol could increase volatilization of one of these species within a specific pH range. The similarily, however, between the ^{203}Hg lost in both acidic and basic media suggests that the disappearance of mercury is due to loss of mercury in the inorganic form. This inorganic mercury could be an intrinsic contaminant or from CH_3-Hg bond cleavage in the incubation vessel. If the methyl group is split from mercury and oxidized to carbon dioxide, it would then be lost from an acid medium and retained in the basic medium as bicarbonate, whereas if methane was formed it would be lost from both acid and basic environments.

Phosphate buffered saline lost ^{203}Hg both in the presence and absence of ethanol, 6 and 9% respectively, whilst only in the presence of ethanol did any loss (11%) of ^{14}C occur. The ^{203}Hg loss was fairly rapid in the first 6 - 10 h of both experiments and insignificant thereafter.

The largest and most sustained mercury-203 and carbon-14 losses were from the high salt environments, 0.5 M sodium chloride and sea water. In both media the loss was continuous over the 96 h period and relatively more ^{203}Hg was lost (20% ethanol absent, 10% ethanol present) than ^{14}C (16% ethanol absent, 0% ethanol present) in sodium chloride solutions. Continued loss of label from sodium chloride and sea water could be explained by either a salting out effect increasing the volatilization of mercury or by chemical interactions which either increase volatilization or, more likely, precipitate the mercury or increase the adsorption/absorption process.

The time course for mercury loss is interesting, in that many solutions lost mercury over the first few hours while only a few showed a prolonged disappearance rate. This would suggest that the early loss could be due to either: (1) Precipitation or volatilization of a small amount of the stock mercury not in the methylated form, (2) adsorption and/or absorption of methylmercury onto the glass of the staining dishes or aerator tubing, or (3) combination and/or breakdown of methylmercury with impurities found in trace amounts in the solutions used, producing a more volatile reaction product.

Organomercury Compounds in Natural Waters

The two main techniques that have been used for the determination of mercury in natural water samples are atomic absorption spectrophotometry and gas chromatography.

Atomic Absorption Spectroscopy. Kalb[3180] used concentrated nitric acid to decompose organomercury compounds in water samples prior to estimation by flameless atomic absorption spectroscopy. Stannous chloride was used to liberate elementary mercury, which is then vaporised by passing a stream of air (1360 ml per min) through the solution. The air stream passes over silver foil, where mercury is retained by amalgamation and other volatile substances pass out of the system. The foil is heated at 350°C in an induction coil, and the air stream carries the mercury vapour through a cell with quartz windows. The atomic absorption at 253.65 nm is measured, and the mercury concentration (up to 0.02 p.p.m.) is determined by reference to a calibration graph.

Other workers who have made earlier contributions to the determination of organically bound mercury compounds and inorganic mercury compounds by flameless atomic absorption spectroscopy include Baltisherger and Knudson 3181, Bisagni and Lawrence[3182], Frimmel and Winckler[3183], Umezaki and Iwanoto[3184], Chau and Saitoh[3185], Stainton[3186], Carr et al[3187], Fitzgerald et al[3188], Watling and Watling[3189], who carried out a literature survey on this subject (34 references). Chart and Boster[3190], Pierce et al[3192], Magos and Clarkson[3192], Garey et al[3193], Workers at the National Institute of Drug Abuse[3194], Done et al[3195] and Misra et al[3196].

Doherty and Dorsett[3197] analysed environmental water samples by separating the total organic and inorganic mercury by electrodeposition for 60 to 90 min on a copper coil in 0.1 M nitric acid medium and then determined it directly by flameless atomic-absorption spectrophotometry[3198, 3199]. The precision and accuracy are within ± 10% for the range 0.1 to 10 parts per 10^9. The sensitivity is 0.1 part per 10^9 (50 ml sample).

The Water Research Council[3200] has described a method for the determination of mercury in nonsaline waters and effluents. This method is not described here in detail as it is readily available. All forms of mercury in non-saline waters, effluents and sludges are converted to inorganic mercury using prolonged oxidation with potassium permanganate[3201]. Solid samples require a more prolonged and vigorous oxidation to bring the mercury completely into solution in the inorganic form using a modification of the Uthe digestion procedure[3202]. The inorganic mercury is determined by the flameless atomic absorption spectrophotometric technique using a method similar to that described by Osland[3203]. Acid stannous chloride is added to the sample to produce elemental mercury:

$$Hg^{2+} + Sn^{2+} \rightarrow Hg^{o} + Sn^{4+}$$

The mercury vapour is carried by a stream of air or nitrogen into a gas cuvette placed in the path of the radiation from a mercury hollow cathode lamp and the absorption of this radiation at 253.7 nm by the mercury vapour is measured (see Figure 174). Many of the potential interferences in the atomic absorption procedure are removed by the preliminary digestion/ oxidation procedure. The most significant group of interfering substances in volatile organic compounds which absorb radiation in the ultraviolet. Most of these are removed by the pretreatment procedure used and the effect of any that remain are overcome by pre-aeration. Substances which are reduced to the elemental state by stannous chloride and then form a stable compound with mercury may cause interference; e.g. selenium, gold, palladium and platinum. The effects of various anions, including bromide and iodide were studied. These are not likely to be important interferers. Excellent performance characteristics methods are presented for this method.

Sampling techniques are described in detail including methods of cleaning sample bottles and fixing the sample with a solution of potassium dichromate in nitric acid.

Simpson and Nickless[3204] have described a rapid dual channel method of cold-vapour atomic-absorption spectroscopy for determining mercury in natural waters. A detection limit of 12.5 ng per litre is claimed.

Abo-Rady[3205] has described a method for the determination of total inorganic plus organic mercury in nanogram quantities in water natural, also fish, plants and sediments. This method is based on decomposition of organic and inorganic mercury compounds with acid permanganate, removal of excess permanganate with hydroxylamine hydrochloride, reduction to metallic mercury with tin and hydrochloric acid, and transfer of the liberated mercury in a stream of air to the spectrometer. Mercury was determined by using a closed, recirculating air stream. Sensitivity and reproducibility of the "closed-system" were better, it is claimed, than those of the "open-system". The coefficient of variation was 13.7% for water, 1.9% for fish, 4.9% for plant and 5.6% for sediment samples.

Lutze[3206] has described a flameless atomic absorption method for deter-

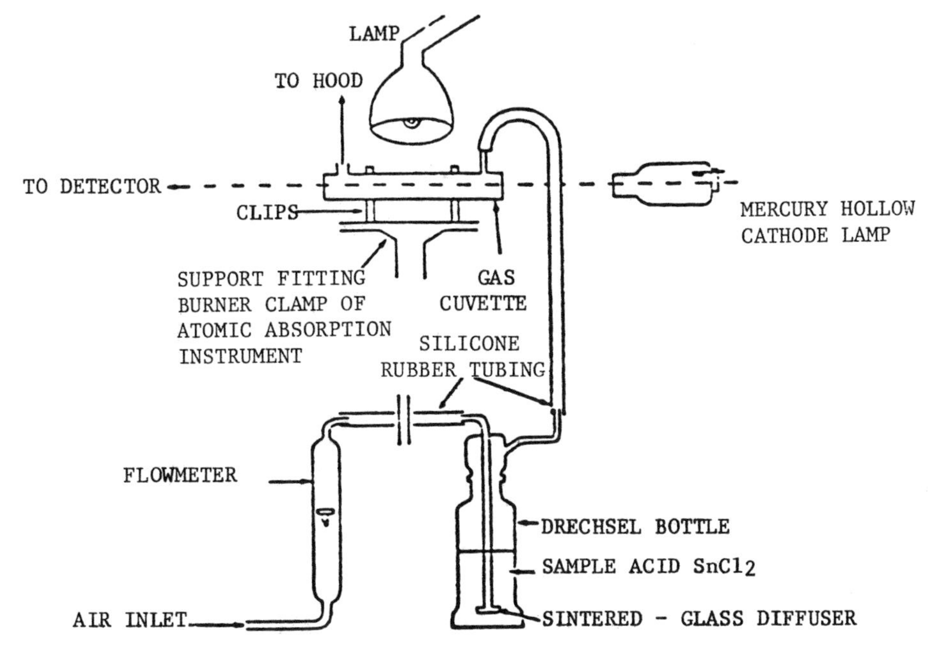

Figure 174 - Details of sample vaporizer.

mining mercury in surface waters and sediments and Grantham[3207] has applied the method to waters, soil, foodstuffs and effluents.

Graf et al[3208] used sulphuric acid acidified potassium permanganate to decompose organically bound mercury, prior to reduction with stannous chloride and determination of the evolved mercury by atomic absorption spectroscopy.

Goulden and Anthony[3209] described how the sensitivity of an automated cold-vapour atomic absorption method for mercury can be improved by equilibriating the reduced sample with a small volume of air at 90°C, to achieve a detection limit of 1 ng mercury per litre. Inorganic mercury, aryl mercury compounds, and alkyl mercury compounds, can be distinguished by changing the chemical reduction system.

Kiemenseij and Kloosterboer[3210] have described an improvement on the Goulden and Afghan[3211] photochemical decomposition of organomercury compounds in the ppb range in natural water prior to determination by cold vapour atomic absorption spectrophotometry. The decomposition of the organomercurials is carried out by means of ultraviolet radiation of a suitable wavelength from small low-pressure lamps containing either Hg, Cd or Zn or a mixture of these metals in their cathodes. These lamps have the strongest lines respectively at 254, 229 and 214 nm. The formed inorganic mercury is determined in the usual way by cold vapor atomic absorption after reduction of divalent mercury to mercury vapour. Determinations with and without irradiation make possible separate determination of total and inorganic mercury, respectively, in about 20 minutes.

The recovery of mercury from aqueous solutions of methylmercury-chloride as a function of the time of irradiation for a number of light sources was studied. The efficiency of destruction increases in the order Hg < Cd < Zn although the relative intensities of the lamps were in the opposite order, namely Hg:Cd:Zn = 6:2:1.

The same order Hg < Cd < Zn was observed in the case of diphenylmercury. This compound even has a slightly lower extinction coefficient at 214 nm (Zn) than at 229 nm (Cd). This shows that for both methylmercurychloride and diphenylmercury, the quantum efficiency for photodecomposition increases with decreasing wavelength, in accord with what might be expected. Both absorbance and quantum yield, therefore, favour the use of short wavelength light sources. In natural waters, on the other hand, the background absorption of dissolved organic compounds which increases towards the uv favours the use of the more intense long wavelength sources.

The cadmium lamp offers a good compromise. If the heating of the sample is not considered as disadvantageous, the more powerful combined Zn-Cd-Hg lamp offers the best results.

Comparison of the photochemical with a wet chemical method (Table 243) showed that the results of prolonged irradiation compare well with the results obtained after complete wet-chemical destruction.

Table 243 - Comparison of photochemical and wet chemical decomposition of mercury compounds in an acidified natural water sample (River Waal)

Sample treatment		Hg found, μg/l.	
Unirradiated		0.31	
Irradiated	10 min	1.01	1.00
(ZnCdHg lamp)	30 min	1.15	1.11
	30 min	1.05	1.12
Stored with $KMnO_4$	2% $KMnO_4$	0.98 (0.06)	1.02 (0.03)
	4% $KMnO_4$	1.04 (0.08)	0.97
	4% $KMnO_4$	1.07	1.01
Stored with $KMnO_4$	2% $KMnO_4$	1.00 (0.24)	1.06 (0.23)
partly evaporated and rediluted	4% $KMnO_4$	1.07 (0.27)	1.06 (0.31)

Farey et al[3212] have discussed ultraviolet photochemical systems for the decomposition of organomercury compounds prior to analysis by cold vapour atomic fluorescence spectroscopy. These workers compared the effectiveness of a bromination treatment for the liberation of mercury from organomercury compounds with a pre-treatment procedure involving oxidation with a permanganate-sulphuric acid mixture recommended by workers at the Water Research Centre, U.K.[3200]. The basis of the bromination technique is that a bromate-bromide reagent in hydrochloric acid reagent is used to generate bromine which quantitatively cleaves both alkyl and aryl mercury compounds to inorganic mercury bromide.

Recoveries of inorganic mercury from distilled water spiked with phenylmercury (II) chloride, thiomersal, ethylmercury (II) chloride, methylmercury (II) chloride, phenylmercury (II) acetate and p-tolymercury (II) chloride are greater than 95%.

When identical conditions of treatment were used on 50 ml samples of tap water and various river waters and sewage effluents, all with added methylmercury (II) chloride there were similar recoveries after 5 min, although the recoveries after a bromination period of 1 min were lower. However, when 2 ml of bromate - bromide solution were added, higher recoveries were obtained after 1 min, indicating that the increase in the amount of brominating reagent accelerated the reaction. The best recoveries were obtained after treatment for 5 min with either 1 or 2 ml of the brominating reagent.

Table 244 shows a comparison of results obtained on a sewage effluent sample by the above method and a method involving digestion with acid permanganate reagent developed by the Water Research Council[3200]. It can be seen that the inorganic mercury value obtained after bromination for 5 min is similar to that found following the acid permanganate pre-treatment. Moreover, the value obtained after a bromination period of 15 min was higher, showing a greater recovery of mercury in this instance.

Table 244 - Total mercury determined by atomic-fluorescence spectroscopy in Crossness sewage works final effluent.
Pre-treatment and analysis performed on four separate aliquots of the sample

	Mercury found $\mu g l^{-1}$	
Pre-treatment	Mean	Standard deviation
$KMnO_4$ - H_2SO_4 80°C, 8 h[4]	1.65	0.20
5 ml of HCl + 2 ml of $KBrO_3$ - KBr		
(i) 5 min	1.56	0.14
(ii) 15 min	1.91	0.13

Farey et al[3212] claim that their pre-treatment compares favourably with an established permanganate - sulphuric acid method. An advantage of the technique is that it can easily be carried out while sampling on-site. The sample is collected in glass bottles containing hydrochloric acid and the bromate - bromide solution is added. A bromination reaction time is then provided from the collection of the sample to the analysis in the laboratory and this is far in excess of that necessary to decompose the organic mercury. In addition as aqueous mercury (II) solutions are stabilised by strong oxidising agents, the oxidising conditions so created will help to preserve the inorganic mercury formed.

Cold-vapor atomic absorption spectrometry with a reduction-aeration technique has been widely applied to the determination of mercury in natural water samples[3213-3217]. Reproducible values could be obtained down to levels of 0.5 μg Hg 1^{-1} both in sea and fresh waters. This sensitivity, however, is not enough to monitor the background values of mercury in unpolluted areas. Consequently, a preconcentration step is necessary, which also separates the mercury from interfering substances. Fitzgerald et al[3217] reported a cold-trap preconcentration technique for the determination of trace amounts of mercury in water. Kramer and Neidhart[3218] determined ppb (μg 1^{-1}) levels of mercury by using an aniline-sulfur resin for the selective enrichment of mercury from surface waters. Chau and Saitoh[3213] reported a method for the determination of submicrogram amounts of mercury in lake water based on dithizone extraction. Preconcentration of mercury prior to the measurement has also been achieved by amalgamation with noble metals[3216,3219,3222]. The origin of organic mercury in fresh water ecology remains obscure in some respects, and there is a need to quantify mercury in both organic and inorganic forms in various environmental waters.

Minagarda et al[3224] describes the use of chelate resin preconcentration for the simultaneous determination of traces of inorganic and organic mercury in fresh water. This resin[3225] contains dithiocarbamate groups which bind strongly with mercury but not with alkali and alkaline earth metals. Both forms of mercury can be collected from pH 1 to 11. Collected mercury is readily eluted with a slightly acidic aqueous 5% thiourea solution. The resin can then be reused. This preconcentration method offers the advantages of using large volumes of water sample and of determining organic and inorganic mercury at ng 1^{-1} levels. The detection limit of the method is 0.2 ng 1^{-1} for both forms of mercury.

The apparatus used for preconcentration consisted of a column (15 mm i.d. 5 cm long) for the resin and a 20.1 high density polyethylene bottle as reservoir for the samples. The 20 - 50 mesh wet dithiocarbamate-treated resin was packed in the column. Each polyethylene bottle was cleaned by soaking in (1 + 9) nitric acid for 2 days and then rinsed thoroughly with distilled-deionized water before use.

The mercury vapor concentration meter, manufactured by Nippon Jarrell-Ash model AMD-F2, had a double-beam, dual-detector system. The inlet of the gas cell was connected by plastic tubing successively to a U-tube (10 x 200 mm, Pyrex glass) containing calcium chloride (6 - 20 mesh) as a water absorbent, a Quickfit 30-ml test-tube (as a reaction vessel) fitted with a Dreschsel bottle head with a sintered-glass (porosity 2) bubbler, and an air pump. When recordings were made the output of the mercury vapor concentration meter was connected to a stripchart recorder (SR 6201). The operating conditions were as follows: air flow rate, 1.01 min^{-1}; source, mercury lamp (253.7 nm); slit, 100 μm; gas cell, 20 x 200 mm quartz tube; scale expansion, x 10; recorder, full scale 10 m V, chart speed 5 mm min^{-1}.

River water and other fresh waters were sampled in a 20.1 high-density polyethylene bottle which was rinsed three times with the water sampled before the sample was taken. The sample was adjusted to pH 2 with concentrated nitric acid, 1 mg of $HAuCl_4$ being added as preservative[3226]. Samples of water should be analyzed within one week of collection to avoid losses of mercury by adsorption and vaporization.

For calibration, known concentrations of mercury standards and 10 ml 30% (w/v) potassium hydroxide were placed in the reaction vessel, and the volume was diluted to 20 ml with the aqueous 5% solution of thiorea. Air was passed through immediately after addition of reductant as described below, and calibration curves were obtained from peak absorption measurements. The resin column was not used in the calibration.

A 20-1 polyethylene sample bottle containing the unfiltered fresh water was placed above the chelating resin column, and connected to it with polyethylene tubing. The sample was allowed to flow through the column at ca. 30 ml min^{-1}. Cleaned resin was used for each experiment.

After the collection, 30 ml of the acidic 5% thiourea solution served to elute total and inorganic mercury. Inorganic and total mercury were determined separately in two 10 ml aliquots of this effluent.

For the determination of inorganic mercury, the 10 ml aliquot of well-mixed effluent was placed in the reaction vessel, 10 ml of 30% (w/v) potassium hydroxide was added followed by 2 ml of the tin (II) chloride solution, and the air flow was started immediately. This mixture was allowed to react for 30 s, during which time the mercury vapor generated passed through the quartz gas cell. Peak height was used for measurements.

For the determination of total mercury, the same procedure was used, except for reduction of the sample, and with the tin(II) chloride - cadmium chloride mixture (10% - 1%), instead of tin (II) chloride alone. The peak heights were again measured. The total mercury minus the inorganic mercury gives an estimate of the organic mercury.

The total blank was determined by carrying out the complete procedure of analysis with 201 of distilled-deionized water. Five replicate measurements gave mean blanks of 0.07 ± 0.05 ng l^{-1} and 0.15 ± 0.06 ng l^{-1} for inorganic and total mercury, respectively.

The precision of the method was evaluated by five analyses of a river water sample. The average mercury contents and the standard deviations were 9.5 ± 0.43 ng l^{-1} and 15.2 ± 0.36 ng l^{-1} for inorganic and organic mercury, respectively.

The chelate resin preconcentration technique was used to measure inorganic and organic mercury in natural waters of the Akita area of Japan. The samples were not filtered so that the concentrations reported include the contribution of acid-leachable mercury associated with particulate matter. Some of this mercury, solubilized by acid treatment, may be in an organic form, and is therefore included in the total mercury measurements. Thus, the particulate matter was not analyzed for mercury. Figure 175 shows some typical responses for standards, a sample and a blank.

The concentrations of inorganic, organic and total mercury found in this study are given in Table 245. Significantly, 35 - 60% of the mercury present in river and lake waters exists as organic compounds or in association with organic matter. About 30 to 60% of the mercury was in this form. The identity of the chemical species making up the organic mercury portion was not established.

Table 245 - Mercury fractions of river, lake waters in the Akita area of Japan (samples were collected in June and July 1978)

Location and type	No, of samples	Mean Hg concentration (ng l^{-1})		
		Total	Inorganic	Organic
Ormono river	3	15.5	8.8	6.7
Taihei river	5	21.1	9.0	12.1
Tazawa lake	3	12.4	6.5	5.9
	2	13.8	8.8	5.5

The calibration graphs obtained for Hg^{2+} and CH_3Hg^+ were identical and are linear over the range 0 - 12 ng.

Possible interferences are other ions, amino acids and naturally-occurring chelating agents which could affect the preconcentration, desorption and reduction steps. No interference was produced in the determination of 0.1 µg of mercury (II) by the presence of at least 1000 µg of each of

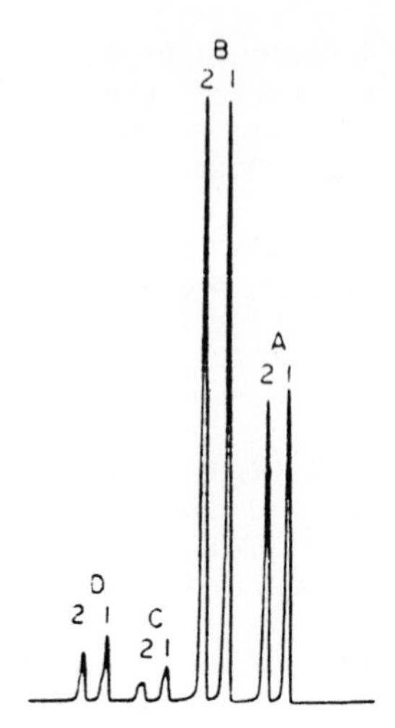

Figure 175 - Resin chromatography of methylmercury compounds (A) inorganic, (B) total.

the following ions or substances added to 5 - 1 aliquots of river water: Cr^{3+}, Mg^{2+}, Na^{+}, K^{+}, Ca^{2+}, Ni^{2+}, Cu^{2+}, Pb^{2+}, Cd^{2+}, Au^{3+}, Fe^{3+}, Al^{3+}, Zn^{2+}, PO_4^{3-}, Cl^{-}, CO_3^{2-}, NO_3^{-}, SO_4^{2-}, silicate, cysteine and humic acid.

The accuracy of the method was tested by analyzing river water samples spiked with known amounts of Hg^{2+} and CH_3Hg^{+}. As shown in Table 246, the accuracy of the method was satisfactory.

Table 246 - Results for river water spiked with mercury (II) chloride and methylmercury chloride[a]

Hg species	Hg added	Hg found	Recovery
	(ng)	(ng)[b]	(%)
Hg^{2+}	5	4.6	92
CH_3Hg^{+}	5	4.6	91
Hg^{2+}	10	9.1	91
CH_3Hg^{+}	10	9.3	93
Hg^{2+}	20	18.6	93
CH_2Hg^{+}	20	18.1	91
Hg^{2+}	100	95	95
CH_3Hg^{+}	100	91	91

[a] Each spike was added to 20 l of sample water
[b] Each result represents the mean of three values after subtraction of the blank.

Minagawa et al[3224] used this method to estimate inorganic and organic mercury in unfiltered Japanese rain water samples; (Table 247). Only about 6% of the mercury present in rain water exists as organic compounds or in association with organic matter.

Yamagami et al[3227] also applied chelating resins (dithiocarbamate type) to the determination of ppb of mercury in water. The samples are adjusted to pH 2.3 and passed through a column packed with 5 g of the resin, at a flow

Table 247 - Mercury fractions of rain waters in the Akita area of Japan (samples were collected in June and July 1978)

Location and type	No. of samples	Mean Hg concentration ($ng\ l^{-1}$)		
		Total	Inorganic	Organic
Akita city	2	10.8	10.1	0.7
rain water	3	15.2	14.3	0.9

rate of 50 ml per min. The resin is then heated under reflux with concentrated nitric acid, and the mercury is determined by atomic absorption spectrophotometry, using the reduction-aeration technique. The method is relatively simple and inexpensive and the detection limit is 10 ng mercury in water samples as large as 10 litres.

Gas Chromatography. This technique has found applications in the determination of organomercury compounds in water and trade effluents.

Nishi and Horimoto[3228,3229] determined trace amounts of methyl, ethyl, and phenyl mercury compounds in river waters and industrial effluents. In this procedure, the organomercury compound present at less than 0.4 ng per litre in the sample (100 - 500 ml) is extracted with benzene (2 x 0.5 vol. relative to that of the aqueous solution). The benzene layer is then back-extracted with 0.1% L-cysteine solution (5 ml) and recovered from the complex by extracting with benzene (1 ml) in the presence of hydrochloric acid (2 ml) and submitted to gas chromatography using a stainless-steel column (197 cm x 3 mm) packed with 5% of diethylene glycol succinate on Chromosorb W (60 - 80 mesh) with nitrogen as carrier gas (60 ml per minute) and an electron-capture detector. The calibration graph is rectilinear for less than 0.1 µg of mercury compound per ml of the cysteine solution. This method is capable of determining mercury down to 0.4 µg per litre for the methyl and ethyl derivatives and 0.86 µg per litre for the phenyl derivative.

The above method has been modified[3229] for the determination of methylmercury (II) compounds in aqueous media containing sulphur compounds that affect the extractions of mercury. The modified method is capable of handling samples containing up to 100 mg of various organic and inorganic sulphur compounds per 100 ml. The aqueous test solution (150 ml) containing 100 mg of methylmercury ions per 100 ml is treated with hydrochloric acid until the acid concentration is 0.4%, then 0.3 to 1 g of mercuric chloride is added (to displace methylmercury groups bonded to sulphur), and the mixture is filtered. The filtrate is treated with aqueous ammonia in excess to precipitate the unconsumed inorganic mercury which is filtered off; this filtrate is made 0.4% in hydrochloric acid and extracted with benzene. The benzene solution is shaken with 0.1% L-cysteine solution, the aqueous phase is acidified with the concentrated hydrochloric acid and then shaken with benzene for 5 minutes and this benzene solution is analysed by gas chromatography as described above.

Ealy et al[3230] have discussed the determination of methyl, ethyl and methoxymercury IIhalides in water and fish. The mercury compounds were separated from the samples by leaching with M-sodium iodide for 24 hours and then the alkylmercury iodides were extracted into benzene. These iodides were then determined by gas chromatography of the benzene extract on a glass column packed with 5% of cyclohexane - succinate on Anakron ABS (70 - 80 mesh) and operated at 200°C with nitrogen (56 ml/min) as carrier gas and electron capture detection. Good separation of chromatographic peaks were obtained for the mercury compounds as either chlorides, bromides or iodides. The extraction recoveries were monitored by the use of alkyl mercury compounds, labelled with 208 Hg.

Cappon and Crispin Smith[3231] have described a method for the extraction, clean-up and gas chromatographic determination of organic (alkyl- and aryl-) and inorganic mercury in biological materials, Methyl-, ethyl-, and phenylmercury are first extracted as the chloride derivatives. Inorganic mercury is then isolated as methylmercury upon reaction with tetramethyltin. The initial extracts are subjected to thiosulfate clean-up, and the organomercury species are isolated as the bromide derivatives. Total mercury recovery ranges between 75 and 90% for both forms of mercury, and is assessed by using appropriate ^{203}Hg-labelled compounds for liquid scintillation spectrometric assay. Specific gas chromatographic conditions allow detection of mercury concentrations of 1 ppb or lower. Mean deviation and relative accuracy average 3.2 and 2.2%, respectively. These workers were concerned with the determination of different inorganic mercury and organomercury species (alkyl and aryl) in a variety of media including water river sediments and fish.

Another application of gas chromatography to natural water analysis is that of Longbottom[3232] who uses a Coleman 50 Mercury Analyser System, as a detector. A mixture of dimethyl, diethyl-, dipropyl- and dibutyl-mercury (1 ng of each) was separated on a 6 ft column packed with 5% of DC-200 and 3% of QF-1 on Gas-Chrom Q and temperature programmed from 60° to 180° at 20° per min. The mercury detector system was used after the column effluent had passed through a flame ionization detector; the heights of the resulting four peaks were related to the percentages of mercury in the compounds.

Dressman[3233] also used the Colman 50 system in his determination of dialkylmercury compound in river waters. These compounds were separated in a glass column (1.86 metres x 2 mm) packed with 5% of DC-200 plus 3% of QF-1 on Gas-Chrom Q (80 to 100 mesh) and temperature programmed from 70° to 180° at 20° per min., with nitrogen as carrier gas (50 ml per min). The mercury compound eluted from the column was burnt in a flame ionisation detector, and the resulting free mercury was detected by a Colman Mercury Analyzer MAS-50 connected to the exit of the flame ionisation instrument; down to 0.1 ng of mercury could be detected. River water (1 litre) was extracted with pentane-ethyl ether (4:1) (2 x 60 ml). The extract was dried over sodium sulphate evaporated to 5 ml and analysed as above.

Zarnegar and Mushak[3234] treated the water sample with an alkylating or arylating reagent. The benzene extract was examined by electron capture gas chromatography. The best alkylating or arylating reagents were pentacyano (methyl) cobaltate[111] and tetraphenylborate. Inorganic and organic mercury could be determined sequentially by extracting and analysing two aliquots of sample, of which only one had been treated with alkylating reagent. The limits of detection were about 10 to 30 ng.

Mushak et al[3235] have described a gas chromatographic method for the determination of inorganic mercury in water, urine and serum. The inorganic mercury in the sample is reacted with lithium pentafluorobenzenesulphinate arylating reagent which converts inorganic mercury to arlymercury compounds. The arylmercury compounds as well as any other organomercury compounds present in the original sample are then determined by a technique based on that described by Westoo (see later) involving gas chromatography on columns of 10% of Dexsil-300 on Anakrom SD (70 - 80 mesh) and of Durapak Carbowax

400 on Porasil F (80 - 100 mesh). The recoveries and precision (standard deviations) were for water 70.5% (6.8), urine 81.4% (10.5) and serum 51.0% (9.4). The limit of detection of inorganic mercury achieved in this method was 20 ng per ml of sample.

Jones and Nickless[3236,3237] have devised methods for the determination of inorganic mercury based on conversion to an organo-mercury compound with arene sulphinites[3236] and formation of the trimethylsilyl derivatives[3237] both of which are ameneable to gas chromatography.

Miscellaneous Methods. Mercury in water samples can exist in inorganic or organic forms or both. The preliminary section is concerned with methods for the preliminary degradation of organomercury compounds in the sample to inorganic mercury preparatory to analysis. This is necessary because the normal methods of reducing inorganic mercury compounds to mercury with reagents such as stannous chloride do not work with organomercury compounds and hence organomercury compounds are not included in such determinations. Owing to the conversion of Hg^{2+} to CH_3Hg^+ in natural water and owing to the presence of mercury in a large number of organic pollutants, it is often observed that a high percentage of the mercury is present in the form of organic compounds. Some organic mercurials like CH_3HgCI and $(CH_3)_2$ Hg may be reduced by a combination of cadmous chloride and stannous chloride but this method requires large quantities of reductants and the use of strong acid and strong alkali[2328].

Organic mercury compounds can be decomposed by heating with strong oxidizing agents such as potassium dichromate or nitric acid - perchloric acid, followed by reduction of the formed divalent mercury to mercury vapour[3239,3240]. Both methods are rather time-consuming and not very suitable for automation. Potassium persulphate has also been used to aid the oxidation of organomercury compounds to inorganic mercury and this forms the basis of an automated method[3241].

Goulden and Afghan[3242] have used ultraviolet irradiation as a means of decomposition, following the original proposal of Armstrong, Williams and Strickland[3243]. After the photochemical oxidation, the formed inorganic mercury is reduced to metallic mercury in the usual way by stannous chloride. This method reduces the consumption of oxidizing agents and thus diminishes considerably the risk of contamination; it also leads to shorter analysis times. Determinations with and without irradiation enable the separate determination of total and inorganic mercury, respectively.

Bennett et al[3244] showed that acid-permanganate alone did not recover three methyl mercuric compounds, while the addition of a potassium persulfate oxidation step increased recoveries to 100%. El-Awady et al[3245] confirmed the low recoveries of methylmercury by acid-permanganate. They showed that only about 30% of methylmercury could be recovered by this method, while the use of potassium persulfate produced complete recovery.

Umezaki and Iwamoto[3246] differentiated between organic and inorganic mercury in river water samples. They used the reduction-aeration technique described by Kimura and Miller[3247]. By using stannous chloride solution in hydrochloric acid only inorganic mercury is reduced whereas stannous chloride in sodium hydroxide medium and in the presence of cupric copper reduces both organic and inorganic mercury. The mercury vapour is measured conventionally at 254 nm. Ions that form insoluble salts or stable complexes with Hg^{11} interfere.

Fitzgerald[3248] reported a coldtrap preconcentration technique for the determination of trace amounts of mercury in water. Kramer and Niedhart [3249] determined ppb ($\mu g l^{-1}$) levels of mercury by using an aniline-sulfur resin for the selective enrichment of mercury from surface waters. Chau and Saitoh[3250] reported a method for the determination of submicrogram amounts of mercury in lake water based on dithizone extraction. Preconcentration of mercury prior to the measurement has also been achieved by amalgamation with noble metals[3251-3255].

Radiochromatographic assay has been used[3256] as the basis of a method for determining inorganic mercury and methylmercury in river water.

Ke and Thibert[3257] have described a kinetic microdetermination of down to 0.05 µg per ml of inorganic and organic mercury in river water and sea water. Mercury is determined by use of the iodide - catalysed reaction between Ce^{IV} and As^{111} which is followed spectrophotometrically at 275 nm.

Van Ettekoven[3258] has described a direct semi-automatic scheme based on ultraviolet light absorption for the determination of total mercury in water and sewage sludge. The full determination time is about 10 min. The lower limit of detection of mercury in water is 0.03 µg per litre and 0.2 ppm (mg per kg dry matter) in sewage sludge.

Matsunaga et al[3259] have discussed possible errors caused prior to measurement of mercury in natural water and sea water.

Potentiometric titration with standard solutions of dithiooxamide at pH 5-6 has been used to estimate less than 100 µg mercury in water samples [3260]. The precision in the range 0.05 - 1.0 ppm mercury is about 4 per cent. The first derivative should be used for end-point determination. A wide variety of ions can be tolerated but silver, copper and chloride interfere, and must be separated in a preliminary step.

Braun et al[3261] showed that polyurethane foam loaded with diethylammonium diethyldithiocarbamate is suitable for concentration of trace amounts of organic mercury and inorganic mercury from potable water samples prior to analysis. Organomercury compounds studied included phenylmercury and methylmercury. The polyurethane discs were then analysed by X-ray fluorescence[3262] spectrometry.

Becknell et al[3263] first converted the organomercury to mercuric chloride using chlorine. The mercury was then concentrated by removal as $HgCl_4{}^{2-}$ by passing the sample solution (500 ml) adjusted to be 0.1 M in hydrochloric acid through a paper filter-disc loaded with SB-2 ion exchange resin. The paper, together with a mercury standard, is then heate-sealed in Mylar bags and irradiated for 2 hr in a thermal-neutron flux of $\sim$ 1.3 x 10^{13} neutrons per sq. cm per sec., after which the concentration of mercury in the sample is determined by the comparison methods using the 77 - keV x-ray photo-peak from the decay of ^{197}Hg.

Kudo et al[3264] measured the proportions of methylmercury to total mercury in river waters in Canada and Japan. In all cases methylmercury was found to represent about 30 per cent of the total mercury. A similar proportion was observed in sewage works effluent.

Glockling[3265] uses tables and graphs to discuss the applications of organometallic compounds, particularly organosilanes, organolead compounds, organomercurials, and organotin compounds, and their potential to cause pollution. A method is outlined for studying the degradation of a compound and determining the toxicity of its intermediates.

Organomercury Compounds in Potable Waters

The proposed World Health Organisation limit for total mercury in potable water has been set at 1 $\mu g\ l^{-1}$ [3266,3267]. A method capable of rapid and reliable measurement of mercury at levels of one tenth of this limit is required should this limit be adopted as a legal restriction for potable waters.

Stary and Prasilova[3268,3269] have described a very selective radiochemical determination of phenylmercury and methylmercury[3270,3271]. These analytical methods are based on the isotope exchange reactions with the excess of inorganic mercury-203 or on the exchange reactions between phenylmercury and methylmercury chloride in the organic phase and sodium iodide-131 in the aqueous phase. The sensitivity of the methods (0.5 p.p.b. in 5 ml sample) is not sufficient to determine organomercurials in natural waters. Subsequently, Stary et al[3272] developed a preconcentration-radioanalytical method for determining down to 0.01 p.p.m. of methyl and phenyl mercury and inorganic mercury using 100 - 500 ml samples of potable or river water. Extraction chromatography and dithizone extraction were the most promising methods for the concentration of organomercurials in the concentration range 0.01 to 2 p.p.b. The dithizone extraction method was used for the preconcentration of inorganic mercury.

Jackson and Dellar[3273] have described a photolysis - cold vapour flameless atomic absorption method for determining down to 0.01 $\mu g\ l^{-1}$ total mercury in potable and natural waters and sewage effluents. Photolysis converts organomercury compounds to the inorganic state and these workers showed the advantages of this approach over chemical oxidation methods for converting organic mercury to inorganic mercury.

Table 248 shows results obtained by this procedure for potable water samples spiked with 0.3 and 2 $\mu g\ l^{-1}$ mercury.

The reproducibilities for inorganic mercury and methylmercury are similar. At the 0.33 $\mu g\ l^{-1}$ level the 95% confidence interval is approximately 0.1 $\mu g\ l^{-1}$ and at the 2.00 $\mu g\ l^{-1}$ level it is approximately 0.30 $\mu g\ l^{-1}$. The limit of detection based on the variation of results when estimating low levels is 0.1 $\mu g\ l^{-1}$ for a single estimation. Hence levels of about 0.1 $\mu g\ l^{-1}$ may be estimated for 95% confidence intervals to the nearest 0.1 $\mu g\ l^{-1}$ up to 0.5 $\mu g\ l^{-1}$ and over this figure to the nearest 20%.

Jackson and Dellor[3273] emphasise that to obtain results for concentrations of mercury present in water at sampling time the container and water itself must be stabilized to avoid loss or gain of mercury. Acidic potassium dichromate is believed to be the best preservative (Feldman[3274]). Any particulate matter present in a water is likely to adsorb dissolved mercury. Before analysis a decision must be made whether total or dissolved mercury concentration is the figure required.

Atomic Absorption Spectroscopy. Oda and Ingle[3275] have described a procedure for the determination of organomercury and inorganic mercury in potable water. They describe a speciation scheme for ultratrace levels of mercury in which inorganic and organomercury are selectively reduced by stannous chloride and sodium borohydride respectively. The volatilized elemental mercury is determined by cold vapor atomic absorption. The detection limits for inorganic and organomercury species are in the 0.003 - 0.005 ppb range and both types of mercury can be determined in a 1 ml sample

Table 248 - Recovery (%) of mercury from 150 ml of sample by Photolytic/cold vapour AA method

Mercury present	Inorganic mercury (as nitrate)		Methylmercury (as chloride)		Phenylmercury (as acetate)	
	0.33 μg l^{-1}	2.00 μg l^{-1}	0.33 μg l^{-1}	2.00 μg l^{-1}	033 μg l^{-1}	2.00 μg l^{-1}
(a) Potable waters						
A Hardness 30 mg l^{-1}	107,106	102,99	101.85	99.98	100	91
B Hardness 150 mg l^{-1}	98,114	99,102	95,100	104,105	93	94
C Hardness 250 mg l^{-1}	111,96	99,94	100,97	97,101	100	95
D Hardness 358 mg l^{-1}	100,103	99,100	102,102	96,102	97	88
Mean recovery %	104	99	98	100	98	92
Reproducibility %	6.3	2.5	5.7	3.3	-	-
Reproducibility μg	0.003	0.007	0.003	0.010	-	-
Reproducibility μg l^{-1}	0.021	0.050	0.019	0.066	-	-
95% confidence interval ± μg l^{-1}	0.058	0.14	0.053	0.18		

in about 3 min. This procedure is much faster than other procedures because no time-consuming sample extraction, sample decomposition, or chromatographic separation steps are required. The detection limits are superior to other techniques because an optimized cold vapor atomic absorption apparatus is used. The accuracy of organomercury determination is better than in many procedures because the organomercury concentration is not determined by difference which is difficult if most of the total mercury is inorganic mercury.

The cold vapor atomic absorption apparatus has been described by Hawley and Ingle[3276], Christman et al[3277] and Ada and Ingle[3278].

Spiked and unspiked samples of tap water were tested with this speciation procedure (Table 249).

Table 249 - Analysis of water

Sample	amt of Hg(II), ppb		
	inorganic	organic	total
test solution 1[a]	1.03	1.00	2.03
tapwater	0.006	0.003	0.009
tapwater	0.005	0.003	0.008
tapwater, spiked[b]	0.035	0.034	0.069
tapwater, spiked[b]	0.033	0.032	0.065
test solution 2[c]	1.02	0.034	1.05
test solution 3[d]	0.029	0.99	1.02

[a] 1.0 ppb HgCl, and CH_3HgCl carried through KOH digestion procedure.
[b] Spiked with 0.030 ppb $HgCl_2$ and 0.030 ppb CH_3HgCl.
[c] 1.0 ppb $HgCl_2$ and 0.030 ppb CH_3HgCl
[d] 0.030 ppb $HgCl_2$ and 1.0 ppb CH_3HgCl.

As shown by the test solution in Table 249, there was no apparent breakdown of CH_3Hg^+ in the procedure.

The tapwater was tested, untreated, as it flowed out of the faucet which was left running for a few minutes before sampling. Parts-per-trillion levels of each species of mercury were evident. The upper limit for mercury in water is 2 ppb as established by the Safe Drinking Water Act [3279]. The calculated concentrations of mercury in the tapwater are near or at the detection limit with relative standard deviations of 19 - 34%. The 0.030 ppb spikes of inorganic - organomercury mixtures resulted in quantitative recovery of the mercury in tapwater.

The effect of a large excess (factor of 33) of one mercury species in solution with respect to the other was examined by running mixtures of 0.03 ppb Hg^{2+} and 1.00 ppb CH_3Hg^+, and 1.0 ppb Hg^{2+} and 0.03 ppb CH_3Hg^+. A factor of 10 scale expansion was used for 0.03 ppb of either mercury species to ensure a reasonable pen deflection for the lower mercury concentrations. The results obtained with synthetic test samples 2 and 3 in Table 249 showed quantitative recovery within experimental error (± 10%) of both species in each case. Relative standard deviations were 2 and 8% for 1 ppb Hg^{2+} and CH_3Hg^+, respectively, and 11 and 16% for 0.030 ppb Hg^{2+} and CH_3Hg^+, respectively. This indicates that a 33-fold excess of one mercury species does not interfere with the determination of the other mercury species. Determination of organomercury when inorganic mercury is in a 33-fold excess would be very difficult by procedures in which organomercury is found by the difference between total and inorganic mercury.

The calibration sensitivities and detection limits achieved for inorganic and organomercury species are as follows: inorganic mercury (0.031 AU/ppb, 0.003 ppb), CH_3HgCl and CH_3CH_2HgCl (0.026 AU/ppb, 0.003 ppb), and C_6H_5HgCl (0.015 AU/ppb, 0.004 ppb). The calibration curves were linear from the detection limit to at least 3 ppb (no significant deviation from linearit was obvious). The detection limit is defined as the concentration yielding a peak absorbance twice the standard deviation of the blank which was calculated as one-fifth of the peak-to-peak base line noise (typically (2 - 4) $\times 10^{-4}$ AU).

The methyl- and ethylmercury slopes are about 84% that of the inorganic mercury slope while the phenylmercuric chloride slope is about 50% of the inorganic mercury slope. Thus a separate calibration curve must be prepared for each mercury species or the inorganic mercury calibration slope can be multiplied by a suitable calibration factor.

Oda and Ingle[3275] carried out preservation study of the breakdown of a 1.0 ppb CH_3HgCl solution caused by 1.0% (v/v) nitric acid alone, 0.01% (w/v) potassium dichromate alone, and a mixture of 1.0% (v/v) nitric acid and 0.1% (w/v) potassium dichromate which are used as preservatives for total mercury was carried out. Measurements of inorganic and methylmercury content using the speciation procedure developed were made within hours of preparation and after 1,3, and 8 days of standing in 100 mL glass volumetric flasks at room temperature. The results were compared to those obtained with an unpreserved 1 ppb CH_3HgCl solution. It has been previously found that 0.01% (w/v) $K_2Cr_2O_7$ and 5.0% (v/v) nitric acid were most effective in preventing loss in Hg^{2+} solutions at parts per billion concentrations. For these studies, the concentration of nitric acid was reduced to 1.0% to minimize decomposition. However, even with the lowered acid concentration, about 20% of the methylmercury was observed to be converted to inorganic mercury (the form easily reducible by stannous chloride) under these conditions in slightly over a day. The total amount of mercury (inorganic and organic) in solution remained fairly constant over a 3-day period with an approximate 25 ± 8% loss over a period of 8 days. Comparison of decomposition induced by 0.01% potassium dichromate alone and 1.0% nitric acid alone to that caused by the combination of the two reagents indicates that the major factor appears to be the presence of nitric acid. Nitric acid alone converts almost half of the methylmercury to mercuric ions in just about 3 days and losses in terms of total mercury from the solution amount to about 26 ± 5% over 8 days. The potassium dichromate is not nearly as destructive although about 15% of the CH_3Hg^+ is decomposed in 3 days while maintaining more than 90% effectiveness in retaining total mercury for more than a week.

It should be noted that in preparing any of the test solutions, the acid or dichromate was diluted to 50 - 75 ml with water before addition of the CH_3HgCl solution so that the organomercury compound was never in direct contact with the concentrated preservation reagent. The unpreserved methylmercury retained its concentration remarkably well over a 3-day period and as expected a minimum of methylmercury breakdown was observed over that time. However, losses of about 33 ± 12% of the total mercury concentration after 8 days was noted, again as expected since no preservatives were present. The presence of inorganic mercury at the end of the test period may be partially attributed to photon-induced decomposition since methylmercury is somewhat sensitive to light.

This study indicates that where speciation of mercury is the primary objective, the use of nitric acid should be avoided to minimize decomposition or if added, the analysis must be run as soon as possible, preferably within hours of the addition.

However, the extended periods of preservation, acid and dichromate should be used and a determination of total mercury can be obtained with a fair amount of accuracy since losses would be minimized. However, the original speciation information for the sample is no longer determinable unless analysis is carried out immediately after preservation.

Organomercury compounds in sea and coastal water. Fitzgerald and Lyons 3280 have described flameless atomic absorption methods for determining organic mercury compounds respectively in coastal and sea waters. Fitzgerald and Lyons[3289] used ultraviolet light in the presence of nitric acid to decompose the organomercury compounds. In this method two sets of 100 ml samples of natural water are collected in glass bottles and then adjusted to pH 1.0 with nitric acid. One set of samples is analysed directly to give inorganically bound mercury, the other set is photo-oxidised by means of ultra-violet radiation for the destruction of organic material and then analyzed to give total mercury. The element is determined by a flameless-atomic absorption technique, after having been collected on a column of 1.5% of OV-17 and 1.95% of QF-1 on Chromosorb W-HP (80 - 100 mesh), cooled in a liquid-nitrogen bath and then released by heating the column, The precision of analysis is 15%. It was found that up to about 50% of the mercury present in river and coastal waters was organically bound or associated with organic matter.

Agemian and Chau[3281] have described an automated method for the determination of total dissolved mercury in fresh and saline waters by ultraviolet digestion and cold vapour atomic absorption spectroscopy. A flow-through ultra-violet digester is used to carry out photo-oxidation in the automated cold vapour atomic adsorption spectrometric system. This removes the chloride interference. Work was carried out to check the ability of the technique to degrade seven particular organomercury compounds. The precision of the method at levels of 0.07 μg/l, 0.23 μg/l, and 0.55 μg/l mercury was ± 6.0%, ± 3.8% and ± 1.0% respectively, The detection limit of the system is 0.02 μg/l.

Agemian and Chan[3282] showed that organomercurials could be decomposed by ultraviolet radiation and that the rate of decomposition of organomercurials increased rapidly in the presence of sulfuric acid and with increased surface area of the ultraviolet irradiation. They developed a flow-through ultraviolet digestor which had a delay time of 3 min to carry out the photooxidation in the automated system. The ultraviolet radiation has no effect on chloride. The method, therefore, can be applied to both fresh and saline waters without the chloride interference.

Millward and Bihan[3282] studied the effect of humic material on the determination of mercury by flameless atomic absorption spectrometry. In both sea and fresh water association between inorganic and organic entities takes place within 90 min at pH values of 7 or above, and the organically-bound mercury was not detected by an analytical method designed for inorganic mercury. The amount of detectable mercury was related to the amount of humic material added to the solutions. However, total mercury could be measured after exposure to ultraviolet radiation under strongly acid conditions.

Table 250 (column e) shows that complete recovery of seven organomercurials is obtained by using the ultraviolet digestor in the presence of sulfuric acid. The systems provide similar recoveries throughout the working range of the calibration curve. The table shows that recoveries by ultraviolet oxidation and sulphuric acid (column e) are complete and comparable to the permanganate-persulfate oxidation method (column e). Under the conditions of analysis, the ultraviolet radiation has no effect on chloride so that chloride ion behaves as an inert constituent of the sample. Analysis of synthetic mercury solutions of the seven compounds in distilled water and synthetic seawater (about 3% w/v sodium chloride) gave similar recoveries, proving that there was no chloride interference.

Levels of sulphide up to 100 μg l^{-1} as S^{2-} did not when the uv digestor was used have any interference effect on mercury determinations, whilst when the digestion was not used a level of 1 μg l^{-1} S^{1-} reduced mercury recoveries down to 50%.

Table 250 - Recoveries of organomercury compounds for different oxidation methods in the automated system[a]

5 μg/l of Hg as organic compound	Method (% recovery = 5)[b] a	b	c	d	e
1. Phenylmercuric acetate	74	80	95	61	102
2. Phenylmercuric nitrate	73	75	95	71	98
3. Diphenylmercury	65	92	84	100	91
4. Dithylmercuric chloride	41	46	89	46	98
5. Ethylmercuric chloride	81	88	88	98	95
6. Methoxyethylmercuric chloride	70	75	94	96	93
7. Ethoxyethylmercuric chloride	85	91	91	85	95

a, H_2SO_4; b, H_2SO_4 + 4% (w/v)$K_2Cr_2O_7$; c,H_2SO_4 + 0.5% (w/v)$KMnO_4$ + 0.5% (w/v) $K_2S_2O_8$; d, UV oxidation e, H_2SO_4 + UV oxidation.

An official method, developed in the U.K.[3283] is suitable for determining in saline, sea and estuarine waters of dissolved inorganic mercury and those organomercury compounds which form dithizonates. In this method inorganic mercury is extracted from the acidified saline water as its dithizonate into carbon tetrachloride. Organomercury compounds may also be extracted by the carbon tetrachloride, but not all these compounds form dithizonates and those which do not may not be determined by this method. In general organo-mercury compounds of the type R-Hg-X in which X is a simple anion form dithizonates, whereas the type R_1-Hg-R_2 does not. Monomethyl mercury ion is extracted though it only appears to have a transient existence in aerobic saline water. The dithizonates are decomposed by the addition of hydrochloric acid and sodium nitrite and the mercury or organomercury compound returned to the aqueous phase. Some organomercury compounds may not be completely re-extracted into the aqueous phase. The mercury in this aqueous phase is determined by the stannous chloride reduction - atomic absorption spectroscopic technique described earlier. The method is based on that used at the Department of Oceanography, University of Liverpool[3284]. This method has a range of application up to 100 μg/l. Standard deviations are 1.3 at the 0.0 ug/l level and 1.15 at the 5.0 mg/l level. The detection limit is 4 ug/l. No perceptable interference was encountered from ions normally present in estuarine and sea water.

The combined effect of the commonly present ions in estuarine and sea waters at the concentrations normally encountered in these waters is less than 1 ng/l at a mercury concentration of 30 ng/l.

Yamamato et al[3285] evolved a technique amalgamation of methylmercury in sea water onto gold followed by atomic absorption spectrophotometry for the determination of picogram quantities of the organomercury compound. Samples of sea water, ground water and river water collected in Japan are analysed for methylmercury and total mercury. Methylmercury is extracted with benzene and concentrated by a succession of three partitions between benzene and cysteine solution. Total mercury is extracted by wet combustion of the sample with sulphuric acid and potassium permanganate. The proportion of methylmercury to total mercury in the coastal sea water sampled was around 1 per cent.

Graphite furnace atomic absorption spectrophtometry has also been applied to the determination of trace levels of divalent mercury in inorganic and organomercury compounds in sea water. Filippelli[3286] has described a technique in which mercury is first preconcentrated using the ammonium tetramethylenedithiocarbamate (ammonium pyrrolidinedithiocarbamate) - chloroform system and then determined by graphite furnace atomic - absorption spectrometry. The technique is capable of detecting mercury (ID in the range 5 - 1500 ng in 2.5 ml of chloroform extract and can be adapted to detect subnanogram levels.

Gas chromatography - helium microwave induced plasma emission spectrometry. The atmospheric pressure helium microwave - induced plasma emission spectrometry has been used as an element - selective detector for gas chromatography of organomercury[3287,3288] compounds in sea water. Chiba et al[3289,3290] used atmospheric pressure helium microwave - induced plasma emission spectrometry with the cold vapor generation technique combined with gas chromatography for the determination of methylmercury chloride, ethylmercury chloride and dimethylmercury in sea water following a 500 - fold preconcentration using a benzene cysteine extraction technique. The analysis system[3291,3292] consisted of a Shimadzu GC-6A gas chromatograph a chemically deactivated four-way valve for solvent ventilation, a heated transfer tube interface, a Beenakker-type TM_{010} microwave resonance cavity, and an Ebert-type monochromator (0.5 m focal length).

The dual column gas chromatograph was equipped with a thermal conductivity detector. The interface between the gas chromatograph and the discharge tube of the microwave induced plasma detector is constructed from a chemically deactivated four-way valve and a heated transfer tube[3291.3292]. The gas chromatographic columns and optimum operating conditions are summarized in Table 251. As is seen in Table 251, the DEGS column was used for the measurement of methylmercury chloride and ethylmercury chloride, and the OV-17 column for that of dimethylmercury. The former column was treated, with dimethylsilane and potassium bromide in order to deactivate the surface.

The TM_{010} microwave cavity is constructed from pure copper metal. The microwave generator, which provides 20 - 200 W of microwave power at 2.45 GHz, is run at 75 W forward power. The width and the height of both monochromator entrance and exit slits are 10 μm and 1 mm, respectively. A photomultiplier tube with low dark current and high gain over a wide wavelength region is used as a detector. The measurement of mercury is carried out at the 253.7 nm mercury line.

Table 251 - Operating conditions for the gas chromatograph

	exptl conditions for determination of CH_3HgCl and C_2H_5HgCl	$(CH_3)_2Hg$
column	Pyrex, 1 m x 3 mm i.d.	Pyrex, 3 m x 3 mm id
column packing	15% DEGS[a] on 80/100 mesh Chromosorb W	3% OV-17 on 80/100 mesh Uniport HP
column temp, °C	160	70
injector temp, °C	180	130
detector oven temp, °C	180	130
transfer tube temp, °C	190	140
carrier gas	helium	helium
carrier gas flow rate, mL/min	80	80
detector	katharometer	katharometer

[a] DEGS = diethylene glycol succinate

The flow rate of carrier helium gas is adjusted at 80 mL/min for both columns, and then the plasma is ignited with a Tesia coil. About 30 min later, the plasma and the temperature of the gas chromatograph stabilize. The adjustments of wavelength and observation position in the plasma are performed as follows. A proper concentration of methylmercury chloride standard (1 - 2 µL) is injected very slowly into the OV-17 column, and a broad mercury peak appears. During the appearance of the peak, wavelength and observation position (both vertical and horizontal) are adjusted quickly. This procedure is repeated two or three times for exact wavelength and observation position adjustment. This adjustment has to be performed daily before the measurement.

The extraction procedure for the ultratrace levels of alkylmercury in seawater is presented in Figure 176. This technique was devised for 500-fold preconcentration of alkylmercury with reference to the Westoo extraction procedure.

The chromatograms were detected with the thermal conductivity and the microwave induced plasma detectors in series. When the microwave induced plasma was used as a detector, the emission signals were monitored at 253.7 nm, and the solvent was vented through a four-way valve before reaching the microwave induced plasma detector.

A typical chromatogram obtained under optimum conditions is shown in Figure 177. As can be seen from Figure 177 (lower curve), only the peaks of methylmercury and ethylmercury chlorides corresponding to 10 µg/L are observed with the microwave induced plasma detector (retention time of each compound was 260 s and 470 s, respectively). These peaks are adequately separated from each other. On the other hand, in the chromatogram obtained with the thermal conductivity detector, only a peak corresponding to the benzene solvent was observed, while the peaks for the mercury compounds could not be detected.

The column packed with OV-17 used for the measurement of dimethylmercury was optimized in a similar manner in terms of carrier gas flow rate and column temperature.

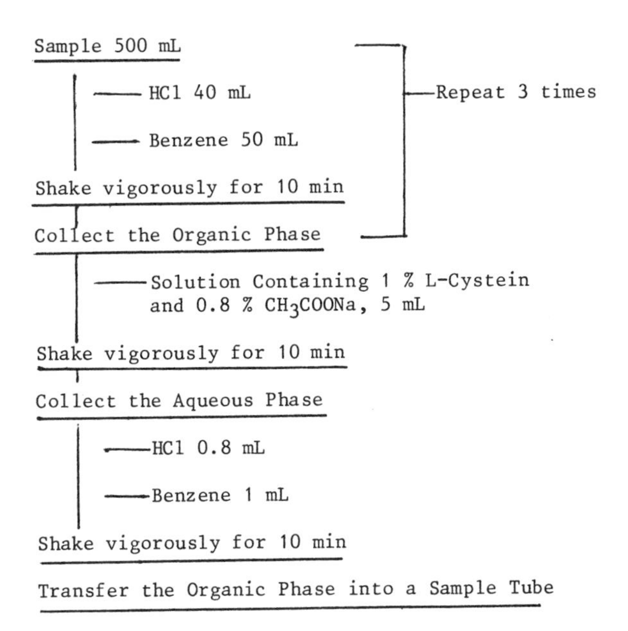

Figure 176 - Extraction procedure for alkylmercury compounds in seawater sample.

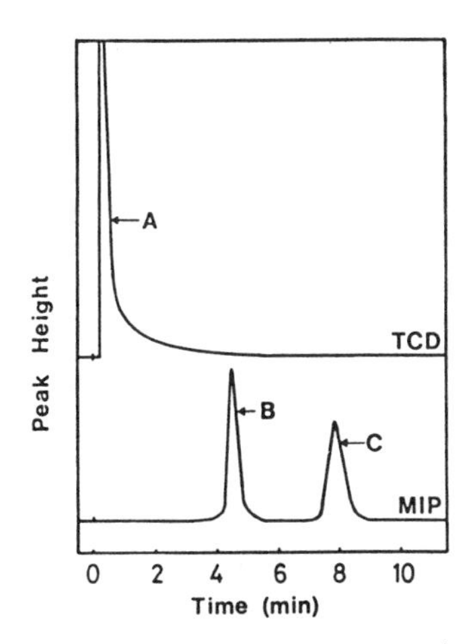

Figure 177 - Gas chromatograms detected by TCD (upper) and MIP (lower); (A) benzene, (B) CH_3HgCl (1 µg of Hg/L), (C), C_2H_5HgCl (1 µg of Hg/L).

Detection limit and standard deviation data obtained for three organomercury compounds without preconcentration are shown in Table 252.

The total extraction efficiency involved in the three stages of the oxidation of methyl mercury chloride from sea water by the cysteine -

benzene extraction technique was reproducible at 2% for a 500 total concentration, giving a detection limit of 0.4 ng/l and a relative standard deviation of 6% at the 20 ug/l level (Table 253).

Table 252 - Analytical figures of merit in the determination of alkylmercury compounds by the GC/MIP system

Compound	detection limit µg/L	rel std. dev,[a] %	dynamic range, decades
methylmercury chloride	0.09 (0.02 pg/s)	2.0	5
ethylmercury chloride	0.12 (0.02 pg/s)	2.0	5
dimethylmercury	0.40 (0.03 pg/s)	3.0	4.5

[a] Measured with 1 µg/L of mercury for each compound.

Table 253 - Analytical characteristics of the solvent extraction technique for methylmercury determination

sample treatment	sample volume injected into GC, µL	detection limit,ng/L	extraction efficiency %	rel std dev, %
without extraction	4	90		2.0
with extraction	4	0.4	42	6.0

Application of the procedure to a sample of sea water taken from Aburatsubo Bay, Tokyo, gave a value of 2 µg/l methylmercury chloride, about 2.3% of the total mercury compounds present.

Miscellaneous Methods. Sipos et al[3293] used subtractive differential pulse voltametry at a twin gold electrode to determine that mercury levels in sea water samples taken from the North Sea.

Fish frequently have 80 - 100% of the total mercury in their bodies in the form of methylmercury regardless of whether the sites at which they were caught were polluted with mercury or not (Holden[3294]). Methylmercury in the marine environment may originate from industrial discharges or be synthesized by natural methylation processes[3295]. Fish do not themselves methylate inorganic mercury (Pennacchioni[3296]); (Pentreath[3297]) but can accumulate methylmercury from both seawater (Pentreath[3297]) and food (Pentreath[3297,3298]). Methylmercury has been detected in seawater only from Minamata Bay, Japan (Egawa and Tajima[3299]) an area with a history of gross mercury pollution from industrial discharge. It has been found in some sediments but at very low concentrations, mainly from areas of known mercury pollution. It represents usually less than 1% of the total mercury in the sediment and frequently less than 0.1% (Andren and Harriss[3300]); Olsen and Cooper[3301]; Windom et al[3302]; Bartlett et al[3304]. Micro-organisms within the sediments are considered to be responsible for the methylation (Olsen and Cooper[3301]; Shin and Krenkel[3305]) and it has been suggested that methylmercury may be released by the sediments to the seawater, either in dissolved form or attached to particulate material and thereafter rapidly taken up by organisms (Jernelov[3306]; Gillespie[3307]; Langley[3308]; Shin and Krenkel[3305]).

Davies et al[3304] set out to determine the concentrations of methylmercury in seawater samples much less polluted than Minamata Bay, viz the Firth of Forth, Scotland. They described a tentative bioassay method for determining methylmercury at the 0.06 $\mu g\ l^{-1}$ level. Mussels from a clean environment were suspended in cages at several locations in the Firth of Forth. A small number were removed periodically, homogenized, and analysed for methylmercury by solvent extraction gas chromatography, as described by West[3309].

The rate of accumulation of methylmercury was determined, and by dividing this by mussel filtration rate, the total concentration of methylmercury in the seawater was calculated.

The methylmercury concentration in caged mussels increased from low levels (less than 0.01 $\mu g\ g^{-1}$) to 0.06 - 0.08 ug g^{-1} in 150 days giving a mean uptake rate of 0.4 ng $g^{-1}\ d^{-1}$, i.e. a 10 g mussel accumulated 4 ng d^{-1}. The average percentage of total mercury in the form of methylmercury increased from less than 10% after 20 days to 33% after 150 days. This may be compared with analyses of natural intertidal mussels from the area, in which the proportion of methylmercury was higher in mussels of lower than 0.10 $\mu g\ g^{-1}$ than of higher total mercury concentration.

Davies et al[3304] calculated that the total methylmercury concentration in the seawater at 60 pg dm^{-3} (0.06 ng dm^{-3}), i.e. 0.1 - 0.3% of the total mercury concentration, as opposed to less than 32 ng dm^{-3} methylmercury found in Minamata Bay, Japan. These workers point out that a potentially valuable consequence of this type of bio-assay is that it may be possible to obtain estimates of the relative abundance of methylmercury at different sites by the exposure of 'standardised' mussels as used in their experiment, in cages for controlled periods of time and by the comparision of the resultant accumulations of methylmercury.

Various other workers[3310-3314] have reported on the levels of total mercury in seawaters. Generally, the levels are less than 0.2 $\mu g\ l^{-1}$ with the exception of some parts of the Mediterranean where additional contributions due to man made pollution are found[3315-3319].

Stoeppler and Matthes[3320] have made a detailed study of the storage behaviour of methylmercury and mercuric chloride in seawater. They recommend that samples spiked with inorganic and/or methylmercury chloride be stored in carefully cleaned glass containers acidified with hydrochloric acid to pH 2.5. Brown glass bottles are preferred. Storage of methylmercury chloride should not exceed 10 days.

Lampear and Bartha[3321] studied the effects of sea salt anions on the formation and stability of methylmercury. The effect of different anions in seawater on the formation of methylmercury was investigated. Methylation was reduced by sulphide under anaerobic conditions and by bicarbonate under both aerobic and anserobic conditions; other anions had no significant effect. In the dark monomethylmercuric chloride was chemically stable in the presence of all the anions tested.

Sewage and Trade Effluents

Takeshita[3322] has used thin-layer chromatography to detect alkylmercury compounds and inorganic mercury in sewage. The dithizonates were prepared by mixing benzene solution of the alkylmercury compounds and a 0.4% solution of dithizone. When a green coloration was obtained, the solution was shaken with sulphuric acid followed by aqueous ammonia and washed with water. The benzene solution was evaporated under reduced pressure and the dithizonates, dissolved in benzene, were separated by reversed-phase chromatography on layers on corn starch and Avicel SF containing various proportions of liquid paraffin. Aqueous solutions of ethanol and of 2-methoxyethanol were used as developing solvents. The spots were observed in daylight. The detection limit was from 5 to 57 ng (calculated as organomercury chloride) per spot.

Itsuki and Kamuro[3323] determined organomercury compounds in waste water by heating the sample with a 2:1 mixture of nitric and hydrochloric acids (10 ml) and 30% hydrogen peroxide (2 ml) at 90°C for one hour (followed by the addition of 50% ammonium citrate solution (5 ml), diamino cyclohexane-tetracetone (5% v/v in 2% sodium hydroxide) (5 ml) and 10% hydroxylamine hydrochloride solution (1 ml) followed by pH adjustment to pH 3 to 4 with aqueous ammonia. The solution is shaken, 5 mg/ml 1,1,1 trifluoro-4-(2-thienyl)-4-mercaptobut-3-en-2-one in benzene (10 ml) added, the benzene layer washed with 0.1 N borate pH 11 (50 ml) and the solution evaluated spectrophotometrically at 365 nm.

Thin-layer chromatography has been used[3324] to evaluate organomercury compounds in industrial waste water. C_1 - C6 h-alkylmercury chlorides were separated on layers prepared with silica gel (27.75 g) plus sodium chloride (2.25 g in 60 ml water) using as development solvent cyclohexane-acetone-28% aqueous ammonia (60:40:1). The R_f values decrease with increasing C-chain length and phenylmercury acetate migrated between the C_1 and C_2 compounds. The spots are detected by spraying with dithizone solution in chloroform. Water samples (100 to 200 ml) were treated with hydrochloric acid (to produce a concentration of 0.1 N to 0.2 M) and with potassium permanganate until a pink colour persists, then shaken (x 3) with chloroform (one-third the volume of the aqueous layer) for 3 min. The combined extracts are shaken with 0.1 to 0.2 N aqueous ammonia (3 x 20 ml); the aqueous solution neutralised to p-nitrophenol with hydrochloric acid and adjusted to 0.2 N in hydrochloric acid and the organomercury compounds extracted with chloroform (4 ml). The chloroform extract usually recovered about 95% of the organomercury compounds and was in a suitable form for thin-layer chromatography.

Murakami and Yoshinaga[3325] determined organomercury compounds in industrial wastes by a spectrophotometric procedure using dithizone. The sample (100 ml) is neutralized to p-nitrophenol, and hydrochloric acid added to give an acid concentration of 0.1 to 0.2 N. The solution is shaken with chloroform (one-third the volume of the sample). 6% potassium permanganate solution is added (until the mixture is pink, then 0.2 ml in excess), shake the mixture for 3 min, and remove the chloroform phase. Add 10% hydroxylammonium chloride solution to decolorise the aqueous phase and repeat the extraction, with chloroform twice. Wash the combined chloroform extracts with 0.1 hydrochloric acid (3 x 50 ml) then extract mercury with 0.1 to 0.5 N-aqueous ammonia (2 x 20 ml) filter the aqueous solution and determine mercury using dithizone.

Carpenter[3326] has reviewed the application of flameless atomic absorption spectroscopy to the determination of mercury in paper mill effluents. Thiosulphate oxidation is recommended as a means of converting organomercury compounds to inorganic mercury. Carpenter concludes that the practical limit of detection of mercury in effluents using this technique is 1 part in 109.

ORGANOGERMANIUM COMPOUNDS

Although earlier workers[3327-3329] have reported that organogermanium compounds are absent in natural waters, Hambrick et al[3330] observed germanium compounds in some natural waters which are reduced and trapped similarly to $Ge(OH)_4$, but which elute from chromatographic packings after GeH_4[3331]. They suggested that these peaks are unidentified methylgermanium species by analogy with previous observations for arsenic, antimony, and tin[3327,3331-3335]. Further work confirmed the presence of methylgermanium in natural waters and led them to modify their technique in order to optimize the recovery of the methylgermanium species. The technique used by Hambrick et al[3330] was a modification of the method developed earlier for inorganic germanium by Andreae and Froelich[3328].

Inorganic and methylgermanium species were determined in aqueous matrix at the parts-per-trillion level by a combination of hydride generation, graphite furnace atomization, and atomic absorption spectrometry. The germanium species were reduced by sodium borohydride to the corresponding gaseous germanes and methylgermanium hydrides, stripped from solution by a helium gas stream, and collected in a liquid-nitrogen-cooled trap. The germanes were released by rapid heating of the trap and enter a modified graphite furnace at 2700°C. The atomic absorption peak was recorded and electronically integrated. The absolute detection limits are 155 pg of Ge for inorganic germanium (Ge_i), 120 pg of Ge for monomethylgermanium (MMGe), 175 pg of Ge for dimethylgermanium (DMGe), and 75 pg of Ge for trimethylgermanium (TMGe). The precision of the determination ranges from 6% for TMGe to 16% for MMGe.

A typical chromatogram is shown in Figure 178 of the separation of germane (GeH_4) monomethylgermane, dimethylgermane and trimethylgermane on a column comprising 15% .OV3 on Chromosorb W-AW-DMCS.

The sensitivities of the system for the different germanium species, as determined by peak heights and peak areas, are given in Table 254.

The method was applied to marine, freshwater and rainwater samples (Table 255). The major germanium species in seawater is monomethylgermanium. Trimethylgermanium is not found in seawater. Total germanium values for seawater are slightly lower than those of El Wardani[3336,3337] (50 ng L^{-1}), Johnson and Braman[3338] (42 ng L^{-1}), Burton et al[3339] (60 ng L^{-1}) and Braman and Tompkins[3327] (79 ng L^{-1}).

Table 254 - Sensitivities and Detection limits.

	peak height			peak area	
Compound	Sensitivity, A/ng	abs. detection limit, pg	concn detection limit (250 mL), pg L^{-1}	sensitivity (A s)/ng	abs. detection limit, pg
Ge_i	0.00986	155[a]	620	0.0178	195[a]
MMGe	0.00830	120[b]	480	0.0293	
DMGe	0.00565	175[b]	710	0.0149	
TMGe	0.0135	75[b]	300	0.0426	

[a] Based on twice the standard deviation of replicate blank determinations (n = 6).
[b] Based on twice the standard deviation of the base line noise (0.0005 A).

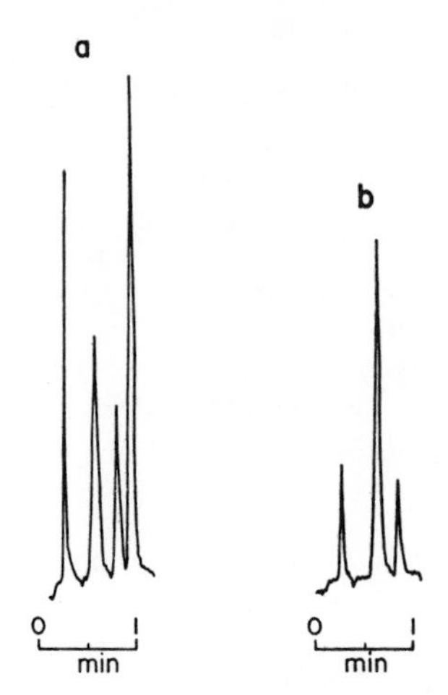

Figure 178 - Chromatograms show the peaks of (a) germane, monomethylgermane, dimethylgermane, and trimethylgermane resulting from reduction of a standard solution containing 1 ng of Ge_i, 1.85 ng of Ge as MMGe, 2.33 ng of Ge as DMGe, and 2.27 ng of Ge as TMGe and (b) germane, monomethylgermane, and dimethylgermane resulting from reduction of 100 mL of deep-ocean seawater from a hydrographic station in the Sargasso Sea (North Atlantic Ocean).

ORGANOTIN COMPOUNDS

For environmental studies, sensitive and species-specific methods are required to determine the volatile tetraalkyltin of R_4Sn type and the polar and solvated alkyltin $R_nSn^{(4-n)+}$ species in solution.

Methods have been described for the conversion of these compounds to their volatile hydrides which can then be separated and detected in an element-specific detector[3340,3341]; an alternate method is to further alkylate the alkyltin species with a selected Grignard reagent to convert them to tetraalkyltin forms which have lower boiling points. Butyltin (IV) compounds have been determined by gas chromatography mass spectrometry after methylation[3342] and by gas chromatography after pentylation[3343].

Atomic absorption spectrometry. Chau et al[3344] have described an improved extraction procedure for the polar methyltin compounds and the use of the gas chromatography atomic absorption spectroscopy system for the determination of their butylated derivatives including inorganic Sn(IV). In this method the highly polar and solvated methyltin, dimethyltin, trimethyltin and Sn(IV) species are extracted into benzene containing tropolone from water saturated with sodium chloride. These compounds are butylated in the extract to the tetramethylbutyltins, Me_nSnBu_{4-n}, (methylbutyl magnesium chloride) which have sufficient volatility to be separated and analyzed by the gas chromatography atomic absorption spectroscopy system. Large volumes of water sample can be handled. Under normal laboratory conditions, detection limit of 0.04 µg/L can be achieved with 5 L of water sample. Absolute detection limit of the gas chromatography atomic absorption spectroscopy for tin is 0.1 ng. Volatile organotin compounds such as tetramethyltin and methyltin hydrides can also be analyzed by this method.

These workers studied the effect of various acids in the extraction of

Table 255 - Typical concentration ranges of germanium species in natural waters

Sample location	no. of analyses	concentration, ng L^{-1}				
		Ge_i [a]	MMGe [b]	DMGe [b]	TMGe	total Ge
Atlantic, Sargasso Sea	11	0.4 - 1.8	21.9 - 24.5	10.0 - 11.5	0	32.2 - 37.8
Bering Sea, Geosecs 219	21	2.2 -12.3	18.2 - 20.3	8.0	0	28.4 - 40.6
rain water, Tallahassee, FL	5	0.5 - 0.7	0	1.0 - 5.9	0.2 -1.0	1.7 - 7.3
Peace River and Charlotte Harbor Esturary FL	30	0.7 - 7.1	0 - 18.6	0 - 10.0	0	7.1 - 29.3
Ochlockonee River and Bay Estuary, FL	20	2.4 - 5.7	0 - 19.5	0 - 12.0	0	5.7 - 33.9
Tejo River and Estuary, Lisbon, Portugal	11	0.6 - 7.4	0 - 19.6	0 - 6.6	tr	7.4 - 26.8

[a] Ge_i increases in concentration with increasing water depth in the oceans and with decreasing salinity in estuaries. Thus the ranges reported here reflect natural gradients in surface and deep-sea water and along the salinity gradient of estuaries.

[b] MMGe and DMGe are vertically homogeneous in the oceans. Thus the ranges for each ocean reflect analytical variability. In estuaries, MMGe and DMGe vary from nondetectable in the riverine freshwater to near the seawater values at the ocean end. Thus these ranges reflect a gradient in the estuary.

methyltin compounds from water using tropolone in benzene. After the mixture was shaken for ca. ½ h, the benzene layers were separated and butylated with 1 mL of butyl magnesium chloride reagent in 10 mL glass-stoppered micro Erlenmeyer flasks with stirring for ca. 10 min. The mixture was washed with 5 mL of 1 sulphuric acid to destroy the excess Grignard reagent. The organic phase was separated, and dried with anhydrous sodium sulphate. The mixture now contained the butyl-derivatized methyltins $MeSnBu_3$, Me_2SnBu_2, Me_3SnBu, and Bu_4Sn ready for analysis by the gas chromatographic atomic absorption technique. The standard methyltbutyltin compounds for evaluation of the absolute recovery of the method were prepared by adding 25 μL of each of the aqueous standard solutions (1 μg μL^{-1}) to 5 mL of tropolone solution, and, after shaking, carrying through the butylation and cleanup procedures as described.

The results (Table 256) indicated that acidification with hydrochloric acid and hydrobromic acid enhanced the recovery of Sn(IV) species, probably due to prevention of hydrolysis and adsorption of Sn(IV) compounds on container walls. Although the use of hydrobromic acid to enhance extraction of butyltin compounds has been reported, it was not found beneficial for overall extraction of methyltin species. It was found, on the contrary, that both hydrobromic acid and hydrochloric acid suppressed the extraction of the Me_3Sn^+ and Me_2Sn^{2+} species. Acetic acid did improve the recovery of Me_2Sn^{2+} and $MeSn^{3+}$ but did not satisfactorily recover Me_3Sn^+ and Sn(IV). Sulfuric acid had no beneficial effect on the extraction of the four tin species. Low pH is therefore not desirable for the recovery of all the methyltins.

Table 256 - Effect of acids on the extraction of methyltin and Sn(IV) compounds from water[a]

	% recovery in acid				
	HBr	HCl	H_2SO_4	HOAc	nil
Me_3Sn^+	6	7	0	0	0
Me_2Sn^{2+}	8	6	5	98	90
$MeSn^{3+}$	85	83	18	97	91
Sn(IV)	102	106	12	23	47

[a] Water, 100 mL; Sn compounds, 25 μg; 5 mL 0.1% tropolone-benzene; concentrated acids used, 10 mL: H_2SO_4, 5 mL; extraction efficiency in %.

Since the methyltine are highly polar and solvated, extraction in the presence of salts may improve their recoveries. Chau et al[3344] tried the effect of adding various amounts of sodium chloride to 100 ml of water containing the four tin species on recovery. After subsequent butylation, the absolute recovery was found satisfactory for Me_2Sn^{2+}, $MeSn^{3+}$, and Sn(IV) but was only 75% for the Me_3Sn^+ species even with 40 g of sodium chloride used to achieve a saturated salt solution (Table 257). The Me_3Sn^+ species was only extracted in the presence of sodium chloride. The recovery, although not quantitative, was consistent as evidenced in the calibration curve and in the reproducibility tests.

Chau et al[3344,3345] found that the introduction of hydrogen to the quartz furnace was necessary to elevate the furnace temperature to ca. 900°C and to enhance atomization of the methylbutyltin derivatives. It was also found that introduction of air further enhanced the sensitivity. Heating the transfer line to 165°C was necessary to give sharp peaks. Although this temperature was well above the boiling points of the tin derivatives, no decomposition of the alkyltin compounds was noted at the transfer line.

Precision at the 250 µg/l level of Me_3Sn^+, Me_2Sn^{2+}, $MeSn^+$ and Sn(IV) in lake water was acceptable as reflected in standard deviations, of 5.41, 8.6, 7.3 and 11%, respectively. Table 258 indicates good overall recovery in the method.

Figure 179 shows a typical GC-AAS chromatogram of the butyl derivatized methyltin and Sn(IV) species.

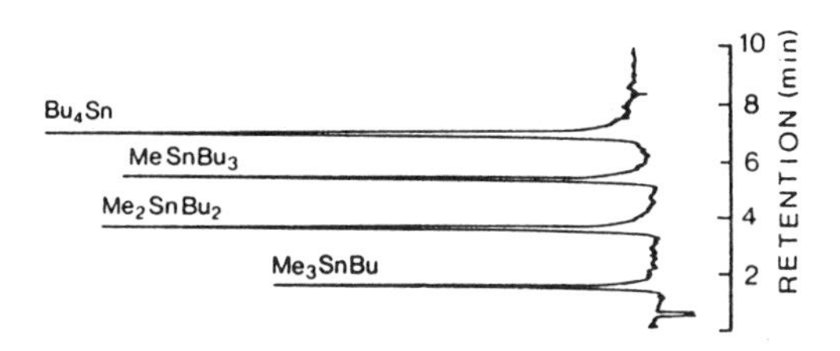

Figure 179 - GC-AAS chromatograms of methyltin and Sn(IV) species after butylation. Each peak represents ca. 8 ng of Sn.

Table 257 - Effect of NaCl on the extraction of methyltin and Sn(IV) compounds from water[a]

	% recovery with NaCl						
	0	10 g	20 g	30 g	40 g	40 g[b]	40 g[c]
Me_3Sn^+	0	10	31	61	75	60	54
Me_2Sn^{2+}	90	95	105	103	97	33	9
$MeSn^{3+}$	87	108	108	107	105	97	100
Sn(IV)	56	110	105	107	101	98	112

[a] Water, 100 mL; Sn compounds, 25 µg; 5 mL 0.1% tropolone-benzene; extraction efficiency in %.
[b] With 2 mL of HCl.
[c] With 10 mL of HCl.

The only tetravalent elements that could be coextracted by tropolone, and similarly butylated to the tetraalkyl derivatives are the Ge(IV) and Pb(IV) species. Tetraalkyllead compounds, however, do not give any signals in the AA detection system for tin analysis at the 224.6 nm spectral line. There should not be any worry of interference from Ge(IV) in natural waters nor is its spectral interferences expected.

Tetramethyltin and other volatile organotin compounds including methyltin hydrides, Me_nSnH_{4-n}, have been found in tin contaminated harbours, and tetramethyltin formation has been observed through methylation processes when trimethyltin compounds were incubated with sediments. All these volatile tin compounds can be directly analyzed by the present gas chromatography atomic absorption spectroscopic system without derivatization. The only requirement is the sampling technique modication. Tetramethyltin is volatile and is accumulated in the headspace of a closed culture system. Evacuation of the headspace gases into a cold trap can quantitatively recover tetramethyltin. In open systems such as the environment, purging a water sample with an inert gas can remove any residual tetramethyltin including the organotin hydrides. Such technique has been used in studying the biogenesis of organotins. In their methylation studies, Chau et al[3344] trapped the headspace in a U-trap packed with 3% OV-1 on Chromosorb W at - 160°C and subsequently mounted the trap through a four-way valve to the gas chromatography atomic absorption spectroscopic system for analysis of Me_4Sn. With other instrument parameters unchanged, the temperature used was isothermal at 30°C. Some results obtained by this procedure are tabulated in Table 259 to illustrate the application of this method. The sampling sites are either harbours, marinas, or heavily industralized areas. The results obtained are higher than data previously obtained for natural waters[3340,3341]

Table 258 - Recovery of methyltin and Sn(IV) compounds from water[a]

amount of Sn compound added, μg	recovery, %			
	Me_3Sn^+	Me_2Sn^{2+}	$MeSn^{3+}$	Sn(IV)
5	91	100	85	100
10	95	98	100	95
25	95	100	117	102
50	99	102	106	108
100	100	96	100	100

[a] Distilled water, 100 mL

Table 259 - Analysis of methyltin and Sn(IV) species in lake waters[a] ($\mu g\ L^{-1}$)

Location	Me_3Sn^+	Me_2Sn^{2+}	$MeSn^{3+}$	Sn(IV)
Port Maitland	nd	0.14	0.37	0.13
Mitchell Bay 1	nd	0.10	0.35	0.98
Mitchell Bay 2	nd	0.22	0.53	0.18
Toronto Harbour	nd	0.29	0.96	0.54
Port Dover	nd	0.16	0.61	0.14
Kingston Harbour	nd	0.40	1.22	0.49

[a] Surface water taken from locations in Ontario. Sample size, 5 - 8 L; nd, not detector.

Gas chromatography. Soderguist and Crosby[3346] have developed a method for the simultaneous determination of triphenyltin hydroxide and its possible degradation products tetraphenyltin, diphenyltin oxide, benzenestannoic acid, (and inorganic tin) in water. The method is rapid (one sample set per hour), sensitive to less than 0.01 g/ml for most of the tin species and exhibits no cross-interferences between the phenyltins. The phenyltins are detected by electron capture gas-liquid chromatography after conversion to their hydride derivatives, while inorganic tin is determined by a procedure which responds to tin(IV) oxide as well as aqueous tin (IV).

Soderquist and Crosby[3346] found that the nonvolatile compounds hydroxyoxyphenyl stannane ($PhSnO_2H$) oxodiphenyl stannane (PhSnO) and hydroxytriphenyl stannane (Ph_3SnOH) upon conversion to their hydrides by lithium aluminium hydride resulted in derivatives with excellent GLC properties, high response to electrol-capture detection, and none of the attendant column stability problems encountered with other derivatives.

$Ph_3Sn_{aq}^{+} \longrightarrow Ph_2Sn_{aq}^{+2} \longrightarrow PhSn_{aq}^{+3} \longrightarrow Sn_{aq}^{+4}$; $PhSn_{aq}^{+3} \longrightarrow SnO_{2aq}$ — Possible degradation

$Ph_3Sn_{aq}^{+} \downarrow Ph_3SnH, Ph_2SnH_2, PhSnH_3$ EC-GLC ; $SnO_{2aq} \downarrow$ Sn-PCV complex Colorimetry — Analysis

The basis for this method thus involves extraction of the phenyltin species from water followed by their quantitation as phenyltin hydrides by electron capture gas chromatography and analysis of the remaining aqueous phase for inorganic tin (Sn^{4+} plus SnO_2) by colorimetry.

Soderquist and Crosby[3346] used a dual column/dual detector Varian model 2400 gas chromatograph equipped, on one side, with a flame-ionization detector (FID) and a 0.7 m by 2 mm (i.d.) glass column containing 3% OV-17 on 60 / 80 mesh Gas Chrom Q. Column, injector and detector temperatures were 265, 275 and 300°C, respectively; carrier gas (nitrogen) flow rate was 25 ml/min. Tetraphenyltin eluted within 8 min under those conditions. The second side of the chromatograph was equipped with a tritium EC detector and a 1.1 m by 2 mm (i.d.) glass column containing 4% SE-30 on 60 / 80 mesh Gas Chrom Q. The injector and detector temperatures were 210°C the carrier gas (nitrogen) flow rate was 20 ml / min, and column temperatures which elulted the following compounds within 6 min were: Triphenylstannane (PhSnH) (190°C), diphenylstannase (Ph_2SnH_2) (135°C), phenylstannane ($PhSnH_3$) (45°C). Combined gas-liquid chromatography / mass spectrometry (GC-MS) was performed on a Finnigan Model 1015 utilizing a 1.0 m by 2 mm (i.d.) glass column containing 3% OV-17 on 60 / 80 mesh Gas Chrom Q.

The 200 ml sample in a 250 ml separatory funnel was mixed with 5 ml of acetate buffer, the mixture extracted with two 15 ml portions of dichloromethane, and the pooled extract divided into three equal portions, each of which was concentrated to about 0.1 ml in a screw-capped test tube at less than 40°C under a gentle stream of nitrogen.

To one of the concentrates (EX-1, Table 260) was added 5 ml of hexane followed by 0.5 ml of lithium aluminium hydride solution. After 2 - 3 min, the mixture was diluted with hexane, about 0.5 ml of water carefully added, the phases were mixed, and the hexane phase was analyzed by electron-capture gas chromatography for Ph_3SnH, Ph_2SnH_2, and $PhSnH_3$. A standard curve was prepared for each of the hydrides with the ng/µl hexane standard solutions, generally in the 0.2 - 2.0 ng range. Quantitation was done by comparison of sample peak heights to the standard curve.

To the second dichloromethane concentrate (EX-2) was added 0.50 ml of sulphuric acid and the dichloromethane removed from the mixture with a vigorous stream of nitrogen. The tube was sealed and heated at 100°C in a water bath for 20 min. After cooling, 4.0 ml of citric acid solution was added and the sample analyzed for inorganic tin.

To the third dichloromethane concentrate (EX-3) was added 1 ml of hexane, the contents were concentrated under nitrogen to about 0.1 ml, and then diluted back to 1.0 ml with hexane. The sample was transferred to a Florisil micro-column (prepared by packing a disposable Pasteur piplet with 0.35 g of 60/100 mesh Florosil held with a small glass wool plug and rinsing with two 5 ml portions of hexane before use) and eluted with hexane. The first 2.5 ml of eluate was collected, concentrated to 0.1 - 0.5 ml, and analysed by FID-GLC for tetraphenyltin. Quantitation was accomplished by comparison of sample peak heights to the tetraphenyltin standard curve in the 10 - 50 ng range.

Table 260 - Flow diagram for analytical procedure

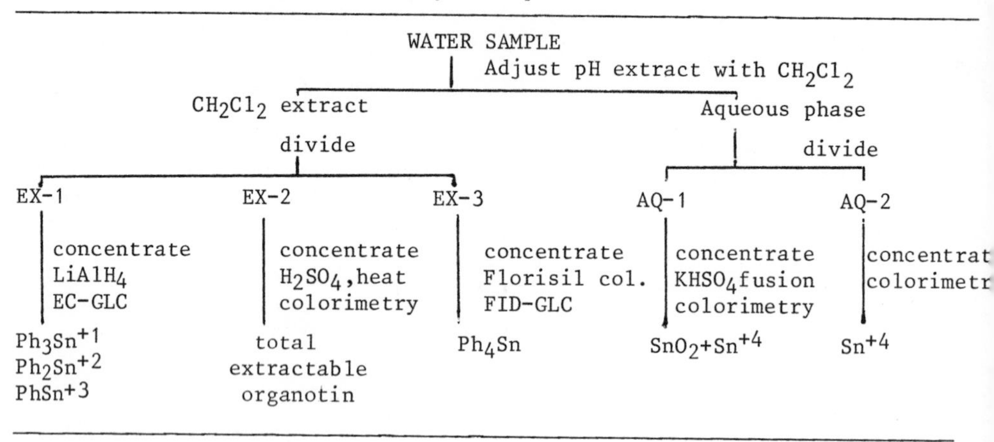

The hydride derivatives differ widely in volatility and could not be simultaneously detected with any columns examined. Since gas chromatography at different temperatures was required, it was found most efficient to analyse all of a series of derivatized samples, blanks, and standards for one of the hydrides before equilibrium of the chromatograph at a new temperature. Typical chromatograms of the hydride standards near the limit of detectability are shown in Figure 180.

Whilst recoveries of tri- and diphenyltin compounds were good those of monophenyltin compounds were in the range 11 to 81%.

Soderquist and Crosby[3346] checked the recovery of their reduction - gas chromatographic procedure for the determination of organotin compounds by spiking distilled water samples (200 ml) with 50 µl of the fortification standard to give the organotins at 0.050 µg/ml; an unfortified (blank) sample also was analyzed. All the organotins except $PhSn^{3+}$ were successfully recovered (Table 261). Analysis for total extractable organotin, on a mole basis, resulted in slightly more tin than the sum of the specific hydride analyses and in the absence of any intentionally-added inorganic tin, some inorganic tin still was found (parenthetic values in Table 261); both observations support the notion that $PhSn^{3+}$ is unstable in water. There was no cross-interference between any of the phenyltin compounds when analysis was made via the hydride procedure for any one organotin in a sample containing a ten-fold excess of each of the other organotins. While some of the added tin species were not recovered intact, the sum of the recoveries of procedures EX-2 plus AG-1 (0.207 mol Sn^{4+}) was 92% of the total added tin (0.224 mol) - an acceptable overall accountability.

The sensitivity of the proposed method for each of the tin species examined is summarized in Table 262. These limits could be decreased either by increasing the sample size, avoiding some of the sub-analyses, or both.

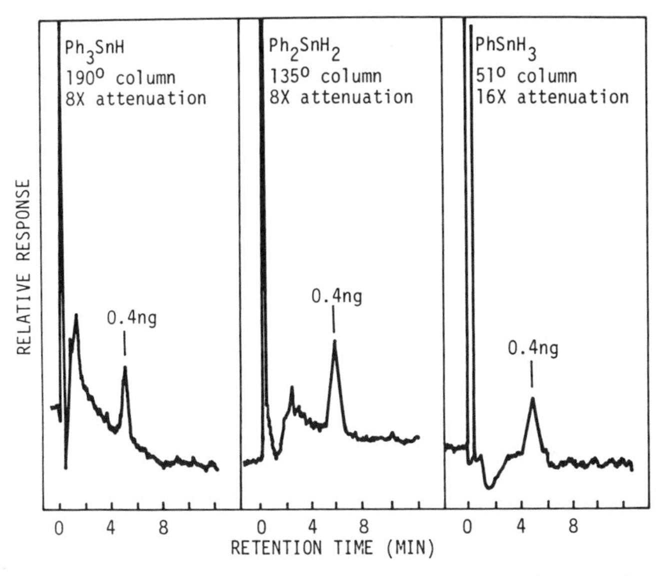

Figure 180 - Typical chromatograms of the hydride derivatives.

While the method has mainly been used for samples generated during laboratory studies (e.g., photodecomposition), none of the natural water samples analyzed have contained materials which interfered with the determination of any of the tin compounds of interest. For example, a 1 L rice-field water sample was analyzed for Ph_3Sn^+ at 1.0 µg/l with 60% recovery.

Table 261 - Recovery of tin from water

species	procedure[a]	fortified samples µg/ml added	fortified samples µg/ml found	blank sample µg/ml found
Ph_4Sn	EX-3	0.050	0.035 ± 0.002[b]	0.003
Ph_3Sn^{1+}	EX-1	0.050	0.054 ± 0.001[b]	0.002
Ph_2Sn^{2+}	EX-1	0.050	0.031 ± 0.003[b]	0.006
$PhSn^{3+}$	EX-1	0.050	0.005[b]	0.005
total extractable organotins (as Sn)	EX-2	0.082	0.050 ± 0.005[b]	0.008

[a] Refers to sub-analyses given in procedure and Table 260.
[b] Average and average deviation of four replicates.
[c] Average and average deviation of three replicates.
[d] Parenthetic values are from single samples to which no inorganic tin was added.

Smith[3347] has discussed the determination of tin in organisms and in water.

Methods for the determination of nanogram amounts of methyltin compounds and inorganic tin in natural waters and human urine have been described by Braman and Tomkins[3348]. In this method the tin compounds in aqueous solution at pH 6.5 are converted by sodium borohydride to the corresponding volatile hydride, SnH_4, CH_3SnH_3, $(CH_3)\ SnH_2$, and $(CH_3)_3SnH$, by reaction with sodium borohydride. These are scrubbed from solution, cryogenically trapped on a U-tube, and separated upon warming. Detection limits are approximately 0.01 ng as Sn when using a hydrogen-rich, hydrogen-air flame emission type detector (SnH band) of a type having considerably lower detection limits than any previously reported[3349-3353]. Average tin recoveries ranged from 96 - 109% for seawater and from 83 - 108% for human urine samples for six samples analyzed to which were added 0.4 to 1.6 ng of methyltin compounds and 3 ng inorganic tin. Re-analysis of analyzed samples shows that all methyltin and inorganic is removed in one analysis procedure.

Table 262 - Method sensitivity

Species	method	minimum detectable amount	method sensitivity -g/ml[a]
Ph_4Sn	FID-GLC	5.0 ng as Ph_4Sn	0.015
Ph_3Sn^{1+}	EG-GLC	0.2 ng as Ph_3SnH	0.003
Ph_2Sn^{2+}	EG-GLC	0.2 ng as Ph_2SnH_2	0.003
$PhSn^{3+}$	EG-GLC	0.2 ng as $PhSnH_3$	0.003
total extractable organotins	colorimetry	1.0 μg as Sn	0.01
Sn^{4+}	colorimetry	1.0 μg as Sn	0.007
SnO_2 + Sn^{4+}	colorimetry	1.0 μg as Sn	0.007

[a] For 200 ml samples

Figure 181 shows the apparatus used by Braman and Tomkins[3348] which consists of a sample reaction chamber, U-trap, the flame emission type detector, and conventional type photometric readout and recording system. Inorganic and methyltin compounds in aqueous solution in the reaction chamber are reduced to stannane or the corresponding methylstannanes by treatment with sodium borohydride solution buffered at pH 6.5. Helium carrier gas scrubs the volatile stannanes out of solution and into the liquid nitrogen-cooled U-trap where they are frozen out. Upon removal of the liquid nitrogen and warming, the stannanes are separated and carried into the detector.

Figure 182 gives details of the quartz burner and burner housing design. The burner is constructed of quartz tubing. The air and helium entrained sample enter the burner assembly through 6 mm o.d. tubing and then pass into the 4 mm o.d. tubing which carries the helium-air-sample mixture to the burner tip and concurrently into the excitation zone. The burner tip is contained within 8 mm o.d. tubing, into which the hydrogen enters, jacketing the excitation zone within a hydrogen-rich atmosphere, just the reverse of usual flame burners. In this air-hydrogen flame configuration, sulphur (S_2) and hydrocarbon band (CH,C_2) excitation is much reduced and occurs at the flame tip while the tin,SnH molecular band emission is much increased. The SnH excitation does not occur at the tip of the burner, but above the air-hydrogen flame interface. Partial advantage can be taken of the physical separation of the SnH band emission from the flame cone to optically isolate the SnH emission. Thus, the SnH band is observed in the presence of much less background emission than in other flame detectors.

This separation is probably responsible for the good selectivity of this detector for tin, relative to sulphur, phosphorus, and various organics and also for its sensitivity. The best detection limits were obtained when

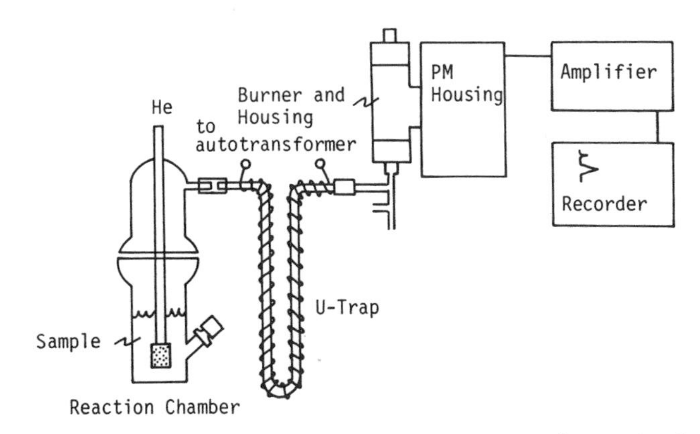

Figure 181 - Apparatus arrangements for tin analysis.

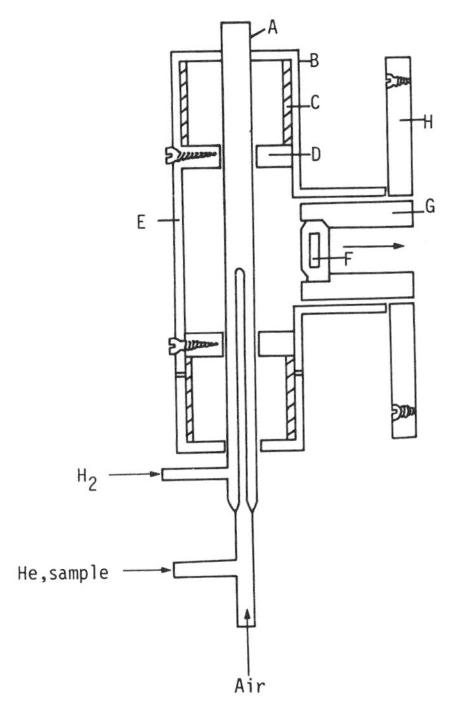

Figure 182 - Quartz burner and housing. A, Quartz burner; B, PVC cap; C, PVC tubing; D, mounting ring; E, PVC T-joint, 1.25 inch; F, filter and holder; G, PVC coupling.

the flame cone was located 2 cm below the centre of the filter. The burner housing shown in Figure 182 is constructed of a 1.25 inch poly (vinyl-chloride) (PVC) T-joint, 1.25 inch PVC pipe, two 1.25 inch PVC end-caps and a 1.25 to 1 inch PVC reducing coupling. The PVC reducing couple was cut such that one end would fit tightly into an opening drilled into a 0.375 inch thick mounting plate. This plate and couple was fitted next to the PM tube module. Within the housing, the reducing coupling holds the filter in place. The burner is held within the housing by two collars and secured by two set screws. It was painted with 2 - 3 coats of flat-black paint to reduce light transmission.

The burner, burner housing, and band-pass interference filter assembly were directly mounted on a Heath Company photo-multiplier module, with a Hamamatsu (middlesex, N.J.) R-818 photomultiplier tube. The R-818 tube is used in the 250 to 800 nm range and displays a wavelength of maximum response near 500 mm. The photomultiplier output was amplified by a Heath Company electrometer. The amplified signal was then read on a Linear Instruments Corp. integrating strip chart recorder. A band-pass interference filter, 610 nm, was used for wavelength selectivity. It was found that a hydrogen to air ratio between 2 and 3, with a helium to air ratio between 1 and 2, produced a flame highly sensitive to tin emission, yet resistant to flame perturbation and long-term base-line drift. Gas optimum flow rates found for the burner used were 183 ml/min, He 127 ml/min air, and 283 ml/min H_2.

Similarly, carrier gas flow rate had an effect on response and Braman and Tomkins[3348] showed that based on an estimated burner active volume of 1 cm^3 and an optimal flow rate of 130 ml/min, the residence time of methylstannane within the flame was 0.5 seconds.

Braman and Tomkins[3348] used a column consisting of fully packed 30 cm U-tubes of silicone oil type, OV-3, on Chromosorb W to separate the stannanes. They noted a resolution of two peaks within the dimethystannane signal. Similar results were observed during the analysis of natural waters containing dimethyltin compounds. This may be due to the formation of stable bypyramidal geometric isomers of dimethylstannane but this was not verified. While the reasons for the observed effect are not known, the split signal is quantitative for dimethyltin dichloride and produced no problems during analysis.

Component resolution depends upon U-trap warming rates. The separation depicted in Figure 183 results from a gradual warming of the trap to room temperature. At higher warming rates, the sharpness of the response signals improves at the expense of signal separation. Selection of the optimum U-trap warming rate which is controlled by the autotransformer is best done using standards.

Table 263 shows the linear response curve data obtained by Braman and Tomkins[3348] for tin (IV), methyltin, dimethyltin and trimethyltin compounds. Since a 100 ml sample volume was used for natural water analyses, this places the calculated concentration detection limits between approximately 0.07 and 0.2 ng L^{-1} depending upon the compound detected. Precision of the method averaged ± 5% relative over the range of the response curve.

Braman and Tomkins[3348] did not observe any interferences in this method by organics in the determination of organotin compounds in water or urine samples. They did, however, observe that organic arsenic III compounds in seawater caused a positive interference effect due to the emission of

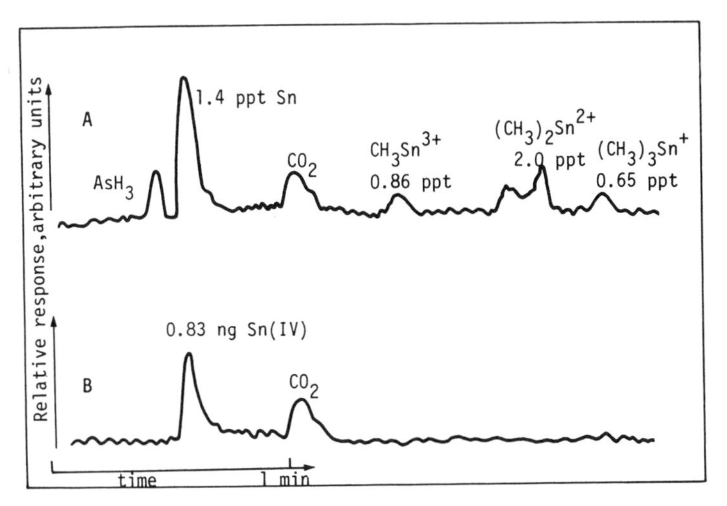

Figure 183 - Environmental sample analysis and blank. A, Environmental analysis, Old Tampa Bay; B, Typical blank.

Table 263 - Response curve data for tin compounds[a]

Compound	concentration range, ng(as Sn)	Samples	Slope,M (± S.D.)	Intercept[b] (±S.D.)	Detection limit pg(as Sn)
Tin	1.5 - 26.	7	6.6(± 0.24)	(0.026(±1.0)	1.3
Methyltin	0.12 - 5.8	8	18(± 0.57)	0.31(± 1.0)	1.6
Dimethyltin	0.084 - 4.4	6	13 (± 0.89)	2.1 (± 1.0)	0.65
Trimethyltin	0.63 - 5.5	7	8.5 (± 0.48)	0.81 (± 0.99)	0.92

a The operational C_L for inorganic tin is limited by the uncertainty of the inorganic tin blank, approximately 40 pg.

the As_2^+ molecular band at 611.5 ng. Inorganic arsenic (III) is reduced to arsine at pH 6.5. It is, nevertheless, separated from stannane on the OV-3 column and is not an interference. The arsenic (III) peak can be eliminated by oxidation of arsenic (III) to arsenic (V), by addition of a few drops of sodium thiosulphate solution to dispel excess iodine. Neither arsenic (V) nor the methylarsenic acids are reduced to corresponding arsines at pH 6.5.

Certain metal ions, Ag^+, Cu^{2+}, Hg^{2+}, Ni^{2+}, MoO_4^{2-}, and Pb^{2+} at 2 ppm in analyzed solutions were found to reduce the complete removal of stannane. Seawater did not inhibit recovery of stannane or Sb^{3+} did not interefere at 20 ppm while Fe^{3+}, BiO_3^-, Cd^{2+}, S^{2-}, VO_3^{2-}, and Zn^{2+} did not interefere at 2 ppm.

In Table 264 are presented some results obtained by this procedure when applied to seawater,estuarine water, surface water and rainwater samples.

Maguire et al[3354,3355] have described a method for the determination of butyltin species including bis(tri-n-butyltin) oxide and some of its dealkylation products in lake and river waters by gas chromatography with flame photometric detection.Mass spectra could be obtained with about 25 ug of the derivatives.

Table 264 - Analysis of saline and estuarine water samples[a]

Sample	tin (IV) ngL^{-1}	%	methyltin ngL^{-1}	%	dimethyltin ngL^{-1}	%	trimethyltin ngL^{-1}	%	Total tin ngL^{-1}
saline waters									
Gulf of Mexico, Sarasota	62.	73.	15.	18.	7.0	8.3	0.98	1.2	85
Gulf of Mexico, Fort Desota	2.2	60.	ND[b]		0.74	20.	0.71	20.	3.6
Gulf of Mexico, St. Petersburg	4.5	54.	0.62	7.4	3.2	39.	ND		8.3
estuarine surface waters									
Sarasota Bay	5.7	47.	3.3	27.	2.0	16.	1.1	9.1	12.
Tampa Bay	3.3	27.	8.0	66.	0.79	6.5	ND		12.
McKay Bay	20.	88.	ND		2.2	9.6	0.45	2.0	23.
Hillsborough Bay	ND		ND		1.8	71.	0.71	29.	2.5
Hillsborough Bay, Seddon Channel; North	12.	86.	0.74	5.3	0.91	6.6	0.35	2.5	14.
fresh waters[c]									
Lake Eckles, Tampa	10.	52.	0.99	4.9	1.2	5.8	7.6	38.	20.
Lake Carroll, Tampa; inlet	7.7	65.	ND[b]	-	0.96	8.1	3.3	28.	12.
Lake Carroll, Tampa; South	6.0	58.	ND	-	0.80	7.7	3.6	34.	10.
rain water and tap water samples[a]									
rain (8/12/77, p.m. after a steady rain)	6.4	65.	1.2	12.	1.7	17.	0.61	6.1	9.9
rain (8/12/77, p.m. after all night rain)	2.7	44.	2.1	34.	0.83	14.	0.45	7.4	6.1
rain (8/31/77, p.m. last rain)	6.7	60.	3.7	33.	0.88	7.9	ND	-	11.

[a] This set of values not used in computing the average.
[b] ND, less than 0.01 ngL^{-1} for methyltin compounds of 0.3 ngL^{-1} for inorganic tin.
[c] Data are average of duplicate analyses.

Jackson et al[3356] carried out gas chromatographic speciation studies of methylstannanes in the Chesapeake Bay area using trap sampling with a tin selective detector. The sampler is coupled automatically to a gas chromatograph equipped with a commercial flame photometric detector, modified for tin-specific detection by use of an interference filter. The method allows non-destructive speciation and detection of both hydrophilic and hydrophobic organotin species in aqueous systems.

Mueller[3357] detected tributyltin compounds at trace levels in water and sediments using gas chromatography with flame photometric detection and gas chromatography - mass spectrometry.

The tributyltin compounds are first converted to tributylmethyltin, and then analysed using capillary gas chromatography with flame photometric detection and gas chromatography-mass spectrometry. Tributyltin was found in samples of river and lake water, and sediment, and these results demonstrated that the technique has detection limits of less than 1 ppt for water and 0.5 ppb for sediment.

ORGANOLEAD COMPOUNDS

Atomic absorption spectrophotometry. Jonghe et al[3358] have developed a method for the determination in water of traces of triethyllead compounds without interference from mono-, di- and tetralakyllead compounds and inorganic lead.

After enrichment of the sample by a fast vacuum distillation technique and saturation of the residual volume with sodium chloride, the analytes are extracted in chloroform. By incorporation of specific purification steps, interference from other forms of organic and inorganic lead is completely eliminated. The final chloroform extract is treated with a sulfuric acid solution in order to transfer the trialkyllead compounds present back into an aqueous solution. The analysis is completed by graphite furnace atomic absorption spectrometry. Under normal laboratory conditions, a detection limit of 0.02 μg can be achieved with 1 L samples. In contrast to earlier approaches where other forms of organic and inorganic lead may give rise to serious interferences, this determination is highly specific for trialkyllead compounds even in the presence of up to 100 $\mu g\ L^{-1}$ of inorganic lead salts.

For environmental applications a 1 L volume is advised. After shaking for 1 min with hexane (1 mL per 100 mL of water), the sample is filtered on a Type RA Milliprefilter of 1.2 μm pore size. The water is brought into a rotary evaporator and evaporated under vacuum at 60°C until a residual volume of ca. 15 mL remains. Thereupon the sample is quantitatively transferred to a 100 mL separating funnel and the volume is adjusted to 25 mL with distilled water. About 8 g of sodium chloride is added and if necessary the pH is brought to below 10. After this, the sample is extracted twice with 25 mL portions of chloroform (shaking for 1 min). The extracts are combined and 1 mL of a 10^{-3} M dithizone in chloroform solution is added. Next, this organic phase is shaken for about 2 min with an aqueous solution consisting of 15 mL of an ammonium citrate/ammonia buffer and 35 mL of 0.1 M EDTA. The chloroform layer is separated off, shaken first with distilled water to remove traces of EDTA and buffer and then for 1 min with 5 mL of a 0.1 N sulfuric acid solution. Analysis of this sulfuric acid extract by means of graphite furnace atomic absorption spectrometry allows the quantitation of the trialkyllead initially present.

For practical reasons, the extraction described above should not be carried out with more than 25 mL of water. Since the detection limit directly depends on the volume of sample, however, the analysis of larger volumes is desirable. The procedure includes therefore a concentration step. It was found that in a rotary evaporator 1 L samples can be evaporated to less than 10 mL without loss of the analytes. After the enrichment, the volume is adjusted to 25 mL for quantitation of the analysis. Under these conditions, the detection limit of the method, defined as three times the standard deviation of the blank, is as low as 0.02 μg/L.

Different levels of trialkyllead were spiked to 1 L of distilled water and carried through the entire procedure. Typical results are shown in Table 265. For a total of ca. 30 synthetic samples analyzed, the overall recovery amounts to 87 ± 4 and 92 ± 5%, for, respectively, trimethyllead and triethyllead, taking into account an average recovery factor of 90%, results will hence be accurate within 5%. Moreover, the results summarized in Table 266 suggest that the method can also be applied successfully to samples in which mixtures of trialkyllead species are present.

Table 265 - Recovery of trimethyllead and triethyllead from 1 L of distilled water (two consecutive chloroform extractions)

amt of TriAL added, ng of Pb	amt of TriAl found, ng of Pb	recovery, %
	TriML	
43	40	93
87	78	90
174	140	80
261	232	89
350	284	81
435	367	84
4349	4066	93
43488	39444	91
	TriEL	
82	76	93
163	143	88
244	213	87
325	284	87
3253	3083	95
32530	20458	94

Table 266 - Recovery of mixtures of trimethyllead and triethyllead from distilled water (two consecutive chloroform extractions)

amt of TriAl added, ng of Pb			amt of TriAl found ng of Pb	recovery %
TriMl	TriEL	total		
336	0	336	293	87
252	70	322	312	97
168	139	307	297	97
84	209	293	275	94
0	278	278	263	95

It was found that in this method up to 1000 µg of inorganic lead and up to 100 µg of dialkyllead and tetraalkyllead compounds (in the sample portion taken) can be tolerated without exceeding the limit of detection.

Jonghe et al[3358] added trialkyllead to a number of lake and river water samples and did not observe any detectable difference in recovery as compared to distilled water. Similarly, the recovery remained the same when the sample was filtered before the analysis or when 0.5 mL of hydrochloric acid was added to avoid adsorption losses on the container walls. Some problems were experienced with the phase separation in the sodium chloride /chloroform extraction for natural water samples, but in general adding a few milliliters of methanol destroyed the emulsion which sometimes developed; addition of methanol did not change the trialkyllead recovery.

It was found that in lake water the concentration of trimethyllead remains fairly constant during 24 hours in daylight whilst the concentration of triethyllead had reduced practically to zero under similar conditions. In the dark, the concentration of both species remained practically unaltered for several weeks.

Gas Chromatography. Chau et al[3359] have applied gas chromatography atomic absorption to the determination of tetraalkyllead compounds in water, sediments and fish samples in high lead areas. Of some 50 fish samples analyzed, only one sample was found to contain detectable amounts (0.26 g/g) of tetramethyllead in the fillet. Since there is no known tetraalkyllead industry and tetramethyllead is not used in gasoline in this area, the source of tetramethyllead is not yet known. The possibility that it comes from in-vivo lead methylation in the sediment or in the fish cannot be totally disregarded.

Chau et al[3360] have described a simple and rapid extraction procedure to extract the five tetraalkyllead compounds (Me_4Pb, Me_3EtPb, Me_2Et_2Pb, $MeEt_3Pb$, and Et_4Pb) from water, sediment, and fish samples. The extracted compounds are analyzed in their authentic forms by a gas chromatographic - atomic adsorption spectrometry system. Other forms of inorganic and organic lead do not interfere. The detection limit for water (200 ml) is 0.50 µg/l.

The gas chromatographic - atomic absorption system used by Chau et al (used without a sample injection trap) for this procedure is described elsewhere[3359].

The extract was injected directly into the column injection port of the chromatograph. Instrumental parameters were identical as previously described. A Perkin-Elmer Electrodeless Discharge Lead Lamp was used; peak areas were integrated with an Autolab-Minigrator (Spectra-Physics, Calif.).

The procedures used by Chau et al[3360] are outlined below.

Water analysis. Place 200 ml of water and 5 ml of high purity hexane in a 250 ml separatory funnel. Shake rigorously for 30 min in a reciprocating shaker. Let stand for about 20 min for phase separation. Drain off approximately 195 ml of the water and transfer the remaining mixture into a 25 ml tube with a Teflon-lined cap. Without separating the phases, inject a suitable aliquot, 5 - 10 µl of the hexane, to the GC-AAS system.

In Table 267 are tabulated recoveries obtained by the procedure for five tetraalkyllead compounds from lake water samples. The spiked samples were equilibrated for approximately 1 h and processed as described in the procedures for water, the recoveries averaged about 89%. Sediment is a much more complex matrix and the recoveries of the five compounds ranged 81 - 85% at the concentration level of 2 - 3 ppm.

Table 267 - Recovery of tetraalkyllead compounds from water samples[a]

Compound	Added µg	Found µg	Recovery %
Me_4Pb	10.00	8.78	87.8 ± 3
Me_3EtPb	13.15	11.80	89.7 ± 4
Me_2Et_2Pb	14.30	12.50	87.4 ± 3
$MeEt_3Pb$	10.15	9.08	89.5 ± 4
Et_4Pb	14.20	12.82	90.3 ± 7
		average	88.9 ± 7

[a] Four determinations for each sample.

Individual triaalkyllead species have been determined by gas chromatography with an element specific detector[3361-3363].

Several methods for the determination of tetraalkyllead have been proposed which depend on a combination of gas chromatography with selective detectors. There have been few reports on the direct determination of trialkyllead salts[3364] and none for dialkyllead salts. Forsythe and Marshall's[3365] approach to the determination of these salts was to further alkylate them with a Grignard reagent to convert them to their tetraalkyl analogues prior to analysis. Their paper describes a method for the isolation of trialkyl- and dialkyllead salts at the low parts per billion level from water and their determination, after derivatization, by capillary column gas chromatography with electron capture detection.

Extraction procedure. Ammoniacal buffer (pH 8.5, 10 mL) was added to the sample (60 mL) which was then extracted three times with 0.005% (w/v) dithizone (10 mL) in 50% benzene/hexane. The organic extracts were combined, reduced in volume to 0.5 mL, and derivatized directly.

Derivatization. Anhydrous tetrahydrofuran (2 mL) and 0.5 mL of phenylmagnesium bromide (3 M) were added to the concentrated extracts under nitrogen. The solution was stirred for 30 min at room temperature then transferred quantitatively to a centrifuge tube (15 mL). The volume was adjusted to 10 mL with water and then extracted three times with hexane (3 mL). Centrifugation followed each extraction to hasten phase separation. In some determinations, the combined hexane extracts were washed once with water (5 mL) and twice with 30% H_2O methyl cyanide (2 mL) and then diluted to 15 mL with hexane. In other determinations, the hexane extracts were washed once with water (5 mL) diluted to 15 mL with hexane, and then washed three times with methylcyanide (1 mL). The hexane was readjusted to 15 mL and analyzed directly.

A feature of the gas chromatographic apparatus, to avoid sample decomposition, was the provision of an all-glass insert and modified injector, which increased internal volume and decreased metal surface area.

Forsythe and Marshall[3365] used a 30 m fuzed silicon DB-1 column (J and W Scientific Co.) as this gave a superior separation of alkylleads from coextractives. Helium was chosen as carrier gas over nitrogen because it increased resolution of the mixture and nitrogen was chosen as makeup gas relative to 5% argon methane because it resulted in more stable detector operation. Nitrogen doped with 10 ppm oxygen caused increased detector response time. These data were used to predict retention times of $EtMe_2PbPh$, $Et_2MePbPh$, and $EtMePbPh_2$. Calibration curves were generated for each of the alkylphenyllead standards using chromatographic conditions described above. A linear increase in detector response was observed with increasing analyte concentrations (range 4 - 500 pg) of each of the four alkylphenyllead standards.

Recoveries of dialkyl and trialkyllead compounds obtained from water by this procedure were consistently high in the 0 - 20 ppb lead range.

Gas chromatography/mass spectrometry identified the following mixed alkylphenyl leads, $EtMe_2PbPh$, $Et_2MePbPh$, and $EtMePbPh_2$ as well as biphenyl and terphenyls in the crude reequilibration reaction mixtures. Forsythe and Marshall[3365] were not able to detect any of these "mixed" alkylleads during recovery trials of alkylphenyllead standards or during recovery trials using trialkyllead chlorides or dialkyllead chlorides.

Because recoveries from water or phosphate buffer were consistently high and because there were no significant peaks in the chromatograms which could have been attributed to transalkylation products, it is concluded that transalkylation was not a major source of loss in the methods.

Estes[3366] described a method for the measurement of triethyl- and trimethyllead chloride in tap water, using fused silica capillary column gas chromatography with microwave excited helium plasma lead specific detection. Element specific detection verified the elution of lead species, a definite advantage to the packed column method reported. The method involved the initial extraction of trialkllead ions from water as described by Bolanowska 3367,3368 into benzene, which was then vacuum reduced to further concentrate the compounds. Direct injection of the vacuum concentrated benzene solutions into the gas chromatography - microwave excited helium plasma system gave detectability of triethyllead chloride at the 30 ppm level and trimethllead chloride at the parts per million level, but the method was time-consuming and only semiquantitative.

The equipment used by Estes et al[3366] featured a gas chromatograph interfaced to a microwave induced and sustained atmospheric pressure helium plasma (9C-HED) for element selective and sensititive detection. It incorporates a chemically deactivated, low-volume, valveless fluidic logic gas switching interface, to vent large quantities of eluent solvent which would disrupt the helium discharge as sustained by the TM_{010} cylindrical resonance cavity. The inertness and venting characteristics of the interface are outstanding. The detection system features a low-resolution scanning monochromator with a quartz retractor plate background corrector directly after the entrance slit to improve selectivity ratios of elements whose emission wavelengths occur in the cyanogen background region. The plasma can be used as a nonselective universal organic compound detector.

Estes et al[3366] demonstrated the good resolution and great inertness of this system by applying it to the determination of chemically active and thermally sensitive trialkyllead chlorides.

The GC-MED operating parameters for lead are shown in Table 268. In order to avoid thermal decomposition of trialkyllead compounds at active sites on the apparatus, the quartz injection port liner, quartz discharge tube, and the quartz interface tubing were removed and silanized by passing 100% dichlorodimethylsilane through the tubing. The tubing was placed in an airtight container flushed with nitrogen and allowed to react for 30 min. The tubes were washed with a large quantity of "spectrograde" methanol in order to quench any remaining chlorosilane bonds and dried at 250°C for 1 h while being flushed with helium. This was followed by helium flushing at room temperature overnight. It is quite evident from examination of the gas chromatograms that the differences in triethyllead chloride response from the "at plasma" and "venting" column positions was negligible. Thus, the loss of triethyllead chloride when utilizing the nonsilanized quartz interface tubing was due to the chemically active quartz surface.

Table 268 - GC-MED operating parameters for lead

Paramaters	for lead
column	
packing material	SP-2100 WCOT fused silica Carbowax pretreated
dimensions	12.5 m x 300 μm o.d. x 200 μm i.d.
injection split	100 to 1
carrier gas flow rate (helium)	1 mL/min
Temperatures	
column or program	140 - 165°C
injector	180°C
transfer block	180°C
interface oven	180°C
total plasma flow rate	125 mL/min
PMT tube & voltage	RCA 1P28, 700 V
entrance & exit slit	
widths	25 μm
height	12 mm
microwave input power	54 W
wavelength obsd	405.8 nm
picoammeter time constant	0.10 s

In an application of this procedure, five-hundred-milliliter samples of tap water were spiked with trimethyllead chloride and/or triethyllead chloride in the range from 10 ppb to 10 ppm. The mixture was stirred for 10 min with a magnetic stirring bar and the sodium chloride concentration was then raised to 300 mg/mL. If the pH was not between 5 and 7, it was adjusted with hydrochloric acid or sodium hydroxide. The resulting mixture (500 mL) was extracted with 50 mL of "spectrograde" benzene in a 1 L separatory funnel and allowed to separate for 20 min. The benzene was drawn off and placed in a 100 mL round-bottom flask (14/20 ground glass joint). The volume of benzene was vacuum reduced to between 0.25 and 4 mL. A sample of the final benzene volume (2 - 5 μL) was analyzed with the gas chromatograph - microwave emission detector system. The total extraction and vacuum reduction required approximately 2.5 h.

The trimethyllead chloride is more volatile than triethyllead chloride and elutes long before the discharge has returned to equilibrium. Therefore, although trimethyllead chloride is at a higher level than triethyllead chloride, the latter compound gives the stronger signal. The detection limit is at the 10-30pb level of triethyllead chloride in tap water under the chromatographic and extraction - vacuum reduction procedures utilized. The extraction - vacuum reduction procedure was determined to be approximately

50% efficient for triethyllead chloride but probably less efficient for the trimethyl compound The loss of sample is probably as a result of decomposition of the trialkyllead chlorides on the glass walls of the round-bottom flask during benzene vacuum reduction, since the compounds will not sublime.

Estes et al[3369] have also reported the n-butyl Grignard derivatization of the trialkyllead ions extracted as the chlorides from spiked tap water and industrial plant effluent. A precolumn trap enrichment technique is substituted to replace solvent extract vacuum reduction. Final measurement of the lead compounds, now as n-butyltrialkylleads, is undertaken with the gas chromatograph microwave emission detector system. Precolumn Tenax trap enrichment of the derived trialkylbutylleads enables determination to low parts per billion levels to be carried out. Also investigated are extraction efficiencies and injection split ratios onto a fused silica capillary column.

If the aqueous sample was not at approximately pH 7, it was adjusted to be so with hydrochloric acid or sodium hydroxide. No less than 500 mL of the plant effluent was adjusted to pH 7. A 100 ml aliquot was saturated with sodium chloride (ca. 36 - 37 g of sodium chloride/100 mL) and extracted with 5 mL of spectrograde benzene. The benzene was drawn off with a dry syringe and placed in a clean dry test tube and the salt saturated aqueous portion was again extracted with 5 mL of benzene. The benzene fractions were combined.

A 5 mL portion of the benzene extract was placed in a 10 mL conical bottomed centrifuge test tube and 1 mL of 2.0 M n-butylmagnesium bromide Grignard reagent in tetrahydrofuran was added. The test tube was tightly stoppered and was allowed to stand for 1 h. After n-butylation, 250 μL of a tetra-n-butyllead in benzene-internal-reference solution (0.316 mg/mL) was added. Excess Grignard remained in the mixture which could be stored at 0°C or lower without decomposition for at least 1 day, if the subsequent analysis could not be performed immediately.

The direct quantitative gas chromatographic measurement of trimethyllead chloride or triethyllead chloride suffers from two major difficulties. (i) Both compounds are thermally unstable and tend to decompose even at the lowest possible injection port temperatures (ca. 160 - 170°C) required to give complete and rapid volatilization. (ii) Both compounds are very chemically reactive giving some tailing of chromatographic peaks even with the most inert chromatographic column available. Thus, direct quantitative gas chromatographic measurement of trialkyllead compounds is difficult.

If the trialkyllead compounds can be converted to tetraalkylleads, quantitative determination is feasible. The use of n-butylmagnesium bromide Grignard reaction for trimethyl and triethyllead chlorides appeared to be promising. Tetra-n-butyllead can be used as an internal reference which will not interfere with speciation of methyl or ethyl tetra- or trialkyllead compounds, and should mark the termination of the gas chromatograph microwave emission detector lead specific analysis, since it should be the last tetra-alkyllead compound to elute.

To determine derivatization efficiency, Estes et al[3369] made direct injections of benzene standards corresponding to the extraction of 100 mL portions of tap water containing trimethyl or triethyllead chlorides at the 10 ppb, 100 ppb, or 1 ppm level, (5 mL) of benzene were employed for each extraction. Comparison injections were made separately of 0.1 μL samples of tetraethyllead standards in benzene (5 ppm, 50 ppm, and 500 ppm). Assuming the 100% extraction from tap water occurred for both chlorides, then 95 ± 8% derivatization/n-butylation efficiency was obtained.

It was then necessary to determine the injection split ratios, n-$BuMe_3Pb$/n-Bu,Pb and n-$BuEt_3Pb$/n-Bu_4Pb since analyte and internal reference tetraalkylleads are of different volatility and size. Injection port splitting differs due to a dilution effect caused by varying release from the Tenax trap and perhaps molecular size split selectivity. In addition, the longer retained tetra-n-butyllead suffers band broadening (partially a result of the Tenax trap retention) with an accompanying loss in detectability. The broadening contribution from desorption could not be readily differentiated.

For the range of benzene n-butylated trimethyl- and triethyllead chloride standards coinjected with tetrabutyllead, the tetraalkyllead analyte to internal reference split ratios were established as n-$BuMe_3Pb$/n-Bu_4Pb = 12.0 ± 1.0 and n-$BuEt_3Pb$/n-Bu_4Pb = 4.79 ± 0.50 based on lead content. Concentration range was from 5 to 500 ppm.

Over the spiked tap water concentration range investigated, the extraction efficiency of trimethylleadchloride was 5.7 ± 0.6% and of triethyllead chloride was 93 ± 12%, a not unexpected result in view of the more ionic nature of the former compound.

Detection limits were 35.0 ppb (based on a 2 x noise signal) for trimethyllead chloride and 5.6 ppb for triethyllead chloride. In addition, the detectability of both compounds could be improved by the use of a larger precolumn Tenax trap.

Other organolead compounds elute at detectable levels after the compounds of interest but before the internal reference. These compounds could result from decomposition of impurities in the trimethyl- to triethyllead chloride, thermal redistribution products occurring from reaction in the injection port, impurities in the internal reference, or organoleads in the tap water. A tap water blank was carried through the analysis procedure and to ensure detection of organoleads other than the internal reference, three 30 µL samples were placed on the same trap before desorption. The additional organolead compounds were not seen in these experiments. Injection port thermal redistribution of alkyllead compounds is unlikely to occur; hence it seems likely that the unidentified lead peaks are derivatized species perhaps of decomposition impurities (i.e. di- and monoethyl and -methylleads) in the trialkyllead chloride standards.

Estes et al[3369] also applied their procedure to an industrial plant effluent containing trialkyllead compounds. Gas chromatographic microwave emission detector analysis for trimethyl- and triethyllead ions in the effluent showed the presence of the latter at a 19.0 ± 4.0 ppb level. Single 30 µL samples, however, contained insufficient trimethylbutyllead for detection, and the multiple sample preconcentration technique was required.

The procedure described by Estes[3369] offers several advantages for the simultaneous determination of trimethyl- and triethyllead ions in aqueous media. It is rapid and reproducible; it does not require the most strictly deactivated fused silica columns, since the chemically active trialkyllead chlorides are quantitatively converted to relatively chemically inert n-butyltrialkylleads. Thus, other capillary columns can be utilized with little effect on the quantitation, because the tetraalkylleads are so amenable to gas chromatography. The precolumn trap enrichment procedure eliminates the need to destroy excess Grignard reagent (to prevent column degradation) and the need for solvent extract vacuum reduction. Elimination of the vacuum reduction step solves the problems of trialkyllead chloride decomposition on the walls of the vacuum vessel and the loss of analyte compound due to volatilization at reduced pressure. The use of an n-butyl Grignard allows the simultaneious speciation of all organoleads of the methyl and ethyl alkyls (i.e. tetraalkyltrialkyl, dialkyl, and monoalkyl) which might be extracted and derivatized.

Chau et al[3370] described a technique in which dialkyl lead and trialkyl lead are quantitatively extracted into benzene from aqueous solution, following their chelation with dithiocarbamate. Using a Grignard reagent they are then converted to their butyl derivatives for gas chromatography/ atomic absorption spectrometry. A detection limit of 0.1 ug lead per litre can be achieved with one litre of water. Other metals coextracted with the chelating agent do not interfere. Molecular covalent tetraalkyllead species, if present in the sample, are also extracted and quantified simultaneously.

Brief details of this procedure are given below; to 1 L of water are added 50 ml of aqueous 0.5 M sodium diethyldithiocarbamate, 50 g of sodium chloride and 50 ml of benzene, and the mixture is shaken for 30 min. The benzene phase is then carefully evaporated in a rotary evaporator to 4.5 ml in a 10 ml centrifuge tube to which 0.5 ml of butyl Grignard reagent is added. The mixture is gently shaken for 10 min, and washed with 5 ml of 0.5 M sulfuric acid to destroy the excess of Grignard reagent. About 2 - 3 ml of the organic phase is pipetted into a small vial and dried with anhydrous sodium sulfate. Appropriate amounts (5 - 10 µl) are injected into the g.c. - a.a.s. system.

Extraction efficiencies of alkyllead compounds are tabulated in Table 269.

Table 269 - Extraction efficiency of alkyllead compounds

Medium	Extraction efficiency (%)[b]				
	Me_2Pb^{2+}	Me_3Pb^+	Et_2Pb^{2+}	Et_3Pb^+	Pb(II)
NaCl(sat.)	10(3)	40(4)	14(4)	95(3)	0
NaCl(sat.) + KI (40g)	45(3)	100(5)	58(6)	112(8)	5(4)
Tropolone	17(4)	25(4)	20(7)	15(7)	20(6)
NaDDTC	109(8)	97(6)	105(9)	94(5)	94(8)
NaDDTC + NaCl(sat.)	98(5)	100(7)	97(6)	98(6)	93(5)

[a] Distilled water, 100 ml; lead compounds, 20 µg each species; volume of benzene in all cases, 5 ml; tropolone (0.5% in benzene), 5 ml; NaDDTC (0.5 M), 5 ml.

[b] Average of two results with average deviation in parentheses.

NaDDTC = sodium diethyldithiocarbamate.

A chromatogram showing the separation of five tetraalkylleads, four ionic alkyllead and Pb^{2+} species in a synthetic sample is illustrated in Figure 184.

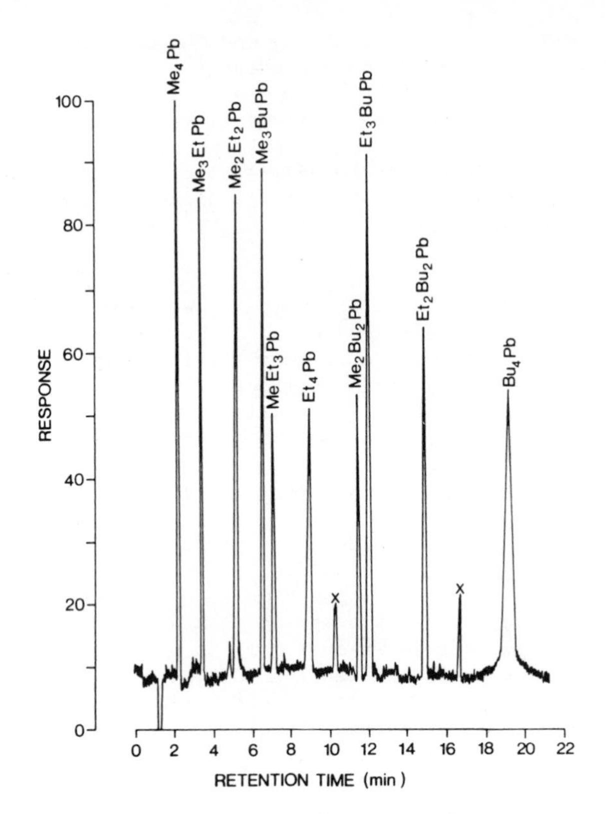

Figure 184 - Gas chromatography - atomic absorption spectrometry of five tetraalkyllead compounds (10 ng each); four butyl derivatives of dialkyl- and trialkyllead (8 ng each) and Pb^{2+} (15 ng). x, unidentified lead compounds.

Polarography. Direct polarography has been used to determine trialkyllead compounds in water[3371], Colombini et al[3372] have described a method for the electrochemical speciation and determination of alkyllead compounds in natural waters and seawater. The method is based on selective organic-phase extraction coupled with differential pulse polarography. The analytical procedure was found to be reliable for the individual detection and determination of organolead complexes at trace levels.

Bond et al[3373] examined the interferences occuring in the stripping volammetric determination of trimethyllead in seawater by polarography and mercury 199 and lead 207 nuclear magnetic resonance spectrometry. NMR and electrochemical data show that Hg(II) reacts with $[(CH_3)_3Pb]^+$ in seawater. Consequently, anodic stripping voltammetric methods for determining $[(CH_3)_3Pb]^+$ and inorganic Pb(II) may be unrelaible.

Miscellaneous. Solvent extraction followed by spectrophotometry have been used to determine trialkyllead compounds in water[3374-3378]. Individual trialkyllead species have been determined in water by thin-layer chromatography[3379,3380]. Jarvie et al[3381] have studied the reactions between trimethyllead chloride and sulphides in aqueous systems, and factors affecting these reactions. It is concluded that the formation of tetramethyllead from trimethyllead compounds in natural waters is due to chemical reactions, and not to biomethylation.

Glockling[3382] has devised ways of studying the degradation of organolead compounds and determining the toxicity of their breakdown products.

ORGANOARSENIC COMPOUNDS

Atomic absorption spectrometry. Various workers have used volatile hydride generation with sodium borohydride followed by various separation and detection systems for the determination of organoarsenic species. Separation of the volatile arsines is achieved by gas chromatography[3383] or sequential volatilization[3384,3385]. The detection systems are microwave emission spectrometry, electron capture and flame ionization detection[3383], dc discharge emission[3386,3377], atomic absorption spectrometry[3383,3388] and neutron activation analyses[3389].

Edmonds and Francesconi[3390] have reported that alkylated arsenicals methylarsonic acid and dimethylarsinic acid occuring universally in the environment may be estimated directly by vapor generation atomic absorption spectrometry without prior digestion. Sodium borohydride treatment produces methylarsine, and dimethylarsine, respectively, which are swept directly into a hydrogen - nitrogen entrained air flame by the excess hydrogen generated by hydrolysis of the sodium borohydride. These methylated arsines are estimated in a manner identical to the arsine produced following acid digestion or dry ashing. The calibration curves and instrument response are directly comparable and are dependent only on the quantity of arsenic entering the flame.

These workers used a Varian Model 1 000 atomic absorption spectrometer fitted with a Varian Model 64 arsenic determination apparatus with the following instrumental parameters: Wavelength, 193.7 nm; spectral band width, 1.0 nm; lamp current, 7 mA. Gas flow rates were: hydrogen, 20 L min^{-1}; nitrogen 20 L min^{-1}. Estimations were carried out as described by Duncan and Parker[3391] with similar calibration curves and instrumental response. Detection limits were 500 ng for inorganic arsenic and 1 µg for monomethylarsonic acid and dimethylarsinic acid. Aqueous solutions of methylarsonic acid and dimethylarsinic acid were adjusted to 2% in hydrochloric acid before addition of the sodium borohydride solution.

Standard mixtures of sodium arsenate, methylarsonic acid, and dimethylarsinic acid were treated with sodium borohydride and the mixed arsines generated trapped in a glass bead packed tube (200 mm x 25 mm) at - 180°C. The cooling agent was removed and the trap allowed to warm slowly in the laboratory atmosphere. A valve assembly kept the trap sealed and was released periodically with a simultaneous flow of nitrogen through the trap into the flame. The valve was opened for 5 s each min. The instrument response was recorded on a chart moving at 2 mm min^{-1} and resembled, in outline, a gas chromatographic trace with arsine peaking at 3 min, methylarsine at 8 min, and dimethylarsine at 13 min.

Arsine itself, originating by the reduction of inorganic arsenic, and methyl arsines, originating by the reduction of organoarsenic compounds, have different responses in colorimetric versus atomic absorption methods of analysis. The colorimetric diethyldithiocarbamate method is much more responsive to arsine than to methyl arsines, whilst vapour generation atomic absorption spectrometry is equally responsive to both forms of arsenic, depending only on the weight of arsenic entering the flame. This explains the difference in the results obtained by Uthe[3392] and Penrose[3393] in the determination of organoarsenic compounds. These authors each compared the efficiency of wet ashing (oxidizing acid) and dry ashing (magnesium oxide/ magnesium nitrate). Whereas Uthe[3392] found wet ashing gave a slightly higher recovery than dry ashing (both approach 100%), Penrose[3393] obtained good recovery for dry ashing but less than 5% for wet ashing. The difference lies in their methods of estimating the arsine generated after the reductive step. Penrose used a colorimetric method whereas Uthe used vapor generation atomic absorption spectrometry. Therefore around 100% recovery would have been achieved if no oxidation had occurred under the digestion conditions.

Pensso and Inguin[3394] have described a method for the determination of dimethylarsinic acid in sea water in the sub ppm range by electrothermal atomic absorption spectrometry after preconcentration of a strong cation exchange column and elution with ammonia.

Gas chromatography. Lussi-schlatter and Brandenberger[3395,3396] have reported a method for inorganic arsenic and phenylarsenic compounds based upon gas chromatography with mass specific detection after hydride generation with head space sampling. However, methylarsenic species were not examined.

Odanaka et al[3397] have recently reported that the combination of a GC-MID (gas chromatography with multiple ion detection)system and hydride generation heptane cold trap (HG-HCT) technique is useful for the quantitative determination of inorganic, monomethyl-, dimethyl-, and tri- methylarsenic oxide (TMA = O) compounds, and this approach is applicable to the analysis of environmental and biological samples.

In this method arsine and methylarsines produced by sodium borohydride reduction are collected in n-heptane (-80°C) and then determined by GC-MID. The limit of detection for a 50 mL sample was 0.2 - 0.4 ng ml^{-1} of arsenic for arsenic compounds. Relative standard deviations ranged from 2% to 5% for distilled water replicates spiked at the 10 ppb level. Recoveries of all four arsenic species from river water ranged from 85% to 100% (Table 270).

Table 270 - Analysis of standard arsenic[a]

standard	generated arsines	arsine species, %
DSMA	arsine	0.08
	methylarsine	99.8
	dimethylarsine	0.05
	trimethylarsine	0.03
DMAA	arsine	0.07
	methylarsine	0.05
	dimethylarsine	99.8
	trimethylarsine	0.06
TMA = O	arsine	0.05
(trimethyl-	methylarsine	0.05
arsine	dimethylarsine	0.08
oxide)	trimethylarsine	99.8

[a] Reduction samples containing 20 µg of standard arsenic.

Aqueous samples (1 - 50 mL) that had been previously neutralized with hydrochloric acid and/or sodium bicarbonate, were placed in the reaction vessel and were diluted to 60 mL with water. Then 6 mL of 4 N hydrochloric acid and 2 mL of methanol were added. The cold trapping system was then connected and the carrier gas (helium) was allowed to pass through the system at 100 - 150 mL/min for 1 min to flush out any air. The reduction was initiated by injecting 3 mL of 10% sodium borohydride solution through the rubber septum into the aqueous sample. The volatile arsines were collected in a n-heptane (3 - 5 mL) cold trap for 2 min. The low temperature was maintained by submerging the trap in a dry ice-acetone (- 80°C) bath. The helium flow as continued for 1 min to ensure complete generation and trapping of the arsines. After the collection of the arsines, n-heptane (5 µL) was injected into the gas chromatograph/mass spectrometer, for quantitative determination.

The following ions were characteristic as intense ions in the mass spectra of arsines: arsine, m/z 78 M+, 76 (M - 2)+, methylarsine: m/z 92 M+. 90 (M - 2)+, 76 [(M - CH_3) - 1]+, dimethylarsine, m/z 106 M+, 90 [(M - CH_3) - 1]+, trimethylarsine, m/z 120 M+, 105 [(M - CH_3) - 1]+, 103 [(M - CH_3) - 2]+. To achieve simultaneous measurement and to assess the specificity of the analysis, for instance, the m/z 76, 78, 89 and 90 were monitored for arsine, methylarsine, and dimethylarsine and/or m/z 90, 103, 105, and 106 for alkylarsines as methylarsine, dimethylarsine, and trimethylarsine. However, simultaneous determination of all four arsenicals could not be done at one injection because of a limited range of detectable mass spectra in the multiple ion detection system used.

The collection efficiency of arsenic species, especially in dimethylarsinic acid and trimethylarsine oxide (TMA = O) was found to be highly dependent on the concentration of hydrochloric acid and sodium borohydride. However, Table 270 indicates that even if the rearrangement occurs, the degree of that is very little under the conditions of the optimized procedure.

The calibration plots for each arsenic compound are linear from 0.2 ng of As mL^{-1} to 2000 ng of As mL^{-1} in aqueous solutions (50 mL). The absolute detection limit per injected sample is 30 pg of As for arsine (m/z 78) and trimethylarsine (m/z 120) and 20 pg of As for methylarsine (m/z 90) and dimethylarsine (m/z 90). When arsines which have been generated by reduction of a 50 mL sample are collected in 3 mL of heptane and 5 µL is used for analysis, the relative detection limits are 0.4 ng mL^{-1} of arsenic for arsenate and trimethylarsine oxide and 0.2 ng mL^{-1} for dimethylarsinic acid and methylarsonic acid (sodium salt).

The reproducibility of the method for distilled water samples containing 0.5 µg of As for each arsenic species was quite good. The relative standard deviation values were 5% for arsenate, 2% for methylarsonic acid sodium salt and dimethylarsinic acid, and 5% of trimethylarsine oxide. Recoveries of inorganic arsenic and mono di and tri methylarsenic compounds were in the range 83 to 110% in spiking experiments on river water.

Parris et al[3398] have studied in detail the chemical and physical considerations that apply in the determination of trimethyl arsine in gases using an atomic absorption spectrophotometer with a heated graphite tube furnace as a detector for a gas chromatograph. 5 ug arsenic could be detected by this technique.

Andreae[3383] described a method for the sequential determination of arsenate, arsenite, mono-, di- and trimethyl arsine, monomethylarsonic and dimethylarsinic acid, and trimethylarsine oxide in natural waters with detection limits of several ng/L. The arsines are volatilized from the sample by gas stripping; the other species are then selectively reduced to the corresponding arsines and volatilized. The arsines are collected in a cold trap cooled with liquid nitrogen. They are then separated by slow warming of the trap or by gas chromatography, and measured with atomic absorption, electron capture, and/or flame ionization detectors. He found that these four arsenic species all occurred in natural water samples.

The apparatus for the volatilization and trapping of the arsines (Figure 185) is constructed from Pyrex glass, with Teflon stopcocks and tubing, and with Nylon Swagelok connectors. The sample trap consists of a 6 mm o.d. Pyrex U-tube of ca. 15 cm length, filled with silane treated glass wool. The interior parts of the six-way valve which interfaces the volatilization system with the gas chromatograph are made of Teflon and stainless steel.

The gas chromatography (Hewlett-Packard Research Chromatograph HP 5750B) is equipped with a ^{63}Ni electron capture detector mounted in parallel with a flame ionization detector and an auxiliary vent by the use of a column effluent splitter. The separation is performed on a 4.8 mm o.d. 6 m long stainless steel column, packed with 16.5% silicone oil DC-550 on 80 - 100 mesh Chromosorb WAW DMCS. The helium carrier gas flow rate is 80 mL/min.

The electron capture detector (Hewlett-Packard 2-6195, with a ^{63}Ni source) is operated in the constant pulse mode with a pulse interval of 50 µs. The atomic absorption detection system consists of a Varian AA5 with a hollow cathode arsenic lamp; the standard burner head is replaced by a 9 mm i.d. quartz burner cuvette (Figure 186)[3399].

Isolation of the Arsenic Species. The sample, 1 - 50 mL, is introduced into the gas stripper with a hypodermic syringe through the injection port. Any volatile arsines in the sample are stripped out by bubbling a helium stream through the sample. Then 1 mL of the Tris buffer solution for each 50 mL sample is added, giving an initial pH of about 6. Into this solution 1.2 mL of 4% sodium borohydride solution is injected while continuously stripping with helium. After about 6 - 10 min, the As(III) is converted to arsine and stripped from the solution. the pH at the end of this period is about 8. Then 2 mL of 6 N hydrochloric acid is added, which brings the pH to about 1. The addition of three aliquots of 2 mL of 4% sodium borohydride solution during 10 min reduces As(V), monomethylarsonic acid, dimethylarsinic acid and trimethylarsine oxide to the corresponding arsines which are swept out of the solution by the helium stream coming from the reaction vessel and the evolved hydrogen.

It was found necessary to dry the helium gas stream coming from the reaction vessel. Effective drying was obtained with a 28 cm long Pyrex U-tube, 7 mm i.d., immersed in a dry ice-isopropyl alcohol bath.

For direct detection by the atomic absorption technique, the stripping gas stream can be used to carry the sample into the burner. The arsines are isolated by immersing the gas trap in liquid nitrogen and released by slowly warming it up to room temperature. This slow warming results in a sequential separation of the arsines on the basis of their boiling points.

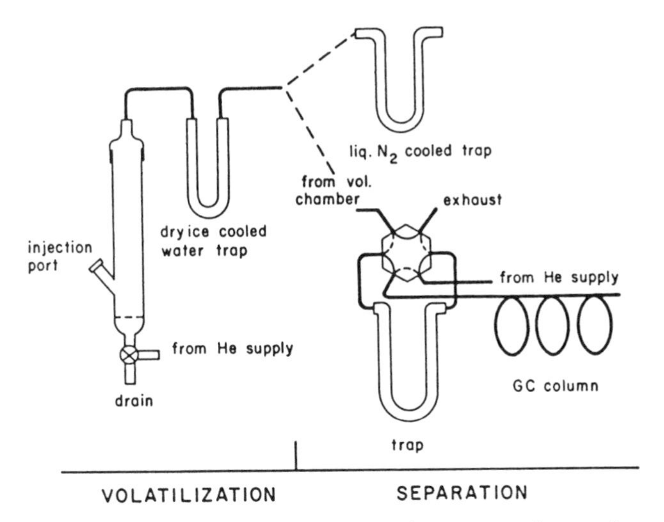

Figure 185 - Apparatus for the volatilization, trapping and separation of the arsines.

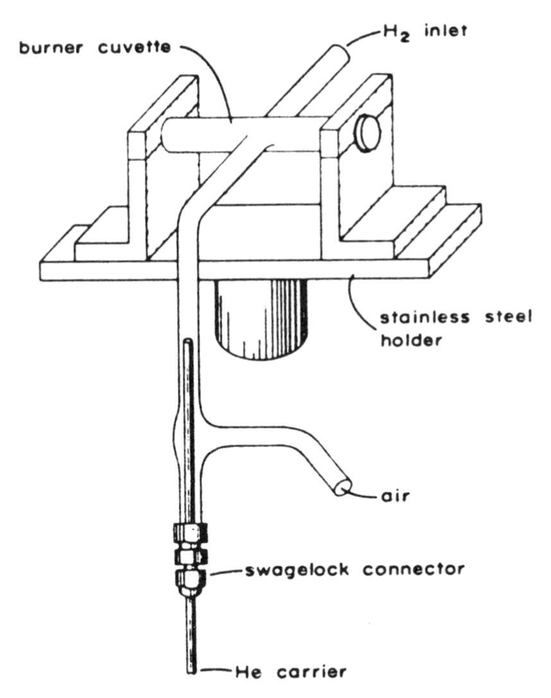

Figure 186 - Quartz cuvette burner head. The cylindrical base of the cuvette holder fits into the standard AA burner socket.

If the gases are to be separated by gas chromatography, the sample trap is attached to a six-way valve, which allows it to be switched from the stripping gas stream of the generating apparatus to the carrier stream of the gas chromatograph (Figure 185). The sample is first trapped at liquid nitrogen temperature from the stripping gas, then the valve is switched to the carrier stream and the arsines are rapidly evaporated by immersing the trap in hot water. Three different detection devices were investigated: a flame ionization detector, an electron capture detector, and an atomic absorption detector. Any combination of them can be attached to the outlet of the gas chromatographic column.

The modular approach is highly efficient in the analysis of natural waters with a high variability of arsenic species composition. Whereas the inorganic forms of arsenic could always be determined by the relatively rapid boiling point-separation - atomic absorption detection technique (about 20 min per sample for As(III) and As(V), the slower and somewhat more cumbersome gas chromatographic method with electron capture detection allows the determination of the organic arsenic species in the low parts per trillion range often found in natural waters. For samples with relatively high concentrations of the organic forms, e.g. urine, a configuration using gas chromatographic separation and detection by atomic absorption or flame ionization detection allows working in the high nanogram to microgram range, thus avoiding some of the problems of work in the low nanogram - subnanogram range.

The methods outlined above were applied to a number of natural water samples.

Sample treatment and storage. If stored in air-tight containers, the free arsines are stable in solutions for a few days. They are slowly oxidized by traces of air to the corresponding acids. From untreated samples, the methylated acids are lost measureably after a period of about three days, depending on the initial concentrations. They are stable indefinitely if the sample is made 0.05 N in hydrochloric acid. Arsenite is slowly oxidized to arsenate; in samples below 0.05 ppb As(III) a loss of arsenite becomes detectable after about one week. Acidification of the sample increases the oxidation rate, and arsenite loss can be detected after one day. If the samples are stored in a freezer below - 15°C or under dry ice, an initial loss of arsenite corresponding to about 0.02 ppb is experienced. The sample then remains unchanged with prolonged storage.

Ion exchange chromatography. Early work on the determination of organoarsenic compounds by ion-exchange chromatography was discussed by Reay and Asher[3400].

Ion-exchange chromatography has been used to achieve separations of monomethylarsonate, dimethylarsinite and tri and pentavalent arsenic. Various workers[3401-3404] have described methods which separated the inorganic arsenic from each of the organic species using ion-exchange chromatography. Here, further inorganic speciation relies on redox-based colorimetry[3405]; both the accuracy and precision suffer from the low As(III) /As(V) and As(total)/P ratios normally encountered in the environment. Henry and Thorpe determined these four arsenicals by coupling a digestion and reduction scheme with ion-exchange chromatography[3406]. However, the utility of this technique for routine environmental analysis is limited, since the implementation time is substantial. This method also relies on estimating As(V) by arithmetic difference.

Grabinski[3407] has described an ion-exchange method for the complete separation of the above four arsenic species on a single column containing

Table 271 - Arsenic species concentrations in natural waters, ppb As

Locality and sample type	AS(III)	As(V)	MMAA	DMAA
Seawater, Scripps Pier, La Jolla, Calif.				
5 Nov. 1976	0.019	1.75	0.017	0.12
11 Nov. 1976	0.034	1.70	0.019	0.12
Seawater, San Diego Trough				
Surface	0.017	1.49	0.005	0.21
25 m below surface	0.016	1.32	0.003	0.14
50 m below surface	0.016	1.67	0.003	0.004
75 m below surface	0.021	1.52	0.004	0.002
100 m below surface	0.060	1.59	0.003	0.002
Sacramento River, Red Bluff, Calif.	0.040	1.08	0.021	<0.004
Owens River, Bishop, Calif.	0.085	42.5	0.062	0.22
Colorado River, Parker, Ariz.	0.114	1.95	0.063	0.051
Colorado River, Slugh, near Topcock, Calif.	0.085	2.25	0.13	0.31
Saddleback Lake, Calif.	0.053	0.020	<0.002	0.006
Rain, La Jolla, Calif.				
10 Sept 1976	<0.002	0.180	<0.002	0.024
11 Sept 1976	<0.002	0.094	<0.002	<0.002

both cation- and anion-exchange resins. Flameless atomic absorption spectrometry with a deuterium arc background correction is used as a detection system for this procedure. This detection system was chosen because of its linear response and lack of specificity for these compounds combined with its resistance to matrix bias in this type of analysis. The elution sequence was as follows: 0.006 M trichloroacetic acid (pH 2.5), yielding first As(III) and then monomethylarsonate; 0.2 M trichloroacetic acid yielding As(V); 1.5 M NH_4OH followed by 0.2 M trichloroacetic acid yielding dimethylarsinite. Detection was by flameless atomic absorption spectrometry. Arsenic recoveries (full procedure) ranged from 97% to 104% for typical lake water samples; more erratic but still acceptable recoveries (96% to 107%) were obtained from arsenic contaminated sediment interstitial water. The overall analytical detection limit was 10 ppb (original sample mixture) for each individual arsenic species. Relative standard deviations ranged from 0.7% to 1.3% for lake water and distilled deionized water replicates spiked at the 500 ppb level.

In order to demonstrate the practical utility of their technique, Grabinski[3407] spiked 0.500 µg of each arsenic species into filtered (0.45 µM) Lake Mendota water (sampled in Madison, WI) and distilled, deionized water. Arsenic recoveries for the entire procedure averaged 104%, 100%, 97% and 99% for As(III), monomethylarsonate, As(V) and dimethylarsinite, respectively. Relative standard deviations for replicate determinations ranged from 0.7% for As(III) and dimethylarsinite to 1.3% for As(V). A sample chromatogram of Lake Mendota water is shown in Figure 187.

Interstitial water from sediment sampled at the mouth of the Menominee River near Marinette,WI, was also analyzed for each arsenic species. The water was contaminated with arsenic (primarily monomethylarsonate) and had considerable air exposure. The results of this analysis are presented in Table 272.

Aggett and Kudwani[3408] point out that although speciation by methods

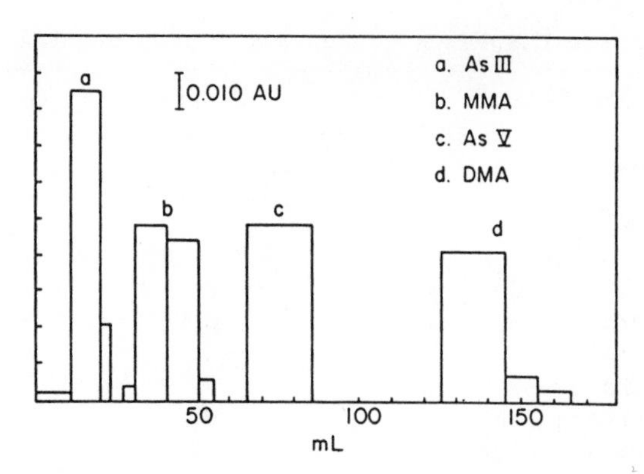

Figure 187 - Sample chromatogram of spiked Lake Mendota Water.

Table 272 - Arsenic speciation from spiked and unspiked contaminated sediment interstitial water

	species			
	As(III)	MMMA	As(V)	DMA
interstitial water, ng	0	2278	300	18
spike, ng	300	300	300	300
interstitial water + spike, ng	310	2576	620	330
recovery of spike, ng	310	298	320	312
recovery of spike, %	103	99	107	104

based on hydride generation is attractive for those systems to which it is applicable it is of course limited to the determination of those species which can be converted into volatile hydrides. Fortunately, the four species most commonly considered to be of environmental importance at the present time, i.e. arsenate, arsenite, monomethylarsonic acid and dimethylarsinic acid are amenable to this form of analysis. However, conventional ion-exchange and ion-chromatographic methods appear to possess the potential advantage that it should be possible to extend or modify them to include the analysis of additional environmentally important arsenic species should that become necessary.

They report the development and application of a relatively simple anion-exchange method for the speciation of arsenate, arsenite, monomethylarsonic acid and dimethylarsinic acid. As these four arsenic species are weak acids the dissociation constants of which are quite different (Table 273) it seemed that separation by anion-exchange chromatography was both logical and possible. However, the first two published ion-exchange methods for speciation of arsenic[3402,3403] both used cation-exchange chromatography. The mechanism for these separations has not been established and attempts to use them to separate inorganic arsenic(III) and inorganic arsenic(V) were not successful. Subsequently, Henry and Thorpe[3406] and other workers[3404,3407] have published methods that require the use of both cation- and anion-exchange columns.

Aggett and Kadwani[3408] employed a two-stage single column anion-exchange method using hydrogen carbonate and chloride as eluate anions. These species appear to have no adverse effects in subsequent analytical procedures. Its

Table 273 - Dissociation constants of arsenic species

Acid	pk_{a1}	pk_{a2}	pk_{a3}
Arsenic acid	2.25	7.25	12.30
Arsenious acid	9.23		
Monomethylarsonic acid	4.26	8.25	
Dimethylarsinic acid	6.25		

successful application is dependent on careful control of pH. It has been applied to the speciation of arsenic in samples obtained during studies on the fate of arsenic released into the Waikato River from geothermal sources in its catchment. For developmental purposes analyses were performed by hydride generation atomic-absorption spectroscopy, although any extension to more general speciation would require the use of a more general technique for analysis, such as graphite furnace atomic absorption spectroscopy or inductively coupled plasma atomic emission spectroscopy.

Preliminary experiments were conducted with various combinations of resins and eluates in the pH range 5 - 7. In these the only promising separations were obtained with carbon dioxide - hydrogen carbonate as the eluate.

The behaviour of arsenic (III) and dimethylarsinic acid on SRA 70 resin (BDH Chemicals Ltd. U.K.) as a function of the pH of the carbon dioxide - hydrogen carbonate buffer is shown in Figure 188. Neither monomethylarsonic acid nor arsenic (V) was eluted by 250 ml of eluate at 1 - 4 ml per min in the pH range 4.8 - 6.4. These observations are understandable in terms of the dissociation constants of the respective acids and the nature of the buffer solution used for elution.

Arsenic(III) exists as an undissociated molecule over the pH range studied and as a consequence is eluted rapidly in a manner independent of pH. At lower pH the elution of dimethyl-arsinic acid overlaps that of arsenic (III) but as the pH is raised there is greater tendency for dimethylarsinic acid to be retained by the resin, presumably a consequence of dissociation. This is reversed somewhat in the region of pH 6.0 - 6.4. The cause of this is believed to be the increase in hydrogen carbonate concentration in the eluent in this pH region. In practice, the hydrogen carbonate concentration increased from about 6×10^{-3} mol l^{-1} at pH 5.4 to about 6×10^{-2} mol l^{-1} at pH 6.4.

The results showed that separation of arsenic(III) and dimethylarsinic acid by elution with carbon dioxide - hydrogen carbonate was satisfactory in the pH range 5.2 - 6.0.

Elution of monomethylarsonic acid was accelerated and satisfactory separation from arsenic(V) achieved using saturated aqueous carbon dioxide (pH 4.0 - 4.2) containing 10 g l^{-1} of ammonium chloride. Thus complete separation of the four species was obtained by eluting first with 150 ml (30 x 5 ml aliquots) of carbon dioxide - hydrogen carbonate buffer at pH 5.5 ± 0.3 followed by elution with a further 150 ml (30 x 5 ml aliquots) of carbon dioxide - ammonium chloride buffer solution at pH 4.0 - 4.2. Although the pH range for successful separation of arsenic(II) from dimethylarsinic acid appears small, no problems were encountered in meeting that specification during subsequent speciation procedures.

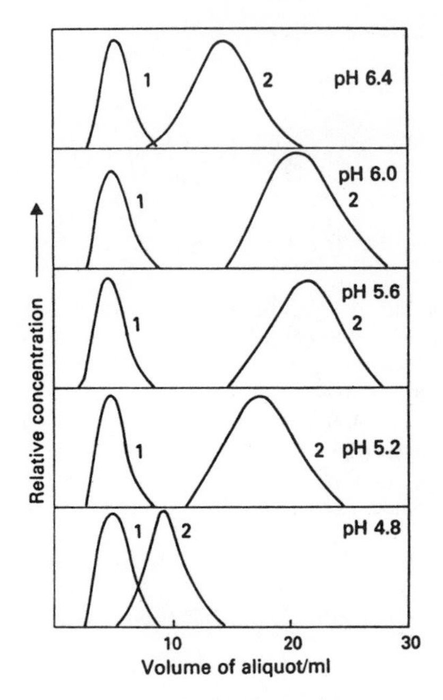

Figure 188 - Separation of arsenic(III) and dimethylarsinic acid on SRA70 by elution with carbon dioxide - hydrogen carbonate as a function of pH. 1, arsenic(III); 2, dimethylarsinic acid.

Initial investigation into quantitative aspects revealed a significant problem, i.e. that arsenic (III) was oxidised to arsenic (V) during elution. This was indicated by the 70 - 80% recoveries for arsenic (III) associated with the elution of low concentrations of arsenic in aliquots 35 - 45 when arsenic (III) was eluted on its own. Although the cause of this problem was not positively identified the problem was removed when resins were treated with nitric acid (1 mol 1^{-1}) and ethylenediaminetetraacetic acid (0.1 mol 1^{-1}, pH 5) before use. Recoveries obtained with resin treated in this way are shown in Table 274 and a chromatogram of a synthetic mixture of the four species is shown in Figure 189.

Table 274 - Recoveries of arsenic species on treated SRA 70

Species	Mean recovery, %*
Arsenic(III)	97.0
Arsenic(V)	98.1
Monomethylarsonate	98.8
Dimethylarsinate	94.8

* Four samples analysed.

The application of these methods to interstitial waters of lake sediments, and also species of lakeweed revealed no methylated arsenic species were found only As(III) and As(IV) in the 0.5 μg 1^{-1} range.

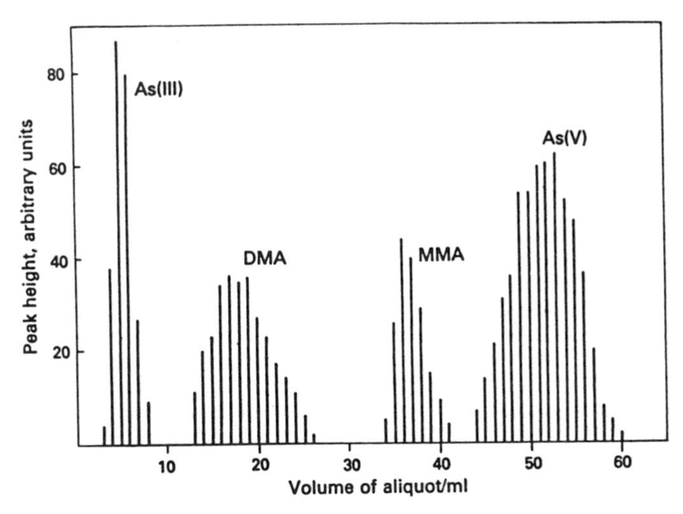

Figure 189 - Separation of a synthetic mixtures of arsenic(III) (0.5 μg), dimethylarsinic acid (0.5 μg), monomethylarsonic acid (0.5 μg) and arsenic(V) (1 μg) on SRA 70.

Only one of a number of solid phase lakeweed samples speciated appeared to contain a significant fraction of monomethylarsonic acid and none gave any indication of the presence of dimethylarsinic acid.

Miscellaneous. Numerous papers[3401,3402,3404,3407,3409-3432] have been published discussing the use of sodium borohydride for reducing inorganic arsenic compounds to arsine preparatory to its determination by atomic absorption spectrometry and other means such as inductively couples plasma emission spectrometry. These papers did discuss the application of these techniques to organoarsenic compounds. Gifford and Bruckenstein[3436] generated the hydrides of As(III), Sn(II) and Sb(III) by sodium borohydride reduction and separated them on a column of Poropak Q. Detection was at a gold gas porous electrode by measurement of the respective electrooxidation currents. Detection limits were, (5 ml samples) As(III), 0.2 ppb; Sn(II), 0.8 ppb; and Sb(III), 0.2 ppb. Most of the classical procedures for decomposing any organoarsenic compounds present in samples prior to the determination of total inorganic arsenic incorporate some mode of wet or dry digestion to destroy any organically bound arsenic, in addition to any other organic constituents present in the sample.

Probably the most frequently used method of digestion incorporates the use of nitric and sulfuric acids. Kopp[3437] used this digestion method and experienced 91 to 114% recovery of arsenic trioxide added to deionized water and 86 to 100% recovery of the compound added to river water. Evans and Bandemer[3438] recovered 87% of the arsenic trioxide added to eggs. By modifying the above digestive method by the addition of perchloric acid, Caldwell et al[3439] observed 80 to 90% arsenic recovery with o-nitrobenzene arsonic acid, 85 to 94% arsenic recovery with o-arsanilic acid, and 76.7% arsenic recovery with disodium methylarsenate.

Two uncertainties seem to arise when reviewing digestive methods using nitric and sulfuric acid. First, the addition of inorganic arsenic to an organic matrix and subsequent recovery of all the inorganic arsenic added is not definite proof of total recovery of any organoarsenicals present.

Secondly, the choice of o-nitrobenzene arsonic acid and o-arsanilic acid seems unfortunate since both compounds represent arsenic attached to an aromatic ring which is atypical of cacodylic acid and disodium methylarsonate, two widely used organoarsenicals.

Aside from nitric and sulfuric acid, a relatively simple digestive method employing 30% hydrogen peroxide in the presence of sulfuric acid was reported by Kolthoff and Belcher[3440] and subsequently used by Dean and Rues[3441] to determine arsenic in triphenylarsine.

Armstrong et al[3442] observed that organic matter in seawater could be oxidized to carbon dioxide on exposure to sufficient ultraviolet radiation from a medium pressure mercury arc vapor lamp. This approach has also been used to decompose organoarsenicals for o-arsanilic acid, 97% for sodium cacodylate, and 108% for arsenazo giving 111% recovery.

Stringer and Altrap[3443] appliied the Dean and Rues[3441] sulphuric acid - hydrogen peroxide and the ultraviolet[3442] decomposition methods to the determination of organoarsenic compounds in waste water. Arsenic determinations were performed by either the silver diethyldithiocarbamate colorimetric procedure described by Kopp[3437] or the arsine - atomic absorption method described by Manning[3444]. The organoarsenicals investigated were disodium methanearsonate, dimethylarsinic acid, and triphenylarsine oxide. All the digestive methods gave quantitative arsenic recoveries for the three organoarsenic compounds when added to wastewater samples. The ultraviolet photodecomposition proved to be an effective digestive technique, requiring a 4 h irradiation to decompose a primary settled raw wastewater sample containing spiked quantities of the three organoarsenicals.

The percent recoveries obtained by the digestion procedure using 5 mL of 30% hydrogen peroxide and 5 mL of sulfuric acid with silver diethyldithiocarbamate analysis and atomic absorption analysis are given in Table 275. The recovery results obtained by this digestion method applied to the aeration feed and settling basis effluent samples spiked with each of the three organoarsenicals equivalent to 5 μg of arsenic are presented in Table 276.

The effects of ultraviolet irradiation as a function of time for triphenylarsine oxide, disodium methanearsonate, and dimethylarsinic acid are illustrated in Figure 190. The extent of arsenic recovery using photooxidation in conjunction with high sensitivity arsenic analysis when applied to the aerator feed and settling basin effluent samples is shown in Table 277.

A chemical analysis of the aerator feeding and settling basin effluent samples is provided in Table 278 in order to illustrate the overall water quality of the samples used in this study.

The digestive method employing hydrogen peroxide and sulfuric acid with analysis by silver diethyl dithiocarbamate gave extremely good reduction in both organic and total carbon content.

The settling basin effluent sample showed 102.8% recoveries for both disodium methanearsonate and triphenylarsine oxide. An 88.6% recovery of dimethylarsinic acid added to settling basin effluent indicated that 2 h irradiation was insufficient for all of this compound to photodecompose in the presence of other organic matter. The total organic carbon content was reduced from an initial value of 12 mg/L carbon to 6 mg/L carbon - a 50% reduction.

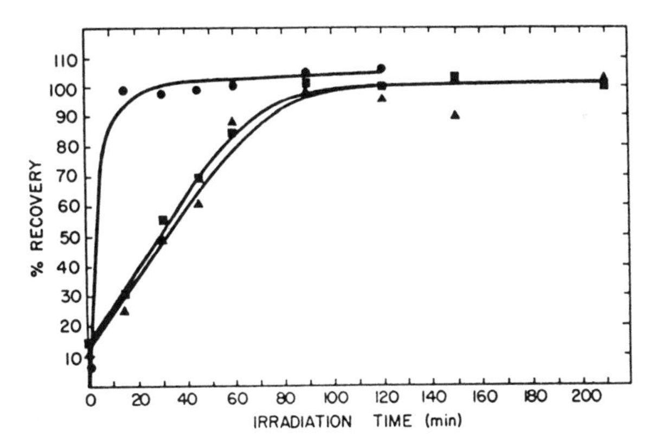

Figure 190 - Recovery of arsenic after ultraviolet exposure as a function of time. ●, triphenylarsine oxide; ■ , disodium methanearsenate; ▲ , dimethylarsinic acid.

Table 275 - Recovery of arsenic from triphenylarsine oxide, disodium methanearsonate, and dimethylarsinic acid employing wet digestion with 5 mL of 30% hydrogen peroxide and 5 mL of sulfuric acid with analyses by AgDDC and atomic absorption

compound	recovery, %[a]	
	AgDDC	atomic absorption
triphenylarsine oxide	101.0 ± 5.2(4)	100.8(2)
disodium methanearsonate	103.5 ± 1.2(4)	100.8(2)
dimethylarsinic acid	99.9 ± 2.3(3)	99.2(2)

[a] Each sample contained 5 µg As. Number in parentheses indicates the number of samples run.

Table 276 - Recovery of arsenic from triphenylarsine oxide, disodium methanearsonate, and dimethylarsinic acid spiked into AF and SBE sample with wet digestion employing 15 mL of 30% hydrogen peroxide and 5 mL of sulfuric acid with analysis by AgDDC[a]

sample	recovery, %[b]	
	sample 1	sample 2
100 mL AF	(9.7 µg As/L)	(6.9 µg As/L)
100 mL AF + triphenylarsine	92.4	94.1
100 mL AF + disodium methanearsonate	90.4	89.1
100 mL AF + dimethylarsinic acid	96.0	90.2
100 mL SBE	(22.7 µg As/L)	(18.5 µg As/L)
100 mL SBE + triphenylarsine oxide	101.1	100.6
100 mL SBE + disodium methanearsonate	98.1	104.4

(continued)

Table 276 (Continued)

100 mL SBE + dimethylarsinic acid	89.4	96.6

[a] All 5 μg weights are as arsenic.
[b] Corrected for amount of arsenic initially present.

Table 277 - Recovery of arsenic from triphenylarsine oxide, disodium methanearsonate, and dimethylarsinic acid spiked into aerator feed (AF) and settling basin effluent (SBE) with digestion by UV and analysis by high sensitivity atomic absorption

compound	recovery, %[a]		
	aerator feed sample 1	aerator feed sample 2	setting basin effluent
triphenylarsine oxide	110.0	100.0	102.8
disodium methanearsonate	100.0	102.4	102.8
dimethylarsinic acid	100.0	109.8	88.6

[a] Corrections were made for arsenic present in the samples.

Table 278 - Chemical analysis of the aerator feed sample and the settling basin effluent sample taken at Dallas

parameter[a]	samples	
	aerator feed	settling basin
pH @ 25°C	7.3	7.7
total alkalinity as $CaCo_3$	216.0	160.0
ammonia nitrogen	12.7	1.93
organic nitrogen	12.8	3.70
nitrite-nitrate nitrogen	0.2	1.7
nitrite nitrogen	0.01	0.08
total phosphorus as P	5.5	6.5
chemical oxygen demand	222.2	42.65
total organic carbon as C	57.0	12.0
total inorganic carbon as C	50.9	33.0
total carbon as C	107.0	45.0
suspended solids	132.	21.

[a] All values reported in mg/L units.

There were several observations made concerning the arsenic analysis by both the silver diethyldithiocarbamate technique and the high sensitivity atomic absorption procedure during the course of this investigation. Turbidity problems were initially encountered in the silver diethyldithiocarbamate colorimetric procedure.

Following the generation and collection of arsine into the chloroform solution containing l-epedrine and silver diethyldithiocarbamate, a precipitate would form which would seriously interfere with the absorbance measurements. The apparent cause of this problem was traced to the silver diethyldithiocarbamate. Of three different lots of silver diethyldithiocarbamate on hand, all three lots were distinctively different in color. The colors ranged from brown to pale green with the brown compound apparently responsible for the difficulty.

In conclusion, it is felt that the data support the premise that the recovery of arsenic from the organoarsenic compounds studied - triphenylarsine oxide, disodium methanearsonate, and dimethylarsenic acid - is complete by wet digestion using hydrogen peroxide - sulfuric acid and by photodecomposition using ultraviolet light.

Results with arsenic recoveries ranged from a low of 98.5 to a high of 104.9%. This same digestive method when applied to a primary settled raw sewage sample gave arsenic recoveries ranging from 89.1 to 96.0%. Arsenic recoveries of 89.4 to 104.4% were experienced from an activated sludge effluent sample. The same digestive method employing the high sensitivity arsenic analysis by atomic absorption resulted in arsenic recoveries of 99.2 to 100.8%.

The hydrogen peroxide-sulfuric acid digestion seemed to provide the most consistent and complete recoveries of any of the wet digestive procedures previously examined. Previous studies using nitric-sulfuric acid, with and without ammonium oxalate, revealed arsenic recoveries ranging from 80.2 to 105.9% and from 100.5 to 110.5%, respectively. This digestion procedure coupled with the silver diethyldithiocarbamate colorimetric analysis resulted in quantitative arsenic recoveries from wastewater samples. The hydrogen peroxide - sulfuric acid digestion combined with the high sensitivity arsenic analysis gave acceptable recoveries of arsenic and should be considered a viable technique.

The most encouraging digestive technique examined during the course of this investigation was the photodecomposition approach using ultraviolet light. The time-dependent study of the digestive effects of ultraviolet irradiation of the three organoarsenicals resulted in some interesting observations. A 15 min exposure of triphenylarsine oxide resulted in greater than 99% photodecomposition of the compound. The monoalkylated arsenic compound reacted much slower than triphenylarsine oxide, requiring 2 h for complete decomposition.

Several methods for the quantitative determination of the chemical forms of arsenic have been reported. These methods included various chromatographic techniques; thin-layer chromatography[3445-3447]; thin-layer electrophoresis[3452]; paper[3448,3449]; and high pressure liquid chromatography[3450,3451].

ORGANOANTIMONY COMPOUNDS

Andreae et al[3453] have described a method for the determination of methylantimony species, Sb(III) and Sb(V) in natural waters by atomic absorption spectrometry with hydride generation. Some results are also reported for estuary and sea waters.

Talmi and Norwell[3454] have studied the application of the gas chromatography-microwave plasma detector technique to the analysis of organo-antimony compounds in environmental samples[3397].

ORGANOCOPPER COMPOUNDS

Baceum et al[3455] have studied the application of reversed phase high performance liquid chromatography with molecular and atomic absorption detectors to the separation and quantification of organocopper speciation in soil-pore waters. Polar dissolved organic compounds and associated copper complexes are separated using either a single Hypersil ODS column or two Hypersil ODS columns and a Hamilton PRP 1 column in series. Quantification was achieved using ultraviolet detectors for the organic molecular species and graphite furnace atomic-absorption spectrometry for the copper. As well as the high relative molecular mass compounds, such as polysaccharides, peptides, lipids and humic substances, there is a wide range of low relative molecular mass metabolites produced in soils by micro-organisms and plant roots. Reversed-phase columns will retain polar compounds most effectively when their ionisation is suppressed. As many of the polar organic ligands present in pore waters are acidic, e.g., citric acid, the eluent system chosen to suppress ionisation was also acidic (0.02% V/V orthophosphoric acid, pH 2.6). A less acidic eluent was utilised for less polar compounds. Ammonium formate solution (0.01 M, pH 6.1) was found to be suitable because it gave a low signal to noise ratio for GFAAS analysis. This eluent was used for substances with higher retention values than the citric acid-copper complex.

One soil-pore water examined contained two recognisable polar dissolved organo copper compounds (Figure 191) (a) as revealed by absorbance measurement at 215 nm on the eluate.

Analysis of soil-pore waters using the single-column high performance liquid chromatographic system interfaced with graphite furnace atomic absorption spectroscopic system showed that the copper was not always associated with the same polar dissolved organic compounds in the same proportions (Figure 192). In the majority of pore waters association of the copper with citric acid and neighbouring eluting compounds was found (Figure 192 a).

A number of the more extreme distributions found in the soil-pore waters examined and the copper levels are quantified in Table 279. Citric acid-copper association was found in all the samples but the association of copper with allantoin, formic and lactic acids occurred rarely.

ORGANOSILICON COMPOUNDS

Mahone et al[3456] have described a method for the quantilative and quantitative characterisation of water organosilicon compounds substances such as silanols or silanol-functional materials are converted to trimethylsilylated derivatives which can be determined by gas liquid chromatography. The method gives good accuracy and precision in the ppm range and with suitable precautions can be extended to the low ppb range.

Wanatabe et al[3457] give details of a method for the determination of trace amounts of siloxanes in water. The method involves extraction into petroleum ether, evaporation to dryness, dissolution of the residue in methyl isobutyl ketone, and determination by inductively coupled plasma emission spectrometry. The detection limit is 0.01 ug per ml for organosilicones.

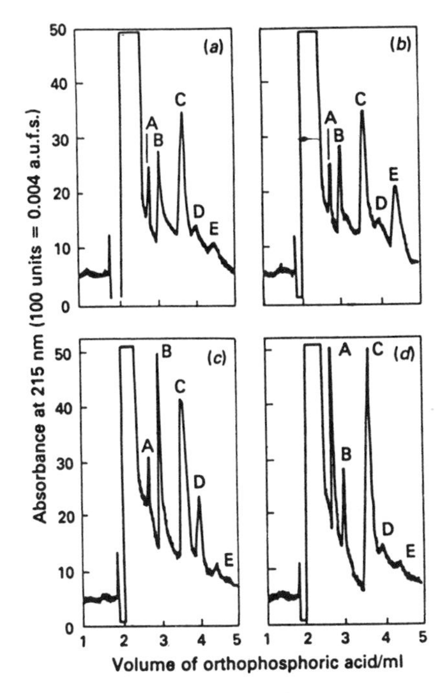

Figure 191 - UV absorbance signals of various PDOCs separated from a soil pore water on a Hypersil 5 µm ODS column with a 0.02% V/V orthophosphoric acid elution system (a) with identification and quantification by co-injections of (b) citric acid (1 µg) (E); (c) formic acid (1 µg) (B) and lactic acid (1 µg) (D); and (d) allantoin (0.05 µg) (A) and malic acid (0.3 µg) (C).

Table 279 - Amount of copper in soil-pore waters as determined by GFAAS after HPLC separation using a hypersil 5 µm ODS column

Eluent 0.02% V/V orthophosphoric acid, flow-rate 0.1 ml min^{-1}; for chromatographic separations.

HPLC peak	Amount of copper in soil-pore water					
	Sample 7(a) / µg l^{-1}	Percentage of total	Sample 7(b) / µg l^{-1}	Percentage of total	Sample 7(c) / µg l^{-1}	Percentage of total
Solvent front	64	53	13	21	13	30
2-Ketoglutamic acid-cystine					5	12
Allantoin					2	5
Formic acid			19	30		
Malic acid	5	4	20	32	16	37
Lactic acid	13	11				
Citric acid	39	32	11	17	7	16

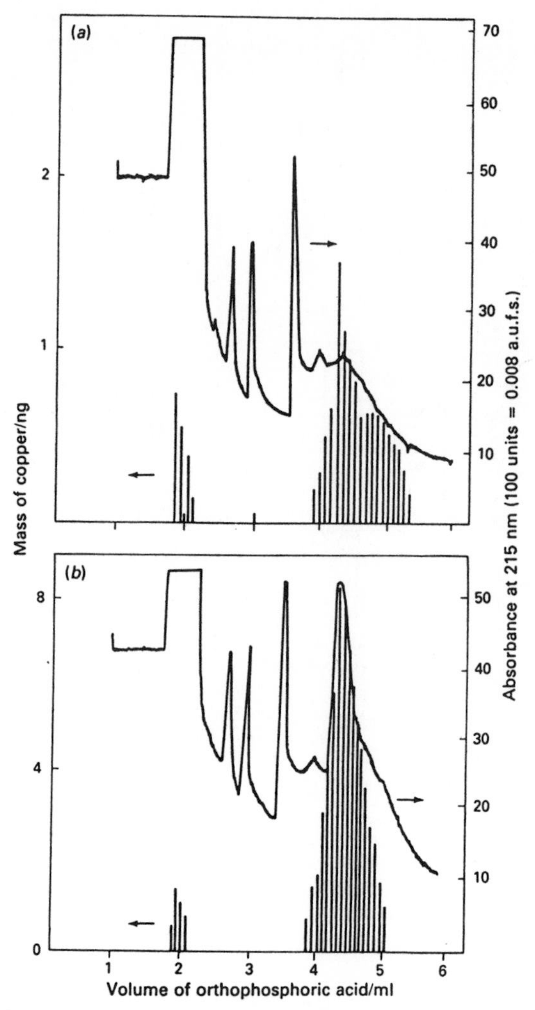

Figure 192 - UV absorbance and copper distribution patterns of a soil pore water using (a) a Hypersil 5 µm ODS column with a 0.02% V/V orthophosphoric acid elution system with flow-rate 0.1 ml min^{-1} and the HPLC - GFAAS interface; (b) an aliquot of the same pore water spiked with 50 ng of copper and 3 µg of citric acid.

C. ORGANOMETALLIC COMPOUNDS IN SEDIMENTS

Organomercury Compounds

In lakes, streams and rivers, mercury can collect in the bottom sediments, where it may remain for a long time. It is difficult to release this mercury from the matrixes for analysis. Mercury is also found in soil as a result of applications of mercury containing compounds, or sewage contaminated with traces of mercury. Much concern has been expressed in recent years concerning the contamination of the environment by mercury compounds, both organic mercury originating in industrial effluents and organic mercury originating as fungicides, seed dressings, etc.

Atomic absorption spectrophotometry. Earlier work on the determination of total mercury in river sediments also includes that of Oskander et al[3458] and Craig and Morton[3459]. Oskander applied flameless atomic absorption to a sulphuric acid nitric acid digest of the sample following reduction with potassium permanganate, potassium persulphate and stannous chloride. A detection limit of one part in 10^9 is claimed for this somewhat laborious method. Langmyhr and Aamodt[3460] determined nanogram amounts of organic mercury in river sediments using cold vapor atomic absorption spectrometry. Craig and Morton[3459] found 2.2 $\mu g\ l^{-1}$ mean total mercury level in samples of bottom deposits from the Mersey Estuary.

A method[3461] has been described for the determination of down to 2.5 ppb alkyl mercury compounds and inorganic mercury in river sediments. This method uses steam distillation to separate methyl mercury in the distillate and inorganic mercury in the residue. The methylmercury is then determined by flameless atomic absorption spectrophotometry and the inorganic mercury by the same technique after wet digestion with nitric acid and potassium permanganate[3462]. These workers considered the possible interference effects of clay, humic acids and sulphides, all possible components of river. sediment samples on the determination of alkylmercury compounds and inorganic mercury.

The well known adsorptive properties of clays for alkylmercury compounds does not cause a problem in the above method. The presence of humic acid in the sediment did not depress the recovery of alkylmercury compounds by more than 20%.

In the presence of metallic sulphides in the sediment sample the recovery of alkylmercury compounds decreased when more than 1 mg of sulphur was present in the distillate. The addition of 4 N hydrochloric acid, instead of 2 N hydrochloric acid before distillation completely eliminated this effect giving a recovery of 90 to 100%.

The method described above was applied to river sediment samples spiked with between zero and 0.06 -g g^{-1} methylmercury and 0.6 $\mu g\ g^{-1}$ mercuric chloride. Results indicated the presence in the original sediment of about 0.02 $\mu g\ g^{-1}$ methylmercury and 9 $\mu g\ g^{-1}$ inorganic mercury.

Jurka and Carter[3463] have described an automated determination of down to 0.1 $\mu g\ l^{-1}$ total mercury in river sediment samples. This method is based on the automated procedure of El Awady[3464] for the determination of total mercury in waters and wastewaters in which potassium persulfate and sulfuric acid were used to digest samples for analysis by the cold vapour technique. These workers proved that the use of potassium permanganate as an additional oxidizing agent was unnecessary.

Aromatic organic compounds such as benzene, which are not oxidized in the digestion, absorb at the same wavelength as mercury. This represents a positive interference in all cold vapour methods for the determination of mercury. For samples containing aromatics, i.e., those contaminated by some industrial wastes, a blank analysis must be performed, and the blank results must be subtracted from the sample results. The blank analysis is accomplished by replacing the potassium persulphate reagent and the stannous chloride reagent with distilled water, and re-analyzing the sample.

This automated procedure was estimated to have a precision of 0.13 to 0.21 μg Hg kg^{-1} at the 1 mg Hg kg^{-1} level with standard deviations varying from 0.011 to 0.02 mg Hg kg^{-1}, i.e. relative standard deviations of 8.4 to 12%. At the 17.2 to 32.3 mg Hg kg^{-1} level in sediments recoveries in methyl mercuric chloride spiking studies were between 85 and 125%. The detection limit for the automated method is dependent upon the weight of sample taken for analysis. It is 0.1 μg Hg/l in the aqueous samples. The results for the automated method are routinely reported to a lower limit of 0.1 mg/kg which corresponds to a dry sample weight of 0.25 g.

As Umezaki and Iwamoto[3465] have reported that organic mercury can be reduced directly with stannous chloride in the presence of sodium hydroxide and copper(II), the determination of organic mercury can be simplified, particularly if the reagent used for back-extraction does not interfere with the reduction of organic mercury. Matsunga and Takahasi[3466] found that back-extraction with an ammoniacal glutathione solution was satisfactory.

In their proposed method, contamination only from the ammoniacal glutathione solution is expected. However, any inorganic mercury in this solution will be adsorbed on the glass container walls with a half-life about 2 d, i.e., the blank value becomes zero if the solution is left to stand for more than a week. This method for mercury in sediments does not distinguish between the different forms of organomercury. Results are calculated as methylmercury.

In the method 10 ml 2 H hydrochloric acid is added to 10.2 g sediment and the mixture left 2 days[3467]. Filter the samples through a glass filter, and wash with 10 ml of 2 M hydrochloric acid. Extract organic mercury from 20 ml of the filtrate into 40 ml of benzene by shaking for 3 min, and discard the aqueous layer. Add 20 ml of 3 - 10^{-4}% glutathione in 0.1 M ammonia solution, and back-extract organic mercury into the aqueous solution by shaking for 2 min.

To a gas washing bottle, add 150 ml of water, 10 ml of 10 M sodium hydroxide, 2 ml of 1000 ppm copper solution and 5 ml of 5% tin(II) chloride dihydrate solution. Pass nitrogen gas at a flow rate of 1.41 min^{-1} for 6 min to eliminate any mercury in the reagent solutions. Then add the aqueous back-extract from the sample. Concentrate[3468] the mercury on 1.5 g of gold granules (about 1 mm diameter) packed in a glass tube (4 mm i.d.) by passing nitrogen gas for 6 min. Heat the gold granules in a boat to 500°C in a furnace for 2 min, and measure the absorbance at 253.7 um using an atomic absorption spectrophotometer with a 1.5 cm dia x 20 cm quartz tube by passing nitrogen gas at a flow rate of 1.21 min^{-1}.

In this method, the relative standard deviations were 4.1 and 10.2% at the 50 and 5 ng absolute level of methylmercury(II) added to 20 ml of 2 M hydrochloric acid. The final recovery of methylmercury was 86%.

This method was applied to the determination of organic mercury in unpolluted sediments. Organic mercury in two sediment samples was 0.22 ± 0.03 and 0.43 ± 0.03 ng Hg g^{-1} (dry weight).

Gas Chromatography. Andrew and Harris[3469] have reported a methylmercury concentration of 0.02 - 0.1 ng Hg g^{-1} in unpolluted sediments by using a gas chromatograph with an electron capture detector.

Ealy et al[3470] determined methyl, ethyl and methoxyethyl mercury compounds in sediments be leaching the sample with sodium iodide for 24 hours and then extracting the alkylmercury iodides into benzene. These iodides are then determined by gas chromatography of the benzene extract on a glass column packed with 5% of cyclohexylenedimethanol succinate on Anskrom ABS (70 - 80 mesh) and operated at 200°C with nitrogen (56 ml per min) as carrier gas and electron-capture detection (^3H foil). Good separation of chromatographic peaks is obtained for the mercury compounds as either chlorides, bromides or iodides.

Batti et al[3471] determined methyl mercury in river sediments from industrial and mining areas.

Gas chromatographic methods have been described for the determination of alkyl mercury compounds in sediments[3472,3473].

Bartlett et al[3474] observed unexpected behaviour of methyl mercury containing river Mersey sediments during storage. They experienced difficulty in obtaining consistent methyl mercury values; supposedly identical samples analysed at intervals of a few days gave markedly different results. They therefore followed the levels of methyl mercury in selected sediments over a period, to determine if any change was occurring on storage. They found that the amounts of methyl mercury observed in the stored sediments did not remain constant; initially there is a rise in the amount of methyl mercury observed, and then, after about ten days, the amount present begins to decline to levels which in general only approximate to those originally present. They have observed this phenomenon in nearly all of the Mersey sediments samples they examined. It was noted that sediments sterilised, normally by autoclaving at approximately 120°C, did not produce methyl mercury on incubation with inorganic mercury suggesting a microbiological origin for the methyl mercury. A control experiment was carried out in which identical samples were collected and homogenised. Some of the samples were sterilised by treatment with an approximate 4% w/w solution of formeldehyde. Several samples of both sterilised and unsterilised sediments were analysed at intervals and all of the samples were stored at ambient room temperature (18°C) in the laboratory. There is a difference in behaviour between the sterilised and unsterilised samples. This work suggests that the application of laboratory-derived results directly to natural conditions could, in these cases, be misleading; analytical results for day 10 if extrapolated directly might lead to the conclusion that natural methylmercury levels and rates of methylation are much greater than in fact they really are. Work in this area with model or laboratory systems needs to be interpreted with particular caution.

Bartlett et al[3474] used the method of Uthe et al[3475] for determining methyl mercury. Sediment samples of 2 - 5 g were extracted with toluene after treatment with copper sulphate and an acidic solution of potassium bromide. Methyl mercury was then back-extracted into aqueous sodium thiosulphate. This was then treated with acidic potassium bromide and copper sulphate following which the methyl mercury was extracted into pesticide grade benzene containing approximately 100 µg dm^{-3} of ethyl mercuric chloride as an internal standard. The extract was analysed by electron capture gas chromatography using a Pye 104 chromatography equipped with a nickel 65 detector. The glass column (1 m x 0.4 cm) was packed with 5% neopentyl glycol adipate on Chromosorb G (AW-DMCS). Methyl mercury was measured by comparing the peak heights with standards of methyl mercuric chloride made up in the ethyl mercury-benzene solution. The results were calculated as nanograms of methylmercury per gram of dry sediment. The detection limit was 1 - 2 ng g^{-1}.

Miscellaneous. Using their technique (Kimura and Miller[3476]) collected and determined vapours produced by the decomposition in soil of phenyl- and alkylmercury compounds. They found that the air above soil containing phenylmercury acetate contained mercury vapour and traces of phenylmercury acetate.

Ethylmercury acetate produced about equal amounts of mercury vapour and an uncharacterized volatile ethylmercury compound, whilst methylmercury chloride and methylmercury dicyanamide both produced an uncharacterized methylmercury compound plus some mercury vapour.

Several investigators have liberated mercury from soil and sediment samples by application of heat to the samples and collection of the released mercury on gold surfaces. The mercury was then released from the gold by application of heat or by absorption in a solution containing oxidizing agents[3477,3478].

Bretthaur, Moghissi, Snyder and Mathews described a method in which samples were ignited in a high-pressure oxygen-filled bomb[3479]. After ignition, the mercury was absorbed in a nitric acid solution. Pillay, Thomas Sondel and Hyche used a wet-ashing procedure with sulfuric acid and perchloric acid to digest samples[3480]. The released mercury was precipitated as the sulfide. The precipitate was then redigested using aqua regia.

Feldman digested solid samples with potassium dichromate, nitric acid, perchloric acid and sulfuric acid[3481]. Bishop, Taylor and Neary used aqua regia and potassium permanganate for digestion[3482]. Jacobs and Keeney oxidized sediment samples using aqua regia, potassium permanganate, and potassium persulfate[3483]. The approved U.S. Environmental Protection Agency digestion procedure requires aqua regia and potassium permanganate as oxidants[3484].

These digestion procedures are slow and often hazardous because of the combination of strong oxidizing agents and high temperatures. In some of the methods, mercuric sulfide is not adequately recovered. The oxidizing reagents, especially the potassium permanganate, are commonly contaminated with mercury, which prevents accurate results at low concentrations.

Kimuna and Miller[3485] have described a procedure for the determination of organomercury (methyl mercury, ethyl mercury and phenylmercury compounds) and inorganic mercury in soil. In this method the sample is digested in a steam bath with sulphuric acid (1.8 N) containing hydroxyammonium sulphate, sodium chloride and, if high concentrations of organic matter are present, potassium dichromate solution. Then, 50% hydrogen peroxide is added in portions with vigorous mixing. Allow sufficient time for the peroxide to decompose after each addition. The decomposition of hydrogen peroxide being exothermic, the temperature will rise gradually to about 150°C. The addition of peroxide is discontinued after the solution turns blue green, or in the absence of chromium, light yellow, and residual peroxide is allowed to decompose with the use of low heat after the condenser has been washed down with water. When the decomposition of peroxide has apparently ceased, add 5% permanganate slowly to give a 5 ml excess while the temperature is maintained. The permanganate is added in 5 ml portions, or less, until the mixed colour persists for 15 minutes. Cool the sample and add 20 ml of reducing solution consisting of hydroxyammonium sulphate dissolved in sodium chloride.

Air is then passed through this solution in the airator apparatus and the volatilized elemental mercury collected in an absorber solution consisting potassium permanganate in dilute sulphuric acid, followed by stannous chloride solution. To the absorber solution is added a reducing solution consisting of hydroxyammonium sulphate dissolved in sodium chloride. To determine mercury in the absorber solution transfer a suitable aliquot or the entire sample containing not more than 10 μg mercury to a separatory funnel and dilute to 50 ml with the 1.8 N sulphuric acid solution. Add 3.5 ml of dithizone solution (containing 11 mg/l) and shake for 1 minute. Transfer the chloroform phase to a 13 x 100 mm test-tube, allowing any residual water to adhere to its walls. Then transfer to a 1 cm square cuvette and measure the excess of dithizone at 605 mμ. Read as soon as possible after shaking with the mercury solution.

Kimura and Miller[3485] demonstrated (Table 280) that mercury in several organic forms can be digested and aerated from unfiltered soil digests. For samples of 10 g of soil cores containing 5 μg mercury or less, the standard deviations of a single determination were 0.12, 0.15 and 0.23 μg, respectively, using 2 cm cylindrical optical cells.

Table 280 - Mercury recovery

Sample	Hg added (μg)	Total Hg found(μg)	Added Hg recovered (μg)
	(Added as $HgCl_2$)		
10 g Puyallup sandy loam soil (air dried and screened, organic matter 6% air-dry basis)	0.00	1.66	-
	0.56	2.17	0.51
	7.77	9.41	7.75
	111.0	111.1	110.1
	(Added as methylmercury chloride)		
	0.40	2.15	0 49
	4.00	5.55	3.89
	(Added as ethylmercury chloride)		
	0.38	2.08	0.41
	7.57	9.19	7.53
	(Added as $HgCl_2$)		
1.9 x 2.4 cm Turf cores (18 - 22 g as sampled)	0.00	1.35	-
	1.00	2.18	0.83
	5.00	6.46	5.11
	100.0	97.45	96.1

A similar digestion procedure to that of Kimura and Miller[3485] has been discussed in a report prepared by the Metallic Impurities in Organic Matter Sub-Committee of the Society for Analytical Chemistry, London[3486]. In this procedure the sediment/soil is wet-oxidized with dilute sulphuric acid and nitric acids in an apparatus in which the vapour from the digestion is condensed into a reservoir from which it can be collected or returned to the digestion flask as required. The combined oxidized residue and condensate are diluted until the acid concentration is 1 N and nitrate is removed by addition of hydroxylammonium chloride with boiling. Fat is removed from the cooled solution with carbon tetrachloride and the liquid is then extracted with a solution of dithizone in carbon tetrachloride. The extract is shaken with 0.1 N hydrochloric acid and sodium nitrite solution and, after treatment of the separated aqueous layer with hydroxylammonium chloride a solution of urea and then EDTA solution are added to prevent subsequent extraction of copper. The liquid is then extracted with a measured small amount (0.5 - 0.2 ml) of a 0.01% solution of dithizone in carbon tetrachloride. When mercury extraction is complete, the combined extracts are made up to 4 ml and filtered through glass wool. The extinction of the filtrate is measured at 485 nm against a blank solution and referred to a calibration graph. The results are reliable down to 0.1 parts/10^6 mercury calculated on the dried sample weight.

Kimura and Miller[3487] also describe the following methods for the determination in soil samples of extractable organic mercury, total mercury and extractable ionic mercury.

Extractable phenyl- and alkylmercury compounds. Phenyl- and alkylmercury compounds are extracted from about 1 g soil by shaking for 2 h with 25 ml 0.1 M phosphate pH 8 buffer containing 6 mg thiomalic acid, added just prior to use, and analysed after dilution of a 5 ml aliquot of the centrifuged extract with 5 ml water, and acidification with 5 ml 9 N hydrochloric acid containing 150 mg hydroxylammonium chloride. The final determination is made by the dithizone microprocedure of Miller and Polley[3488]. Diphenyl- and dialkylmercury compounds are extracted from 1 g soil by shaking for 2 h with 10 ml chloroform and analysed by cleaving the disubstituted mercurial to give an aryl- or alkylmercury salt, using 9 N or 12 N hydrochloric acid, followed by the dithizone microprocedure[3489].

Ionic mercury. Ionic mercury is extracted from about 1 g soil by shaking for 2 h with each of two 25 ml portions of 2 M sodium chloride. The combined centrifuged and filtered (using 1 M sodium chloride for washing) extract is analysed by the procedure of Polley and Miller[3489].

Total mercury. Total mercury is determined in the soils containing phenylmercury acetate and or ethylmercury acetate using the method described by Polley and Miller[3489]. Total mercury is determined in soils containing methylmercury chloride and methylmercury dicyanamide by the method described by Kimura and Miller[3485].

Kimura and Miller[3476] have also studied the decomposition of organic fungicides in soil to mercury vapour and to methyl- or ethylmercury compounds and devised methods for the determination of these compounds in the vapours liberated from the soil sample. The mixed vapours of mercury and organomercury compounds is passed successively through bubblers containing a carbonate-phosphate solution to absorb organic mercury and through an acidic potassium permanganate solution to absorb inorganic mercury vapour. In both cases the mercury in the scrubber solution is determined photometrically at 605 nm with dithizone. The method is capable of determining 10 µg or more of organic mercury/1000 l air in the presence of mercury vapour.

In this procedure the separation of metallic mercury and organic mercury vapour is based on the passage of the metallic mercury vapours through a solution of sodium carbonate and dibasic sodium phosphate and its quantitative capture in acid permanganate solution (Table 281). Vapours of ethyl- and methylmercury chloride, in the 100 - 1000 µg range, are 95% - 99% retained in a single carbonate - phosphate absorber (Table 282). This retention is attributed to the formation of the extremely water-soluble methyl- and ethylmercury hydroxide and phosphate, and the foaming characteristics of the carbonate solution.

Table 281 - Retention and distribution of metallic mercury vapour in an aeration train composed of one carbonate - phosphate absorber and two acid permanganate absorbers. (Mercury in μg)

Trial	First absorber carbonate-phosphate: Oxidation by H_2O_2 method	First absorber carbonate-phosphate: acid dichromate washings	Second absorber acid permanganate: without further oxidation	Second absorber acid permanganate: Further oxidation by H_2O_2 method	Third absorber acid permanganate: Without further oxidation	Third absorber acid permanganate: Further oxidation by H_2O_2 method
I	0	0	1.860	1.960	0	4
II	0	0	1.780	1.780	0	3
III	0	0	1.730	1.700	0	4
IV	1	0	1.660	1.670	2	7

Table 282 - Distribution of methylmercury chloride (MMC) and ethylmercury chloride (EMC) after aeration using carbonate - phosphate and acid permanganate absorbers

Residue*	First Absorber # Na_2CO_3 - Na_2HPO_4	Second Absorber * Acid permanganate	Total found (μg)	Original (μg)
		methylmercury chloride		
8	975	15	998	1000
13	770	9	792	800
9	383	6	398	400
2	189	5	196	200
3	94	4	101	100
		ethylmercury chloride		
3	984	9	996	1000
2	393	4	399	400
2	193	4	199	200
1	96	3	100	100

* Determined as mercury and calculated as MMC or EMC.
# Determined directly as MMC or EMC.

Organotin Compounds

Mueller[3490] has recently described a gas chromatographic method for the determination of tributyltin compounds in sediments. The tributyltin compounds are first converted to tributylmethyltin, and then analysed using capillary gas chromatography with flame photometric detection and gas chromatography - mass spectrometry. Tributyltin was found in samples of river and lake water, and sediment, and these results demonstrated that the technique has detection limits of less than 1 ppt for water and 0.5 ppb for sediment.

Organolead Compounds

The application of a combination of gas chromatography and atomic absorption spectrometry to the determination of tetraalkyllead compounds has been studied by Chau et al[3491] and by Segar[3492]. In these methods the gas chromatography flame combination showed a detection limit of about 0.1 µg Pb. Chau et al have applied the silica furnace in the atomic absorption unit and have shown that the sensitivity limit for the detection of lead can be enhanced by three orders of magnitude. They applied the method to the determination of tetramethyllead in sediment systems.

The system used by these workers consisted of a Microtek 220 gas chromatograph and a Perkin-Elmer 403 atomic absorption spectrophotometer. These instruments were connected by means of stainless steel tubing (2 mm o.d.) connected from the column outlet of the gas chromatograph to the silica furnace of the a.a.s. (Figure 193). A 4-way valve was installed between the carrier gas inlet and the column injection port so that a sample trap could be mounted, and the sample could be swept into the g.c. column by the carrier gas. The recorder (10 mV) was equipped with an electronic integrator to measure the peak areas, and was simultaneously actuated with the sample introduction so that the retention time of each component could be used for identification of peaks.

The furnace was constructed from silica tubing (7 mm i.d., 6 cm long) with open ends (Figure 194). The lead compounds separated by g.c. were introduced to the centre of the furnace through a side-arm. Hydrogen gas was introduced at the same point at a flow rate of 1.35 ml min^{-1}; the burning of hydrogen improved the sensitivity. The furnace was wound with 26-gauge Chromel wire to give a resistance of about 5 ohms. The voltage applied to the furnace was about 20 Va.c. regulated by a variable transformer so that the furnace temperature with the hydrogen burning was about 1000°C. The silica furnace was mounted on top of the a.a.s. burner and aligned to the light path.

The sample trap was a glass U-tube (6 mm dia., 26 cm long) packed with 3% OV-1 on Chromosorb W, which was immersed in a dry ice-methanol bath at ca -70°C as described by Chau et al[3491]. A known amount of gaseous sample was drawn through the trap by a peristaltic pump operated at 130 - 150 ml min^{-1}. After sampling, the trap was mounted to the 4-way valve and heated to ca. 80 - 100°C by a beaker of hot water, and the adsorbed compounds were swept into the g.c. column.

Liquid samples can be directly injected to the column through the injection port, without a sample trap.

Instrument Parameters. Glass column, 1.8 m long, 6 mm dia., packed with 3% OV-1 on Chromosorb W 80 - 100 mesh; carrier gas, 70 ml nitrogen min^{-1}; injection port temperature, 150°C; temperature program, initial 50°C for 2 min, programmed at 15°C min^{-1}, until 150°C; sample trap temperature, 80 - 100°C.

Lead line, 217 nm; lamp current, 8 mA; spectral band width, 0.7 nm; scale expansion 4 x (0.25 A full scale); furnace gas, 135 ml H_2 min^{-1}. The deuterium background corrector was used.

The relative standard deviation was in the range 10 - 15% at the 5 ng level (as Pb). When the absorbances were plotted against lead concentrations, each of the five tetraalkyl compounds gave similar calibration curves; the response was linear up to at least 200 ng Pb, above which overlapping of the peaks occurred. If only one compound was present (e.g. tetramethyllead), the plot was linear up to at least 2000 ng. For

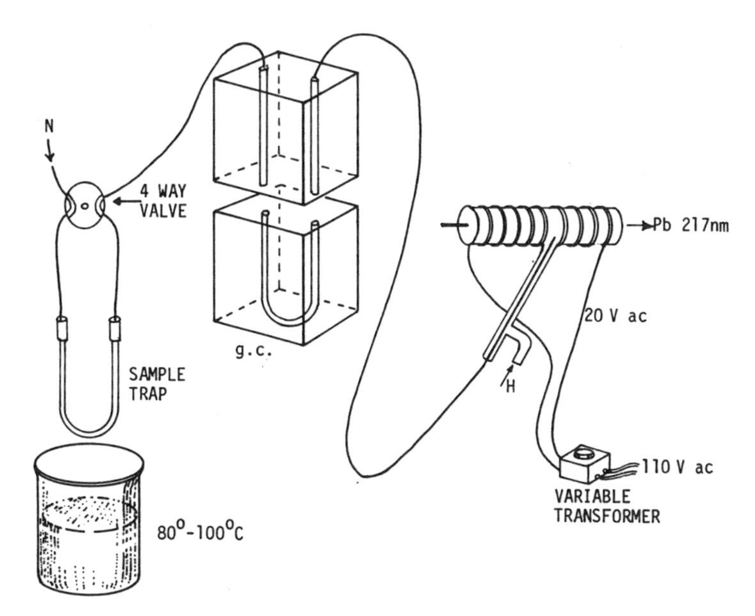

Figure 193 - Schematic diagram showing the interfacing of the g.c. - a.a.s. system.

determinations at the microgram level, the flame a.a.s. technique[3491] is more suitable. Figure 195 illustrates a typical recorder tracing of a mixture of the five tetraalkyllead compounds.

Solvents such as chloroform, carbon tetrachloride, hexane and benzene gave absorption signals because of their non-specific absorption at the lead resonance line. Although these solvent peaks generally emerged well before the lead compounds, the use of the background corrector is recommended to eliminate these potential interferences.

Chau et al[3360] have described a simple and rapid extraction procedure to extract the five tetraalkyllead compounds from sediment. The extracted compounds are analysed in their authentic forms by a gas chromatographic - atomic absorption spectrometry system. Other forms of inorganic and organic lead do not interfere. The detection limits for sediment (5 g) is 0.01 µg/g. The gas chromatographic-atomic absorption system used by Chau[3359] is discussed in the section on water analysis. The procedure is outlined below;

Place 5 g of wet sediment, 5 ml of EDTA reagent, (0.1 M, 37 g Na_2EDTA $2H_2O$/1, disperses fish and sediment homogenates in a suspension to provide better extraction and produce clarified organic phase), and 5 ml of hexane in a 25 ml test tube, with a Teflon-lined screw cap. Shake rigorously in a reciprocating shaker for 2 h. Centrifuge the sample for 10 min at 2000 xg. Inject a suitable aliquot, 5 - 10 µl, of the hexane extract to the gas chromatographic atomic absorption system.

Determination of the ionic forms of alkyllead compounds is difficult because of the incomplete extraction of the dimethyl and trimethyl species from sample matrices. Recently a chelation extraction method followed by

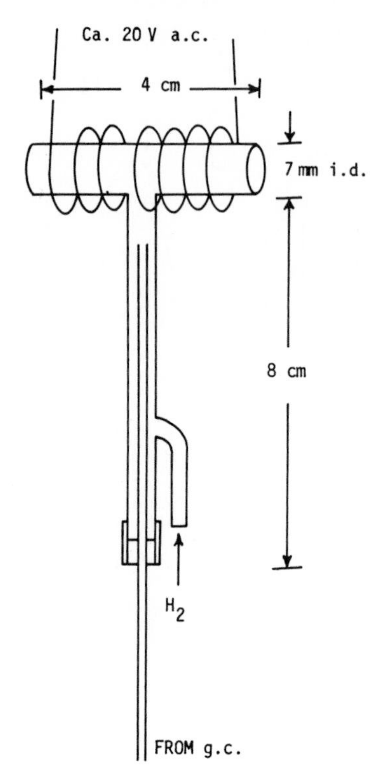

Figure 194 - Silica Furnace.

derivatization to their butyl homologues has overcome all the previous difficulties to achieve quantitative extraction of the dialkyl- and trialkyllead (R = Me, Et) from water samples at nanogram levels[3493].

In the determination of these ionic alkyllead compounds in biological tissues, difficulties are further compounded by their strong affinity with protein and lipid matrices. There has been a dearth of information on the occurrence of these compounds in biological samples mainly because of the lack of suitable methodology. Up to the present time, there have only been relatively few methods[3493-3495] dealing with the ionic alkyllead compounds in biological samples.

Chau et al[3496] have described the use of a tissue solubilizer to digest biological samples and the optimum conditions for extraction of these compounds from biological as well as sediment samples for alkyllead speciation analyses. Analyses of some environmental samples revealed for the first time the occurrence of dialkyl- and trialkyllead in fish, macrophytes, and sediment in areas of lead contamination.

The various alkyllead species and lead(II) are isolated quantitatively by chelation extraction with sodium diethyldithiocarbamate, followed by n-butylation to their corresponding tetraalkyl forms, $R_nPbBu_{(4-n)}$, and Bu_4Pb, respectively (R = Me, Et), all of which can be determined by a gas chromatography using an atomic absorption detector. The method determines simult-

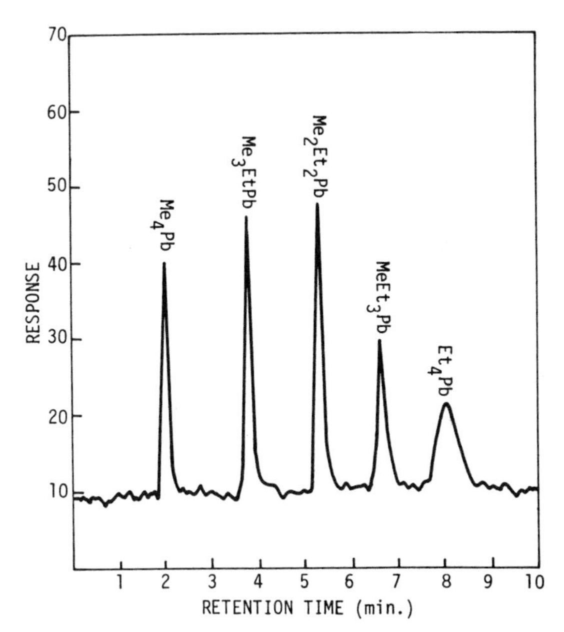

Figure 195 - Recorder tracings for a mixture of five tetraalkyllead compounds. Each peak represents ca. 5 ng of the compound expressed as lead.

aneously the following species in one sample: tetraalkyllead (Me_4Pb, Me_3EtPb, Me_2Et_2Pb, $MeEt_3Pb$, Et_4Pb); ionic alkyllead (Me_2Pb^{2+}, Et_2Pb^{2+}, Me_3Pb^+, Et_3Pb^+); Pb^{2+}. Detection limits expressed for Pb are 7.5 ng/g and 15 ng/g respectively, for biological and sediment samples.

In this method a dried sediment (1 - 2 g) or wet (5 g) sediment sample was extracted in a capped vial with 3 mL of benzene after addition of 10 mL of water, 6 g of sodium chloride, 1 g of potassium iodide, 2 g of sodium benzoate, 3 mL of sodium diethyldithiocarbamate, and 2 g of course glass beads (20 - 40 mesh) for 2 h in a mechanical shaker. After centrifugation of the mixture, a measured aliquot (1 mL) of the benzene was butylated using 0.2 ml n-butyl magnesium chloride, with occasional mixing for 10 min. The mixture was washed with 2 ml of sulphuric acid (1 N) to destroy excess Grignard reagent. The organic layer was separated in a capped vial and dried with anhydrous sodium sulphate. Suitable aliquots were injected into the gas chromatograph.

Chau et al[3496] found that in spiking experiments on sediment both the diethyl and triethyl species were recovered at satisfactory levels (Table 283)

Average recovery varied from 94% for triethyllead and 111% for trimethyllead in the range 1 - 20 µg of alkyllead compounds spiked to 1 g of sediment. Deplicated analyses showed an average standard deviation of 4% for trimethyllead and triethyllead and 15% for the dialkyllead compounds.

Organoarsenic Compounds

Odanake et al[3497] have reported the application of gas chromatography with multiple ion detection after hydride generation with sodium borohydride

Table 283 - Recovery and reproducibility of alkyllead and lead(II) compounds from sediment

amt of Pb added, μg	recovery,[b] %				
	Me_3Pb	Et_3Pb	Me_2Pb	Et_2Pb	Pb(II)
1	113(9)	73(14)	103(6)	104(15)	
5	111(3)	86(2)	116(2)	94(5)	
10	122(4)	106(1)	114(3)	85(1)	
20	99(1)	111(4)	118(2)	89(3)	
av	111	94	113	93	
% rel std dev (n = 6) at 5 μg/g level	4	4	14	15	9[c]

[a] Sediment, 1 g; spiked compounds expressed as Pb.
[b] Average of two results with average deviation in parenthesis.
[c] The sediment contained 71 μg/g of Pb(II) which was used to evaluate the reproducibility. No Pb(II) was added to sample.

to the determination of mono and dimethyl arsenic compounds, trimethyl arsenic oxide and inorganic arsenic in soil and sediments; this work is discussed more fully in the section on organoarsenic compounds in water. Recoveries in spiking experiments were 100 to 102% (mono and dimethyl arsenic compounds and inorganic arsenic) and 72% (trimethyl arsenic oxide).

Organosilicon Compounds

Wanatake et al[3498] give details of a method for the determination of trace amounts of siloxanes in sediments and sewage sludge. The method involves extraction into petroleum ether, evaporation to dryness, dissolution of the residue in methyl isobutyl ketone, and determination by inductively coupled plasma emission spectrometry. The detection limit is 0.01 ug per ml for organosilicons.

D. ORGANOMETALLIC COMPOUNDS IN PLANTS AND CROPS

Organomercury Compounds

Gutenmann and Lisk[3499] suggested an alternate method of overcoming mercury losses during decomposition of mercury containing organic materials by adopting the Schoniger oxygen flask combustion technique. They determined mercury in apples by first drying the apple tissue on cellophane, in vacuo. overnight. The dry material is then burnt in an oxygen-filled flask and the combustion products absorbed in 0.1 N hydrochloric acid. Mercury is extracted with dithizone and determined spectrophotometrically. Recovery of 0.3 - 0.6 parts/10^6 mercury in apples by this procedure averaged 83.6%.

Langryhn and Aamodt[3500] have described a cold vapour atomic absorption spectrometric method for the determination of nanogram amounts of mercury in aquatic organisms.

Houpt and Compaan[3501] used emission spectrographic analysis for the identification of traces of organic matter containing halogens and mercury isolated from fish, eggs and grass by gas chromatography. They transferred the gas chromatographic fractions sequentially, through a heated stainless-steel capillary tube, to a silica tube (3 mm i.d.) in which they were submitted to a hf discharge (2.45 MHz) in helium at 10 torr. The emission spectrum arising from the fragmentation, ionization and excitation of the organic molecule was then analysed with the aid of two monochromators, the intensities of the required analytical lines being measured photo-electrically. One monochromator was focused on a characteristic line, e.g. the 247.86 nm carbon line (as a chromatographic detector) and, when the intensity of this line is a maximum for any one fraction detected in the discharge-tube, a 10 sec sweep over the range 200 - 600 mm was made by the other monochromator. Examination of the resulting complete spectrograms revealed the presence or absence of phosphorus, sulphur, chlorine, bromine, iodine and mercury. Houpt and Compaan[3501] report for a mixture (in dekalin) of bromo- and iodo-ethanes, iodopropane, chloroform and thiophen (containing 10 ng each of chlorine, bromine, iodine and sulphur). This method permits the determination of 5 pg of methylmercury in plant and biological samples.

Organomercury compounds are used extensively as fungicides in agriculture and horticulture and Tatton and Wagstaffe[3502] have described gas chromatographic procedures for the determination of these substances in apples, potatoes and tomatoes. In general they studied organomercury salts including nitrates, sulphates, acetates halides and dicyanimide. The following types of salts were examined, methylmercury, ethylmercury, phenylmercury, tolymercury and methoxyethylmercury and ethoxyethylmercury.

Tatton and Wagstaffe[3502] found it was convenient to convert the organomercury compounds to their dithizonates for gas chromatography. They found that, in general, the more polar phases such as Carbowax 20 M and ethylene glycol adipate, on Chromosorb G, gave good separations but had a distinct tendency to produce tailing peaks on the chromatograms. By far the most satisfactory column consisted of 2% of polyethylene glycol succinate on Chromosorb G.

Typical retention times for this column are given in Table 284. The dithizonates of the various alkyl- and alkoxyalkyl-mercury compounds have fairly short retention times but are clearly separated from one another. Sensitivity is good and the system can easily detect 0.05 ng of these compounds. By contrast the arly-mercury dithizonates had relatively long retention times with peaks that were correspondingly broader at the base. The peaks corresponding to phenylmercury dithizonate and tolymercury dithizonate were also slightly asymetrical; this type of peak appears to be an inherent characteristic of the arylmercury dithizonates, for which no obvious reason could be found. It is very marked on some types of column. Stationary phases such as Apiezon L, Silicone GE SE-52, Cyanosilicone GE XE-60, Carbowax 1500M and ethylene and diethylene glycol succinates, on Chromosorb W, G or Q as support, all showed this feature to some extent. Teflon, 40 - 60 mesh, was probably the best support but has certain intrinsic disadvantages. Direct "on column" injection tended to minimise this effect and was used by Tatton and Wagstaffe[3502]. Nevertheless, excellent reproducibility of these peaks for the arylmercury dithizonates was obtained on the polyethylene glycol succinate column referred to above and 1 ng of these compounds could be readily detected. A shorter column, containing only 1% of polyethylene glycol succinate, specifically for the arlymercury dithisonates, was also useful in that shorter retention times were obtained together with narrower peaks on the chromatogram. Typical retention times obtained by use of this column are also given in Table 284. This system would readily detect 0.5 ng of these arlymercury compounds.

Table 284 - Typical GLC retention times for organomercury dithizonates

(1) 2% polyethyleneglycol succinate on Chromosorb G (acid-washed, DMCS-treated, 60 - 80 mesh) in glass columns 1.5 m long, 3 mm i.d.; carrier gas, nitrogen.

Dithizonate	Column Temperature (°C)				
	140	150	160	170	180
Methylmercury	3.8	2.8	2.2	1.6	1.2
Ethylmercury	6.6	4.6	3.6	2.7	2.0
Ethoxyethylmercury	17.0	11.6	8.7	6.2	4.9
Methoxyethylmercury	17.4	12.0	8.7	6.2	4.9
Tolymercury	-	-	-	29.0	19.5
Phenylmercury	-	-	-	42.0	27.0

(11) 1% polyethyleneglycol succinate on Chromosorb G (acid-washed, DMCS-treated, 60 - 80 mesh) in glass columns, 1.2 m long, 3 mm i.d.; carrier gas, nitrogen.

Dithizonate	Column Temperature (°C)	
	170	180
Tolylmercury	6.4	3.2
Phenylmercury	10.0	5.0

The method was applied to potatoes, tomatoes and apples and gave recoveries of 85 - 95% for samples spiked with 1.0, 0.1 and 0.01 ppm of methyl-, ethyl and ethoxy ethylmercury as their chlorides, and 5 and 0.5 ppm of phenyl- and tolyl-mercury acetates.

In this method five grams of chopped peel of apples or potatoes, or 5 g of the macerated fruit in the case of tomatoes, are macerated with a mixture of 10 ml of propan-2-ol and 5 ml of alkaline cysteine hydrochloride solution (1% aqueous solution adjusted to pH 8.0 by the addition of 5 N ammonia solution). After allowing the liquor to settle, the clear layer is decanted and the extraction repeated twice more with further portions of extractant solutions. The combined extracts are then centrifuged at 2500 rmp for 5 min. The clear liquor is separated, diluted with 700 ml of 4% sodium sulphate solution and the solution washed with three 50 ml portions of diethyl ether. The organomercurials are then extracted from the aqueous solution using three 25 ml portions of a 0.005% solution of dithizone in diethyl ether. The combined extracts are then dried by passage through a short column of granular anhydrous sodium sulphate and concentrated to a suitable volume, usually 5 ml, in a Kuderna-Danish evaporator. The final solution is injected on to the first of the gas chromatographic columns described in Table 284 operating at 180°C. The shorter column described in Table 284 should also be used if arylmercury compounds are present.

Ealy et al[3503] have discussed the determination of methyl, ethyl and methoxyethyl mercury II halides in environmental samples such as aquatic systems, seeds and fish.

The mercury compounds were separated from the samples by leaching with

M sodium iodide for 24 hours and then the alkylmercury iodides were extracted into benzene. These iodides were then determined by gas chromatography of the benzene extract on a glass column packed with 5% of cyclohexylene-succinate on Anakrom ABS '70 - 80 mesh) and operated at 200°C with nitrogen (56 ml per min) as carrier gas and electron-capture detection. Good separation of chromatographic peaks were obtained for the mercury compounds as either chlorides, bromides or iodides. The extraction recoveries were monitored by the use of alkylmercury compounds labelled with 208 Hg.

Organotin Compounds

Chronopotentiometry (Nangniot and Martens[3504]) has also been applied to the determination of triphenyltin acetate at very low concentrations in plant material. In this method a hanging-drop electrode is used at which the ions are reduced in a pre-electrolysis step at - 0.7 v the silver-silver chloride saturated potassium chloride electrode for 5 min; the potential is then increased gradually to - 0.1 V and the anodic diffusion current is registered at about - 0.45 V. The sample for analysis is obtained by extraction of plant material with chloroform; the extract is washed with 0.1 N potassium hydroxide and 0.5 N potassium tartrate, then mineralized with sulphuric acid - nitric acid and the residue is dissolved in 5 N hydrochloric acid. About 0.8 μg tin per ml can be determined by this method.

Vogel and Deshusses[3405] have described a method for estimating triphenyltin acetate residues in vegetables in which as little as 25 μg of triphenyltin acetate can be determined with a precision of ± 5%. The sample (10 to 25 g) is extracted with chloroform in the presence of sodium hydroxide, the extract is purified by shaking with alkaline tartrate solution, and the solvent is evaporated. The residue after oxidation is dissolved in water aluminium trichloride solution is added, and tin and aluminium are coprecipitated as hydroxides. The precipitate is dissolved in dilute hydrochloric acid, ammonium chloride solution is added, and the polarogram is recorded. Lead interferes and causes high results.

Gauer et al[3506] have described a gas chromatographic method for the determination of the residues of tricylohexylhydroxystannane and its dicyclohexyl metabolite on strawberries, apples and grapes that have been treated with Plictran miticide. Crop samples were treated with aqueous hydrobromic acid to form bromo-derivatives of the organotin compounds, and these derivatives were extracted into benzene. When the residue levels were less than 1 p.p.m., the derivative solution was cleaned-up on a column of silica gel. The derivatives were determined by g.l.c. at 200°C on a glass column (60 cm x 4 mm) packed with 2% of OV-225 on Chromosorb G AW-DMCS (100 - 120 mesh) or at 100° on a column (120 cm x 4 mm) packed with 0.5% of OV-225 on glass beads (100 mesh) with helium (30 ml per min) as carrier gas. Background interference was minimised by use of a halide-sensitive Coulson detector. Recovery of 1 p.p.m. of added tricyclohexylhydroxystannane was 80 to 95%; that of 0.1 p.p.m. was 78 to 89%. Conditions are also described for the gas chromatographic determination of cyclohexylstannoic acid, another possible degradation product of Plictran.

Organolead Compounds

Binnie and Hodges[3507] have described a method for the determination of ionic lead species in marine forms. The technique is based on solvent extraction followed by differential pulse anodic stripping volametry.

Organoarsenic Compounds

Sachs et al[3508] have described methods for the identification of cacodylic acid, sodium methane arsonate and several inorganic arsenic compounds, arsenites and arsenates in plant tissues.

Samples of bean seedlings were first digested with nitric acid to avoid losses of arsenic as volatile chlorides during charring, then H_2SO_4 was added and heating was continued to fuming. The inorganic As^V was then determined with silver diethyldithiocarbamate. There was a rectilinear relationship between extinction at 540 nm and the weight of arsenic in the sample in the range 0.6 to 20 µg. The herbicides cacodylic acid, sodium methanearsonate, sodium arsenate and sodium arsenite were identified by paper chromatography of aqueous extracts with propanol - aqueous NH_2 (7:3), isopropyl alcohol - H_2O - acetic acid (100:30:1), isopropyl alcohol - H_2O (7:3) and methanol - mM-aqueous NH_3 (4:1) as solvents.

Odanaka et al[3509] have reported the application of gas chromatography with multiple ion detection after hydride generation with sodium borohydride to the determination of mono, di-and methyl arsenic compounds, trimethylarsenic oxide and inorganic arsenic in plant materials. This work is discussed more fully in the section on organoarsenic compounds in water. Recoveries in spiking experiments were 92 to 103%.

Whyte and Englar[3510] have described methods for the analysis of inorganic and total arsenic in several species of marine algae. Arsenic trichloride formation and distillation was used for inorganic arsenic and acid-oxidative digestion for total arsenic.

Organomagnesium Compounds

Various workers have discussed the analysis of chlorophyll. Bruinsina [3411] discusses various methods for the extraction and determination of chlorophyll in plant materials. He suggests acetone homogenization of the plant material followed by filtration and dilution with water to a concentration of 80% of acetone. The chlorophyll is then estimated photometrically by evaluation of three wavelengths to compensate for background interefer-ence.

Strell et al[3512] examined water soluble chlorophylls as follows: After removal of copper by acid treatment, and of insoluble matter, the chlorophyll derivatives were fractionated by extraction from ether with graded concentrations of hydrochloric acid. The fractions, after further similar purification, and in some cases after chromatography, were examined chemically and spectroscopically.

Kondo and Mori[3513] determined chlorophyll in dentrifice by the so-called quantitative bridge method. The method was studied using a mixture of butanol, ethanol and ammonium chloride (1%) (2:1:1). Two parallel lines (2 mm apart, both 60 mm long) are slit with a sharp blade on chromatographic paper; melted paraffin wax is smeared outside the slits so that the solution can ascent only through the narrow bridge ("quantitative bridge"). Tooth paste or powder (10 - 30 mg) is dissolved in 6 N hydrochloric acid and extracted with ether (10 - 20 ml); the ether layer is evaporated to dryness and dissolved in butanol (1 ml). This solution is spotted at a point 60 mm from one end of the bridge and developed at 25°C. The logarithm of the length (5 - 10 mm) of the green pattern on the bridge is proportional to the concentration of chlorophyll (2 to 20 μg/ml). The overall error of determination was ~ 4.5%.

E. ORGANOMETALLIC COMPOUNDS IN BIOLOGICAL MATERIALS

Organomercury Compounds in Fish

Atomic absorption spectrometry. Amrstrong et al[3514] and Uthe et al [3515,3516], also Hatch and Ott[3517] and Goulden and Afghan[3518] have developed a semi-automated procedure for the determination of mercury in fish and animal tissue based on flameless atomic absorption spectroscopy. Using Technicon Auto Analyser Equipment, samples can be analysed at the rate of 30 samples/h with a recovery of 95%. (SD ± 3% to ± 8%).

In this procedure the sample (0.1 - 0.5 g) is placed in the bottom of a 30 ml Kjeldahl flask, making sure that no sample will be above the acid level during digestion. Add 1.0 ml nitric acid (sp. gr. 1.42) and 4.0 ml sulphuric acid (sp. gr. 1.84). Incubate flasks in shaking water bath at 58°C or until a clear solution is obtained. Remove samples from bath, cool in ice, then add slowly 15 ml 6% potassium permanganate solution; close flask with stopper, allow to stand overnight. The reaction with permanganate is vigorous, with evolution of oxygen. If the permanganate is added very slowly to the warm digest, oxidation is rapid, being complete in 30 minutes. Add, dropwise, 30% hydrogen peroxide until the precipitate just redissolves. Make solution to 25.0 ml and mix.

Mercury is then estimated on the Autoanalyser using the hydroxylamine-stannic sulphate reduction system.

Results obtained in recovery experiments carried out by Uthe et al[3516] in the semi-automated determination of nanogram levels of organically bound mercury in fish is shown in Table 285. The recovery is 94%.

The precision of routine determinations, estimated from duplicate determinations, using analyses of several batches of fish in which the batches had low or high mercury content is shown in Table 286 in which it is seen that standard deviation increases with increasing mercury content. Relative standard deviation ranges from ± 3% to ± 8%.

Collett et al[3519] steam distilled the fish sample and determined alkyl-mercury compounds by cold vapour atomic absorption spectroscopy.

The applications of flameless atomic absorption spectroscopy to the determination of total mercury in fish have been dealt with by Shultz[3520], Stainton[3521], Kopp et al[3522], the U.S. Environmental Protection Agency [3523] and Hendzel and Jamieson[3524].

The determination of mercury in solid environmental samples such as fish requires low temperature preparation techniques to prevent loss of organomercury compounds. Armstrong and Uthe[3525] used a sulfuric and nitric acid digestion at 58°C followed by permanganate oxidation to extract mercury from fish tissue.

The Analytical Methods Committee[3526,3527] described the use of hydrogen peroxide and sulfuric acid for the destruction of organic matter, and it has been applied[3528] to mercury in biological materials. This digestion system

Table 285 - Recovery of mercury added to whitefish muscle.

Sample wt.	In tissues*	Added	Hg(ng) Found	Recovered	Recovery %
a. Mercuric chloride, whitefish sample					
0.214	-	-	13	-	-
0.271	-	-	15	-	-
0.227	13	100	111	98	98
0.146	8	200	204	196	98
0.260	15	100	201	186	93
0.246	14	300	299	285	95
0.290	17	300	315	298	99
0.236	14	400	364	350	88
0.274	16	400	395	379	95
0.174	10	500	498	488	98
0.163	10	500	418	408	82
b. Methyl mercuric chloride, whitefish sample					
0.318	-	-	15	-	-
0.238	-	-	10	-	-
0.349	-	-	15	-	-
0.227	-	-	10	-	-
0.310	14	93	98	84	90
0.252	11	93	101	90	97
0.322	14	166	186	172	92
0.293	13	186	194	181	97
0.349	15	279	294	279	100
0.329	14	372	367	353	95
0.352	16	372	345	329	95

* Calculated

Table 286 - Precision of routine mercury determination of fish

Sample	Mean mercury content of batch (20 Fish) p.p.m.	Standard deviation p.p.m.	Relative standard deviation(%)
Whitefish	0.10	± 0.008	± 8
Pike	0.28	± 0.014	± 5
Pike	4.94	± 0.292	± 6
Pike	6.12	± 0.298	± 5
Pike	8.96	± 0.621	± 7
Walleye	9.00	± 0.254	± 3

can be coupled with permanganate-persulfate oxidation to recover completely organomercury compounds from fish. This sample preparation technique has been adapted to automated reduction and determination by a.a.s.[3529] of mercury by atomic absorption spectroscopy.

Agemian and Cheam[3530] developed a procedure for the simultaneous extraction of organomercury and organoarsenic compounds from fish tissues.

Methyl mercury compounds have been specifically dealt with by various workers[3531-3535]. Shum et al[3536] carry out a toluene extraction of the fish, then treat the extract with dithizone to form methylmercury dithizonate which is then determined in amounts down to 0.08 μg Hg g^{-1} fish sample by graphite furnace atomic absorption spectroscopy.

Oda and Ingle[3537] have described a procedure for the determination of organomercury and inorganic mercury in tuna fish, hair and urine. They describe a speciation scheme for ultratrace levels of mercury in which inorganic and organomercury are selectively reduced by stannous chloride and sodium borohydride respectively. The volatilized mercury is determined by cold vapour atomic absorption spectroscopy.

The detection limits for inorganic and organomercury came in the 0.003 - 0.005 ppb range and both types of mercury can be determined in a 1 ml sample in about 3 min.

The samples were first digested by heating in 2 dram capped vials at 90°C for 15 - 30 minutes, urine (1 ml) or hair (20 - 60 mg) or tuna (0.1 - 0.3 g) with 2.5 ml 10 M potassium hydroxide. After digestion, the resultant solutions were centrifuged to separate remaining particulates and the supernatant was decanted into a 100 mL volumetric flask. The vials were washed three to four times with 1% (w/v) sodium chloride with centrifuging before each decant. When the sample had been transferred, 7.5 mL of concentrated nitric acid and 1 mL of 1.0% (w/v) potassium dichromate were added and the remainder of the volume was diluted with 1.0% (w/v) sodium chloride.

The results obtained are tabulated in Table 287.

Table 287 - Analysis of hair, urine and tuna

	Amount of HgII p.p.b.		
Sample	Inorganic	Organic	Total
urine	3.2	1.1	4.3
urine	2.9	0.80	3.7
hair	2.1×10^3	1.9×10^3	4.0×10^3
hair	2.3×10^3	2.0×10^3	4.3×10^3
tuna	35	4.1×10^2	4.5×10^2
tuna	39	4.4×10^2	4.8×10^2

Total concentration of mercury in hair (relative standard deviations of 6 - 11%) was fairly consistent with existing data[3538] which indicates mercury concentrations in hair ranging from 1 to 25 ppm for samples from rural to industrial areas. About half of the mercury found was in the organic form concentrated by the body with the other half probably in the form of externally adsorbed Hg^{2+}.

Tuna samples showed about 0.5 ppm total mercury contamination which is approximately normal for canned tuna[3539]. This is also the upper limit allowed by the United States Food and Drug Administration. The organomercury to total mercury ratio is also in agreement with previous studies[3540,3541] with about 92% in the tuna appearing as organomercury (probably mainly methyl-mercury) and relative precisions of 18 and 7% for inorganic and organomercury, respectively.

The mercury in the urine sample was in the parts per billion range which is not unusual[3542]. The relative precision for the urine measurements was about 5 and 10% for inorganic and organomercury, respectively. The inorganic form predominates as expected since most organomercury that is introduced to the body is adsorbed or broken down before excretion.

Hanamura et al[3543] applied thermal vapourization and plasma emission spectrometry to the determination of organomercury compounds and inorganic mercury in fish. Solid samples of 250 mg or more are used to avoid problems with sample heterogenelty. The precision of characteristic appearance temperatures is ± 2°C. The single electrode atmosphere pressure microwave plasma system is extremely tolerant to the introduction of water, organic solvents and air. The measurement system contained a repetition wavelength scan device to allow background correction. The plasma temperature was 5500 K. The single electrode microwave plasma emission allows very sensitive detection of both metals and nonmetals and the higher power minimizes sample matrix interferences. The need for high sensitivity is apparent when one considers the requirement of a low temperature ramp (~ 15°C/min) over the temperature range of 25°C to 450°C; the low rate of temperature rise is needed to achieve higher resolution in the vaporization separation process. The analytical results consist of the emission signal of the analyte species recorded as a function of the temperature of the sample, resulting in a plot similar to differential thermal analysis. Because analyte species in the solid sample vaporize at different temperatures, each species produces a peak at a temperature characteristic of the analyte species and the sample type.

The furnace vaporizer is shown schematically in Figure 196. The sample is held in a quartz crucible; as the sample is heated, the volatile constituents are swept by the carrier gas flowing past the indented top of the quartz furnace tube. An alumel-chromel thermocouple (B+S gauge 28) is used to monitor the temperature of the quartz sample crucible. The temperature of the furnace, which was increased at a rate of 15°C/min for most samples, was determined by a 250 W heating coil controlled by a linear temperature programming unit taken from a commercial gas chromatograph.

The 2540 MHz microwave plasma torch, which was operated with argon, nitrogen or helium at 500 W, is schematically shown in Figure 197. The plasma torch and power generator system was similar to the one described by Murayama[3544,3545]. The plasma operating characteristics are given in Table 288.

Table 288 - Operating characteristics of experimental system

Helium plasma	He flow rate, 3.6 L/min sample vapor carrier flow, 0.4 L/min
Argon plasma	Ar flow rate, 3.6 L/min sample vapor carrier flow, 0.4 L/min (argon passed through the water gas wash bottle to become saturated in water vapor producing a more stable plasma)
Nitrogen plasma	N_2 flow rate, 3.6 L/min sample vapor carrier flow, 0.4 L/min
Monochromator	slit width, 0.03 mm (spectral band-pass = 0.05 nm) slit height, 2 mm scanning speed (repeated scan modes), 2.0 nm/min
Observation area	centre of plasma - just above the Pt electrode (1-2 mm)
Magnetron	current, 200 mA supply voltage, 2 kV

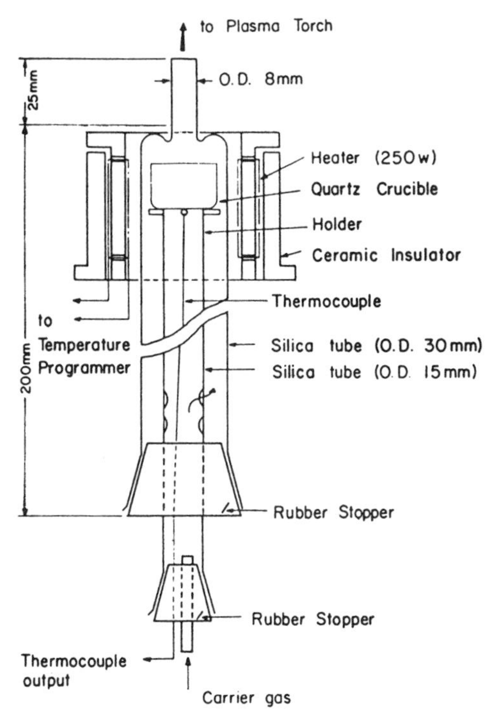

Figure 196 - Schematic diagram of evolved gas furnace.

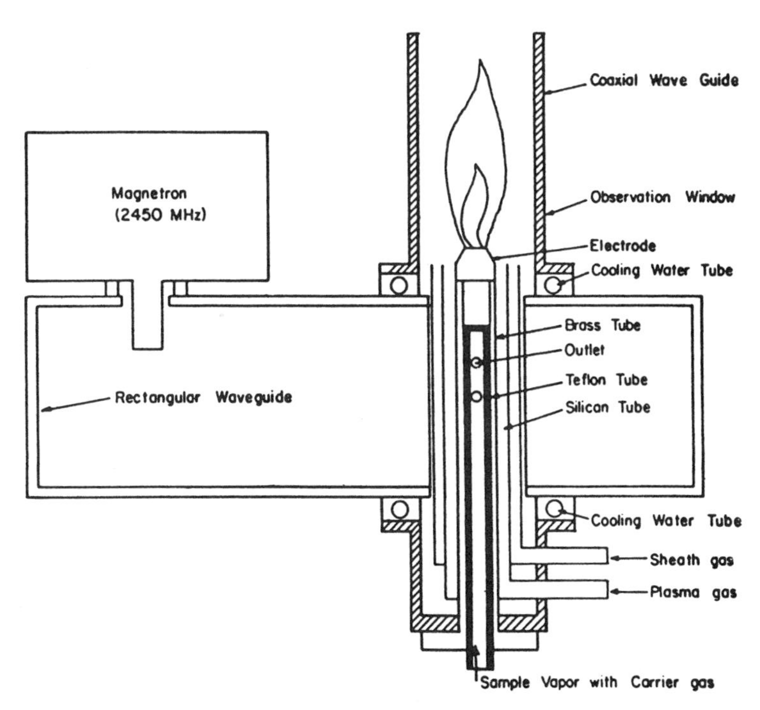

Figure 197 - Schematic diagram of microwave single electrode plasma torch assembly.

The vaporized molecular species, introduced into the microwave induced plasma, appear as peaks at characteristic temperatures dependent primarily upon the molecular form and secondarily upon the sample composition. The identification power of the thermal vaporization (evolved gas)/microwave emission detection is evident by the results in Table 289.

Table 289 - Characteristic temperatures (peak values) for Hg compounds

molecular species	sample	characteristic temp[a] °C
$HgCl_2$	freeze dried fish	72[c] (104)[b]
$HgCl_2$	pure	75
HgCl	pure	97
HgO	pure	89, 133
$HgBr_2$	pure	69
HgI_2	pure	74
CH_3HgCl	pure	33

[a] Characteristic temperatures determined by extrapolation of the linear portion of the low temperature wing of the peak to the base line. Values have a precision of ± 2°C.
[b] Direct.
[c] Cold trapped.

In Table 290 the detection limits amount of analyte producing a signal-to-noise ratio of 3 for several nonmetals (hydrogen, oxygen and nitrogen) and mercury, are given for three plasma types (helium, argon and nitrogen) with the evolved gas analysis/microwave emission spectrometric system.

Table 290 - Detection limits (in μg) for several nonmetals and mercury with the evolved gas analysis/microwave emission spectrometric system

		plasma gas		
	wavelength nm	helium	argon	nitrogen
hydrogen	656.28	0.4 (0.01)[a]	5 (0.08)	20 (0.3)
oxygen	777.19	0.8 (0.004)	4 (0.07)	4 (0.07)
nitrogen	746.53	5 (0.03)	240 (4)	
carbon	247.86	0.7 (0.001)	4 (0.07)	10 (0.2)
mercury	253.65	0.1 (0.001)	0.006 (0.0001)	0.007 (0.001

[a] Values in parentheses are in ng/s.

The evolved gas analysis/microwave plasma emission spectrometer was used for the meaurement of carbon, hydrogen, nitrogen, oxygen and mercury in orchard leaves (NBS-SRM-1517) and in tuna fish (NBS-RM-50). A separate sample was measured for each element, and the plots of emission intensity vs. sample volatilization temperature are given. In the case of tuna fish standard, the first peak is a combination of adsorbed water and nitrogen and since a carbon peak does not occur, the mercury peak would appear to be metallic mercury. The second peak certainly is an organomercury compound since mercury, carbon, oxygen and hydrogen appear (200 - 300°C). The third peak (300 - 400°C) contains carbon nitrogen and some hydrogen (Figure 198).

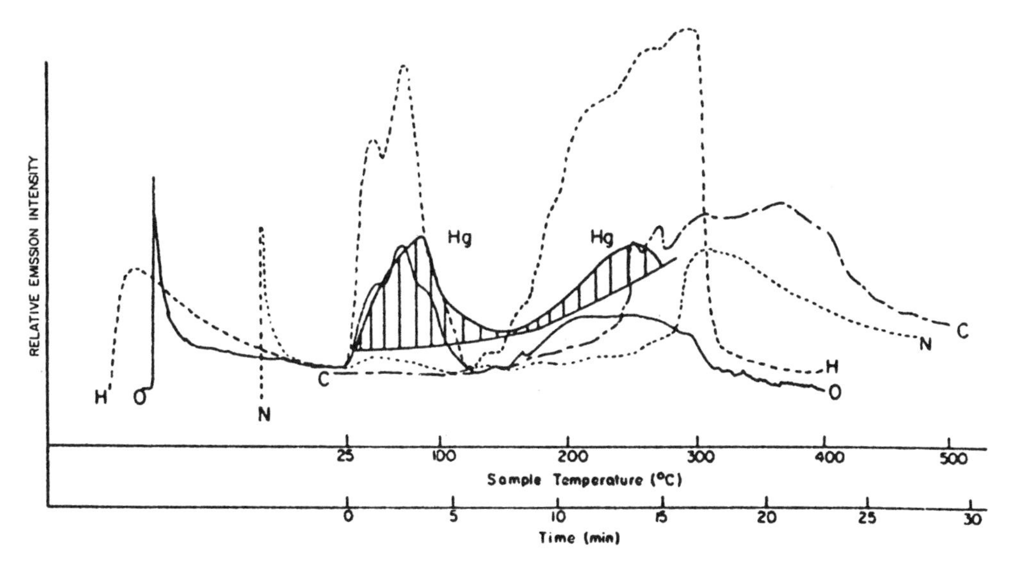

Figure 198 - Evolved gas analysis/plasma emission spectrometer application to measurement of C, H, N, O, and Hg in tuna fish (NBS-RM-50): He plasma for C. H. N. and O; Ar plasma for Hg; He plasma; He plasma gas flow rate, 3.6 L/min; He carrier gas flow rate, 0.4 L/min; microwave power, 400 W for C, H, N. and O; all flows the same for He or Ar plasma and carrier gas but microwave power is 500 W for an Ar plasma. Sample weights were as follows: 252 mg for H, 258 mg for O, 117 mg for N, 110 mg for C, and 504.5 mg for Hg. Emission wavelengths were the same as those in Table 290 except for Hg, 253.7 nm. The sample heating rate is 15°C/min.

Gas chromatography. In view of the comparatively high mercury content of fish found (Westermark[3546]) in Swedish lakes and rivers, (Westhoo[3547-3549]; Westhoo et al[3550]) embarked on an extensive survey of the nature and the concentrations of mercury in fish from these waters. Westhoo points out that several authors (e.g. Polley and Miller[3551]; Gage[3552]) have described methods for the determination of organic mercury compounds, but these methods either do not separate different compounds, e.g. methylmercury from phenylmercury compounds, or are designed for mercury contents higher than those usually met with in foods.

He describes a combined gas chromatographic and thin-layer chromatographic method (Westhoo[3547,3548]), for the identification and determination of methylmercury compounds in fish, in animal foodstuffs, egg-yolk, meat and liver. He (Westhoo[3550]) has also used a combination of gas chromatography and mass spectrometry to identify and determine methylmercury compounds in fish.

To extract organically bound mercury from muscle tissue of fish, Westhoo homogenized the fish with water and acidified with concentrated hydrochloric acid (1/5 of the volume of the suspension). Organomercuric compounds were then extracted in one step with benzene using the method described by Gage[3452]. Methylmercury, either originally present or added to the fish, could be extracted, though with difficulty, also when only a small amount of acid was added, e.g. at pH 1. From an aliquot of the benzene solution organomercury could be extracted with ammonium or sodium hydroxide solution, saturated with sodium sulphate for elimination of lipids. The yields were low and variable, but could be improved as described below.

Several workers have found that a cleanup procedure is necessary to remove fatty acids and amino acids, which could otherwise poison the column. The cleanup is achieved by adding to the organic phase a reagent, such as sodium sulphide[3552], cysteine[3547-3549], sodium thiosulphate[3453] or glutathione[3554], which forms a strong water-soluble alkylmercury complex to extract the mercury complex into the aqueous phase. A halide is added to the aqueous phase, and the alkylmercury halides formed are re-extracted into an organic phase. Aliquots of this phase are finally injected into the gas chromatograph.

Uchida et al[3555] have shown that the mercury compounds in the shellfish that caused the Minimata disease (Japan) was methyl(methylthio)mercury. Westhoo concluded that it is reasonable to assume that methylmercury, if present in Swedish fish, should at least to some extent be a methylthio derivative. The Hg - S bond is stronger than Hg - NH or Hg - OH bonds. Accordingly it prevents the formation of these bonds, which should be produced by the ammonium hydroxide solution and increase the solubility in water. Any methylthio group present should therefore be removed before the extraction with alkali.

Distillation of the benzene extract at reduced pressure at room temperature or at ordinary pressure at 80°C to 1/10 of the original volume removed the factor that prevented an acceptable extraction by ammonium or sodium hydroxide solution (probably methanethiol and perhaps hydrogen sulphide). After the distillation and subsequent extraction with ammonium hydroxide solution the extract was acidified with hydrochloric acid, and the organomercury compound was extracted once with benzene. After drying with anhydrous sodium sulphate, the benzene solution was ready for gas chromatography, and after concentration, also for thin-layer chromatography.

In the above procedure about 30% of the methylmercury was lost, mainly by unfavourable partition coefficients. In a model experiment of the benzene extraction of methylmercury from a hydrochloric acid solution, for instance, 14% of the methylmercury was left in the water layer. The losses by partition are, however, characteristic of the compounds involved and reproducible. Consequently they can be included in the calibration curve, thus disturbing the results only slightly. The yields can be increased by repeated extractions, but good results are obtained with the above simple procedure. The calibration curve is based on the partition laws for methylmercury chloride, though some methylmercaptide and perhaps sulphide are probably present in fish. However, when hydrogen sulphide or methanethiol was added (30 μg per 5 μg mercury as methylmercury) to the aqueous phase before the first extraction, when preparing the calibration curve, the 5 μg point was unaltered. Large amounts of these sulphur compounds disturbed the analysis because they were not completely removed by the distillation.

When known amounts of methylmercury dicyandiamide were added to saltwater fish (frozen cod, Gadus morrhua, or haddock, Gadus aeglefinus), 82% - 95% of the additions were recovered.

Westhoo[3547-3550] used an electron capture detector and 60 in x 1/8 in stainless steel columns filled with Carbowax 1500 (10%) on Teflon 6 or in Chromosorb W, acid washed DMCS. Nitrogen was used as carrier gas and column temperatures were 130 - 145°C. He identified methylmercury chloride in pike caught in the Baltic ocean at concentrations between 0.07 and 4.4 mg per kg of fish.

Westhoo[3547] pointed out that if methylmercury attached itself to a sulphur atom by reaction with a thiol or hydrogen sulphide then the non-volatile HgS compound produced would not be included in the determination. More recently (Westhoo[3548]) he has developed a modification to this method, to render it applicable to a wider range of foodstuffs (fish, egg-yolk and white, meat and liver) by binding interfering thiols in the benzene extract of the sample to mercuric ions added in excess or, by extracting the benzene extract with aqueous cysteine to form the cysteine-methylmercury complex.

Westhoo[3550] discussed the identification and determination of methylmercury compounds in benzene extracts of fish by the use of combined gas chromatographic mass spectrometric analysis and also by using a standard gas chromatograph with an electron capture detector for detecting organic halogen compounds.

Westhoo et al[3550] reported results obtained by gas chromatography with electron capture and with mass spectrometric detection on a range of samples of fish. (Table 291). Total mercury was also determined on these samples by neutron activation analysis. Results obtained by the three methods agree within ± 10% of the average value.

Table 291 - Comparison between results for mercury levels in fish flesh, determined by combination gas chromatograph - mass spectrometer, gas chromatograph with electron capture detector, and activation analysis

	Methylmercury, mg Hg/kg fish flesh		Total Hg,mg/kg fish flesh
	GLC-mass spectrometric measurement of $^{202}Hg^{+}$	Gas chromatography with electron capture detector	Activation analysis
Pike 1	0.14	0.17	Not determined
Pike 2	0.55	0.54	0.59
Pike 3	2.53	2.57	2.70
Pike 4	0.43	0.41	0.39
Pike 5	0.49	0.55	0.54
Pike 6	0.75	0.66	0.63
Pike 7	0.72	0.70	0.66
Perch 8	3.19	3.29	3.12

Kampe and McMahon[3556] used the Westhoo procedure to determine methylmercury in fish. Their procedure involved the partitioning of methylmercury chloride in benzene and gas chromatographic analysis with electron capture detection. Down to 0.02 ppm of methylmercury chloride can be detected in a 10 g sample.

Longbottom et al[3557] improved the Westhoo cleanup procedure by replacing cysteine with the more stable sodium thiosulphate when forming the methylmercury adduct. For the gas chromatography of methylmercury iodide, these workers recommend the use of a 63 Ni electron capture detector, as it does not form an amalgam at 280°C, the temperature at which it is used. This method was used to detect down to 0.01 µg of methylmercury per g of fish, 0.001 µg per g of sediment and 0.01 µg per litre of water.

Shaefter et al[3558] determined methylmercury in fish.

Fabbriani et al[3559] have reviewed methods for the determination of organomercury compounds in canned fish products. They describe a method in which methylmercury is converted into a mercaptide complex[3560] followed by gas chromatography. All the samples of canned tuna examined contained inorganic and methylmercury, 25% of samples contained more than 1 ppm of total mercury and 20% contained more than 0.5 ppm of organic mercury. The ratio of organic to inorganic mercury ranged from 36 - 93%. Contents of mercury in sardines and mackerel were low (less than 0.3 ppm of total mercury and less than 0.21 ppm of methylmercury).

Newsome[3561] has described a method for the determination of methyl mercury in fish and cereal grain products in which fish fillet (10 g) is homogenized for 10 min with N hydrobromic acid - 2 N potassium bromide (60 ml) and filtered through glass wool. The residue and glass wool are homogenized with a further 60 ml of the same solution. The combined filtrates are extracted twice with benzene, emulsions being broken by shaking with sodium sulphate. The combined benzene layers are extracted with a cysteine acetate solution (8 ml), an aliquot (5 ml) of which is acidified with 48% hydrobromic acid (1 ml) and extracted with benzene. Flour or ground oats (10 g) is homogenized with benzene (50 ml) and 90% formic acid (5 ml), the mixture is filtered, the aliquot of the filtrate (30 ml) is transferred to a column (10 cm x 1.6 cm) of silicic acid and the column is percolated with benzene under pressure (1 ml per min). The 15 - 55 ml fraction of the percolate is extracted with cysteine acetate solution (6 ml) and treated as before. The benzene extracts are submitted to gas chromatography on a glass column (40 cm x 4 mm) packed with 2% of butanediol succinate on Chromosorb W (AW-DMCS) (100 - 120 mesh) operated at 120°C with nitrogen as carrier gas (80 - 100 ml per min) and a ^{3}H-foil electron-capture detector. The sensitivity of the method is in the range 0.01 - 0.90 ppm and the mean recovery generally exceeds 95%.

Bache and Lisk[3562] determined methylmercury compounds in fish by chromatography on a 2 ft glass column of Chromosorb 101 or 20% 1:1 OV-17/QF-1. Detection of the separated organomercury compounds was achieved by measurement of the emission spectrum of the 2537°A atomic mercury line which gave a linear response over the range of 0.1 - 100 μg of injected methylmercury chloride. Average recoveries of methylmercury chloride in fish were 62% at the 0.3 ppm level.

Uthe et al[3563] have described a rapid semi-micro method for determining methylmercury in fish and crustacean and aquatic mammal tissue. The procedure involves extracting the methylmercury into toluene as methylmercury bromide, partitioning the bromide into aqueous ethanol as the thiosulphate - complex, re-extracting methylmercury iodide into benzene, and gas chromatography on a glass column (4 ft x 0.25 in) packed with 7% Carbowax 20M on Chromosorb W and operated at 170°C with nitrogen as carrier gas (60 ml per min) and electron-capture detection. Down to 0.01 ppm of methylmercury in a 2 g sample could be detected. A comparison of the results with those obtained by atomic absorption (which gave total mercury content) indicated that all the fish samples examined contained mercury, more than 41% of the mercury as methylmercury.

Zalenko and Kosta[3564] have described a method for the determination of methylmercury at the 1 in 10^{9} level in biological tissue, both human and fish, which involves volatilisation of methylmercury cyanide, formed by the reaction of methylmercury ions with hydrogen cyanide released in the reaction of potassium ferrocyanide with sulphuric acid at 75°C on filter-paper treated with cysteine in a micro-diffusion cell, conversion into methylmercury chloride with hydrogen chloride, extraction into benzene and determination by gas chromatography with electron-capture detection. Two methods are described, the first for samples containing more than 30 ng of methylmercury per g involves extraction into 1 ml of benzene, the second (for lower concentrations) involves extraction into 0.15 ml of benzene in a special enclosed extractor. Gas chromatography employing electron-capture detection has been used quite extensively as a sensitive technique for analysing organomercury[3566,3567] compounds in biological materials. Blair et al[3565] has described a procedure based on reductive combustion in a flame ionization detector of a conventional GC followed by the cold vapor atomic absorption detection of mercury vapour for the determination of organomercurials in bacterial respirant gases. This technique has also been discussed by Longbottom[3568].

Bye and Paus[3570] have described atomic absorption - gas chromatographic procedure for the determination of alkylmercury compounds in fish tissue. These workers point out that when direct gas chromatographic methods are used in the determination of alkylmercury compounds, interferences are often a problem, especially with the electron-capture detector, which is very sensitive to other halogen compounds. Some workers have solved this problem by utilizing atomic absorption as a specific mercury detector. Gonzales and Ross[3570] burned the effluent in an oxygen atmosphere and led the gases through a MAS 50 mercury analyzer (an atomic absorption instrument). Longbottom[3571] cooled the gases from the flame ionization detector and led the gases through such a mercury analyzer, but reported that it was less sensitive than the electron-capture detector for dialkyl mercury compounds. Bye and Paus[3570] solved this problem by leading the effluent from the gas chromatographic column through a steel tube in a furnace at a temperature at which the organic mercury molecules are cracked. The products are then led through a 10 cm quartz cuvette placed in the beam from a hollow-cathode lamp in an atomic absorption spectrometer. Bye and Paus[3569] state that for many of the earlier methods, the calibration curves are obtained from measurements of peaks from pure standard solutions of organic mercury compounds. They doubt the correctness of such a procedure, because it does not take into account the fact that appreciable amounts of mercury may be lost during the many extraction steps used in the analysis, especially in work with small samples and small volumes and state that a standard addition procedure should be used for calibration, and the standard organic mercury solution should be added as early as possible in the procedure. Their method is a modification of that of Longbottom et al[3572] and for their method they claim greater precision than that achieved in earlier methods due to improvements in the extraction sequence.

A schematic diagram of the apparatus is shown in Figure 199. A Perkin-Elmer model 800 gas chromatograph was used. The following operating conditions were satisfactory: column, 10% SP2300 on Chromosorb W 80 - 100 mesh; oven temperature 145°C; inlet temperature 200°C; carrier gas, nitrogen at a pressure of 3.5 k Pa cm^{-2} measured at the g.c. inlet; flow rate, 90 ml min^{-1}.

The furnace was made by winding a nichrome resistance wire around a quartz tube 6 cm long, 4 mm o.d., 2 mm i.d. This unit was placed inside another quartz tube (7 cm long, 8 mm o.d., 6 mm i.d.). The nichrome wire coil was connected to a 0-230-V- Variac transformer. The circuit was equipped with voltmeter and ammeter. The furnace was operated at about 10 V and 2.3 A; the temperature was then about 620°C. From the end of the stainless steel tube, a PVC tube led to the inlet device (Figure 200) for the cuvette.

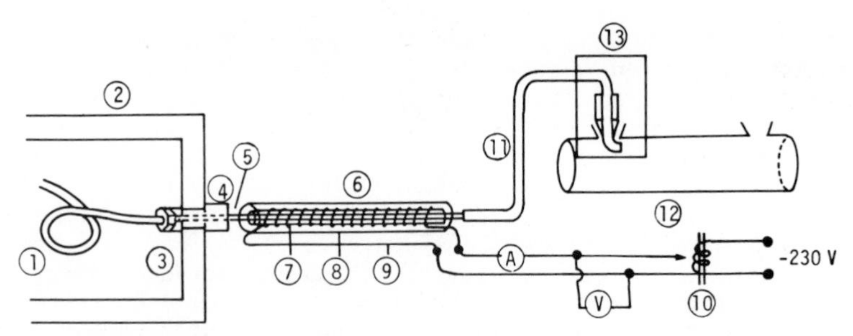

Figure 199 - The complete apparatus for measuring organic mercury compounds in gas chromatographic effluents. (1) column, stainless steel; (2) gas chromatograph; (3) g.c. oven; (4) Teflon joint; (5) stainless steel tube; (6) electrical furnace; (7) inner quartz tube; (8) outer quartz tube; (9) resistance wire; (10) Variac transformer, (11) PVC tube; (12) 10 cm quartz cuvette; (13) inlet device for the cuvette.

To homogenize the sample, a Bellco No. 1977 12 ml graduated tissue grinder was used (Figure 200).

The Perkin-Elmer model 303 atomic absorption spectrometer was run at the 254 nm mercury line. Deuterium background correction was essential. The signals were recorded on a Perkin-Elmer 159 chart recorder. Three weighed 0.5 g portions of frozen fish were transferred to three tissue grinders, 0.5 g of 1 M copper sulphate solution was added to each, and 50- and 100-1 of the standard mercury solution were added to two of the samples. To each sample 2 ml of bromine reagent were added, and the samples were homogenized well.

Potassium bromide (360 g) was dissolved in 700 ml of water. Concentrated sulfuric acid (110 ml) was added to 100 ml of water. After cooling to room temperature, the solutions were mixed and made up to 1 litre with water. After homogenization, the rod was lifted carefully from the solution and rinsed with water until the total volume of sample, reagents and water was exactly 7 ml. Toluene (3.5 ml) was pipetted into each solution, the grinders were shaken for 2 minutes, and the mixture was centrifuged for 10 minutes at 2000 rpm.

A 3 ml portion of each toluene phase was pipetted into one of the three centrifuge tubes, with a separate pipette for each sample. To the residue in each grinder, 3.5 ml of toluene was added and the operation repeated, the 3.0 ml portion of each toluene phase being transferred, with the same pipettes as before, to the appropriate centrifuge tube. To each tube, 2.0 ml of 5 x 10^{-3} M sodium thiosulphate solution was added. These were shaken for 2 mins and centrifuged for 15 mins. Graduated pipettes (2 ml), equipped with Peleus balloons, were inserted through the organic phases into the lower aqueous phases, and as much aqueous phase as possible (but equal volumes of 1.6 - 1.9 ml) was taken from each tube, starting with the tube in which the cloudy layer between the phases was thickest. These volumes were transferred to three centrifuge tubes, and the same volumes of 3 M potassium iodide solution were added to each. From that moment the solutions had to be kept in the dark, Benzene (1.0 ml) was added to each tube - the tubes were shaken vigorously for 2 mins and centrifuged for 10 mins. Each benzene phase was transferred to a 3 ml "Microflex" tube with screw cap and septum, contain-

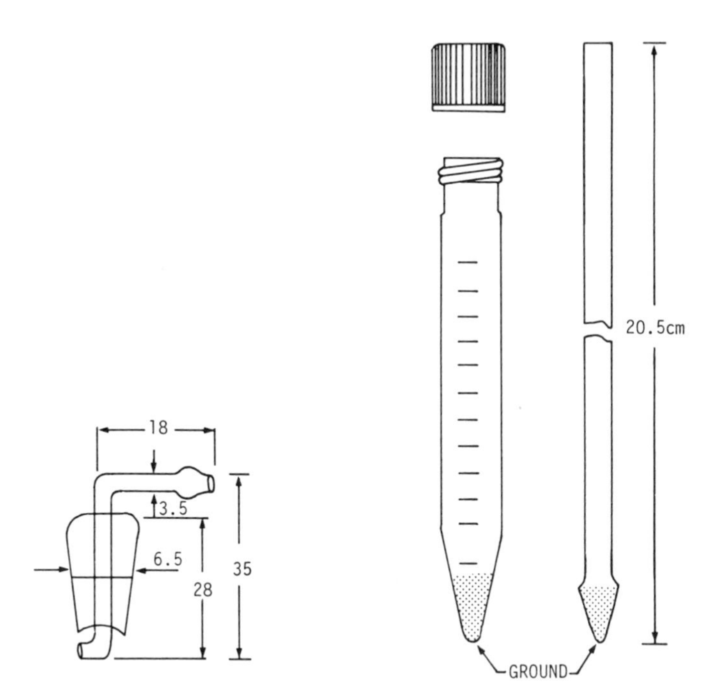

Figure 200 - Inlet device for the cuvette. Measurements in mm. Graduated tissue grinder, "Bellco" No. 1977, 12 ml.

ing a few crystals of anhydrous sodium sulphate. Suitable aliquots (5 - 25 μl) of the solutions were injected into the gas chromatograph and measured on the atomic absorption instrument optimized at 254 nm with deuterium background correction and suitable scale expansion.

Bye and Paus[3570] detected methylmercury (not ethyl- or phenylmercury) in a fish sample. A typical example of the peaks obtained is shown in Figure 201. The calibration curves showed linear ranges up to 10 ppm Hg for methylmercury and ethylmercury chlorides in mixtures. From these graphs, the fish samples were found to contain 2.2 ppm of mercury as methylmercury.

Liquid chromatography. Liquid chromatography has been applied to organomercury compounds employing a differential pulse electrochemical detector [3573] and a graphite furnace atomic absorption detector[3574]. These detection techniques can permit the specific determination of inorganic mercury or methylmercury with very high sensitivity. However, an improvement in sensitivity of the detection techniques inevitably requires complicated and elaborate preliminary chemical procedures - digestion, separation, and cleanup steps prior to detection - for an accurate and precise determination. Because the amount of mercury to be determined is smaller, its quantitative separation from the sample is more difficult.

Liquid chromatography using differential pulse electrochemical detection has been used to determine organomercury cations in tuna fish and shark meat (MacCrehan and Dunst[3575]).

Previous work had shown that many important organometallic compounds of

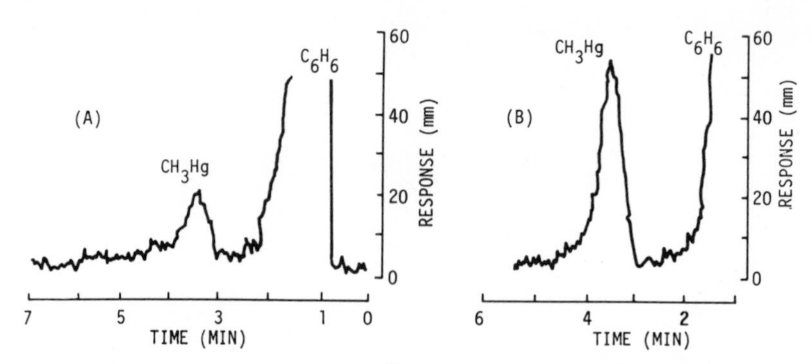

Figure 201 - Peaks obtained from (a) the fish samples, and (b) the sample with 2.32 ng Hg added as CH_3HgCl. Ethylmercury and phenylmercury peaks did not appear in the sample chromatogram, therefore peaks for these compounds after standard addition are not shown.

mercury, tin, lead, arsenic and antimony are reduced at potentials below - 1.1 V[3576,3577]. An amperometric mode of detection had been used to monitor the reverse-phase separation of such species as methyl-, ethyl-, and phenylmercury and also trimethyl-, and triethyllead. The selectivity of the elecrochemical detection approach can be further enhanced by the use of a differential pulse waveform[3578,3579].

MacCrehan and Durst[3577] used a gold electrode because the hydrogen overvoltage on gold amalgam is greater (0.80 V vs. 0.10 V) than on a platinum amalgam. The gold amalgamated mercury electrode provides a negative potential range to about - 1.2 V for amperometric detection (at pH 5.5) and gives reproducible results.

Another important consideration is the purification of the chromatographic solvent of reducible species. Oxygen can be removed to very low levels by purging the solvent reservoir continuously with ultrapure nitrogen (O_2 < 0.5 ppm). Other reducible impurities are removed by electrolytic reduction at a mercury pool. In order to maintain the purity of the solvent, it is necessary to enclose the entire chromatograph in a nitrogen-purged box to avoid re-entry of oxygen through Teflon components in the pump, sampling valve, and detector cell.

The final consideration in reductive electrochemical detection is the type of waveform applied to the cell. The simplest approach is to use amperometry where a constant potential is applied to the working electrode and the resulting current is monitored. In this approach, all species with reduction potentials below that applied will give a response. Amperometric detection provides excellent detection limits (about 5×10^{-9} mol/L) for use with a solid electrode and has good selectivity for easily reduced analytes. However, when the analytes of interest have reduction potentials approaching - 1.0 V, the number of possible interfering electroactive species becomes larger. The differential pulse mode of detection offers a substantial increase in selectivity over amperometry. In Figure 202, for example, electroactive species B will have a large current response when the potential is pulsed from the applied potential E to $E + \Delta E$. The differential readout of the current measured before and at the end of the pulse will provide a signal proportional to the concentration of B only, A and C will

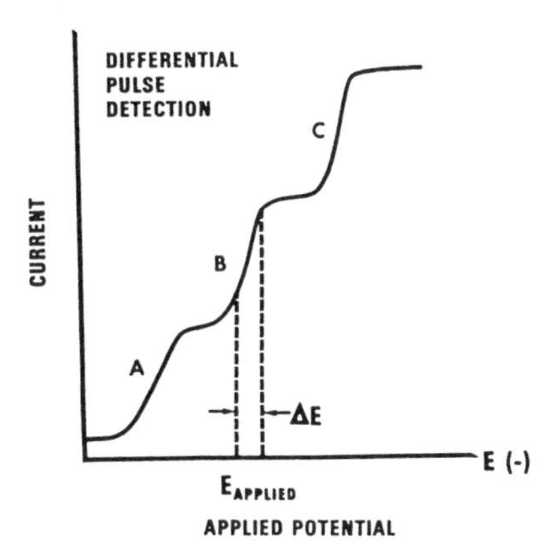

Figure 202 - Hydrodynamic voltammetry of three reducible, coeluting species.

not be detected at potential E with the pulse height (ΔE) employed in the figure.

The apparatus used has been described in detail by MacCrehan et al[3577]. Alkaline hydrolysis is followed by acidification with hydrochloric acid. The organomercury cations can then be extracted from the aqueous solution with toluene as the neutral chloride complexes. The procedure employed in this work for the fish sample preparation for the determination of methylmercury follows the recommendation of the Analytical Methods Committee of the Chemical Society[3580] with some modification. The sample and reagent amounts were reduced by 4/5. The aqueous back-extraction solution used was 0.01 mol/L disodium thiosulfate buffered to pH 5.5 with 0.05 mol/L ammonium acetate. This extraction solution was compatible with the chromatographic separation, and the determination was performed directly on this aqueous extract after filtering through a 0.2 μm syringe filter. The procedure eliminates the final toluene extraction.

In all cases, a standard additions procedure was used for the determination with known amounts of diluted CH_3Hg^+ solution added to the solid material before the hydrolysis step. The recovery was checked by comparison to a standard curve and found to be about 95%. Various intereference effects on the determination of organomercury compounds and how they are overcome are discussed by these workers.

Table 292 shows the results obtained when the method was applied to standard NBS fish samples.

Table 292 - Methylmercury content of fish samples

sample	mercury species in μg/g			
	$MeHg^+$	$EtHg^+$	$PhHg^+$	total Hg
RM 50 Albacore Tuna	0.93 ± 0.1	nd[a]	nd	0.95 ± 0.1 (20)
Japanese shark paste	8.41 ± 0.1	nd	nd	7.4 (22)

a nd = not detected

For the methylmercury determination in these two materials, duplicate

1.00 g samples were taken and also two other duplicates were spiked with known amounts of methylmercury.

The sample chromatograms were characterized by a single response for methylmercury with a high signal-to-noise ratio. Ethyl- and phenylmercury were not detected in these samples. The identity of the peak corresponding to methylmercury was verified by comparison of the potential-current response function to a standard. The standard additions curves were drawn for quantitative analysis. A linear result was obained for each.

The results obtained (see Table 292) for the methylmercury content of the fish samples were in fairly close agreement to the total mercury (as measured by alternate technique such as atomic absorption and neutron activation analysis). This high proportion of methylmercury to total mercury in tissues is consistent with the results of other workers.

Thin-layer chromatography. Thin-layer chromatography was carried out by Westhoo[3547], either on the original methylmercury chloride containing fish extract or on derivatives prepared from this extract, such as the dithizonate, bromide, iodide or cyanide. Light petroleum- diethylether (70:30) was used as developing solvent, using aluminium oxide on silica gel plates. Separated organomercury compounds were detected with a saturated solution of Michlers thioketone in ethanol.

Methylmercury dithizonate and phenylmercury dithizonate could be separated from each other in the fish extracts by thin-layer chromatography on aluminium oxide (limit of detection: 0.2 μg). Methylmercury cyanide, chloride, bromide and iodide were separated by thin-layer chromatography on silica gel (limit of detection of the chloride and bromide: 0.02 μg on spraying with an ethanol solution of Michler's thioketone).

Miscellaneous. Stuart[3581] used 203 Hg labelled methylmercuric chloride for in-vivo labelling of fish to study the efficacy of various wet ashing procedures.

Yamanaka and Ueda[3582] determined ethylmercury; originating as man-made pollution, in fish found in Japanese waters.

Organomercury Compounds in other Biological Materials

Druzhinin and Kishtsin[3583] have described a method for the determination of mercury in biological tissue based on electrolytic deposition of mercury from solution. The tissue is dissolved by digestion with a mixture of sulphuric acid and nitric acid. A portion of the digest (5 ml) is diluted to 100 ml and saturated sodium acetate (1- ml) and formaldehyde (formalin) solution (2 drops) added. The solution is transferred to an electrolyer vessel and electrolysed at 80°C for 45 - 60 min using a zinc anode and a graphite or copper cathode and passing carbon dioxide or hydrogen through the solution during this operation. Mercury is then dissolved off the cathode in concentrated nitric acid (5 ml) diluted to 10 ml with water and 5% potassium permanganate is added dropwise until a pink colour is obtained. Excess permanganate is then destroyed by the addition of ferrous sulphate solution or hydrogen peroxide. To this solution is added ferric alum solution (2 ml/100 ml test solution) and then the solution is titrated with 0.1 or 0.01 N ammonium thiocyanate. It is claimed that this method will determine mercury in the 1 - 6 mg range in 100 ml with an error of 4.3% (maximum).

Various workers have described methods for the determination of mercury in tissues. Cafruny[3584], for example, determined mercury in urine and kidney by forming a complex in aqueous trichloroacetic acid between potassium bromide and the mercury in certain organic mercurials. After adjustment to pH 5.0 with formate buffer solution, the mercury in this form is extracted with dithizone solution in chloroform for spectrophotometric determination at 475 nm. The procedure can be used to detect as little as 1 μg of mercury and was used for the study of urinary excretion after oral administration of organic mercurial diuretics and of inorganic mercury.

Miller and Lillis[3585] have described methods for the determination of phenylmercury acetate in urine, kidney, liver, muscle, spllen and brain. For urine, the sample (5 - 10 ml containing 5 - 20 μg of phenylmercury acetate) is refluxed with N sodium hydroxide (20 ml), cooled, and 5% potassium permanganate solution added, (12.5 or 25 ml according to the sample volume) with thorough mixing. Then 30% hydroxylamine sulphate:aqueous ammonia (1:1) (12.5 or 25 ml) and 30% ammonium sulphamate solution (5 ml) are added. The mixture is cooled and sufficient 12 N hydrochloric acid added under the surface of the liquid with vigorous swirling to lower the pH to not greater than 1. After further cooling, the solution is shaken well with purified chloroform (11 ml) for one min, the chloroform layer washed with N hydrochloric acid for 45 s and the washings rejected. The chloroform layer is washed for 30 s with 0.3 N acetic acide (25 ml) and a 0.0033% solution of dithizone in purified chloroform (1 ml). The chloroform layer is separated, diluted with chloroform to 11 ml and the extinction measured at 620 nm.

For kidney, liver, muscle or spleen, 1 g of the sample is treated as described above, except that the chloroform layer is clarified by passage through a column of Hyflo Supercel before reaction with dithizone. For brain the procedure is similar, but more vigorous treatment is necessary. Results obtained by these methods are accurate to within approximately ± 1 microgram of the amount of phenylmercury compound present.

Gage[3586] has described a method for the trace determination of phenylmercury acetate in biological material. In this method an acidified aliquot of a 5% aqueous homogenizate of the tissue is extracted with benzene (20 ml). A 15 ml portion of the extract is shaken with 1% aqueous sodium sulphide solution and the evaporated aqueous layer is oxidized with potassium permanganate. To the oxidized solution, decolorized with hydroxyl-ammonium chloride solution, are added urea and EDTA; the pH is adjusted to 1.5. Chloroform is added and the solution is titrated with dithizone solution until the colour of the chloroform layer is intermediate between the orange of the mercury complex and the green of the dithizone solution. The same volume of dithizone solution is added to a control solution, which is then titrated with a standard mercury solution. Recoveries of added phenylmercury salts and methylmercury salts from rat tissues and rat urine were low by up to 15% by this procedure, but concentrations down to 1 ppm can be measured. Inorganic mercury does not interfere.

Ashley[3587] has reviewed procedures for the determination of micro amounts of mercury in biological materials involving destruction of organic matter, use of dithizone for mercury extraction and the avoidance of the simultaneous extraction of copper. Ashley describes an absorptiometric method for determining the mercury-dithizone complex and also a photometric method in which the separated mercury is volatilized and its concentration in the vapour determined by means of a detector cell with a monochromatic light source, the output of a photo-cell being measured and referred to calibration measurements.

Miller and Wachter[3588] and others (Honskova[3589]; Fabre et al[3590]) have used a procedure based on reduction and digestion with sulphuric acid for the determination of low concentrations of mercury in biological materials.

Neutron activation analysis (Sjosfraud[3591]) has been very successfully applied to the determination of mercury in the parts per billion range. This technique does not, of course, distinguish between inorganic mercury and organically bound mercury.

In the method developed by Westhoo[3547], described earlier, for the identification and determination of methylmercury in fish by gas chromatography, the methylmercury was extracted with benzene from a homogenate of the fish acidified with hydrochloric acid. It was then taken up into ammonium hydroxide solution and finally re-extracted into benzene after acidification with hydrochloric acid. The extraction with alkali was incomplete unless the benzene extract was previously concentrated by distillation. The distillation procedure was assumed to remove volatile thio compouunds binding part of the methylmercury and preventing its uptake into ammonia. Any methylmercury attached to a sulphur atom of non-volatile compounds giving rise to alkali-insoluble methylmercury salts at the purification stage would not be determined.

When, however, small amounts of methylmercury dicyanidiamide (less than 0.05 mg/kg) were added to meat, liver or egg yolk and analysed according to the above method, the methylmercury was completely lost in liver and egg yolk, and only partly recovered from meat. After addition of 10 mg/kg of methylmercury to meat or liver, most of it was recovered from meat, but only 5% from liver. Such a failure of the procedure can be expected, if the methylmercury in the neutralized extracts from these foodstuffs is firmly attached, exclusively or to a considerable extent, to thiol groups of non-volatile compounds, but only if the methylmercury salts formed are insoluble in alkali solutions. Model experiments showed, in fact, that after the addition of excess methanethiol or thiophenol to methylmercury chloride in benzene, an extraction with 2 N aqueous ammonia or with sodium hydroxide did not extract the mercury compound from the benzene layer.

Westhoo[3547-3550] developed two method involving, respectively, the use of mercuric chloride and cysteine acetate binding agents for determining methylmercury chloride in fish, egg white, meat and liver, without interference from thiols.

In Table 293 are shown methylmercury chloride contents obtained in some foods by the mercuric chloride method.

In a further paper Wethoo[3549] examined problems associated with the determination of methylmercury salts in egg yolk and white with low methylmercury content, liver, aquaria sediments and sludge, also bile, kidney, blood, meat and moss, in many of which mercury could not be accurately determined by the mercuric chloride method or the cysteine method. By combining these two methods, however, he was able to obtain good results with these various types of samples.

Excess mercuric ions were added to an aqueous liver suspension containing known amounts of a methylmercury salt. The analysis was performed according to the cysteine acetate modification. More than 100% of the methylmercury was recovered. When the acidified liver suspension containing mercuric ions was kept at room temperature overnight, the recovery increased. This indicated a synthesis of methylmercury ions from mercuric ions by the liver under the conditions used. Thus, this combined mercuric ion-cysteine procedure for analysis of methylmercury could not be applied to liver. Some results obtained without addition of methylmercury compounds are seen in Table 294.

For egg yolk with low content of methylmercury the cysteine acetate procedure gave less than 90% recovery[3548], with the combined method using cysteine and mercuric ions the recovery of methylmercury salt decreased almost to zero. But for sediments in aquaria and sludge, which similarly could not be analysed by the original cysteine acetate modification, the combined method gave good results. (Table 295).

Table 293 - Some methylmercury contents in foods, analysed by mercuric chloride procedure (A)

Foods	µg of total mercury/g of food	Methylmercury in foods mg Hg/kg	% of total Hg
Meat (ox)	0.074	0.068	82
Meat (hen)	0.051	0.037	73
Meat (hen)	0.023	0.017	74
Liver (pig)	0.130	0.095	73
Liver (pig)	0.140	0.095	68
Liver (pig)	0.096	0.075	78
Egg Yolk	0.010	0.005	50
Egg Yolk	0.010	0.009	90
Egg White	0.023	0.020	87
Egg White	0.025	0.019	76
Egg White	0.012	0.011	92
Egg White	0.025	0.024	96
Egg White	0.012	0.011	92

Table 294 - Methylmercury compounds analysed by combined mercuric ion-compounds cysteine acetate procedure

Sample	Total Mercury mg/kg	mg of Hg/kg Reaction time 0.5 h	Reation time 20 h
Ox Liver	0.006	0.029	0.045
Ox Liver, boiled	0.006	0.031	0.047

Table 295 - Methylmercury compounds found after addition of methylmercury dicyanamide

Sample	Total mercury mg/kg	Methylmercury compounds found mg of Hg/kg		Methylmercury compounds found after addition of methylmercury (0.1 mg of Hg/kg)			
		Cysteine acetate procedure	Combined procedure	Cysteine acetate procedure mg of Hg/kg	Cysteine acetate procedure % recovery	Combined procedure mg of Hg/kg	Combined procedure % recovery
Egg yolk	0.004	0.000	0.000	0.082	82	0.001	1
Sediment	0.063	0.000	0.017	0.004	4	0.109	92
Sludge	0.52	0.01	0.041	0.028	18	0.125	84

A second attempt to improve the recovery in the cysteine acetate method involved a precipitation of the proteins in liver by molybdic acid. This increased the recovery of added methylmercury salt to about 90%. In egg yolk with a low content of methylmercury compounds, however, neither molybdic acid nor phosphomolybdic acid improved the results.

Fujiki[3592] pointed out that in the presence of divalent sulphide ion, methylmercury compounds form bis (dimethylmercury) sulphide. This is insoluble in aqueous cysteine acetate or in aqueous glutathione which in the Westhoo procedure is added to extract the organomercury compounds from benzene solution in order to free them from interfering thiols. By the addition of cuprous chloride, the bis (dimethylmercury) compounds are converted into methylmercury chloride; this is extracted into benzene and then into aqueous glutathione. After acidification of the aqueous phase with hydrochloric acid, methylmercury chloride is re-extracted into benzene and determined by gas chromatography on a column (40 cm x 4 mm) packed with 25% of poly (diethylene glycol succinate) on Celite (60 - 80 mesh) or 10% of poly (butanediol succinate) on Chromosorb W (60 - 80 mesh) with nitrogen as carrier gas and electron-capture detection.

Gas chromatography has been extesively used to separate mixtures of organomercury compounds prior to their determination. Gas chromatography employing electron-capture detection with prior cleanup, has been used quite extensively as a sensitive technique for analyzing organomercury compounds in biological materials[3593,3594], urine[3596], hair[3597], sediment [3595] and blood[3598]. Ealy et al[3599] described a gas chromatographic method for the determination of methyl, ethyl and methoxyethyl mercury halides in organic sediments in aquatic systems, seeds and fish.

The mercury compounds are separated from the samples by leaching with M sodium iodide for 24 hr, and then extracting the alkylmercury iodides into benzene. These iodides are then determined by gas chromatography of the benzene extract on a glass column packed with 5% of cyclohexylenedimethanol succinate on Anakrom ABS (70 to 80 mesh) and operated at 200°, with nitrogen (56 ml per min) as carrier gas and electron-capture detection (31.1 foil). Good separation of chromatographic peaks is obtained for the mercury compounds as either chlorides, bromides or iodides. The extraction recoveries are monitored by the use of alkylmercury compounds labelled with ^{203}Hg.

Perhaps the most frequently used approach for the analysis of the cationic organometals is derivitization with a halogen and subsequent gas chromatography with electron capture detection[3600,3601]. Detection limits with this approach are quite good, 2 ng/g for CH_3HgBr, for example. However, the relatively poor selectivity of this detection approach requires rather extensive sample "cleanup" before analysis. Gas chromatography has been widely used for methylmercury with an electron capture detector[3602-3606]. Westhoo[3547-3590] pointed out that any methylmercury that remained attached to sulphydryl groups in tissue protein would not be determined.

Uthe et al[3607] later described the use of acidic sodium bromide reagent in the extraction of methylmercury as the bromide. Copper(II) ions were added to mask any free sulphydryl groups present and displace mercury bound to sulphur. Cappon and Smith[3608] introduced an initial treatment of samples with 40% sodium hydroxide solution and urea to improve extraction of methylmercury by uncoiling protein chains and hence exposing more tissue-bound methylmercury for eventual acid cleavage.

Gas chromatographic detectors[3609-3611] based upon flame ionization, electron capture, thermal conductivity, and photoionization are generally nonselective and rely totally upon the separation power of gas chromatography. Recently Fuwa et al[3613] has published a critical review of sepectochemical methods for chemical speciation using element specific detectors with various chromatographic techniques. Atomic spectrometric detectors are inexpensive, sensitive, precise, elemental selective, and simple to use. Atomic spectrometric devices[3614-3616] based upon atomic absorption in flames[3615,3616] and furnaces, atomic fluorescence in flames and furnaces [3615], atomic emission in flames[3615], arcs and plasmas[3614,3616,3618], have been used with great success to detect elements including mercury in separated molecular species.

The approach which shows considerable promise for speciation of molecular compounds (inorganic and metallo-organic) in solid biological and environmental materials is evolved gas analysis/microwave induced emission detection. Evolved gas analysis involves volatilization of the molecular species as a function of temperature. This technique was first proposed by Hanamura[3621] and later developed by Bauer and Natusch[3619] and Mitchell et al[3620]. Bauer and Natusch[3619] developed an evolved gas analysis/microwave emission spectrometer for identifying trace inorganic compounds in solid samples. The solid samples were heated from 25 to 1000 °C at 140°C/min; and the molecular components were vaporized into a low power atmospheric pressure He microwave induced plasma. Both metals and nonmetals could be measured. These workers indicated some difficulties with the efficiency of mass transfer of the evolved vapors to the microwave plasma and the lower power (< 100 W) and microwave coupling method led to problems associated with sensitivity to matrix effects influencing the plasma background. The Mitchell et al[3620] approach involved a furnace-microwave plasma system to determine organic carbon by differential vaporization. The Mitchell approach had the same problems as the Bauer and Natusch approach. Both of these systems also did not allow the vaporization of sample of sufficient size, e.g. > 250 mg to reduce problems due to sample heterogeneity as recommended by the NBS certification of standard reference materials[3622].

The microwave cavity emission detection has been utilized by Taluri [3623] and Reamer et al[3624]. However, the gas chromatographic separation requires thermally stable, strong complexes of the cationic organometallics to be made by derivatization before analysis.

Gas chromatography with a microwave emission spectrometric detector [3525], has been used for the determination of methylmercury.

Cappon and Crispin Smith[3526] have described a method for the extraction, cleanup and gas chromatographic determination of organic (alkyl- and aryl-) and inorganic mercury in biological materials. Methyl-, ethyl-, and phenylmercury are first extracted as the chloride derivatives. Inorganic mercury is then isolated as methylmercury upon reaction with tetramethyltin. The initial extracts are subjected to thiosulfate cleanup, and the organomercury species are isolated as the bromide derivatives. Total mercury recovery ranges between 75 and 90% for both forms of mercury, and is assessed by using appropriate ^{203}Hg-labelled compounds for liquid scintillation spectrometric assay. Specific gas chromatographic conditions allow detection of mercury concentrations of 1 ppb or lower. Mean deviation and relative accuracy average 3.2 and 2.2%, respectively. These workers were concerned with the determination of different inorganic mercury and organomercury species (alkyl and aryl) in a variety of media (blood grain, faeces, fish, hair, milk, sediment, soft tissue, urine and water).

Whole samples, aqueous homogenates, or alkaline digests of samples can be used for analysis. Biological materials that lend themselves to the three methods of sample preparation are listed below:

Preparation	Samples
Whole sample	Blood, grain, sediment, urine, water
Aqueous homogenate	Faeces, fish, sediment, soft tissue
Alkaline digest (45% sodium hydroxide=	Blood, faeces, fish, grain, hair, soft tissue

The analytical procedure is schematically outlined in Figure 203. Urea aids in enhancing both ^{203}Hg incubation and organic mercury recovery in the initial extraction step by uncoiling proteins in the sample matrix and exposing mercury-binding sulfhydryl sites for acid cleavage. The first benzene extraction results in 85 - 90% recovery of organic mercury. A second extraction removes any remaining organic mercury, and is necessary for samples containing inorganic and methylmercury, since the isolation and quantitation of inorganic mercury is based upon its conversion to methylmercury.

Cappon and Crispin Smith[3626] spiked the sample with $C_2H_5\,^{203}Hg\,Cl$ or $C_6H_5\,^{203}Hg\,Cl$ in order to check on organomercury recoveries through the whole procedure. A Packard Model 7401 gas chromatograph equipped with a Model 810 electron capture detector was used by Cappon and Crispin Smith [3626]. The detector employs a 150 mCi ^3H foil and is operated in the d.c. mode. The instrument operating conditions are presented in Table 296, along with corresponding retention times for methyl-, ethyl-, and phenyl-mercuric bromide.

Table 296 - Gas chromatography

Column:	Coiled glass, 1.22 m length, 4 mm i.d.		
Packing:	1.5% OV-17 + 1.95%, QF-1 on 80/100 Chromosorb W-HP (Alltech Associates)		
Instrument settings:			
Temperature, °C	Inlet	Column	Detector
MeHg, EtHg	130	110	150
PhHg	185	180	185
Carrior flow rate:	120 cm^3/min., nitrogen		
Detector settings:	Sensitivity: 3 x 10^{-9}A full-scale (AFS)		
	Suppression current: 1 - 2 x 10^{-7} A		
	Potential: 5 V		
Retention times, min.	MeHgBr: 0.6	EtHgBr: 1.6	PhHgBr: 2.0

For the given instrument settings, the minimum detectable organomercury bromide concentration is 0.02 ng/10 µl, or 2 ppb RHgBr. Expressed as mercury, this represents minimum detectable levels of 1.36, 1.30 and 1.12 ppb for methyl-, ethyl-, and phenylmercury, respectively. For 2 g sample (whole) containing methylmercury, assuming 1 ml of final extract and 80% recovery, this translates to a sample concentration of 0.85 ppb Hg. If 2 g of a sample containing methylmercury is used to prepare 10 ml of aqueous homogenate or alkaline digest and a 1 ml aliquot is analyzed, assuming equal extract volume and recovery, the mercury concentration giving the same instrument response would be 8.5 ppb. The minimum detectable concentration can be further lowered by increasing detector sensitivity but at the expense of increased baseline noise level.

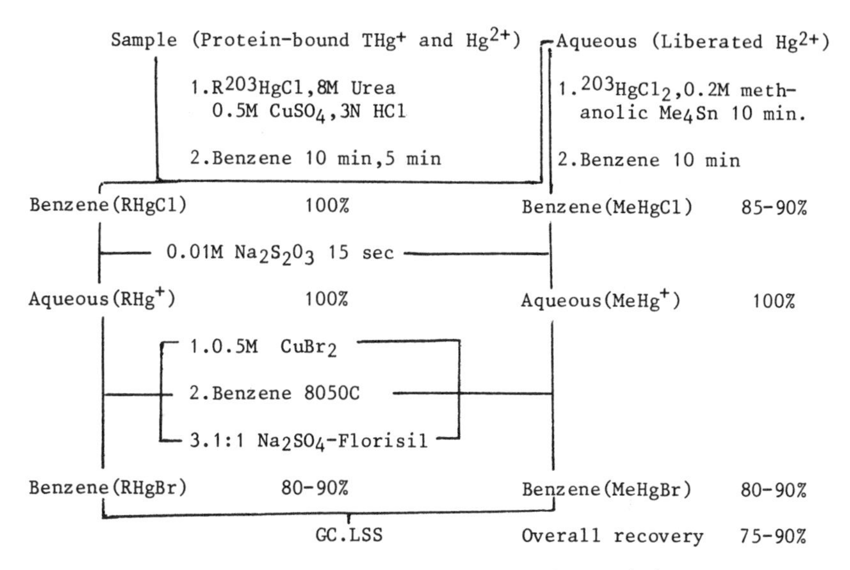

Figure 203 - Outline of HCl-Me_4Sn procedure.

In general, benzene extracts from this procedure are quite clean as seen in Figure 204. The organomercury peaks are well resolved from any contaminants, and any late peaks elute within 3 minutes from sample injection. No attempts were made to identify any contaminant peaks present on the chromatograms. However, such peaks may arise from other electron - capturing organic components present in and extracted from the original sample matrix (e.g., chlorinated pesticides from grain or PCB residues from fish) rather than from the reagents or the extraction solvent.

Cappon and Crispin Smith[3626] evaluated the accuracy and precision of the method by analyzing different sample types fortified with mercuric chloride and methylmercuric chloride (Table 297). Mean deviation and relative accuracy averaging 3.2 and 2.2% respectively, were observed. They also cross checked results obtained by their method and by an atomic absorption procedure. Results obtained on samples by both methods are given in Table 298. There is good agreement between the two methods for samples methyl-, ethyl-, and inorganic mercury, and this is expressed in terms of gas chromatographic / atomic absorption ratios.

Methylmercury compounds present in tissue is often very tightly bound to protein and, as such, is not easily amenable to extraction techniques applied prior to analysis. Various methods have been studied for releasing the organomercury compounds from the tissue prior to analysis for mercury compounds.

Callum et al[3627] have described the use of the proteolytic enzyme subtilision Carlsberg type A for breakdown of human and animal tissues prior to the release of methylmercury. This enzyme has a very high, non-specific proteolytic activity, which gives excellent breakdown of protein. It has been shown to be especially suitable for the release of drugs known to be bound to proteins. The yields obtained are greater than those found using the more conventional acid hydrolysis[3628,3629].

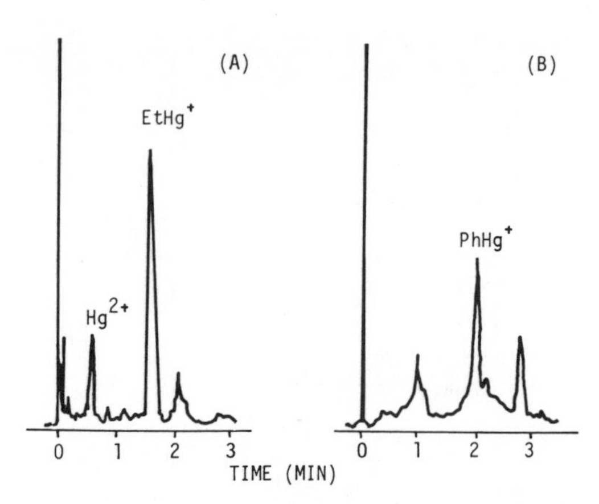

Figure 204 - (A) Chromatogram of heart sample (alkaline digest) containing 1.32 ppm ethylmercury and 0.07 ppm inorganic mercury. (B) Chromatogram of egg yolk sample (aqueous homogenate) containing 0.63 ppm phenylmercury.

Table 297 - Analytical and recovery data

Sample	Added MeHgCl[a] (ng/g)	Mean[b] (ng/g)	% Mean deviation	% Relative accuracy[c]	% Recovery (average)
Blood	139.2	141.4	1.3	1.6	79.8
Faeces	12.2	12.4	4.9	2.4	86.3
Liver	196.1	203.6	2.8	3.8	85.0
	Added $HgCl_2$[a]				
Blood	198.0	194.6	3.4	1.7	83.4
Faeces	22.4	21.9	1.3	1.3	81.5

[a] Whole sample. All samples were anlyzed as alkaline digests.
[b] Mean values are the average of three analyses. All have been corrected for recovery.
[c] Percent relative accuracy of the mean.

Finely chopped tissue (1 g) was homogenised with 15 ml of 1.0 M tris (hydroxymethyl aminomethane (Tris)-hydrochloric acid buffer solution (pH 8.5) then incubated with 1.0 m of subtilisin Carsberg Type A (Nova Enzyme Products Ltd., Windsor) for 1 h at 50°C in water-bath with continuous stirring. Then 2 ml of 40% m/V sodium hydroxide solution and 1 ml of 1% m/V cysteine hydrochloride solution were added and the samples stirred for 5 min at 50°C. When cool, 1 ml of 0.5 M copper(II) bromide and 10 ml of acidic sodium bromide (31 g of sodium bromide in 70 ml of water + 22 ml of 50% sulphuric acid) was added. The methylmercury (II) bromide was then extracted with two 5 ml portions toluene. In each extraction the mixtures were shaken for 2 min then centrifuged at 6.00° for 10 mins. The two toluene extracts were removed and combined and the methylmercury was extracted twice with 1 ml of ethanolic sodium thiosulphate solution (a 1 + 1 mixture 95% ethanol and 0.005 M sodium thiosulphate solution). During each extraction the solute were vortex mixed and centrifuged at 4000 g for 2 min. The lower aqueous layers were removed and combined then 0.5 ml of 3 M potassium iodide was added to these combined aqueous extracts followed by 0.5 ml of benzene (pesticide grade; distilled in glass) containing ethylmercury(II) iodide as an internal standard (50 ng ml^{-1}); this internal standard concentration could be increased if the methylmercury concentration was high. These solutions were shaken and then centrifuged at 3000 g for 1 min. Standard solutions of methylmercury(II) iodide were prepared in the benzene containing the internal standard. Samples (5 µl) were analyzed by gas chromatography with electron capture detection.

Table 298 - GC-AA Intercomparison Study

Sample	GC	ppm Hg AA as MeHg	GC/AA
Fish	1.10	1.06	1.04
Hair	266.2	282.9	0.98
Muscle	0.27	0.70	1.03
		as EgHg	
Blood	0.72	0.77	0.94
Kidney	0.66	0.68	0.97
		as inorganic	
Blood	0.59	0.57	1.47
Fish	0.08	0.07	1.14
Sediment	0.17	0.19	0.89

A 150 cm x 4 mm i.d. glass column packed with 5% m/m ethylene glycol adipate poly (Hi-EFF-2AP) on 80 - 100 mesh Gas-Chromb Q was used, with a carrier gas (argon - methyl flow rate of 60 ml min^{-1}. The operating temperatures were: injection port, 175°C; column 155°C; and nickel-63 electron capture detector, 225°C. The results were calculated by comparison of peak-area ratios of methylmercury to ethylmercury in the samples standards.

Sample extracts that gave many interfering peaks when chromatographed could be cleaned up using Florisil PR column. The column consisted of a Pasteur pipette loosely filled with glass-wool on top of which were placed, in ascending order, a 5 mm layer of anhydrous sodium sulphate, a 50 mm layer of Florisil PR (60 - 100 mesh) and finally another 5 mm of anhydrous sodium sulphate. The column was pre-washed with benzene, then 0.4 ml the extract to be cleaned was allowed to drain through the column. The column was washed first with 0.4 ml and then 1.0 ml of benzene. Finally, a further 2.0 ml of benzene were run through the column and collected for injection into the gas chromatograph. 0.4 ml volume of the standard mixture prepared previously was treated in the same way and the 2 ml fraction collected.

The method described was used to analyse a variety of methylmercury solutions and tissue samples (Table 299). The results obtained for the tissue analysis were compared with those obtained using other methods of analysis (Table 300) involving extraction with acidic sodium bromide[3630].

Albacore tuna and fish homogenate analysis by the more conventional method of Uthe et al[3530] for methylmercury determination in fish gave significantly lower levels than those found when the enzymatic sample breakdown prior to extraction was carried out (Table 300). This finding indicates that the release of methylmercury from tissue is enhanced by the enzymatic sample breakdown.

Table 299 - Analysis of water and dry biological tissue (1 g samples)

Material	No. of determinations	Methylmercury/ng Added	Found (mean value)	Range	Coefficient of variation %	Mean recovery %
Water	6	-	-	-	-	-
	6	37.4	36.3	35.1-37.7	2.3	97
	6	128	124	121 - 128	2.5	96
Albacore tuna	11	-	912	812 - 1000	5.5	-
Bovine liver	5	-	15.4	14.7-16.8	4.7	-
Whole blood	6	-	10.9	9.95-11.6	5.1	-
Fish homogenate	6	-	346	329 - 363	3.6	-

Table 300 - Analysis of dry albacore tuna and dry fish homogenate (1 g samples)

Material	Method	No. of determinations	Methylmercury/ng mean	range	Coefficient variation %
Albacore tuna	Uthe et al	12	693	661 - 734	3.7
	This work	11	912	812 - 1006	5.5
Fish homogenate	Uthe et al	6	57	43 - 69	15.4
	This work	6	346	329 - 362	3.6

Substoichiometric analysis proposed by Suzuki[3631,3632] and by Ruzicka and Stary[3633,3634] is a very attractive analytical method based on isotope dilution analysis. This method offers an accurate and precise determination of trace amounts of elements by measurement of raioactivity alone without corrections for chemical yield. A number of substiochiometric methods have been developed for the determination of about 40 elements in various types of samples[3635,3636]. This technique has been applied to the determination of organomercury compounds in environmental samples (discussed later).

Kande and Suzuki[3657] have applied substiochiometric isotope dilution analysis to the determination of inorganic mercury and organically bound mercury in hair. Thionalide (thioglycolic-β-aminonaphthalide) was used as the extracting agent.

A finely chopped hair sample was placed into a 50 mL flask containing 1.98 μg $^{203}Hg(I)$, and 3 ml of sulphuric acid (1:1) and 2 ml of 30% hydrogen peroxide were added. The flask was connected to a reflux condenser and then heated gently until the solution was clear and pale yellow (usually for about 30 min). After the flask was cooled in an ice-bath, the digest solution was transferred into a separatory funnel with 5 ml of distilled deionized water and shaken for 5 min with two 5 ml portions of methyl isobutyl ketone to extract Hg(II). The combined methyl isobutyl ketone extract was washed with two 5 ml portions of water and shaken for 1 min with 5 ml of ethylenediaminetetraacetic acid (EDTA) solution of pH 6.5 to back-extract Hg(II). The acidity of the aqueous solution was adjusted to 0.1 M in sulphuric acid and then the substoichiometric extraction of Hg(II) was carried out with 5 ml of 1 x 10^{-6} M thionalide in chloroform. The organic phase was washed with 0.1 M sulphuric acid solution and then a 3.0 ml aliquot of the organic phase was pipetted into a polyethylene test tube for measurement of its radioactivity.

In the procedure for methylmercury, a finely chopped hair sample was put into a 50 ml conical beaker, and 1.37 µg of ^{203}Hg/methylmercury and 20 ml of hydrochloric acid were added. The beaker was allowed to stand for 5 h with occasional shaking. The hair was removed by filtration and then the hydrochloric acid extract was shaken for 5 min with two 10 ml portions of benzene to extract the methylmercury. Methylmercury in the combined benzene extract was back-extracted into 5 ml of 0.02 M sodium sulphite solution of pH 6.5 and then the substoichiometric extraction was carried out with 5 ml of 5 x 10^{-7} M thionalide in chloroform. A 3.0 ml aliquot of the organic phase was pipetted into a polyethylene test tube for measurement of its radioactivity.

The precision of substiochiometric analysis has mainly two components; one associated with substoichiometric separation and the other associated with radioactivity measurements. As the latter component can be improved by use of a radiosotope with a high specific activity, the precision depends primarily on the former component.

From the activities in the substoichiometric region a reproducibility with a relative standard deviation of 2.4% was obtained for 1 to 5 µg Hg(II). Further, the reproducibility was examined for the single substiochiometric extraction from five test solutions of the same component and also for five successive substoichiometric extractions from the same test solution. As only a small portion of the element is extracted in substoichiometry, the extraction can be carried out repeatedly from the same aqueous solution. Satisfactory reproducibility was obtained for both the single extraction from a 0.1 M sulphuric acid solution containing 1.4 µg Hg(II) and the successive extraction from a 0.1 M sulphuric acid solution containing 4.2 µg Hg(II) with 2.5 x 10^{-9} thionalide, with relative standard deviations of 1.7 and 1.2% respectively.

The results in Table 301 show the effect of diverse metal ions. Though metals such as Cu^{2+}, Ag^{+}, As^{3+}, Bi^{3+} and Pd^{2+} have been reported to be precipitated quantitatively in acidic solution with thionalide, interference with the substoichiometric extraction of Hg(II) was observed only for Pd^{2+}.

The capillary column in the gas chromatographic method developed by Forsythe and Marshall[3638] for the determination of di- and trialkyl mercury compounds in water, described earlier, has also been applied to the determination of these compounds in whole egg samples. Alkyllead salts (R_3Pb+ and R_2Pb^{2+}, R = Me or Et) are recovered from whole eggs by complexometric extraction with dithizone. The dithizonates are phenylated and speciated by capillary column GC with electron capture detection. The method is sensitive to low parts per billion levels of lead salts in 2.5 g egg homogenate. At these levels methyllead salts (but not ethyllead salts) interact strongly with the sample matrix. Treatment of the matrix with lipases and proteases releases them.

Whole Egg Hydrolysis Procedure. Whole egg homogenate was incubated at 37°C for 24 hr in 60 ml of 5% ethanol/0.1 M phosphate buffer (pH 7.5) containing 30 mg of Lipase Type III and 30 mg of Protease Type XIV. This technique was found to be 7,2 ± 9% effective after 24 hr relative to classical acid hydrolysis.

Extraction Procedure. Absolute ethanol (15 - 22 ml) and ammoniacal buffer (pH 9,5, 10 ml) were added to the sample. The mixture was extracted three times with 0.05% (w/v) dithizone (10 ml) in 50% benzene/hexane. The organic extracts were combined, centrifuged and back-extracted three times with 10 ml of nitric acid (0.15 M). The aqueous washes were combined, neutralized with sodium hydroxide, and further basified with 5 mL ammoniacal buffer (pH 9.5). The alkyllead salts were recovered by extracting the aqueous phase three times with 0.01% dithizone (10 ml) in 50% benzene/hexane. These washes were combined and centrifuged and the organic layer was reduced in volume to 0.5 mL.

Table 301 - Effect of diverse ions on substoichiometric extraction[a]

ion[b]	activity of extract, cpm
none	10 915
Ag^{+}	10 359
Mn^{2+}	10 350
Cu^{2+}	10 807
Zn^{3+}	11 042
Cd^{2+}	10 715
Pd^{2+}	295
Pb^{2+}	10 842
Cr^{3+}	10 954
Fe^{3+}	10 931
As^{3+}	10 941
Bi^{3+}	10 762

[a] Aqueous solution, 0.1 M H_2SO_4; thionalide, 2.5×10^{-9} mol.
[b] Amount of each ion added, 0.1 mg.

Gas chromatography was carried out on a 30 m fused silicon DB-1 column (J & W Scientific Co.) as described earlier. This method was developed to speciate lead in Herring Gull (Larus argentatus) eggs. This species is considered to be an excellent biological indicator of pollution in that it occupies a position high on the aquatic food chain and has a quite diverse diet. Increased salt concentration improved the efficiency of extraction of alkyllead salts by organic solvents (benzene, ethyl acetate, or methyl isobutyl ketone). Alkali and alkaline earth halides from several sources, however, contained appreciable quantities of inorganic lead. The possibility of transalkylation of alkyllead salts[3639] precluded heat treatment in any proposed method, hence, the use of solvent extraction with complexing agents. This would also avoid the need for column cleanup. The hydrolysis when incorporated into the determination sequence had no substantial effect on the mean recovery of ethyllead salts; however, precision was considerably improved. In contrast the mean recovery of methyllead salts was dramatically improved. These results suggest that the methyllead salts are rapidly bound up by the egg matrix. The determinations were commenced 15 min after each sample was spiked. The enzymatic pretreatment apparently releases the bound analytes.

Organotin Compounds

A variety of seashells have been analysed for organotin compounds using the gas chromatographic procedure described by Braman and Tompkins [3348] (see section on organotin compounds in water).

Analytical results for these seashells and domestic chicken egg samples are given in Table 302. The average total tin content was between 1 - 2 ng g^{-1} and methylated tin compounds were detected in the seashells. The higher concentration of tin in the seashells relative to the water in which they were found would indicate the presence of a bioaccumulation process. Also, during the analyses of the seashells, several other small unidentified signals were noted to follow the methylated tin compounds. Identification of these compounds was not done.

Table 302 - Analysis of shell samples[a]

Seashells	tin (IV) ng g^{-1}	%	methyltin ng g^{-1}	%	dimethyltin ng g^{-1}	%	trimethyltin ng g^{-1}	%	total tin ng g^{-1}
white coral	0.73	73	0.20	20	0.21	12	ND[b]	-	1.0
miscellaneous shells									
sample 1	ND	-	0.23	85	0.038	14	ND	-	0.27
sample 2	ND	-	0.45	92	0.038	7.8	ND	-	0.49
sample 3	1.3	87	0.20	13	ND	-	ND	-	1.5
sample 4	0.94	85	0.12	11	0.059	5.4	ND	-	1.1
average	0.59	67	0.24	27	0.051	5.8	ND	-	0.88
domestic chicken egg shell	2.9	97	ND	-	0.083	2.8	ND	-	3.0

[a] Data are averages of duplicates.
[b] ND, less than 0.01 ng g^{-1} for methyltin compounds, less than 0.05 ng g^{-1} for inorganic tin.

Gas chromatographic methods have been described for the determination of tetraalkyl[3640] and trialkyl tins[3641] in biological materials. Unfortunately, these methods are not easily applicable to the determination of the dialkyl tin homologues because of their adsorption and decomposition during chromatography.

Spectrofluorimetry has been applied to the determination of triphenyltin compounds. Coyle and White showed that 3-hydroxyflavone could be used to determine submicrogram amounts of inorganic tin[3642]; and then Vernon used the reagent to determine triphenyltin compounds in potatoes[3643]. On the basis of this procedure, Blunden and Chapman spectrofluorometrically determined triphenyltin compounds in water[3644]. Further, they showed that chloride ions quenched the fluorescence but that on shaking with aqueous sodium acetate solution a stable complex was formed, although the instability to light of the triphenyltin chloride-3-hydroxyflavone complex had been initially pointed out by Aldridge and Cremer[3645].

Arakawa et al[3646] have shown that Morin (2',3,4',5,7-pentahydroxyflavone) can be used as a fluorescence reagent for organotin, especially dialkyltin compounds. Although quercetin and 3-hydroxyflavone are similar to Morin in structure, they are unsuitable because of their sensitivity and instability. Morin produces a green fluorescence with various organotin compounds in organic solvent. The reagent is especially sensitive to dialkyltin compounds. The excitation and emission spectra show peaks at ca. 415 nm and ca. 495 nm, respectively, for each alkyltin-Morin complex and at ca. 405 nm and ca. 520 nm for the triphenyltin-Morin complex. The maximum fluorescence requires a ratio of 3 to 9 mol of Morin for 1 mol of dialkyl- and triphenyltin and a 6 - 12 to 1 molar ratio for trialkyltin. Detection limits are 1×10^{-9} M for dialkyltins, 1×10^{-7} M for monoalkyltin, 5×10^{-7} M for trialkyltins, and 1×10^{-7} M for triphenyltin. The fluorometric procedure can be used for the determination of individual organotin, especially a dialkyltin compound in biological samples. Recoveries of organotin added to various tissues at the 1.0 - 100 nmol level ranged from 91.0 to 99.7% depending upon the organotin species.

The general analytical procedure is outlined below. To 3 mL of n-hexane solution containing 0.001 - 1.0 nmol/mL of dialkyltin or 1.0 - 100 nmol/mL of monoalkyltin or 1.0 - 1000 nmol/mL of trialkyltin or 1.0 - 100 nmol/mL of triphenyltin, add 0.5 mL of a 0.005% solution of Morin in absolute ethanol. After mixing thoroughly, measure the fluorescence intensity at ca. 495 nm, using an excitation wavelength of ca. 415 nm. A reagent blank should be run concurrently.

Morin alone had very little fluorescence. However, the reagent produced a strong green fluorescence with various organotin compounds in organic solvent. Although the fluorescent intensity varied only slightly depending upon the organosolvent species, i.e., benzene, toluene, hexane, ether, ethyl acetate, chloroform, and ethanol, n-hexane was the most satisfactory as a medium for this test because of the greater solubility and stability of the complexes. Ethanol was suitable for some methyltin compounds, however, because of their polarity. The excitation and emission spectra for each alkyltin-Morin complex show peaks at ca. 415 nm and ca. 495 nm respectively. Similarly both spectra for triphenyltin - Morin complex show peaks at ca. 405 nm and ca. 520 nm. The formation of the organotin - Morin complexes progressed very rapidly at room temperature and the fluorescence intensities remained constant for hours. Particularly, the dialkytin complexes were stable over a number of hours. Less than 10% of acetic acid in the reaction solution had no effect on the fluorescence intensity readings, although the presence of hydrochloric and sulfuric acids resulted in lower fluorescent readings. A quenching of the fluorescence by chloride ions and the instability of the complexes to light were not found under the conditions of the test. For a maximum fluorescence, the Morin concentration should be at least in a 3 - 9 to 1 molar ratio to each dialkyl- and tri-phenyltin, and in a 6 - 12 to 1 molar ratio to each trialkytin.

Dialkyltin compounds produced a much stronger fluorescence than other organotin compounds with Morin. For 1 x 10^{-6} M of each organotin compound the relative fluorescence intensity was 10.2 for $BuSnCl_3$, 42.5 for Me_2SnCl_2, 99.2 for Et_2SnCl_2, 99.4 for Pr_2SnCl_2, 51.7 for Bu_2SnCl_2, 2.2 for $Et3SnCl$, 2.8 for Pr_3SnCl, 1.6 for Bu_3SnCl, and 12.7 for Ph_3SnCl at the same instrument setting. The concentration detection limits for dialkyltin compounds were in the 10^{-8} to 10^{-9} M range, at which other organotin compounds could not be detected. This large difference in fluorescent intensities among different organotin - Morin complexes appears to be dependent on the valence state of the metal.

Organolead compounds such as di- and triethyllead and organosilane compounds such as di- and monomethylsilane did not interfere at 1 x 10^{-3} M under the conditions used for the determination of organotin. Other organometal compounds such as dimethyl arsenide and methyl- and ethyl- mercury chlorides did not fluoresce at all. Although aluminium(III), zinc (II), tin (IV), magnesium (II) and cadmium (II) produced a strong fluorescence, and manganese (II) , selenium (IV) and mercury (II) produced a very weak fluorescence with Morin in water solutions, these organometallic compounds did not interfere at 1 x 10^{-3} M under the conditions of the organotin determinations. Arsenic (III or V), lead (II), chromium (III or VI), copper (II), and iron (II or III) did not fluoresce even in water solutions.

In general, organotin compounds in environmental and biological samples were isolated by extraction with n-hexane or ethyl acetate. An aqueous sample containing 0.01 - 1.0 nmol of dialkyltin, 1.0 - 100 nmol of monoalkyltin, 0.01 - 1.0 µmol of trialkyltin, or 1.0 to 100 nmol of triphenyltin compound in up to 100 mL of solution was directly shaken with 10 mL of n-hexane in a separatory funnel for 10 min, and the two layers were allowed to separate. A tissue sample (2) weighing between 1.0 and 5.0 g (wet weight) was first homogenized in 10 ml of normal saline solution. Hydrochloric acid (8 mL) was carefully added to the homogenate, and the contents were mixed thoroughly. After the mixture was allowed to stand for 5 min, ethyl acetate (20 mL) and sodium chloride (2 g) were further added and shaken for 10 min. The extraction procedure was repeated twice; the recovery from double extractions was about 98%. The combined ethyl acetate layers were then concentrated under reduced pressure at about 20°C to 0.5 - 1.0 mL. Loss of organotin compounds through the concentration procedure was not seen. n-Hexane (10 mL) was added to the concentrated solution and the precipitate produced was removed by centrifugation. By this replacement of the extraction solution with n-hexane, ethyl acetate soluble and n-hexane insoluble substances are eliminated.

Each n-hexane layer was directly subjected to the fluorometric procedure for the determination of individual organotin compounds after being concentrated to a suitable volume and being made up to 3 mL with n-hexane finally.

For the analysis of organotin mixtures, the individual organotin compound must be preseparated from each other by an appropriate chromatography technique for subsequent determination by the fluorometric procedure.

In Table 303 are tabulated the recoveries of organotin compounds from various biological materials, as obtained by this procedure.

Organolead Compounds

Atomic absorption spectrometry. Sirota and Uthe[3647] have described a fast sensitive atomic absorption procedure for determining tetraalkyllead compounds in biological materials such as fish tissue. Tissue homogenates were extracted by shaking with a benzene/aqueous EDTA solution, a measured portion of the benzene was removed, and, after digestion, the residue was defatted if necessary. The resultant Pb^{2+} was determined by flameless atomic absorption spectroscopy using a heated graphite atomizer. Using a sample weight of 5 g, 10 ppb of lead as PbR, can be determined with a relative standard deviation of 5%. No other forms of lead that were tested, e.g. PbR_3X, PbR_2X_2, were found to partition into the benzene layer under these conditions.

Procedure - Benzene (10 mL) and EDTA reagent (10 mL 0.4% w/v. PH_{6-7}) were added to a homogenized tissue sample (5 g) in a 50 mL glass centrifuge tube fitted with a screw cap, and shaken on a wrist action shaker fitted with extension clamps to give 2½ - 3 inch strokes for 10 min, after which the phases were separated by centrifugation at 2200 rpm for 30 min to give two layers.

An accurately measured 3 mL portion of the benzene layer was transferred to a 50 mL calibrated Folin-Wu digestion tube and the contents were acidified with 3 mL of concentrated nitric acid. The benzene layer was evaporated under a stream of nitrogen (high purity) at room temperature, and the residue was then digested for at least 2 h at 80 - 90°C until evolution of large amounts of nitrogen oxides ceased. The sample was then made up to 10 mL with glass distilled water and shaken with approximately 2 mL of hexane. After removal of the hexane layer, the aqueous phase was analyzed by the method

Table 303 - Recovery of organotin compound added to human urine and rat tissues

	% recovery[a]							
	$BuSnCl_2$ (10 nmol)[b]	Et_2SnCl_2 (1.0 nmol)	Pr_2SnCl_2 (1.0 nmol)	Bu_2SnCl_2 (1.9 nmol)	Et_3SnCl (100 nmol)	Pr_3SnCl (100 nmol)	Bu_3SnCl (100 nmol)	Ph_3SnCl (10 nmol
human urine	93.2±0.6	96.7±0.3	97.1±0.7	99.3±0.4	93.5±0.8	94.7±0.8	91.3±0.6	94.8±0.8
rat organ								
liver	93.7±0.5	98.9±0.7	98.3±0.5	99.1±0.7	94.7±0.7	96.1±0.6	93.2±0.9	94.6±0.9
kidney	93.8±0.6	98.6±0.7	98.2±0.8	98.2±0.6	94.2±0.6	95.9±0.9	94.3±0.8	93.9±0.7
spleen	94.2±0.6	99.0±0.5	99.1±0.5	98.9±0.6	95.2±0.7	97.0±0.9	93.8±0.6	94.7±0.6
brain	92.0±0.7	97.8±0.9	97.5±0.9	98.6±0.6	92.9±0.5	93.4±0.8	93.4±1.0	92.8±0.9
thymus	95.9±0.5	99.2±0.6	98.9±0.4	99.7±0.5	95.6±0.8	97.1±0.8	95.2±0.8	96.9±0.7

[a] Data are average and average deviation of the five replicates.
[b] Amount of compound added.

Table 304 - Recovery of tetraalkyllead compounds from cod liver homogenate.

Compound	Amount added µg Pb	Amount added, ppb	Total Pb present prior to spike µg	Total Pb found after pike µg	Amount of spike found µg	Recovery %
Tetramethyllead	0.10	20	0.25	0.38	0.13	130
	0.10	20	0.27	0.40	0.13	130
	0.50	100	0.06	0.575	0.515	103
	0.50	100	0.06	0.625	0.565	113
Tetraethyllead	0.10	20	0.14	0.21	0.07	70
	0.10	20	0.16	0.26	0.10	100
	0.50	100	1.056	1.548	0.492	98
	0.50	100	0.053	0.65	0.542	119
	0.50	100	0.045	0.42	0.375	75

of standard additions and lead concentration calculated following linear regression analysis of peak heights using a programmable calculator.

The recovery and selectivity of the method was evaluated by adding known amounts of different lead compounds to previously analyzed tissue samples. The results obtained are summarized in Table 304 and indicate a satisfactory recovery and selectivity for tetraalkyllead compounds. Various marine tissues were sampled for total lead and tetraalkyllead. Results are summarized in Table 305. Di- and tri-substituted alkylleads were also evaluated in this system and the results are shown in Table 306.

Table 305 - Concentrations of total lead and tetraalkyllead in various marine tissues

Tissue		Concentration total Pb,ppm	Concentration PbR_4,ppm	% tetraalkl lead of total lead
Frozen cod (liver homogenate)		0.39 ± 0.04	0.037 ± 0.003	9.5
Large, freshly killed cod			0.010 ± 0.001	
(liver homogenate)		0.52 ± 0.05	0.125 ± 0.005	24
Small, freshly killed cod	A	0.21 ± 0.04[a]	0.028	13.3
(2 separate lobes nalyzed)	B		0.444	20.9
Lobster digestive gland				
(homogenate)		0.20 ± 0.02	0.162 ± 0.004	81
Frozen mackerel muscle				
(homogenate)		0.14 ± 0.02	0.054 ± 0.005	38.6
Flounder meal		5.34 ± 1.02	4.79 ± 0.32	89.7

[a] For total lead determination, both lobes blended.

Table 306 - Extraction of di- and tri-substituted alkyllead compounds into benzene phase

Compound	Percent extraction into benzene phase	
	After 1st extration	After 2nd extraction
Diethyllead dichloride	7.1	0
Trimethyllead acetate	0.4	0

As can be seen from Table 306, there was no evidence of extraction of lead into the benzene phase after the second extraction, and it is postulated that the substituted alkyl compounds were contaminated with small amounts of tetraalkyllead, which were extracted into the benzene phase in the first extraction.

Ionic lead, as both the chloride and nitrate, also remained in the aqueous phase. All levels of both total lead and tetraalkyllead in the marine samples analyzed were quite variable, and the percent of lead present as the tetraalkyl species, ranged from 9.5% in a cod liver to 89.7% in a flounder meal. The relatively high concentration of lead compounds in flounder meal is a result of the large water loss, approximately 75%, during production of this meal.

Gas chromatography. Hayakawa first described the determination of trialkyllead ion as the chloride at biological and environmental trace levels with gas chromatography[3648]. Nanogram levels were claimed eluted from a packed polyester column.

A procedure[3649] for determining tetraalkyllead compounds in fish samples employs vacuum extraction of the tetraalkyllead into a cold trap under liquid nitrogen, followed by solvent extraction of the condensate for gas chromatographic determination. In this method, tetraalkyllead compounds have been found in fish and mussels. The presence of tetraethyllead in aquatic organisms may indicate that the alkyllead compounds are not immediately metabolized by living organisms and may remain in their authentic forms in the living tissues for a long time.

Several analytical methods for the determination of trialkyllead compounds have been reported: (i) the separation of triethyllead ions as the benzoate from liver and identification by infrared spectrometry[3650] (ii) the separation of trialkyllead compounds from rat blood, urine, brain, liver, and kidney via a laborious multiple extraction separation procedure with final dithizone complexation of the decomposed organoleads and colorimetric determination of the lead dithizone complex[3651-3653] and (iii) extraction of Et_3Pb in brain[3654] with subsequent lead measurement by aas.

Chau et al[3360] have described a simple and rapid extraction procedure to extract the five tetraalkyllead compounds (Me_4Pb), Me_3EtPb, Me_2Et_2Pb, $MeEt_3Pb$ and Et_4Pb), from fish samples. The extracted compounds are analyzed in their authentic forms by a gas chromatographic atomic absorption spectrometry system. Other forms of inorganic and organic lead do not interfere. The detection limit for fish samples (2 g) is 0.025 µg/g. The gas chromatographic atomic absorption system used by Chau[3359] is discussed in the section on water analysis.

The procedure is outlined below: Homogenize fish tissue in a Hobart grinder and a Polytron homogenizer. Place 2 g of the fish homogenated with 5 ml of EDTA reagent, and 5 ml of hexane in a 25 ml test tube with a Teflon-lined screw cap. Shake rigorously for 2 h in a reciprocating shaker. Centrifuge to facilitate phase separation. Carefully withdraw a suitable aliquot 5 - 10 µl, of the hexane phase and inject to the GC-AAS system.

Recovery experiments carried out by Chau et al showed that benzene, hexane, and octanol gave the most satisfactory recovery of tetraalkyllead compounds (Table 307).

Table 307 - Extraction of tetraalkyllead compounds from fish tissue by different solvents [a]

solvent	averaged recovery, %
hexane	80.0
cyclohexane	54.0
octanol	90.0
butyl acetate	55.0
methylisobutyl ketone	30.0
chloroform	57.0
benzene	78.0

[a] Fish homogenate 2 g; EDTA 5 ml; solvent 5 ml.

In Table 308 are shown results obtained in measurements of the accumulation of tetramethyllead in rainbow trout. The trout after exposure to water containing 3.5 μg/l tetramethyllead for different periods of time were found to contain tetramethyllead. Preliminary results showed that this compound was mainly concentrated in the lipid layer of the tissues.

Table 308 - Accumulation of tetramethyllead in rainbow trout

			concentration of Me_4Pb in		
exposure, day	wt. of fish, g	fish, alive or dead	water averaged μg/l	fish μg/g wet wt.	concn factors[a]
1	0.1211	dead	3.46	0.43	124
2	0.3661	dead		1.08	312
	0.7982	dead		2.00	578
3	0.4116	dead		1.32	382
	0.6300	dead		2.09	604
7	1.3045	alive		2.94	850
	1.5466	alive		3.23	934
	0.8100	alive		2.25	650
	0.4926	alive		1.73	500

[a] Concentration factor = Concentration of Me_4Pb in fish/concentration of Me_4Pb in water.

Chau et al[3655] have described a method for the determination of dialkyllead and trialkyllead compounds in fish. This method involves use of a tissue solubilizer to digest the sample followed by chelation extraction with sodium diethyl dithiocarbamate, followed by n-butylation using methyl magnesium chloride to their corresponding tetraalkyl forms, $R_nPbBu_{(4-n)}$, and R_4Pb respectively (R = methyl and ethyl). The method determines simultaneously in one sample; tetraalkyllead, ionic alkyllead (R_2Pb^{2+} and R_3Pb^+) and divalent inorganic lead, all of which are determined by gas chromatography using an atomic absorption detector.

In this method the fish samples were homogenized a minimum of five times in a commercial meat grinder. About 2 g of the homogenized paste was digested in 5 mL of tetramethylammonium hydroxide solution in a water bath at 60°C for 1 - 2 h until the tissue had completely dissolved to a pale yellow solution. After cooling, the solution was neutralized with 50% hydrochloric acid to pH 6 - 8. The mixture was extracted with 3 mL of benzene for 2 h in a mechanical shaker after addition of 2 g of sodium chloride and 3 mL of sodium diethyldithiocarbamate. After centrifugation of the mixture, a measured amount (1 mL) of the benzene was transferred to a glass-stoppered vial and butylated with 0.2 mL of n-butyl magnesium chloride with occasional mixing for ca. 10 min. The mixture was washed with 2 mL of sulphuric acid (1 N) to destroy the excess Grignard reagent. The organic layer was separated in a capped vial and dried with anhydrous sodium sulphate. Suitable aliquots (10 - 20 μL) were injected to the gas chromatographic atomic absorption system for analysis.

The recoveries of trialkyllead and dialkyllead species at different levels obtained by this procedure are shown in Table 309. The relative low recovery of dimethyllead is in agreement with the results of other investigators[3656-3658] who concluded that the dialkyllead remained bound to the tissue or existed in solution as a complex unextractable in solvent. Chau et al noticed that there was a large Pb(II) peak in the fish sample containing spiked dimethyllead but such was not found in the standard which was run in parallel but without the sample. They attributed such low recovery to the decomposition of dialkyllead in the fish matrix. The decomposition reaction of dialkyllead to Pb(II) in aqueous medium had been documented [3658]. Diethyllead, however, did not decompose significantly and was recovered near quantitative levels.

Table 309 - Recovery and reproducibility of alkyllead and lead(II) compounds from fish[a]

Amount of Pb added, μg	recovery, [b] %				
	Me_3Pb	Et_3Pb	Me_2Pb	Et_2Pb	Pb(II)
1	72 (5)	102 (5)	79 (4)	93 (0)	
5	88 (4)	88 (3)	89 (5)	103 (2)	
10	93 (2)	98 (2)	56 (10)	92 (2)	
20	91 (2)	81 (2)	62 (6)	114 (2)	
av	86	92	71	101	
% rel std dev (n=6) at 5 μg/g level	15	7	18	20	14[c]

[a] Fillet, 2 μg; spiked compounds expressed as Pb.
[b] Average of two results with average deviation in parentheses.
[c] The fish fillet contained 142 ng/g of Pb(II) which was used to evaluate the reproducibility. No Pb(II) was added to sample.

The average recovery obtained by this method in spiking experiments in fish samples varied from 71% for dimethyllead to 101% for diethyllead in the range 1 - 20 μg of spike per 2 g of fish fillet. The precision of the method was also evaluated by replicate analysis of samples of fillet spiked with 5 μg of each of the alkylleads. The reproducibility varied from 6.5% for triethyllead to 20% for diethyllead expressed in relative standard deviation at this level. For the first time, the occurrence of triethyl- and diethyllead compounds was detected in fish samples and in other environmental materials (Table 310). The method is very useful in studying the degradation of tetraalkyllead and the pathways of alkyllead in the environment.

Chau et al[3659] have described a method for the determination of dialkyllead, trialkyllead, tetraalkyllead and inorganic lead in biological samples. After sample digestion with a tissue solubilizer, ten organolead and Pb(II) species are isolated by chelation extraction and butylated for determination by gas chromatography with atomic absorption spectrometric detection.

Organoarsenic Compounds

Gas-phase methods have been described for the separation and identification of inorganic species as well as for methyl arsonic acid (MMA) and dimethylarsinic acid (DMAA)[3660-3668]. Braman and Foreback succeeded in separating and quantitatively determining As(III), As(V), methyl arsonic acid and dimethyl arsinic acid by a procedure based on sodium borohydride reduction of the arsenicals to the corresponding arsines with a dc helium plasma emission spectrometer as the detector[3670]. Talmi and Bostick improved the method by achieving the arsine separation with gas chromatography[3671]. Although the method is highly sensitive, it has some disadvantages. It cannot discriminate all arsenic compounds in the biological samples because the same arsines can be produced by sodium borohydride reduction from different organoarsenicals. Furthermore, difficulty lies in the conversion step: collection efficiency is incomplete for very volatile arsine (bp -55°C), or two reduction steps are necessary to discriminate As(III) and As(V).

Ion chromatography can separate these arsenicals in the liquid phase. Ion exchange resins have been employed for the separation of As(III), As(V), methyl arsonic acid and dimethyl arsinic acid in biological, water and soil samples[3672,3673]. Stockton and Irgolic separated As(III), As(V), arsenobetaine, and arsenocholine by high performance liquid chromatography using a reversed-phase ion suppression technique[3673]. Ion exchange high performance liquid chromatography seemed applicable to the separation of these arsenicals with better resolution than the conventional ion exchange chromatography using resin.

Because a selective and sensitive detection is necessary after the separation, atomic absorption spectrometry has been used for this purpose [3667]. Dc plasma and microwave helium plasma atomic emission spectrometry have been employed for gas-phase detection[3660,3668,3669]. For liquid phase, graphite furnace Zeeman effect atomic absorption method has been used with the automated sampler[3673]. Inductively coupled argon plasma emission spectrometry seems another choice as a high performance liquid chromatography detector because it has high sensitivity for arsenic, low chemical interference, and wide dynamic range.

Schwedt and Ruessel[3676] have described a method for the gas chromatographic determination of arsenic (as triphenylarsine) in biological material. In this method the dry sample is burnt in a Schoniger flask containing 3 N hydrochloric acid, the products are washed out with 3 N hydrochloric acid and aqueous potassium iodide and aqueous sodium bisulphite are added to the solution which is then extracted with diethyldithiocarbamate solution in dichloromethane. The extract is evaporated and the residue is stirred for 30 min with diphenylmagnesium solution in ethyl ether. After addition of dilute sulphuric acid the separated ether phase is evaporated and the residue is treated with mercapoacetic acid solution. After being set aside for 20 min, the solution is chromatographed on a glass column (2 metres x 3 mm) packed with 5% of terephthalic acid-treated Carbowax 20 M on Gas-Chrom Q (80 - 100 mesh) and operated at 220°C with nitrogen, helium or argon as carrier gas and flame ionization detection, respectively. Down to 2 ppm of arsenic in the sample could be determined by this procedure.

Morita et al[3674] have described a method of speciation and quantitative analysis of organoarsenic compounds in biological samples using high performance liquid chromatography with inductively coupled argon plasma atomic emission spectrometric detection. High performance liquid chromatography is used to separate mixtures of arsenic compounds on anion and cation exchange columns using phosphate buffer. The inductively coupled argon plasma atomic emission spectrometric detector is used as a selective detector by observing arsenic emissions at 193.6 nm. The detection limit is 2.6 ng As/s.

With anion exchange chromatography, the best separation was achieved with Nagel-N-$(CH_3)_3$ column packing and with 0.06 M phosphate buffer (Figure

Table 310 - Analysis of environmental samples (St. Lawrence River, near Maitland, Ontario)[a]

Sample	Me_4Pb	Me_3EtPb	Me_2Et_2Pb	$MeEt_3Pb$	Et_4Pb	Me_3Pb^+	Me_2Pb^{2+}	Et_2Pb^+	Et_2Pb^{2+}	Pb^{2+}
carp	137	-[b]	-	-	780	2735	362	906	707	1282
	-	-	96	142	7475	162	-	1216	1310	4133
pike	-	-	-	169	1018	215	-	-	-	1040
	-	-	-	146	1125	205	-	53	-	1187
white sucker	-	-	-	-	4384	196	-	3433	4268	3477
	-	-	-	293	2948	95	-	2171	2196	3610
small mouth bass	-	-	57	187	1204	-	-	223	92	254
	-	-	71	252	1834	-	-	660	275	305
sediment	-	-	-	142	1152	-	-	187	22	10000
	-	-	-	-	329	-	-	-	-	5582
macrophytes, mixed										
surface	-	-	-	-	68	-	-	132	-	4327
4 m deep	-	38	1501	3613	16515	-	-	558	113	59282

[a] Data expressed in ng/g as Pb, wet weight; whole fish for fish samples.
[b] Not detectable.

205). A slight tailing is seen at the As(III) and dimethyl arsinic acid (DMAA) peaks. A weak ion exchanger, μ-Bondapak-NH_2, separated these arsenicals as well. However, its resolution was inferior to the strongly basic ion exchange Nagel-$N(CH_3)_3$.

With a cation exchange chromatography (Nagel-SO_3H-10), these arsenicals were separated. Since arsenate is a stronger acid than arsenite, As(V) is more negatively changed than As(III) at the neutral region employed here. Arsenobetaine may be neutral or weakly positive. These ionic characteristics may explain the elution sequence of arsenobetaine, arsenite, and arsenate: AB → As(III) → As(V) in anion exchange chromatography, and As(III), As(V) → AB in cation exchange chromatography. At the same time, however, hydrophobic ineraction may need to be considered.

Dimethyl arsinic acid eluted later than monomethyl arsonic acid on both columns, indicating the affinity of methyl groups on As to alkyl groups of the column packing. In ion suppression reversed-phase chromatography, dimethyl arsinic acid was eluted later than monomethyl arsonic acid.

As(III) was oxidized to As(V) during sample storage. This tendency was especially marked in dilute solution. When the mixed solution of As(III) and As(V) (7 μg/mL As each) was stored at room temperature for 4 weeks, As (III) disappeared and, instead, twice the amount of As(V) appeared on the chromatogram. The mixed solution at 70 μg/mL As each gave the residual As(III) at the 20% level during the same period. However, stock solutions which contained 0.7% As(III) did not show deterioration during 4 weeks. Therefore, standard mixture solutions should be prepared just before high performance liquid chromatographic analysis.

Morita et al[3674] pointed out that with the argon/hydrogen flame and 193.7 nm line, the sensitivity of atomic absorption spectrometry was about 1/20th of that obtained using inductively coupled argon plasma - atomic emission spectrometry as the detector in conjunction with high performance liquid chromatography. Therefore, atomic absorption spectrometry can be applied only for relatively concentrated samples. Arsenic was also monitored with dc plasma atomic emission spectrometry. The sensitivity of this technique was about one-fifth of that obtained using the inductively coupled plasma technique.

Hanamuka et al[3675] applied thermal vaporization and plasma emission spectrometry to the determination of organoarsenic compounds in fish.

Odanaski et al[3397] have reported the application of gas chromatography with multiple ion detection after hydride generation with sodium borohydride to the determination of mono, dimethyl arsenic compounds, trimethyl arsenic oxide and inorganic arsenic in soil and sediments; this work is discussed more fully in the section of organoarsenic compounds in water. Recoveries in spiking experiments were 87 to 103%. These experiments were part of a study of the fate of arsenicals in some laboratory animals.

Arsenate was orally administered at the 1 mg As/kg level to rats, mice, and hamsters. Urine samples were collected at 24 h intervals and were analyzed by this method.

Table 311 shows the results of arsenic species in the urine from rats, mice, and hamsters after a single oral administration of arsenate (1 mg of As/kg). Inorganic, monomethylated, dimethylated and trimethylated arsenic compounds were detected as urinary metabolites. The former three products

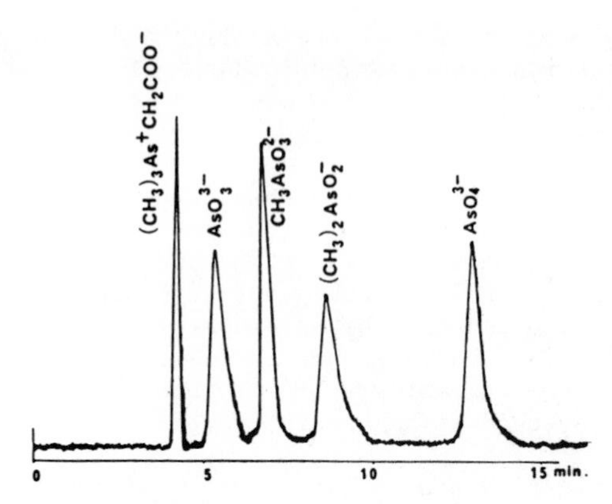

Figure 205 - Separation of five arsenicals by anion exchange chromatography. Each arsenical containing 350 ng of As was injected.

have been previously reported, however, the last one was detected as a new metabolite. These data indicate that inorganic arsenic may be biomethylated to monomethylated, dimethylated, and even trimethylated compounds in these animals.

Maher[3676] used ion-exchange chromatography to separate inorganic arsenic and methylated arsenic species in marine organisms and sediments. The method determines monomethylarsenic, and dimethylarsenic. The procedure involves the use of solvent extraction to isolate the arsenic species which are then separated by ion-exchange chromatography, and determined by arsine generation.

Organosilicon Compounds

Wanatabe et al[3677] give details of a method for the determinations of trace amounts of siloxanes in biological materials. The method involves extraction into petroleum ether, evaporation to dryness, dissolution of the residue in methyl isobutyl ketone, and determination by inductively coupled plasma emission spectrometry. The detection limit is 0.01 ug per ml for organosilicone.

Horner et al[3678] describe two methods for the visible and infrared spectroscopic determination of trace amounts of silicones in biological materials. In one method the sample is decomposed by wet oxidation and the residues are fused with sodium carbonate. The melts are dissolved in hydrochloric acid, and the silicon is determined by the molybdenum blue procedure. In cases where it is desired to identify the type of silicone present, an infrared technique is used. Preliminary separation of the silicone from the sample by solvent extraction may or may not be necessary, depending on its concentration and on the degree of interference by other sample constituents. Recoveries ranged from 86 - 98% for the colorimetric method on a ground-beef sample, and from 79 - 89% for the technique when applied to pineapple juice.

F. ORGANOMETALLIC COMPOUNDS IN AIR

Organomercury Compounds

Many methods have been developed for the determination of organomercury compounds in air. Many of the older methods did not detect organo-

Table 311 - Urinary metabolites from three species of animals after single oral administration of arsenate (1 mg of As/kg)

animal tested	hours postdose	% of dose recovered				
		Inorg. As	Monomethylated Arsonic - As	Dimethylated Arsinic-As	Trimethyl ated Arsenic As	Total As
rat	0 - 24	12.9	0.36	1.47	0.10	14.83
	24 - 48	0.30	0.03	0.34	0.18	0.85
	48 - 72	0.02	0.01	0.15	0.04	0.22
	total	13.22	0.40	1.96	0.32	15.90
mouse	0 - 24	17.1	0.70	20.4	<0.01	38.2
	24 - 48	0.58	0.23	0.09	1.45	11.35
	48 - 72	0.02	0.01	0.99	0.05	1.07
	total	17.70	0.94	30.48	1.50	50.62
hamster	0 - 24	16.4	1.63	9.78	<0.01	27.81
	24 - 48	0.87	1.27	12.0	0.24	14.38
	48 - 72	0.03	0.07	2.15	0.18	2.43
	total	17.30	2.97	23.93	0.42	44.62

mercury compounds such as toxic methylmercury(II) and dimethylmercury compounds or mercury(II) chloride compounds in air, unless these were converted to elemental mercury prior to analysis. Therefore, many indirect methods have been developed in which an enrichment step on efficient adsorbers is used. A large volume of air can be sampled and the mercury is determined by resonance absorption of the 253.7 nm wavelength line after desorption. The adsorber can be a liquid such as an acidified solution of potassium permanganate[3699] or a solid such as activated charcoal[3680], silver[3682], or gold[3681]. These adsorbers collect not only elemental mercury but also the inorganic mercury(II) and organic RHgX and R_2Hg compounds.

Christie et al[3683] describe procedures capable of determining mercury at levels down to 10 mg/m^3 in air. In one method the air sample (500 ml) is passed at 50 l/min through the apparatus described containing 0.75 g of active carbon, and the carbon, freed from moisture by a stream of dry air, is then removed for determination of mercury. In a second method air (500 l) is drawn at 33.3 l/min, through a glass-fibre pad treated with cadmium acetate and sodium sulphide. The pad is then removed for determination of mercury. The active carbon or the cadmium sulphide pad is ignited in the apparatus of Sargent et al[3684] the absorbent replacing the iodized carbon and the mineral wool being omitted. The colour produced on selenium sulphide test paper is compared with a colour chart[3685]. Both methods are applicable to the determination of ethylmercury (chloride and phosphate), diphenylmercury and methylmercury dicyandiamide. The first method only is applicable to diethylmercury.

Polley and Miller[3686] have described a method suitable for the determination in air samples or soil volatiles of amounts of methyl- and ethylmercury chlorides down to 1 - 5 μ in 50 - 100 ml of sample solution. An alternate method developed by these workers (Miller and Polley[3687]) is best used for the determination of amounts above 10 μ of methyl- and ethylmercury chloride in 25 ml of the carbonate-phosphate absorber solution previously mentioned by these workers (Kimura and Miller[3688]). Large aqueous sample volumes are not deleterious in this method as they are in the direct method of analysis mentioned above. This is because of a favourable distribution coefficient of p-tolyl mercury chloride between the chloroform and water phases. The Miller and Swanberg[3689] method is suitable for the determination of below 30 μg of alkyl-mercury compounds in sample sizes of up to 100 ml of carbonate - phosphate absorber solution.

Methods have been described for the determination of organomercury dusts and vapours in air. Hamilton and Ruthven[3690] draw the sample of air through a furnace at 800°C in which the organomercury compounds are decomposed to metallic mercury, and finally through an u.v. spectrophotometer in which total mercury (original organo- and metallic mercury) is determined. A two-way stopcock permits the sample to by-pass the furnace; metallic mercury only is then measured, and the organomercury is calculated by difference.

Dumarew et al[3691] have applied the Coleman mercury analyser system to the determination of volatile organomercury compounds in air. In this method nanogram amounts of mercury are trapped on gold coated sand and the desorbed volatile mercury compounds determined by atomic absorption spectrophotometry.

Field samples were collected by drawing air through the absorber with a small pump (KNF, No. 5 ANE membrane pump) at flow rates of 2.5 - 3.5 l min^{-1}. A prefilter (Whatman GF/A, 13 mm diameter) is used to retain particulates and solid mercury compounds. The volume of air is measured by a calibrated dry gas meter. The volume sampled depends on the expected mercury concentrations. The loaded samplers can be stored for several days without change in mercury content[3692].

The mercury was desorbed by heating the absorber to 800°C and sweeping the products by a carrier gas (0.1 l/min) stream into an atomic absorption spectrophotometer.

Gold-coated sand is the only one of the three collectors tested which is capable of collecting the total content of volatile mercury compounds in ambient air. Gold is not affected by sulfur compounds and so can be used almost indefinitely. The gold absorber retains 1 - 3 μg of mercury per gram before breakthrough occurs. The capacity of the collector depends on the rate of mercury collection and the size of the carrier granules; the value cited is a lower limit because the injections used contained mercury at higher concentrations than are found in ambient air.

Using this method several air samples were taken from areas with different degrees of contamination. The results of the analyses summarized in Table 312 include only the volatile mercury content in ambient air.

Ballantine and Zoller[3693] developed a mercury collection method that is compatible with a chromatographic method of analysis and is capable of detecting levels of organic mercury in the atmosphere as low as 0.1 ng/m^3. Chromosorb 101 is used as a collection substrate for methylmercuric chloride and dimethylmercury. The method involves direct elution of the organic mercury compounds from the collection substrate onto a gas chromatographic column prior to detection with a microwave plasma detector. Methylmercury chloride is collected at ambient temperatures, and dimethylmercury is collected by use of a cryogenic trap at - 80°C. Collection efficiencies are, respectively, 95 ± 3% and 98 ± 2%.

Table 312 - Results of field sample analysis

	Sample size (m^3)	Mercury content ($ng\ m^{-1}$)
Location A (university laboratories)		
Polarography lab	0.005	2616
Purification room	0.005	1610
(ventilated)	0.005	1448
	0.005	1498
office	0.020	205
Location B ("uncontaminated" building)		
Lab	0.300	158
	0.250	151
Office	0.250	99
Room (recent mercury spillate)	0.175	789
Computer room	0.250	54
Reactor hall	0.250	56
Location C (outdoor urban environment)		
Day 1	3000	7.3
2	3000	7.4
3	3000	7.0
4	3000	7.4
5	3000	6.7

Samples and standards were analyzed by using the gas chromatograph/ microwave plasma detector system. The gas chromatograph had been modified so that samples or standards collected on suitable substrates could be thermally desorbed directly onto the gas chromatographic column. The plasma emission detector has the advantage of being extremely selective for mercury when set for the 253.7 nm Hg emission line[3694].

Total volatile mercury samples were collected on gold-coated glass beads prepared by dissolving gold metal or auric chloride in aqua regia. Glass beads that had been etched with dilute hydrofluoric acid were added to the solution, and the volume of the mixture was reduced by heating. The resulting glass bead/gold chloride slurry was packed into a glass tube and heated to 350 - 400°C while flushing the tube with hydrogen. Total volatile mercury samples were collected by drawing air through blanked gold-coated glass bead tubes. Blanked quartz fiber filters were placed in front of the collection tubes to remove particulate matter from the sampling stream.

The optimum conditions for carrying out the gas chromatographic analysis are tabulated in Table 313.

For the analysis of methyl mercury chloride, sample tubes of Chromosorb 101 were heated to 200°C for 45 min at a flow rate of 85 mL/min and the eluting methylmercury was re-collected onto tubes containing 5% FFAP on Gas Chrom Q. These tubes were subsequently heated to 175°C and the methylmercury eluted directly onto the gas chromatographic column. Samples and standards of methyl mercury chloride analyzed in this manner showed an elution volume for methyl mercury chloride of 460 ± 15 mL at a column temperature of 165°C.

Table 313 - Analytical system parameters for the analysis of MMC. DMM and total volatile mercury

parameter	System MMC	DDM	total volatile Hg
flow rate, mL/min	85 ± 5	75 ± 5	60 ± 5
oven temp, °C	165 ± 5	same	same
column (1) (1 m x 6 mm od)	5% FFAP on Gas Chrom Q (80/100 mesh)	Chromosorb 101 (60/80 mesh)	empty glass column
collection tube (T1)	Chromosorb 101 (60/80) (2-4 cm)	Chromosorb (4 - 6 cm)	gold-coated glass beads
preconcn tube (T2)	5% FFAP on Gas Chrom Q (9 cm) Tenax GC (60/80 mesh) (5 cm)	Chromosorb 101 (5 cm)	gold-coated glass beads
desorption temp °C	200	90	350
	Detector		
microwave power	26 W (forward) < 1 W (reflected)	slit width carrier gas pressure at outlet	110 µm argon 755 ± 5 mmHg
monochromator setting	253.7 nm		
quartz capillary	1 mm i.d. 6 mm o.d.		

During the collection of dimethylmercury, the drying tube was less than 100% efficient, with some condensation appearing in the Chromosorb 101 tube. While not enough condensation occurred to impede airflow during sampling, water in the collection tubes extinguished the plasma when the sample was eluted onto the gas chromatographic column. For the analysis of dimethylmercury, Chromosorb 101 collection tubes were heated to 90°C and eluted samples were re-collected on clean, blanked Chromosorb 101 tubes.

When all the water had passed through the system and the plasma was reignited, the Gas Chrom Q collection tubes were heated to 90°C and the sample eluted directly onto the gas chromatograph. Resulting peak areas showed good reproducibility (± 3%). The retention volume for dimethylmercury at 165°C was 406 ± 17 mL/min.

The system was also used for the analysis of total volatile mercury. Mercury species collected on gold-coated glass beads were analyzed by replacing the gas chromatographic column with an empty glass column.

Organomercury in air contents of samples taken from a Power Plant are shown in Table 314.

Major disadvantages of this technique are the necessity of long collection times (3 - 12 h) when studying background levels, and relatively long analyses (approximately 1 h) for methylmercury chloride samples. The significant advantages over previous methods are the simplification of sample elution directly onto the gas chromatographic column which minimizes the possibilities of contamination and the ability to positively identify two of the most important organic forms of mercury present in the atmosphere.

Table 314 - Mercury in air data from Chalk Point Power Plant, MD.

date	site 1 (intake), ng of Hg/m^3		site 2 (canal), ng of Hg/m^3	
	as methyl-mercuric chloride	as non-methyl-mercuric chloride	as methyl-mercuric chloride	as non-methyl-mercuric chloride
3/24/82 day	2.9	3.3	7.3	19
night	1.04	1.3	4.7	26
3/25/82 day	1.4		4.0	16
night	0.74	1.13	2.1	1.5
3/26/82 day	2.1	6.2	2.5	16.4
av, day	2.0 ± 0.9	5 ± 2	5 ± 2	19 ± 5
av, night	0.9 ± 0.2	1.2 ± 0.1	2.1	1.5

Schroeder et al[3694] determined mercury in the atmosphere using a method of selective preconcentration followed by pyrolysis and cold vapor atomic fluorescence detection to determine different mercury species collected from the atmosphere. Other studies have been performed by using sequential specific absorption tubes which separate different chemical forms of mercury by selective collection[3695]. In these latter studies, mercury compounds were thermally desorbed and re-collected on gold surfaces prior to elution into an emission detector. While these methods represented a significant advance in atmospheric mercury sampling by achieving the separation of volatile species of mercury, the analytical methods prevented positive identification of the compounds by converting all forms to elemental mercury prior to detection. Chromatographic substrates have been used successfully for the collection of organics and of organic mercury[3695-3700]. By logical extension a chromatographic method of analysis would permit the positive identification of organic mercury compounds by comparison of sample elution times/volumes with standard compounds.

The "electrostatic accumulation furnace for electrothermal atomic spectrometry" (EAFEAS) technique[3701] has been successfully used for the precise, simple and fast determination of mercury in air samples, It has also been shown that the electrostatic accumulation furnace can be easily operated with a collection efficiency of approximately 100% with particulate matter as well as with mercury vapour. The high capture efficiency of mercury atoms, even if not unexpected[3704] represents a remarkable indication of the potentialities of the proposed method and suggests that gaseous compounds other than mercury vapour could also be tested.

Organomanganese Compounds

Bykhovskaya[3705] described a method for the determination of cyclopentadienylmanganese tricarbonyl vapour in air. The carbonyl compound is absorbed in a fluidised bed of silica gel (0 - 5 mm), which permits complete absorption at high rates of flow of air (5 - 7 litres per min). It is recovered by treatment with ethanol, or decomposed with nitric acid and the manganese is determined polarographicially or photometrically.

Kawamura and Matsumoto[3706] have described a method for the determination of small amounts of hydrogen selenide in air. The sample (not less than 50 litres) is passed at 2 litres per min through an absorber containing 20 ml of 40% hydrogen bromide solution with 18% of free bromine; the absorbent is made up to 50 ml and a 25 ml portion is treated with 10% hydroxylammonium chloride solution to destroy unchanged bromine. The colourless solution is neutralised with aqueous ammonia, the pH is adjusted to 2.5 with 2.5 M formic acid, and selenium in this solution is determined photometrically by the 3,3' diaminobenzidine method.

Coe et al[3707] have described a gas chromatographic atomic absorption technique for the determination of methylcyclopentadienylmanganese at the μgm^3 level in air samples. The method involves trapping of MMT in a small segment of gas chromatographic column and then determination by gas chromatography with an electrothermal atomic absorption detector. The detection limit of the procedure is 0.05 ng m^{-3}.

A Perkin-Elmer 603 atomic absorption spectrometer was used with a Perkin-Elmer HGA 2100 furnace. A deuterium arc background corrector was employed. A Pye gas chromatograph (Series 104) was interfaced to the graphite furnace using a tantalum connector as previously described[3708]. A glass chromatographic column (2.3 m long, 6 mm o.d.) was packed with 3% OV-1 on high-performance Chromosorb W (80/100 mesh). The gas from the chromatograph was transferred to the furnace through teflon-lined aluminium tubing.

The gas chromatographic set up is illustrated in Figure 206. (A) connects to a nitrogen cylinder. (B) is the sample oven containing the sample trap and 4-way valve, (C) is the gas chromatograph connected through a standard injection valve to the column. (D) is the graphite furnace. The operating conditions for the gas chromatograph and the graphite furnace were as shown in Table 315.

Samples of air were collected in teflon-lined aluminium U-tubes (30 cm long, 3 mm o.d.) packed with 3% OV-1 on Chromosorb W (80/100 mesh). These tubes were placed in a water - ice cooling bath. Air entered through an air filter and was pumped through the U-tube trap at about 70 ml min^{-1} using a vacuum pump. The length of sampling time and the average flow rate, checked frequently during sampling, were used to compute sample volume.

Coe et al[3707] took air samples of up to 15 m^3 at a variety of locations on the streets of Toronto. No organo manganese was detected in any sample (detection limit 0.05 ng m^{-1}). This compares with 14 ng m^{-3} for total tetraalkyllead compounds found in street air in similar locations. Samples were then collected in an underground car park. In a few of these samples organomanganese was detected.

It is interesting to speculate as to why the organomanganese levels in air are so much lower than those of tetraalkyllead when it is likely that a large fraction of the vehicles are using "unleaded" gasoline containing organomanganese. To this end, a cursory study of organomanganese in air was attempted.

Organomanganese, liquid and vapor, was injected into clear 2.1 glass bottles kept in the light and in the dark. Samples were then taken of the contained air and isooctane rinsings of the bottle walls at various intervals. Results of these studies suggest that organomanganese is quickly decomposed in air, more quickly in the light than in the dark.

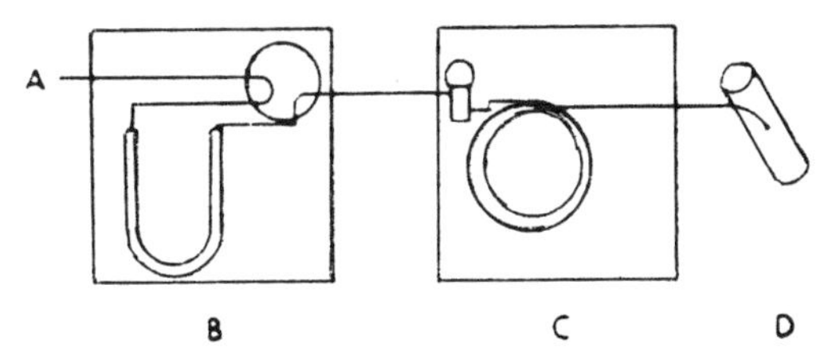

Figure 206 - Gas chromatograph system.

Table 315 - Instrumental operating conditions

Gas chromatography		Atomic absorption spectrometry	
Carrier gas flow rate	80 ml min^{-1}	Ash 300°C	120 s (dry cycle)
Oven temperature program	115°C isothermal	Atomize 1800°C	240 s (char cycle)
Injection port temperature	150°C	Internal gas flow	0
Outlet and transfer tube temperature	150°C	External gas flow (N_2)	60 ml min^{-1}
		Wavelength	279.5 nm
		Slit 4	(0.7 nm)
		Background correction mode,	
		Scale expansion (x 5)	

Organonickel Compounds

Nickel tetracarbonyl is one of the most dangerous chemicals known. It exhibits acute toxicity, carcinogenicity, teratogenicity, and can be produced spontaneously in unsuspecting environments whenever carbon monoxide contacts an active form of nickel. The toxicological and carcinogenic data on nickel carbonyl has led to the lowest eight-hour coverage allowable exposure of any chemical. The Occupational Safety and Health Administration has set this level at 0.5 mg/m^3 or 1 ppb[3709] which has dictated a need for ultra methods of analysis for nickel carbonyl in ambient air.

Recently the instrumental methods, infrared spectrophotometry (IR)[3710] Fourier transform IR (FTIR)[3710,3711] plasma chromatography[3712] and chemiluminescence[3713], have been used to analyze for nickel carbonyl. All of these methods have the advantage of direct measurement with detection limits of about 1 ppb and are also adaptable to process stream monitoring. The methods based on the FTIR and the plasma chromatography methods have been compared in real sample analyses and agree within a few percent. The chemiluminescent analysis for nickel carbonyl demonstrates a detection limit of 0.01 ppb with a linearity over four order of magnitude. Therefore, it is suitable for industrial applications from the practical and economic standpoint.

In order to meed Federal Standards, some automobile manufacturers in the U.S.A. have added thermal and catalytic reactors to gasoline engine exhaust systems. These catalytic reactors may contain a combination of a variety of metals in active forms on an inert support with a significant surface area. It is reasonable to expect that during normal operation these metals will be emitted from the exhaust system in the form of metal complexes or metal particulate matter. The complexity of exhaust gases, particulate matter, and the range of catalytic converter operating conditions during normal automobile operation suggest that compounds may be formed within the catalytic exhaust system.

Campara and Risky[3714] have studied the possibility of the formation of nickel carbonyl by the reaction of carbon monoxide (present in exhaust gases in the range 1 - 15%) with nickel containing catalysts (Monel) under typical automotive operating conditions.

Chemical ionization mass spectrometry was used to determine trace levels of nickel carbonyl in a typical exhaust gas mixture. Also a catalytic flow reactor system was designed to study the formation of nickel carbonyl where the reacted gas was analyzed for nickel carbonyl. This reactor approximated the conditions found in an automotive exhaust catalytic converter system. These workers found various positive ions which can be attributed to nickel carbonyl and its fragments in this mass spectrum. The effect of pressure on the intensity of both the reactant gas ions and the nickel carbonyl was studied in an attempt to find the optimum condition for the quantification of nickel carbonyl. Using these data, the minimum detectable limit was found to be 10 ppb for nickel carbonyl. This methodology was used to monitor nickel carbonyl in the effluent from a model reactor for a catalytic controlled automobile. Based on these studies, no measureable quantities of nickel carbonyl can be expected from catalytic controlled automobiles.

Organotin Compounds

Luskina and Syavtsillo[3715] have described procedures for the determination of tetraethyltin and tetrabutyltin in air and in water.

Jeltes[3716] has discussed the determination of bis (tributyltin) oxide in air by atomic absorption spectroscopy and pyrolysis gas chromatography.

Organolead Compounds

Atomic absorption spectroscopy. Various workers[3717-3720] have applied this technique to the determination of organolead compounds in air. Flame photometry has also been used[3721]. The area of organometallics in general and of tetraalkyllead (TAL) in particular is of great interest. Methods such as direct atomic absorption spectrometry[3719,3720], flame photometry[3721] are unsuitable for the determination of TAL in environmental samples because of the lack of specificity and/or sensitivity or because of the long time required to analyse a single air sample.

Torsi and Palmisano[3722,3734] have described a procedure for sampling air using a battery powered field sampler and capturing organolead compounds by electrostatic capture and subsequently determining them by electrothermal atomic absorption spectrometry. Down to 100 pg tetraalkyllead (as lead) can be determined by this procedure.

The electrostatic accumulation furnace (Figure 207) is composed of a brass block which screws on to C fixing the stainless-steel sheet B (0.2 mm thickness) between A and C. The brass blocks C and E are screwed on to a glass-reinforced PTFE ring D. The ring material can be substituted with high-temperature ceramic material such as Aremco ceramic, Type 502 - 1100. Radial holes I are machined in the brass blocks C and E to facilitate the purging and cooling of the furnace. The graphite tube G can be fixed

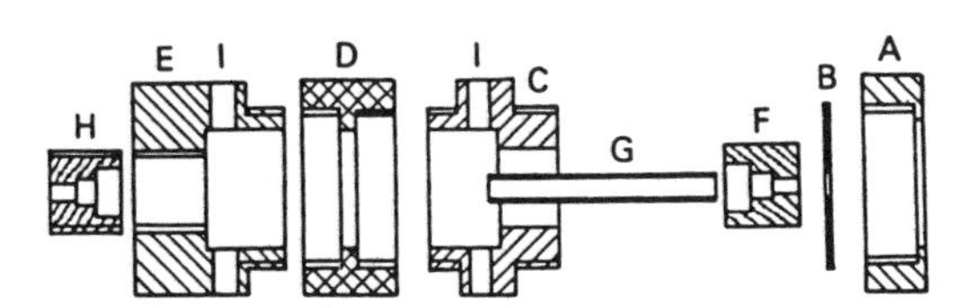

Figure 207 - Exploded view of the electrostatic accumulation furnace. For a detailed description see text.

tightly on to the furnace axis by means of the graphite rings F and H, because the ring F can alide inside C against B, while the ring H is screwed on to E. To obtain better data reproducibility using different furnaces a torque wrench has been used in screwing H, producing in this way a reproducible force between F and H. The stainless steel sheet B, acting as a spring, allows the expansion of the graphite tube G without unduly increasing the thrust against F and H. In this way the possibilities of breaking the furnace are practically eliminated.

The electrostatic precipitator (Figure 208) serves to accommodate the furnace and to introduce and evaporate efficiently a TAL solution under a stream of air without contamination from (inorganic) lead associated with particulate matter. In Figure 208 A is the body (PTFE) of the furnace container cup, B is a Pyrex tube, C is a tightly fixed silicone-rubber septum and D is a Gelman type filter-holder accommodating a Gelman Ga-6 0.10 μm Metricel membrane filter E, which ensures the practical removal of lead associated with the particulate matter. (The volume of air typically sucked was 300 - 400 cm^3 and did not show a detectable signal for lead). F is a PTFE tip machined in the cup A, 10 mm long, 2 mm o.d. and 1 mm i.d., which protrudes about 5 mm inside the graphite tube G when the cup A is screwed on to the furnace container H. This arrangement will bring the tetraalkyllead vapours well inside the graphite tube, thus minimising the amount of analyte absorbed by diffusion at the entrance and in the first section of the graphite tube. I is a tungsten tapered wire mounted precisely along the axis of the furnace and L is a spring contact that ensures the electrical connection with the furnace G via the metal block M.

A known volume of air was sucked through the membrane E (in order to exclude lead from the particles) with the voltage on. The flow-rate (approximately 0.8 $cm^3\ s^{-1}$ corresponding to an average linear velocity of approximately 10 cm s^{-1}) and the potential applied (1.8 - 2.0 kV) were in the range of maximum capture efficiency[3723].

Gas chromatography. Gas chromatography with electron-capture detection [3724,3725], gas chromatography with catalytic hydrogenation pre-derivatisation and flame-ionisation detection[3726,3727] and gas chromatography with microwave plasma detection[3728] are unsuitable for tetraalkyllead determination in environmental samples because of the lack of specificity and/or sensitivity or because of the long time required for the analysis of a single air sample. Most of the methods with high sensitivity and low detection limits presently available[3729,3734,3736,3737] for the determination of tetraalkyllead in air samples are based on a sampling step, often performed by cryogenic trapping, followed by an analysis step by gas chromatography combined with atomic-absorption spectrometric (both flame and electrothermal) detection. These methods determine the tetraalkyllead compounds present but are time consuming, require a gas chromatograph on line

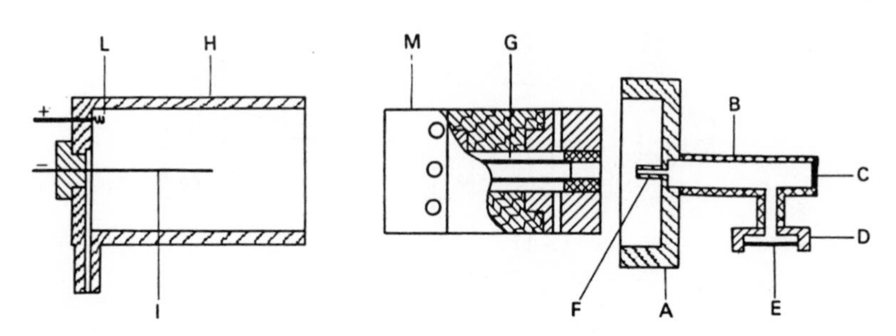

Figure 208 - Schematic view of the electrostatic precipitator. For a detailed description see text.

with an atomic-absorption spectrometer and have a relatively complex sample collection step.

A critical review concerning the chemical and instrumental problems associated with the measurement of tetraalkyllead compounds in air can be found in reference 3735.

Boettner and Dallas[3738] have compared the sensitivities of the electron capture, thermal, conductivity, argon ionization and flame ionization detectors to the chromatographic determination of organolead and aliphatic chlorine compounds in the atmosphere. They used a Wilkins Hi-Fi Model 600 Chromatograph with both hydrogen flame and electron capture detectors, a Beckmann Model GC-2A Chromatograph with a thermal conductivity detector and a Research Specialities Model 600 Chromatograph with an argon ionization detector. All separations were made on 1/8 in or 1/4 in x 6 ft stainless steel columns. The compounds studied and the column coatings used are tabulated in Table 316. The sensitivities of the various detectors to the different compounds examined are shown in Table 317.

Several observations can be made from the results presented in Table 317. The sensitivities of the thermal conductivity and the argon ionization detectors are independent of the molecular weight or the number of chlorine atoms in the chlorinated compounds, but the flame detector decreases slightly in sensitivity with increasing numbers of chlorine atoms. The electron capture detector was found to have its greatest response to the chlorinated compounds at 10 volts. With the electron capture detector the sensitivity was dependent in rather a complex manner on the molecular weight and the number of chlorine atoms in the chlorine compound.

In the case of the two alkyllead compounds examined, no large gain in sensitivity of the electron capture detector over the ionization detectors was realized, and the sensitivity gain was only a factor of two. The sensitivity values given for lead tetramethyl and lead tetraethyl in air are about the same as reported by Bonelli and Hartmann[3739] for these compounds in gasoline. Here again the greatest sensitivity was attained at 10 volts, which is somewhat lower than the findings of Bonelli and Hartmann[3739] and Dawson [3740].

The conclusions reached by Boettner and Dallas[3738] concerning the electron capture detector were as follows:-
That for analysis of volatile chlorinated aliphatic hydrocarbons, this detector is no more sensitive than the ionization detectors for those

Table 316 - Column coatings and supports for the separation of chlorinated aliphatic and lead alkyl compounds using gas chromatographs with various detectors.

	Thermal Conductivity	Argon Ionization	Flame Ionization	Electron Capture
Methyl chloride	C	A	A	A
Dichloromethane	C	A	C	B
Chloroform	C	A	A	B
Carbon tetrachloride	A	A	A	B
Ethyl chloride	A	A	A	A
1,2 Dichloroethane	A	A	A	B
1,1,1 Trichloroethane	C	A	A	B
1,1,2 Trichloroethane	C	A	C	B
1,1,2,2 Tetrachloroethane	C	A	C	E
1,2 Dichloropropane	C	A	C	C
1,2,3 Trichloropropane	C	A	C	E
Chloroethylene	A	A	A	A
1,2 Dichloroethylene cis	A	A	A	A
1,2 Dichloroethylene trans	A	A	C	B
Trichloroethylene	C	A	A	B
Tetrachloroethylene	D	A	C	B
Lead Tetramethyl	D	G	D	F
Lead Tetraethyl	D	G	F	F

A - 20% Carbowax 600 on C-22 Firebrick
B - 5% Silicone 550 and 5% Ucon (water insoluble) on Chromosorb P
C - 10% Silicone 500 on C-22 Firebrick
D - 20% Carbowax 20 M on C-22 Firebrick
E - 10% Silicone SE 30 on Chromosorb W
F - 5% Silicone SE 3- on Chromosorb W
G - 5% Silicone 550 on Anakrom ABS

compounds having one or two chlorine atoms. For those compounds having three or four chlorine atoms, the electron capture detector is from 100 to 1,000 times more sensitive than the ionization detectors. For the two alkyl lead compounds tested, the electron capture detector gives little improvement in sensitivity but its discrimination toward the lead substituted compounds as compared with unsubstituted hydrocarbons makes it a preferable detector for analyzing mixtures of these two types of compounds.

Chau et al[3741,3742] found that the recovery of the five lead compounds from air in a cold trap containing OV-1 at 80°C was 94 - 105%. This was done without precondensing moisture.

Cantuti and Cartoni[3743] described a procedure for the direct collection of tetraethyllead from polluted air and its chromatographic determination at ppm levels by using electron capture detection. The air under test was passed through a sampling tube containing the material used for packing the chromatographic column. When equilibrium conditions had been established, the sample was desorbed by flushing the tube (heated at 130°C) with carrier gas, and was injected directly into a glass column (1 metre x 0.3 mm) packed with 10% of silicone rubber SE-52 on Chromosorb P (80 to 100 mesh), operated at 80°C (with electron-capture detection and pure nitrogen 30 ml per min) as carrier gas. The method is sensitive to down to 0.1 ppm of lead tetraethyl in air. Their technique was later modified and improved in

Table 317 - Limits of detection for chlorinated aliphatic and lead alkyl compounds using gas chromatographs with various detectors.

1. Methyl chloride	1.2	2.0	3.0	8.5
2. Dichloromethane	4.2	5.0	1.3	8.6
3. Chloroform	6.0	4.3	20.	0.08
4. Carbon tetrachloride	4.8	5.0	20.	0.002
5. Ethyl chloride	1.4	6.0	1.6	11.
6. 1,2 Dichloroethane	3.4	4.1	13.	13.
7. 1,1,1 Trichloroethane	2.6	5.2	6.0	0.03
8. 1,1,2 Trichloroethane	2.8	4.0	8.6	0.07
9. 1,1,2,2 Tetrachloroethane	5.0	8.0	16.	0.008
10. 1,2 Dichloropropane	0.9	5.5	8.8	23.
11. 1,2,3 Trichloropropane	2.8	3.8	4.0	0.07
12. Chloroethylene	0.2	1.9	2.2	2.3
13. 1,2 Dichloroethylene cis	4.0	6.5	2.6	13.
14. 1,2 Dichloroethylene trans	2.2	3.5	2.5	8.4
15. Trichloroethylene	2.5	10.	8.5	0.02
16. Tetrachloroethylene	3.2	5.3	21.	0.003
17. Lead tetramethyl	2.5	13.	2.5	1.5
18. Lead tetraethyl	3.3	33.	6.6	3.3

order to apply it to the analysis of city air[3744]. In this way detection limits of about 0.05 and 0.5 $\mu g/m^3$ were reported for the determination of tetraethyllead and tetramethyllead. Since both methods use electron capture detection, they lack the specificity required for environmental samples, as these contain a variety of other compounds with high electron affinity.

By use of cooled gas chromatographic column packing material for the collection, the tetraalkyllead compounds, tetramethyllead, trimethyllead, dimethyldiethyllead, methyltriethyllead, and tetraethyllead) have also been determined with gas chromatography/atomic absorption spectrometry. However, the low analytical sensitivity combined with a very slow air sampling rate (130 - 150 mL/min) made this procedure unsuitable for the analysis of ambient air[3745]. In later modifications, the sensitivity was greatly improved [3746,3748] but the determination of tetraalkyllead compounds in street air still required collection times of 16 h or more. Other reports also mentioned the possibilities of the gas chromatography atomic absorption spectrometry technique for the analysis of tetraalkyllead compounds in the air without entering into details about the method used for sampling[3748] or analysis [3749,3750]. In the latter work only samples of polluted atmospheres with organic lead concentrations of 0.1 $\mu g/m^3$ or higher were adequate in order to separate the different alkyllead components.

A method described by Kaveeskoy[3751] involving gas chromatography - mass spectrometry for the determination of alkyllead species in air which is reasonably rapid and certainly more rapid than many of the earlier methods. Hayakawa[3752] has described a procedure for the determination of triethyllead ion in street dust.

Harrison et al[3753] trapped organic lead compounds from street air and eluted them directly into the flame of an atomic absorption spectrometer, thus determining total organic lead. In such a study it would be an advantage to employ furnace atomization since organic lead exists in air at very low levels and the furnace can give a detection limit gain of up to 3 orders of magnitude.

Robinson et al[3754] have carried out a detailed study of atomization processes in carbon furnace atomizers during the development of a gas chromatography-furnace atomic absorption combination for the determination of organolead compounds in gasoline and air. The carbon furnace atomizer is attached directly to the gas chromatograph column. The atomizer described exhibits high sensitivity and eliminates many of the problems involved with interferences encountered with furnace atomization. Robinson et al[3754] point out that the previously described gas chromatographic - atomic absorption methods utilizing carbon atomizers[3755-3762] for the determination of organolead compounds are unwieldy in many respects. In addition, many of the problems involved with commercial carbon atomizers persist when they are used in g.c. - a.a.s. combination systems. The commercial systems are certainly capable of performing analyses at very high levels of analytical sensitivity and precision; but the development of reliable quantitative procedures is much more difficult. Their study of the furnace atomization step revealed the problems involved in obtaining accurate quantitative data.

The atomization step with a furnace or rod atomizer is very difficult to examine; the formation of atoms is a very fast process. The number of atoms formed depends on many variables and the population never reaches a steady state. Consequently, to control the percentage of atoms formed (atomization efficiency) and the total number of free atoms entering the light path during optical measurement, it is vital to control all the variables which affect the rate and the degree of atomization. The difficulty involved in effective control can be more readily understood when the entire atomization process is considered. Many sample atomizers involve three steps: (1) evaporation of the solvent at a relatively low temperature; (2) ashing of the sample at an increased temperature; and (3) atomization at a relatively high temperature.

When carbon rod or a tube is used, the liquid sample diffuses into the carbon to a lesser or greater extent depending on the physical form of the carbon. If the carbon is impervious and the surface is shiny, as is the case when pyrolytic carbon is deposited on the surface the liquid sample will not penetrate the carbon; atomization will then be relatively rapid. However, if the carbon is not impervious, then the liquid sample soaks into the carbon and the subsequent evaporation, ashing and atomization steps are slowed down because the components of the solution must first re-emerge from the carbon.

The atomization and ashing steps both depend on the chemical form of the sample. Extensive studies by Ebdon et al[3763] indicated that the chemical interferences with carbon atomizers can lead to serious errors. With commercial atomizers, the chemical interferences can be overcome to some extent by rigidly controlling the atomization step, with regard to timing, rate of heating and temperatures. However, it is quite impossible to control these steps completely since the normal sample is often not completely characterized. It is much easier to obtain accurate and reproducible results from 'pure' solutions, than from real samples.

A further problem relates to the variation of the electrical resistance of the carbon rod or tube with use. Thus, with a standardized atomization programme, the temperatures achieved will differ somewhat as the carbon ages, although the time and voltage remain the same, and this will affect the atomization rate.

With these atomizers, the free atoms are swept through the light path in a very brief period of time. Short, sharp peaks are frequently recorded and the peak heights are usually measured. Equipment in which the peak area is integrated is more accurate because peak areas correspond more to the total number of atoms than to the maximum number of atoms produced. A problem frequently arises with coupling a furnace atomizer to equipment purchased for flame atomizers. The latter equipment usually utilizes amplifiers and read-out systems with slow response times which are totally unsuitable for use with furnace atomizers. Such equipment can give answers which are more a measure of response time than of metal concentration.

Despite the above variables, it is possible to obtain precise measurements. The reproducibility may be very good for a given instrument under strictly controlled conditions. However, when actual samples are run, the rates at which the samples diffuse into and out of the carbon rod change, the rates of decomposition vary because of chemical interferences, and quite erroneous answers may be obtained even though every precaution has been taken to match samples and the standards used. The processes never reach equilibrium. Under these non-identical conditions, which might be compared with the conditions encountered in gravimetric analysis before the days of thermogravimetric analysis, highly reproducible data are possible, but the production of accurate, reliable results is a very different matter. These considerations lead Robinson et al[3754] to the development of the atomizer with different processes described below.

A schematic diagram of the atomizer developed by Robinson et al[3754] is shown in Figure 209. The atomizer is left hot at all times when in use. The effluent from the gas chromatograph enters the base of the atomizer where the gaseous sample is decomposed and the atomization takes place. The atoms flow into the cross-piece, which is in the optical light path. The advantage of the process is that the peak of the solvent used is quite separate from the peak of the metal-bearing component on the gas chromatograph. The gas chromatograph separates the metal-bearing components from the rest of the material, which eliminates much of the problems encountered in the solvent evaporation step and other matrix effects. Decomposition is fairly rapid, although several seconds elapse from the time that the sample enters the carbon atomizer before the atoms reach the optical light path. This permits chemical decomposition to take place and virtually eliminates chemical interference, which is usually caused by varying rates of atomization from different compounds rather than by prevention of decomposition. Even if the rate varies, decomposition is virtually complete before the free atoms enter the light path.

The peak height is a product of the metal concentration and the resolution of the g.c. column. The definition of sensitivity as that concentration which results in 1% absorption is therefore unrealistic. Sensitivity measurements were based on peak area data.

The atomizer is just a few cm long (Figure 209) and can be fitted into almost any commercial a.a.s. unit. Fast electronics are not necessary because the metal concentration is measured as a gas chromatographic peak which takes several seconds to evolve. The atomizer overcomes many of the difficulties involved in commercial furnace carbon atomizers.

The sensitivity of the equipment was shown to be of the order of 10^{-10}g. Robinson et al[3754] undertook studies on the effect of evaporation on the concentrations of lead components in the anti-knock materials. Under some circumstances, tetramethyllead tended to evaporate more rapidly than the

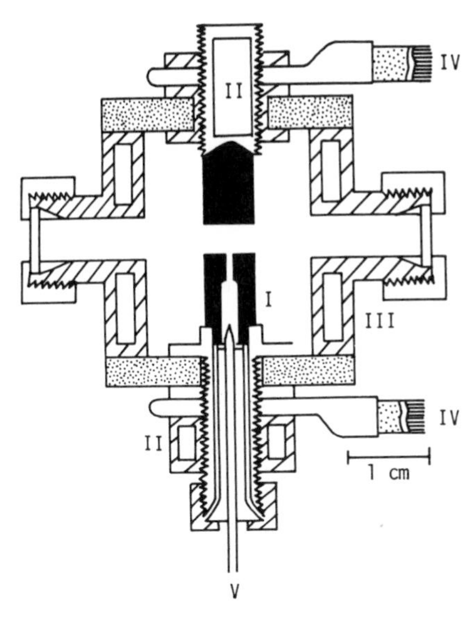

Figure 209 - Schematic diagram of atomizer. I, carbon resistance atomization chamber. (II) water-cooled electrodes. (III) water-cooled housing. (IV) power supply (welding cables). (V) Pyrex capillary transfer line from g.c. to atomizer.

other forms of lead; this is not surprising since it is the most volatile of the lead components. Robinson et al[3754] also studied the applicability of their technique to the determination of organolead compounds in the atmosphere. In this work 1 m3 of air was pulled through a cryogenic trap. The trapped material was then put into the gas chromatograph under conditions suitable for tetraethyllead or other organic lead compounds. The concentrations of TEL were so small that in 9 cases out of 10 none was detected; in the 10th case some lead was detected but the compound could not be identified. These results confirm that the tetraalkyllead concentrations in the atmosphere are very low.

Reamer et al[3764] have discussed the applicability of a gas chromatograph coupled with a microwave plasma detector for the determination of tetraalkyllead species in the atmosphere. The tetraalkyllead species are collected by a cold trap. The volatile lead species are concentrated within an organic solvent, separated by a gas chromatographic column and determined by an MPD system which measures the emission intensity of the lead 405.78 nm spectral line. Previous workers have used an acidic solution of hydrochloric acid[3768,3766], activated charcoal[3620], Apiezon L on a silanized universal support[3767], the chromatographic support OV-1 and silicone rubber on Chromosorb P[3768], to collect alkyllead compounds from the atmosphere. Reamer et al[3666] utilized a cold trap containing SE-52 on Chromosorb P at - 80°C for the collection at atmospheric alkyllead compounds. Reamer et al used a Matronic gas chromatograph for the separation and introduction of alkyllead compounds into the detector. The detector consisted of a microwave plasma system which was basically the same as that described by McCormack et al[3769] with some instrumental modifications. The system used was comprised of a transparent quartz capillary which contained the effluent from the gas chromatograph (see Figure 210). The capillary was contained within a 3/4 wave cylindrical microwave cavity. The argon plasma within the capillary

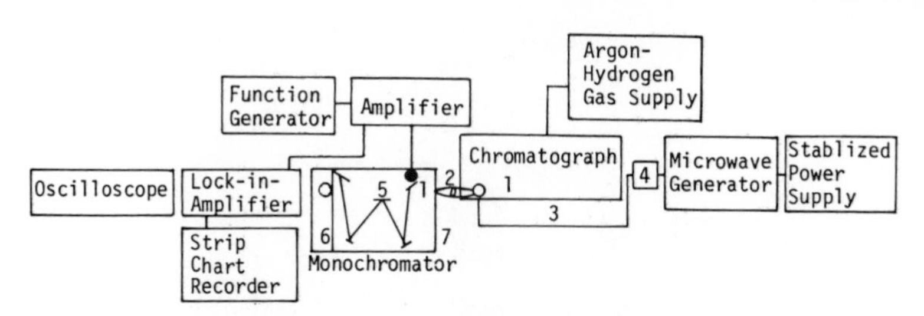

Figure 210 - Block diagram of the GC-MPD system. (1) microwave cavity and plasma. (2) quartz lens. (3) sample injection port. (4) reflected power meter. (5) Diffraction grating. (6) photomultiplier power supply. (7) vibrating plate.

was generated by a 200-W Microtron microwave generator. The signal was viewed at tight angles to the quartz capillary and was focused on the entire area of the entrance slit. A background correction system employed a wavelength - molulation technique. The GC-MPD operating conditions are summarized below.

GC column : 3% OV-1 on 80/100 mesh Chromosorb-W (1.8 x 31 mm)
quartz capillary : 1.0 mm i.d., 6.0 mm o.d.
carrier gas : argon, 1% hydrogen (1% hydrogen included to avoid mercury effects)
carrier gas flow rate : 22 ml/min
column temperature : 80°C
injection temperature : 130°C
microwave power : 125 W
monochromator settings :
slit height 12 mm
spectram bandpass : 0.22 nm
wavelength : Pb, 4)5.78 nm
modulation interval : 0.38 nm
photomultiplier tube : Jarrell-Ash R106.

When using the wavelength - modulation mode for background correction, the tetraalkyllead calibration curves extended from the low pg range to the low ng range with the following detection limits : tetramethyllead, 6 pg: trimethylethyllead 10 pg; dimethyldiethyllead 23 pg; methyltriethyllead 35 pg; and tetraethyllead 40 pg. Figure 211 shows a representative chromatogram of an ambient air sample, illustrating the degree of separation of the five tetraalkyllead species.

The absorption tubes used by Reamer et al[3764] consisted of Teflon tubing 9.5 mm o.d. x 25 cm long filled with 10% SE-52 on Chromosorb-P (non-acid washed) approximately 2.8 g. The absorbent was pretreated by heating to 150°C for 6 h while passing ultra-pure nitrogen through it. During collection, the absorption tube was maintained in a dry ice-methanol slurry at - 80°C and was preceded by a 47 mm Teflon filter holder with a Nuclepore filter of 0.2 m pore radius. These workers showed that 0.07 µg (0.01%) of tetraethyllead was retained by the Nuclepore filters and 1.7 µg (0.3%) by Millipore filters when 650 µg of tetraethyllead was passed through the filters. A twin-piston vacuum pump was used for the air collection with a flow meter to measure the volume of air sampled, (0.6 m^3 sample size). A flow rate of approximately 0.3 m^3/h was used. A high vacuum Tempest pump (= 15 m^3/h) was also used simultaneously with the absorption system to

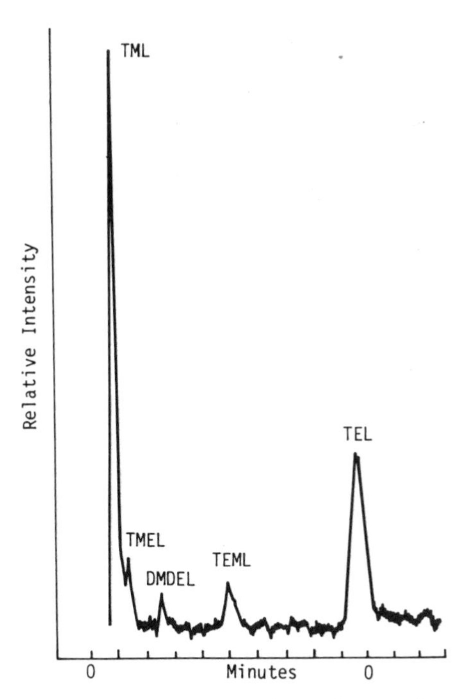

Figure 211 - Chromatogram of an air sample collected near a highway. Sample collected for 2 h.

collect particles on 110 mm Whatman 40 filter paper. All Teflon parts were cleaned in nitric acid and rinsed in distilled water prior to use. After sampling was completed, the absorption tubes were sealed and packed in dry ice until analyzed. After a 2 h sampling period, the filters were folded, placed in plastic bags and polyethylene vials, and stored in dry ice for transport to the laboratory. Using this system Reamer et al[3764] showed that collection efficiencies for tetraethyllead were between 84% and 100% at the 150 ng of tetraethyllead level.

The absorbent from the sampling tube was transferred to a 50 ml round-bottom flask which was attached to a miniature freeze-drying system. The sample was dried for 12 h, resulting in the quantitative removal of the tetraalkyllead compounds and water present in the sample. The gases were quantitatively trapped cryogenically at liquid nitrogen temperature. The trap was removed from the system and slowly warmed. The container walls were rinsed with distilled water to ensure quantitative retention of the analyte. The organic lead compounds were extracted with 200 µl of hexane by shaking for 10 min (extraction efficiency ranged from 83 to 95%). Sample analysis involved injecting a 1- to 5-µl aliquot of the hexane layer into the gas chromatograph with a 10 µl microsyringe. Calibration curves were prepared after each set of sample injections, to minimize the effect of slight fluctuations with the plasma.

Table 318 gives the concentrations of the five tetraalkyllead compounds found by Reamer et al[3764] in some environmental samples.

Table 318 - Concentrations of tetraalkyl and particulate leads from various sampling sites [a]

Sampling Site	species of tetraalkyllead ng/m³					total tetra-alkyl leads ng/m³	total partic-ulate lead ng/m³	% tetra-alkyl leads
	TML	TMEL	DMDEL	MTEL	TEL			
vehicle A exhaust	240	0.5	0.5	0.5	140	380	17 500	2.2
vehicle B exhaust	650	0.5	0.5	0.5	380	1030	55 000	1.9
vehicle C (cold)	435	25	107	2	28	585		
vehicle C (hot)	270	8	9	3	11	298		
Baltimore						57	12 200	0.47
Harbor							19 220	
Tunnel						80	21 600	0.37
7/19/76						130	19 600	0.66
						72	18 400	0.39
						122	14 700	0.83
Baltimore	21	0.5	11	6	36	74	12 600	0.59

Harrison et al[3770] trapped organic lead compounds from street air and eluted them directly into the flame of an atomic absorption spectrometer, thus determining total organic lead. In such a study it would be an advantage to employ furnace atomization since organic lead exists in air at very low levels and the furnace can give a detection limit gain of up to 3 orders of magnitude.

Radzuik et al[3771] have described a sensitive gas chromatographic - atomic absorption method for the determination of individual alkyllead compounds in air. Gas chromatography - mass spectrometry was used to identify the separated organolead compounds.

A Perkin-Elmer 603 atomic absorption spectrophotometer used was equipped with a deuterium background corrector, and HGA 2100 graphite furnace and a Perkin-Elmer Model 56 recorder. The radiation source was a Perkin-Elmer electrodeless discharge lamp operated at 10 W. A Pye gas chromatograph (Series 104) was interfaced to the graphite furnace with a tantalum connector machined from a 6.4 mm diameter rod. The glass chromatographic column (150 cm long, 0.6 cm o.d.) was packed with 3% OV-101 on Chromosorb W (80 - 100 mesh). The effluent was transferred to the furnace by Teflon-lined aluminium tubing (3 mm o.d.= heated electrically to 80°C.

The gas chromatographic - mass spectrometry system consisted of a Varian Series 1400 gas chromatograph and a Finnigan quadrupole spectrometer Model 1015-C. The chromatographic column was identical to the above.

The adsorption tubes for air samples were U-shaped Teflon-lined aluminium tubes (30 cm long, 3 mm o.d.) packed with 3% OV-101 on Chromosorb W (80 - 100 mesh). Moisture was condensed from air by using glass U-tubes at - 15°C. The air was sampled with a peristaltic pump.

The atomic absorption spectrometer was operated, with background correction, at the 283.3 nm lead resonance line, with a spectral bandwidth of 0.7 nm. The graphite furnace was heated continuously for 20 min periods at 1500°C by using the charring stage at the longest time setting. For optimal gas flow from the gas chromatograph, the quartz windows were removed from the furnace assemblies. The graphite furnace was operated with an internal flow of 40 ml min^{-1}. Because of the high carrier gas flow, the sensitivity was unaffected by the internal gas flow. The above optimal operating conditions were arrived at while using a gas chromatograph oven temperature of 150°C and a carrier gas flow rate of 140 ml min^{-1}.

Air and exhaust were sampled directly for organic lead compounds with four parallel traps maintained at - 72°C in a dry ice - methanol bath. Water was precondensed in U-tubes at - 15°C (sodium chloride in crushed ice). The flow through the traps averaged 70 ml min^{-1} over periods from 30 min (exhaust) to 18 h. Each trap was kept in dry ice until it was attached to a 4-way valve installed between the carrier gas inlet and the injection port of the gas chromatograph. The trap was immersed in boiling water. After 1 min, the volatilized fraction of the sample was introduced to the gas chromatographic column by diverting the carrier gas through the 4-way valve. The gas chromatographic oven was programmed from 50 to 200°C with a heating rate of 40°C min^{-1}. Mixed alkyl lead standards were added to blank traps which were run under identical conditions for calibration.

Air samples were introduced to the gas chromatographic - mass spectrometric system as described above. The chromatographic conditions were also identical to those used in the g.c. - a.a.s. analysis.

The detection limit of this method was found to be about 40 pg of lead for each compound, based on peak-height measurements. For a 70-1 air sample, 0.5 ng m^{-3} of each compound could be detected.

Radzuik et al[3771] showed that total atmospheric alkyl lead averaged 14 ng Pb m^{-3}. Vehicular exhaust fumes are an insignificant contributor to this total. Tetraethyllead, the only alkyl lead compound used in southern Ontario gasoline, is unstable in air. Besides decomposing, it reacts to give other alkyl lead compounds, which can be determined by the technique described. Evaporation of gasoline is almost exclusively the source of alkyl lead compounds in street air.

De Jonghe et al[3772,3773] have described a sampling system for the analysis by gas chromatography - atomic absorption spectrometry of alkyllead compounds in air. This method, when compared with many other published procedures is relatively rapid. Sampling periods of 1 h or less proved to be sufficient, even for the determination of alkyllead species in relatively non-polluted air. The major difficulty in collecting the compounds from air samples on GC column packing material is the condensation of moisture in the trap. Ice condensation on the column material leads to clogging of the pores and a sharp decrease in the air flow rate. The volume of air that can be sampled is therefore limited. Several previous workers pass the air before it enters the adsorption tube through a predeposition trap, to condense the excess of atmospheric water. By use of sampling apparatus with an empty U-tube at - 15°C, maximal reported volumes are about 75 L as water condensation on the adsorption tube is not completely prevented. An empty impinger held at - 78°C is still not capable of extending the sampling volumes above 200 L.

Jonghe et al[3772,3773] used a large U-tube filled with glass beads to achieve a more efficient water condensation trap. In this way the predeposition of water much improved as a result of a better cooling efficiency of the air. From preliminary experiments it appeared that this predeposition trap was really effective, only if temperatures of about -80°C or lower are used. However, at these low temperatures, a substantial fraction of the tetraalkylleads is retained also especially the less volatile species, whereas at higher temperatures also a rapid obstruction of the chromatographic adsorption tube occurs. Provided the trap is cooled down to sufficiently low temperatures to retain also the more volatile species dimethyldiethyllead, trimethylethyllead and tetramethyllead, it could therefore be used for the direct collection of the lead alkyl compounds. This allows much higher air flow rates than is possible with chromatographic adsorption tubes.

The air to be analyzed was passed at a flow rate of about 6 L/min for 1 h through a two-component collection system (Figure 212(a)). The first stage was a 47 mm Nuclepore membrane filter (0.4 μm) to remove the lead-containing particulates. The second stage was a cryogenic sampling trap for the collection of the volatile tetraalkyllead compounds. It consists of a U-shaped Pyrex tube (50 cm long by 25 mm i.d.) filled with glass beads of 4 mm diameter and immersed in a liquid nitrogen - ethanol slush bath at - 130°C. After sampling is completed, the U-tube remains in the slush bath until analyzed.

After the sample was obtained, the trapped alkyllead compounds were thermally desorbed from the large U-tube and transferred to a short adsorption tube (Figure 212 (b)) by connecting the sampling tube, still immersed in the slush bath, with a short glass column (26 cm long by 6 mm o.d. and 2 mm i.d.) packed with 0.2 g of 3% OV-101 on 100/120 mesh Gaschrom Q and kept in liquid nitrogen. While air was passed at a flow rate of 1 L/min, the U-tube was removed from the slush bath and allowed to warm slowly in air and then in a water bath at 60°C. With this treatment a rather constant air flow through the desorption system can be maintained, but near the end of the operation, when appreciable amounts of water start to evaporate out of the trap and condense on top of the adsorption tube, the flow rate decreases and desorption stops.

The adsorption tube was then attached to the four-port valve installed between the carrier gas inlet and the injection port of the gas chromatograph (Figure 212 (c)). The tube was immersed in a hot water bath at ca. 90°C, and the trapped sample was swept into the gas chromatograph by the carrier gas. Simultaneous with this injection, the gas chromatograph oven temperature program was initiated and the graphite furnace brought to 2000°C.

The reproducibility of measurements achieved by this method was better than 2% at the nanogram level. Detection limits defined as the ratio of three times the standard deviation of the noise and the sensitivity are shown in Table 319. They could be improved by extending the sampling period, although this depends on the relative air humidity as the cryogenic trap becomes ultimately saturated with ice. In many practical situations, the volume of air sampled could well be increased from 400 L to 1 m^3, which decreases the detection limits to well below 100 pg/m^3 per

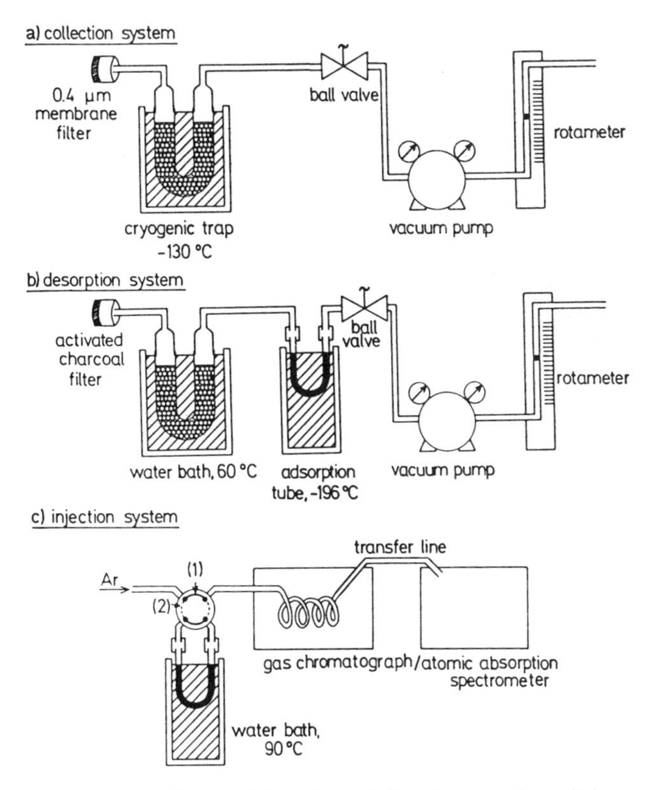

Figure 212 - Apparatus for collection (a), desorption (b), and injection (c) of tetraalkyllead compounds.

Table 319 - Response of GC/AAS system for tetraalkyllead compounds

Species	Retention time min	absolute detection limit, ng	detection limit for 1 h sampling, ng/m^3
TML	1.0	0.04	0.1
TMEL	1.7	0.06	0.2
DMDEL	2.7	0.06	0.2
MTEL	3.6	0.07	0.2
TEL	4.6	0.09	0.3

To demonstrate the analytical capability and sensitivity of this method De Jonghe[3772,3773] collected samples at sites which were relatively unpolluted. The concentrations measured at these locations are summarized in Table 320. The first site is an air-conditioned laboratory in which the gas chromatograph - atomic absorption apparatus is located and where the different experiments mentioned above were performed. The others are two office rooms in the same building, an outside location near the building, located in a residential area, and finally, a location in a shopping centre in the same area. The inside air corresponds closely with the outside air, whereas the shopping area is somewhat more affected by lead emissions from the automobile traffic.

Table 320 - Tetraalkyllead concentrations (ng/m^3) as determined with two identical sampling systems

Sampling site (date)	vol of air, L	alkyllead concn, ng/m^3					Total TAL, ng/m^3
		TML	TMEL	DMDEL	MTEL	TEL	
laboratory (02/01/80)	358	3.6	1.9	1.2	0.4	3.6	10.7
	330	4.0	2.5	1.5	0.2	4.1	12.3
office room 1 (02/05/80)	369	5.3	3.7	0.4	<0.2	<0.3	9.4
	378	5.9	3.3	0.4	<0.2	<0.3	9.6
office room 2 (02/11/80)	321	2.1	0.5	<0.2	<0.2	<0.3	2.6
	347	1.8	0.4	<0.2	<0.2	<0.3	2.2
residential area (02/04/80)	361	4.8	2.5	1.0	0.4	<0.3	8.7
	355	4.6	2.6	0.9	0.4	<0.3	8.5
shopping area (02/06/80)	208	22.7	4.3	1.3	0.7	3.0	32.0
	211	20.6	4.5	1.6	0.8	2.8	30.3

High performance liquid chromatography. Several researchers have used atomic absorption methods as detectors for high performance liquid chromatography[3774-3776]. These methods generally have many advantages over other techniques. However, interference caused by the eluent was still a problem, because ashing temperatures had to be kept as low as possible to avoid evaporation of organometallic compounds. Strictly speaking, furnace ashing should not be used for organometallic compounds because of their volatile nature. To overcome this problem Koizumi et al[3777] designed a new furnace that allows the atomic absorption detection of volatile organolead compounds eluting from a high performance liquid chromatographic column. They demonstrated the operation of this furnace by analyzing the eluent of a high pressure liquid chromatograph utilizing Zeeman atomic absorption spectrometry. The method was applied to the determination of lead in automobile exhaust.

The Zeeman atomic absorption technique has features that make it attractive for determinations of this kind[3778]. It can measure accurate values of atomic absorption signals with 500 - 1000 times larger background absorption[3779-3781]. The stability of the base line is also a desirable feature for long time measurement with high performance liquid chromatography[3782]. However, at times, there is a problem in the determination of organometallic compounds, even though the Zeeman technique provides excellent background correction. For example, the absorption signal of lead in alkyllead compounds cannot be observed with the conventional graphite furnace because highly volatile alkyllead escapes from the cuvette before atomization.

There are also problems that arise when the flame is used for atomization of alkyllead in gasoline which require special treatment of the sample to obtain accurate results. These problems arise because tetramethyllead produces a different absorbance value than does the same amount of tetraethyllead[3783]. It is much more difficult to get the same absorbance response from the various species in conventional furnace atomic absorption spectrometry than in flame atomic absorption spectrometry. Therefore, a different furnace is necessary to atomize the metal in organometallic compounds and for speciation studies in conjunction with high performance liquid chromatography.

One way to minimize the different response of different species is to dissociate the organometallic compound as completely as possible into individual atoms by heating to high temperatures. Conventional furnaces are not very effective because the gas temperature of the furnace is much lower than the wall temperature and the sample vapor easily diffuses out of the hot part of the cuvette before reaching a high temperature[3784,3785].

Figure 213 shows the cross section of a high temperature furance designed by Koizuma et al[3777]. This furnace consists of several separate parts: the sample cup, the thermal converter and reactor of porous graphite, a narrow hole (which at high temperatures also acts as a thermal converter) and the absorption cell. The sample vapor flows through the thermal converter and its temperature is raised sufficiently to decompose the compound and to atomize the metal. After that, the sample vapor is carried to the absorption cell for Zeeman atomic absorption spectrometry measurement.

In the case of the present furnace, the center portion of the cuvette is heated to a high temperature first because of the small heat capacity and the large electrical resistance. Conductive heating causes the porous graphite to be heated next and, finally, the tantalum cup is heated. A few seconds after the current is turned on, the temperature difference between

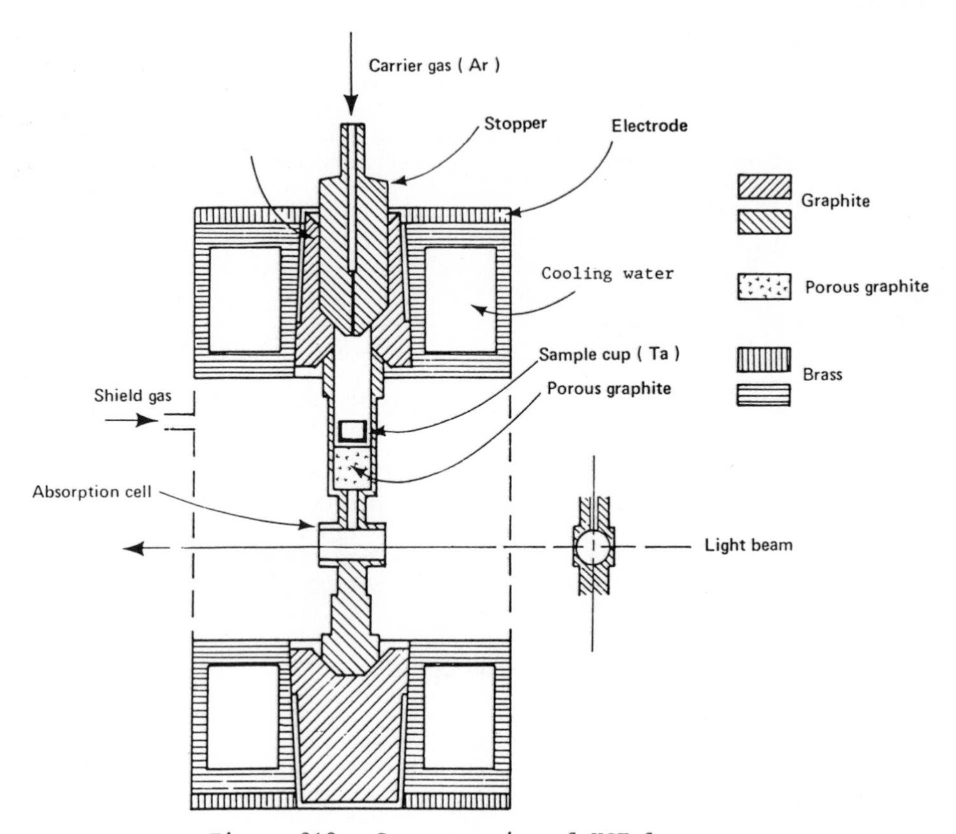

Figure 213 - Cross section of HGT furnace.

the three sections begins to decrease. Vaporized sample in the cup follows the flowing argon gas through the porous graphite where its temperature is raised by coming into intimate contact. It then passes through a small hole, the walls of which have the highest temperature. Because of intimate contact with these surfaces, the gas temperature becomes equal to the wall temperature before passing into the absorption cell. Using this equipment it is easy to raise the temperature of the cuvette to 2800°C by supplying about 5 kw of electrical power.

The system used to separate alkyllead compounds is described in Table 321. Methyl alcohol was used as the eluent. The pressure was about 30 kg/cm^2, and the flow rate was 0.67 mL/min. A sample of 10 μl was injected into the high performance liquid chromatograph while the flow was stopped. A 10 μL aliquot from each 250 μL portion of column effluent was intermittently introduced into the furnace.

To determine organomercury compounds in automotive exhaust gases, the gas was collected in a polyethylene bag and was forced to flow through the furnace carrier gas inlet port at a flow rate of 90 cm^3/min. The lead concentration in the exhaust gas was determined from the area under the absorption signal to be 0.3 ng/cm^3. Hence, this furnace is capable of on-line monitoring of trace elements in gas or ambient air.

Table 321 - Instrumentation for HPLC-ZAA system

HPLC
high performance liquid chromatograph, Hitachi M633, 0-350 kg/cm^3, 0.36 3.6 mL/min
column, Hitachi, 2.5 x 500 mm
resin, Hitachi Gel No. 3010
Furance
graphite, Ultra Carbon 0.5 in diameter
porous graphite, RVC 100 PPI porosity grade
furnace power supply, reactor controlled, 20 V, 700 A
ZAA spectrophotometer
light source, magnetically confined lamp (dc + rf (50 MHz))
magnet, permanent, 12 kg
variable retardation plate, 0 ~ $\lambda/2$, 30 Hz
polarizer, Rochon prism (quartz, optical contact)
monochromator, Hitachi M100 spectrophotometer
photomultiplier, Hamamatsu T.V., YA 7122
chopper, Bulova L2C, 1.0 KHz
Electronics
lock-in amplifier (including log convertor, AGC)
recorder, Honeywell Electronik 17

Miscellaneous. The American Conference of Government Industrial Hygienists[3786] has recommended values for maximum allowable concentrations of tetraethyllead and inorganic lead of 0.75 mg lead per 10 m^3 and 2.0 mg lead per 10 m^3 respectively.

Skalicka and Cejka[3787,3789] have compared three methods for the determination of tetraethyllead in air. The polarographic procedure involves trapping the tetraethyllead from the air sample in fuming nitric acid. A second method involves the use of the Unijet lead-in-air analyser. The preferred method involves spectrophotometry. In this method air is passed at 3 l per min through a tube containing finely crystallized iodine. The iodine is then dissolved in acidified potassium iodide solution and the solution is treated with potassium iodide-ammonium citrate-sodium sulphite-aqueous ammonia solution. The lead is extracted into a chloroform solution of dithizone and the extinction of the filtered extract is measured at 530 nm. This procedure is sensitive enough to detect down to 0.7 µg of lead in 5 ml of the final test solution. Snyder et al[3788] have also discussed the spectrophotometric determination of tetraethyllead in air using dithizone as the chromogenic reagent.

Moss and Browett[3789] have described a method for the determination of tetraethyllead in air in which the air sample is passed through an acidic solution of iodine monochloride to produce dialkyllead ions;

$$R_4Pb \rightarrow PbR_3^+ \rightarrow PbR_2^{2+} \rightarrow Pb^{2+}$$

These workers describe manual and automatic procedures for the determination of the amount of tetraalkyllead collected. The manual method involves reaction of the dialkyllead ions with dithizone at high pH and matching the colour of the dialkyllead dithizonate with a standard disc. In the automatic procedure, the dialkyllead is converted to the inorganic state before reaction with dithizone and colorimetric measurement as lead dithizonate.

The method measures lead-in-air concentrations down to 0.1 mg of lead per 10 m^3 of air, with sampling periods of at least 8 hours. A modified method based on a sampling period of 30 min, and having a sensitivity of 0.3 mg of lead per 10 m^3, is also described.

In Figure 214 are shown the absorption spectra of the orange coloured diethyllead dithizonate and the dimethyllead dithizonate complexes produced, respectively, by the reaction of dithizone with the products of the reaction of tetraethyllead and tetramethyllead and iodine monochloride.

Inorganic lead compounds may be present as dust in atmospheres that are monitored for tetraalkyllead vapour. Most of these compounds are soluble in the acid solution of iodine monochloride and would, therefore, interfere in the subsequent determination of dialkyllead ions derived from tetraethyllead or tetramethyllead. Filtration of the air under test, prior to contact with the iodine monochloride reagent, will avoid the possibility of this interference. Moss and Browett[3789] also describe how the solids collected on the filter may be analysed for lead content to give more complete results on the total toxic hazard of the atmosphere under test. Manual and automatic methods for determining tetraethyllead and tetraethyllead and elemental lead in the air are described.

The apparatus used by Moss and Browett[3789] for the filtration of the air sample is shown in Figure 215. To collect an air sample a Gelman glass-fibre filter paper, of 1 in diameter, type E, is placed in a Gelman open filter-holder which is then attached to the inlet of a Drechsel-type scrubber 100 ml of iodine monochloride reagent is measured into the scrubber which is then placed in the sampling position. The outlet of the scrubber is connected via a second scrubber containing sodium thiosulphate solution to a gas meter that is connected to a suitable vacuum system. Air is drawn through the filter and the iodine monochloride reagent at a rate of 3 ft^3 per h for a period of 8, 12 or 24 hours. The scrubber is disconnected from the sampling train and the total volume of air that has passed through it is noted.

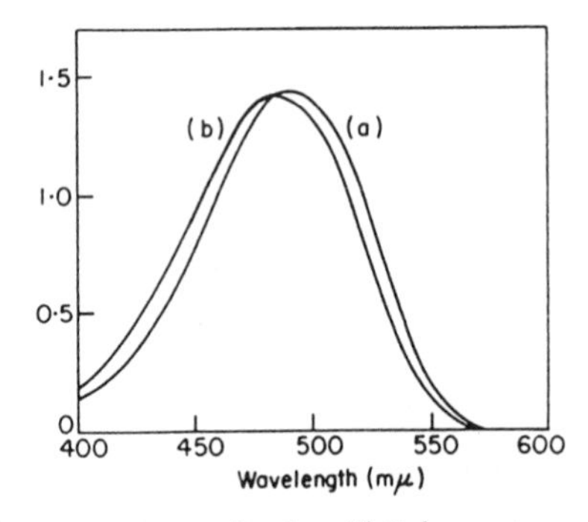

Figure 214 - Absorption spectra of the dithizonates of the products of reaction of iodine monochloride with tetraethyllead and with tetramethyllead at room temperature; curve A diethyllead dithizonate; curve B, dimethyllead dithizonate, in amounts equivalent to 100 μg of lead.

Organosilicon Compounds

Trichlorosilane (Frivoruchko[3790]) and alkyl chlorosilanes (Penegud and Kozlova[3791]) have both been estimated in air using methods based on the formation of molybdenum blue when organosilicon compounds react with ammonium molybdate. In the method for determining alkylchlorosilanes

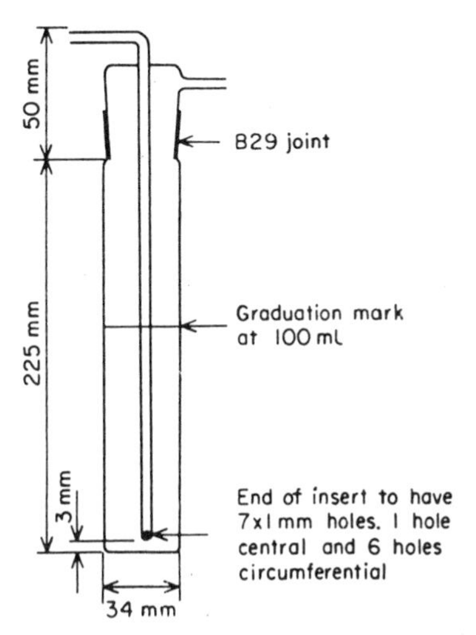

Figure 215 - Drechsel-type scrubber used to contain iodine monochloride reagent.

(Penegud and Kozlova[3791]) the air is passed through a sulphuric acid adsorbent, then the solution is evaporated, and the residue fuzed with potassium carbonate - sodium carbonate mixture prior to solution of the melt in dilute sulphuric acid and production of molybdenum blue by the addition of ammonium molybdate tartaric acid and ascorbic acid.

Krivoruchko[3792] described several methods for the estimation of nitrogen containing organosilicon compounds in air. In one method the compound is absorbed in dilute sulphuric acid and, after wet oxidation, the silicon is determined colorimetrically as ammonia using Nesslers reagent after sulphuric acid hydrolysis of the organosilicon compound. In a further method the organosilicon compound is oxidized with potassium bromate and the liberated bromine determined using the eosin method.

REFERENCES

1. D.F.Hagen, D.G.Biechler, W.D.Leslie and D.E.Jordan Anal.Chim.Acta, 41, 557 (1968).
2. T.R.Crompton, Chemical Analysis of Organometallic Compounds, Vol. V, Academic Press, London, New York, San Fransisco, 20, 115-122 (1977).
3. E.Wannier and A.Ringbom, Anal Chem., 12, 308 (1955).
4. F.Nydahl, Talanta,4, 1017 (1960).
5. G.S.Shvindherman and E.N.Zavadovskaya, Zavod. Lab., 31, 32 (1965).
6. C.Hennart, Chim. Anal., 43, 283 (1961).
7. D.E.Jordan and W.D.Leslie, Anal. Chim. Acta., 50, 161 (1970).
8. R.Belcher, B.Crossland and T.R.F.W.Fennell, Talanta, 17, 639 (1970).
9. M.P.Stukova and T.V.Kirillova, Zh. Anal. Khim., 21, 1236 (1966).
10. M.F.Smith and R.A.Bayer, Anal. Chem., 35, 1098 (1963).
11. N.E.Gel'man and I.I.Bryushkova, Zh. Anal. Khim., 19, 369 (1964).
12. M.O.Korshun, Anal. Abstr., 6, 2625 (1959).
13. E.L.Head and J.Holley, Anal.Chem., 28, 1172 (1956).
14. T.R.Crompton, Chemical Analysis of Organometallic Compounds, Vol. V, Academic Press, London, New York, San Fransisco, 20, 124-135 (1977).
15. E.Bonitz, Chem. Ber., 88, 742 (1955).
16. K.Ziegler, Justus Liebigs Ann. Chem., 589, 91 (1954).
17. T.R.Crompton and V.W.Reid, Analyst (London), 88, 713 (1963).
18. T.R.Crompton, Anal. Chem., 39, 1464 (1967).
19. R.Dijkstra and E.A.M.Dahmen, Z. Anal. Chem., 181, 399 (1961).
20. Stauffer Chemical Company, Anderson Chemical Division, Exclusive Sales Agent, Weston, Michigan, USA, Alkyls Bulletin, Triethylaluminium Analytical Methods,T.68 5-4 and T.68 5-5 (1959).
21. B.J.Phillip, W.L.Mundry and S.C.Watson, Anal. Chem., 45, 2298 (1973).
22. R.Z.Lioznova and M.L.Genusov, Zavod. Lab., 26, 945 (1960).
23. W.P.Neumann, Justus Liebigs Ann. Chem., 629, 23 (1960).
24. T.R.Crompton, Chemical Analysis of Organometallic Compounds, Vol. V, Academic Press, London, New York, San Fransisco, 20, 186-199 (1977).
25. T.R.Crompton, Chemical Analysis of Organometallic Compounds, Vol. V, Academic Press, London, New York, San Fransisco, 20, 206-213 (1977).
26. T.R.Crompton, Analyst (London), 86, 652 (1961).
27. A.I.Vogel, Elementary Practical Organic Chemistry, Part 111, Quantitative Organic Analysis, Longmans, Green & Co., (Fig. XIV,3.3) 655 (1958).
28. V.W.Reid and R.K.Truelove, Analyst (London), 77, 325 (1952).
29. V.W.Reid and D.G.Salmon, Analyst (London), 80, 704 (1955).
30. E.A.Bondarevskaya, S.V.Syavtsillo and R.N.Potsepkina, Zh. Anal. Khim., 14, 501 (1959); Chem. Abstra., 54, 9611f (1959).
31. E.A.Bondarevskaya, S.V.Syavtsillo and R.N.Potsepkina, Tr. Kom. Anal. Khim. Akad. Nauk SSSR, 13, 178 (1963).
32. G.A.Belova, A.R.Kozlova and T.N. Boltunova, Prom. Sintet. Kauch., 4, 19 (1967).
33. S.Mitev, N.Damyanov and P.K.Ajasyan, Zh. Anal. Khim., 28, 821 (1973).
34. T.R.Crompton, Chemical Analysis of Organometallic Compounds, Vol. V, Academic Press, London, New York, San Fransisco, 20,222-232 (1977).
35. T.R.Crompton, Chemical Analysis of Organometallic Compounds, Vol. V, Academic Press, London, New York, San Fransisco, 20, 232-235 (1977).
36. T.R.Crompton, Chemical Analysis of Organometallic Compounds, Vol. V, Academic Press, London, New York, San Fransisco, 20, 235-238 (1977).

37. J.Jenik, Chem. Listy, 55, 509 (1961).
38. M.Juracek and J.Jenik, Sb. Ved. Prac. Vysoke Skoly Chem. Technol. Pardubice, 105 (1959).
39. Y.Kinoshita, J. Pharm. Soc. Jap., 78, 315 (1958).
40. H.Bieling and K.H.Thiele, Z. Anal. Chem., 145, 105 (1955).
41. A.P.Terent'ev, M.A.Volodina and E.G.Fursova, Dokl. Akad. Nauk SSSR, 109, 851 (1966).
42. K.M.Orr and B.B.Sithole, Analyst, 107, 1212 (1982).
43. M.Corner, Analyst (London), 84, 41 (1959).
44. R.Belcher, A.M.G.Macdonald and T.S.West, Talanta, 1, 408 (1958).
45. W.Merz, Mikrochim Acta, 640 (1959).
46. K.Eder, Mikrochin, Acta, 471 (1960).
47. W.H.Guttermann, S.F.John, J.E.Barry, D.L.Jones and E.D.Lisk, J. Agric. Food Chem., 9, 50 (1961).
48. R.Puschel and Z.Stefanac, Mikrochim. Acta, 6,1108 (1962).
49. Z.Stefanac, Mikrochim. Acta., 6, 1115 (1962).
50. R.Belcher, A.M.G.Macdonald, S.E.Phang and T.S.West, J. Chem. Soc., 2044 (1965).
51. A.D.Wilson and D.T.Lewis, Analyst (London), 88, 510 (1963).
52. H.Kashiwagi, Y.Tukamoto and M.Kan, Ann. Rep. Takeda Res. Lab., 22, 69 (1962).
53. G.Bahr, H.Bieling and K.H.Thiele, Z. Anal. Chem., 143, 103 (1954).
54. R.Pietsch, Z. Anal. Chem., 144, 353 (1955).
55. C.DiPietro and W.A.Sassaman, Anal. Chem., 36, 2213 (1964).
56. A.Steyermark, Quantitative Organic Microanalysis, 2nd Ed. Academic Press, New York, London, 367 (1961).
57. M.M.Tuckerman, J.H.Hodecker, B.C.Southworth and K.D.Fliescher, Anal. Chim.Acta., 21, 463 (1959).
58. M.Jean, Anal. Chim. Acta., 14 172 (1956).
59. V.J.Neulenhoff, Pharm. Weekbi., 100, 409 (1968).
60. M.Juracek and J.Jenik, Collect. Czech. Chem. Commun., 20, 550 (1955).
61. M.Jutacek and J.Jenik, Collect. Czech. Chem. Commun., 21, 890 (1956).
62. J.S.Edmonds and K.A.Francesconi, Anal. Chem., 48, 2019 (1976).
63. D.E.Fleming and G.A.Taylor, Analyst, 102, 101 (1977).
64. R.B.Denyszyn, P.M.Grohe, D.E.Wagoner, Anal. Chem., 50, 1094 (1978).
65. A.G.Howard, M.H.Arbab-Zavar, Analyst, 106, 213 (1981).
66. M.Juracek and J.Jenik, Sb. Ved. Prac. Vysoke Skoly Chem. Technol. Pardubice, 105 (1959).
67. L.Maier, D.Seyferth, F.G.A.Stove and B.G.Rochow, J. Am. Chem. Soc., 79, 5884 (1957).
68. H.Trutnovsky, Mikrochim. Acta, 3, 499 (1963).
69. G.Bahr, H.Bieling and K.H.Thiele, Z. Anal. Chem., 145, 105 (1955).
70. A.P.Terent'ev, M.A.Volodina and E.G.Fursova, Dokl. Akad. Nauk SSSR. 169, 851 (1966).
71. E.L.Head and E.Holley, Anal. Chem., 28, 1172 (1956).
72. R.Schmitz, Dtsch. Apoth.-Zig., 100, 693 (1960).
73. R.C.Rittner and R.Culmo, Anal. Chem., 34, 673 (1962).
74. S.Mizukami and T.Ieka, Michrochem. J., 7, 485 (1963).
75. G.Kainz and G.Chromy, Mikrochim. Acta., 16, 1140 (1966).
76. F.Martin and A.Floret, Chim. Anal., 41, 181 (1959).
77. D.G.Shaheen and R.S.Braman, Anal. Chem., 33, 893 (1961).
78. I.Dunstan and J.V.Griffiths, Anal. Chem., 33, 1598 (1961).
79. S.Kato, K.Kimura and Y.Tsuzuki, J.Chem. Soc. Jap. (Pure Chem. Sect.) 83, 1039 (1962).
80. R.H.Pierson, Anal.Chem., 34, 1642 (1962).
81. R.D.Strahm and M.F.Hawthorne, Anal. Chem., 32, 530 (1960).
82. P.Arthur and W.P.Donahoo, U.S. Atom. Energy Comm. Rep., CCC-1024-TR-221 (1957).

83. J.A.Kuck and E.C.Grim, Microchem. J., 3, 35 (1959).
84. I.M.Kolthoff, PH and Electrotitations, Wiley, New York, 120-122 (1931).
85. H.Allen and S.Tannenbaum, Anal. Chem., 31, 265 (1959).
86. J.J.Bailey and D.G.Gehring, Anal. Chem., 33, 1760 (1961).
87. T.M.Shanina, N.E.Gel'man and V.S.Mikhailovskaya, Zh. Anal. Khim., 22, 732 (1967).
88. S.K.Yasuda and R.N.Togers, Microchem. J., 4, 155 (1960).
89. S.I.Obtemperanskaya and V.N.Likhosherstova, Vestn. Mosk. Univ., 1, 57 (1960).
90. M.Corner, Analyst (London), 84, 41 (1959).
91. R.Wickbold, Angew. Chem., 69, 530 (1957).
92. R.Wickbold and F.Nagel, Angew. Chem., 71, 405 (1959).
93. B.E.Buell Anal. Chem., 30, 1514 (1958).
94. T.Yoshizaki, Anal. Chem., 35, 2177 (1963).
95. A.Shah, A.A.Qadri and R.Rehana, Pak. J. Sci. Ind. Res., 8, 282 (1965).
96. M.Giorgini and A.Lucchesi, Ann. Chim. (Rome), 54, 832 (1964).
97. L.Erdev, E.Gegus and E.Kocsis, Acta. Chim. Acad. Sci. Hung., 7, 343 (1955).
98. N.C.Malysheva, L.P.Starchik, I.S.Panidi and Ya, M.Paushkin, Zh. Anal. Khim., 18, 1367 (1963).
99. J.M.Thoburn, Diss. Abstr., 15, 3-4 (1955).
100. N.G.Nadeau, D.M.Oaks and R.D.Buxton, Anal. Chem., 33, 341 (1961).
101. H.Noth and H.Beyer, Chem. Ber., 93, 928 (1960).
102. G.E.Ryschkewitsch and E.R.Birnbaum, Inorg. Chem., 4. 575 (1965).
103. H.C.Kelly, Anal. Chem., 40, 240 (1968).
104. R.C.Rittner and R.Culmo, Anal. Chem., 35, 1268 (1963).
105. P.Arthur, R.Annino and W.P.V.Donahoo, Anal. Chem., 29, 1852 (1957).
106. R.F.Putnam and H.W.Myers, Anal. Chem., 34, 486 (1962).
107. S.A.Greene and H.Pust, Anal. Chem., 30, 1039 (1958).
108. A.F.Zhigach, E.B.Kazakova and R.A.Kigel, Zh. Anal. Khim., 14, 746 (1959).
109. I.Lysyj and R.C.Greenough, Anal. Chem., 35, 1657 (1963).
110. D.G.Newman and O.Tomlinson, Mikrochim. Ichnoanal. Acta., 1023 (1964).
111. E.A.Kalinovskaya and L.S.Sil'vestrova, Zavod. Lab., 34, 30 (1968).
112. J.C.Booker and T.L.Isenhour, Anal. Chem., 41, 1705 (1969).
113. O.Palade, Zh. Anal. Khim., 21, 377 (1966).
114. S.A.Bartkiewicz and E.A.Hammatt, Anal. Chem., 36, 833 (1964).
115. M.A.H.Sharif, Pak. J. Sci. Ind. Res., 1, 160 (1958).
116. O.Kidson and O.Eskew, N.Z. J. Sci. Technol., B21, 178 (1940).
117. R.K.Mitra, P.C.Bose, G.K.Ray, B.Mukerji, Indian J. Pharm., 24, 152 (1962).
118. D.S.Kinnory, E.Kaplan, Y.J.Oester and A.A.Imperato, J.Lab. Clin. Med., 50, 913 (1957).
119. E.C.Abd, A.A.Raheem, M.M.Dokhana, Z. Anal. Chem., 189, 389 (1962).
120. O.Menyharth, Acta. Chim. Hung., 41, 195 (1964).
121. V.Brustier, A.Castaigne, E.de Montety, A.Anselem, Am. Pharm. Franc., 22, 373 (1964).
122. G.Parissakis, P.B.Issopoulos, Pharm. Acta., 40, 653 (1965).
123. J.Konechy and J.Talgyessy, Chemicke Zvesti, 20, 692 (1966).
124. T.V.Reznitskaya and E.I.Burtsera, Zh. Anal. Khim., 21, 1132 (1966).
125. K.L.Cheng, Michrochem. J., 7, 100 (1963).
126. W.L.Wuggatzer and J.E.Christian, J. Am. Pharm. Assoc. Sci. Ed., 44, 645 (1955).
127. G.Sebor, G.I.Lang, D.Kolihova and O.Weisser, Analyst, 107, 1350 (1982).
128. T.Arany and A.Erden, Magy. Kem. Lab., 20. 164 (1965).
129. V.A.Klimova, M.O.Korshun and E.G.Bereznitskaya, Zhur. Anal. Khim., 11, 223 (1956); Anal. Abstra., 3, 3663 (1956).
130. M.P.Brown and G.W.A.Fowles, Anal. Chem., 30, 1689 (1958).

131. V.A.Klimova and N.D.Vitalini, Zh. Anal. Khim., 19, 1254 (1963); Anal. Abstra., 10, 4716 (1964).
132. R.P.Kreshkov and E.A.Kucharev, Zavod. Lab., 32, 558 (1966).
133. M.Schulunz and A.Koster-Pflugmacher, Z. Anal. Chem., 232, 93 (1967).
134. R.S.Braman and M.A.Tompkins, Anal. Chem., 50, 1088 (1978).
135. D.T.Burns and D.Dadgar, Analyst April,452 (1982).
136. H.Pieters and W.J.Buis, Mickrochem. J., 8, 383 (1964).
137. H.H.Anderson, J. Org. Chem., 21, 869 (1956).
138. V.A.Klimova, K.S.Zabrodina and N.L.Shitikova, Izv Akad Nauk SSSR, Ser. Khim, 1. 178 (1965).
139. J.S.Fritz, G.H.Schlenk, Anal. Chem., 31, 1808 (1959).
140. P.Dostal, J.Cermak, B.Novotna, Collection Czech. Chem. Commun., 30, 34 (1965).
141. J.A.Magneson, Anal. Chem., 35, 1487 (1963).
142. A.Berger and J.A.Magnuson, Anal. Chem., 36, 1156 (1964).
143. J.A.Magnusson and E.W.Knaub, Anal. Chem., 37, 1607 (1965).
144. T.R.F.W.Fennell and J.R.Webb, Talanta, 9, 795 (1962).
145. E.A.Terent'eva and T.M.Malolina, Zh. Anal. Khim., 19, 353 (1964).
146. H.M.Rosenberg and C.Riber, Microchem. J., 6, 103 (1962).
147. J.Jenik and F.Renger, Collect. Czech. Chem. Commun., Engl. Ed., 29, 2237 (1964).
148. J.Decombe and J.P.Ravoux, Bull. Soc. Chim. Fr., 17, 260 (1964).
149. L.V.Myshlyaeva, V.V.Krasnoshchekov, T.G.Shatunova and I.V.Sedova, Tr. Mosk. Khim. Tekhnol. Inst., 49, 178 (1965); Ref. Zh. Khim., 19GD, Abstr. 18G254 (1966).
150. R.Renger and J.Jenik, Sb. Ved. Prac. Vysoke Skoly Chem. Technol. Pardubice, 1, 55 (1963).
151. E.N.Kuchkarev, Tr. Mosk. Khim. Tekhnol. Inst., 48, 51 (1965); Ref. Zh. Khim., 19GD, Abstra. 11G230 (1966).
152. J.Fischl, Clin. Chim. Acta., 4, 686 (1959).
153. D.G.Nathan, T.G.Gabuzda and F.H.Gardner, J. Lab. Clin. Med., 62, 511 (1963).
154. J.Fine, J. Clin. Pathol., 14, 561 (1961).
155. S.Natelson and B.Shied, Clin. Chem., 7, 115 (1961).
156. L.C.Willemsens and G.J.M.Van der Kerk, Investigations in the Fields of Organolead Chemistry, Institute for Organic Chemistry TNO, Utrecht, Published by the International Lead-Zinc Research Organization, New York, 87 (1965).
157. R.P.Kreshkov and E.A.Kucharev, Zavod. Lab., 32, 558 (1966).
158. F.B.Ashbel, A.M.Parshina, M.S.Goizman, L.T.Zhizhina and K.M.Zuptsova, Zavod. Lab., 31, 1062 (1965).
159. G.R.Sirota and J.F.Uthe, Anal. Chem., 49, 823 (1977).
160. J.W.Robinson, Anal. Chim. Acta., 24, 451 (1961).
161. D.J.Trent, At. Absorpt. Newsl., 4, 348 (1965).
162. R.M.Dagnall and T.S.West, Talanta, 11, 1553 (1964).
163. H.W.Wilson, Anal. Chem., 38, 920 (1966).
164. A.C.Menzies, Anal. Chem., 32, 905 (1960).
165. C.S.Rann, The Element No. 9, Aztec Instruments, Westport, Conn., USA (1965).
166. R.A,Mostyn and A.F.Cunningham, J. Inst. Pet. 53, 101 (1967).
167. E.J.Moore and J.R.Glass, Microchem. J., 10, 148 (1966).
168. K.Campbell and R.Moss, J. Inst. Pet., 53, 521, 194 (1967).
169. P.R.Ballinger and I.M.Whittemore, Proceedings of the American Chemical Society Division of Petroleum Chemistry, Atlantic City Meetings, 13, 133 Sept. 1968 (1968).
170. B.Kolb, G.Kemnner, F.H.Schleser and E.Wiedeking, Z. Anal. Chem., 221, 166 (1966).
171. Americal Society for Testing and Materials, ASTM Standards on Petroleum Products and Lburicants, Methods D1269-53T (1963) and D1269-61 (1961).

172. Institute of Petroleum (1966) Standards for Petroleum and its products 25th edition.
(a) Lead in gasoline 1 P 96/64, Part 1, Section 1, P.374.
(b) Tetraethyllead in gasoline, 1 P 116/64, Part 1, Section 1, P.442.
(c) Separation of TEL and TML in gasoline, 1 P 188/66 T, Part 1, Section 11, P.837.
Also, 1965 Book of ASTM Standards, Part 17.
(a) ASTM D 526/61 (1 P 96/64) P.272.
(b) ASTM D 526/61 (1 P 188/66T) P.849.
(c) Lead anti-knock in gasolines, ASTM D1269/61, P.578.
173. W.Meine, Erdol Kohle, 8, 711 (1955).
174. W.Linne and H.D.Wulfken, Erdol Kohle, 10, 757 (1957).
175. J.Van Rysselberg and R.Leysen, Nature (London), 189, 478 (1961).
176. T.Okada, T. Ueda and T.Kohzuma, Bunko Kenkyu, 4, 30 (1956); Chem.Abs., 53, 65866 (1959).
177. P.Buncak, Ropa Uhlie, 8,148 (1966).
178. F.W.Lamb, O.O.Niebylski and E.W.Kiefer, Anal. Chem., 27, 129 (1955).
179. C.Campo, Chim. Anal., 45, 343 (1963).
180. L.Vajta and M.Moser, Period. Polytech., 9. 275 (1965).
181. E.L.Gunn, Appl. Spectrosc., 19, 99 (1965).
182. S.Szeiman, Proc. VII Hungarian Symp. Emission Spectrogr., Pecs, Hungary, August 1964, pp 564-565; Magy, Kem Foly., 70, 511 (1964).
183. G.S.Smith, Chem. Ind. (London), 22, 907 (1963).
184. G.D.Christofferson and B.Y.Beach, Colloq. Spectros. Int. 19th, Lyons, 3, 492 (1961).
185. A.Ferro and C.P.Galotto, Ann. Chim. (Rome), 45, 1234 (1955).
186. J.F.Brown and R.J.Weir, J. Sci. Instrum., 33, 222 (1956).
187. G.Calingaert, F.W.Lamb, H.L.Miller and G.E.Noakes, Anal. Chem., 22, 1238 (1950).
188. H.K.Hughes and F.P.Hochgesang, Anal. Chem., 22, 1248 (1950).
189. M.Farkas and P.Fodor, Magy. Kem. Foly., 69, 407 (1963).
190. F.B.Ashbel, A.M.Parshina, A.M.Goizman, M.S.Zhizhina and K.M.Kuptsova, Zavod Lab., 31, 1062 (1965)
191. H.E.Howard, W.C.Ferguson and L.R.Snyder, Anal. Chem., 32, 1814 (1960).
192. American Society Testing Materials (1956) and (1961) ASTM Standards on Petroleum Products and Lubricants - methods D 526-56 (1956) and D526-61 (1961).
193. American Society for Testing and Materials, ASTM Standards on Petroleum Products and Lubricants, Methods D526-56 (1956) and D526-61 (1961).
194. American Society for Testing Materials, Materials Research and Standards Committee D2, 2, 494 (1962).
195. G.Nagypataki and Z.Tamasi, Magy. Kem. Lapja, 17, 140 (1962).
196. J.Krotky, Paliva, 36, 124 (1956).
197. J.M.Lopez, Colloq. Spectrosc. Int., 7th Sept. 1958, University of Rev. Univ. Mines. Mentall. Mec., 15, 299 (1959).
198. L.C.Willemsens, Organolead Chemistry, Institute for Organic Chemistry TNO, Utrecht, Published by the International Lead-Zinc Research Organization, New York, (1964) p. 69.
199. D.Colaitis and M.Lesbre, Bull. Soc. Chim. Fr., 1069 (1952).
200. R.N.Meals, J. Org. Chem., 9, 211 (1944).
201. A.S.Zabrodina and U.P.Miroshino, Vestn. Mosk. Univ., 2,195 (1957).
202. E.L.Head and C.E.Holley, Anal. Chem., 28, 1172 (1956).
203. J.B.Honeycutt (Ethyl Corp.), US. Pat., 3 059 036 Oct. 16th (1962).
204. H.L.Johnson and R.A.Clark, Anal. Chem., 19, 869 (1947).
205. E.C.Juenge and D.Seyferth, J. Org. Chem., 26, 563 (1961).
206. H.Sinni and D.Aumann, Makromol. Chem., 57, 105 (1962).
207. H.Gilman, Bull. Soc. Chim. Fr., 1356 (1963).

208. R.Reimschneider and K.Petzoldt, Z. Anal. Chem., 176, 401 (1960).
209. R.Reinschneider and K.Petzoldt, Z. Anal. Chem., 193, 193 (1963).
210. T.Mitsui, K.Yoshikawa and Y.Sakai, Microchem. J., 7, 160 (1963).
211. T.Mitsui, O.Yamamoto and K.Yoshikawa, Mikrochim. Acta, 521 (1961); Anal. Abstra., 9, 174 (1962).
212. R.Belcher, D.Gibbons and A.Sykes, Mikrochem. Mitrochim. Acta., 40, 76 (1952).
213. A.Sykes, Mikrochim. Acta., 1155 (1956).
214. T.Sudo and D.Shinoe, Bunseki Kagaku, 4, 88 (1955).
215. V.Pechanec and J.Horacek, Collect. Szech. Chem. Commun., 27, 239 (1962).
216. F.Martin and A.Floret, Bull. Soc. Chim. Fr., 4, 610 (1960).
217. A.M.G.Macdonald, Ind. Chem., 35, 33 (1959).
218. A.M.G.Macdonald, Analyst (London), 86, 3 (1961).
219. B.C.Southworth, J.H.Hodecker and K.D.Fleischer, Anal. Chem., 30, 1152 (1958).
220. P.Gouveneur and W.Hoedeman, Anal. Chim. Acta., 30, 519 (1964).
221. R.Wickbold, Z. Anal. Chem., 152, 262 (1956).
222. H.Roth and W.Beck in Quantitative Organische Mikroanalyse (Ed. F.Pregl and H.Roth) 7th ed. Springer-Verlag, Vienna, 184 (1958).
223. R.Donver, Z. Chem., 5, 466 (1965).
224. C.Vickers and J.V.Wilkinson, J. Pharm. Pharmacol., 13, 156T (1961).
225. C.A.Johnson and C.Vickers, J. Pharm. Pharmacol., 11, 218r (1959); Anal. Abstr., 7, 329 (1960).
226. K.Kamen, Chem. Listy., 47, 1008 (1953).
227. K.Shigeo, Microchem. J., 8, 79 (1964).
228. H.Romanowski, Chem. Anal. (Warsaw), 11, 1027 (1966).
229. M.Wronski, Chem. Anal. (Warsaw), 5, 601 (1960).
230. J.C.Merodio, An. Asoc. Quim. Argent., 49, 225 (1961).
231. D.Pirtea and I.Albescu. Acad. R.P.R. Stud. Cercet. Chim., 7,137 (1959).
232. H.F.Walton and H.A.Smith, Anal. Chem., 28, 406 (1956).
233. G.Spacu and G.Suciu, Z. Anal. Chem., 78, 244 (1929).
234. L.Manger, Kem. Ind., 14, 317 (1965).
235. A.Dondrio and C.Molina, Inf. Quim. Anal., 19, 77 (1965).
236. T.Pelczar and J.Weyers, Diss. Phar. (Krakow), 13, 243 (1961).
237. B.Hetnarski and K.Ketnarska, Roczn. Chem., 34, 457 (1960).
238. A.Kondo, Bunseki Kagaku, 10, 658 (1961).
239. H.Takehara, T.Takeshita and I.Hara, Bunseki Kagaku, 15, 332 (1966).
240. U.H.Chambers, F.H.Cropper and H.Crossley, J. Sci. Food Agric., 7, 17 (1956).
241. C.A.Johnson, Mfg. Chem., 36, 72 (1965).
242. F.A.J.Armstrong and J.F.Uthe, Atom. Absorpt. Newsl., 10, 101 (1971).
243. J.F.Uthe, F.A.J.Armstrong and M.P.Stainton, J. Fish Res. Board Can., 27, 805 (1970).
244. J.F.Uthe, F.A.J.Armstrong and K.C.Tam, J. Assoc. Off. Anal. Chem., 54, 866 (1971).
245. D.T.Burns, F.Glocking, V.B.Mahale and W.J.Swindan, Analyst (London), 103, 985 (1978).
246. I.M.Davies, Anal. Chim. Acta., 102, 189 (1978).
247. K.Minagania, Y.Takizawa and I.Kifune, Anal. Chim. Acta., 115, 103 (1980).
248. R.Bye, P.E.Paus, Anal. Chim. Acta., 107, 169 (1970).
249. R.Dumarey, R.Heindryck, R.Dams and J.Hoste, Anal. Chim. Acta., 107, 159 (1979).
250. G.J.C.Shum, H.E.Freeman and J.F.Uthe, Anal. Chem., 51, 414 (1979).
251. A.I.Lebedeva and K.S.Kramer, Izv. Akad. Nauk SSSR, Otd. Khim. Nauk, 7, 1305 (1962).
252. M.O.Korshun and N.S.Sheveleva, Zh. Anal. Khim., 7,104 (1952).
253. A.I.Lebedeva and E.F.Fedorova, Zh. Anal. Khim., 16, 87 (1961).

254. A.A.Abramyan and A.A.Kocharyan, Izv. Akad. Nauk Arm. SSR. Khim. Nauk, 20, 515 (1967).
255. I.F.Holmes and A.Lander, Analyst (London), 90, 307 (1965).
256. A.Verdino, Mikrochemie, 6, 5 (1928).
257. A.Verdino, Mikrochemie, 9, 123 (1931).
258. F.Hernlet, Mikrochemie, Pregl Festschrift, 154 (1929).
259. M.Furter, Mikrochemie, 9, 27 (1931).
260. G.Ingram, J. Soc. Chem. Ind. (London), 61, 112 (1942).
261. G.Ingram, J. Soc. Chem. Ind. (London), 58, 34 (1939).
262. L.Synek and M.Vecera, Chim. Listy, 51, 1551 (1957).
263. M.Vecera, M.Vojtech and L.Synek, Collect. Czech. Chem. Commun., 25, 93 (1960).
264. Y.A.Gawargious and A.M.G. Macdonald, Anal. Chim. Acta., 27, 119 (1962).
265. R.Belcher and G.Ingram, Anal. Chim. Acta., 4, 119 (1950).
266. V.Pechanec and J.Horacek, Collect. Czech. Chem. Commun., 27, 232 (1962).
267. V.Pechanec, Collect. Czech. Chem. Commun., 27, 2976 (1962).
268. V.Pechanec, Collect. Czech. Chem. Commun., 27, 2009 (1962).
269. V.Pechanec, Collect. Czech. Chem. Commun., 27, 1817 (1962).
270. V.Pechanec, Collect. Czech. Chem. Commun., 27, 1702 (1962); Anal. Abstra., 10, 1054 (1963).
271. V.Pechanec and J.Horacek, Chim. Anal., 46, 457 (1964).
272. A.A.Abramyan and R.A.Megroyan, Arm. Khim. Zh., 21, 115 (1968).
273. M.O.Korshun, N.S.Sheveleva and N.E.Gel'man, Zhur. Anal. Chim., 15, 99 (1960).
274. M.O.Korshun and N.E.Gel'man "New methods of elementary microanalysis" State University Publication, Moscow, Leningrad, USSR (1949).
275. A.I.Lebedeva and N.A.Nikolaeva, Izv. Akad. Nauk SSSR. Ser. Khim., 10, 1867 (1965).
276. A.I.Lebedeva and N.A.Nikolaeva, Zh. Anal. Khim., 18, 984 (1963).
277. C.E.Oda, J.D.Ingle, Anal. Chem., 53, 2305 (1981).
278. A.B.Farag, M.E.Attla and H.N.A.Hassan, Anal. Chem., 52, 2153 (1980).
279. T.T.Gorsuch, "Treatise on Analytical Chemistry", I.M.Kolthoff, P.J. Elving., Eds; Interscience: New york (1965); Vol. 12 Part 11, p.317-324.
280. M. Boetius, Prakt. Chem., 151, 279-306 (1938).
281. R.Belcher, D.Gibbons and A.Sykes, Mikrochim. Acta., 40, 76-103 (1952).
282. L.T.Hallet, Ind. Eng. Chem. Anal. Ed., 14, 956-993 (1942).
283. M.O.Korshun and E.V. Lavrovskaya, Zh. Anal.Khim., 3, 322-328 (1948): Chem. Abstra., 43, 8955d (1949).
284. A.Verdino, Mikrochemie, 6, 5-12 (1928).
285. T.F.Holmes and A.Lauder, Analyst (London), 90, 307-308 (1965).
286. R.Belcher and G.Ingram, Anal. Chim. Acta., 4, 118-129 (1950).
287. V.Pechanec, J.Horacek, Collect. Czech Chem. Commun., 27, 232-238 (1962).
288. A.L.Lebedeva and K.Sh.Kramer, Izy. Akad. Nauk SSSR, Ser. Khim., 1305-1307 (1962); Chem. Abstra., 57, 15805f (1962).
289. V.Grignard and A.Abelmann, Bull. Soc. Chim. Fr., 19, 25-27 (1916).
290. G.J.Ingram, Soc. Chem. Ind., London, 61, 112-115 (1942).
291. M.O.Korshun, N.S.Sheveleva and N.E.Gel'man, Zh. Anal. Khim., 15, 99-103 (1960): Chem. Abstr., 54, 13946d (1960).
292. M.O.Korshun and V.A.Kilmova, Zh. Anal. Khim., 2, 274-280 (1947): Chem. Abstra., 43, 6936h (1949).
293. V.A.Kilmova and T.A.Antipova, Zh. Anal. Khim., 16, 465-468 (1961): Chem. Abstra., 56, 5402i (1962).
294. M.Vecera, F.Vojtech and L.Synek, Collect. Czech. Chem. Commun., 25, 93-96 (1960).
295. D.T.Burns, F.Glockling, V.B.Mahale and W.J.Swindall, Analyst (London) 103, 985 (1978).

296. G.Constantinides, G.Arich and C.Lomi, Chim. Ind. (Milan), 41, 861(1959).
297. C.W.Dwiggins and H.N.Dunning, Anal. Chem., 32, 1137 (1960).
298. M.P.Stukova, I.I.Kashiricheva and A.A.Lapshova, Zhr. Anal. Khim., 22, 1110(1967).
299. L.J.Kanner, E.D.Slesin and L.Gordon, Talanta, 7, 288 (1961).
300. F.R.Hartley and J.L.Wagner, J. Organomet. Chem., 42, 477 (1972).
301. F.E.Beamish and J.Dale, Ind. Eng. Chem. Anal. Ed., 10, 697 (1938).
302. J.D.Burton and J.P.Riley, Analyst (London), 80, 391 (1955).
303. W.J.Kirsten and M.E.Carlsson, Microchem. J., 4, 3 (1960).
304. W.I.Stephen, Ind. Chem., 37, 86 (1961).
305. T.Salvage and J.P.Dixon, Analyst (London), 90, 24 (1965).
306. P.M.Saliman, Anal. Chem., 36, 112 (1964).
307. R.A.Apodacu, Quim. Ind. (Bilbao), 9, 167 (1962).
308. A.J.Christopher and T.R.F.W.Fennell, Microchem. J., 12, 593 (1967).
309. T.R.F.W.Fennell and J.R.Webb, Microchem. J., 10, 456 (1966).
310. H.Levine, J.J.Rowe and F.S.Grimardi, Anal. Chem., 27, 258 (1955).
311. T.R.F.W.Fennell, M.W.Roberts and J.R.Webb, Analyst (London), 82,639 (1957).
312. O.B.Michelsen, Anal. Chem., 29 60 (1957).
313. C.Whalley, Hydrogen Peroxide as an Analytical Reagent. In "Analytical Chemistry 1962) (P.W.West, A.M.G.Macdonald and T.S.West Editions) pp 397-404, Elsevner, Amsterdam (1963).
314. R.P.Taubinger and J.R.Wilson, Analyst, 90, 429 (1965).
315. R.Belcher and A.M.G.Macdonald, Talanta, 1, 185 (1958).
316. R.Belcher and A.M.G.Macdonald, S.E.Phang and T.S.West, J.Chem. Soc., 3, 2044 (1965).
317. M.Kan and H.Kashiwagi, Bunseki Kagaku, 10, 789 (1961).
318. R.Belcher and A.L.Godbert, Analyst (London, 66, 184 (1941).
319. L.Ubaldini and F.C.Maitan, Chim. Ind. (Milan), 37, 871 (1955).
320. J.Horacek, Collect. Czech. Chem. Commun., 27, 1811 (1962).
321. A.Nara, Y.Urushibata and N.Oc, Bunseki Kagaku, 12, 294 (1963).
322. M.Tanaka and S.Kanamori, Anal. Chim. Acta., 14, 263 (1956).
323. C.DiPietro, R.E.Kramer and W.A.Saisaman, Anal. Chem., 54, 586 (1962).
324. F.L.Schaeffer, J.Forg and P.L.Kirk, Anal. Chem., 25, 343 (1953).
325. R.Belcher, A.M.G.Macdonald, S.E.Phang and T.S.West, J. Chem. Soc., 2044 (1965).
326. C.A.Rush, S.St.Cruickshank and E.J.H.Rhodes, Mikrochim. Acta., 858 (1956).
327. H.N.Wilson, Analyst (London), 76, 65 (1951).
328. R.M.Lincoln, A.S.Carney and E.C.Wagner, Ind. Eng. Chem. Anal. Ed., 13, 358 (1941).
329. A.Kondo, Buneseki Kagaku, 9, 416 (1960).
330. R.Vasiliev and G.Anastasecu, Rev. Chim. (Bucharest), 13, 558 (1962).
331. F.S.Malyukova and A.D.Zaitseva, Plast. Massy, (1966); Ref. Zh. Khim., 19 GD, Abstr. 21G216 (1966).
332. H.Buss, H.W.Kohlshlutter and M. Preiss, Z. Analyt. Chem., 214, 106 (1965).
333. F.F.Hoffman, L.C.Jones, O.E.Robbins and F.F.Alsbert, Anal. Chem., 30, 1334 (1958).
334. R.May, Anal. Chem., 31, 308 (1959).
335. M.Maruyama and K.Hasegawa, Ann. Rep. Takamine Lab., 13, 173 (1961).
336. Ya.A.Mandel'baum, A.F.Gropov and A.L.Itskova Zh. Anal. Khim., 20, 873 (1965).
337. M.O.Korshun, E.A.Terent'eva and V.A.Klimova, Zh. Anal. Khim, 9, 275 (1954).
338. A.A.Abramyan, R.S.Sarkisyan and M.A.Balyou, Izv. Akad. Nauk Arm. SSR. Khim. Nauk., 14, 561 (1961).
339. M.Juracek and J.Jenik, Chem. Listy, 51, 1312 (1957).

340. W.Schoniger, Mikrochim. Acta., 52 (1963).
341. F.D.Fleischer, B.C.Southworth, J.H.Hodecker and M.M.Tuckerman, Anal. Chem., 30, 152 (1958).
342. M.Jean, Anal. Chim. Acta., 14, 172 (1956).
343. L E.Cohen and F.W.Czech, Chemist-Analyst, 47, 86 (1958).
344. U.Bartels and H.Hayme, Chem. Tech (Berlin), 11, 156 (1959).
345. E.Meier, Mikrochim. Acta., 70 (1961).
346. K.Zeigler, H.Till and H.Schindbauer, Mikrochim. Acta., 1114 (1963).
347. H.Y.Yu and H.I.Sha, Chem. Bull. (Peking), 557 (1965).
348. T.L.Hunter, Anal. Chim. Acta., 35, 398 (1966).
349. D.E.Ott and F.A.Gunther, Bull. Environ. Contam. Toxicol., 1, 90 (1966).
350. N.E.Gel'man and T.M.Shanina, Zh. Anal. Khim., 17, 998 (1962).
351. J.E.Barney, J.G.Bergmann and W.G.Tuskan, Anal. Chem., 31, 1394 (1959).
352. L.L.Farley and R.A.Winkler, Anal. Chem., 35, 772 (1963).
353. D.W.Brite, Anal. Chem., 27, 1815 (1955).
354. C.M.Welch and P.W.West, Anal. Chem., 29, 874 (1957).
355. A.P.Terent'ev, M.A.Volodina and E.G.Fursova, Dokl. Akad. Nauk SSSR, 169, 851 (1966).
356. V.N.Kotova, Zh. Anal. Khim., 22, 1239 (1967).
357. B.M.Luskina, A.P.Terent'eva and S.V.Syavtsillo, Zh. Anal. Khim., 17, 639 (1962).
358. A.P.Terent'ev, S.V.Syavtsillo and B.M.Luskina, Zh. Anal. Khim., 16, 83 (1961); Anal. Abstr., 8, 3309 (1961).
359. M.N.Chumachenko and V.P.Burlaka, Izv. Akad Nauk SSSR, Otd. Khim, Nauk 10, 5 (1963).
360. S.Ostrowski, R.Piekos and A.Radecki, Chem. Anal. (Warsaw), 10, 531 (1965).
361. E.A.Terent'eva and N.N.Smirnova, Zavod. Lab., 32, 924 (1966).
362. I.Lysyj and J.E.Zarembo, Microchem. J., 2, 245 (1958).
363. J.Korbl, Chem. Listy, 49, 858, 862, 1532 (1955).
364. J.Korbl and K.Blabolil, Chem. Listy, 49, 1664 (1955).
365. J.Korbl and P. Komers, Chem. Listy, 50, 1120 (1956).
366. J.Korbl and R.Pribl, Chem. Listy, 50, 232, 236 (1956).
367. J.Binkowski and B.Babranski, Chem. Anal. (Warsaw), 9, 515 (1964).
368. J.Binkowski and B.Babranski, Chem. Anal. (Warsaw), 9, 777 (1964).
369. J.Binkowski and M.Vecera, Mikrochim. Acta., 842 (1965).
370. M.O.Korshun and E.A.Terent'eva, Dokl. Akad. Nauk SSSR, 100, 707 (1955).
371. D.S.Galanos and V.M.Kapoulas, Anal. Chim. Acta., 34, 360 (1966).
372. B.D.Holt, Anal. Chem., 37, 751 (1965).
373. R.D.Hinkel and R.Raymond, Anal. Chem., 25, 470 (1953).
374. W.H.Jones, Anal. Chem., 25, 1449 (1953).
375. I.J.Oita and H.S.Conway, Anal. Chem., 26, 600 (1954).
376. J.Unterzaucher, Analyst (London), 77, 584 (1952).
377. I.Sheft and J.J.Katz, Anal. Chem., 29, 1322 (1957).
378. G.Tolg, Z. Anal. Chem., 194, 20 (1963).
379. E.Debal, Chim. Anal., 45, 66 (1963).
380. C.Eger and J.Lipke, Anal. Chim. Acta., 20, 548 (1959).
381. G.Stark, Z. Anorg. Chem., 70, 173 (1911).
382. J.I.Hoffman and G.E.F.Lundell, J. Res. Nat. Bur. Stand., 3, 581 (1929).
383. B.C.Saunders in Phosphorus and Fluorine, Cambridge University Press, Cambridge, pp 206-211 (1960).
384. W.M.Hoskins and C.A.Ferris, Ind. Eng. Chem., Anal. Ed., 8, 6 (1936).
385. W.D.Armstrong, J.Am. Chem. Soc., 55,741 (1933).
386. H.H.Willard and O.B.Winter, Ind. Eng. Chem. Anal. Ed., 5. 7 (1933).
387. S.Sass, N.Beitsch and C.U.Morgan, Anal. Chem., 31, 1970 (1959).
388. B.I.Stepanov and G.I.Migachev, Zavod. Lab., 32, 416 (1966).
389. E.L.Gefter, Zavod. Lab., 29, 419 (1963).
390. M.P.Strukova and T.V.Kirkllova, Zh. Anal. Kihm., 21, 1236 (1966); Anal. Abstr., 15, 3998 (1968).

391. V.V.Mikhailov and T.I.Tarasenko, Zavod. Lab., 33, 1380 (1967).
392. V.I.Kuznetsov and N.N.Basargin, USSR Patent 165566 (1964); Chem,Abstra. 62, 4623a (1965); Anal. Abstr., 13, 4801 (1966).
393. J.S.Fritz and S.S.Yamamura, Anal. Chem., 27, 1461 (1955).
394. B.Jaselskis and S.F.Vas, Anal. Chem., 36, 1965 (1964).
395. A.M.G.Macdonald and W.I.Stephen, J. Chem. Educ., 39, 528 (1962).
396. R.R.Balodis, A.Comerford and C.E.Childs, Microchem. J., 12, 606 (1967).
397. K.L.Cheng, Anal. Chem., 33, 783 (1961).
398. Determination of Sulphate, Technological Service Bulletin 6470-E-G-5, Arthur H.Thomas Co., 1958, California.
399. R.Levin, Israel J. Chem., 3, 11 (1966).
400. D.J.Driscoll, D.A.Clay, C.H. Rogers, R.H.Jungers and F.E.Butler, Anal. Chem., 50, 767 (1978).
401. F.Kasler, M.Tierney, Anal. Chim., 51, 1070 (1979).
402. F.R.Hartley and L.M.Venanzl, unpublished results.
403. E.B.Sandell, Colorimetric Metal Analysis, Interscience, New York, 726 (1959).
404. M.G.Reed and A.D.Scott, Anal. Chem., 33, 773 (1961).
405. A.S.Zabrodina and M.R.Bagreeva, Vestn. Mosk. Univ., 187 (1958).
406. N.Kunimine, H.Nakamaru and M.Nakamaru, J. Pharm. Soc. Jpn., 83, 59 (1963).
407. K.L.Cheng, Chemist Analyst, 45, 67 (1956); Anal. Abstr., 4. 1189 (1957).
408. Mejer and N.Shaltiel, Mikrochim. Acta., 580 (1960).
409. Z.Stefanac and Z.Rakovic, Mikrochim. Ichnoanal. Acta., 81 (1965).
410. W.Ihn, G.Huse and P.Neuland, Mikrochim. Acta., 628 (1962).
411. Kh, Ya, Huse and L.A.Lipp, Zh. Anal. Khim., 21, 1266 (1966).
412. S.Barabas and W.C.Cooper, Anal. Chem., 28, 129 (1956); Anal. Abstr., 3, 1694 (1956).
413. L.Barcza, Acta. Chim. Hung., 47, 137 (1966).
414. L.Barcza, Acta. Chim. Hung., 45, 23 (1965); Anal. Abstr., 13, 6189 (1966).
415. A.S.Zabrodina and M.R.Bagreeva, Vestn.Mosk. Univ., 187 (1958); Anal. Abstr., 6, 2197 (1959).
416. A.S.Zabrodina and A.P.Khiystova, Vestn. Mosk. Univ., 69 (1960).
417. A.Kondo, Bunseki Kagaku, 6, 583 (1957).
468. D.Dingwall and W.D.Williams, J. Pharm. Pharmacol., 13, 12 (1961).
469. H.Bieling, W.Wagenknecht and E.M.Arndt, Z. Anal. Chem., 201, 419 (1964).
470. W.J.Kelleher and M.J.Johnson, Anal. Chem., 33, 1429 (1961).
471. J.W.Shell, J. Pharm. Sci., 51, 731 (1962).
472. Z.Stefanac, Bull, Sci. Yugosl., 8, 136 (1963).
473. A.S.Zabrodina and S.Ya.Levina, Vestn. Mosk. Univ., 181 (1957).
474. R.Huls and M.Renson, Bull Soc. Chim. Belg., 65, 696 (1956).
475. R.Belcher, D.Gibbons and A.Sykes, Mikrochem., 76, (1952).
476. A.Sykes, Mikrochim. Acta., 1155 (1956).
477. J.A.McHard, P.C.Servais and H.A.Clark, Anal. Chem., 20, 325 (1948).
478. H.Wilson, Analyst, 74, 243 (1949).
479. M.Jean, Chem. Anal., 37, 125 (1955).
480. I.A.Voinovitch, Chim. Anal., 40, 332 (1958).
481. A.Ringbom, P.E.Ahlers and S.Sutonen, Anal. Chim. Acta., 20, 78 (1959).
482. R.F.Milton, J. Appl. Chem., 1, 126 (1951).
483. R.F.Milton, Analyst, 76, 431 (1951).
484. L.Halzapfel and G.Gottschalk, Anal. Chem., 142, 115 (1954).
485. A.J.Christopher and T.R.F.Fennell, Talanta, 12, 1003 (1965).

486. J.H.Wetters and R.C.Smith, Anal. Chem., 41, 379 (1969).
487. H.J.Horner (1965) in "Treatise of Analytical Chemistry" Part 11 Vol 12 p 287-8, Wiley Interscience, New York, NY.(1965)
488. T.Urin, T.Wade and A.Hanyo, J. Chem. Soc. Japan Ind. Chem. Sect., 60, 1274 (1957).
489. T.M.Shanina, N.E.Gel'man and L.M.Kiparenko, Zh. Analit. Khim., 20, 118 (1965).
490. Z.Sir and R.Komers, Coll. Czech. Chem. Commun., 21, 873 (1956).
491. A.P.Kreshbov, S.V.Syavtsillo and U.T.Shemyatenkova, Zavod Lab., 22, 1425 (1956).
492. B.Smith, Acta. Chem. Scand., 11, 558 (1957).
493. L.V.Myshlyaeva, V.V.Krishoshchekov and I.V.Sedova, Zhur. Khim., 19GDE (2); Abstr., No. 2G184 (1965).
494. L.V.Myshlyaeva and I.V.Sedova, Zav. Lab., 34, 263 (1968).
495. I. Ya Chulkov, USSR Patent 121,963 (18/8/59) (1959).
496. A.P.Terent'eva, S.U.Syavtsillo and B.M.Luskina, Zhur. Anal. Khim., 16, 83 (1961).
497. E.L.Fogel'son, Zavod. Lab., 23, 1427 (1957).
498. R.B.Lew and W.Oyung, Anal. Chem., 36, 1587 (1964).
499. A.P.Kreshkov, Myshlyaeva and V.V.Krasnoshekekov, Plant Massy,(12) 51 (1962).
500. A.P.Kreshkov, L.V.Myshlyaeva, O.B.Khachaturyan and V.V.Krasnoshekekov, Zhur. Anal. Khim., 18, 1375 (1963).
501. A.P.Kreshkov and L.V.Myshlyeava, Zavod. Lab., 29, 924 (1963).
502. V.Popov, A.Deryabin, L.V.Myshlyaeva and V.V.Krasnoschchekov (1964) Trudy. Mosk. Khim. Tekhnol. Inst., 46, 7 (1965). Zhur. Khim. 19GDE 7 (5) Abstr. No. 5G214.
503. L.Kalman and A.Vago, Magyor Kem Foly, 64, 123 (1958).
504. L.V.Myshlyaeva, V.V.Krasnoshchekov, Trudy. Mosk. Khim. Tekhnol. Inst. 1m DI Mendeleeva (42) 178 (1964). Zhur. Khim. 19GDE (11) Abstr. No. 11G220.
505. A.P.Kreshkov and N.I.Gludiva, Trudy Mosk. Khim. Tekhnol. Inst., 18, 73 (1957), Zhur. Khim. Abstra. No. 1276.
506. M.P.Brown and G.W.A.Fowles, Anal. Chem., 30, 1689 (1958).
507. R.Riverchon and Y.Legrand, Chim. Analyst, 47, 194 (1965).
508. L.Govett, Analyt. Chim. Acta., 25, 69 (1961).
509. A.P.Kieshkov, Ya Mikhailenko and E.A.Kuchkarev, Zavod. Lab., 30, 555 (1964).
510. J.Radall and P.D.Hunt, Anal. Chem., 30, 1280 (1958).
511. F.L.Chau in "Proceedings of the Society for Analytical Chemistry Conference", Nottingham, pp 89-101, W.Heffer and Sons Ltd. Cambridge U.K.
512. T.R.F.W.Fennell and J.R.Webb, Talanta, 2, 389 (1959).
513. T.R.F.W.Fennell, N.W.Roberts and J.R.Webb, Analyst, 82, 639 (1957).
514. R.Belcher and A.L.Godbert, Analyst, 66, 194 (1941).
515. T.R.F.W.Fennell and J.R.Webb, Talanta, 2, 105 (1959).
516. H.R.Shell and H.Craig, Anal. Chem., 26, 996 (1954).
517. S.Harel, E.R.Horman and A.Talmi, Anal. Chem., 27, 1144 (1955).
518. T.R.F.W.Fennell and J.R.Webb, Talanta, 11, 1323 (1964).
519. O.Schwarzkopf, R.Henleim, Actas de Congresso XVth International Congress of Pure and Applied Chemistry (Analytical Chemistry) 1956, Ramos Afono. and (1957) Moita, Lld, Lisbon, Vol. 1 p.301 (1957).
520. T.R.F.W.Fennell, Chem. Ind., 1404 (1955).
521. J.A.Brabson, H.C.Mattraw, G.E.Maxwell, A.Darrow and M.P.Needham, Anal. Chem., 20, 504 (1948).
522. S.Greenfield, Analyst, 84, 380 (1959).
523. H.N.Wilson, Analyst, 74, 243 (1949).
524. K.A.Andrianov, L.M.Khanonashvili and T.V.Vasil'eva, Zhur. Anal. Khim., 16, 738 (1961).

525. S.Ostrowski, R.Pickos and A.Radecki, Chemia. Analit., 10, 531 (1965).
526. L.V.Myshlyaeva, T.G.Shatunova, Trudy, Mosk. Khim. Tekhnol. Inst., (48) 48 (1966). Zhur. Khim., 19GD (10) Abstr. No. 10G236.
527. I.Chulkov Ya and L.A.Solov'ena, Vestn. Tekhnol. Ekon. Inform. Gos. Kom. Sov. Min. SSSR (1960) po Khimic (10) 32 Ref (1961). Zhur. Khim., 21, Abstr. No. 21D40.
528. A.P.Kreshkov, L.V.Myshlyaeva, E.A.Kucharev and T.G.Shatunova, Lakotz-rasoch. Mater ikh Primenenie (24) Abstra. No. 24G257 (1966).
529. E.A.Terent'eva, Zavod. Lab., 28, 807 (1962).
530. O.Willard and O.Dean, Anal. Chem., 22,1264 (1950).
531. E.A.Terent'eva, B.M.Luskina and S.V.Syavtsillo, Zhur. Anal. Khim., 16, 635 (1961).
532. E.A.Terent'eva, S.V.Syavtsillo and B.M.Luskina, Anal. Khim. Akad. Nauk. SSSR, 13, 3 (1963). Ref: Zhur. Khim. 19GDE (4) Abstra. No. 4G239 (1964).
533. A.P.Terent'eva, B.M.Luskina and S.V.Syavtsillo, Anal. Abstra., 8, 3309 (1961).
534. B.M.Luskina, A.P.Terent'eva and N.A.Gradkova, Zhur. Analit. Khim., 19, 1251 (1964).
535. B.M.Luskina, A.P.Terent'eva and N.A.Gradskova, Zhur. Analit. Khim., 20, 990 (1965).
536. E.A.Terent'eva and N.N.Smirnova, Zavod. Lab., 32, 924 (1966).
537. S.Kohama, Bull. Chem. Soc. Japan, 36, 380 (1963).
538. V.A.Klimova, M.O.Forshun and E.G.Bereznitskaya, Zhur. Anal. Khim., 11, 223 (1956).
539. V.A.Klimova and E.G.Bereznutskaya, Zhur. Anal. Khim., 11, 292 (1956).
540. V.A.Klimova and E.G.Bereznitskaya, Zhur. Anal. Khim., 12, 424 (1957).
541. H.Lammer, Chem. Tech., Berlin, 4, 491 (1952).
542. W.F.Gilliam, H.A.Lieblafsky, A.F.Winslow , J. Amer. Chem. Soc., 76, 918 (1954).
543. E.W.Balis, H.A.Lieblafsky and L.B.Bronk, Ind. Eng. Chem. Anal. Ed., 17, 56 (1945).
544. E.G.Rochow and W.F.Gilliam, J. Amer. Chem. Soc., 63, 798 (1941).
545. E.G.Rochow, J. Amer. Chem. Soc., 70, 2170 (1958).
546. T.Aranyi and A.Erdey, Magy. Kem. Lab., 20, 164 (1965).
547. Y.A.Gawargious and A.M.G.Macdonald, Anal. Chim. Acta., 27,300 (1962).
548. J.Korbl and R.Komers, Chem. Listy., 50, 1120 (1956). Anal. Abstra., 3, 373 1384 (1956).
549. K.Uhle, Z. Analyt. Chem., 231, 194 (1967).
550. N.Platonov, Trudy Komiss. Anal. Khim. Akad. Nauk. SSSR., 13, 15 (1963). Zhur. Khim., 19GDE (1) Abstra. No. 19236 (1964).
551. H.B.Bradley, Anal. Chem., 27, 2021 (1955).
552. A.Radecki, Chem. Anal., Warsaw, 8, 607 (1959).
553. A.Radecki and R.Pickos, Roczn. Chem., 33, 57 (1959).
554. H.Kautasky, G.Fritz, H.P.Siebel and D.Siebel, Z. Anal. Chem., 147, 327 (1955).
555. E.A.Bondarevskaya, V.M.Kuznetskova and S.A.Syavtsillo, Zhur. Anal. Khim., 16, 472 (1961).
556. E.A.Bondarevskaya,A.P.Kreshkov, S.A.Syavtsillo and V.M.Kuznetsova, Trudy Komis. Anal. Khim. Akad. Nauk. SSSR., 13, 24 (1963). Ref: Zhur Khim, 19GDE (1) Abstra. No. 1G235 (1964).
557. V.A.Klimova, M.D.Vitalina, Izv. Akad. Nauk. SSSR. Odtel Khim. Nauk (12) 2245 (1962).
558. J.A.Magnuson, P.W.Baillargeon, Anal. Chim. Acta., 32, 156 (1965).
559. J.A.Magnuson, Chem. Anal., 53, 15 (1964).
560. Z.Sir and R.Komers, Chem. Listy., 50, 162 (1956).
561. S.V.Syavtsillo, U.T.Shemyatenkova and A.M.Neshumova, Zavod. Lab., 24, 287 (1958).

562. O.Watt, Chem. Rev., 46, 317 (1960).
563. L.H.Sommer, E.Dorfman, G.M.Goldberg and F.C.Whitmore, J. Am. Chem. Soc., 68, 488 (1946).
564. F.Hirata and T.Takiguchi, J. Chem. Soc. Japan. Ind. Chem. Sect., 60,
565. T.A.Khudayakova, E.I.Pavel'eva and R.V.Kozlov, Zhur. Prikl. Khim. Lening, 38, 2002 (1965).
566. I.Chulkov Ya, USSR Patent No. 118,224 (29/2/59) (1959).
567. L.V.Myshlyaeva, Trudy. Mosk. Khim. Tekhnol. Inst. in D.I.Mendeleev (25) 29 (1957) Ref: Zhur. Khim. Abstr. No. 24900 (1958).
568. L.V.Myshlyaeva, Ya Mikhailenko Yu, V.V.Krasnoshchekov and E.A.Kuchkarev Trudy. Khim. Moskow. Khim. Tekhnol. Inst. (44) 139 (1963) Ref: Zhur. Khim. 19GDE (20) Abstr. No. 20G153 (1964).
569. T.Takiguchi, J. Chem. Soc. Japan. Ind. Sect., 61, 1236 (1958).
570. L.V.Myshlyaeva, Trudy Moskov Khim. Tekhnol Inst. 1 m, D.I.Mendeleev (25) 29 (1957) Ref. Zhur. Khim. Abstra. No. 24900 (1958).
571. T.Takiguichi, Analyst, 83, 482 (1958).
572. T.Takiguichi, J. Chem. Soc. Japan, Ind. Chem. Sect., 61, 587 (1955).
573. T.R.F.W.Fennell, Chem. Ind., 1404 (1955).
574. R.Belcher and A.M.G.Macdonald, Mikrochim Acta, 510 (1957).
575. G.M.Kline in "Analytical Chemistry of Polymers" p 372 Interscience, New York (1959).
576. C.R.N.Strouts, J.H.Gilfillan and H.N.Wilson in "Analytical Chemistry - The Working Tools",Vol. 1, p.417 Oxford University Press (1965).
577. B.Smith, Acta. Chem. Scand., 11, 558 (1957).
578. P.Brown and A.L.Smith, Anal. Chem., 30, 549 (1958).
579. J.S.Fritz and G.H.Schlenk, Anal. Chem., 31,1808 (1959).
580. J.A.Magnuson, Anal. Chem., 35, 1487 (1963).
581. P.Postal, J.Cermiak and P.Novotna, Coll. Czeck. Chem. Commun. (Engl. Ed.), 20, 34 (1965).
582. A.P.Terent'eva, E.A.Bondarevskaya and T.V.Kirillova, Zhur. Analit. Khim., 22, 1242 (1967).
583. A.P.Kreshkov and V.A.Bork, Zhur. Anal. Khim., 12, 764 (1957).
584. A.P.Terent'eva and E.A.Bondarevskaya, Zhur. Analit. Khim., 20, 249 (1965).
585. E.Kreshkov and P.Nessonova, Zhur. Anal. Khim., 4, 220 (1949).
586. S.V.Syavtsillo and E.A.Bondarevskaya, Zhur. Anal. Khim., 11, 613 (1956).
587. V.A.Klimova, K.S.Zabrodina and L.L.Shitikova, Izv. Akad. Nauk, SSSR. Ser. Khim., 1, 178 (1965).
588. G.D.Nessonova and E.K.Pogosyants, Zavod. Lab., 24, 953 (1958).
589. R.A.Bourrique, Chem. Anal., 43, 40 (1954).
590. A.Colson, Analyst, 58, 594 (1933).
591. W.Kirsten and E.Rogozinsky, Anal. Abstra., 2, 3080 (1955).
592. V.A.Drozdov and E.G.Vlasova, Trudy, Komiss, Anal. Khim. Akad. Nauk, SSSR 13, 187 (1963). Zhur. Kihm., 19GDE (19) Abstra. No. 19 G191 (1963).
593. A.P.Kreshkov, V.A.Drozdov and E.G.Vlasova, USSR Patent, 137,299 (10/4/61) (1961).
594. F.O.Guenther, Anal. Chem., 30, 1118 (1958).
595. J.F.Lees and R.T.Lobeck, Analyst, 88, 782 (1963).
596. J.F.Hampton, C.W.Lacefield and J.F.Hyde, Inorg. Chem., 4, 1659 (1956).
597. K.Damm and W.Noll, Kolloid., 158, 97 (1958).
598. J.Haslam and H.A.Willis in "Identification and Analysis of Plastics" pp 256-7, Van Nostrand, New York (1965).
599. G.M.Kline in "Analytical Chemistry of Polymers" Part 1, Analysis of monomers and polymeric materials, Interscience, New York, 9, 373 (1959).
600. G.R.Lucas and R.W.Martin, J. Amer. Chem. Soc., 74, 5225 (1952).

601. L.H.Sommer and L.J.Tyler, J. Am. Chem. Soc., 76, 1030 (1954).
602. K.Damm, D.Belitz and W.Noll, Angew. Chem., 76, 273 (1964).
603. H.Gilman and L.S.Miller, J. Am. Chem. Soc., 73, 2367 (1951).
604. W.T.Grubb, J. Am. Chem. Soc., 76, 3408 (1954).
605. R.C.Smith and G.E.Kellum, Anal. Chem., 39,338 (1967).
606. G.H.Barnes and N.E.Daughenbough, Anal. Chem., 35, 1308 (1963).
607. J.S.Fritz, G.H.Schlenk, Anal. Chem., 31, 1808 (1959).
608. P.Dostal, J.Cermak and P.Novotna, Coll. Czech. Chem. Commun. (Eng. Ed.), 30, 34 (1965).
609. J.A.Magnuson, Anal. Chem., 35, 1487 (1963).
610. J.A.Magnuson and R.J.Cerri, Anal. Chem., 38, 1088 (1966).
611. G.E.Kellum and K.L.Uglum, Anal. Chem., 39, 1623 (1967).
612. F.O.Guenther, Anal. Chem., 30, 1118 (1958).
613. R.Fuchs, O.Moore, D.Miles and H.Gilman, J. Org. Chem., 21, 1113 (1952).
614. R.O.Sauer, J. Am. Chem. Soc., 66, 1707 (1944).
615. J.F.Hyde, J. Amer. Chem. Soc., 75, 2166 (1953).
616. W.Noll, K.Damm, W.Krauss and U.Fette, Seifen Farbe U. Lack, 65, 17 (1959).
617. A.Berger and J.A.Magnuson, Anal. Chem., 36, 1156 (1964).
618. K.K.Kunder and M.N.Das, Anal. Chem., 31, 1358 (1959).
619. G.Fritz and H.Burdt, Z. Anorg. Chem., 317, 35 (1962).
620. A.P.Kreshkov, V.A.Bork, M.I.Aparsheva, Plast. Massy (4) 63 (1965). Ref. Zhur. Khim., 19GDE (21) Abstr. No. 21G196 (1965).
621. A.P.Terent'eva, E.A.Bondarevskaya and T.V.Kirillova, Zavod. Lab., 33, 156 (1967).
622. N.E.McDougall, Appl. Spectros., 19, 196 (1965).
623. V.A.Bork, A.P.Kreshkov and L.A.Shvyrkova, USSR Patent 141,334 (6/10/61) (1961).
624. A.P.Kreshkov, V.A.Bork and L.A.Shvyrkova, Zhur. Anal. Khim., 28,151 (1962).
625. A.P.Kreshkov and V.A.Bork, Zhur. Anal. Khim., 17, 359 (1962).
626. V.A.Bork and L.A.Shvyrkova, Trudy Kommiss, Anal. Khim. Akad. Nauk, SSSR, 13, 148 (1963). Ref. Zhur. Khim. 19GDE (22) Abstr. No.22G159 (1963).
627. J.Cermak and P.Dostal, Coll. Czech. Chem. Commun., 28, 1384 (1963).
628. A.Schwarz, Z. Anal. Chem., 115, 161 (1939).
629. F.S.Kipping and J.E.Sands, J.Chem. Soc., 119, 848 (1921).
630. F.S.Malyukova and L.I.Surkova, Plast. Massy (12) 51 (1965). Ref. Zhur. Khim. 19G7 (10) Abstr. No. 10G237 (1966).
631. A.P.Kreshkov, V.A.Bork and M.Aparsheva, Zavod. Lab., 30, 1208 (1964).
632. A.P.Kreshkov, V.A.Bork and M.Aparsheva, Izv. Vysshucheb. Zavod. Khim., Khim. Tekhnol. 7, (5) 742 (1964). Ref. Zhur. Khim., 19GD(11) Abstr. No. 11G212 (1965).
633. Y.Y.Milhailenko, L.P.Senetskaya and E.G.Kutyrina, Trudy Kommiss. Anal. Khim., Akad. Nauk SSSR, 13, 383 (1963). Ref. Zhur. Khim. 19GDE (20) Abstr. No. 20G168 (1963).
634. H.Waledziak, Chemia. Analit., 10, 579 (1965).
635. A.P.Kreshkov, U.T.Shemyatenkova, S.V.Syavtsillo and N.A.Palamarchuk, Khur. Anal. Khim., 15, 635 (1960).
636. M.G.Voronkov and V.T.Shemyatenkova, Izv. Akad. Nauk. SSSR, (1) 178 (1961).
637. F.S.Kipping and L.L.Lloyd, J.Chem. Soc., 79, 449 (1901).
638. C.A.Burkhard and F.J.Norton, Anal. Chem., 21, 304 (1949).
639. H.S.Booth, Chem. Rev., 41, 97 (1947).
640. W.H.Perlson, T.J.Brice and J.H.Simons, J. Am. Chem. Soc., 67, 1769 (1945).
641. H.S.Booth and M.L.Freedman, J. Amer. Chem. Soc., 72, 2847 (1950).

642. A.P.Kreshkov, V.A.Drozdov and E.G.Vlasova, Izv. Vyssh. Ucheb. Zavedenii i Khim. Tekhnol., 3, (1) 85 (1960).
643. A.P.Freshkov and V.A.Drozdov, Dokl. Acad. Nauk SSSR, 131, (6) 1345 (1960). Ref. Zhur. Khim. (18) Abstr. No. 73,199 (1960).
644. A.P.Kreshkov, Trudy Mosk. Khim. Tekhnol. Inst. Im D.I.Mendelieva (32) 333 (1961). Ref. Khur. Khim. (8) Abstr. No. 8D211 (1962).
645. A.P.Kreshkov, V.A.Bork and H.I.Aparsheva, Zhur. Anal. Khim., (18), 10, 1149 (1963).
646. A.P.Kreshkov, V.A.Drozdov and E.G.Vlasova, Zavod. Lab., 26, 1080 (1960).
647. A.P.Kreshkov, V.A.Drozdov and E.G.Vlasova, Izv. Vyssh. Ucheb. Zavedenic Khim. Tekhol., 3, (1) 80 (1960). Ref. Zhur. Khim. (18) Abstr. No. 73, 203 (1960).
648. A.P.Kreshkov, V.A.Drozdov and R.R.Tarasyants, Zavod. Lab., 30, 413 (1964).
649. A.P.Kreshkov, V.A.Drozdov, Dokl. Akad. Nauk. SSSR, 131 (6) 1345 (1960) Ref. Zhur. Khim (18) Abstr. No. 73,199 (1960).
650. A.P.Kreshkov, Trudy Mosk. Khim. Tekhnol. Inst. Im. D.I.Mendeleeva, (32) 333 (1961). Ref. Zhur. Khim. (8) Abstr. No. 8D211 (1962).
651. M.Farnsworth and J.Pekola, Anal. Chem., 31, 410 (1959).
652. D.Dunn and T.Norris, Australia, Commonw. Dept. Supply, Defence Standard Lab. Rep., No. 269, 1964 , 21,pp.
653. I.G.M.Campbell, G.W.A.Fowles and L.A.Nixon, J. Chem. Soc., 1398 (1964).
654. R.Geyer and H.J.Seidlitz, Z. Chem., 4, 468 (1964).
655. G.Tagliavini, Studi Urbinan, Fac. Farm., 10, 39 (1967).
656. R.Reverchon, Chim. Anal. (Paris), 47, 70 (1965).
657. V.Chromy and J.Vrestal, Chem. Listy., 60, 1537 (1966).
658. L.V.Myshlaeva and T.G.Maksimova, Zh. Anal. Khim., 23, 1584 (1968).
659. S.Genda, K.Morikawa and T.K.Kegaku, To Kogyo (Osaka), 43, 265 (1969).
660. F.Guenther, R.Geyer and D.Stevenz, Neue Hutte, 14, 563 (1969).
661. R.P.Kreshkov and E.A.Kucharev, Zavod. Lab., 32, 558 (1966).
662. P.Ochsenhein, Kunststoffe, 58, 366 (1968).
663. J.C.Maire, Ann. Chim. (Paris), 6, 969 (1961).
664. C.Mohr and G.Z.Stock, Anal. Chem., 221, 1 (1966).
665. R.Geyer and H.T.Seidlitz, Z. Chem., 7, 114 (1967).
666. H.Gilman and W.B.King, J. Am. Chem. Soc., 51, 1213 (1929).
667. K.A.Kocheshkov, Ber. Dtsch. Chem. Ges., 62, 1659 (1926).
668. J.G.A.Luijten and G.J.M. Van der Kerk, in Investigations in the Field of Organotin Chemistry, Tin Research Institute, Greenford, Middlesex p. 84 (1966).
669. S.Kohama, Bull. Chem. Soc. Jap., 36, 830 (1963).
670. S.Kohama, Bull. Chem. Soc. Jap., 36, 830 (1963).
671. M.P.Brown and G.W.A.Fowles, Anal. Chem., 30, 1689 (1958).
672. G.J.M.Van der Kerk, J.G.Noltes and J.G.A.Luijten, J. Appl. Chem., 7, 366 (1957).
673. G.Fritz and H.Scheer, Z. Anorg. Allg. Chem., 331, 151 (1964).
674. H.Gilman and S.D.Rosenberg, J. Am. Chem. Soc., 75, 3592 (1953).
675. P.I.Seinorklin, Hyg. Sanit., 31, 270 (1966).
676. E.Kroller, Dtsch. Pettenkofer Inst., 56, 190 (1960).
677. R.Bock, S.Borback and H.Oeser, Angew. Chem., 70, 272 (1958).
678. P.Nangniot and P.H.Martens, Anal. Chim. Acta., 24,276 (1961).
679. K.Hartel, Agric. Vet. Chem., 3, 19 (1962).
680. J.Herokand H.Gotte, in The Use of Radioisotopes in Animal Biology and the Medical Sciences, Symposium, Mexico City, 21st Nov. - 1st Dec.
681. J.Herok and H.Gotte, in The Use of Radioisotopes in Animal Biology and the Medical Sciences, Symposium, Mexico City, 21st Nov. - 1st Dec. 1961 Vol. 11, 4, Academic Press, London , New York, 1961 p.177.
682. K.Burger, Z.Lebensm. Unters. Forsch., 114, 1 (1961).

683. H.J.Hardon, A.F.H.Bessemer and H.Brunink, Dtsch. Lebensm. Rundsch., 58, 349 (1962).
684. G.A.Lloyd, C.Otaci and F.T.Last, J. Sci. Food Agric., 13, 353 (1962).
685. J.Herok and H.Gotte, Int. J. Appl. Radiat. Isotop., 14, 461 (1963).
686. J.Bruggemann, K.Barth and K.H.Neisar, Zentralbi. Veterinaermed., Reihe, 11, 4 (1964).
687. K.Burger, Z. Anal. Chem., 192, 280 (1962).
688. J.Markland and F.C.Shenton, Analyst (London), 82, 43 (1957).
689. T.D.Holmes and I.F.Storey, Plant Pathol., 11, 139 (1962).
690. L.H.Adcock and W.G.Hope, Analyst (London), 95, 868 (1970).
691. H.B.Corbin, J. Assoc. Off. Anal. Chem., 53, 140 (1970).
692. P.Rivett, J. Appl. Chem., 15, 469 (1965).
693. L.Chromy and Z.Uhaez, J. Oil Colour Chem. Assoc., 51, 494 (1968).
694. R.Warchol. J. Oil Colour Chem. Assoc., 53, 121 (1970).
695. I.R.McCallum, J. Oil Colour Chem. Assoc., 52, 434 (1969).
696. P.N.Vijan and C.Y.Chan, Anal. Chem., 48, 1788 (1976).
697. V.F.Hodge, S.I.Seidel and E.D.Goldberg, Anal. Chem., 51, 1256 (1979).
698. S.Kojima, Analyst (London), 104, 550 (1979).
699. I.L.Marr, J. Anwar Analyst, 107. 260 (1982).
700. M.O.Andreae and J.T.Byrd, Anal. Chim. Acta., 156, 147 (1984).
701. T.Burns, D.Dadgar and M.Harriott, Analyst, 109, 1099 (1984).
702. D.Colaitis and M.Lesbre, Bull. Soc. Chim. Fr., 19, 1069 (1952).
703. F.C.Silbirt and W.R.Kirner, Ind. Eng. Chem. Anal. Ed., 8, 353 (1936).
704. U.S.Bazalitskaya and M.K.Dzhamletdinova, Zavod. Lab., 33, 427 (1967).
705. K.A.Kocheshkov, Ber. Dtsch. Chem. Ges., 61, 1659 (1928).
706. M.P.Strukova and I.I.Kashiricheva, Zh. Anal. Khim., 24, 1244 (1969).
707. G.J.M.Van der Kerk and J.G.A.Luijten, J. Appl. Chem., 7, 369 (1957).
708. H.Matsuda and S.Matsuda, J. Chem. Soc. Jap. Ind. Chem. Sect., 64, 539 (1961).
709. V.Chromy and L.Srp. Chem. Listy, 61, 1509 (1967).
710. E.W.Abel and D.B.Brady, J. Chem. Soc., 1192 (1965).
711. J.Jenik and F.Renger, Coll. Czech. Chem. Commun. (English Ed.) 29, 2237 (1964).
712. B.M.Luskina, E.A.Terent'eva and N.A.Gradskova, Zhur. Anal. Khim., 20, 990 (1965).
713. E.A.Terent'eva and M.V.Bernatskaya, Zhur. Anal. Khim., 21, 870 (1966).
714. E.A.Terent'eva, Anal. Abstra., 13, 1306 (1966).
715. S.Kohoma, Bull. Chem. Soc. Japan, 36, 830 (1963).
716. E.E.Kotlyar and T.N.Nazarchuk, Zhur. Anal. Chim., 16, 631 (1961).
717. E.Niki and K.Matsato, Japan Analyst, 9, 324 (1960).
718. Y.I.Chulkov, USSR Patent, 128, 190 28/4/60 (1960).
719. Y.I.Chulkov, Zavod. Lab., 26, 550 (1960).
720. T.R.Crompton, Chemical Analysis of Organometallic Compounds, Vol. V, Academic Press, London, New York. San Fransisco, 1977, Chapter 21, pp 378-386.
721. D.J.David, Analyst (London), 83, 655 (1958).
722. G.Westoo, Analyst (London), 88, 287 (1963).
723. H.J.Gudzinowicz and V.J.Suciano, J. Assoc. Off. Anal. Chem., 49, 1 (1966).
724. W.U.Malik, R.Hague and S.P.Verma, Bull. Chem. Soc. Jpn., 36, 746 (1963).
725. T.R.Crompton, Analyst (London), 86, 652 (1961).
726. T.R.Crompton, Analyst, 91, 374 (1966).
727. T.R.Crompton and V.W.Reid, Analyst, 88, 713 (1963).
728. B.J.Phillip, W.L.Mundry and S.C.Watson, Anal. Chem., 45, 2298 (1973).
729. E.A.Terent'eva and M.V.Bernatskaya, Zh. Anal. Khim., 19, 876 (1964).
730. E.A.Terent'eva and N.N.Smirnova, Zavod. Lab., 32, 924 (1966).
731. K.Ziegler and H.Gellert, Justus Liebigs. Ann. Chem., 629, 20 (1960).

732. S.A.Bartkiewicz and W.J.Robinson, Anal. Chim. Acta., 20, 326 (1959).
733. T.R.Crompton, Analyst (London), 91, 374 (1966).
734. E.Bonitz, Chem. Berichte.,88, 742 (1955).
735. C.A.Uraneck, J.E.Burleigh and J.W.Cleary, Anal. Chem., 40, 327 (1968).
736. Razuvaev and S.I.Graevsky, Dokl. Akad. Nauk SSSR, 128, 309 (1959), (English translation, p 747).
737. D.F.Hagen and W.D.Leslie, Anal. Chem., 35, 814 (1963).
738. J.H.Mitchen, Anal. Chem., 33, 1331 (1961).
739. D.E.Jordan, Anal. Chem., 40, 2150 (1968).
740. G.W.Heunisch, Anal. Chem., 44, 741 (1972).
741. D.E.Jordan and W.D.Leslie, Anal. Chim. Acta., 50, 161 (1970).
742. E.Bonitz, Chem. Ber., 88, 742 (1955).
743. A.I.Graevskii, S.SH.Shchegal and Z.S.Smalian, Dokl. Akad. Nauk SSSR, 119, 101 (1958).
744. E.Bonitz and W.Hubner, Z. Anal. Chem., 186, 206 (1962).
745. T.R.Crompton, Anal. Chem., 39, 268 (1967).
746. L.M.Shtifman, S.V.Syaztillo and G.G.Larikova, Trudy, Komiss. Anal. Khim. Nauk. SSSR, 13, 325 (1963).
747. M.Farina, M.Donati and M.Raazzini, Ann. Chim. (Rome), 48, 501 (1958).
748. L.Nebbia and B.Pagani, Chim. Ind. (Milan), 44, 383 (1962).
749. M.Uhniat and T.Zawada, Chem. Anal. (Warsaw), 9, 701 (1964).
750. V.P.Mardykin, E.I.Kvasyuk and P.N.Gaponik, Zh. Prikl. Khim. (Leningrad), 42, 947 (1969).
751. E.G.Hoffman and W.Tornau, Z. Anal. Chem., 188, 321 (1962).
752. E.G.Hoffman and W.Tornau, Z. Anal. Chem., 186, 231 (1962).
753. M.Dimbat and G.A.Harlow, Anal. Chem., 34, 450 (1962).
754. W.L.Everson and E.M.Ramirez, Anal. Chem., 37, 806 (1965).
755. W.L.Everson, Anal. Chem., 36, 854 (1964).
756. W.P.Neumann, Justus Liebigs Ann. Chem., 629, 23 (1960).
757. M.Wronski, Chem. Anal. (Warsaw), 8, 299 (1963).
758. R.Montequi, A.Doadrio and C.Serrano, An. Rl. Soc. Esp. Fis. Quim. Ser. B., 53, 447 (1957).
759. H,Flaschka and F.Sadek, Chem. Anal. (Warsaw), 47, 30 (1958).
760. A.Schneer and H.Hartmann, Magy, Kem. Foly., 67, 309 (1961).
761. A.J.Krol, L.B.Eddy and D.R.Mackey, U.S. Atom. Energy Comm. Rep., CCC-1024-TR-239, 1957, 12 pp.
762. A.F.Zhigach, E.B.Kazakova and R.A.Kigel, Zh. Anal. Chim., 14, 746 (1959).
763. G.A.Guter and G.W.Schaeffer, J. Am. Chem. Soc., 78, 3346 (1956).
764. R.S.Braman, D.D.DeFord, T.N.Johnson and L.J.Kuhns, Anal. Chem., 32, 1258 (1960).
765. A.E.Messner, Anal. Chem., 30, 547 (1958).
766. L.J.Kuhns, R.S.Braman and J.E.Graham, Anal. Chem., 34, 1700 (1962).
767. M.I.Fauth and C.F.McNercy, Anal. Chem., 32, 91 (1960).
768. A.Heyrovsky, Z. Anal. Chem., 173, 301 (1960).
769. A.Heyrovsky, Collect. Czech. Chem. Commun., 26, 1305 (1961).
770. F.F.Crane, Anal. Chim. Acta., 16, 370 (1957).
771. C.A.Uraneck, Anal. Chem., 40, 327 (1968).
772. O.Menyharth, Acta. Chim. Hung., 41, 195 (1964).
773. R.Iwanaga, Bull. Chem. Soc. Jap., 35, 247 (1962).
774. O.Boos, Science, 117, 603 (1953).
775. J.W.Collat and S.L.Tackett, J. Electroanal. Chem., 4, 59 (1962).
776. J.C.Chapin, Am. Digest. Rep., 53, 164 (1964).
777. H.H.Anderson, J. Org. Chem., 21, 869 (1956).
778. V.A.Klimova, K.S.Zabrodina and N.L.Shitikova, Izy. Akad. Nauk. SSSR, Ser. Khim., 1, 178 (1965).
779. J.S.Fritz and G.H.Schlenk, Anal. Chem., 31, 1808 (1959).
780. P.Dostal, J.Cermak and B.Novotna, Collect. Czech. Chem. Commun., 30, 34 (1965).

781. J.A.Magnuson, Anal. Chem., 35, 1487 (1963).
782. A.Berger and J.A.Magnuson, Anal. Chem., 36, 1156 (1964).
783. J.A.Magnuson and E.W.Knaub, Anal. Chem., 37, 1607 (1965).
784. R.R.Jay, Anal. Chem., 36, 667 (1964).
785. K.Abel. Biochim. Biophys. Acta., 101, 286 (1963).
786. A.Majerova and V.Porbusky, Soudri Lek., 6, 81 (1958).
787. J.G.Mason and M.Rosenblum, J. Am. Chem. Soc., 82, 4206 (1960).
788. M.Peterlik and K.Schogl, Z. Anal. Chem., 195, 113 (1963).
789. B.P.Nikol'shif, M.S.Zakhar'evskii and A.A.Pendin, Zh. Anal. Khim., 19, 1407 (1964).
790. D.M.Knight and R.C.Schlitt Anal. Chem., 37, 470 (1965).
791. A.Grunwald, Erdol Kohle, 6, 550 (1953).
792. E.I.Uvarova and N.M.Vanyardina, Zavod. Lab., 26, 1097 (1960).
793. K.Katsumi, Y.Taguchi and S.Eguchi, Bunseki Kagaku, 12, 435 (1963).
794. J.J.Russ and W.Reeder, Anal. Chem., 29, 1331 (1957).
795. A.S.T.M. Standards on Petroleum Products and Lubricant Methods DS26-56 (1956) and DS26-61 (1961), American Soc. for Testing and Materials, Washington, D.C.
796. Y.K.Chau, P.T.S.Wong and P.D.Goulden, Anal. Chim. Acta, 85, 421 (1976).
797. J.W.Robinson, E.C.Kiesel, J.P.Goodhead, R.Bliss and R.Marshall, Anal. Chim. Acta., 92, 321 (1977).
798. T.M.Vickrey, G.V.Harrison and G.J.Ramelow, Atomic Spectroscopy 1 (4), 116 (1980.
799. D.R.Scott and L.E.Holboke, Anal. Chem., 55, 2006 (1983).
800. G.Torsi and F.Palmisono, Analyst, 108, 1318 (1983).
801. S.T.Holding and J.M.Palmer, Analyst, 109, 507 (1984).
802. K.H.Braun, Chem. Tech., (Berlin), 10, 159 (1958).
803. O.I.Milner and G.F.Shipman, Anal. Chem., 26, 1222 (1954).
804. M.Brandt and R.H.Van Der Berg, Anal. Chem., 31, 1921 (1959).
805. DIN51781, Erdol Kohle, 12, 987 (1959).
806. A.Blumenthal, Mitt. Geb. Lebensmitteluntes, Hyg., 51, 159 (1960).
807. J.Mendes Cipriano, Rev. Port. Quim., 5, 129 (1963).
808. A.Fernandez Perez, A.Peralonso and F.G.Regalado, Inf. Quim. Anal., 20, 79 (1966).
809. J.Hurtado de Mendoza Riquelme, Inf. Quim. Anal., 18, 27 (1964).
810. L.G.Escolar and M.P.Castro, Inf. Quim. Anal., 18, 66 (1964).
811. R.Gelius and K.R.Pressner, Chem. Tech. (Leipzig), 15, 290 (1963).
812. V.S.Dimitrievskii, Khim. Technol. Topliv. Masel., 3, 59 (1958).
813. H.Saori, Shoseki Giho, 2, 182 (1958).
814. E.I.Uvarova and N.M.Vanyarkina, Zavod. Lab., 26, 1097 (1960).
815. H.Gilman and L.S.Miller, J. Am. Chem. Soc., 73, 2367 (1951).
816. A.F.Clifford and R.R.Olsen, Anal. Chem., 32, 544 (1960).
817. G.Tagliavini, Chim. Ind. (Milan), 39, 902 (1957).
818. L.Newman, J.F.Philip and A.R.Jensen, Ind. Eng. Chem. Anal. Ed., 19, 451 (1947).
819. U.Belluco, G.Tagliavini and R.Barbieri, Ric. Sci., 30, 1675 (1960).
820. S.Imura, K.Fukutaka and Y.Takahiko, Bunseki Kagaku, 16, 1351 (1967).
821. G.Pilloni and G.Plazzogna, Anal. Chem. Acta., 35, 325 (1966).
822. G.Pilloni, Farmacotd. Prat., 22, 666 (1967).
823. R.M.Dagnall, T.S.West and P.Young, Talanta, 12, 583 (1965).
824. G.Pilloni, Anal. Chim. Acta., 37, 497 (1967).
825. R.Barbieri, U.Belluco and G.Tagliavini, Ric. Sci., 30, 1671 (1960).
826. G.Plazzogna and G.Pilloni, Anal. Chim. Acta., 37, 260 (1967).
827. G.Pilloni, Corsi. Semin. Chem., 9, 98 (1968).
828. P.W.Swanson and P.H.Daniels, J. Inst. Petrol., 39, 487 (1953).
829. M.S.Jovanovic, J.Tomic, Z.Masic and M.Dragojevic, Chemia.Analit., 11. 479 (1966).
830. G.Pilloni and G.Plazzogna, Ric. Sci. Parte 2, Sez. A., 34, 27 (1964).

831. G.Pilloni, Farmaco. Ed. Prat., 22, 666 (1967).
832. G.Tagliavini, U.Bulluco and L.Ricoboni, Ric. Sci. Parte 2, Sez. A., 31, 338 (1961).
833. H.L.Kies and G.Charlot, in Modern Electronalytical Methods, Elsevier, Amsterdam, 14 (1958).
834. G.Tagliavini, Anal. Chim. Acta., 34, 24 (1966).
835. G.Tagliavini, Anal. Chim. Acta., 34. 26 (1966).
836. A.L.Gol'dshtein, N.P.Lapisova and I.M.Shtifman, Zh. Anal. Khim., 17, 143 (1962).
837. N.P.Lapisova and A.L.Gol'dshtein, Zh. Anal. Khim., 16, 508 (1961).
838. K.Ziegler, F.Croismann, H.Kleiner and Schafer, Justus Liebigs, Ann, Chem., 31, 473 (1929).
839. H.Gilman, P.D.Wilkinson, W.P.Fishel and C.H.Meyers, J. Am. Chem. Soc., 45, 150 (1923).
840. H.Gilman, H.Haubine, J. Am. Chem. Soc., 66, 1515 (1944).
841. Yu N.Baryshnikov and G.I.Vesnovskaya, Zh. Anal. Khim., 19, 1128 (1964).
842. H.Gilman and F.K.Cartledge, J. Organomet. Chem., 2, 447 (1964).
843. P.F.Collins, C.W.Kamienski, E.L.Esmay and R.B.Ellestad, Anal. Chem., 33, 468 (1961).
844. C.W.Kamienski and D.L.Esmay, J. Org. Chem., 25, 115 (1960).
845. W.L.Everson, Anal. Chem., 36, 854 (1964).
846. K.C.Eberly, J. Org. Chem., 26, 1309 (1961).
847. D.E.Applequist and A.H.Peterson, J.Am. Chem. Soc., 83, 862 (1961).
848. D.E.Applequist and D.F.O'Brian, J. Am. Chem. Soc., 85, 743 (1963).
849. H.J.S.Winkler, A.W.P.Jarvie, D.J.Peterson and H.Gilman, J. Am. Chem. Soc., 83, 4089 (1961).
850. H.Gilman, F.K.Cartledge and S.Y.Sim, J. Organomet. Chem., 1, 8 (1963).
851. C.Tamborski, F.E.Ford and E.J.Soloski, J. Org. Chem., 28, 181 (1963).
852. H.Gilman, Bull. Soc. Chim. Fr., 1356 (1963).
853. S.J.Crystal and R.S. Bly, J. Am. Chem. Soc., 83, 4027 (1961).
854. I.M.Kolthoff and E.B.Sandell, Textbook of Quantitative Inorganic Analysis, Macmillan, New York, 573 (1949).
855. R.R.Turner, A.G.Alterman and T.C.Cheng, Anal. Chem., 42, 1835 (1970).
856. R.L.Eppley and J.A.Dixon, J.Organomet. Chem., 8, 176 (1967).
857. H.Gilman, R.A.Klein and H.J.S.Winkler, J. Org. Chem., 26, 2474 (1961).
858. American Society for Testing and Materials, Standard Method for Assay of n-Butyllithium Solutions, ASTM Designation E233-68 (1968).
859. Foote Mineral Co., Technical Data Bulletin, No. T.D.109 (December 1961).
860. R.A.Finnigan and H.W.Kutta, J. Org. Chem., 30, 4138 (1965).
861. H.W.Kutta, MSc. Thesis, Ohio State University (1964).
862. W.N.Smith, Foote Mineral Co., Exton, Pa., USA.
863. C.A.Uraneck, J.E.Burleigh, and J.W.Cleary, Anal. Chem., 40,327 (1968).
864. I.M.Kolthoff and W.E.Harris, Ind. Eng. Chem. Anal. Ed., 18, 161 (1946).
865. R.Adams, Org. React., 6, 353 (1951).
866. W.L.Everson, Anal. Chem., 36, 854 (1964).
867. S.C.Watson and J.F.Eastham. J. Organomet. Chem., 9, 165 (1967).
868. A.F.Clifford and R.R.Olsen, Anal. Chem., 31, 1860 (1959).
869. L.Kniel, The Determination of Active Metal Content in Alkali Metal Dispersions, CD-2817, Office of Rubber Reserve, USA (1952).
870. S.C.Watson and J.F.Eastham. J.Organomet. Chem., 9, 165 (1967).
871. R.A.Ellison, R.Griffin and F.N.Kotsonis, School of Pharmacy, University of Wisconsin, USA, personal communication.
872. D.Setferth and M.A.Weiner, J. Am. Chem. Soc., 83, 3583 (1961).
873. H.L.Johnson and R.A.Clark, Anal. Chem., 19, 869 (1947).
874. E.C.Juenge and D.Seyferth, J. Org. Chem., 26, 563 (1961).
875. A.F.Clifford and R.R.Olsen, Anal. Chem., 32, 544 (1960).
376. J.Jordan and T.G.Alleman, Anal. Chem., 29, 9 (1957).

877. S.C.Watson and J.F.Eastham, Anal. Chem., 39, 171 (1967).
878. M.Dimbat and G.R.Harlow, Anal. Chem., 34, 450 (1962).
879. Jolibois, C. R. Acad. Sci., 155, 213 (1912).
880. H.Gilman, P.D.Wilkinson, W.P.Fishel and C.H.Meyers, J. Am. Chem. Soc., 45, 150 (1923).
881. D.Mitter, J. Am. Chem. Soc., 41, 287 (1919).
882. H.Gilman and C.H.Meyers, Rec. Trav. Chim. Pays-Bas., 45, 314 (1926).
883. J.Reich, Bull. Soc. Chim. Fr., 33, 1414 (1923).
884. S.Champtier and R.Kullman, Bull. Soc. Chim. Fr., 693, 155 (1949).
885. R.D'Hollander and M.Anteunis, Bull. Soc. Chim. Belg., 72, 77 (1963).
886. H.Gilman and C.H.Myers, J. Am. Chem. Soc., 45, 159 (1923).
887. H.Gilman, E.A.Zoellner and J.B.Dicky, J. Am. Chem. Soc., 51, 1576 (1929).
888. M.Kharash and D.Reinmuth, Grignard Reactions of Non-metallic substances, Prentice Hall, New Yord (1954).
889. J.F.Eastham and S.C.Watson, J.Organomet. Chem., 9, 165 (1967)
890. A.F.Clifford and R.R.Oben, Anal. Chem., 32, 544 (1960).
891. Yu N.Barysknikov and A.A.Kvasou, Anal. Chem., 19, 117 (1964).
892. C.A.Unaneck, J.E.Burleigh and J.W.Cleary, Anal. Chem., 40, 327 (1968).
893. F.J.Langmyhr and J.Aamodt, Anal. Chim. Acta., 87, 483 (1976).
894. S.Honamura, B.W.Smith and J.D.Winefordner, Anal. Chem., 55, 2026 (1983).
895. R.Dumarez, R.Heindyck, R.Davis and J.Hoste, Anal. Chim. Acta., 107, 159 (1979).
896. B.Sjosfraud, Anal. Chem., 36, 814 (1964).
897. N.Iritani, K.Ozawa and H.Hoshida, J. Pharm. Soc., Jpn., 80, 1008 (1960).
898. K.K.Kundu and M.N.Das, Sci. Cult., 23, 660 (1958).
899. K.A.Connors and D.R.Swanson, J. Pharm. Sci., 53, 432 (1964).
900. L.Dragusin, Rev. Roum. Chim., 12, 1235 (1967).
901. L.Dragusin and N.Totir, Stud. Cercet. Chim., 13, 947 (1965).
902. W.Wronski, Chem. Anal. (Warsaw), 7, 1011 (1962).
903. H.Lanbie, Bull. Soc. Pharm. (Bordeaux), 96, 65 (1957).
904. I.Dragusin and N.Totir, Rev. Chim. (Bucharest), 15, 112 (1964).
905. I.Dragusin and N.Totir, Stud. Cercet. Chim., 13, 947 (1965).
906. I.Dragusin and A.German, Rev. Chim. (Bucharest), 14, 352 (1963).
907. J.C.Gage, Analyst (London), 86, 457 (1961).
908. E.Hoffman and A.Saracz, Z. Anal. Chem., 214, 428 (1965).
909. D.V.Carter and G.Sykes, Analyst (London), 83, 536 (1958).
910. I.G.Druzhinin and P.S.Kislitsin, Tr. Inst. Khim. Akad, Nauk. Kirk. SSR., 8, 21 (1957).
911. B.Hetnarski and K.Hetnarska, Bull. Acad. Polon, Sci. Ser. Chim. Geol. Geogr., 7, 645 (1959).
912. F.H.Merkle and C.A.Discher, J. Pharm. Sci., 51, 117 (1962).
913. R.W.Wilkinson and T.F.Williams, United Kingdom Atomic Energy Research Establishment Report AERE R-3258 (1960).
914. U.A.Sant, H.Sankar, Anal. Chim. Acta., 19, 202 (1958).
915. A.W.Ashbrook, Anal. Chim. Acta., 24, 504 (1961).
916. A.N.Pukhonto, A. Ya Zhavoronkova, E.I.Moiseeva and V.F.Smirnov, Zhur. Anal. Khim., 20, 372 (1965).
917. D.A.Bernhart and K.H.Rattenberg, Anal. Chem., 28, 1765 (1956).
918. M.Geiger and R.Furer, Zhur. Anal. Chem., 174, 401 (1960).
919. K.Braundsorf, Pharm. Zentralh. Deut., 93, 222 (1954).
920. Y.C.Lee and G.Ting, Anal. Chim. Acta., 106, 373 (1979).
921. A.Yarden and C.Eger, Bull. Res. Council. Israel. A7, 80 (1958).
922. Q.E.Thompson, J. Amer. Chem. Soc., 83, 845 (1961).
923. R.Siegfried, Anal. Chem., 49, 1584. (1977).
924. A.P.Kreshkov, V.A.Drozdov and N.A.Kolchina, Zhur. Anal. Khim., 19, 1177 (1964).
925. A.P.Kreshkov, V.A.Drozdov and N.A.Kolchina, Zavod. Lab., 31, 160 (1965).

926, T.Jasinki, A.Madro and T.Madro, Chem. Anal. (Warsaw), 10, 929 (1965).
927. S.Sass, I.Master, P.M.Davis and N.Beitsch, Anal. Chem., 32, 285 (1960).
928. L.K.Beach and S.Sass, Anal. Chem., 33, 901 (1961).
929. L.K.Beach, U.S.Patent Application SN 738,241 May 27th 1958, filed to issue 4/7/61 (1958).
930. R.Sass, N.Beitsch and A.Morgan, Anal. Chem., 31, 1970 (1959).
931. S.T.Ross and D.B.Denney, Anal. Chem., 32, 1896 (1960).
932. J.S.Fritz and G.S.Hammond in "Quantitative Organic Analysis" Chapter 3 Wiley, New York (1957).
933. C.A.Streuli,Anal. Chem., 31, 1652 (1959).
934. C.A.Streuli, Anal. Chem., 32, 985 (1960).
935. G.Fritz and G.Z.Pappenburg, Anorg. Chem., 331, 147 (1964).
936. D.C.Wimer, Anal. Chem., 30, 2060 (1958).
937. D.C.Wimer, Anal. Chem., 34, 873 (1962).
938. W.A.Henderson, C.A.Streuli and S.A.Buckler, in "Abstracts 137th meeting American Chemical Society", Cleveland, Ohio, April 1960 p.47 (1960).
939. V.Batora and Z.Vasela, Sb. Prac. Vyczk. Ust. Agrochem. Technol. Bratizlava, 1, 85 (1961).
940. U.M.Shul'man, S.V.Larionov and L.A.Padol'skaya, Zhur. Anal. Khim., 22, 1165 (1967).
941. H.Bode and W.Arnswald, Z. Anal. Chem., 185, 99 (1961).
942. P.R.E.Lewkowitsch, J. Inst. Petroleum, 48, 217 (1962).
943. P.R.E.Lewkowitsch, Chem. and Ind., 27, 1241 (1962).
944. Ya I.Yamaleev and M.V.Pal'yanova, Zavod. Lab., 33, 1381 (1967).
945. E.L.Gefter, Zavod. Lab., 24, 691 (1958).
946. A.P.Kreshkov, V.A.Drozdov and R.R.Tarasyants, Plast. Massy, 4, 57 (1963).
947. R.Huls and M.Renson, Bull. Soc. Chim. Belg., 65, 696 (1956).
948. C.A.Uraneck, J.E.Burleigh and J.W.Cleary, Anal. Chem., 40, 327 (1968).
949. O.Foss and S.R.Svendsen, Acta. Chem. Scand., 8, 1351 (1954).
950. A.Kotarski, Chem. Anal. (Warsaw), 10, 541 (1965).
951. A.Kotarski, Chem. Anal. (Warsaw), 10, 321 (1965); Anal. Abstr., 13, 4805 (1966).
952. L.M.Shtifman, V.V.Lastovich and L.G.Kuryakova, Zvod. Lab., 29, 546 (1963).
953. R.R.Turner, A.G.Alteron and T.C.Cheng, Anal. Chem., 42, 1835 (1970).
954. B.L.Pepe, R.Rivarola-Barbieri, Analyt, Chem., 40, 432 (1968).
955. A.P.Kreshkov, V.A.Drozdov and D.G.Vlasova, Izv. Uyssh. Ucheb. Zavedenii i Khim. Tekhnol., 3, 85 (1960): Ref. Zhur. Khim., 18, Abstr. No. 73,200 (1960).
956. A.P.Kreshkov and V.A.Drozdov, Dokl. Akad. Nauk SSSR., 131, 1345 (1960): Ref. Zhur Khim., 18, Abstr. No. 73,199 (1960).
957. A.P.Kreshkov, Trudy Mosk. Khim. Tekhnol. Inst. Im. D.I.Mendeleeva, 32. 333 (1961). Ref. Zhur. Khim., 8, Abstra. No. 8D211 (1962).
958. A.P.Kreshkov, V.A.Bork and M.I.Aparsheva, Zhur. Anal. Khim., 18, 10, 1149 (1963).
959. A.P.Kreshkov, V.A.Drozdov and E.G.Vlasova, Zavod. Lab., 26, 1080 (1960).
960. A.P.Kreshkov, V.A.Drozdov and E.G.Vasova, Igv. Vyssh. Ucheb. Zavedenii. Khim. i Khim. Tekhnol., 3, 80 (1960). Ref. Zhur. Khim., 18, Abstr. No. 73,203 (1960).
961. E.G.Kreshkov, V.A.Drozdov and R.R.Tarasyants, Zavod. Lab., 30, 413 (1964).
962. R.R.Tarayants, Trudy. Moskov. Khim. Tekhnol. Inst., 44, 149 (1963). Ref. Zhur. Khim., 19GDE (20), Abstr. No. 20G155 (1964).
963. V.A.Drozdov and E.G.Vlasova, Trudy, Komiss. Anal. Khim. Akad. Nauk. SSSR., 13, 187 (1963). Ref. Zhur. Khim., 19GDE (19) Abstr. No. 19G191 (1963).

964. H.Gilman, R.A.Klein and H.J.S.Winkler, J. Org. Chem., 26, 2474 (1961).
965. J.Efer, D.Quaas and W.Spichale, Z. Chem., 5, 390 (1965).
966. K.Burger, Z. Lebensm. Unters. Forsch., 114,1 (1961).
967. K.G.Bergner, U. Audt and D.Mack, Dsch. Lebensm. Rundsch., 63, 180 (1967).
968. R.Geyer and H.J.Seiditz., Z. Chem., 4, 468 (1964).
969. E.Friebe and H.Kelker, F. Z. Anal. Chem., 192, 267 (1963).
970. B.G.Kushlefsky and A.Ross, Anal. Chem., 34, 1666 (1962).
971. B.Kushlefsky, I.Simmons and A.Ross, Inorg. Chem., 2, 187 (1963).
972. H.Gilman and L.S.Miller, J. Am. Chem. Soc., 73, 2367 (1951).
973. H.H.Anderson, Anal. Chem., 34, 1340 (1962).
974. W.P.Neumann and R.Sommer, Justus Liebigs. Ann. Chem., 701, 28 (1967).
975. H.Owoki, H.Maeda and N.Wada, Nagoyashi Kogyo Kenkyusho Kenkyu Hakoku, 24, 92 (1960).
976. G.N.Freeland and R.M.Hoskinson, Analyst (London), 95, 579 (1970).
977. H.Gilman, F.K.Cartledge and S.Y.Sim, J. Organomet. Chem., 1, 8 (1963).
978. G.Tagliavini and P.Zanella, Anal. Chim. Acta., 40, 33 (1968).
979. V.Gutman and F.Mairinger, Z. Anorg. Allg. Chem., 289, 279 (1957).
980. G.Plazogna and G.Pilloni, Anal. Chim. Acta., 37,260 (1967).
981. G.Pilloni, Corsi. Semin. Chim., 9, 98 (1968).
982. L.Magon, R.Portanova, A.Cassol and G.Rizzardi, Ric. Sci., 38, 782 (1968).
983. G.Tagliavini, Anal. Chim. Acta., 34, 24 (1966).
984. U.Belluco, G.Tagliavini and R.Barbieri, Ric. Sci., 30, 1675 (1960).
985. H.L.Kies and G.Charlot, in Modern Electrchemical Methods, Elsevier, Amsterdam, 14 (1958).
986. L.Haasova and M.Pribyl, Z. Anal. Chem., 249, 35 (1970).
987. F.Magno and G.Pilloni, Anal. Chim. Acta., 41, 413 (1968).
988. G.Tagliavini and G.Plassogna, Ric. Sci., Parte, 2, Sez. A., 2,356 (1962).
989. H.Gilman, F.K.Cartledge and S.Y.Sim., J. Organometallic Chem., 1, 8 (1963).
990. T.R.Crompton, Analyst (London), 91, 374 (1966).
991. K.Novak, Chem. Prum., 12, 551 (1962).
992. P.T.Makarov and K.F.Pershina, Khim. Tekhnol. Topl. Masel., 10, 62 (1963).
993. O.Lorenz, E.Echte and U.Kantsch, Gummi, 9, 300 (1956).
994. W.L.Everson and E.M.Ramirez, Anal. Chem., 37, 812 (1965).
995. T.E.Bonitz, Chem. Ber., 88, 742 (1955).
996. W.P.Neumann, Angew. Chem., 69, 730 (1957).
997. W.P.Neumann, Justus Liebigs Ann. Chem., 629, 23 (1960).
998. W.P.Neumann, Dissertation, University of Geissen)1959).
999. J.H.Mitchen, Anal.Chem., 33, 1331 (1961).
1000. C.N.Wadelin, Talanta, 10, 97 (1962).
1001. K.Ziegler and H.Gellert, Justus Liebigs. Ann. Chem., 629, 20 (1960)
1002. D.F.Hagen and W.D.Leslie, Anal. Chem., 35, 814 (1963).
1003. G.Bruekner and M.Parkany, Magy. Kem. Foly., 68,164 (1962).
1004. J.W. Cavett, J. Assoc. Off. Agric. Chem., 39, 857 (1956).
1005. C.E.Stringer and M.Attrep, Anal. Chem., 51, 731 (1979).
1006. P.Malastesta and A.Lorenzini, Ric. Sci., 28, 1874 (1958).
1007. R.M.Fournier, Mein, Poudres., 40, 386 (1958).
1008. N.Z.Bruja, Rev. Chim. (Bucharest), 17, 359 (1966).
1009. M.M.Auerbach and W.W.Haughtaling, Drug Stand., 28, 115 (1960).
1010. W.H.Hill, L.J.Kuhns, J.M.Merrill, B.J.Palm, J.Seals and U.Urquiza, Am. Ind. Hyg. Assoc. J., 21, 231 (1960).
1011. L.J.Kuhns, R.H.Forsyth and I.Masi, Anal. Chem., 28, 1750 (1956).
1012. H.G.Offner, Anal. Chem., 37,370 (1965).
1013. W.H.Hill and M.S.Johnston, Anal. Chem., 27, 1300 (1955).

1014. R.S.Braman, D.D.De Ford, T.N.Johnston and L.J.Kuhns, Anal. Chem., 32, 1258 (1960).
1015. E.A.Pfitzer and J.M.Seals, Am. Ind. Hyg. Assoc. J., 20, 329 (1959).
1016. R.S.Braman and T.N.Johnston, Talanta, 10, 810 (1963).
1017. W.H.Hill, J.M.Merrill and R.H.Larsen, U.S.Atomic Energy Comm., Rep. CCC-1024-TR-228, 13 (1957).
1018. D.L.Hill, E.I.Gipson and J.F.Heacock, Anal. Chem., 28, 133 (1956).
1019. W.H.Hill, J.M.Merrill and B.J.Palm, U.S. Atomic Energy Comm., Rep. CCC-1024-TR-223, 10 (1957).
1020. D.Thierig and F.Umland, Z. Anal. Chem., 215, 24 (1966).
1021. A.H.Sholoway and J.R.Messer, Anal. Chem., 36, 433 (1964).
1022. G.R.Friston, L.Bennett and W.G.Bert, Anal. Chem., 31, 1696 (1959).
1023. J.G.Heathcote and P.J.Duff, Analyst (London), 79, 727 (1954).
1024. H.L.Newark and M.Leff, Drug Stand., 25, 177 (1957).
1025. C.F.Bruening, W.L.Hall and O.L.Kline, J. Am. Pharm. Assoc. Sci. Ed., 47, 15 (1958).
1026. C.F.Bruening and O.L.Kline, J. Pharm. Soc., 50, 537 (1961).
1027. J.Bayer, Chimia., 15, 555 (1961).
1028. M.Covello and O.Schettino, Ann. Chim. (Rome), 52, 1135 (1962).
1029. H.Brink, Pharm. Weekbi., 97, 505 (1962).
1030. J.Bayer, Gyogyszereszet, 8, 21 (1964).
1031. J.Bayer, Pharmazie, 19, 602 (1964).
1032. G.Asensi, Quim. Ind. (Bilbao), 12, 141 (1965).
1033. D.Monnier, Y.Ghaliounglu and R.Saba, Anal. Chim. Acta., 28, 30 (1963). Anal. Abstr., 10, 5410 (1963).
1034. T.Rahandraha, M.Chanez and M.Sagot-Masson, Ann. Pharm. Fr., 22, 663 (1964).
1035. R.K.Mitra, P.C.Bose, G.K.Ray and B.Mukerji, Indian J. Pharm., 24, 152 (1962).
1036. N.E.Dowd, A.M.Killard, H.J.Pazdera and A.Ferrari, Ann. N.Y. Acad. Sci., 130, 558 (1965).
1037. A.Sezerat, Ann. Pharm. Fr., 22, 159 (1964).
1038. A.Dominquez, G.Oller and M.Oller, Galenica Acta., 14, 157 (1961).
1039. G.O.Rudkin and R.J.Taylor, Anal. Chem., 34, 1155 (1952): Anal. Abstr., 24, 1155 (1952).
1040. G.Parissakis and P.B.Issopoulos, Pharm. Acta. Helv., 40, 653 (1965).
1041. E.C.Abe, A.A.Raheem and M.M.Dokhana, Z. Anal. Chem., 189, 389 (1962).
1042. V.Brustier, A.Castaigne, E. de Montety and A.Anselem, Ann. Pharm. Fr., 22, 373 (1964).
1043. V.N.Danilova, Ukr. Khim. Zh., 30, 651 (1964).
1044. A.K.Babko and N.M.Lukovskaya, Zh. Anal. Khim., 17, 50 (1962); Anal. Abstr., 9, 4078 (1962).
1045. A.K.Babko and N.M.Lukovskaya, Zavod. Lab., 29, 4104 (1963); Anal. Abstr., 11, 1300 (1964).
1046. W.U.Malik and S.I.Ahmed, Indian J. Chem., 2, 247 (1964).
1047. W.U.Malik and R.Haque, Z. Anal. Chem., 189, 179 (1962); Anal. Abstr., 10, 1100 (1963).
1048. W.U.Malik and S.I.Ahmed, J. Am. Oil. Chem. Soc., 42, 451 (1963).
1049. W.J.Mader and R.G.Johl, J. Am. Pharm. Assoc. Sci. Ed., 44, 577 (1955).
1050. E.L.Smith, J.L.Martin, R.J.Gregory and W.H.C.Shaw, Analyst (London), 87, 183 (1962).
1051. G.Bellomonte and E.Cingolani, R.C. 1st. Sup. Sanit., 26, 1050 (1963).
1052. W.U.Malik and R.Haque, Z. Anal. Chem., 189, 179 (1962).
1053. A.G.Kempton, M.Greenberger and A.M.Kaplan, Am. Digest Rep., 51, 19 (1952).
1054. E.Hoffman, A.Saracz and B.Z.Burstyn, Anal. Chem., 215, 101 (1966).
1055. E.Hoffman, A.Saracz and Z.Bursztyn, Z. Anal. Chem., 208, 431 (1965): Anal. Abstr., 13, 3672 (1966).
1056. British Standards Institution, BS 3769: 1964, 20 pp.

1057. V.G.Hiatt, J. Assoc. Off. Agric. Chem., 47, 253 (1964).
1058. S.U.Kreingol'd and E.H. Bozhevov'nov, Trudy Kom. Analit. Khim., 16, 194 (1968).
1059. T.N.Gladyshevskaya, Yas. Kvyatkovskaya and B.A.Sobchuk, Ukr. Biokhim Zh., 29, 317 (1957): Ref. Zh. Khim., Abstract. No. 10, 852 (1958).
1060. V.V.Popov and B.A.Sobshuk, Lab. Delo, 3, 19 (1957).
1061. O.O.Clegg, Br. Med. J., 2, 329 (1942).
1062. J.F.Scaife, Analyst (London), 80, 562 (1955).
1063. E.J.van Kampen and W.G.Zijlstra, Clin. Chim. Acta., 6, 538 (1961).
1064. E.J.van Kampen and W.G.Zijlstra, Clin. Chim. Acta., 7, 147 (1962).
1065. N.Ressler, N.A.Nelson and I.M.Smith, J. Lab. Clin. Med., 54, 304 (1959).
1066. W.G.Zijlstra and E.J.van Kampen, Clin. Chim. Acta., 7, 96 (1962).
1067. W.G.Zijlstra and E.J.van Kampen, Clin. Chim. Acta., 5, 719 (1960).
1068. O.Stadic, J. Biol. Chem., 41, 237 (1920).
1069. L.Tentor, G.Vivaldi and A.M.Salvati, Clin. Chim. Acta., 14, 276 (1966).
1070. W.Pilz, I.Johann and E.Steizl, Z. Anal. Chem., 215, 260 (1966).
1071. W.Pilz, I.Johann and E.Steizl, Z. Anal. Chem., 215, 105 (1966): Anal. Abstr., 14, 1627 (1967).
1072. G.Vanzetti and A.Nordeschi, J. Lab. Clin. Med., 67, 116 (1966).
1073. H,V.Connerly and A.R.Briggs, Clin. Chem., 8, 151 (1962).
1074. A.Kumlien, K.Paul and S.Ljungberg, J. Clin. Lab. Invest., 12, 381 (1960).
1075. P.C.Elwood and A.Jacobs, Br. Med. J., 1, 20 (1966).
1076. British Standards Institution, BS 3420; (1961).
1077. W.Steffens, Naturwissenschaften., 49, 109 (1962).
1078. M.Harboe, Scand. J. Clin. Lab. Invest., 11, 66 (1959).
1079. G.E.Hanks, M.Cassell and H.Chaplin, J. Lab. Clin. Med., 56, 486 (1960).
1080. G.V.Derviz and N.K.Byalko, Lab. Delo., 8, 461 (1966).
1081. T.R.Johnson, J. Lab. Clin. Med., 53, 495 (1959).
1082. W.Pollmann, Klin. Wochenschr., 44, 789 (1966).
1083. K.B.McCall, Anal. Chem., 28, 189 (1956).
1084. A.Martin and M.Zade-Oppen, Acta. Soc. Med. Ups., 65, 249 (1960).
1085. A.F.Beau, Am. J. Clin. Pathol., 38, 111 (1962).
1086. J.M.McKenzie, P.R.Fowler and V.Fiorica, Anal. Biochem., 16, 139 (1966).
1087. N.Chamori, R.J.Henry and O.J.Golub, Clin. Chim. Acta., 6, 1 (1961).
1088. O.Malloy and O.Evelyn, J. Biol. Chem., 119, 481 (1937).
1089. W.E.Huckabee, J. Lab. Clin. Med., 46, 486 (1955).
1090. G.G.Nahas, J. Appl. Physiol., 13, 147 (1958).
1091. W.G.Zijlstra, Klin. Wochenschr., 34, 384 (1956).
1092. H.E.Refsum and S.L.Sveinssen, Scand. J. Clin. Lab. Invest., 8, 67 (1956).
1093. R.G.Martinek, J. Am. Med. Technol., 21, 42 (1966).
1094. O.Wolff, Ann. Med. Leg., 221 (1947).
1095. T.P.Whitehead and S.Worthington, Clin. Chim. Acta., 6, 346 (1961).
1096. F.Vogt-Lorentzen, Scand. J. Clin. Lab. Invest., 14, 648 (1962).
1097. L.Heilmeyer, in Spectrometry in Medicine, Adam Hilger, London (1943).
1098. H.Hartmann, Ergeba Physiol., 39, 413 (1937).
1099. B.Steinmann, Archw. Exp. Pathol. Pharm., 191, 237 (1939).
1100. P.L.Drabkin, in Medical Physics, Vol. 11,Chicago Book Publishers, Chicago (1950).
1101. B.T.Commins and P.J.Lawther, Br. J. Ind. Med., 22, 139 (1965).
1102. R.Richterich, in Clinical Chemistry, Academic Press, New York (1969).
1103. R.A.Knight, D.C.Ephraim and F.E.Payne, W. G. Pye Unicam. Anal. News, Vol. 3 (Dec 1973).

1104. A.A.Christman and E.L.Randall, J. Biol. Chem., 102, 595 (1933).
1105. R.Wennesland, Acta Physiol. Scand., 1, 49 (1940).
1106. C.H.Gray and H.Sandiford, Analyst (London), 71, 107 (1946).
1107. M.T.Ryan, J.Nolan and E.J.Conway, Biochem. J., 42, 94 (1948).
1108. L.A.Williams, R.A.Linn and B.Zak, Am. J. Clin. Pathol., 34, 334 (1960).
1109. I.I.Datsenko, Lab. Delo., 9, 8 (1963).
1110. W.G.Zijlstra and C.J.Moller, Clin. Chim. Acta., 2, 237 (1957).
1111. G.E.Martin, J.I.Munn and L.Biskup, J. Assoc. Off. Agric. Chem., 43, 743 (1960).
1112. J.C.Kaplan, Rev. Fr. Etud. Clin. Biol., 10, 856 (1965).
1113. G.V.Derviz, Lab. Delo., 9, 527 (1966).
1114. T.Leahy and R.Smith, Clin. Chem., 6, 148 (1960).
1115. O.Evelyn and O.Malloy, J. Biol. Chem., 126, 655 (1960).
1116. O.Crosby, U.S. Armed Forces Med. J., 4, 693 (1954).
1117. J.Sendroy, H.A.Collison and H.J.Mark, Anal. Chem., 27, 1641 (1955).
1118. R.S.Brief, R.S.Ajemain and R.G.Conter, Am. Ind. Hyg. Assoc. J., 28, 21 (1967).
1119. A.B.Densham, P.A.A.Beale and R.Palmer, J. Appl. Chem., 13, 576 (1963).
1120. United Kingdom Atomic Energy Authority, Production Group Chemical Services Technical Department, Windscale, UKAEA Rep., PG 483 (W) 1963, 9 pp.
1121. H.Irving and J.J.Cox, J. Chem. Soc., 1470 (1961).
1122. R.Barbieri, G.Tagliavini and U.Belluco, Ric. Sci., 30, 1963 (1960).
1123. S.R.Henderson and L.J.Snyder, Anal. Chem., 33, 1172 (1961).
1124. J.E.Cremer, Br. J. Ind. Med., 16, 191 (1959).
1125. W.N.Aldridge and J.E.Cremer, Analyst (London), 80, 37 (1957).
1126. S.R.Henderson and L.J.Snyder, Anal. Chem., 31, 2113 (1959).
1127. M.E.Griffing, A.Rozek, L.J.Snyder and S.R.Henderson, Anal. Chem., 29, 190 (1957).
1128. G.Pilloni and G.Plazzogna, Anal. Chim. Acta., 35, 325 (1966).
1129. G.Pilloni, Farmac, Ed. Prat., 22, 666 (1967).
1130. R.M.Dagnall, T.S.West and P.Young, Talanta, 12, 583 (1965).
1131. G.Pilloni, Anal. Chim. Acta, 37, 497 (1967).
1132. L.Roncucci, G.Faraglia and R.Barbieri, J. Organomet. Chem., 1, 427 (1964).
1133. D.Blake, G.E.Coates and J.M.Tate, J. Chem. Soc., 756 (1961).
1134. D.L.Alleston and A.G.Davies, J. Chem. Soc., 2050 (1962).
1135. T.Tanaka, M.Komuna, Y.Kawasaki and R.Okawara, J. Organomet. Chem., 1, 484 (1964).
1136. M.Yasuda and R.S.Tobias, Inorg. Chem., 2, 207 (1963).
1137. Y.Kawaskai, T.Tanaka and R.Dkawara, Bull. Chem. Soc. Jap., 27, 903 (1964).
1138. W.Bolanowska, Chem. Anal. (Warsaw), 12, 121 (1967).
1139. L.Newmann, J.F.Phillip and A.R.Jensen, Ind. Eng. Chem. Anal. Ed., 19, 451 (1947).
1140. V.A.Smith, W.E.Dilaney, W.J.Tanag and J.C.Bailie, Anal. Chem., 22, 1230 (1950).
1141. D.Loroue and G.Paul, Fr. Tech. Pet. Bull. Assoc., 127, (1963).
1142. D.Loroue and G.Paul, Rev. Inst. Fr. Pet. Ann. Combust. Liq., 17, 830 (1962).
1143. W.K.Rudnevskii, Yyshinskii Izv. Akad. Nauk SSSR, Ser. Fiz., 23, 1228 (1959).
1144. R.De Hollander and M.Anteunis, Bull. Soc. Chim. Belg., 72, 77 (1963).
1145. C.Blomberg, A.D.Vrengdenkill and P.Vink, Recl. Trav. Chim..Pays-Bas Belg., 83, 662 (1964).
1146. H.Gilman, P.D.Wilkinson, W.P.Fishel and C.H.Meyers, J. Amer. Chem. Soc., 45, 150 (1923).

1147. H.Gilman, E.A.Soeller and J.B.Dicky, J. Amer. Chem. Soc., 51, 1576 (1929).
1148. A.D.Vrengdenhill and C.Blomberg, Recl. Thav. Chim., Pays-Bas. Belg., 82, 453 (1963).
1149. H.Flaschka, Mikrochemie, 39, 38 (1952).
1150. M.Pedinelli, Chem. et Ind., 41, 1180 (1959).
1151. V.L.Miller and D.Polley, Anal. Chem., 26, 1333 (1954).
1152. E.Hoffman, Z. Anal. Chem., 174, 48 (1960).
1153. S.Ishikuro and K.Yokata, Chem. Pharm. Bull. Jap., 11, 939 (1963).
1154. M.Iguchi, A.Nishiyama and Y.Nugase, J. Pharm. Soc. Jap., 80, 1437 (1960).
1155. E.J.Cafruny, J. Lab. Clin. Med., 57, 468 (1961).
1156. V.L.Miller and D.Lillis, Anal. Chem., 30, 1705 (1958).
1157. E.Hoffman, Z. Anal. Chem., 182, 193 (1961).
1158. Y.Kimura and V.L.Miller, Anal. Chim. Acta., 27, 325 (1962).
1159. J.Story, K.Kratzer and J.Prasilova, Anal. Chim. Acta., 100, 643 (1978).
1160. Metallic Impurities in Organic Matter Sub-Committee of Analytical Methods Committee, Analyst (London), 90, 515 (1965).
1161. M.G.Ashley, Analyst (London), 84, 124 (1959).
1162. T.Etchu, Bull. Agric. Chem. Insp. Stn., 7, 17 (1967).
1163. T.Etchu, Bull. Agric. Chem. Insp. Stn., 7, 21 (1967).
1164. T.Etchu, Bull. Agric. Chem. Insp. Stn., 7, 25 (1967).
1165. D.Polley and V.L.Miller, Anal. Chem., 24, 1622 (1952).
1166. J.Stary, K.Kratzer and J.Prasilova, Anal. Chim. Acta., 100, 627 (1978).
1167. I.Iwantscheff, Das. Dithizon und Seine Anwendung in der Nikro and Sprin analyse, Verlag Chemie, Weinheim (1958).
1168. T.D.Waugh, H.F.Walton and J.A.Laswick, J. Phys. Chem., 59, 395 (1955).
1169. G.Schwarzenbach and M.Scheilenberg, Hely. Chim. Acta., 48, 28 (1965).
1170. M.Vancea and A.German, Rev. Chim., 19, 58 (1968).
1171. K.Ozawa and S.Egoshira, Bunseki Kagaku, 11, 506 (1962).
1172. A.Eldridge and T.R.Sweet, Anal. Chem., 28, 1268 (1956).
1173. I.M.Kolthoff and E.B.Sandell, Text Book of Inorganic Quantitative Analysis, 3rd ed., Macmillan, New York,, 461 (1952).
1174. G.Gran, Sven. Papperstidn., 53, 234 (1950).
1175. D.Polly and V.L.Miller, J. Argric. Food Chem., 2, 1030 (1954).
1176. G.Rentsch, Z. Anal. Chem., 178, 100 (1960).
1177. J.E.McCarley, R.S.Saltzman and R.H.Osborn, Anal. Chem., 28, 880 (1956).
1178. V.S.Fikhtengol'ts and N.P.Kozlov, Zavod. Lab., 23, 917 (1957).
1179. H.Suszuki and K.Oishi, Bunseki Kagaku, 12, 1011 (1963).
1180. R.S.Brief, F.S.Venable and R.S.Aiemian, Am. Ind. Hyg. Assoc. J., 26, 72 (1965).
1181. A.B.Denshaw, P.A.A.Beale and R.Palmer, J. Appl. Chem., 13, 576 (1963).
1182. United Kingdom Atomic Energy Authority (1963) Production Group Chemical Services Dept., Windscale, United Kingdom, UKAED Report PG-483 (w) 91 p (1963).
1183. E.Raboz, W.C.Hess, R.R.Ni Nella and W.Cevallos, J. Lab. Clin. Med., 52, 158 (1958).
1184. A.K.R.McDowell, J. Dairy Res., 25, 192 (1958).
1185. K.E.Richardson and N.E.Tolbert, J. Biol. Chem., 236, 1285 (1961).
1186. A.J.Fudge and G.C.Hutton, Atomic Energy Research Establishment, Harwell, United Kingdom, AERE Report C/R.2384 (1957).
1187. S.Sass and J.Cassidy, Anal. Chem., 28, 1968 (1956).
1188. H.D.Baerstein, Ind. Eng. Chem. Anal. Ed., 15, 251 (1943).
1189. F.C.Whitmore in "Organic Chemistry", p.731, Van Nostrand, New York (1937).
1190. B.C.Saunders and B.P.Stark, Tetrahedron, 4, 197 (1958).
1191. S.I.Bass, Zhur. Anal. Khim., 17, 113 (1962).

1192. D.Serafin, A.Janik and F.Maciejowski, Chem. Anal. (Warsaw), 9, 65 (1964).
1193. H.J.Brandt, Anal. Chem., 33, 1390 (1961).
1194. P.A.Giang, W.F.Barthel and S.A.Hall, J. Agr. Food Chem., 2, 1281 (1954).
1195. R.M.Fournier, Chem. et Ind., 76, 246 (1956).
1196. B.Gehanf, J.Epstein, G.B.Wilson, B.Witten, S.Sass, V.E.Bauer and W.H.C.Rueggeberg, Anal. Chem., 29, 278 (1957).
1197. S.Sass, W.D.Ludemann, B.Witten, V.Fischer, A.J.Sisti and J.I.Miller, Anal. Chem., 29, 1346 (1957).
1198. D.J.Marsh and E.Neale, Chemi and Ind. (London), 494 (1956).
1199. J.Goldenson, Anal. Chem., 29, 877 (1957).
1200. J.Kolmerton and J.Epstein, Anal. Chem., 30, 1536 (1958).
1201. S.R.Dickmann and R.H.Bray, Ind. Eng. Chem. Anal. Ed., 12, 665 (1940).
1202. C.J.Anderson and R.A.Keeler, Anal. Chem., 26, 213 (1954).
1203. J.W.O'Laughlin, F.W.Sealock and C.V.Banks, Anal. Chem., 36, 224 (1964).
1204. J.W.O'Laughlin, C.V.Banks, U.S. Atomic Energy Comm. Report IS-737 (1963).
1205. J.Epstein, A.Koblin in "Division of Analytical Chemistry Symposium on Air Pollution, 130th meeting", American Chemical Society, Atlantic City, U.S.A., September (1956).
1206. A.Koblin J.Epstein, Chemical Warefare Laboratories, Army Chemical Centre, Maryland, U,S.A., presented at Air Pollution Symposium, Atlantic City Meeting (130th) of American Chemical Society, September (1956).
1207. E.Neale and B.J.Perry, Analyst, 84, 226 (1959).
1208. M.Mascero and M.Perini, Chem. Ind. (Milan), 37, 945 (1955).
1209. S.A.Kerfedt, Acta. Chem. Scand., 13, 1479 (1959).
1210. A.Gomori, J. Lab. Clin. Med., 27, 955 (1942).
1211. M.A.Petrov, Yu Yu Lur'e, Zavod. Lab., 29, 416 (1963).
1212. J.Kovac and E.Paulinyova, Chem. Prumysl., 13, 582 (1963).
1213. S.Akerfeldt, Acta. Chem. Scand., 13, 1479 (1959).
1214. M.Garnier and A.Wakli, Lebanese Pharm., 5, 39 (1957).
1215. R.K.Morton, Biochem. J., 70, 134 (1958).
1216. O.Heva and D.A.Rappoport, T. Biol. Chem., 149, 47 (1943).
1217. J.J.Boltralik and H.Noll, Anal. Biochem., 1, 269 (1960).
1218. O.Bublitz and O.Kennedy, J. Biol. Chem., 211, 951 (1954).
1219. M.Kawamura and K.Matsumoto, Bunseki Kagaku, 14, 789 (1965).
1220. K.L.Cheng, Chemist Analyst, 45, 67 (1956); Anal. Abstr., 4, 1189 (1957).
1221. A.J.Bawd, D.T.Burns and A.G.Fogg, Talanta, 16, 719 (1969).
1222. B.L.Pepe and R.Barbieri-Rivarola, Anal. Chem., 40, 432 (1968).
1223. M.F.Borisov, E.Y.Gutt-saet, L.M.Shtifman and L.A.Efremova, Vesty. Tekh. i Edon, Inform. Nauch-Issled. Inst. Tekhe-Ekon. Issled. Gos. Kom. Khim. i Neft. Prom. Pri. Gosplane. SSSR., 10, 51 (1963); Zhur. Khim., 19GDE (18) No. 18G171 (1964),
1224. K.K.Popkov and S.L.Lulchuk, USSR Pat. 129,385 (5/6/60) (1960).
1225. S.U.Kreingol'd and E.A.Bozheval'nov, Trudy. Kom. Analyt. Khim., 16, 194 (1968).
1226. M.Myamoto, Japan Analyst, 12, 233 (1963).
1227. A.Schneer, T.Holmes and T.Szekely, Z. Anal. Chem., 182, 178 (1961).
1228. S.Fugiwara and H.Narasaki, Anal. Chem., 36, 206 (1964).
1229. W.N.Aldridge and J.E.Cremer, Nature (London), 178, 1306 (1956).
1230. W.N.Aldridge and J.E.Cremer, Analyst (London), 82, 37 (1957).
1231. E.B.Sandell, Colorimetric Determination of Trace Metals, Interscience, New Yor, London, 90 (1944).
1232. R.Barbieri, G.Tagliavini and U.Belluco, Ric. Sci., 30. 1963 (1960).

1233. H.Irving and J.J.Cox, J. Chem. Soc., 1470 (1961).
1234. R.T.Skeel and C.E.Bricker, Anal. Chem., 33, 428 (1961).
1235. G.L.Snable, Senior Thesis, Princeton University (1959).
1236. G.Pilloni and G.Plazzogna, Anal. Chim. Acta., 35, 325 (1966).
1237. G.Pilloni. Anal. Chim. Acta., 37, 497 (1967).
1238. G.Pilloni, Farmaco, Ed. Prat., 22, 666 (1967).
1239. R.S.Tobias, I.Ogrins and B.A.Nevett, Inorg. Chem., 1, 638 (1962).
1240. R.S.Tobias and M.Yasuda, J. Phys. Chem., 68, 1820 (1964).
1241. R.S.Tobias and M.Yasuda, Can. J. Chem., 42, 781 (1964).
1242. R.S.Tobias, Organomet. Chem. Rev., 1, 93 (1966).
1243. A.Corsino, I. May-Ling Yih, Q.Fernando and H.Frieser, Anal. Chem., 34, 1090 (1962).
1244. D.Betteridge, Q.Fernando and H.Freiser, Anal. Chem., 35, 294 (1963).
1245. T.Tanaka, M.Komura, Y.Kawasokai and R.Okawara, J. Organomet. Chem., 1, 484 (1964).
1246. L.Roncucci, G.Faraglia and R.Barbieri, J. Organomet. Chem., 1, 427 (1964).
1247. D.Blake, G.E.Coates and J.M.Tate, J. Chem. Soc., 756 (1961).
1248. D.L.Alleston and A.G.Davies, J. Chem. Soc., 2050 (1962).
1249. M.Yasuda and R.S.Tobias, Inorg. Chem., 2, 207 (1963).
1250. Y.Kawasaki, T.Tanaka and R.Okawara, Bull. Chem. Soc. Jap., 27, 903 (1964).
1251. A.Cassol and T.Magon, J. Inorg. Nucl. Chem., 27, 1297 (1965).
1252. W.N.Aldridge and B.W.Street, Analyst,London, 106, 60 (1981).
1253. Y.Arakawa, O.Wada and M.Manabe, Anal. Chem., 55, 1901 (1983).
1254. M.Frankel, D.Wagner, D.Gertner and A.Zilha, Israel J. Chem., 4, 183 (1966).
1255. A.H.Chapman, M.W.Duckworth and J.W.Price, Br. Plast., 32, 78 (1959).
1256. R.Sawyer, Analyst (London), 92, 569 (1967).
1257. O.R.Klimmer and I.U.Nebel, Arzneim.-Forsch., 10, 44 (1960).
1258. G.Gras and J.Castel. Soc. Pharm. Montpellier, 25, 178 (1965).
1259. J.H.Adamson, Analyst (London), 87, 597 (1962).
1260. H.J.Hardon, H.Brunink and E.W. Van der Pol, Analyst (London), 85, 847 (1960).
1261. H.J.Hardon, A.F.H.Bessemer and H.Brunik, Dtsch. Lebensm.-Rundsch., 58, 349 (1962).
1262. S.J.Blunden and A.H.Chapman, Analyst (London), 103, 1266 (1978).
1263. J.W.Zijp, IRI Proc., 6, 108 (1959).
1264. M.C.Kerssen and P.Riepma, Tijdschr. Plantenziekten, 65, 27 (1959).
1265. G.C.Pimental, Anal. Chem., 17, 882 (1949).
1266. J.R.Urwin and P.J.Reed, J. Organomet. Chem., 15, 1 (1968).
1267. C.Blomberg, A.D.Vrengdenhil and P.Vink, Rec. Trav. Chim. Pays-Bas., 83, 662 (1964).
1268. A.D.Vrengdenhil and C.Blomberg, Rec. Trav. Chim. Pays-Bas., 82, 453 (1963).
1269. D.Hollander and M.Anteunis, Bull. Soc. Chim. Belg., 72, 77 (1963).
1270. H.Flaschka, Mikrochemie, 39, 38 (1952).
1271. O.Datta and O.Mitten, J. Am. Chem. Soc., 41, 287 (1919).
1272. B.G.Gowenlock, Anal. Abstra., 1, 3020 (1954).
1273. B.G.Gowenlock. J. Chem. Soc., 1454 (1955).
1274. G.A.Hamilton and A.D.Ruthven, Lab. Pract., 15, 995 (1966).
1275. D.N.Hague and R.H.Prince, J. Chem. Soc., 4, 690 (1965).
1276. W.Drenth, M.J.Janssen, G.J.M.Van der Kerk and J.A.Viiegenthart, J. Organomet. Chem., 2, 265 (1964).
1277. C.W.N.Cumper, A.Meinikoff and A.I.Vogel. J. Chem. Soc. A., 242, (1966).
1278. V.S.Griffiths and G.A.W.Derwish, J. Mol. Spectrose., 3, 165 (1959).

1279. R.H.Wismall and C.P.Smythe, J. Chem. Phys., 9, 352 (1941).
1280. G.Schomberg,Dissertakion,University of Aachem, Aachem. (1956).
1281. E.G.Hoffman, Z. Elektrochem., 64, 616 (1960).
1282. P.H.Lewis and R.E.Rundle, J. Chem. Phys., 21, 986 (1953).
1283. K.S.Pitzer and R.K.Sheline, J. Chem. Phys., 16, 552 (1948).
1284. A.P.Gray and A.B.Callear, Edgecombe F.H.C., Canad. J. Chem., 41, 1502 (1963).
1285. A.P.Gray, Canad. J. Chem., 41, 1511 (1963).
1286. L.O.Brockway and R.R.Davidson, J. Amer. Chem. Soc., 63, 3287 (1941).
1287. N.R.Davidson and H.C.Brown, J. Amer. Chem. Soc., 64, 316 (1942).
1288. M.P. Groenewege, Z. Physich.Chem.(Frankfurt), 18, 147 (1958).
1289. G.P.Van der Kelen and M.A.Hermann, Bull. Soc. Chim. Belges., 65, 350 (1956).
1290. W.Klemperer, J. Chem. Phys., 24, 353 (1956).
1291. R.P.Bell and H.C.Languet-Higgins, Proc. Roy. Soc. London, 183, 357 (1945).
1292. V.A.Grosse and J.M.Mavity, J. Org. Chem., 5, 106 (1940).
1293. R.N.Badger and L.R.Zumwalt, J. Chem. Phys., 6, 711 (1938).
1294. G.Herzliery, in "Infrared and Raman Spectra of Polyatomic Materials" D.Van Nostrand, New York, p.419 (1947).
1295. M.P.Groenewege, J.Smidt and H.de Vries, J. Am. Chem. Soc., 82, 4425 (1960).
1296. E.G.Hoffman and G.Schomberg, Z. fur Elecktrochemie., 61, 1101 (1957).
1297. E.G.Hoffman and G.Schomberg, Z. fur Elecktrochemie., 61, 1110 (1957).
1298. R.L.Hudson, Anal. Chem., 29, 1895 (1957).
1299. J.V.Bell, J.Heisler, H.Tannerbaum and J.Goldenson, Anal. Chem., 25, 1720 (1953).
1300. D.V.Guertin, S.E.Wiberley, W.H.Bauer and J.Golderson, J. Phys. Chem., 60, 1018 (1956).
1301. C.G.Barraclough, D.C.Bradley, J.Lewis and I.M.Thomas, J. Chem. Soc. (B) 2601 (1961).
1302. V.A.Zeiter and C.A.Brown, J. Phys. Chem., 61, 1174 (1957).
1303. R.C.Wilhoit, J.R.Burton, Fu-Tien Kuo, Sui-Rang Huang and A.Vignesnel, J. Inorg. Nucl. Chem., 24, 851 (1962).
1304. L.M.Brown and K.S.Mazdiynasi, Anal. Chem., 41, 1243 (1969).
1305. C.T.Lynch, K.S.Mazdiynasi, J.S.Smith and W.J.Crawford, Anal. Chem., 36, 2332 (1964).
1306. F.A.Scott, J.Golderson, S.E.Wiberley and J.Bauer, Phys. Chem., 58, 61 (1954).
1307. N.V.Sutton and H.Schneider, Mikrochim. J., 9, 209 (1965).
1308. O.Yamamoto, Bull. Chem. Soc. Japan., 35, 619 (1962).
1309. K.W.F.Kohlraush and J.Wagner, Z. Physik. Chem.,1352, 185 (1952).
1310. R.E.Rundle, J. Chem. Phys., 21, 986 (1953).
1311. H.P.Groenewege, Rev.Universelle Mines, 9 Serie, 15, 461 (1959).
1312. E.G.Hoffman, Zhur. Anal. Khim., 170, 176 (1959).
1313. S.Broustein, B.C.Smith, G.Erlich and A.W.Laubergayer, J. Amer. Chem. Soc., 82, 1000 (1960).
1314. N.Muller and D.E.Prichard, J. Amer. Chem. Soc., 82, 248 (1960).
1315. J.Smidt, J.Groenewage and H.de Vries, Rec. Trav. Chim., 81, 729 (1962).
1316. V.J.Shiner, D.Whittaker and V.P.Fernandes, J. Amer. Chem. Soc., 85, 2318 (1963).
1317. D.C.Bradley in "Metal Alkoxides - Advances in Chemistry Series," Vol. 23 - American Chemical Society, Washington D.C., p.10 (1959).
1318. D.E.O'Reilly, J. Chem. Phys., 32, 1007 (1960).
1319. G.V.Bortnikov, E.N.Vyankin, E.N.Gladyshev and V.S.Andreevickev, Zavodskaya Lab., 35, 1445 (1969).

1320. R.C.Mehrotra, J. Indian Chem. Soc., 21, 157 (1956).
1321. G.W.Hennische, Anal. Chem., 46, 1018 (1974).
1322. L.M.Brown and K.S.Mazdiynasi, Anal. Chem., 41, 1243 (1969).
1323. W.Fieggan, H.Gerding and N.N.Nihberling, Rec. Trav. Chim., 87, 377 (1968).
1324. H.G.Nadeau and D.M.Oakes, Anal. Chem., 32, 1480 (1960).
1325. H.W.Myers and R.F.Putnam, Anal. Chem., 34, 664 (1962).
1326. L.V.McCarty, G.C.Smith and R.S.McDonald, Anal. Chem., 26, 1027 (1954).
1327. L.J.Kuhns, R.S.Braman and J.E.Graham, Anal. Chem., 34, 1700 (1962).
1328. O.O.Le Sachs and M.P.Garrigues, Anal. Chem., 43, 174 (1961).
1329. D.R.Mackey, J.W.Brogan, H.J.Birch and A.E.Weber, U.S. Atomic Energy Commission Report CCC-1024-TR-232, 1, 13pp (1957).
1330. J.J.Kaufman, W.S.Koski, L.J.Kuhns and S.S.Wright, J. Amer. Che. Soc., 85, 1369 (1963).
1331. J.J.Kaufman and W.S.Koski, J. Amer. Chem. Soc., 78, 5774 (1956).
1332. J.J.Kaufman and W.S.Koski, J. Chem. Phys., 24, 403 (1956).
1333. H.J.Hrostowski and G.C.Pimental, J. Amer. Chem. Soc., 76, 998 (1954).
1334. N.V.Sutton and H.Schneider, Mikrochem. J., 9, 209 (1965).
1335. W.S.Koski, J.J.Kaufman and P.C.Lauterbur, J. Amer. Chem. Soc., 79, 2382 (1957).
1336. W.S.Koski and J.J.Kaufman, J. Chem. Phys., 24, 221 (1956).
1337. W.S.Koski, J.J.Kaufman and C.F.Pachuki, J. Amer. Chem. Soc., 81, 1326 (1959).
1338. J.J.Kaufman, W.S.Koski, L.J.Kuhns and R.W.Law, J. Amer. Chem. Soc., 84, 4198 (1962).
1339. C.O.Wilson and I.Shapiro, Anal. Chem., 32, 78 (1960).
1340. J.C.Booker and T.L.Isenhour, Anal. Chem., 41, 1705 (1969).
1341. B.Ruzicka, Chemia. Analit., 12, 1093 (1967).
1342. J.H.Benyon, R.A.Saunders and A.E.Williams, Appl. Spectrosc., 17, 63 (1963).
1343. K.Nakamura and H.Orii, Anal. Chem., 52, 532 (1980).
1344. O.Bayer, Annual Infrared Spectroscopy Institute, Canuis College. pp 151-176, Buffalo, New York (1962); Plenum Press, New York (1963).
1345. M.P.Brown, R.Okawara and E.G.Rochow, Spectrochim. Acta., 16, 595 (1960).
1346. I.G.Shafran and A.V.Kuraeva, Trudy. Uses. Nauchno-issled. Inst. Khim. Reakt., 28, 96 (1966); Ref. Zhur. Khim., 19GD 1967 (6) Abstr. No. 6G134.
1347. L.Vercernyes and L.Hangos, Z. Analyt. Chem., 208, 407 (1965).
1348. N.V.Larin, G.G.Devyatykh and L.L.Agafonov, Zh. Analit. Khim., 22, 285 (1967).
1349. G.G.Devyatykh, N.V.Lavin and L.L.Agafonov, Trudy Kom. Analit. Khim., 16, 206 (1968). Ref. No. Zh. Khim., 19GD (14) (1968).
1350. W.E.Winter, B.Curnette and S.E.Whitcomb, Spectrochim. Acta., 15, 1085 (1959).
1351. O.Lippencott and O.Nelson, Spectrochim. Acta., 10, 307 (1958).
1352. A.F.Reid, D.E.Scaife and P.C.Wailes, Spectrochim. Acta., 20, 1257 (1964).
1353. J.Lewis and R.G.Wilkins in "Modern Co-ordination Chemistry" p.355 Interscience, New York (1960).
1354. G.Wilkinson and F.A.Cotton in "Progress in Inorganic Chemistry" p.21 Vol. 1 (Edition F.A.Cotton), Interscience, New York (1959).
1355. V.E.Fischer in "Advances in Inorganic Chemistry and Radiochemistry", Vol. 1 (Edition, H.J.Emeleus, H,J.Sharpe) Academic Press, New York (1959).
1356. G.Wilkinson, F.A.Cotton and J.M.Birmingham, J. Inorg. Nucl. Chem., 2, 95 (1955).

1357. K.Nakamoto in "Infrared Spectra of Inorganic and Co-ordination Compounds", Wiley, New York (1963).
1358. E.B.Wilson, J.C.Decins and P.C.Cross in "Molecular Vibrations", p.331, McGraw-Hall, New York (1955).
1359. J.P.Pellegrini, Township O'Hara, A.Country and I,J.Spilners, U.S. Patent 3,350,434 (31/10/67) (1967).
1360. I.J.Spilners and J.P.Pellegrini, Applications of Mass Spectrometry and N.M.R. to structure problems in ferrocene chemistry. Cleavage of ferrocene by aluminium chloride. Presented at the American Chemical Society 153rd National Meeting, Organic Division, Miami Beach, Florida, U.S.A. (1967).
1361. V.V.Korshak, S.I.Sosin and V.P.Alexeeva, Dokl. Akad. Nauk. SSSR., 132. 360 (1960).
1362. I.J.Spilners, J. Organometal. Chem., 11, 381 (1968).
1363. S.I.Goldberg, J. Chem. Soc., 43, 554 (1966).
1364. M.Rosenblum, N.Danieli, R.W.Rish and V.Schalter, J. Amer. Chem. Soc., 85, 316 (1963).
1365. J.D.Dunitz, L.E.Orgel and A.Rish, Acta. Cryst., 9, 373 (1956).
1366. J.R.Leto, F.A.Cotton and J.S.Waugh, Nature, 180, 978 (1957).
1367. M.B.Laing and K.N.Trueblood, Acta. Cryst., 19,373 (1965).
1368. K.L.Rinehart, A.K.Frerichs, P.A.Kittle, L.F.Westmas, D.H.Gustafson, R.L.Pruett, J.E.McMahon, J. Amer. Chem. Soc., 82, 4111 (1960).
1369. I.C.Paul, Chem. Commun., 12, 377 (1966).
1370. H.P.Fritz in "Advances in Organometallic Chemistry" Vol. 1 (F.G.A.Stone and R.West Editions), Academic Press, New York, p.267 (1964).
1371. P.L.Parson, M.A.Sandhu and W.E.Watts, J. Chem. Soc., 3C, 251 (1966).
1372. G.G.Dvoryantseva, S.L.Portrova, Y.N.Sheinker, L.P.Yur'eve and A.N. Nesmeyanov, Dokl. Akad. Nauk. SSSR., 169, 1083 (1966).
1373. G.G.Dvoryantseva, S.L.Portrova, Y.N.Sheinber, L.P.Yur'eva and A.N. Nesmeyanov, Dokl. Akad. Nauk. SSSR., 160, 1075 (1965).
1374. L.Friedmann, A.P.Irsa and G.Wilkinson, J. Amer. Chem. Soc., 77, 3689 (1955).
1375. R.I.Reed and F.M.Talyza, App. Spectry., 17, 124 (1963).
1376. E.Schumacker and R.Taubenest, Helv. Chim. Acta., 47, 1525 (1964).
1377. N.Moaz, A.Mandelbaum and M.Cass, Tetrahedron Letters, 25, 2087 (1965).
1378. C.Cordes, K.L.Rinehart, Abstracts. No. S37, 150th Meeting American Chemical Society, Atlantic City, New Jersey, U.S.A., Sept. (1965).
1379. P.W.Slocum, R.Lewis and G.J.Mains, Chem. Ind. (London),2095 (1966).
1380. H.Egger, Monatsch. Chem., 97,602 (1966).
1381. J.Muller and L.D'Or, J. Organometal. Chem., 10, 313 (1967).
1382. D.T.Roberts, W.F.Little and M.M.Bursey, J. Amer. Chem. Soc., 89, 6156 (1967).
1383. D.T.Roberts, W.F.Little and M.M.Bursey, J. Amer. Chem. Soc., 89, 4917 (1967).
1384. D.T.Roberts, W.F.Little and M.M.Bursey, J. Amer. Chem. Soc., 90, 973 (1968).
1385. F.L.Hedberg and H.Rosenberg, J. Amer. Chem. Soc., 91, 1258 (1969).
1386. R.B.King, Canad. J. Chem., 47, 559 (1969).
1387. D.J.Clancy and I.J.Spilners, Anal. Chem., 34, 1839 (1962).
1388. D.J.Spilners and J.Pellegrini, J. Org. Chem., 30, 3800 (1965).
1389. D.J.Spilners and J.Pellegrini, 153rd National Meeting of the American Chemical Society, Miami Beach, Florida, U.S.A. No. 0112, April (1967).
1390. I.J.Spilners and J,Pellegrini, Application of Mass Spectrometry and N.M.R. to structure problems in ferrocene chemistry, cleavage of ferrocene by aluminium chloride. Presented at the Americal Chemical Society 153rd National Meeting, Organic Division, Miami Beach, Florida, U.S.A. (1967).

1391. K.I.Rinehart, P.A.Kittle and F.A.Ellis, J. Amer. Chem. Soc., 82, 2082 (1960).
1392. H.R.Schulten, P.B.Monkhouse and R.Miller, Anal. Chem., 54, 654 (1982).
1393. W.L.Baun, Anal. Chem., 31, 1308 (1959).
1394. P.F.Eiland and R.Pepinsky, J. Amer. Chem. Soc., 74, 4971 (1952).
1395. T.J.Kealy and P.L.Pauson, Nature, 168, 1039 (1951).
1396. O.De Bruyne, Fermentation, 1,28 (1964).
1397. D.A.Moringstar, G.Z.Williams and P.Smintorinen, Amer. J. Clin. Path., 46, 603 (1966).
1398. F.J.Bajer in "Progress in Infrared Spectroscopy" Vol. 11 based on lectures from the 6th and 7th Annual Infrared Spectroscopy Institutes held at Canisins College, Buffalo, New York (1962) and (1963). Plenum Press, New York pp 151-176 (1964).
1399. R.P.Curry, Anal. Chem., 31,959 (1959).
1400. H.E.Howard, W.C.Ferguson and L.R.Snyder, Anal. Chem., 32, 1814 (1960).
1401. G.L.Bate, D.S.Miller and J.L.Kulp, Anal. Chem., 29, 84 (1957).
1402. T.J.Ulrych and R.D.Russell, Geochim. Cosmoch. Acta., 28, 455 (1964).
1403. J.R.Urwin and P.J.Reed, J. Organometal. Chem., 15, 1 (1968).
1404. H.Susi and H.E.Rectar, Anal. Chem., 30, 1933 (1958).
1405. H.Gauthier, R.Goupil, G.Mangoney, S.Didelot and R.Merlier, Chim. Anal., 44, 12 (1962).
1406. J.E.Campana and T.H.Risby, Anal. Chem., 52, 468 (1980).
1407. N.N.Shesterikov, Y.A.Pentin, and E.G.Teterin, USSR Patent 142,809 (28/12/61) (1961).
1408. Y.A.Pentin, E.G.Teterin and N,V.Shesterikov, Zhur. Anal. Khim., 17, 239 (1962).
1409. United Kingdom Atomic Energy Authority, Production Group. Technical Dept., Winscale, Cumberland, U.K. UKAEA Report No. 344/W1 (1962).
1410. J.Lorenz and A.Kaiser, Z. Anal. Chem., 215, 438 (1966).
1411. J.Ferraroin "Developments in Applied Spectroscopy" Vol. 11 p. 89 Proceedings of the 13th Annual Symposium on Spectroscopy, Chicago, U.S.A. April - May 1962; distributed by Plenum Press, New York (1963).
1412. J.L.Occolowitz and G.L.White, Anal. Chem., 35, 1179 (1963).
1413. I.E.O'Neill and M.A.Pringner, Anal. Chem., 49, 588 (1977).
1414. R.Greenhalgh and J.N.Shoolery, Anal. Chem., 50, 2039 (1978).
1415. A.L.Smith, J.A.McHard, Anal. Chem., 31, 1174 (1959).
1416. G.Munro, J.H.Hunt and L.R.Rowe, Anal. Chem., 51, 311 (1979).
1417. C.A.Burkhard and E.H.Winslow, J. Amer. Chem. Soc., 72, 3276 (1950).
1418. C.Eaborn, J. Chem. Soc., 3, 148 (1953).
1419. H.Gilman and G.E.Dunn, J. Amer. Chem. Soc., 72, 2178 (1950).
1420. K.Bowden and E.A.Brande, J. Chem. Soc., 1068 (1952).
1421. A.Pozefsky and M.E.Greenoble, Drug and Cosmetic Ind., 80, 752 (1957).
1422. P.Brown and A.L.Smith, Anal. Chem., 30, 1016 (1958).
1423. A.L.Smith and N.C.Angelotti, Spectrochim. Acta., 10, 124 (1972).
1424. G.M.Kline (Editor) in "Chemical Analysis of High Polymers" Chapter 14, Interscience, New York (1959).
1425. K.P.Bradley and W.J.Potts, Appl. Spectroscopy, 12, 77 (1958).
1426. D.N.Ingebrigtson and A.L.Smith, Anal. Chem., 26, 1765 (1954).
1427. U.Scheidt, Appl. Spectroscopy, 7, 75 (1953).
1428. D.C.Harms, Anal. Chem., 25, 1141 (1953).
1429. N.Wright and M.J.Hunter, J. Amer. Chem. Soc., 69, 803 (1947).
1430. C.W.Young, P.C.Servais, C.C.Curme and H.J.Hurder, J. Amer. Chem. Soc., 70. 3758 (1948).
1431. D.C.Smith, J.M.French and J.J.O'Neill in U.S.Naval Research Laboratory Office of Technical Services - Report 2746 (January 1946).

1432. Midland Silicones Ltd. Publications A3-2 7 pp "The Analysis of Organosilicon Compounds, (February 1957).
1433. D.Hummel, Farbe. Lach., 62, 529 (1956).
1434. I.W.Fishl and Young, Appl. Spectroscopy, 10 213 (1956).
1435. D.Hummel, Kunst.Ofle., 46, 442 (1956).
1436. R.E.Kagarise and L.A.Weinberger, U.S.Naval Research Laboratory Office of Technical Services, Report p.13, 111438 (May 11th 1954).
1437. A.L.Smith, L.H.Brown, L.J.Tyler and M.J.Hunter, Ind. Eng. Chim., 49, 1903 (1957).
1438. T.Yonemoto and K.Senzok, Bull. Electrotech. Lab. (Tokyo), 18, 428 (1954).
1439. W.H.T.Davidson and G.R.Bates in "Proceedings of the Third Rubber Technology Conference, London", p.281 (1954).
1440. W.Ecknig, H.Kriegsiman, H.Rotzsche, H.Jancke and K.Witke, Proceedings on Application of Physicochemical Methods of Chemical Analysis", Budapest, Hungary, April 1966. Abstracts of Plenary Lecture, Vo. 111 476 pp, Section V, Spectrochemical Analysis p.11-12, 12, (1966).
1441. L.J.Bellamy in "Infrared Spectra of Complex Molecules" (1st edition) Chapter 20, Methuen & Co. London (1954).
1442. H.Gilman (Editor, in "Organic Chemistry" vol. 111 p. 143-151, Wiley, New York (1953).
1443. A.P.Kreshkov, Yu Ya Mikhailenko and G.F.Yakimovich, Zhur. Anal. Khim., 9, 208 (1954).
1444. A.L.Smith and J.A.McHard, Anal. Chem., 31, 1174 (1959).
1445. A.L.Smith, Spectro. Chim. Acta., 16, 87 (1960).
1446. W.D.Reents, D.I.Woods and A.M.Mujsce, Anal. Chem., 57, 104 (1985).
1447. A.G.Sharkess, R.A.Friedel and S.Langer, Anal. Chem., 29,770 (1957).
1448. V.H.Dibelier, J. Res. Nat. Bureau Standards, 49, 235 (1952).
1449. V.H.Dibelier,F.L.R.Mohler and M.L.Reese, J. Chem. Phys., 21,180 (1953).
1450. N.A.Sokolov, K.A.Andrianov and S.M.Akimova, Zhur. Obshchei. Khim., 25, 675 (1955). J. Gen. Chem., (USSR), 25, 647 (1955).
1451. G.W.Bethke and M.K.Wilson, J. Chem. Phys., 26, 1107 (1957).
1452. M.I.Batnev, V.A.Panamarenko, A.D.Matseeva and A.D.Snegova, Izvest. Akad. Nauk. SSSR. Odel. Khim. Nauk 1420, Consultants Bureau Translation p.1457.
1453. R.Ulbrich, Z. Natur forschung, 9b, 380 (1954).
1454. Y.P.Bazhulin, Y.P.Yegorov and V.F.Mironov, Doklady Akad. Nauk SSSR., 88, 647 (1953).
1455. P.A.Bazhulin, Y.P.Yegorov and V.F.Mironov, Doklady Akad. Nauk SSSR., 92, 515 (1953).
1456. J.Goubeau, H.Siebert and M.Winterweb, Z. Anorg. Chem., 259, 240 (1949).
1457. H.Murata and S.Hayashi, J. Chem. Phys., 19, 1217 (1951).
1458. J.Gourbeau and R.Warneke, Z. Anorg. Chem., 259, 233 (1949).
1459. H.Batnev, A.D.Petrov, V.A.Panamorenko and A.D.Matveeva Izvest. Akad. Nauk. SSSR, Odtel Khim. Nauk, 1070 (Consultants Bureau Translation p.1087).
1460. H.Murata, R.Okawara and T.Watase, J. Chem. Phys., 18, 1308 (1950).
1461. L.Sanidan, Bull. Soc. Chim. France, 411, (1953).
1462. H.Mataka, Sci. Ind. (Osaka), 30, 164 (1956).
1463. C.C.Cerato, J.L.Laver and H.C.Beachell, J. Chem. Phys., 22, 1 (1954).
1464. H.Murata, J. Chem. Phys., 19, 659 (1951).
1465. Y.M.Slabodin, Y.E. Shmulyakovskii and K.A.Rzhedzinskaya, Doklady Akad. Nauk, SSSR., 105, 958 (1955).
1466. D.C.Smith, J.M.French and J.J.O'Neill, in US Naval Research Laboratory Office of Technical Services Report No. 2746 (January 1946).

1467. W.J.Currie and G.W.Harrison, J. Org. Chem., 23, 1219 (1958).
1468. L.Goodman, R.M.Silverstein and J.N.Schoolery, J. Amer. Chem. Soc., 78, 4493 (1956).
1469. E.Schnell and E.G.Rochow, J. Amer. Chem. Soc., 78, 4178 (1956).
1470. E.Schnell and E.G.Rochow, J. Inorg. and Nuclear Chem., 6, 303 (1958).
1471. G.R.Halzman, P.C.Lauterbur, J.Anderson and W.Koth, J. Chem. Phys., 25, 172 (1956).
1472. E.G.Rochow and H.G.Le Clair, J. Inorg. and Nuclear Chem., 1, 92 (1955).
1473. H.Kriegsmann, Z. Anorg. Allgem. Chem., 299, 138 (1959).
1474. N.Wright and M.J.Hunter, J. Amer. Chem. Soc., 69, 803 (1947).
1475. C.W.Young, P.C.Servais, C.C.Currie and M.J.Hunter, J. Amer. Chem. Soc., 70, 3758 (1948).
1476. A.L.Smith, J. Chem. Phys., 21, 1997 (1953).
1477. C.W.Young, J.S.Koeler and D.S.McKinney, J. Amer. Chem. Soc., 69, 1410 (1947).
1478. S.Kaye and S.Tannbaum, J. Org. Chem., 18, 1750 (1953).
1479. M.C.Harvey, H.W.Negergall and J.S.Peake, J. Amer. Chem. Soc., 76, 4555 (1954).
1480. H.Westermark, Acta. Chem. Scand., 9, 947 (1955).
1481. L.J.Bellamy in "Infrared Spectra of Complex Molecules" (2nd edition) Chapter 20, Methuen and Co., London.
1482. G.D.Oshesky and F.F.Bentley, J. Amer. Chem. Soc., 79, 2057 (1957).
1483. J.W.Curry, J. Amer. Chem. Soc., 78, 1686 (1956).
1484. R.G.Scott and K.C.Frisch, J. Amer. Chem. Soc., 73, 2599 (1951).
1485. M.Kanazachi, Bull. Chem. Soc. Japan, 26, 493 (1953).
1486. K.C.Frisch, P.A.Goodwin and R.E.Scott, J. Amer. Chem. Soc., 74, 4584 (1952).
1487. E.R.Schull, R.A.Thursack and C.M.Birdsall, J. Chem. Phys., 24, 147 (1956).
1488. D.L.Bailey and A.N.Pines, Ind. Eng. Chem., 46, 2363 (1954).
1489. L.D.Nasiak and H.W.Post, J. Org. Chem., 24, 489 (1959).
1490. L.J.Bellamy in "Infrared Spectra of Complex Organic Molecules" (2nd edition), Chapter 3 (1958).
1491. R.E.Richards and H.W.Thompson, J. Chem. Soc., 124 (1949).
1492. M.C.Harvey, H.W.Negergill and J.S.Peake, J. Amer. Chem. Soc., 76, 4555 (1954).
1493. J.G.Noltes, M.C.Henry and M.S.Janssen, Chem. and Ind. (London) 298 (1959).
1494. H.Kriegsmann and K.H.Schowtka, Z. Physik. Chem. (Leipzig), 209, 261 (1958).
1495. L.Spaltier, D.C.Priest and C.W.Harris, J. Amer. Chem. Soc., 77, 6227 (1955).
1496. R.D.Kross and V.A.Fassel, J. Amer. Chem. Soc., 77, 5858 (1955).
1497. P.D.Whitten, J. Chem. Soc., 1350 (1956).
1498. E.N.Beck, W.H.Dandt, H.J.Fletcher, H.J.Hunter and A.J.Barry, J. Amer. Chem. Soc., 81, 1256 (1959).
1499. H.A.Clark, A.F.Gordon, C.W.Young and H.J.Hunter, J. Amer. Chem. Soc., 73, 3798 (1951).
1500. V.A.Young, R.B.DuVall and N.Wright, Anal. Chem., 23, 709 (1951).
1501. N.B.Colthup, J. Opt. Soc. Amer., 40, 397 (1950).
1502. G.R.Wilson, G.M.Hutzel and A.G.Smith, J. Org. Chem., 24,381 (1959).
1503. J.W.Gilkey and L.J.Tyler, J. Amer. Chem. Soc., 73, 4982 (1951).
1504. R.West and E.G.Rochow, J. Org. Chem., 18, 303 (1953).
1505. W.A.Snowacka and T.Bierwacka, Chem. Anal. Warsaw, 9, 303 (1964).
1506. T.Burnacka and A.Wokroj, Chemia. Analit., 10,1233 (1965).
1507. J.H.Lady, G.M.Bower, R.E.Adams and F.P.Byre, Anal. Chem., 31, 1100 (1959).
1508. A.D.Grant and A.L.Smith, Anal. Chem., 30, 1016 (1958).

1509. W.Fish'l and I.G.Young, Appl. Spectroscopy, 10, 213 (1956).
1510. T.Kubota and T.Takamura, Bull. Chem. Soc. Japan, 33, 70 (1960).
1511. T.Urin and T.Hakamada, J. Chem. Soc. Japan, Ind. Chem. Sect., 62, 1421 (1959).
1512. T.Urin, S.Tanaka and T.Wada, J. Chem. Soc. Japan Ind. Chem. Sect., 62, 1577 (1959).
1513. T.Urin, S.Tanaka and H.Yamamato, J. Chem. Soc. Japan, Ind. Chem. Sect., 62, 1581 (1959).
1514. O.Kuratani, Chem. Soc. Japan Pure Chem. Sect., 73, 576 (1952).
1515. M.Margoshes and V.A.Fassel, Anal. Chem., 27, 351 (1955).
1516. L.A.Efremova and K.K.Papkov, Zavod. Lab., 29, 708 (1963).
1517. E.R.Shull, Anal. Chem., 32, 1627 (1960).
1518. M.Kakudo, P.N.Kasai and T.Watase, J. Chem. Phys., 21, 1894 (1953).
1519. M.Kakudo, P.N.Kasai and T.Watase, J. Chem. Phys., 21, 167 (1953).
1520. K.Damm, W.Noll and Z.Kolloid, 158, 97 (1958).
1521. V.I.Kastochkin, M.F.Shos'takovskei, U.I.Zil'berbrau and D.A.Kochkin, Izvest. Akad. Nauk. SSSR. Ser. Fiz., 18, 726 (1954).
1522. H.Kriegsmann, Z. Anorg. Chem., 299, 78 (1959).
1523. I.Ya Ryskin, M.G.Voronkov, Zhur. Fiz. Khim., 30, 2275 (1956).
1524. S.W.Kantor, J. Amer. Chem. Soc., 75, 2712 (1953).
1525. Ya I.Tyskin, Optika i Spekroskopiya, 4, 532 (1958).
1526. H.A.Benesi and A.C.Jones, J. Phys. Chem., 63, 179 (1959).
1527. G.J.Young, J. Colloid Sci., 13, 67 (1958).
1528. M.Folman and D.J.C.Yates, J. Phys. Chem., 63, 183 (1959).
1529. R.S.McDonald, J. Amer. Chem. Soc., 79, 850 (1957).
1530. V.E.Lygin, Vestrik Hosk. Univ. Ser. Mat. Mekh. Astrien. Fiz. Khim., 13, 223 (1958).
1531. R.S.McDonald, J. Phys. Chem., 62, 1168 (1958).
1532. F.Simon and H.O.McMahon, J. Chem. Phys., 20, 905 (1952).
1533. A.P.Kreshkov, Yu Ya Mikhailenko and G.F.Yakimovich, Zhur. Anal. Khim. 9, 208 (1954).
1534. R.Okawara Bull. Chem. Soc. Japan, 31, 154 (1958).
1535. R.Forneris and E.Funck, Z. Elektrochem., 62, 1130 (1958).
1536. P.Brown and A.L.Smith, Anal. Chem., 30, 549 (1958).
1537. H.Kriegsmann and K.Light, Z. Elektrochem., 62, 1163 (1958).
1538. A.A.Stuart, C.L.Lan and H.Breederveld. Rec. Trav. Chim., 74, 747 (1955).
1539. A.P.Kreshkov, Yu Ya Mikhailenko and G.F.Yakimovich, Zhur. Fiz. Khim., 28, 537 (1954).
1540. M.Hayashi, Nippon Kagaku. Zasshi., 79, 436 (1958).
1541. M.O.Bulanin, B.N.Dolgov, T.A.Speranskaya and N.P.Kharitnov, Zhur. Fiz. Khim., 31, 1321 (1957).
1542. A.N.Lazarev, M.G.Voronkov, Optika i Spekroskopiya, 4, 180 (1958).
1543. A.N.Lazarev, Optika i Spekroskopiya, 4, 805 (1958).
1544. H.Breederveld and H.I.Watermann, Rec. Trav. Chim., 73, 871 (1954).
1545. T.Tanaka, Bull. Chem. Soc., Japan, 31, 762 (1958).
1546. R.Okamara, Bull. Chem. Soc. Japan, 31, 154 (1958).
1547. L.J.Bellamy in "Infrared Spectra of Complex Molecules" (2nd edition) p.174, Methuen & Co. (London), (1958).
1548. F.C.Lanning and M.Moore, J. Org. Chem., 23, 288 (1958).
1549. R.A.Khwel'nitskil, A.A.Polyakova and A.A.Petrov, Trudy Komiss. Anal. Khim. Akad. Nauk. SSSR., 13, 482 (1963). Ref: Zhur. Khim. 19GD, (21) Abstract No. 21G188 (1963).
1550. A.L.Smith and N.C.Angelotti, Spectrochim. Acta., 15, 412 (1959).
1551. L.Kaplan, J. Amer. Chem. Soc., 76, 5880 (1954).
1552. E.A.V.Ebsworth, M.Onyszchuk and N.Sheppard, J. Chem. Soc., 1453 (1958).
1553. R.Kniseley, V.A.Fassel and E.Conrod, Spectrochim. Acta., 15, 651 (1959).

1554. H.Z.Kriegsmann, Elektrochem., 62, 1033 (1958).
1555. H.Kriegsmann and K.H.Showatka, Z. Physik. Chem. (Leipzig), 209, 261 (1958).
1556. R.N.Dixon and N.Sheppard, J. Chem. Phys., 23, 215 (1955).
1557. C.Newman, J.K.O'Loane, S.R.Polo and M.K.Wilson, Chem. Phys., 25, 855 (1956).
1558. H.Westermark, Acta. Chem. Scand., 9, 947 (1955).
1559. R.West and E.G.Rochow, J. Org. Chem., 18, 303 (1953).
1560. M.H.Herchenson in "Infrared Absorption Spectra - Index for 1945 - 1957) Academic Press, New York, London (1959).
1561. D.F.Ball, P.L.Goggin, D.C.McKean and L.A.Woodward, Spectrochim. Acta., 16, 11 (1960).
1562. H.Z.Kriegsmann, Elekrochem., 61, 1088 (1957).
1563. N.Wright and M.S.Hunter, J. Amer. Chem. Soc., 69, 803 (1947).
1564. R.C.Lord, D.W.Robinson and W.C.Schumb, J. Amer. Chem. Soc., 78, 1327 (1956).
1565. R.E.Richards and H.W.Thompson, J. Chem. Soc., 124 (1949).
1566. C.W.Young, P.C.Servais, C.C.Currie and M.J.Hunter, J. Amer. Chem. Soc., 70, 3758 (1948).
1567. H.Kriegsmann, Z. Anorg. Allgen. Chem., 298, 223 (1956).
1568. H.Kriegsmann, Z. Elekrochem., 62, 1033 (1958).
1569. A.J.Barry, N.H.Davdt, S.S.Domincone and J.W.Gilkey, J. Amer. Chem. Soc., 77, 4248 (1965).
1570. D.C.McKean, Spectrochim. Acta., 13, 38 (1958).
1571. P.J.Launer and M.E.Grenoble, Anal. Chem., 32, 441 (1960).
1572. Y.J.Yamamato, J. Chem. Soc., Japan, Ind. Chem. Sect., 64, 1464 (1961).
1573. F.A.Anderson and B.Bok, Acta. Chem. Scand., 8, 738 (1954).
1574. R.C.Collins and J.R.Nielsen, J. Chem. Phys., 23, 351 (1955).
1575. M.Hayashi, Nippon, Kagaku Zasshi., 78, 1472 (1957).
1576. E.A.Jones, J.S.Kirby-Smith, P.J.H.Woltz and A.H.Nielson, J. Chem. Phys., 19, 242 (1951).
1577. R.Okawara, Bull. Chem. Soc., Japan, 31, 154 (1958).
1578. A.L.Smith, J. Chem. Phys., 21, 1997 (1953).
1579. T.G.Gibian and D.S.McKinney, J. Amer. Chem. Soc., 73, 1431 (1951).
1580. R.E.Richards, H.W.Thompson, J. Chem. Soc., 124 (1949).
1581. A.L.Smith, J. Chem. Phys., 21, 1997 (1953).
1582. L.Burnelle and J.Duchesne, J. Chem. Phys., 20, 1324 (1952).
1583. K.Shimazu and H.Murata, Bull. Chem. Soc., Japan, 32, 46 (1959).
1584. M.C.Tobin, J. Amer. Chem. Soc., 75, 1788 (1953).
1585. R.Forneris and E.Funck, Z. Elekrochim., 62, 1130 (1958).
1586. H.Kriegsmann, Z. Elektrochem., 61, 1088 (1957).
1587. M.E.Grenoble and P.J.Launer, Tenth Annual Symposium on Spectroscopy, Chicago (1959).
1588. P. Brown and A.L.Smith, Anal. Chem., 30, 1016 (1958).
1589. H.R.Linton and E.R.Nixon, Spectrochim. Acta., 12, 41 (1958).
1590. D.W.Mayo, H.E.Opitz and J.S.Peake, J. Chem. Phys., 23, 1344 (1955).
1591. H.E.Opitz, J.C.Peake and W.H.Nebergall, J. Amer. Chem. Soc., 78, 292 (1956).
1592. A.L.Smith and S.A.McHard, Anal. Chem., 31, 1174 (1959).
1593. L.A.Efremova and K.K.Popkov, Zavod. Lab., 29, 708 (1963).
1594. M.E.Grenoble and P.J.Launer, Appl. Spectroscopy, 14, 85 (1960).
1595. A.Aulinger and W.Reerink, Z. Anal. Chem., 197, 24 (1963).
1596. C.A.Hirt, Anal. Chem., 33, 1786 (1961).
1597. C.C.Cerato, J.L.Laver and H.C.Beachell, J. Chem. Phys., 22, 1 (1954).
1598. E.A.V.Ebsworth, J.R.Hall, M.J.Mackillop, D.C.McKean, N.Sheppard and L.Woodward, Spectrochim. Acta., 13, 202 (1958).
1599. H.Kriegsmann and W.Foster, Z. Anorg. Allgem. Chem., 298, 212 (1959).

1600. D.W.Robinson, J. Amer. Chem. Soc., 80, 5924 (1958).
1601. J.J.McBride and H.C.Beachell, J. Amer. Chem. Soc., 74, 5247 (1952).
1602. M.Prober, J. Amer. Chem. Soc., 78, 2274 (1956).
1603. H.R.Linton and E.R.Nixon, J. Chem. Phys., 28, 990 (1958).
1604. H.R.Linton and E.R.Nixon, Spectrochim. Acta., 10, 299 (1958).
1605. P.J.Launer and A.S.Crouse, Appl. Spectroscopy, 15, 118 (1961).
1606. E.A.V.Ebsworth, R.Taylor and L.A.Woodward, Trans. Faraday Soc., 55, 211 (1959).
1607. H.J.Emeleus, A.G. MacDiarmid and A.G.Maddock, J. Inorg. and Nuclear Chem., 1, 194 (1955).
1608. H.R.Linton and E.R.Nixon, J. Chem. Phys., 29, 291 (1958).
1609. E.A.Jones, J.S.Kirby-Smith, P.J.H.Woltz and A.H.Nielson, J. Chem. Phys., 19, 242 (1951).
1610. J.Goubeau and J.Rushing, Z. Anorg. Allgem. Chem., 294, 96 (1958).
1611. A.G.MacDiarmid and A.G.Maddock, J. Inorg. Nuclear Chem., 1, 411 (1955).
1612. H.Murata and K.Shimazu, J. Chem. Phys., 23, 1968 (1955).
1613. W.S.Tatlock and E.G.Rochow, J. Org. Chem., 17, 1555 (1952).
1614. V.A.Zeitler and C.A.Brown, J. Amer. Chem. Soc., 79, 4618 (1957).
1615. V.A.Zeitler and C.A.Brown, 61, 1174 (1957).
1616. W.H.Keeber and H.W.Post, J. Org. Chem., 21, 509 (1956).
1617. H.R.Linton and H.R.Nixon, Spectrochim. Acta. 15, 146 (1959).
1618. W.P.Newmann, H.Niermann and R.Sommer, Annalen., 653, 164 (1962).
1619. W.P.Newmann, H.Niermann and R.Sommer, Annalen, 659, 27 (1962).
1620. E.Friebe and H.Kelker, Anal. Chem., 192, 267 (1962).
1621. L.Domange and J.Guy, Ann. Pharm. France, 16, 161 (1958).
1622. E.A.Gasilovich, D.N.Shigorin and N.V.Komarov, Trudy Komis Po. Spektrosopii, Akad. Nauk. SSSR., 3, 70 (1964). Chem. Abstr. 65, 593f(1966).
1623. N.I.Shergina, N.I.Golovanova, N.Y.Kamarov and V.K.Misyaras, Primen. Mal. Spektros. Khim. Sb. Dokl. Sb. Soveshch, 3rd Krasroyarsk. USSR 93, (1964) Chemical Abstr. 69, 14227(1968).
1624. N.I.Shergina, N.I.Golovanova, R.G.Mirshkov, V.M.Vlasov, Izv. Akad. Nauk. SSSR, Ser. Khim., 1378 (1967). Chemical Abstr., 68, 38836 (1968).
1625. M.F.Shostakovskii, N.I.Sherlina, N.I.Gologanava, N.V.Komarov, E.I.Brodskaya and V.K.Misyanas, Zhur. Obschch. Khim., 35, 1768 (1965).
1626. J.Medelsohn, A.Marehand and J.Volade, J. Organometallic Chem., 6, 25 (1966).
1627. R.A.Cummins, Austral. J. Chem., 18, 98 (1965).
1628. H.G.Kuivila in "Advances on Organometallic Chemistry" vol. 1 p.47 Academic Press, London and New York (1964).
1629. Yu P. Egorov, Teor i Eksperim, Khim. Akad. Nauk. Ukr. SSSR., 1, 30 (1965).
1630. K.Kawakami, T.Saito and R.Okawara, J. Organometallic Chem., 8, 377 (1967).
1631. A.K.Sawyer and J.E.Brown, J. Organometallic Chem., 5, 438 (1966).
1632. E.W.Randall, J.J.Ellmer and J.Zuckermann, J. Inorg. Nuclear Chem. Letters, 1, 109 (1966).
1633. K.Sisido and S.Kozima, J. Org. Chem., 29, 907 (1964).
1634. K.Kawakami and R.Okawara, J. Organometallic Chem., 6, 249 (1966).
1635. Y.Tanaka and T.Morikawa, Japan Analyst, 13, 753 (1964).
1636. T.Tanaka, Y.Matsumura, R.Okawana and Y.Musya, Bull. Chem. Soc. Japan, 41, 1497 (1968).
1637. P.Taimsalu and J.L.Wood, Trans. Faraday Soc., 59, 1754 (1963).
1638. R.A.Cummins Austral. J. Chem., 18, 985 (1965).
1639. W.Steingross and W.Zeil, J. Organometallic Chemistry, 6, 109,464 (1966).

1640. I.R.Beattie and G.P.McQuillan, J. Chem. Soc.,1519 (1963).
1641. R.J.H.Clark, A.G.Davies and R.J.Duddenphalt, J. Chem. Soc. (C) 1829 (1968).
1642. R.Okawara and K.Yasuda, J. Organometallic Chem., 1, 356 (1964).
1643. P.P.Simons and W.A.G.Graham, J. Organometallic Chem., 8, 479 (1967).
1644. H.C.Clark and J.H.Tsai, Inorg. Chem., 5, 1407 (1966).
1645. P.Taimsalu and J.L.Wood, Spectrochim. Acta., 20, 1043 (1964).
1646. D.H.Lohmann, J. Organometallic Chem., 4, 382 (1965).
1647. R.Okawara and M.O'hara, J. Organometallic Chem., 1, 360 (1964).
1648. F.K.Butcher, W.Gerrard, E.F.Mooney, R.G.Rees, H.A.Willis, A.Anderson and H.A.Gebbie, Organometallic Chem., 1, 431 (1964).
1649. H.C.Clark and R.G.Goel, J. Organometallic Chem., 7, 263 (1967).
1650. R.S.Tobias, Organometallic Chem. Rev., 1, 93 (1966).
1651. T.Tanaka, Inorganica Chim. Acta., 1,217 (1967).
1652. J.P.Clark and C.J.Wilkins, J. Chem. Soc.,(A)871 (1966).
1653. R.S.Tobias and C.E.Friedline, Inorg. Chem., 4, 215 (1965).
1654. W.R.McWhinnie, R.Poller, J.N.R.Ruddick and M.Thevarasa, J. Chem. Soc., (A) 2327 (1969).
1655. R.J.H.Clark and C.S.Williams, Spectrochim. Acta., 21, 1861 (1965).
1656. H.Kriegsmann and S.Pauly, Z. Anorg. Allgem. Chem., 330. 275 (1964).
1657. R.A.Cummins and J.Austral, Chem., 16, 985 (1963).
1658. R.C.Poller in "The Chemistry of Organotin Compounds", Section 13, Logas Press Ltd. (1970).
1659. E.A.Gastilovich, D.N.Shigorin and N.V.Komorov, Trudy. Komis. Po Spekroskopic. Akad. Nauk. SSSR., 3, 70 (1964). Chemical Abstr., 65, 5930 (1966).
1660. D.H.Brown, A.Mohammed and D.W.A.Sharp, Spectrochim. Acta., 21, 1013 (1965).
1661. R.C.Poller, Spectrochim. Acta., 22, 935 (1966).
1662. T.S.Srivastava, J. Organometallic Chem., 10, 373 (1967).
1663. H.Kriegsmann and H. Geissler, Z. Anorg. Allgem. Chem., 323, 170 (1963).
1664. H.Keissler and H.Kriegsmann, J. Organometallic Chem., 11, 85 (1968).
1665. R.C.Poller and D.L.B.Toley, J. Chem. Soc., (A). 1578 (1967).
1666. T.Tanaka, M.Komura, Y.Kawasokai and R.Okawara, J. Organometallic Chem., 1, 484 (1964).
1667. H.Schumann, P.Jutzi, A.Roth, P.Schwake and E.Schauer, J. Organometallic Chem., 10, 71 (1967).
1668. J.C.Maire and R.Quaki, Helv. Chim. Acta., 51, 1151 (1968).
1669. H.Kriegsmann,H.Hoffmann and S.Pischtschau, Z. Anorg. Allgem. Chem., 315; 283 (1962).
1670. E.Friebe, H.Kelker and Fresenius, Z. Anal. Chem., 192, 267 (1963).
1671. B.Kushlefsky, L.Simmons and A.Ross, Inorg. Chem., 2, 187 (1963).
1672. R.C.Poller, J. Inorg. Nucl. Chem., 24, 593 (1962).
1673. N.N.Vyshinskii and N.K.Rudnevskii, Opt. Spectrosc. (USSR), 10, 421 (1961).
1674. R.A.Cummins and J.V.Evans, Spectrochim. Acta., 21, 1016 (1965).
1675. A.Marchand, J.Mendelsohn and J.Valade, CR. Acad. Sci., Paris, 259, 1737 (1964).
1676. K.Yasuda, H.Matsumoto and R.Okawara, J. Organometall. Chem., 6, 528 (1966).
1677. K.Yasula and R. Okawara, J. Organometall. Chem., 3, 76 (1965).
1678. Y.Kawaski, T.Tanaka and R.Okawara, Spectrochim. Acta., 22, 1571 (1966).
1679. H.Schmidbauer and H.Hussek, J. Organometall. Chem., 1, 244 (1964).
1680. J.P.Clark, V.M.Langford and C.J.Wilkins, J. Chem. Soc. (A) 792 (1967).

1681. Y.Kawasaki, M.Hori and O.O.Uenake, Bull. Chem. Soc., Japan, 40, 2463 (1967).
1682. Yu P.Egorov, Teor i Eksperimi Khim. Akad. Nauk. Ukr, SSR., 1, 30 (1965). Chemical Abstr., 63, 7773 (1965).
1683. A.K.Sawyer and J..E.Brown, J. Organometallic Chem., 5, 438 (1966).
1684. H.Schumann and M.Schmidt, J. Organometall. Chem., 3, 485 (1965).
1685. A.Finch, R.C.Poller and D.Steele, Trans. Faraday Soc., 61, 2628 (1965).
1686. R.C.Poller and J.A.Spillman, J. Chem. Soc. (A), 1024 (1966).
1687. H.M.Gager, J.Lewis and M.J.Ware, Chem. Commun., 616 (1966).
1688. N.A.D.Carey and H.C.Clark, Chem. Commun., 292 (1967).
1689. E.W.Randall, J.J.Ellmer and S.Zuchermann, J. Inorgan. Nucl. Chem., Letters, 1, 109 (1966).
1690. K.Sisido and S.Kozima, J. Org. Chem., 29, 907 (1964).
1691. R.E.Hester and K.Jones, Chem. Commun., 317 (1966).
1692. G.Engelhardt, P.Reich and H.Schumann, Z. Naturforsch, 22B, 392 (1967).
1693. H.Schumann, P.Jutzi, A.Roth, P.Schwabe and E.Schauer, J. Organometall. Chem., 10, 71 (1967).
1694. M.L.Maddox, S.L.Stafford and H.D.Kaesz in "Applications of N.M.R. to the study of Organometallic Chemistry", Academic Press, London and New York (1965).
1695. K.Kawakami and R.Okawara, J. Organometallic Chem., 6, 249 (1966).
1696. P.P.Simone and W.A.G.Graham, J. Organometallic Chem., 8, 479 (1967).
1697. H.Schmidbauer and H.Hussek, J. Organometallic Chem., 1, 244 (1964).
1698. E.W.Abel and D.B.Brady, J. Organometallic Chem., 11, 145 (1968).
1699. H.D.Kaesz, J. Amer. Chem. Soc., 83, 1514 (1961).
1700. J.Lorberth and H.Vakrenkamp, J. Organometallic Chem., 11, 111 (1968).
1701. G.P.Van der Kelen, Nature (London), 193,1069 (1962).
1702. A.G.Davies and T.N.Mitchell, J. Organometallic Chem., 6, 658 (1966).
1703. A.B.Envin and D.J.Seyferth, J. Amer. Chem. Soc., 89, 952 (1967).
1704. D.J.Seyferth and A.B.Envin, J. Amer. Chem. Soc., 89, 1468 (1967).
1705. W.H.Atwell and D.R.Weyenberg, J. Org. Chem., 32, 885 (1967).
1706. Van der Berghe and Van der Kelen, J. Organometallic Chem., 6, 515 (1966).
1707. Van der Berghe, Van der Vondel, Van der Kelen, Inorganica Chim. Acta, 1, 97 (1967).
1708. P.P.Simone and W.A.G.Graham, J. Organometallic Chem., 10, 457 (1967).
1709. O.J.Scherer and P.Hornig, J. Organometallic Chem., 8, 465 (1967).
1710. I.Ruidisch and M.J.Schmidt, Organometallic Chem., 1, 160 (1963).
1711. V.G.Das Kumar and O.W.Kitching, J. Organometallic Chem., 10, 59 (1967).
1712. D.J.Patmore and W.A.G.Graham, Inorg. Chem., 6, 981 (1967).
1713. M.Wieber and M.Schmidt, J. Organometallic Chem., 2, 129 (1964).
1714. H.M.McGrady and R.S.Tobias, J. Amer. Chem. Soc., 87, 1909 (1965).
1715. H.R.H.Patil and W.A.G.Graham, Inorg. Chem., 5, 1401 (1966).
1716. M.R.Kula, E.Amberger and K.K.Mayer, Chem. Berichte, 98, 634 (1965).
1717. P.T.Narasimhan and M.T.Rogers, J. Chem. Phys., 34, 1049 (1961).
1718. T.Tanaken, Y.Matsumura, R.Okawara and Y.Musya, Bull. Chem. Soc., 41, 1497 (1968).
1719. L.Verdonck, Van der Kelen, Ber. Bunsenges Physik. Chem., 69, 478 (1965).
1720. R.C.Poller in "The Chemistry of Organotin Compounds", Section 13, Logos Press Ltd. (1970).
1721. L.Verdonck, G.P.Van der Kelen, J. Organometallic Chem., 11, 491 (1968).
1722. L.Verdonck, G.P.Van der Kelen and Z.Eeckhaut, J. Organometallic Chem., 11, 487 (1968).

1723. L.Verdonck and G.P.Van der Kelen, J. Organometallic Chem., 5, 532 (1966).
1724. H.P.Fritz and C.G.Kreiter, J. Organometallic Chem., 1, 323 (1964).
1725. M.P.Simonnin, J. Organometallic Chem., 5, 155 (1966).
1726. A.J.Leusink, H.A.Budding and J.W.Marsman, J. Organometallic Chem., 9, 285 (1967).
1727. M.Delmos, J.C.Maire and J.Santamaria, J. Organometallic Chem., 16, 405 (1969).
1728. D.Seyferth and C.G.Vaughan, J. Organometallic Chem., 1, 138 (1963).
1729. L.G.Vaughan and D.Seyferth, J. Organometallic Chem., 5, 295 (1966).
1730. H.J.Hardon, H.Brunink and E.W.Van der Pol. Analyst, 85,847 (1960).
1731. H.S.Gutowsky, J. Chem. Phys., 17, 128 (1949).
1732. D.R.J.Boyd, H.W.Thompson and R.L.Williams, Discussion of the Faraday Society No. 9 (1950).
1733. H.D.Kaesz and F.G.A.Stone, Spectrochim. Acta., 15, 360 (1959).
1734. M.Ishimori and T.Tsuruta, Die Makromoleculare Chemie, 64, 190 (1963).
1735. W.Gordy and C.G.McCormick, J. Amer. Chem. Soc., 78, 3243 (1956).
1736. P.T.Nasasinhar and M.T.Rogers, J. Amer. Chem. Soc., 82, 5983 (1960).
1737. W.B.Lyttle and H.N.Rexroad, Proc. West Virginia Acad. Sci., 31-32, 190 (1959-1960).
1738. W.L.Everson and E.M.Ramirez, Anal. Chem., 37, 812 (1965).
1739. H.Matschiner and A.Tzschach, Z. Chem., 5, 144 (1965).
1740. M.V.Alekseeva, Inform. Meted. Materily Gas Nauch. Issledovatel Sanit. Inst., 5, 16 (1954); Ref. Zh. Khim., 40,375 (1955).
1741. H.M.Factory Inspectorate, Ministry of Labour, Methods for the Detection of Toxic Substances in Air, No. 9, Arsine, 2nd ed. H.M. Stationery Office, London (1966).
1742. V.Vasak, Collect. Czech. Chem. Commun., 24, 3500 (1959).
1743. F.T.Henry and T.M.Thorpe, Anal. Chem., 52, 80 (1980).
1744. R.K.Elton and W.E.Geiger, Anal. Chem., 50, 712 (1978).
1745. A.Watson Analyst (London), 103, 332 (1978).
1746. S.W.Lee and T.C.Meranger, Anal. Chem., 53, 130 (1981).
1747. A.Watson and G.Svehia, Analyst (London), 100, 489 (1975).
1748. A.Watson and G.Svehia, Analyst (London), 100, 573 (1975).
1749. R.Elton and W.E.Geiger, Jr. Anal. Left., 9, 665 (1976).
1750. R.C.Bess, K.J.Irogolic, J.E.Flannery and T.H.Ridgway, Anal. Lett., 9, 1091 (1976).
1751. R.C.Bess, K.J.Irogolic, J.E.Flannery and T.H.Ridgway, Anal. Lett., 9,
1752. D.J.Myers and J.Osteryoung, Anal. Chem., 45, 267 (1973).
1753. G.Forsberg, J.W.O'Laughlin and R.C.Megargle, Anal. Chem., 47, 1586 (1975).
1754. Standard Methods for the Examination of Water and Wastewater, American Public Health Association, 13th ed., 1971.
1755. J.H.Lowry, R.B.Smart and K.H.Mancy, Anal. Lett., 10, 979 (1977).
1756. H.C.Marks and J.R.Glass, J. Am. Water Works Assoc., 34, 1227 (1942).
1757. W.Stumm, J. Boston Soc. Civl. Eng., 45, 68 (1958).
1758. J.H.Lowry, R.B.Smart and K.H.Mancy, Anal. Chem., 50, 1303 (1978).
1759. A.F.Finders and T.De Vries, Anal. Chem., 28, 209 (1956).
1760. J.D.L.Holloway, F.C.Senftleber and W.E.Geiger, Anal. Chem., 50, 1010 (1978).
1761. R.Geyer and W.Gliem, Z. Chemie Lp3, 1, 64 (1967).
1762. J.B.Headridge, M.Ashraf and H.L.H.Dodds, J. Electroanal, Chem., 16, 114 (1968).
1763. A.M.Hartley and R.E.Visco, Anal. Chem., 35, 1871 (1963).
1764. R.Engst, W.Schnaak and H.Waggon, Z. Anal. Chem., 222, 388 (1966).
1765. J.F.Price and R.P.Baldwin, Anal. Chem., 52, 1940 (1980).
1766. P.W.Swanson and P.H.Daniels, J. Inst. Pet., 39, 487 (1953).
1767. W.Hubis and R.O.Clark, Anal. Chem., 27, 1009 (1955).

1768. V.Sedivec and V.Flek, Prac. Lek., 10, 270 (1958).
1769. P.Nangniot, Chim. Anal., 47, 592 (1965).
1770. M.S.Jovanovic, J. Tomic, Z.Masic and M.Dragojevic, Chem. Anal. (Warsaw), 11, 479 (1966).
1771. ASTM Proc., 52, 365 (1952).
1772. ASTM, Standards on Petroleum Products and Lubricants, Methods D1269-53T (1953) and D1269-61 (1961), American Society for Testing Materials.
1773. I.A.Prashinskii and M.K.Frolova, Tr. Vses. Nauchn.-Issled. Inst. Pererab Nefti Gaza Polycheniyu Iskusstv. Zhidkogo Topliva, 6, 181 (1957).
1774. J.Smelik, J. Prakt. Chem., 5, 9 (1954).
1775. M.Roschig and H.Matschiner, Chem. Tech. (Berlin), 19, 103 (1967).
1776. L.N.Vertyulina and I.A.Korschunov, Khim. Nauka. Prom., 4, 136 (1959).
1777. J.E.DeVries, A.Lauw-Zecha and A.Pellecer, Anal. Chem., 31, 1995 (1959).
1778. R.I.McCallum, J. Oil Colour Chem. Assoc., 52, 434 (1969).
1779. G.Costa, Ann. Chim. (Rome), 41, 207 (1951).
1780. V.F.Toropova, M.K.Saikina, Shornik Statei. Obshchei. Khim. Akad. Nauk. SSSR., 1, 210 (1953).
1781. W.U.Malik and R.Hague, J. Amer. Oil Chem. Soc., 41, 41 (1964).
1782. I.Dragusin and N.Totir, Stud. Cercet. Chim., 13, 955 (1965).
1783. T.M.Hopes, J. Assoc. Off. Anal. Chem., 49, 840 (1966).
1784. G.Radulescu, I.Badilescu and A.Gilici, Rev. Chim. (Bucharest), 15, 164 (1964).
1785. N.Shirota, M.Kotakersori and H.Handa, Ann. Rep. Takamine Lab., 9, 198 (1957).
1786. H.Sato. Bunseki Kagaku, 6, 166 (1957).
1787. H.Sato, Japan Analyst, 6, 84 (1957); Anal. Abstr., 5, 705 (1958).
1788. M.Kotakemori and H.Henda, Ann. Rep. Takamine Lab., 8, 231 (1956).
1789. T.M.Hopes, J. Assoc. Off. Agric. Chem., 48, 585 (1965).
1790. T.Medwick in "Pharmaceutical Analysis" (Editors T.Higuchi, E.Brockmann-Haussen), Interscience, New York, Chapter X11 (1961).
1791. W.U.Malik and P.Hague, Nature (London), 194, 863 (1962).
1792. G.Bontempelli, B.Corain and F.Magro, Anal. Chem., 49, 1005 (1977).
1793. R.F.Makens, H.H.Vaughan and R.R.Chelberg, Anal. Chem., 27, 1062 (1955).
1794. D.C.Tse and T.Kuwana, Anal. Chem., 50, 1315 (1978).
1795. J.O.Bronstad and H.O.Friestad, Analyst (London), 101, 820 (1976).
1796. P.Nangniot, Anal. Chim. Acta., 31, 166 (1964).
1797. H.Sohr, Chem. Zvesti, 16, 316 (1962).
1798. W.Stricks and R.G.Meuller, Anal. Chem., 36, 40 (1964).
1799. V.Sh.Tsveniashvili, S.I.Zhdanov and Z.V.Todres, Z. Anal. Chem., 224, 389 (1967).
1800. H.Jehring and H.Mehner, Z. Chem., 3, 34 (1963).
1801. H.Jehring and H.Mehner, Z. Chem., 4, 273 (1964).
1802. H.Mehner and J.Jehring, Z. Chem., 3. 472 (1963).
1803. H.Mehner, H.Jehring and H.Kriegsmann, 3rd Analytical Conference, Budapest, 24-29 August (1970).
1804. H.Mehner, Dissertation (1967) Humboldt, University, Berlin (1967).
1805. H.Jehring and H.Mehner, Z. Anal. Chem., 1, 136 (1967).
1806. R.Geyer and H.J.Seidlitz, Z. Chem., 7, 114 (1967).
1807. H.Mehner, H.Jehring and H.Kriegsmann, J. Organomet. Chem., 15, 97 (1968).
1808. H.Mehner, H.Jehring and H.Kriegsmann, J. Organomet. Chem., 15, 107 (1968).
1809. H.Jehring, H.Mehner and H.Kriegsmann, J. Organomet. Chem., 17, 53 (1969).

1810. J.Heyrovsky and R.Kalvoda, Akademic Veriag, Berlin, (1960).
1811. R.Geyer and P.Rotermund, Acta. Chim. Acad. Hung., 59, 201 (1969).
1812. G.Costa, Gazz. Chim. Ital., 80, 42 (1950).
1813. G.Costa, Ann. Chim. (Rome), 41, 207 (1951).
1814. V.F.Toropova and M.K.Saikina, Sb. Stat. Obshch. Khim. Akad. Nauk SSSR, 1, 210 (1953).
1815. R.B.Allen, Diss. Abstr., 20, 897 (1959).
1816. H.Jehring and H.Mehner, Z. Chem., 3, 472 (1963).
1817. J.Lorberth and H.Noth, Chem. Ber., 98, 969 (1965).
1818. H.Kriegsmann and S.Pischtschan, Z. Anorg. Allg. Chem., 308, 212 (1961).
1819. K.A.Koreschkov, Ber. Dtsch. Chem. Ges., 62, 996 (1929).
1820. D.Seyferth, Naturwissenschaften, 34 (1957).
1821. V.A.Bork and P.I.Selivokhin, Plast. Massy, (10), 60 (1969).
1822. M.Devaud, C. R. Acad. Sci. Ser. C., 262, 702 (1966).
1823. M.Devaud, P.Souchay and M.Person, J. Chim. Phys. Physiocochim. Biol., 64, 646 (1967).
1824. M.Devaud, C. R. Acad. Sci. Ser. C., 263, 1269 (1966).
1825. M.Devaud, J. Chim. Phys. Physiocochim. Biol., 63, 1335 (1966).
1826. M.Devaud, J. Chim. Phys. Physicochim. Biol., 64, 791 (1967).
1827. M.Devaud and P.Souchay, J. Chim. Phys. Physicochim. Biol., 64, 1778 (1967).
1828. A.Vanachayangkul and M.D.Morris, Anal. Lett., 1, 885 (1968).
1829. M.Devaud and S.Laviron, Rev. Chim. Miner., 5, 427 (1968).
1830. V.N.Flerov and U.M.Tyurin, Zh. Obshch. Khim., 38, 1669 (1968).
1831. M.D.Morris, J. Electroanal. Chem. Interfacial Electrochem., 16, 569 (1968).
1832. T.L.Shkorbatova, O.A.Kochkin, L.D.Sirak and T.V.Khavalit, Zh. Anal. Khim., 26, 1521 (1971).
1833. A.Vogel and J.Deshusses, Helv. Chim. Acta., 47, 181 (1964).
1834. S.Gorbach and R.Bock, Z. Anal. Chem., 163, 429 (1958).
1835. V.D.Bezuglyi, E.A.Preobrazhenskaya and V.N.Dmitrieva, Zh. Anal. Khim., 19, 1033 (1964).
1836. D.A.Kochkin, T.L.Shkorbatova, L.D.Pegusova and N.A.Voronkov, Zh. Obshch. Khim., 39, 1777 (1969).
1837. R.Geyer and H.T.Seidlitz, Z. Chem., 4, 468 (1964).
1838. L.Haasova, M.Pribyl and Z.Fresenius, Z. Anal. Chem., 249, 35 (1970).
1839. C.J.Svehla and F.Glockling, Anal. Chim. Acta., 117, 193 (1980).
1840. B.Fleet and N.B.Fouzder, J. Electroanal. Chem., 63, 59 (1975).
1841. N.B.Fouzder and B.Fleet in "Polorography of molecules of Biological Significance" (Editor W.F.Smythe), Academic Press, London and New York, p.268 (1979).
1842. A.M.S.Alam, J.M.Martin and P.Kapsa, Anal. Chim. Acta., 107, 391 (1979).
1843. P.Longi and R.Mazzocchi, Chim. Ind. (Milano), 48, 718 (1966).
1844. G.V.Bortnikov, E.N.Vyankin, E.N.Gladyshev and V.S.Andreevichev, Zavodskaya Lab., 35, 1445 (1969).
1845. L.M.Brown and K.S.Mazdiynasi, Anal. Chem., 41, 1243 (1969).
1846. R.Lioznova and M.L.Genusov, Zavod. Lab., 26, 945 (1960).
1847. D.F.Hagen, D.G.Biechler, W.D.Leslie and D.E.Jordan, Anal. Chem. Acta, 41, 557 (1966).
1848. H.Kawaguchi, T.Sakamoto and O.Mizuike, Talanta, 20, 231 (1973).
1849. R.M.Pagnall, T.S.West and P.Whitehead, Analyst, 98, 647 (1973).
1850. C.Feldman and D.A.Batistoni, Anal. Chem., 49, 2215 (1977).
1851. B.J.Gudzinowicz and H.F.Martin, Anal. Chem., 34, 648 (1962).
1852. B.J.Gudzinowicz and J.L.Driscoll, J. Gas Chromatogr., 1, 25 (1963).
1853. G.E.Parris, W.R.Blair and T.E.Brinkman, Anal. Chem., 49, 378 (1977).

1854. Y.Odanaka, N.Tsuchiya, O.Matano and S.Goto, Anal. Chem., 55, 929 (1983).
1855. Y.Talmi and D.T.Bostick, Anal. Chem., 47, 2145 (1975).
1856. Y.Talmi and V.E.Norvell, Anal. Chem., 47, 1510 (1975).
1858. J.M.Ammons and J.L.Bricker, Anal. Chem., 49, 621 (1977).
1859. R.S.Braman and N.Dynako, Anal. Chem., 40. 95 (1968).
1860. J.D.Defreese and H.V.Nalmstadt, Paper No. 26, presented before the Federation of Analytical Chemistry and Spectroscopic Societies, Indianapolis, Indiana, November (1975).
1861. N.F.Ives and L.Guiffrida, J. Assoc. Off. Anal. Chem., 53, 973 (1970).
1862. R.E.Weston and B.B.Wheals, and M.J.Kenselt, Analyst, 96, 601 (1971).
1863. J.Guchi, A.Nishyama and Y.Nagase, J. Pharm. Soc., Japan, 80, 1408 (1960).
1864. A.D.Zorin, G.G.Devyatykk, Y.Ya Dudorov and A.M.Amel'chenko, Zhur. Anorg. Khim., 9, 2525 (1964).
1865. M.Covello,G.Ciampa and E.Ciamillo, Farmaco Ed. Pract., 22, 218 (1967).
1866. A.D.Molodyk, G.V.Bondar and L.N.Morozorva, Zav. Lab., 38, 129 (1972).
1867. P.R. Gifford and S.Bruckenstein, S. Anal. Chem., 52, 1028 (1980).
1868. G.Schomberg, R.Koster and D.Henneberg. Z. Anal. Chem., 170, 285 (1959).
1869. R.Koster and G.Bruno, Annaley der Chemie., 629, 89 (1960).
1870. L.J.Kuhns, R.S.Braman and J.E.Graham, Anal. Chem., 34,1700 (1962).
1871. G.R.Seely, J.P.Oliver and D.M.Ritter, Anal. Chem., 32,1993 (1959).
1872. T.D.Parsons, M.B.Silverman and D.M.Ritter, J. Am. Chem. Soc., 79, 5091 (1957).
1873. J.A.Semlyen and C.S.G.Phillips, J. Chromatography, 18, 1 (1965).
1874. A.T.James and A.J.P.Martin, Analyst,(London), 77, 915 (1952).
1875. T.C.L.Chang and C.Karr, Anal. Chim. Acta., 21, 474 (1959).
1876. T.C.L.Chang and C.Karr, Anal. Chim. Acta., 24, 343 (1961).
1877. C.S.G.Phillips, P.Powell and J.A.Semlyen, J. Chem. Soc., 1202 (1963).
1878. P.Powell, J.A.Semlyen, R.E.Blofield and C.S.G.Philipps, J. Chem. Soc., 280, (1954).
1879. C.S.G.Phillips, P.Powells, J.A.Semplyen and P.L.Timms, Z. Anal. Chem., 197, 202 (1963).
1880. J.J.Kaufman, J.E.Todd and W.S.Koski, Anal. Chem., 29, 1032 (1957).
1881. W.S.Koski, P.C.Maybury and J.J.Kaufman, Anal. Chem., 26, 1992 (1954).
1882. F.J.Sawinski and I.H.Suffet, J. Chromat. Sci., 9, 632 (1971).
1883. A.D.Zorin, I.A.Frolov, N.T.Karabanov, V.M.Kedyarkin, V.V.Balabanov, T.S.Kuznetsova and A.N.Gurtyanov, Zhur. Anal. Khim., 25, 389 (1970).
1884. H.W.Myers and R.F.Putnam, Anal. Chem., 34, 664 (1962).
1885. W.S.Pappas and J.G.Milliou, Anal. Chem., 40, 2176 (1968).
1886. V.V.Brazhnikov and K.I.Sakodynskii, Zhue. Prinklad. Khim. (Leningrad), 43, 2247 (1970).
1887. I.A.Zelyaev, G.G.Devyatykh and N.K.Aglinlov, Zhur. Anal. Chem., 24, 1081 (1969).
1888. J.Dazard and H.Mongeot, Bull. Soc. Chim. (France), 1, 51 (1971).
1889. A.Di Lorezo, J. Chromat., 76, 207 (1973).
1890. N.Toshiyuki and T.Matsuda, J. Chem. Soc. (Japan) Pure Chem. Sect., 86, 965 (1965).
1891. P.J.Wood and I.R.Suddiqui, Carbohydrate Research, 19, 283 (1971).
1892. C.J.W.Brooks and D.J.Harvey, J. Chromatogr., 521, 193 (1971).
1893. C.F.Poole and E.D.Morgan, Private communication.
1894. R.F.Putnam and H.W.Myers, Anal. Chem., 34, 486 (1962).
1895. S.A.Greene and H.Pust, Anal. Chem., 30, 1039 (1958).
1896. A.F.Zhigach, E.B.Kakova and R.A.Kigel, Z. Anal. Chem., 14, 746 (1959).

1897. I.Lysyj, and R.C.Greenough, Anal. Chem., 35, 1657 (1963).
1898. H.Veening, W.E.Bockman and D.M.Wilkinson, J. Gas Chromatography, 5, 248 (1967).
1899. H.Veening and J.K.F.Huber, J. Gas Chromatography, 6, 326 (1968).
1900. H.Veening, J.Graver, D.B.Clark and B.R.Willeford, Anal. Chem., 41, 1655 (1969).
1901. H.Veening, J.S.Keller and B.R.Willeford, Anal. Chem., 43, 1516 (1971).
1902. W.J.A.Van der Heuvel, J.S.Keller, H.Veening and B.R.Willeford, Analyt. Lett., 3, 279 (1970).
1903. P.M.Krueger and J.A.McCloskey, Anal. Chem., 41, 1930 (1969).
1904. C.Segard, B.Rogues, C.Pommier and G.Guichon, Anal. Chem., 43, 1146 (1971).
1905. T.L.Mironova, D.A.Vyakhirev and A.M.Golubeva, Zh. Obsch. Khim., 44, 330 (1974).
1906. W.R.Jackson and W.B.Jennings, J. Chem. Soc., (B), 1221 (1969).
1907. V.A.Devyatykh, V.A.Umilin, Yu B.Zverev and I.A.Frolov, Izv. Akad. Nauk. SSSR. Ser. Khim.,(2), 247 (1969).
1908. D.E.F.Gracey, W.R.Jackson, W.B.Jennings and T.R.Mitchell, J. Chem. Soc., (B), 1197 (1969).
1909. D.E.F.Gracey, W.R.Jackson, C.H.Mullen and N.Thompson, J. Chem. Soc., (B) , 1197 (1969).
1910. C.Segard, C.Pommier, B.Pierre Rogues and G.Ginchon, J. Organometallic Chemistry, 77, 49 (1974).
1911. F.A.Serravalto and T.H.Risky, J. Chromatogr. Sci., 12, 585 (1974).
1912. M.T.Yarotovski, E.I.Kozlov, E.A.Obol'wikova and O.I.Volkova and G.I. Samokhval'ov, J. Gas Chromatogr., 6, 520 (1968).
1913. R.M.Pagnell, T.S.West and P.Whitehead, Analyst,(London),98,647(1973).
1914. K.K.Fukin, V.G.Rezchikov, T.S.Kuznetsova and I.A.Frolov, Zav. Lab., 39, 993 (1973).
1915. E.W.Abel, G.Nickless and F.H.Pollard, Proc. Chem. Soc. (London), 288 (1960).
1916. C.S.G.Phillips and P.L.Timms, Anal. Chem., 35, 505 (1963).
1917. F.H.Pollard, G.Nickless and P.C.Uden, J. Chromatogr., 14, 1 (1964).
1918. J.A.Semlyen, G.R.Walker, R.E.Blofield and C.S.G.Phillips, J. Chem. Soc.,4948 (1964).
1919. J.A.Semlyen, C.S.G.Phillips, J. Chromatogr., 18, 1 (1965).
1920. P.L.Timms, C.C.Simpson and C.S.G.Phillips, J. Chem. Soc., 1467 (1964).
1921. A.D.Snegova, L.K.Markov, V.A.Panomarenko and W.D.Zelinsky, Institute of Organic Chemistry, USSR., Academy of Sciences, Moscow, Zhur. Anal. Khim., 19, 610 (1960).
1922. G.Garzo, J.Fekete and M.Blazo, Acta. Chim. Acad. Sci., (Hungary), 51, 359 (1967).
1923. G.N.Bortnikov, N.S.Vyazankin, N.P.Nikulina and I. Ya Yashin, Izv. Akad. Nauk. SSSR, Ser. Khim., 1, 21 (1970).
1924. N.S.Yvazankin, G.N.Bortnikov, I.A.Niguinova, A.V.Kiselev, Ya I.Yashin, A.N.Egorochin and V.F.Mironov, Izv. Akad. Nauk. SSSR. Ser. Khim., 1, 186 (1969).
1925. G.N.Bortnikov, M.N.Bochkarev, N.S.Yyanankin, S.K.R.Ratushvaya and Ya I.Yashin, Izv. Akad. Nauk. SSSR. Ser. Khim., 4, 851 (1971).
1926. V.V.Brazhnikov and K.I.Sakodynskii, J. Chromatogr., 64, 157 (1972).
1927. I.A.Frolov, Trudy Khim. Tckhnol (Gor'kii), 1, (15), 107 (1966).
1928. V.M.Gorbackev, G.V.Trent'yakov, Zav. Lab., 32, 796 (1966).
1929. V.V.Brazhnicov and S.I.Sakodynskii, Zh. Prikl. Khim., 43, 2247 (1970).
1930. G.Parissakis, D.Vranti-Piscou and J.Kontoyannakos, J. Chromatogr., 52, 461 (1970).
1931. D.Vranti-Piscou, J.Kontoyannakos and G.Parissakis, J. Chromatogr. Sci., 9, 499 (1971).

1932. G.G.Devyatykh, A.D.Zorin, A.M.Amel'chenko, S.B.Lyakhmanov and A.E. Ezheleva, Dokl. Akad. Nauk. SSSR., 156, (5), 1105 (1964).
1933. M.Fiser, Ustav. Jad. Fyz. Cesk. Akad. Ved. (Rep),2917 (1972).
1934. R.D.Kadeg and G.D.Christian, 1st Chem. Congr. of the North Amer. Continent, Mexico City (Dec, 1975).
1935. B.Iatridis and G.Parissakis, Anal. Chem., 49, 909 (1977).
1937. K.Tanikawa and K.Arakawa, Chem. Pharin, Bull,, Tokyo, 13, 926 (1965).
1938. A.N.Nesmeyanov, L.P.Yur'eva, N.S.Kochetkova and S.V.Vitt., Izv. Akad. Nauk. SSSR. Ser. Khim., 3 560 (1966).
1939. O.E.Ayers, T.C.Smith, J.D.Burnett and B.W.Pouder, Anal. Chem., 38, 1606 (1966).
1940. C.Pommier and G.Guichon, Chromatographia., 2, 346 (1969).
1941. K.Yamakawa, N.Ishibashi and K.Arakawa, Chem. Pharim. Bull., 13, 926 (1965).
1942. K.Yamakawa, N.Ishibashi and K.Arakawa, Chem. Pharim, Bull., 12, 119 (1964).
1943. R.A.Benkeser and J.L.Barch, J. Am. Chem. Soc., 86, 890 (1964).
1944. R.A.Benkeser, Y.Nagai and J.Hooz, J. Am. Chem. Soc., 86, 3742 (1964).
1945. Y.Nagai, J.Hooz and R.A.Benkeser, Bull. Chem. Soc. Japan, 36, 482 (1964).
1946. Y.Nagai, J.Hooz and R.A.Benkeser, 17th Annual Meeting Chem. Soc. Japan, Abstr. papers 1G 19 Tokyo, (April 1964).
1947. E.J.Forbes, M.K.Sultan and P.C.Uden, Analt. Lett., 5, 927 (1972).
1948. P.C.Uden, D.E.Henderson, F.P.DiSanzo, R.J.Lloyd and T.Tetu, 174th National Meeting American Chemical Society, Chicago, 111. Abstr. Anal. 27, (August 1977).
1949. H.H.Hill and W.A.Aue, J. Chromatogr., 74, 311 (1972).
1950. W.W.Parker, G.Z.Smith and R.L.Hudson, Anal. Chem., 33, 1170 (1961).
1951. W.W.Parker and R.L.Hudson, Anal. Chem., 35, 1334 (1963).
1952. J.E.Lovelock and N.L.Gregory in "Gas Chromatography International Symposium", 3, 219 (1962).
1953. E.J.Bonelli and H.Hartmann, Anal. Chem., 35, 1980 (1963).
1954. H.J.Dawson, Anal. Chem., 35, 542 (1963).
1955. W.A.Aue and H.H.Hill, J. Chromatogr., 74, 319 (1972).
1956. H.H.Hill and A.W.Aue, J. Chromatogr., 74, 311 (1972).
1957. Y.K.Chau, P.T.S.Wong and H.Saitoh, J. Chromatogr. Sci., 14, 162 (1976).
1958. D.T.Coker, Anal. Chem., 47, 386 (1975).
1959. D.C.Reamer, W.H.Zoller and T.C.O'Haver, Anal. Chem., 50, 1449 (1978).
1960. H.J.Dawson, Jr., Anal. Chem., 35, 542 (1963).
1961. E.A.Boettner and F.C.Dallos, J. Gas Chromatogr., 3 (6), 190 (1965).
1962. N.L.Soulages, Anal. Chem., 38, 28 (1966).
1963. W.N.Parker, G.Z.Smith and R.L.Hudson, Anal. Chem., 33, 1170 (1961).
1964. W.N.Parker and R.L.Hudson, Anal. Chem., 35, 1334 (1963).
1965. A.Laveskag, Second International Clean Air Congress Proceedings, Washington D.C., p.549 (Dec. 1970).
1966. J.E.Lovelock and A.Zlatkis, Anal. Chem., 33, 1958 (1961).
1967. J.E.Lovelock, Anal. Chem., 33, 162 (1961).
1968. H.J.Dawson, Anal. Chem., 35, 542 (1963).
1969. E.Barrall and P.Ballinger, J. Gas Chromatography, 1, 7 (1963).
1970. E.Bonnelli and H.Hartmann, Anal. Chem., 35, 1980 (1963).
1971. E.W.Abel, G.Nickless and F.H.Pollard, Proc. Chem. Soc. (London), 288 (1960).
1972. M.D.DePuis and H.H.Hill, Anal. Chem., 51, 292 (1979).
1973. W.A.Aue and H.H.Hill, Jr,, J. Chromatogr., 74, 319 (1972).
1974. W.A.Aue and H.H.Hill, Jr., Anal. Chem., 45, 729 (1973).
1975. H.H.Hill, Jr., and W.A.Aue, J. Chromatogr. Sci., 12, 541 (1974).
1976. H.H.Hill, Jr. and W.A.Aue, J. Chromatogr., 122, 515 (1976)
1977. H.H.Hill, Jr. Ph.D. Thesis, Dalhousie University, Halifax, N.S., Canada, (1975).

1978. D.C.Reamer, W.H.Zoller and T.C.O Haver, Anal. Chem., 50, 1449 (1978).
1979. B.D.Quimby, P.C.Uden and R.M.Barnes, Anal. Chem., 50, 2112 (1978).
1980. A.J.McCormack, S.C.Tong and W.D.Cooke, Anal. Chem., 37, 1470 (1965).
1981. H.Kawaguchi, T.Sakamoto and O.Mizuike, Talanta, 20, 321 (1973).
1982. N.M.Karayannis and A.H.Corwin, J. Chromatogr., 47, 247 (1970).
1983. Y.Talmi, Anal. Chim. Acta., 74, 107 (1975).
1984. R.M.Pagnell, T.S.West and P.Whitehead, Analyst,(London), 98, 647 (1973).
1985. D.T.Bostick and Y.Talmi, J. Chromatogr. Sci., 15, 163 (1977).
1986. F.A.Serravallo and T.H.Risky, J. Chromatogr. Sci., 12, 585 (1974).
1987. Y.Talmi and W.W.Audren, Anal. Chem., 46, 2122 (1974).
1988. Y.Talmi and D.T.Bostick, Anal. Chem., 47, 2145 (1975).
1989. Y.Talmi and V.E.Norvell, Anal. Chem., 47, 1510 (1975).
1990. C.A.Bache and D.J.Lisk, Anal. Chem., 39, 786 (1967).
1991. W.R.McLean, D.L.Stanton and G.E.Penketh, Analyst,(London), 98,432 (1973).
1992. R.M.Dagnall, T.S.West and P.Whitehead, Anal. Chem., 44, 2074 (1972).
1993. D.C.Reamer, W.H.Zoller and T.C.O'Haver, Anal. Chem., 50. 1449 (1978).
1994. R.Moss and E.V.Browell, Analyst, 9, 428 (1966).
1995. S.A.Estes, P.C.Uden and R.M.Barnes, Anal. Chem., 53, 1336 (1981).
1996. S.A.Estes, P.C.Uden and R.M.Barnes, Anal. Chem., 54, 2402 (1982).
1997. N.L.Soulages, Anal. Chem., 38, 28 (1966).
1998. N.L.Sculages, Anal. Chem., 39, 1340 (1967).
1999. L.E.Green, Hewlett Packard Facts and Methods, 8, 4 (1967).
2000. K.Kramer, Erdol Kohle, 19, 182 (1966).
2001. V.Svab, Nafta (Zagreb), 20, 557 (1969).
2002. P.M.Mutsaacs and J.E.Van Steen, J. Inst. Petrol, 58, 102 (1972).
2003. F.H.Pollard, G.Nickless and P.C.Uden, J. Chromatogr., 14, 1 (1964).
2004. M.Pressler,V.Martin and J.Janak, J. Chromatogr., 59, 429 (1971).
2005. H.H.Hill and W.A.Aue, J. Chromatogr., 74, 311 (1972).
2006. P.R.Ballinger and L.M.Whitemore, Proc. of Amer. Chem. Soc.,Division of Petroleum Chemistry, Atlantic City Meeting, 13, 130 (September 1968).
2007. B.Kolb, G.Kemner and F.H.Schleser, Wiedeking, Meeting of the German Chemical Society, the Austrian Society for Microchemistry and Analytical and General Chemistry, Rindau, Germany, (April 1966). Ref: Z. Analyt. Chem., 221, 116 (1966).
2008. Y.K.Chau, P.J.S.Wong and P.D.Goulden, Anal. Chim. Acta., 85, 421 (1976),
2009. J.W.Robinson, E.L.Kiesel, J.P.Goodbread, R.Bliss and R.Marshall, Anal. Chim. Acta., 92, 321 (1977).
2010. B.Radzuik, Y.Thomassen, J.C.Van Loon and Y.K.Chau, Anal. Chim. Acta., 105, 255 (1979).
2011. B.Kolb, G.Kemmner, F.H.Schleser, E.Wiedeking, Z. Anal. Chem., 221, 166 (1966).
2012. T.Kator, and R.Nagagawa, Bull. Inst. Environ. Sci. Technol., 1, 19 (1974).
2013. Y.K.Chau, P.T.S.Wong and H.Saitohn, J. Chromatogr. Sci., 14, 162 (1976).
2014. D.T.Coker, Anal. Chem., 47, 386 (1975).
2015. D.A.Segar, Anal. Lett., 7, 89 (1974).
2016. J.W.Robinson, Vidarreta, D.K.Wolcott, J.P.Goodbread and E.Kiesel, Spectrosc. Lett., 8, 491 (1975).
2017. R.Bye, P.E.Paus, R.Solberg and Y.Thomassen, Atomic Absorption Newsletter, 17, 131 (1978).
2018. B.Radzuik, Y.Thomassen, L.R.P.Butler, J.C.Van Loon and Y.K.Chau, Anal. Chim. Acta., 108, 31 (1979).
2019. De Jon'ghe, D.Chakraborti and F.Adams, Anal. Chim. Acta., 115, 89 (1980).
2020. D.S.Forsythe and W.D.Marshall, Anal. Chem., 57, 1299 (1985).

2021. R.G.Anderson, J. Org. Chem., 23, 750 (1958).
2022. B.Bartocha, Z. Naturforsch., 14b, 809 (1959).
2023. D.Seyferth and M.A.Weiner, J. Am. Chem. Soc., 83, 3583 (1961).
2024. R.Waack and M.A.Doran, J. Am. Chem. Soc., 85, 1651 (1963).
2025. H.L.Johnson and R.A.Clark, Anal. Chem., 19, 869 (1947).
2026. E.C.Juenge and D.Seyferth, J. Org. Chem., 26, 563 (1961).
2027. W.S.Leonhardt, R.C.Morrison and C.W.Kamienski, Anal. Chem., 38,466 (1966).
2028. P.F.Collins, C.W.Kamienski, D.L.Esmay and R.B.Ellestad, Anal. Chem., 33, 468 (1961),
2029. D.Bernstein, Z. Anal. Chem., 182, 321 (1961).
2030. H.O.House and W.L.Respers, J. Organometal. Chem., 4, 95 (1965).
2031. R.D'Hollander and M.Anteunis, Bull. Soc. Chim. Belg., 72, 77 (1963).
2032. H.O.House and W.L.Respers, J. Organometal. Chem., 4, 95 (1965).
2033. M.A.Molinari, J.Lombardo, O.A.Lires and G.J.Videla, An. Assoc. Quim. Argentina, 48, 223 (1960).
2034. L.V.Giold, C.A.Hollingsworth, D.H.McDaniel and J.H.Wotiz, Anal. Chem., 38, 1156 (1961).
2035. A.Wowk and S.Di Giovanni, Anal. Chem., 38, 742 (1966).
2036. N.M.Karayannis and A.H.Corwin, J. Chromatography, 47, 247 (1970).
2037. G.N.Turkel'taub, B.M.Luskina and S.V.Syavtsillo, Khim. Tekhnol. Topliv. Masel., 12, 58 (1967).
2038. P.C.Uden, R.M.Barnes and P.Di Sanzo, Anal. Chem., 50, 852 (1978).
2039. M.Coe and J.C.Van Loon, Anal. Chim. Acta., 120, 171 (1980).
2040. K.Brodersen and U.Z.Schlenker, Anal. Chem., 182, 421 (1961).
2041. S.Nishi and Y.Horimoto, Japan Analyst, 20, 16 (1971).
2042. R.C.Dressman, J. Chromato. Sci., 10, 472 (1972).
2043. R.C.Dressman, J. Chromato. Sci., 10, 468 (1972).
2044. G.L.Baughman, M.H.Carter, N.L.Wolf and R.G.Zepp, J. Chromatogr., 76, 471 (1973),
2045. K.Teremoto, M.Kita'atake, M.Tanabe and Y.Noguchi, J. Chem. Soc. Ind. Chem. Sect., 70, 1601 (1967).
2046. H.Hey, Z. Analyt. Chem., 256, 361 (1971).
2047. J.A.Rodrigues-Vasques, Talanta, 25, 299 (1978).
2048. C.A.Bache and D.J.Lisk, Anal. Chem., 43, 950 (1971).
2049. C.A.Bache and D.J.Lisk, Anal. Chem., 43, 951 (1971).
2050. C.A.Bache and D.J.Lisk, Anal. Chem., 43, 1950 (1971).
2051. W.E.L.Grossman, J.Eng and Y.C.Tong. Anal. Chim. Acta., 60. 447 (1972).
2052. Y.Talmi, Anal. Chim. Acta., 74, 107 (1975).
2053. V.A.Umilin and I.N.Tyutyaev, Izv. Akad. Nauk. SSSR. Ser. Khim., 6, 1262 (1972).
2054. H.Veening, N.J.Graven, D.C.Clark and B.R.Willeford, Anal. Chem., 41, 1655 (1969).
2055. A.Davis, A.Roadi, J.G.Michalovic and H.M.Joseph, J. Gas Chromatogr., 1, 23 (1963).
2056. C.E.Legate and H.D.Burnham, Anal. Chem., 32, 1042 (1960).
2057. C.J.Hardy, J. Chromatogr., 13, 372 (1964).
2058. A.Apelblat and A.Hornik, J. Chromatogr., 24, 175 (1966).
2059. A.Apelblat and A.Hornik, in "Proc. Int. Conf. Solvent Extraction Chemistry", Gothenburg (1966) (J.Rydberg, D.Dryssen J.O.Lilenzin editors), pp 296-304 North Holland, Amsterdam (1966).
2060. A.Apelblat and A.Hornik, Trans. Faraday Soc. No. 529, 63, 185 (1967).
2061. A.Apelblat, J. Inorg. Nucl. Chem., 31,483 (1969).
2062. A.Apelblat, J. Inorg. Nucl. Chem., 32, 3647 (1970).
2063. F.Strache, Dtsh. Lebensmith. Rdsch., 56. 173 (1960).
2064. M.H.Cambell, Anal. Chem., 38, 237 (1966).

2065. United Kingdom Atomic Energy Authority UKAEA Report Pg 755 W (1966).
2066. A.J.Moffat and R.D.Thompson, J. Inorg. Nucl. Chem., 16, 365 (1961).
2067. K.D.Berlin, T.H.Austin, M.E.Nagahushanam, J.Peterson, J.Calvert, L.A.Wilson and D.Hopper, J. Gas Chromatogr., 3, 256 (1965).
2068. L.S.Lewis and H.W.Patton in "Gas Chromatography" (J.V.Coates et al editors), p. 149, Academic Press, New York, London (1958).
2069. S.H.Shipotofsky, U.S.Atomic Energy Commission Report TID 6437 (1960).
2070. R.Feinland, J.Sass and S.A.Buckler, Anal. Chem., 35.,920 (1963).
2071. Y.C.Lee and G.Ting, Anal. Chim. Acta., 106, 373 (1979).
2072. A.De Rosi, W.Gerrard and E.F.Mooney, Chem. Ind. (London) 1449 (1961).
2073. B.T.Gudinowicz and R.H.Campbell, Anal. Chem., 33, 1510 (1961).
2074. B.T.Gudinowicz and R.H.Campbell, Anal. Chem., 33, 521 (1961).
2075. J.B.Plumb and C.E.Griffin, J. Org. Chem., 28, 2908 (1963).
2076. R.Dulon, G.Quesnel and M. de Bottom, Bull. Soc. Chim. (France), 9, 1340 (1959).
2077. T.F.Williams, R.W.Wilkinson and T.Rigg, Nature, 179, 540 (1957).
2078. R.L.Grob and G.L.McCrea, Anal. Lett., 1, 55 (1967).
2079. S.A.Buckler, J. Amer. Chem. Soc., 84, 3093 (1962).
2080. K.D.Berlin, T.H.Austin and K.Stone, J. Amer. Chem. Soc., 86, 1787 (1964).
2081. K.D.Berlin and M.E.Nagabhushanam. Chem. and Ind. (London),974 (1964).
2082. S.H.Shipotofsky and H.C.Moser, Anal. Chem., 33, 521 (1961).
2083. F.G.Stamford, J. Chromatogra., 4, 419 (1960).
2084. Y.Abe, Bunseki. Kagaku, 9, 795 (1960).
2085. T.J.Mon, R.D.Dresdner and J.A.Young, J. Amer. Chem. Soc., 81, 1020 (1959).
2086. A.C.Chapman, N.L.Paddock, H.D.Paine, H.T.Seale and D.R.Smith, J. Chem. Soc., 3608 (1960).
2087. F.G.R.Gimblett, Chem. and Ind. (London), 12, 365 (1958).
2088. H.Rotzsche, R.Stahlberg and E.Steger, J. Inorg. Nucl. Chem.., 28, 687 (1966).
2089. E.Steger and R Stahlberg, Z. Anorg. Allg. Chem., 326, 243 (1964).
2090. B.K.Schulte and L.W.Shive, Anal. Chem., 54, 2392 (1982).
2091. J.E.Lovelock and S.R.Lipsky, J. Amer. Chem. Soc., 82, 431 (1960).
2092. J.E.Lovelock, Anal. Chem., 33, 162 (1961).
2093. H.Egan, E.W.Hammond and J.Thompson, Analyst, (London), 85,177 (1960).
2094. E.S.Goodwin, R.Goulden and J.G.Reynolds, Analyst, 86. 697 (1961).
2095. F.K.Kawahara, J.J.Lichtenberg and J.W.Eichelberger, J. Wat. Pollut. Control Fed., 39, 446 (1967).
2096. C.Crasso, A.Giachetti and G.Bernardi, Ric. Sci., 36. 1083 (1966).
2097. J.Hrivnack and I.Pastorek, Coll. Czeck. Chem. Commun., 31, 3402 (1966).
2098. M.Horiguchi, M.Ishida and N.Higosake, Chem. Phar,. Bull. Tokyo, 12, 1315 (1964).
2099. J.Hrivnak, V.Batora and Z.Vasela, Chemicke Zvesti., 20. 600 (1966).
2100. J.S.Thornton and C.A.Anderson, J. Agr. Food Chem., 14, 143 (1966).
2101. C.H.Hartmann, Aerograph Research Notes, 1-6 (1966).
2102. J.Kanazawa, H.Kubo, Sator. Agr. Biol. Chem., 29, 56 (1965).
2103. D.F.McCaulley, J. Assoc. Offic. Agr. Chem., 48, 659 (1965).
2104. D.C.Abbot, N.T.Crosby and J.Thompson,. "Proceedures of the Society for Analytical Chemistry Conference, Nottingham, United Kingdom", pp 121-124, W.Heffer and Son, Cambridge, (1965).
2105. J.Kanazawa and T.Kawahara, J. Agr. Chem. Soc., Japan, 40, 178 (1966).
2106. L.Guiffrida, N.F.Ives and D.C.Bostwick, J. Asso. Office. Anal. Chem., 49, 8 (1966).
2107. A.Kamen, J. Gas Chromatogr., 3, 336 (1965).
2108. S.S.Brody and J.E.Chaney, J. Gas Chromatogr., 4, 42 (1966).
2109. H.P.Burchfield, J.W.Rhoades and R.J.Wheeler, J. Agr. Food Chem., 13, 511 (1965).

2110. H.P.Burchfield, D.E.Johnson, J.W.Rhodes and R.J.Wheeler, J. Gas Chromatogr., 3, 28 (1965).
2111. N.F.Ives and L.Guiffrida, J. Ass. Offic. Anal. Chem., 50, 1 (1967).
2112. I.B.Rubin and C.K.Bayne, Anal. Chem., 51, 541 (1979).
2113. A.N.Nesmeyenov, L.P.Yur'eva, S.N.Kochetkova and S.V.Vitt, Akad-Nauk. SSSR. Ser. Khim., 31, 560 (1966).
2114. K.Ballshmitter, J. Chromato. Sci., 8. 496 (1970).
2115. C.S.Evans and C.M.Johnson, J. Chromatography, 21, 202 (1966).
2116. G.E.Parris, W.R.Blair and F.E.Brinkman, Anal. Chem., 49, 378 (1977).
2117. Y.K.Chau, P.T.S.Wong and P.D.Goulden, Anal. Chem., 47, 2279 (1975).
2118. Y.Talmi and W.W.Audren, Anal. Chem., 46, 2122 (1974).
2119. G.N.Bortnikov, N.N.Bockkarev, N.S.Yyazankin, S.K.R.Ratushnaya and Ya I.Yashin, Izv. Akad. Nauk. SSSR. Ser. Khim., 4, 851 (1971).
2120. V.Vlazakova, J.Benes and J.Parizek, Radiochem. Radioanalyt. Lett., 10, 251 (1972).
2121. S.Nakashima and K.Toei, Talanta, 15, 1475 (1968).
2122. Y.Shimoishi and K.Toei, Talanta, 17, 165 (1970).
2123. C.S.G.Phillips and P.L.Timms, Anal. Chem., 35, 505 (1963).
2124. P.L.Timms, C.C.Simpson and C.S.G.Phillips, J. Chem. Soc., 1467, (1964).
2125. J.A.Semlyen, R.G.Walker, R.E.Bloefield and C.S.G.Phillips, J. Chem. Soc., 4948 (1964).
2126. F.Feher and H.Strack, Naturwissenschaften, 50, 570 (1963).
2127. F.H.Pollard, G.Nickless and P.C.Uden, J. Chromatogr., 19, 28 (1965).
2128. F.H.Pollard, G.Nickless and P.C.Uden, J. Chromatogr., 14, 1 (1964).
2129. G.Garzo, J.Fekete and H.Blazo, Acta. Chim. Akad. Sci., (Hungary), 51, 359 (1967).
2130. F.H.Polland, G.Nickless and D.B.Thomas, J. Chromatogr., 22, 286 (1966).
2131. J.A.Semnlyn and C.S.G.Phillips, J. Chromatogr., 18, 1 (1965).
2132. J.Bohemen, S.H.Langer, R.H.Perrett and J.J.Runnell, J. Chem. Soc., 2444 (1960).
2133. J.A.Semlyen, G.R.Walker, R.E.Blofield and C.S.G.Phillips, J. Chem. Soc.,4948 (1964).
2134. L.B.Peetre and B.E.F.Smith, J. Chromatogr., 90, 41 (1977).
2135. L.B.Peetre, O.Ellren and B.E.F.Smith, J. Chromatogr., 88, 295 (1974).
2136. L.B.Peetre, J. Chromatogr., 88, 311 (1974).
2137. A.Di Lorenzo, J. Chromatogr., 76, 207 (1973).
2138. A.D.M.Hailey and G.Nickless, J. Chromatogr., 49, 187 (1970).
2139. M.Wurst and J.Churacek, Coll. Czeck, Chem. Commun., 36, 3497 (1971).
2140. N.S.Nametkin, V.G.Berezkin, N.Y.Vanyukova and V.M.Vadvin, Nettekhiyma, 4, 137 Ref: Zhur. Khim. 1GGDE (24) Abstr. No. 24G220 (1964).
2141. A.D.Snegova, L.K.Markov, V.A.Panomorenko and W.D.Zelinsky, Institute of Organic Chemistry, USSR Academy of Sciences, Moscow, Zhur. Anal. Khim., 19, 610 (1960).
2156. G.N.Bortnikov, N.S.Vyazankin, N.P.Nikulina, Yashin YaI, Izv. Akad. Nauk. SSSR. Ser. Khim., 1, 21 (1964).
2157. D.Bersadchi and V.Stefan, Rev. Chem. (Bucharest), 15, 224 (1964).
2158. N.A.Palamarchuk, S.V.Syavtsillo, N.M.Turkeltaub, V.T.Shemyatenkova, Trudy Komiss, Anal. Khim. Akad. Nauk. SSSR. 13 277 Ref. Zhur. Khim. 19GE(20) Abstr. No. 20G169 (1963).
2159. Y.Abe, Bunseki Kagaku, Japan Analyst, 9, 795 (1960).
2160. M.Wurst and R.Dusek, Coll. Czech. Chem. Commun., 26, 2022 (1961).
2161. G.Fritz and D.Ksinsik, Z. Anorg. Chem., 304, 241 (1960).
2162. H.Rotzsche, Z. Anorg. Chem., 324, 197 (1963).
2163. M.Wurst and R.Dusek, Coll. Czech. Chem. Commun., 27, 23-1 (1962).
2164. Vronti Piscou and Parissakis, J. Chromat. Sci., 9, 499 (1971).
2165. P.K.Basu, C.J.King and S.Lynn, J. Chromat. Sci., 10, 479 (1972).

2166. N.K.Aglinlov, V.V.Luchinkin and G.G.Devyatykh, Zhur. Analyt. Khim. 23, 951 (1968).
2167. K.Friedrich, Chemistry and Industry, 2. 47 (1957).
2168. J.Joklik, Coll. Czech. Chem. Commun., 26, 2079 (1961).
2169. K.Kawazumi, S.Lataoka and K.Maruyama, J. Chem. Soc. Japan Ind. Chem. Sect., 64, 784 (1961).
2170. N.M.Turkeltaub, N.A.Palamarchuk. V.T.Shemyatenkova and S.V.Syavtsillo, Plast. Massy (4) 51 Ref. Zhur. Khim.,(23) Abstr. No. 23D135 (1961).
2171. T.Garzo, F.Till and I.Till, Magyar Kem Ely., 68, 327 (1962).
2172. T.Oiwa, M.Sato, Y.Miyakava and I.Miyazaki, J. Chem. Soc. Japan Pure Chem. Sect., 84, 409 (1963).
2173. B.Lengyel, G.Garzo and T.Szekely, Acta. Chim. Acad. Sci. Hung., 37, 37 (1963).
2174. H.Rotzsche, Z. fur Anorg. und Allgemeine. Chemie., 328, 79 (1964).
2175. G.V.Avdonin, Z.I.Alekseeva, V.N.Detinova, V.D.Merkulov, L.A.Nechaeva, N.A.Palamarchuk, V.E.Syavtsillo Tenina and S.G.Yagodina, Plast. Massy., (3) 56 Anal. Abs. 15 (6) 3387 (1968).
2176. J.Cermak and J.Franc, Colln. Czech. Chem. Commun., 30, 3278 (1965).
2177. E.V.Sivitsova and V.B.Kogan, Ogorodnikov S. K. Khur. Prikl. Khim. Leningrad, 38, 2609 (1965).
2178. A.N.Popov, V.M.Gorbackev and E.I.Torgova, Izv. Sib. Otdel, Akad. Nauk. SSSR. (11), Ser. Khim. Nauk. (3) 17 Ref. Zhur. Khim. 19GD (13) (1967) Abstr. No. 13G250.
2179. N.M.Turkeltaub, N.A.Palamarchuk (1970) Trudy Nauchno-issled. geol. Razv. Neft. Inst., (64) 143 - Anal. Abstr. (1971) 20 (6) 3932.
2180. A.A.Kreshkov, V.A.Drosdov, N.D.Rumyantseva and V.F.Andronov (1971) Plast. Massy (9) 65 Ref. Zhur. Khim. (1972) 19GD(2) Abstr. No. 2G229.
2181. G.Garzo and F.Till, Talanta, 10, 583 (1963).
2182. J.Joklik and V.Bazant, Chem. Listy, 53, 277 (1959).
2183. A.A.Ainshtein and T.I.Shulyat'eva,Zhur. Anal. Khim., 27, 816 (1972). Anal. Abstr. 77 (8) 56183 (1972).
2184. K.R.Burson and C.T.Kenner, Anal. Chem., 41, 870 (1969).
2185. J.Frank and M.Wurst (1960) Gus. Kromatografaya, Moskow, Akad. Nauk SSSR., 289. Ref: Zhur. Khim. (13) Abstr. No. 13D222 (1961).
2186. J.Frank and M.Wurst, Coll. Czech. Chem. Commun., 25, 701 (1960).
2187. M.Wurst, Coll. Czech. Chem. Commun., 34, 3297 (1969).
2188. M.Wurst, Coll. Czech. Chem. Commun., 30, 2038 (1965).
2189. M.Wurst, Anal. Abstr., 12, 5229 (1965).
2190. D.Knauz, G.Gorzo, P.Gomory and L.Telgdi, Magyar Kem. Foly., 70, 119 (1964).
2191. S.A.Volkov, Sakodynskii Atomnaya Energuiya, 17, 70 (1964).
2192. T.Gabor and J.Takacs, Periodica Polytech., 10, 341 (1966).
2193. J.H.Taylor, J. Gas Chromatography, 6, 557 (1968).
2194. E.A.Kirichenki and B.A.Markov, Trudy Mosk. Khim. Technol. Inst., 71, 143 (1972). Ref: Zhur. Khim. 19GD (4) Abstr. No. 4G244 (1973).
2195. G.W.Heunish, Anal. Chem. Acta., 48, 405 (1969).
2196. C.R.Thrash, J. Gas Chromatography, 2, 390 (1964).
2197. M.Wurst, Mikrochim. Acta., 379,(1966).
2198. J.H.Taylor, J. Gas Chromatography, 6, 557 (1968).
2199. O.Ellren, I.B.Peetre and B.E.F.Smith, J. Chromat., 88, 295 (1974).
2200. L.B.Peetre, J. Chromatography, 90, 35 (1974).
2201. O.Ellren, I.B.Peetre and B.E.F.Smith, J. Chromat., 93, 383 (1974).
2202. L.S.Ettre and J.Billeb, J.Chromatogr., 30, 112 (1967).
2203. C.B.Moore and H.A.Dewhurst, J. Org. Chem., 27, 693 (1962).
2204. M.Wurst, Coll. Czech. Chem. Commun., 29, 1458 (1964).
2205. M.Wurst, Anal. Abstr., 10, 2760 (1963).
2206. B.M.Luskina, G.N.Turkel'taub and S.V.Syavtsillo, Zavod Lab., 33, 1496 (1967).
2207. K.Otto and M.Doubek, Chem. Prumsyl., 10, 476 (1960).

2208. H.Rotzsche and H.Rosler, Report of Conference on Modern Methods of Analysis of Organic Compounds,Munich, October 26-29 (1960). Also - Anal. Chem., 181, 407 (1961).
2209. J.B.Carmichael, D.J.Gordon and C.E.Ferguson, J. Gas Chromatography, 4, 347 (1966).
2210. J.Franc, K.Polacek and F.Mikes, Coll. Czech. Chem. Commu., 32, 2242 (1967).
2211. G.U.Heylmun and J.E.Pikula, J. Gas Chromatography, 3, 266, (1965).
2212. K.R.Burson and C.T.Kenner, J. Chromatographic Sci., 7,63 (1969).
2213. N.T.Ivanova, S.V.Syavtsillo and L.D.Prigozhina, Zav. Lab., 39, 1455 (1973).
2214. G.Alexander and G.Garzo, Chromatographia, 7, 190 (1974).
2215. G.Alexander and G.Garzo, Anal. Abstr., 23, 2589 (1972).
2216. M.Wurst, F.Loember and H.Kelker, Chromatographia, 6, 359 (1973).
2217. M.Wurst, F.Loember and H.Kelker, Anal. Abstr., 23, 390 (1972).
2218. M.Wurst, Chemicky Prum., 22,. 124 (1972).
2219. G.N.Turkel'taub and B.M.Luskina, J. Analyt. Chem. USSR., 26, 2010 (1971) also Zhur. Anal. Khim., 26, 2243 (1971).
2220. F.Sutter and P.Fuchs, Arzneimittel - Forsch, 22, 553 (1972).
2221. M.Tochner, I.A.Magnuson and L.Z.Soderman, J. Chromat. Sci., 7, 740 (1969).
2222. F.H.Pollard, G.Nickless and P.C.Uden, J. Chromat., 11, 312 (1963).
2223. G.N.Bortnikov, M.N.Bochkarev, N.S.Vyazankin, S.K.R.Ratushnaya, Ya I. Yashin, Izv. Akad. Nauk. SSSR. Ser. Khim., 4, 851 (1971).
2224. E.D.Smith, J. Chromat. Sci., 10, 34 (1972).
2225. G.N.Bortnikov, N.S.Vasileiskaya, L.V.Gorbunova, N.P.Nikulina, Ya I. Yoshin, Izv. Akad,. Nauk. SSSR. Ser. Khim., 3, 686 (1970).
2226. G.N.Bortnikov, N.S.Vyazankin, E.N.Gladyshev and Ya I.Yashin, Izv. Akad. Nauk. SSSR. Ser. Khim., 7, 1661 (1970).
2227. N.S.Yashin, A.N.Egorochkin and V.F.Mironov, Izv. Akad. Nauk. SSSR. Ser. Khim., 1, 186 (1969).
2228. G.N.Bortnikov, A.V.Kiselev, N.S.Yyasankin, Ya I.Yashin, Chromatographia, 4, 14 (1971).
2229. E.V.Sivtsova, Zhur. I. Prikl. Khim. Leningrad., 45, 201 (1972).
2230. A.D.M.Hanley and G.Nickless, J. Chromatography, 49, 180 (1970).
2231. J. Franc, Plast. Liwoty. Kanenk., 4, 249 (1967).
2232. N.M.Turkel'taub, S.A.Aunshtein and B.V.Kuznetsov (1961) Khim. Technil Topliv. Masel., 12, 44 Ref. Zhur. Khim., 11(ii) Abstr. No. 11D170 (1962).
2233. N.M.Turkel'taub, S.A.Shemyatenkova, S.A.Ainstein and S.V.Syavtsillo (1963) Trudy Komiss. Anal. Khim. Akad. Nauk. SSSR., 13, 284. Ref: Zhur. Khim. 19GDE (20) Abstr. No. 20G167 (1963).
2234. G.Garzo and F.Till, Talanta, 10, 583 (1963).
2235. D.Fritz, G.Garzo, T.Szekely and F.Till, Acta. Chem. Hung.,45, 301 (1965).
2236. G.Garzo and D.Fritz, Gas Chromatography. Proceedings of the 6th International Symposium on Gas Chromatography and Associated Techniques, Rome, Italy (September 1966).
2237. M.Dressler, V.Martinu and J.Janek, J. Chromat., 59, 429 (1971).
2238. D.T.Bostick and Y.Talmi, J. Chromatogr. Sci., 15, 163 (1977).
2239. T.G.Crowley, V.A.Fassel and R.N.Kniseley, Spectro. Chim. Acta., Part B 23, 771 (1968).
2240. D.Fritz, G.Garzo, T.Szekely and F.Till, Acta. Chim. Hung., 45, 301 (1965).
2241. B.Lengnel, G.Garzo, D.Fritz and F.Till, J. Chromatogr., 24, 8 (1966).
2242. R.W.Morrow, J.A.Dean, W.D.Shults and M.R.J.Guerin, J. Chromatogr. Sci., 7, 572 (1969).
2243. H.H.Hill and W.A.Aue, J. Chromatogr., 122, 515 (1976).

2244. H.H.Hill and W.A.Aue, J. Chromatogr., 140, 1 (1971).
2245. M.A.Osman, H.H.Hill, M.W.Holdren and H.H.Wetberg, Anal. Chem., 51, 1286 (1979).
2246. M.A.Osman, H.H.Hill, M.W.Holdren and H.H.Wetberg, Anal. Chem., 51, 1289 (1979).
2247. K.J.Slatkavitz, L.D.Hoey, P.C.Uden and R.M.Barnes, Anal. Chem., 57, 1846 (1985).
2248. B.Lengyel, G.Garzo, D.Fritz and F.Till, J. Chromatogr., 24, 8 (1966).
2249. J.Franc and J.Dvoracek, J. Chromatography, 14, 340 (1964).
2250. J.Franc and K.Placek, J. Chromatography, 48, 295 (1970).
2251. C.R.Thrash, D.L.Viosinet and K.E.Williams, J. Gas Chromatography, July 248 (1965).
2252. D.F.Helm and E.Malk, J. Am. Chem. Soc., 59, 60 (1937).
2253. C.E.Waring, Trans. Faraday Soc.,London, 36. 1142 (1940).
2254. G.Fritz and B.Raabe, Z. Anorg. Allgem. Chem., 229, 232 (1959).
2255. G.Fritz, J.Grobe, Z. Anorg. Allgem. Chem., 315, 157 (1962).
2256. G.Fritz, J.Bahl, D.Grobe, D.Aulinger and W.Reerink, Z. Anorg. Allgem. Chem., 312, 301 (1961).
2257. G.Fritz and R. Fortschr,Chem. Forsch., 44, 459 (1963).
2258. G.N.Bortnikov, H.N.Bochkarev, N.S.Yyazankin, S.K.R.Ratushnaya and Ya I.Yashin, Izv. Akad. Nauk. SSSR. Ser. Khim., 4, 851 (1971).
2259. R.K.Ingham, S.D.Rosenberg and H.Gilman, Chem. Rev., 60, 459 (1960).
2260. J.G.A.Luijten and G.J.M.Van der Kerk, Investigations in the field of Organotin Chemistry. Tin Research Inst., Greenford, Middlesex, United Kingdom, p.84 (1966).
2261. F.H.Pollard, G.Nickless and D.J.Cooke, J. Chromatography, 13, 48 (1964).
2262. F.H.Pollard, G.Nickless and D.J.Cooke, J. Chromatography, 17, 472 (1965).
2263. J.Boheman, S.H.Langer, R.H.Perrett and J.J.Purnell, J. Chem. Soc., 2444 (1960).
2264. F.F.Eggersten and S.Groennings, Anal. Chem., 30, 20 (1958).
2265. A.James, A.J.P.Martin, J. Biochem., 50, 679 (1952).
2266. A.Liberti (1958) in "Gas Chromatography 1958" ed. H.Desty, Butterworths, London p.214.
2267. C.H.Orr and J.E.Callen, Ann. N.Y.Acad. Sci., 72, 649 (1950).
2268. E.C.Ormerod and R.P.W.Scott, J. Chromatography, 2, 65 (1959).
2269. G.A.Howard and A.J.P.Martin, Biochem., J, 46, 532 (1950).
2270. A.Kwantes and G.W.A.Rynders (1958) in "Gas Chromatography 1958" Ed. H. Desty, Butterworths, London, p.125.
2271. A.V.Kiselev, D.H.Everett and F.S.Stove in "The Structure and Properties of Porous Materials" volume X Colston Papers, Butterworths London, p.257 (1958).
2272. H.W.Kohlschutter, P. Best and G.Wirzing, Z. Anorg. Chem., 285 336 (1956).
2273. S.H.Langer, S.Connell and I.Wender, J. Org. Chem., 23, 50 (1958).
2274. S.H.Langer, O.Pantages and I.Wender, Chem. and Ind. (London),1664 (1958).
2275. R.H.Perrett and J.H.Purnell, J. Chromatography, 7, 455 (1962).
2276. R.G.Ackman, R.D.Burgler, J.C.Spos and P.H.Odense, J. Chromatography, 9,531 (1962).
2277. J.Novak and J.Janak, J. Chromatography, 4, 249 (1960).
2278. J.Stuve (1958) in "Gas Chromatography 1958) ed. D.H.Desty, Butterworths, London, p.178.
2279. A.J.P.Martin and A.T.James, Biochem. J., 63, 138 (1956).
2280. G.H.M.Van der Kerk, J.C.Noltes, J. App. Chem. (London), 9, 106 (1959).
2281. F.H.Polland, G.Nickless and P.C.Uden, J. Chromatogr., 14, 1 (1964).
2282. R.C.Putman and H.Put, J. Gas Chromatography, 3, 2 (1965).

2283. R.C.Putnam, H.Put, J. Gas Chromatography, 3, 160 (1965).
2284. U.Proesch and H.J.Zoepfl, Z. Chem., 3, 97 (1963).
2285. E.W.Abel, G.Nickless and F.H.Polland, Proc. Chem. Soc. (London), 228 (1960).
2286. K.Hoppner, U.U.Prosch and H.Wieglab, Z. Chem., 4, 31 (1964).
2287. K.Hoppner and H.J.Zoepfl, Abh. Deutsch. Akad. Berlin. Lk. Chem. Geol. Biol., 393 (1966).
2288. A.Wowk and S.Dogiovanni, Anal. Chem., 38, 742 (1966).
2289. V.A.Umilin and Yu N.Tsinvoi, Izv. Akad. SSSR. Ser. Khim. 1409, (1962).
2290. Y.Jitsu, N.Kudo, K.Sato and T.Teshinia, Japan Analyst, 18, 169 (1969).
2291. D.A.Vyakhirev and O.P.Chereshnya, Trudy Khim. Tekhnol. (Gorkii) (i) 121 (1971) Ref. Zhur. Khim. 19GD (1972).
2292. V.A.Chernoplekova, K.I.Sakodynskii and V.M.Sakharov, Zhur. fiz. Khim., 46, (6) 1502 (1972) Ref. Zhur. Khim. 19GD 1972 (21) Abstr. No. 21G175 (1972).
2293. E.S.Selikhova and V.A.Umilin (1969) Trudy Khim. Tekhnol (Gorkii) 1 (22) 121 Ref. Zhur. Khim. 19GD (8) Abstr. No. 8G266 (1970).
2294. D.J.Cooke, G.Nickless and F.H.Pollard, Chem. Ind. (London), 1493, (1963).
2295. G.G.Devyatyk, V.A.Umilin and U.N.Tsinovoi, Trudy Khim. Tekhnol, 82 (1968).
2296. H.D.Nelson, Doctorial Dissertation, University of Utrecht, 29 (1967).
2297. K.Jeugen and J.Figge, J. Chromatogr., 109, 89 (1975).
2298. V.V.Brazhnikov and K.I.Sakodynskii, J. Chromatogr., 66, 361 (1972).
2299. W.O.Gauer, J.N.Selber and D.G.Crosby, J. Agric. Food Chem., 22, 252 (1974).
2300. H.Geissler and H.Kriegsmann, Z. Chemie. Lpz., 4, 354 (1964).
2301. B.L.Tonge, J. Chromatography, 19, 182 (1965).
2302. G.E.Parris, W.R.Blair and F.E.Brinkman, Anal. Chem., 49, 378 (1977).
2303. R.D.Steinmeyer, A.F.Fentiman and E.J.Kahler, Anal. Chem., 37, 520 (1965).
2304. H.Geissler and H.Kriegsmann, Third Analytical Conference 24-29 August Budapest, Hungary (1970).
2305. J.Franc, M.Wurst and V.Moudry, Collect. Czech. Chem. Commun., 26, 1313 (1961).
2306. W.Gerrard, E.F.Mooney and R.G.Rees, J. Chem. Soc., 740, (1964).
2307. V.Chromy, A.Groagova, O.Pospichal and K.Jurak, Chem. Listy., 60, 1599 (1966).
2308. W.P.Neumann and G.Burkhardt, Liebigs. Ann. Chem., 663, 11 (1963).
2309. Z.Hladky, Z. Chem., 11, 423 (1965).
2310. R.A.Keller, J. Chromatography, 5, 225 (1961).
2311. H.D.Kaesz, F.L.Stafford and F.G.A.Stone, J. Am. Chem. Soc., 82, 6232 (1960).
2312. H.D.Kaesz, J.R.Phillips and F.G.A.Stone, J. Am. Chem. Soc., 82, 6228 (1960).
2313. P.Barrell and A.E.Platt, Trans. Faraday Soc., London, 66, 2286 (1970).
2314. G.N.Bortnikov, E.Vyanoukin, E.N.Gladyshov and Ya I.Yashin, Izv. Akad. Nauk SSSR. Ser. Khim., 7, 1661 (1970).
2315. Y.N.Bortnikov, E.N.Yyankin, E.N.Gladyshev and V.S.Andreeviche, Zav. Lab., 35, 1445 (1969).
2316. M.Dressler, V.Martina and J.Janak, J. Chromatogr., 59,429 (1971).
2317. H.H.Hill and W.A.Aue, J. Chromatogr., 74, 311 (1972).
2318. H.H.Hill and W.A.Aue, Anal. Abstr., 24, 1304 (1973).
2319. D.R.Hansen, T.J.Gilfoil and H.H.Hill, Anal. Chem., 53, 857 (1981).
2320. P.Longi and R.Mazzochi, Chimica. Ind., Milano, 48,718 (1966).
2321. N.M.Karayannis and A.H.Corwin, J. Chromatogr., 47, 242 (1970).

2322. S.G.Perry, J. Gas Chromatography, 93, March (1964).
2323. C.E.Legate and N.D.Burnham, Anal. Chem., 32, 1042 (1960).
2324. V.M.Krasikova, A.N.Kaganova and V.D.Lohkov, J. Anal. Chem. USSR (Eng. Transl.), 26, 1458 (1971).
2325. V.H.Krasikova, A.N.Kaganova and V.D.Labkov, Zhur. Anal. Khim., 26, 1635 (1971).
2326. D.D.Schleuter and S.Siggia, Anal. Chem., 49, 2343 (1977).
2327. D.D.Schleuter, Ph. D. Dissertation, University of Massechusetts, Amherst, Massechusetts, (1976).
2328. V.M.Krasikova and A.N.Kaganova, Zhur. Analit. Khim., 25, 409 (1970).
2329. V.M.Krasikova and A.N.Kaganova, J. Anal. Chem. USSR., 25, 1212 (1970).
2330. G.W.Heylum, R.L.Bajaiski and H.B.Bradley, J. Gas Chromatogr., 2, 300 (1964).
2331. E.R.Bissell and D.R.Fields, J. Chromatogr. Sci., 10, 164 (1972).
2332. C.L.Hanson and R.C.Smith, Anal. Chem., 44, 1571 (1972).
2333. S.P.Evdokimova, N.A.Isakova and V.F.Evdokinov, J. Analyt. Chem. USSR., 26, 704 (1971).
2334. J.Franc and F.Mikes, Coll. Czech, Chem. Commun., 31, 363 (1966).
2335. G.Garzo, T.Szekely, J.Tamas and K.Ujsaski, Anal. Chim. Sci. Hungary, 69, 273 (1971).
2336. D.C.Tse and T.Kuwana, Anal. Chem., 50, 1315 (1978).
2337. J.R.Marquant and R.L.Belford, Anal. Chem., 50, 656 (1978).
2338. T.M.Vickrey, H.E.Howell, G.V.Harrison and G.J.Ramelow, Anal. Chem., 52, 1743 (1980).
2339. B.I.Anvaer, Yu Drugov, Zh. Analit. Khim., 26, 1180 (1971).
2340. A.G.Hamlin, G.Iveson and T.R.Phillips, Anal. Chem., 35, 2037 (1963).
2341. J.F.Ellis, C.W.Forrest and P.Allen, Anal. Chim. Acta., 22,27 (1960).
2342. H.Shinolara, N.Asakura and S.Tsujimuras, J. Nucl. Sci. Technol., 3, 373 (1966).
2343. F.M.Zado and R.S.Juvet, Anal. Chem., 38, 569 (1966).
2344. R.S.Juvet and R.L.Fischer, Anal. Chem., 38, 1860 (1966).
2345. R.L.Fischer, Dissert. Abstr., 27B, 3809 (1967).
2346. W.S.Pappas and J.G.Milliou, Anal. Chem., 40, 2176 (1968).
2347. J.G.Milliou, W.S.Pappas and C.N.Weber, J. Chromat. Sci., 7, 182 (1969).
2348. O.Pitak, Chromatographia, 3, 29 (1970).
2349. R.M.Dagnall, D.R.Deans, B.Fleet and T.H.Risby, Talanta, 18, 155 (1971).
2350. R.M.Dagnall, B.Fleet, T.H.Risby and D.R.Deans, Analyt. Letters, 4, 497 (1971).
2351. R.Aubeau, G.Balndenet and F.Leccia, Chromatographia, 5, 240 (1972).
2352. N.N.Aleinikov, D.N.Sokolov, L.K.Golubeva, B.L.Korsumskii and F.I. Dubovitskii, Izv. Akad. Nauk. SSSR. Ser. Khim., 11, 2614 (1973).
2353. H.Freiser, Anal. Chem., 31, 1440 (1959).
2354. W.J.Bierman and H.Gesser, Anal. Chem., 32, 1525 (1960).
2355. J.Tadmor, J. Inorg. Nucl. Chem., 23, 158 (1961).
2356. J.Tadmor, Research Council of Israel Bulletin, Chem. Soc., 10A, No. 3, 17 (Sept. 1961).
2357. E.Stumpp, Z. Analyt. Chem., 242, 225 (1968).
2358. H.J.Becker, J.Chevallier and J.Spitz, Z. Analist. Chem., 247, 301 (1969).
2359. T.S.Zvarova and I.Zvara, J. Chromato., 44, 604 (1969).
2360. O.Pitak, Chromatographia., 2, 304 (1969).
2361. T.S.Zvarova and I.Zvara, J. Chromat., 49, 290 (1970).
2362. D.Parissakis, D.Vranti-Piscou and J.Kontoyannakos, J. Chromat., 52, 461 (1970).
2363. V.V.Brazhnikov and K.I.Sakodynskii, Zh. Prikl. Khim. Leningr., 43, 2247 (1970).

2364. N.K.Aglinlov, M.V.Zueva and G.G.Devyatykh, Zhur. Anal. Khim., 24, 1220 (1969).
2365. D.Vranti-Piscou and J.Kantoyannakos, J. Chromato. Sci., 9, 499 (1971).
2366. G.Parissakis, D.Vranti-Piscou and Kontoyannakos, Z. Analit. Chem., 254, 188 (1971).
2367. I.Tohyama and K.Otozai, Z.Analyt. Chem., 262, 346 (1972).
2368. N.P.Kuznetsova, V.A.Fedorov and A.A.Yushkin, Z. Analist. Chem., 27, 821 (1972).
2369. V.V.Lebedev, A.D.Molodyk, L.N.Morozova and N.M.Korenchuk, Zavod. Lab., 38, 525 (1972).
2370. B.Iatridis and G.Parissakis, Anal. Chim. Acta., 89, 347 (1977).
2371. R.E.Sievers, B.W.Ponder, M.L.Morris and R.W.Moshier, Inorg. Chem., 2, 693 (1963).
2372. R.W.Moshier and R.E.Sievers, in Gas Chromatography of Metal Chelates, PergamonPress, London, pp 46 and 92 (1965).
2373. P.C.Uden and D.E.Henderson, Analyst, 102, 889 (1977).
2374. R.A.Minear and C.M.Palesh, Papers presented before the Division of Water, Air and Waste Chemistry at the 165th National Meeting of the American Chemical Society, Dallas, Texas, USA (April 1973). Prepr. Div. Water Air Waste Chem. Am. Chem. Soc. pp 13-18 (1973).
2375. R.P.Durbin, Doctoral Dissertation, University of Illinois, (1966) Diss. Abstr., 27, 710B (1966).
2376. D.R.Gere and R.W.Moshier, J. Gas Chromatog., 6, 89 (1968).
2377. M.Lederer, Nature, 176, 462 (1955).
2378. A.A.Dustwald, Doctoral Dissertation Purdue University (1958); Diss. Abstr., 20, 52 (1959).
2379. W.V.Floutz, Master's thesis, Purdue University (1959).
2380. W.W.Brandt, Gas Chromatography (1960). R.P.W.Scott (Ed.), Butterworths London, 305-306 (1960).
2381. W.W.Brandt and J.E.Heveran, 142nd Meeting of the American Chemical Society, Atlantic City, New Jersey, Sept. 9-14 (1962).
2382. W.W.Brandt, Proc. 3rd Gas Chromatographic Process Symposium, Edinburgh, pp 505 (1960).
2383. W.W.Brandt in "Gas Chromatography 1960" edited by R.P.W.Scott, Butterworths, Washington, USA., 305 (1960).
2384. J.Janak, 3rd Gas Chromatographic Process Symposium, Edinburgh, 306 (1960).
2385. J.Janak, in "Gas Chromatography 1960" edited by R.P.W.Scott, Butterworths, Washington, USA. 306 (1960).
2386. M.L.Morris and R.W.Moshier, Inorg. Chem., 2, 693 (1963).
2387. R.E.Sievers, S.W.Ponder and R.W.Mosher, 141st National Meeting, American Chemical Society, Washington D.C., USA., March 24th (1962).
2388. R.D.Hill, Master's Thesis, University of Manitoba (1962).
2389. R.D.Hill and H.Gesser, J. Gas Chromatog., 1st Oct. (1963) 11.
2390. R.G.Melcher, Master's Thesis, Purdue University (1961).
2391. J.E.Heveran, Master's Thesis, Purdue University (1962).
2392. W.J.Biermann and H.Gesser, Anal. Chem., 32, pp 1525-2526 (1960).
2393. W.G.Baldwin, Master's Thesis, University of Manitoba (1961).
2394. A.A.Duswalt, Dissertation Abstracts, 20, 52 (1959).
2395. T.Fuginaga, T.Kiwamoto and Y.Ono, J. Chem. Soc. Japan, Pure Chem. Sect., 86, 1294 (1965).
2396. T.Fuginaga, T.Kiwamoto and Y.Ono, Japan Analyst, 12, 1199, Chem. Abstr., 60 6209 (1964).
2397. R.E.Sievers, B.W.Ponder and R.W.Moshier, 141st National Meeting of the Amer. Chem. Soc., Washington D.C. March 24 (1962).
2398. K.Yamakawa, K.Tanikawa and K.Arakawa, Chem. Pharm. Bull.(Tokyo), 11 11, 1405 (1963). Chem. Abstr., 60,7464 (1964).

2399. R.E.Sievers, R.W.Moshier and M.L.Morris, Inorg. Chem., 1, 966 (1962).
2400. R.E.Sievers, B.W.Ponder, M.L.Morris and R.W.Moshier, Inorg. Chem., 2, 693 (1963).
2401. J.P.Cullman, R.L.Marshall, W.L.Young and S.D.Coldby, Inorg. Chem., 1, 704 (1962).
2402. R.E.Sievers, R.W.Moshier and M.L.Morris, Inorg. Chem., 1, 966 (1962).
2403. R.E.Sievers, 16th Annual Summer Symposium, American Chemical Society, Tuscan, Arizona, Chemical Engineering News, 41, (1/6/63).
2404. R.E.Sievers, Dayton Gas Chromatography - a discussion group meeting, Dayton, Ohio, USA., (November 1961).
2405. J.E.Schwarberg, Masters Thesis, Perdue University of Dayton, Ohio, USA., (1964).
2406. J.E.Schwarberg, R.W.Moshier and J.H.Walsh, Talanta, 11, 1213 (1964).
2407. R.W.Moshier, J.E.Schwarberg, R.E.Sievers and M.L.Morris, American Chemical Society 14th Conference on Analytical Chemistry and Applied Spectroscopy, Pittsburgh, Pennsylvania, USA (March 5th 1963).
2408. R.W.Moshier and J.E.Schwarberg, XXth International Congress of Pure and Applied Chemistry, Moscow, USSR (July 1965).
2409. R.W.Moshier and R.E.Sievers in "Gas Chromatography of Metal Chelates", Pergammon Press Ltd., Oxford UK., pp23 and 36 (1965),
2410. J.A.Stokeley, Masters Thesis, Oak Ridge University, Diss. Abstr., 27, 1388 B (1966).
2411. R.E.Sievers, 16th Annual Summer Symp. ACS. Div. Anal. Chem., Tucson, Arizona (June 1963).
2412. Anonymous, Chem. Eng. News, 41, (July 1 1963).
2413. Q.Fernando, H.Freiser and E.N.Wise, Anal. Chem., 35, 1994 (1963).
2414. J.E.Schwarberg, R.W.Moshier and J.H.Walsh, Talanta, 11, 1213-1224 (1964).
2415. J.E.Schwarberb, Masters Thesis, University of Dayton (1964).
2416. H.Veening, W.E.Bachman and D.M.Wilkinson, J. Gas Chromatog., 5, 248 (1967).
2417. P.Jacquelot and G.Thomas, Bull. Soc. Chim. Fr., 3167 (1970).
2418. P.Jacquelot and G.Thomas, Bull. Soc. Chim. Fr., 702 (1971).
2419. K.Tanikawa, K.Hirano and K.Arakawa, Chem. Pharm. Bull.(Tokyo), 15, 915 (1967).
2420. W.G.Scribner, W.J.Treat, J.D.Weiss and R.W.Moshier, Anal. Chem., 37, 1136 (1965).
2421. H.Veening and J.F.K.Huber, J. Gas Chromatog., 6, 326 (1968).
2422. W.G.Scribner, J.D.Weis and R.W.Moshier, 148th National Meeting of the Amer. Chem. Soc., Chicago, Ill., (Sept. 1964).
2423. R.W.Moshier and J.E.Schwarberg, Talanta, 13, 445 (1966).
2424. G.P.Morie and T.R.Sweet, Anal. Chem., 37, 1552 (1965).
2425. G.P.Morie, Doctoral Dissertation, Ohio State University, Diss. Abstr. 27, 1386 B-1387 B (1966).
2426. G.P.Morie and T.R.Sweet, Anal. Chim. Acta., 34, 314(1966).
2427. W.D.Ross, R.E.Sievers and G.Wheeler, Anal. Chem., 37, 598(1965).
2428. W.D.Ross and R.E.Sievers, Gas Chromatography (1966), A.B.Littlewood (Ed.), The Institute of Petroleum, London (1967) pp 272-282.
2429. W.D.Ross and R.E.Sievers, Talanta, 15, 87 (1968)
2430. W.D.Ross and R.E.Sievers, Symposium on Inorganic Gas Chromatography, 156th National ACS Meeting, Atlantic City, New Jersey, (Sept. 1968).
2431. J.Savory, P.Mushak and F.W.Sunderman, J. Chromatog. Sci., 7, 674 (1969).
2432. W.D.Ross and R.E.Sievers, Anal. Chem., 41, 1109 (1969).
2433. M.L.Taylor, E.L.Arnold and R.E.Sievers, Anal. Letters, 1, 735-757 (1968).
2434. C.Gentry, C.Houin, P.Malherbe and R.Schott, Gas Chromatography (1968) C.J.A.Harbourne (Ed.), The Institute of Petroleum, London (1969), pp 142-147; Anal. Chem., 43, 235 (1971).

2435. M.L.Taylor and E.L.Arnold, Anal. Chem., 43, 1328 (1971).
2436. W.R.Wolf, M.L.Taylor, B.M.Hughes, T.O.Tierman and R.E.Sievers, Anal. Chem., 44, 616 (1972).
2437. L.C.Hansen, W.Scribner, T.W.Gilbert and R.E.Sievers, Anal. Chem., 43, 349 (1971).
2438. K.J.Eisentraut, D.J.Griest and R.E.Sievers, Anal. Chem., 43, 2003 (1971).
2439. T.A.Gosink, Anal. Chem., 47, 165 (1975).
2440. R.C.Fay and T.S.Piper, J. Am. Chem. Soc., 85, 500 (1963).
2441. Anonymous, Chem. Eng. News, 40, 50 (April 1962).
2442. E.W.Berg and F.R.Hartlage, Anal. Chim. Acta., 54, 46 (1966).
2443. R.Fontaine, Doctorial Dissertation, Paris (1972).
2444. W.D.Ross, Anal. Chem., 35, 1596 (1963).
2445. M.L.Morris, R.W.Moshier and R.E.Sievers, Inorg. Chem., 2, 411 (1963).
2446. W.C.Butts, C.V.Banks, Anal. Chem., 42, 133 (1970).
2447. K.Arakawa and K.Tanikawa, Japan Analyst, 16, 812 (1967).
2448. W.D.Ross and G.Wheeler, Anal. Chem., 36, 266 (1964).
2449. M.L.Morris, R.W.Moshier and R.E.Sievers, Inorg. Synth., 9, 28-30 (1967).
2450. R.W.Moshier, J.E.Schwarberg, M.Morris and R.E.Sievers, 14th Pittsburgh Conf. on Anal. Chem. and Appl. Spectroscopy, Pittsburgh, Pa., March (1963); Progr. Abstra., p.58.
2451. R.E.Sievers, R.W.Moshier and M.I.Morris, Inorg. Chem., 1, 966-967 (1962).
2452. R.G.Linck, W.D.Ross, G.Wheeler and R.E.Sievers, Unpublished results.
2453. R.W.Moshier and R.E.Sievers, unpublished results.
2454. R.G.Linck and R.E.Sievers, 148th National Meeting of the Amer. Chem. Soc., Chicago III, September 1964; Ref. (1), p. 131.
2455. K.J.Eisentraut and R.E.Sievers, J. Inorg. Nucl. Chem., 29, 1931 (1967).
2456. R.Belcher, A.W.L.Dudeney and W.I.Stephen, J. Inorg. Nucl. Chem., 31, 625 (1969).
2457. W.L.Stephen, I.J.Thompson and P.C.Uden, Chem. Communs., 269 (1969).
2458. K.J.Eisentraut and R.E.Sievers, J. Am. Chem. Soc., 87, 5254 (1965).
2459. Anonymous, Chem. Eng. News, 43, 39-40 (Nov. 22, 1965).
2460. J.E.Sicre, J.T.Dubois, K.J.Eisentraut and R.E.Sievers, J. Am. Chem. Soc., 91, 3476 (1969).
2461. R.W.Moshier and D.R.Gere, 152nd ACS Meeting, New York (1966).
2462. T.Shigematsu, M.Matsui and K.Utsunomiya, Bull. Chem. Soc. Japan, 41, 763 (1968).
2463. M.Tanaka, T.Shono and K.Shinta, Anal. Chim. Acta., 43, 157 (1968).
2464. T.Shigematsum, M.Matsui and K.Utsunomiya, Bull. Chem. Soc. Japan, 42, 1278 (1969).
2465. P.C.Uden and C.R.Jenkins, Talanta, 16, 893 (1969).
2466. M.Tanaka, T.Shono and K.Shinra, Nippon Kagaku Zasshi, 89, 669-672 (1969).
2467. R.Belcher, J.R.Najer, W.I.Stephen, I.J.Thompson and P.C.Uden, Anal. Chim. Acta., 50, 423 (1970).
2468. J.Savory, P.Mushak and F.W.Sunderman in "Advances in Chromatography" edition A, Zlatis, Preston Technical Abstracts, Evanston, USA, p. 181 (1960).
2469. J.Savory, P.Mushak, N.O.Roszel and F.W.Sunderman, Federation Proceedings, (Biochem.) Abstracts, 52nd Meeting, Atlantic City, New Jersey, USA. p. 777 (April 15-20 1968).
2470. W.G.Scribner and A.M.Kotechki, Anal. Chem., 37, 1304 (1965).

2471. W.G.Scribner,M.J.Borchers and W.J.Treat, Anal. Chem., 38, 1779 (1966).
2472. C.Gentry, C.Houim and R.Shott, Preprints of 7th International Symposium on Gas Chromatography and its Exploitation, Copenhagen, Denmark, paper 9 (June 1968).
2473. R.E.Sievers, J.W.Connolly and W.D.Ross, J. Gas Chromatog., 5. 241-247 (1967).
2474. W.D.Ross, N.G.Scribner and R.E.Sievers, Gas Chromatography (1970) R. Stock (Ed.), The Institute of Petroleum, London p. 381 (1971).
2475. R.Fontaine, D.Santoni, C.Pommier and G.Guichon., Chromatographia, 3, 532-533 (1970).
2476. C.S.Springer, D.W.Meek and R.E.Sievers, Inorg. Chem., 6, 1105-1110 (1967).
2477. R.Belcher, A.Khalique and W.I.Stephens, Anal. Chim. Acta., 110, 503 (1978).
2478. T.Shigematsu, M.Matsui and K.Utsunomiya, Bull. Inst. Chem. Res., Kyoto Univ., 46, 256-261 (1968).
2479. W.I.Stephen, I.J.Thompson and P.C.Uden, Chem. Commun., 269-270 (1969).
2480. R.S.Barratt, R.Belcher, W.I.Stephen and P.C.Uden, Anal. Chim. Acta, 59 59-73 (1972).
2481. R.Belcher, W.I.Stephen, I.J.Thompson and P.C.Uden, J. Inorg. Nucl., Chem., 33, 1851 (1971).
2482. R.Belcher, W.I.Stephen, I.J.Thompson and P.C.Uden, Chem. Communs. 101G (1970).
2483. R.S.Barratt, R.Belcher, W.I.Stephen and P.CUden, Anal. Chim. Acta., 59, 59 (1972).
2484. N.M.Karayannis, A.H.Corwin, E.W.Baker, E.Klesper and J.A.Walter, Anal. Chem., 40, 1736-1739 (1968).
2485. T.Dono, Y.Ishihara, K.Saito and T.Nakazawa, Japan Analyst, 15, 181-182 (1966).
2486. K.Ballschmiter, J. Chromatog. Sci., 8, 496-498 (1970).
2487. M.Miyazaki, T.Imanari, T.Kunugi and Z.Tamura, Chem. Pharm. Bull. (Tokyo), 14, 117-120 (1966).
2488. K.Utsunomiya, Anal. Chim. Acta., 59, 147-151 (1972).
2489. P.Jacquelot and G.Thomas, C.R. Acad. Sci. Paris, Series C, 272, 448-450 (1971).
2490. P.Jacquelot and G.Thomas, J. Chromatog., 66, 121-128 (1972).
2491. E.Bayer, H.P.Muller and R.E.Sievers, Anal. Chem., 43, 2012-2014 (1971).
2492. R.Belcher, M.Pravica, W.I.Stephen and P.C.Uden, J. Chem. Soc., D. 41-42 (1971).
2493. R.F.Sieck, J.J.Richard, K.Iversen and C.V.Banks, Anal. Chem., 43, 913-917 (1971).
2494. R.Belcher, A.W.L.Dudeney and W.I.Stephen, J. Inorg. Nucl. Chem., 31, 625-631 (1969).
2495. R.Belcher, J.R.Mayer, R.Perry and W.I.Stephen, Anal. Chim. Acta., 45, 305-309 (1969).
2496. R.Belcher, J.Majer, R.Perry and W.I.Stephen, J. Inorg. Nucl. Chem., 31, 471-478 (1969).
2497. J.E.Schwarberg, R.E.Sievers and R.W.Moshier, Anal. Chem., 42, 1828-1830 (1970).
2498. W.I.Stephen, Proc. Soc. Analyt. Chem., 9, 137 (1972).
2500. S.Dilli, A.M.Maitra, J. of Chromatography, 254, 133 (1983).
2501. R.Belcher, K.Blessel, T.J.Cardwell, M.Pravica, W.I.Stephen and P.C.Uden, J. Inorg, Nucl. Chem., 35, 1127 (1973).
2502. R.Belcher, R.J.Martin, W.I.Stephen, D.E.Henderson, A.Kamalizad and P.C.Uden, Anal. Chem., 45, 1197 (1973).
2503. P.C.Uden, D.E.Henderson and A.Kamalizad, J. Chromatogr. Sci., 12, 591 (1974).

2504. T.J.Cardwell and P.S.McDonough, Inorg. Nucl. Chem. Lett., 10, 283 (1974).
2505. P.C.Uden and B.A.Waldman, Anal. Lett., 8, 91 (1975).
2506. G.D'Ascenzo and W.W.Wendlandt, J. Therm. Anal., 1, 423 (1969); J. Inorg. Nucl. Chem., 32, 2431 (1970).
2507. E.H.Daughtrey, Jr. A.W.Fitchett and P.Mushak, Anal. Chim. Acta., 79, 199 (1975).
2508. R.Belcher, K.Blessel, T.J.Cardwell, M.Pravica, W.I.Stephen and P.C.Uden, J. Inorg. Nucl. Chem., 35, 1127 (1973).
2509. R.Belcher, D.E.Henderson, A.Kamalizad, R.J.Martin, W.I.Stephen and P.C.Uden, Anal. Chem., 45, 1197 (1973).
2510. P.C.Uden, D.E.Henderson and A.Kamalizad, J. Chrom. Sci., 12, 591 (1974).
2511. P.C.Uden and D.E.Henderson, Analyst, 102, 889 (1977).
2512. R.Belcher, A.Khalique and W.I.Stephen, Anal. Chim. Acta., 100, 511 (1978).
2513. S.Dilli and E.Patsalides, J. Chromatography, 130, 251 (1977).
2514. A.Khalique, W.I.Stephen, D.E.Henderson and P.C.Uden, Anal. Chim. Acta., 101, 117 (1978).
2515. T.J.Cardwell and D.J.Desarra, Anal. Chim. Acta., 85, 415 (1976).
2516. Y.Shimoishi and K.Toei, Anal. Chim. Acta., 100, 65 (1978).
2517. T.Shigamatson, M.Matsui and K.Utsuromiya, Bull. Inst. Chem. Res. KyotoIlniv, 46, 256 (1968).
2518. H.Kawaguchi, T.Sakamoto and A.Mizuike, Talanta, 20, 321 (1973).
2519. R.Belcher, C.R.Jenkins, W.J.Stephen and P.C.Uden, Talanta, 17, 455 (1970).
2520. R.E.Sievers, K.J.Eisentraut, 15th American Chemical Society Meeting, Atlantic City Meeting, USA (September 1968).
2521. K.Tanikawa, H.Ochi and J.Arakawa, Japan Analyst, 19, 1669 (1970).
2522. R.Belcher, J.R.Majer, W.I.Stephen, I.J.Thompson and P.C.Uden, Anal. Chim. Acta., 50, 423 (1970).
2523. W.R.Wolf, R.E.Sievers and G.H.Brown, Inorg. Chem., 11, 1995 (1972).
2524. R.Belcher, C.R.Jenkins, W.I.Stephen and P.C.Uden, Talanta, 17, 455 (1970).
2525. R.E.Sievers, K.J.Eisentraut, D.W.Meek and C.S.Springer, Proc. of the 9th Int. Conf. on Coordination Chemistry, St. Moritz, Switzerland, p.479 (1966).
2526. C.S.Springer, Doctorial Dissertation, Ohio State University, Ohio, USA. (1967).
2527. R.W.Moshier and J.E.Schwarberg, Talanta, 13, 445 (1966).
2528. R.Fontaine, B. Santoni and C.Pommier, Guichon G. Anal. Chim. Acta. 62, 337 (1972).
2529. W.D.Ross, W.G.Scribner and R.E.Sievers, Reprint of 8th Int. Symp. on Gas Chromatography, Ballsridge, Dublin, Ireland, (September 1970).
2530. P.Jacquelot and G. Thomas, J. Chromatogr., 66, 121 (1972).
2531. W.C.Butts, Dissert. Abstr., 29 B(2), 506B (1968).
2532. R.S.Barratt, R.Belcher, W.I.Stephen and P.C.Uden, Anal. Chim. Acta., 59, 59 (1972).
2533. A.Kito, Y.Mijake, H.Kabayashi and K.Uero, Japan Analyst, 20. 1363 (1971).
2534. C.S.Springer, D.W.Meek and R.E.Sievers, Inorg. Chem., 6, 1105 (1967).
2535. K.Utsunomiya, Anal. Chim. Acta., 59, 147 (1972).
2536. C.A.Burgelt and S.S.Fritz, Anal. Chem., 44, 1738 (1972).
2537. R.S.Barratt, R.Belcher, W.I.Stephen and P.C.Uden, Anal. Chim. Acta., 57, 447 (1971).
2538. N.Kanayannis and A.H.Corwin, J. Chromatogr., 47, 247 (1971).
2539. N.M.Karayannis, A.H.Corwin, E.W.Baker, E.Klesper and J.A.Walker, Anal. Chem., 40, 1736 (1968).

2540. N.M.Karayannis and A.H.Corwin, J. Chromatogr. Sci., 8, 251 (1970).
2541. R.F.Sieck, J.J.Richard, K.Iverson and C.V.Banks, Anal. Chem., 43, 913 (1971).
2542. P.Jacquelot and G.Thomas, Bull. Soc. Chim. France, 2, 702 (1971).
2543. E.Klesper, A.H.Corwin and D.A.Turner, J. Org. Chem., 27, 700-701 (1962).
2544. J.C.Giddings, W.A.Manwaring and M.N.Myers, Science, 154, 146-148 (1966).
2545. S.T.Sie and G.W.A.Rijnders, Anal. Chim. Acta., 38, 31-44 (1967).
2546. N.M.Karayannis and A.H.Corwin, Anal. Biochem., 26, 34-50 (1968).
2547. N.M.Karayannis and A.H.Corwin, J. Chromatography, 47, 247-256 (1970).
2548. R.S.Juvet and F.Zado in "Advances of Gas Chromatography, vol. 1", Chapter 8, Marcel Dekker, New York, USA.
2549. R.W.Moshier and R.E.Sivers in "Gas Chromatography of Metal Chelates" Pergamon Press Ltd. Oxford, U.K. (1965).
2550. R.E.Sievers, B.W.Ponder, M.L.Morris and R.W.Moshier, Inorg. Chem., 2, 693 (1963).
2551. R.E.Sievers, J.W.Connolly and W.D.Ross, J. Gas Chromatography, 5, 241 (1967).
2552. R.E.Sievers, K.J.Eisentraut, C.S.Springer and D.W.Meck, Advances in Chemistry, 71, 141 (1967).
2553. R.E.Sievers, J.W.Connolly, A.S.Hilton, M.F.Richardson, J.E.Schwarberg and W.D.Ross, 155th American Chemical Society Meeting, San Fransisco, California, USA (April 1968).
2554. C.S.Springer, D.W.Meek and R.E.Sievers, Inorg. Chem., 6, 1105 (1967).
2555. K.J.Eisentraut and R.E.Sievers, J. Amer. Chem. Soc., 87, 5254 (1965).
2556. H.Veening, W.E.Beckman and D.M.Wilkinson, J. Gas Chromatography, 5, 248 (1967).
2557. H.Veening and J.K.F.Huber, J. Gas Chromatography, 6, 326 (1968).
2558. D.K.Albert, Anal. Chem., 36, 2034 (1964).
2559. R.E.Sievers, B.W.Ponder, M.L.Morris and R.W.Moshier, Inorg. Chem., 2, 693 (1963).
2560. W.D.Ross, Anal. Chem., 35, 1596 (1963).
2561. R.S.Barrett, Proc. Soc. Analyt. Chem., 9, 86 (1972).
2562. D.K.Albert, Anal. Chem., 36, 2034-2035 (1964).
2563. T.Fujnaga and Y.Ogino, Bull. Chem. Soc. Japan, 40, 434 (1967).
2564. R.S.Juvet and R.P.Durbin, J. Gas Chromat. 1, 14 (1963).
2565. R.S.Juvet and R.P.Durbin, Anal. Chem., 38, 565 (1966).
2566. R.S.Juvet and F.M.Zado, Anal. Chem., 38, 569 (1966).
2567. W.R.Wolf, M.L.Taylor, B.M.Hughes, T.O.Tiernan and R.E.Sievers, Anal. Chem., 44, 616 (1972).
2568. A.J.McCormack, S.C.Tong and W.D.Cooke, Anal. Chem., 37, 1470 (1965).
2569. C.A.Bache and D.J.Lisk, Anal. Chem., 38, 1757 (1966).
2570. C.A.Bache and D.J.Lisk, Anal. Chem., 43, 951 (1971).
2571. J.K.Runnels and J.H.Gibson, Anal. Chem., 39, 1398 (1967).
2572. R.M.Dagnall, B.L.Sharp and T.S.West, Nature, Phys. Sci., 235, 65 (1972).
2573. R.M.Dagnall, T.S.West and P.Whitehead, Anal. Chim. Acta., 60, 25 (1972).
2574. M.S.Black and R.E.Sievers, Anal. Chem., 48, 1872 (1976).
2575. H.Kawaguchi, T.Sakamoto and T.Mizuike, Talanta, 20, 321 (1973).
2576. F.A.Serravallo and T.H.Risby, J. Chromatogr. Sci., 12, 585 (1974).
2577. R.M.Dagnall, T.S.West and P.Whitehead, Analyst, 98, 647 (1973).
2578. M.S.Black and R.F.Browner, Anal. Chem., 53, 249 (1981).
2579. M.S.Black, M.B.Thomas and R.F.Browner, Anal. Chem., 53, 2224 (1981).
2580. V.I.Mishin and S.L.Dobychin, Zb. Prikl. Khim., 43, 1584-1585 (1970).
2581. W.D.Ross and R.T.Jefferson, J. Chromatog., 8, 386 (1970).
2582. J.F.K.Huber, J.C.Kraak and H.Veening, Anal. Chem., 44, 1554 (1972).

2583. P.C.Uden, I.E.Bigley and F.H.Walters, Anal. Chem., 100, 555 (1978).
2584. N.M.Karayannis and A.H.Corwin, J. Chromatog. Sci., 8, 251-256 (1970).
2585. P.C.Uden, I.E.Bigley and E.Imogene, Anal. Chem., 94, 29 (1977).
2586. E.Gaetani, C.F.Laureri, A.Magnia and G.Parolari, Anal. Chem., 48, 1725 (1976).
2587. T.Honjo, R.Honda and T.Keba, Anal. Chem., 49, 2246 (1977).
2588. J.M.Greenwood, H.Veening and B.R.Wileford, J. Organomet. Chem., 38, 345-348 (1972).
2589. J.F.K.Huber, J.C.Kraak and H.Veening, Chem. Commun., 1305 (1969).
2590. R.E.Graf and C.P.Lillya, J. Organomet. Chem., 47, 413-418 (1973).
2591. J.F.K.Huber, J.C.Kraak and H.Veening, Anal. Chem., 44, 1554-1559 (1972).
2592. E.Gaetani, C.F.Laureri and A.Mangia, Abstracts, 12th National Meeting of the Societa Chimica Italiana, Cagliari, September 1975 pp 68-70.
2593. P.C.Uden and F.H.Walters, Anal. Chim. Acta., 79, 175-183 (1975).
2594. D.E.Henderson, R.Chafter and F.P.J.Novak, J. Chromatogr. Sci., 19, 79 (1981).
2595. L.K.Hanson, M.Gouterman, J.C.Hanson, J. Am. Chem. Soc., 95, 4822-4829 (1973).
2596. A.D.Adler, F.R.Longo and V.Varadi, Inorg. Synth., 16, 213-220 (1976).
2597. R.H.Felton and H.Linschitz, J. Am. Chem. Soc., 88, 1113-1116 (1966).
2598. P.E.Wei, A.H.Corwin and R.Arellano, J. Org. Chem., 27, 3344-3346 (1962).
2599. T.C.Chu and E.J.H.Chu, J. Biol. Chem., 212, 1-7 (1955).
2600. T.C.Chu and E.J.H.Chu, J. Chromatogr., 26 475-478 (1976).
2601. D.W.Lamson, A.F.W.Coulson and T.Yonetani, Anal. Chem., 45, 2273-2276 (1973).
2602. Y.Yamamoto, M.Yamamotom S.Ebisul, T.Takagi, T.Hashimoto and M.Izuhara, Anal. Lett., 6, 451-460 (1973).
2603. K.Saltoh, M.Satoh and N.Suzuki, J. Chromatogr., 92, 291-297 (1974).
2604. K.Saitoh and N.Suzuki, J. Chromatogr., 109, 333-339 (1975).
2605. N.Suzuki, K.Saitoh and M.Shibukawa, J. Chromatogr., 138, 79-87 (1977).
2607. N.Suzuki and K.Saitoh, Bull. Chem. Soc., Jpn., 50, 2907-2910 (1977).
2608. K.Saitoh and N.Suzuki, Bull. Chem. Soc., Jpn., 51, 116-120 (1978).
2609. H.Noda, K.Saitoh and N.Suzuki, J. Chromatogr., 168, 250-254 (1979).
2610. A.G.Ogston, Trans. Faraday Soc., 54, 1754-1757 (1958).
2611. K.Saitoh and N.Suzuki, Anal. Chem., 52, 30-32 (1980).
2612. T.C.Laurent and J.Killander, J. Chromatogr., 14, 317-330 (1964).
2613. N.Suzuki, J.Suzuki and J.Koichi, J. Chromatogr., 177, 166 (1979).
2614. K.Saitoh and J.Suzukin, J. Chromatogr., 92, 371 (1974).
2615. M.Saito, R.Kuroda and M.Shibukawa, Anal. Chem., 55, 1025 (1983).
2616. K.Saitoh, M.Kabayashi and N.Suzuki, Anal. Chem., 53, 2309 (1981).
2617. M.Sato and T.Kwan, Chem. Pharm. Bull., 20, 840-841 (1972).
2618. K.S.Hui, B.A.Davis and A.A.Baulton, J. Chromatogr., 115, 581-586 (1975).
2619. G.P.Morie and T.S.Sweet, Anal. Chem., 34, 314 (1966).
2620. R.E.Sievers, G.Wheeler and W.D.Ross, Anal. Chem., 38, 306 (1966).
2621. R.E.Sievers, J.W.Connolly and W.D.Ross, J. Gas Chromatography, 5, 241 (1967).
2622. W.D.Ross and R.E.Sievers, 156th American Chemical Soc. Meeting, Atlantic City, New Jersey, USA (Sept. 1968).
2623. J.Savoury, P.Mushak, N.O.Roszel and F.W.Sunderman, Federation Proceedings (Biochem), Abstracts, 52nd Meeting, Atlantic City, New Jersey, USA,(April 15-20 1968) p.777.
2624. R.S.Juvet and R.L.Fischer, Anal. Chem., 38, 1860 (1966).

2625. K.Arakawa and K.Tamkawa, Bunseki Kagakin, 16, 812 (1967).
2626. K.Tankkawa, Anal. Abstr., 15, 5857 (1968).
2627. K.Tanikawa, K.Hirano and K.Arakawa, Chem. Pharm. Bull, Tokyo, 15, 915 (1967).
2628. R.C.Fay and T.S.Piper, Inorg. Chem., 3, 348 (1964).
2629. G.Gentry, C.Houin and R.Schott, Preprints of 7th International Symposium on Gas Chromatography and its Exploitation, Copenhagen, Denmark paper G (June 1968).
2630. W.D.Ross and R.E.Sievers, 6th International Symposium on Gas Chromatography, Rome, Italy, (Sept. 1966).
2631. R.E.Sievers, K.J.Eisentraut, D.W.Meek and C.S.Springer, Proc. of the 9th International Conference on Coordination Chemistry, St. Mortiz, Switzerland (1966) p.479.
2632. C.S.Springer, Doctorial Dissertation, Ohio State University, Ohio. USA (1967).
2633. P.Jacquelot and G.Thomas, Bull. Soc. Chim. France, 8-9, 3167 (1970).
2634. T.Shigmatsu, T.Uchiike, T.Aoki and M.Matsui, Bull. Inst. Chem. Res. Kyoto University, 51, 273 (1973).
2635. R.E.Sievers, B.W.Ponder, M.I.Morris and R.W.Moshier, Inorg. Chem., 2, 693 (1963).
2636. T.Fujunaga, T.Tuwamato and Y.Ono, Japan Analyst, 12, 1199 (1963).
2637. R.A.Palmer, R.C.Fay and T.S.Piper, Inorg. Chem., 3, 875 (1964).
2638. R.W.Mosher, R.E.Sievers, in "Gas Chromatography of Metal Chelates", p23, 36, Pergamon Press Ltd., Oxford, U.K. (1965).
2639. R.S.Juvet and F.Zado in "Advances of Gas Chromatography, vol. 1" Chapter 8 - Marcel Dekker New York, USA. (1966).
2640. C.Kutal and R.E.Sievers, Inorganic Chem., 13, 897 (1974).
2641. S.Yeh and C.Ke, J. Chim. Chem. Soc., Taipei, 20, 129 (1973).
2642. G.H.Booth and W.J.Darby, Anal. Chem., 43, 831 (1971).
2643. J.Savory, P.Mushak and F.W.Sunderman, J. Chromat. Sci., 7, 674 (1969).
2644. J.Savory, P.Mushak and F.W.Sunderman, Fed. Proc., 27, 777 (1968).
2645. J.Savory, M.T.Glenn and J.A.Allstrom, J. Chromat. Sci., 10. 247 (1972).
2646. J.Savory, P.Mushak, N.O.Rozzel and F.W.Sunderman, Federation Proceedings (Biochem) Abstracts, 52nd Meeting, Atlantic City, New Jersey U.S.A., p. 777 (April 15-20 1968).
2647. J.Savory, P.Mushak, F.W.Sunderman, R.H.Estes and N.O.Roszel, Anal. Chem., 42, 294 (1970).
2648. L.C.Hansen, W.G.Scribner, T.W.Gilbert and R.B.Sievers, Anal. Chem., 43, 349 (1971).
2649. J.Savory, P.Mushak and P.W.Sunderman in "Advances in Chromatography Ed. Zlatkis A., Preston Technical Abstracts Co. Evanston, Illinois, U.S.A. p.181-186 (1969).
2650. M.Muyazaki and H.Kaneko, Chem. Pharm. Bull., Tokyo, 18, 1933 (1970).
2651. D.N.Sokolov, G.N.Nesterenko and L.K.Golubeva, Zav. Lab., 39, 939 (1973).
2652. O.Gentry, Anal. Abstr., 21, 4032 (1971).
2653. W.D.Ross and R.E.Sievers, Envir. Sci. and Technology, 6, 155 (1972).
2654. M.H.Noweir and J.Cholack, Envir. Sci. Technol., 3, 927 (1969).
2655. G.Kaiser, E.Grallath, P.Tschoepel and G.Toelg, Z. Analyt. Chem., 259, 257 (1972).
2656. M.L.Taylor and E.L.Arnold, Anal. Chim., 43, 1328 (1971).
2657. J.K.Foreman, T.A.Gough and E.A.Walker, Analyst, 95, 797 (1970).
2658. P.C.Uden and C.R.Jenkins, Talanta, 16, 893 (1969).
2659. C.Gentry, C.Houin and P.Malherbe, Anal. Chem., 43, 235 (1971).
2660. V.I.Mishin and S.L.Dobychin, Zh. Prikl. Khim., Leningr., 43, 1584 (1970).

2661. M.L.Taylor, E.L.Arnold and R.E.Sievers, Anal. Letters, 1, 735 (1968).
2662. W.D.Ross and R.E.Sievers, Talanta, 15, 87 (1968).
2663. W.G.Scribner, M.J.Borchers and W.J.Treat, Anal. Chem., 38, 1779 (1966).
2664. W.J.Biermann and H.Geisser, Anal Chem., 32, 1525 (1960).
2665. C.Hista, J.P.Meiserby, R.F.Reschke, D.H.Fredericks and W.D.Cooke, Anal. Chem., 32, 880 (1960).
2666. R.D.Hill, Sc., Thesis University of Manitoba , Winnipeg. Canada, (1962).
2667. A.A.Duswalt, Doctorial Dissertation Perdue University, U.S.A. (1958).
2668. W.V.Floutz, MSc. Thesis, Perdue University, U.S.A. (1959).
2669. R.S.Melcher, MSc. Thesis, Perdue University, U.S.A. (1961).
2670. R.E.Sievers, B.W.Ponder, M.L.Morris and R.W.Moshier, Inorg. Chem., 2, 693 (1963).
2671. W.W.Brandt in "Gas Chromatography 1960" p. 305 ed. R.P.W.Scott, Butterworths, Washington, U.S.A. (1960).
2672. J.Janek in "Gas Chromatography 1960" p. 306 ed. R.P.W.Scott, Butterworths, Washington, U.S.A. (1960).
2673. J.E.Heveran, MSc. Thesis, Perdue University, U.S.A. (1962).
2674. W.W.Brandt and J.E.Heveran, 142nd National Meeting, American Chemical Society, Atlantic City, New Jersey, U.S.A., (September 9-14, 1962).
2675. W.G.Baldwin, MSc. Thesis, University of Manitoba, Winnipeg. Canada, (1961).
2676. M.L.Taylor and F.L.Arnold, Anal. Chem., 43, 1328 (1971).
2677. W.D.Ross, Anal. Chem., 35, 1596 (1963).
2678. W.D.Ross, R.E.Sievers and G.Wheeler, Anal. Chem., 37, 598 (1965).
2679. R.E.Sievers, 16th Annual Summer Symposium, American Chemical Soc., Tuscan, Arizona, Chemical Engineering News, Vol. 41 (1/6/63).
2680. R.E.Sievers, 16th Annual Summer Symposium, Americal Chemical Soc., Tuscan, Arizona, Chemical Engineering News, Vol. 41 (1/6/63).
2681. J.E.Schwarberg, Masters Thesis, Perdue University of Dayton, Ohio, U.S.A. (1964).
2682. A.E.Sievers, 16th Annual Summer Symposium, Americal Chemical Soc., Tuscan, Arizona, Chem. Eng. News, 41 (1.6.63).
2683. J.Stary and E.Hladky, Anal. Chim. Acta., 28, 227 (1963).
2684. R.W.Moshier and R.E.Sievers in "Gas Chromatography of Metal Chelates" Pergamon Press Ltd., Oxford, U.K. (1965).
2685. T.Fujinaga, T.Kuwamoto and S.Murai, Talanta, 18, 429 (1971).
2686. T.Fujinaga, T.Kuwamoto and S.Murai, Talanta, 18, 433 (1971).
2687. W.D.Ross and R.E.Sievers, Talanta, 15, 87 (1968).
2688. R.W.Moshier and J.E.Schwarberg, Talanta, 13, 445 (1966).
2689. G.P.Morie and T.S.Sweet, Anal. Chem., 34, 314 (1966).
2690. R.D.Hill and H.Gesser, J. Gas Chromatography, 1, 11 (1963).
2691. R.E.Sievers, J.W.Connolly and W.D.Ross, J. Gas Chromatography (1967).
2692. R.E.Sievers, J.W.Connolly and W.D.Ross, J. Gas Chromatography, 241 (May 1967).
2693. M.Tanaka, T.Shono and K.Shinra, Nippon Kagakin Zasshi, 89, 669 (1968).
2694. W.D.Ross, Anal. Chem., 35, 1596 (1963).
2695. M.Tanaka, S.Toshiyuki and K.Shinra, J. Chem. Soc. Japan, Pure Chem. Sec., 89, 669 (1968).
2696. R.W.Moshier and R.E.Sievers, in "Gas Chromatography of Metal Chelates" pp 23 and 36, Pergamon Press Ltd. Oxford, U.K. (1965).
2697. J.C.Bailer and D.H.Busch, in "Chemistry of the Coordination Compounds" p 42, Reinhold Publications Corp. New York, U.S.A. (1956).
2698. K.J.Eisentraut and R.E.Sievers, J. Am. Chem. Soc., 87, 5254 (1965).
2699. K.J.Eisentraut and R.E.Sievers, U.S.Patent No. 3,453,319, Chem. Abstr. 71, 103712 (1969).
2700. W.C.Buets and C.V.Banks, Anal. Chem., 42, 133 (1960).

2701. R.Belcher, R.J.Njager, R.Perry and W.I.Stephen, J. Inorg. Nucl. Chem., 31, 471 (1969).
2702. T.Shigmatsu, M.Matsui and K.Utsumova, Bull. Chem. Soc. Japan, 42, 1278 (1969).
2703. K.Utsunomiya and T.Shigematsu, Anal. Chim. Acta., 58, 411 (1972).
2704. K.Utsunomiya and T.Shigimatsu, Bull. Chem. Soc. Japan, 45, 303 (1972).
2705. C.Burgett and J.S.Fritz, J. Chromatogr., 77, 265 (1973).
2706. H.Veening and J.F.K.Huber, J. Gas Chromatogr., 6, 326 (1968).
2707. R.S.Juvet and R.P.Durbin, Anal. Chem., 38. 565 (1966).
2708. R.S.Juvet and R.P.Durbin, J. Gas Chromatography, 1, 14 (1963).
2709. J.E.Schwarberg, R.E.Sievers and R.W.Moshier, Anal. Chem., 42, 1828 (1970).
2710. W.R.Wolf, M.L.Taylor, B.M.Hughes, T.O.Tiernan and R.E.Sievers, Anal. Chem., 44, 616 (1972).
2711. R.E.Sievers, Chem. and Engineering News, 41, 41 (1963).
2712. G.P.Morie and T.S.Sweet, Anal. Chem., 37, 1552 (1965).
2713. R.W.Moshier and D.R.Gere, p.1332, Abstracts Winter Meeting, American Chem. Soc., Phoenix, Arizona, U.S.A. (January 1966).
2714. R.E.Sievers, K.J.Eisentraut, D.J.Griest, M.F.Richardson, W.R.Wolf, W.D.Ross, N.M.Frew and T.L.Isenhour, Variations in contents in lunar fines compared with crystalline rocks, Proc. Second Lunar Sci. Conf., 2, 1451 (1971).
2715. K.J.Eisentraut, M.S.Black, D.J.Griest and R.E.Sievers, Earth and Planetary Science Letters, 15, 169 (1972).
2716. K.J.Eisentraut, D.J.Griest and R.E.Sievers, Anal. Chem., 43, 2003 (1971).
2717. W.R.Wolf, Anal. Chem., 48, 1717 (1976).
2718. E.L.Arnold and B.L.Dold, Anal. Chem., 50, 1708 (1978).
2719. T.J.Cardwell, D.J.Desarro and P.C.Uden, Anal. Chim. Acta., 85, 415 (1976).
2720. K.Utsunomiya, Bull. Chem. Soc. Japan, 44, 2688 (1971).
2721. R.Belcher, A.Khalique and W.L.Stephen, Anal. Chim. Acta., 100, 503 (1978).
2722. P.C.Uden, I.E.Bigley and F.H.Walters, Anal. Chim. Acta., 100, 555 (1978).
2723. T.J.Cardwell, D.Caridi and M.S.Loo, J. Chromatography, 351, 331 (1986).
2724. T.J.Cardwell and D.Caridi, J. Chromatography, 288, 357 (1984).
2725. R.C.Fay and T.S.Piper, J. Am. Chem. Soc., 84, 2303 (1962).
2726. R.C.Fay and T.S.Piper, J. Am. Chem. Soc., 85, 500 (1963).
2727. J.F.K.Huber, J.C.Kraak and H.Veening, Anal. Chem., 44, 1554 (1972).
2728. P.C.Uden and F.H.Walters, Anal. Chim. Acta., 79, 175 (1975).
2729. P.C.Uden , D.M.Parees and F.H.Walters, Anal. Lett., 8, 795 (1975).
2730. E.Gaetani, C.F.Laurieri, A.Mangis and G.Parolari, Anal. Chem., 48, 1725 (1976).
2731. P.C.Uden and I.E.Bigley, Anal. Chim. Acta., 94, 29 (1977).
2732. E.Steger and R.Stahlberg, Z. Anorg. Allg. Chem., 326, 243 (1964).
2733. C.J.Cohen, W.R.Betker, D.M.Wasleski and J.C.Cavagnol, J. Agr. Food Chem., 14, 314 (1966).
2734. N.T.Crosby and E.Q.Laws, Analyst, 89, 319 (1961).
2735. J.Derkosch, H.Jansch, R.Leutner and F.X.Mayer, Chonatsch. Chem.,85, 684 (1954).
2736. D.F.McCaulley and J.W.Cook, J.Ass. Offic. Agr. Chem., 43, 710 (1960).
2737. G.Stanescu, O.Radulescu and M.Keul, Rev. Chim. (Bucharest), 15, 416 (1964).
2738. T.T.White and G.G.McKinley, J. Ass. Offic. Agr. Chem., 44, 589 (1961).
2739. J.Kovac, Chem. Zvesti., 10, 222 (1956).
2740. D.A.Giang and R.L.Caswell, J. Agr. Food Chem., 5, 753 (1957).

2741. A.J.Blotchy, G.T.Hansen, R.Laura, O.Buencamino and E.P.Rack, Anal. Chem., 57, 1937 (1985).
2742. J.M.Vobetsky, V.D.Nefedov and E.N.Sinotova, Zh. Obshch. Khim., 33, 4023 (1963).
2743. K.Berei, J. Chromatogr., 20, 406 (1965).
2744. K.Berei and L.Vasana, Magy. Kem. Foly., 73, 313 (1967).
2745. D.T.Haworth and A.F.Kardis, J. Chromatogr., 27, 302 (1967).
2746. S.Hermanek, J.Plesck and V.Gregor, Collect. Czech. Chem. Commum., 31. 1281 (1966).
2747. J.Huber, J.Ruckbeil and R.Kiessig, Pharm. Zentralhalle Dtschl., 102, 783 (1963).
2748. L.Cima and R.Mantovan, Farmaco, Ed. Pract., 17, 473 (1962).
2749. T.Ono, Bitamin, 30,280 (1964).
2750. M.Covello and O.Schettino, Farmaco, Ed. Pract., 19, 38 (1964).
2751. L.Cima, C.Levarato and R.Mantovan, Farmaco. Ed. Pract., 21, 244 (1966).
2752. Ya.G.Popova, K.Popov and M.Ilieva, J. Chromatogr., 24, 263 (1966).
2753. Ya.G.Popova, K.Popov and M.Ilieva, J. Chromatogr., 21, 164 (1966).
2754. T.Sassaki, J. Chromatogr., 24, 452 (1966).
2755. C.Maglitto, L.Gianotti and C.Maltarei, Bull. Lab. Chim. Provinciali., 15, 354 (1964).
2756. K.Schloegl, H.Pelousek and A.Mohae, Monatsh. Chem., 92, 533 (1961).
2757. R.E.Bozak and O.Fukuda, J. Chromatogr., 26, 501 (1967).
2758. L.C.Willemsens and G.J.M.Van der Kerk, in Investigations in the Field of Organolead Chemistry, International Lead-Zinc Research Organization New York, p.84 (1957).
2759. R.Donner and K.Lohs, J. Chromatogr., 17, 349 (1965).
2760. H.Reuter and H.Hank, Pharm. Zentrahl., Deut., 104, 323 (1965).
2761. G.Neubert, J. Chromatogr., 20, 342 (1965).
2762. J.Rabincwicz, B.Boehler and G.Weher, Helv. Chim. Acta., 49, 590 (1966).
2763. B.Berai, J. Agr. Food Chem., 13, 373 (1965).
2764. D.Braun, Kunststoffe, 52, 2 (1962).
2765. J.E.Barney, J. Chromatogr., 20, 334 (1965).
2766. C.S.Hanes and F.A.Isherwood, Nature, 164, 1107 (1949).
2767. P.J.Bunyan, Analyst, 89, 615 (1964).
2768. T.A.Taulli, Anal. Chem., 39, 1901 (1967).
2769. H.Hecker and F.Hein, Z. Anal. Chem., 174, 354 (1960).
2770. E.E.Nifant'ev, Zh. Obshch. Khim., 35, 1980 (1965).
2771. H.Schindbauer and F.Mitterhofer, Z. Anal. Chem., 221, 394 (1966).
2772. K.Berai and L.Vasana, Magy. Kem. Foly., 73, 313 (1967).
2773. H.Petschik and E.Steger, J. Chromatogr., 9, 307 (1962).
2774. H.Petschik and E.Steger, J. Chromatogr., 31, 369 (1967).
2775. J.Haukiewicz and K.Studniarski, Chem. Anal. (Warsaw), 10, 941 (1965).
2776. T.A.Mastryukova, T.B.Saktarova and M.I.Kabachnik, Izv. Akad. Nauk, SSSR. Ser. Khim., 12, 2211 (1963).
2777. J.Kovac, J. Chromatogr., 11, 412 (1963).
2778. L.Gelder, Erdol. Kohle., 18, 545 (1965).
2779. V.Amormino and E.Cingolani, Annali. Ist. Sup. Sanita., 2, 545 (1966).
2780. H.Wagner, Fette. Seif. Anstrichm., 63, 1119 (1961).
2781. P.Melchoirri, F.Maffei and J.J.Siesto, Farmaco Ed. Prat., 19, 610 (1964).
2782. N.A.Smart and A.R.C.Hill. J. Chromatogr., 30, 626 (1967).
2783. D.P.Braithwaite, Nature, 200, 1011 (1963).
2784. I.Tadic-Pejkovie and S.L.Vitorvic, Hrana. Ishana, 9, 29 (1968).
2785. D.Pantovic, Hrana, Ishana, 8, 769 (1967).
2786. W.A.Steller and A.N.Curry, J. Ass. Offic. Agr. Chem., 47, 645 (1964).
2787. B.Bazzi, R.Sarti, G.Carale and M.Radice, Soc. Gen. pen l'min. e. Chem. Milan, 12 (1963).

2788. D.M.Wasleski, J. Agr. Food Chem., 14, 156 (1966).
2789. F.Grimmer, W.Spichale, R.Kliche and D.Quass, J. Chromatogr., 22, 316 (1966).
2790. R.C.Blinn, J. Ass. Offic. Agr. Chem., 47, 641 (1964).
2791. M.T.H.Ragab, J. Ass. Offic. Anal. Chem., 50, 1088 (1967).
2792. C.N.Stanley, J. Chromatogr., 16, 467 (1964).
2793. M.Salame, J. Chromatogr.. 16, 476 (1964).
2794. J. Baumler, and S.Rippstein, Helv. Chim. Acta., 44, 1162 (1961).
2795. R.S.Bandurski and B.Axelrod, J. Biol. Chem., 193, 405 (1951).
2796. I.M.Hais and K.Macek in "Paper Chromatography" p.646 Academic Press, New York and London (1963).
2797. N.H.Doman and Z.S.Kagan, Biokhimija., 17, 719 (1952).
2798. T.K.H.Otter, Nature, 176, 1078 (1955).
2799. J.W.Cook, J. Ass. Offic. Anal. Chem., 38, 150 (1955).
2800. T.Wood, Nature, 176, 175 (1955).
2801. J.J.Menn, W.R.Erwiss and H.T.Gordon, J. Agr. Food Chem., 5, 601 (1957).
2802. H.F.MacRae and W.P.McKinley, J. Ass. Offic. Agr. Chem., 44, 207 (1961).
2803. H.E.Wade and D.M.Morgan, Nature, 171, 529 (1953).
2804. R.R.Watts, J. Ass. Offic. Agr. Chem., 48, 1161 (1965).
2805. Y.Takeshi, Ann. Rep. Sankyo Res. Lab., 16, 104 (1964).
2806. K.Uhle, Z. Chemie. Lipz., 7, 236 (1967).
2807. A.J.Bawd, D.T.Burns and A.Fogg, Talanta, 16, 719 (1969).
2808. M.Vobetsky, V.D.Nefedov and E.N.Sinotova, Zh. Obshch. Khim., 33, 4023 (1963).
2809. M.Vobetsky, V.D.Nefedov and E.N.Sinotova, Zh. Obshch. Khim., 35, 1684 (1965).
2810. H.Huber and J.Wimmer, Kunststoffe, 58, 786 (1968).
2811. H.Seidler, H.Waggon, M.Haertig and W.J.Uhde, Nahrung, 13, 257 (1969).
2812. M.Turler and O.Hogl, Mitt. Lebensmitt. Hyg. Bern., 52, 123 (1961).
2813. K.Burger, Z. Anal. Chem., 192, 280 (1962).
2814. G.Neubert, Z. Anal. Chem., 203, 265 (1964).
2815. D.Helberg, Dtsch. Lebensm-Rundsch., 62, 178 (1966).
2816. D.Helberg, Dtsch. Lebensm-Rundsch., 63, 69 (1967).
2817. R.F.Van der Heide, Z. Lebensm-Unters.-Forsch., 124, 348 (1964).
2818. K.Figge, J. Chromatogr., 39, 84 (1969).
2819. B.Herold, K.H.Droege and Z.Fresenius, Anal. Chem., 245, 295 (1969).
2820. Y.Jitsu, N.Kudo and T.Sugujama, Noyaku Seison Gijutsu., 17, 17 (1967).
2821. D.Braun and H.T.Heimes, F. Z. Anal. Chem., 239, 6 (1968).
2822. V.D.Nefedov, V.E.Khuravlev, N.G.Molchanova and N.N.Kalinina, Zh. Obshch. Khim., 38, 1219 (1968).
2823. H.Wieczorek, Dtsch. Lebensm.-Rundsch., 65, 74 (1969).
2824. P.P.Otto, H.M.J.C.Creemers and J.G.A.Luijten, J. Labelled Comp., 2, 339 (1966).
2825. E.S.Kosmatyi, L.L.Bulic and G.V.Gaurilova, Fiziol. Biokhim. Kul't Rast., 4, 317 (1972).
2826. L.C.Mitchell, J. Assoc. Off. Agric. Chem., 42,684 (1959).
2827. A.Castiglioni, Z. Anal. Chem., 161, 40 (1958).
2828. C.Cardini, G.Cavina, E.Cingolani, A.Marioni and C.Vicari, Farmaco, Ed. Prat., 17, 583 (1962).
2829. M.Covello and O.Schettino, Ann. Chim. (Rome), 52, 1135 (1962).
2830. J.Bayer, J. Chromatogr., 8, 123 (1962).
2831. A.Sauciuc, L.Jonescu and M. Albu-Budai, Revta. Chim., 18, 237 (1967).
2832. J.L.Martin and W.H.C.Shaw, Analyst (London), 88, 292 (1963).
2833. M.Covello and O.Schettino, Farmaco, Ed. Prat., 20, 581 (1965).
2834. I.V.Konova, N.M.Neronov, N.D.Jerusalimskii and A.I.Borisova, Mikrobiologiya, 28, 490 (1959).
2835. N.D.Jerusalimskii, J.V.Konova and N.M.Neronova, Mikrobiologiya, 28, 433 (1959).

2836. T.D.Miles, A.C.Delasant and J.C.Barry, Anal. Chem., 33, 685 (1961).
2837. A.N.DeBelder, E.J.Bourne and J.B.Pridham, Chem. Ind. (London), 996 (1959).
2838. C.A.J.Goldberg, Clin. Chem., 5, 446 (1959).
2839. T.C.Chu and E.J.H.Chu, J. Biol. Chem., 212, 1 (1955).
2840. D.G.Hill-Cottingham and C.P.Lloyd-Jones, J. Sci. Food. Agric., 12, 69 (1961).
2841. R.Barbieri, U.Belluco and G.Tagliavini, Ric. Sci., 30, 1671 (1960).
2842. R.Barbieri, U.Belluco and G.Tagliavini, Ann. Chim. (Rome), 48, 940 (1958).
2843. H.Schafer, Z. Anal. Chem., 180, 15 (1961).
2844. M.Pedinelli, Chim. Ind. (Milan), 44, 651 (1962).
2845. J.N.Bartlett and G.W.Curtis, Anal. Chem., 34, 80 (1962).
2846. V.L.Miller, D.Polley and C.J.Gould, Anal Chem., 23, 1286 (1951).
2847. J.Kanazawa, K.Koyama, M.Aya and R.Sato, J. Agric. Chem. Soc. Jap., 31, 872 (1957).
2848. J.Kanazawa and R.Sato, Bunseki Kagaku, 8. 322 (1959).
2849. K.Sera, M.Kanda, A.Murakami, Y.Sera, Y.Kondo and T.Yanagi, Kumamato Med. J., 15, 38 (1962).
2850. K.Broderson, Schlenker, Z. Anal. Chem., 182, 421 (1961).
2851. V.P.Shvedov and S.P.Rosyanov, Trudy Leningrad. Tecknol Inst. Im. Lensoveta, 55, 55 (1961). Ref. Zhur. Khim. 5, Abstract No. 5D205 (1962).
2852. N.Cyjeticanin and J.Cvoric, Bull. Inst. Nucl. Sci., "Boris Kidrich" Belgrade, 13, 35 (1962).
2853. V.P.Shvedov and S.P.Rosyanov, Zhur. Anal. Khim., 14, 507 (1959).
2854. N.I.Gabov and A.I.Shafiev, Zhur. Anal. Khim., 21, 1107 (1966).
2855. E.Cerri, C.Cesarano and F.Gadda, Energia. Nucleare., 4, 405 (1957).
2856. C.S.Hanes and F.A.Isherwood, Nature, 164, 1107 (1949).
2857. C.J.Hardy and D.Scargill, J. Inorg. Nucl. Chem., 10, 323 (1959).
2858. H.E.Wade and D.M.Morgan, J. Biochem., 60,264 (1955).
2859. F.W.Plapp and J.E.Canida, Anal. Chem., 30, 1622 (1958).
2860. J.P.Crowther, Anal. Chem., 26, 1383 (1954).
2861. E.Tyskiewicz, Anal. Biochem., 3, 164 (1962).
2862. A.Eggleston and M.Hems, Biochem. J., 52, 156 (1952).
2863. J.L.Drummond and A.J.Blair, Unpublished work, Chemistry Division, Dounreay Experimental Reactor Establishment, United Kingdom Atomic Energy Authority (1959).
2864. L.H.Towle and R.S.Farrand, United States Atomic Energy Commission Document T ID.6186 (1960).
2865. H.A.Moule and S.Greenfield, J. Chromatogr., 11, 77 (1963).
2866. V.C.Runeckles and G.Krotov, Arch. Biochem. Biophys., 80, 94 (1959).
2867. E.Gerlach, E.Weber and H.J.Doring, Arch. Exp. Path. Pharmak., 226, 9 (1955).
2868. G.R.Havey and T.C.Loughhead, Chem. and Ind., 25, 702 (1955).
2869. R.Donnel and K.Lohs, J. Chromatogr., 17, 349 (1965).
2870. L.Macho, Chem. Zvesti., 11, 175 (1957).
2871. E.S.Rorem, Nature, 183, 1739 (1959).
2872. T.K.H.Otter, Nature, 176, 1078 (1955).
2873. H.T.Gordon, L.N.Werum and W.W.Thornbury, J. Chromatogr., 13, 272 (1964).
2874. R.Sokolawska, Rocza. Panst. Zakl. Hig., 17, 1 (1966).
2875. N.I.Gabov and A.I.Shafiev, Zhur. Anal. Khim., 21, 1107 (1966).
2876. T.Weil, Helv. Chim. Acta., 38, 1274 (1955).
2877. M.Halmann and L.Kugel, Bull. Res. Council. Israel, 10, 124 (1961).
2878. N.I.Gabov, Zh. Anal. Khim., 22, 814 (1967).
2879. A.Siuda, Nukleonika, 10, 459 (1965).
2880. H.W.Gordde, K.G.Blume and H.Halzer, Biochim. Biophys. Acta., 62, 1 (1962).

2881. S.Sato, Japan Analyst, 9, 471 (1960).
2882. J.P.Vigne, R.L.Taban and J.Fondarai, Bull. Soc. Chim. France, 10, 1282 (1955).
2883. J.P.Vigne, R.L.Taban and J.Fondarai, Bull. Soc. Pharm., Marseille, 5, 49 (1956).
2884. J.P.Vigne, R.L.Taban and J.Fondarai, Bull. Soc. Pharm., Marseille, 5, 55 (1956).
2885. Y.Nagai and Y.Kimura, Nature, 181, 1730 (1958).
2886. F.E.G.Hanrap, Analyst, 85, 452 (1960).
2887. F.E.G.Hanrap, Nature, 182, 876 (1958).
2888. W.S.Cowgill, J. Asso. Offic. Agr. Chem., 16. 614 (1955).
2889. B.Axelrod and R.S.Bandurski, J. Biol. Chem., 193, 405 (1951).
2890. P.J.Heald, J. Biochem., 63, 235 (1956).
2891. R.M.Bock, L.Nan-Sing, S.A.Morell and S.H.Lipton, Arch. Biochem. Biophys., 62, 253 (1956).
2892. P.Cerletti and V.Siliprandi, Ric. Sci., 25, 2084 (1955).
2893. E.Gerlach and H.J.Doring, Naturwissenschaften, 42, 344 (1955).
2894. R.M.C.Dawson and J.Eichberg, Biochem. J. 96, 634 (1965).
2895. R.F.Witter, G.U.Marinetti, L.Heicklin and M.A.Cottone, Anal. Chem., 30, 1624 (1958).
2896. J.Kovac, Chem. Zvesti, 11, 162 (1957).
2897. W.Paulus and H.J.Mallach, Arzneim.-Forsch., 6, 766 (1956).
2898. J.W.Cook, J. Ass. Offic. Agr. Chem., 37, 987 (1954).
2899. A.Cortes and D.R.Gilmore, T. Chromatogr., 19, 450 (1965).
2900. H.Waggon, D.Spranger and H.Ackerman, Nahrung., 7, 612 (1963).
2901. I.Sato, Kumamato. Med. J., 14, 1 (1961).
2902. R.B.Marsh, R.L.Metcalf and T.R.Fukuto, J. Agr. Food Chem., 2, 732 (1954).
2903. J.J.Menn, M.E.Eldfrawi and H.T.Gordon, J. Agr. Food Chem., 8, 41 (1960).
2904. B.Bazzi, R.Santi, G.Canale and M.Radice, Soc. Gen. per l'Min. e. Chem. Milan, 12 (1963).
2905. K.Yamashita, Kumamato Med., 14, 13 (1961).
2906. P.Caspeieri and H.H.Keppler, J. S. Agr. Chem. Inst., 12, 26 (1959).
2907. R.Muller, G.Ernst and H.Schoch, Mitt. Lebensmitt. Hyg. Bern., 48, 152 (1957).
2908. O.Jucker, Mitt. Lebensmitt. Hyg. Bern., 49, 299 (1958).
2909. M.E.Getz, J. Ass. Offic. Agr. Chem., 45, 393 (1962).
2910. A.Irudayasamy and A.R.Natarjan, Analyst, 90, 503 (1965).
2911. J.Zadroninska, Roczn. Panst. Zakl. Hyg., 16, 53 (1965).
2912. R.Anliker and R.E.Menzer, J. Agr. Food Chem., 11, 291 (1963).
2913. M.C.Kinley and P.Johal, J. Ass. Offic. Agr. Chem., 46, 840 (1963).
2914. J.W.Cook, J. Ass. Offic. Agr. Chem., 38, 150 (1955).
2915. Y.Takeshi, Ann. Rep Sankyo Res. Lab., 16, 104 (1964).
2916. T.Wood, Nature, 176, 175 (1955).
2917. R.R.Watts, J. Ass. Offic. Agr. Chem., 48, 1161 (1965).
2918. J.J.Menn, W.R.Erwin and H,T.Gordon, J. Agr. Food Chem., 5, 601 (1957).
2919. R.Fischer and N.Otterbech, Sci. Pharm., 27, 1 (1959).
2920. W.P.McKinley and S.I.Read, J. Ass. Offic. Agr. Chem., 46, 467 (1962).
2921. N.H.Smart and A.R.C.Hill, J. Chromatogr., 30, 626 (1967).
2922. W.P.McKinley, Proc. Canad. Soc. Forensic Sci., 2, 433 (1964).
2923. K.R.Millar, J. Chromatogr., 21, 344 (1966).
2924. J.Scala and H.H.Williams, J. Chromatogr., 15, 546 (1964).
2925. G.Fritz and D.Wick, Z. Anorg. Chem., 342, 130 (1966).
2926. D.J.Williams and J.W.Price, Analyst (London), 85, 579 (1960).
2927. D.J.Williams and J.W.Price, Analyst (London), 89, 220 (1964).
2928. O.A.Reutov, O.A.Putsyna and M.F.Turchinskii, Dokl. Akad. Nauk. SSSR., 139, 146 (1961).

2929. Y.Tanaka and T.Morikawa, Bunseki Kagaku, 13, 753 (1964).
2930. J.Gasparic and A.Cee, J. Chromatogr., 8, 393 (1962).
2931. J.Franc, M.Wurst and V.Moudry, Collect. Czech. Chem. Commun., 26, 1313 (1961).
2932. R.Barbieri, U.Belluco and G.Tagliavini, Ann. Chim. (Rome), 48, 940 (1958).
2933. A.Cassol. L.Magon and R.Barbieri, J. Chromatogr., 19, 57 (1965).
2934. A.Cassol and R.Barbieri, Ann. Chim. (Rome), 55, 606 (1965).
2935. B.Visintin, A.Pepe and S.A.Guiseppe, Ann. Ist. Sup. Sanita, 1, 767 (1965).
2936. K.Burger, Z. Anal. Chem., 192,280 (1962); Anal. Abstr., 10, 3742 (1963).
2937. R.F.Elton and W.E.Geiger, Anal. Chem., 50, 712 (1978).
2938. F.T.Henry and T.M.Thorpe, Anal. Chem., 52, 80 (1980).
2939. D.G.Iverson, M.A.Anderson, T.R.Halm and R.R.Starforth, Environ. Sci. Technol., 13, 1491 (1979).
2940. L.D.Hansen, B.E.Richter, D.K.Rollins, D.K.Lamb and D.J.Eatough, Anal. Chem., 51, 633 (1973).
2941. M.Morita, T.Uchiro and K.Fuwa, Anal. Chem., 53, 1806 (1981).
2942. G.R.Ricci, L.S.Shepard, G.Colovos and N.H.Hester, Anal. Chem., 53, 610 (1981).
2943. D.Monnier, Y.Ghalioughi and R.Salia, Anal. Chim. Acta., 28, 30 (1963).
2944. F.Gstirner and S.K.Baveja, Arch. Pharm. (Weinheim), 298, (2); Mitt. Dtsch. Pharm. Ges., 35, 29 (1965).
2945. L.Klotz, Pharm. Zentralhalle Dtsch., 104, 393 (1965).
2946. B.Petrangeli, Bull. Chim. Farm., 105, 770 (1966).
2947. R.Huttenrauch and L.Klotz, J. Chromatogr., 12, 464 (1963).
2948. J.Pawelkiewicz, W.Walerych, W.Friedrich and K.Bernhauer, J. Chromatogr. 3, 359 (1960).
2949. M.Rosenblum and R.B.Woodword, J. Am. Chem. Soc., 80, 5443 (1958).
2950. C.J.Muller and C.Pik, Clin. Chim. Acta., 7, 92 (1962).
2951. R.T.Jones and W.A.Schroeder, J. Chromatogr., 10, 421 (1963).
2952. T.H.J.Huisman and H.K.Prins, J. Lab. Clin. Med., 46, 255 (1955).
2953. P.C.Van der Schaaf and T.H.J.Huisman, Rec. Trav. Chim. Pays-Bas, 74, 563 (1955).
2954. S.Raymond and M.Nakamichi, Anal. Biochem., 3, 23 (1962).
2955. F.J.Gulter, E.A.Peterson and H.A.Sober, Arch. Biochem. Biophys., 80. 353 (1959).
2956. T.H.J.Huisman and H.K.Prins, Nature, 175, 903 (1955); Anal. Abstr., 2, 2498 (1955).
2957. T.H.J.Huisman and H.K.Prins, Clin. Chim. Acta., 2, 307 (1957).
2958. H.K.Prins, J. Chromatogr., 2, 445 (1959).
2959. E.Awad, B.Cameron and L.Kotite, Nature (London), 198, 1201 (1963).
2960. T.C.S.Ruo, M.L.Selucky and O.P.Strauzz, Anal. Chem., 49, 1761 (1977).
2961. T.C.S.Ruo, M.L.Selucky and O.P.Strauzz, Anal. Chem., 49, 1764 (1977).
2962. J.D.Messman and T.C.Rains, Anal. Chem., 53, 1632 (1981).
2963. C.Botre, F.Cacace and H.Cozzani, Anal. Lett., 9, 825-830 (1976).
2964. T.C.S.Ruo, M.L.Selucky and O.P.Strausz, Anal. Chem., 49, 1761-1765 (1977).
2965. H.Kolzumi, R.D.McLaughlin and T.Hadeishi, Anal. Chem., 51, 387-392 (1979).
2966. T.M.Vickrey, H.E.Howell, G.V.Harrison and G. J.Ramelow, Anal. Chem., 52, 1743-1746 (1980).
2967. T.M.Vickrey, T.E.Howell, G.V.Harrison and G.J.Ramelow, Anal. Chem., 52, 1743 (1980).
2968. T.M.Vickrey, H.E.Howell, G.V.Harrison and G.J.Ramelow, Anal. Chem., 1744 (1980).

2969. A.Mangia, G.Parolari, E.Gratana and C.F.Lamreci, Anal. Chim. Acta., 92, 111 (1977).
2970. W.A.MacCrehan and R.A.Durst, Anal. Chim., 50. 2108 (1978).
2971. W.A.MacCrehan, Anal. Chem., 53, 74 (1981).
2972. H.Koizumi and R.D.McLaughlin, Hadeishe Anal. Chem., 51, 387 (1979).
2973. H.Koizumi, T.Tadeishi and R.McLaughlin, Anal. Chem., 50, 1700 (1978).
2974. W.Holak Analyst, 107, 1457 (1982).
2975. D.R.Jones and S.E.Manahan, Anal. Chem., 48, 502-505 (1976).
2976. D.M.Fraley, D.A.Yates, S.E.Manahan, D.Stalling and J.Petty, Appl. Spectros, 35, 525-531 (1981).
2977. D.M.Fraley, D.A.Yates and S.E.Manahan, Anal. Chem., 51, 2225-2229 (1979).
2978. F.E.Brinckman, W.R.Blair, K.L.Jewell and W.P.Iverson, J. Chromatogr. Sci., 15, 493-503 (1977).
2979. H.Kolzumi, T.Hadeishi and R.McLaughlin, Anal. Chem., 50, 1700-1701 (1978).
2980. W.A.McCrehan, Anal. Chem., 53, 74-77 (1981).
2981. W.A.McCrehan and R.A.Durst, Anal. Chem., 50, 2108-2112 (1978).
2982. S.Inoue, S.Hoshi and M.Sasaki, Bunseki Kagaku, 31, E243-E246 (1982).
2983. N.Haring and K.Ballschmiter, Talanta, 27, 873-879 (1980).
2984. J.C.Van Loon, B.Radziuk, N.Kohn, J.Lichwa, F.J.Fernandez and J.D. Kerber, Absorpt. Newsl., 16, 79-83 (1977).
2985. B.R.Willeford and H.Veening, J. Chromatogr., 251,61-88 (1982).
2986. M.Fujita and E.Takrabatake, Anal. Chem., 55, 454 (1983).
2987. W.A.MacCrehan, Anal. Chem., 53, 74 (1981).
2988. E.Klesper, A.H.Corwin and D.A.Turner, J. Org. Chem., 27, 700 (1962).
2989. Schnitzerling and Schunter, J. Chromatogr., 20, 621 (1965).
2990. R.H.A.Blimmer and W.J.N. Bunch, J. Chem. Soc., 292 (1929).
2991. W.D.Kumler and J.J.Eiler, J. Amer. Chem. Soc., 65, 2355 (1943).
2992. D.F.Peppard, G.W.Mason, J.L.Maier and W.J.Driscoll. J. Inorg. Nucl. Chem., 4, 334 (1957),
2993. D.E.Peppard, J.R.Ferraro and G.W.Mason, J. Inorg. Nucl. Chem., 21, 1156 (1958).
2994. D.S.Stewart and H.W.Crandell, J. Amer. Chem. Soc., 73, 1377 (1951).
2995. F.W.Plapp and J.E.Casida, Anal. Chem., 30, 1622 (1958).
2996. C.E.Higgins and W.H.Baldwin, J. Org. Chem., 21, 1156 (1956).
2997. H.Schmitz and G.Walpurger, Angew. Chem., 71, 549 (1959).
2998. F.Jakob, K.C.Park, J.Ciric and W.Rieman, Talanta, 8, 431 (1961).
2999. W.Rieman and R.Sargent in "Ion-exchange in Physical Methods of Chemical Analysis" (W.G.Beal editor) vol. IV, Academic Press, New York and London (1961).
3000. K.D.Quinlan and A.M.Desesa, Anal. Chem., 27, 1616 (1955).
3001. C.J.Barton, Anal. Chem., 20, 1068 (1948).
3002. G.Misson, Chem. Ztg., 32, 633 (1908).
3003. J.Kolmerton and J.Epstein, Anal. Chem., 30, 1536 (1958).
3004. A.Varon, F.Jakob, K.C.Park, J.Ciric and W.Rieman, Talanta, 9, 573 (1962).
3005. R.B.Lew, H.Gard and F.Jakob, Talanta, 10, 911 (1963).
3006. R.B.Lew and F.Jakob, Talanta, 10, 322 (1963).
3007. E.Cerrai and F.Gadda, Nature, 183, 1528 (1959).
3008. O.Cesaranal and C.Lepseky, J. Inorg. Nucl. Chem., 14, 276 (1960).
3009. R.M.Wheaton and W.C.Bauman, Ann. New York Acad. Sci., 57, 159 (1953).
3010. R.M.Wheaton and W.C.Bauman, Ind. Eng. Chem., 45, 228 (1953).
3011. L.K.Beach and S.Sass, Anal. Chem., 33, 901 (1961).
3012. A.G.Caldwell and C.A.Black, Proc. Soil Sci. Amer., 22, 290 (1958).
3013. E.Mehta, Proc. Soil Sci. Amer., 18, 443 (1954).
3014. D.A.Rappoport and P.T.Chau, Biochim. Biophys. Acta., 38, 156 (1960).
3015. R.K.Morton, J. Biochem., 70, 134 (1958).

3016. C.Stolzer and A.Simon, J. Chromatogr., 9, 224 (1962).
3017. S.Hirotoshi, Y.Kakimoto, T.Nakajima, A.Kanazawa and I.Sano, Nature, 207, 1197 (1965).
3018. W.Paulus and H,J.Mallach, Arzneim-Forsch 6, 766 (1956).
3019, W.Paulus and H.J.Mallach, Arzneim-Forsch, 6, 636 (1956).
3020. R.Fischer and N.Otterbeck, Sci. Pharm., 27, 1 (1959).
3021. E.Sandi, Z. Anal. Chem., 167, 241 (1959).
3022. R.C.Double, J. Ass. Offic. Agr. Chem., 47, 693 (1964).
3023. B.Bazzi, R.Santi and M.Radice, Soc. Gen. per l'Min. e. Chem. Milan, 12 (1963).
3024. C.L.Dunn, J. Agr. Food Chem., 6, 203 (1958).
3025. C.A.Anderson, J.M.Adams and D.MacDougall, J. Agr. Food Chem., 7, 256 (1959).
3026. L.F.Dupee, K.Gardner and P.Newton, Analyst, 85, 177 (1960).
3027. P.Tittarelli and A.Mascherpa, Anal. Chem., 53, 1466 (1981).
3028. J.M.Preston, W.Karasek and S.H.Kim, Anal. Chem., 49, 1746 (1977).
3029. D.C.Paschal, R.Birknell and D.Dresbach, Anal. Chem., 49, 1551 (1977).
3030. J.L.Martin and L.M.Cummins, Anal. Biochem., 15, 530 (1966).
3031. G.Fritz and D.Wick, Z. Anorg. Allgerm. Chem., 342, 130 (1966).
3032. I.G.Sarishvili, M.V.Solielovskii, K.P.Grinevich and P.Z.Sorobin, Zh. prikl. Khim. Leningr., 40, 2062 (1967).
3033. J.Nagy, Brandt and O.Petriko, Periodica Polytech., 10, 443 (1966).
3034. D.T.Burns , F.Glockling and M.Harriott, Analyst, 106, 921 (1981).
3035. R.H.Fish and J.J.Komlenic, Anal. Chem., 56, 510 (1984).
3036. J.A.Owen and C.Got, Clin. Chim. Acta., 2, 588 (1957).
3037. F.Rappaport and M.Rabinovitz, Clin. Chim. Acta., 4, 535 (1959).
3038. E.R.Huehns and A.O.Jakubovic, Nature (London), 186, 729 (1960).
3039. P.Fessas and A.Karaklis, Clin. Chim. Acta., 7, 133 (1962).
3040. A.M.M.Zade-Oppen, Scand. J. Clin. Lab. Invest., 15, 491 (1963).
3041. A.M.M.Zade-Oppen, Scand. J. Clin. Lab. Invest., 15, 331 (1963).
3042. R.C.Bartlett, Clin. Chem., 9, 325 (1963).
3043. R.C.Bartlett, Clin. Chem., 9, 317 (1963); Anal. Abstr., 11, 713 (1964).
3044. J.A.Owen, H.J.Silberman and C.Got, Nature (London), 182, 1373 (1958).
3045. C.B.Laurell, Scand. J. Clin. Lab. Invest., 11, 18 (1959).
3046. T.G.Ferris, R.E.Easterling and R.E.Budd, Clin. Chim. Acta., 8, 792 (1963).
3047. W.Dobryszycka, J.Moretti and M.F.Jayle, Bull. Soc. Chim. Biol., 45, 301 (1963).
3048. M.F.Jayle, Bull. Soc. Chim. Biol., 33, 876 (1951).
3049. C.A.J.Goldberg, Clin. Chem., 3, 1 (1957).
3050. V.J.Yakulis, P.Heller, A.M.Josephson and L.Singer, Am. J. Clin. Pathol., 34, 28 (1960).
3051. C.A.J.Goldberg, Clin. Chem., 4, 484 (1958).
3052. C.A.J.Goldberg and A.C.Ross, Clin. Chem., 6, 254 (1960).
3053. W.B.Gratzer and G.H.Beaven, Clin. Chim. Acta., 5, 577 (1960).
3054. R.L.Engle, A.Markev, J.H.Pert and K.R.Woods, Clin. Chim. Acta., 6, 136 (1961).
3055. M.Aksay and S.Erdem, Clin. Chim. Acta., 12, 696 (1965).
3056. K.Aksay and S.Erdem, Turk Tip Cemiy. Mecm., 31, 593 (1965).
3057. R.N.Ibbotson and B.A.Crompton, J. Clin. Pathol., 14, 164 (1961).
3058. T.R.Johnson and O.N.Barrett, J. Lab. Clin. Med., 57, 961 (1961).
3059. H.S.Friedmann, Clin. Chim. Acta., 7, 100 (1962).
3060. E.Afonso, Clin. Chim. Acta., 7, 545 (1962).
3061. A.S.Pinfield and D.O.Rodgerson, Clin. Chem., 12, 883 (1966).
3062. T.G.Ferris, R.E.Easterling and R.E.Budd, Nature (London), 208, 1103 (1965).

3063. O.Smithies, Biochem. J., 61, 629 (1955).
3064. J.L.Graham and B.W.Grunbaum, Am. J. Clin. Pathol., 39, 567 (1963).
3065. R.O.Briere, T.Golias and J.G.Batsakis, Am. J. Clin. Pathol., 44, 695 (1965).
3066. H.R.Marti, Experientia, 17, 235 (1961).
3067. L.A.Lewis, Clin. Chem., 12, 596 (1966).
3068. F.W.Sunderman, Am. J. Clin. Pathol., 40, 227 (1963).
3069. B.Zak, F.Volini, J.Briski and L.A.Williams, Am. J. Clin. Pathol., 33, 75 (1960).
3070. B.Zak, E.M.Eggers, T.L.Jarkowski and L.A.Williams, J. Clin. Med., 54, 288 (1959).
3071. C.D.McDonald and T.H.J.Huisman, Clin. Chim. Acta., 8, 639 (1963).
3072. V.N.Tomplins, Am. J. Clin. Pathol., 25, 1430 (1955).
3073. W.Leyko and R.Gondko, Biochim. Biophys. Acta., 77, 500 (1963).
3074. M.Guistiniani, G.Faraglia and R.Barbieri, J. Chromatogr., 15, 207 (1964).
3075. B.S.Van der Leiden, Anal. Biochem., 8, 1 (1964).
3076. C.S.Hanes and F.A.Isherwood, Nature, 164, 1107 (1949).
3077. V.C.Runeckles and G.Krotov, Arch. Biochem. Biophys., 70, 442 (1957).
3078. P.J.Heald, Biochem. J., 63, 235 (1956).
3079. A.Siuda, Nukleonika, 10, 459 (1965).
3080. M.Halmann and L.Kugel, Bull. Res. Council, Israel, 10, 124 (1961).
3081. G.Foraglia, L.Roncucci, B.Fioroni, P.Lassandro and R.Barbieri, Ric. Sci. Riv., 37, 986 (1967).
3082. P.A.Krenkle, W.D.Burrows and R.S.Reimers, Crit. Rev. Environ. Control, 3, 303-362 (1973).
3083. J.F.Uthe and F.A.J.Armstrong, Toxicol. Environ. Chem. Rev., 2, 45-77 (1974).
3084. W.P.Ridley, L.J.Dizikes and J.M.Wood, Science, 197, 329-332 (1977).
3085. N.Imura, E.S.S.Pan, K.N.J.Kim and T.K.T.Ukita, Science, 172, 1248-1249 (1971).
3086. J.M.Wood, F.Scott Kennedy and G.G.Rosen, Nature (London), 220, 173-174 (1968).
3087. S.Jensen and A.Jernelov, Nature (London), 223, 753-754 (1969).
3088. W.P.Ridley, L.J.Dizikes and J.M.Wood, Science, 197, 329 (1977).
3089. J.S.Thayer, J. Organometal Chem., 76, 265 (1974).
3090. F.E.Greifenstein, M.DeVault, J.Yashitake and J.E.Gajewski, Anaesth. Analg. Curr. Res., 37, 283 (1958).
3091. B.M.Davies and H.L.Beech, J. Mental Sci., 106, 912 (1960).
3092. K.Bachmann, Talanta, 29, 1 (1982).
3093. G.Westhoo, Acta. Chem. Scand., 20, 2131-2137 (1966).
3094. G.Westhoo, Acta. Chem. Scand., 21, 1790-1800 (1967).
3095. G.Westhoo, Acta. Chem. Scand., 22, 2277-2280 (1968).
3096. C.A.Bache and D.J.Lisk, Anal. Chem., 43,950-952 (1971).
3097. P.Jones and G.Nickless, Analyst (London), 103, 1121-1126 (1978).
3098. D.L.Collett, D.E.Fleming and G.A.Taylor, Analyst (London), 105, 897-901 (1980).
3099. J.Vostal, "Mercury in the Environment" CRC Press: Cleveland, Ohio, (1972).
3100. S.Jensen and A.Jernelov, Nature (London), 223, 173-174 (1969).
3101. J.M.Wood, C.G.Rosen and S.F.Kennedy, Nature (London),220, 173-174 (1968).
3102. S.Jensen and A.Jernelov, Nature, 223, 753 (1969).
3103. A.Jernelov, Limnol. Oceanogr., 15, 958 (1970).
3104. D.G.Langley, J. Water Pollution Control Fed., 45, 44 (1973).
3105. C.Huey, F.E.Brinckman, S.Grim and W.P.Iverson, Proc. Int. Conf. Transp. Persistent Chem. Aquat. Ecosyst., 73-78 (1974).
3106. Y.K.Chau, P.T.S.Wong, O.Kramar and G.A.Bengert, Proc. Int. Conf. Heavy Met. Engiron., 641-644 (1981).

3107. H.E.Guard, A.B.Cobet and W.M.Coleman, 111 Science,213, 770-771 (1981).
3108. F.E.Brinckman, J.A.Jackson, W.R.Blair, G.J.Olson and W.P.Iverson, Ultratrace speciation and biogenesis of methyltin transport species in estuarine waters. Trace Met. Seawater, NATO Adv. Res. Inst., in press.
3109. J.E.Cremer, Br. J. Ind. Med., 16, 191-193 (1959).
3110. A.W.Bolanowska, Br. J. Ind. Med., 25, 203-205 (1968).
3111. A.W.Bolanowska, W. Chem. Anal. (Warsaw), 12, 121-129 (1967).
3112. J.R.Grove, in "Lead in the Marine Environment": M.Branica, Z.Kosrad; Eds., Pergamon Press, Oxford, 45-52 (1980).
3113. K.Hayakawa, Jpn. J. Hyg., 26, 526-535 (1972).
3114. C.Botre, E.Mailizia, P.Melchlorri, A.Stacchini, G.Tiravanti and C. deZorsl, Proc. Eur. Soc. Toxicol, 18 (1977).
3115. P.Grandjean and T.Nielson, Residue Rev., 72, 97-120 (1979).
3116. P.T.S.Wong, Y.K.Chau and P.L.Luxon, Nature (London), 253, 263-264 (1975).
3117. U.Schmidt and F.Huber, Nature (London), 259, 157-158 (1976).
3118. K.Reisinger, M.Stoeppier and H.W.Nurnberg, Nature (London), 291, 228-230 (1981).
3119. A.W.P.Jarvie and A.P.Whitmore, Environ. Tech. Lett., 2, 197-204 (1981).
3120. K.Hayakawa, Nippon Eiselgaku. Zasshi., 26, 526 (1972).
3121. G.R.Sirota and J.F.Uthe, Anal. Chem., 49, 823-825 (1977).
3122. Y.K.Chau, P.T.S.Wong, O.Kramer, G.A.Bengert, R.B.Cruz, J.O.Kinrade, J.Lye and J.C.Van Loon, Bull. Environ. Contam. Toxicol., 24, 265-269 (1980).
3123. R.M.Harrison and R.Perry, Atmos. Environ., 11, 847-852 (1977).
3124. W.R.A.De Joughe and F.C.Adams, Atmos. Environ., 14, 1177-1180 (1980).
3125. D.C.Reamer, W.H.Zoller and T.C.O'Haver, Anal. Chem., 50, 1449-1455 (1978).
3126. E.Rohbock, H.W.Georgii and J.Muller, Atmos. Environ., 14, 89-98 (1980).
3128. R.M. Harrison, J. Environ. Sci. Health, Part A, A11 419-423 (1976).
3129. J.Ahmad, Y.K.Chau, P.T.S.Wong, A.J.Carty and L.Taylor, Nature (London) 287, 70-71 (1981).
3130. K.Reisinger, M.Stoeppier and W.H.Nurnberg, Nature (London), 291, 228-230 (1981).
3131. P.J.Craig, Environ. Technol. Lett., 1, 17-20 (1980).
3132. A.W.P.Jarvie, R.N.Markall and H.R.Potter, Nature (London), 255, 217-218 (1975).
3133. Y.K.Chau and P.T.S.Wong, "Lead in the Marine Environment"; M.Branica, and Z.Konrad, Eds.; Proceedings of the International Expert Discussion on Lead Occurrence, Fate and Pollution in the Marine Environment, Yugoslavia, (1977); Pergamon Press: New York (1980).
3134. P.T.S.Wong, Y.K.Chau and P.L.Luxon, Nature (London) 253 263-264 (1975).
3135. U.Schmidt and F.Huber, Nature (London), 259, 157-158 (1976).
3136. J.S.Thayer, "Occurrence of Biological Methylation of Elements in the Environment"; American Chemical Society; Washington D.C. (1978) ACS Advances in Chemistry Series No. 182, pp 188-205.
3137. J.E.Cremer, Br. J. Ind. Med., 16, 191-199 (1959).
3138. W.Bolanowska and J.M.Wisnlewska-Knypie, Biochem. Pharmacol., 20, 2108-2110 (1971).
3139. P.Grandjean and T.Nielsen, Residue Rev., 72, 97-148 (1979).
3140. W.Bolanowska, Br. J. Ind. Med., 25, 203-208 (1968).
3141. K.H.Byington, D.A.Yates and W.A.Mullins, Toxicol. Appl. Pharmacol. 52, 379-385 (1980).
3142. P.Grandjean and T.Nielsen, Residue Rev., 72, 97-148 (1979).
3143. J.R.Grove in "Lead in the Marine Environment"; M.Branica and Z.Konrad Eds. Pergamon Press: Oxford, pp 45-52 (1980).

3144. National Academy of Sciences, "Lead: Airborn Lead in Perspective", National Academy of Sciences, Washington D.C. (1972).
3145. J.L.Webb, "Enzyme and Metabolic Inhibitors", Academic Press, New York (1966) Vol. 3, Chapter 6.
3146. R.E.Stauffer, J.W.Ball and E.A.Jenne, "Chemical Studies of Selected Trace Elements in Hot Spring Drainages of Yellowstone National Park," Geological Survey Professional Paper 1044-F; U.S.Government Printing Office: Washington, DC (1980).
3147. E.A.Crecellus, Limnol. Oceanogr., 20, 441 (1975).
3148. M.O.Andreae, Anal. Chem., 49, 820 (1977).
3149. F.Challenger, C.Higgenbottom and L.Ellis, J. Chem. Soc., 1, 95 (1933).
3150. P.T.S.Wong, Y.K.Chaw, L.Luton, G.A.Bengut and D.J.Swaine, "Methylation of Arsenic in the Aquatic Environment", Conference Proceedings on Trace Substances in Environmental Health-X1; Hemphill, University of Missouri (1977).
3151. B.C.McBride and R.S.Wolfe, Biochemistry, 10, 4312 (1971).
3152. R.S.Braman and C.C.Foreback, Science, 182, 1247 (1973).
3153. J.S.Edmonds and K.A.Francesconi, Nature (London), 265, 436 (1977).
3154. R.S.Braman, "Arsenical Pesticide", E.A.Woolson; Ed. American Chemical Society, Washington DC (1975): Ser. 7, pp 108-123.
3156. D.L.Johnson and R.S.Braman, Deep Sea Research, 22, 503 (1975).
3157. R.S.Braman and L.L.Justen, Deep Sea Research, 22, 506 (1975).
3158. D.P.Cox, "Arsenical Pesticide", E.A.Woolson, Ed: American Chemical Society, Washington, DC (1975); Ser. 7, pp 81-96.
3159. D.W.Von Endt, P.C.Kearney and D.D.Kaufman, J. Agric. Food Chem., 16, 17 (1968).
3160. E.A.Woolson, J.H.Axley and P.C.Kearney, Soll Sci. Am. Proc., 35, 101 (1971).
3161. E.A.Woolson, Weed Sci., 25, 412 (1977).
3162. J.U.Lakso and S.A.Peoples, J. Agric. Food Chem., 23, 674 (1975).
3163. S.M.Charbonneau, G.K.H.Tam, F.Bryce, Z.Zawidzka and E.Sandi, Toxicol. Lett., 3, 107 (1979).
3164. Y.Odanaka, O.Matano and S.Goto, J. Agric. Food Chem., 26, 505 (1978).
3165. Y.Odanaka, O.Matano and S.Goto, Bull. Environ. Contam. Toxicol., 24, 452 (1980).
3166. E.A.Crecellus, EHP. Environ. Health Perspect, 19, 147 (1977).
3167. G.K.H.Tam, S.M.Chabonneau, F.Bryce, C.Pomroy and E.Sandi, Toxicol. Appl. Pharmacol., 50, 319 (1979).
3168. A.C.Chapman, Analyst (London), 51, 548 (1926).
3169. G.Lunde, Int. Rev. Gesamten Hydrobiol. 52, 265 (1967).
3170. G.Lunde, Nature (London), 244, 186 (1969).
3171. J.S.Edmonds, K.A.Francesconi, J.R.Cannon, C.L.Raston, B.W.Skelton, and A.H.White, Tetrahedron Lett. 18, 1543 (1977).
3172. M.O.Andreae, Anal. Chem., 49, 820 (1977).
3173. G.K.H.Tam, S.M.Charbonneau, F.Bryce and G.Lacroix, Bull. Environ. Contam. Toxicol., 21, 371 (1979).
3174. K.H.Tam, S.M.Charbonneau, F.Bryce and G.Lacroix, Anal. Biochem., 86, 505 (1978).
3175. J.U.Lasko and S.A.Peoples, J. Agric. Food Chem., 23, 674 (1975).
3176. K.May, K.Reisinger, R.Flucht, M.Stoeppler, Sanderdlruch aus Von Wasser, 55, 63 (1980).
3177. K.Olson, Anal. Chem., 49, 23 (1977).
3178. W.D.Burrows and P.A.Krenkel, Anal. Chem., 46, 1613 (1974).
3179. W.D.Burrows and P.A.Krenkel, Environ. Sci. Technol., 7, 1127 (1973).
3180. G.W.Kalb, Atom. Absorp. Newsletter, 84, 9 (1970).
3181. R.J.Baltisherger and C.L.Knudson, Anal. Chim. Acta., 265, 73 (1974).
3182. J.J.Bisagni and A.W.Lawrence, Environmental Science and Technology, 850, 8 (1974).

3183. F.Frimmel and H.A.Winckler, Zeitschrift fur Wassen and Abwassen, 8, 67 (1975).
3184. Y.Umezaki and K.Iwamoto, Japan Analyst, 20, 173 (1971).
3185. Y.K.Chau and H.Saitoh, Environ. Sci. Technol., 4, 839 (1970).
3186. M.P.Stainton, Anal. Chem., 43, 625 (1971).
3187. R.A.Carr, J.B.Hoover and P.W.Wilkniss, Deep-Sea Res., 19, 747 (1972).
3188. W.F.Fitzgerald, W.B.Lyons and C.D.Hunt, Anal. Chem., 46, 1882 (1974).
3189. R.J.Watling, H.R.Watling and S.A.Water, 1, 113 (1975).
3190. L.D.Chait and R.D.Bolster, Commun. Pschopharmacal, 2, 351 (1978).
3191. W.O.Pierce, T.C.Lamorreaux, F.M.Urry, L.Kopjak and B.S.Finkle, J, Anal. Toxicol., 2, 26 (1978).
3192. L.Magos and T.W.Clarkson, J. Assoc. Off. Anal. Chem., 55, 986-971 (1972).
3193. H.E.Garey, L.A.Weisberg and R.G.Heath, J. Psychedelic Drugs, 9, 280 (1977).
3194. National Institute on Drug Abuse, Res. Monogr., National Institute on Drug Abuse, Rockville, Md. Vol. 21 (1978).
3195. A.K.Done, R.Aronow and J.M.Miceli, National Institute on Drug Abuse Res. Monogr.,National Institute on Drug Abuse: Rockville, Md., Vol. 21, p. 210 (1978).
3196. A.L.Misra, R.B.Pontani and J.Bartolomeo, Res. Commun. Chem. Pathol. Pharmacol., 24, 431 (1979).
3197. P.E.Doherty and R.S.Dorsett, Anal. Chem., 43,1887 (1971).
3198. O.Brandenbergen and O.Bader, Anal. Abstr., 15, 5883 (1968).
3199. O.Brandenberger and O.Bader, Anal. Abstr., 17, 617 (1969).
3200. Department of the Environment and National Water Council (U.K.), H.M.Stationary Office, London, 23 pp (pt. 22 Abenv.), (1978).
3201. S.H.Omang, Anal. Chim. Acta., 52, 415 (1972).
3202. J.E.Uthe, F.A.J.Armstrong and M.P.Stainton, J. Fish Research Board, Canada, 27, 805 (1970).
3203. R.Osland, Pye Limean Spectrovision, 11, No. 24 (1970).
3204. W.R.Simpson and G.Nickless, Analyst (London), 102, 86 (1977).
3205. M.D.K.Abo-Rady, Frezenius Zeitschrift fur Analylische Chemise,187, 299 (1979).
3206. R.L.Lutze, Analyst (London), 104, 979 (1979).
3207, D.L.Grantham, Laboratory Practice, 294, 27 (1978).
3208. E.Graf, L.Polos, L.Bezar and E.Pungor, 79, 471 (1973).
3209. P.D.Goulden and P.H.J.Anthony, Anal. Chim. Acta., 120, 129 (1980).
3210. A.M.Kiemeneiu and J.GKloosterboer, Anal. Chem., 48, 545 (1976).
3211. P.D.Goulden and B.K.Afghan, Tech. Bull. Inland Waters Branch Department of Energy, Mines and Resources, Ottawa, Canada.
3212. B.J.Farey, L.A.Nelson and M.G.Rolph, Analyst (London),103 ,656 (1978).
3213. Y.K.Chan and H.Saitoh, Environ. Sci. Technol., 4, 839 (1970).
3214. Y.Umezaki and K.Iwamoto, Jpn. Anal., 20, 173 (1971).
3215. M.P.Stainton, Anal. Chem., 43, 625 (1971).
3216. R.A.Carr, J.B.Hoover and P.W.Wilkniss, Deep-Sea Res., 19, 747 (1972).
3217. W.F.Fitzgerald, W.B.Lyons and C.D.Hunt, Anal. Chem., 46, 1882 (1974).
3218. H.J.Kramer and B.Neidhard, J. Radioanal. Chem., 37, 835 (1977).
3219. M.J.Fishman, Anal. Chem., 42, 1462 (1972).
3220. V.I.Muscat, T.J.Vickers and A.Andrery, Anal. Chem., 44, 218 (1972).
3221. K.Matsunaga, Mizushori-gijutau, 15, 431 (1974).
3222. J.Olafsson, Anal. Chim. Acta., 68 207 (1974).
3223. Y.Nanba, S.Sekine and K.Matsuda, Sumitomo Kagaku, Toku, 19, 1974-1 (1974).
3224. K.Minagawa, Y.Takizawa and I.Kufune, Anal. Chim. Acta., 115, 103 (1980).
3225. Y.Nauba, S.Sekine and K.Matsuda, Sumitomo Kagaku, Toku, 19, 1974-1 (1974).

3226. S.L.Law, Science, 174, 285 (1971).
3227. E.Yamagami, S.Tateishi and A.Hashumato, Analyst (London), 105, 491. (1980).
3228. S.Nishi and H.Horimoto, Japan Analyst, 17, 1247 (1968).
3229. S.Nishi and H.Horimoko, Japan Analyst, 19, 1646 (1970).
3230. J.Ealy, W.D.Schultz and O.A.Dean, Anal. Chim. Acta., 64, 235 (1974).
3231. C.J.Capon and V.Crispin Smith, J. Anal. Chem., 49, 365 (1977).
3232. J.E.Longbottom, Anal. Chem., 44, 1111 (1972).
3233. R.C.Dressman, J. Chromat. Sci., 10, 472 (1972).
3234. P.Zarnegar and P.Mushak, Anal. Chim. Acta., 69, 389 (1974).
3235. P.Mushak, F.E.Tibetts, P.Zarneger and G.B.Fisher, J. Chromatogr., 87, 215 (1973).
3236. P.Jones and G.Nickless, J. Chromatogr., 76, 285 (1973).
3237. P.Jones and G.Nickless, J. Chromatogr., 89, 207 (1974).
3238. L.Magos, Analyst (London), 96, 847 (1971).
3239. Y.Kimura and V.L.Miller, Anal. Chim. Acta., 27, 325 (1962).
3240. T.C.Rains and O.Menis, J. Assoc. Off. Anal. Chem., 55, 1339 (1972).
3241. Environmental Protection Agency, Methods for Chemical Analysis of Water and Waste Water. EPA publication no. EPA-625/6-74-003 P 118 US., E.P.A. Office of Technology Transfer. Washington DC 20460 (1972).
3242. P.D.Goulden and B.K.Afghan, Tech-Bull. Inland Waters Branch Department of Energy, Mines and Resources, Ottawa, Canada.
3243. F.A.J.Armstrong, P.M.Williams and J.D.Strickland, Nature (London), 211, 481 (1966).
3244. T.B.Bennett Jr., J.H.McDaniel and R.N.Hemphill, "Advances in Automated Analysis, Technical International Congress", Vol. 8, Mediad. Inc., Tarrytown, N.Y. (1972).
3245. A.A.El-Awaday, R.B.Miller and M.J.Carter, Anal. Chem., 48, 110 (1976).
3246. Y.Umezaki and K.Iwamoto, Japan Analyst, 20, 173 (1971).
3247. O.Kinuma and O.Miller, Anal-Abstr., 10, 2943 (1963).
3248. W.F.Fitzgerald, W.B.Lyons and C.D.Hunt, Anal. Chem., 46, 1882 (1974).
3249. H.J.Kramer and J.Neid Lamb. Radio-Anal. Chem., 37,835 (1977).
3250. Y.K.Chan and H.Saitoh, Environmental Science and Technology, 4, 839 (1970).
3251. M.J.Fishman, Anal. Chem., 42, 1462 (1972).
3252. V.I.Muscat, T.J.Vickers and A.Andrery, Anal. Chem., 44, 218 (1972).
3253. K.Matsunaga, Mizushori-gijutsu, 15, 431 (1974).
3254. J.Olafsson, Anal. Chim. Acta., 68, 207 (1974).
3255. P.Jones and S.G.Nickless, Analyst, 103, 1121 (1978).
3256. M.C.Culhen and E.T.McGuinness, Analyt. Biochem., 42, 455 (1971).
3257. P.J.Ke and T.R.Thibert, Mikrochim. Acta., 3, 417 (1973).
3258. K.G.Van Ettebover, 13, 326 (1980).
3259. K.Matsunaga, S.Konishi and M.Nishimura, Environmental Science and Technology, 13, 63 (1979).
3260. S.Rubel, Anal. Chim. Acta., 115, 343 (1980).
3261. T.Braun, M.N.Abbas, L.Bakos and A.Elek, Anal. Chim. Acta., 131, 311 (1981).
3262. T.Braun, M.N.Abbas, Tokosk. Szakefalvi-Nagy Z. Anal. Chim. Acta., 160, 277 (1984).
3263. D.F.Becknell, R.H.Marsh and W.Allie, Anal. Chem., 43, 1230 (1971).
3264. A.Kudo, H.Nagase and Y.Ose, Water Research, 16, 1011 (1982).
3265. F.Glockling, Queens Uninversity of Belfast, Analytical Procedings, 17, 417 (1980).
3266. World Health Organization International Standards for Drinking Water, W.H.O., Geneva, (1970).
3267. Official Journal of the European Community (1975) Proposal for a Council directive relating to the quality of water for human consumption, 18, C214 2-17.

3268. J.Stary and J.Prasilova, Radiochem. Radionala. Letters, 24, 143 (1976).
3269. J.Stary and J.Prasilova, Radiochem. Radionala. Letters, 26, 33 (1976).
3270. J.Stary and J.Prasilova, Radiochem. Radional. Letters, 26, 193 (1976).
3271. J.Stary and J.Prasilova, Radiochem. Radional. Letters, 27, 51 (1976).
3272. J.Stary, B.Havlik, J.Prasilova, K.Kratzer and J.Hannsova, International J. Environment and Chemistry, 5, 89 (1978).
3273. F.Jackson and D.Dellar, Water Research, 13, 381 (1979).
3274. C.Feldman, Anal. Chem., 46, 99 (1974).
3275. C.E.Oda and J.D.Ingle, Anal. Chem., 53, 2305 (1981).
3276. J.E.Hanley and J.D.Ingle, Anal. Chem., 47, 719 (1975).
3277. D.R.Christman and J.D.Ingle, Anal. Chim. Acta., 86, 53 (1976).
3278. C.E.Oda and J.D.Ingle, Anal. Chem., 53, 2030 (1981).
3279. M.S.Quimby-Hunt, Am. Lab. (Fairfield, Comm.)., 10, 17 (1978).
3280. W.F.Fitzgerald and W.B.Lyons, Nature (London), 242, 452 (1973).
3281. H.Agemian and A.S.Y.Chau, Anal. Chem., 50, 13 (1978).
3282. G.E.Millward and A.I.Bihan, Water Research, 12, 979 (1978).
3283. Department of the Environment and National Water Council U.K., H.M. Stationary Office, London, 23 pp (1978).
3284. D.Gardner and J.P.Riley, J. Cons. Out. Explor. Mer., 35, 202 (1974).
3285. J.Yamamoto, Y.Kaneda and Y.Hijaka, Int. Journal of Environmental Analytical Chemistry, 1, 16 (1983).
3286. M.Filippelli, Analyst, 109, 515 (1984).
3287. K.Tanabe, H.Haraguchi and K.Fuwa, Spectrochim. Acta., Part B, 36B, 633-639 (1981).
3288. K.Chiba, K.Yoshida, K.Tanabe, M.Ozaki, H.Haraguchi, J.D.Winefordner and K.Fuwa, Anal. Chem., 54, 761-764 (1982).
3289. K.Tanabe, K.Chiba, H.Haraguchi and K.Fuwa, Anal. Chem., 53, 1450 (1981).
3290. K.Chiba, K.Yoshida, K.Tanabe, H.Haraguchi and K.Fuwa, Anal. Chem., 55, 450 (1983).
3291. K.Tanabe, H.Haraguchi and K.Fuwa, Spectrochim. Acta., Part B, 36B 633 (1981).
3292. K.Chiba, K.Yashida, K.Tanabe, M.Ozaki and M.Haroguchi, J.D.Winefordner and K.Fuwa, Anal. Chem., 54, 761 (1982).
3293. L.Sipos, H.W.Nurnberg, P.Valenta and M.Brancia, Anal. Chim. Acta., 115, 25 (1980).
3294. A.V.Holden, J. Food Techno., 8, 1 (1973).
3295. H.N.S.O. Central Unit on Environmental Pollution, Poll. Pap. 10 : 92 (1976).
3296. A.Pennacchioni, R.Marchetti and G.R.Gaggino, J. Environ. Qual., 5, 451 (1976).
3297. R.J.Pentreath, J. Pleuronectes platessa L.J. Exp. Mar. Biol. Ecol., 25, 103 (1976).
3298. R.J.Pentreath, Pleuronectes platessa L.J.Exp. Mar. Biol. Ecol., 25, 25-51 (1976).
3299. H.Egawa and S.Tajima, Proc. 2nd U.S. Japan Experts Meeting, Oct. 1976 Tokyo, Japan (1977).
3300 A.W.Andren and R.C.Harriss, Nature, 245, 256 (1973).
3301. B.H.Olsen and R.C.Cooper, Water. Red., 10, 113 (1976).
3302. H.Windom, W.Gardner, J.Stephens and F.Taylor, Est. Coast. Mar. Sci., 4, 579 (1976).
3304. I.M.Davies, W.C.Graham and S.M.Pirie, Marine Chemistry, 7, 111 (1979).
3305. E.Shin and P.A.Krenkel, J. Water Poll. Cont. Fed., 48, 473 (1976).
3306. A.Jernelov, Limnol. Oceanogr., 15, 958 (1979).
3307. D.C.Gillespie, J. Fish Res. Board Can., 29, 1035 (1972).
3308. D.G.Langley, J. Water Poll. Cont. Fed., 49, 44 (1973).

3309. G.Westoo, Acta. Chem., Scand., 22, 2277 (1968).
3310. R.Chester, D.Gardner, J.P.Riley and J.Stoner, Mar. Poll. Bull., 2, 28 (1973).
3311. A.Renzoni, E.Bacci and L.Falciai, Rev. Intern. Oceanogr., 32, 31 (1973).
3312. J.Olafsson, Anal. Chim. Acta., 68, 207 (1974).
3313. W.F.Fitzgerald, W.B.Lyons and C.D.Hunt, Anal. Chem., 46, 1882 (1974).
3314. R.A.Fitzgerald, D.C.Gordon Jr. and R.E.Cranston, Deep-Sea Res., 21, 139 (1974).
3315. Y.Thibaud, Science et Peche, Bull. Inst. Peche Marit., 209, 1 (1971).
3316. G.Cumont, G.Viallex, H.Lelievre and P.Bobenrieth, Rev. Intern. Oceanogr. Med., 26, 95 (1972).
3317. A.Renzoni and F.Baldi, Acqua and Aria.,597 (1975).
3318. F.Stoeppler, F.Backhaus, W.Matthes, M.Bernhard and B.Schulte, Proc. Verb. XXVth. Congress and Plenary Assembly of ICSEM, Split (1976).
3319. M.Stoeppler, M.Berhard, F.Backhaus and E.Schulte, Mar. Poll. Bull, submitted.
3320. M.Stoeppler and W.Natthas, Anal. Chim. Acta., 98, 389 (1978).
3321. G.Coupean and R.Bartha, Bull. of Environmental Contamination and Toxicology, 31, 486 (1983).
3322. R.Takeshita, H.Akagi, M.Fujita and Y.Sakegami, J. Chromatography, 51, 283 (1970).
3323. K.Itsuki and H.Komuro, Japan Analyst, 19, 1214 (1970).
3324. T.Murakami and T.Yoshinaga, Japan Analyst, 20, 1145 (1971).
3325. T.Murakami and T.Yoshinaga, Japan Analyst, 20, 989 (1971).
3326. W.L.Carpenter, N.C.A.S.I. Stream Improvement Tech. Bull. No. 263 (1972).
3327. R.S.Braman and M.A.Tompkins, Anal. Chem., 51, 12 (1979).
3328. M.O.Andreae and P.N.Froelich, Anal. Chem., 53, 287 (1981).
3329. P.M.Froelich and M.O.Andreae, Science, 213, 205 (1981).
3330. G.A.Hambrick, P.M.Froelich, O.A.Meinrate and B.L.Lewis, Anal. Chem., 56, 421 (1984).
3331. G.A.Hambrick and P.N.Froelich, Trans. Am. Geophys, Union, 63, 71 (1982).
3332. M.O.Andreae, Anal. Chem., 49, 820 (1977).
3333. M.O.Andreae, in "Trace Metals in Sea Water", C.S.Wong, E.Boyle, K.W. Bruland and J.D.Burton, Eds.: Plenum, New York (1983).
3335. M.O.Andreae, J.Asmode, P.Foster and L.Van't dack, Anal. Chem., 53, 1766 (1981).
3336. S.A.El Wardani, Geochim. Cosmochim. Acta., 13, 5 (1957).
3337. S.A.El Wardani, Geochim. Cosmochim. Acta., 15, 237 (1958).
3338. D.L.Johnson and R.S.Braman, Deep-Sea Res., 22, 503 (1975).
3339. J.D.Burton, F.Culkin and J.P.Riley, Geochim. Cosmochim. Acta., 10, 151 (1959).
3340. R.S.Braman and M.A.Tompkins, Anal. Chem., 51, 12-19 (1979).
3341. V.F.Hodge, S.L.Seidel and E.D.Goldberg, Anal. Chem., 51, 1256-1259 (1979).
3342. H.A.Meinema, T.Burger-Wiersma, G.Versluis-de Haan and E.Ch.Gevers, Environ. Sci. Technol., 12, 288-293 (1978).
3343. R.J.Maguire and H.Hunealt, J. Chromatogr., 209, 288-293 (1981).
3344. Y.K.Chau, P.T.S.Wong and G.A.Bengert, Anal. Chem., 54, 246 (1982).
3345. Y.K.Chau, P.T.J.Wong and P.D.Goulden, Anal. Chim. Acta., 85, 421 (1976).
3346. C.J.Soderquist and D.G.Crosby, Anal. Chem., 50, 1435 (1978).
3347. J.D.Smith, Nature, 103, 225 (1970).
3348. R.S.Braman and M.A.Tomkins, Anal. Chem., 51, 12 (1979).
3349. R.Herman and C.T.J.Alkemade "Chemical Analysis by Flame Photometry" Interscience Publishers, New York and London (1963).

3350. W.R.S.Garton, Proc. Phys. Soc., 64, 591 (1951).
3351. W.W.Watson and R.Simon, Phys. Rev., 55, 358 (1939).
3352. R.M.Dagnall, K.C.Thompson and T.S.West, Analyst, 93, 518 (1968).
3353. W.A.Aue and H.H.Hill, J. Chromatography, 70, 158 (1972).
3354. R.J.Maguire and H.Runeault, J. of Chromatography, 209, 458 (1981).
3355. R.J.Maguire, Y.J.Chau, G.A.Bengert, E.J.Hale, P.T.S.Wong and O.Kramero, Environmental Science and Technology, 16, 698 (1982).
3356. J.A.Jackson, W.R.Blair, F.E.Brinckman and W.P.Iverson, Environmental Science and Technology, 16,110 (1982).
3357. M.D.Mueller, Fresenius Zeitschrift fur Analylishe Chemie, 317, 32 (1984).
3358. W.R.De Jonghe, W.E.Van Mol and E.C.Adams, Anal. Chem. 55, 1050 (1983).
3359. Y.K.Chau, P.T.S.Wong and P.D.Goulden, Anal. Chim. Acta., 421, 85 (1976).
3360. Y.K.Chau, P.T.S.Wong, G.A.Bergest and O.Kramer, Anal. Chem., 51, 186 (1979).
3361. W.Bolanowska, J.Piotrowski and H.Garcynski, Arch. Toxikol., 22, 278-282 (1967).
3362. R.M.Harrison, J. Environ. Sci. Health, Part A, A11, 6, 417-423 (1976).
3363. T.Hielsen, K.A.Jensen and P.Grandjean, Nature (London), 274, 602-603 (1978).
3364. S.A.Estes, P.C.Uden and R.M.Barnes, Anal. Chem., 53, 1336-1340 (1981).
3365. D.S.Forsythe and W.D.Marshall, Anal. Chem., 55, 2132 (1983).
3366. S.A.Estes, P.C.Uden and R.M.Barnes, Anal. Chem., 53, 1336 (1981).
3367. A.W.Bolanowska, Br. J. Ind. Med., 25, 203 (1968).
3368. A.W.Bolanowska, Chem. Anal. (Warsaw), 12, 121 (1967).
3369. S.A.Estes, P.C.Uden and R.M.Barnes, Anal. Chem., 54, 2402 (1982).
3370. Y.K.Chau, P.T.B.Wong and O.Kramar, Anal. Chim. Acta., 146, 211 (1983).
3371. D.J.Hodges and F.C.Noden, Presented at the International Conference on Heavy Metals in the Environment, London pp 408-411 (1979).
3372. M.P.Colombini,R.Fuoco and P.Popoff, Science of the Total Environment, 37, 61 (1984).
3373. A.M.Bond, J.R.Bradbury, P.J.Hanna, G.N.Howell, H.A.Hudson and S. Strother, Anal. Chem., 56,2392 (1984).
3374. W.N.Aldridge and B.W.Street, Analyst (London), 106, 60-68 (1981).
3375. J.E.Cremer, Br. J. Ind. Med., 16, 191-199 (1959).
3376. S.R.Henderson and L.J.Snyder, Anal. Chem., 33, 1172-1180 (1961).
3377. G.Roderer, J. Environ. Sci. Health, Part A, A17, 1-20 (1982).
3378. U.Schmidt and F.Huber, Anal. Chim. Acta., 98, 147-149 (1978).
3379. C.D.Stevens, C.J.Feldhake and R.A.Kehoe, J. Pharmacol. Exptl. Ther., 128, 90 (1960).
3380. H.H.Potter, A.W.P.Tarvie and R.N.Markall, Water Pollution Control (Maidstone) U.K., 123 (1977).
3381. A.W.P.Jarvie, A.P.Whitmore, R.N.Markall and H.R.Potter, Environmental Pollution (Series B), 6, No. 1 69 (1983).
3382. F.Glockling, Queens University of Belfast, Analytical Proceedings, 17, 417 (1980).
3383. M.O.Andreae, Anal. Chem., 49, 820 (1977).
3384. R.S.Braman, D.L.Johnson, C.C.Foreback, J.M.Ammons and J.L.Bricker, Anal. Chem., 49, 621 (1977).
3385. E.A.Crecelius, Anal. Chem., 50, 826 (1978).
3386. Y.Talmi and D.T.Bostick, Anal. Chem., 47, 2145 (1975).
3387. R.S.Braman and C.C.Foreback, Science, 182, 1247 (1973).
3388. A.U.Shalkh and D.E.Tallman, Anal. Chim. Acta., 98, 251 (1978).
3389. S.Gohda, Bull. Chem. Soc. Japan, 48, 1213 (1975).
3390. J.S.Edmonds and K.A.Francesconi, Anal. Chem., 48, 2019 (1976).
3391. L.Duncan and C.R.Parker, Varian Techtron, Palo Alto, Calif. "Technical Topics" (1974).

3392. J.F.Uthe, H.C.Freeman, J.R.Johnston and P.Michalik, J. Assoc. Offic. Anal. Chem., 57, 1363 (1974).
3393. W.R.Penrose, R.Black and M.J.Hayward, J. Fish Res. Bd. Can., 32, 1275 (1975).
3394. J.A.Persson and K.Irgum, Anal. Chim. Acta., 138, 111 (1982).
3395. B.Lussi-Schlatter and H.Brandenberger, "Advances in Mass Spectrometry in Biochemistry and Medicine", Sepctrum Publications New York (1976) Vol. 2 pp231-248.
3396. B.Lussi-Schlatter and H.Brandenberger, Z.Klin. Chem. Biochem., 12, 224 (1974).
3397. Y.Odanaka, N.Tsuchlya, O.Matano and S.Goto, Anal. Chem., 55, 929 (1983).
3398. G.E.Parris, W.R.Blair and F.E.Brinkmann, Anal. Chem., 49, 378 (1977).
3399. Y.K.Chau, P.T.S.Wong and P.D.Goulden, Anal. Chem., 47, 2279 (1975).
3400. P.F.Reay and C.J.Asher, Anal. Biochem., 78, 557 (1977).
3401. D.G.Iverson, M.A.Anderson , T.R.Holm and R.R.Stanforth, Environ. Sci. Technol., 13, 1491 (1979).
3402. M.Yamamoto, Soil Sci. Soc. Am. Proc., 39, 859 (1975).
3403. D.E.Dietz and M.E.Perez, Anal. Chem., 48, 1088 (1976).
3404. G.E.Pacey and J.A.Ford, Talanta, 28, 935 (1981).
3405. D.L.Johnson and M.E.A.Pilson, Anal. Chim. Acta., 58, 289 (1972).
3406. F.T.Henry and T.M.Thorpe, Anal. Chem., 52, 80 (1980).
3407. A.A.Grabinski, Anal. Chem., 53, 966 (1981).
3408. J.Aggett and R.Kadwani, Analyst, 108, 1495 (1983).
3409. W.L.Jolly, J, Am. Chem. Soc., 83, 335 (1961).
3410. W.Holak, Analyt. Chem., 41, 1713 (1969).
3411. E.F.Dalton and A.J.Malanoski, Atom. Absorption Newsl., 10, 92 (1971).
3412. F.J.Fernandez and D.C.Manning, Atom. Absorption Newsl., 10, 86 (1971).
3413. D.C.Manning, Atom. Absorption Newsl., 10, 123 (1971).
3414. F.J.Fernandez, Atom. Absorption Newsl., 12, 93 (1973).
3415. F.J.Schmidt and J.L.Royer, Analyt. Lett., 6, 17 (1973).
3416. Kan Kwok-Tai, Analyt. Lett., 6, 603 (1973).
3417. P.N.Vijan and G.R.Wood, Atom. Absorption Newsl., 13, 33 (1974).
3418. R.K.Skogerbore and A.P.Bejnink, Anal. Chim. Acta., 94, 297 (1977).
3419. R.D.Kadeg and G.D.Christian, Anal. Chim. Acta., 88, 117 (1977).
3420. M.Thompson, S.J.Paklavanpour G.F.Welton, Analyst, 103, 568 (1978).
3421. R.C.Chu, G.P.Barron and P.A.W.Baumgarner, Analyt. Chem., 44,1476 (1972).
3422. M.Lansford, E.M.McPherson and M.J.Fishman, Atom. Absorption Newsl., 13, 103 (1974).
3423. R.S.Braman, L.L.Justan and C.C.Foreback, Analyt. Chem., 44, 2195 (1972).
3424. K.C.Thompson and D.R.Thomerson, Analyst, 99, 595 (1974).
3425. A.E.Smith, Analyst, 100, 300 (1975).
3426. M.Bedard and J.D.Kerbyson, Can. J. Spectrosc., 21, 64 (1976).
3427. H.D.Fleming and R.G.Ide, Analytica Chim. Acta., 83, 67 (1976).
3428. P.N.Vijan, A.C.Rayner, D.Sturgiss and G.R.Wood, Analytica Chim. Acta., 82, 329 (1976).
3429. P.N.Vijan and G.R.Wood, Talanta, 23, 89 (1976).
3430. S.Greenfield, I.L.Jones, H.McD.McGeachin and P.B.Smith, Analytica Chim. Acta., 74, 225 (1975).
3431. R.H.Wendt and V.A.Fassel, Analyt. Chem., 37, 920 (1965).
3432. J.Aggett and A.C.Aspell, Analyst, 101, 341 (1976).
3433. A.G.Howard and M.H.Arbab-Zavar, Analyst, 105, 338 (1980).
3434. F.E.Brinkman, K.L.Jewett, W.P.Iverson, K.J.Irgolic, K.C.Ehrhardt and R.A.Stockton, J. Chromatogr., 191, 31 (1980).
3435. G.R.Ricci, L.S.Shepard, G.Colovos and N.E.Hester, Anal. Chem., 53, 610 (1981).

3436. P.R.Gifford and S.Bruckenstein, Anal. Chem., 52, 1028 (1980).
3437. J.F.Kopp, Anal. Chem., 45, 1789 (1973).
3438. R.J.Evans and S.L.Bandemar, Anal. Chem., 26, 595 (1954).
3439. J.S.Caldwell, R.L.Lishka and E.F.McFarren, J. Am. Water Works Assoc., 65 731 (1973).
3440. I.M.Kolthoff and R.Belcher, "Volumetric Analysis", Vol. 3, Interscience Publishers, New York, pp 511-13 (1957).
3441. J.A.Dean and R.E.Rues, Anal. Lett., 2, 105 (1969).
3442. P.A.J.Armstrong, P.M.Williams and J.D.H.Strickland, Nature (London), 211, 418 (1966).
3443. C.E.Stringer and M.Attrep, Anal. Chem., 51, 731 (1979).
3444. D.C.Manning, At. Absorpt. Newsl., 10, 6 (1971).
3445. D.W.Von Endt, P.C.Kearney and D.D.Kaufman, J. Agric. Food Chem., 16, 17 (1968).
3446. Y.Odanoka, O.Manato and S.Goto, J. Agric. Food Chem., 26, 505 (1978).
3447. H.Abe, K.Anma, K.Ishikawa and K.Akasaki, Bunseki Kagaku, 29. 44 (1980).
3448. V.Mikelukova, J. Chromatogr., 34, 284 (1968).
3449. R.M.Sach, F.B.Anastasia and W.A.Walls, Proc. Northeast. Weed Control Conf., 24, 316 (1970).
3450. E.A.Woolson and N.Aharonson, J. Assoc. Off. Anal. Chem., 63, 523 (1980).
3451. R.Iadevaia, N.Aharonson and E.A.Woolson, J. Assoc. Off. Anal. Chem., 63, 742 (1980).
3452. P.F.Reay and G.Asher, J. Anal. Biochem., 78, 557 (1977).
3453. M.D.Andreae, J.F.Asmode and P.Foster, Van't Dock Anal. Chem., 53, 1766 (1981).
3454. Y.Talmi, and V.E.Norwell, Anal. Chem., 47, 1510 (1975).
3455. L.Brown, S.J.Haswell, M.M.Rhead, P.O'Neill and C.C.Bancroft, Analyst, 108, 1511 (1983).
3456. L.G.Mahone, P.J.Garner, R.R.Buch, T.H.Lane, J.F.Tatera, R.C.Smith and C.L.Frye, Environmental Toxicology and Chemistry, 2, 307 (1983).
3457. N.Wanatabe, Y.Yasuda, E.Kato, T.Nakamura, R.Funasaka, K.Shimokawa, E.Sato and Y.Ose, Science of the Total Environment, 34, No. 1/2 169 (1984).
3458. I.K.Israndee, J.K.Syers, L.W.Tabobs, D.R.Keaney and J.T.Gilmour, Analyst (London), 97, 388 (1972).
3459. P.J.Craig and S.F.Norton, Nature (London), 261, 125 (1976).
3460. F.J.Langmyre and J.Aamodt, Anal. Chim. Acta., 87, 483 (1976).
3461. Official Methods of Analysis of AOAC (11th edition), 418, (1970).
3462. H.Nagase, T.Sato, T.Ishikawa and K.Mitani, International Journal of Environmental Analytical Chemistry, 7, 261 (1980).
3463. A.M.Jurka and M.J.Carter, Anal. Chem., 50, 91 (1978).
3464. A.A.El-Awady, R.B.Miller and M.J.Carter, Anal. Chem., 48, 110 (1976).
3465. Y.Umezaki and K.Iwamato, Japan Analyst, 20, 173 (1971).
3466. K.Matsunaga and S.Takahashi, Anal. Chim. Acta., 87, 487 (1976).
3467. K.Irukayama, M.Fukiki, S.Tajima and S.Omori, Jpn. J. Public Health, 19, 25 (1972).
3468. M.Nishimura, K.Matsunaga and S.Konishi,. Jpn. Anal., 24, 655 (1975).
3469. A.W.Andren and R.C.Harris, Nature (London), 245, 256 (1973).
3470. J.A.Ealy, W.D.Shultz and J.A.Dean, Anal. Chim. Acta., 64, 235 (1973).
3471. R.Balti, R.Magnaval and E.Lanzola, Chemosphere, 4, 13 (1975).
3472. J.E.Longbottom, R.C.Dreisman and J.J.Lichtenbert, J. Assoc. Anal. Chem., 56, 1297 (1973).
3473. S.E.Longbottom, Private Communication.
3474. P.D.Bartlett, P.J.Craig and S.F.Morton, Nature, 267, 606 (1977).
3475. J.F.Uthe, J.Solomon and B.Grift, J. Assoc. Analyt. Chem., 55, 583 (1977).

3476. Y.Kimura and V.L.Miller, Anal. Chem., 32, 420 (1960).
3477. P.C.Leong and H.O.Ong, Anal. Chem., 43, 940 (1971).
3478. D.H.Anderson, J.H.Evans, J.J.Murphy and W.W.White, Anal. Chem., 45, 1511 (1971).
3479. E.W.Bretthaur, A.A.Moghissi, S.S.Snyder and N.W.Matthews, Anal. Chem., 46, 445 (1974).
3480. K.K.S.Pilley, C.C.Thomas, J.A.Sondel and C.M.Hyche, Anal. Chem., 43, 1419 (1971).
3481. S.Feldman, Anal. Chem., 46, 1606 (1974).
3482. J.N.Bishop, L.A.Taylor and B.O.Henry, "The Determination of Mercury in Environment Samples", Ministry of the Environment, Ontario, Canada (1973).
3483. L.W.Jacobs and D.R.Keeney, Environ. Sci. Technol., 8, 267 (1976).
3484. "Methods for Chemical Analysis of Water and Wastes", U.S.Environmental Protection Agency, Cincinnati, Ohio, p.134-138 (1974).
3485. Y.Kimura and V.L.Miller, Anal. Chem., 27, 253 (1962).
3486. Metallic Impurities in Organic Matter. Sub-committee (of the Analytical Methods Committee) of the Society for Analytical Chemistry (London), Report by. Analyst, 90, 515 (1965).
3487. Y.Kimura and V.L.Miller, Agric. and Food Chem., 12, 253 (1964).
3488. V.L.Miller and D.Polley, Anal. Chem., 26, 1333 (1954).
3489. D.Polloy and V.L.Miller, Anal. Chem., 27, 1162 (1955).
3490. M.D.Mueller, Fresenius Zeitschrift fur Analytische Chemie, 317, 32 (1984).
3491. Y.K.Chau, P.T.S.Wong and H.Saitoh, J. Chromatogr. Sci., 162, 14 (1976).
3492. D.A.Segar, Anal. Lett., 7, 89 (1974).
3493. Y.K.Chau, P.T.S.Wong and O.Kramar, Anal. Chim. Acta., 146, 211-217 (1983).
3494. J.E.Cremer, Br. J. Ind. Med., 16, 191-199 (1959).
3495. W.Bolanowska, J.Piotrowski and H.Garczynchi, Arch. Toxikol., 22, 278-282 (1967).
3496. Y.K.Chau, P.T.S.Wong, G.A.Bengert and J.L.Dunn, Anal. Chem., 56, 271 (1984).
3497. Y.Odanaka, N.Tsuchlya, O.Matano and S.Goto, Anal. Chem., 55, 929 (1983).
3498. N.Wanatabe, Y.Yasuda, E.Kato, T.Nakmura, R.Funasaka, K.Shimokana, E.Sato and Y.Ose, Science of the Total Environment, 34, No. 1/2 169 (1984).
3499. W.H.Gutenmann and D.Lisk, J. Agr. Food Chem., 8, 316 (1960).
3500. F.J.Langmyhr and J.Aamodt, Anal. Chim. Acta., 87, 483 (1976).
3501. D.M.Houpt and H.Compaan, Analysis, 1, 27 (1972).
3502. J.O.G.Tatton and P.J.Wagstaffe, J. Chromatography, 44, 284 (1969).
3503. J.Ealy, W.D.Schultz and D.A.Dean, Anal. Chim. Acta., 64, 235 (1974).
3504. P.Nangniot and P.H.Mortens, Anal. Chim. Acta., 24, 276 (1961).
3505. A.Vogil and J.Deshusses, Helv. Chim. Acta., 47, 181 (1964).
3506. W.O.Garner, J.N.Seiber and D.G.Crosby, J. Agric. Food Chem., 22, 252 (1974).
3507. S.E.Birnie and D.J.Hodges, Environmental Technology Letters, 2, 433 (1981).
3508. R.M.Sachs, J.L.Michael, F.B.Anastasia and W.A.Wells, Weed Science, 19, 412 (1971).
3509. Y.Odanaka, N.Tsuchlya, O.Matano and S.Goto, Anal. Chem., 55, 929 (1983).
3510. J.N.C.Whyte and J.R.Kuglar, Botanica Marina, 26, 159 (1983).
3511. J.Bruinsina, Photochem. Photobiol., 2, 241 (1963).
3512. M.Strell, A.Kalojanoff and F.Zuther, Anzneimmittel Forsch, 5, 640 (1955).

3513. S.Kondo and I.Mori, J. Pharm. Soc. Japan, 75, 519 (1955).
3514. F.A.J.Armstrong and J.F.Uthe, Perkin Elmer Ltd., Atomic Absorption Newsletter, 10. 101 (1971).
3515. J.F.Uthe, F.A.J.Armstrong and M.P.Stainton, J. Fish Res. Bd. Can., 27, 805 (1970).
3516. J.F.Uthe, F.A.J.Armstrong and K.C.Tam, J. Ass. Off. Anal. Chem., 54, 866 (1971).
3517. W.R.Hatch and W.L.Ott, Anal. Chem., 40, 2085 (1968).
3518. P.D.Goulden and B.K.Afghan, Technical Bulletin No. 27 Inland Waters Branch, Department of Energy, Mines and Resources, Ottawa, Canada.
3519. D.L.Collett, J.E.Fleming and G.E.Taylor, Analyst, 105, 897 (1980).
3520. C.D.Shultz, D.Clear, J.E.Pearson, J.B.Rivers and J.W.Hylin, Bulletin of Environmental Contamination and Toxicology, 15, 230 (1976).
3521. M.P.Stainton, Anal. Chem., 43, 625 (1971).
3522. J.F.Kopp, M.C.Longbottom and L.B.Lobring, J. Amer. Water Works Assoc., 64, 20 (1972).
3523. U.S.Environmental Protection Agency, Mercury in water - provisional method, Analytical Quality Control Laboratory, Cincinnati, Ohio, (1972) - Environmental Protection Agency, Mercury in fish, provisional method. Analytical Quality Control Laboratory, Cincinnati, Ohio, (1972).
3524. M.R.Hendzel and D.H.Jamieson, Anal. Chem., 48, 926 (1976).
3525. F.A.J.Armstrong and J.R.Uthe, At. Absorpt. Newsl., 10, 101 (1971).
3526. Analytical Methods Committee, Analyst, 92, 403 (1967).
3527. Analytical Methods Committee, Analyst, 101,62 (1976).
3528. M.T.Friend, C.A.Smith and D.Wishart, At. Absorpt. Newsl., 16, 46 (1977).
3529. H.Agemian and A.S.Y.Chau, Anal. Chim. Acta., 75, 297 (1975).
3530. H.Agemian and V.Cheam. Anal. Chim. Acta., 101, 193 (1978).
3531. K.I.Aspila and J.M.Carron, "Inter-laboratory Quality Control Study No. 18 - Total Mercury in Sediments", Report Series, Inland Waters Directorate Water Quality Branch, Special Services Section, Department of Fisheries and Environment, Burlington, Ontario, Canada.
3532. Analytical Methods Committee, Chemical Society, London, Analyst (London), 102, 769 (1977).
3533. I.M.Davies, Anal. Chim. Acta., 102, 189 (1978).
3534. P.Jones and G.Nickless,Analyst (London), 103, 1120 (1978).
3535. R.Capelli, C.Fezia and A.Franchi, Zanicchi. Analyst (London), 104, 1197 (1979).
3536. G.T.C.Shum, H.C.Freeman and J.F.Uthe, Anal. Chem., 51, 414 (1979).
3537. C.E.Oda and J.D.Ingle, Anal. Chim., 53, 2305 (1981).
3538. S.A.Katz, Am. Lab. (Fairfield Comm.), 11, 44 (1979).
3539. L.R.Kamps, R.Carr and H.Miller, Bull Envir. Contam. Toxicol., 8, 273 (1972).
3541. J.B.Rivers, J.E.Pearson and C.D.Shutz, Bull. Environ. Contam. Toxicol., 8, 257 (1972).
3542. I.M.Trakhtenberg, "Chronic Effects of Mercury on Organisms" (translated from Russian), Geographic Health Studies Program of the J.E.Fogarty International Centre for Advanced Study in the Health Sciences, U.S. Department of Health, Education and Welfare Public Health Service National Institutes of Health (1974).
3543. S.Hanamura, B.W.Smith and J.D.Winefordner, Anal. Chem., 55, 2026 (1983).
3544. S.Murayama, J. Appl. Phys., 39, 5478-5484 (1968).
3545. S.Murayama, H.Matsuno and M.Yamamoto, Spectrochim. Acta. Part B, 23B, 513-520 (1968).
3546. T.Westermark, Kvicksilverfrageni Sverige, 25, 25 (1964 ars naturresursutredrung, Jordbruks-departementet, Kvicksilverkonferensen (1965).

3547. G.Westhoo, Acta. Chem. Scand., 20. 2131 (1966).
3548. G.Westhoo, Acta. Chem. Scand., 21, 1790 (1967).
3549. G.Westhoo, Acta. Chem. Scand., 22, 2277 (1968).
3550. G.Westhoo, B.Johansson and R.Ryhage, Acta. Chem. Scand., 24, 2349 (1970).
3551. D.Polley and V.L.Miller, J. Agric. Food Chem., 2, 1030 (1954).
3552. J.C.Gage, Analyst, 86, 457 (1961).
3553. J.E.Longbottom, R.C.Dressman and J.C.Lichtenberg, J. Assoc. Off. Anal. Chem., 56, 1297 (1973).
3554. K.Sumino, Kobe J. Med. Sci., 14, 115 (1968).
3555. M.Uchida, K.Hirakawa and T.Inoue, Kumamoto Med. J., 14, 181 (1961).
3556. L.R.Kampe, B.McMahon, J. Ass. Off. Analyt. Chem., 55, 590 (1972).
3557. J.E.Longbottom, R.C.Dressman and J.E.Lichtenberg, J. Assoc. Off. Analy. Chem., 56, 1297 (1973).
3558. M.L.Shaeffer, U.Rhea, J.T.Peelar, C.H.Hamilton and J.E.Campbell, J. Agric. Food Chem., 23, 1079 (1975).
3559. A.Fabbrini, G.Modi, L.Signorelli and G.Simiani, Boll. Laboratory Chim. Proc., 22, 339 (1971).
3560. V.Sumino and O.Kobe, J. Med. Sci., 14, 115 131 (1968).
3561. W.H.Newsome, J. Agric. Food Chem., 19, 567 (1971).
3562. C.A.Bache and D.J.Lisk, Anal. Chem., 43, 950 (1971).
3563. J.F.Uthe, J.Solomon and B.Griff, J. Ass. Off. Analy. Chem., 55, 585 (1972).
3564. V.Zalenko and L.Kosta, Talanta, 20, 115 (1973).
3565. L.Frishbein, Chromatogr. Rev., 13, 83 (1970).
3566. P.Mushak, Environ. Health Perspect., 4, 55 (1973).
3567. W.Blair, W.P.Iveson and F.E.Brinkmann, Chromosphere, 3. 167 (1974).
3568. J.E.Longbottom, Anal. Chem., 44, 111 (1972).
3569. R.Bye and P.E.Paus, Anal. Chim. Acta., 107, 169 (1979).
3570. J.G.Gonzales and T.R.Ross, Anal. Lett., 5, 683 (1972).
3571. J.E.Longbottom, Anal. Chem., 44, 1111 (1972).
3572. J.E.Longbottom, R.C.Dressman and J.C.Lichtenberg, J. Assoc. Off. Anal. Chem., 56, 1297 (1973).
3573. J.K.Baker, R.E.Skelton and C.Ma, J. Chromatogr., 168, 417 (1979).
3574. L.S.Rosenberg and H.V.Vunakis, Res. Commun. Chem. Pathol. Pharmacol, 25, 547 (1979).
3575. W.A.MacCrehan and R.A.Durst, Anal. Chem., 50, 2110 (1978).
3576. S.G.Mairanovskii, Russ. Chem. Rev., 45, 298 (1976).
3577. W.A.MacCrehan, R.A.Durst and J.M.Bellama, Anal. Lett., 10, 1175 (1977).
3578. D.G.Swartzfagar, Anal. Chem., 48, 2189 (1976).
3579. W.A.MacCrehan, R.A.Durst and J.M.Bellama in "Trace Organic Analysis: A New Frontier in Analytical Chemistry", Nati. Bur. Stand. Spec. Publ., in press.
3580. Analytical Methods Committee, Society for Analytical Chemistry, London, Analyst, 769 (1977).
3581. D.C.Stuart, Anal. Chem., 96, 83 (1978).
3582. S.Hamanaka and K.Ueda, Bulletin of Environmental Contamination and Toxicology, 14, 409 (1975).
3583. J.G.Druzhinin, P.S.Kislitin, Tandy Inst. Khim. Akad. Nauk. Kirk. SSR., 8, 21 (1957).
3584. E.J.Calfruny, Analyst, 57, 468 (1961).
3585. V.L.Miller and D.Lillis, Anal. Chem., 30, 1705 (1958).
3586. J.C.Gage, Analyst, 86, 457 (1961).
3587. M.G.Ashley, Analyst, 84, 692 (1959).
3588. V.L.Miller and L.E.Wachter, Anal. Chem., 22, 1312 (1957).
3589. M.Hanscova, Proc. Lek., 5, 339 (1953).
3590. R.Fabre, R.Truhart and C.Bondere, C.R.M. Acad. Sci., 246, 2086 (1958).
3591. B.Sjosfraud, Anal. Chem., 36, 814 (1964).

3592. M.Fujika, Japan Analyst, 19, 1507 (1970).
3593. L.Fishbein, Chromatogr. Rev., 13, 83-162 (1970).
3594. P.Mushak, Environ. Health Perspect., 4, 55-60 (1973).
3595. J.E.Longbottom, R.C.Dressman and J.J.Lichtenberg, J. Assoc. Off. Anal. Chem., 56, 1297-1303 (1973).
3596. R.T.Ross and J.G.Gonzalez, Bull. Environ. Contam. Toxicol., 10, 187-192 (1973).
3597. T.Giovanoli-Jakubczak, M.R.Greenwood, J.C.Smith and T.W.Clarkson, Clin. Chem. (Winston-Salem, N.C.), 20, 222-229 (1974).
3598. R.Von Burg, F.Farris and J.C.Smith, J. Chromatogr., 97, 65-80 (1974).
3599. J.A.Ealy, W.D.Shultz and J.A.Dean, Anal. Chim. Acta., 64, 238 (1973).
3600. C.J.Cappon and J.C.Smith, Anal. Chem., 49, 365 (1976).
3601. Analytical Methods Committee, Analyst (London), 769 (1977).
3602. National Institute for Drug Abuse. Res. monogram. National Institute on Drug Abuse, Rockville Hd. Vol. 21 (1978).
3603. R.C.Gupta, I. Lu, G.Oel and G.D.Lundberg, Clin. Toxicol., 6, 611 (1975).
3604. J.A.Marshman, M.P.Ramsay and E.M.Sellers, Toxicol. Appl. Pharmacol., 35, 129 (1976).
3605. R.Saferstein, J.J.Manura, T.A.Brettell and P.K.De, J. Anal. Toxicol. 2, 245 (1978).
3606. D.C.K.Lin, A.F.Fentiman, R.L.Foltz, R.D.Forney and I.Sunshine, Biomed. Mass. Spectrom., 2, 206 (1975).
3607. J.F.Uthe, J.Solomon and B.Grift, J. Assoc. Off. Anal. Chem., 55, 583 (1972).
3608. C.J.Cappon and J.C.Smith, Bull. Environ. Contam. Toxicol., 19, 600 (1978).
3609. Y.Taimi, Anal. Chim. Acta., 74, 107-117 (1975).
3610. M.D.DuPuis and H.H.Hill, Anal. Chem., 51, 292-295 (1979).
3611. K.Toel and Y.Shimoishi, Talanta, 28, 967-972 (1981).
3612. D.M.Fraley, D.A.Yates, S.E.Manahan, D.Stalling and J.Petty, Anal. Spectrosc., 35, 525-531 (1981).
3613. K.Fuwa, H.Haraguchi, M.Morita and J.C.Van Loon, J. Spectrosc. Soc. Jpn., 31, 289 (1982).
3614. J.C.Van Loon, Anal. Chem., 52, 955A (1980).
3615. K.Z.Hammu Backmann, Anal. Chem., 306, 183 (1981).
3616. J.C.Van Loon, B.Radzuik, N.Kahn, J.Lichwa, F.J.Fernandez and J.D. Kerber, At. Abs. Newsl., 16, 79 (1977).
3617. C.F.Bauer and D.F.S.Natusch, Anal. Chem., 53, 2020-2027 (1981).
3618. D.G.Mitchell, K.M.Aldous and E.Carell, Anal. Chem., 49, 1235-1238 (1977).
3619. C.F.Bauer and D.F.S.Natusch, Anal. Chem., 53, 2020 (1981).
3620. D.G.Mitchell, K.M.Aldous and E.Carelli, Anal. Chem., 49, 1235 (1977).
3621. S.Hanamura, NBS Spec. Publ. (U.S.), 422, 621-624 (1976).
3622. NBS Certificate for Standard Reference Materials; SRM 1566-1575; National Bureau of Standards: Washington D.C.
3623. Y.Taimi, Anal. Chim. Acta., 74, 107 (1975).
3624. D.C.Reamer, T.C.O'Haver and W.H.Zoller in "Methods and Standards for Environmental Measurement", Nati. Bur. Stand. Spec. Publ., 464, 609 (1977).
3625. A.E.Wilson and E.F.Domino, Biomed. Mass Spectrom., 5, 112 (1978).
3626. C.J.Cappon and V.Crispin Smith, J. Anal. Chem., 49, 365 (1977).
3627. G.I.Callum, M.M.Ferguson and J.M.A.Leniham, Analyst, 106, 1010 (1981).
3628. M.D.Osselton, J. Forensic Sci. Soc., 17, 189 (1977).
3629. M.D.Osselton, M.D.Hammond and P.J.Twitchett, J. Pharm. Pharmacol., 29, 460 (1977).
3630. J.F.Uthe, J.Solomon and B.Grift, J. Assoc. Off. Anal. Chem., 55, 583 (1972).

3631. A.K.Done, R.Aronow, J.N.Micell and D.C.K.Lin in "Management of the Poisoned Patient", B.H.Rumack and A.R.Temple, Eds. Princeton Scientific Press, Princeton, N.J., 72-102 (1977).
3632. D.C.K.Lin, R.L.Foltz, A.K.Done, R.Aronow, E.Arcinue and J.N.Micell, "Quantitative Mass Spectrometry in Life Sciences", A.P.DeLeenheer and R.R.Roncucci, Eds. Elsevier, Amsterdam, pp 121-129.
3633. L.K.Wong and K.Beiman, Clin. Toxicol., 9, 583 (1976).
3634. C.R.Martin and H.Freiser, Anal. Chem., 52, 562 (1980).
3635. T.Goina, S.Habal and L.Rosenberg, Farmacia (Bucharest), 26, 141 (1978).
3636. T.Higuchi, C.Illian and J.Tossounian, Anal. Chem., 42, 1674 (1970).
3637. Y.Kanda and N.Suzuki, Anal. Chem., 52, 1672 (1980).
3638. D.S.Forsythe and W.D.Marshall, Anal. Chem., 55, 2132 (1983).
3639. G.Calingaert, H.Beathey and H.Soroos, J. Amer. Chem. Soc., 62, 1099 (1940).
3640. Y.Arakawa, O.Wada, T.H.Yu and H.Iwal, J. Chromatogr., 207, 237 (1981).
3641. Y.Arakawa, O.Wada, T.H.Yu and H.Iwal, J. Chromatogr., 216, 209 (1981).
3642. C.F.Coyle and C.E.White, Anal. Chem., 29. 1486 (1957).
3643. F.Vernon, Anal. Chim. Acta., 71, 192 (1974).
3644. S.J.Blunden and A.H.Chapman, Analyst (London) (short papers) 103, 1266 (1978).
3645. W.N.Aldridge and J.E.Cremer, Analyst (London), 82. 37 (1957).
3646. Y.Arakawa, O.Wada and M.Manabe, Anal. Chem., 55, 1901 (1983).
3647. G.R.Sirota and J.F.Uthe, Anal. Chem., 49, 823 (1977).
3648. K.Hayakawa Japan J. Hygiene, 26, 377 (1971).
3649. M.S.Epstein and T.C.O'Haver, Spectrochim. Acta. Part B., 30, 135 (1975).
3650. H.R.Potter, A.W.P.Jarvie and R.N.Markall, Water Pollution Control (Maidstone, U.K.), 76, 123 (1977).
3651. S.A.Listes, P.C.Uden and R.M.Barnes, Anal. Chem., 53, 1336 (1981).
3652. W.R.A.De Jonghe and F.C.Adams, Talanta, 29, 1057-1067 (1982).
3653. R.M.Harrison and D.P.H.Laxen, Nature (London), 275, 738-740 (1978).
3654. J.W.Robinson, E.L.Kiesel and A.L.Rhodes, J. Environ. Sci. Health, Part A, A14, 65 (1979).
3655. Y.K.Chau, P.T.S.Wong, G.A.Bengert and J.L.Dunn, Anal. Chem., 56, 271 (1984).
3656. J.E.Cremer, Br. J. Ind. Med., 16, 191 (1959).
3657. J.E.Birnie and D.J.Hodges, J. Environ. Technol. Lett., 2 433 (1981).
3658. A.W.P.Jarvie, R.N.Markall and H.R.Potter, Environ. Res. 25, 241 (1981).
3659. Y.K.Chau, P.T.S.Wong, G.A.Bengert and J.L.Dunn, Anal. Chem., 56, 271 (1984).
3660. M.O.Andreae, Anal. Chem., 49, 820 (1977).
3661. J.Tadmor, J. Gas Chromatgr., 2, 385-389 (1964).
3662. D.Vranti-Piscou, J.Kontoyannakos and G.Parissakis, J. Chromatogr. Sci., 9, 499-501 (1971).
3663. W.C.Butts and W.T.Rainey, Anal. Chem., 43, 538-542 (1971).
3664. F.Roy, J. Gas Chromatogr., 6, 245-247 (1968).
3665. B.J.Gudzinowicz and J.L.Driscol, J. Gas Chromatogr., 1, 108-113 (1972).
3666. L.Johnson, K.Gerhardtdt and W.Aue, Sci. Tot. Environ., 1, 108 (1972).
3667. J.S.Edmond and K.A.Francesconi, Anal. Chem., 48, 2019 (1976).
3668. L.R.Overby, S.F.Bocchieri and R.C.Frederickson, J. Assoc. Off. Anal. Chem., 48, 17-22 (1965).
3669. E.A.Dietz Jr. and M.E.Perez, Anal. Chem., 48, 1088-1092 (1976).
3670. R.S.Braman and C.C.Foreback, Science, 182, 1247-1249 (1973).
3671. Y.Taimi and D.T.Bostick, Anal. Chem., 47, 2145-2150 (1975).
3672. M.Yamamoto, Soil Sci. Soc. Am. Proc., 39, 859-861 (1975).

3673. R.A.Stockton and K.J.Irgolic, Int. J. Environ. Anal. Chem., 6, 313-319 (1979).
3674. M.Morita, T.Uchiro and K.Fuwa, Anal. Chem., 53, 1806 (1981).
3675. S.Hanamura, B.W.Smith and J.D.Winefordner, Anal. Chem., 55, 2026 (1983).
3676. G.Schwedt and H.A.Ruessil, Chromatographia, 5, 242 (1972).
3677. N.Wanatabe, Y.Yasuda, E.Kato, T.Nakamura, R.Funasaka, K.Shimokana, E.Sato and Y.Ose, Science of the Total Environment, 34, No. 1/2, 169 (1984).
3678. H.J.Horner, J.A.Weiler and N.C.Angelotti, Anal. Chem., 32, 858 (1960).
3679. Y.Kimura and V.L.Miller, Anal. Chim. Acta., 27, 325 (1962).
3680. Y.Matsumura, Atmos. Environ., 8, 1321 (1974).
3681. S.J.Long, D.R.Scott and R.J.Thompson, Anal. Chem., 45, 2227 (1973).
3682. J.D.Aubort, H.Rollier and A.Ramuz, Trav. Chim. Aliment. Hyg., 68, 155 (1977).
3683. A.A.Christie, A.J.Dundsor and B.S.Marshall, Analyst, 92, 185 (1967).
3684. O.O.Sargeant, Anal. Abstra., 4, 2814 (1957).
3685. H.M.S.O. (1957). Methods for the Detection of Toxic Substances in Air, Booklet No. 13, Mercury and compounds of mercury. Her Majesty's Stationery Office, London.
3686. D.Polley and V.L.Miller, Anal. Chim., 27, 1162 (1955).
3687. V.L.Miller and D.Polley, J. Agr. Food Chem., 2, 1030 (1954).
3688. Y.Kimura and V.L.Miller, Anal. Chem., 32, 420 (1960).
3689. V.L.Miller and F.Swanberg, Anal. Chem., 29, 391 (1957).
3690. G.A.Hamilton and A.D.Ruthven, Lab. Pract., 15, 995 (1966).
3691. R.Dumarev, R.Heindryckx, R.Dams and J.Hoste, Anal. Chim. Acta., 107, 150 (1979).
3692. R.S.Braman and D.L.Johnson, Environ. Sci. Technol., 8, 996 (1974).
3693. D.S.Ballantine and W.H.Zoller, Anal. Chem., 561, 1288 (1984).
3694. W.E.L.Grossman, J.Eng. and Y.C.Tong, Anal. Chim. Acta., 60, 447-449 (1972).
3695. W.H.Schroeder and R.Jackson, Proceedings of the APCA Speciality Conference on Measurement and Monitoring of Non-Criteria (Toxic) Contaminants in Air, held at Chicago, IL. March 22-24 1983; APCA Speciality Conference Proceedings SP-50, pp 91-100.
3696. R.S.Braman and D.L.Johnson, Environ. Sci. Technol., 8, 996-1003 (1974).
3697. B.A.Soldano, P.Bien and P.Kwan, Atomos. Environ., 9, 941-944 (1975).
3698. Y.Takizawa, K.Minagawa and M.Fugil, Chemosphere, 10, 801-809 (1981).
3699. B.Versino M.De Groot and F.Geiss, Chromatographia, 7, 302-304 (1974).
3700. W.A.Aue and P.M.Tell, J. Chromatogr., 62, 15-27 (1971).
3701. G.Torsi, E.Desimoni, F.Palmisano and L.Sabbatini, Anal. Chem., 53, 1035 (1981).
3702. G.Torsi, E.Desimoni, F.Palmisano and L.Sabbatini, Analyst, 107, 96 (1982).
3703. G.Torsi, F.Palmisano, E.Desimoni and R.Rinaldi, Ann. Chim., 72, 365 (1982).
3704. O.M.G.Newman and D.J.Palmer, Nature (London), 526, 275 (1978).
3705. M.S.Bykhovskaya, Zavod Lab., 29, 667 (1963).
3706. M.Kawahara and K.Matsumoto, Japan Analyst, 14, 789 (1965).
3707. M.Coe, R.Cruz and J.C.Van Loon, Anal. Chim. Acta., 120, 171 (1980).
3708. B.Radziuk, Y.Thomassen, Y.K.Chau and J.C.van Loon, Anal. Chim. Acta., 105, 255 (1979).
3709. Fed. Regist., USA, 37, 22141 (1972).
3710. R.S.McDowell, Am. Ind. Hyg. Assoc. J., 32, 621 (1971).
3711. A.W.Mantz, Appl. Spectrosc., 30, 539 (1976).

3712. R.F.Werniund and M.J.Cohen, Res./Dev., 32, July 1975.
3713. D.H.Stedman and D.A.Tammaro, Anal. Lett., 9, 81 (1976).
3714. J.E.Campana and T.H.Risby, Anal. Chem., 52, 468 (1980).
3715. B.M.Luskina and S.V.Syavtsillo, Nov. Obh. Prom. Sarit. Khim., 186 (1969).
3716. R.Jeltes, Ann. Occup. Hyg., 12, 203 (1969).
3717. S.Hancock and A.Slater, Analyst, 100, 422 (1975).
3718. W.De Jonghe and F.Adams, Anal. Chim. Acta., 108, 21 (1979).
3719. G.Thilliez, Anal. Chem., 39, 427 (1967).
3720. B.Kolb, G.Kemmner, F.H.Schesler and E.Wiedeking, Fresenius Z. Anal. Chem., 221, 166 (1966).
3721. P.M.Mutsaars and J.E.Van Steen, J. Inst. Pet., London, 58, 102 (1972).
3722. G.Torsi and F.Palmisano, Analyst, 108, 1318 (1983).
3723. G.Torsi, E.Desimoni, F.Palmisano and L.Sabbatini, Analyst, 107, 96 (1982).
3724. V.Cantuti and G.Cartoni, J. Chromatogr., 32, 641 (1968).
3725. H.Tausch, Ber. Oesterr, Studienges, Atomenerg., SGAE No. 2636. (1979).
3726. N.L.Soulages, Anal. Chem., 38, 28 (1966).
3727. N.L.Soulages, J. Gas Chromatogr., 6, 356 (1968).
3728. D.O.Reamer, W.H.Zoller and T.C.O'Haver, Anal. Chem., 50, 1449 (1978).
3729. W.R.A.De Jonghe, D.Chakraborti and F.C.Adams, Anal. Chem., 52, 1974 (1980).
3730. W.R.A.De Jonghe, D.Chakraborti and F.C.Adams, Anal. Chim. Acta., 115, 89 (1980).
3731. Y.K.Chau, P.T.S.Wong, G.A.Bengert and O.Kramar, O. Anal. Chem., 51, 186 (1970).
3732. D.Chakraborti, S.G.Jiang, P.Surkiju, W.R.A.De Jonghe and F.C.Adams, Anal. Proc., 18, 347 (1981).
3733. "ASTM Standards on Petroleum Products and Lburicants" American Soc. for Testing and Materials, Philadelphia, Method D 526-61 (1967).
3734. D.T.Coker, Ann. Occup. Hyg., 21, 33 (1978).
3735. W.R.A.De Jonghe and F.C.Adams, Talanta, 29, 1057 (1982).
3736. B.Kolb, G.Kemmner, F.H.Schleser, E.Wiedeking and Fresenins, Z. Anal. Chem., 221, 166 (1966).
3737. Y.K.Chau, P.T.S.Wong and H.J.Saitoh, Chromatogr. Sci., 14, 162 (1976).
3738. E.A.Boettner and F.C.Dallas, J. Gas Chromatography, June 1965, 190 (1965).
3739. E.J.Bonelli and H.Hartmann, Wilkins Aerograph Research Notes, Autumn Issue (1963).
3740. H.J.Dawson, Anal. Chem., 35, 542 (1963).
3741. Y.K.Chau, P.T.S.Wong, G.A.Bengert and O.Kramer, Anal. Chem., 186, 51 (1979).
3742. Y.K.Chau, P.T.S.Wong and H.Saitoh, J. Chromatogr. Sci., 162, 14 (1976).
3743. V.Cantuti and G.P.Cartoni, J. Chromatogr., 32, 641-647 (1968).
3744. H.Tausch, Oesterr. Studienges. Atomenerg.,SGAE No. 2636 (1976).
3745. R.M.Harrison and R.Perry, Atomos.Environ., 11, 847-852 (1977).
3746. Y.K.Chau, P.T.S.Wong and P.D.Goulden, Anal. Chim. Acta., 85, 421-424 (1976).
3747. B.Radziuk, Y.Thomassen, J.C.Van Loon, and Y.K.Chau, Anal. Chim. Acta, 105, 255-262 (1979).
3748. J.W.Robinson, E.L.Kiesel, J.P.Goodbread, R.Bliss and R.Marshall, Anal. Chim. Acta., 92, 321-328 (1977).
3749. E.Rohbock and J.Miller, Microchim. Acta., 1, 423-434 (1979).
3750. E.Rohbock, H.W.Georgil and J.Muller, Atmos. Environ., 14, 89-98 (1980).
3751. A.Laveskog, Proc. Int. Clean Air Congress, 2nd 549 (1970).

3752. K.Hayakawa, Japan J. Hygiene, 20, 377 (1971).
3753. R.M.Harrison, R.Perry and D.H.Slater, Atmos. Environ., 8, 1187 (1974).
3754. J.W.Robinson, E.L.Kiesel, J.P.Goodbread, R.Bliss and R.Marshall, Anal. Chim. Acta., 92, 321 (1977).
3755. D.A.Segar, Anal. Lett., 7, 89 (1974).
3756. Y.K.Chau, P.T.S.Wong and P.D.Goulden, Anal. Chem., 2279, 47 (1975).
3757. J.W.Robinson, L.E.Vidaurreta, D.K.Wolcott, J.P.Goodbread and E.Kiesel, Spectrosc. Let., 8, 491 (1975).
3758. D.A.Segar, Anal. Lett., 7, 89 (1974).
3759. G.Gonzalez and R.T.Ross, Anal. Lett., 5, 683 (1972).
3760. D.T.Coker, Anal. Chem., 47, 386 (1975).
3761. H.Hey, Z. Anal. Chem., 256, 361 (1971).
3762. B.Kolb, G.Kemner, F.H.Schleser and E.Wiedeking, Z. Anal. Chem., 221, 166 (1966).
3763. L.Ebdon, G.F.Kirkbright and T.S.West, Anal. Chim. Acta., 58, 38 (1972).
3764. D.C.Reamer, W.H.Zoller and T.C.O'Haver, Anal. Chem., 50, 1449 (1978)/
3765. L.J.Snyder and S.R.Henderson, Anal. Chem., 33, 1175 (1961).
3766. L.J.Snyder, Anal. Chem., 39, 591 (1967).
3767. R.M.Harrison, R.Perry and D.H.Slater, Atmos. Envir., 8, 1187 (1974).
3768. V.Cartuti and G.P.Cortoni, J. Chromatogr., 32, 641 (1968).
3769. A.J.McCormack, S.C.Tong and W.D.Cooke, Anal. Chem., 37, 1470 (1965).
3770. R.M.Harrison, R.Perry and D.H.Slater, Atmos. Environ., 8, 1187 (1974).
3771. B.Radziuk, Y.Thomassen, J.C.Van Loon and Y.K.Chau, Anal. Chim. Acta., 105, 255 (1979).
3772. W.De Jonghe, D.Chakraborti and F.Adams, Anal. Chim. Acta., 115, 89 (1980).
3773. W.R.A.De Jonghe, D.Chakraborti and F.C.Adams, Anal. Chem., 52, 1974 (1980).
3774. J.C.Van Loon, B.Radziuk, N.Kahn, J.Lichwa, F.J.Fernandez and J.D. Kerber, At.Absorp. Newsl., 16, 79 (1977).
3775. D.R.Jones and S.E.Manaham, Anal. Chem., 48, 502 (1976).
3776. F.E.Brinckman, W.R.Blair, K.L.Jewett and W.P.Iverson, J. Chromatogr. Sci., 15, 493 (1977).
3777. H.Koizumi, R.D.McLaughlin and T.Hadeishi, Anal. Chem., 51, 387 (1979).
3778. H.Koizumi, T.Hadeishi and R.D.McLaughlin, Anal. Chem., 50, 1700 (1978).
3779. T.Hadeishi, Appl. Phys. Lett., 21, 438 (1972).
3780. T.Hadeishi and R.D.McLaughlin, Anal. Chem., 48, 1009 (1976).
3781. H.Kolzumi and K.Yasuda, Spectrochim. Acta. Part B, 31, 237 (1976).
3782. H.Kolzumi, K.Yasuda and M.Katayama, Anal. Chem., 49, 1106 (1977).
3783. J.W.Robinson, Anal. Chim. Acta., 24, 451 (1961).
3784. R.E.Sturgeon and C.L.Chakrabarti, Spectrochim. Acta. Part B., 32, 231 (1977).
3785. R.E.Sturgeon and C.L.Chakrabarti, Anal. Chem., 49, 1100)1977).
3786. American Conference of Governmental Hygienists Archs. Envir. Health, 9, 545 (1964).
3787. B.Shalicko, M.Cejka, Ropa Uhlie, 16, 246 (1966).
3788. L.J.Snyder, W.R.Barnes and J.V.Takos, Anal. Chem., 20, 772 (1948).
3789. R.Moss and E.V.Browett, Analyst, (London),91, 428 (1966).
3790. F.D.Krivoruchko, Zavod. Lab., 27, 290 (1961).
3791. E.A.Pereguid and N.P.Kozlova, J. Anal. Chem. USSR., 9, 47 (1954).
3792. F.D.Krivoruchko, Zavod. Lab., 29, 927 (1963).
3793. W.S.Pappas and J.G.Millou, Anal. Chem., 40, 2176 (1968).
3794. G.N.Bortnikov, A.V.Kiselev, N.S.Vyankin and Y.I.Yashin, Chromatographia, 4, 14 (1971).
3795. E.J.Forbes, M.K.Sutan and P.C.Uden, Analyst. Lett., 5, 927 (1972).
3796. H.Matsuda and A.Matsuda, J. Chem. Soc. Japan Ind. Chem. Sect., 63, 1960 (1960).

INDEX